Netter's Gastroenterology

2nd edition

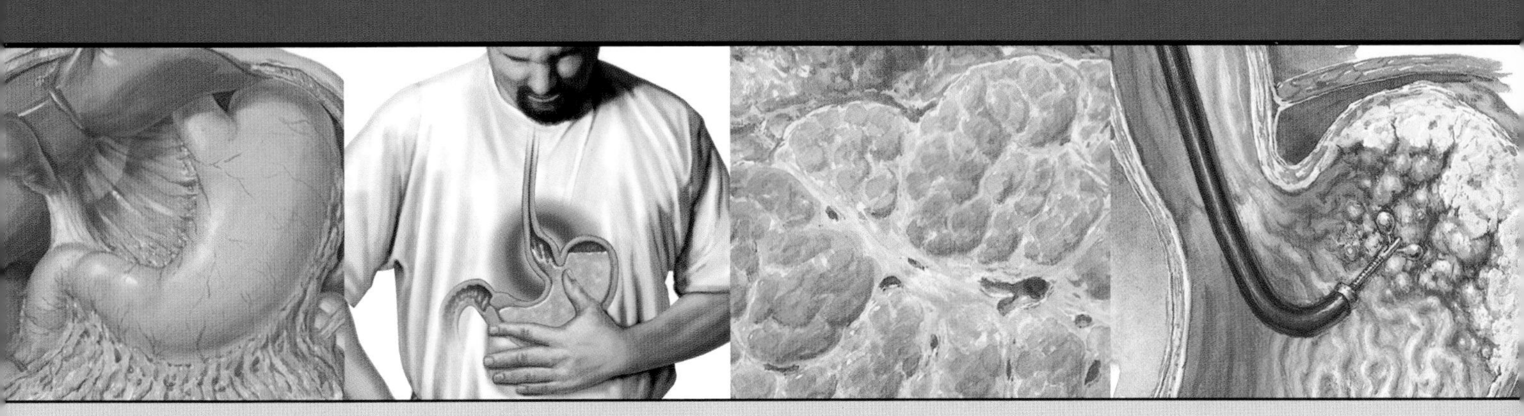

MARTIN H. FLOCH, MD, FACP, MACG, AGAF
NEIL R. FLOCH, MD, FACS
KRIS V. KOWDLEY, MD, FACP
C. S. PITCHUMONI, MD, FRCP, FACP, MACG, MPH
RAUL J. ROSENTHAL, MD
JAMES S. SCOLAPIO, MD

Illustrations by Frank H. Netter, MD

CONTRIBUTING ILLUSTRATORS
Carlos A. G. Machado, MD
John A. Craig, MD
Kip Carter, MS
David Mascaro, MS
Steven Moon
Kristen Wienandt Marzejon
Mike de la Flor

SAUNDERS

SAUNDERS
ELSEVIER

1600 John F. Kennedy Blvd.
Ste 1800
Philadelphia, PA 19103-2899

NETTER'S GASTROENTEROLOGY ISBN: 978-1-4377-0121-0

Notice

Knowledge and best practice in this field are constantly changing. As new research and experience broaden our knowledge, changes in practice, treatment and drug therapy may become necessary or appropriate. Readers are advised to check the most current information provided (i) on procedures featured or (ii) by the manufacturer of each product to be administered, to verify the recommended dose or formula, the method and duration of administration, and contraindications. It is the responsibility of the practitioner, relying on their own experience and knowledge of the patient, to make diagnoses, to determine dosages and the best treatment for each individual patient, and to take all appropriate safety precautions. To the fullest extent of the law, neither the Publisher nor the Editors assume any liability for any injury and/or damage to persons or property arising out of or related to any use of the material contained in this book.

The Publisher

Library of Congress Cataloging-in-Publication Data
Netter's gastroenterology / edited by Martin H. Floch . . . [et al.]; illustrations by Frank H. Netter; contributing illustrators, Carlos A.G. Machado . . . [et al.]—2nd ed.
p. ; cm.
Includes bibliographical references and index.
ISBN 978-1-4377-0121-0 (hardcover : alk. paper) 1. Gastroenterology. 2. Gastrointestinal system—Diseases. I. Floch, Martin H. II. Netter, Frank H. (Frank Henry), 1906-1991. III. Title: Gastroenterology.
[DNLM: 1. Gastrointestinal Diseases. WI 140 N474 2010]
RC801.N48 2010
616.3′3—dc22

2009030539

Netter Director: Anne Lenehan
Editor: Elyse O'Grady
Developmental Editor: Marybeth Thiel
Publishing Services Manager: Patricia Tannian
Senior Project Manager: Sarah Wunderly
Design Manager: Steve Stave
Illustrations Manager: Karen Giacomucci
Marketing Manager: Jason Oberacker
Editorial Assistant: Julie Goolsby

Printed in China

Last digit is the print number: 9 8 7 6 5 4 3 2 1

In today's hectic world, writing scientific text takes place during what most people would consider leisure time. We have forsaken time with our families to compile this book. Hence, we would like to dedicate it completely to our wives and children.

To our wives—Gladys Floch, Prema Pitchumoni, Bonnie Kowdley, Robin Floch, Sima Rosenthal, and Elizabeth Scolapio.

To our children—Jeffrey Aaron Floch (in memoriam), Craig Lawrence Floch, Lisa Susanne Adelmann, Neil Robert Floch, Sydney Floch, Jake Floch, Sheila Pitchumoni, Shoba Pitchumoni, Suresh Pitchumoni, Sara Kowdley, Noam Rosenthal, Dana Rosenthal, Julia Scolapio, and Anthony Scolapio.

Contributors

Editor
Martin H. Floch, MD
Clinical Professor of Medicine
Section of Digestive Diseases
Yale University School of Medicine
New Haven, Connecticut

Associate Editors
Neil R. Floch, MD
Director, Minimally Invasive and Bariatric Surgery
Norwalk Hospital
Norwalk, Connecticut

Kris V. Kowdley MD, FACP
Director, Center for Liver Disease
Virginia Mason Medical Center
Seattle, Washington

C.S. Pitchumoni, MD, MPH
Clinical Professor of Medicine, Robert Wood Johnson Medical School
Chief, Division of Gastroenterology, Hepatology, and Clinical Nutrition
Saint Peter's University Hospital
New Brunswick, New Jersey

Raul J. Rosenthal, MD
Director, The Bariatric Institute
Division of Minimally Invasive Surgery
Cleveland Clinic Florida
Weston, Florida

James S. Scolapio, MD
Director of Nutrition
Department of Gastroenterology
Mayo Clinic
Jacksonville, Florida

Preface

Anyone who has studied medicine in the past 25 years knows the tremendous value of the illustrations of the late Dr. Frank H. Netter. Students and teachers alike recognize his illustrations as the criteria by which all others are judged and consider them timeless classics because they "teach rather than intimidate." Perhaps some thus far unimagined technical innovation will one day permit the development of teaching tools that surpass the clarity and educational value of Dr. Netter's illustrations, but even with the present electronic revolution that day has yet to come, and new generations of students worldwide continue to discover for themselves his remarkable talent.

I have published two previous books in my career—one is among the first texts on the small intestine, and the other is among the first on nutrition in gastrointestinal disease. Although I pledged I would never again undertake such arduous work, the need for a concise yet comprehensive volume covering the basic concepts and the latest developments in the field, along with the opportunity to use the famous Netter collection to illustrate it, overcame my resistance. The result is *Netter's Gastroenterology*. It was a success, and now we have finished the second edition.

My charge from the publisher was to write a text aimed at the generalist, understandable by the student, and of interest to the gastroenterologist. The approach of this book is different from that of traditional textbooks, which rely primarily on the written word. I sought to cover the field of gastroenterology and nutrition and to provide an efficient and meaningful learning experience to my audience by balancing the visual and the verbal, staying true to the philosophy of Dr. Netter by using the power of illustration to teach while providing essential information in the text. I selected approximately 300 of Dr. Netter's best illustrations and, with my coauthors, wrote text to illuminate and expand on the concepts the illustrations present. In a traditional textbook, the text is written first, and illustrations are then created to accompany it. In this case, my coauthors and I wrote with the illustrations before us, and our goal was to forge text and pictures into a seamless whole. In this second edition, we have tried to update the narrative and modify the illustrations where necessary. Because medicine is a constantly advancing science, we called on the talents of medical artists Kip Carter, David Mascaro, Steven Moon, Mike de la Flor, and Kristen Wienandt Marzejon to create new illustrations in the Netter style and to update others where appropriate.

The book is organized into 10 sections that correspond to the organs of the gastrointestinal system and to special topics within that system. These sections have been further divided into 272 concise, condition-oriented chapters, including introductory chapters on anatomy and physiology. The text provides core information about the "clinical picture," diagnosis, treatment and management, course and prognosis, and, when applicable, prevention and control of each condition. With the increased observation of eosinophilic esophagitis, Neil Floch has added a chapter on that subject. With the advances in the treatment of obesity, this edition adds chapters on bariatric surgery and complications of that surgery by Drs. Neil Floch and Raul Rosenthal. I believe the result is a thorough look at almost all clinical topics in the field.

Although I felt comfortable writing the text for the sections on the stomach, small intestine, colon, and infectious diseases, I invited colleagues to undertake the other areas according to their interests and expertise. Dr. C. S. Pitchumoni, an outstanding scholar in the field of pancreatic diseases working at St. Peter's University Hospital in New Jersey, agreed to write and update the text for the sections on the pancreas and the gallbladder. Dr. Kris Kowdley of the University of Washington, who has now assumed the leadership in liver disease at the Virginia Mason Medical Center and is recognized as a world authority in liver diseases, agreed to write and update the section on the liver. Dr. Neil Floch, who spent endless hours as a surgeon mastering the techniques of minimally invasive surgery and the Nissan fundoplication procedure and who continues as Head of Minimally Invasive Surgery at Norwalk Hospital in Connecticut, agreed to write and update the section on the esophagus. Dr. Raul Rosenthal, who is so active as a mentor and surgical practitioner at the Cleveland Clinic, in collaboration with his fellows in training, accepted the task of writing the sections on the abdomen. To complete the book, Dr. James Scolapio of the Mayo Clinic in Jacksonville again wrote the section on nutrition for this edition. I am grateful to all these colleagues for their hard work and dedication. The staff of Elsevier, Linda Belfus, Elyse O'Grady, Marybeth Thiel, and Sarah Wunderly, has been of immeasurable help in arranging for the book, editing the text and combining the text and illustrations into the finished form. We are forever thankful to them. Without them, this book and now its second edition would not have been published.

I hope our readers will enjoy this second edition as they did the first and find it helpful as part of their education and an ongoing reference.

Martin H. Floch, MD

About the Artists

Frank H. Netter, MD

Frank H. Netter was born in 1906 in New York City. He studied art at the Art Student's League and the National Academy of Design before entering medical school at New York University, where he received his MD degree in 1931. During his student years, Dr. Netter's notebook sketches attracted the attention of the medical faculty and other physicians, allowing him to augment his income by illustrating articles and textbooks. He continued illustrating as a sideline after establishing a surgical practice in 1933, but he ultimately opted to give up his practice in favor of a full-time commitment to art. After service in the United States Army during World War II, Dr. Netter began his long collaboration with the CIBA Pharmaceutical Company (now Novartis Pharmaceuticals). This 45-year partnership resulted in the production of the extraordinary collection of medical art so familiar to physicians and other medical professionals worldwide.

In 2005, Elsevier, Inc. purchased the Netter Collection and all publications from Icon Learning Systems. There are now over 50 publications featuring the art of Dr. Netter available through Elsevier, Inc. (in the US: www.us.elsevierhealth.com/Netter, and outside the US: www.elsevierhealth.com)

Dr. Netter's works are among the finest examples of the use of illustration in the teaching of medical concepts. The 13-book *Netter Collection of Medical Illustrations*, which includes the greater part of the more than 20,000 paintings created by Dr. Netter, became and remains one of the most famous medical works ever published. *The Netter Atlas of Human Anatomy*, first published in 1989, presents the anatomical paintings from the Netter Collection. Now translated into 16 languages, it is the anatomy atlas of choice among medical and health professions students the world over.

The Netter illustrations are appreciated not only for their aesthetic qualities, but, more important, for their intellectual content. As Dr. Netter wrote in 1949, ". . . clarification of a subject is the aim and goal of illustration. No matter how beautifully painted, how delicately and subtly rendered a subject may be, it is of little value as a *medical illustration* if it does not serve to make clear some medical point." Dr. Netter's planning, conception, point of view, and approach are what inform his paintings and what makes them so intellectually valuable.

Frank H. Netter, MD, physician and artist, died in 1991.

Learn more about the physician-artist whose work has inspired the Netter Reference collection: http://www.netterimages.com/artist/netter.htm

Carlos A. G. Machado, MD

Carlos Machado was chosen by Novartis to be Dr. Netter's successor. He continues to be the main artist who contributes to the Netter collection of medical illustrations.

Self-taught in medical illustration, cardiologist Carlos Machado has contributed meticulous updates to some of Dr. Netter's original plates and has created many paintings of his own in the style of Netter as an extension of the Netter collection. Dr. Machado's photorealistic expertise and his keen insight into the physician-patient relationship informs his vivid and unforgettable visual style. His dedication to researching each topic and subject he paints places him among the premier medical illustrators at work today.

Learn more about his background and see more of his art at: http://www.netterimages.com/artist/machado.htm

Contents

SECTION IV

Editor: Martin H. Floch

SMALL INTESTINE

SECTION V

Editor: Martin H. Floch

COLON, RECTUM, AND ANUS

SECTION VI

Editor: Martin H. Floch

INFECTIOUS AND PARASITIC DISEASES OF THE ALIMENTARY TRACT

SECTION VII

Editor: C. S. Pitchumoni

PANCREAS

SECTION VIII

Editor: C. S. Pitchumoni

GALLBLADDER AND BILE DUCTS

SECTION IX

Editor: Kris V. Kowdley

LIVER

SECTION X

Editor: James S. Scolapio

NUTRITION AND GASTROINTESTINAL DISEASE

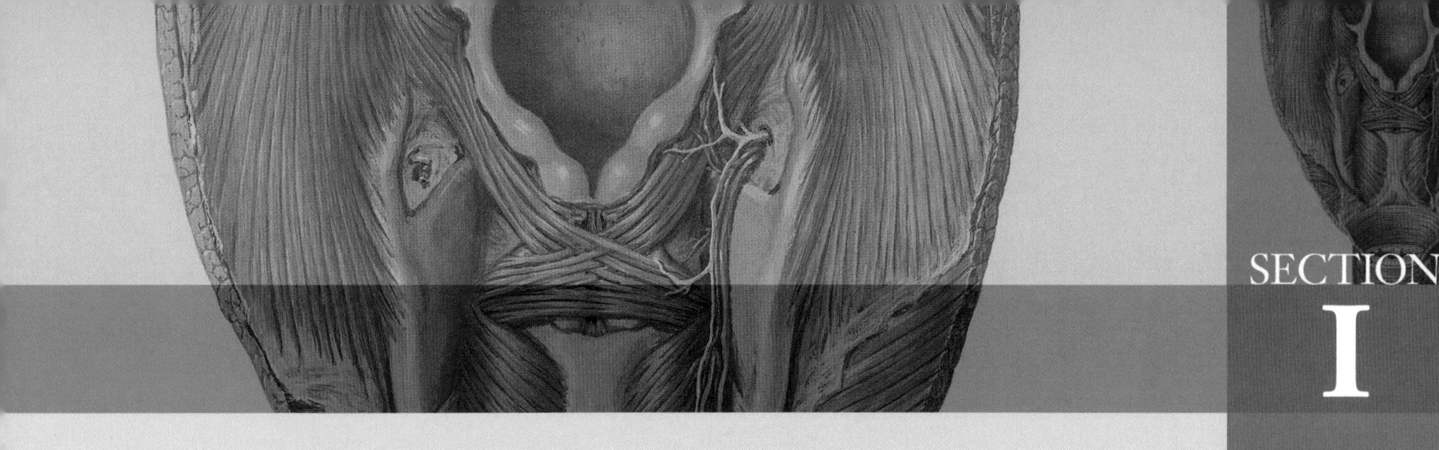

SECTION I

Esophagus

Topographic Relations of the Esophagus

1

Neil R. Floch

There is a smooth transition from the end of the pharynx, at the level of the cricoid cartilage and the sixth cervical vertebra (C6), to the esophagus (**Figs. 1-1** and **1-2**). On average, the esophagus is 40 cm (16 inches) long from the upper incisor teeth to the cardia of the stomach, but it may be as long as 43 cm in tall persons or in those with long trunks. The esophagus is divided, with the first part extending 16 cm from the incisors to the lower border of the cricopharyngeus muscle and the rest extending 24 cm.

The aortic arch crosses the esophagus from the left side and is located 23 cm from the incisors and 7 cm below the cricopharyngeus muscle; 2 cm below this level, the left main bronchus crosses in front of the esophagus. The lower esophageal sphincter (LES) begins 37 to 38 cm from the incisors. The esophageal hiatus is located 1 cm below this point, and the cardia of the stomach is yet lower. In children the dimensions are proportionately smaller. At birth the distance from the incisor teeth to the cardia is approximately 18 cm; at 3 years, 22 cm; and at 10 years, 27 cm.

Like a "good soldier," the esophagus follows a left-right-left path as it marches down the anteroposterior curvature of the vertebral column. It descends anterior to the vertebral column, through the lower portion of the neck and the superior and posterior mediastinum. The esophagus forms two lateral curves that, when viewed anteriorly, appear as a reverse S: the upper esophagus has a convex curve toward the left, and the lower esophagus has a convex curve toward the right. At its origin, the esophagus bends ¼ inch (0.6 cm) to the left of the tracheal margin. It crosses the midline behind the aortic arch at the level of the fourth thoracic vertebra (T4). The esophagus then turns to the right at the seventh thoracic vertebra (T7), after which it turns sharply to the left as it enters the abdomen through the esophageal hiatus of the diaphragm, to join the cardia of the stomach at the gastroesophageal (GE) junction.

The esophagus is composed of three segments: cervical, thoracic, and abdominal. Anterior to the cervical esophagus is the membranous wall of the trachea. Loose areolar tissue and muscular strands connect the esophagus and the trachea, and recurrent laryngeal nerves ascend in the grooves between them. Posterior to the esophagus are the longus colli muscles, the prevertebral fascia, and the vertebral bodies. Although the *cervical esophagus* is positioned between the carotid sheaths, it is closer to the left carotid sheath. The thyroid gland partially overlaps the esophagus on both sides.

The *thoracic esophagus* lies posterior to the trachea. It extends down to the level of the fifth thoracic vertebra (T5), where the trachea bifurcates. The trachea curves to the right as it divides, and thus the left main bronchus crosses in front of the esophagus. Below this, the pericardium separates the esophagus from the left atrium of the heart, which lies anterior and inferior to the esophagus. The lowest portion of the thoracic esophagus passes through the diaphragm into the abdomen.

On the left side of the esophageal wall, in the upper thoracic region, is the ascending portion of the left subclavian artery and the parietal pleura. At approximately the level of T4, the arch of the aorta passes backward and alongside the esophagus. Below this, the descending aorta lies to the left, but when that vessel passes behind the esophagus, the left mediastinal pleura again comes to adjoin the esophageal wall. On the right side, the parietal pleura is intimately applied to the esophagus, except when, at the level of T4, the azygos vein intervenes as it turns forward.

In the upper thorax, the esophagus lies on the longus colli muscle, the prevertebral fascia, and the vertebral bodies. At the eighth thoracic vertebra (T8), the aorta lies behind the esophagus. The azygos vein ascends behind and to the right of the esophagus as far as the level of T4, where it turns forward. The hemiazygos vein and the five upper-right intercostal arteries cross from left to right behind the esophagus. The thoracic duct ascends to the right of the esophagus before turning behind it and to the left at the level of T5. The duct then continues to ascend on the left side of the esophagus.

A small segment of *abdominal esophagus* lies on the crus of the diaphragm and creates an impression in the underside of the liver. Below the tracheal bifurcation, the esophageal nerve plexus and the anterior and posterior vagal trunks adhere to the esophagus.

As the esophagus travels from the neck to the abdomen, it encounters several indentations and constrictions. The first narrowing occurs at the cricopharyngeus muscle and the cricoid cartilage. The aortic arch creates an indentation on the left side of the esophagus, and the pulsations of the aorta may be seen during esophagoscopy. Below this point, the left main bronchus creates an impression on the left anterior aspect of the esophagus. The second narrowing occurs at the LES.

Although the esophagus is described as a "tube," it is oval and has a flat axis anterior to posterior with a wider transverse axis. When the esophagus is at rest, its walls are approximated and its width is 2 cm, but it distends and contracts depending on its state of tonus.

ADDITIONAL RESOURCES

Cameron JL, editor: *Current surgical therapy*, ed 6, St Louis, 1998, Mosby, pp 1-74.

Gray H, Bannister LH, Berry MM, Williams PL, editors: *Gray's anatomy: the anatomical basis of medicine and surgery*, New York, 1995, Churchill Livingstone.

Peters JH, DeMeester TR: Esophagus and diaphragmatic hernia. In Schwartz SI, Shires TG, Spencer FC, editors: *Principles of surgery*, ed 7, New York, 1999, McGraw-Hill, pp 1081-1179.

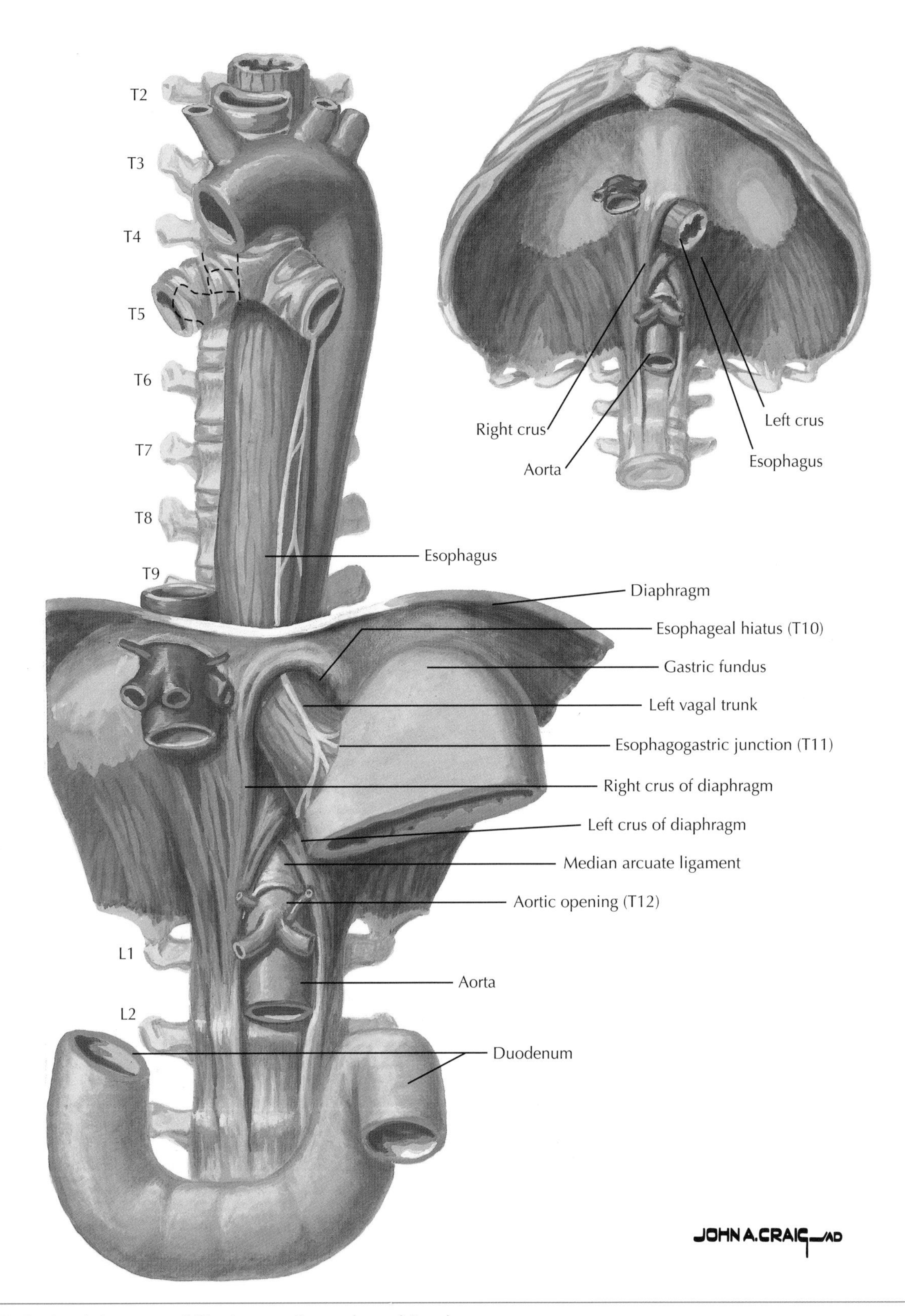

Figure 1-1 *Regional Anatomy of Diaphragm, Stomach, and Esophagus.*

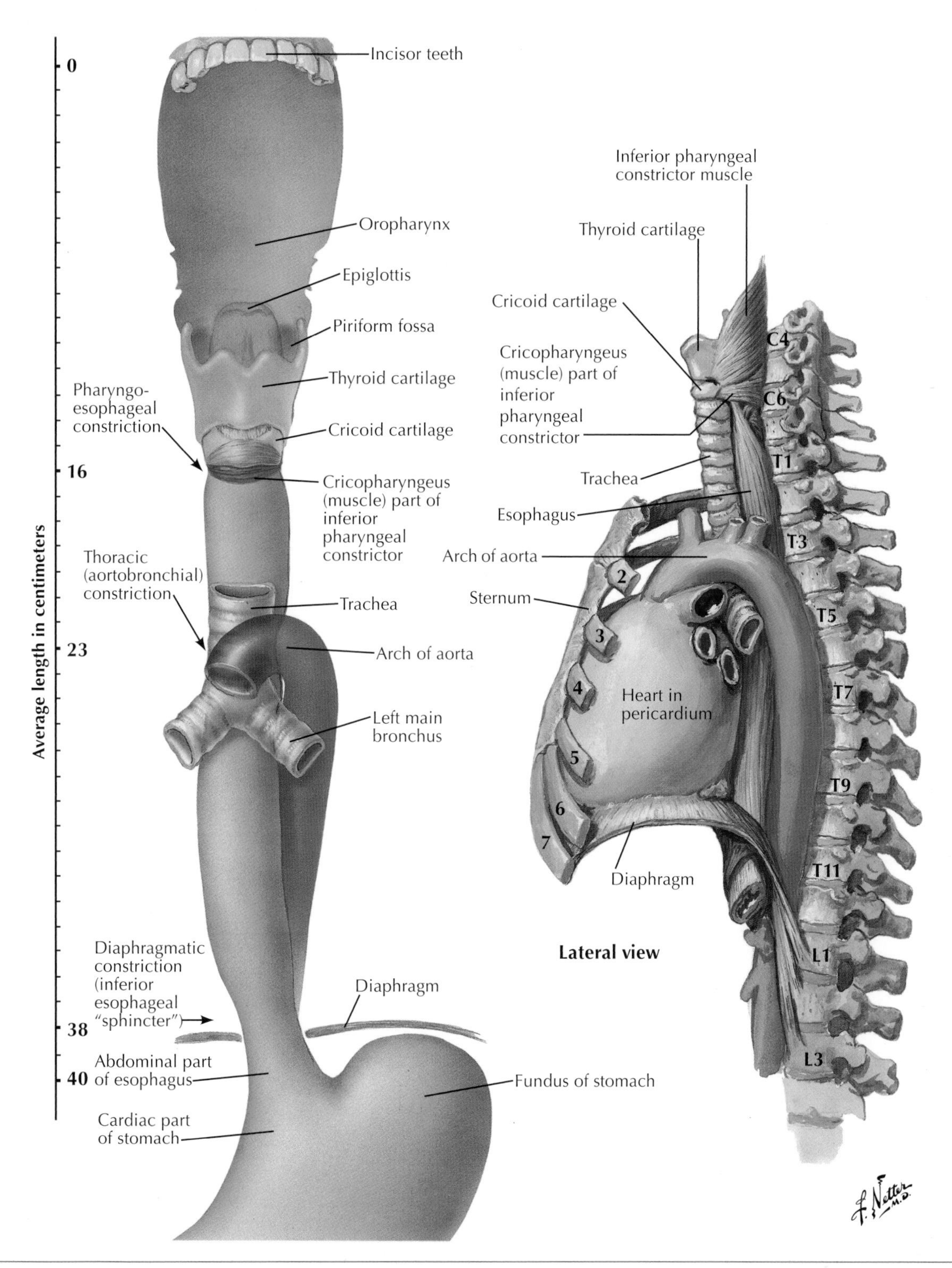

Figure 1-2 *Topography and Constrictions of Esophagus.*

Musculature of the Esophagus

2

Neil R. Floch

The esophagus is composed of outer longitudinal and inner circular muscle layers (**Figs. 2-1** and **2-2**). On the vertical ridge of the dorsal aspect of the cricoid cartilage, two tendons originate as they diverge and descend downward around the sides of the esophagus to the dorsal aspect. These tendons weave in the midline of the ventral area, creating a V-shaped gap between the two muscles known as the V-shaped area of Laimer. This gap, or bare area, exposes the underlying circular muscle. Located above this area is the cricopharyngeus muscle. Sparse longitudinal muscles cover the area, as do accessory fibers from the lower aspect of the cricopharyngeus muscle.

In the upper esophagus, longitudinal muscles form bundles of fibers that do not evenly distribute over the surface. The thinnest layers of muscle are anterior and adjacent to the posterior wall of the trachea. The longitudinal muscle of the esophagus receives fibers from an accessory muscle on each side that originates from the posterolateral aspect of the cricoid cartilage and the contralateral side of the deep portion of the cricopharyngeus muscle. As the longitudinal muscle descends, its fibers become equally distributed and completely cover the surface of the esophagus.

The inner, circular, muscle layer is thinner than the outer longitudinal layer. This relationship is reversed in all other parts of the gastrointestinal (GI) tract. In the upper esophagus, the circular muscle closely approximates the encircling lower fibers of the cricopharyngeus muscle. The upper esophageal fibers are not circular but elliptical, with the anterior part of the ellipse at a lower level of the posterior part. The ellipses become more circular as the esophagus descends, until the start of its middle third, where the fibers run in a horizontal plane. In one 1-cm segment, the fibers are truly circular. Below this point, the fibers become elliptical once again, but they now have a reverse inclination; that is, the posterior part of the ellipse is located at a lower level than the anterior part. In the lower third of the esophagus, the fibers follow a spiral course down the esophagus. The elliptical, circular, and spiral fibers of this layer are not truly uniform and parallel but may overlap and cross, or they may even have clefts between them. Some fibers in the lower two thirds of the esophagus pass diagonally or perpendicularly, up or down, joining fibers at other levels. These branched fibers are 2 to 3 mm wide and 1 to 5 cm long and are not continuous.

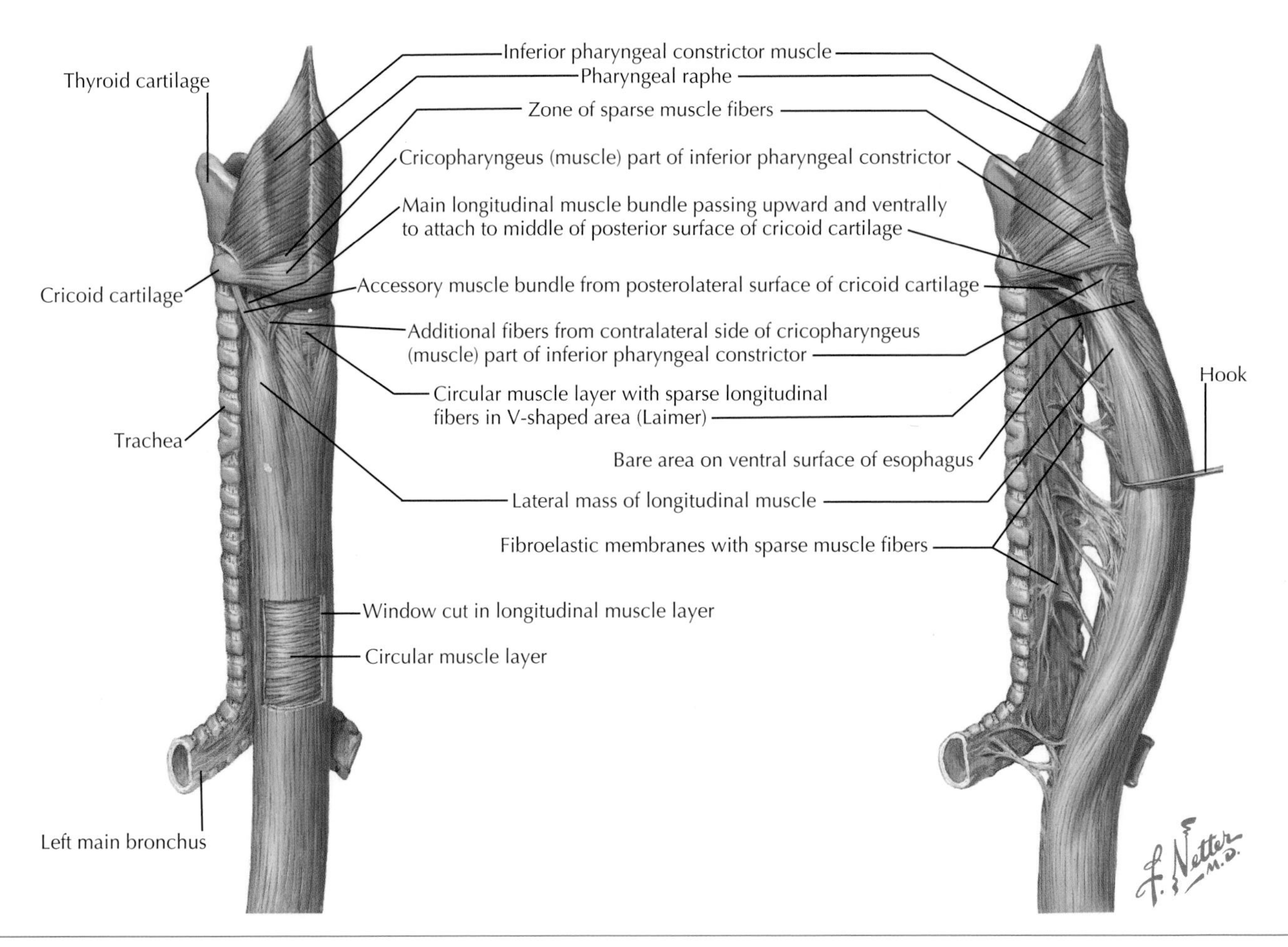

Figure 2-1 *Musculature of the Esophagus.*

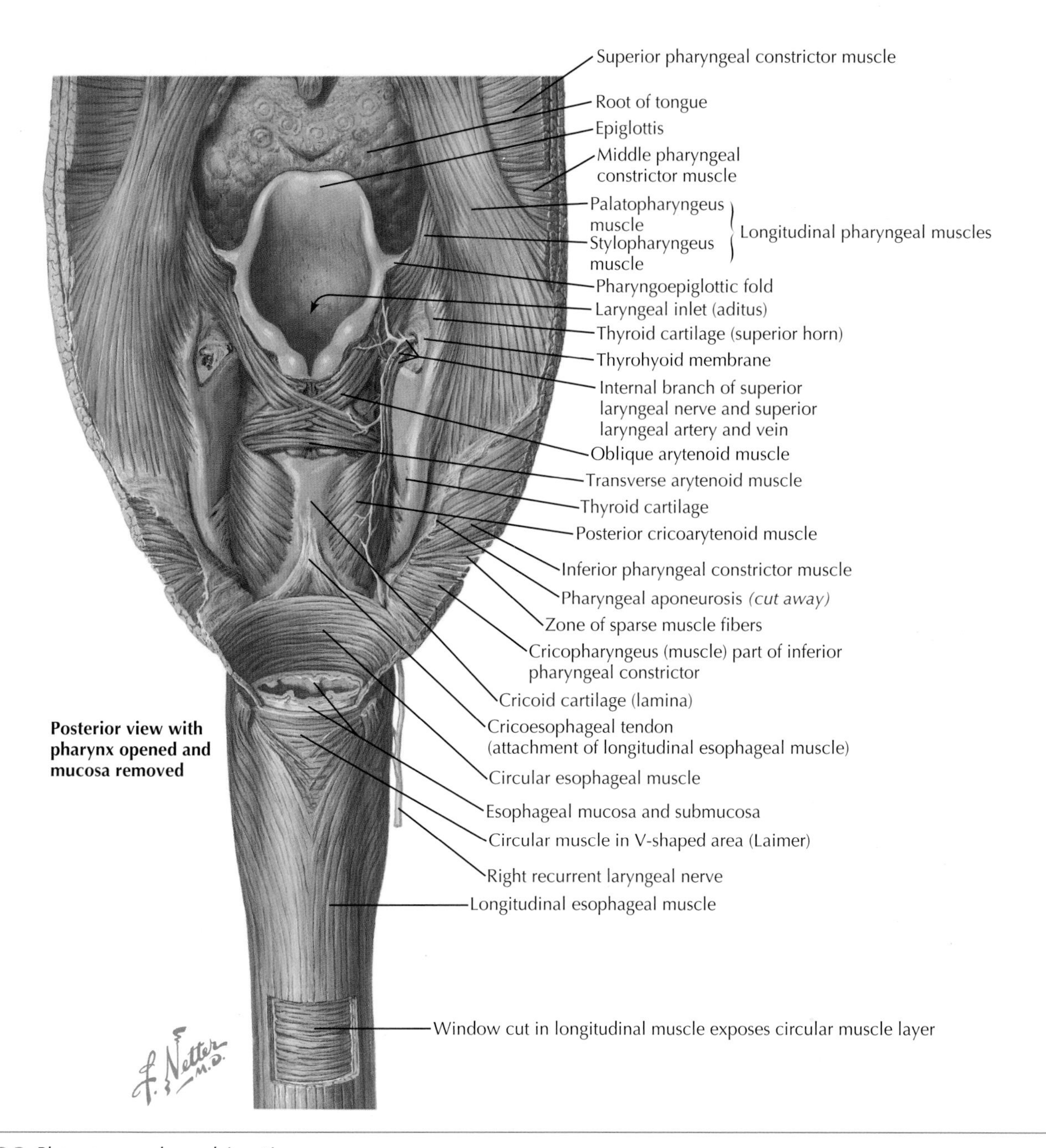

Figure 2-2 *Pharyngoesophageal Junction.*

The cricopharyngeus muscle marks the transition from pharynx to esophagus. It is the lowest portion of the inferior constrictor of the pharynx and consists of a narrow band of muscle fibers that originate on each side of the posterolateral margin of the cricoid cartilage. The cricopharyngeus then passes slinglike around the dorsal aspect of the pharyngoesophageal (PE) junction. Upper fibers ascend and join the median raphe of the inferior constrictor muscle posteriorly. Lower fibers do not have a median raphe; they pass to the dorsal aspect of the PE junction. A few of these fibers pass down to the esophagus. The cricopharyngeus functions as a sphincter of the upper esophagus. Muscle tone of the esophageal lumen is greatest at the level of the cricopharyngeus, and relaxation of this muscle is an integral part of the act of swallowing. There is a weak area between the cricopharyngeus and the main part of the inferior constrictor where Zenker diverticula are thought to develop.

The upper 25% to 33% of the esophagus is composed of striated muscle, whereas the lower or remaining portion is smooth muscle. Within the second fourth of the esophagus is a transitional zone where striated muscle and smooth muscle are present. The lower half contains purely smooth muscle. Between the two muscular coats of the esophagus, a narrow layer of connective tissue is inserted that accommodates the myenteric plexus of Auerbach.

ADDITIONAL RESOURCES

Gray H, Bannister LH, Berry MM, Williams PL, editors: *Gray's anatomy: the anatomical basis of medicine and surgery*, New York, 1995, Churchill Livingstone.

Peters JH, DeMeester TR: Esophagus and diaphragmatic hernia. In Schwartz SI, Shires TG, Spencer FC, editors: *Principles of surgery*, ed 7, New York, 1999, McGraw-Hill, pp 1081-1179.

Arterial Blood Supply of the Esophagus

3

Neil R. Floch

The blood supply of the esophagus is variable (**Fig. 3-1**). The inferior thyroid artery is the primary supplier of the cervical esophagus; esophageal vessels emanate from both side branches of the artery and from the ends of the vessels. Anterior cervical esophageal arteries supply small branches to the esophagus and trachea. Accessory arteries to the cervical esophagus originate in the subclavian, common carotid, vertebral, ascending pharyngeal, superficial cervical, and costocervical trunk.

Arterial branches from the bronchial arteries, the aorta, and the right intercostal vessels supply the thoracic esophagus. Bronchial arteries, especially the left inferior artery, distribute branches at or below the tracheal bifurcation. Bronchial artery branches are variable. The standard—two left and one right—occurs in only about 50% of patients. Aberrant vessel patterns include one left and one right in 25% of patients, two right and two left in 15%, and one left and two right in 8%. Rarely do three right or three left arteries occur.

At the tracheal bifurcation, the esophagus receives branches from the aorta, aortic arch, uppermost intercostal arteries, internal mammary artery, and carotid artery. Aortic branches to the thoracic esophagus usually consist of two unpaired vessels. The cranial vessel is 3 to 4 cm long and usually arises at the level of the sixth to seventh thoracic vertebrae (T6-T7). The caudal vessel is longer, 6 to 7 cm, and arises at the level of T7 to T8. Both arteries pass behind the esophagus and divide into ascending and descending branches. These branches anastomose along the esophageal border with descending branches from the inferior thyroid and bronchial arteries, as well as with ascending branches from the left gastric and left inferior phrenic arteries. Right intercostal arteries, mainly the fifth, give rise to esophageal branches in approximately 20% of the population.

The abdominal esophagus receives its blood supply from branches of the left gastric artery, the short gastric artery, and a recurrent branch of the left inferior phrenic artery. The left gastric artery supplies cardioesophageal branches either through a single vessel that subdivides or through two to five branches before they divide into anterior and posterior gastric branches. Other arterial sources to the abdominal esophagus are (1) branches from an aberrant left hepatic artery, derived from the left gastric, an accessory left gastric from the left hepatic, or a persistent primitive gastrohepatic arterial arc; (2) cardioesophageal branches from the splenic trunk, its superior polar, terminal divisions (short gastrics), and its occasional, large posterior gastric artery; and (3) a direct, slender, cardioesophageal branch from the aorta, celiac, or first part of the splenic artery.

With every resection surgery, areas of devascularization may be induced by (1) excessively low resection of the cervical segment, which always has a supply from the inferior thyroid; (2) excessive mobilization of the esophagus at the tracheal bifurcation and laceration of the bronchial artery; and (3) excessive sacrifice of the left gastric artery and the recurrent branch of the inferior phrenic artery to facilitate gastric mobilization. Anastomosis around the abdominal portion of the esophagus is usually copious, but sometimes it is limited.

ADDITIONAL RESOURCES

Gray H, Bannister LH, Berry MM, Williams PL, editors: *Gray's anatomy: the anatomical basis of medicine and surgery*, New York, 1995, Churchill Livingstone.

Peters JH, DeMeester TR: Esophagus and diaphragmatic hernia. In Schwartz SI, Shires TG, Spencer FC, editors: *Principles of surgery*, ed 7, New York, 1999, McGraw-Hill, pp 1081-1179.

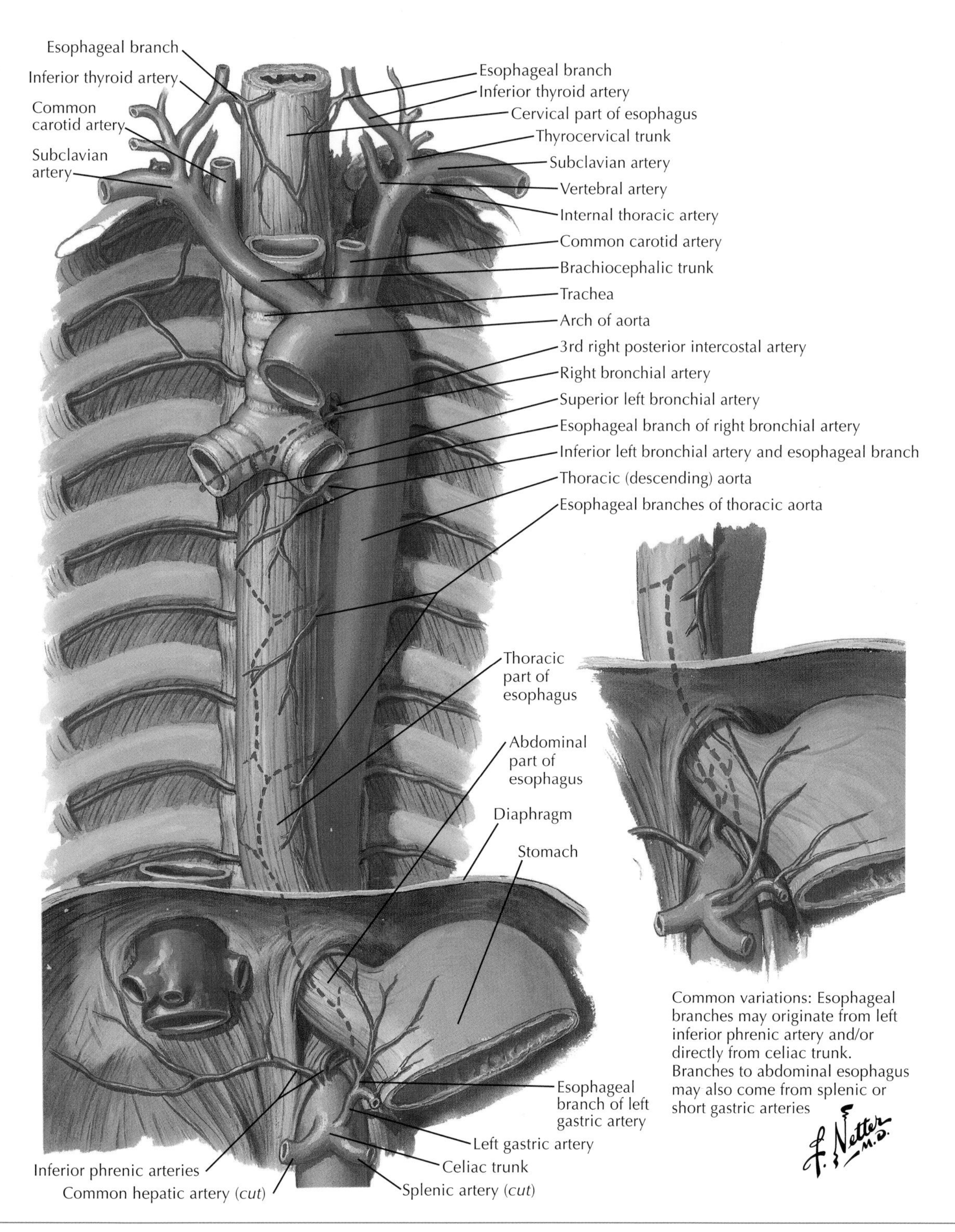

Figure 3-1 *Arteries of the Esophagus.*

Venous Drainage of the Esophagus

4

Neil R. Floch

Venous drainage of the esophagus begins in small tributaries that eventually empty into the azygos and hemiazygos veins (**Fig. 4-1**). Drainage begins in a submucosal venous plexus that exits externally to the surface of the esophagus. Tributaries from the cervical periesophageal venous plexus drain into the inferior thyroid vein, which empties into the right or left brachiocephalic (innominate) vein, or both. Tributaries from the thoracic periesophageal plexus on the right side join the azygos, the right brachiocephalic, and occasionally the vertebral vein; on the left side, they join the hemiazygos, the accessory hemiazygos, the left brachiocephalic, and occasionally the vertebral vein. Tributaries from the short abdominal esophagus drain into the left gastric (coronary) vein of the stomach. Other tributaries are in continuity with the short gastric, splenic, and left gastroepiploic veins. They may also drain to branches of the left inferior phrenic vein and join the inferior vena cava (IVC) directly or the suprarenal vein before it enters the renal vein.

The composition of the azygos system of veins varies. The *azygos vein* arises in the abdomen from the ascending right lumbar vein, which receives the first and second lumbar and the subcostal veins. The azygos may arise directly from the IVC or may have connections with the right common iliac or renal vein. In the thorax, the azygos vein receives the right posterior intercostal veins from the fourth to the eleventh spaces and terminates in the superior vena cava (SVC). The highest intercostal vein drains into the right brachiocephalic vein or into the vertebral vein. Veins from the second and third spaces unite in a common trunk, the right superior intercostal, which ends in the terminal arch of the azygos.

The *hemiazygos vein* arises as a continuation of the left ascending lumbar or from the left renal vein. The hemiazygos receives the left subcostal vein and the intercostal veins from the eighth to the eleventh spaces, and then it crosses the vertebral column posterior to the esophagus to join the azygos vein.

The accessory hemiazygos vein receives intercostal branches from the fourth to the eighth intercostal veins, and it crosses over the spine and under the esophagus to join the hemiazygos or the azygos vein. Superiorly, the accessory hemiazygos communicates with the left superior intercostal that drains the second and third spaces and ends in the left brachiocephalic vein. The first space drains into the left brachiocephalic or vertebral vein. Often the hemiazygos, the accessory hemiazygos, and the superior intercostal trunk form a continuous longitudinal channel with no connections to the azygos. There may be three to five connections between the left azygos, in which case a hemiazygos or an accessory hemiazygos is not formed. If the left azygos system is very small, the left venous drainage of the esophagus occurs through its respective intercostal veins. Connections between left and right azygos veins occur between the seventh and ninth intercostal spaces, usually at the eighth.

At the gastroesophageal (GE) junction, branches of the left gastric coronary vein are connected to lower esophageal branches so that blood may be shunted into the SVC from the azygos and hemiazygos veins. At the GE junction, blood may also be shunted into the splenic, retroperitoneal, and inferior phrenic veins to the caval system. Retrograde flow of venous blood through the esophageal veins leads to dilatation and formation of varicosities. Because the short gastric veins lead from the spleen to the GE junction of the stomach, thrombosis of the splenic vein may result in esophageal varices and fatal hemorrhage.

ADDITIONAL RESOURCES

Gray H, Bannister LH, Berry MM, Williams PL, editors: *Gray's anatomy: the anatomical basis of medicine and surgery*, New York, 1995, Churchill Livingstone.

Peters JH, DeMeester TR: Esophagus and diaphragmatic hernia. In Schwartz SI, Shires TG, Spencer FC, editors: *Principles of surgery*, ed 7, New York, 1999, McGraw-Hill, pp 1081-1179.

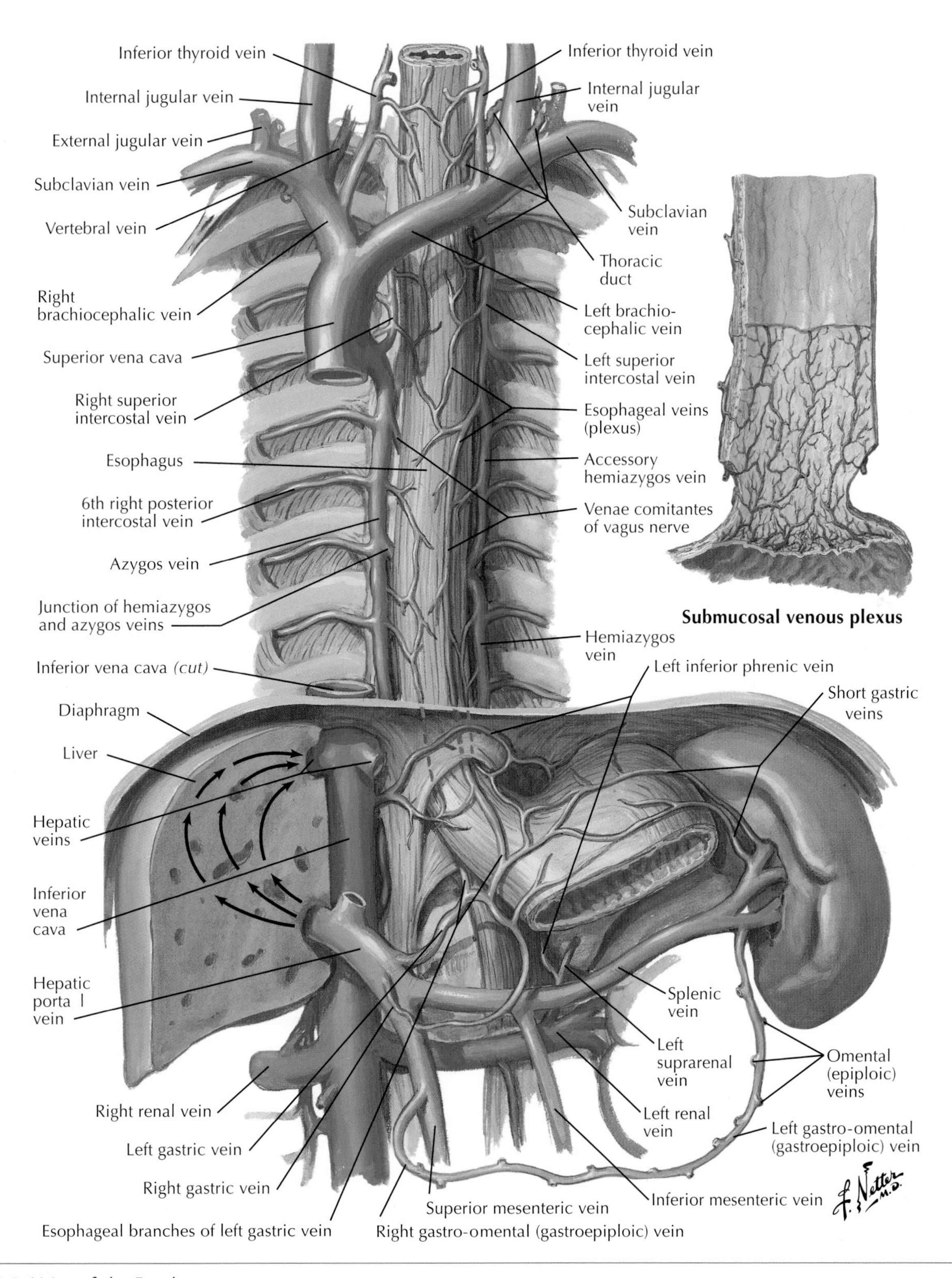

Figure 4-1 *Veins of the Esophagus.*

Innervation of the Esophagus: Parasympathetic and Sympathetic

5

Neil R. Floch

The esophagus is supplied by a combination of parasympathetic and sympathetic nerves (**Fig. 5-1**). Constant communication occurs between efferent and afferent fibers that transmit impulses to and from the vessels, glands, and mucosa of the esophagus.

Anterior and posterior vagus nerves carry parasympathetic efferent fibers to the esophagus, and afferent fibers carry them from the esophagus. These parasympathetic fibers terminate in the dorsal vagal nucleus, which contains visceral efferent and afferent cells. The striated muscle of the pharynx and upper esophagus is controlled by parasympathetic fibers that emanate from the nucleus ambiguus. Vagus nerves intermingle with nerve fibers from the paravertebral sympathetic trunks and their branches such that the nerves in and below the neck are a combination of parasympathetic and sympathetic.

In the neck, the esophagus receives fibers from the recurrent laryngeal nerves and variable fibers from the vagus nerves, lying posterior to and between the common carotid artery and the internal jugular vein in the carotid sheath. On the right side, the recurrent laryngeal nerve branches from the vagus nerve and descends, wrapping itself around the right subclavian artery before it ascends in the esophageal-tracheal groove. On the left side, the recurrent laryngeal nerve branches from the left vagus nerve, descends and wraps around the aortic arch, and ascends between the trachea and the esophagus.

In the superior mediastinum, the esophagus receives fibers from the left recurrent laryngeal nerve and both vagus nerves. As the vagus nerves descend, small branches intermingle with fibers from sympathetic trunks to form the smaller anterior and the larger posterior pulmonary plexuses. Below the mainstem bronchi, the vagus nerves divide into two to four branches that become closely adherent to the esophagus in the posterior mediastinum. Branches from the right and left nerves have anterior and posterior components that divide and then intermingle to form a mesh nerve plexus, which also contains small ganglia.

At a variable distance above the esophageal hiatus, the plexus reconstitutes into one or two vagal trunks. As the vagus enters the abdomen, it passes an anterior nerve, which is variably embedded in the esophageal wall, and a posterior nerve, which does not adhere to the esophagus but lies within a layer of adipose tissue. Small branches from the plexus and the main vagus enter the wall of the esophagus. Variations in the vagal nerves and plexuses are important for surgeons performing vagotomy because there may be more than one anterior or posterior vagus nerve.

Sympathetic preganglionic fibers emanate from axons of intermediolateral cornual cells, located in the fourth to sixth thoracic spinal cord segments (T4-T6). Anterior spinal nerve roots correspond to the segments containing their parent cells. They leave the spinal nerves in white or mixed rami communicans and enter the paravertebral sympathetic ganglia. Some fibers synapse with cells in the midthoracic ganglia and travel to higher and lower ganglia in the trunks. Axons of the ganglionic cells have postganglionic fibers that reach the esophagus. Afferent fibers travel the same route in reverse; however, they do not relay on the sympathetic trunks, and they enter the spinal cord through the posterior spinal nerve roots. Afferent nerve perikaryons are located in the posterior spinal nerve root ganglia.

The pharyngeal plexus innervates the upper esophagus. As the esophagus descends, it receives fibers from the cardiac branches of the superior cervical ganglia, but rarely receives them from the middle cervical or vertebral ganglia, of the sympathetic trunks. Fibers may also reach the esophagus from the nerve plexus that travels with the arterial supply.

In the upper thorax, the stellate ganglia supply esophageal filaments called ansae subclavia, and the thoracic cardiac nerves may be associated with fibers from the esophagus, trachea, aorta, and pulmonary structures.

In the lower thorax, fibers connect from the greater thoracic splanchnic nerves to the esophageal plexus. The greater splanchnic nerves arise from three to four large pathways, and a variable number of smaller rootlets arise from the fifth to tenth thoracic ganglia and the sympathetic trunks. The roots pass in multiple directions across the sides of the thoracic vertebral bodies and discs to form a large nerve. On both sides, the nerve enters the abdomen through the diaphragm by passing between the lateral margins of the crura and the medial arcuate ligament.

In the abdomen, the nerves branch into the celiac plexus. The lesser and least thoracic splanchnic nerves end primarily in the aortorenal ganglia and the renal plexuses, respectively. Filaments from the terminal part of the greater splanchnic nerve and from the right inferior phrenic plexus reach the abdominal portion of the esophagus.

ADDITIONAL RESOURCES

Gray H, Bannister LH, Berry MM, Williams PL, editors: *Gray's anatomy: the anatomical basis of medicine and surgery*, New York, 1995, Churchill Livingstone.

Peters JH, DeMeester TR: Esophagus and diaphragmatic hernia. In Schwartz SI, Shires TG, Spencer FC, editors: *Principles of surgery*, ed 7, New York, 1999, McGraw-Hill, pp 1081-1179.

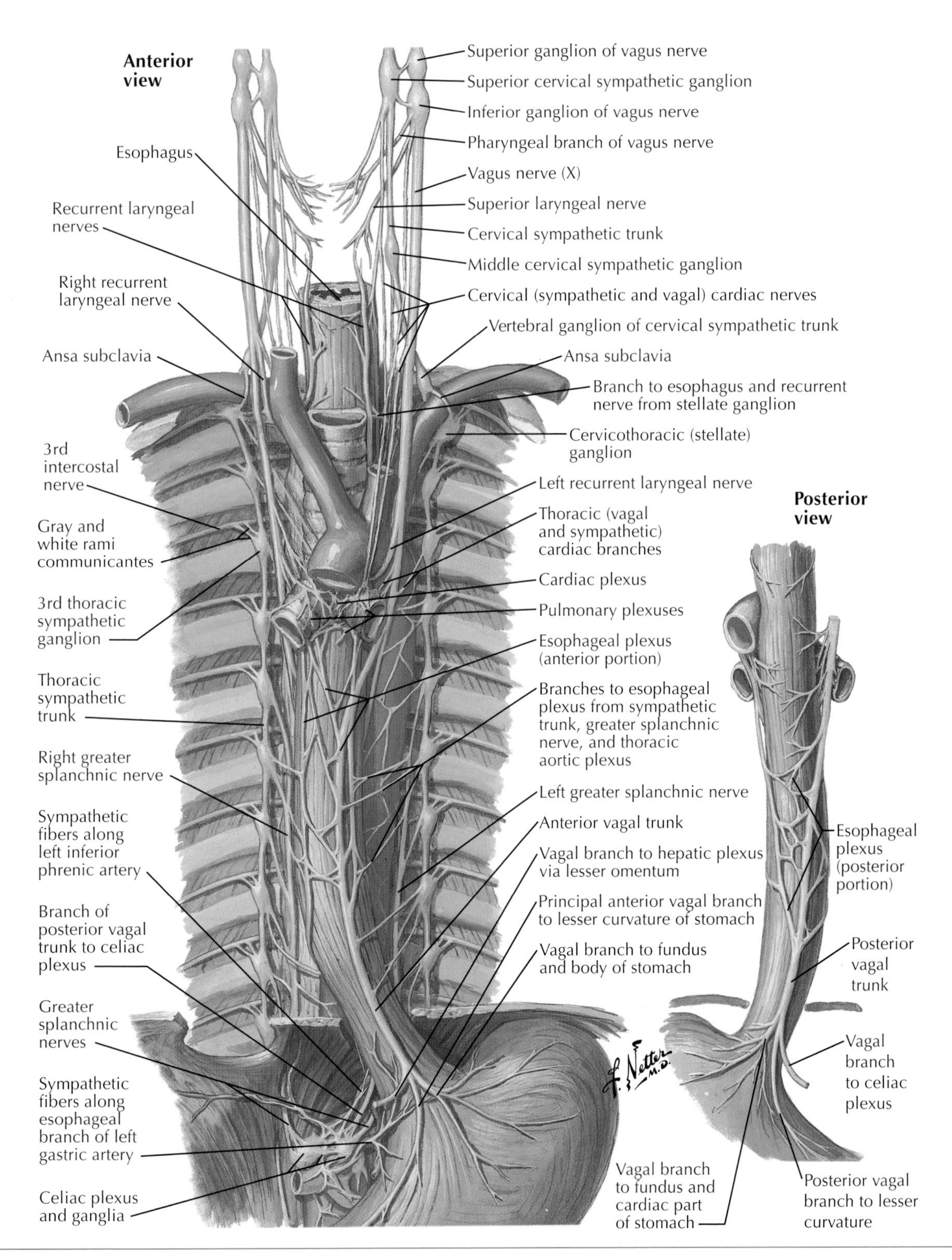

Figure 5-1 *Nerves of the Esophagus.*

Intrinsic Innervation of the Alimentary Tract

6

Neil R. Floch

Enteric plexuses that extend from the esophagus to the rectum control the gastrointestinal (GI) tract (**Fig. 6-1**). Numerous groups of ganglion cells interconnect in a network of fibers between the muscle layers. Synaptic relays are located in the myenteric plexus of Auerbach and the submucosal plexus of Meissner. The Meissner plexuses are coarse and consist of a mesh of thick, medium, and thin bundles of fiber, which represent the primary, secondary, and tertiary parts. The thin plexus is delicate.

Subsidiary plexuses appear in other areas covered by peritoneum. Enteric plexuses vary in pattern in different parts of the alimentary tract. They are less developed in the esophagus and are more developed from the stomach to the rectum. Ganglion cells also are not uniformly distributed; they are at their lowest levels in the Auerbach plexus and the esophagus, increase in the stomach, and reach their highest levels in the pylorus. Distribution is intermediate throughout the small intestine and increases along the colon and in the rectum. Cell population density in Meissner plexus parallels that in Auerbach plexus.

The vagus nerve contains preganglionic parasympathetic fibers that arise in its dorsal nucleus and travel to the esophagus, stomach, and intestinal branches. The proportion of efferent parasympathetic fibers is smaller than that of its sensory fibers. Vagal preganglionic efferent fibers have relays in small ganglia in the visceral walls; the axons are postganglionic parasympathetic fibers. Gastric branches have secretomotor and motor functions to the smooth muscle of the stomach, except for the pyloric sphincter, which is inhibited. Intestinal branches function similarly in the small intestine, cecum, appendix, and colon, where they are secretomotor to the glands and motor to the intestinal smooth muscle and where they inhibit the ileocecal sphincter.

Enteric plexuses contain postganglionic sympathetic and preganglionic and postganglionic parasympathetic fibers, afferent fibers, and intrinsic ganglion cells and their processes. Sympathetic preganglionic fibers have already relayed in paravertebral or prevertebral ganglia; thus the sympathetic fibers in the plexuses are postganglionic and pass through them and their terminations without synaptic interruptions. Afferent fibers from the esophagus, stomach, and duodenum are carried to the brainstem and cord through the vagal and sympathetic nerves, but they form no synaptic connections with the ganglion cells in the enteric plexuses.

Except for interstitial cells of Cajal, two chief forms of nerve cells, types 1 and 2, occur in the enteric plexuses. Interstitial cells of Cajal are pacemaker cells in the smooth muscles of the gut and are associated with the ground plexuses of all autonomic nerves. Type 1 cells are multipolar and confined to Auerbach plexus, and their dendrites branch close to the parent cells. Their axons run for varying distances through the plexuses to establish synapses with type 2 cells, which are more numerous and are found in Auerbach and Meissner plexuses. Most type 2 cells are multipolar, and their longer dendrites proceed in bundles for variable distances before they ramify in other cell clusters. Many other axons pass outwardly to end in the muscle, and others proceed inwardly to supply the muscularis mucosae and to ramify around vessels and between epithelial secretory cells; their distribution suggests that they are motor or secretomotor in nature.

Under experimental conditions, peristaltic movements occur in isolated portions of the gut, indicating the importance of intrinsic neuromuscular mechanisms, but the extrinsic nerves are probably essential for the coordinated regulation of all activities. Local reflex arcs, or axon reflexes, may exist in the enteric plexuses. In addition to types 1 and 2 multipolar cells, much smaller numbers of pseudounipolar and bipolar cells can be detected in the submucosa and may be the afferent links in local reflex arcs.

In megacolon (Hirschsprung disease), and possibly in achalasia, the enteric plexuses apparently are undeveloped or have degenerated over a segment of alimentary tract, although the extrinsic nerves are intact. Peristaltic movements are defective or absent in the affected segment, indicating the importance of the intrinsic neuromuscular mechanism.

ADDITIONAL RESOURCES

Gray H, Bannister LH, Berry MM, Williams PL, editors: *Gray's anatomy: the anatomical basis of medicine and surgery*, New York, 1995, Churchill Livingstone.

Peters JH, DeMeester TR: Esophagus and diaphragmatic hernia. In Schwartz SI, Shires TG, Spencer FC, editors: *Principles of surgery, ed 7*, New York, 1999, McGraw-Hill, pp 1081-1179.

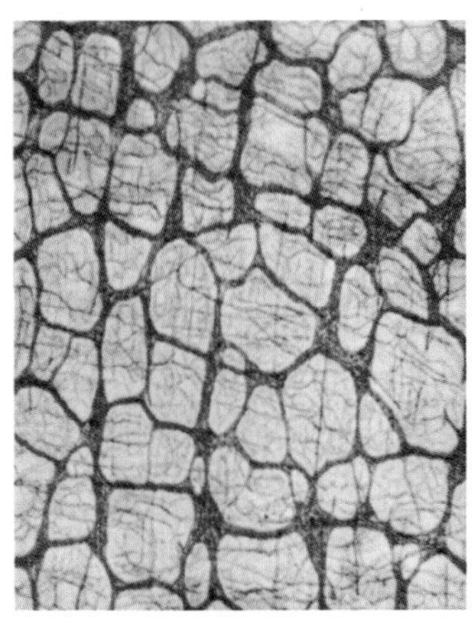

1. Myenteric plexus (Auerbach) lying on longitudinal muscle coat. Fine tertiary bundles crossing meshes (duodenum of guinea pig. Champy-Coujard, osmic stain, ×20)

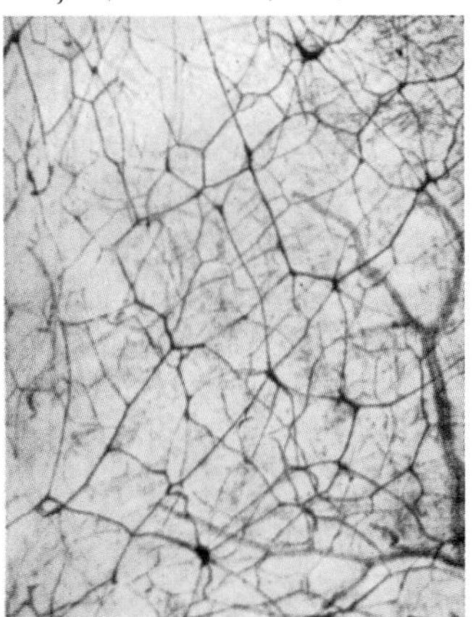

2. Submucous plexus (Meissner) (ascending colon of guinea pig. Stained by gold impregnation, x20)

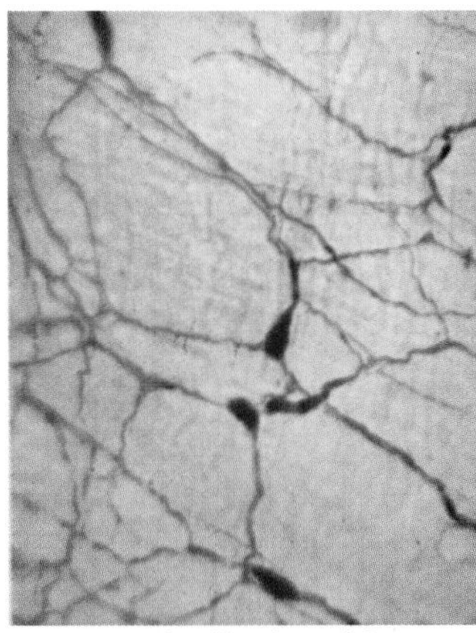

3. Interstitial cells of Cajal forming part of dense network between muscle layers (descending colon of guinea pig. Methylene blue, x375)

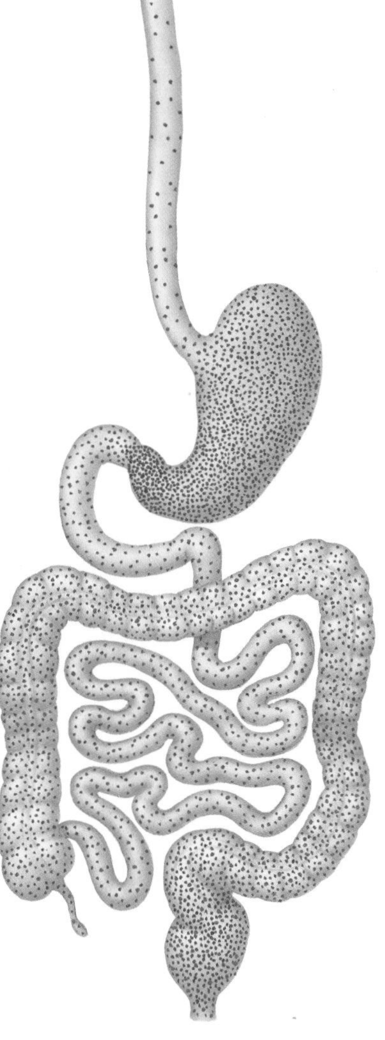

Relative concentration of ganglion cells in myenteric (Auerbach) plexus and in submucous (Meissner) plexus in various parts of alimentary tract (myenteric plexus cells represented by maroon, submucous by blue dots)

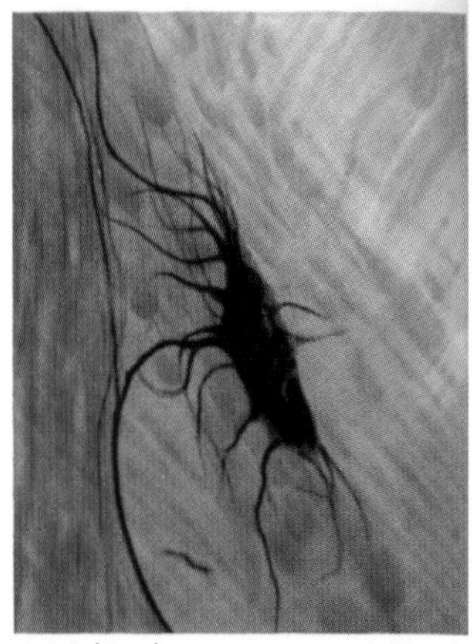

4. Multipolar neuron, type I (Dogiel), lying in ganglion of myenteric (Auerbach) plexus (ileum of monkey. Bielschowsky, silver stain, x375)

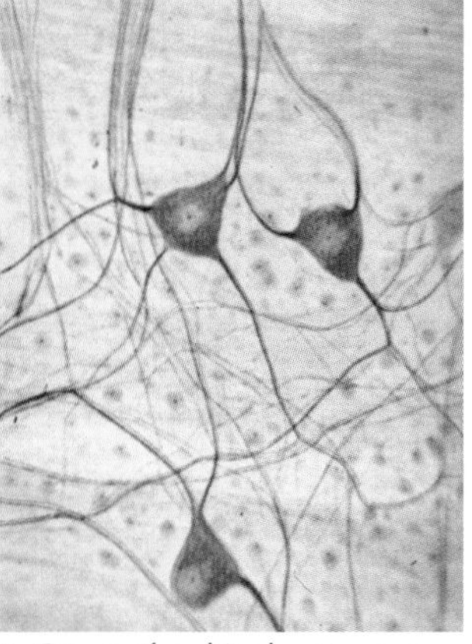

5. Group of multipolar neurons, type II, in ganglion of myenteric (Auerbach) plexus (ileum of cat. Bielschowsky, silver stain, x200)

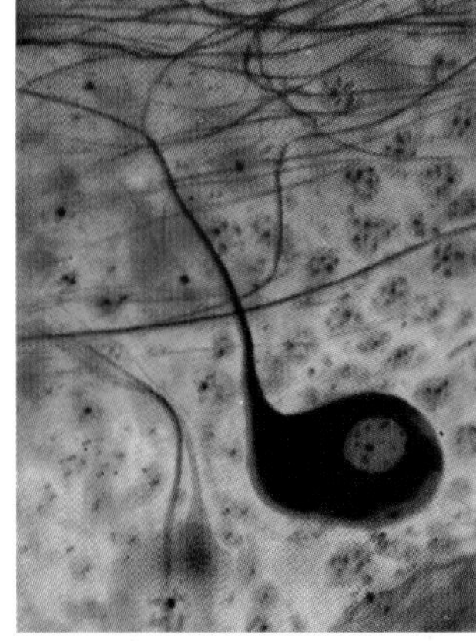

6. Pseudounipolar neuron within ganglion of myenteric plexus (ileum of cat. Bielschowsky, silver stain x375)

Figure 6-1 *Enteric Plexuses.*

Histology of the Esophagus

Neil R. Floch

7

Esophageal layers include the mucosa, submucosa, muscularis externa, and adventitia (**Fig. 7-1**). The esophageal mucosa ends abruptly at the gastroesophageal (GE) junction, where columnar epithelia with gastric pits and glands are found. The esophageal epithelium is 300 to 500 µm thick, nonkeratinized, stratified, and squamous and is continuous with the pharyngeal epithelium. Tall papillae rich in blood and nerve fibers assist in anchoring the tissue to its base. The epithelial layer is constantly renewed by mitosis as cuboidal basal cells migrate, flatten, and slough in 2 to 3 weeks.

The barrier wall of the esophagus functions well with the aid of mucus-producing glands that protect against mechanical invasion. However, this protection is limited. Repeated exposure of acid and protease-rich secretions from the stomach may occur during episodes of GE reflux and may cause fibrosis of the esophageal wall. Patients with nonerosive reflux disease (NERD) have evidence of increased cell permeability, which may contribute to their symptoms but does not exhibit visible damage. Exposure may also cause metaplastic epithelial cell changes consistent with Barrett esophagus. In the most serious cases, neoplastic changes may occur. A competent GE sphincter should prevent significant acid exposure.

With its lymphoid aggregates and mucous glands, especially near the GE junction, the lamina propria is supportive. Two types of glands reside in the esophagus. The cardiac glands are at the proximal and distal ends of the esophagus. Their ducts do not penetrate the muscularis mucosae, and their branched and coiled tubules are located in the lamina propria rather than in the submucosa. The other glands, the esophageal glands proper, produce mucus and are located throughout the esophagus.

The muscularis mucosae is composed primarily of sheets of longitudinal muscle that aid in esophageal peristalsis. It loosely adheres to both the mucosa and the muscularis as it invades the longitudinal ridges of the esophagus. Muscularis mucosae contain blood vessels, nerves, and mucous glands. The muscularis externa is approximately 300 µm thick and is composed of an outer longitudinal and an inner circular layer, as described previously.

ADDITIONAL RESOURCES

Gray H, Bannister LH, Berry MM, Williams PL, editors: *Gray's anatomy: the anatomical basis of medicine and surgery*, New York, 1995, Churchill Livingstone.

Peters JH, DeMeester TR: Esophagus and diaphragmatic hernia. In Schwartz SI, Shires TG, Spencer FC, editors: *Principles of surgery*, ed 7, New York, 1999, McGraw-Hill, pp 1081-1179.

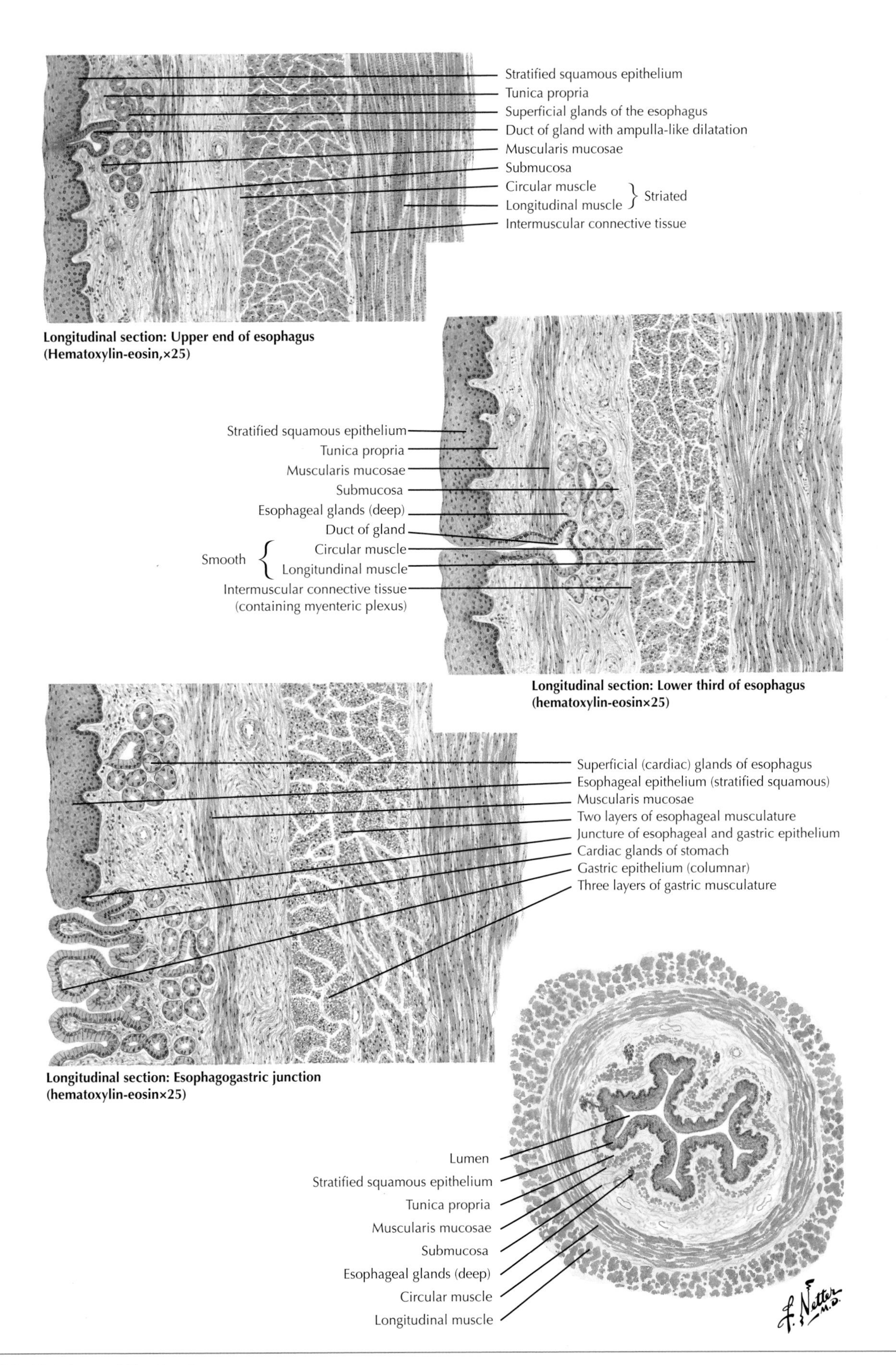

Figure 7-1 *Histology of the Esophagus.*

Gastroesophageal Junction and Diaphragm

8

Neil R. Floch

The sphincter mechanism of the gastroesophageal (GE) junction prevents retrograde flow of gastric contents into the lower esophagus while allowing deposition of a food bolus from the esophagus to the stomach (**Figs. 8-1** and **8-2**). The lower esophageal sphincter (LES) mechanism is a combination of functional contractions of the diaphragm, thickening of the circular and longitudinal muscles of the esophagus, an intraabdominal-esophageal component, gastric sling muscles, and the angle created by the entry of the esophagus into the abdomen through the diaphragm. Proper functioning of the LES mechanism depends on all its muscular components and the complex interaction of autonomic nerve inputs. Failure of this sphincterlike mechanism results in the symptoms of gastroesophageal reflux disease (GERD) with reflux and regurgitation of gastric contents. Physical damage, including esophagitis, ulcers, strictures, Barrett esophagus, and esophageal carcinoma, may develop.

At the GE junction, the Z line, indicating the transition from squamous to columnar gastric mucosa, is easily recognized by the color change from pale to deep red and texture change from smooth to rugose. The Z line is located between the end of the esophagus and the level of the hiatus and diaphragm. In some patients, the gastric mucosa may extend several centimeters proximally, into the esophagus.

Toward the distal esophagus, the circular and longitudinal muscles gradually thicken and reach their greatest width 1 to 2 cm above the hiatus. These characteristics define the location of the LES, which is capable of tonic contraction and neurologically coordinated relaxation. Manometry reveals a high-pressure zone in the distal 3 to 5 cm of the esophagus, with a pressure gradient between 12 and 20 mm Hg.

Pressure magnitude and sphincter length are important for maintaining the competency of the valve. The intraabdominal portion of the esophagus is important for the antireflux mechanism. The intrathoracic esophagus is exposed to −6 mm Hg of pressure during inspiration through 6 mm Hg of pressure within the abdomen, for a pressure difference of 12 mm Hg. *Sliding hiatal hernia* is defined as the lower esophagus migrating into the chest, where the pressure is −6 mm Hg. In this situation, negative pressure resists the LES remaining tonically closed.

The longitudinal muscle of the esophagus continues into the stomach to form the outer longitudinal muscle of the stomach. The inner circular or spiral layer of the esophagus divides at the cardia to become the inner oblique layer and the middle circular layer. Inner oblique fibers create a sling across the cardiac incisura, and the middle circular fibers pass horizontally around the stomach. These two muscle layers cross at an angle and form a muscular ring known as the collar of Helvetius and thought to be a component of the complex LES.

Muscle fibers of the hiatus usually arise from the larger right crus of the diaphragm, not from the left crus. Fibers that originate from the right crus ascend and pass to the right of the esophagus as another band, originating deeper than the right crus, ascending and passing to the left of the esophagus. The bands cross scissorslike and insert ventrally to the esophagus, into the central tendon of the diaphragm. Fibers that pass to the right of the esophagus are innervated by the right phrenic nerve, whereas right crural fibers, which pass to the left of the esophageal hiatus, are innervated by a branch of the left phrenic nerve.

In some patients, an anatomic variation may be found by which fibers from the left crus of the diaphragm surround the right side of the esophageal hiatus. Rarely, the muscle to the right of the esophageal hiatus originates entirely from the left crus, and fibers surrounding the left of the hiatus originate from the right crus. The ligament of Treitz originates from the fibers of the right crus of the diaphragm.

The diaphragm independently contributes to sphincter function. As the crura contract, they compress the esophagus. This action is most exaggerated during deep inspiration, when the diaphragm is in strong contraction and the passage of food into the stomach is impeded. The LES mechanism is exaggerated by the angulation of the esophagus as it connects to the stomach at the angle of His. How much this angulation contributes is not clearly defined.

Phrenicoesophageal and diaphragmatic esophageal ligaments connect the multiple components of the sphincter as the esophagus passes through the hiatus. The phrenicoesophageal ligament arises from the inferior fascia of the diaphragm, which is continuous with the transversalis fascia. At the margin of the hiatus, the phrenicoesophageal ligament divides into an ascending leaf and a descending leaf. The ascending leaf passes through the hiatus, climbs 1 to 2 cm, and surrounds the mediastinal esophagus circumferentially. The descending leaf inserts around the cardia deep to the peritoneum. Within the intraabdominal cavity formed by the phrenicoesophageal ligament is a ring of dense fat. The phrenicoesophageal ligament fixates the esophagus while allowing for respiratory excursion, deglutition, and postural changes. Its role in the closure of the sphincteric mechanism is unclear.

Resting LES pressure is maintained by a complex interaction of hormonal, muscular, and neuronal mechanisms. The muscular sphincter component functions with coordinated relaxation and contraction of the LES and the diaphragm. Its action may be observed during deglutition as it relaxes and tonically closes to prevent the symptoms of reflux and regurgitation. As the muscle groups contract externally, the mucosa gathers internally into irregular longitudinal folds.

When a swallowed bolus of food reaches the LES, it pauses before the sphincter relaxes and enters the stomach. The mechanism depends on the specialized zone of esophageal circular smooth muscle and possibly the gastric sling. At resting state, the LES is under tonic contraction. During swallowing, these muscles relax, the sphincter opens, and the food bolus empties

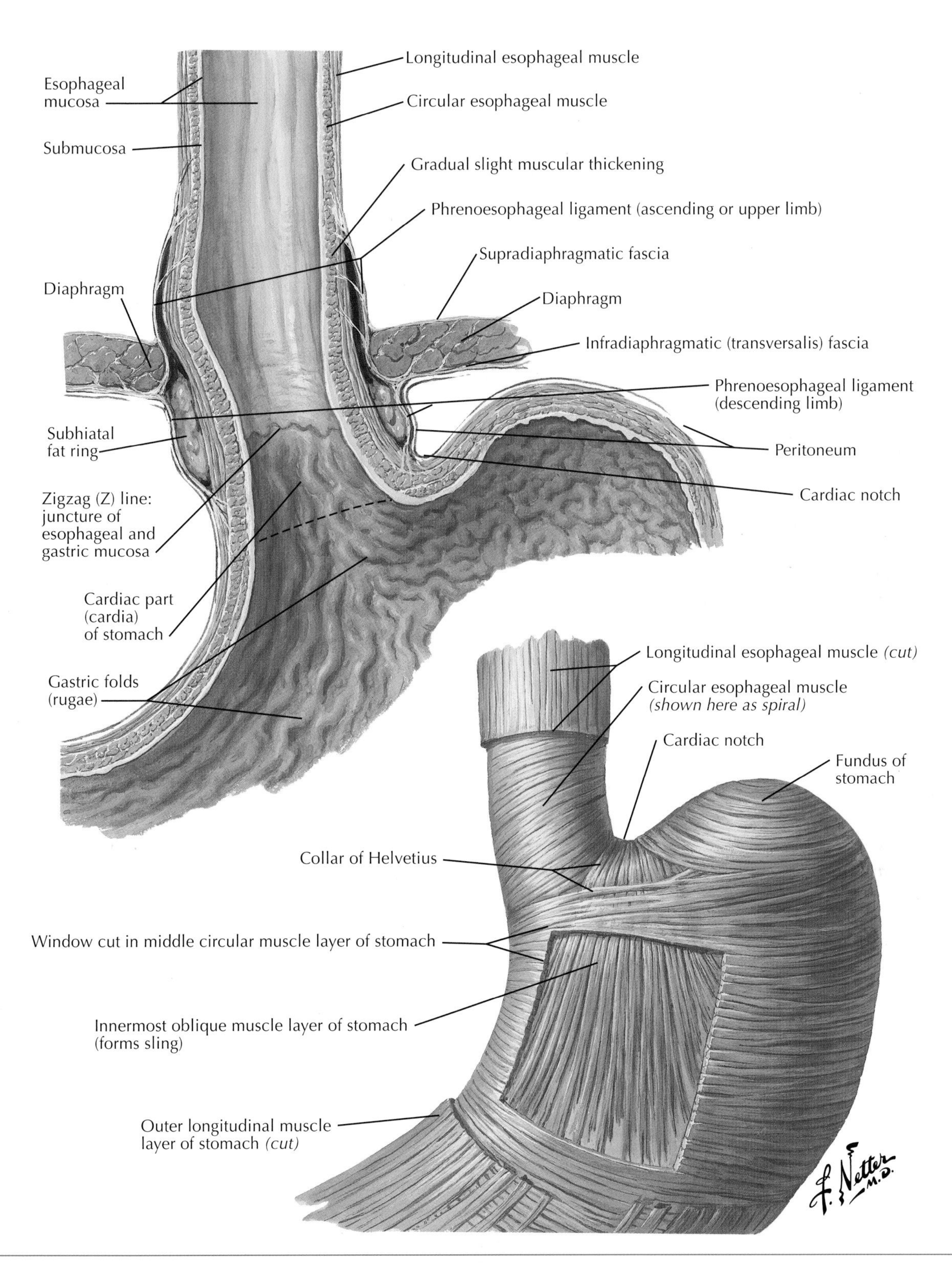

Figure 8-1 *Gastroesophageal Junction.*

into the stomach. Conversely, during vomiting, the LES relaxes to emit fluid into the esophagus.

The diaphragm contributes an external, sphincterlike function through the right crus of the diaphragm, which is attached by the phrenicoesophageal ligament. Manometry and electromyographic studies reveal that fibers of the crura contract around the esophagus during inspiration and episodes of increased intraabdominal pressure. In patients with hiatal hernia, the diaphragmatic component is no longer functional.

The muscular component is only partially responsible for the resting LES pressure. Parasympathetic, sympathetic, inhibitory, and excitatory autonomic nerves innervate the intramural plexus of the LES. Resting pressure decreases after administration of atropine, supporting the presence of a cholinergic neural com-

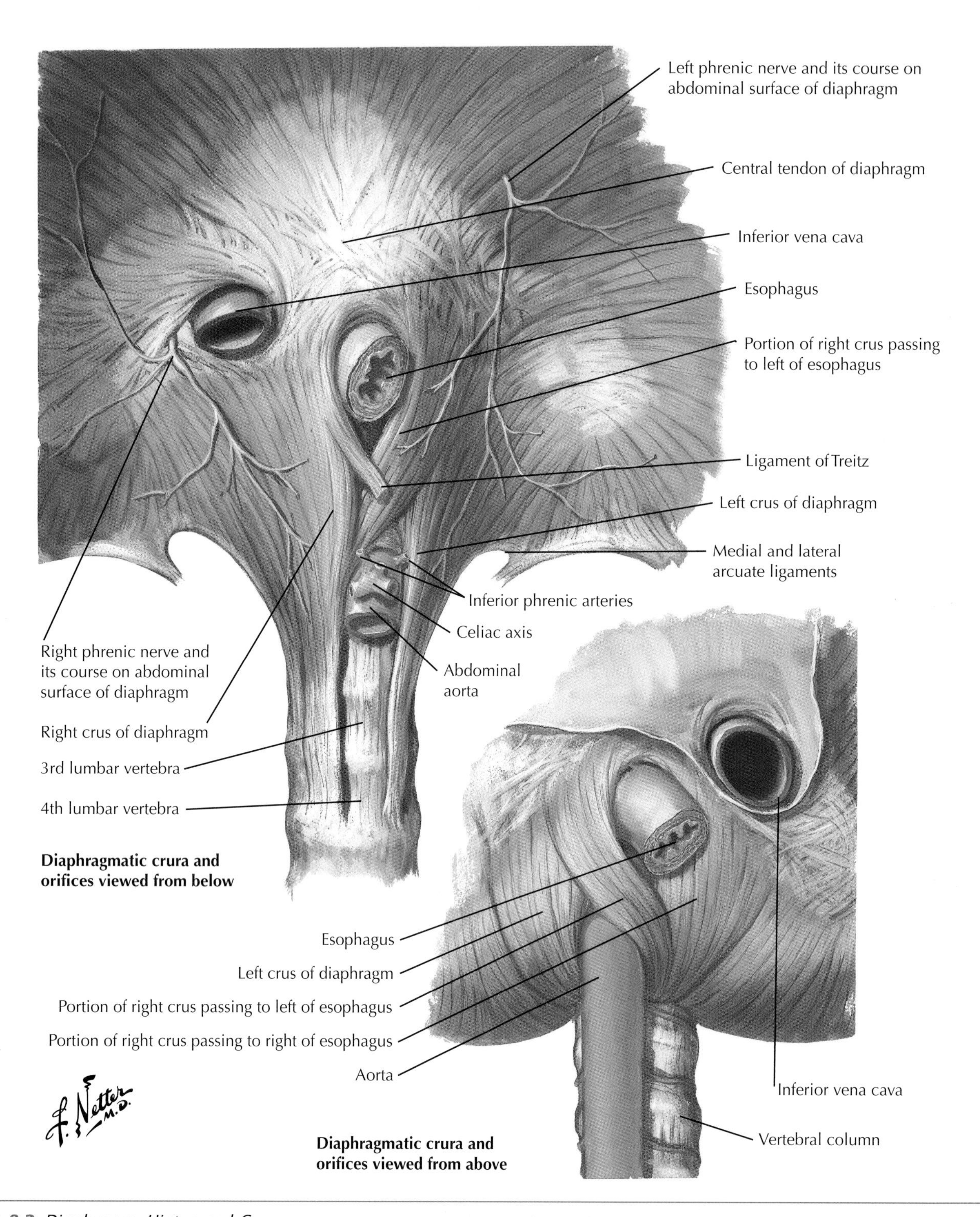

Figure 8-2 *Diaphragm: Hiatus and Crura.*

ponent. Cell bodies of the inhibitory nerves are located in the esophageal plexus, and the vagus nerves supply the preganglionic fibers. These nerves mediate sphincter relaxation in response to swallowing. Evidence suggests that nitric oxide controls relaxation through the enteric nervous system.

ADDITIONAL RESOURCES

Cameron JL, Peters JH, editors: Gastroesophageal reflux disease. In *Current surgical therapy,* ed 6, St Louis, 1998, Mosby, pp 33-46.

Gray H, Bannister LH, Berry MM, Williams PL, editors: *Gray's anatomy: the anatomical basis of medicine and surgery,* New York, 1995, Churchill Livingstone.

Peters JH, DeMeester TR: Esophagus and diaphragmatic hernia. In Schwartz SI, Shires TG, Spencer FC, editors: *Principles of surgery,* ed 7, New York, 1999, McGraw-Hill, pp 1081-1179.

Deglutition

Neil R. Floch

9

Swallowing, once initiated, becomes a reflex response (**Figs. 9-1** and **9-2**). Although a continuous process, deglutition is divided it into three stages—oral, pharyngeal, and esophageal—and may be observed by cineradiography and manometry. Deglutition requires the physiologic ability to (1) prepare a bolus of suitable size and consistency, (2) prevent dispersal of this bolus during the phases of swallowing, (3) create differential pressure that propels the bolus in a forward direction, (4) prevent food or liquid from entering the nasopharynx or larynx, (5) pass the bolus rapidly through the pharynx to limit the time respiration is suspended, (6) prevent gastric reflux into the esophagus during free communication between the esophagus and the stomach, and (7) clear residual material from the esophagus. Failure of these mechanisms leads to difficulty with swallowing and may lead to regurgitation of gastric contents into the esophagus and possibly into the pharynx.

The oral phase of deglutition follows mastication. The food bolus in the mouth breaks down into smaller pieces with the assistance of saliva. The tongue pushes the bolus posteriorly into the oropharynx as it simultaneously closes the nasopharynx with the help of the soft palate, fauces, and posterior wall of the oropharynx to prevent food from being pushed through the nose. Afterward, a peristaltic wave propels the bolus distally. Paralysis of the soft palate may occur in patients after a cerebrovascular accident (stroke) and cause regurgitation into the nasopharynx.

When the bolus enters the oropharynx, the hyoid bone elevates and moves anteriorly. Concomitantly, the larynx elevates, moves forward, and tilts posteriorly, pulling the bolus inward as the anteroposterior diameter of the laryngopharynx increases. This action causes the epiglottis to move under the tongue, tilt backward, and overlap the opening of the larynx to prevent aspiration of the food. Depression of the epiglottis may not completely close the larynx, and small particles of food may infringe on the opening. A liquid bolus may be split by the epiglottis and travel on each side of the larynx through the piriform recesses, rejoining behind the cricoid cartilage. The pharyngeal mechanism of swallowing occurs within 1.5 seconds.

At the same time, the upper esophageal sphincter (UES) closes as the tongue moves backward and the posterior pharyngeal constrictors contract. In the hypopharynx, pressure increases from 15 mm Hg to a closing pressure of 30 to 60 mm Hg. A pressure difference then develops between the hypopharynx and the midesophagus, creating a vacuum effect that, with the help of peristalsis, pulls the food from the hypopharynx into the esophagus during relaxation of the cricopharyngeus muscle. The 30-mm Hg closing pressure prevents reflux of food back into the pharynx. When the bolus reaches the distal esophagus, pressure in the UES returns to 15 mm Hg.

Passage of the food bolus beyond the cricopharyngeus muscle signifies the completion of the pharyngeal phase and the beginning of the esophageal phase. Hyoid bone, larynx, and epiglottis return to their original positions, and air reenters the trachea. The peristaltic wave begins in the oropharynx and continues into the esophagus, propelling the food in front of it. Sequential, coordinated contractions in middle and distal esophageal smooth muscles function to propel the food down to the lower esophageal sphincter (LES). In its travels, the bolus moves from an area with intrathoracic pressure of –6 mm Hg to an area with intraabdominal pressure of +6 mm Hg.

Peristaltic contractions may range from 30 to 120 mm Hg in a healthy person. The average wave peaks in 1 second, remains at that peak for 0.5 second, and subsides for 1.5 seconds. The total rise and fall of each wave proceeds for 3 to 5 seconds. A primary peristaltic contraction, initiated by swallowing, travels down the esophagus at a rate of 2 to 4 cm/sec, reaching the LES approximately 9 seconds after the initiation of swallowing. If swallowing is rapidly repeated, the esophagus remains relaxed; a wave develops only after the ending movement of the pharynx.

Efferent vagal nerves that arise in the medulla control esophageal peristalsis. When the esophagus is distended, a wave is initiated with the forceful closure of the UES and contracts down the esophagus. This phenomenon is a secondary contraction and occurs without movement of the mouth or pharynx. Secondary peristalsis is a dependent, local reflex that attempts to remove any food substance that remains in the esophagus after primary contraction is complete. The propulsive force of the esophagus is not very strong. Normal contractions of the esophageal muscles and relaxation of the inferior esophagus are necessary for efficient deglutition. So-called tertiary waves, which occur particularly in elderly persons and in patients with hiatal hernia, are nonperistaltic, repetitive, ringlike contractions at multiple levels in the distal half of the esophagus, usually during stages of incomplete distention. A patient with a large hiatal hernia lacks the ability for distal fixation and adequate food propulsion.

In the resting state, the LES divides the esophagus from the stomach and functions as a pressure barrier with a 12-mm Hg gradient. The LES is a thickening of muscle fibers that performs a sphincterlike action, although no distinct sphincter exists. Tonically, the LES remains closed, preventing gastroesophageal (GE) reflux. With the onset of swallowing, the peristaltic wave creates a transient peak behind the bolus and stops in the terminal esophagus. The LES then relaxes through a reflex mechanism. It does not relax completely until the pressure immediately proximal is great enough to overcome the LES pressure. The esophagus immediately proximal to the LES functions as a collecting area in which pressure builds after the peristaltic wave and the bolus is temporarily delayed.

After the bolus enters the stomach, LES pressure increases temporarily before it returns to a resting state. The UES returns to its resting pressure. The bolus does not completely clear the esophagus; rather, small amounts may remain, especially if a

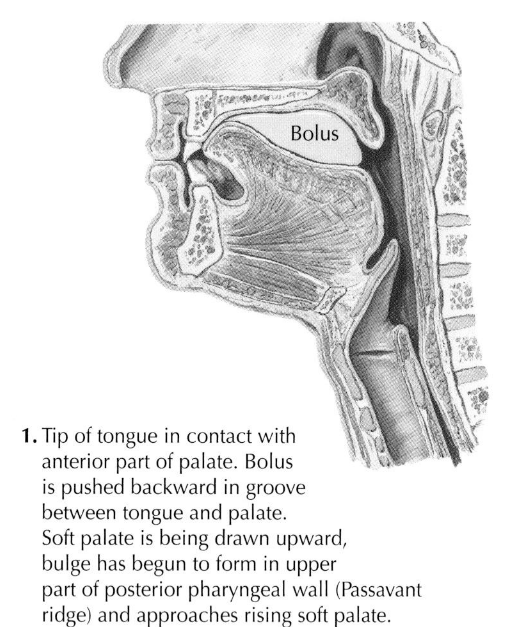

1. Tip of tongue in contact with anterior part of palate. Bolus is pushed backward in groove between tongue and palate. Soft palate is being drawn upward, bulge has begun to form in upper part of posterior pharyngeal wall (Passavant ridge) and approaches rising soft palate.

2. Bolus lying in groove on lingual dorsum formed by contraction of genioglossus and transverse intrinsic musculature of tongue.

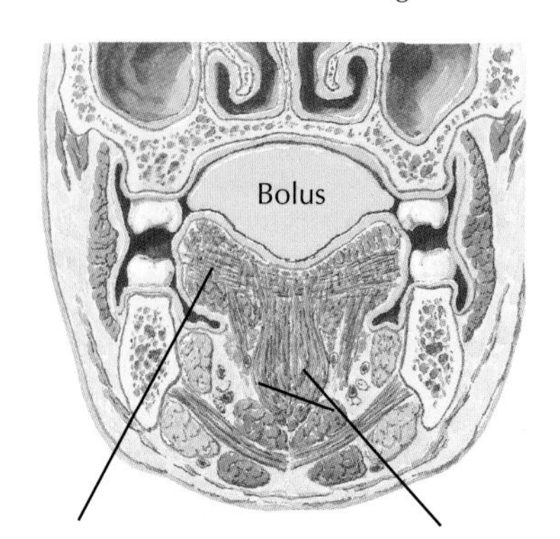

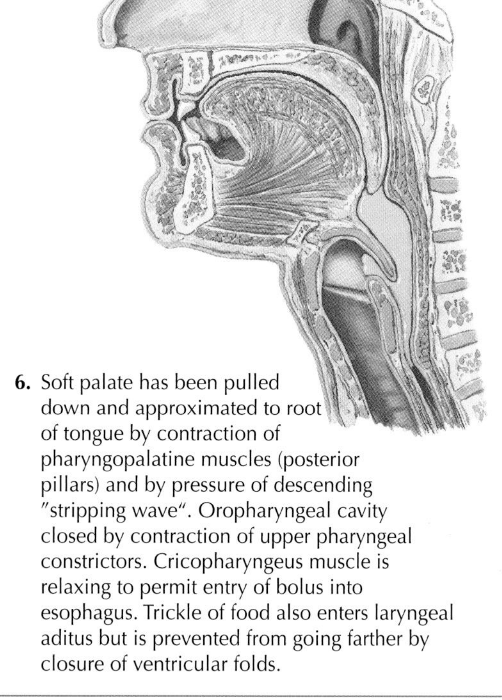

6. Soft palate has been pulled down and approximated to root of tongue by contraction of pharyngopalatine muscles (posterior pillars) and by pressure of descending "stripping wave". Oropharyngeal cavity closed by contraction of upper pharyngeal constrictors. Cricopharyngeus muscle is relaxing to permit entry of bolus into esophagus. Trickle of food also enters laryngeal aditus but is prevented from going farther by closure of ventricular folds.

7. Laryngeal vestibule is closed by approximation of aryepiglottic and ventricular folds, preventing entry of food into larynx (coronal section: AP view).

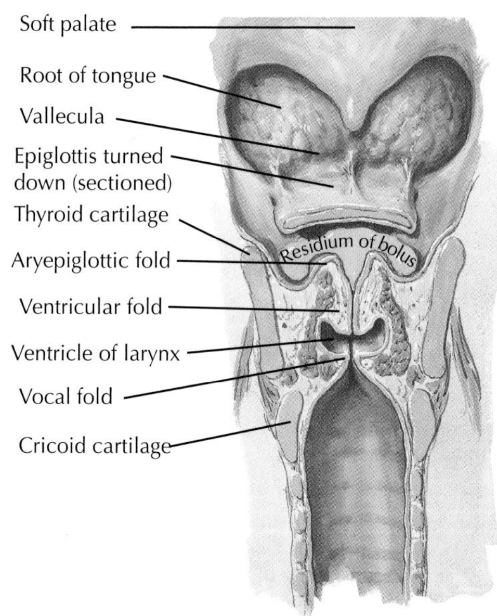

Figure 9-1 *Deglutition: Oral and Pharyngeal.*

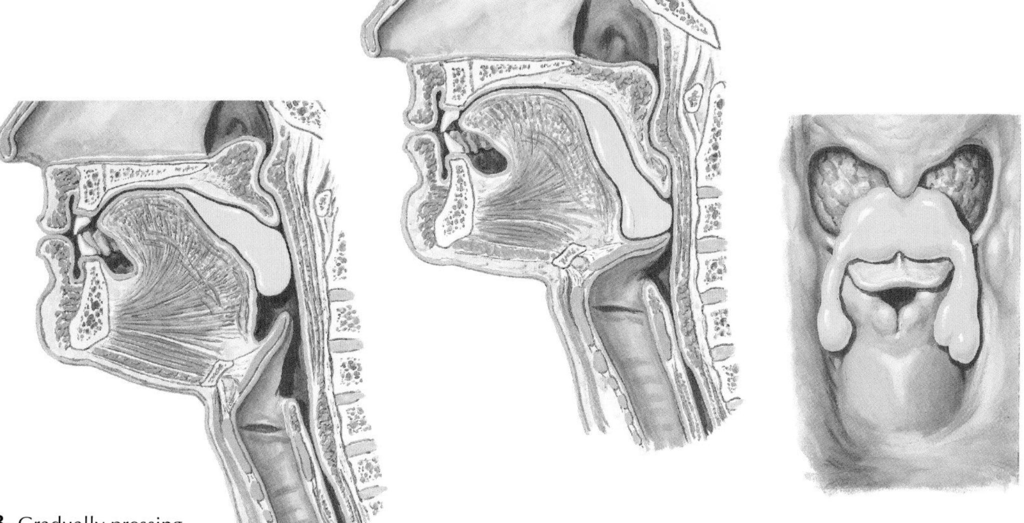

4. Bolus has reached vallecula; hyoid bone and larynx move upward and forward. Epiglottis is tipped downward. "Stripping wave" on posterior pharyngeal wall moves downward.

3. Gradually pressing more of its dorsal surface against hard palate, tongue pushes bolus backward into oral pharynx, soft palate is drawn upward to make contact with Passavant ridge, closing off nasopharynx, receptive space in oral pharynx forms by slight forward movement of root of tongue, contraction of stylopharyngeus and upper pharyngeal constrictor muscles draws pharyngeal wall upward over bolus

5. Epiglottis is tipped down over laryngeal aditus but does not completely close it. Bolus flows in two streams around each side of epiglottis to piriform fossae. Streams will then unite to enter esophagus. Trickle of food may enter laryngeal aditus (viewed from behind).

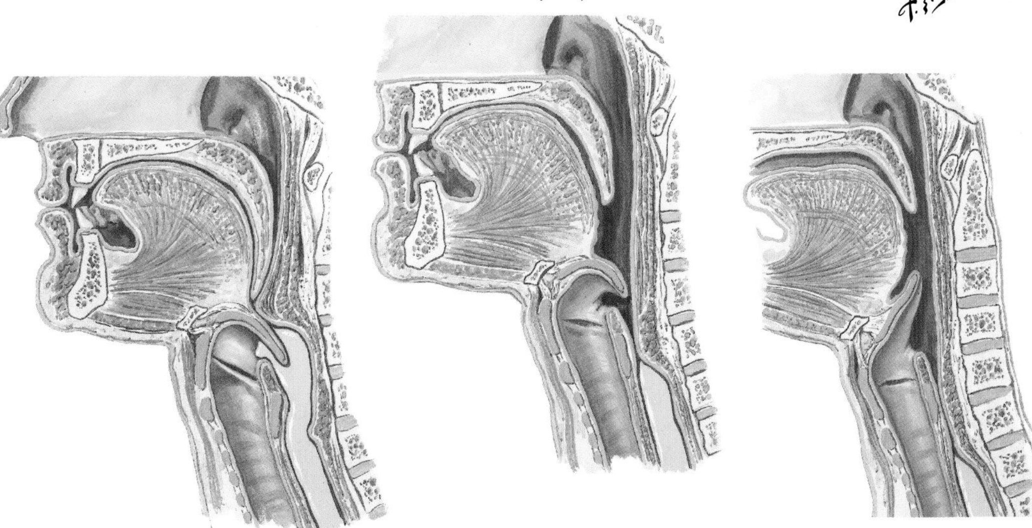

9. "Stripping wave" has passed pharynx. Epiglottis is beginning to turn up again as hyoid bone and larynx descend. Communication with nasopharynx has been reestablished.

8. "Stripping wave" has reached vallecula and is pressing out last of bolus. Cricopharyngeus muscle has relaxed and bolus has largely passed into esophagus.

10. All structures of pharynx have returned to resting position as "stripping wave" passes down into esophagus, pushing bolus before it.

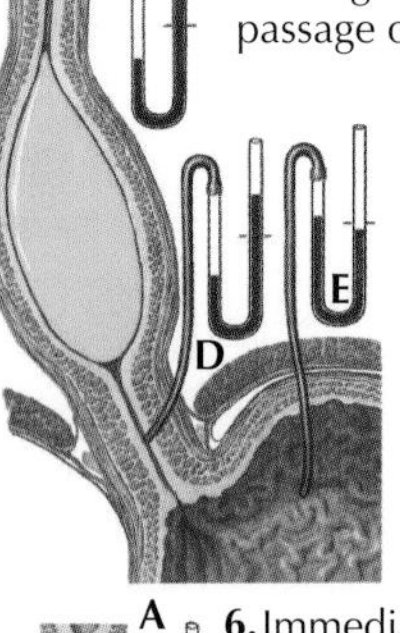

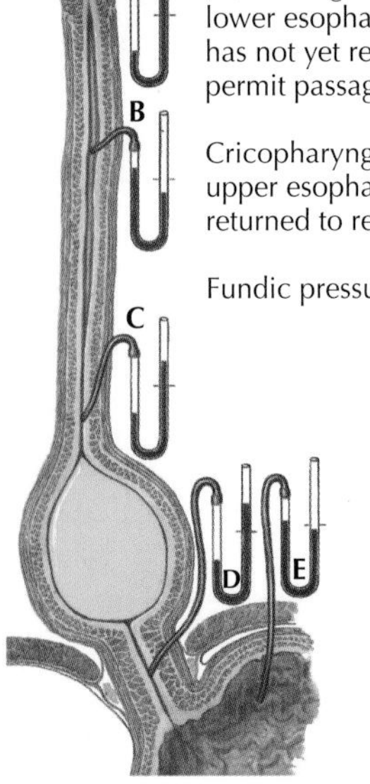

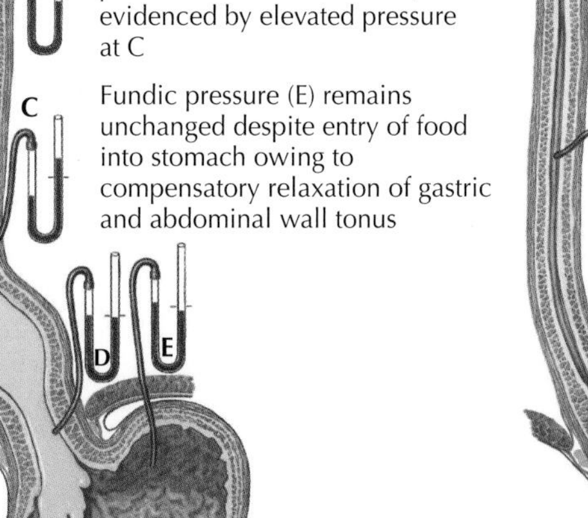

Figure 9-2 *Deglutition: Esophageal.*

person consumes thick food or swallows in the recumbent position.

If a pharyngeal swallow does not result in peristalsis of the esophagus, relaxation of the LES results in the reflux of gastric contents that cannot be propelled back into the stomach. Vagal function is responsible for relaxation of the LES. Therefore, preventing reflux requires a functioning LES and stomach and an esophagus capable of peristalsis.

ADDITIONAL RESOURCES

Gray H, Bannister LH, Berry MM, Williams PL, editors: *Gray's anatomy: the anatomical basis of medicine and surgery,* New York 1995, Churchill Livingstone.

Peters JH, DeMeester TR: Esophagus and diaphragmatic hernia. In Schwartz SI, Shires TG, Spencer FC, editors. *Principles of surgery,* ed 7, New York, 1999, McGraw-Hill, pp 1081-1179.

Neuroregulation of Deglutition

10

Neil R. Floch

Swallowing is controlled by the cortical area located in the inferior portion of the precentral gyrus, near the insula (**Fig. 10-1**). Efferent connections are made by the hypothalamus with the medulla, where a deglutition center is located near the ala cinerea and the nuclei of cranial nerve X. This medullar deglutition center coordinates the nerves and muscles involved in the act of swallowing.

Sensory impulses ready the swallowing center through afferent fibers from the mucosa of the mouth, soft palate, tongue, fauces, pharynx, and esophagus. Stimulation of the anterior and posterior tonsils and of the sides of the hypopharynx may stimulate swallowing as a reflex; but swallowing may also be initiated voluntarily. The glossopharyngeal nerve, the superior laryngeal branches to the vagus, and the pharyngeal branches to the vagus serve as the afferent sensory nerves of the pharynx. These nerves initiate a reflex reaction and regulate the response of the muscle groups that control breathing, positioning of the larynx, and movement of the bolus into the esophagus. The voluntary component of swallowing is completed when sensory nerves on the pharynx detect the food bolus. After this, swallowing is an involuntary process.

Afferent sensory fibers travel through cranial nerves V, IX, and X into their respective nuclei, after which fibers travel to the swallowing center in the medulla. Data are coordinated in the deglutition center and are stimulated by these nerves, which facilitate the act of swallowing by emitting impulses in a delicately timed reflex sequence through cranial nerves V, VII, X, XI and XII. Fibers from cranial nerves V, X, and XII innervate the levator muscles of the soft palate. Cranial nerve X travels to the constrictor muscles of the pharynx. Cervical and thoracic spinal nerves innervate the diaphragm and intercostal muscles. Cranial nerves V and XII travel to the extrinsic muscles of the larynx. Fibers from cranial nerve X control the intrinsic muscles of the larynx and the musculature of the esophagus. Cranial nerves VII and XI and cervical motor neurons C1 to C3 are also involved. The entire coordinated event occurs over 0.5 second. Swallowing may be initiated by several different impulses. The output, however, always follows the same sequence of coordinated events.

These events may be altered after a cerebrovascular accident (stroke). Efferent nerves through the vagus, cranial nerve X, and recurrent laryngeal nerves activate the cricopharynx and upper esophageal muscles. Innervation is required for the cricopharynx to relax as the pharyngeal constrictors contract concomitantly. Damage to these nerves may result in aspiration.

The bolus is propelled down the esophagus by peristaltic muscle contractions controlled by the vagus nerves. In addition to the recurrent laryngeal nerve, the cervical esophagus receives an additional efferent supply either from a pharyngoesophageal nerve arising proximal to the nodose ganglion or from an esophageal nerve.

Visceral afferent nerves from the upper five or six thoracic sympathetic roots convey nerve stimuli that result from esophageal distention, chemical irritation, spasm, or temperature variations. These impulses travel to the thalamus and then to the inferior portion of the postcentral gyrus. After this, stimuli are interpreted as sensations of pressure, burning, gas, dull ache, or pain in the tissues innervated by the somatic nerves from the corresponding spinal levels. These connections explain why pain from esophageal disease may be referred to the middle or to either side of the chest, to the sides of the neck, or to the jaws, teeth, or ears. The similarity between atypical chest pain and referred pain of cardiac origin is controversial and complicates the differential diagnosis of cardiac and esophageal disease. Distention, hypertonus, or obstruction of the distal esophagus may give rise to difficulty in swallowing and to reflex contraction of the upper esophageal sphincter, with the resultant sensation of a lump, known as globus, at the level of the suprasternal notch.

ADDITIONAL RESOURCES

Gray H, Bannister LH, Berry MM, Williams PL, editors: *Gray's anatomy: the anatomical basis of medicine and surgery,* New York, 1995, Churchill Livingstone.

Peters JH, DeMeester TR: Esophagus and diaphragmatic hernia. In Schwartz SI, Shires TC, Spencer FC, editors: *Principles of surgery,* ed 7, New York, 1999, McGraw-Hill, pp 1081-1179.

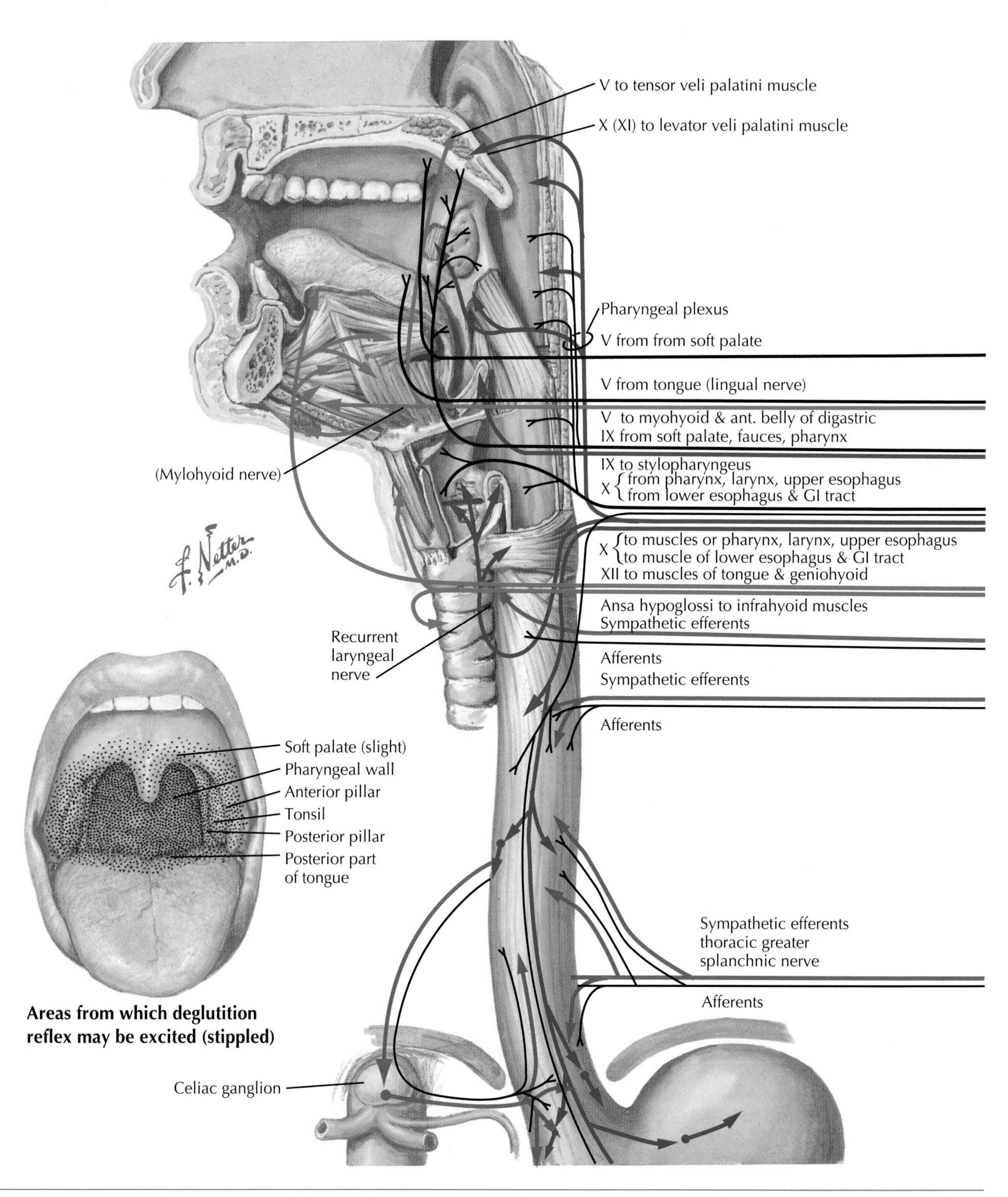

Figure 10-1 *Neuroregulation of Deglutition.*

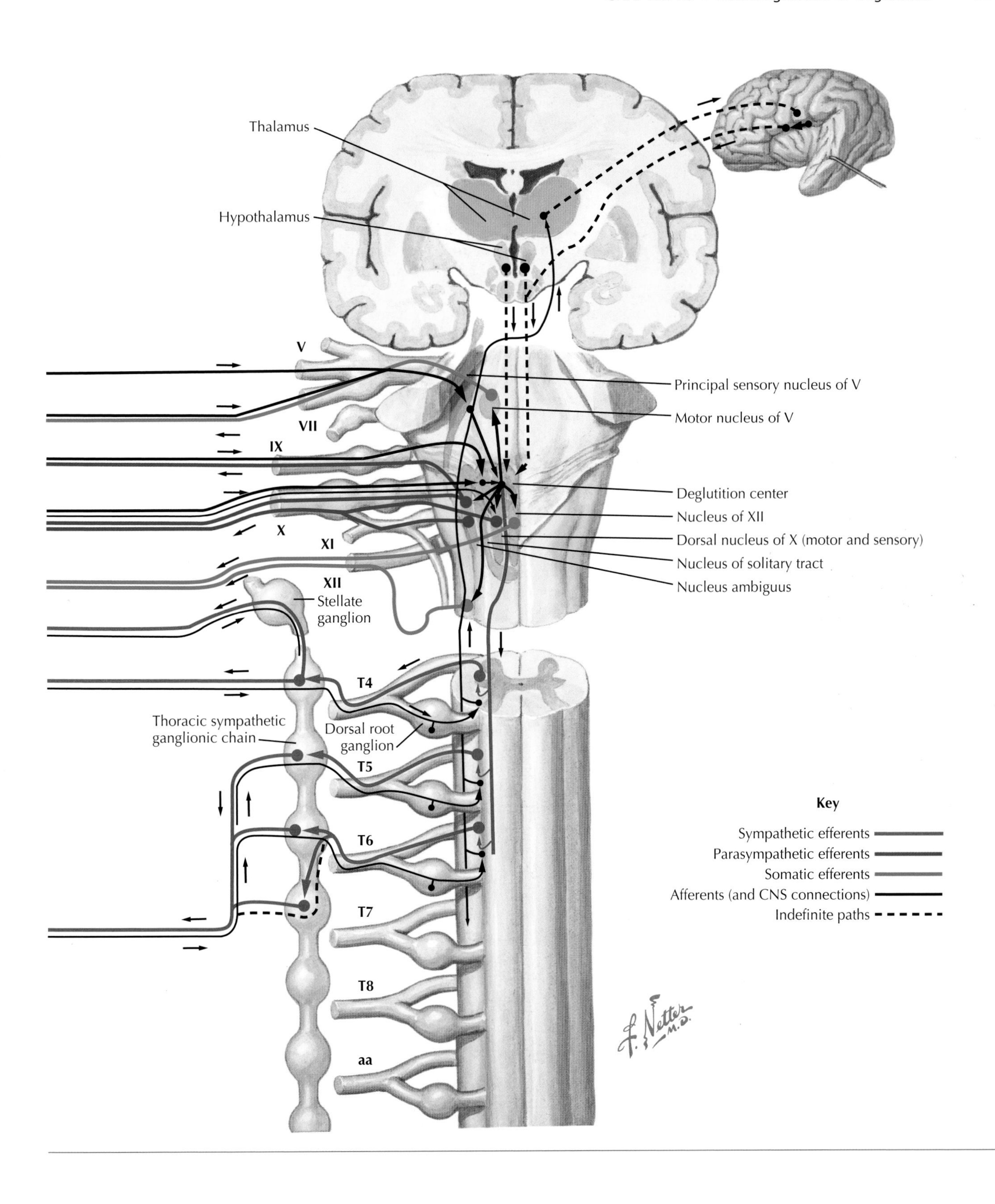
Thalamus
Hypothalamus
V
VII
IX
X
XI
XII
Stellate ganglion
Principal sensory nucleus of V
Motor nucleus of V
Deglutition center
Nucleus of XII
Dorsal nucleus of X (motor and sensory)
Nucleus of solitary tract
Nucleus ambiguus
T4
Thoracic sympathetic ganglionic chain
Dorsal root ganglion
T5
T6
T7
T8
aa
Key
Sympathetic efferents
Parasympathetic efferents
Somatic efferents
Afferents (and CNS connections)
Indefinite paths
F. Netter M.D.

Congenital Anomalies of the Esophagus

11

Neil R. Floch

The most frequently encountered anomaly in the newborn is *esophageal atresia* (EA), which occurs with or without *tracheoesophageal fistula* (TEF) (**Fig. 11-1**). The incidence of EA is 1 in 4500 live births, with no gender predilection. Infants born with EA have a 95% chance of survival.

Anatomic classification of EA comprises five categories. The most common category is proximal pouch with distal fistula *(A1, A2)*, which occurs in 85% of patients. The upper esophagus ends at the level of the second thoracic vertebra (T2), leaving a gap of 1 to 2 cm. The lower esophagus enters the trachea at the carina. Over time, the upper esophagus dilates as swallowing ends in a blind pouch. The distal esophagus is of normal caliber but tapers proximally to 3 to 4 mm at its tracheal communication. The second most common EA category is long-gap esophageal atresia *(B)*, which develops in 8% of patients. In this EA, a fibrous cord connects the proximal and distal parts of the esophagus. Occasionally, no cord exists, and the esophagus ends in two pouches. Isolated H-type TEF *(D)*, which can exist anywhere along the posterior wall of the trachea in a normal esophagus, occurs in 4% of patients. EA with distal and proximal fistulae (C) may develop in 6% of patients. Rare variants occur in 2% to 3% of patients; for example, congenital atresia caused by a stenosing web may develop anywhere in the esophagus of a patient with a normal trachea (E).

The pathogenesis of EA remains controversial. In the fetus, the tracheoesophageal septum is a single tube of mesoderm that divides into the esophagus and the lung bud between the fourth and twelfth week of development. The laryngotracheal groove forms the floor of the gut. The esophageal lumen closes as it is filled with epithelial-lining cells. After this, vacuolation occurs, and the lumen is reestablished. An early traumatic event may result in failure of the mesoderm to separate or differentiate during growth of the lung and esophageal components, resulting in reabsorption of a portion of the esophagus. If vacuoles fail to coalesce, a solid core of esophageal cells remains, resulting in atresia. An abnormal esophagus forms, with or without pulmonary communication. Recent evidence indicates that a tracheal fistula may develop as a trifurcation of the trachea that grows and connects to the stomach bud.

Congenital anomalies of the esophagus are frequently associated with organ anomalies. Certain anomalies are incompatible with life unless they are surgically corrected. The most common associated syndrome is the VATER/VACTERL (vertebral, anorectal, cardiac, tracheoesophageal, renal, and limb abnormalities) syndrome, which occurs in 46% of patients; 15% of infants have two or more components of the syndrome.

CLINICAL PICTURE

Shortly after birth, the infant with a congenital abnormality of the esophagus experiences respiratory distress, tachypnea, coughing, and choking. Excessive salivation and drooling occur, along with regurgitation and cyanosis with feedings. Cardiac murmur or cyanosis may be noted. Symptoms are worse after feedings. In infants with EA, the obstruction occurs 10 to 12 cm from the mouth. H-type TEF may be diagnosed in older children with pneumonia and reactive airway disease.

DIAGNOSIS

Diagnosis of a congenital esophageal anomaly in the prenatal period may be possible with ultrasound. Abdominal radiographs may confirm the diagnosis if air/fluid is found in the mediastinum or if a nasogastric tube in the chest has become curled. TEF may indicate air in the stomach. Barium esophagraphy, upper endoscopy, and bronchoscopy are the best diagnostic tests for TEF. The VATER syndrome may be detected using radiography, renal ultrasound, or echocardiographic studies.

TREATMENT AND MANAGEMENT

An infant with a congenital abnormality should be kept in an incubator that provides oxygen or ventilation and has controlled temperature and humidity. A rubber catheter should be introduced into the pharynx for suction, and the infant should be placed in a slight Trendelenburg position to facilitate the aspiration of mucus. Aspiration pneumonia is treated with antibiotics. Intravenous fluids should be given, and total parenteral nutrition should be maintained if possible. Premature infants may receive surfactants.

In the past, surgery was performed as soon as the patient was stabilized after birth. Current procedures include closure of the fistula, if present, and anastomosis of the esophageal ends. A temporary gastrostomy feeding tube may be inserted. Newborns who undergo surgery within the first 48 hours tend to do better, but the timing of repair depends on the patient's condition.

In patients with VATER, the most severe problem is corrected first. Bronchoscopy is performed at surgery, and the incision is made on the side opposite the aortic arch. In patients with dysphagia secondary to webs located at the cricopharyngeal folds, dilatation with bougie may be well tolerated.

COURSE AND PROGNOSIS

Although long-term outcomes are favorable, children with EA and TEF encounter many difficulties after initial surgery and later in life. In the first 5 years, almost 50% of children acquire gastroesophageal reflux disease, and 45% have dysphagia. At least 25% of children with EA and TEF contract pulmonary infections. Many experience developmental delays, but these usually resolve as the children grow. Support groups have been successful at helping children mature and helping parents through their children's difficult years.

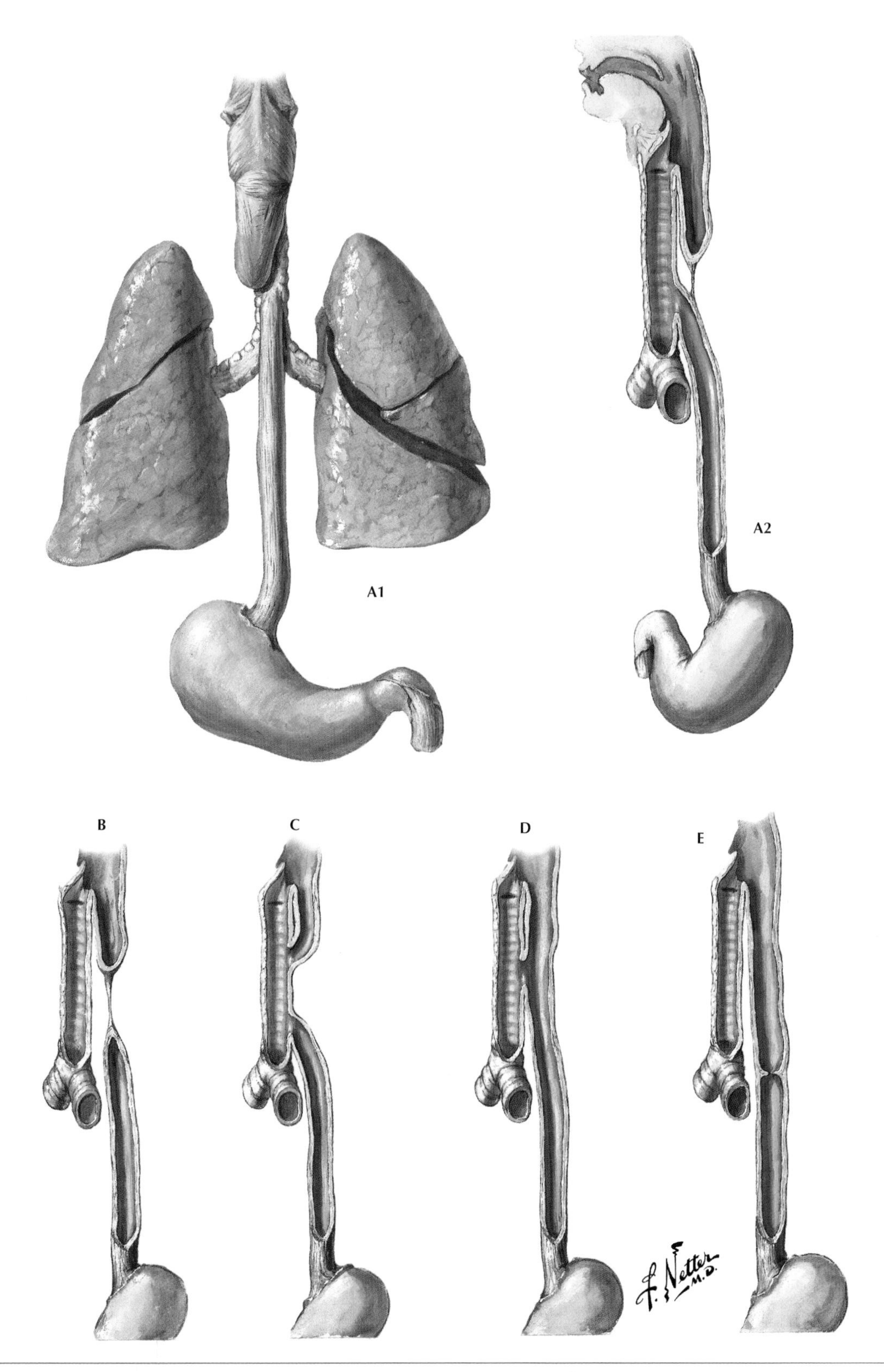

Figure 11-1 *Congenital Anomalies.* **A1, A2,** *Proximal pouch with distal fistula;* **B,** *long-gap esophageal atresia,* **C,** *EA with distal and proximal fistulae,* **D,** *Isolated H-type TEF,* **E,** *congenital atresia caused by stenosing web.*

ADDITIONAL RESOURCES

Little DC, Rescorla FJ, Grosfeld JL, et al: Long-term analysis of children with esophageal atresia and tracheoesophageal fistula. I, *Pediatr Surg* 38:852-856, 2003.

Peters JH, DeMeester TR: Esophagus and diaphragmatic hernia. In Schwartz SI, Shires TG, Spencer FC, editors: *Principles of surgery*, ed 7, New York, 1999, McGraw-Hill, pp 1081-1179.

Spilde TL, Bhatia AM, Marosky JK, et al: Complete discontinuity of the distal fistula tract from the developing gut: direct histologic evidence for the mechanism of tracheoesophageal fistula formation, *Anat Rec* 267:220-224, 2002.

Schatzki Ring

12

Neil R. Floch

In 1953, Schatzki first reported a circumferential stricture, or ring, at the gastroesophageal (GE) junction. It is now believed that up to 18% of patients undergoing routine upper endoscopy exhibit this characteristic. No consensus exists on the cause, location, or significance of the ring. It has been found in patients with sliding hiatal hernia and in whom the GE junction has migrated proximally. The ring may form from an infolding of tissue at the GE junction (**Fig. 12-1**). Of patients with a Schatzki ring, 65% have reflux, 50% erosive esophagitis, and 25% a nonspecific dysmotility disorder.

Histologically, the lower esophageal (Schatzki) ring marks the abrupt change from squamous esophageal cells to columnar gastric cells. The ring consists of connective tissue and muscularis mucosa. Over time, it may progress to form a stricture. Differential diagnosis includes a congenital web, gastroesophageal reflux disease, or carcinoma-induced strictures. Eosinophilic esophagitis and reflux may play a role in the development of Schatzki ring. Some evidence indicates that a ring may have a protective effect from Barrett esophagus. A Schatzki ring may be a rare cause of swallow syncope.

CLINICAL PICTURE

Patients may report symptoms of dysphagia or odynophagia after swallowing meat, bread, or hard vegetables. Analysis of the data by Schatzki indicates that decreasing the ring diameter by 1 mm results in a 46% increase in the incidence of dysphagia. Patients may present with food or a pill that has lodged at the site of the ring.

DIAGNOSIS

A Schatzki ring is diagnosed by barium esophagram, which reveals two protrusions that resemble pencil tips at the GE junction. Esophagogastroscopy may reveal the ring within a sliding hiatal hernia. Infrequently, a ring may be distended beyond 13 mm. Swallowing a marshmallow bolus results in impaction in 75% of patients at barium esophagraphy. Manometry usually reveals high-amplitude contractions.

TREATMENT AND MANAGEMENT

Patients should chew food thoroughly to prevent impaction. Esophagoscopy with bolus extraction is the simplest measure to relieve obstruction. However, glucagon administration has successfully reduced spasm and allowed an obstructed object to pass. Balloon or bougie dilatations are equally effective for patients with chronic dysphagia from Schatzki ring. Bougienage is generally effective, but relapse is common. The ring may be incised if repeat dilatations are ineffective. Dilatation after incision is performed if there is further failure.

ADDITIONAL RESOURCES

DiSario JA, Pedersen PJ, Bichis-Canoutas C, et al: Incision of recurrent distal esophageal (Schatzki) ring after dilation, *Gastrointest Endosc* 56:244-248, 2002.

Gawrieh S, Carroll T, Hogan WJ, et al: Swallow syncope in association with Schatzki ring and hypertensive esophageal peristalsis: report of three cases and review of the literature, *Dysphagia* 20(4):273-277, 2005.

Johnson AC, Lester PD, Johnson S, et al: Esophagogastric ring: why and when we see it, and what it implies—a radiologic-pathologic correlation, *South Med J* 85:946-952, 1992.

Marshall JB, Kretschmar JM, Diaz-Arias AA: Gastroesophageal reflux as a pathogenic factor in the development of symptomatic lower esophageal rings, *Arch Intern Med* 150:1669-1672, 1990.

Nurko S, Teitelbaum JE, Husain K, et al: Association of Schatzki ring with eosinophilic esophagitis in children, *J Pediatr Gastroenterol Nutr* 39(1):107, 2004.

Pezzullo JC, Lewicki AM: Schatzki ring, statistically reexamined, *Radiology* 228:609-613, 2003.

Scolapio JS, Pasha TM, Gostout CJ, et al: A randomized prospective study comparing rigid to balloon dilators for benign esophageal strictures and rings, *Gastrointest Endosc* 50:13-17, 1999.

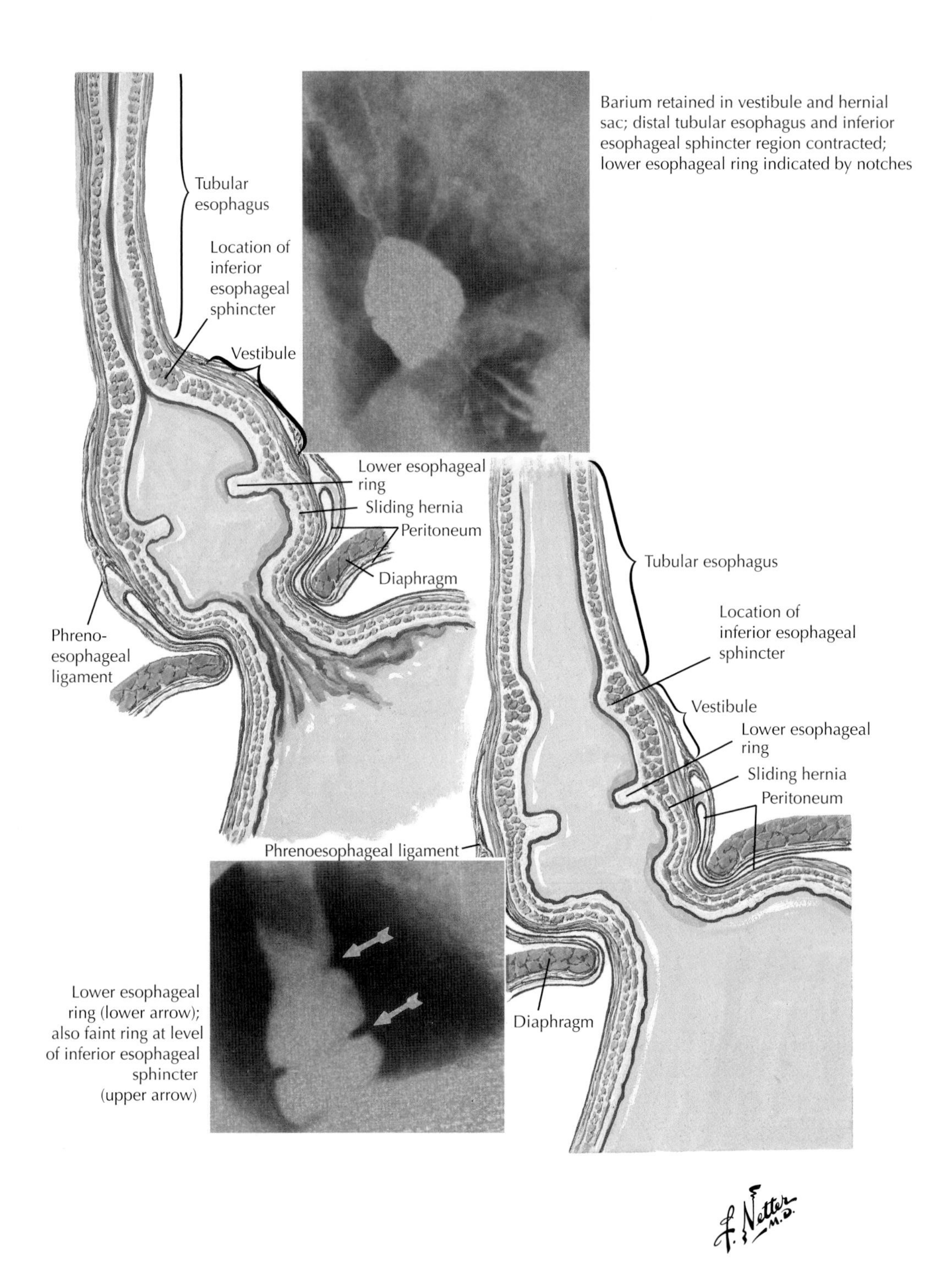

Figure 12-1 *Schatzki Esophageal Ring Formation.*

Plummer-Vinson Syndrome

Neil R. Floch

13

The disease named after two Americans, physician Henry Stanley Plummer and surgeon Porter Paisley Vinson, usually occurs in edentulous, premenopausal, married women and rarely in men (**Fig. 13-1**). Plummer-Vinson syndrome (PVS) develops over months to years, manifests in the fourth to fifth decades of life, and is more common in Scandinavian countries than in the United States. Because PVS is a risk factor for developing squamous cell carcinoma of the esophagus and hypopharynx, it is considered a premalignant process.

Dysphagia symptoms of PVS are caused by the hallmark finding of a weblike structure that originates on the posterior wall of the cervical esophagus between the hypopharynx and 1 to 2 cm below the cricopharyngeal region. At its root, the web is usually thick. It becomes thinner as it protrudes inward, and it may have the consistency of paper. The cause of the web's formation is unknown, but genetic factors and nutritional deficiencies may play a role.

CLINICAL PICTURE

Dysphagia, iron-deficiency anemia, and weakness are the most common symptoms of PVS. Dysphagia of solids occurs frequently, but dysphagia of liquids is rare. Odynophagia may also be present. Oral symptoms are common, and patients complain of glossitis or burning of the tongue and oral mucosa. Possible atrophy of lingual papillae produces a visually smooth and shiny glossal dorsum. Patients may have stomatitis with painful cracks in the angles of a dry mouth.

Atrophic mucosa may involve the esophagus and the hypopharynx. Patients with PVS may also have achlorhydria, brittle fingernails (which may indicate vitamin deficiency), and splenomegaly (33% of patients). Anemia may result in hemoglobin levels that are 50% of normal values.

DIAGNOSIS

Barium esophagraphy reveals a fibrous web under the cricopharyngeus muscle, seen as a filling defect below the level of the cricoid cartilage in the esophagus. The web may involve the entire circumference of the esophagus and is thought to be the cause of dysphagia. Serum tests may reveal hypochromic microcytic anemia, consistent with iron-deficiency anemia. Biopsy of mucosa should demonstrate epithelial atrophy and submucosal chronic inflammation, as well as possible epithelial atypia or dysplasia.

TREATMENT AND MANAGEMENT

Treatment of PVS is primarily aimed at correcting the iron-deficiency anemia. Patients should receive iron supplementation, as well as foods high in iron content. With treatment, symptoms such as dysphagia, as well as oral and tongue pain, usually resolve. Iron supplementation usually resolves the anemia.

Dilatation of the esophageal web may be necessary. Only a small amount of pressure ruptures a web, so introducing an endoscope is usually therapeutic because it reestablishes a normal passage through the esophagus.

COURSE AND PROGNOSIS

Iron replacement therapy reverses anemia, and strictures are almost always dilated successfully. Unfortunately, malignant lesions of the oral mucosa, hypopharynx, and esophagus may be observed in as many as 100% of patients with PVS on long-term follow-up.

ADDITIONAL RESOURCES

Novacek G: Plummer-Vinson syndrome, *Orphanet J Rare Dis* 1:36, 2006.

Peters JH, DeMeester TR: Esophagus and diaphragmatic hernia. In Schwartz SI, Shires TC, Spencer FC, editors: *Principles of surgery*, ed 7, New York, 1999, McGraw-Hill, pp 1081-1179.

Plummer HS: Diffuse dilatation of the esophagus without anatomic stenosis (cardiospasm): a report of ninety-one cases, *JAMA* 58:2013-2015, 1912.

Vinson PP: A case of cardiospasm with dilatation and angulation of the esophagus, *Med Clin North Am* 3:623-627, 1919.

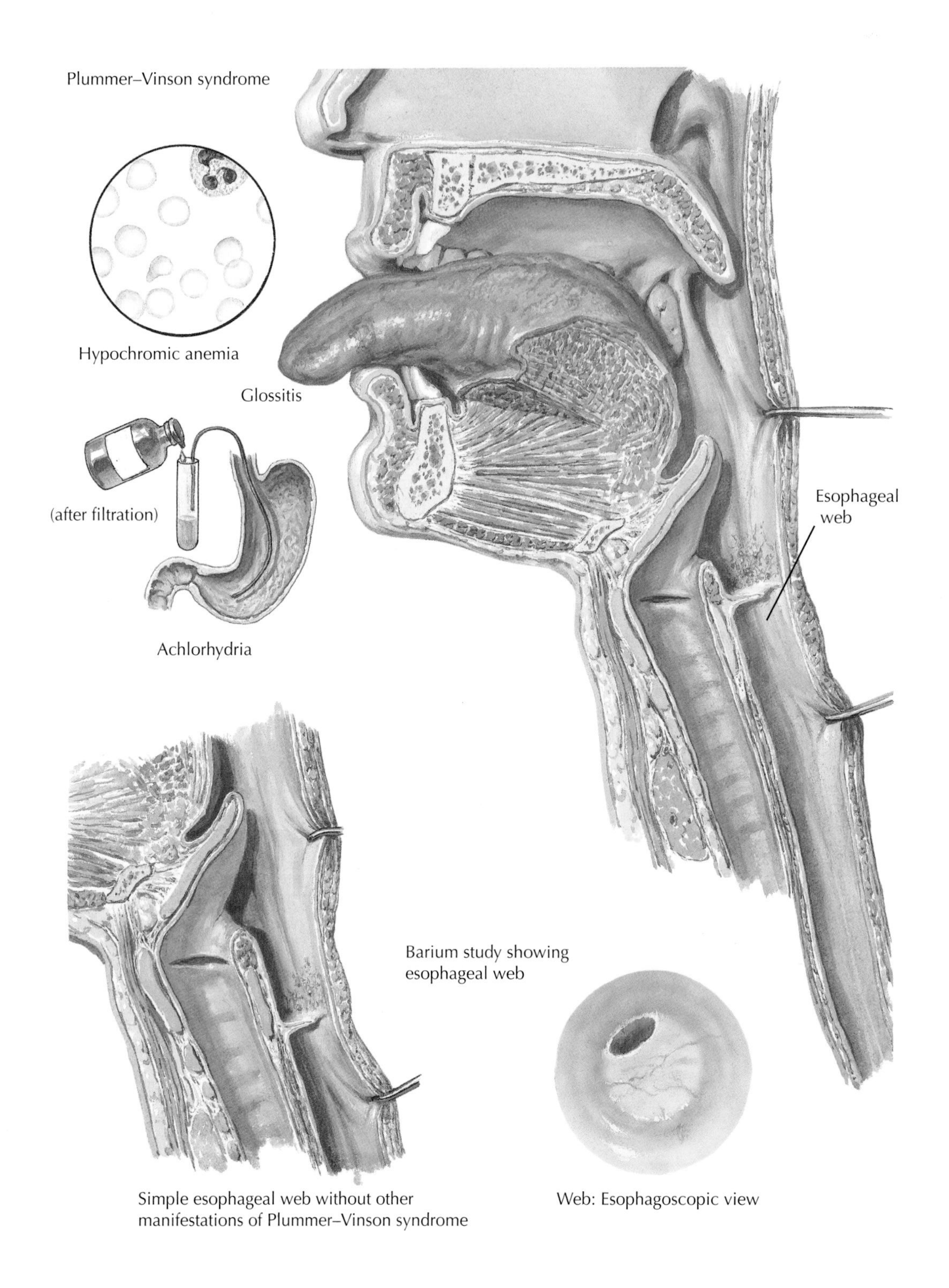

Figure 13-1 *Plummer-Vinson Syndrome.*

Esophageal Motility Disorders

14

Neil R. Floch

The connection between unexplained chest pain and esophageal spasm was first discovered by William Osler in 1892. Since then, multiple esophageal motility disorders have been encountered in clinical practice, with a wide range of symptoms, manometric findings, and responses (**Fig. 14-1**). These disorders vary from minimal changes to extensive radiologic and manometric abnormalities. The etiology of motility disorders has yet to be clearly defined.

Esophageal motility disorders have been best classified into four categories according to manometric findings (Spechler and Castell, 2001), as follows:

1. Inadequate lower esophageal sphincter (LES) relaxation is indicative of achalasia and its variants (see Chapter 15).
2. Uncoordinated esophageal contractions indicate the presence of diffuse esophageal spasm.
3. Hypercontractility disorders include "nutcracker" esophagus, *high-amplitude peristaltic contraction* (HAPC), and *hypertensive lower esophageal sphincter* (HLES).
4. Hypocontraction occurs in ineffective esophageal motility.

Diffuse esophageal spasm (DES) is a disease of the esophageal body characterized by rapid wave progression down the esophagus. It has an incidence of 1 in 100,000 population per year. DES is unique in that it is distinguished by a nonperistaltic response to swallowing, and it may be closely related to achalasia. Significant overlap exists between DES and other esophageal spasm disorders; their clinical presentations are similar, their pathologic basis is undetermined, and their management is almost the same.

Nutcracker esophagus (NE) was first diagnosed in the 1970s. HLES is an uncommon manometric abnormality found in patients with dysphagia and chest pain that is sometimes associated with gastroesophageal reflux disease (GERD).

Ineffective esophageal motility (IEM) is a manometrically defined disorder associated with severe GERD, obesity, respiratory symptoms, delayed acid clearance, and mucosal injury. IEM may occur secondary to other diseases, including alcoholism, diabetes mellitus, multiple sclerosis, rheumatoid arthritis, scleroderma, and systemic lupus erythematosus.

Unfortunately, pathologic distinction between these disorders is usually not helpful because muscles and neural plexuses cannot be properly biopsied. The degree of increase in muscle mass may be an important determinant of the type and severity of esophageal motor dysfunction. The LES and esophageal muscles are thickest in patients with achalasia, thicker in patients with esophageal spasm disorders, and least thick in patients with DES and NE. In certain studies, no specific change in ganglion cells, vagus nerve, or disease progression has been found. However, a nerve defect is suspected because many patients may be sensitive to cholinergic stimulation.

CLINICAL PICTURE

Classic symptoms of esophageal motility disorders include chest pain (80%-90% of patients), dysphagia (30%), and heartburn (20%). Dysphagia of liquids and solids indicates a functional disorder of the esophagus; dysphagia of solids alone indicates a physical lesion. Very hot or cold liquids and stress may exacerbate dysphagia. The pain is usually retrosternal and frequently radiates to the back. Patients describe a pain more severe than angina that is intermittent and variable from day to day. It may last from minutes to hours. Usually, a disparity exists between symptoms and manometric findings, and the chest pain may be unrelated to the dysmotility. Anxiety and depression are common in these patients. Stress, loud noises, and ergonovine maleate may stimulate muscular contractions. The cause may be a sensory abnormality, and psychiatric illness may alter patients' sensory perception.

Patients with DES complain of chest pain and dysphagia. The pain may be associated with eating quickly or drinking hot, cold, or carbonated beverages. Anxiety is common.

Patients with NE or HAPC usually present with chest pain; dysphagia is present in only 10%. There is a 30% incidence of associated psychiatric disorders. Patients with IEM present with typical symptoms of heartburn and reflux and rarely dysphagia. Patients with HLES have dysphagia (71%) and chest pain (49%). Other common symptoms are regurgitation (75%) and heartburn (71%).

DIAGNOSIS

A diagnosis of esophageal motility disorder is made using multiple tests. First, barium esophagraphy typically is used to evaluate for nonpropulsive contractions, which indicate "corkscrew" esophagus; unfortunately, NE and other spastic disorders present minimal findings. Endoscopy is not diagnostic but may reveal associated disorders, such as hiatal hernia, reflux esophagitis, and strictures.

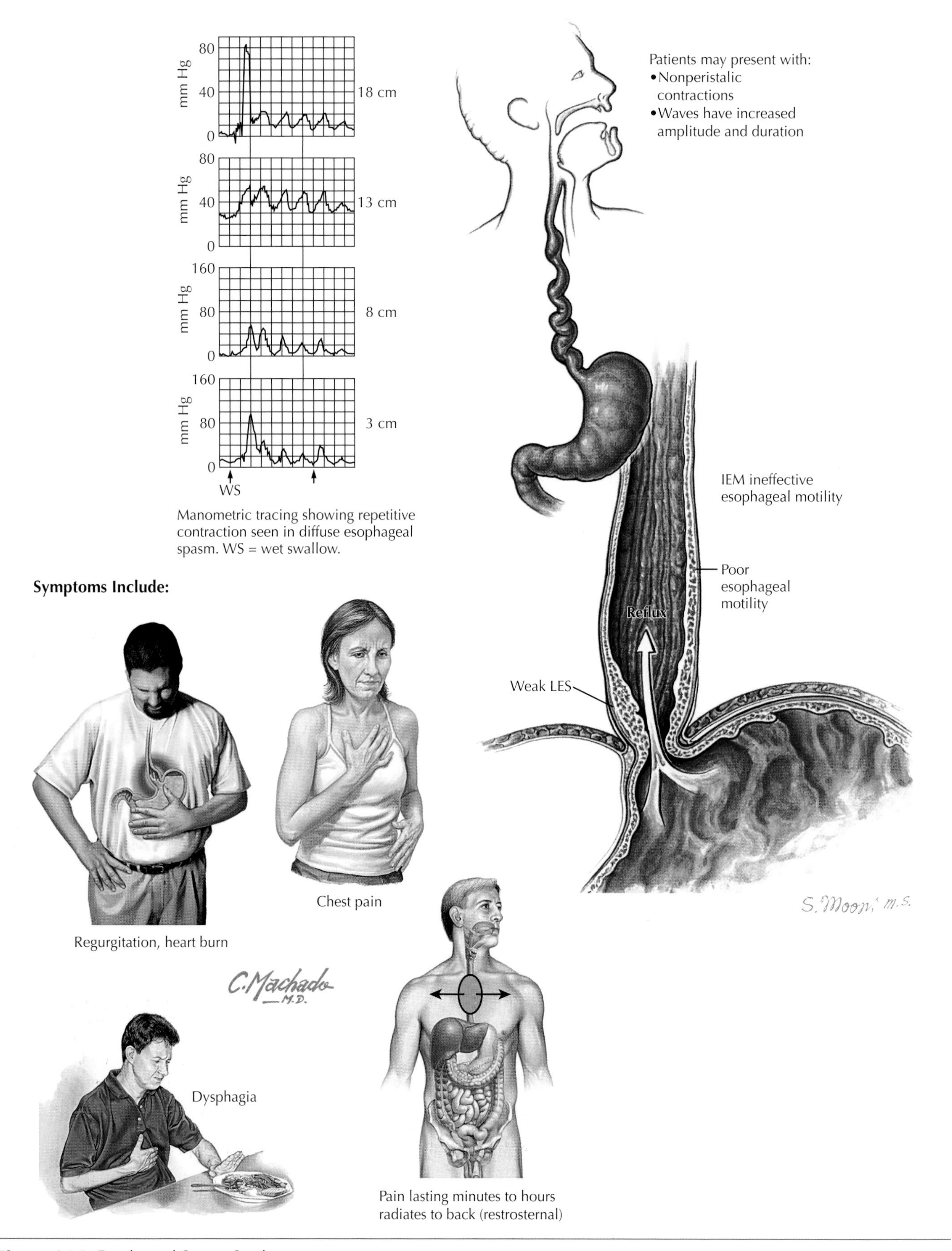

Figure 14-1 *Esophageal Spasm Syndromes.*

Manometry is the definitive test for evaluating esophageal motility disorders, but symptoms correlate poorly with findings. Most findings are present in the distal esophagus and at the LES, where smooth muscle is located. Patients may have a combination of nonperistaltic contractions after most swallows, waves of increased amplitude and duration, or frequent multi-peak waveforms.

In DES patients, an esophagram is usually normal but may reveal a corkscrew appearance with segmentation. The diagnosis of DES is made by manometry, which shows aperistalsis in more than 30% of wet swallows, 20% of contractions that are simultaneous, and amplitudes greater than 30 mm Hg in the distal three fifths of the esophagus. DES patients rarely have repetitive contractions or LES dysfunction. Vigorous achalasia can easily be confused with DES. Recently, computed tomography (CT) has been found to be sensitive in detecting esophageal wall thickening in the distal 5 cm of the esophagus of DES patients, showing promise as a diagnostic test. Degeneration of esophageal vagal branches may be seen on biopsy.

The remaining esophageal motility disorders are diagnostically nonspecific. Patients may have combinations and degrees of increased amplitude, wave duration, and triple-peaked contractions. Double-peaked waves may be present, but these may also be found in healthy patients.

Both NE and HAPC are characterized by abnormally elevated contractions, with peaks greater than 180 mm Hg on manometry despite normal LES. These disorders may be related to DES.

Hypertensive LES is defined as a resting LES greater than 45 mm Hg, with normal relaxation of the sphincter. Intrabolus pressure, a manometric measure of outflow obstruction, is significantly higher in patients with HLES. Residual pressure measured during LES relaxation induced by a water swallow is also significantly higher than in healthy persons. Hypermotility of the LES may be present in conjunction with other muscular disorders, and 33% of patients may have elevated LES pressures with poor relaxation.

Ineffective esophageal motility is associated with GERD and is characterized by multiple low-amplitude waves in the lower esophagus that are less than 30 mm Hg. Also, nontransmitted contractions are present and ineffective at propelling food through a normal LES. Esophagraphy may confirm these findings. Endoscopy should be performed to exclude malignancy or associated disorders. In patients with GERD and respiratory symptoms, 30% to 50% have IEM, and 75% of IEM patients and 25% of HLES patients have an abnormal DeMeester score on 24-hour pH monitoring. Recent studies suggest that evaluation of the distal esophagus with impedance manometry may be helpful in the diagnosis and differentiation of esophageal motility abnormalities.

TREATMENT AND MANAGEMENT

Medical therapy for esophageal motility disorders is limited, and all disorders are treated similarly. These disorders rarely progress and are not known to be fatal. Treatment focuses on symptom reduction and begins with reassurance, because many disorders may have a psychiatric component. These patients' most frequent complaint is pain, which may be related to GERD and not the motor disorder; therefore, treatment should also include proton pump inhibitors (PPIs). Treatment directed at resolving GERD will cure the spasm. It is unknown whether IEM is the cause or the effect of GERD, but resolving reflux helps improve IEM. Unfortunately, effective motility medications for IEM are no longer available.

Trials should be performed with isosorbide nitrate and calcium channel blockers because these agents are successful in relaxing muscle, but studies show no significant benefit. Sildenafil lowers LES pressure and spastic contractions of the esophagus in healthy people as well as in patients with NE, HLES, or achalasia, and these effects may last for 8 hours. Patients with HLES and NE may benefit from sildenafil, but adverse effects are a limiting factor.

Tricyclic antidepressants, which have proven benefit for chest pain, have produced the most success with motility disorders. Recently, a 70% incidence of concomitant psychiatric disorders has been observed. Trazodone at 100 to 150 mg daily decreases stress and symptoms, and imipramine is also beneficial. Antidepressants such as sertraline may also be used. Up to 75% of patients may experience prolonged remission of symptoms.

If symptoms include dysphagia or regurgitation accompanied by poor LES relaxation, treatment is similar to that for achalasia, including smooth muscle relaxants and bougie dilatation (beneficial in 40% of patients with severe manometric abnormalities). Botulinum toxin A (Botox) injections show promise as a medical treatment.

Surgical therapy is reserved for patients in whom medical intervention has failed and symptoms of dysphagia and chest pain have remained severe. Surgical intervention includes a long myotomy from the arch of the aorta across the LES, with an added antireflux procedure to address severe dysmotility. Thoracoscopy, the preferred technique for long myotomy and a viable alternative to open surgery, provides effective relief for spastic disorders, reduces surgical trauma, decreases hospital stay, and speeds recovery. Myotomy with partial fundoplication for isolated HLES relieves dysphagia and chest pain, suggesting a primary sphincter dysfunction. Because medication for IEM is used only to treat associated reflux, patients may be offered partial fundoplication to treat GERD, with expected relief of reflux in 79% of patients.

COURSE AND PROGNOSIS

Open surgical therapy has been successful in only 50% of patients and only when dysmotility is clearly associated with dysphagia. Thoracoscopic myotomy for NE and DES has resulted in a good or excellent result in 80% of patients, compared with a good to excellent result in only 26% of patients treated with medication or dilatation. Minimally invasive surgery offers patients with NE and DES the best opportunity to become asymptomatic. Patients with IEM should undergo laparoscopic partial fundoplication for relief of reflux.

ADDITIONAL RESOURCES

Achem SR: Treatment of spastic esophageal motility disorders, *Gastroenterol Clin North Am* 33(1):107-124, 2004.

Balaji NS, Peters JH: Minimally invasive surgery for esophageal motility disorders, *Surg Clin North Am* 82:763-782, 2002.

Blonski W, Hila A, Vela MF, Castell DO: An analysis of distal esophageal impedance in individuals with and without esophageal motility abnormalities, *J Clin Gastroenterol* 42(7):776-781, 2008.

Cameron JL, editor: *Current surgical therapy*, ed 9, St Louis, 2008, Mosby, pp 1-80.

Goldberg MF, Levine MS, Torigian DA: Diffuse esophageal spasm: CT findings in seven patients, *AJR Am J Roentgenol* 191(3):758-763, 2008.

Mittal RK, Kassab G, Puckett JL, Liu J: Hypertrophy of the muscularis propria of the lower esophageal sphincter and the body of the esophagus in patients with primary motility disorders of the esophagus, *Am J Gastroenterol* 98:1705-1712, 2003.

Shakespear JS, Blom D, Huprich JE, Peters JH: Correlation of radiographic and manometric findings in patients with ineffective esophageal motility, *Surg Endosc* 18:459-462, 2004.

Smout AJ: Advances in esophageal motor disorders, *Curr Opin Gastroenterol* 24(4):485-489, 2008.

Spechler SJ, Castell DO: Classification of oesophageal motility abnormalities, *Gut* 49(1):145-151, 2001.

Tamhankar AP, Almogy G, Arain MA, et al: Surgical management of hypertensive lower esophageal sphincter with dysphagia or chest pain, *J Gastrointest Surg* 7:990-996 (discussion 996), 2003.

Watson DI, Jamieson GG, Bessell JR, et al: Laparoscopic fundoplication in patients with an aperistaltic esophagus and gastroesophageal reflux, *Dis Esophagus* 19(2):94-98, 2006.

Achalasia

Neil R. Floch

15

With an incidence of 1 to 6 per 100,000 population in North America, achalasia is the most common motor disorder of the esophagus (**Fig. 15-1**). It affects both genders equally and usually occurs in persons 20 to 40 years of age. The traditional form, characterized by extensive esophageal dilatation, aperistalsis, and thickened lower esophageal sphincter (LES) that does not relax to baseline pressure, affects 75% of patients with achalasia. The remaining 25% of patients have "vigorous" achalasia. Compared with patients with traditional achalasia, those with vigorous achalasia seek treatment at an earlier stage of disease and have higher muscle contraction amplitude, minimal esophageal dilatation, higher LES pressure, and prominent tertiary contractions.

Patients with achalasia lack ganglion cells in the myenteric plexus of Auerbach in the distal esophagus. Degeneration of the vagal motor dorsal nucleus and destruction of the vagal nerve branches have been observed.

The etiology of achalasia is becoming clearer. Achalasia is now believed to be an immune-mediated inflammatory disease in which esophageal neurons are destroyed by herpes simplex virus type 1 (HSV-1) reactive T cells present in the LES muscles. Myenteric antiplexus antibodies are present in 100% of women and 67% of men with achalasia. The response occurs after HSV-1 infection and is believed to have a genetic predisposition.

Secondary achalasia results from Chagas disease, and pseudoachalasia results from malignancy, infiltrative disorders, diabetes, and other causes.

CLINCAL PICTURE

Almost all patients with achalasia have dysphagia of solids, and 66% have dysphagia of liquids. Patients initially feel heaviness or constriction in the chest when under stress. Food itself causes some stress, eventually resulting in obstruction. Retrosternal chest pain may occur in up to 50% of patients but improves over time. Patients eventually become afraid to eat as symptoms of dysphagia, chest pain, and regurgitation of food develop. Regurgitation of undigested food occurs in 60% to 90% of patients. Most patients maintain their nutritional status with little weight loss. Pneumonia is common in elderly patients from the regurgitation and aspiration of food.

Neither the severity nor the total number of achalasia-related symptoms correlates with the severity of radiographic findings. Although the most common symptom, dysphagia is the initial symptom in only 39% of patients. Heartburn occurs in 25% to 75% of patients. During barium esophagraphy or swallowing, symptoms increase. Slow eating occurs in 79% of patients, regurgitation occurs in 76%, and 60% engage in characteristic movements such as arching the neck and shoulders, raising the arms, standing and sitting straight, and walking.

DIAGNOSIS

Barium esophagraphy shows dilatation of the distal esophagus, aperistalsis, and poor relaxation of the LES. There is a classic bird-beak appearance as a dilated portion of the LES esophagus tapers to a point. Fluoroscopy may visualize spasms in the esophagus as it attempts to empty its contents through the LES. Epiphrenic diverticula are often associated with achalasia.

Diagnosis is made through manometric evaluation, which reveals simultaneous low-amplitude contractions of the esophageal body that do not propagate. The LES narrows to 2 cm, and the LES relaxes incompletely. The esophagus responds with increased activity to parasympathetic agents. An acetylcholine analog, methacholine chloride (Mecholyl), causes exaggerated esophageal body and LES contractions that indicate achalasia.

Esophagoscopy is performed to rule out malignancy and other diseases that are a part of the differential diagnosis and to evaluate the mucosa before any procedure is undertaken. Endoscopic findings may include dilatation and atony of the esophageal body and LES closing that is difficult to open; a pop may be heard as the scope passes through the LES. Small particles of food may be retained early and large amounts retained late in the disease process. Inspissated food particles may adhere to the thickened mucosa, causing leukoplakia. The esophagus may also become elongated before dilatation.

TREATMENT AND MANAGEMENT

Therapy consists of medications, local injections, pneumatic dilatation, and surgery and is directed at palliation of symptoms and prevention of complications. Medications have had limited success in relieving achalasia symptoms. Nitrates, such as amyl nitrite or sublingual isosorbide, can enhance esophageal emptying and relieve symptoms in up to 70% of patients. Calcium channel blockers have been used to relax the LES, but studies show no significant benefit.

Botulinum toxin A (Botox) inhibits acetylcholine release from the nerve endings within the myenteric plexus and at the nerve-muscle junction. It decreases LES pressure in patients with achalasia and has limited adverse effects. It is initially successful in 30% to 75% of patients after circumferential injection into the LES. Unfortunately, the results last only 6 to 9 months on average. Only 50% of patients respond for more than 1 year, and 70% experience relapse at 2 years. Long-term success is highest in elderly patients and in patients with LES pressure that exceeds normal by only 50%. Injections must therefore be repeated several times. Botulinum toxin A injection is a good option for elderly, debilitated patients who are not candidates for more invasive procedures, as well as for patients who prefer this option.

Pneumatic dilatation is another option that is less invasive than surgery. Dilatation with a 50-French dilator provides temporary relief for only 3 days. Forceful dilatation with a balloon is more successful because the circular muscles must be torn to achieve long-term relief. The balloon creates pressure to 300 mm Hg for 1 to 3 minutes and distention to a diameter of 3 cm. After dilatation, a meglumine diatrizoate (Gastrografin) swallow is performed, and the patient is observed for 6 hours before discharge. The most severe complication after dilatation

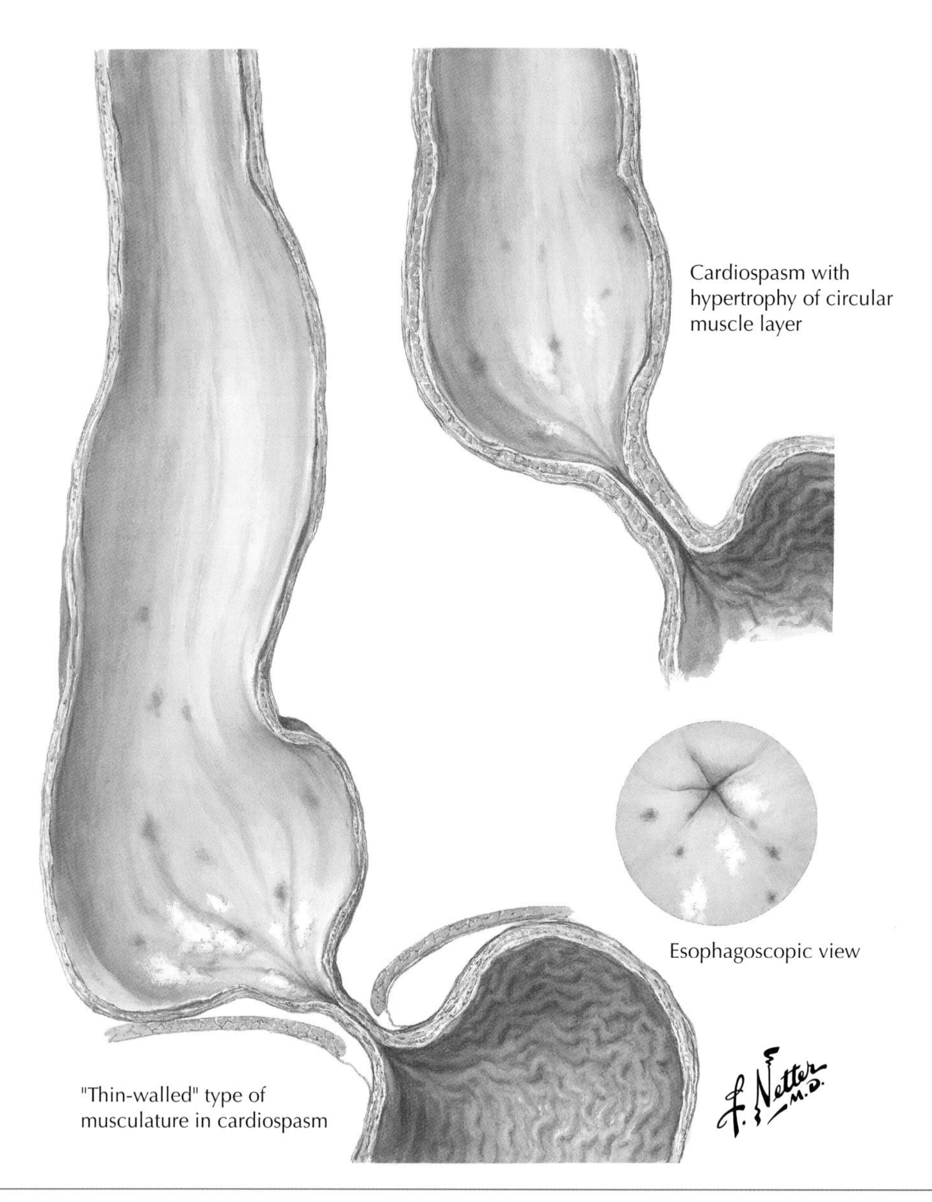

Figure 15-1 *Achalasia (Cardiospasm or Achalasia Cardiae).*

is esophageal perforation, which occurs in 3% of patients. Small tears with free flow of contrast back into the esophagus may be treated conservatively. If there is free flow into the mediastinum, emergency thoracotomy is indicated. Surgery is usually necessary for 50% of perforations. Symptom relief is successful in 55% to 70% of achalasia patients with an initial dilatation and up to 93% with multiple dilatations. Symptom relief is the traditional measure used to assess treatment success. The timed barium study (TBS) is also used to assess esophageal emptying and correlates with a successful outcome in patients undergoing pneumatic dilatation. Poor esophageal emptying can be seen on barium esophagraphy in almost 30% of achalasia patients reporting complete symptom relief after pneumatic dilatation; 90% of these patients experience failed treatment within 1 year.

In the past, surgery performed by thoracotomy had an 80% to 90% success rate. However, minimally invasive procedures such as left thoracoscopic myotomy and laparoscopic myotomy are now the preferred methods, with the latter gaining greatest acceptance. Surgery is indicated in patients younger than 40 years or in those who have recurrent symptoms after botulinum type A treatment or pneumatic dilatation. It is also indicated in those who are at high risk for perforation from dilatation because of diverticula, previous gastroesophageal (GE) junction surgery, or tortuous or dilated esophagus. Myotomy involves the division of all layers of muscle down to the mucosa, with extension of at least 1 cm onto the stomach. A postoperative esophagram showing excellent initial esophageal clearance correlates well with a very good clinical outcome after esophageal myotomy.

Thoracoscopic myotomy without fundoplication produces excellent response in 85% to 90% of patients. Compared with the open procedure, this approach results in shorter average duration of surgery, less blood loss, less need for postoperative narcotic analgesics, and quicker recovery to normal activity. Median hospital stay is 3 days. Gastroesophageal reflux disease

(GERD) may develop in 60% of patients after thoracoscopic myotomy. The most significant complication is postoperative reflux, and dysphagia persists in approximately 10% of patients.

Laparoscopic myotomy is the procedure of choice for patients with achalasia. It results in all the same benefits as thoracoscopy with less dysphagia and less reflux. A partial fundoplication, Toupet or Dor, should be performed to prevent reflux. Median hospital stay is only 2 days.

The incidence of GE reflux in patients who underwent esophageal myotomy alone was 64%, but 27% in those who had myotomy and antireflux procedure. At 15 years after surgery, 11% of patients will develop esophagitis, and more than 40% will have reflux with a partial fundoplication. Good to excellent long-term results were seen in approximately 90% of patients at 3-year follow-up and 75% to 85% after 15 years. Resting pressures of the esophageal body and LES are lower after surgery. Esophageal transit improves in postoperative patients but is still slower than in healthy controls. Persistent postoperative dysphagia may occur in up to 5% and may be treated with dilatation or repeat surgery. Approximately 2% of patients develop esophageal cancer after surgery.

COURSE AND PROGNOSIS

Achalasia can now be more specifically classified. Type II patients are most likely to respond to all therapies, such as Botox in 71%, pneumatic dilatation in 91%, and Heller myotomy in 100%, compared with only 56% overall response rate in type I patients and 29% in type III patients. If left untreated, esophagitis may develop from stasis of retained food; 30% of patients may aspirate esophageal contents. Coughing attacks and pulmonary infections may also occur. Carcinoma may develop in 2% to 7% of patients. Currently, there is no recommended surveillance program for malignancy.

Surgery for achalasia can fail in 10% to 15% of patients, more frequently in those with previous endoscopic procedures, longer duration of symptoms, severely dilated esophagus, and very low LES pressure. Recurrent dysphagia after myotomy should first be treated with pneumatic dilatation. If this fails, laparoscopic reoperation for achalasia is safe and feasible and the procedure of choice. It is performed in approximately 5% of patients with recurrent or persistent dysphagia. Repeat surgery improves symptoms in more than 85% of patients. The surgeon's experience and recognizing the cause for failure of the original surgery are the most important factors in predicting outcome.

In primary treatment of achalasia, dilatation is superior to botulinum type A treatment in the short term; clinical remission rates at 4 months are approximately 90% (dilatation) versus 40% (Botox). Also, myotomy is more reliable in reducing LES pressure than pneumatic dilatation. Good or excellent relief of dysphagia is obtained in 90% of myotomy patients (85% after thoracoscopic, 90% after laparoscopic). Mortality is rare. After documentation that laparoscopic treatment outperforms balloon dilatation and botulinum type A injection, the laparoscopic Heller myotomy has created a shift in practice and has become the preferred treatment for achalasia.

Over their lifetime, patients with achalasia develop many complications that warrant therapy, but their life expectancy and eventual cause of death are no different than in the average population.

ADDITIONAL RESOURCES

Bansal R, Nostrant TT, Scheiman JM, et al: Intrasphincteric botulinum toxin versus pneumatic balloon dilation for treatment of primary achalasia, *J Clin Gastroenterol* 36:209-214, 2003.

Boeckxstaens GE: Achalasia: virus-induced euthanasia of neurons? *Am J Gastroenterol* 103(7):1610-1612, 2008.

Camacho-Lobato L, Katz PO, Eveland J, et al: Vigorous achalasia: original description requires minor change, *J Clin Gastroenterol* 33:375-377, 2001.

D'Onofrio V, Annese V, Miletto P, et al: Long-term follow-up of achalasic patients treated with botulinum toxin, *Dis Esophagus* 13:96-101 (discussion 102-103), 2000.

Gorecki PJ, Hinder RA, Libbey JS, et al: Redo laparoscopic surgery for achalasia, *Surg Endosc* 16:772-776, 2002.

Oezcelik A, Hagen JA, Halls JM, et al: An improved method of assessing esophageal emptying using the timed barium study following surgical myotomy for achalasia, *J Gastrointest Surg*, October 2008 (Epub).

Ortiz A, de Haro LF, Parrilla P, et al: Very long-term objective evaluation of Heller myotomy plus posterior partial fundoplication in patients with achalasia of the cardia, *Ann Surg* 247(2):258-264, 2008.

Pandolfino JE, Kwiatek MA, Nealis T, et al: Achalasia: a new clinically relevant classification by high-resolution manometry, *Gastroenterology*, July 2008 (Epub).

Schuchert MJ, Luketich JD, Landreneau RJ, et al: Minimally invasive esophagomyotomy in 200 consecutive patients: factors influencing postoperative outcomes, *Ann Thorac Surg* 85(5):1729-1734, 2008.

Esophageal Diverticula

16

Neil R. Floch

Diverticula of the esophagus may be classified according to cause (pulsion or traction), location (pharyngoesophageal, midesophageal, or epiphrenic), or wall component (full thickness [true diverticula] or mucosa/submucosa [pseudodiverticula]) (**Fig. 16-1**).

CRICOPHARYNGEAL DIVERTICULA

Zenker, or pharyngoesophageal, diverticula occur 10 times more often than other esophageal diverticula; 80% to 90% of cases occur in men, and the average age is 50 years. Zenker diverticula develop as the mucosa and submucosa of the hypopharynx herniate between the inferior constrictor and the cricopharyngeus muscles in the posterior midline. The developing sac becomes stretched over time as it protrudes to the left, posterior to the esophagus, and anterior to the prevertebral fascia. Evidence suggests that patients with Zenker diverticula have more scar tissue, and that degenerated muscle fibers of the cricopharyngeus have a smaller opening and increased hypopharyngeal bolus pressure during swallowing. Changes in the morphology of the unique fiber orientation of the cricopharyngeus muscle may impair its dilatation and are thought to be caused by progressive denervation of the muscle.

Clinical Picture

Initially, patients may have the sensation of a lump in the throat and may accumulate copious amounts of mucus. Dysphagia to liquids and eventually dysphagia of solids may occur. Patients may regurgitate undigested food when coughing; some may develop pneumonia. Patients may also report foul-tasting food, halitosis, and nausea. As the disease progresses, obstruction may result in significant weight loss and malnutrition.

Diagnosis

Examination may reveal fullness under the left sternocleidomastoid muscle, with resultant gurgling on compression. Barium esophagraphy reveals the size, location, and degree of distention of the diverticulum. Esophagoscopy reveals a wide mouth pouch that ends blindly. The opening of the esophagus is pushed anteriorly and kinked by the diverticulum. Endoscopy demonstrates the presence of two lumens above the cricopharyngeus muscle. Manometry may reveal dysmotility of the upper esophageal sphincter (UES) and may differentiate dysphagia secondary to a recent cerebrovascular accident (stroke).

Treatment and Management

Treatment is surgical through an endoscopic or external cervical approach and should include a cricopharyngeal myotomy. Surgery is indicated for patients with moderate to severe symptoms and especially for those with a history of aspiration pneumonia or lung abscess. Surgery has been associated with significant morbidity because of the procedure itself and the poor medical condition of most of these patients.

Course and Prognosis

Open surgery for Zenker diverticulum includes diverticulectomy, invagination, diverticulopexy, and myotomy. Morbidity ranges from 3% for myotomy to 23% for diverticulectomy with myotomy. Significant improvement occurs in 92% of patients; 6% experience recurrence with diverticulectomy, and 21% have recurrence with invagination. Open techniques result in better symptomatic relief than endoscopic staple diverticulostomy (ESD), especially in patients with small diverticula. Resection without myotomy is initially effective but may result in recurrence or fistulae in the long term.

The ESD procedure is a minimally invasive or endoscopic approach. It may be performed in up to 85% of patients with Zenker diverticulum, although a large diverticulum with redundant mucosa is a risk factor for recurrence. A linear stapler is placed with one blade in the esophagus and the other in the diverticulum as the stapler is fired across the cricopharyngeus muscle. ESD is a safe, effective procedure with a high level of patient satisfaction. It is performed with diathermy, lasers, or staplers through a rigid esophagoscope or by diathermy through a flexible endoscope. The morbidity rate is 2% to 13% with staplers and 26% with lasers. Symptoms improve in 91% to 99% of patients after diathermy. The recurrence rate is 12%, but it is as high as 64% in some studies. Generally, patients recover and return to their normal diets quickly, and complication and mortality rates are lower than with open procedures. When comparing ESD with other endoscopic procedures, duration of surgery and mortality rate are similar, but fewer complications and quicker convalescence occur with ESD. It is safer than laser division.

Small diverticula may be treated by diverticulectomy, with or without myotomy. Large diverticula may be treated by all methods. Patients younger than 60 years old or those with very large diverticula should undergo diverticulectomy. Elderly patients with multiple comorbidities should be treated through ESD.

PULSION DIVERTICULA

Epiphrenic or pulsion diverticula usually occur singly and are located in the distal 10 cm of the esophagus. Multiple diverticula are found in persons with scleroderma. They occur equally on the left and right sides at an incidence of less than 1 in 100,000. They usually range in size from 3 to 10 cm.

There is a high prevalence (up to 100% of patients) of primary motility disorders in patients with epiphrenic diverticula. Diverticula may be associated with achalasia, diffuse esophageal spasm (DES), nutcracker esophagus (NE), and other nonspecific dysmotility disorders (see Chapter 14). They are believed to occur secondary to dyscoordination of muscular contractions that cause the inner mucosa to protrude through the outer esophageal muscle and to a high resting lower esopha-

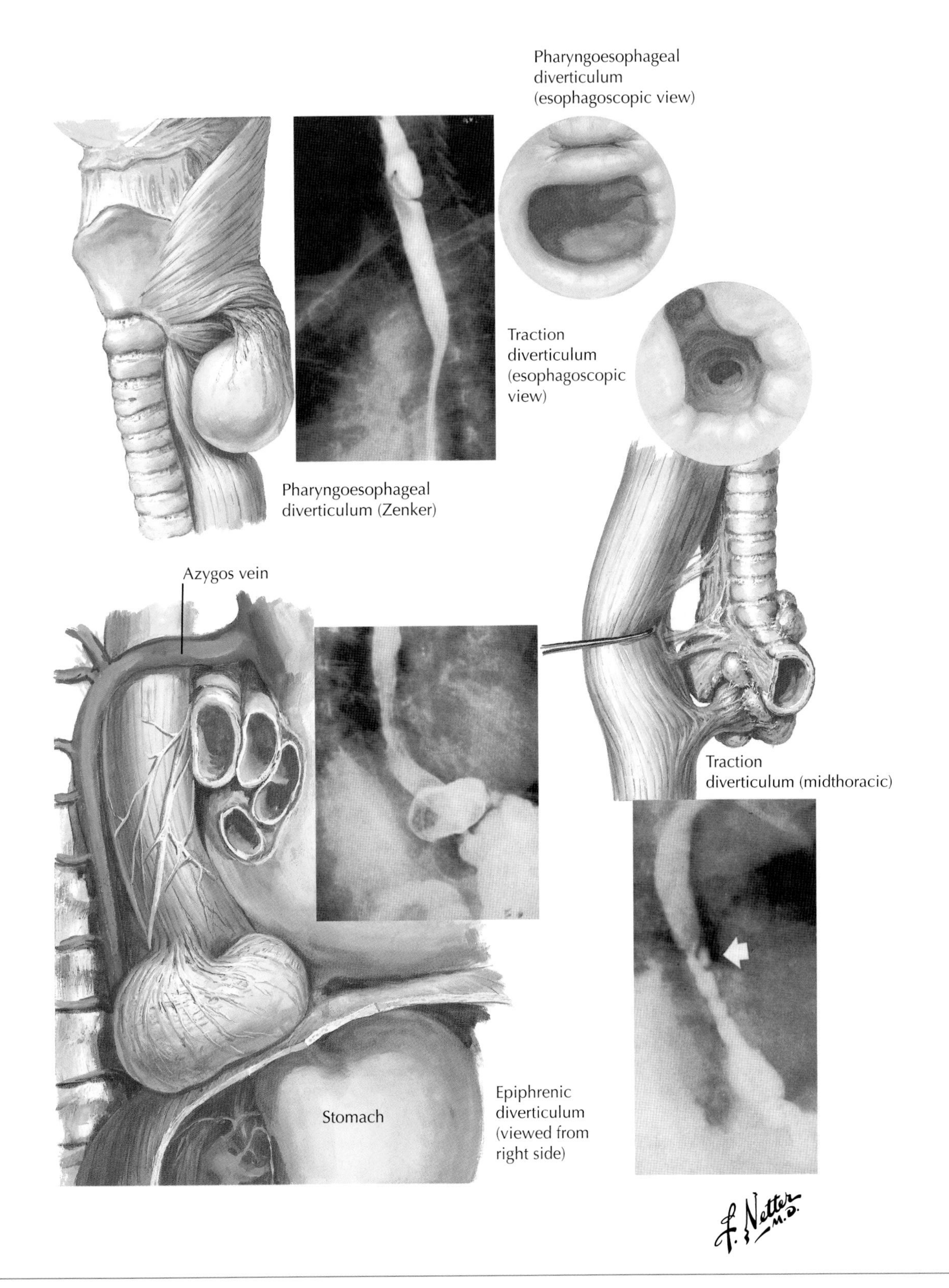

Figure 16-1 *Esophageal Diverticula.*

geal sphincter (LES) pressure with resultant increased intraluminal pressure. Patients usually have associated hiatal hernia with reflux that may result from poor esophageal clearance caused by dysmotility. Distal esophageal diverticula also have been associated with reflux strictures and other lesions. Earlier literature has categorized diverticula according to location and not by cause. Midesophageal diverticula are usually pulsion diverticula that develop secondary to motility disorders.

Clinical Picture

Most diverticula are asymptomatic or cause only minimal dysphagia or regurgitation. Symptoms usually relate to the size of the diverticula. Primary symptoms are dysphagia in approximately 25% of patients, dysphagia and regurgitation in 50%, and pulmonary symptoms in 25%. The usual duration of the primary symptoms before presentation is 10 years. In more than one third of patients, these symptoms are severe, and lethal aspiration is a risk. Halitosis may occur from the retention of food contents in the lesion, and chest pain may result from an associated motility disorder. If the contents of the pouch become infected, the pouch can rupture, resulting in bronchopulmonary complications such as bleeding or sepsis. Symptoms of midesophageal diverticula are similar to those for epiphrenic diverticula, except that reflux is usually not present.

Diagnosis

An esophageal diverticulum is easily visualized during barium esophagraphy. Videoesophagraphy may add further benefit. Endoscopy should be performed to evaluate any coexistent abnormalities or to obtain a biopsy specimen. Manometry is used to determine the esophageal body function and LES pressure and usually indicates that diverticula are the result of motility disorders. An esophageal motor disorder is diagnosed through motility testing in approximately 90% of patients. When diagnosis is difficult, 24-hour ambulatory motility testing may be used and may clarify the diagnosis in almost 100% of patients. Underlying disorders are achalasia in 17% to 43%, hypertensive LES in 14%, DES in 24%, NE in 10%, and nonspecific motor disorder in 10% to 66% of patients.

Treatment and Management

Treatment is limited to surgery and should be reserved for those with severe symptoms and lesions measuring 5 cm or larger; 50% to 75% of patients with diverticula undergo surgery. Treating the underlying motor disorder is the main therapeutic goal, whereas diverticulectomy is performed for large diverticula. Surgical approach is by thoracotomy, thoracoscopy, or laparoscopy. Some centers now recommend concomitant use of endoscopy. If a diverticulum is excised at its base, muscle is closed over the area, and myotomy is performed on the opposite side, at the same level.

Small diverticula are inverted and oversewn. Whether the diverticulum should be surgically resected or suspended depends on its size and proximity to the vertebral body. Usually, midesophageal diverticula are adjacent to the spine and may be suspended. In these cases, myotomy is performed from the neck of the diverticulum to below the LES. Long myotomy is performed for patients with motility disorders, and its length is tailored according to manometry results. Myotomy of the LES should be performed to prevent breakdown of the staple line and rupture of the esophagus, caused by the same intraluminal pressure that initially gave rise to the diverticulum. Partial or total fundoplication is also performed to prevent reflux.

Midesophageal diverticula are treated with thoracotomy or thoracoscopy. Patients with moderate to severe symptoms undergo surgery. Diverticula are removed, and myotomy is performed. Because the LES is not divided, fundoplication is not performed.

Course and Prognosis

Results are good to excellent in 90% to 100% of surgical patients followed long term after resection or imbrication of the diverticula. The most common morbidity is caused by suture leakage. Good results are indicated by resolution of symptoms, weight gain, and no clinical recurrence. Approximately 50% of patients who do not undergo myotomy have less favorable results. Results for thoracoscopy and laparoscopy approach those for open techniques, but with less morbidity. Approximately 66% of patients who do not undergo surgery remain symptomatic or become symptomatic.

TRACTION DIVERTICULA

Traction diverticula were first discovered in patients with tuberculosis and mediastinal lymph nodes. Currently, tuberculosis and histoplasmosis infections are the usual cause, although other etiologies, such as sarcoidosis, have been reported.

Traction diverticula result from inflammation of paratracheal and subcarinal lymph nodes that adhere to and scar the esophagus. Adhesion pulling results in a diverticulum, usually in the midesophagus. Traction diverticula are an outpouching of all the esophageal layers. Pulsion diverticula may also occur in the midesophagus but are caused by dysmotility.

Clinical Picture

Most midesophageal diverticula are asymptomatic and are discovered incidentally. Symptomatic patients report chest pain, odynophagia, and regurgitation. Evaluation should be conducted to determine the presence of an esophageal motility disorder to distinguish it from pulsion diverticula. If dysmotility is not present, a traction or congenital diverticulum should be suspected. Rarely, a patient will have a bronchoesophageal fistula with symptoms of coughing and aspiration of food.

Diagnosis

Traction diverticula of the midesophagus are usually incidental findings on barium swallow or upper endoscopy. Barium esophagraphy reveals poorly demarcated diverticula. Endoscopy may also be helpful in the diagnosis.

Treatment and Management

Most patients with traction diverticula are not treated. If symptoms are severe, thoracotomy is performed, the diverticulum is removed, and the opening is sewn. No myotomy is necessary.

Course and Prognosis

If left untreated, some lesions may erode or extend into the adjacent lung or bronchial arteries and may result in clinical symptoms such as pneumonia or gastrointestinal bleeding.

ADDITIONAL RESOURCES

Anselmino M, Hinder RA, Filipi CJ, Wilson P: Laparoscopic Heller cardiomyotomy and thoroscopic esophageal long myotomy for the treatment of primary esophageal motor disorders, *Surg Laprosc Endosc* 3:437-441, 1993.

Evrard S, Le Moine O, Hassid S, Deviere J: Zenker's diverticulum: a new endoscopic treatment with a soft diverticuloscope, *Gastrointest Endosc* 58: 116-120, 2003.

Kelly KA, Sare MC, Hinder RA: *Mayo Clinic gastrointestinal surgery*, Philadelphia, 2004, Saunders, p 49.

Klaus A, Hinder RA, Swain J, Achem SR: Management of epiphrenic diverticula. I, *J Gastrointest Surg* 7:906-911, 2003.

Melman L, Quinlan J, Robertson B, et al: Esophageal manometric characteristics and outcomes for laparoscopic esophageal diverticulectomy, myotomy, and partial fundoplication for epiphrenic diverticula. *Surg Endosc*, September 2008 (Epub).

Nehra D, Lord RV, DeMeester TR, et al: Physiologic basis for the treatment of epiphrenic diverticulum, *Ann Surg* 235:346-354, 2002.

Schima W, Schober E, Stacher G, et al: Association of midoesophageal diverticula with oesophageal motor disorders: videofluoroscopy and manometry, *Acta Radiol* 38:108-114, 1997.

Zaninotto G, Portale G, Costantini M, et al: Long-term outcome of operated and unoperated epiphrenic diverticula, *J Gastrointest Surg* 12(9):1485-1490, 2008.

Foreign Bodies in the Esophagus

17

Neil R. Floch

More than 100,000 cases of ingested foreign bodies occur in the pediatric population each year. Although most are accidental, intentional ingestion starts in adolescence. Children are often exposed to random household objects, and they often swallow coins, particularly those ages 2 to 5 years. Children also swallow toy parts, jewels, batteries, sharp objects (needles, pins, fish or chicken bones), metal objects, food, seeds, plastic material, magnets, buttons, nuts, hard candy, and jewelry, which can become lodged in the esophagus. Even a safety pin can become impacted in the esophagus of an infant or a small child. Batteries represent less than 2% of foreign bodies ingested by children. Ingestion of multiple magnets can cause esophageal obstruction and perforation.

Foreign bodies become entrapped as frequently in adults as in children **(Fig. 17-1)**. In adults, the foreign body most often entrapped is food, usually meat (33%). Hasty eating may result in people swallowing chicken or fish bones. A large proportion (30%-38%) of these people has an underlying esophageal disease, most often peptic stricture. Underlying strictures obstruct such foods as seeds and peas, which usually pass unobstructed into the stomach.

Tacks, pins, and nails held between the lips may be swallowed and may attach to the esophageal wall or descend into the stomach and beyond. Of the 40% to 60% that become lodged in the esophagus, ingested objects are then found above the cricopharyngeus in 57% to 89% of patients, at the level of the thoracic esophagus in approximately 26% of patients, and at the gastroesophageal junction in 17% of patients.

CLINICAL PICTURE

Symptoms caused by foreign bodies lodged in the esophagus depend on the object's size, shape, consistency, and location. About 50% of patients have symptoms at the time of ingestion, such as retrosternal pain, choking, gagging, or cyanosis. They may drool, and dysphagia may occur in up to 70% and vomiting in 24%. Patients also report odynophagia, chest pain, and interscapular pain.

Infants are unable to express their discomfort or locate the sensation of pain; they may have vague symptoms, making diagnosis difficult. Retching, difficulty swallowing, and localized cervical tenderness may be the only ways to confirm obstruction.

DIAGNOSIS

Radiopaque substances, such as metallic objects, chicken or fish bones, or clumps of meat, can readily be recognized on x-ray film. Nonradiopaque objects, such as cartilaginous and thin fish bones, may be seen on computed tomography (CT) or during esophagoscopy.

TREATMENT AND MANAGEMENT

Treatment of foreign bodies depends on the type of object, its location, and the patient's age and size. In general, esophageal foreign bodies require early intervention because of the risk of respiratory complications and esophageal erosion or perforation.

Passage occurs naturally in 50% of all foreign body ingestions. Small, smooth objects and all objects that have passed the duodenal sweep should be managed conservatively by radiographic surveillance and stool inspection.

Spontaneous passage of coins in children occurs in 25% to 30% of cases without complications; therefore, these patients should be observed for 8 to 16 hours, especially with distally located coins. Spontaneous passage of coins is more likely in older, male patients, especially when they become lodged in the distal third of the esophagus. If coins do not pass, esophageal bougienage or endoscopic removal may be required.

When considering most objects, esophageal bougienage entails the lowest complication rate and the lowest cost. Esophagoscopy should almost always be performed because it is diagnostic and therapeutic. Emergency flexible endoscopy is the most effective method for removing foreign bodies from the esophagus. It is successful in 95% to 98% of patients and results in minimal morbidity. Innovative methods such as loop basket, suction retrieval, suture technique, double-snare technique, and combined forceps/snare technique for long, large, and sharp foreign bodies, along with newer equipment such as retrieval nets and specialized forceps, may be necessary if removing the object is difficult.

Management of sharp foreign bodies has a higher complication rate than for other foreign bodies, from less than 1% to 15% to 35%, except for straight pins, which usually do not cause significant problems unless multiple pins are ingested. Ingested batteries that lodge in the esophagus require urgent endoscopic removal even in the asymptomatic patient because of the high risk of burns and possible death. Batteries that are 2 cm or larger are especially likely to become lodged.

Patients must be anesthetized. Approximately 90% can tolerate conscious sedation; the rest require general anesthesia. The pressure from a large mass in the esophagus against the trachea may cause asphyxia, necessitating tracheotomy before the object can be removed, especially in children.

Food often accumulates above an entrapped object and must be removed by forceps. Maximal dilatation of the esophageal wall allows visualization of the foreign body. Sharp or pointed objects (e.g., nails, pins, bristles) may become embedded in the esophageal wall, with only their tips visible, and must be retrieved using endoscopic forceps.

On occasion, magnets are used to localize a metallic foreign body and position it so that it can be removed. Magill forceps enable quick, successful, and uncomplicated removal of coins in children, especially coins lodged at or immediately below the level of the cricopharyngeus muscle. Proximal dilatation using an oral side balloon is safe and effective for removing sharp foreign bodies from the esophagus, avoiding surgery and possible perforation; it is successful in 95% of patients.

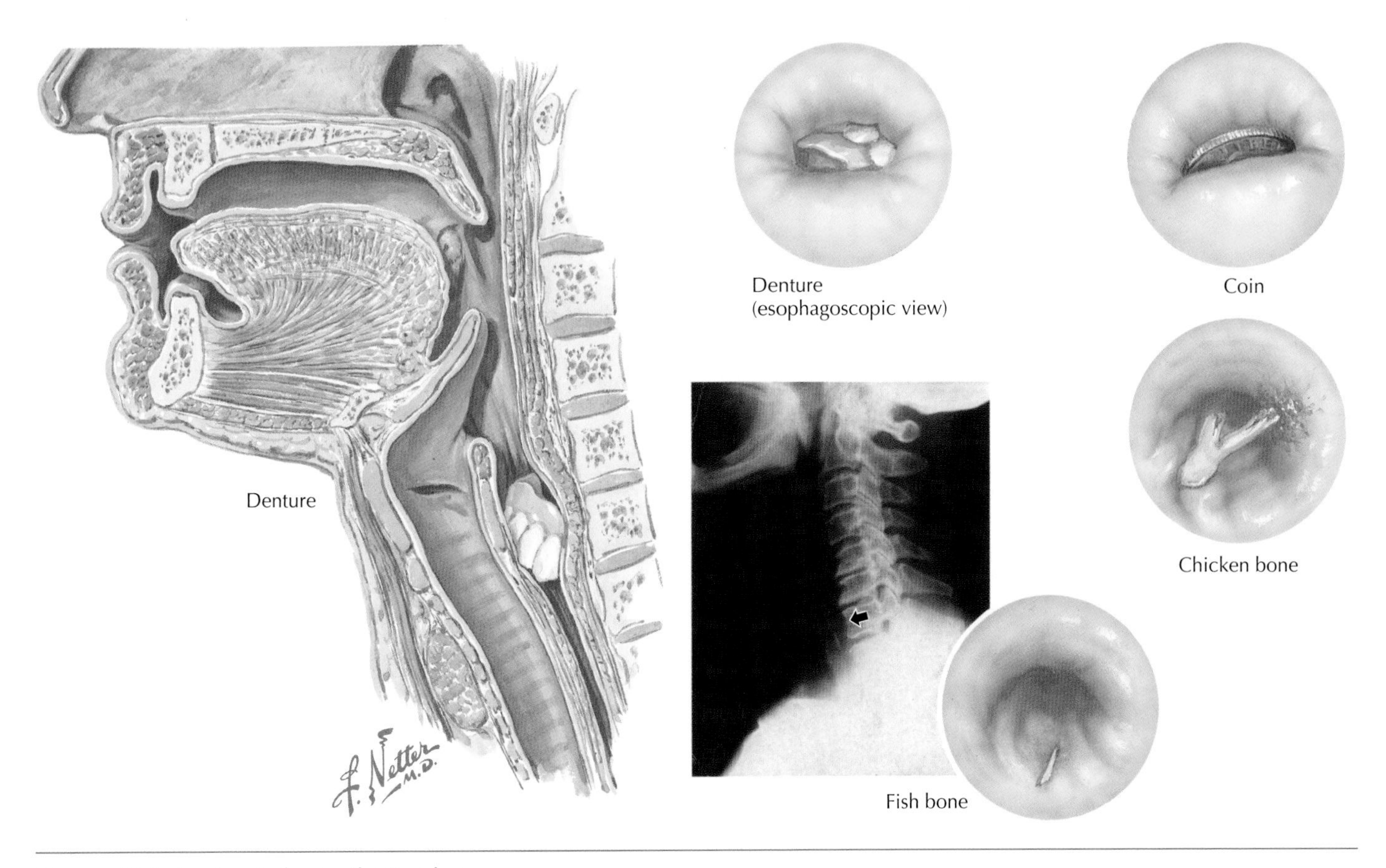

Figure 17-1 *Foreign Bodies in the Esophagus.*

Surgical treatment is unavoidable for the 1% of patients from whom an object cannot be retrieved by endoscopy. These objects usually are lodged in the cervical esophagus. If the esophagus has been perforated, conservative treatment or surgery is performed. Conservative treatment is successful in patients with perforation but no abscess or significant contamination. These patients are treated immediately with broad-spectrum antibiotics and are not permitted food or liquids, receiving either enteral feeding or total parenteral nutrition until healing is documented by meglumine diatrizoate (Gastrografin) swallow.

In the presence of a cervical abscess or mediastinitis, the patient should undergo exploratory surgery, and the area should be surgically drained. Surgical treatment of perforation includes cervical mediastinotomy or thoracotomy and drainage. Esophageal perforation may cause death. Successful therapy for perforation depends on the size of the injury, its location, the time elapsed between rupture and diagnosis, the patient's underlying medical condition, and whether sepsis has developed.

ADDITIONAL RESOURCES

Athanasstadi K, Cerazounis M, Metaxas E, Kalantzi N: Management of esophageal foreign bodies: a retrospective review of 400 cases, *Eur J Cardiothorac Surg* 21:653-656, 2002.

Janik JE, Janik JS: Magill forceps extraction of upper esophageal coins, *J Pediatr Surg* 38:227-229, 2003.

Jeen YT, Chun HJ, Song CW, et al: Endoscopic removal of sharp foreign bodies impacted in the esophagus, *Endoscopy* 33:518-522, 2001.

Kay M, Wyllie R: Pediatric foreign bodies and their management, *Curr Gastroenterol Rep* 7(3):212-218, 2005.

Lam HC, Woo JK, van Hasselt CA: Esophageal perforation and neck abscess from ingested foreign bodies: treatment and outcomes, *Ear Nose Throat J* 82:786, 789-794, 2003.

Mosca S, Manes C, Martino R, et al: Endoscopic management of foreign bodies in the upper gastrointestinal tract: report on a series of 414 adult patients, *Endoscopy* 33:692-696, 2001.

Waltzman ML, Baskin M, Wypij D, et al: A randomized clinical trial of the management of esophageal coins in children, *Pediatrics* 116(3):614-619, 2005.

Yardeni D, Yardeni H, Coran AG, et al: Severe esophageal damage due to button battery ingestion: can it be prevented? *Pediatr Surg Int* 20(7): 496-501, 2004.

Caustic Injury of the Esophagus

18

Neil R. Floch

Each year in the United States, 34,000 people ingest caustic substances (**Fig. 18-1**), leading to tissue destruction through liquefaction or coagulation reactions. The severity of destruction depends on the type, concentration, and amount of substance ingested, as well as the time and intent of ingestion. Ingesting caustic substances is the most common toxic exposure in children and is almost always accidental. In 60% of all cases, caustic substance ingestion is suicidal, and in 40% it is accidental. Adults usually ingest substances in attempts at suicide. Solid crystal lye was the substance most often used for suicide attempts until 1960, when liquid oven cleaners superseded it. Liquid oven cleaners cause more distal esophageal burns. Severe injury almost always occurs after liquid oven cleaner ingestion, but occurs in only 25% of patients after lye ingestion. In the acute stage, perforation and necrosis may occur.

Chemical burns result from strong acid and alkali exposure, and their severity depends on the time of exposure to these radicals. At a pH greater than 11, alkali causes liquefaction necrosis, whereas acid causes coagulation necrosis. Alkali ingestion leads to esophageal stricture formation more often than

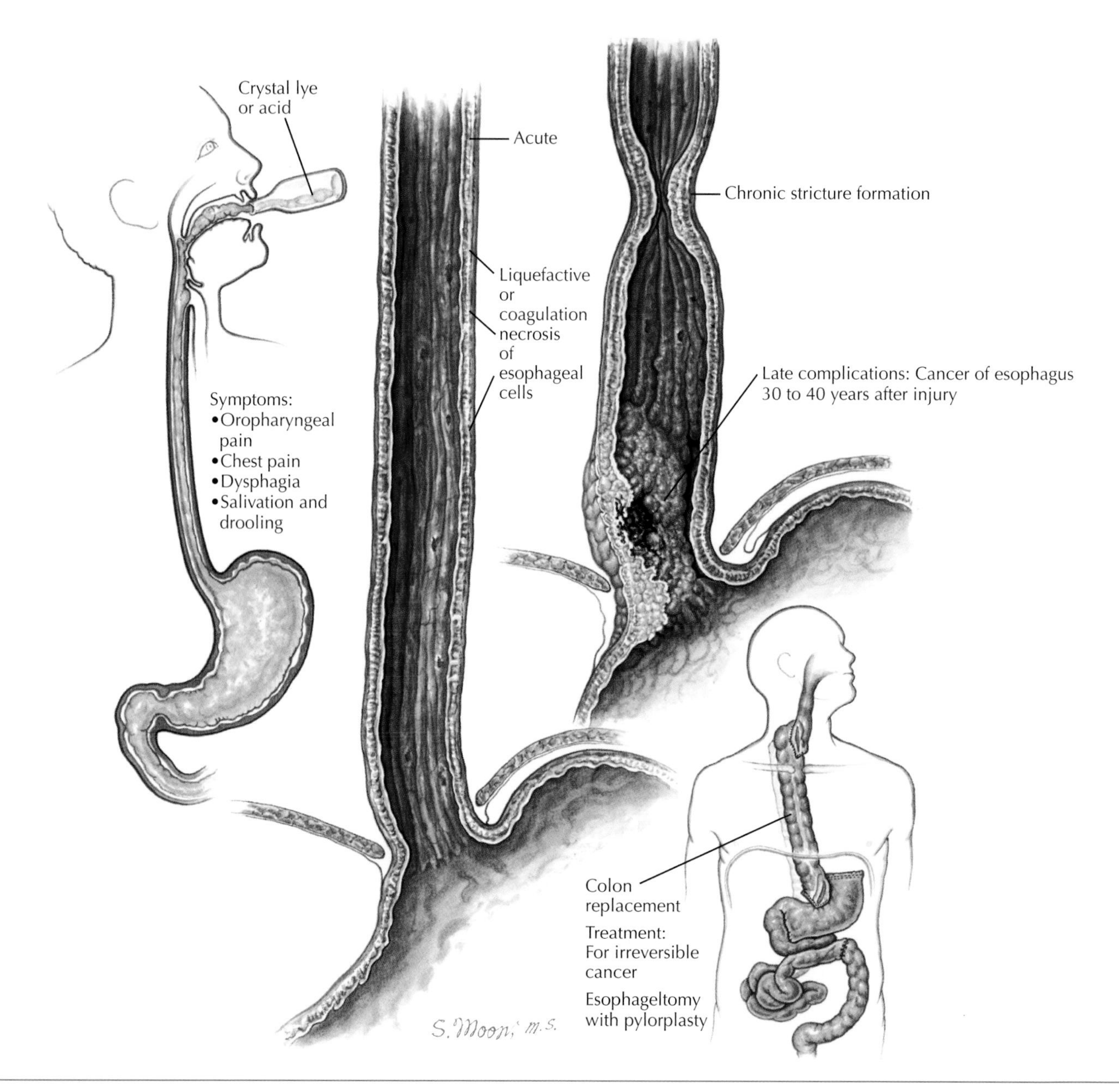

Figure 18-1 *Caustic Injury.*

acid exposure. The natural areas of esophageal approximation (e.g., at cricopharyngeus, aortic arch, and lower esophageal sphincter) are most frequently affected. Cell death is complete by 4 days, and 80% of scars are formed within 60 days.

CLINICAL PICTURE

Symptoms of caustic substance ingestion include oropharyngeal pain, chest pain, dysphagia, salivation, and drooling. Hoarseness and stridor may indicate a need for intubation. The presence of additional symptoms and signs suggests a more severe injury, which warrants more aggressive management.

DIAGNOSIS

Before any diagnostic examination can be conducted, a detailed history of the type and quantity of ingested material must be taken. Plain radiographs of the abdomen will reveal a pneumothorax, pneumomediastinum, perforated viscus, or pleural effusion. Once the patient is stabilized, the clinician performs laryngoscopy to examine the vocal cords and endoscopy to assess the degree of esophageal and gastric damage, regardless of the presence or absence of symptoms. Endoscopy should be performed within the first 12 to 24 hours. Procedure-related perforation is rare.

TREATMENT AND MANAGEMENT

After staging is complete, oral intake should be restricted for the patient with moderate to severe burns, and the patient should receive intravenous fluids, antibiotics, and total parenteral nutrition. If oral intake is restricted for several weeks, high-protein and hypercaloric feedings should be administered through a jejunostomy tube. The patient should remain under observation unless there are signs of perforation or transmural necrosis that require immediate esophagectomy. Water or milk should not be given, vomiting should be prevented, and no nasogastric tube should be placed. In the patient with minor burns, high-dose corticosteroids may improve the prognosis and prevent the formation of esophageal strictures. Corticosteroids have not shown benefit in the treatment of caustic injuries or in the prevention of esophageal strictures, but rather increase the risk for other complications.

Patients who survive beyond several weeks should undergo esophagoscopy to reassess for esophageal strictures. Strictures, a severe complication of caustic ingestion, develop in 5% to 47% of patients and are accompanied by severe esophagitis. Strictures are most common in patients with second-degree and third-degree burns. Severe endoscopic lesions, involvement of the entire length of the esophagus, hematemesis, and increased levels of serum lactate dehydrogenase (LDH) are risk factors for fibrotic strictures.

Once stenosis occurs, the strictures are dilated. Earlier treatment results in better outcomes. Initially, chronic strictures may necessitate serial dilatation so that patency can be established. Some patients require repeat dilatation to maintain adequate lumen diameter. For severe strictures, the lumen may be restricted to 2 to 3 mm. Severe strictures may require esophagectomy. If the patient's condition deteriorates, surgery may be indicated. Esophagectomy with colon interposition is usually performed, although gastric tube pull-up and small-bowel interposition are alternative options.

COURSE AND PROGNOSIS

In the acute phase of caustic substance ingestion, approximately 90% of patients have esophagitis, and 75% experience progression to stenosis. Predictors of stricture formation are lesions extending the entire length of the esophagus, hematemesis, and elevated serum LDH levels. Strictures are mild in approximately 15% of patients, moderate in 60%, and severe in 25%. During the acute phase, 1% of patients die. During the chronic phase, 1.4% die, and 5% of patients have perforation, 1% fistula, and 1% brain abscess.

Dilatation is successful in 60% to 80% of patients. Early complications of surgery include graft ischemia (10%), anastomotic leak (6%-10%), proximal strictures (5%), small-bowel obstruction (2%), and death (1%). Late complications include stenosis requiring dilatation (50%), graft stenosis (1%), and bile reflux requiring surgical diversion (2%). Swallowing function is excellent in 24% of patients, good in 66%, and poor in 10%. Surgical revision is required in 4%. Overall mortality is 4%, predictors of which are increased age, strong acid ingestion, and elevated white blood cell count.

Progression to cancer of the esophagus occurs in 2% of patients. Esophageal carcinoma may develop 30 to 40 years after injury. Although there is an increased incidence of esophageal carcinoma in patients who ingested a caustic substance, regular endoscopic screening has not been advocated in the past.

ADDITIONAL RESOURCES

Bernhardt J, Ptok H, Wilhelm L, Ludwig K: Caustic acid burn of the upper gastrointestinal tract: first use of endosonography to evaluate the severity of the injury, *Surg Endosc* 16:1004, 2002.

Boukthir S, Fetni I, Mazigh Mrad S, et al: High doses of steroids in the management of caustic esophageal burns in children, *Arch Pediatr* 11:13-17, 2004.

Erdogan E, Emir H, Eroglu E, et al: Esophageal replacement using the colon: a 15-year review, *Pediatr Surg Int* 16:546-549, 2000.

Hamza AF, Abdelhay S, Sherif H, et al: Caustic esophageal strictures in children: 30 years' experience, *J Pediatr Surg* 38:828-833, 2003.

Katzka DA: Caustic injury to the esophagus, *Curr Treat Options Gastroenterol* 4:59-66, 2001.

Kukkady A, Pease PW: Long-term dilatation of caustic strictures of the oesophagus, *Pediatr Surg Int* 18:486-490, 2002.

Nunes AC, Romaozinho JM, Pontes JM, et al: Risk factors for stricture development after caustic ingestion, *Hepatogastroenterology* 49:1563-1566, 2002.

Ramasamy K, Cumaste VV: Corrosive ingestion in adults. I, *Clin Gastroenterol* 37:119-124, 2003.

Esophageal Rupture and Perforation

19

Neil R. Floch

Traumatic perforation represents 75% of esophageal injuries, with spontaneous rupture of the esophagus less common; however, both are surgical emergencies (**Fig. 19-1**).

Mechanisms of perforation include iatrogenic injuries, trauma, malignancy, inflammation, and infection. Iatrogenic injuries can occur from endoscopy with dilatation, ultrasound, ablation, resection, and endoscopic antireflux procedures. Nasogastric tubes, endotracheal or Sengstaken-Blakemore tubes, and bougies, as well as neck or chest surgery, may also cause esophageal injury. Trauma can result from penetration, blunt injury, foreign bodies (e.g., coins, pins), and food (e.g., fish or chicken bones). Barotrauma can occur from seizure, weightlifting, or Boerhaave syndrome. Caustic alkaline or acid injury may also cause esophageal damage. Perforation may result from malignancies or inflammatory processes such as Crohn's disease and gastroesophageal reflux with ulcers. Infection is always a possibility as well.

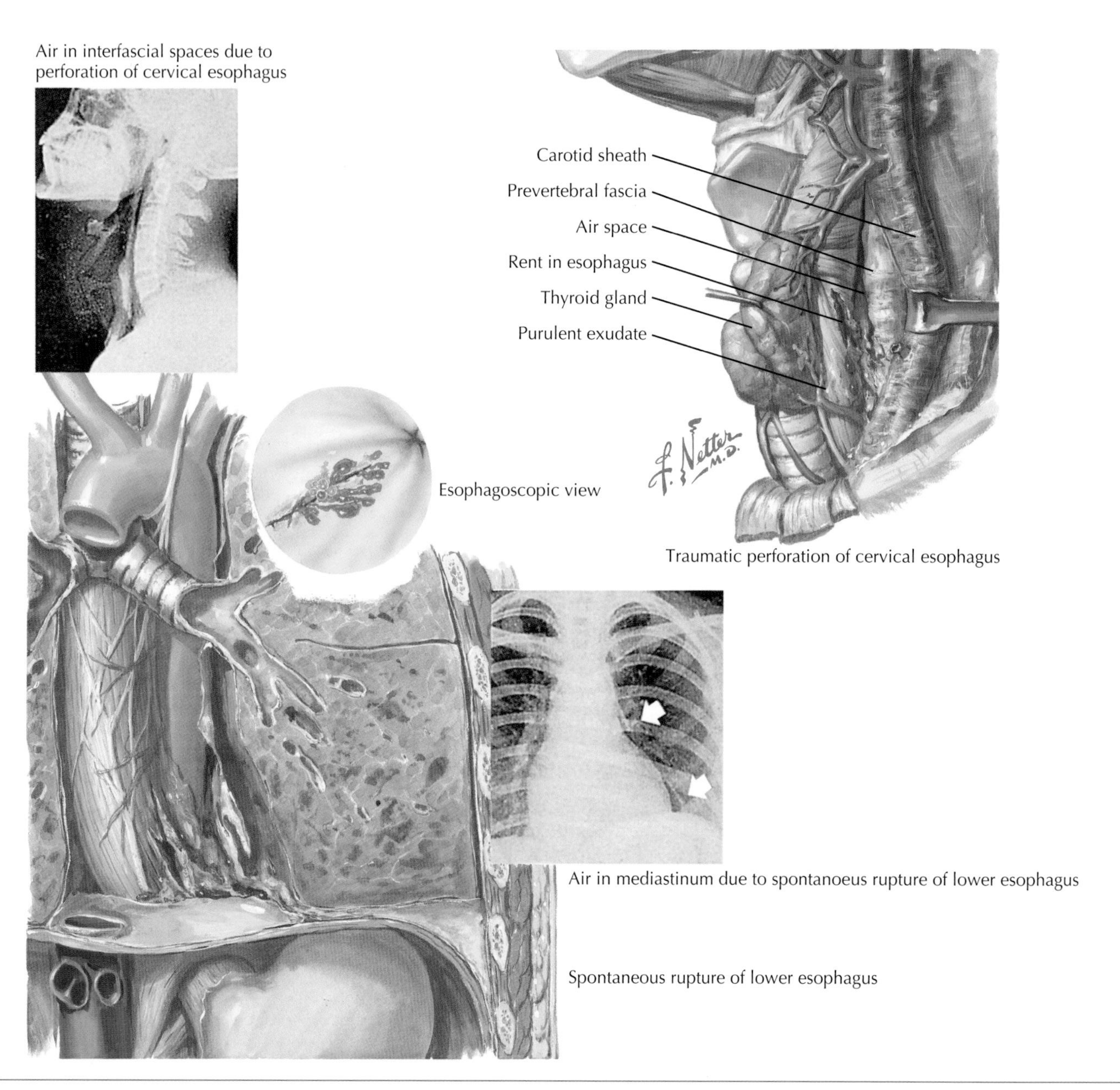

Figure 19-1 *Rupture and Perforation of the Esophagus.*

Approximately 70% of perforations occur on the left side of the esophagus, 20% occur on the right side, and 10% are bilateral. Most patients are 40 to 70 years of age, and 85% are men.

Esophageal perforation usually occurs at narrow areas of the anatomy and at points weakened by benign or malignant disease. Perforation of the cervical esophagus through endoscopy is likely in areas of blind pouches, such as a Zenker diverticulum or the pyriform sinus. It is common in elderly persons who have kyphosis and are unable to open their mouths completely because of muscle contracture. The endoscopist typically is immediately aware of the perforation because bleeding occurs and the anatomy is difficult to discern. Overall, the distal third of the esophagus is the most common site of perforation because it is also the most frequent location for tumors and inflammation. Patients with evidence of a malignancy at the time of esophagogastroduodenoscopy may have as high as a 10% incidence of perforation.

Boerhaave syndrome, or spontaneous rupture of the esophagus, occurs from barotrauma with violent coughing, vomiting, or weightlifting or from the Heimlich maneuver. A sudden pressure transfer of 150 to 200 mm Hg across the gastroesophageal junction causes damage. Spontaneous rupture occurs in the distal or lower third of the esophagus on the posterolateral wall and results in a 2- to 3-mm linear tear, frequently on the left side of the chest and in alcoholic patients. Penetrating trauma is more likely to cause rupture than blunt trauma. Tearing may occur during misidentification of the retroesophageal space during laparoscopy or with improper passage of a bougie.

With only a sparse connective tissue barrier and no adventitia, the esophagus has limited defenses. Once it is ruptured, infection migrates diffusely and rapidly. The mortality rate from perforation is high because the anatomy of the esophagus enables direct communication with the mediastinum, allowing the entry of bacteria and digestive enzymes and leading to sepsis, mediastinitis, empyema, and multiorgan failure.

CLINICAL PICTURE

Symptoms are determined by the location and size of the perforation and by the interval between injury and discovery. Diagnosis is difficult in most patients because 50% have atypical histories. Often, however, patients with esophageal injury have an acute attack or "ripping" chest, back, and epigastric pain. Crepitus may be palpated, and hematemesis, fever, and leukocytosis may develop. Patients with cervical injuries frequently have dysphagia and odynophagia, which increases with neck flexion. Thoracic perforations cause not only substernal chest pain but also epigastric pain. Substernal pain, cervical crepitus, and vomiting affect 60% of patients with spontaneous rupture from barotrauma. Patients with abdominal perforations have epigastric, shoulder, and back pain. Fever, dyspnea, cyanosis, sepsis, shock, and eventually multiorgan failure may develop with increasing contamination of the mediastinum and chest.

DIAGNOSIS

Chest radiographs are obtained first in patients with esophageal injury but have limited sensitivity and specificity. An hour after the incident, the chest radiograph may show air under the diaphragm or subcutaneous or mediastinal emphysema in 40% of patients. Pneumothorax may be seen in 77% of patients, in which case the pleura must also have been injured. Pleural effusion then develops.

Meglumine diatrizoate (Gastrografin) esophagraphy is performed next because the material used is better tolerated if leaked into the mediastinum. If no leak is found, a barium study is performed because it has 90% sensitivity for finding a small leak. Patients at risk for aspiration should have a barium swallow, given that Gastrografin may cause pulmonary edema. Studies are performed in the right lateral decubitus position. Computed tomography can confirm the diagnosis by revealing extraluminal air, periesophageal fluid, esophageal thickening, or extraluminal contrast. Esophagoscopy may demonstrate small bruises or tears and has not been shown to worsen the clinical situation.

TREATMENT AND MANAGEMENT

Treatment goals are to prevent contamination, control infection, maintain nutrition, and restore continuity of the esophagus. All treatment begins with intravenous fluids and broad-spectrum antibiotics. Distal injury may be treated using a nasogastric tube. Treatment depends on the location of the injury and presence of any underlying disease. Cervical esophageal injuries may be treated conservatively, but midthoracic and distal injuries are usually treated surgically. Most patients seek treatment in the first 24 hours and may undergo primary surgical closure, with or without cervical esophagotomy. Patients with associated malignancy, long-segment Barrett esophagus, or chronic strictures will require esophagectomy. Other patients, with severe reflux, dysphagia, aspiration, or severely dilated esophagus from achalasia, may be considered for esophagectomy. Intraabdominal perforations may do well with repair and Nissen fundoplication.

Surgery is not performed if patients seek treatment late, have minimal symptoms, do not have sepsis, or are in a poor medical condition. Patients who have intramural perforation after balloon dilatation may also be treated conservatively. Best results are achieved in patients who have normal white blood cell counts, free communication of the injury with the esophagus, no fever, and no sepsis. They may be treated with antibiotics and total parental nutrition, and no food may be allowed by mouth for 1 to 2 weeks. Another option is endoscopic clipping of a mucosal defect, which can be performed with a double-lumen endoscope. Percutaneous drainage or closure may be performed to treat cervical rupture, if diagnosed early. If severe soilage occurs, patients should undergo esophagectomy with delayed reconstruction, because they will do better than with drainage alone. The chest may be drained or thoracotomy performed with T-tube placement. For patients with life-threatening illness, excision of the esophagus is performed.

Covered metallic stents may be used to seal perforations in patients with distal esophageal perforation. Large-diameter stents are placed, thoracostomy tubes drain pleural cavities, and antibiotics are administered. Complete sealing occurs in 80% of patients; no further therapy is necessary except for eventual removal of the stent.

Treatment is shifting toward the possibility of primary esophageal repair of nonmalignant esophageal perforations that present at any time.

COURSE AND PROGNOSIS

Treatment outcome depends on comorbidities, the interval between diagnosis and treatment, the cause and location of injury, and the presence of esophageal disease. Survival rates are 92% for patients with thoracic perforations closed primarily within 1 day of injury and 30% to 35% for patients with thoracic perforations discovered after 24 hours. In the past, results included 10% to 25% mortality if the perforations were treated within the first 24 hours and 40% to 60% mortality if treated after 48 hours. The mortality rate is highest, 67%, in patients with spontaneous rupture of the esophagus. Recent studies indicate lower overall mortality of 3.8% and morbidity of 38%.

ADDITIONAL RESOURCES

Cameron JL, editor: *Current surgical therapy*, ed 9, St Louis, 2008, Mosby, pp 16-20.

Duncan M, Wong RK: Esophageal emergencies: things that will wake you from a sound sleep, *Gastroenterol Clin North Am* 32:1035-1052, 2003.

Gupta NM, Kaman L: Personal management of 57 consecutive patients with esophageal perforation, *Am J Surg* 187:58-63, 2004.

Kollmar O, Lindemann W, Richter S, et al: Boerhaave's syndrome: primary repair vs. esophageal resection—case reports and meta-analysis of the literature, *J Gastrointest Surg* 7:726-734, 2003.

Port JL, Kent MS, Korst RJ, et al: Thoracic esophageal perforations: a decade of experience, *Ann Thorac Surg* 75:1071-1074, 2003.

Rubesin SE, Levine MS: Radiologic diagnosis of gastrointestinal perforation, *Radiol Clin North Am* 41:1095-1115, 2003.

Zubarik R, Eisen G, Mastropietro C, et al: Prospective analysis of complications 30 days after outpatient upper endoscopy, *Am J Gastroenterol* 94:1539-1545, 1999.

Zumbro GL, Anstadt MP, Mawulawde K, et al: Surgical management of esophageal perforation: role of esophageal conservation in delayed perforation, *Am Surg* 68:36-40, 2002.

Esophageal Varicosities

20

Neil R. Floch

Varicosities occur secondary to portal hypertension and are defined as a dilatation of various alternative pathways when cirrhosis obstructs the portal return of blood (**Fig. 20-1**). Varicosities occur most often in the distal third but may occur throughout the esophagus. Varices of the esophagus are a less common cause of upper gastrointestinal hemorrhage, but the consequences of bleeding are an ever-impending threat to life. *Acute variceal hemorrhage* is the most lethal complication of portal hypertension. The median age of these patients is 52 years, and 73% are men. The most common cause for portal hypertension, affecting 94% of patients, is cirrhosis. The most common causes of cirrhosis are alcoholism (57%), hepatitis C virus (30%), and hepatitis B virus (10%).

Mortality rates from the initial episode of variceal hemorrhage range from 17% to 57%. Larger vessels bleed more frequently. Hospitalizations for acute bleeding from esophageal varices have been declining in recent years, believed to be a result of more active primary and secondary prophylaxis. Bleeding occurs when the tension in the venous wall leads to rupture, and shock may occur. Occasionally, the bleeding may stop spontaneously, but more often the bleeding will recur. Thrombocytopenia and impaired hepatic synthesis of coagulation factors both interfere with hemostasis.

CLINICAL PICTURE

Cardinal symptoms of esophageal varicosities are recurrent hematemesis and melena. Patients with acute variceal bleeding have hemodynamic instability (61%), tachycardia (22%), hypotension (29%), and orthostatic hypotension (10%).

DIAGNOSIS

To prevent a first variceal hemorrhage, patients with cirrhosis should undergo endoscopy so that the large varices that cause hemorrhage can be detected and treated. Endoscopy should be performed when the patient's condition is stable. The risk of initiating bleeding from the varices is negligible. Screening should be performed for patients with low platelet counts, splenomegaly, or advanced cirrhosis. Endoscopy should also be performed for any patient who has hemorrhage of unexplained cause. In 25% of patients with varices that bleed, the cause is something other than varices. Esophageal varices are believed to be the cause of bleeding if no other source of bleeding is found. Other causes include gastric or duodenal ulcers, gastritis, Mallory-Weiss tear, and gastric varices.

At endoscopy, the varices are blue, round, and surrounded by congested mucosa as they protrude into the lumen of the distal esophagus. They are soft and compressible, and an esophagoscope can be passed easily beyond them. Erosion of the superficial mucosa, with an adherent blood clot, signifies the site of a recent hemorrhage. On establishing the presence of esophageal varices, the clinician should also search for gastric varicosities, because surgical treatment may need to be modified if these have developed.

Only 40% of varicosities can be seen on radiographs. A typical finding is a "honeycomb" formation produced by a thin layer of barium surrounding the venous protrusion that does not constrict the lumen. Endoscopic color Doppler ultrasonography is a useful modality for obtaining color flow images of esophageal varices and their hemodynamics. Capsule endoscopy is now being studied as a possible screening tool for esophageal varices; it has a sensitivity and specificity of 84% and 88%, respectively. Recently, 64-row multidetector computed tomography (CT) portal venography reliably displayed the location, morphology, origin, and collateral types of esophageal varices, showing promise as a diagnostic tool. CT was found to have 90% sensitivity and 50% specificity in finding esophageal varices. It also has the benefit of detecting extraluminal pathology that cannot be seen by endoscopy.

TREATMENT AND MANAGEMENT

Variceal management encompasses three phases: (1) prevention of initial bleeding, (2) management of acute bleeding, and (3) prevention of rebleeding. Treatment includes pharmacologic, endoscopic, and radiologic shunting and surgery. Once large varices are identified, patients should begin β-blocker therapy, such as propranolol, which reduces portal pressure and variceal blood flow and decreases risk of bleeding by 50%. Adding isosorbide mononitrate further reduces recurrent bleeding. Hepatic venous pressure measurements are used to monitor the success of this combination pharmacologic therapy, shown to be superior to sclerotherapy and possibly superior to band ligation. A recent meta-analysis showed that a combination of endoscopic and pharmacologic therapy reduces overall and variceal rebleeding in cirrhosis more than either therapy alone.

If β-blockers are not tolerated or are contraindicated, or if patients are at high risk for bleeding, endoscopic band ligation is preferred over sclerotherapy because of fewer complications and lower cost. Surveillance of varices, with potential rebanding, should be repeated every 6 months.

Bleeding requires simultaneous control, resuscitation, and prevention/treatment of complications. Medical treatment of bleeding with vasopressin, terlipressin, somatostatin, or octreotide is started. These medications stop the bleeding in 65% to 75% of patients, but 50% will bleed again within a week. Vasopressin is a posterior pituitary hormone that constricts splanchnic arterioles and reduces portal flow and pressure. Prophylactic intravenous antibiotics should also be started. Endoscopy is performed to diagnose and treat hemorrhage.

Definitive therapy is first performed with sclerotherapy or band ligation, which is successful in 90% of patients. Varices are injected with sclerosing solutions to stop acute bleeding. Repeated injections will cause variceal obliteration and may

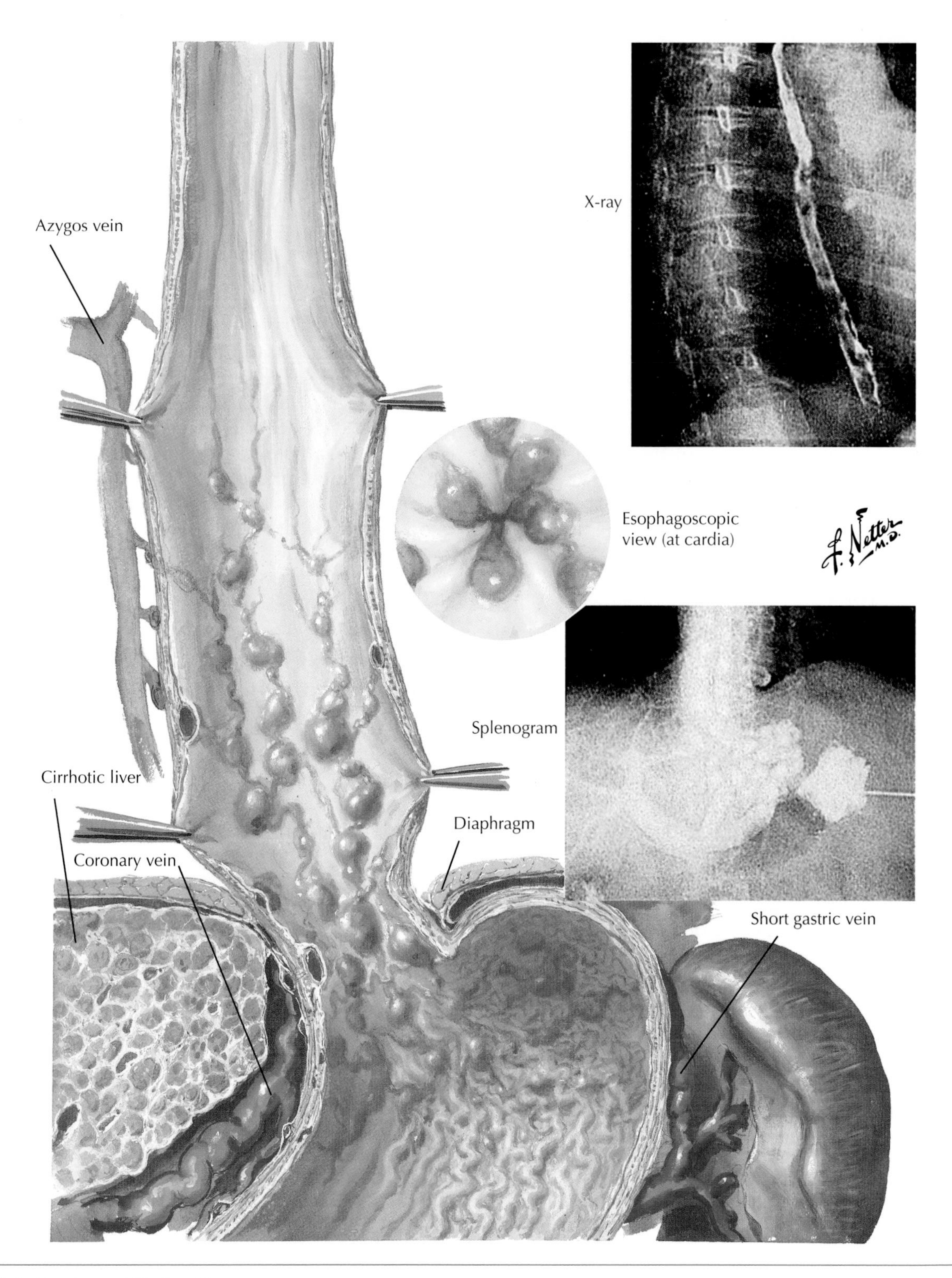

Figure 20-1 *Esophageal Varicosities.*

prevent recurrent bleeding. However, recurrence is common before complete obliteration, and esophageal strictures typically develop.

Endoscopic band ligation results in fewer strictures and ulcers than sclerotherapy and faster eradication. Rebleeding is less frequent with ligation than with sclerotherapy (26% vs. 44%), but number of blood transfusions, duration of hospital stay, and mortality risk are comparable.

When bleeding is under control, endoscopic ligation and sclerotherapy are repeated every 1 to 2 weeks until the varices are eradicated. This technique has the fewest complications and the lowest incidence of recurrence. Surveillance is performed at 3- to 6-month intervals to detect and treat any recurrence. Patients who have two or more rebleeding episodes should be considered for surgery or transplantation.

Balloon tamponade is used as a bridge to definitive therapy in 6% of patients when hemostasis is not achieved. Connected balloons in the stomach and the esophagus compress the varices. Bleeding stops in 80% to 90% of patients, but unfortunately, 60% of them have recurrences. Complications such as aspiration and esophageal rupture may also occur. A new method involves the use of a self-expanding stent to stop acute bleeding from esophageal varices; initial studies reveal no method-related mortality or complications.

If medical and endoscopic therapies fail, transjugular intrahepatic portosystemic shunt (TIPS) is the procedure of choice for emergency bleeding. TIPS should be reserved for patients who have poor liver function. It can be performed in 90% of patients but is used in only 7%. The mortality rate with TIPS is low. Bleeding may recur in 15% to 20% of patients over 2 years. Patients must be followed closely because the shunt may occlude in up to 50% of cases within 18 months.

Shunt procedures are not the modality of choice because they result in a high rate of complications compared with medical therapy. Shunts are used now in less than 1% of patients. Emergency bleeding may be controlled with a central portacaval shunt or with combined esophageal transaction, gastric devascularization, and splenectomy in patients hopeful for liver transplantation. Emergency shunt surgery carries a 50% mortality risk and is rarely undertaken.

Surgical shunts should be used to prevent rebleeding in patients who do not tolerate, or who are noncompliant with, medical therapy and who have relatively preserved liver function. Portal decompression procedures create a connection between the high-pressure portal and the low-pressure systemic venous systems. Nonselective shunts include portacaval anastomoses and TIPS, which decompress the entire portal system. Selective shunts, such as the distal splenorenal shunt, only decompress esophageal varices. Shunt surgery does not improve survival and may result in hepatic encephalopathy. Elective shunt procedures are avoided in candidates for liver transplantation but may be performed in those with Child A and B cirrhosis. Liver transplantation is the best therapy for patients with Child C cirrhosis and is performed in only 1% of patients.

COURSE AND PROGNOSIS

Acute variceal hemorrhage occurs more often in patients with Child B and C cirrhosis. Endoscopic banding is the most common single endoscopic intervention. Early rebleeding occurs in 13% of patients within a week. Although medical therapy, banding, and sclerotherapy are still used frequently for rebleeding, balloon tamponade is necessary in 17%, TIPS in 15%, and surgical shunting in 3% of patients. Early complications after acute variceal bleeding include esophageal ulceration (2%-3% of patients), aspiration (2%-3%), medication adverse effects (0%-1%), dysphagia and odynophagia (0-2%), encephalopathy (13%-17%), and hepatorenal syndrome (2%). The prognosis for patients with bleeding esophageal varices depends directly on liver function. Overall short-term mortality rates after acute bleeding are 10% to 15%. However, in patients with cirrhosis who have variceal bleeding, mortality risk is as high as 60% at 1 year.

ADDITIONAL RESOURCES

Comar KM, Sanyal AJ: Portal hypertensive bleeding, *Gastroenterol Clin North Am* 32:1079-1105, 2003.

De Franchis R, Eisen GM, Laine L, et al: Esophageal capsule endoscopy for screening and surveillance of esophageal varices in patients with portal hypertension, *Hepatology* 47(5):1595-1603, 2008.

Jamal MM, Samarasena JB, Hashemzadeh M, et al: Declining hospitalization rate of esophageal variceal bleeding in the United States, *Clin Gastroenterol Hepatol* 6(6):689-695 (quiz 605), 2008.

Laine L, el-Newihi HM, Migikovsky B, et al: Endoscopic ligation compared with sclerotherapy for the treatment of bleeding esophageal varices, *Ann Intern Med* 119:1-7, 1993.

Perri RE, Chiorean MV, Fidler JL, et al: A prospective evaluation of computerized tomographic (CT) scanning as a screening modality for esophageal varices, *Hepatology* 47(5):1587-1594, 2008.

Sorbi D, Gostout CJ, Peura D, et al: An assessment of the management of acute bleeding varices: a multicenter prospective member-based study, *Am J Gastroenterol* 98:2424-2434, 2003.

Zaman A: Current management of esophageal varices, *Curr Treat Options Gastroenterol* 6:499-507, 2003.

Zehetner J, Shamiyeh A, Wayand W, et al: Results of a new method to stop acute bleeding from esophageal varices: implantation of a self-expanding stent, *Surg Endosc* 22(10):2149-2152, 2008.

Gastroesophageal Reflux Disease

Neil R. Floch

21

Gastroesophageal reflux disease (GERD) is a common, lifelong condition that requires long-term treatment. Accounting for 75% of diseases that occur in the esophagus, GERD entails the reflux of gastric and duodenal contents through the lower esophageal sphincter (LES) into the esophagus to cause symptoms or injury to esophageal, oropharyngeal, or respiratory tissues. GERD cannot be diagnosed by symptoms alone because patients with similar presentation may have achalasia, diffuse esophageal spasm, gastritis, cholecystitis, duodenal ulcer, esophageal cancer, or coronary artery disease. Patients may also have atypical symptoms and may consult several physicians before the correct diagnosis is established. Esophagogastroduodenoscopy (EGD) may reveal esophagitis, but only 90% of esophagitis is secondary to reflux. Many patients may have *nonerosive reflux disease* (NERD), which shows no sign of esophageal inflammation. The best way to determine the presence of GERD is to use 24-hour pH testing (**Fig. 21-1**).

The pathophysiology of GERD is complex and not completely understood. The antireflux mechanism depends on proper function of the esophageal muscle, LES, and stomach. Reflux develops when LES pressure drops, as occurs with gastric distention, which shortens the LES length. Over time, transient lower esophageal sphincter relaxations (TLESRs) become more common, and the valve becomes permanently damaged, resulting in manifestations of GERD. The esophageal muscle works to clear the lumen of both acid and duodenal contents. Poor luminal clearance increases the exposure time, allowing previously healthy epithelium to become damaged tissue. The com-

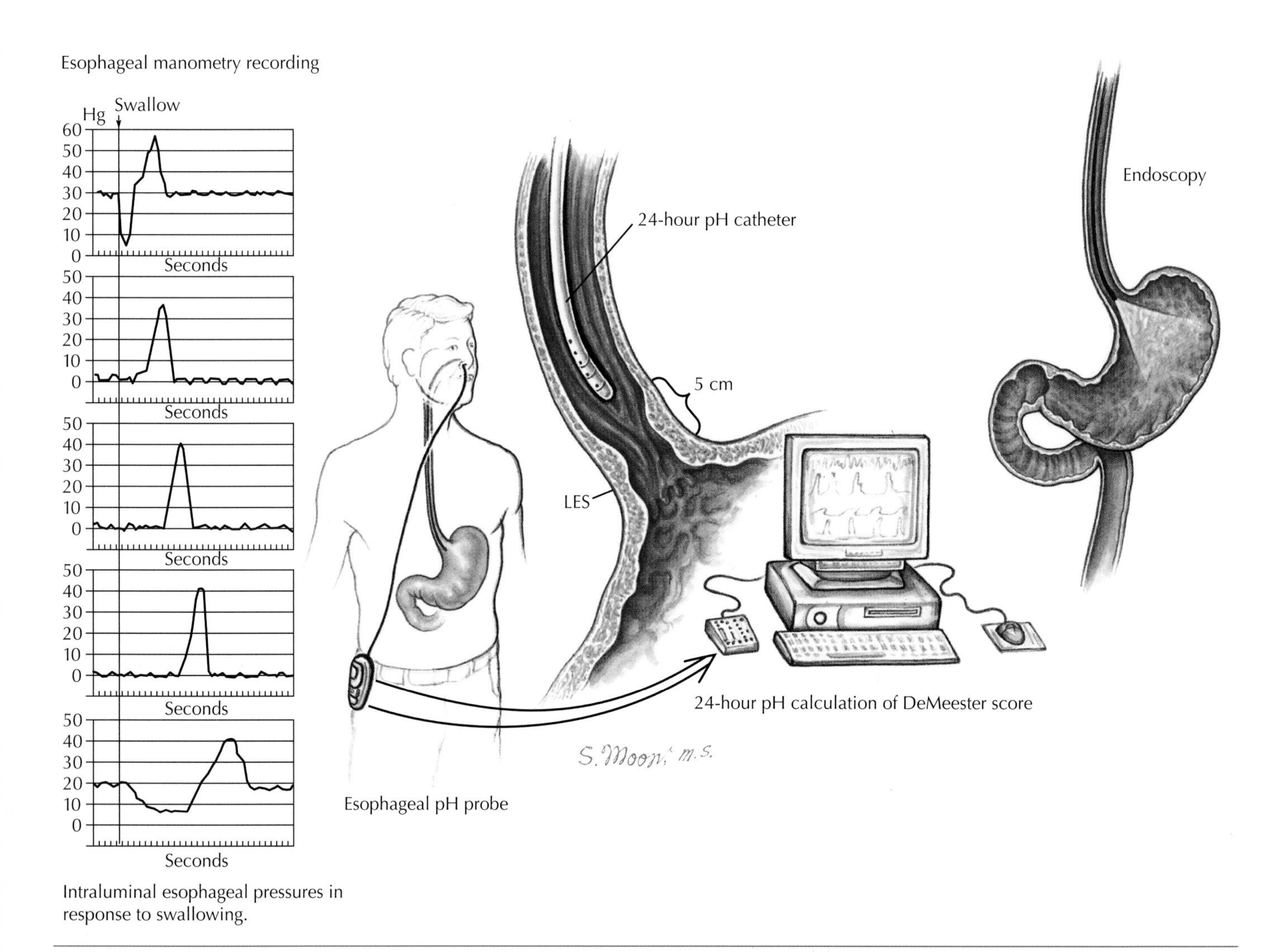

Figure 21-1 *Esophageal Tests.* Graph from Waters PF, DeMeester TR. *Med Clin North Am* 65:1238, 1981.

position of the reflux fluid and the susceptibility of the esophagus, oropharynx, and respiratory structures to damage affect the pathogenesis of the symptoms and possible lesions.

Although GERD is chronic and usually nonprogressive, complications may include peptic esophageal erosion, ulceration, stricture, Barrett esophagus, and esophageal adenocarcinoma. Progression from one complication to another is not clearly established across the GERD continuum, although there is a clear progression from Barrett esophagus to esophageal adenocarcinoma. Evidence suggests that patients with NERD may be less susceptible to complications. Recent studies reveal that *Helicobacter pylori* eradication leads to more resilient GERD. *Duodenogastroesophageal reflux* (DGER) involves not only acid reflux but also the retrograde flow of duodenal contents into the esophagus. The presence of bile, pepsin, and pancreatic enzymes in addition to acid indicate a more destructive atmosphere and therefore more severe disease. NERD is the most common type of GERD. These patients have typical reflux symptoms but have no visible mucosal changes. Only 50% of patients with NERD have abnormal 24-hour pH monitoring. The histopathologic feature found in the NERD patients is dilated intercellular spaces within the squamous cell epithelium. This ultrastructural abnormality is detected on transmission electron microscopy and light microscopy.

CLINICAL PICTURE

Heartburn (pyrosis) is the main symptom of GERD. It is a burning sensation in the chest or epigastrium caused by stomach acid, which rises into the esophagus. Of American adults, 44% experience heartburn monthly, 18% weekly, and 5% to 10% daily. Typical symptoms of GERD also include reflux of acid, regurgitation of food, epigastric abdominal pain, dysphagia, odynophagia, nausea, bloating, and belching. Recent data support that being overweight, or even moderate weight gain among persons of normal weight, may cause or exacerbate symptoms of reflux.

Atypical or extraesophageal symptoms include noncardiac chest pain, choking, laryngitis, coughing, wheezing, difficulty breathing, sore throat, hoarseness, asthma, and dental erosions. GERD is present in the 50% of patients who have atypical chest pain and negative results on coronary angiography. GERD is linked to asthma and chronic cough, and it is found in 80% of persons with asthma. Physiologic changes caused by asthma and chronic cough cause airway inflammation and may promote acid reflux. This involves nerve reflexes, cytokines, inflammatory and neuroendocrine cells, and occasionally tracheal aspiration of refluxed gastric contents.

Reflux symptoms are responsible for almost one third of otolaryngeal disorders. Patients with extraesophageal reflux (EER) have increased amounts of laryngeal reflux despite an adequate esophageal clearance mechanism. TSLERs may be the mechanism. The ciliated epithelium of otolaryngeal structures is more susceptible to damage from refluxate, which can occur from fewer and briefer episodes. The active pepsin in EER disease contributes to laryngeal lesions and eustachian tube dysfunction.

The severity of symptoms is not a reliable indicator of the severity of erosive esophagitis. Chronic abnormal gastric reflux results in erosive esophagitis in 50% of patients, but GERD patients may also be asymptomatic.

DIAGNOSIS

Administering the proper therapy requires determining the presence of complications and the cause of GERD. Diagnostic testing should be done for patients with persistent symptoms who are already receiving therapy; those with recurrent symptoms, weight loss, dysphagia, or gastrointestinal bleeding; and those at risk for complications of esophagitis, as indicated by stricture formation, Barrett esophagus, and adenocarcinoma. Diagnosis depends on a combination of radiologic, pathologic, physiologic, and endoscopic findings. Tests are selected based on the information needed and may include esophageal pH monitoring, impedance testing, acid provocation tests, modified barium swallow, and endoscopy.

Endoscopy is the preferred method to diagnose reflux or hiatal hernia, to grade esophagitis, and to obtain a biopsy sample of the esophagus to rule out Barrett esophagus or cancer. Among the classification systems used to grade disease severity, the Los Angeles Classification is the most widely accepted. Up to 50% of patients with GERD have no endoscopic evidence of esophagitis (NERD). Compared with patients who have erosive esophagitis (75%) and Barrett esophagus (93%), patients with NERD (45%) were significantly less likely to have abnormal pH findings.

In 24-hour pH testing for GERD, a probe is placed 5 cm above the LES to obtain pH readings. The test measures real-time acid exposure and the ability of acid to clear the esophagus, correlating symptoms with acid exposure. Six determinants are used to calculate a DeMeester score: total time of reflux, upright time, supine time, number of episodes, number of episodes longer than 5 minutes, and longest episode. Any patient with a score greater than 14.72 is considered positive; the sensitivity and specificity of the test is 96%. Although 24-hour pH monitoring is the most sensitive and specific test for GERD, 25% of patients with GERD-compatible symptoms have a normal pH test. This test must be performed before surgery if the patient has no signs of GERD on EGD (i.e., patient has NERD). The Bravo pH monitoring system (Medtronic, North Shoreview, Minn) is an endoscopically placed device that measures 24-hour pH without the need for a nasogastric tube. It is a more comfortable option for patients.

Multichannel intraluminal impedance (MII) is a new technique to assess the movement of substances in the esophagus based on differences in their conductivity to an alternating current. MII reacts to the electrical charges within the esophageal mucosal, submucosal, and muscular layers and to any other material within the esophagus that produces a change. Electrical *impedance* is the converse of conductivity and decreases from air to mucosal lining, to saliva, to swallowed material, and finally to refluxed gastric contents (lowest impedance). Impedance increases and decreases depending on the material encountered. Using multiple impedance detection sites on a single catheter reveals the direction of bolus movement. Combining MII with esophageal manometry or 24-hour pH on the same catheter expands the diagnostic tools for evaluation of esophageal function. Combined MII-pH allows detection of all types of gastroesophageal refluxate: acid, nonacid, liquid, mixed, and air. In combined MII-pH, the pH sensor is used simply to characterize whether the refluxate is acid or nonacid based. The MII technology is not a replacement for current manometry or pH techniques, but rather a complementary procedure that expands the diagnostic potential of esophageal function testing.

Recent studies reveal that 60% of reflux episodes are not conventional and can be detected only by impedance changes, not by 24-hour pH testing. More than 98% of reflux events detected by a decrease in pH to less than 4 were detected by impedance changes. Liquid-only reflux occurs in approximately 35% of patients, mixed liquid and gas reflux in 36%, and gas reflux in 27%. Liquid is confined to the distal esophagus in approximately 30% of patients, reaches the midesophagus in 60%, and reaches the proximal esophagus in 11%. Additional information provided by impedance technology is likely to have a major impact on the clinical management of patients with GERD.

Manometry is used to determine LES pressure and motility of the esophageal body. A catheter with lateral side holes is connected to a transducer to measure LES pressure, esophageal body pressure, and LES length. The data are useful before surgery. Ambulatory manometry over a 24-hour period provides 100 times more data and is therefore more helpful in the diagnosis of esophageal body disorders. Conditions such as scleroderma must be ruled out. A complete absence of peristalsis and a hypotensive LES are characteristic of scleroderma. A Toupet (partial) fundoplication is required to avert postoperative obstruction.

Videoradiography records the act of swallowing, which may then be observed at several speeds. This technique is helpful during the pharyngeal phase of swallowing by identifying structural abnormalities of the esophagus, such as ulcers, strictures, paraesophageal hernia, masses, reflux, and obstruction. Simple barium esophagraphy may also reveal esophageal disease, but it is not as sensitive as cineradiography because it cannot detect spontaneous reflux in 60% of patients. However, when reflux is found on barium swallow, it is specific and is almost always confirmed by 24-hour pH testing.

Use of 24-hour ambulatory bile monitoring is best for determining the presence of DGER. Not only can acid injure the esophagus, but pepsin, bile, and pancreatic juices may also cause damage. Bile serves as a marker for duodenal substances and can be detected by its light wavelength using an indwelling spectrophotometer probe. Monitoring is useful for identifying patients at risk for esophageal injury and therefore candidates for surgery.

The standard acid reflux test, formerly the Bernstein test, is performed by placing a pH probe 5 cm above the LES and injecting 300 mL of 0.1 N hydrochloride (HCl) into the stomach with a manometry probe. Four maneuvers are performed in four different positions, giving 16 recordings. More than two reflux episodes is a positive finding. This test is helpful for patients receiving long-term proton pump inhibitor (PPI) therapy whose 24-hour pH values may be inaccurate. In several studies, PPIs were found 40 days after the dose was taken.

Gastric emptying, as measured with a radionucleotide-tagged meal, is helpful in determining delayed emptying. Solids and liquids may be measured simultaneously with different markers. Pictures are taken at 5-minute intervals for 2 hours.

TREATMENT AND MANAGEMENT

Medical therapy cannot resolve abnormal LES function. Therefore, the medical treatment of GERD centers on suppression of intragastric acid secretion. Goals of treatment are to provide effective symptomatic relief, prevent symptom relapse, achieve healing of esophageal damage, and prevent complications of esophagitis.

Lifestyle and dietary changes are the first steps for treatment because of their low cost and simplicity; these include elevating the head of the bed, modifying the size and composition of meals, consuming low-fat foods, and avoiding coffee, wine, tomato, chocolate, and peppermint. These changes should be continued even if more potent therapies are added.

Medical treatment begins with a step-up approach as an H_2-receptor antagonist (H_2RA) is prescribed for 8 weeks. If symptoms do not improve, patients are changed to a PPI. Titration to the lowest effective medication type and dosage should be performed in all patients. H_2RA is taken on demand, whereas PPIs are taken 30 to 60 minutes before the first meal of the day. PPI therapy is the most successful medical treatment. Of the PPI medications, esomeprazole at 40 mg once daily is more effective than standard doses of lansoprazole, omeprazole, pantoprazole, or rabeprazole in patients with symptoms of GERD. For patients with erosive esophagitis, esomeprazole has demonstrated higher healing rates and more rapid, sustained resolution of heartburn than omeprazole or lansoprazole after up to 8 weeks of once-daily treatment. Although healing of the esophageal mucosa is achieved with a single dose of any PPI, symptoms are difficult to control in more than 80% of patients.

An estimated 30% of GERD patients who require a PPI once daily will fail treatment, most often patients with NERD. Suggested mechanisms include weakly acidic reflux, duodenal gastric reflux (DDGR), visceral hyperalgesia, delayed gastric emptying, psychologic comorbidity, and functional bowel disease. Available diagnostic modalities provide limited information on the cause of failure. Current treatment relies on increasing doses of PPIs. The pathophysiology of PPI failure should provide alternative therapeutic options in the future.

If symptom control fails or symptoms return after medication is discontinued, endoscopy establishes the diagnosis. For patients who have erosive esophagitis, as identified on endoscopy, a PPI is the initial treatment of choice. These patients should undergo 24-hour pH testing and bile probe testing in selected centers to determine the severity of reflux. Patients who have supine reflux, poor esophageal contractility, erosive esophagitis, Barrett esophagus, or defective LES are predicted to do poorly with medication and are at high risk for complications of GERD. These patients should be offered the option of surgery.

The best medical treatment for DGER is PPI therapy, which decreases the level of gastric acidity and the volume of gastric fluid available for esophageal reflux. Adding γ-aminobutyric acid (GABA) receptor agonist baclofen may further reduce DGER in patients not responding to PPIs. Prokinetic agents may also alleviate symptoms by promoting increased gastric emptying. In patients with refractory disease, a Roux-en-Y diversion or duodenal-switch procedure may be helpful.

Antireflux surgery, including open and laparoscopic versions of Nissen and Toupet fundoplication, are as effective as PPI therapy and should be offered to patients with DGER as an alternative to medication for chronic reflux with recalcitrant symptoms. Surgery should be performed if symptoms fail to resolve while the patient is taking medication or if symptoms develop during drug therapy or recur after medication is stopped. Similarly, surgery should be performed if the patient is noncompliant, has lifelong PPI dependence, has experienced complications despite medication, or has recurrent strictures, pulmonary symptoms, severe esophagitis, symptomatic Barrett's esophagus, or symptomatic paraesophageal hernia.

Best results from surgery occur in patients who are young, have typical GERD symptoms, have abnormal pH study findings, and show good response to PPI therapy, but these patients are the best candidates for medical therapy, as well. Results after 10 years reveal at least a 90% success rate. Side effects include bloating (20.5%), diarrhea (12.3%), regurgitation (6.4%), heartburn (5.8%), and chest pain (4.1%); 27.5% of patients reported dysphagia, and 7% required dilatation. Although 14% of patients continue receiving PPI therapy, 79% of these patients are treated for vague abdominal or chest symptoms with unclear indications. The surgical outcome for NERD patients is similar to that for patients with erosive esophagitis; therefore, surgery is effective regardless of the endoscopic appearance of the esophageal mucosa.

Endoluminal transoral procedures offer an outpatient therapy option that is less invasive than laparoscopic fundoplication. The Stretta system uses radiofrequency energy to cauterize the gastroesophageal junction and causes thickening of the muscle and therefore reduced compliance of the sphincter. The Bard Endocinch procedure involves an "overtube" placed over an endoscope that allows tissue 0.5 to 2.5 cm below the squamocolumnar junction to be "sucked" into the tube and sutured. A total of four sutures are placed. The Full-Thickness Plicator (NDO Medical, Mansfield, Mass) is composed of a gastroscope and a suturing device that takes full-thickness, serosa-to-serosa bites and deploys two prettied 2-0 polypropylene sutures, two expanded polytetrafluoroethylene bolsters, and two titanium retention bridges. Despite multiple studies with all these devices, resolution of symptoms, improvement of GERD, and cessation of PPIs have reached 50% at best. Many long-term studies have not been promising. Although these procedures are less invasive and induce fewer complications than antireflux surgery, their success rates are significantly lower. The role of these procedures in the treatment of reflux has yet to be determined and may depend on advanced modifications and techniques with the possibility of improved results.

COURSE AND PROGNOSIS

Antacids result in symptom relief in 20% of GERD patients but have minimal effect on pH (acidity) and no effect on healing. H_2RA therapy results in symptom relief in 40% to 70% of patients and healing in 20% to 50%. Remission is maintained in only 25% to 40%. Higher and more frequent doses may improve symptoms minimally. All H_2RAs are similar in efficacy, and adverse effects are uncommon and mild. PPIs have the best acid-blocking effect, alleviating symptoms in 90% of patients and promoting healing in 80% to 90%. Once-daily omeprazole (20 mg) has a greater acid-blocking effect than twice-daily ranitidine (150 mg). Even so, up to 70% of patients do not have adequate nocturnal control of gastric acid secretion with omeprazole (20 mg) twice daily. GERD is a chronic, relapsing disease; long-term maintenance therapy is safe and necessary to relieve symptoms, prevent complications, and prevent recurrence in 40% to 50% patients.

Despite its usefulness, pH testing cannot definitively establish a causative relationship between GERD and extraesophageal symptoms. Therefore, effective treatment resulting in significant improvement or remission of extraesophageal symptoms is the best evidence of GERD's pathogenic role. Extraesophageal symptoms usually require more prolonged and aggressive antisecretory therapy than typical GERD.

Since its advent in 1991, laparoscopic Nissen fundoplication has become the "gold standard" for the treatment of severe GERD. Multiple trials comparing surgical fundoplication and PPI therapy reveal similar effectiveness in controlling GERD and its symptoms. Longer studies reveal an advantage of surgery that is eliminated when PPI dosage is increased.

ADDITIONAL RESOURCES

Bammer T, Hinder RA, Klaus A, Klingler PJ: Five- to eight-year outcome of the first laparoscopic Nissen fundoplications, *J Gastrointest Surg* 5:42-48, 2001.

Fass R: Proton-pump inhibitor therapy in patients with gastro-oesophageal reflux disease: putative mechanisms of failure, *Drugs* 67(11):1521-1530, 2007.

Jacobson BC, Somers SC, Fuchs CS, et al: Body-mass index and symptoms of gastroesophageal reflux in women, *N Engl J Med* 354(22):2340-2348, 2006.

Kahrilas PJ: GERD pathogenesis, pathophysiology, and clinical manifestations, *Cleve Clin J Med* 70(suppl 5):S4-S19, 2003.

Lundell L, Attwood S, Ell C, Fiocca R, et al: Comparing laparoscopic antireflux surgery with esomeprazole in the management of patients with chronic gastro-oesophageal reflux disease: a 3-year interim analysis of the LOTUS trial, *Gut* 57(9):1207-1213, 2008.

Martinez SD, Malagon IB, Garewal HS, et al: Non-erosive reflux disease (NERD)–acid reflux and symptom patterns, *Aliment Pharmacol Ther* 17:537-545, 2003.

Napierkowski J, Wong RK: Extraesophageal manifestations of GERD, *Am J Med Sci* 326:285-299, 2003.

Richter JE: Diagnostic tests for gastroesophageal reflux disease, *Am J Med Sci* 326:300-308, 2003.

Richter JE: Duodenogastric reflux–induced (alkaline) esophagitis, *Curr Treat Options Gastroenterol* 7:53-58, 2004.

Spechler SJ: Clinical manifestations and esophageal complications of GERD, *Am J Med Sci* 326:279-284, 2003.

Tutuian R, Castell DO: Management of gastroesophageal reflux disease, *Am J Med Sci* 326:309-318, 2003.

Esophagitis: Acute and Chronic

22

Neil R. Floch

Acute esophagitis may have numerous causes, of which gastroesophageal reflux disease (GERD) is the most common (**Fig. 22-1**). *Chronic* esophagitis occurs more frequently and results from multiple episodes of acute inflammation. Of patients undergoing esophagogastroduodenoscopy (EGD), 14% have esophagitis, and most are men. Hiatal hernia is present in 79% to 88% of patients with active reflux esophagitis. The incidence of reflux esophagitis is rapidly increasing; in one study, it had doubled over 10 years. In Belgium the incidence of erosive esophagitis (EE) rose dramatically and then stabilized with a sixfold increase in the use of proton pump inhibitors (PPIs).

Esophagitis is believed to be caused not only by acid but also by the reflux of bile, enzymes, pepsin, and pancreatic juices. Acid-induced esophagitis may induce hyperresponsive longitudinal smooth muscle contraction and impairment of circular smooth muscle contractility, which may lead to chronic complications.

Esophagitis may occur from pills that remain in the esophagus for an extended period, causing irritation. Opportunistic infections of the esophagus are a common cause of morbidity in patients with human immunodeficiency virus (HIV) infection and may reflect the severity of the underlying disease. Less frequent causes of esophagitis include swallowing acid or basic household materials, severe vomiting, irritation by feeding tubes or suction catheters, and candidal or other infectious diseases.

CLINICAL PICTURE

Only 50% of patients with endoscopic evidence of esophagitis have typical reflux symptoms of GERD. Heartburn, the most common symptom in patients with esophagitis, is present in only 28% of those with endoscopic evidence. Other symptoms are dysphagia (19%), acid regurgitation (18%), odynophagia (6%), nausea, vomiting, and belching. Older age, male gender, severe symptoms, and presence of a hiatal hernia are independent risk factors for severe esophagitis. Patients with HIV-related diseases have symptoms associated with the specific etiology.

DIAGNOSIS

Barium esophagraphy and EGD have low but comparable rates of accuracy for detecting reflux esophagitis, with sensitivities of 35% and 39% and specificities of 79% and 71%, respectively. Esophagoscopy reveals congestion, erythema, and edema of the mucosa, as well as pinpoint hemorrhages. Endoscopic biopsy is the best way to detect reflux esophagitis. Microscopy reveals epithelial necrosis, erosions, small cell infiltration, and hypertrophy of muscle fibers. Esophagitis is usually located between the gastroesophageal junction and 10 cm above. Manometry may reveal ineffective esophageal motility, found to be independently associated with EE.

TREATMENT AND MANAGEMENT

Goals of treatment for EE are to heal lesions, relieve symptoms, and prevent relapse. Daytime and nighttime esophageal pH must be kept above 4 for as long as possible. In the past, short-term treatment with both PPIs and H_2 blockers was effective in healing EE, but PPIs have proved to be better, and therefore H_2 blockers are no longer used as a primary treatment. Multiple prospective trials indicate that esomeprazole has a higher rate of action, lower interpatient variability, and more prolonged action in achieving esophageal healing in the subset of patients with esophagitis. "On-demand" maintenance therapy is not effective in treating EE. Numerous PPI studies show no differ-

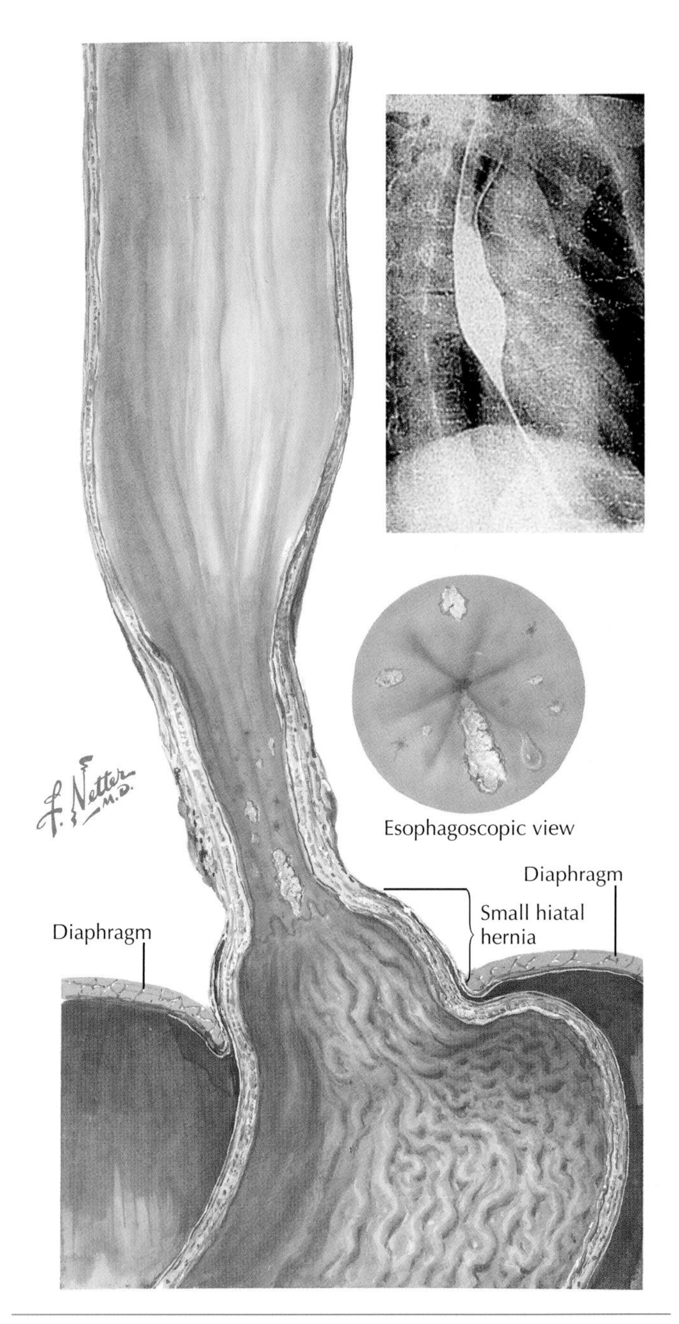

Figure 22-1 *Acute and Chronic Esophagitis.*

ence in effectiveness with intravenous versus oral, or with intake of pills versus oral granular suspension, in treatment of EE.

In maintenance therapy, only PPIs reduce symptoms, as well as the incidence of and interval to relapse, making PPIs the recommended medical therapy for the long-term management of EE. Maintenance PPI therapy is also cost-effective.

If maintenance therapy is not initiated, most patients relapse within 1 year. Relapse increases the severity of esophagitis and the risk for complications such as Barrett esophagus and adenocarcinoma. Poor compliance is the main reason for failure and relapse, followed by nonacid reflux, especially in patients with regurgitation or cough that persists despite treatment.

Antireflux surgery is effective in relieving symptoms and healing EE. It is performed after medical treatment has failed or as an alternative to long-term maintenance. The effectiveness of newer modalities of endoscopic treatment is not yet known.

"Pill esophagitis" is treated acutely with sucralfate, after which patients are instructed in the proper timing and use of water when swallowing pills.

COURSE AND PROGNOSIS

All PPIs resolve esophagitis in 89% to 93% of patients at 8 weeks, although resolution is faster and more common in patients taking esomeprazole. Maintenance success at 6 to 12 months varies from 82% to 93% with most PPIs.

Laparoscopic antireflux surgery (LARS) resolves persistent GERD symptoms and maintains resolution regardless of the endoscopic appearance of the esophageal mucosa. Multiple studies show at least similar effectiveness of long-term, continuous medical therapy and surgery, although data at up to 7 years suggest a benefit to surgery. Long-term surgical problems include increased gas-bloat and use of PPIs. Patients may become poorly compliant or may have nonacid reflux. Complications of esophagitis include multiple small superficial ulcerations, larger flat ulcerations, and fibrous tissue formation leading to strictures.

Improved medical therapeutic response and remission may depend on the development of new PPI isomers, potassium-competitive acid blockers, and inhibitors of transient LES relaxation.

ADDITIONAL RESOURCES

Coron E, Hatlebakk JG, Galmiche JP: Medical therapy of gastroesophageal reflux disease, *Curr Opin Gastroenterol* 23(4):434-439, 2007.

Edwards SJ, Lind T, Lundell L: Systematic review: proton pump inhibitors (PPIs) for the healing of reflux oesophagitis—a comparison of esomeprazole with other PPIs, *Aliment Pharmacol Ther* 24(5):743-750, 2006.

Fornari F, Callegari-Jacques SM, Scussel PJ, et al: Is ineffective oesophageal motility associated with reflux oesophagitis? *Eur J Gastroenterol Hepatol* 19(9):783-787, 2007.

Katz PO, Ginsberg GG, Hoyle PE, et al: Relationship between intragastric acid control and healing status in the treatment of moderate to severe erosive oesophagitis, *Aliment Pharmacol Ther* 25(5):617-628, 2007.

Lundell L, Miettinen P, Myrvold HE, et al: Seven-year follow-up of a randomized clinical trial comparing proton-pump inhibition with surgical therapy for reflux oesophagitis, *Br J Surg* 94(2):198-203, 2007.

Okamoto K, Iwakiri R, Mori M, et al: Clinical symptoms in endoscopic reflux esophagitis: evaluation in 8031 adult subjects, *Dig Dis Sci* 48:2237-2241, 2003.

Pandolfino JE: Gastroesophageal reflux disease and its complications, including Barrett's metaplasia. In Feldman M, Friedman LS, Sleisenger MH, editors: *Gastrointesinal and liver disease,* ed 7, Philadelphia, 2002, Saunders, pp 599-622.

Wells RW, Morris GP, Blennerhassett MG, Paterson WG: Effects of acid-induced esophagitis on esophageal smooth muscle, *Can J Physiol Pharmacol* 81:451-458, 2003.

Esophageal Ulcers

23

Neil R. Floch

Esophageal ulcers are mucosal defects that have distinct margins (**Fig. 23-1**). They are found in1% of patients undergoing esophagogastroduodenoscopy (EGD). In 66% of patients, the cause is gastroesophageal reflux disease (GERD), resulting from prolonged contact between squamous epithelial cells and gastric refluxate containing acid, pepsin, bile, and pancreatic juices. Drug-induced ulcers account for 23% of all esophageal ulcers and are usually caused by nonsteroidal antiinflammatory drugs (NSAIDs) that have prolonged direct contact with the esophageal mucosa. Ulcers may also be a complication of medications such as doxycycline.

Less prevalent, infectious causes of esophageal ulcer include *Candida*, *Mycobacterium tuberculosis*, *Actinomyces*, herpes simplex virus (HSV), and cytomegalovirus (CMV). Infections may result from caustic injury, marginal ulceration, foreign bodies, and variceal banding, as well as unknown etiologies. Patients with human immunodeficiency virus (HIV) have a higher incidence of infectious ulcers from CMV (45%), idiopathic causes (40%), *Candida* esophagitis (27%), and HSV (5%).

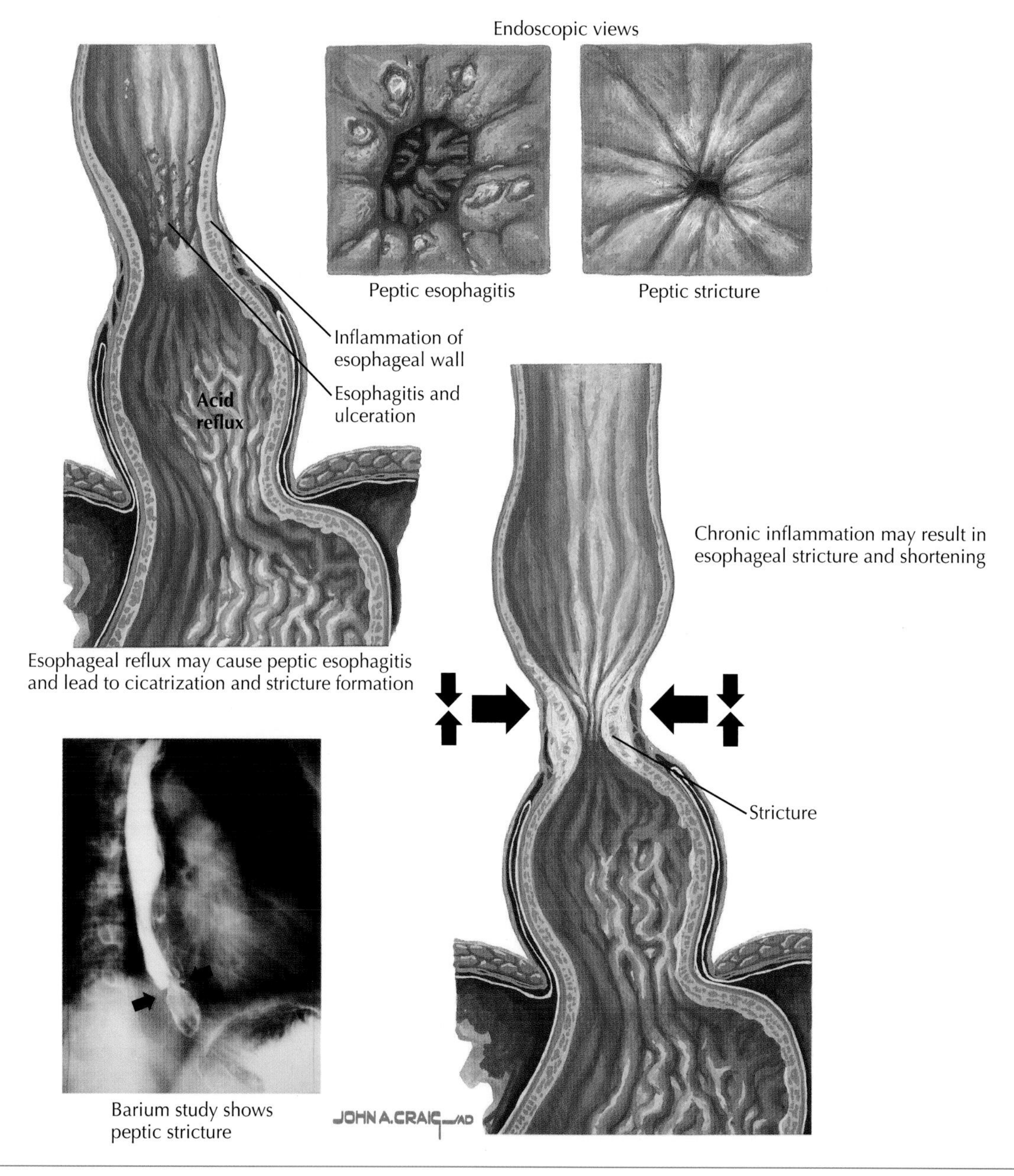

Figure 23-1 *Complications of Peptic Reflux (Esophagitis and Stricture).*

Esophageal ulcers may be complicated by hemorrhage, perforation, and fistulization into the airway. Ulcers may lead to fibrous tissue formation and collagen production (strictures). Healing may occur with intestinal epithelium. This metaplastic process results in Barrett's esophagus; ulceration occurs in 46% of patients with Barrett's esophagus. Since the advent of proton pump inhibitor (PPI) therapy, esophageal ulcers occur less frequently.

CLINICAL PICTURE

Symptoms are rarely different from those in patients with GERD. Most patients have substernal chest pain and may have dysphagia; others may be asymptomatic. The most common sign of esophageal ulcer is anemia; one third of patients may present with acute gastrointestinal (GI) bleeding. In patients with Barrett's esophagus and ulcers, 24% present with active GI bleeding. Melena occurs in 40% of patients, and melena and hematemesis occur concomitantly in another 40%. Fifty percent of patients have orthostatic hypotension, and 8 in 10 patients require blood transfusion.

Bleeding ulcers are associated with NSAIDs in 50% of patients, hiatal hernia in 60%, and esophagitis in 40%. Drug-induced ulcers are usually located in the midesophagus, near the aortic arch, at an area of natural esophageal tapering where pills may become temporarily lodged. Only 13% of ulcers occur in the distal esophagus. Midesophageal ulcers have a greater tendency to hemorrhage than ulcers at the gastroesophageal junction; this may reflect the cause. Strictures occur in 12.5% and esophageal perforation in 3.4% of patients.

DIAGNOSIS

Barium esophagraphy or endoscopy establishes the diagnosis of esophageal ulcer. Both studies may show evidence of GERD, such as overt reflux. Barium esophagraphy may reveal the position of the ulcer, which may be posterior in 69% of patients, lateral in 17%, and anterior in 14%. Nine of 10 ulcers are within 4 cm of the lower esophageal sphincter. Esophagraphy may also reveal hiatal hernias, mucosal nodularity, and strictures, each in 40% of cases. Esophagraphy can make optimal determinations at an average depth of 5 mm.

Endoscopy is the best study to establish a diagnosis. Location, visual characteristics, and biopsy results at esophagoscopy elucidate the cause of the ulcer. A chronic GERD ulcer may be well demarcated, may have undetermined edges and a crater of granulation tissue, and may be covered with a yellow-gray membrane. Esophagitis is usually adjacent to the GERD ulcer and has signs of inflammation, congestion, edema, and superficial erosions. At the site of these changes is a narrowing secondary to segmental spasms. NSAID ulcers have normal surrounding mucosa. Drug-induced ulcers are larger and shallower than GERD-induced ulcers, but both range from 2.75 to 3.0 cm. Biopsy should be performed to exclude the presence of Barrett esophagus and malignancy. Biopsy during EGD is integral to the diagnosis of ulcers in patients with HIV.

TREATMENT AND MANAGEMENT

Ulcer healing is a repair process that involves inflammation, cell proliferation, reepithelialization, formation of granulation tissue, and angiogenesis, as well as cell communication and matrix and tissue remodeling that eventually develop into a scar. These processes are under control of cytokines, growth factors, and transcription factors stimulated by injury to the esophageal lining.

In patients with GERD, uncomplicated, previously untreated esophageal ulcers should be treated with PPI therapy. Currently, the most clinically effective medication for healing erosive esophagitis and later maintenance therapy is esomeprazole. With a drug-induced esophageal ulcer, healing occurs if the ulcer is recognized early. The medication should be discontinued and the patient instructed to swallow pills in the upright position in the future and to drink a glass of water each time. Antacids and H_2 blockers are the fastest-acting therapy, and PPIs allow optimal acid blockade. In patients with HIV, medical therapy focuses on the specific cause of the ulcer. Infectious causes are treated by eradication with the appropriate antimicrobial agent.

Although acute bleeding frequently necessitates blood transfusion, most bleeding stops without endoscopic therapy. Endoscopic hemostasis for esophageal bleeding from ulcers may be required as emergency therapy in 4% of patients. Emergency surgery is reserved for esophageal stricture and perforation in 8% of patients. Elective laparoscopic fundoplication may be necessary for patients whose ulcers fail to heal over the long term.

Future treatments to improve ulcer healing may include the use of stem cells and tissue engineering. Local gene therapy with VEGF + Ang1 and/or SRF cDNAs has shown the ability to accelerate and improve the quality of esophageal ulcer healing.

COURSE AND PROGNOSIS

Nonsurgical therapy is successful in 92% of patients with GERD- and drug-induced ulcers. Follow-up endoscopy indicates that NSAID-induced ulcers heal in 3 to 4 weeks. The healing rate in treated HIV patients is 98%. Strictures complicate GERD-induced esophageal ulcers, but not drug-induced esophageal ulcers. Esophageal dilatation is an effective treatment for most strictures associated with esophageal ulcers. Death from ulcers is rare, but 2% of patients die from acute hemorrhage or perforation.

ADDITIONAL RESOURCES

Higuchi D, Sugawa C, Shah SH, et al: Etiology, treatment, and outcome of esophageal ulcers: a 10-year experience in an urban emergency hospital, *J Gastrointest Surg* 7:836-842, 2003.

Murphy PP, Ballinger PJ, Massey BT, et al: Discrete ulcers in Barrett's esophagus: relationship to acute gastrointestinal bleeding, *Endoscopy* 30:367-370, 1998.

Raghunath AS, Green JR, Edwards SJ: A review of the clinical and economic impact of using esomeprazole or lansoprazole for the treatment of erosive esophagitis, *Clin Ther* 25:2088-2101, 2003.

Spechler SJ: Clinical manifestations and esophageal complications of GERD, *Am J Med Sci* 326:279-284, 2003.

Sugawa C, Takekuma Y, Lucas CE, Amamoto H: Bleeding esophageal ulcers caused by NSAIDs, *Surg Endosc* 11:143-146, 1997.

Tarnawski AS: Cellular and molecular mechanisms of gastrointestinal ulcer healing, *Dig Dis Sci* 50(suppl 1):S24-S33, 2005.

Wolfsen HC, Wang KK: Etiology and course of acute bleeding esophageal ulcers, *J Clin Gastroenterol* 14:342-346, 1992.

Eosinophilic Esophagitis

24

Neil R. Floch

Eosinophilic esophagitis (EOE) is a chronic inflammatory disorder propagated by interleukin-5 (IL-5) and unrelated to gastroesophageal reflux disease (GERD). Formerly a rare disease initially described in children and young men, EOE has been diagnosed more frequently in the past 10 years. According to current estimates, EOE has an annual incidence of 10 per 100,000 in children and teenagers and 30 per 100,000 in the adult population. EOE has a male/female ratio of 3:1. It leads to structural esophageal alterations but does not impact the nutritional state and has no malignant potential. EOE is distinguished by the presence of eosinophilic infiltration of the esophageal mucosa of at least 15 eosinophils per high-power field (hpf) in a patient without a previously identified cause of eosinophilia.

The pathogenesis of EOE is not completely understood, but clinical evidence and basic science support that it is an immune-mediated disease initiated by allergens that are inhaled or consumed. Exposure to the allergens with resultant sensitization may be a genetically acquired predisposition. Foods that are most allergenic include corn, chicken, wheat, beef, soy, eggs, and milk. The pathologic process may entail the activation of eosinophils, mast cells, and lymphocytes with the resultant release of molecules that trigger the onset of symptoms.

CLINICAL PICTURE

Eosinophilic esophagitis is suspected in adults with symptoms of progressive and persistent dysphagia and food impaction. EOE should also be suspected in children with feeding intolerance and GERD symptoms. EOE may have signs and symptoms similar to GERD, but EOE often continues despite prolonged treatment with proton pump inhibitors (PPIs). Interestingly, a history of extensive allergies has been found in more than 50% of patients.

A recent analysis of 24 studies revealed the presence of dysphagia in 93% of EOE patients, food impaction in 62%, heartburn in 24%, and peripheral eosinophilia in 31%. Other symptoms include chest pain, dyspepsia, nausea/vomiting, odynophagia, abdominal pain, and weight loss.

DIAGNOSIS

Although normal in 7% of patients with EOE, endoscopy shows a "feline" or corrugated esophagus in 55% of patients, proximal strictures in 38%, linear furrows in 33%, and diffusely narrowed esophagus in 10% (**Fig. 24-1**). Other features include adherent white plaques (16%) and friable mucosa that shreds easily. All these characteristics, as well as dysphagia, odynophagia, heart-

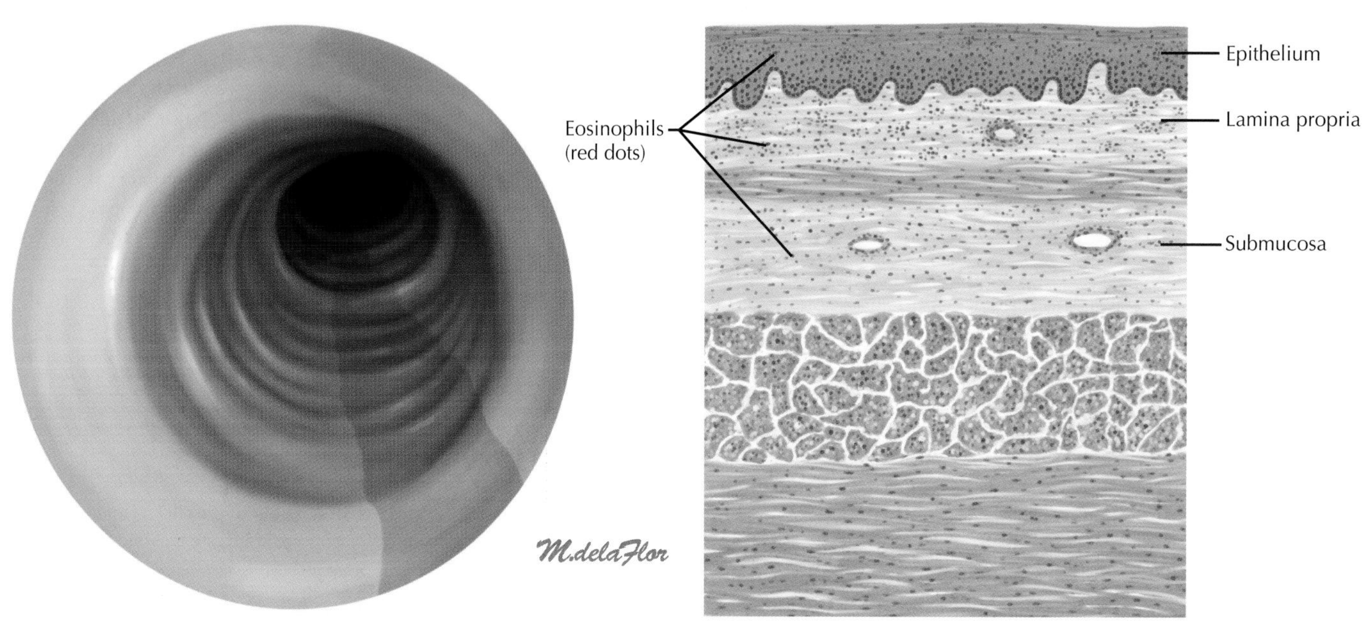

Endoscopic view demonstrates characteristic rings seen in the esophagus with eosinophilic esophagitis

Cross sectional microscopic view of the esophagus demonstrates the infiltration of all layers of the esophagus with eosinophils. The infiltrate is diagnosed most frequently by endoscopic biopsy so it is seen in the biopsy specimen in the epithelium and lamina propria.

Figure 24-1 *Eosinophilic Esophagitis.*

burn, and chest pain in the presence of a normal-appearing esophagus, should warrant a biopsy on endoscopy.

Diagnosis is established by the finding of 15 or more eosinophils/hpf on microscopy of a mucosal biopsy. Also, 97% of these patients have mucosal furrows. At least five biopsies should be performed in the distal esophagus (5 cm above the gastroesophageal [GE] junction) and proximal esophagus (at least 15 cm above the GE junction). At least four biopsies should have sensitivity near 100%. The patient also must have normal gastric and duodenal biopsies. Patients should undergo endoscopic biopsy after 6 to 8 weeks of treatment with a twice-daily PPI or a negative DeMeester score on 24-hour pH monitoring. Esophageal manometry is not a diagnostic modality with EOE but will reveal evidence of an esophageal motility disorder in 40% of patients.

TREATMENT AND MANAGEMENT

Patients with suspected EOE should first receive at least 4 to 8 weeks of PPI therapy, to exclude the presence of acid reflux. The decision is then made whether to treat with pharmacologic or dietary methods. Dilatation is reserved for patients with severe dysphagia from strictures. It is a safe therapy that rarely results in perforation, although superficial mucosal tears can occur in one third of dilatations. Most patients will need two dilatations to achieve symptomatic relief.

The administration of corticosteroids results in symptomatic improvement in more than 95% of patients with EOE. Systemic corticosteroids may be used in the acute setting, but symptoms may recur when stopped. Corticosteroids such as oral prednisone, topical/swallowed fluticasone spray, and swallowed budesonide mixed in a sucralose suspension have improved clinical symptoms and histologic findings. These therapies are more effective on a chronic basis to abate symptoms. Adverse effects include growth retardation, bone abnormalities, and adrenal suppression. Currently, steroids that are swallowed and directly cover the squamous mucosa are the best treatment option for EOE. Cromolyn sodium may offer some benefit. Anti–IL-5 antibodies have shown promise in reducing clinical symptoms as well as blood and esophageal eosinophils and may lead to the development of future therapies.

An elemental diet leads to complete healing and resolution of symptoms in patients with EOE, but the reintroduction of foods leads to return of symptoms. Therefore, treatment must be based on a balance between food exclusion and patient tolerance and compliance of diet. After skin and patch tests, three options exist: removal of foods that react to the skin test, removal of the foods most often responsible, or use of an elemental diet. The patient follows the diet for 2 months, after which endoscopy is repeated with biopsy. If the biopsy is normal, foods are reintroduced. If abnormal, an elemental diet is implemented. Reintroduction of food starts with the least allergenic foods, then slow introduction of more allergenic foods. If foods are associated with symptoms, they are stopped. This method has resulted in a socially acceptable diet in almost 70% of patients.

COURSE AND PROGNOSIS

Long-term treatment of EOE focuses on symptomatic control and mucosal healing. Currently, topical steroids and dietary restriction are the most successful options to achieve this goal. Concomitant use of PPIs is believed to treat secondary acid reflux. Future therapies such as anti–IL-5 antibodies show significant promise.

Evidence-based guidelines for the management of EOE are not currently available. In adults, no randomized trials have demonstrated the efficacy of any particular treatment, and no prospective studies have described the natural history of EOE after treatment.

ADDITIONAL RESOURCES

Furuta GT, Liacouras CA, Collins MH, et al: Eosinophilic esophagitis in children and adults: a systematic review and consensus recommendations for diagnosis and treatment, *Gastroenterology* 133(4):1342-1363, 2007.

Furuta GT, Lightdale CJ: Eosinophilic esophagitis, *Gastrointest Endosc Clin North Am* 18:1, 2008.

Helou EF, Simonson J, Arora AS: Three-year follow-up of topical corticosteroid treatment for eosinophilic esophagitis in adults, *Am J Gastroenterol,* June 2008 (Epub).

Lucendo AJ, Castillo P, Martín-Chávarri S, et al: Manometric findings in adult eosinophilic oesophagitis: a study of 12 cases, *Eur J Gastroenterol Hepatol* 19(5):417-424, 2007.

Sgouros SN, Bergele C, Mantides A: Eosinophilic esophagitis in adults: a systematic review, *Eur J Gastroenterol Hepatol* 18(2):211-217, 2006.

Benign Esophageal Stricture

Neil R. Floch

25

Strictures occur more frequently in men and are most common in elderly white patients. Esophageal strictures develop in 10% to 15% of patients with gastroesophageal reflux disease (GERD) (**Fig. 25-1**) and in 13% of patients with esophageal ulcers. GERD accounts for almost 70% of all esophageal strictures. Less common causes of strictures include ingestion of caustic substances, Barrett esophagus, mediastinal irradiation, ingestion of drugs, malignancy, surgical resection line, congenital esophageal stenosis, skin diseases, and pseudodiverticulosis.

In reflux esophagitis, acid and pepsin secretions eventually erode the mucosa of the esophagus, causing replacement with fibrous tissue, which eventually contracts and results in a lumen as narrow as 2 to 3 mm. Severe strictures form less frequently since the advent of proton pump inhibitor (PPI) therapy. In general, GERD strictures are associated with severe esophagitis or Barrett's esophagus. They occur at the squamocolumnar junction. As intestinal metaplasia advances to the proximal esophagus in Barrett, the stricture follows.

CLINICAL PICTURE

Patients report varying symptoms of dysphagia, odynophagia, regurgitation, and chest pain. Dysphagia begins with solid foods and advances to liquids as the stenosis becomes severe. Painful swallowing (odynophagia) develops as food irritates the mucosa overlying the strictured area. The patient's inability to ingest proper amounts of food results in weight loss and poor nutrition.

DIAGNOSIS

Clinical history suggests the diagnosis, and a combination of endoscopy and barium esophagraphy confirms stricture. Usually, barium esophagraphy shows a variable segment of narrowed esophagus. The margins are smoothly tapered, not jagged as found in patients with malignancy. Esophagogastroduodenoscopy allows direct visualization, and biopsy confirms a benign stricture. The esophagus is rigid, and the endoscope may meet resistance as it advances. In severe cases, a pediatric endoscope may be used to pass through the lumen. In GERD, the active reflux of acid may be observed above the level of the lesion. Twenty-four–hour pH monitoring should be performed to distinguish GERD-induced strictures from drug-induced strictures in 45% of patients. Peptic strictures must also be differentiated with a Schatzki ring, or weblike narrowing, thought to be related to reflux and found at the squamocolumnar junction (see Chapter 12).

TREATMENT AND MANAGEMENT

Repeated bougie dilatation with either rigid dilators or balloons is the treatment of choice for strictures. Bougie dilatation of GERD-related strictures results in resolution of symptoms in 75% of patients. Multiple topical postdilatation applications of mitomycin C show promise in decreasing dilatations and increasing their intervals, with overall improved results; however, further trials are needed. The underlying cause of reflux must be treated chronically with aggressive PPI therapy.

Surgery is indicated when recurring strictures require frequent dilatations or when medical therapy fails or is impractical. Surgical fundoplication should be performed within 2 years of diagnosis to resolve the underlying cause of reflux. Laparoscopic repair may be performed with good results and minimal complications. A recent study of 200 medical patients and surgical patients concluded that resecting peptic strictures is rarely indicated.

COURSE AND PROGNOSIS

In 30% to 40% of patients with benign stricture, symptoms will recur within 1 year. Patients with nonpeptic strictures and narrow strictures have the highest rates of recurrence. In GERD-related strictures, continued heartburn and hiatal hernia are the strongest predictors for failure of PPI therapy. When drug-induced strictures are resolved, heartburn does not need to be treated. Drug-induced injury may occur in a patient with an underlying GERD-induced stricture, causing pills to become lodged and resulting in further injury. These strictures may not respond to dilatation.

In a Mayo Clinic study, dilatations decreased from 5.3 per patient 26 months before surgery to 1.8 per patient 25 months after surgery. After laparoscopic fundoplication for dysphagia and strictures, the overall satisfaction rate is 88% to 91%, with a 10% recurrence rate for dysphagia. Laparoscopic surgery results in a good clinical outcome with minimal complications and a good quality of life.

ADDITIONAL RESOURCES

Bonavina L, DeMeester TR, McChesney L, et al: Drug-induced esophageal strictures, *Ann Surg* 206:173-183, 1987.

Kelly KA, Sare MG, Hinder RA: *Mayo Clinic gastrointestinal surgery*, Philadelphia, 2004, Saunders, p 49.

Klingler PJ, Hinder RA, Cina RA, et al: Laparoscopic antireflux surgery for the treatment of esophageal strictures refractory to medical therapy, *Am J Gastroenterol* 94:632-636, 1999.

Olson JS, Lieberman DA, Sonnenberg A: Practice patterns in the management of patients with esophageal strictures and rings, *Gastrointest Endosc* 66(4):670-675 (quiz 767, 770), 2007.

Rosseneu S, Afzal N, Yerushalmi B, et al: Topical application of mitomycin C in oesophageal strictures, *J Pediatr Gastroenterol Nutr* 44(3):336-341, 2007.

Said A, Brust DJ, Gaumnitz EA, Reichelderfer M: Predictors of early recurrence of benign esophageal strictures, *Am J Gastroenterol* 98:1252-1256, 2003.

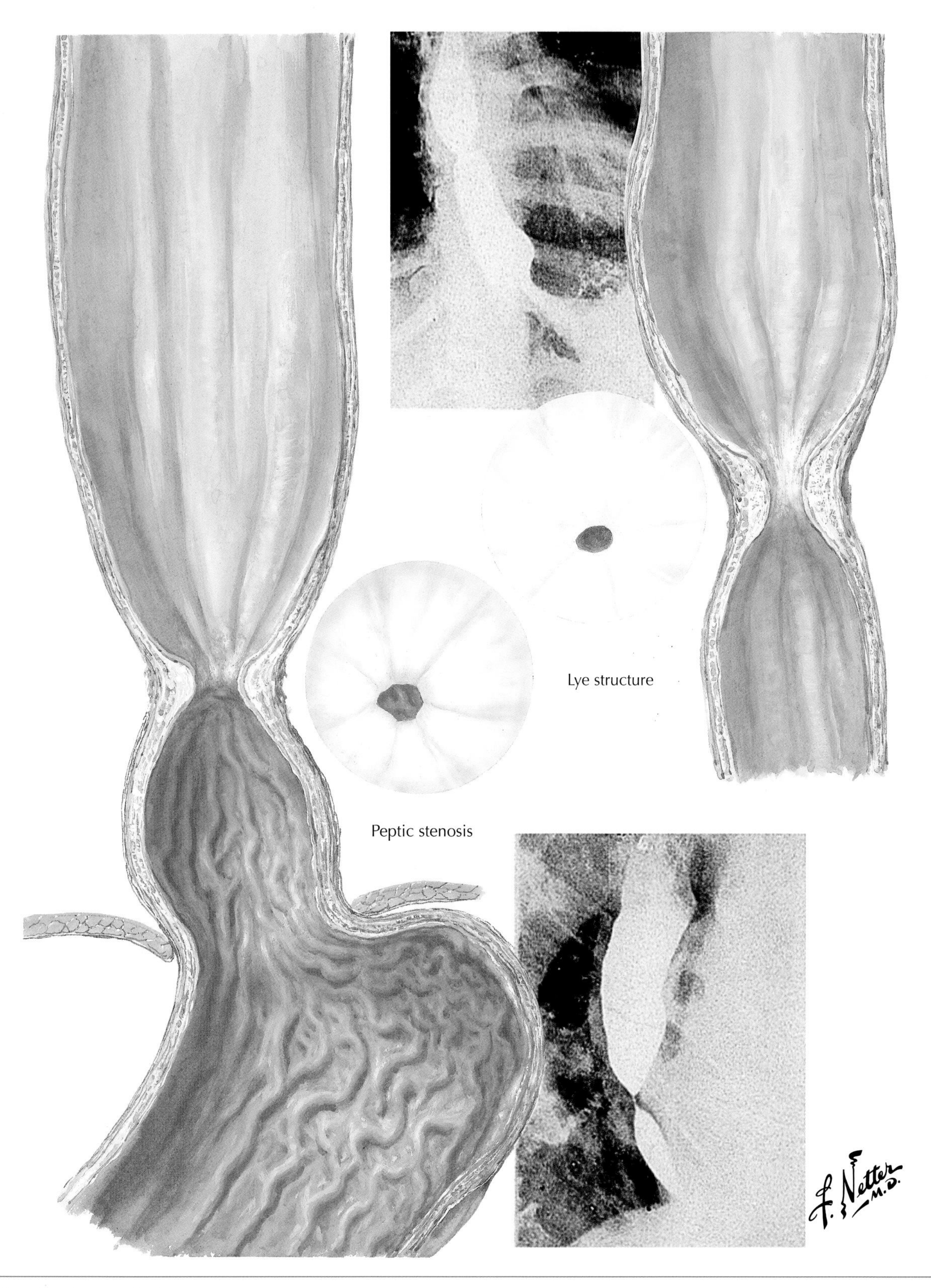

Figure 25-1 *Esophageal Stricture.*

Sliding and Paraesophageal Hiatal Hernias Types 1, 2, and 3

26

Neil R. Floch

The distinction among sliding, true paraesophageal, and mixed paraesophageal hernias has now been defined (**Figs. 26-1** and **26-2**). Also, although laparoscopy has replaced thoracotomy and laparotomy since 1991 as the standard treatment approach, with new synthetic material now available, hernia repair with mesh is challenging classic laparoscopic paraesophageal hernia repair.

In North America, hiatal hernias develop in 10% to 50% of the population. Average age for patients with a sliding hernia is 48 years and for a paraesophageal hernia, 61 years. There are four types of hiatal hernias. Type 1 accounts for 85% of all hernias. It develops when the gastroesophageal (GE) junction slides above the diaphragm. Of the remaining hernias, 14% are type 2, or pure paraesophageal, hernias. These develop when the gastric fundus herniates into the chest, lateral to the esophagus, but the GE junction remains fixed in the abdomen. Type 3, or mixed paraesophageal, hernia accounts for 86% of the remaining hernias. They develop with movement of the lower esophageal sphincter (LES) and the fundus into the chest. Type 4 hernia is a subset of type 3 and contains not only the entire stomach, but also other viscera, such as the omentum, colon (13%), spleen (6%), and small bowel. Patients with type 4 hernias may have bowel obstruction; 50% seek emergency treatment, and 25% experience major complications. *Parahiatal hernia* is movement of the stomach through a diaphragmatic defect separate from the hiatus and accounts for less than 1% of all hiatal hernias. Iatrogenic or postoperative paraesophageal hernia may occur after a previous distal esophageal procedure and accounts for 0.7% of paraesophageal hernias.

Hiatal hernia forms as the phrenicoesophageal membrane, preaortic fascia, and median arcuate ligament become attenuated over time. The pressure differential between the abdomen and the chest creates a vacuum effect during inspiration that pulls on the stomach. The degree of herniation into the posterior mediastinum and the type of volvulus that occurs may depend on the relative laxity of the gastrosplenic, gastrocolic, and gastrohepatic ligaments. As the hiatal hernia becomes larger, two types of volvulus may develop. *Organoaxial* volvulus (longitudinal axis) occurs with movement of the greater curvature of the stomach anterior to the lesser curvature. *Mesenteric axial* volvulus is less common and occurs when the stomach rotates along its transverse axis.

When the GE junction cannot be reduced below the diaphragm, it is considered to be shortened. This phenomenon is believed to occur in patients with chronic gastroesophageal reflux disease (GERD) with resultant transmural inflammation and contraction of the esophageal tube.

CLINICAL PICTURE

Although small, type 1 hiatal hernias may be asymptomatic, most patients complain of typical and atypical symptoms of GERD. Heartburn is the main symptom of GERD, but patients may also complain of acid reflux, regurgitation of food, epigastric abdominal pain, dysphagia, odynophagia, nausea, bloating, and belching. Atypical or extraesophageal symptoms include noncardiac chest pain, choking, laryngitis, coughing, wheezing, difficulty breathing, sore throat, hoarseness, asthma, and dental erosions.

Symptoms of types 2 and 3 paraesophageal hernia differ from GERD symptoms. Although paraesophageal symptoms vary, most series describe dysphagia, chest pain, and regurgitation as the most common. One series defined the symptoms as regurgitation (77%), heartburn (60%), dysphagia (60%), chest pain (52%), pulmonary problems (44%), nausea or vomiting (35%), hematemesis or hematochezia (17%), and early satiety (8%). Asymptomatic patients may constitute 11% of the population, and the hernia may be discovered on routine chest radiography or endoscopy. Questioning may reveal the presence of symptoms in most patients.

Dysphagia may result from compression of the lower esophagus by the adjacent stomach or from twisting of the esophagus by a herniated stomach. Chest pain may be confused with angina, resulting in emergency cardiac evaluation with negative results. Dyspnea may be secondary to loss of intrathoracic volume caused by a large hiatal hernia. Coughing may be a sign of aspiration, which may develop into pneumonia or bronchitis. Symptoms of asthma are severe enough to require bronchodilator therapy in 35% of patients. In 14% of patients with mixed hernia, a pulmonary condition ranging from dyspnea to severe bronchoconstriction may be the only symptom.

Iron-deficiency anemia has been reported in as many as 38% of patients with paraesophageal hernia. Most patients with iron deficiency are unaware of the problem until they experience symptoms such as pallor, palpitations, or dyspnea on exertion. Usually there is no direct evidence of gastrointestinal (GI) bleeding. Cameron ulcers or mucosal ulcerations of the stomach are a cause of anemia. Ischemia and mucosal injury occur secondary to the friction of the stomach moving through the esophageal hiatus during respiration and are diagnosed during endoscopy in 5.2% of patients with paraesophageal hernias. Larger hernias are associated with a higher incidence of ulcers, and 66% of patients have multiple ulcers. Although rare, bleeding is an indication for immediate repair. The patient's condition can usually be stabilized, but transfusion may be necessary. Elective surgery after stabilization is most prudent.

The progression of symptoms gives insight into the changes that occur with hernias. *Postprandial distress,* defined as chest pain, shortness of breath, nausea, and vomiting, occurs in 66% of patients, but eventually most patients have these symptoms as the hernia enlarges. Conversely, as a hernia enlarges, heartburn decreases. Heartburn is less common in type 3 than in type 1 hernia. Although 66% of patients initially have heartburn, the

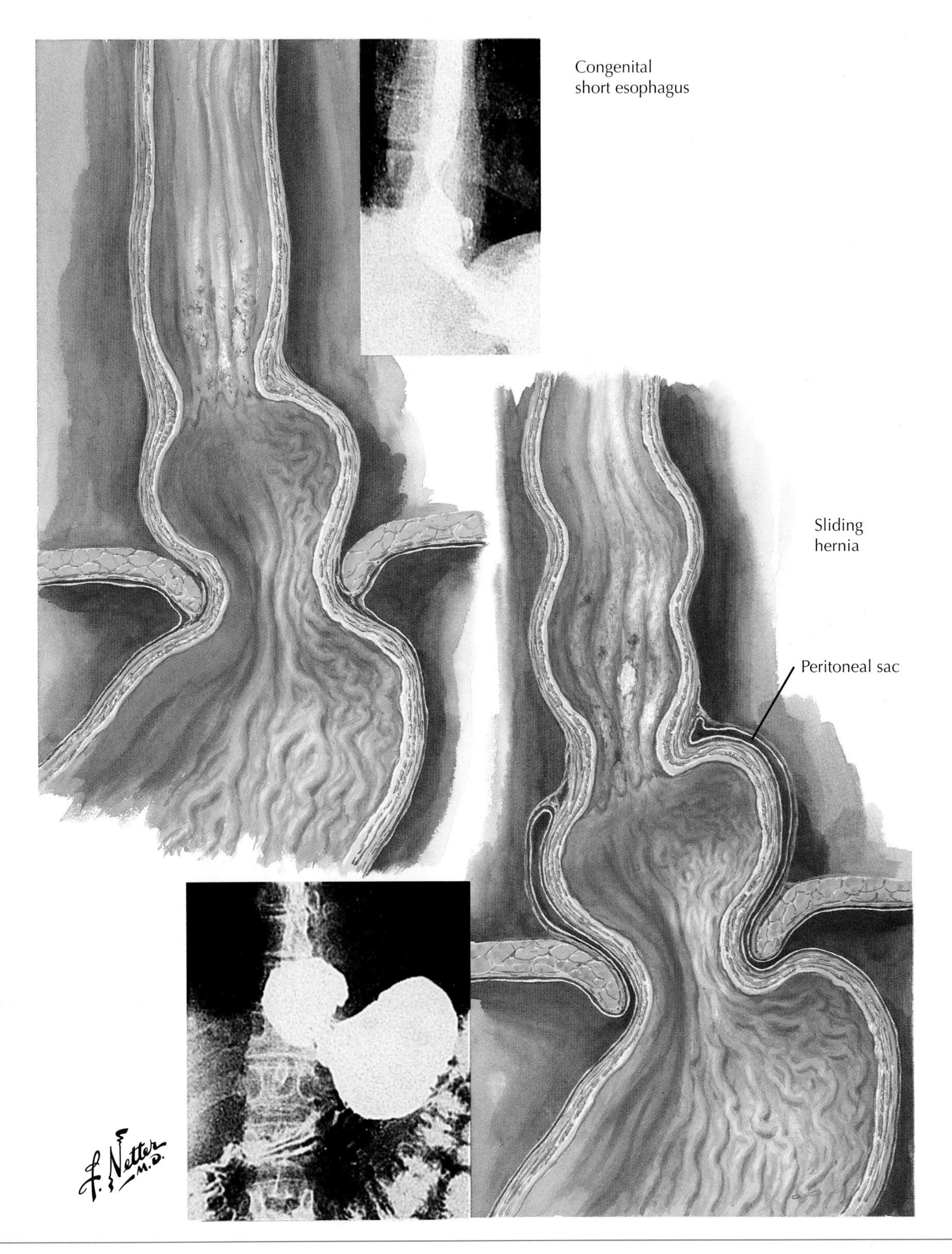

Figure 26-1 *Type I: Sliding Hiatal Hernia.*

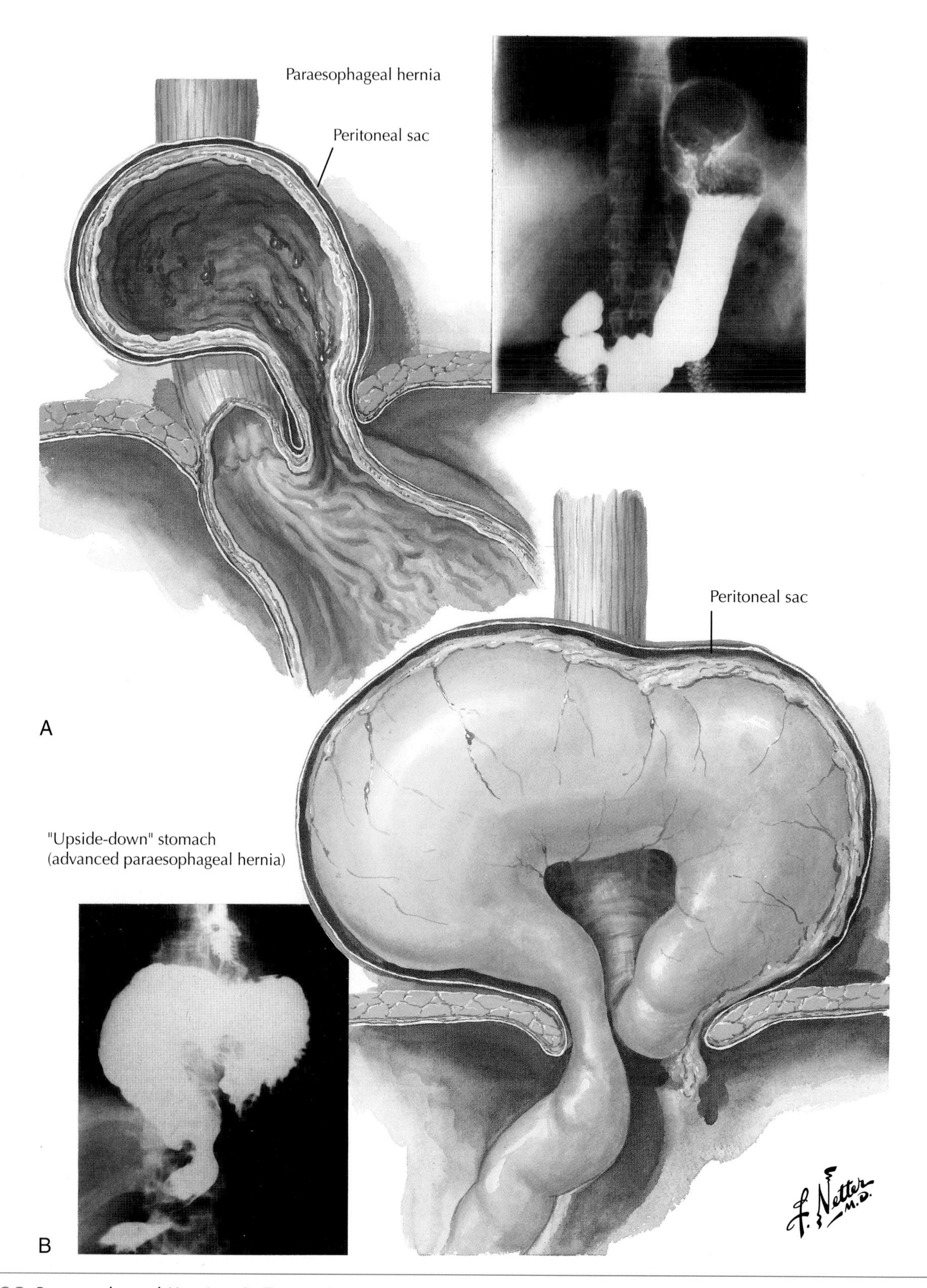

Figure 26-2 *Paraesophageal Hernias.* ***A****, Type II.* ***B****, Type III.*

symptom progresses in only 59% of these patients. As a type 3 hernia enlarges, kinking is thought to occur at the GE junction.

As many as 30% of patients undergo emergency surgery for bleeding, acute strangulation, gastric volvulus, or total obstruction. Recent studies report that 2% to 17% of patients need emergency surgery for acute obstruction or volvulus; the complication rate is 40%. Surgery is performed to treat perforation after strangulation with peritonitis, but mortality is 17%. If gastric necrosis has developed, mortality may reach 50%. Proponents of elective surgery have stressed these data to support early repair.

DIAGNOSIS

A sliding hiatal hernia is rarely seen on routine chest radiographs unless it is large. Computed tomography (CT) may also detect hiatal hernia. Most type 1 hiatal hernias are detected either by barium esophagraphy or by upper endoscopy, the most common method. Unless in an emergent situation, all patients should undergo esophageal manometry to determine the presence of an associated motility disorder before any surgical intervention. Specifically, achalasia should be ruled out (see Chapter 15). Evidence now indicates that patients with disorders such as ineffective esophageal motility or scleroderma may benefit from surgery but should undergo a partial fundoplication. Determination of the presence of acid or bile reflux can be performed with classic 24-hour pH testing, impedance testing, or the Bravo technique.

Chest radiography performed with the patient in the upright position can establish the diagnosis of a paraesophageal hernia by revealing air-fluid behind the heart in 95% of patients. Nasogastric tube placement in the intrathoracic stomach confirms the diagnosis. Paraesophageal hernia can also be easily detected on CT. An upper GI series can establish the diagnosis in almost all patients because it defines the type of hiatal hernia. In a series of 65 patients, 56 (86%) were found on barium swallow or esophagogastroduodenoscopy (EGD) to have a type 3 paraesophageal hernia. Nine (14%) had type 2 paraesophageal hernia. In 21% of patients, more than half of the stomach was in the chest.

A herniated stomach can be intubated using EGD, nasogastric tube, or manometry in approximately 50% of patients. When possible, manometry can assess esophageal body motility, LES pressure, LES length, and total esophageal length. At least 50% of patients with paraesophageal hernias have hypotensive LES. Incompetent LES was found in 56% to 67% of patients, with an average pressure of less than 6 mm Hg. Short intra-abdominal length of the LES combined with a sliding hernia may also contribute to reflux.

The amplitude of peristaltic waves is reduced in 52% to 58% of patients. Poor body motility can result in delayed clearance of refluxed acid that requires partial fundoplication, although some authors advocate floppy Nissen in this situation. A short esophagus may be related to mixed, or type 3, paraesophageal hernia and is believed to result from injury to the esophageal wall secondary to stricturing and fibrosis from reflux. Whether short esophagus is a result or the cause of paraesophageal herniation has yet to be determined.

Twenty-four–hour esophageal pH testing is not a diagnostic test for a paraesophageal hernia but may be helpful in identifying associated GE reflux in 50% to 65% of patients. Type 3 hernias are associated with reflux because of the migration of the LES into the chest. Some patients with paraesophageal hernias may have abnormal 24-hour pH test results but normal LES pressure.

TREATMENT AND MANAGEMENT

Repair of small, type 1 sliding hiatal hernia entails reduction of the hernia sac from the chest and performing either a partial (Toupet) or total (Nissen) fundoplication. Most patients are not treated surgically but with proton pump inhibitor (PPI) medication. Larger type 1 hiatal hernias can become more challenging as more stomach protrudes into the chest. As the esophagus contracts into the chest, so does the proximal cardia, then the fundus. If the fundus moves alongside the esophagus, the hernia is then classified as paraesophageal. The most difficult type 1 sliding hernias involve a shortened esophagus. When the GE junction is unable to be reduced easily, 3 cm below the diaphragm, the technique of extensive mediastinal dissection must be used. This involves dissecting all lateral, anterior, and posterior attachments of the esophagus to the mediastinum, taking care to avoid entering the pleura or disturbing major vessels. Dissection may be necessary up to the level of the bronchial bifurcation. GERD treatment or type 1 hiatal hernia repair results in at least a 90% patient satisfaction rate.

Observation of paraesophageal hernias can result in emergency complications such as incarceration, strangulation, perforation, splenic vessel bleeding, and acute dilatation of the herniated stomach in 20% of patients. A cohort study concluded that watchful waiting is reasonable for the initial management of patients with asymptomatic or minimally symptomatic paraesophageal hernias. Asymptomatic patients at high risk for morbidity after surgery may be observed. Nonsurgical management resulted in 29% mortality, but this rate is now believed to be lower. Asymptomatic patients have lower risk for complications. Symptoms indicate the need for elective repair. Elective surgery carries a zero to 3% mortality rate. In comparison, emergency surgery results in up to a 40% complication rate and a 19% to 40% mortality rate.

Although there is no proof that laparoscopy has changed the indications for paraesophageal hernia repair, patients with comorbidities who undergo laparoscopy may experience the low complication rate, short recovery, and long-term results seen after open surgery. It may be argued that the low morbidity and mortality rates achieved by experienced surgeons should encourage all patients with paraesophageal hernia to undergo laparoscopic hernia repair.

Paraesophageal hernia is a surgical disease and cannot be adequately treated medically. Symptoms of reflux may be reduced with H_2 blockers and PPIs. Before laparoscopy, paraesophageal hernias were repaired by thoracotomy or laparotomy. Open paraesophageal hernia repair has average morbidity of 14% and average mortality of 3%, and length of hospital stay is 3 to 10 days. Major complications include bowel obstruction and splenectomy. The recurrence rate after laparotomy is 11%. Thoracotomy results in 19% morbidity and up to 25% mortality. Reoperation may be necessary in 5% of patients.

Although the operative time is longer for laparoscopy than for open surgery, the results are similar. In addition, laparoscopy involves significantly lower rates of blood loss, intensive care unit stay, ileus, hospital stay, and overall morbidity. The visibility of the hiatus is superior with the laparoscope. The only disadvantage is that laparoscopy can cause decreases in systolic blood pressure and cardiac index, which may be detrimental to patients with poor cardiac function. Overall, laparoscopy is beneficial in the elderly population.

Failing to perform concomitant antireflux surgery results in postoperative reflux in 20% to 40% of patients. Antireflux surgery may improve motility in 50% of patients. Concomitant antireflux surgery produces several other benefits: (1) positive findings on 24-hour pH test; (2) destruction of LES after surgical dissection of the hiatus; (3) incompetence of LES no longer masked by paraesophageal hernia; (4) fundoplication securing the stomach in the abdomen; (5) minimal morbidity added to the procedure; and (6) emergency surgery necessitating a concomitant antireflux procedure because testing cannot be performed.

Short esophagus has an overall incidence of 1.5%. Among patients with paraesophageal hernias, 15% to 20% have short esophagus. The diagnosis of short esophagus is made at surgery if the LES is 5 cm above the hiatus or if the esophagus is difficult to mobilize from the mediastinum. Preoperative indicators of short esophagus include paraesophageal hernias larger than 5 cm, severe esophagitis, strictures, Barrett esophagus, reoperative antireflux surgery, and evidence of poor esophageal body motility.

In the past, the Collis-Belsey procedure has been recommended to treat short esophagus in patients with type 3 paraesophageal hernias. Newer techniques rely on more aggressive dissection. First, the hernia sac is reduced. Extensive dissection enables the esophagus to be mobilized from the chest into the abdomen. Patients with short esophagus who undergo laparoscopic transmediastinal dissection have a 90% success rate for fundoplication, almost equal the rate for patients with normal esophageal length (89%). The advent of laparoscopic transmediastinal dissection has rendered Collis gastroplasty less favorable.

Gastrostomy and gastropexy should be considered for elderly and debilitated patients who have many comorbidities and cannot tolerate extensive surgery. Both procedures secure the stomach, preventing future herniation. Disadvantages are the discomfort and inconvenience of a gastrostomy tube.

If a primary crural closure is not possible, a tension-free mesh repair may be indicated. Most recent reports promote a primary suture closure of the muscle reinforced with an onlay of mesh that has a keyhole opening stapled to the crura. A defect larger than 5 cm is usually reported when mesh is used. Various prosthetic materials, including polyester (Mersilene), polytetrafluoroethylene (PTFE), and polypropylene, have been used for mesh. PTFE may be the most preferable material because it causes the least inflammatory reaction and fewest adhesions and is typically used for ventral hernia repairs where the mesh is exposed to bowel. Polypropylene is easier to staple but causes more inflammatory reaction and therefore carries a higher likelihood of erosion. Other types of materials are now being investigated for repair. Small intestinal submucosa is developed from small-bowel submucosa; it maintains strength while being gradually resorbed and replaced by native host tissue. Human acellular dermal matrix shows promise, as does the use of a patient's own ligamentum teres.

COURSE AND PROGNOSIS

Complications may be divided into intraoperative complications and conversions, postoperative complications, late sequelae, reoperations, and death. Intraoperative complications occur in up to 17% of patients. Esophageal and gastric perforations, tears, and lacerations occur in 11% of patients. Perforation of the esophagus has been related to bougie usage. Excessive bleeding may occur after dissection in the wrong anatomic plane, tearing of the short gastric vessels, or retraction of the liver. Vagal nerve injury is rare but may lead to gastric atony and bezoar formation.

Pleural entry and pneumothorax may occur in 14% of patients. Chest tube placement is rarely indicated because increasing intrathoracic ventilatory pressure at the end of the operation forces carbon dioxide into the abdomen. Rarely of clinical significance is pneumomediastinum or crepitus, which resolves with no sequelae. Acute intraoperative complications may include respiratory acidosis secondary to carbon dioxide exposure or, rarely, pulmonary embolus.

The 3% conversion rate frequently reflects the inability to decrease mediastinal contents caused by mediastinal scarring of a shortened esophagus. Traumatic vessel injury is a common reason for conversion, but other causes may be adhesions and difficulty with exposure. Exposure may be limited in patients who are obese or who have hepatomegaly. Visualization has improved with the advent of 30-degree, 45-degree, and flexible-tip esophagoscopes.

Postoperative complications occur in 3% to 28% of patients. The most serious postoperative complications are pulmonary embolism, myocardial infarction (heart attack), cardiac dysrhythmias, cerebrovascular accident (stroke), and respiratory failure. Other conditions that may develop are pneumonia or pleural effusion, congestive heart failure, deep vein thrombosis, urinary retention, and superficial wound infections. Dysphagia is the most frequent postoperative problem but is considered inherent to the surgery. Dilatation may be required in 6% of patients. Over time, some fundoplications slip, become undone, or migrate to the mediastinum. Postoperative pain is usually limited to incisions, but patients may have left shoulder pain caused by diaphragmatic irritation.

Reoperation rates range from zero to 9%. Early reoperation may be necessary for hernia recurrence, fundoplication slippage, esophageal or stomach perforation, or small-bowel obstruction. Dilatation may be necessary for patients with dysphagia after surgery.

Mortality rates range from zero to 5%. The population of patients with paraesophageal hernias is older and has a higher frequency of comorbidities than patients with type 1 hernias. Late sequelae do occur but are well tolerated. Reflux may develop in patients who have not undergone the antireflux procedure. Frequently, patients have gas-bloat syndrome, which includes bloating, abdominal gas, increased flatus, uncontrolled flatus, belching, and abdominal discomfort. Patients may also experience early satiety, pain after meals, and weight loss.

Operative time averages from 2 to 3 hours. Most patients stay in the hospital for 1 to 5 days. On average, patients return to normal activities within 3 weeks. Displacement of the paraesophageal hernia from the chest results in improved (15%-20%) pulmonary function. On average follow-up at 1.5 years, 92% of patients are satisfied with the surgical result.

Recurrence rates are based on the definition of "recurrence," which can be determined clinically or by barium esophagraphy. Subjective evidence or barium esophagraphy is used to determine recurrence. Asymptomatic recurrence rates may be very high, but they range from zero to 32% when barium esophagraphy is used. Clinically symptomatic recurrences are much lower, as indicated by the zero to 9% reoperation rate. Most patients with recurrences undergo surgery only if they have symptoms, including regurgitation, heartburn, dysphagia, and white saliva. Evaluation usually reveals a sliding herniation of their wrap in about 80% of patients, although a recurrent paraesophageal hernia may occur. Although recurrence is common, reoperation is found to be rarely necessary at 10-year follow-up.

Symptomatic patients may be candidates for paraesophageal hernia repair after extensive evaluation. Laparotomy and thoracotomy are successful approaches but have higher morbidity and longer recovery rates. Laparoscopy is the best approach, if performed by experienced surgeons. An antireflux procedure should be added to prevent reflux. Sac excision, transmediastinal mobilization of the esophagus, primary crural repair, and antireflux surgery may reduce the recurrence of hernia. Gastropexy and mesh repair show promise in reducing recurrence, with results of long-term studies pending.

ADDITIONAL RESOURCES

Diaz S, Brunt LM, Klingensmith ME, et al: Laparoscopic paraesophageal hernia repair, a challenging operation: medium-term outcomes of 116 patients, *J Gastrointest Surg* 7:59-66, 2003.

Ferguson MK: Paraesophageal hiatal hernia. In Cameron JL, editor: *Current surgical therapy*, ed 6, St Louis, 1998, Mosby, pp 51-54.

Floch NR: Paraesophageal hernias: current concepts, *J Clin Gastroenterol* 29:6-7, 1999.

Perdikis G, Hinder RA, Filipi CJ, et al: Laparoscopic paraesophageal hernia repair, *Arch Surg* 132:586-589, 1997.

Peters JH, DeMeester TR: Esophagus and diaphragmatic hernia. In Schwartz SI, editor: *Principles of surgery*, ed 6, New York, 1994, McGraw-Hill, pp 1043-1122.

Stylopoulos N, Gazelle GS, Rattner DW: Paraesophageal hernias: operation or observation?, *Ann Surg* 236(4):492-500, 2002.

White BC, Jeansonne LO, Morgenthal CB, et al: Do recurrences after paraesophageal hernia repair matter? Ten-year follow-up after laparoscopic repair, *Surg Endosc* 22:1107-1111, 2008.

Barrett's Esophagus

Neil R. Floch

27

Barrett's esophagus is defined as metaplasia occurring in any length of epithelium, at any location above the gastroesophageal (GE) junction, that is identified at endoscopy and confirmed by biopsy, and that does not include metaplasia of the gastric cardia (**Fig. 27-1**). Squamous epithelium is replaced by columnar epithelium with goblet cells. Barrett's esophagus originally included gastric fundic type, junctional type, and *intestinal metaplasia* (IM). Fundic-type epithelium has minimal malignant potential and is no longer included in the definition of Barrett's esophagus. Endoscopy cannot accurately distinguish between IM and gastric-type epithelium, so biopsy is necessary for diagnosis.

Evidence supports a strong correlation between Barrett's esophagus and chronic gastroesophageal reflux disease (GERD). Abnormal pH study findings are present in 93% of Barrett's esophagus patients compared with 45% to 75% of all other GERD patients. Barrett's esophagus is twice as prevalent in men as in women. The men are usually white, have chronic heartburn, and are older than 50. Barrett's esophagus is present in 10% to 20% of patients undergoing endoscopy for GERD. It occurs in 3% of patients with weekly heartburn, 5% of patients with daily heartburn, and 0.5% to 2% of asymptomatic adults in the United States.

Barrett's esophagus has the potential to progress to *adenocarcinoma,* the incidence of which has increased dramatically over the past 20 to 30 years, making it the most rapidly increasing cancer in the United States. Barrett's esophagus develops in the presence of persistent GERD, which is an independent risk factor for adenocarcinoma. Patients with GERD have a risk for esophageal adenocarcinoma that is 30 to 60 times greater, at an incidence rate more than 100 times greater, than that of the general population. The prevalence of Barrett's esophagus and adenocarcinoma increases with age and severity of symptoms. A surgical series of esophageal resections shows that adenocarcinoma predominates in older white men.

Carcinoma is the final step in the progression from squamous cells, to IM, to *low-grade dysplasia* (LGD), to *high-grade dysplasia* (HGD), and then to invasive disease. All stages may coexist. Other risk factors for adenocarcinoma in patients with Barrett's esophagus include length of Barrett's epithelium, LGD, and HGD.

Intestinal metaplasia is the most important risk factor for the development of dysplasia and cancer, and most adenocarcinomas of the esophagus and GE junction are accompanied by IM. Short-segment Barrett's esophagus carries a lower risk for dysplasia. Patients with long-segment Barrett's esophagus have greater pH exposure and lower esophageal sphincter (LES) pressure than those with short-segment disease. Recently, attempts were made to differentiate short-segment Barrett's esophagus from IM of the gastric cardia. The distinction between short-segment and long-segment Barrett's esophagus was poorly defined.

Reflux of acid and duodenal contents may contribute to the development of Barrett's esophagus. Bile, in conjunction with acid and pepsin, disrupts the mucosal barrier of the esophagus and causes esophagitis. Acid exposure is associated with the development of columnar mucosa, and bile exposure has been deemed an independent predictor of Barrett's esophagus. Severe duodenogastroesophageal reflux (DGER) occurs after subtotal esophagectomy and pyloroplasty and provides an environment for development of Barrett's metaplasia through a sequence that begins with cardiac epithelium and eventually transforms into IM. In this human model of severe DGER, 33% of patients have esophagitis, 23% have Barrett's esophagus, and another 18% progress from cardiac mucosa to Barrett's esophagus over time. Clearly, acid is not the only cause.

CLINICAL PICTURE

No symptoms or signs distinguish patients with Barrett's metaplasia from those without it. Signs and symptoms of GERD are the same as those for Barrett's esophagus.

DIAGNOSIS

Patients with an extensive history of GERD symptoms are more likely to have Barrett's esophagus and should undergo esophagogastroduodenoscopy (EGD). Patients at higher risk may be male and may have abnormal bile reflux, hiatal hernia larger than 4 cm, defective LES, distal esophageal dysmotility, reflux episodes longer than 5 minutes, and GERD symptoms for more than 5 years.

The diagnosis of Barrett's esophagus requires biopsy of abnormal-appearing mucosa to determine the presence of IM and dysplasia. Squamocolumnar (SC) and GE junctions should be specified. When the SC junction is displaced cranial to the GE junction, Barrett's esophagus is suspected. Gastric folds define the beginning of the stomach. It is difficult to distinguish gastric mucosa and esophagitis from Barrett's esophagus by visualization, but Barrett is typically described as salmon colored. Guidelines suggest that biopsies be taken in four quadrants, beginning 1 cm below the GE junction and extending 1 cm above the SC junction at 2-cm intervals. Endoscopic staining with methylene blue may assist in locating cells with IM in the esophagus.

TREATMENT AND MANAGEMENT

Management includes controlling reflux, healing esophagitis, and detecting dysplasia early. The principles and treatment for Barrett's esophagus are the same as those for GERD, although patients with Barrett's esophagus have worse responses because of more severe disease. How to prevent Barrett's esophagus or stop its progression has not been determined. Treatment involves a combination of endoscopy, medical therapy, surgery, and possibly ablative therapies.

Surveillance is based on the increasing risk for adenocarcinoma, and dysplasia is the most sensitive indicator of the risk for cancer. Surveillance endoscopy is recommended in all patients with Barrett's esophagus to detect dysplasia and to initiate early intervention. These measures attempt to decrease the

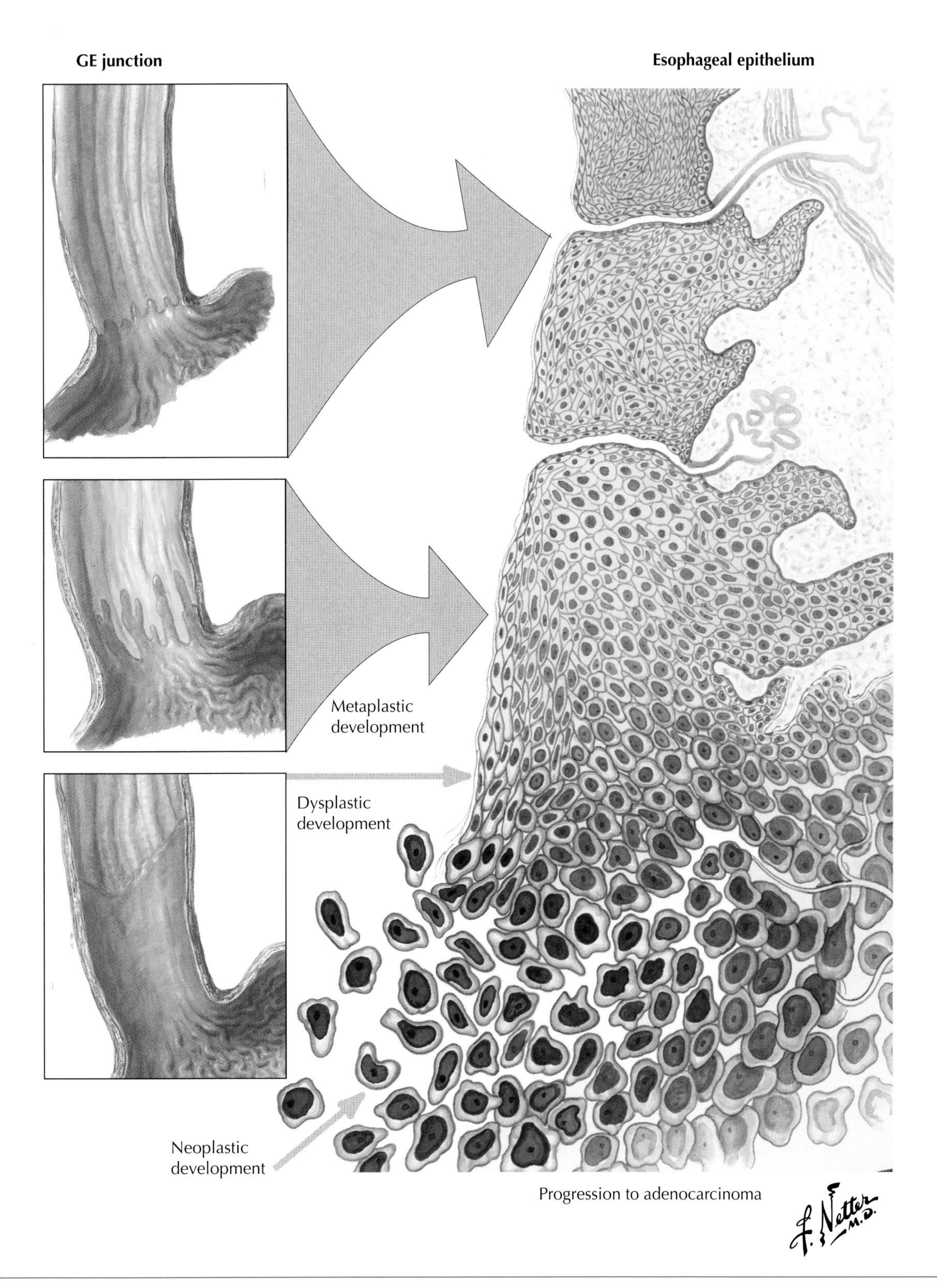

Figure 27-1 *Barrett's Esophagus.*

incidence of esophageal adenocarcinoma and improve patient survival. The result has been detection of adenocarcinoma at an earlier stage compared with cancer detected after symptoms such as dysphagia, but these surveillance measures have not yet affected survival.

Unfortunately, most patients with Barrett's cancer were not found to have a premalignant condition such as dysplasia and were not under surveillance. Asymptomatic patients still account for most patients with adenocarcinoma. Therefore, all patients undergoing EGD should undergo careful examination of the distal esophagus. The goal is to detect dysplasia, which occurs on top of IM. Any level of dysplasia may be located adjacent to frank carcinoma.

Unfortunately, evidence-based data to determine the timing of surveillance endoscopy intervals for screening remain to be defined. The interval at which endoscopy is performed depends on the grade of dysplasia. Any nodule or ulcer on the epithelial surface should undergo careful biopsy. Endoscopy for surveillance may be performed every 3 years if there is no evidence of dysplasia on two consecutive endoscopies.

If LGD is found at endoscopy, patients should be treated intensively with high-dose proton pump inhibitor (PPI) therapy for 3 to 12 weeks, after which repeat biopsy should be performed. Esophagitis will resolve, but dysplasia will not, eliminating confusion regarding the diagnosis at repeat biopsy. If LGD is found after repeat biopsy, repeat endoscopy should be performed at 6-month intervals for 1 year. If LGD has not progressed, annual endoscopy should be performed. The presence of HGD should be confirmed by a second pathologist and warrants aggressive treatment.

In patients with Barrett's esophagus, medical and surgical therapies are effective in controlling reflux symptoms. However, more than 60% of patients continue to have pathologic GERD and abnormally low esophageal pH, despite doses of esomeprazole that control reflux symptoms. Higher doses of PPIs must be used to prevent the development of esophageal adenocarcinoma. Treatment success can be measured only by repeat pH monitoring. Although aggressive PPI therapy is the first-line treatment, no significant clinical evidence supports that acid suppression prevents adenocarcinoma or progression of IM to dysplasia. PPI treatment entails titration of medication dosage to a level that controls symptoms and heals esophagitis.

Heartburn is alleviated in 96% and resolves in 70% of patients with Barrett's esophagus after laparoscopic antireflux surgery (LARS). No differences occur in medication use or symptom control after LARS, but the failure rate is higher in patients with Barrett's esophagus (12%) than in those without it (5%). In 89% of patients with Barrett's esophagus, LARS provides excellent control of esophageal acid exposure.

Antireflux surgery is superior to medical therapy for preventing the development of, and inhibiting the progression to, Barrett's carcinoma. A meta-analysis reports that the risk for adenocarcinoma in patients with Barrett's esophagus is low and decreases by 1.5 cancers per 1000 patient-years, more after antireflux surgery than after medical treatment; however, these findings are not significant. Regression of Barrett's esophagus depends on the length of the columnar-lined esophagus and the time of follow-up after antireflux surgery. Endoscopy and pathology findings reveal complete regression of IM in 33% to 55% of patients with short-segment Barrett's esophagus after LARS. In patients with segments of Barrett's esophagus longer than 3 cm, 20% have disease regression, but 20% have disease progression from IM to dysplasia.

If HGD is found at endoscopy, there are three options for treatment: esophagectomy, intense surveillance, and ablation therapy. Intense surveillance for up to 46 months results in the development of adenocarcinoma in approximately 25% of patients, regression in 25% of patients, and stability in the remaining 50% of patients. The chance for concomitant esophageal cancer is 47% in all patients.

Esophagectomy is the most conservative approach but has high morbidity and mortality rates (3%-10%). Endoscopic therapy may be performed with multiple techniques, including thermal, chemical, and mechanical methods. The goal is to remove all dysplastic epithelium to allow the regrowth of squamous epithelium. Argon beam, laser, electrocautery, and photodynamic therapy have been used. Photodynamic therapy results in a downgrading of dysplasia in 90% of patients, but residual Barrett's esophagus may be found in 58% of patients. Complications of chest pain, nausea, and esophageal strictures may develop.

ADDITIONAL RESOURCES

Bammer T, Hinder RA, Klaus A, et al: Rationale for surgical therapy of Barrett's esophagus, *Mayo Clin Proc* 76:335-342, 2001.

Cossentino MJ, Wong RK: Barrett's esophagus and risk of esophageal adenocarcinoma, *Semin Gastrointest Dis* 14:128-135, 2003.

Dresner SM, Griffin SM, Wayman J, et al: Human model of duodenogastro-oesophageal reflux in the development of Barrett's metaplasia, *Br J Surg* 90:1120-1128, 2003.

Fass R, Sampliner RE: Barrett's oesophagus: optimal strategies for prevention and treatment, *Drugs* 63:555-564, 2003.

Gurski RR, Peters JH, Hagen JA, et al: Barrett's esophagus can and does regress after antireflux surgery: a study of prevalence and predictive features, *J Am Coll Surg* 196:706-712 (discussion 712-713), 2003.

Lee TJ, Kahrilas PJ: Medical management of Barrett's esophagus, *Gastrointest Endosc Clin North Am* 13:405-418, 2003.

Morales TG, Camargo E, Bhattacharyya A, Sampliner RE: Long-term follow-up of intestinal metaplasia of the gastric cardia, *Am J Gastroenterol* 95:1677-1680, 2000.

Peters JH, DeMeester TR: Esophagus and diaphragmatic hernia. In Schwartz SI, Shires TG, Spencer FC, editors: *Principles of surgery*, ed 7, New York, 1999, McGraw-Hill, pp 1081-1179.

Benign Neoplasms of the Esophagus

Neil R. Floch

28

Benign tumors of the esophagus are more common than previously thought, occurring in 0.5% to 8% of the population, as indicated by autopsy studies. Esophageal carcinoma is 50 times more prevalent. Since the advent of computed tomography (CT), tumors have been discovered more frequently (**Fig. 28-1**). Most are asymptomatic and nonepithelial. Categories include mucosal or intraluminal tumors, submucosal tumors, and muscle wall tumors. Mucosal tumors include leiomyoma, gastrointestinal (GI) stromal tumor, squamous papilloma, fibrovascular polyps, retention cysts, and granular cell tumors. Submucosal tumors include lipoma, fibroma, neurofibroma, granular cell tumor, hemangioma, and salivary gland tumors. Fibromas or fibrovascular polyps are located in the upper esophagus, may reach lengths of 7 to 10 cm, and may become freely suspended in the lumen. Periesophageal tissue includes foregut cysts.

Intraluminal leiomyomas are the most common benign tumor of the esophagus, accounting for two thirds of all benign esophageal tumors. The male/female ratio is 2:1; 33% occur in the middle and 56% in the lower third of the esophagus; 80% are intramural. Leiomyomas may extend into the stomach as well. Half the tumors are smaller than 5 cm. They are usually firm, encapsulated, rubbery, and elastic and typically not pedunculated because they are muscular in origin and are covered by the mucosa. In 13% of patients, intraluminal leiomyomas are annular, or completely encircle the esophagus. They usually are isolated but may appear in multiples.

CLINICAL PICTURE

In 15% to 50% of benign esophageal neoplasms, patients are usually asymptomatic. The most common presenting symptoms for leiomyomas are dysphagia (~50%), pain (~50%), weight loss (15%), and nausea or vomiting (12%). Other symptoms include odynophagia, reflux, regurgitation, respiratory symptoms, shoulder pain, atypical chest pain, hiccups, and anorexia. There is little correlation between size and symptoms. However, larger pedunculated tumors may occlude the esophageal lumen, causing dysphagia, or may be aspirated into the trachea. Bleeding into the lumen may occur from ulceration of lesions such as angiomas.

DIAGNOSIS

Large endoluminal tumors may be visualized on barium esophagraphy as a concave mass with smooth borders. CT is 91% sensitive and excellent for smaller lesions. If obstruction has occurred, proximal dilatation of the esophagus may be detected. Endoscopy is most sensitive and may determine the presence, location, and integrity of the mucosa. Normal mucosa over a leiomyoma frequently rules out malignancy. In patients with leiomyoma, biopsy is contraindicated because it may cause infection, bleeding, or perforation. It also increases the risk for mucosal tear at surgical excision. If an ulcer is identified on endoscopy, biopsy should be performed.

The best method of classification is endoscopic ultrasound, which is capable of delineating five layers of the esophageal wall. These layers are detected by alternating hyperechoic and hypoechoic transmissions. The superficial, or inner, layer is hyperechoic and the remaining layers alternate as described; deep mucosa (second layer), submucosa (third layer), and muscularis propria (fourth layer). Periesophageal tissue is seen as the fifth layer. Ultrasound is limited in determining the nature of the tumor and whether it is malignant.

TREATMENT AND MANAGEMENT

Most patients without symptoms could be managed by observation because of the benign nature of the lesions, but this is controversial. Advocates for nonsurgical management argue that the risk for malignant transformation is extremely rare, that slow-growing tumors may be observed, and that the risk for surgery may be more harmful than observation alone. When symptoms develop, the lesion should be removed. It should also be removed if the tumor becomes larger or if mucosal ulceration develops, and it especially should be removed to obtain a definitive diagnosis.

Transthoracic excision by thoracotomy is the most common approach, but lesions may be removed by thoracoscopy, laparoscopy, or hand-assisted laparoscopy. Endoscopic methods may be used to snare and cauterize esophageal polyps. Other endoscopic methods are being developed, but concomitant endoscopy already has a role to ensure adequate esophageal luminal patency after resection. Enucleation of the mass with primary closure is the preferred technique. Benign tumors larger than 8 cm may require esophagectomy with gastric pull-up, using thoracotomy, thoracoscopy, or laparoscopy.

COURSE AND PROGNOSIS

Results of leiomyoma enucleation are excellent, and recurrence is rarely reported. Symptoms resolve with tumor excision. Overall, prognosis is excellent because the lesions are benign. Minimally invasive techniques result in minimal morbidity and rare mortality, and patients generally require hospital stays of 1 to 3 days.

ADDITIONAL RESOURCES

Cameron JL, editor: *Current surgical therapy*, ed 9, St Louis, 2008, Mosby, pp 1-80.

Lee LS, Singhal S, Brinsler CJ, et al: Current management of esophageal leiomyoma, *J Am Coll Surg* 198:136-146, 2004.

Peters JH, DeMeester TR: Esophagus and diaphragmatic hernia. In Schwartz SI, Shires TG, Spencer FC, editors: *Principles of surgery*, ed 7, New York, 1999, McGraw-Hill, pp 1081-1179.

Samphire J, Nafteux P, Luketich J: Minimally invasive techniques for resection of benign esophageal tumors, *Semin Thorac Cardiovasc Surg* 15:35-43, 2003.

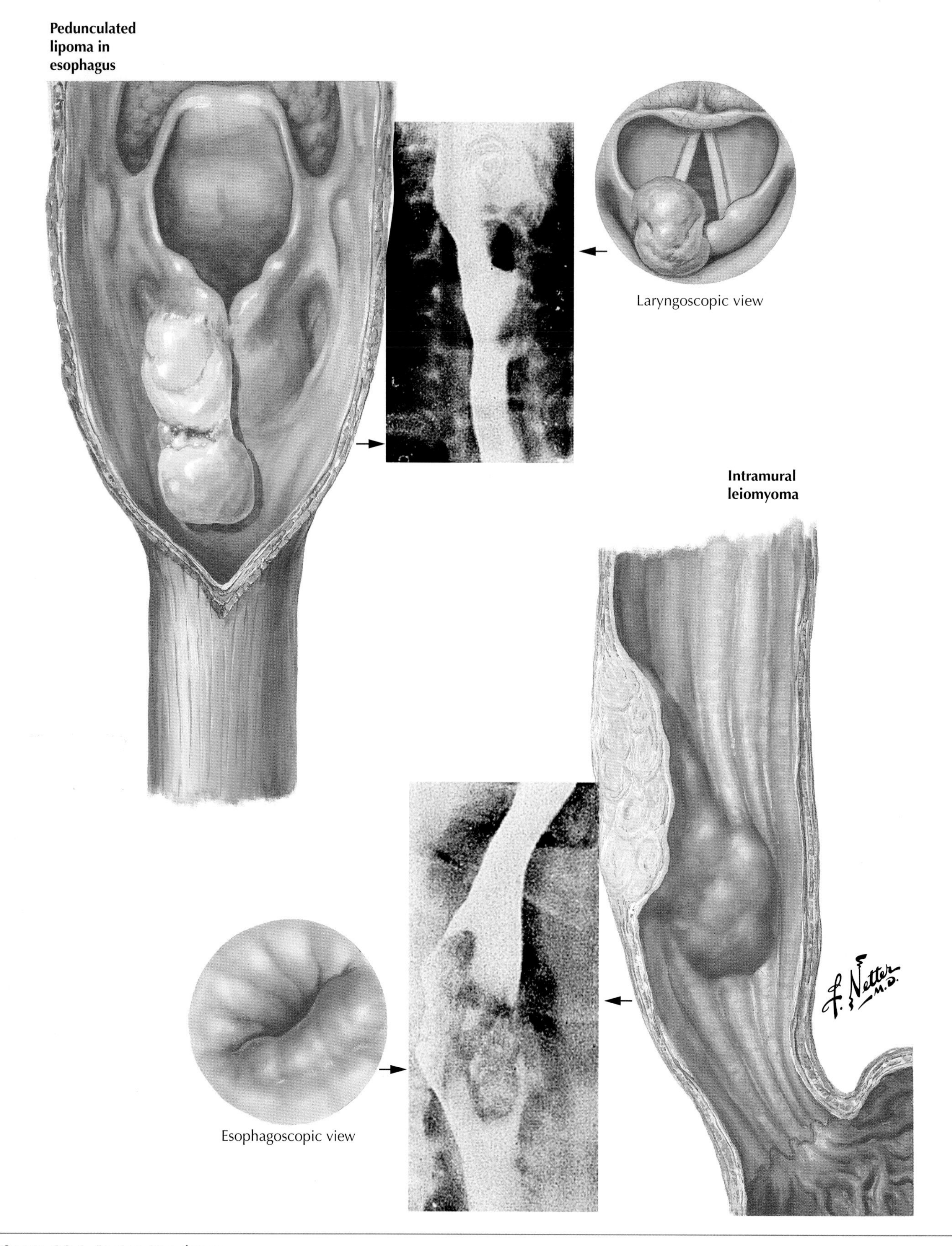

Figure 28-1 *Benign Neoplasms.*

Malignant Neoplasms: Upper and Middle Portions of the Esophagus

29

Neil R. Floch

Four of five carcinomas in the gastrointestinal (GI) tract occur in men, but carcinomas of the upper third of the esophagus are more common in women (**Fig. 29-1**). In the esophagus, the upper third is the *least* common site for carcinomas to occur. A correlation exists between esophageal carcinoma and chronic hypopharyngitis. Patients with Plummer-Vinson syndrome have an increased incidence of malignancy. Tumors are usually *squamous cell carcinoma* and may be anaplastic. (See also Chapter 30.)

CLINICAL PICTURE

Patients frequently have symptoms of hoarseness, dysphagia, and aspiration. Hoarseness usually results from recurrent laryngeal nerve involvement by the tumor. Dysphagia may be the first symptom, and its presence should initiate a thorough evaluation.

DIAGNOSIS

Evaluation should include laryngoscopy and esophagoscopy to evaluate the upper respiratory and GI tracts and to determine the location of the lesion. If a lesion is encountered, biopsy should be performed. Computed tomography (CT) or magnetic resonance imaging (MRI) may show the soft tissue lesion more accurately. Determination of the need for resection or radiation therapy may be made on the basis of this test. Lesions, if any, are ulcerated and fungating.

TREATMENT AND MANAGEMENT

Immediate esophageal reconstruction improves results. Surgical treatment involves laryngoesphagectomy with reconstruction, using a skin graft. The posterior half of the larynx and the cricoid cartilage are often involved and must be removed. Radical dissection is performed to remove lymph nodes in the neck and superior mediastinum.

Aggressive surgical resection yields better results than radiation therapy. Positive margins, invasion, and vocal cord paralysis indicate a worse prognosis. Patients who undergo esophagectomy with gastric pull-up have better palliative responses than those treated with chemotherapy and radiation. Lesions that are not fixed to surrounding tissues should be resected. If nodes are present or if the lesion is in close contact with the cricopharyngeus muscle, preoperative chemotherapy should be administered, followed by surgical resection.

COURSE AND PROGNOSIS

Treatment with surgical excision has produced results that are as poor as for lower esophageal malignancies. Radiotherapy had been reserved for these lesions but has resulted in a high rate of local recurrence with vascular and tracheal erosion, causing dysphagia, bleeding, and aspiration. Patients who undergo complete resection usually die of metastatic disease. Unfortunately, 80% of patients fail after radiation, and 20% require palliation to control local disease.

ADDITIONAL RESOURCES

Netter FH, Som MX, Wolf BS: Diseases of the esophagus. In Netter FH, Oppenheimer E, editors; with Bachrach WH, Michels NA, Mitchell GAG, et al: *The Netter collection of medical illustrations.* Vol 3. Digestive system. I. Upper digestive tract, Teterboro, NJ, 1979, Icon Learning Systems, pp 137-156.

Peters JH, DeMeester TR: Esophagus and diaphragmatic hernia. In Schwartz SI, Shires TG, Spencer FC, editors: *Principles of surgery,* ed 7, New York, 1999, McGraw-Hill, pp 1081-1179.

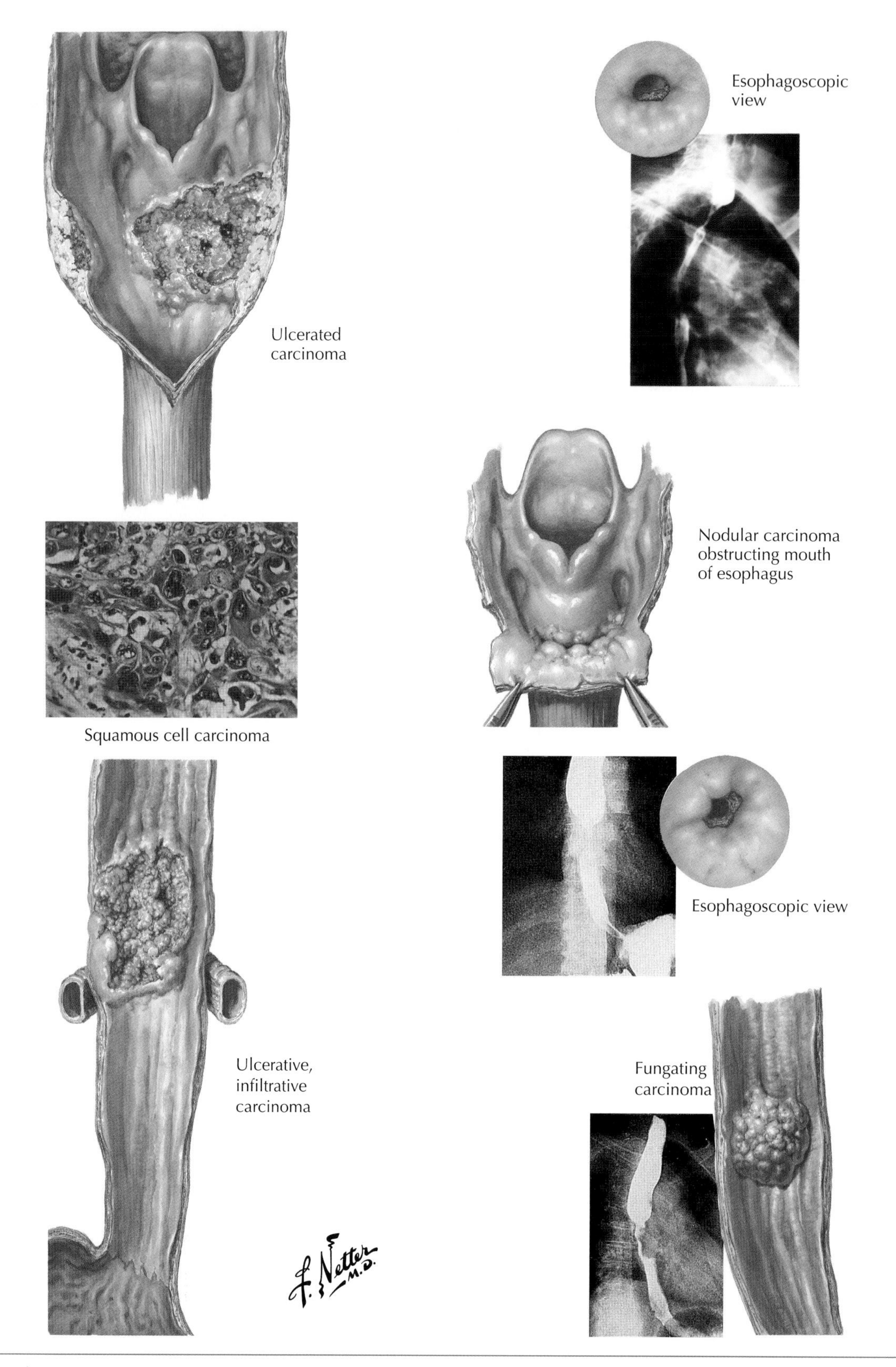

Figure 29-1 *Malignant Tumors: Upper and Middle Portions of the Esophagus.*

Malignant Neoplasms: Lower End of the Esophagus

30

Neil R. Floch

Malignancy of the esophagus accounts for 5% of all gastrointestinal (GI) cancers (**Fig. 30-1**). *Squamous cell carcinoma* is the most common esophageal malignancy worldwide. *Adenocarcinoma* is the most common malignancy in Western countries and has continued its 20-year increase in incidence, especially in white men in the sixth decade of life. The black/white male ratio is 3:10 for adenocarcinoma; the reverse is true for squamous cell cancer. The increase in esophageal cancer has mirrored the increase in adenocarcinoma in patients with Barrett esophagus and has raised concern for risk in these patients. It is not surprising that 85% of cancers occur in the mid-to-distal esophagus. Recommendations of routine surveillance and aggressive surgical treatment for patients with high-grade dysplasia (HGD) have been instituted to discover and treat esophageal cancer early.

Patients with achalasia, caustic strictures, Barrett esophagus, or gastroesophageal reflux disease (GERD) are predisposed to adenocarcinoma; those who drink alcohol, smoke, or have tylosis or Plummer-Vinson syndrome are predisposed to squamous cell cancer. Unfortunately, despite best attempts, only 5% of patients seek treatment while they have local disease. Five-year survival ranges from 16% to 32% of patients.

Adenocarcinoma and squamous cell cancer are the most common malignancies of the esophagus. Adenocarcinoma is thought to develop from metaplastic columnar mucosa or Barrett esophagus, which occurs in the distal esophagus in patients with GERD. Distal esophageal and proximal gastric malignancies are similar. Bile, pancreatic juice, pepsin, and gastric acid may cause a transformation of squamous cells to columnar cells. In time, metaplastic cells may be transformed from dysplastic to malignant. Barrett mucosa is almost always present in those with adenocarcinoma. *Helicobacter pylori* infection may be protective for adenocarcinoma but can cause gastritis, ulcers, and lymphoid tumors. Adenocarcinomas and squamous cell cancers invade the mucosa and the submucosa, spreading quickly up the length of the esophagus.

Other, uncommon lower-end esophageal malignancies include adenosquamous carcinoma, small cell cancer, and lymphoma.

CLINICAL PICTURE

Most patients with tumors have initial symptoms of dysphagia, odynophagia, and weight loss. Presentation with luminal obstruction indicates poor prognosis. As a tumor invades, there may be pain, hoarseness from recurrent laryngeal nerve involvement, superior vena cava syndrome, malignant pleural effusions, hematemesis, or bronchotracheoesophageal fistulae.

DIAGNOSIS

Diagnosis is not limited to determining the presence of malignancy and must include determination of extent. History and physical examination suggest the disease. Confirmation may be made using endoscopy and biopsy; barium esophagraphy to determine mucosal extent; endoscopic ultrasound (US) with or without fine-needle aspiration to determine depth of invasion; computed tomography (CT) or magnetic resonance imaging (MRI) to determine local invasion and lymph node involvement; bronchoscopy to determine invasion of the airway; and positron emission tomography (PET) to detect distant metastases. Diagnostic mediastinoscopy, laparoscopy, or thoracoscopy may determine metastasis to the lymph nodes.

Staging of tumors is the best predictor of survival and is performed by the *tumor-node-metastasis* (TNM) classification (Tables 30-1 and 30-2). Endoscopic US is superior to CT for determining depth of tumor wall invasion, lymph node involvement, and characteristics of adjacent structures. US has the added benefit of facilitating endoscopic-guided fine-needle

Table 30-1 TNM Staging for Esophageal Cancer

Stage	Description
Primary Tumor (T)	
TX	Primary tumor cannot be assessed.
T0	No evidence of primary tumor.
Tis	Carcinoma in situ (into mucosa only).
T1	Tumor invades lamina propria or submucosa.
T2	Tumor invades muscularis propria.
T3	Tumor invades adventitia.
T4	Tumor invades (or is adherent to) adjacent structures.
Regional Lymph Nodes (N)	
NX	Regional lymph nodes cannot be assessed.
N0	No regional lymph node metastases.
N1	Regional lymph node metastases.
Distant Metastases (M)	
MX	Distant metastases cannot be assessed.
M0	No distant metastases.
M1	Distant metastases.

From Kelly KA, Sarr MG, Hinder RA: *Mayo Clinic gastrointestinal surgery,* Philadelphia, 2004, Saunders, p 49.

Table 30-2 Stage Groupings for Esophageal Cancer

Stage	T	N	M
0	Tis	N0	M0
I	T1	N0	M0
IIA	T2-T3	N0	M0
IIB	T1-T2	N1	M0
III	T3	N1	M0
	T4	N1	M0
IV	Any T	Any N	M1

From Kelly KA, Sarr MG, Hinder RA: *Mayo Clinic gastrointestinal surgery,* Philadelphia, 2004, Saunders, p 49.

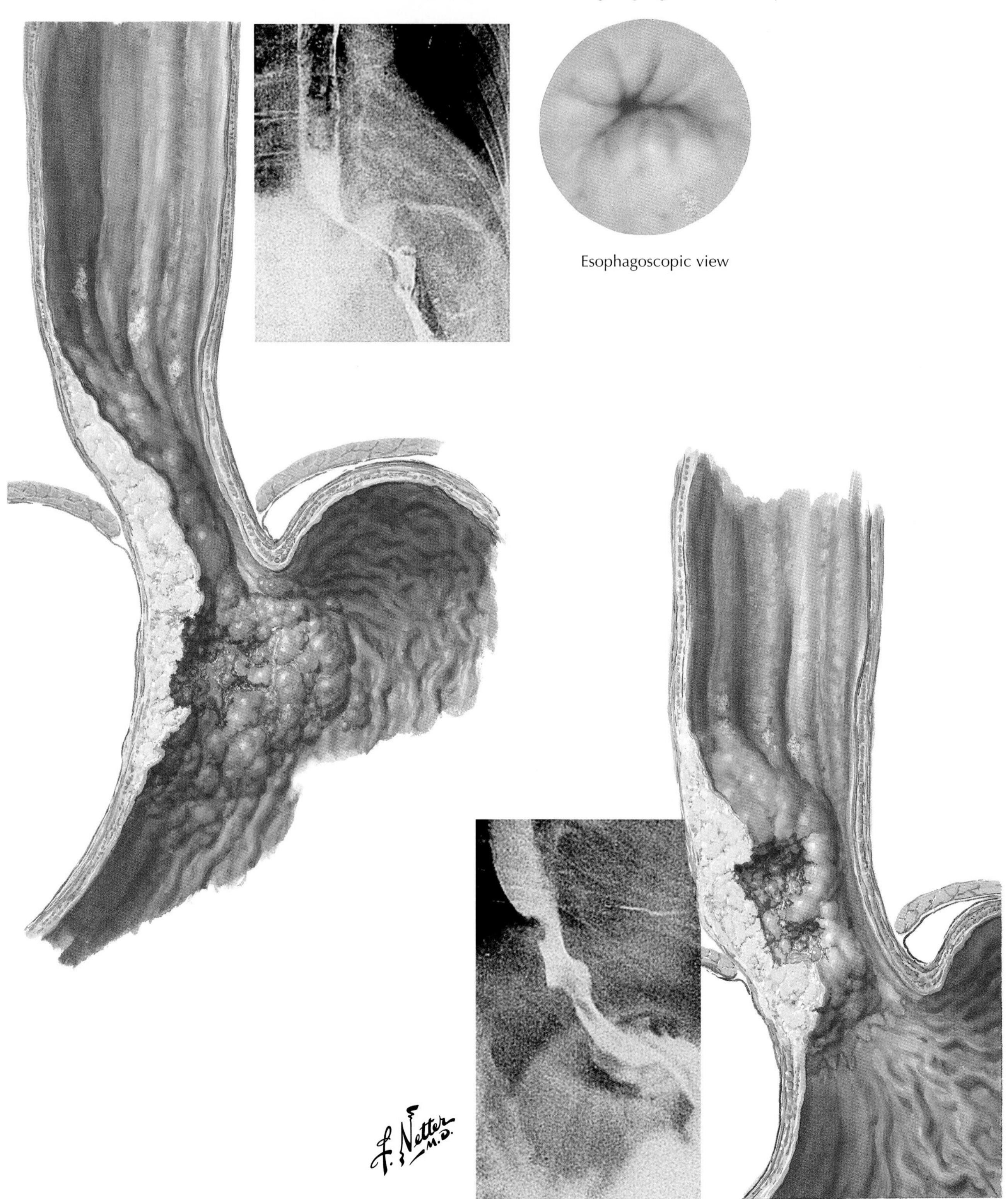

Figure 30-1 *Malignant Tumors: Lower End of the Esophagus.*

aspiration of lymph nodes and mediastinal lesions. US is capable of delineating four layers of the esophageal wall and has 90% and 85% accuracy, respectively, for measuring tumor and lymph node status. Laparoscopy and laparoscopic US may be superior to CT and esophageal US and can avoid noncurative laparotomies in 11% to 48% of patients.

TREATMENT AND MANAGEMENT

Esophageal cancer spreads early and rapidly; two thirds of patients will have lymph node metastases at presentation. Aggressive multimodal therapy is necessary to achieve control and attempt cure. If cure is not possible, palliation with attempts at maintaining nourishment and quality of life are paramount. The TNM classification aids in determining the feasibility of surgical resection and treatment options.

Total esophagectomy to obtain free margins and to clear all mediastinal lymph nodes is the first line of therapy and is the only modality necessary for those with T1-2 N0 M0 disease. The mortality rate associated with surgery is 3%. Techniques include Ivor Lewis esophagogastrectomy and transhiatal esophagectomy, which is recommended for stage I tumors. T1b and T2 tumors are removed through thoracotomy.

Chemoradiation alone is reserved for those in poor medical condition. In patients with T1-2 N1 M0, T3 N0 M0, and T3 N1 M0, chemoradiation is preferably given before surgery but may be given after it. Cisplatin and 5-fluorouracil are the medications of choice at the Mayo Clinic. T4 N0-1 M0 disease is treated with preoperative chemoradiation, which may achieve adequate palliation in itself. Afterward, patients may be restaged. Patients without hematogenous spread or peritoneal implants are candidates for surgical resection and intraoperative radiation.

Palliation of metastatic disease is reserved for those with any T, any N, M1 disease. Chemotherapy is the treatment of choice, but radiation may be added. Endoscopic metal stents, photodynamic tumor ablation with laser, metalloporphyrin, and brachytherapy may be used for palliation of obstruction. Radical surgery to bypass obstruction is rarely performed.

Other tumors of the esophagus are usually treated by surgery. Most patients with small cell carcinoma have metastatic disease at presentation and rarely survive a year. Therapy is palliative, but surgery and chemoradiation may result in a rare cure. Melanomas of the esophagus have a worse prognosis than cutaneous disease because they are discovered late. Surgery is performed but rarely helpful. Results after surgical excision of salivary tumors are worse than after excision of head and neck tumors. Lymphomas develop after direct spread from other organs. Primary lymphomas usually develop in patients with immune disorders. Sarcomas may develop, but not from degeneration of benign tumors, and are removed by surgery. Metastatic lesions of the breast and lung and melanomas are most common and are treated by palliation.

COURSE AND PROGNOSIS

Patient survival depends directly on disease stage. Five-year survival rates for esophageal cancer are 78.9% for stage I, 37.9% for stage IIA, 27.3% for stage IIB, 13.7% for stage III, and 0% for stage IV.

ADDITIONAL RESOURCES

Cameron JL, editor: *Current surgical therapy*, ed 6, St Louis, 1998, Mosby, pp 1-74.

Gee DW, Rattner DW: Gastrointestinal cancer: management of gastrointestinal tumors, *Oncologist* 12:175-185, 2007.

Kelly KA, Sarr MG, Hinder RA: *Mayo Clinic gastrointestinal surgery*, Philadelphia, 2004, Saunders, p 49.

Peters JH, DeMeester TR: Esophagus and diaphragmatic hernia. In Schwartz SI, Shires TG, Spencer FC, editors: *Principles of surgery*, ed 7, New York, 1999, McGraw-Hill, pp 1081-1179.

Van Dam J: Endosonographic evaluation of the patient with esophageal cancer, *Chest* 112(suppl):184S-190S, 1997.

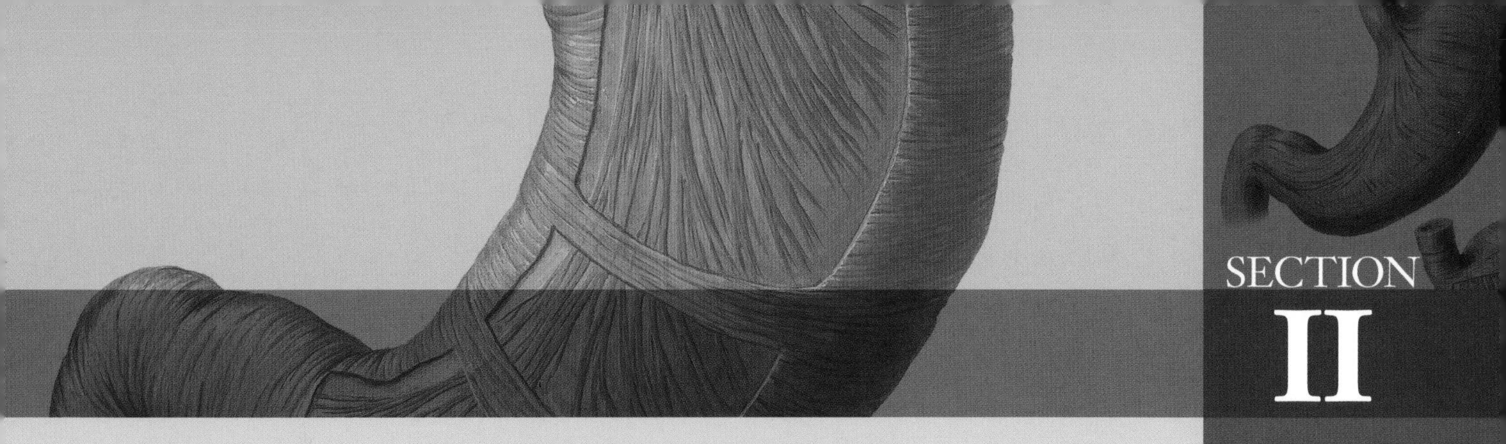

SECTION II

Stomach and Duodenum

Anatomy of the Stomach: Normal Variations and Relations

31

Martin H. Floch

The stomach is a J-shaped reservoir of the digestive tract in which ingested food is soaked in gastric juice containing enzymes and hydrochloric acid and then is released spasmodically into the duodenum by gastric peristalsis. The form and size of the stomach vary considerably, depending on the position of the body and the degree of filling. Special functional configurations of the stomach are of interest to the clinician and radiologist (**Fig. 31-1**).

The stomach has a ventral surface and a dorsal surface that may be vaulted or flattened and that almost make contact when the organ is empty. The stomach also has two borders, the concave *lesser curvature* above on the right and the convex *greater curvature* below on the left. The two join at the *cardia*, where the esophagus enters. The poorly defined cardia is the point of demarcation between both curvatures, whereas on the right the esophagus continues smoothly into the lesser curvature. On the left there is a definite indentation, the incisura cardialis (cardial or cardiac incisure, or notch), that becomes most obvious when the uppermost, hoodlike portion of the stomach (fundus, or fornix) is full and bulges upward. The major portion of the stomach (body, or corpus) blends imperceptibly into the pyloric portion, except along the lesser curvature, where a notch, the incisura angularis (angular incisure) marks the boundary between the corpus and the pyloric portion. The pylorus consists of the pyloric antrum, or vestibule, which narrows into the pyloric canal and terminates at the pyloric valve. External landmarks of the pylorus form a circular ridge of sphincter muscle and the subserosal pyloric vein.

During esophagogastroduodenoscopy, selective views can evaluate almost all these areas. For example, retroflexion of the endoscope permits visualization of the scope entering the stomach. The endoscopist can see the normal mucosa of the gastroesophageal junction as it hugs the scope, forming a fold or flap at the cardiac incisure. The pyloric channel is usually closed, and waves of contractions move aborally from the pylorus and end at the angular incisure of the pyloric antrum.

The stomach is entirely covered with peritoneum. A double layer of peritoneum, deriving from the embryonal ventral mesogastrium, extends on the lesser curvature beyond the stomach known as the *lesser omentum*. It passes over to the porta hepatis and may be divided into a larger, thinner, proximal portion (hepatogastric ligament) and a smaller, thicker, distal portion (hepatoduodenal ligament), which attaches to the pyloric region and to the upper horizontal portion of the duodenum. The free edge of the hepatoduodenal ligament, through which run the portal vein, hepatic artery, and common bile duct (see Chapter 32), forms the ventral margin of the epiploic foramen of Winslow, which gives access to the lesser peritoneal sac (bursa omentalis). The *greater omentum*, a derivative of the embryonal dorsal mesogastrium, passes caudally from the greater curvature and contains, between its two frontal and two dorsal sheets, the inferior recess of the bursa omentalis.

The anterior surface of the stomach abuts the anterior abdominal wall, against the inferior surface of the left lobe of the liver and, to some extent in the pyloric region, against the quadrate lobe of the liver and the gallbladder. Its posterior surface is in apposition with retroperitoneal structures (pancreas, splenic vessels, left kidney, and adrenal gland) from which, however, it is separated by the bursa omentalis. The fundus bulges against the left diaphragmatic dome. On the left, adjacent to the fundus, is the spleen, which is connected to the stomach by the gastrosplenic ligament (also derived from the dorsal mesogastrium).

The four recognized principal functional types of stomach are known as *orthotonic*, *hypertonic*, *hypotonic*, and *atonic*. In the hypotonic and atonic types, the axis of the stomach is more longitudinal, whereas in the orthotonic and particularly the hypertonic types, it is more transverse.

ADDITIONAL RESOURCES

Russo MA, Redel CA: Anatomy, histology, and developmental anomalies of the stomach and duodenum. In Feldman M, Friedman LS, Brandt LJ, editors: *Gastrointestinal and liver disease*, ed 8, Philadelphia, 2006, Saunders-Elsevier.

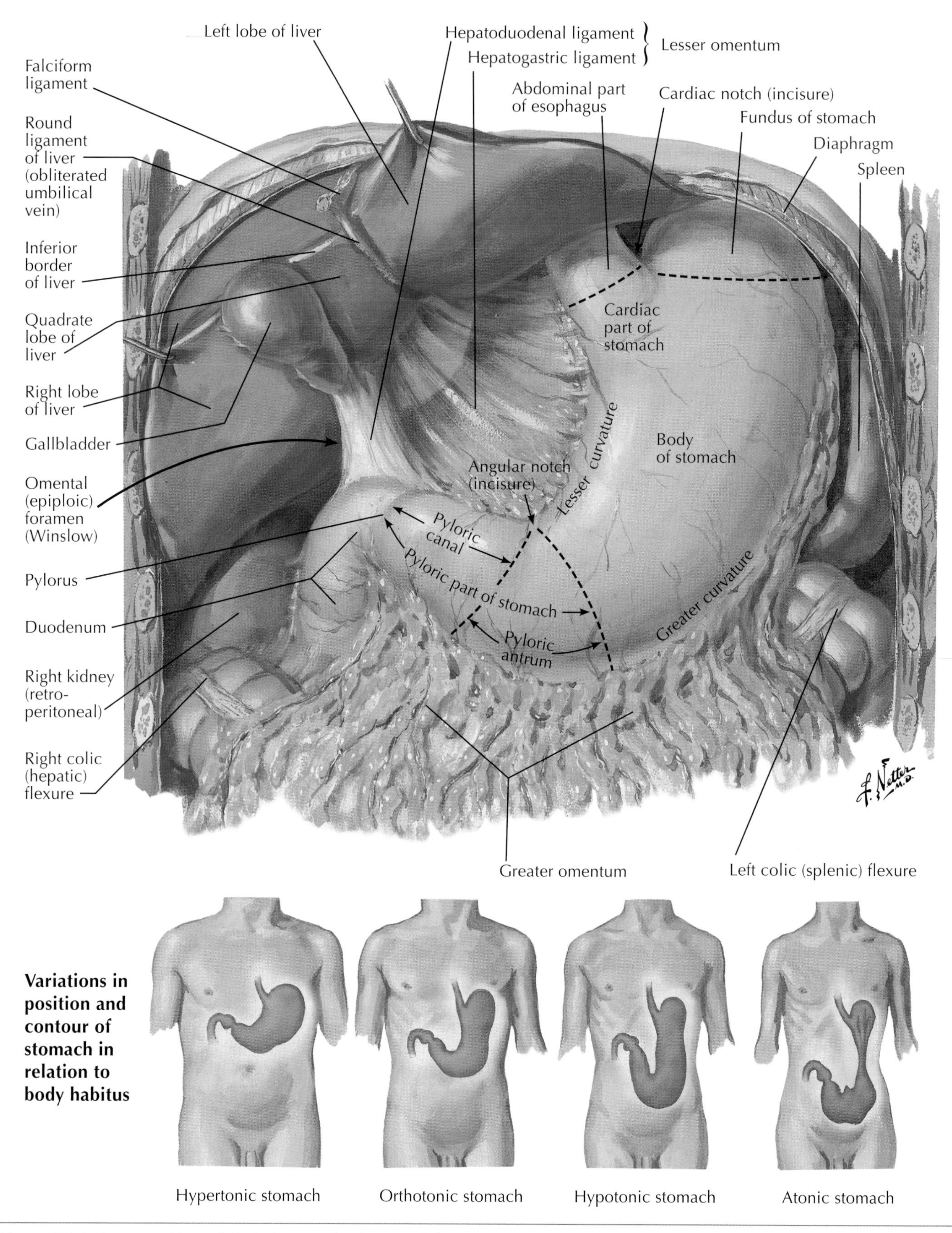

Figure 31-1 *Anatomy, Normal Variations, and Relations of the Stomach.*

Anatomy and Relations of the Duodenum

32

Martin H. Floch

The duodenum, the first part of the small intestine, has a total length of approximately 25 to 30 cm (10-12 inches). It is horseshoe shaped, with the open end facing left, and is divided into four parts (**Fig. 32-1**).

The first part of the duodenum, or the *pars superior*, lies at the level of the first lumbar vertebra (L1) and extends almost horizontally from the pylorus to the first flexure. As a result of its intraperitoneal position, this first duodenal portion is freely movable and can adapt its course according to the filling condition of the stomach. The anterior and superior surfaces of the first half of this duodenal segment are in close relation to the inferior surface of the liver (lobus quadratus) and the gallbladder. The radiographic designation *duodenal bulb* refers to the most proximal end of the pars superior duodeni, which is slightly dilated when the organ is filled and then is more sharply separated from the stomach because of pyloric contraction.

The two layers of peritoneum, which cover the anterosuperior and the posteroinferior surfaces, join together on the upper border of the superior portion of the duodenum and move as the hepatoduodenal ligament cranially toward the liver, forming the right, free edge of the lesser omentum (see Chapter 31). This ligament contains the important triad of the portal vein, hepatic artery, and common bile duct.

The second part of the duodenum, the descending portion, extends vertically from the first to the second duodenal flexure, the latter lying approximately at the level of the third lumbar vertebra (L3). The upper area of this portion rests laterally on the structures of the hilus of the right kidney; medially, its whole length is attached by connective tissue to the duodenal margin of the caput pancreatis (head of pancreas). Approximately halfway its length, the descending portion is crossed anteriorly by the parietal line of attachment of the transverse mesocolon. The common bile duct, together with the portal vein, occupies the start of the hepatoduodenal ligament, a position dorsal to the superior duodenal portion, and continues its course between the descending portion and the pancreatic head to its opening at the major duodenal papilla (Vater).

The third part of the duodenum, the inferior portion, begins at the second flexure. It begins almost horizontally (horizontal part) or sometimes in a slightly ascending direction, until it reaches the region of the left border of the aorta, where it changes direction and curves cranially to pass into the terminal duodenal segment (ascending part). Although the caudal part of the second portion and the second flexure lie over the psoas major of the right side of the body, the third duodenal portion, with its horizontal segment, passes over the vena cava and the abdominal aorta. The superior mesenteric vessels, before entering the root of the mesentery, cross over the horizontal part of the third portion near its transition to the ascending part. During its course, the third portion is increasingly covered by the peritoneum, and a complete intraperitoneal configuration is attained at the duodenojejunal flexure, which is located caudal to the mesocolon transversum at the level of the second lumbar vertebra (L2) or of the disk between L1 and L2.

As the third part of the duodenum courses up to the left of the aorta to reach the border of the pancreas, it is frequently referred to as the fourth part of the duodenum. This fourth part joins the jejunum and is fixed posteriorly by the *ligament of Treitz*, a suspensory muscle of the duodenum. The fourth part of the duodenum then leaves the retroperitoneal area to join the intraperitoneal jejunum. On radiographs, the duodenum usually takes the form of a C, although it may show individual variations, such as a redundant second part or a reversal of curve (see Fig. 32-1).

ADDITIONAL RESOURCES

Russo MA, Redel CA: Anatomy, histology, and developmental anomalies of the stomach and duodenum. In Feldman M, Friedman LS, Brandt LJ, editors: *Gastrointestinal and liver disease*, ed 8, Philadelphia, 2006, Saunders-Elsevier.

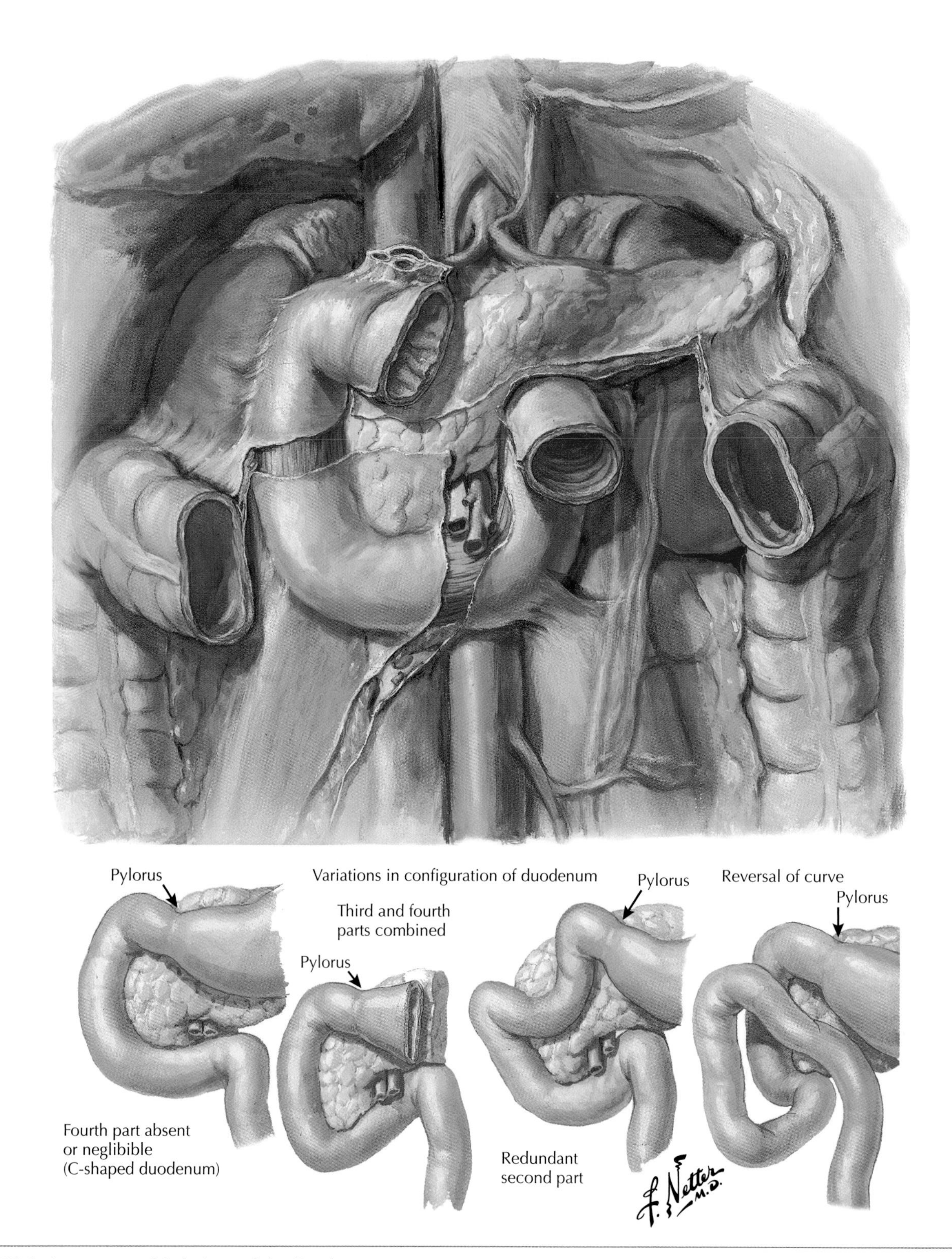

Figure 32-1 *Anatomy and Relations of the Duodenum.*

Mucosa of the Stomach

33

Martin H. Floch

The reddish gray mucous membrane of the stomach, composed of a single surface layer of epithelial cells (tunica propria) and the submucosa, begins at the cardia along an irregular or zigzag line, often referred to as the Z line (**Fig. 33-1**). The mucosa appears as a more or less marked relief of folds, or *rugae*, which flatten considerably when the stomach is distended. In the region of the lesser curvature, where the mucosa is more strongly fixed to the muscular layer, the folds take a longitudinal course, forming what has been called the *magenstrasse* ("stomach street," canalis gastricus). The rugae are generally smaller in the fundus and become larger as they approach the antrum, where they tend to run diagonally across the stomach toward the greater curvature. In addition to these broad folds, the gastric mucosa is further characterized by numerous shallow invaginations, which divide the mucosal surface into a mosaic of elevated areas varying in shape. When viewed under magnification with a lens, these *areae gastricae* reveal several delicate ledges and depressions, the latter known as gastric pits, or *foveolae gastricae*. The glands of the stomach open into the depth of these pits, which have varying widths and lengths.

The gastric epithelium, a single layer of columnar cells at the gastroesophageal junction, is sharply demarcated from the stratified and thicker esophageal mucosa. The *epithelial cells* are mucoid type and contain mucigen granules in their outer portions and an ovoid nucleus at their base.

The glands of the stomach are tubular; three types can be differentiated. The *cardiac glands* are confined to a narrow 0.5- to 4-cm zone in width around the cardiac orifice. They are coiled and are lined by mucus-producing cells. The *gastric, oxyntic,* or *fundic glands* are located in the fundus and over the greater part of the body of the stomach. They are fairly straight, simply branched tubules, with a narrow lumen reaching down almost to the muscularis mucosae. They are lined largely by three types of cells. *Mucoid cells* are present in the neck and differ from the cells of the surface epithelium in that their mucigen granules have slightly different staining qualities and their nuclei tend to be flattened or concave at the cell base. *Chief cells*, or *zymogenic cells*, line the lower half of glandular tubules. They have spheric nuclei and contain strongly light-refracting granules and a Golgi apparatus, the size and form of which vary with the state of secretory activity. Chief cells produce pepsinogen, the precursor of pepsin (see Chapter 40). *Parietal cells* are larger than chief cells and are usually crowded away from the lumen, to which they connect by extracellular capillaries stemming from intracellular canaliculi. Their intraplasmatic granules are strongly eosinophilic and less light refracting than those of the chief cells. Parietal cells produce hydrochloric acid. Histochemical and electron microscope studies have shown the elaborate molecular mechanisms by which hydrogen chloride forms and is secreted as hydrochloric acid within parietal cells and reacts to hormonal, chemical, and neurologic stimuli.

Pyloric glands, the third type of stomach gland, are located in the pyloric region but also spread to a transitional zone, where gastric and pyloric glands are found and which extends diagonally and distally from the lesser to the greater curvature. Tubes of the pyloric glands are shorter, more tortuous, and less densely packed and their ends more branched than in fundic glands. Pits are much deeper in the region of the pyloric glands. These glands are lined by a single type of cell, which resembles, or may be identical to, the mucoid neck cells of the fundic glands.

Specialized endocrine-secreting cells have been identified and are scattered through gastric glands, in the antrum, and in the pylorus. They are fewer in number than chief or parietal cells but are significant in their endocrine and physiologic functions. They secrete into the lumen to affect other endocrine cells or into the circulation for a distal endocrine effect. The *D (delta) cells* secrete somatostatin, which may have a paraendocrine or an endocrine effect. *Enterochromaffin-like* (ECL) cells, or *argentaffin cells* that stain with silver, secrete histamine. Other argentaffin cells that stain with potassium dichromate are called *enterochromaffin* (EC) *cells*, and these contain serotonin. The pylorus also contains a small but significant number of gastrin-secreting cells, called *C cells*. The role of gastrin is discussed in Chapters 38 and 41. *Ghrelin* is secreted by endocrine cells of the pylorus and has a significant effect on appetite and eating behavior.

ADDITIONAL RESOURCES

Date Y, Cojima M, Hosoda H, et al: Ghrelin, growth hormone–releasing associated peptide, synthesized in a distinct endocrine cell type in the gastrointestinal tracts of rats and humans, *Endocrinology* 141:4255-4261, 2000.

Russo MA, Redel CA: Anatomy, histology, and developmental anomalies of the stomach and duodenum. In Feldman M, Friedman LS, Brandt LJ, editors: *Gastrointestinal and liver disease*, ed 8, Philadelphia, 2006, Saunders-Elsevier.

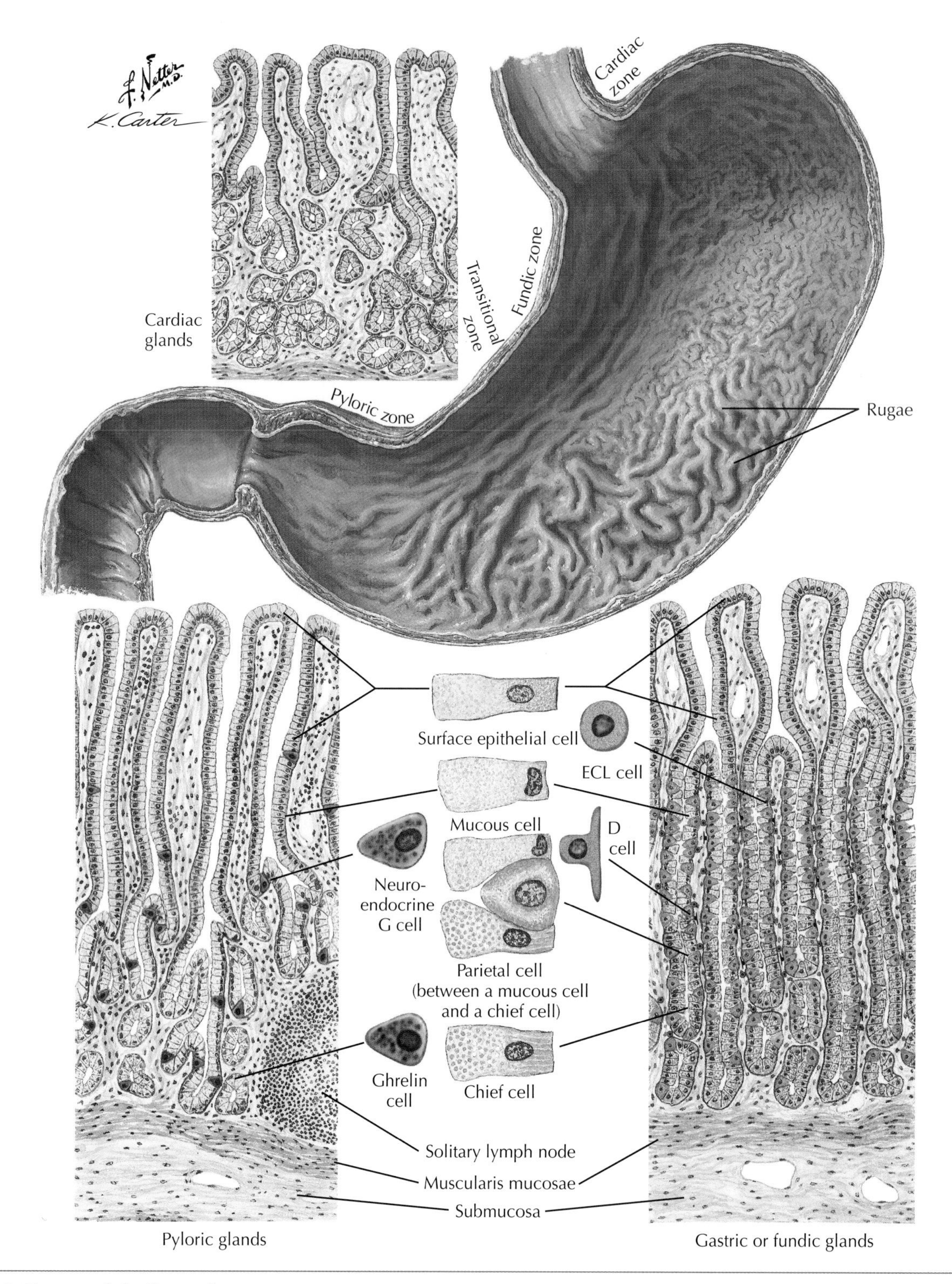

Figure 33-1 *Mucosa of the Stomach.*

Duodenal Mucosa and Duodenal Structures

34

Martin H. Floch

The mucosa of the widened first portion of the duodenum, also known as the bulbus duodeni (duodenal bulb; see Chapter 32), is flat and smooth, in contrast to the more distal duodenal part, which displays the mucosal *Kerckring folds*, as does the entire small intestine (**Fig. 34-1**). These circular folds (plicae), which augment the absorption surface of the intestine, begin in the region of the first flexure and increase in number and elevation in the more distal parts of the duodenum. Kerckring folds do not always form complete circles along the entire intestinal wall; some are semicircular, and others branch out to connect with adjacent folds. Both the mucosa and the submucosa participate in the structure of these plicae, whereas all the other layers of the small intestine, including its two muscular coats, are flat and smooth.

Approximately halfway down the posteromedial aspect of the descending portion of the duodenum, at a distance of 8.5 to 10 cm from the pylorus, is the *papilla of Vater*. The papilla and its relationship to the local anatomy and the anatomic variations are essential to the investigating endoscopist for interpretation of endoscopic retrograde cholangiopancreatography (ERCP) and endoscopic ultrasound of the area. Here the common bile duct (ductus choledochus) and the major pancreatic duct, or duct of Wirsung, open into the duodenum. The common bile duct approaches the duodenum within the enfolding hepatoduodenal ligament of the lesser omentum (see Chapter 31) and continues caudally in the groove between the descending portion of the duodenum and the pancreas (see Section VII). In the posteromedial duodenal wall, the terminal part of the ductus choledochus produces a slight but perceptible longitudinal impression known as the plica longitudinalis duodeni. This fold usually ends at the papilla but occasionally may continue for a short distance beyond the papilla in the form of the so-called frenulum. Small, hoodlike folds at the top of the papilla protect the mouth of the combined bile duct and pancreatic duct.

Numerous variations occur in the types of union of the bile and pancreatic ducts, as illustrated and discussed in Section VII. A small, wartlike, and generally less distinct second papilla, the papilla duodeni minor, is situated approximately 2.5 cm above, and slightly farther medially from, the major papilla. It serves as an opening for the minor pancreatic duct, or duct of Santorini, which is almost always present, despite great variations in development (see also Section VIII).

Except for the first portion of the duodenum, the mucosal surface, which is red in living patients, is lined with villi (see Section IV); these account for its typical velvetlike appearance. The high magnification of videoendoscopes enables endoscopists to determine when villi are flattened. A biopsy specimen is still needed to be certain of villous atrophy.

The duodenal bulb, varying in form, size, position, and orientation, appears in the anteroposterior radiographic projection as a triangle, with its base at the pylorus and its tip pointing toward the superior flexure or the transitional region of the first and second parts of the duodenum. As with the wall of the whole intestinal tract, the wall of the duodenum comprises one mucosal, one submucosal, and two muscular layers and an adventitia, or a subserosa and a serosa, wherever the duodenum is covered by peritoneum. Embryologically, morphologically, and functionally, the duodenum is an especially differentiated part of the small intestine. The epithelium of the duodenal mucosa consists of a single layer of high columnar cells with a marked cuticular border. In the fundus of the crypts, there are cells filled with eosinophilic granules (cells of Paneth) and some cells filled with yellow granules, which have a strong affinity to chromates. The tunica or lamina propria of the mucosa consists of loose connective tissue. Between the mucosa and the submucosa lies a double layer of smooth muscle cells, the fibers of which enter the tunica propria and continue to the tips of villi, enabling the villi to perform a sucking and pumping function.

The submucosa, lying between the mucosal and the muscular layers, allows these two layers to shift in relation to each other. It is made up of collagenous connective tissue, the fibers of which are arranged in the form of a mesh. In this network are embedded the duodenal *glands of Brunner*, characteristic of the duodenum. These are tortuous, acinotubular glands with multiple branches at their ends; breaking through the muscularis mucosae, they open into the crypts. Brunner glands are more numerous and denser in the proximal parts of the duodenum, diminishing in size and density as the duodenum approaches the duodenojejunal junction, although their extension and density vary greatly among individuals.

ADDITIONAL RESOURCES

Date Y, Cojima M, Hosoda H, et al: Ghrelin, growth hormone–releasing associated peptide, synthesized in a distinct endocrine cell type in the gastrointestinal tracts of rats and humans, *Endocrinology* 141:4255-4261, 2000.

Russo MA, Redel CA: Anatomy, histology, embryology and developmental anomalies of the stomach and duodenum. In Feldman M, Friedman LS, Brandt LJ, editors: *Gastrointestinal and liver disease*, ed 8, Philadelphia, 2006, Saunders-Elsevier.

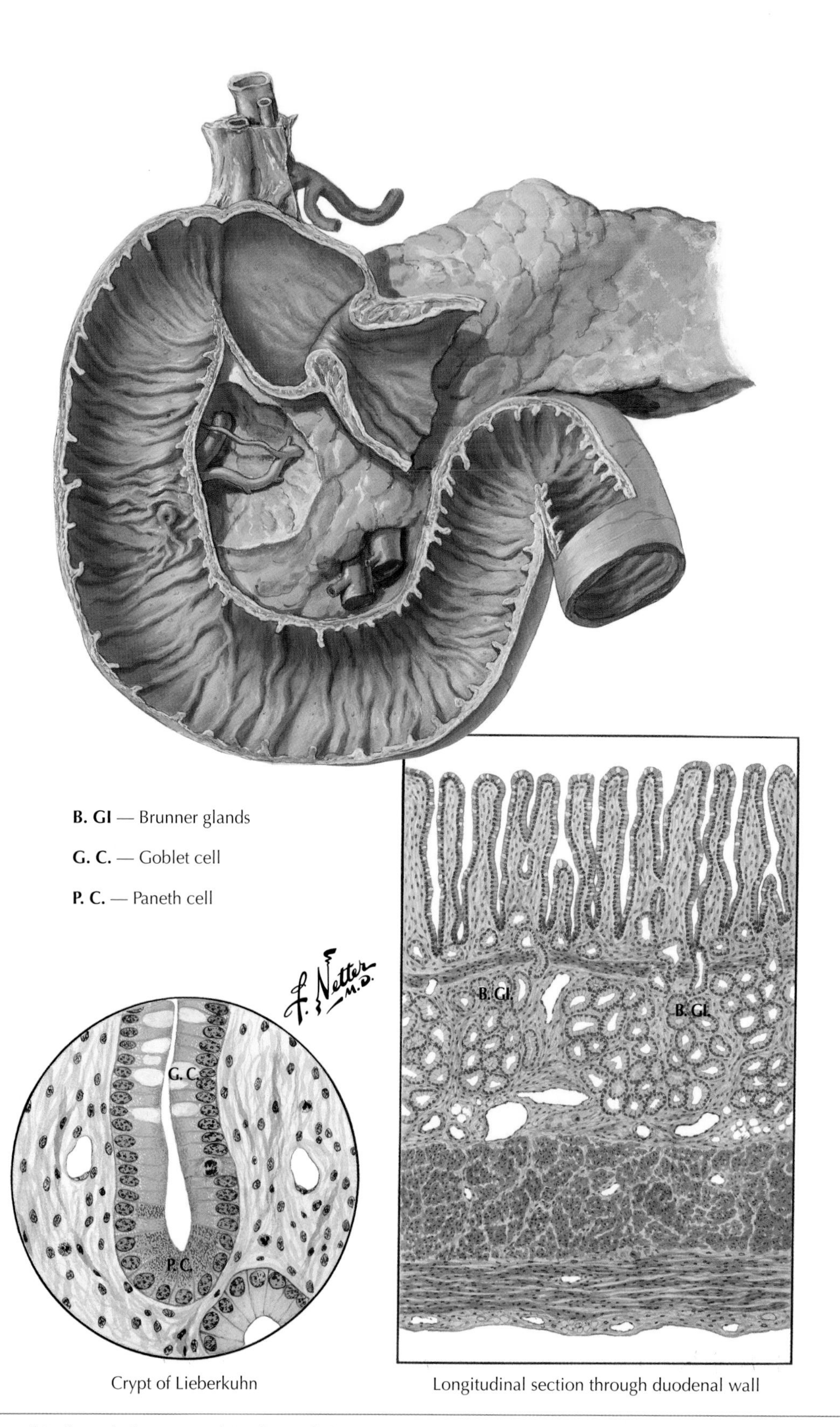

Figure 34-1 *Duodenal Bulb and the Mucosal Surface of the Duodenum.*

Blood Supply and Collateral Circulation of Upper Abdominal Organs

35

Martin H. Floch

Conventional textbook descriptions of the blood supply of the stomach and duodenum and associated organs (e.g., spleen, pancreas) present the misleading concept that the vascular pattern of these organs is relatively simple and uniform. On the contrary, these vascular patterns are always unpredictable and vary in almost all cases. Clinicians should remember this when interpreting angiography and imaging. It is important for the student of gastroenterology to understand the rich collateral circulation in this area of the body. The following are classic descriptions.

Typically, the entire blood supply of the liver, gallbladder, stomach, duodenum, pancreas, and spleen is derived from the *celiac artery;* a small, supplementary portion is supplied by the superior mesenteric artery, inferior pancreoduodenal branch. The celiac varies from 8 to 40 mm in width. When typical and complete, it gives off three branches—hepatic, splenic, and left gastric—constituting a complete trunk, frequently in the form of a tripod.

This conventional description of the celiac artery, with its three branches, occurs in only 55% of the population. In the other 45%, numerous variations occur; the interested reader is referred to classic anatomy texts. Observing a bleeding vessel through endoscopy, surgery, or angiography can be frustrating but requires an open mind and an understanding of the variations in the vascular anatomy.

No other region in the body presents more diversified collateral pathways of blood supply than the supracolonic organs (stomach, duodenum, pancreas, spleen, liver, and gallbladder). Michels identified at least 26 possible collateral routes to the liver alone (**Fig. 35-1**). Because of its many blood vessels and loose arrangement of its extensive connective tissue network, the great omentum is exceptionally well adapted as an area of compensatory circulation, especially for the liver and the spleen, when either the hepatic or the splenic artery is occluded. Through interlocking arteries, the stomach may receive its blood supply from six primary and six secondary sources: the pancreas from the hepatic, splenic, and superior mesenteric arteries; and the liver from three primary sources—celiac, superior mesenteric, and left gastric arteries—and, secondarily, from communications with at least 23 other arterial pathways. In view of the relational anatomy of the splenic artery, it is obvious that most of the collateral pathways to the upper abdominal organs can be initiated through this vessel and its branches and can be completed through communications established by the gastroduodenal and superior mesenteric arteries.

The most important collateral pathways in the upper abdominal organs are as follows:

- *Arcus arteriosus ventriculi inferior.* This infragastric omental pathway is made by the right and left gastroepiploics as they anastomose along the greater curvature of the stomach. The arc gives off ascending gastric and descending epiploic (omental) arteries.
- *Arcus arteriosus ventriculi superior.* This supragastric pathway, with branches to both surfaces of the stomach, is made by the right and left gastrics anastomosing along the lesser curvature. Branches of the right gastric may unite with branches from the gastroduodenal, supraduodenal, retroduodenal superior pancreaticoduodenal, or right gastroepiploic. Branches of the left gastric may anastomose with the short gastrics from the splenic terminals or the left gastroepiploic or with branches from the recurrent cardioesophageal branch of the left inferior phrenic or with those of an accessory left hepatic, derived from the left gastric.
- *Arcus epiploicus magnus.* This epiploic (omental pathway) is situated in the posterior layer of the great omentum below the transverse colon. Its right limb is made by the right epiploic from the right gastroepiploic; its left limb is made by the left epiploic from the left gastroepiploic. Arteries involved in this collateral route include hepatic, gastroduodenal, right gastroepiploic, right epiploic, left epiploic, left gastroepiploic, and interior terminal of the splenic.
- *Circulus transpancreaticus longus.* This important collateral pathway is affected by the inferior transverse pancreatic artery coursing along the inferior surface of the pancreas. By way of the superior or the dorsal pancreatic, of which it is the main left branch, it may communicate with the first part of the splenic, hepatic, celiac, or superior mesenteric, depending on which artery gives rise to the dorsal pancreatic. At the tail end of the pancreas, it communicates with the splenic terminals through the large pancreatic and the caudal pancreatic and at the head of the pancreas with the gastroduodenal, superior pancreaticoduodenal, or right gastroepiploic.
- *Circulus hepatogastricus.* This is a derivative of the primitive, embryonic, arched anastomosis between the left gastric and the left hepatic. In the adult, the arc may persist in its entirety; the upper half may give rise to an accessory left gastric, the lower half to a so-called accessory left hepatic from the left gastric (25%).
- *Circulus hepatolienalis.* An aberrant right hepatic or the entire hepatic, arising from the superior mesenteric, may communicate with the splenic through a branch of the dorsal pancreatic or gastroduodenal or through the transverse pancreatic and caudal pancreatic.
- *Circulus celiacomesentericus.* Through the inferior pancreaticoduodenal, blood may be routed through the anterior and posterior pancreaticoduodenal arcades to enter the gastroduodenal, from which, through the right and left gastroepi-

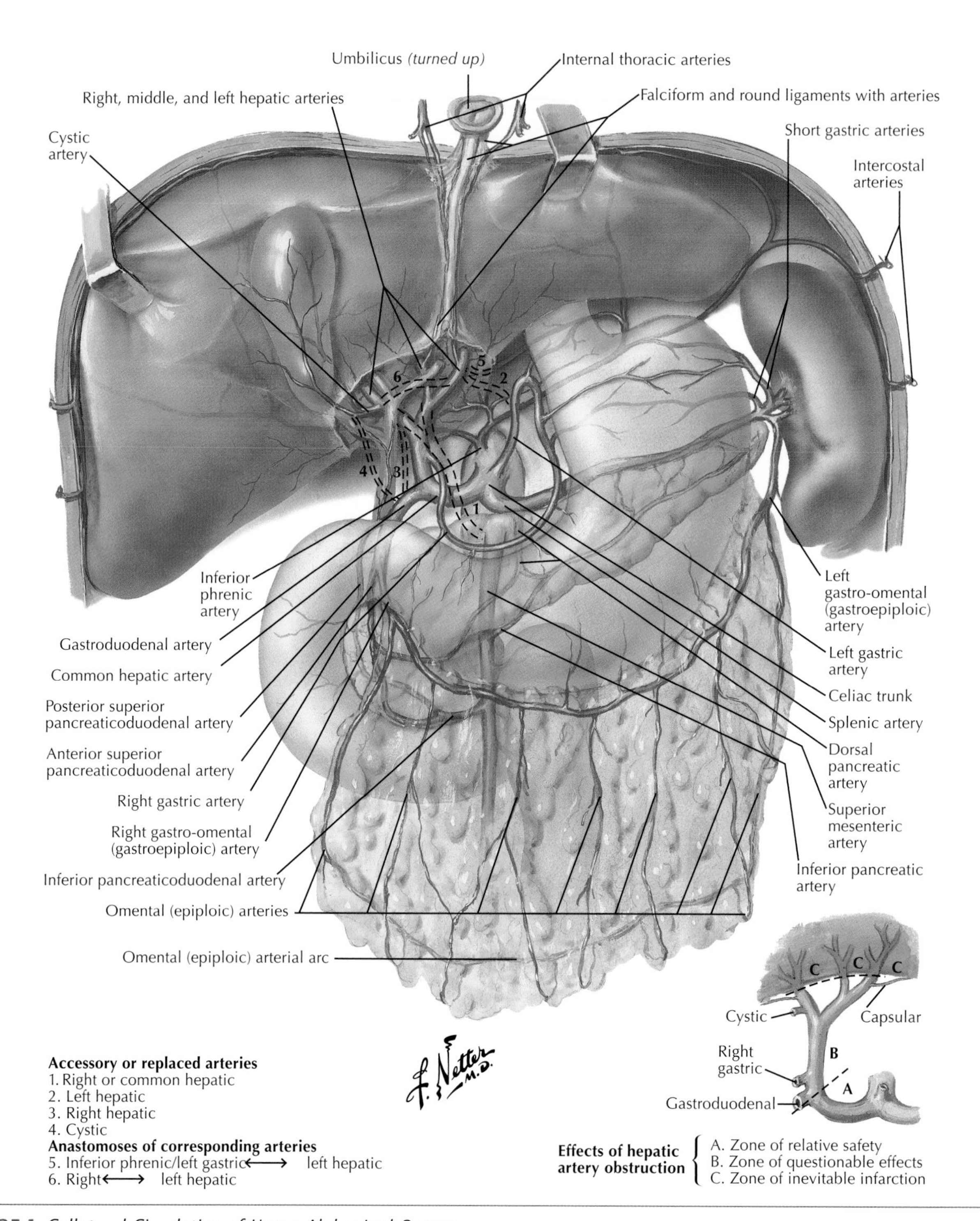

Figure 35-1 *Collateral Circulation of Upper Abdominal Organs.*

ploics, it reaches the splenic, or, through the common hepatic, it reaches the celiac.

- *Circulus gastrolienophrenicus.* This pathway may be affected by a communication between the short gastrics from the splenic terminals and the recurrent cardioesophageal branches of the left inferior phrenic or by a communication between the latter and the cardioesophageal branches given off by the left gastric, its aberrant left hepatic branch, or an accessory left gastric from the left hepatic.

For venous drainage, see Section IX.

ADDITIONAL RESOURCES

Netter FH, Michels NA, Mitchell GAG, Wolf-Heidegger G: Anatomy of the stomach and duodenum. In Netter FH, Ernst Oppenheimer, eds: *The Netter collection of medical illustrations*, Vol 3, *Digestive System I: upper digestive tract*, Teterboro, NJ, 1979, Icon Learning Systems.

Russo MA, Redel CA: Anatomy, histology, and developmental anomalies of the stomach and duodenum. In Feldman M, Friedman LS, Brandt LJ, editors: *Gastrointestinal and liver disease*, ed 8, Philadelphia, 2006, Saunders-Elsevier.

Lymphatic Drainage of the Stomach

Martin H. Floch

36

Lymph from the gastric wall collects in lymphatic vessels, which form a dense subperitoneal plexus on the anterior and posterior surfaces of the stomach (**Fig. 36-1**). The lymph flows in the direction of the greater and lesser curvatures, where the first regional lymph nodes are situated.

On the upper half of the lesser curvature (i.e., portion near cardia) are situated the *lower left gastric* (LLG) *nodes* (lymphonodi gastrici superiores), which are connected with the *paracardial nodes* surrounding the cardia. Above the pylorus is a small group of suprapyloric nodes (not labeled). On the greater curvature, following the trunk of the right gastroepiploic artery and distributed in a chainlike fashion within the gastrocolic ligament, are the *right gastroepiploic* (RGE) *nodes* (lymphonodi gastrici inferiores). From these nodes the lymph flows to the right toward the *subpyloric* (S′pyl) *nodes*, which are situated in front of the head of the pancreas, below the pylorus and the first part of the duodenum. There are a few smaller *left gastroepiploic* (LGE) *nodes* in the part of the greater curvature nearest the spleen.

For purposes of simplification, a distinction can be made among four different draining areas into which the gastric lymph flows, although, in fact, these areas cannot be so clearly separated. The lymph from the upper left anterior and posterior walls of the stomach (region I in the diagram) drains through the lower left gastric and paracardial nodes. From here, the lymphatics follow the left gastric artery and the coronary vein toward the vascular bed of the celiac artery. Included in this system are the *upper left gastric* (ULG) *nodes*, which lie on the left crus of the diaphragm. The LLG nodes, paracardial nodes, and ULG nodes are known collectively as the left gastric nodes.

The pyloric segment of the stomach, in the region of the lesser curvature (region II), discharges its lymph into the *right suprapancreatic* (RS′p) *nodes*, directly and indirectly, through the small suprapyloric nodes. The lymph from the region of the fundus facing the greater curvature (i.e., adjacent to spleen) flows along lymphatic vessels running within the gastrosplenic ligament. Some of these lymphatics lead directly to the *left suprapancreatic* (LS′p; pancreaticolienal) *nodes*, and others lead indirectly through the small *left gastroepiploic* (LGE) *nodes* and through the splenic nodes lying within the hilus of the spleen.

Lymph from the distal portion of the corpus facing the greater curvature and from the pyloric region (region IV) collects in the RGE nodes. From here, the lymph flows to the subpyloric nodes, which lie in front of the head of the pancreas. From the ULG nodes (region 1), RS′p nodes (regions 2 and 4), and LS′p (pancreaticolienal) nodes (region III), the lymph stream leads to the celiac (middle suprapancreatic [MS′p]) nodes, which are situated above the pancreas and around the celiac artery and its branches. From the celiac lymph nodes, the lymph flows through the gastrointestinal (GI) lymphatic trunk to the thoracic duct, in the initial segment of which (i.e., where it arises from various trunks) there is generally a more or less pronounced expansion in the form of the cisterna chyli.

In the region where the thorax borders on the neck, the thoracic duct—before opening into the angle formed by the left subclavian and left jugular veins—receives, among other things, the left subclavian lymphatic trunk. In cases of gastric tumor, palpable metastases may develop in the left supraclavicular nodes (also known as Virchow or Troisier nodes). The lymphatics of the duodenum drain into the nodes that also serve the pancreas.

ADDITIONAL RESOURCES

Netter FH, Michels NA, Mitchell GAG, Wolf-Heidegger G: Anatomy of the stomach and duodenum. In Netter FH, Ernst Oppenheimer, eds: *The Netter collection of medical illustrations*, Vol 3, *Digestive System I: upper digestive tract*, Teterboro, NJ, 1979, Icon Learning Systems.

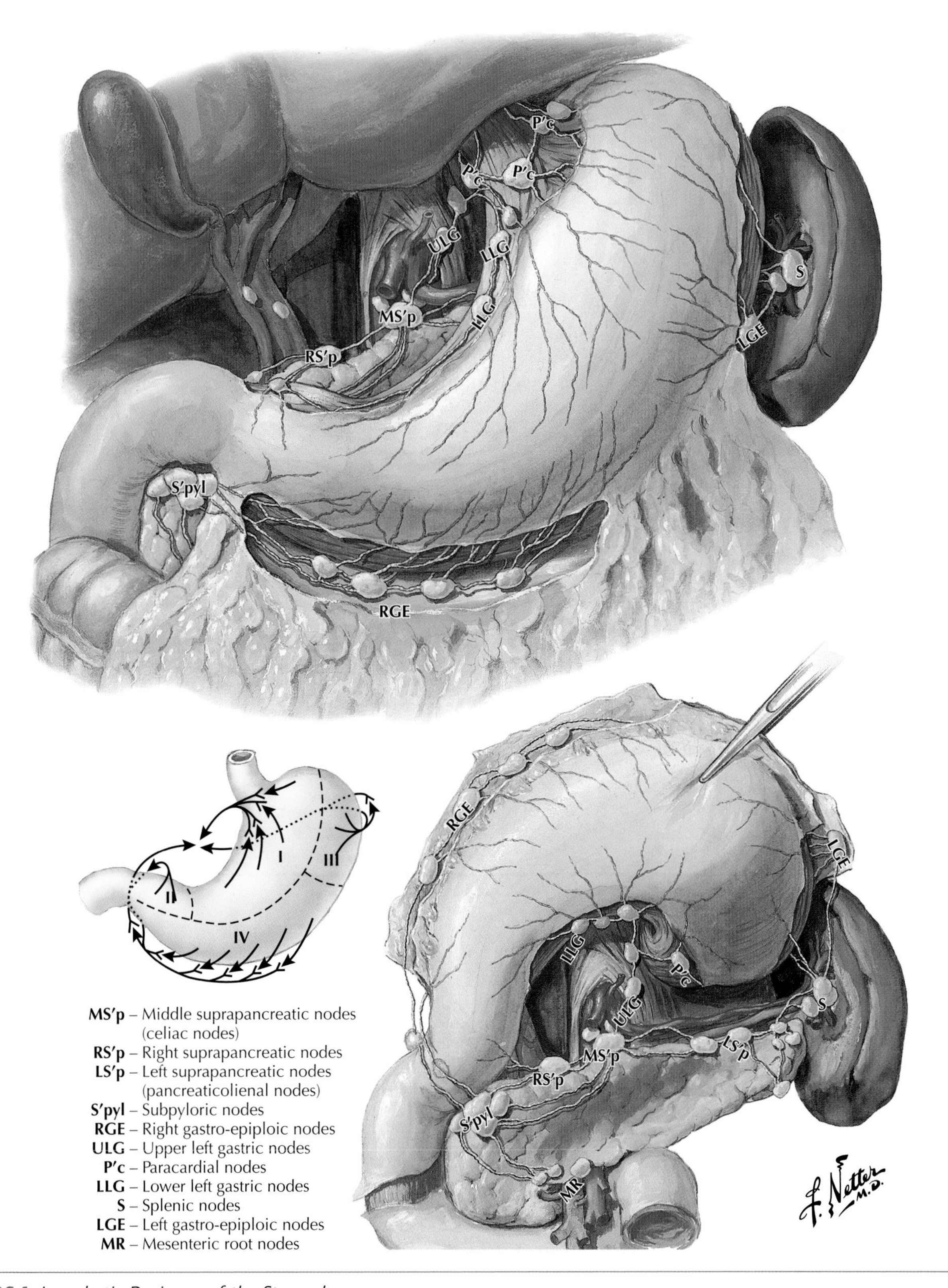

Figure 36-1 *Lymphatic Drainage of the Stomach.*

Innervation of the Stomach and Duodenum

37

Martin H. Floch

This description of the innervation of the stomach and duodenum, although complex and detailed, is important to those who want to understand the common motility disorders of the stomach, such as gastroparesis and dyspepsia.

Sympathetic and parasympathetic nerves that contain efferent and afferent fibers innervate the stomach and the duodenum (**Fig. 37-1**). The sympathetic supply emerges in the anterior spinal nerve roots as *preganglionic fibers*, which are axons of lateral cornual cells located at about the sixth to the ninth or tenth thoracic segments. These fibers are carried from the spinal nerves in rami communicantes, which pass to the adjacent parts of the sympathetic ganglionated trunks, then into the thoracic splanchnic nerves to the celiac plexus and ganglia. Some fibers form synapses in the sympathetic trunk ganglia, but most form synapses with cells in the celiac and superior mesenteric ganglia. The axons of these cells, the *postganglionic fibers*, are conveyed

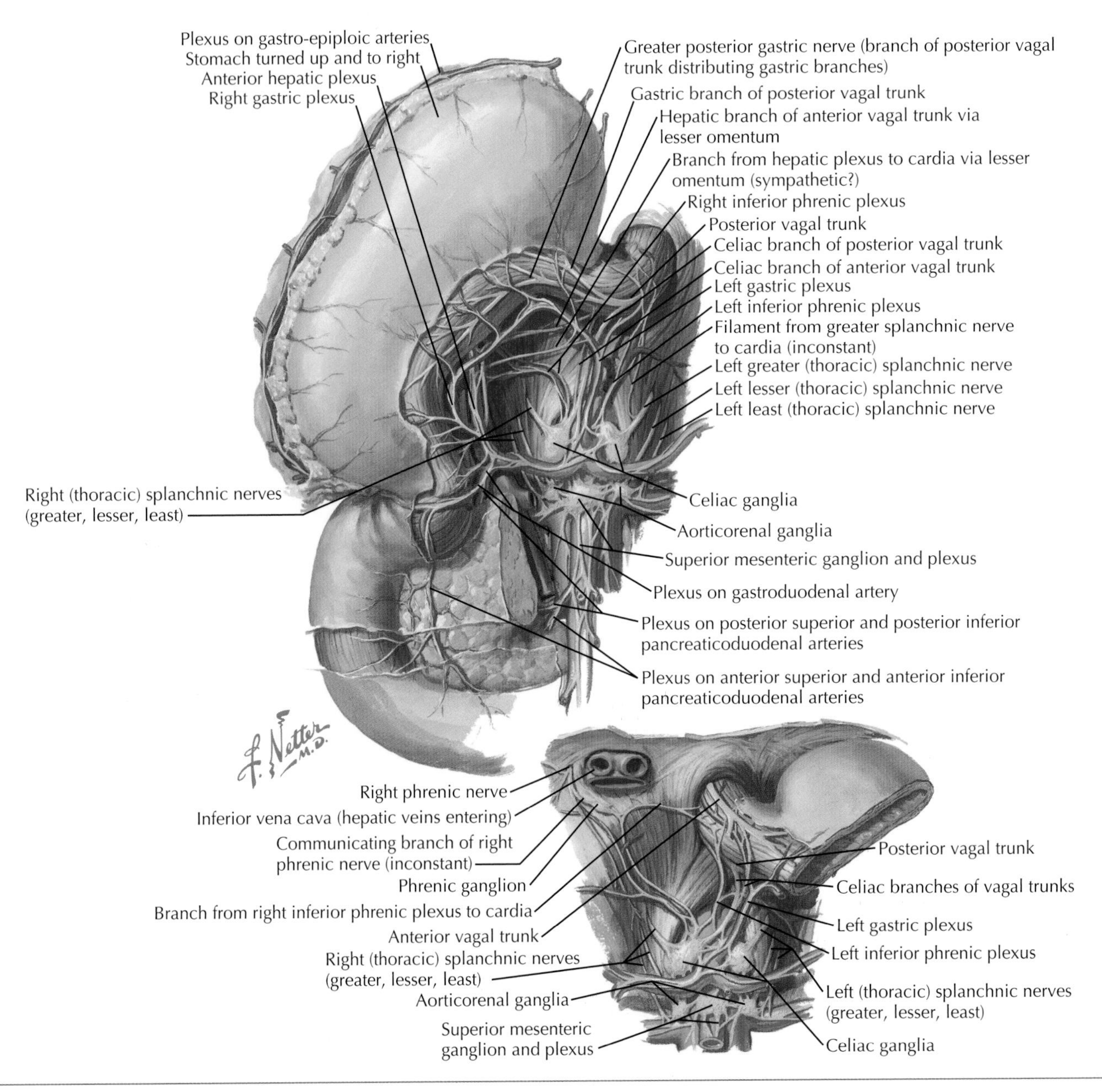

Figure 37-1 *Innervation of the Stomach and Duodenum.*

to the stomach and duodenum in the nerve plexuses alongside the various branches of the celiac and superior mesenteric arteries. These arterial plexuses are composed mainly of sympathetic fibers but also contain some parasympathetic fibers, which reach the celiac plexus through the celiac branches of the vagal trunks.

Afferent impulses are carried in fibers that pursue the reverse route of that just described. However, afferent impulses do not form synapses in the sympathetic trunks; their cytons (perikaryons) are located in the posterior spinal root ganglia and enter the cord through the posterior spinal nerve roots.

The *celiac plexus* is the largest of the autonomic plexuses and surrounds the celiac arterial trunk and the root of the superior mesenteric artery. It consists of right and left halves, each containing one larger celiac ganglion, a smaller aorticorenal ganglion, and a superior mesenteric ganglion, which is often unpaired. These and other, even smaller ganglia are united by numerous nervous interconnections to form the celiac plexus. It receives sympathetic contributions through the greater (superior), lesser (middle), and least (inferior) thoracic splanchnic nerves and through filaments from the first lumbar ganglia of the sympathetic trunks. Its parasympathetic roots are derived from the celiac division of the posterior vagal trunk and from smaller celiac branches of the anterior vagal trunk.

The celiac plexus sends direct filaments to some adjacent viscera, but most of its branches accompany the arteries from the upper part of the abdominal aorta. Numerous filaments from the celiac plexus unite to form open-meshed nerve plexuses around the celiac trunk and the left gastric, hepatic, and splenic arteries. Subsidiary plexuses from the hepatic arterial plexus are continued along the right gastric and gastroduodenal arteries and from the latter along the right gastroepiploic and anterior and posterior superior pancreaticoduodenal arteries. The splenic arterial plexus sends offshoots along the short gastric and left gastroepiploic arteries.

The *superior mesenteric plexus* is the largest derivative of the celiac plexus and contains the superior mesenteric ganglion or ganglia. The main superior mesenteric plexus divides into secondary plexuses, which surround and accompany the inferior pancreaticoduodenal, jejunal, and other branches of the artery.

The *left gastric plexus* consists of one to four nervelets connected by oblique filaments that accompany the artery and supply "twigs" to the cardiac end of the stomach, communicating with offshoots from the left phrenic plexus. Other filaments follow the artery along the lesser curvature between the layers of the lesser omentum to supply adjacent parts of the stomach. They communicate with the *right gastric plexus* and with gastric branches of the vagus.

The *hepatic plexus* also contains sympathetic and parasympathetic efferent and afferent fibers and gives off subsidiary plexuses along all its branches. Following the right gastric artery, these branches supply the pyloric region, and the gastroduodenal plexus accompanies the artery between the first part of the duodenum and the head of the pancreas, supplying fibers to both structures and to the adjacent parts of the common bile duct. When the artery divides into its anterosuperior pancreaticoduodenal and right gastroepiploic branches, the nerves also subdivide and are distributed to the second part of the duodenum, terminations of the common bile and pancreatic ducts, head of the pancreas, and parts of the stomach. The part of the hepatic plexus lying in the free margin of the lesser omentum gives off one or more hepatogastric branches, which pass to the left between the layers of the lesser omentum, to the cardiac end and lesser curvature of the stomach; they unite with and reinforce the left gastric plexus.

The *splenic plexus* gives off subsidiary nerve plexuses around its pancreatic, short gastric, and left gastroepiploic branches, which supply the structures indicated by their names. A filament may curve upward to supply the fundus of the stomach.

The *phrenic plexuses* assist in supplying the cardiac end of the stomach. A filament from the *right* phrenic plexus sometimes turns to the left, posteroinferior to the vena caval hiatus in the diaphragm, and passes to the region of the cardiac orifice. The *left* phrenic plexus supplies a constant twig to the cardiac orifice. A delicate branch from the left phrenic nerve (not illustrated) supplies the cardia.

The parasympathetic supply for the stomach and duodenum arises in the dorsal vagal nucleus in the floor of the fourth ventricle. The afferent fibers also end in the dorsal vagal nucleus, which is a mixture of visceral efferent and afferent cells. The fibers are conveyed to and from the abdomen through the vagus nerves, esophageal plexus, and vagal trunks. The vagal trunks give off gastric, pyloric, hepatic, and celiac branches.

The *anterior vagal trunk* gives off gastric branches that run downward along the lesser curvature, supplying the anterior surface of the stomach almost as far as the pylorus. The pyloric branches (not illustrated) arise from the anterior vagal trunk or from the greater anterior gastric nerve and run to the right between the layers of the lesser omentum, before turning downward through or close to the hepatic plexus to reach the pyloric antrum, pylorus, and proximal part of the duodenum. Small celiac branches run alongside the left gastric artery to the celiac plexus, often uniting with corresponding branches of the posterior vagal trunk.

The *posterior vagal trunk* gives off gastric branches that radiate to the posterior surface of the stomach, supplying it from the fundus to the pyloric antrum. One branch, the *greater posterior gastric nerve*, is usually larger than the others. As on the anterior aspect, these branches communicate with adjacent gastric nerves, although no true posterior gastric plexus exists. The celiac branch is large and reaches the celiac plexus alongside the left gastric artery. Vagal fibers from this celiac branch are distributed to the pylorus, duodenum, pancreas, and so on, through the vascular plexuses derived from the celiac plexus.

ADDITIONAL RESOURCES

Netter FH, Michels NA, Mitchell GAG, Wolf-Heidegger G: Anatomy of the stomach and duodenum. In Netter FH, Ernst Oppenheimer, eds: *The Netter collection of medical illustrations*, Vol 3, *Digestive System I: upper digestive tract*, Teterboro, NJ, 1979, Icon Learning Systems.

Gastric Secretion

38

Martin H. Floch

The stomach produces endocrine and exocrine secretions. The *endocrine* secretions are somatostatin, histamine, gastrin, neuropeptides (gastrin-releasing peptide), calcitonin, pituitary adenylate cyclase–activating polypeptide, and ghrelin. The *exocrine* secretions are water, electrolytes (hydrogen, potassium, sodium, chlorate, bicarbonate), pepsinogen, lipase, intrinsic factor, and mucins. Small amounts of zinc, iron, calcium, and magnesium are also secreted. (Pepsinogen and lipase are activated in acid media to assist in digestion.)

The anatomy and the cells involved in gastric secretion are described in Chapter 33, the influences on secretion in Chapter 39, and the role in digestion in Chapter 40. Gastric secretion varies greatly during the day, from resting periods to active periods while eating (**Fig. 38-1**). Actual secretion is integrated and includes many stimulatory and inhibitory factors. Basal secretion does have a circadian variation. Observations on gastric secretion clearly identify an interdigestive period and a digestive period. The *interdigestive* phase includes the basic secretion and is influenced heavily by emotional factors. Although most experiments have been on animals, human experiments have shown that anger, resentment, hostility, and fear can influence the volume and content of secretion. Secretions clearly are influenced by vagus and neurohumoral stimuli.

The *digestive* period can be divided into three phases: cephalic, gastric, and intestinal. The *cephalic* phase includes the secretory response to all stimuli acting in the region of the brain. These may be unconditional (unlearned) reflexes, such as the secretion to sham feeding in a decorticate animal or the conditioned (learned) reflexes exemplified by the secretory effect of the thought, odor, sight, or taste of food. Conditioned or psychic secretion (Pavlov) is the principal component of the cephalic phase; the copious flow of gastric juice that occurs when appetizing food is masticated amounts to almost half the volume output of the gastric glands. Its presence contributes to the effective initiation and the subsequent efficiency of gastric digestion. The cephalic phase is mediated primarily through the vagus nerve and hormonal stimuli as gastrin release from the antrum.

The *gastric* (second) phase is so named because effective stimuli are within the stomach and are of two types: *mechanical*, from distension of the stomach as a result of the meal, and *chemical*, from secretagogues in foods or hormones that are released in the process of digestion. Hormonal stimulation of the secretion occurs by humoral and paraendocrine methods through receptors and intracellular signal transduction. The most potent stimulators are gastrin, acetylcholine, neurotransmitters such as gastrin-releasing peptide (GRP), and histamine.

The *intestinal* phase begins when chyme enters the duodenum and humoral effects occur. By the time a significant amount of the gastric content has been delivered to the intestine, regulatory mechanisms are already in operation to terminate the digestive period of gastric secretion. When the stomach is filled and absorption begins, satiety sets in, eating ceases, and psychic stimuli are withdrawn. Acidity of pH 1.5 or less acts on the antral mucosa to inhibit the release of gastrin. The production of secretoinhibitory hormones from the antrum results in the withdrawal of humoral and mechanical stimuli of the gastric phase. The main inhibitors of secretion are somatostatin, secretin, and a host of peptides that include corticotropin-releasing factor, β-endorphin, bombesin, neurotensin, calcitonin, calcitonin gene–related peptide, and interleukin-1 (IL-1). Other polypeptides, such as tropin-releasing hormone, peptide-y, and peptide-yy, have action, but their roles are unclear.

The relationship between secretions and food is discussed in Chapter 40. Importantly, a viscous layer of mucus is secreted. The mucous layer, consisting of a glycoprotein, coats the stomach and measures approximately 0.2 to 0.6 mm. Transport occurs across this mucous barrier of hydrogen ions (H^+), which is constantly being digested by pepsin and then replaced as it acts as an interface between the passage of H^+ and the neutralization by bicarbonate. This mucous barrier presumably prevents autodigestion of the stomach.

The role of prostaglandins has some importance in gastric physiology, but their exact effect is not clear. Similarly, the role of ghrelin in secretion has not been fully explained.

ADDITIONAL RESOURCES

Del Valle J, Todisco A: Gastric secretion. In Yamada T, editor: *Textbook of gastroenterology*, ed 4, Philadelphia, 2003, Lippincott–Williams & Wilkins.

Feldman M: Gastric secretion. In Feldman M, Friedman LS, Brandt LJ, editors: *Gastrointestinal and liver disease*, ed 8, Philadelphia, 2006, Saunders-Elsevier, pp 1029-1047.

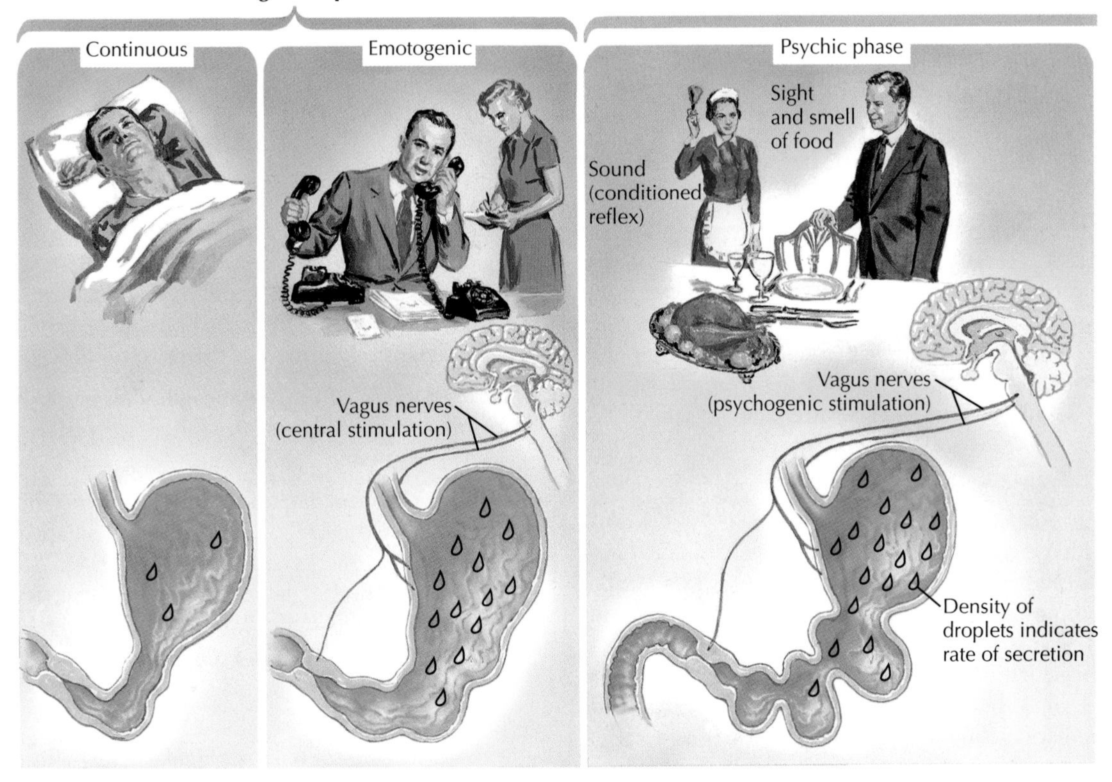

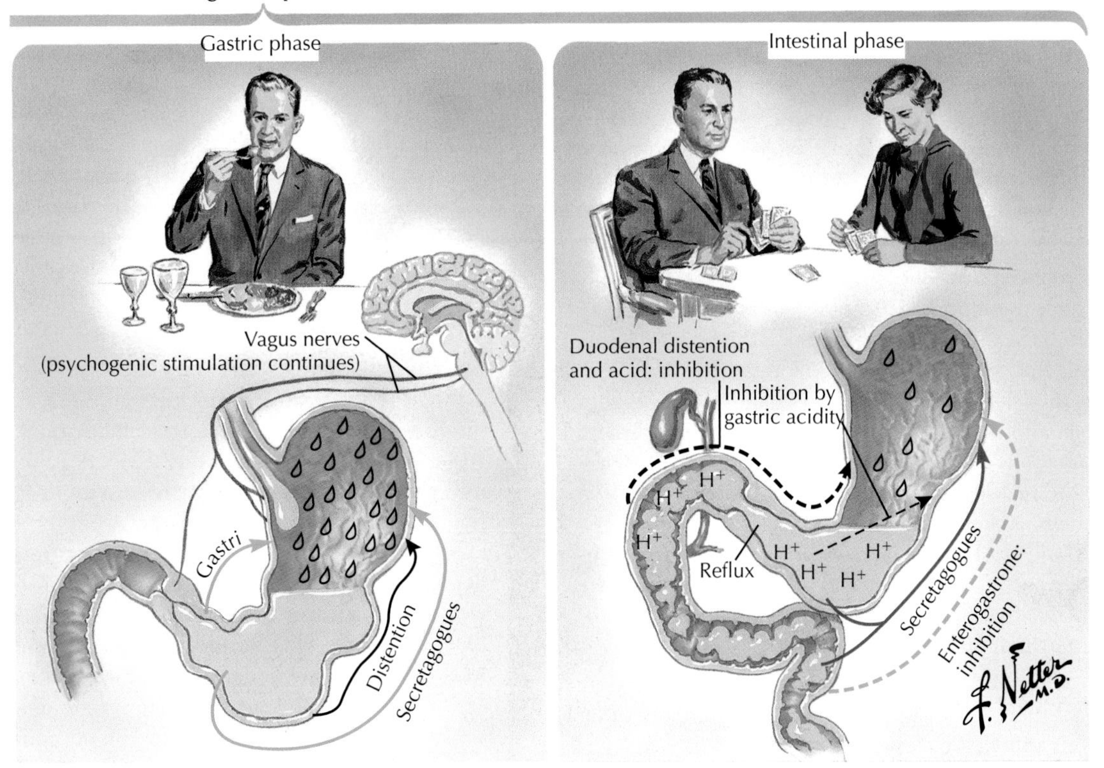

Figure 38-1 *Mechanisms of Gastric Secretion.*

Factors Influencing Gastric Activity

39

Martin H. Floch

Stomach activity is modified by the factors that stimulate and inhibit gastric secretion **(Fig. 39-1).** Emptying of the stomach is affected by many factors; the types of food eaten and the environment in which they are eaten play major roles through direct nerve and hormonal influences. Factors that modify motor and secretory activities of the stomach, usually simultaneously and in the same direction, include the following:

1. *Tonus of the stomach.* The hypertonic, or "steerhorn," stomach is hypermotile and empties relatively rapidly compared with the hypotonic, or "fishhook," type. In addition, the hypertonic stomach secretes more hydrochloric acid (HCl) and, as a corollary, experiences accelerated secretion and diminished intragastric stasis. Barium residue in the stomach 4 or 5 hours after upper gastrointestinal (GI) x-ray examination must be interpreted with consideration that the stomach's inherent tonicity is a factor in its emptying rate.
2. *Character of the food.* A meal sufficiently high in fat to yield an intragastric fat content in excess of approximately 10% empties more slowly and stimulates considerably less acid secretion than does a meal consisting predominantly of protein. The inhibitory effect of fat on gastric secretion is not local but is a primarily a result of enterogastric neural reflexes and hormones, primarily cholecystokinin, after fat has entered the upper intestine.
3. *Starch and protein.* A meal consisting exclusively or mainly of starch tends to empty more rapidly, although stimulating less secretion, than does a protein meal. Thus, other factors being equal, a person may expect to be hungry sooner after a breakfast of fruit juice, cereal, toast, and tea than after bacon, eggs, and milk. The amount of total secretion and of acid content is highest with the ingestion of proteins. However, the relationship of quantity and rate of secretion to its acid or pepsin concentration varies greatly among individuals and in a single person under different conditions. Numerous GI hormones and neural mechanisms are involved in feedback to the stomach. The so-called ileal break occurs when fat enters the ileum.
4. *Consistency of the food.* Liquids, whether ingested separately or with solid food, leave the stomach more rapidly than do semisolids or solids. This does not apply to liquids such as milk, from which solid material is precipitated on contact with gastric juice. With any foods requiring mastication, the consistency of the material reaching the stomach should normally be semisolid, thereby facilitating gastric secretion, digestion, and evacuation. Important exceptions to the general rule that liquids are weak stimulants of gastric secretion are (1) the broth of meat or fish, because of their high secretagogue content, and (2) coffee, which derives its secretory potency from its content of caffeine and of the secretagogues formed in the roasting process.
5. *Mixed meals.* In a mixed meal, liquids empty first and solids empty in two phases. An initial lag phase is followed by linear emptying.
6. *Hunger.* A meal eaten at a time of intense hunger tends to be evacuated more rapidly than normal, apparently in consequence of the heightened gastric tonus. Because hunger results from the depletion of body nutrient stores (see Section X), it is understandable on teleologic grounds that in the hunger state, the body should have some mechanism for hastening the delivery of ingested nutrients into the intestine.
7. *Exercise.* Mild exercise, particularly just after eating, shortens the emptying time of the meal. With strenuous exercise, gastric contractions are temporarily inhibited, then augmented, so that final emptying is not significantly delayed. Secretory activity does not appear to be materially influenced by exercise.
8. *Position.* In some persons, gastric emptying is facilitated when the position of the body is such that the pylorus and the duodenum are in a dependent position, that is, with the person lying on the right side. In the supine position, particularly in infants and adults with cascade stomach, the gastric content pools in the dependent fundic portion, and emptying is delayed. No evidence suggests that secretion is affected by position.
9. *Emotion.* The impairing effect of emotional states on gastric motility and secretion has been well documented by clinical and experimental observations. Evidence indicates that the influence of emotions on gastric activity may be augmentative or inhibitory, depending on whether the emotional experience is of an aggressive (hostility, resentment) or a depressive (sorrow, fear) type, respectively. One point of view holds that it is not the manifest or conscious emotion that determines whether the stomach is stimulated or inhibited, but rather the unconscious or symbolic content of the emotional state, and further, that certain emotions may be accompanied by dissociation in the response among the various components of the gastric secretions.
10. *Pain.* Severe or sustained pain in any part of the body (e.g., kidney stones or gallstones, migraine, sciatica, neuritis) inhibits gastric motility and evacuation by nervous reflex pathways.

Factors Affecting Gastric Emptying

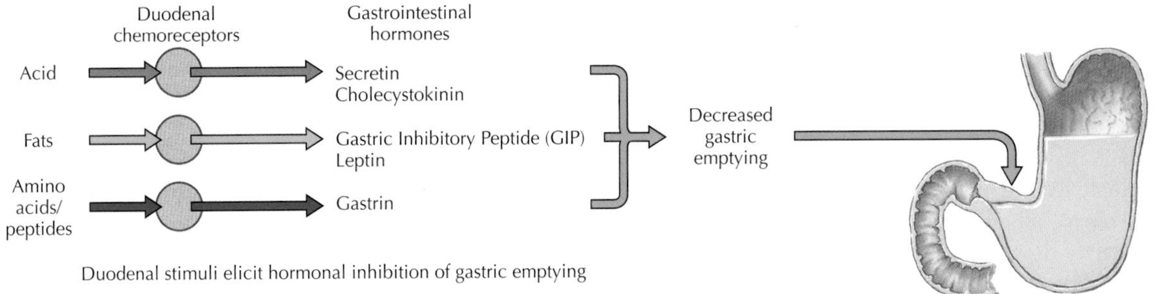

Sequence of Gastric Motility

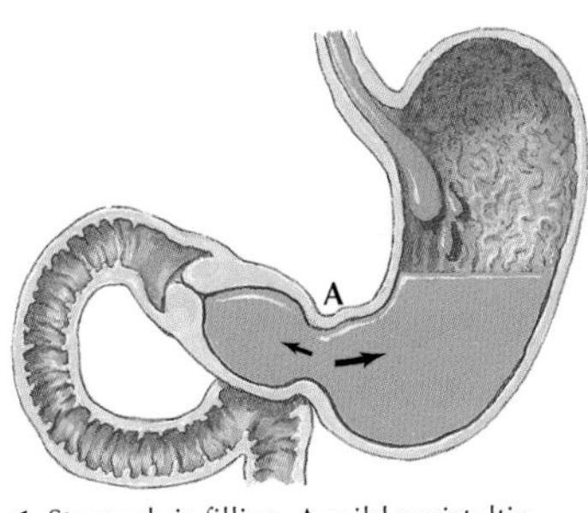

1. Stomach is filling. A mild peristaltic wave (A) has started in antrum and is passing toward pylorus. Gastric contents are churned and largely pushed back into body of stomach

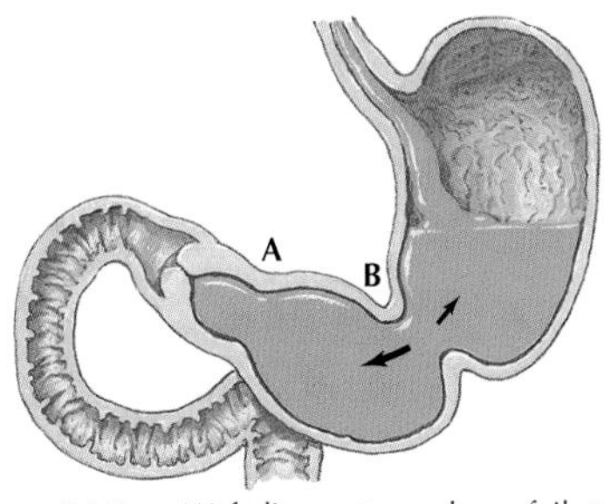

2. Wave (A) fading out as pylorus fails to open. A stronger wave (B) is originating at incisure and is again squeezing gastric contents in both directions

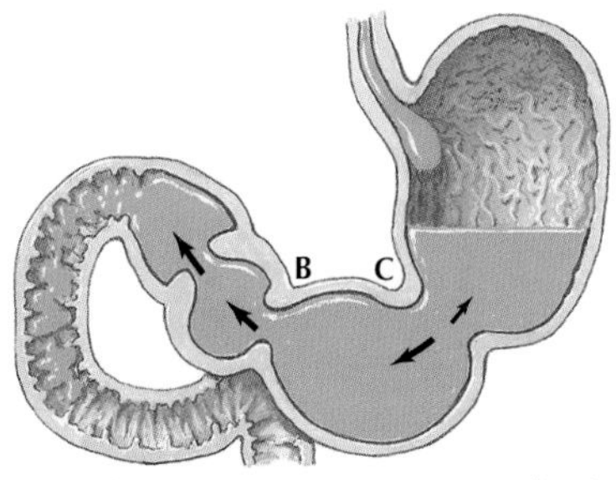

3. Pylorus opens as wave (B) approaches it. Duodenal bulb is filled, and some contents pass into second portion of duodenum. Wave (C) starting just above incisure

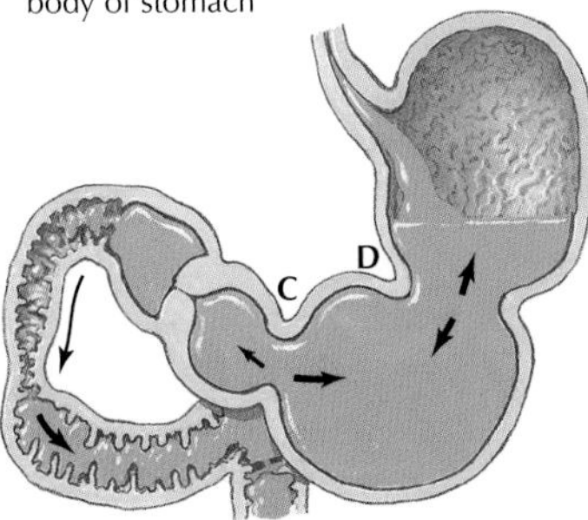

4. Pylorus again closed. Wave (C) fails to evacuate contents. Wave (D) starts higher on body of stomach. Duodenal bulb may contract or may remain filled as peristaltic wave originating just beyond it empties second portion

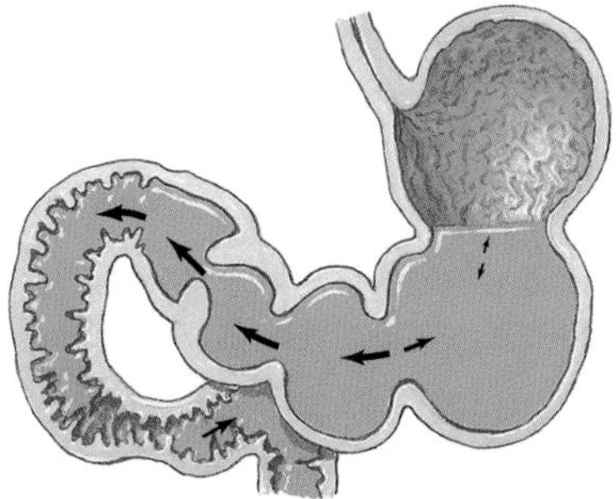

5. Peristaltic waves are now originating higher on body of stomach. Gastric contents are evacuated intermittently. Contents of duodenal bulb area pushed passively into second portion as more gastric contents emerge

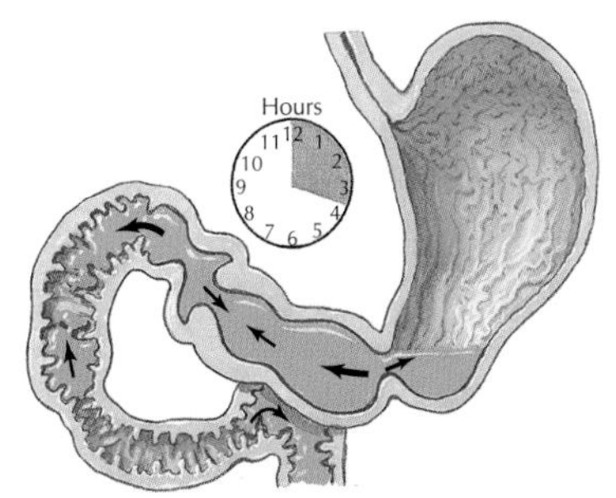

6. 3 to 4 hours later, stomach is almost empty. Small peristaltic wave empties duodenal bulb with some reflux into stomach. Reverse and antegrade peristalsis present in duodenum

Figure 39-1 *Factors Influencing Gastric Activity.*

ADDITIONAL RESOURCES

Collins PJ, Houghton LA: Nutrients and the control of liquid gastric emptying, *Am J Physiol* 276:997, 1999.

Collins PJ, Houghton LA, Read NW, et al: Role of the proximal and distal stomach in mixed solid and liquid meal emptying, *Gut* 32:615-619, 1991.

Liddle RA: Gastrointestinal hormones and neurotransmitters. In Feldman M, Friedman LS, Brandt LJ, editors: *Gastrointestinal and liver disease*, ed 8, Philadelphia, 2006, Saunders-Elsevier, pp 3-25.

Moran TH, Wirth JB, Schwartz, GS, et al: Interactions between gastric volume and duodenal nutrients and the control of liquid gastric emptying, *Am J Physiol* 276:R997, 1999.

Quigley MM: Gastric motor and sensory function, and motor disorders of the stomach. In Feldman M, Friedman LS, Brandt LJ, editors: *Gastrointestinal and liver disease*, ed 8, Philadelphia, 2006, Saunders-Elsevier, pp 999-1028.

Role of the Stomach in Digestion

Martin H. Floch

40

The stomach plays an important role in nutrition; maintaining adequate weight and nutrient intake would be difficult without a stomach. When adequately chewed, food arrives in the stomach, where the motility enables churning and the initial digestive processes (see Chapters 39 and 46). Regulating cephalic, gastric, and intestinal phases of gastric secretion is complicated (see Chapter 38).

Normal gastric secretion is essential for the normal digestion of foods (**Fig. 40-1**). Hydrochloric acid (HCl) is secreted from the parietal cell in a concentration of 0.16 N, but this maximal concentration is quickly diluted by the metabolic activity in the mucus layer and with food. In addition to the normal physiologic regulator mechanisms of gastric secretion, a number of systemic and local effects are unique to HCl secretion. The stimulating effect of the oral administration of sodium bicarbonate ($NaHCO_3$), popularly called "acid rebound," probably results from a combination of factors, including a direct stimulating action on the gastric mucosa, annulment of the antral acid-inhibitory influence, and acceleration of gastric emptying. The "alkaline tide," or decrease in urinary acidity that may occur after a meal, is generally attributed to increased alkalinity of the blood resulting from the secretion of HCl. An alkaline tide is not predictable and is influenced by (1) relative rate of HCl formation and alkaline digestive secretions, mainly pancreatic, with the high $NaHCO_3$ content in the pancreas; (2) rate of absorption of HCl from the gut; (3) neutralizing capacity of the food; (4) respiratory adjustments after the meal; and (5) diuretic effect of the meal.

Pepsin, the principal enzyme of gastric juice, is preformed and is stored in the chief cells as *pepsinogen*. At pH less than 6, pepsinogen is converted to pepsin, a reaction that then proceeds autocatalytically. Pepsin exerts its proteolytic activity by attacking peptid linkages containing the amino groups of the aromatic amino acids, with the liberation principally of intermediate protein moieties and a few polypeptides and amino acids. An accessory digestive function of pepsin is the clotting of milk, which serves to improve its use by preventing too rapid passage, rendering it more susceptible to enzymatic hydrolysis. Anything that mobilizes vagal impulses for the stomach serves as a powerful stimulus for pepsin secretion. Thus, a gastric juice rich in pepsin content is evoked by sham feeding; by hypoglycemia, which stimulates the vagal centers; and by direct electrical stimulation of the vagus nerves.

The mucoid component of gastric juice consists of at least two distinct mucoproteins. The "visible mucus" has a gelatinous consistency and, in the presence of HCl, forms a white coagulum; evidence indicates that it is secreted by the surface epithelium. The "soluble mucus" or "dissolved mucus" appears to be a product of the neck's chief cells and the mucoid cells of the pyloric and cardiac glands. The secretion of soluble mucus is activated primarily by vagal impulses, whereas the secretion of visible mucus occurs principally in response to direct chemical and mechanical irritation of the surface epithelium. Because of its adherent and metabolic properties and its resistance to penetration by pepsin, mucus secretion protects the mucosa of the stomach against damage by various irritating agents, including its own acid, pepsin.

A normal constituent of the gastric juice, but characteristically deficient or absent in patients with pernicious anemia, is *intrinsic factor*. It interacts with vitamin B_{12} to prepare it for absorption in the intestine. An R factor from saliva mixes and binds with vitamin B_{12}. The R factor is cleaved by pancreatic enzyme action when the combined B_{12}–R factor enters the duodenum, and the intrinsic factor from the stomach binds to B_{12} to enable absorption by receptors in the proximal intestine. Large amounts of intrinsic factor usually are secreted to enable adequate amounts to bind with the B_{12} in the intestine.

Salivary amylases may mix with the starch in foods and may have an initial digestive effect, but major carbohydrate digestion occurs in the intestine from the action of pancreatic enzymes. The degree of salivary enzyme activity in the stomach depends on how long the food is masticated and how fast it is swallowed, because salivary amylases are inactivated rapidly in the stomach by peptic action.

Gastric lipases may begin the process of fat digestion and may account for as much as 25% of intraluminal fat digestion. Again, however, this depends on how fast the stomach empties, as well as other factors that affect emptying. Because of the pH and molar-sensitive receptors in the duodenum, a delay in gastric emptying results when the chyme is too acid or hypertonic at the beginning of the intestinal phase of digestion.

In summary, the major digestive activity in the stomach is proteolytic and prepares chyme to pass into the duodenum for orderly digestion and absorption.

ADDITIONAL RESOURCES

Camilleri M: Integrated upper gastrointestinal response to food intake, *Gastroenterology* 131:640-658, 2006.

Farrell JJ: Digestion and absorption of nutrients: an overview. In Feldman M, Friedman LS, Brandt LJ, editors: *Gastrointestinal and liver disease*, ed 8, Philadelphia, 2006, Saunders-Elsevier, pp 2147-2198.

Feldman M: Gastric secretion. In Feldman M, Friedman LS, Brandt LJ, editors: *Gastrointestinal and liver disease*, ed 8, Philadelphia, 2006, Saunders-Elsevier, pp 1029-1043.

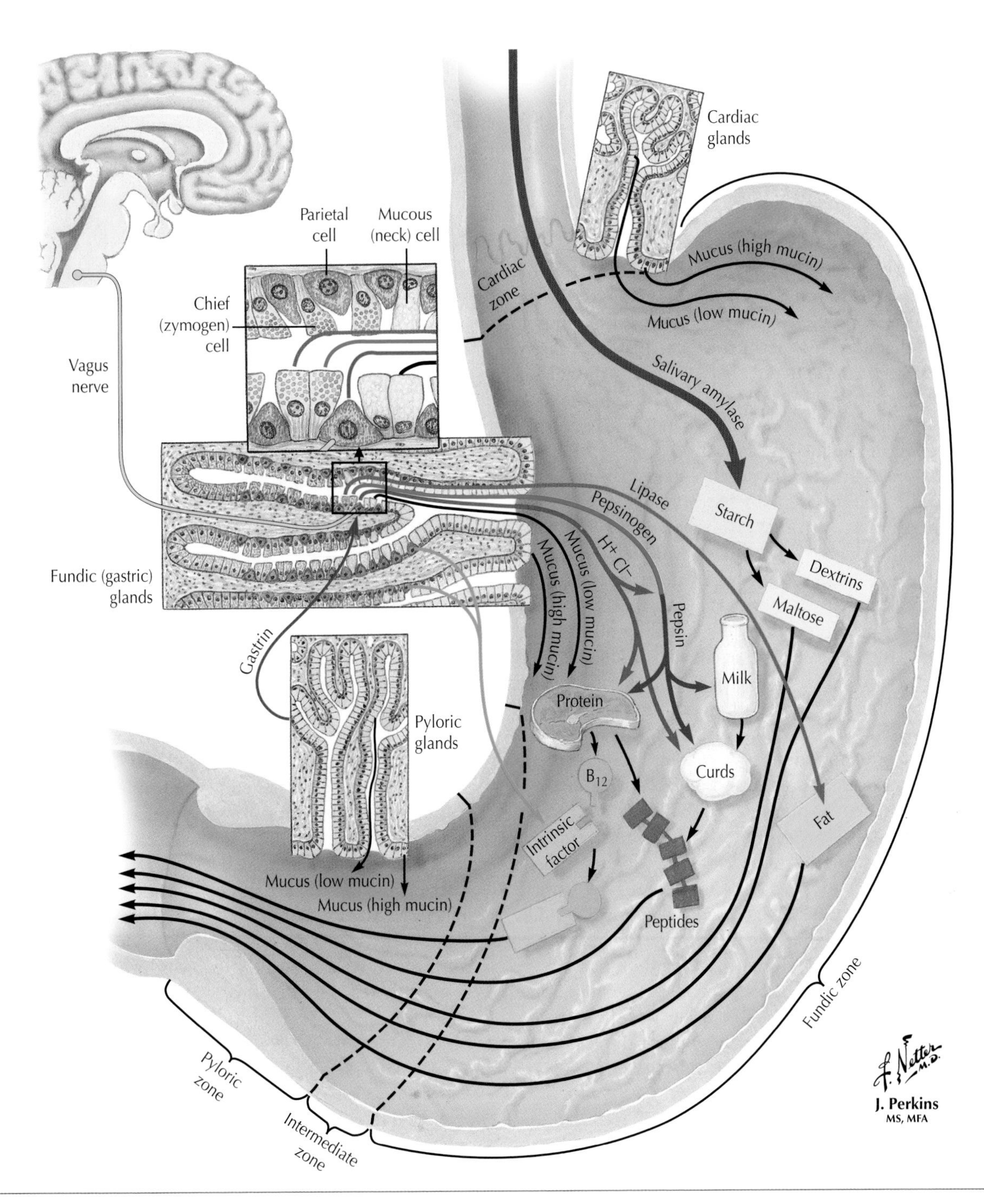

Figure 40-1 *Digestive Activity of the Stomach.*

Gastric Acid Secretion Tests: Hydrochloric Acid and Gastrin

41

Martin H. Floch

Gastric analysis is now rarely used; it was a major test before the development of significant medical therapy for duodenal and gastric ulcer. *Qualitative* gastric analysis is used only in the differential diagnosis of pernicious anemia and gastric atrophy. However, *quantitative* gastric analysis (the classic method is illustrated in **Fig. 41-1**) is still important in the diagnosis and monitoring of Zollinger-Ellison syndrome (ZES) and the multiple endocrine neoplasm syndrome.

The technique seeks to determine, by 1-hour monitoring of basal secretion, the amount of hydrochloric acid (HCl) secreted by the stomach. After an overnight fast, the patient is intubated with a radiopaque tube, and the position is checked by fluoroscopy. The tip of the tube should be in the gastric antrum. Studies have revealed that no more than 5% to 10% of acid secretion is lost by aspirating continually when the tube is positioned correctly. After the tube is placed, the residuum in the stomach is emptied, and then collections are made in separate tubes every 15 minutes. The amount in each tube is measured. Topfer reagent is used to check quickly for acid. Any device that maintains negative pressure can be used to aspirate all the secretions. The patency of the tube should be checked with air every 5 minutes to make sure it is not plugged. The amount of HCl in each tube is calculated using one of two methods: titration with sodium hydroxide or by pH electrode, which measures hydrogen activity. The hydrogen chloride secreted is then calculated and reported as milliequivalents per liter (mEq/L). The total quantity of acid (volume × concentration) can then be reported as milliequivalents per hour. Basal acid output may be zero in approximately one of three people, but the upper limit of normal for the basal acid output is approximately 10 mmol/hr in men and 5 mmol/hr in women. It varies from hour to hour and certainly varies greatly during various phases of gastric activity.

"Maximal acid output" and "peak acid output" are no longer used. They represented the amount of acid after either pentagastrin or histamine stimulation. Neither drug is available in the United States.

Functionally, *gastrin* is the most potent stimulant of acid secretion. The stimulation of acid secretion and the role of gastrin are complex (see Chapters 33, 38, and 40). Specialized G cells of the stomach, along with pyloric and duodenal glands, produce gastrin. It is secreted as preprogastrin, and then by enzymatic action, all the active forms are produced. The two main gastrins are G_{17} and G_{34} (chain lengths of 17 and 34 amino acids). Gastrin stimulates enterochromaffin-like (ECL) cells to release histamine. Histamine stimulates acid secretion by parietal cells and release of other neurotransmitters, such as acetylcholine and gastrin-releasing peptide. Major inhibitors of parietal cell secretion are somatostatin and cholecystokinin. Secretin and other peptides also enter in the inhibitory process, but their roles are less well understood.

Fasting serum gastrin levels may vary greatly. The normal range is 0 to 200 pg/mL. However, medications and other suppressants and stimulants can widen the normal range to as high as 400 pg/mL. A level greater than 1000 pg/mL is usually considered diagnostic of ZES. In some cases, there may still be uncertainty because of chronic severe gastritis. Prolonged use of proton pump inhibitors may also create uncertainty because PPIs increase serum gastrin levels. A combination of a very high fasting serum gastrin level and high levels of gastric acid secretion usually confirm the diagnosis of ZES. However, renal failure can also result in high serum gastrin levels because of poor clearance of the serum gastrin. Therefore, the *secretin provocative test* was designed to confirm the diagnosis of ZES; it is used when the serum gastrin level is less than 1000 pg/mL, or whenever the diagnosis is uncertain. A fasting serum gastrin level is obtained; secretin (2 U/kg) is rapidly injected; and serum gastrin blood levels are obtained at 2, 5, and 10 minutes after injection. In ZES, serum gastrin levels usually rise rapidly after 5 minutes. A 200-pg/mL or greater elevation in gastrin level confirms the diagnosis. The secretin provocative test has a sensitivity of 90% for ZES.

ADDITIONAL RESOURCES

Gregory RA, Tracy J: The constitution and properties of two gastrins extracted from hog antral mucosa, *Gut* 5:103-117, 1964.

Liddle RA: Gastrointestinal hormones and neurotransmitters. In Feldman M, Friedman LS, Brandt LJ, editors: *Gastrointestinal and liver disease*, ed 8, Philadelphia, 2006, Saunders-Elsevier, pp 3-25.

Pisegna JR: The effect of Zollinger-Ellison syndrome and neuropeptide secreting tumors on the stomach, *Curr Gastroenterol Rep* 1:511-517, 1999.

Rehfeld JF: The new biology of gastrointestinal hormones, *Physiol Rev* 78:1087-1108, 1998.

Takasu A, Shimosegawa T, Fukudo S, et al: Duodenal gastrinoma: clinical features and usefulness of selective arterial secretin injection test, *J Gastroenterol* 33:728-733, 1998.

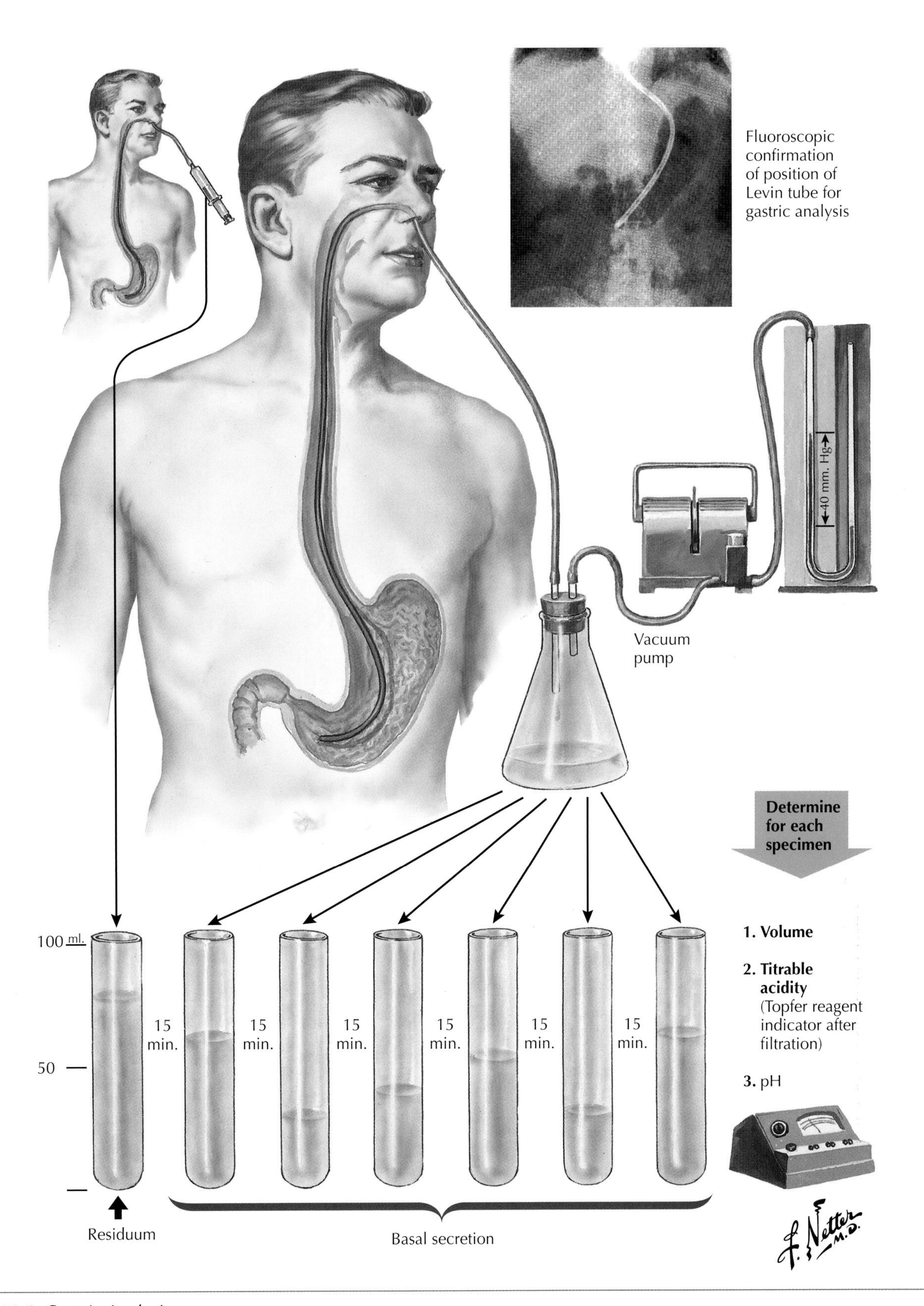

Figure 41-1 *Gastric Analysis.*

Effects of Drugs on Gastric Function

42

Martin H. Floch

Many of the pharmacologic agents widely used in medical therapy adversely affect the upper gastrointestinal (GI) tract (**Fig. 42-1**). Therefore, every patient with symptoms referable to the esophagus, stomach, or duodenum should be questioned carefully regarding the recent use of drugs. Drugs may also adversely affect the liver (see Section IX), pancreas (see Section VII), and other organs. This chapter discusses specific agents and drug categories often implicated in gastric disorders.

SALICYLATES

The primary offenders are salicylates, alone or in combination with other analgesics, antacids, opiates, or steroids. The inflammatory action of salicylates in the stomach of susceptible persons can result in mild dyspepsia to massive hemorrhage. Aspirin is widely used to prevent cardiac disease and polyp formation in the GI tract. The potential for bleeding is dose related, but in some persons, even small doses (81 mg) may lead to bleeding tendencies.

CAFFEINE

Although it is a common component of headache remedies and is responsible for the widespread use of coffee and tea, caffeine is a gastric irritant and a stimulant of gastric secretion and gastric motility. Beverages containing caffeine, which also include most sodas containing cola, have the same effect as the pure xanthine preparation. A cup of coffee contains 100 to 150 mg of caffeine. Teas have even larger amounts. The amount of caffeine in the beverage varies with the brewing process and amount ingested.

Theophylline and its water-soluble salt aminophylline are closely related to caffeine and have similar effects, but are used effectively in bronchospasm.

NONSTEROIDAL ANTIINFLAMMATORY DRUGS

Nonsteroidal antiinflammatory drugs (NSAIDs) are prescribed more frequently worldwide than any other group of medicines. Approximately 25 different NSAIDs are available in the United States. They cause significant rates of morbidity and mortality because of the adverse effects on the GI tract. The most serious complications are bleeding and perforation, which account for almost all associated deaths. The major damage occurs in the stomach and duodenum, but NSAIDs may also affect the small and large intestines. Significant endoscopic evidence indicates that NSAIDs and aspirin directly injure the GI mucosa. However, damage has been observed even when medications are administered intravenously and in enteric-coated preparations, questioning whether the effect is topical or systemic.

The benefit of NSAIDs is decreased cyclooxygenase (COX-1 and COX-2) activity, which in turn decreases the cascade of cytokine formation in inflammation, although NSAIDs also decrease the prostaglandin protection of the GI mucosa. COX-2 appears to cause less, but still significant, GI damage. Common COX-1 inhibitors are diclofenac, ibuprofen, indomethacin, naproxen, and sulindac. Common COX-2 inhibitors are celecoxib and rofecoxib and are used to decrease the side effects of COX-1 agents.

ISONICOTINIC ACID HYDRAZIDE

Isonicotinic acid hydrazide (isonicotinylhydrazine) is used in to treat tuberculosis. When administered in large doses, isonicotine hydrazine and the related drug isoniazid (INH) are gastric secretory stimulants associated with gastric irritation and liver disease.

ANTIBIOTICS

Antibiotics may cause local irritant effects (tetracyclines) or greatly increased motility (erythromycin and clarithromycin). The tetracycline drugs often cause esophageal ulceration. Care must be taken in administration and make sure they are swallowed with adequate amounts of fluid.

CARDIOVASCULAR DRUGS

Digitalis and antihypertensive medications may cause gastric hyperemia. Some may have a central nervous system effect and may cause significant nausea. Alpha and beta blockers may be associated with GI disturbances.

ANTICOAGULANTS AND VASODILATORS

Anticoagulants and vasodilators frequently are used to prevent thromboembolism in cardiac and neurologic disease. However, they may dilate blood vessels in the stomach and the upper GI tract, and they may decrease clotting and result in significant GI hemorrhage. An occult ulcer or tumor of the GI tract may start to bleed when anticoagulation is administered. Once bleeding occurs in a patient taking anticoagulants, the clinician must search for a previously hidden lesion.

ANTICHOLINERGICS

Anticholinergic drugs, both the naturally occurring and the synthetic forms, are used primarily for effects on the GI tract to reduce motility and secretion. Evidence indicates that large doses are needed to decrease gastric secretion, but therapeutic

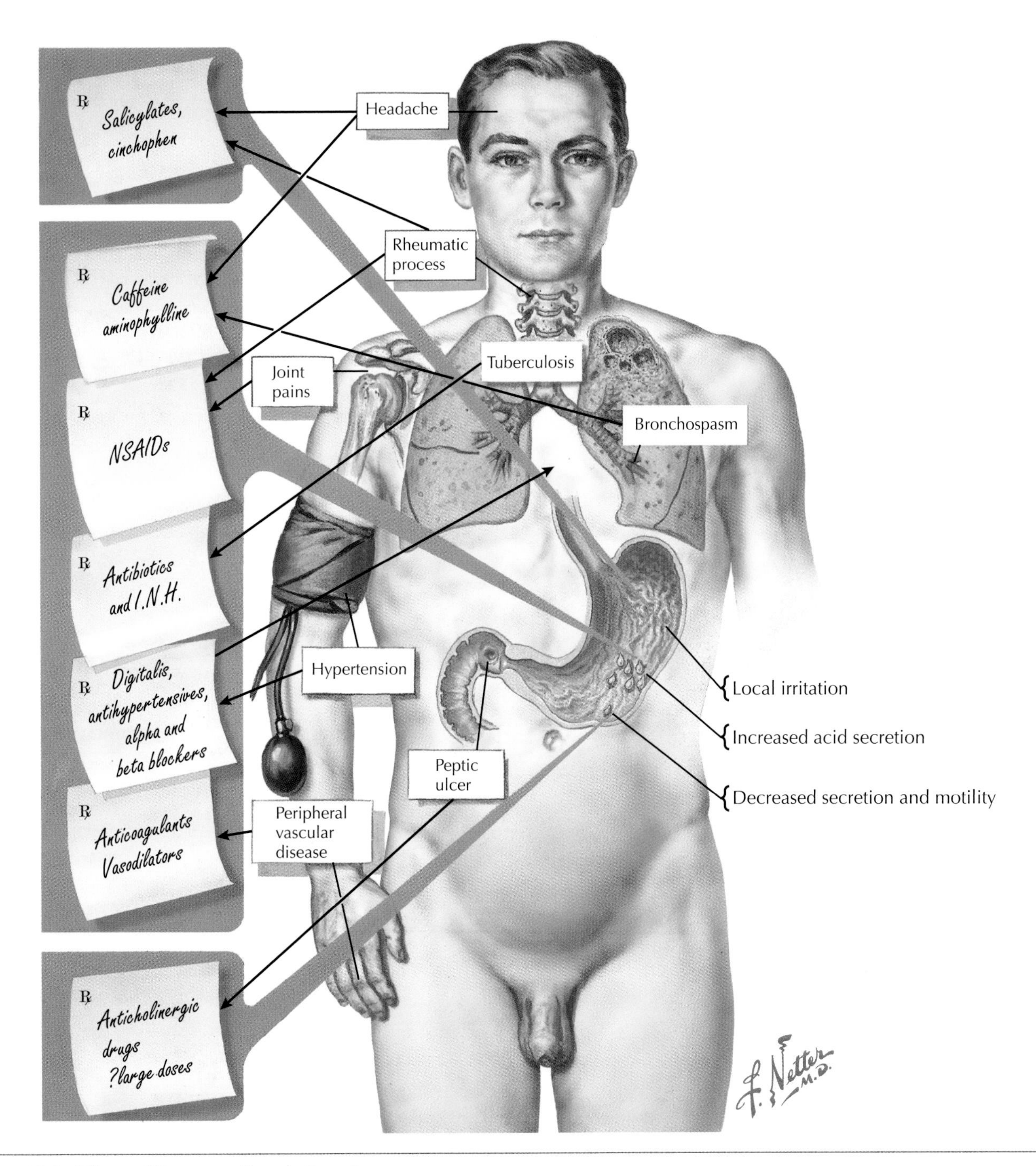

Figure 42-1 *Effects of Drugs on Gastric Function.*

doses can decrease motility and consequently help in some hyperactive states.

ADDITIONAL RESOURCES

Cryer B, Spechler SJ: Peptic ulcer disease. In Feldman M, Friedman LS, Brandt LJ, editors: *Gastrointestinal and liver disease*, ed 8, Philadelphia, 2006, Saunders-Elsevier, pp 1089-1110.

Lanza FL, Royer, CL, Nelson RS: Endoscopic evaluation of the effects of aspirin, buffered aspirin, and enteric-coated aspirin on gastrointestinal duodenal mucosa, *N Engl J Med* 303:136-138, 1980.

Physicians' desk reference, ed 53/54, Montvale, NJ, 2003, Medical Economics Data Production.

Talley NJ, Evans JM, Fleming KC: Nonsteroidal anti-inflammatory drugs and dyspepsia in the elderly, *Dig Dis Sci* 40:1345-1350, 1995.

Upper Gastrointestinal Endoscopy: Esophagogastroduodenoscopy

43

Martin H. Floch

Endoscopic visualization of the esophagus, stomach, or proximal duodenum is an essential procedure in the diagnosis and treatment of diseases of the esophagus, stomach, and duodenum (**Fig. 43-1**). Examining the esophagus and stomach has evolved from using rigid and semirigid instruments to using flexible instruments, first fiberoptic and now videoendoscopic. During the past decade, all endoscopic laboratories converted to videoendoscopy and are adding *endoscopic ultrasound* (EUS) to evaluate the full thickness of the esophagus, stomach, or duodenum and adjacent structures, as well as for direct visualization and histology of the mucosal surface. Further advances in technology now include confocal imaging of the mucosa and staining of mucosa

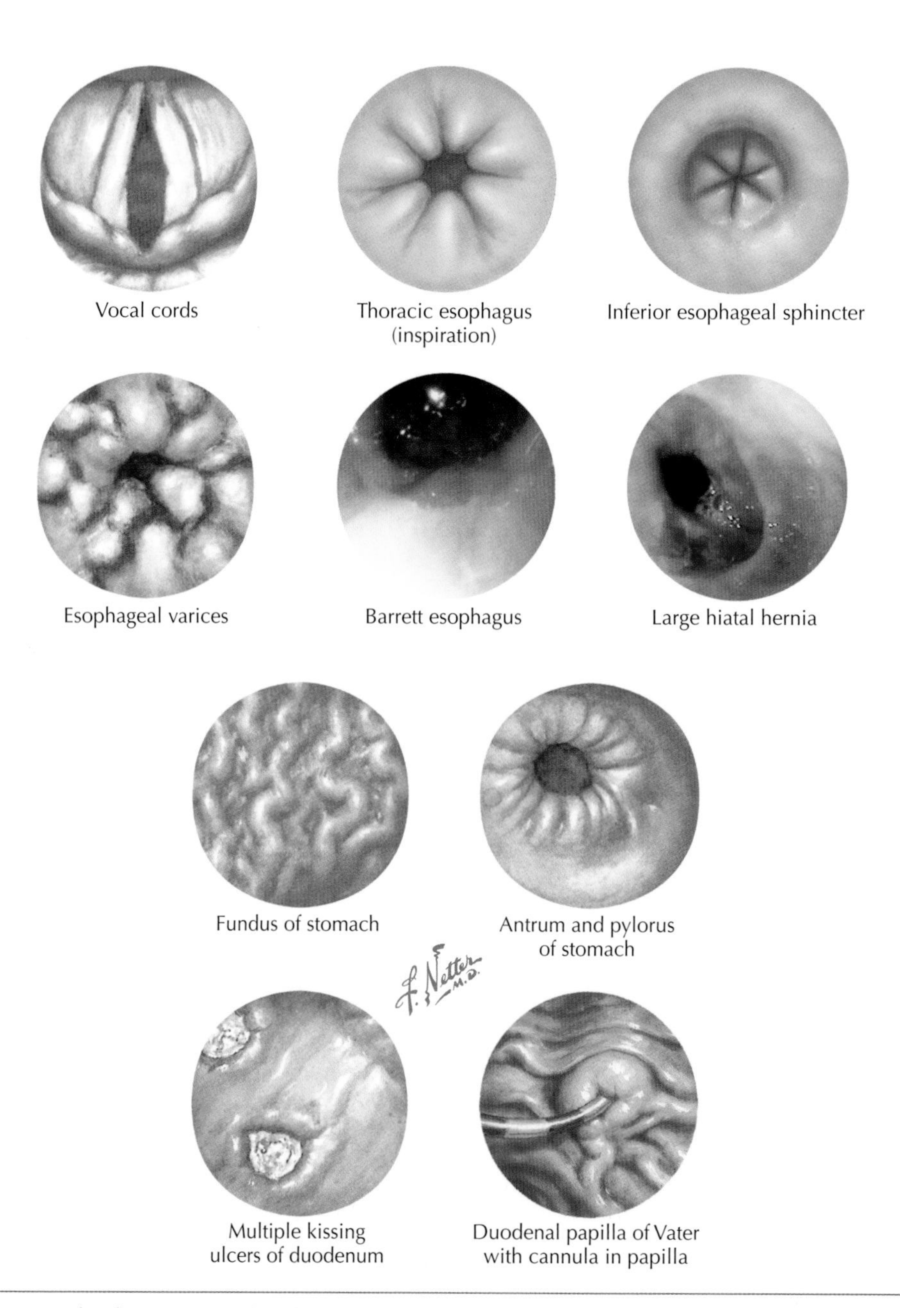

Figure 43-1 *Esophagogastroduodenoscopy and Endoscopic Ultrasound.*

to identify neoplasia. This chapter describes techniques and normal findings; abnormal findings are described in the specific disease chapters.

Although many variations exist, most videoendoscopes now use a color chip that gathers the image at the tip of the endoscope and transmits it through a videoprocessor into a monitor. The image can be preserved and then transmitted to computerized recording systems. Images can be stored easily for reports or archival records.

Most endoscopes include a portal for a biopsy channel. Very thin instruments can be passed transnasally but may not have biopsy capability. Instruments usually are 8 to 10 mm wide, but transnasal instruments may be only 2 to 3 mm wide.

The EUS instruments are wider than other endoscopes but are still easily passable, and they allow direct imaging, biopsy, and ultrasound (US) imaging. There are two types of EUS instruments, and each has a US device built into the tip. The more common type has a US device and an imaging endoscope, but a thin US device can be passed through a large endoscope channel. EUS is a complicated procedure, but it enables better submucosal and full-wall thickness images. It is especially useful for evaluating large folds, submucosal nodules, tumor extensions through the wall, lymph nodes, and associated structures (e.g., pancreas).

Upper endoscopy is used for numerous indications, including dysphasia, gastroesophageal reflux disease (GERD), esophageal (Barrett) or gastric metaplasia, dyspepsia, gastric ulcers, duodenal ulcers, upper gastrointestinal tract bleeding, infection in the esophagus, removal of foreign bodies, caustic injuries, drug-induced injuries; evaluation for esophageal cancer and for all possible premalignant lesions, including mass lesions, esophageal metaplasia (Barrett), achalasia, atrophic gastritis, pernicious anemia; and follow-up after surgery for all malignant lesions.

Although useful for many of these indications, EUS is more complicated, with fewer trained ultrasonographers, than the major screening and diagnostic procedure, video *esophagogastroduodenoscopy* (EGD). After an appropriate indication is identified and the procedure scheduled, EGD usually begins with a local anesthetic throat spray or gargle, followed by injection for sedation. Although drugs used for sedation vary widely worldwide, most protocols include a drug to reduce pain and a drug to induce sedation and produce an amnesic effect. Common drugs include meperidine (50-100 mg) or hydromorphone (2-4 mg) in combination with diazepam or midazolam (titrated at 1-10 mg). Endoscopists may choose to use fentanyl citrate (Sublimaze; 75-100 mg) alone or in combination with midazolam.

Once the patient is appropriately sedated, the endoscope can be passed in several ways through the pharynx. Some endoscopists prefer passing it blindly over the base of the tongue into the upper esophagus. Others insist on passage by direct visualization. Direct visualization permits an examination of the epiglottis and a "peek" at the vocal cords. Once the pharynx is passed, the endoscopist evaluates the esophagus. The upper esophagus, midesophagus, and distal esophagus can be visualized with an injection of air, as in a tubular structure. Contractions are often starlike. Ringlike contractions indicate a motility disturbance. Fixed rings, strictures, polypoid masses, and varices are all described under the particular pathologic condition in other chapters.

After passing through the gastroesophageal (GE) junction (see Section I), the endoscope enters the stomach, and the clinician notes whether any food or significant bilious secretions are retained from duodenal reflux. An injection of air permits evaluation of the fundus, body, and antrum. Normal contractions are seen radiating from the pylorus into the antrum and back in the opposite direction. By retroflexing the instrument along the lesser curvature, the endoscopist can fully evaluate the fundus and the GE junction from below. Hiatal hernias and gastric lesions of this area can be identified.

Staining of the mucosa to differentiate neoplasia is used in some centers but has not gained wide acceptance. Obtaining biopsy specimens from the visualized lesions remains the procedure of choice.

The endoscopist then observes the pyloric channel and, with mild pressure, passes the instrument into the duodenal bulb and then into the second and as far into the third part of the duodenum as needed. The ampulla of Vater can be seen with direct-viewing instruments but is best examined with a lateral-viewing instrument. Masses, abnormal mucosa, and bleeding sites, however, can be thoroughly evaluated with a direct-viewing instrument.

Biopsy specimens can be obtained from any location in the esophagus, stomach, or duodenum. Specimens are usually processed in a fixative medium to be sent for evaluation.

Endoscopic ultrasound has rapidly developed into an important technique in the differential diagnosis of benign and malignant disease. Upper endoscopes are now made for several types of US probes at the end of the scope that provide clear endoscopic and US images. The US image delineates normal layers of the wall of the mucosa, enlarged nodes outside the wall, and extent and type of tumor. Therapeutic and diagnostic EUS has evolved to allow both biopsy of nodes and lesions and drainage of cysts through US guidance.

ADDITIONAL RESOURCES

DiMarino AJ, Benjamin SB: *Gastrointestinal disease: an endoscopic approach*, ed 2, Thorofare, NJ, 2002, Slack.

Emery TS, Carpenter HA, Gostout CJ, Sobin LH: *Atlas of gastrointestinal endoscopy and endoscopic biopsies*, Washington, DC, 2000, Armed Forces Institute of Pathology.

Pech O, Rabenstein T, Manner H, et al: Confocal laser endomicroscopy for in vivo diagnosis of early squamous cell carcinoma in the esophagus, *Clin Gastroenterol Hepatol* 6(1):89-94, 2008.

Rogart JN, Nagata J, Loeser CS, et al: Multiphoton imaging can be used for microscopic examination of intact human gastrointestinal mucosa ex vivo, *J Gastroenterol Hepatol* 6(1):95-101, 2008.

Coated Tongue, Halitosis, and Thrush

44

Martin H. Floch

The tongue is kept clean and normally colored by the cleansing action of saliva, the mechanical action of mastication, the customary oral flora, and adequate nutrition. Consequently, when salivary secretion is insufficient, when the dietary regimen eliminates chewing, when the bacterial flora is altered, or when certain vitamins necessary for the preservation of the normal epithelium are deficient, the normal appearance of the tongue may change. It may become coated with food particles, sloughed epithelial cells, and inflammatory exudates (**Fig. 44-1**). Fungal growths may be deposited on its surface.

Patients at risk for abnormal conditions of the tongue are those whose saliva is diminished by mouth breathing, dehydration, or anticholinergic drugs; those who are comatose and are unable to eat, drink, or rinse the mouth; and those with impaired mobility of the tongue caused by cranial nerve XII paralysis. An exudative oral or pharyngeal inflammatory process or antibiotic therapy that destroys the normal flora may result in an overgrowth of fungi. Hypertrophy of the papillae may give the appearance of a black or hairy tongue, especially in smokers.

Geographic tongue (benign migratory glossitis) is a migratory lesion of unknown cause. It may occur intermittently. Lesions are often irregular and appear as denuded, grayish patches. If lesions persist or any uncertainty exists in the diagnosis, an otolaryngologist should evaluate and biopsy the lesions if necessary. Other tongue lesions of uncertain identity should also prompt a full evaluation.

Fissured tongue is a benign lesion with longitudinal grooves usually considered congenital lingual defects. Again, if the diagnosis is uncertain, otolaryngologic evaluation is indicated.

In patients with pernicious anemia, a varicolored appearance caused by patchy loss of papillae may evolve into geographic tongue, but this does not denote a diagnosis of pernicious anemia. In allergic reactions in the mouth, usually a manifestation of sensitivity to an ingested food, the tongue may swell, and epithelial elements may desquamate and coat the surface.

Unpleasant breath, sometimes imagined, is reported by people who conclude that their sensations of unpleasant taste must be a reflection of, or must be reflected in, breath odor. *Halitosis* is often present, however, brought to a patient's attention by a spouse or other family member. Common causes include infection or neoplasm in the oronasopharyngeal structures, poor oral hygiene, bronchiectasis or lung abscess, cirrhosis with hepatic fetor, gastric stasis inducing aerophagia and eructation, gastroesophageal reflux, and diabetes. Halitosis may also result from absorption of intestinal products and their excretion through the lungs.

The odor of garlic remains on the breath for many hours because garlic is absorbed into the portal circulation and passes through the liver into the general circulation. Volatile oils applied to denuded or even intact skin surfaces are also recognizable on the breath. Enzymatic processes in the intestine in some persons liberate absorbable gases of offensive odor. When introduced rectally, material not normally found in the upper gastrointestinal tract may be recovered from the stomach, which supports the possibility that retrograde passage of odoriferous substances reaches the mouth through the intestine. In a patient with pyloric obstruction, the breath is typically offensive only at eructation. It has also been postulated that substances such as fats, fatty acids, and some end products of fat digestion may cause halitosis, for which a low-fat diet is indicated.

Often, the diligent search for the cause of halitosis uncovers no clues, and recourse must be made to frequent mouth rinsing with antiseptic solutions that contain pleasant-smelling ingredients. Diet manipulation may be helpful in select patients but necessitates individual trials. Manipulation of the enteric flora and use of probiotics may be attempted as well.

Thrush may develop after the use of antibiotics. White or red fibrous lesions appear on the tongue. Thrush is also referred to as *mucocutaneous candidiasis* because of its association with *Candida* species, primarily *Candida albicans*. This organism is part of the normal flora of the tongue but can be disrupted and become infectious after antibiotic therapy or after long-term glucocorticoid therapy. Thrush occurs more frequently in elderly persons, in patients with metabolic disturbances, and in those with autoimmune suppression. Treatment with nystatin, in liquid form or as tablets in 100,000-U doses, is usually effective. Holding the liquid in the mouth or slowly dissolving the tablets three or four times daily for 1 to 2 weeks usually resolves the immediate infection.

ADDITIONAL RESOURCES

Edwards JE: *Candida* species. In Mandell GL, Bennett JE, Dolin R, editors: *Principles and practice of infectious diseases*, ed 6, Philadelphia, 2005, Churchill Livingstone–Elsevier, pp 2933-2957.

Lee SS, Zhang W, Li Y: Halitosis update: a review of causes, diagnoses and treatments, *J Calif Dent Assoc* 35:258-268, 2007.

Mirowski GM, Mark LA: Oral disease and oral-cutaneous manifestations of gastrointestinal and liver disease. In Feldman M, Friedman LS, Brandt LJ, editors: *Gastrointestinal and liver disease*, ed 8, Philadelphia, 2006, Sanders-Elsevier, pp 443-463.

Shulman JD, Carpenter WM: Prevalence and risk factors associated with geographic tongue among U.S. adults, *Oral Dis* 12:381-386, 2006.

Struch F, Schwahn C, Wallaschofski H, et al: Self-reported halitosis and gastroesophageal reflux disease in the general population, *J Gen Intern Med* 23(3):260-266, 2008.

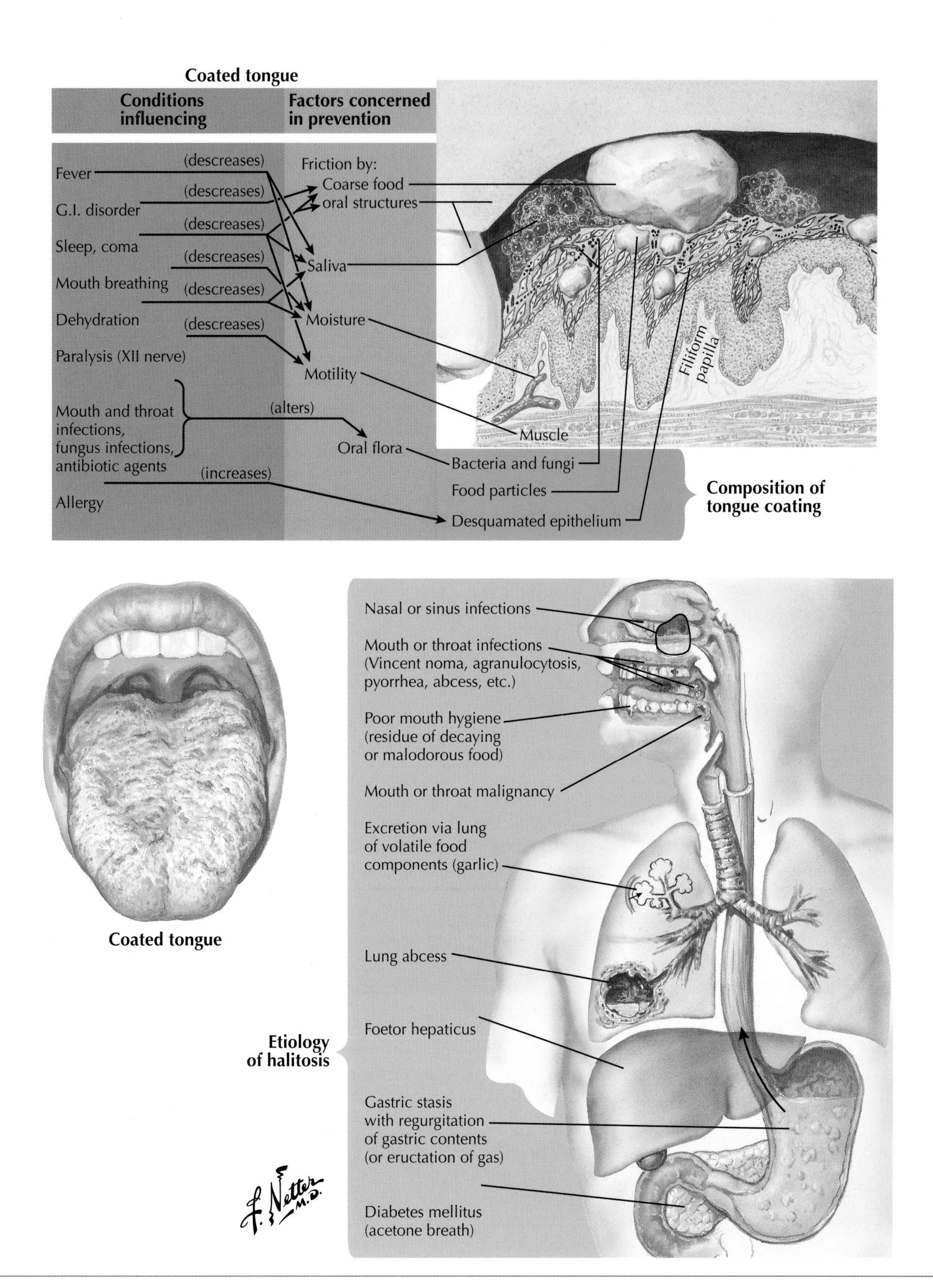

Figure 44-1 *Coated Tongue and Halitosis.*

Aerophagia and Eructation

45

Martin H. Floch

Aerophagia is characterized by excessive swallowing of air that results in repeated belching. Air may be swallowed unconsciously by the patient; when it results in repeated eructation, it becomes a clinical problem (**Fig. 45-1**).

Patients with aerophagia report frequent, uncontrollable belching, or eructation, which often is loud and disturbs family or co-workers. The condition may be acute in onset, but careful history usually reveals it is slow in developing but increases in severity until the patient seeks medical attention. It has been noted in children and may occur at any age. To fulfill the criteria of a functional gastrointestinal (GI) disorder, the condition should have been noted for at least 12 weeks in the year preceding the onset of troublesome, repetitive belching.

PATHOPHYSIOLOGY AND DIAGNOSIS

Eructation is normal during or after a meal, occurring two to six times without significance. Early in life, infants are made to burp with a change of position and then are able to resume a meal interrupted because of stomach distention caused by air swallowed during feeding. Frequent eructation by adults may become a habit.

In the act of belching, the glottis is closed, and the diaphragm and thoracic muscles contract. When the increased intraabdominal pressure transmitted to the stomach is sufficient to overcome the resistance of the lower esophageal sphincter, the swallowed air is eructated.

No diagnostic tests demonstrate normal or abnormal belching. However, in a patient with any symptom associated with belching, the history might indicate that the esophagus or stomach should be evaluated. Patients who are uncomfortable from mild upper abdominal distress may swallow a great amount of air and may have frequent eructation. Upper endoscopy to evaluate for organic disease is important with this symptom complex. The diagnosis is established by observing either the air swallowing or the frequent belching.

DIFFERENTIAL DIAGNOSIS

Aerophagia is now classified as a functional GI disorder. Once the diagnosis is established, the differential is minimal. However, the clinician must be certain that aerophagia is not secondary to upper abdominal discomfort from disease. Esophageal reflux with significant esophagitis, peptic ulcer of the stomach or duodenum, or discomfort from pancreatic or biliary disease in rare cases may cause mild aerophagia. However, other symptoms are apparent, and disease of upper GI organs can easily be evaluated. Often, the classic picture of aerophagia and loud belching confirms the diagnosis.

Persons who consume bicarbonate and patients with aerophagia are basically similar, unless the bicarbonate was taken to relieve the gas pains of peptic ulcer, in which case the carbon dioxide (CO_2) generated in the stomach is belched. The patient obtains relief from the antacid that is not obtained by belching the swallowed air. The relief that follows the ingestion of soda is explained not by deflation of the distended stomach with belching, but rather by neutralization of the acid. Also, some patients buy large amounts of soda or sodium bicarbonate to help them belch not because of aerophagia, but because they are chronic belchers.

Instead of swallowing air, some people are able to suck it in through a relaxed superior esophageal sphincter. This may occur in a patient with emphysema who is "pulling for air," or it may occur deliberately in an accomplished belcher. The same principle of using swallowed air is put to practical use in the development of so-called esophageal speech in patients who have undergone laryngectomy.

TREATMENT AND MANAGEMENT

The rational management of aerophagia and loud belching depends on correction of the underlying disturbance, whether organic or psychologic. Aerophagia is a functional disorder, and its management includes reassurance of the patient, education into the process of air swallowing and eructation, and treatment of any psychiatric component, such as anxiety or depression. Patients with aerophagia are always in some distress, and the physician must reassure them that they have no organic disease, then use pharmacologic therapy or recommend psychologic assistance. The family is frequently upset, so the social situation must be carefully assessed, and family or partners must be involved in the therapeutic regimen of reassurance and therapy.

Although no specific pharmacologic therapy exists, some clinicians have used tranquilizers, whereas others have used antidepressants. Simethicone and activated charcoal are usually ineffective. Again, reassurance, psychotherapy, and behavioral modification may be needed.

ADDITIONAL RESOURCES

Bredenoord AJ, Smout AJ: Physiologic and pathologic belching, *Clin Gastroenterol Hepatol* 5:772-775, 2007.

Bredenoord AJ, Weusten BL, Timmer R, Smout AJ: Psychological factors affect the frequency of belching in patients with aerophagia, *Am J Gastroenterol* 101:2777-2781, 2006.

Castell DO, Richter JE: *The esophagus*, ed 4, Philadelphia, 2004, Lippincott–Williams & Wilkins.

Drossman DA: *The functional gastrointestinal disorders*, ed 2, Lawrence, Kan, 2000, Allan Press, pp 328-330, 556-557.

Hasler WL: Nausea, gastroparesis and aerophagia, *J Clin Gastroenterol* 39:S223-S229, 2005.

Tack J, Talley NJ, Camilleri M, et al: Functional gastroduodenal disorders, *Gastroenterology* 130:1666-1679, 2006.

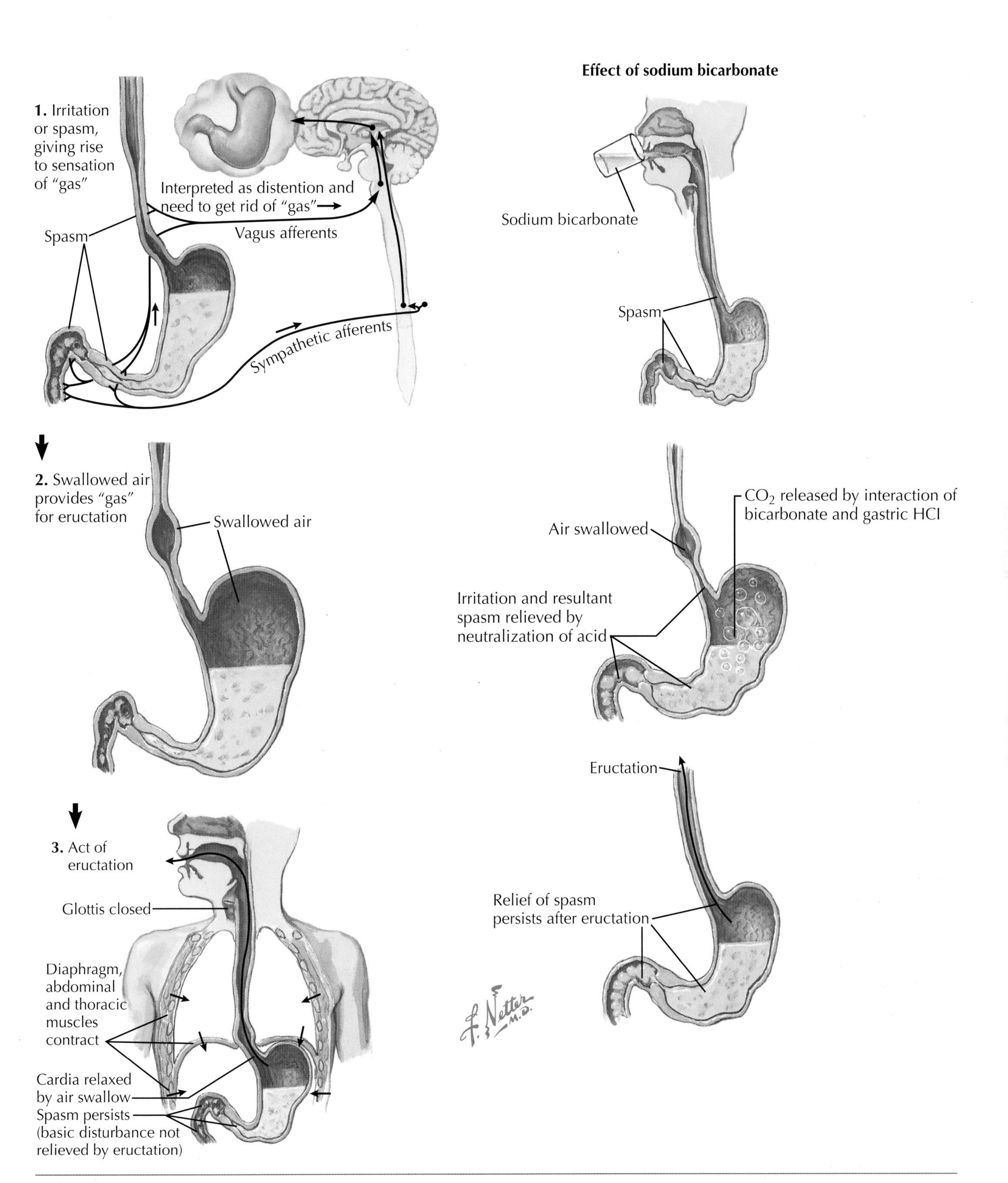

Figure 45-1 *Aerophagia and Eructation.*

Motility of the Stomach

Martin H. Floch

46

Peristalsis usually commences minutes after food reaches the stomach through vagal and splanchnic nerve stimuli. It is first noted in the pyloric portion because of the greater thickness of its musculature, which gives it the strongest triturating (grinding) power (**Fig. 46-1**).

The contractions originate in shallow indentations in the region of the incisura angularis and deepen as they move toward the pylorus. After 5 to 10 minutes, the contractions increase in strength and become progressively more vigorous. The pylorus opens incompletely and intermittently as the waves advance toward it. Most of the material reaching the pyloric portion is forced back into the fundus.

This process continues until some of the content has been reduced to a fluid or semifluid consistency suitable for passing into the small intestine. Evacuation is regulated by the influence of the gastrointestinal (GI) hormones secreted by the stomach and duodenum. Adverse mechanical or physiochemical properties of the chyme (e.g., hypertonicity) or large particles of food give rise to intrinsic or extrinsic nervous and hormonal influences that modify the tone of the pyloric sphincter and the motor activity of the pylorus. Large volumes of food, increased acidity, hypertonicity, large amounts of fat, and concentrated nutrients all slow motility and emptying. Reflexes slow because of fat in the ileum, the so-called ileal brake, and distention of the rectum and colon. The pylorus provides constant resistance to the passage of chyme and blocks the exit of solid particles. By maintaining a narrow orifice, the pylorus filters the gastric contents and helps prevent duodenal reflux.

Antral and duodenal contractions are well synchronized by nerve and hormone influences. Electrophysiologic patterns of gastric motor activity are based on a constant slow-wave pattern. They occur in the stomach at approximately three cycles per minute but do not cause contractions. It is believed that slow waves originate on the greater curvature approximately in the middle of its body. This area is now referred to as the "gastric pacemaker." Electrical signals do not pass the pylorus. Slow waves in the duodenum occur at about 11 to 12 cycles per minute. Electrical impulses of the stomach and the duodenum are clearly separated.

Interstitial cells of Cajal and muscle cells form a sophisticated network that initiate action potentials and begin the process of muscle contraction and peristaltic activity. Muscle activity of the fundus is separate from muscle activity of the antrum and the pylorus. The churning of the chyme occurs during these contractions and relaxations.

The fasting stomach has a basic migrating motor complex that tends to begin and end simultaneously at all sides, whereas in the duodenum and the small bowel, the motor complexes become progressive and migrate aborally. When food swallowing begins, the vagus nerve induces relaxation of the stomach, changing the balance of excitatory and inhibitory tone. Once food enters the stomach, the migrating motor complex pattern is replaced by the fed pattern, which may last 2 to 8 hours. The entire feeding process is under the influence of the vagus nerve and parasympathetic pathways, as well as corticotropin-releasing peptide, cholecystokinin, and other hormonal substances (e.g., vasoactive intestinal peptide, gastrin, somatostatin, dopamine, glucagon, bombesin).

Liquids are disbursed rapidly and have a slower lag period than solids. Solids empty in two phases. First, there is a lag period with slow emptying. Second, as the churning continues, emptying of the mixed chyme that has been exposed to acid and enzymes becomes more rapid. The lag phase for solids lasts approximately 60 minutes.

The pylorus and the coordinated antral pyloric and duodenal activity regulate emptying of the stomach. Once emptying is complete, electrophysiologic activity of the stomach returns to the basic migrating motor complex, awaiting the next feeding.

ADDITIONAL RESOURCES

Parkman HP, Trate DM, Knight LC, et al: Cholinergic effects on human gastric motility, *Gut* 45:346-354, 1999.

Quigley EMM: Gastric motor and sensory function and motor disorders of the stomach. In Feldman M, Friedman LS, Brandt LJ, editors: *Gastrointestinal and liver disease*, ed 8, Philadelphia, 2006, Saunders-Elsevier, pp 999-1028.

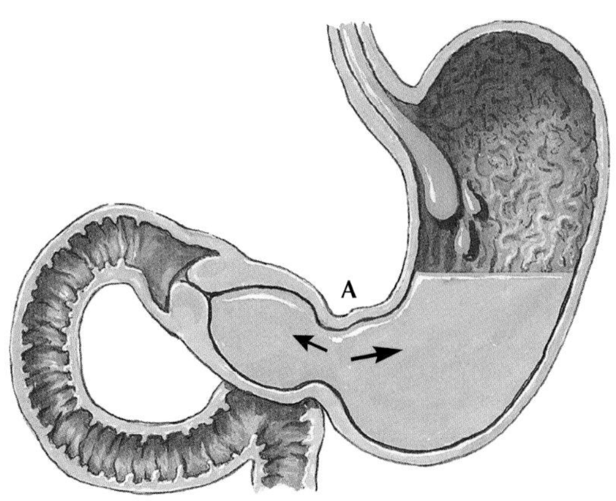

1. Stomach is filling, a mild peristaltic wave (A) has started in antrum and is passing toward pylorus. gastric contents are churned and largely pushed back into body of stomach

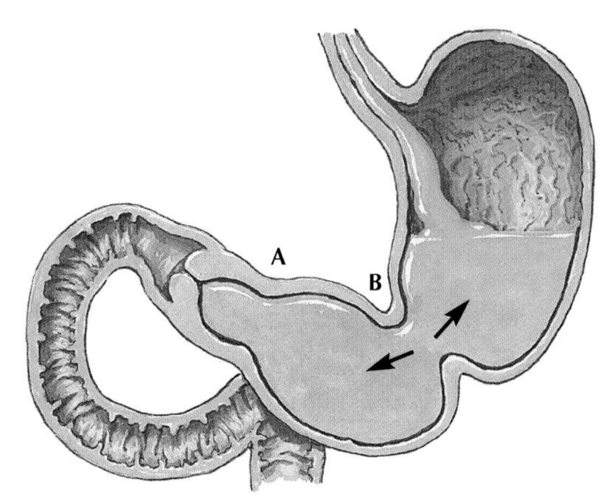

2. Wave (A) fading out as pylorus fails to open. A stronger wave (B) is origination at incisure and is again squeezing gastric contents in both directions

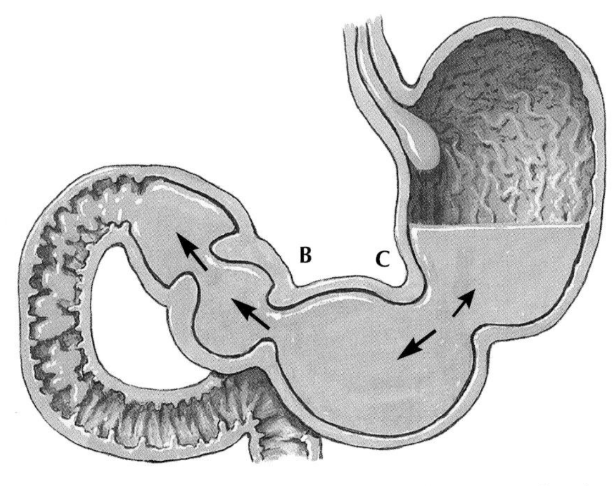

3. Pylorus opens as wave (B) approaches it. Duodenal bulb is filled and some contents pass into second portion of duodenum. Wave (C) starting just above incisure

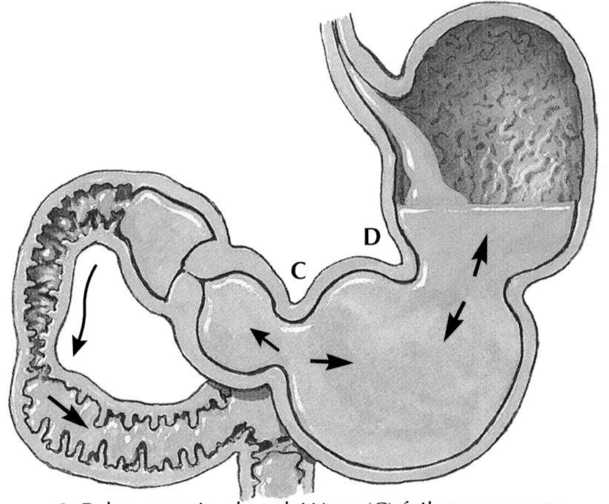

4. Pylorus again closed. Wave (C) fails to evacuate contents. Wave (D) starting higher on body of stomach. Duodenal bulb may contract or may remain filled, as peristaltic wave originating just beyond it empties second portion

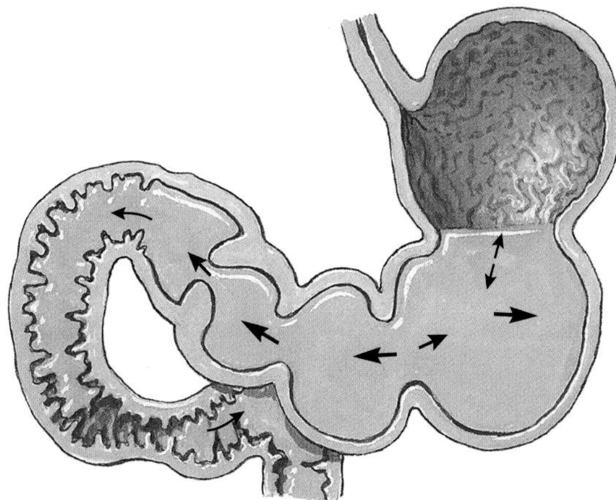

5. Peristaltic waves are now originating higher on body of stomach, gastric contents are evacuated intermittently. Contents of duodenal bulb area pushed passively into second portion as more gastric contents emerge

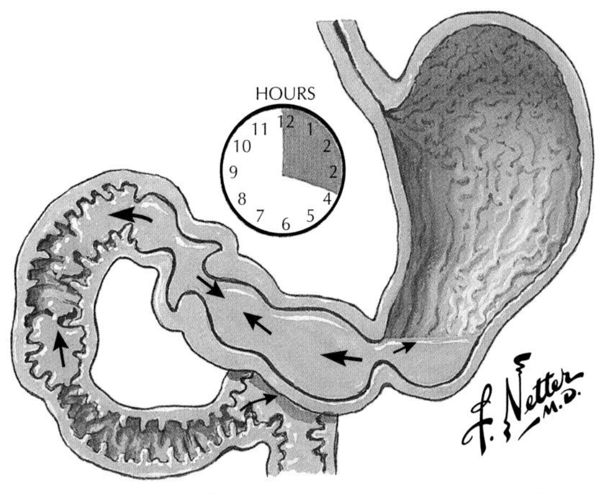

6. 3 to 4 hours later stomach almost empty, small peristaltic wave emptying duodenal bulb with some reflux into stomach. Reverse and antegrade peristalsis present in duodenum

Figure 46-1 *Motility of the Stomach.*

Gastroparesis and Gastric Motility Disorders

47

Martin H. Floch

Gastroparesis is defined as delayed emptying of the stomach. The most common causes of this motility disturbance of the stomach encountered in clinical practice are the association with diabetes mellitus and idiopathic forms (**Fig. 47-1** and **Box 47-1**). Gastroparesis affects persons of almost any age, with no gender predilection.

CLINICAL PICTURE

The presenting symptom of gastroparesis may be bloating, abdominal pain or distention, nausea, or vomiting. The patient may report a history of continued postprandial fullness. Nausea can be persistent and unexplained as the initial symptom. Vomiting may or may not accompany the nausea. The patient may vomit undigested fluid or may experience mild regurgitation of undigested fluid. Anorexia and weight loss may occur when the symptom complex persists over time. There are no hallmark physical findings other than the weight loss if the symptom has persisted.

Acute infectious diseases cause poor gastric emptying but resolve; the diagnosis of gastroparesis is based on chronicity. If gastroparesis symptoms are associated with postsurgical or other trauma or with neurologic disease, the findings of the primary disorder are significant. Patients with diabetes may be unable to control glycemia because of many factors, but irregular gastric emptying may play a role, as well as insulin secretion. In addition, associated findings in diabetes, such as neuropathy or enteropathy, may be present.

Box 47-1 Causes of Gastroparesis

Metabolic
- Diabetes mellitus
- Hypothyroidism
- Pregnancy
- Uremia

Associated gastric/esophageal disease
- Gastroesophageal reflux
- Gastritis
- Atrophic gastritis
- Peptic ulcer disease of stomach
- Acute gastroenteritis

Associated diseases
- Muscular dystrophy
- Parkinson disease
- Scleroderma
- Amyloidosis
- Chronic liver disease
- Idiopathic pseudoobstruction
- Anorexia nervosa
- Postsurgical and other trauma
- Vagotomy

Roux-en-Y surgery
- Head injuries
- Spinal cord injuries
- Medications
- Idiopathic causes

DIAGNOSIS

Of the many ways to document gastric emptying, the most common currently used is scintigraphy. Radioactive tracers are added to liquid and solid foods. Indium-111–diethylenetriamine pentaacetic acid (DTPA)–labeled water or technetium-99m–labeled egg or egg salad is most frequently used for liquid and solid phases. Many institutions use both; others have success using only the labeled solid food.

Breath tests with carbon 13 (^{13}C)–octanoic acid or ^{13}C–acetic acid are used in some institutions, but these require more time than the 1- or 2-hour testing with scintigraphy. Tests vary greatly from institution to institution and require the availability of an accredited nuclear medicine laboratory. Ultrasonography has also been used but requires specialized operator expertise.

Regardless of the method used to diagnose delayed emptying, the finding then must be correlated with a list of associated diseases or deemed "idiopathic" (unknown cause or spontaneous condition). Therefore, it is essential to document any related diseases before treatment. Again, the most common associations are diabetes and idiopathic causes. Diabetes may be subtle and must be discerned; gastroparesis may be the first presentation. The patient who does not have diabetes must be carefully evaluated to rule out associated neurologic disorders before the condition is called idiopathic.

TREATMENT AND MANAGEMENT

Controlling diabetes, if present, is essential. Careful history taking is done to uncover any medication-related cause. Substances that can delay gastric emptying include alcohol, aluminum hydroxide antacids, atropine, β-adrenergic antagonists, calcitonin, calcium channel blockers, dexfenfluramine, diphenylhydromine, glucagon, H_2-receptor antagonists, interleukin-1 (IL-1), L-dopa, lithium, octreotide, opiates, phenothiazine, progesterone, propantheline bromide, sucralfate, synthetic estrogens, tetrahydrocannabinol, tobacco, and tricyclic antidepressants. For patients with idiopathic causes and associated diseases, several medications can be prescribed. However, when

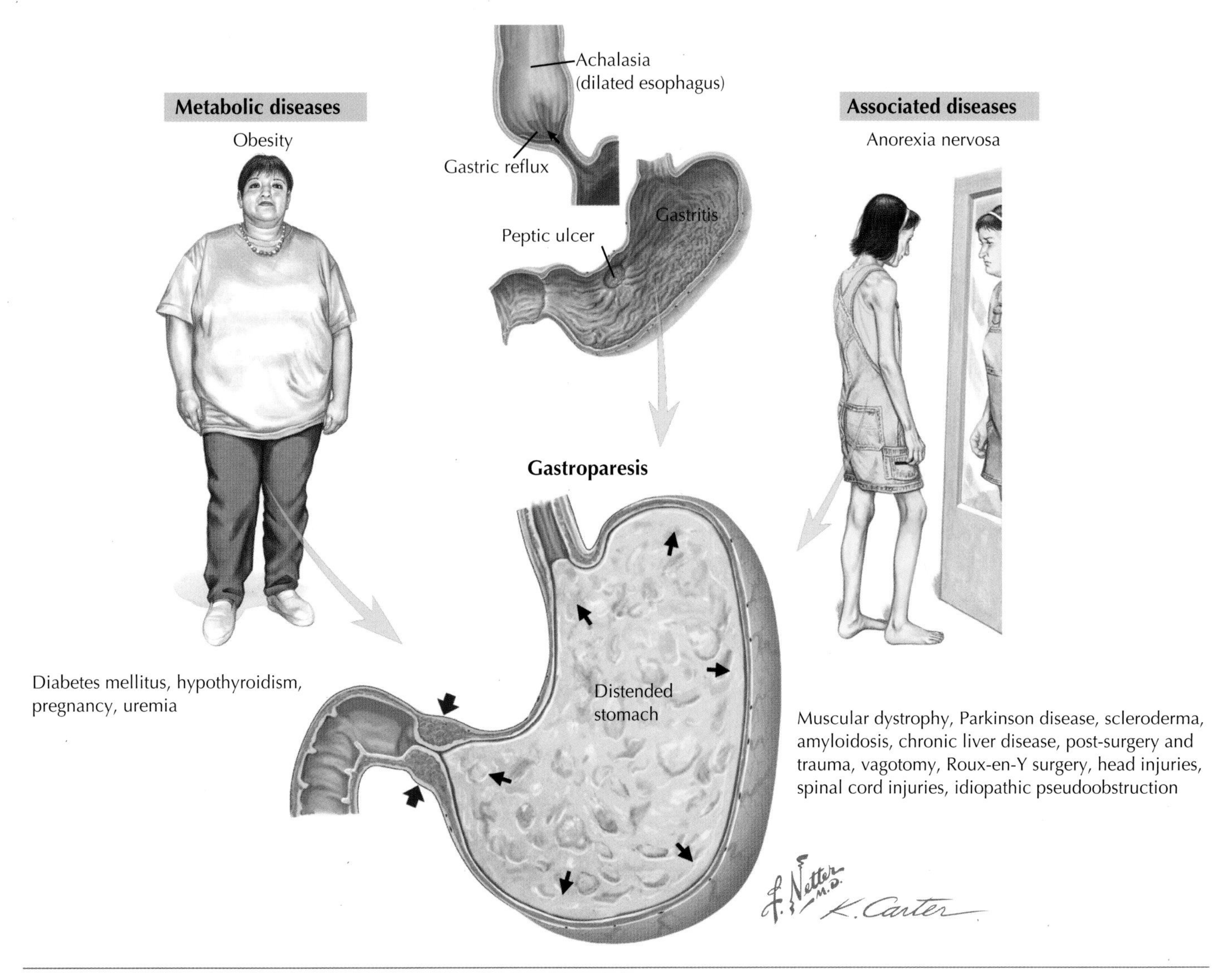

Figure 47-1 *Gastroparesis and Gastric Motility Disorders.*

the diagnosis points to an idiopathic condition, the clinician must keep in mind that it may become a *functional* disorder, and that all forms of therapy used in functional disorders (e.g., reassurance, antianxiety/antidepressant drugs) may be needed.

With recent progress in drug therapy and research into prokinetic agents, the following four categories of drugs are now used in patients with gastroparesis:

1. *Dopamine antagonists.* Domperidone (10-30 mg four times daily) and metoclopramide (5-20 mg four times daily) are dopamine antagonists. Unfortunately, domperidone is only available in the United States in special situations, and metoclopramide, although used frequently, can cause neurologic symptoms with long-term use.
2. *Substituted benzamides.* Cisapride (5-20 mg twice daily) is effective but is unavailable in the United States.
3. *Macrolides.* Erythromycin (50-200 mg four times daily) often may cause pain in women.
4. *Cholinergic agonists.* The use of bethanechol (5-25 mg four times daily) is controversial, but the drug may be helpful in some patients.

Ghrelin, the gastrointestinal hormone that stimulates eating, also has a positive effect on gastric emptying. Although only a few studies have used ghrelin in subjects with gastroparesis, results are promising, and there are research advocates for its use in these patients.

COURSE AND PROGNOSIS

It is important to monitor the patient with gastroparesis and to repeat the gastric-emptying study while the patient is taking medication to determine if the drug is effective. Often, the clinician can correlate the decrease in symptoms with increased gastric emptying. Gastroparesis is chronic but may vary in severity; thus, therapy can be modulated depending on the symptom phase.

Mild cases of gastroparesis may be controlled by prokinetic medication, but patients with severe gastroparesis may require nutrition support and possibly jejunostomy feeding. Weight must be monitored, and when the patient is losing weight and cannot eat sufficiently, nutrition support must be started. Supplemental feedings may control weight loss, but enteral feeding may be necessary in some patients.

Recent experimental therapeutic methods include electronic devices that are wired to the gastric mucosa, with gastric electrical pacing. These techniques have been instituted only in research centers but hold promise for patients who require long-term therapy.

OTHER DISORDERS

Other gastric motility disorders are rare and mainly involve disturbances that can be identified in gastric muscle activity. To identify these abnormalities, sophisticated electrogastrography is necessary. Although available only at a few large university centers, this procedure can identify disturbances in gastric motility and gastric pacing that can cause nausea, vomiting, abdominal pain, anorexia, and weight loss. Gastric pacing disturbances are now experimentally treated with gastric electrical pacing.

ADDITIONAL RESOURCES

Bortolotti M: The "electrical way" to cure gastroparesis, *Am J Gastroenterol* 97:1874-1883, 2002.

Bouras EP, Scolapio JS: Gastric motility disorders: management that optimizes nutritional status, J Clin Gastroenterol 38:549-557, 2004.

Camilleri M: Advances in diabetic gastroparesis, *Rev Gastroenterol Dis* 2:47-56, 2002.

Gaddipati KV, Simonian HP, Kresge KM, et al: Abnormal ghrelin and pancreatic polypeptide responses in gastroparesis, *Dig Dis Sci* 51:1339-1346, 2006.

McCallum RW, Chen JD, Lin Z, et al: Gastric pacing improves emptying and symptoms in patients with gastroparesis, *Gastroenterology* 114:456-461, 1998.

Murray CD, Martin NM, Patterson M, et al: Ghrelin enhances gastric emptying in diabetic gastroparesis: a double-blind, placebo-controlled crossover study, *Gut* 54:1693-1698, 2005.

Owyang C. Hasler WL: Physiology and pathophysiology of the interstitial cells of Cajal: from bench to bedside. VI. Pathogenesis and therapeutic approaches to human gastric dysrhythmias, *Am J Physiol* 283:G8-G18, 2002.

Parkman HP, Hasler WL, Fisher RS: American Gastroenterological Association technical review of the diagnosis and treatment of gastroparesis, *Gastroenterology* 127:1592-1622, 2004.

Pyloric Obstruction and the Effects of Vomiting

48

Martin H. Floch

Pyloric obstruction occurs when the outlet of the stomach narrows to the point of serious interference with gastric emptying (**Fig. 48-1**). In Western countries, tumors are the most common cause of pyloric obstruction in adults. Duodenal ulcer was once a common cause but is now rarely encountered because of the high cure rate of *Helicobacter pylori* and the use of H_2-antagonist and proton pump inhibitor (PPI) therapy for peptic ulcer. It is important to understand the effect of pyloric obstruction, which is vomiting.

Infantile hypertrophic pyloric stenosis is the most common cause of abdominal surgery in the first 6 months of life. The incidence in the United States is approximately 3 in 1000 births.

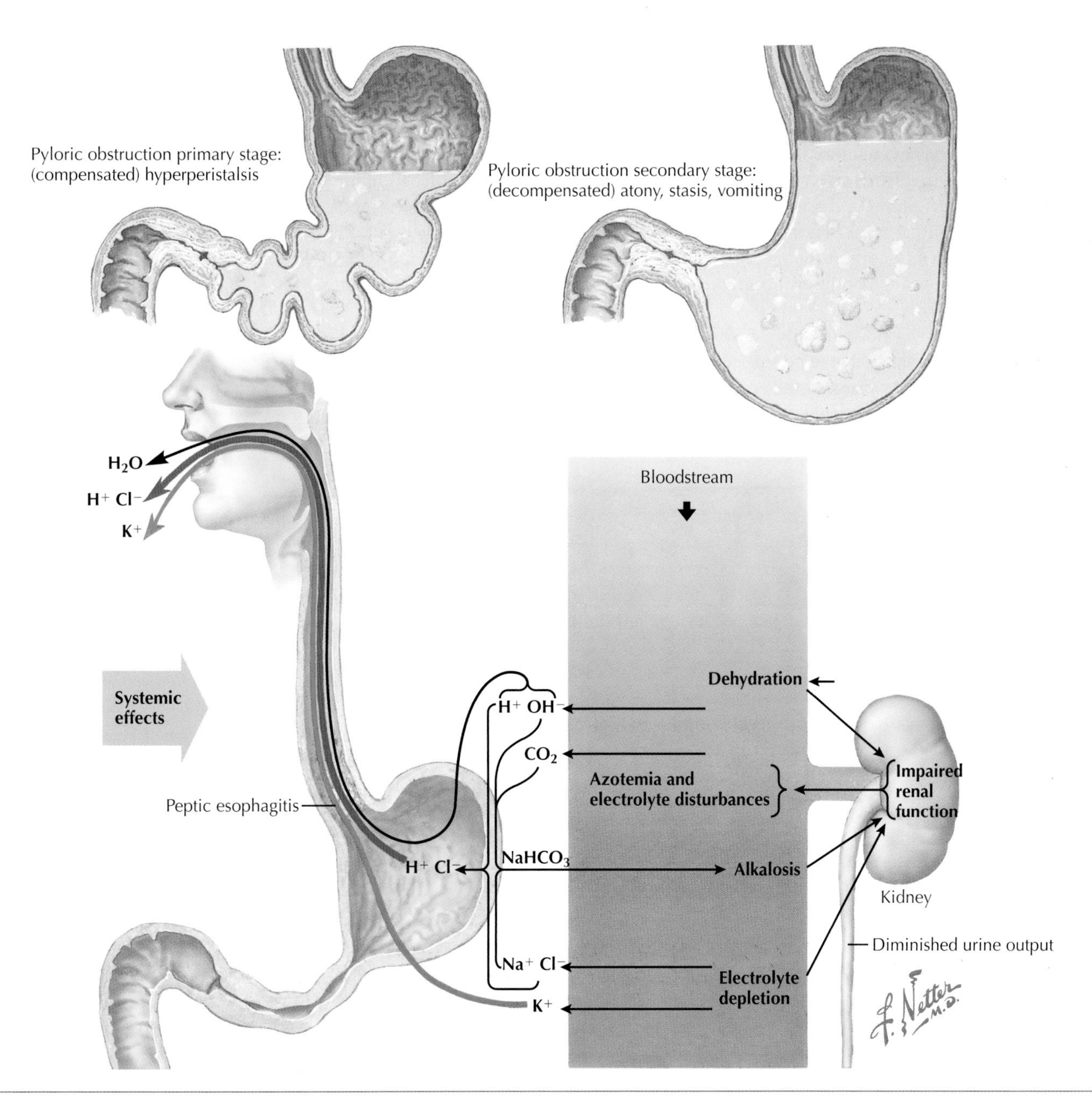

Figure 48-1 *Pyloric Obstruction and the Effects of Vomiting.*

Although rare in adults, hypertrophic pyloric stenosis does occur when missed early in life or when symptoms were not severe in childhood and progressed to diagnosis later in life (see Chapter 50).

CLINICAL PICTURE

When the outlet of the stomach becomes narrowed to the point of interference with gastric emptying, the gastric musculature responds at first with increased peristalsis in an effort to build up sufficient pressure to overcome the resistance at its pyloric end. At this stage, the patient may experience a sensation, or burning, in the epigastrium or left hypochondrium. With persisting obstruction and further stagnation of ingested food and gastric secretion, the stomach begins to dilate; the musculature becomes atonic, and peristaltic activity is minimal. At this stage, the patient reports fullness, vomiting of undigested food consumed many hours earlier, and foul-smelling eructation. If the obstruction is unrelieved, vomiting becomes more frequent and more copious. With so little gastric content now passing into the intestine because of the profound gastric atony, the patient is powerless to keep up with the fluid and electrolytes lost in the vomitus. Dehydration, hypochloremia, hypokalemia, and alkalosis supervene, which in turn affect renal function, with development of oliguria, azotemia, and retention of other electrolytes. Clinically, the patient is weak, anorexic, and drowsy. Unless measures are instituted to correct the metabolic disorder and to relieve the obstruction, the condition progresses to irreversible tissue damage and death.

Pyloric obstruction is not the only cause of vomiting (see Chapter 49), but the diagnosis may be suspected because of the history just described, the pattern of the emesis, and the appearance of the vomitus. In duodenal ulcer, which is the most common cause of pyloric obstruction, the patient usually gives a history of ulcer symptoms. The vomiting is at first intermittent, perhaps 2 or 3 days apart, and the vomitus often contains recognizable particles of food eaten the previous day.

As with excessive vomiting from any cause, the patient has appreciable losses of fluids and hydrogen (H^+), chloride (Cl^-), and potassium (K^+) ions. Because the gastric juice is poor in sodium (Na^+), usually no sodium deficiency occurs, and although Na^+ remains in the blood, bicarbonate (HCO_3^-) substitutes for Cl^-. Loss of K^+ occurs because parietal cells secrete significant amounts of this ion.

Vomiting does not usually occur in uncomplicated ulcer disease, except when the ulcer is located in the pyloric canal. However, many patients with ulcers empty the stomach through vomiting to obtain pain relief.

DIAGNOSIS

Barium contrast imaging or computed tomography can provide a diagnosis of pyloric obstruction, and endoscopic visualization of the pylorus and mucosal biopsy can clarify the cause. The differential diagnosis, as previously indicated, includes benign or malignant tumor and scarring resulting from chronic peptic disease. Rare causes, such as polyp intussusception, usually are more acute in presentation than a chronic obstructive process.

TREATMENT AND MANAGEMENT

Managing the consequences of repeated or excessive vomiting consists of fluid and electrolyte replacement, evacuation of the stomach with adequate drainage, and continuous gastric aspiration with a nasogastric tube for 48 to 72 hours. If the obstruction itself is not relieved, surgery is necessary to reestablish gastrointestinal passage, but only after fluid and electrolyte balance has been restored. The cause of the obstruction is treated after the effects of vomiting are managed. Treatment of tumor obstruction is discussed in Chapter 64, and treatment of peptic disease is discussed in Chapters 55, 56, and 58. Medical treatments depend on the cause, but surgical relief of the obstruction is invariable. In incurable malignant obstruction, stents may be placed to gain temporary relief.

A clinical and physiologic disturbance similar to pyloric obstruction, known as *milk-alkali* (Burnett) *syndrome*, may result from excessive ingestion of a soluble alkali and a rich source of calcium.

COURSE AND PROGNOSIS

Immediate treatment of the vomiting and its metabolic disturbance are usually successful. The long-term course and prognosis for pyloric obstruction and vomiting depend on the cause. If the vomiting was caused by a benign tumor or by scarring from chronic ulcer, the prognosis is usually excellent. If cancer was the cause, the prognosis depends on its type and extent and on the effectiveness of other treatments. However, if cancer causes pyloric obstruction, the prognosis is usually poor.

ADDITIONAL RESOURCES

Malagelada JR, Malagelada C: Nausea and vomiting. In Feldman M, Friedman LS, Brandt LJ, editors: *Gastrointestinal and liver disease*, ed 8, Philadelphia, 2006, Saunders-Elsevier, pp 143-158.

Russo MA, Redel CA: Anatomy, histology, embryology, and developmental anomalies of the stomach and duodenum. In Feldman M, Friedman LS, Brandt JS, editors: *Gastrointestinal and liver disease*, ed 8, Philadelphia, 2006, Saunders-Elsevier, pp 981-998.

Nausea and Vomiting

49

Martin H. Floch

Nausea and vomiting are nonspecific but clinically important symptoms associated with numerous causes. Nausea is variously described as a sick feeling, a tightness in the throat, a sinking sensation, or a feeling of imminent vomiting. It generally precedes vomiting and may be associated with retching when the stomach is empty. Although associated with any disease, acute nausea and vomiting are most often associated with infectious disease, pregnancy, medications (including chemotherapy), postoperative status, and motion sickness. Other common causes include radiation sickness, gastrointestinal (GI) obstruction, hepatitis, metabolic disturbances (e.g., diabetes mellitus, thyroid disease), systemic diseases (e.g., myocardial infarction, renal failure, asthma), Addison disease, and central nervous system (CNS) causes (e.g., brain tumors, stroke, hemorrhage, meningitis). This wide list of associated diseases and causes makes it necessary to understand the nausea/vomiting process and its treatment.

Furthermore, many patients have nausea and vomiting associated with gastric motility disorders or with anorexia nervosa or psychogenic causes, in which the inciting disease or cause is not apparent or diagnosable. Some of these may be considered functional disorders.

CLINICAL PICTURE AND PHYSIOLOGY

Salivation, pallor, tachycardia, faintness, weakness, and dizziness frequently occur concomitantly. Nausea and vomiting may result from disturbances throughout the body and may be precipitated by the following:

- Emotional disturbances
- Intracranial vasomotor and pressure changes
- Unpleasant olfactory, visceral, or gustatory stimuli
- Functional or anatomic alterations in the thoracic and abdominal viscera, including the urogenital tract
- Intense pain in somatic parts
- Exogenous or endogenous toxins
- Drugs (notably opiates)
- Stimulation of the vestibular apparatus (usually by motion)

Impulses from all these sources reach the CNS through the corresponding sensory nerves (**Fig. 49-1**).

The CNS control of vomiting is based in two areas: (1) the *vomiting center*, located in the lateral reticular formation of the medulla, among cell groups governing such related activities as salivation and respiration, and (2) the *chemoreceptor trigger zone*, in a narrow strip along the floor of the fourth ventricle, close to the vomiting center. Functions of these two areas are distinct, although not independent. The vomiting center is activated by impulses from the GI tract and other peripheral structures. The chemoreceptor trigger zone is stimulated by circulating toxic agents and by impulses from the cerebellum; this zone's influence on the vomiting center results in the emetic action.

After irritation in any somatic or visceral area or in any sense organ, impulses travel through their respective sensory nerves to reach the medulla, where they activate the vomiting center. Toxic agents, whether introduced into the body or accumulated endogenously, act on the chemoreceptor trigger zone, through which impulses reach and activate the nearby vomiting center. Before the vomiting threshold is exceeded, impulses passing to the cortex lead to the sensation of nausea. The vomiting center coordinates the discharge of impulses from adjacent neural components to the various structures that participate in the act of vomiting. Salivation, which almost invariably precedes the actual ejection of the vomitus, is stimulated by impulses from the salivary nuclei. Contraction of the intercostal muscles and the diaphragm produces a sharp inspiratory movement and increased intraabdominal pressure, facilitated by contraction of the abdominal muscles. Closure of the glottis forestalls aspiration into the respiratory passages. The pyloric portion of the stomach contracts; the body of the stomach, cardia, esophagus, and cricopharyngeus muscle relax, and the gastric contents are forced out through the mouth and, in vigorous emesis, through the nose as well.

Nausea and vomiting brought on by motion do not require a vertical component; some persons develop the symptoms merely from being rotated. Attempts to resolve the visual disorientation through eye and head movements may result in stimulation of the labyrinth, either directly or by decreased gastric tonus. Visual stimuli are not essential for the development of motion sickness; even blind persons may be susceptible.

Rapid downward motion that comes to a sudden stop, or that is followed by upward motion, causes the abdominal viscera to sag and pull on their attachments. This is the origin of the sinking feeling experienced at the end of a rapid descent in an elevator, or a sudden steep decline in a plane. The sensation does not occur if the subject stands on his head in the elevator, and it is reduced if the subject assumes a horizontal position when the plane is bouncing up and down, because the viscera cannot be displaced as far in the anteroposterior direction as in the craniocaudal direction. Nausea and retching may be induced in a patient under spinal anesthesia by downward traction on the exposed stomach.

Nausea may be difficult to relieve and becomes a serious clinical problem if sufficiently prolonged to interfere with nutrition. *Primary nausea*, or nausea occurring in the postabsorptive state, occasionally accompanies eye strain, myocardial infarction, azotemia, or visceral neoplastic disease, but it is usually of psychologic origin. Protracted vomiting is detrimental not only because of nutrition concerns but also because of electrolyte depletion (see Chapter 48).

If vomiting does not respond to antiemetic drugs, nasogastric suction should be instituted. Correction of a gastric hypotonus may be the factor that brings the condition under control.

DIAGNOSIS

Thorough evaluation must include all possible causes of nausea and vomiting. In the pregnant patient, the pregnancy may be the paramount cause, but GI conditions during pregnancy, such as cholecystitis and appendicitis, must be explored. The large number of causes must therefore be considered in the workup. Once established, the specific cause must be evaluated to determine correct therapy and prognosis. When no cause is found, and the nausea and vomiting fall into the category of "psychogenic" or cyclic vomiting, therapy must include not only pharmacologic but also psychiatric modalities.

Cyclic vomiting syndrome (CVS) is much more common than previously believed. CVS is most often seen in children but now is recognized in adolescents, as well as young and older adults. The classic presentation involves sudden attacks of severe vomiting and retching that subside with acute treatment, often in the emergency room. Migraine headache and abdominal migraine are often associated with CVS. In addition, prodromal symptoms may be identified. The cause and frequency of attacks vary. There are no good evidence-based treatment protocols, and many drugs are used. Specific treatment for CVS, as well as those listed next, in the acute phase includes preventing shock and dehydration and electrolyte loss. Ondansetrone is tried, hydromorphone for pain and, if needed, sedative agents. Removal to a dark area may be helpful and removal of any stimulating actions that are known to precipitate an attack.

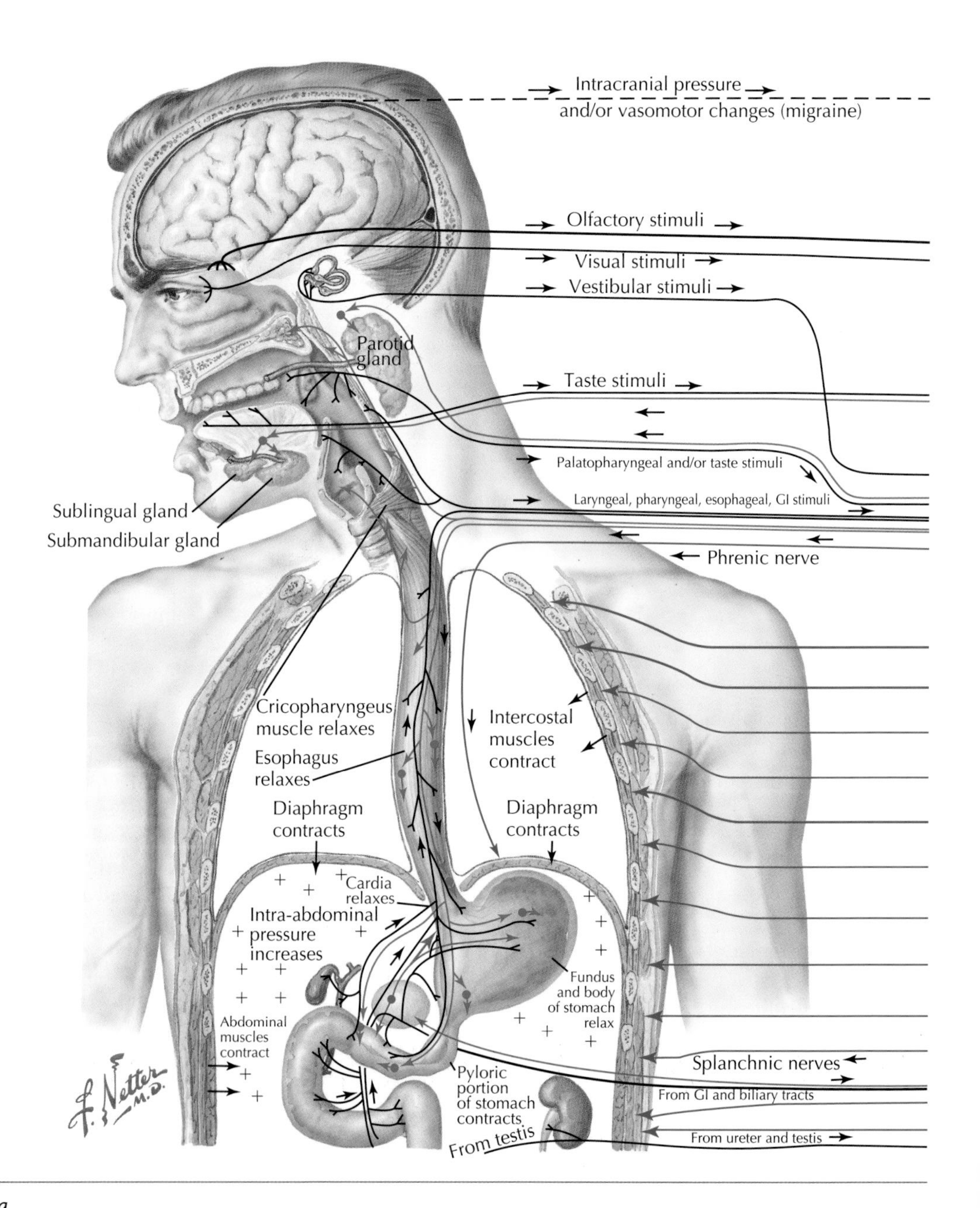

Figure 49-1 *Nausea and Vomiting.*

Prevention of attacks has been tried with cyproheptadine, propranolol, tricyclic antidepressants, and 5-HT_{1d} agonists such as sumatriptan and eletriptan if the patient has migraine symptoms. Management requires an aggressive approach to prevention, and most patients can be helped to live with the disorder.

TREATMENT AND MANAGEMENT

Severe acute vomiting and protracted vomiting may cause significant metabolic and electrolyte disturbances, usually necessitating intravenous treatment for replacement of potassium, sodium, and other electrolytes. Prolonged nausea and vomiting also cause nutritional deficiencies, which must be treated according to duration of the illness (see Section X).

Many drugs are available for pharmacologic therapy (**Table 49-1**). Dosage and frequency of administration depend on the

Table 49-1 Drugs for Therapy of Nausea and Vomiting

Drug Class	Medications
Antihistamine	Meclizine Promethazine Dimenhydrinate
Anticholinergic	Scopolamine Hyoscyamine
Antidopaminergic	Prochlorperazine Chlorpromazine
Prokinetic	Metoclopramide Cisapride Erythromycin Domperidone Bethanechol Octreotide Trimethobenzamide

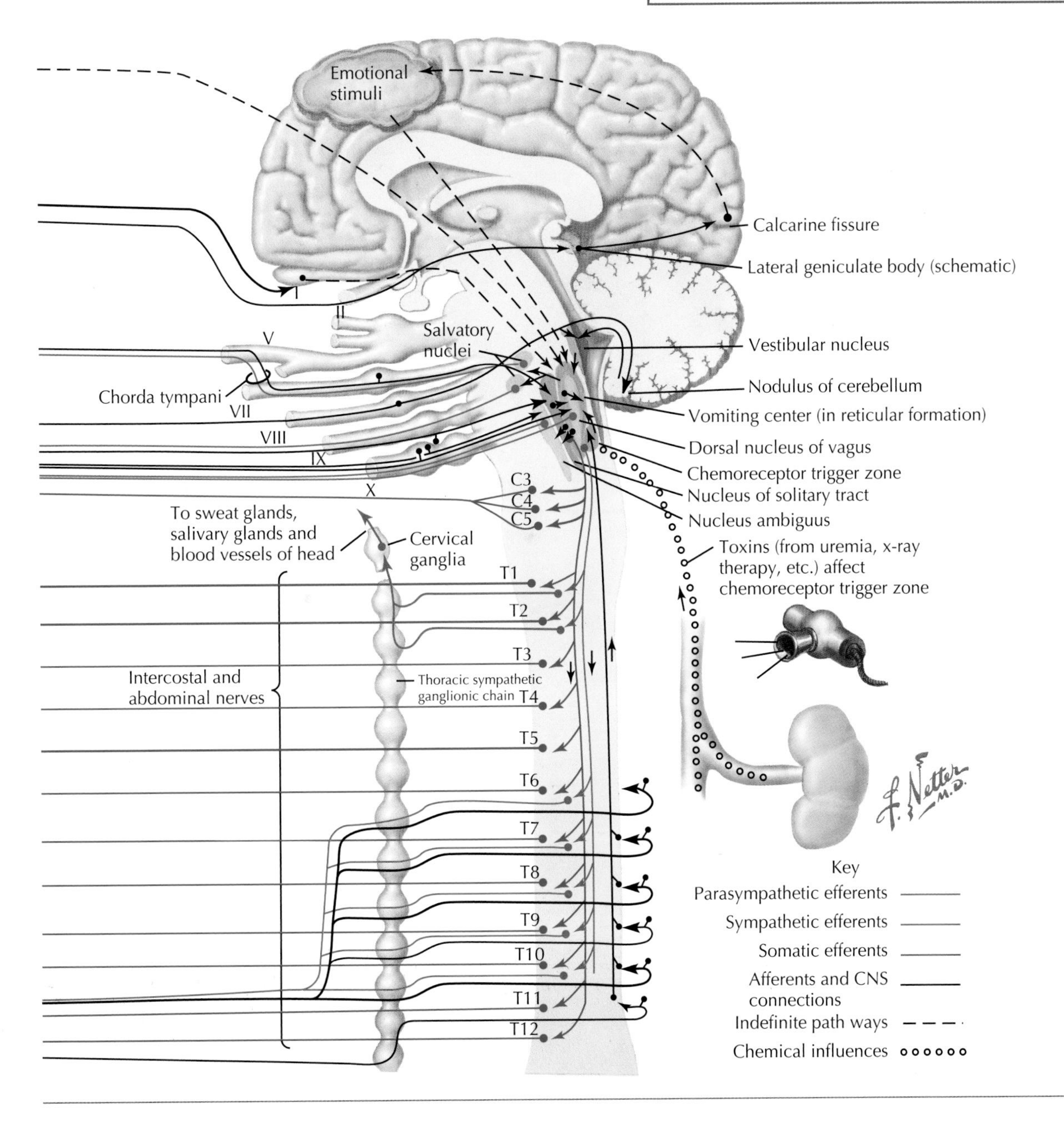

disease process (e.g., motion sickness vs. illness caused by acute disease vs. chemotherapy).

Chemotherapy can cause intensive nausea and recurrent vomiting. Often, oncologists use a combination of drugs, such as the selective 5-HT_3 receptor antagonists ondansetron and granisetron, although these are expensive, or tetrahydrocannabinol, the active ingredient of marijuana. Treating nausea and vomiting during chemotherapy is challenging because it includes nutrition support and symptom relief.

If vomiting is persistent and does not respond to the administration of antiemetic drugs, nasogastric suction should be instituted. Again, correcting gastric hypotonia may be the factor that brings the condition under control.

COURSE AND PROGNOSIS

Course and prognosis depend on the cause of the nausea and vomiting. If the cause is idiopathic, the symptoms can be frustrating and challenging to the patient and physician. However, if the cause is benign and idiopathic, the course is usually benign. Progress may be intermittent, however, and both patient and physician can become frustrated. Psychotherapeutic drugs may be helpful, as well as psychiatric treatment, if necessary.

ADDITIONAL RESOURCES

Alhashimi D, Alhashimi H, Fedorowicz Z: Antiemetics for reducing vomiting related to acute gastroenteritis in children and adolescents, *Cochrane Database Syst Rev* 4:CD005506, 2006.

Malagelada JR, Malagelada C: Nausea and vomiting. In Feldman M, Friedman LS, Brandt JS, editors: *Gastrointestinal and liver disease*, ed 8, Philadelphia, 2006, Saunders-Elsevier, pp 143-158.

Pareek N, Fleisher DR, Abell T: Cyclic vomiting syndrome: what a gastroenterologist needs to know, *Am J Gastroenterol* 102:2832-2840, 2007.

Ramsook C, Sahagun-Carreon I, Kozinetz CA, Moro-Sutherland D: A randomized trial comparing oral ondansetron with placebo in children with vomiting from acute gastroenteritis, *Ann Emerg Med* 39:397-403, 2002.

Wood GJ, Shega JW, Lynch B, Von Roenn JH: Management of intractable nausea and vomiting in patients at the end of life, *JAMA* 298:1196-1207, 2007.

Hypertrophic Pyloric Stenosis

50

Martin H. Floch

Hypertrophic pyloric stenosis is an obstruction in the pylorus caused by hyperplasia of the circular muscle surrounding the pyloric outlet channel (**Fig. 50-1**). It is more common in infants than adults and actually is rare in adults. The incidence is approximately 3 in 1000 live births; boys are affected more often than girls by a ratio of 4:1 to 5:1. The disorder is more common among white persons of northern European descent than among persons of African or Asian descent.

The cause of hypertrophic pyloric stenosis is unknown, but a deficiency of nitric oxide synthetase is suspected. In addition, interstitial cells of Cajal are not seen throughout the pylorus. Fifty percent of identical twins are affected, but the disorder does not follow Mendelian inheritance patterns. Both genetic and environmental factors are thought to be important.

CLINICAL PICTURE

The clinical presentation of hypertrophic pyloric stenosis is different in infants than in adults. The classic infant presentation is vomiting that occurs in the second to sixth weeks of life. Vomiting increases in frequency and severity and is characterized early as occurring suddenly with great force (projectile vomiting). The infant cries, indicating hunger. Because less food is able to pass the pylorus, the infant becomes dehydrated and loses weight. At this stage, metabolic acidosis may become a serious problem. On examination of the infant, the classic "olive" might be felt in the area of the pylorus, and strong peristaltic movements in the stomach may be observed on inspection of the abdomen.

In adults, nausea, vomiting, satiety, and epigastric pain after eating are major symptoms. Physical examination is not helpful because the condition is chronic and the pyloric mass is not easy to palpate. Patients may lose weight if symptoms persist.

DIAGNOSIS

Hypertrophic pyloric stenosis is diagnosed in children based on timing of the presentation and physical examination findings. X-ray examination is important before surgery. The large, dilated stomach is evident; if further evidence is needed, careful barium study can be performed to visualize the narrowed pylorus. Ultrasonography is important because the classic 3-mm sonolucent "doughnut" can be seen in children.

In adults, the clinician first must consider the possibility of stenosis. Ultrasonography is then helpful, but barium contrast reveals the classic narrowed segment. Upper endoscopy is recommended to rule out chronic peptic or malignant disease.

TREATMENT AND MANAGEMENT

In the past, some clinicians preferred a trial of medical treatment with anticholinergic therapy and very soft food for patients with hypertrophic pyloric stenosis. However, the medical therapy had a high failure rate. Pyloromyotomy is the treatment of choice. Ramstedt pyloromyotomy includes a longitudinal incision through the hyperplastic pyloric muscle. Some surgeons prefer resection of the adult pylorus, to rule out malignancy. Although endoscopy can dilate the pylorus, these procedures have failed in as many as 80% of patients within the first 6 months of therapy.

PROGNOSIS

The prognosis is excellent, and once the correct therapy has been applied for hypertrophic pyloric stenosis, patients go on to lead normal lives.

ADDITIONAL RESOURCES

Graadt Van Roggen JF, Van Krieken JH: Adult hypertrophic pyloric stenosis: case report and review, *J Clin Pathol* 51:479-480, 1998.

Safford SD, Pietrobon R, Safford KM, et al: A study of 11,003 patients with hypertrophic pyloric stenosis and the association between surgeon and hospital volume and outcomes, *J Pediatr Surg* 40:967-972, 2005.

Vandiwinden JM, Liu H, de Laet MH, et al: Study of interstitial cells of Cajal in infantile pyloric stenosis, *Gastroenterology* 111:279-288, 1996.

Yamataka A, Tsukada K, Yokoyama-Laws Y, et al: Pyloromyotomy vs. atropine sulfate for infantile hypotrophy pyloric stenosis, *J Pediatr Surg* 35:338-341 (discussion 342), 2000.

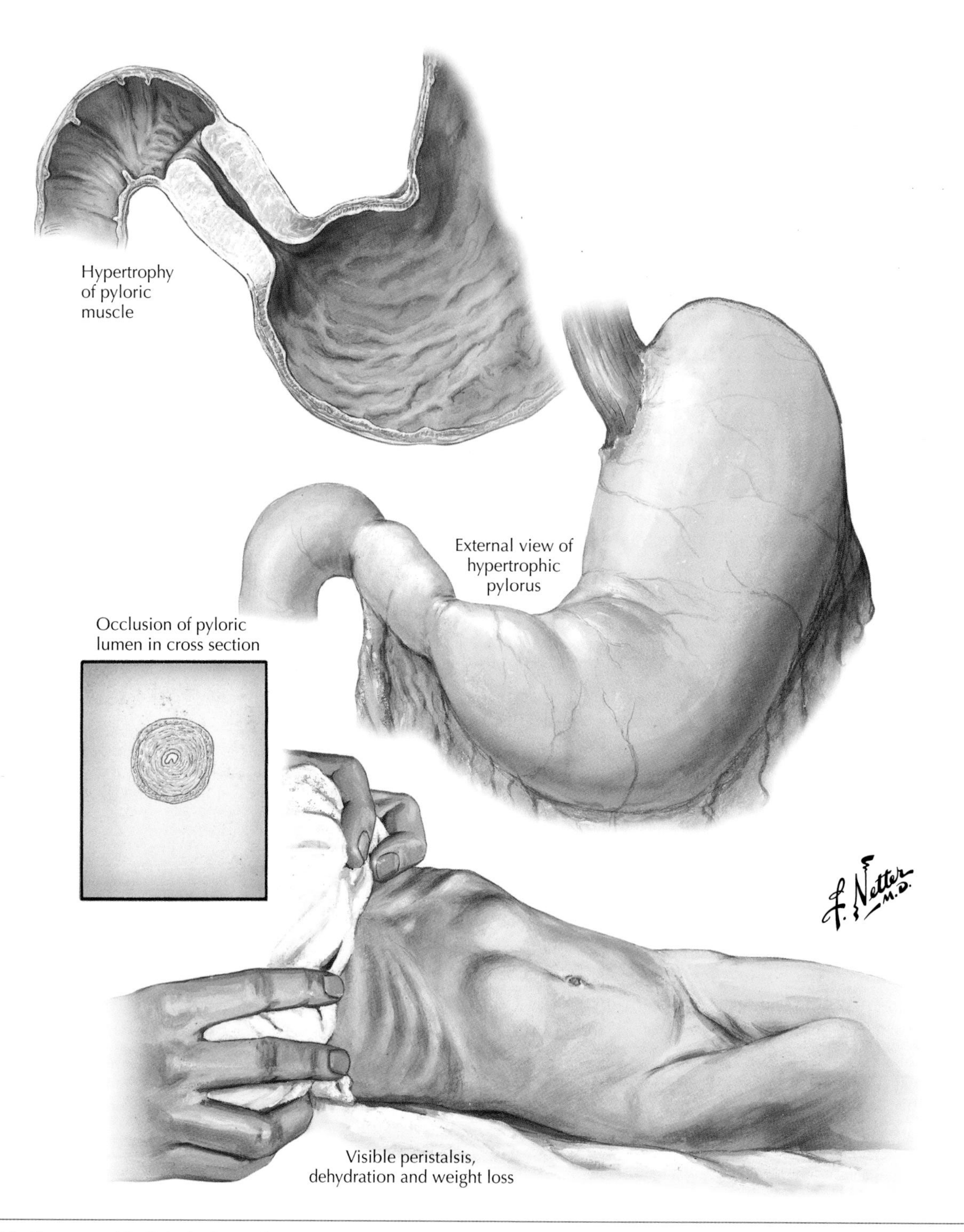

Figure 50-1 *Hypertrophic Pyloric Stenosis.*

Diverticula of the Stomach and Gastroduodenal Prolapse

51

Martin H. Floch

Gastric diverticula are rare and are found in 0.02% of autopsy specimens. Almost all are located on the posterior wall of the cardia and to the left of the esophagus (**Fig. 51-1**). They are thought to be congenital but occur at the structural weakness of the longitudinal muscles on the posterior surface. Usually, the diverticula contain all layers of the muscle wall and are 2 to 3 cm long and 1.2 cm in diameter. Openings are wide, permit free communication with gastric contents, and may be seen endoscopically. Gastric diverticula are best visualized on a retroflexion view. On barium radiography, they can be missed when the stomach is distended but often are seen on the lesser curvature, and they fill and empty regularly.

CLINICAL PICTURE

Diverticula of the stomach are asymptomatic. However, omplications have been reported and resulted in resection. Laparoscopic techniques are used effectively to resect the diverticula.

TREATMENT AND PROGNOSIS

No treatment is needed for diverticula unless the infrequent complication occurs. When bleeding, perforation from the manipulation, or the rare associated malignancy occurs, resection is performed laparoscopically.

Small and most other diverticula are asymptomatic, and prognosis is excellent.

GASTRODUODENAL PROLAPSE

Prolapse of the gastric mucosa into the duodenum probably results from extreme mobility of the antral mucosa and submucosa. The mucosa of the antrum, which normally is thicker than the mucosa of other parts of the stomach and sometimes assumes a cushionlike quality, is pushed through the pyloric ring to lie like a turned-back cuff of a sleeve within the duodenum (see Fig. 51-1). Although a fully developed prolapse is rare, partial prolapse is common but of little or no clinical significance.

Gastroduodenal prolapse is most often a radiologic curiosity, and the duodenal bulb can appear to be filled with a tuberous mass with irregular contours. Diagnosis is easy to make. Occasionally, however, it is difficult to differentiate a prolapse from a polyp or an acute ulcer from marked mucosal edema of the surrounding area. Endoscopists rarely report gastroduodenal prolapse.

The literature contains a report of a rare episode of strangulation of the mucosa with subsequent signs of pyloric obstruction or gastrointestinal bleeding that required surgical correction.

ADDITIONAL RESOURCES

Dickenson RJ, Freeman AH: Gastric diverticula: radiologic and endoscopic features in six patients, *Gut* 27:954-957, 1986.

Fine A: Laparoscopic resection of a large proximal gastric diverticulum, *Gastrointest Endosc* 48:93-95, 1998.

Fork FT, Toth E, Lindstrom C: Early gastric cancer in a fundic diverticulum, *Endoscopy* 30:S2, 1998.

Kim SH, Lee SW, Choi WJ, et al: Laparoscopic resection of gastric diverticulum, *J Laparoendosc Surg Tech* 9:87-91, 1999.

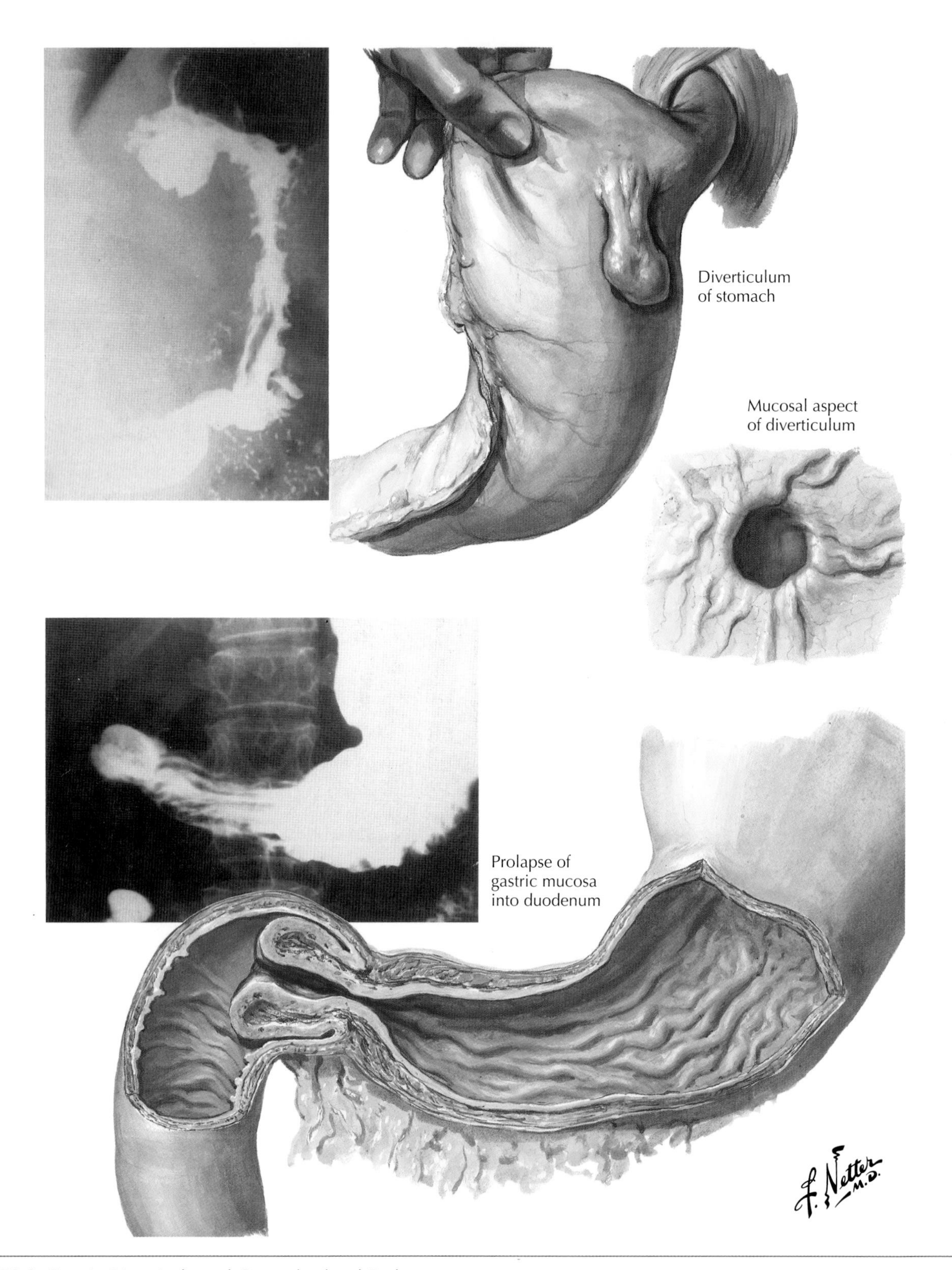

Figure 51-1 *Gastric Diverticula and Gastroduodenal Prolapse.*

Diverticula of the Duodenum

52

Martin H. Floch

A saccular "true" diverticulum can originate from any part of the duodenum (**Fig. 52-1**). It is rare in the first part and usually develops in the second part in the region of the ampulla of Vater. Diverticula have been reported in approximately 6% of barium studies but in as many as 27% of endoscopy studies and in 23% of autopsy evaluations. They have been noted close to the ampulla, and in some cases, the ampulla enters the diverticulum.

EXTRALUMINAL DIVERTICULA

Extraluminal duodenal diverticula are common with an interesting etiology, but debate is ongoing concerning congenital weakness in the duodenal wall and increased internal pressure. In rare cases, diverticula may be multiple. They usually develop on the inner or concave border of the duodenal curve and rarely on the outer border.

Clinical Picture

Approximately 10% of patients with extraluminal diverticula have symptoms. Abdominal discomfort may result when the diverticulum becomes inflamed, particularly from prolonged retention of duodenal content. The resultant diverticulitis can cause pain that radiates the epigastrium or back. Pancreatitis may occur when the ampulla is involved. Diverticula on the lateral wall have been reported to perforate (see Section VII).

Although there is a high incidence of extraluminal diverticula, most patients are asymptomatic. When diverticula are multiple, they can be associated with a malabsorption or bacterial overgrowth syndrome (see Section IV).

Diagnosis

Diagnosis of extraluminal duodenal diverticulum is easily made on barium study or endoscopy. Computed tomography may also make the diagnosis. A simple x-ray film of the abdomen may reveal an air-fluid level in the area of the duodenal sweep that is explained by a diverticulum. When the diverticulum is associated with pancreatic disease, endoscopic retrograde cholangiopancreatography (ERCP) is necessary for full evaluation (see Section VII).

Treatment and Management

Bleeding from the extraluminal diverticulum may be treated endoscopically, but in rare situations, surgical intervention may be necessary. In associated pancreatitis, the ampulla must be evaluated using ERCP, followed by appropriate intervention. Surgery is difficult; pancreatic or biliary surgical intervention may be needed. Therefore, medical and endoscopic treatment is preferred for these patients. Surgery should be performed only in emergencies and only under controlled conditions.

Prognosis

Complications rarely occur, and duodenal surgery is rarely necessary. The prognosis for extraluminal duodenal diverticula is usually excellent unless a severe complication necessitates surgery.

INTRALUMINAL DUODENAL DIVERTICULA

Unlike the prevalent extraluminal type, intraluminal duodenal diverticula are rare. They are congenital abnormalities in which the diverticulum develops within the duodenal wall; occasionally, clinical problems surface in adulthood. They may cause obstruction in the duodenum, with loculation of food particles, and are reportedly associated with pancreatitis.

When patients with intraluminal diverticula present with clinical syndromes, intervention is usually required. Surgical and endoscopic techniques have been successful in opening the intraluminal wall so that there is free passage through the duodenum.

ADDITIONAL RESOURCES

Goelho J, Sousa GS, Lobo DN: Laparoscopic treatment of duodenal diverticula, *Surg Laparosc Endosc* 9:74-77, 1999.

Gore RM, Ghahremani GG, Kirsch MD: Diverticulitis of the duodenum: clinical and radiological manifestations of seven cases, *Am J Gastroenterol* 86:981-985, 1991.

Lobo DN, Balfour TW, Iftikhar SY, et al: Periampullary diverticula and pancreaticobiliary disease, *Br J Surg* 86:588-597, 1999.

Lotveit T, Skar V, Osnes M: Juxtapapillary duodenal diverticula, *Endoscopy* 20:175-178, 1988.

Uomo G, Manes G, Ragozzino A, et al: Periampullary extraluminal duodenal diverticula and acute pancreatitis: estimated etiological association, *Am J Gastroenterol* 91:1186-1188, 1996.

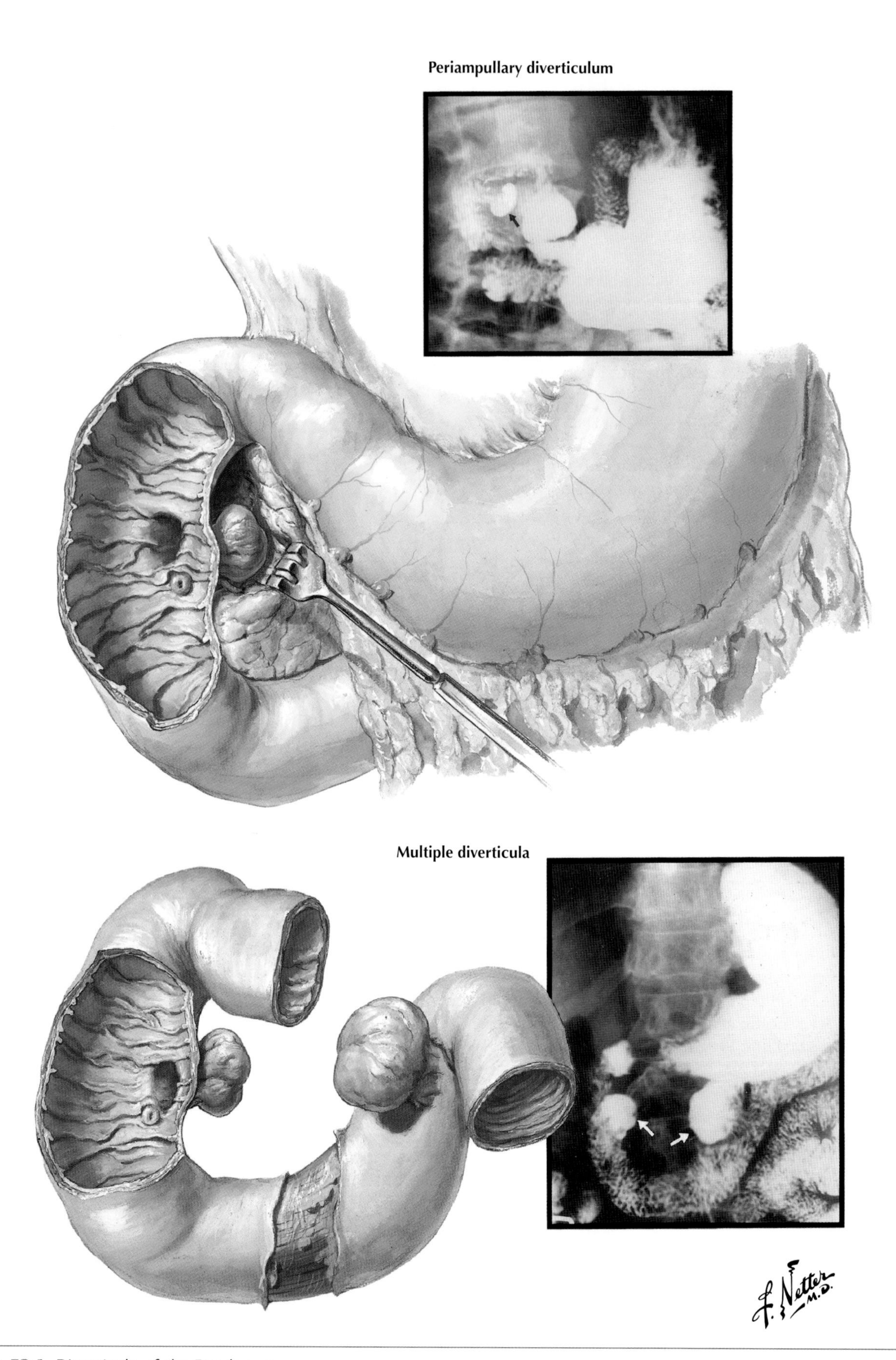

Figure 52-1 *Diverticula of the Duodenum.*

Dyspepsia, Functional Dyspepsia, and Nonulcer Dyspepsia

53

Martin H. Floch

Dyspepsia is pain or discomfort centered in the upper abdomen. Associated disease may cause the symptom. As a *functional* disorder, the term *dyspepsia* is used when the discomfort or pain is chronic, lasts at least 12 weeks during the preceding 12 months, and is accompanied by no evidence of biochemical, metabolic, or organic disease.

Dyspepsia is common. Approximately 25% of adults experience such discomfort, but only 5% seek medical attention. Fewer than half the patients with this type of centered epigastric discomfort have any associated organic disease. Dyspepsia of no organic cause is called *functional dyspepsia*. Therefore, the cause of dyspepsia may be true organic disease, which, when treated, cures the dyspepsia. Without some identifiable pathophysiology, it becomes functional dyspepsia. Patients with associated abnormality usually are also classified as having functional dyspepsia when the abnormality is considered irrelevant.

CLINICAL PICTURE

Patients of all ages have epigastric discomfort. Medical attention is usually sought after the discomfort becomes chronic. Often, the patient has some initial therapy and evaluation, but the treatment is unsuccessful, and it becomes apparent that the discomfort will persist. There may be associated early satiety and loss of appetite, a feeling of fullness, bloating in the upper abdomen, mild nausea, and sometimes even retching without vomiting of food. The degree of associated symptoms varies greatly.

DIAGNOSIS

Because of the discomfort and chronicity of dyspepsia, a workup must ensue. Screening for gastric lesions is essential. Upper endoscopy esophagogastroduodenoscopy (EGD) is preferred, and during the procedure the patient should be evaluated for *Helicobacter pylori*. Because interpretation at endoscopy can vary, it is wise to perform mucosal biopsies of the esophagus and the stomach. Other pertinent evaluation includes a study for gastric emptying, especially if any food is retained in the stomach. If the EGD and biopsy results are negative, ultrasonography should be performed to rule out gallbladder, liver, and pancreatic disease. Depending on the findings, computed tomography may be necessary to rule out gross lesions in the pancreas. Full serum screening should be performed to rule out liver or metabolic disease.

As indicated, 50% of patients will have definite disease, and thus their dyspepsia is caused by disease and is *not* functional. Patients with an identified physiologic abnormality should be treated and the abnormality evaluated. Unfortunately, many of these symptoms persist, and it becomes clear that the abnormality is not the cause. These patients then fall into the category of functional dyspepsia.

Because *H. pylori* is ubiquitous in most societies and because it can cause peptic disease, the recommendation is to treat it before making a diagnosis of functional dyspepsia. Data and studies clearly reveal that as many as 50% of patients may be cured of symptoms after the *H. pylori* infection is treated. However, symptoms persist in many patients, who then fall into the category of functional dyspepsia.

The accompanying algorithm (**Fig. 53-1**) outlines the diagnosis and management of functional dyspepsia.

TREATMENT AND MANAGEMENT

If the cause of the dyspepsia is found and treated, the treatment of that particular entity solves the problem. However, if the diagnosis is functional dyspepsia, the treatment becomes challenging and includes the following:

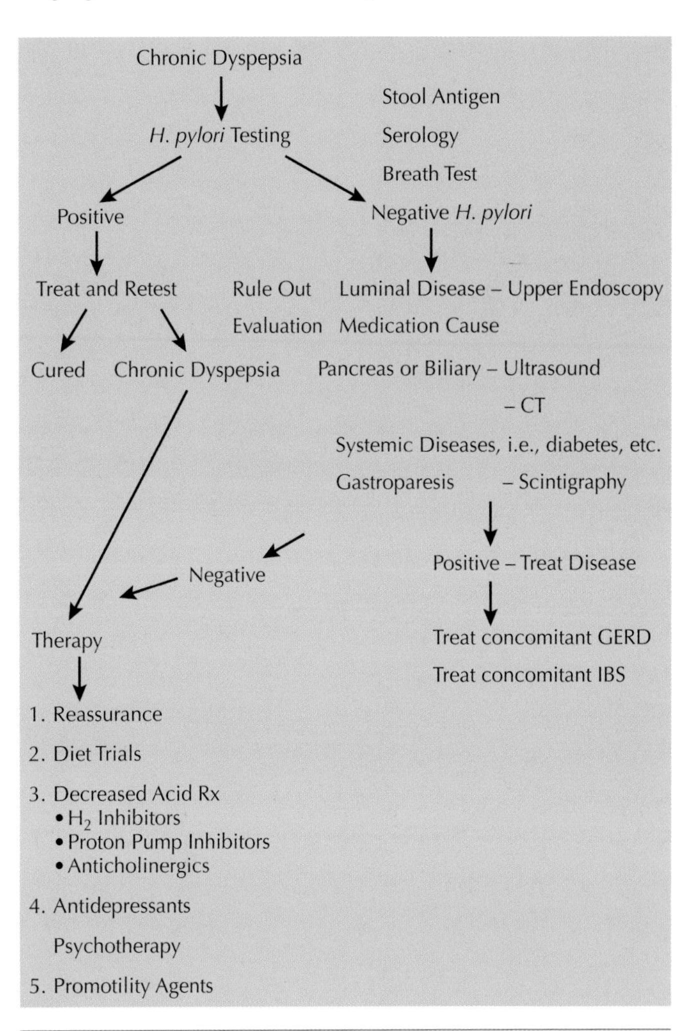

Figure 53-1 *Diagnosis and Treatment of Functional Dyspepsia.*

- *Strong reassurance.* Make sure the patient with dyspepsia understands that he or she has a functional disorder, probably with visceral hypersensitivity. Carefully continue to evaluate dietary stress factors that may aggravate the symptoms, such as caffeine, coffee, alcohol, and spices. Drug therapy may include trials of antisecretory, promotility, and antidepressant agents.
- *Antisecretory agents.* Some patients report some relief using H_2-receptor antagonists or proton pump inhibitors (PPIs).
- *Promotility agents.* Domperidone is available only in some parts of the world. Metoclopramide is more readily available, but long-term use is generally discouraged. These drugs can be effective intermittently in some patients.
- *Antidepressant therapy.* Many dyspeptic patients experience symptom relief once an appropriate antidepressant is used in adequate dosage.

Treating functional dyspepsia is challenging. It requires reassurance and working closely with patients to maintain their confidence.

COURSE AND PROGNOSIS

Patients with dyspepsia have an excellent prognosis, but the disease course varies with periods of increasing and decreasing symptoms. Frequently, patients are satisfied with one or the other forms of therapy: H_2 inhibitor, PPI, or antidepressant. Many require more intensive work and even psychotherapy to manage severe symptoms.

Concomitant disease often includes gastroesophageal reflux disease (GERD), as confirmed on upper endoscopy and other studies. Treating GERD is often helpful. Similarly, some dyspeptic patients have irritable bowel syndrome (IBS) accompanied by recurrent diarrhea or constipation, and treatment for those symptoms often helps their epigastric distress. It is essential that concomitant disease, such as GERD or IBS, be treated simultaneously. This often helps in dyspeptic patients' long-term management.

ADDITIONAL RESOURCES

Camilleri M: Functional dyspepsia: mechanisms of symptom generation and appropriate management of patients, *Gastroenterol Clin North Am* 36:649-664, 2007.

Moayyedi P, Soo S, Deeks J, et al: Eradication of *Helicobacter pylori* for non-ulcer dyspepsia, *Cochrane Database Syst Rev* 1:CD002096, 2003.

Talley NJ: The role of endoscopy in dyspepsia (clinical update), *Am Soc Gastroenterointest Endosc* 15:1-4, 2007.

Talley NJ, Stanghellini V, Heading RC, et al: Functional gastrointestinal disorders. In Drossman DA: *The functional gastrointestinal disorders*, McLean, Va, 2000, Degnon Associates, pp 302-327.

Talley NJ, Vakil N, Moayyedi P: AGA Technical Review on the evaluation of dyspepsia, *Gastroenterology* 129:1756-1780, 2005.

Helicobacter pylori Infection

54

Martin H. Floch

Helicobacter pylori is a gram-negative, spiral, flagellated bacterium that inhabits the mucous layer of the stomach. Warren and Marshall first described *H. pylori* as a pathogen in humans and clearly documented and correlated the organism's association with gastritis and peptic ulceration.

The prevalence of *H. pylori* varies greatly. Approximately 40% of persons in developed countries are affected, and as many as 85% are affected in underdeveloped countries. In all areas, prevalence is associated with low socioeconomic status and advanced age. Most people remain infected for life unless they are treated. Marriage does not appear to be a strong risk factor for acquiring the infection.

Data on how the organism is acquired are controversial, although crowded living conditions and poor hygiene are associated with higher infection rates. Transmission does appear to be based on person-to-person spread (**Fig. 54-1**). However, the exact mode of transmission remains unclear. *Helicobacter heilmanii* colonizes both animals and humans. Although it may be pathogenic, *H. heilmanii* has not been shown to be as prevalent a pathogen as *H. pylori.*

CLINICAL PICTURE

It is now clear that *H. pylori* is a major risk factor for gastritis, peptic ulcer, gastric adenocarcinoma, and gastric lymphoma. Consequently, patients may have epigastric pain and ulceration, bleeding from gastritis or ulceration, or pain, nausea, vomiting, and weight loss from malignancy. It is not clear why *H. pylori* rarely causes diarrhea. Anemia from chronic blood loss may be the only symptom in those who feel little pain. The picture of chronic dyspepsia may unfold over the years.

Helicobacter pylori has been known to cause acute gastritis. In these patients, it may actually cause severe acute achlorhydria, which appears to be self-limiting. Not only can prolonged disease cause ulceration, but it also has been known to cause gastric atrophy and, in association with any metaplasia, appears to result in a high risk for adenocarcinoma.

DIAGNOSIS

The diagnosis of *H. pylori* infection can be made using histologic, serologic, breath, or stool testing.

Histology

Histologic examination is made through endoscopy with biopsy material of the gastric mucosa and appropriate histologic and staining evaluation. Because it is difficult to grow *H. pylori* in culture, cultures are no longer used for the diagnosis of gastric aspirants or gastric tissue. Instead, the rapid urease test (CLO test) can be used by placing gastric biopsy material into a urease medium that changes color when urease from the bacteria metabolizes the urea.

Serology

Serologic tests are sensitive and as specific as histologic biopsy evaluation. Many have been adopted for whole-blood, rapid use in the office. However, only immunoglobulin G (IgG) antibody is reliable. IgA and IgM antibodies are unreliable. Serologic tests are useful because they establish that the patient has had *H. pylori* infection. However, controversy surrounds the rapidity with which the antibody disappears after treatment, and consequently, serology is not a good test to determine whether treatment is effective.

Breath Test

The urea breath test with either carbon 13 or 14 (C_{13} or C_{14}) is accurate. The subject ingests the carbon label, and the test determines whether the carbon is freed by urease activity from the bacteria in the stomach and absorbed, then measures it in expired breath.

Stool Test

The stool antigen test is the latest noninvasive method for diagnosing *H. pylori* infection. Evaluations are as accurate as histologic methods. The stool test can be used easily for monitoring the effectiveness of therapy.

TREATMENT AND MANAGEMENT

Once a diagnosis of a *H. pylori* infection is made, treatment must follow. Many treatment regimens use combinations of bismuth sulfate and numerous antibiotics, including metronidazole, tetracycline, amoxicillin, and clarithromycin. The course varies from 7 to 14 days. The literature attests to the effectiveness of the different regimens, which can be categorized into three types of therapy; double, triple, and quadruple.

Most often used and recommended is *triple therapy*, which includes a proton pump inhibitor (PPI) twice daily, plus amoxicillin (1000 mg) twice daily, plus clarithromycin (500 mg) or metronidazole (500 mg) twice daily. Triple or quadruple therapy using bismuth is equally effective. Triple therapy administers two bismuth tablets four times daily plus tetracycline (500 mg) four times daily and metronidazole (250 mg) three times daily; quadruple therapy uses bismuth and includes a PPI twice daily in addition to the two antibiotics. In trials, clinicians have used variations of these antibiotics with reported success. If a patient develops an allergy or intolerance to one of the drugs, other options are available.

With resistant cases always present and unsuccessful therapy reported in the 20% range, new regimens are constantly being tested. Sequential therapy is new, but initial analysis reveals it may be effective and possibly better than previous therapies. The initial effective regimen consisted of 5 days of amoxicillin

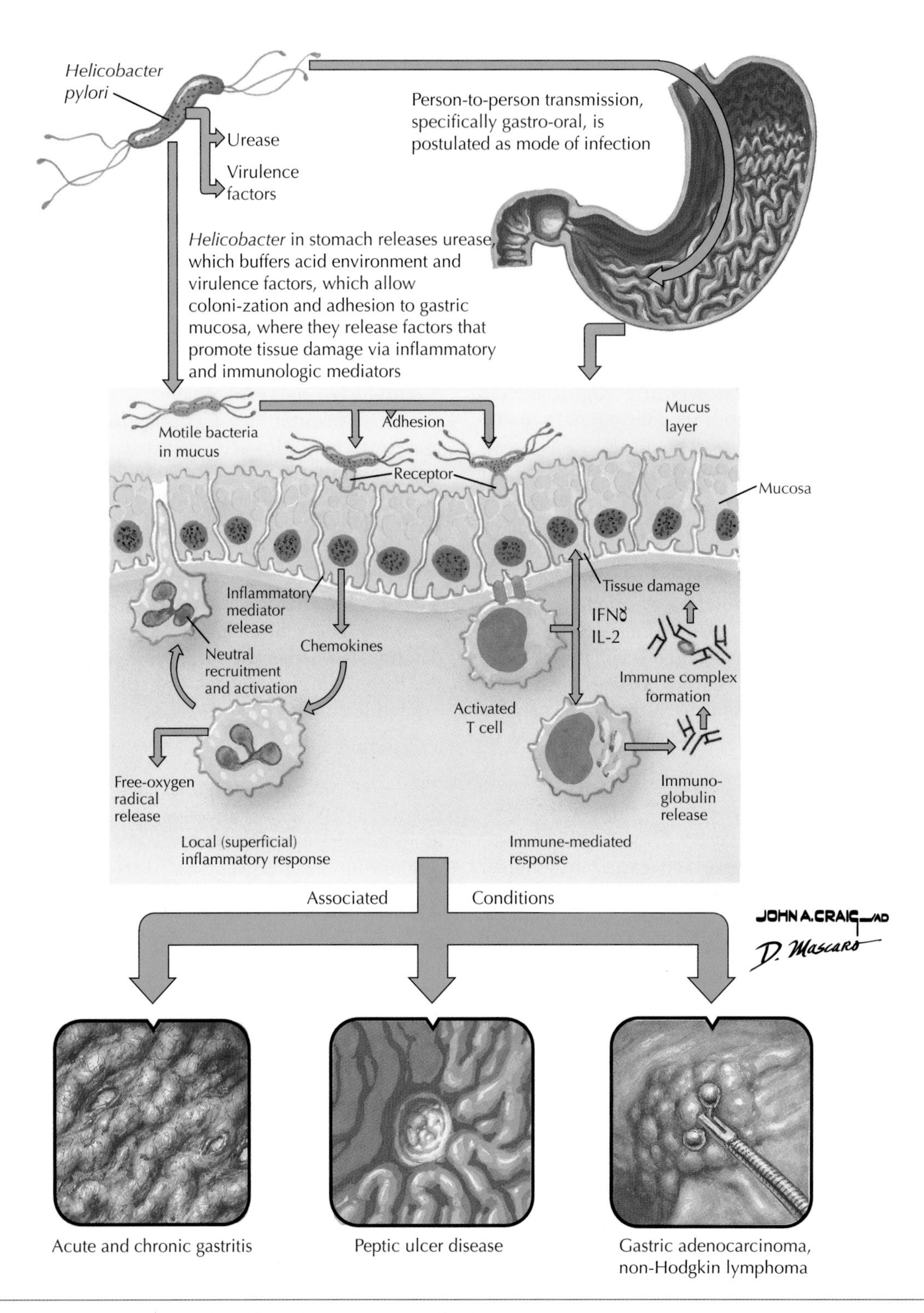

Figure 54-1 *Etiology and Pathogenesis of* Helicobacter pylori *Infection.*

Table 54-1 Treatment and Retreatment Options/ Preferences for *Helicobacter pylori* Infection

Option	**Treatment**
1	P-A-C
2	P-M-C
3	O-B-M-T
4	O-B-F-T
5	P-A-R
6	P3-A3
7	P3-A3-CF
Clinical Program*	**Treatment Preference**
SVZ	1 ↔ 2 → 3
LL	6 → 5 → 7
DG	3 → 4
AA	1 ↔ 2 → 3 → 6
BM	1 ↔ 2 → 7
Treatment Code	**Drug/Dose**
A	Amoxicillin, 1 g bid
A3	Amoxicillin, 1 g tid
M	Metronidazole, 500 mg tid
C	Clarithromycin, 500 mg bid
P or O	Any PPI (omeprazole, 20 mg bid, or equivalent)
P3	Triple-dose PPI (omeprazole, 40 mg tid, or equivalent)
R	Rifabutin, 300 mg bid
F	Furazolidone, 100 mg qid
CF	Ciprofloxacin, 500 mg bid
B	Bismuth citrate, 120 mg qid (De Nol) *or* Bismuth subsalicylate, 250 mg qid (2 Pepto-Bismol tablets qid)
T	Tetracycline, 500 mg qid

Modified from Megraud F, Marshall BJ: *Gastroenterol Clin North Am* 29:759-773, 2000.
*Clinical programs at five academic centers.
PPI, Proton pump inhibitor; *bid,* twice daily; *tid,* three times daily; *qid,* four times daily.

(1 g) plus a PPI twice daily, followed by 5 days of triple therapy with a PPI, clarithromycin (500 mg), and tinidazole (500 mg), all twice daily. These sequential therapies are now in meta-analysis and appear to be more effective than 14 days of triple therapy.

If a patient has carcinoma, treatment of the malignancy is paramount. However, when a lymphoma or mucosa-associated lymphoid tissue (MALT) lesion develops, remission of the lymphoma has been reported if *H. pylori* has been eradicated. Those lesions must be treated in a specific manner (see Chapter 63).

Some organisms are resistant, and failure occurs. It must be emphasized that *H. pylori* requires acid for reproduction; thus, most microbiologists and clinicians believe acid suppression should be part of therapy. Failed therapy occurs in 5% to 10% of treated patients. Failure may result from resistant strains or may be associated with smoking or dense colonization with the *cag*-negative strain. Repeat therapy is often successful, but resistant strains do grow.

Table 54-1 demonstrates the many options for treating *H. pylori.* Option 1 is the most recommended as a beginning, but the other options vary worldwide, with options 3, 6, and 7 often used in repeat therapies. These are the older regimens but are still widely used; newer antibiotics such as levoquin and rifabutin are also used. Most agree that at least 7 days of therapy is required, although some are using shorter courses and others require 14 days.

COURSE AND PROGNOSIS

Effective therapy for *H. pylori* infection is rewarding when patients with chronic gastritis or peptic ulceration are cured of their symptoms and disease. Initial therapy is effective in 70% to 95% of patients, depending on the patient and the regimen. When a strain is resistant, treatment of the associated disease must be continued until rescue therapy can be implemented to eradicate the organism. Attempts to eradicate *H. pylori* should continue, even if it takes years with different antibiotics and extended treatment; to do otherwise is now considered a risk factor for adenocarcinoma.

If a patient has an associated malignancy, course and prognosis depend on the extent of the cancer. *H. pylori* definitely causes MALT) lesions, which may respond to, and may be cured by, *H. pylori* eradication (see Chapter 63).

ADDITIONAL RESOURCES

Blaser MJ: *Helicobacter pylori* and other gastric *Helicobacter* species. In Mandell GL, Bennett JE, Dolin R, editors: *Principles and practice of infectious diseases*, ed 6, Philadelphia, 2005, Churchill Livingstone–Elsevier, pp 2557-2566.

Chey WD, Wong CY: American College of Gastroenterology guideline on the management of *Helicobacter pylori* infection, *Am J Gastroenterol* 102:1808-1825, 2007.

Gisbert JP, Gisbert JL, Marcos S, et al: Empirical rescue therapy after *Helicobacter pylori* treatment failure: a 10-year single-centre study of 500 patients, *Aliment Pharmacol Ther* 27:346-354, 2008.

Graham DY, Malaty HM, Evans DG, et al: Epidemiology of *Helicobacter pylori* in an asymptomatic population in the United States: effective age, race, and socioeconomic status, *Gastroenterology* 100:1495-1501, 1991.

Jafri NS, Hornung CA, Howden CW: Meta-analysis: sequential therapy appears superior to standard therapy for *Helicobacter pylori* infection in patients naïve to treatment, *Ann Intern Med* 148:923-931, 2008.

Megraud F, Marshall BJ: How to treat *Helicobacter pylori:* first-line, second-line, and future therapies, *Gastroenterol Clin North Am* 29:759-773, 2000.

Suerbaum S, Michetti P: *Helicobacter pylori* infection, *N Engl J Med* 347: 1175-1186, 2002.

Warren J, Marshall B: Unidentified curved bacilli in gastric epithelium and acute and chronic gastritis, *Lancet* 1:1273-1275, 1983.

Zullo A, De Francesco V, Hassen C, et al: The sequential therapy regimen for *Helicobacter pylori* eradication: a pooled-data analysis, *Gut* 56:1353-1357, 2007.

Gastritis

Martin H. Floch

55

Gastritis is inflammation of the gastric mucosa, submucosa, or muscularis (**Fig. 55-1**). A gastritis classification proposed in 1991 by an international convention in Sydney, Australia, has not gained support in the past two decades, reflecting the clinical confusion in this area. However, the basic pathologic entity of "inflammation in the mucosa" is considered gastritis. It may be acute or chronic, or it may result in atrophy. Each condition is associated with a clear endoscopic clinical picture.

Gastritis may be chronic, may be associated with disease (e.g., *Helicobacter pylori*, autoimmune), or may be a progressive atrophic form. It may be associated with all forms of infectious disease (viral, bacterial, parasitic, fungal), or it may be granulomatous and associated with chronic disease (e.g., Crohn, tumors). Gastritis may be erosive (often referred to as *reactive*) because of foreign agents such as aspirin, nonsteroidal antiinflammatory drugs (NSAIDs), bile reflux, alcohol, and caffeine. It may be classified as rare entities such as collagenous, lymphocytic, and eosinophilic gastritis (referred to as *distinctive*). A hypertrophic form is known as *Ménétrier disease*, and a post–gastric surgery form is known as *gastritis cystica profunda*. A form now appearing in patients who have had grafts or transplants is known as *graft-versus-host disease*, in which the stomach and other parts of the gastrointestinal (GI) tract are involved.

Recent attempts to classify gastritis according to topography, morphology, and etiology have not changed clinical practice. In the most recent consensus, gastritis was categorized into *nonatrophic*, *atrophic*, and special forms. *H. pylori* gastritis is classified as nonatrophic.

CLINICAL PICTURE

The clinical picture of gastritis can be specific in that patients have abdominal pain, nausea, and anorexia. Patients may report bloating or a burning discomfort in the epigastrium. In severe acute gastritis, patients may vomit and have food intolerance. In chronic gastritis, patients may have anorexia with weight loss. Many academicians believe that gastritis can exist without symptoms; therefore, symptoms often will not be attributed to the mucosal inflammation. Nevertheless, if a cause such as *H. pylori* or an associated disease can be identified and treated, symptoms can be resolved. Gastritis is often part of another disease process.

DIAGNOSIS

The history of onset of symptoms is important. When symptoms are acute and the gastritis is associated with infection, symptoms usually subside within days, and evaluation is unnecessary. The use of NSAIDs must be evaluated. However, when symptoms persist longer than 7 to 14 days, an investigation is necessary. The standard evaluation includes upper GI endoscopy with biopsy to determine the disease process.

When atrophy is present, a test for parietal cell antibodies is indicated. Serum gastrin levels may be elevated if atrophy is diffuse. Evaluation for vitamin B_{12} is necessary.

The most common cause of gastritis is *H. pylori* (see Chapter 54). Finding the organism through endoscopy and biopsy confirms the diagnosis. When present, other organisms can be identified on biopsy, but careful histologic staining must be done to identify chronic infections, such as tuberculosis and fungi. Anisakiasis can be diagnosed on endoscopy; with the increased ingestion of raw fish, *Anisakis* infection should be considered in patients with an appropriate history. Other parasites also may be identified in the stomach.

TREATMENT AND MANAGEMENT

When an infectious agent is identified, such as *H. pylori* or any parasite, treatment for that infectious agent cures the gastritis. Autoimmune diseases and nonspecific gastric diseases are treated symptomatically.

When another disease involving the gastric mucosa is identified, such as Crohn disease or sarcoid, it must be treated. Erosive gastritis is treated by removing its cause, whether alcohol, drugs, or other agents.

During the healing phase of gastritis therapy, acidic and spice-containing foods could further irritate the mucosa and must be removed from the patient's diet. Neutralizing acid is also recommended because the mucosa has many breaks in its barrier and can be invaded by acid. Therefore, it is advisable to use acid suppression therapy (H_2 inhibitors, proton pump inhibitors [PPIs], antacids) as tolerated.

COURSE AND PROGNOSIS

The course of gastritis depends on the cause. It can be chronic, troublesome, and difficult to treat. Most acute forms resolve rapidly. An association with NSAID use must be considered and may be treated by changing the NSAID or adding a PPI to alleviate symptoms if the NSAID is essential therapy. Chronic forms are related to the natural history of an associated disease and must be treated by diet restrictions and antacid therapy. A true atrophic gastritis may be associated with vitamin B_{12} deficiency, which should be evaluated.

ADDITIONAL RESOURCES

Dixon MF, Genta RM, Yardley JH, et al: Classification grading of gastritis: the updated Sydney system, *Am J Surg Pathol* 20:1161-1181, 1996.

Graham DH, Genta RM, Dixon MF: *Gastritis*, Philadelphia, 1999, Lippincott Williams & Wilkins.

Lee EL, Feldman M: Gastritis and other gastropathies. In Feldman M, Friedman LS, Brandt LJ, editors: *Gastrointestinal and liver disease*, ed 8, Philadelphia, 2006, Saunders-Elsevier, pp 1067-1088.

Misiewicz J: The Sydney system: a new classification of gastritis, *J Gastroenterol Hepatol* 6:207-208, 1991.

Rotterdam H: Contributions of gastrointestinal biopsy to an understanding of gastrointestinal disease, *Am J Gastroenterol* 78:140, 1983.

Sipponen P: Update on the pathologic approach to the diagnosis of gastritis, gastric atrophy, and *Helicobacter pylori* and its sequelae, *J Clin Gastroenterol* 32:196-202, 2001.

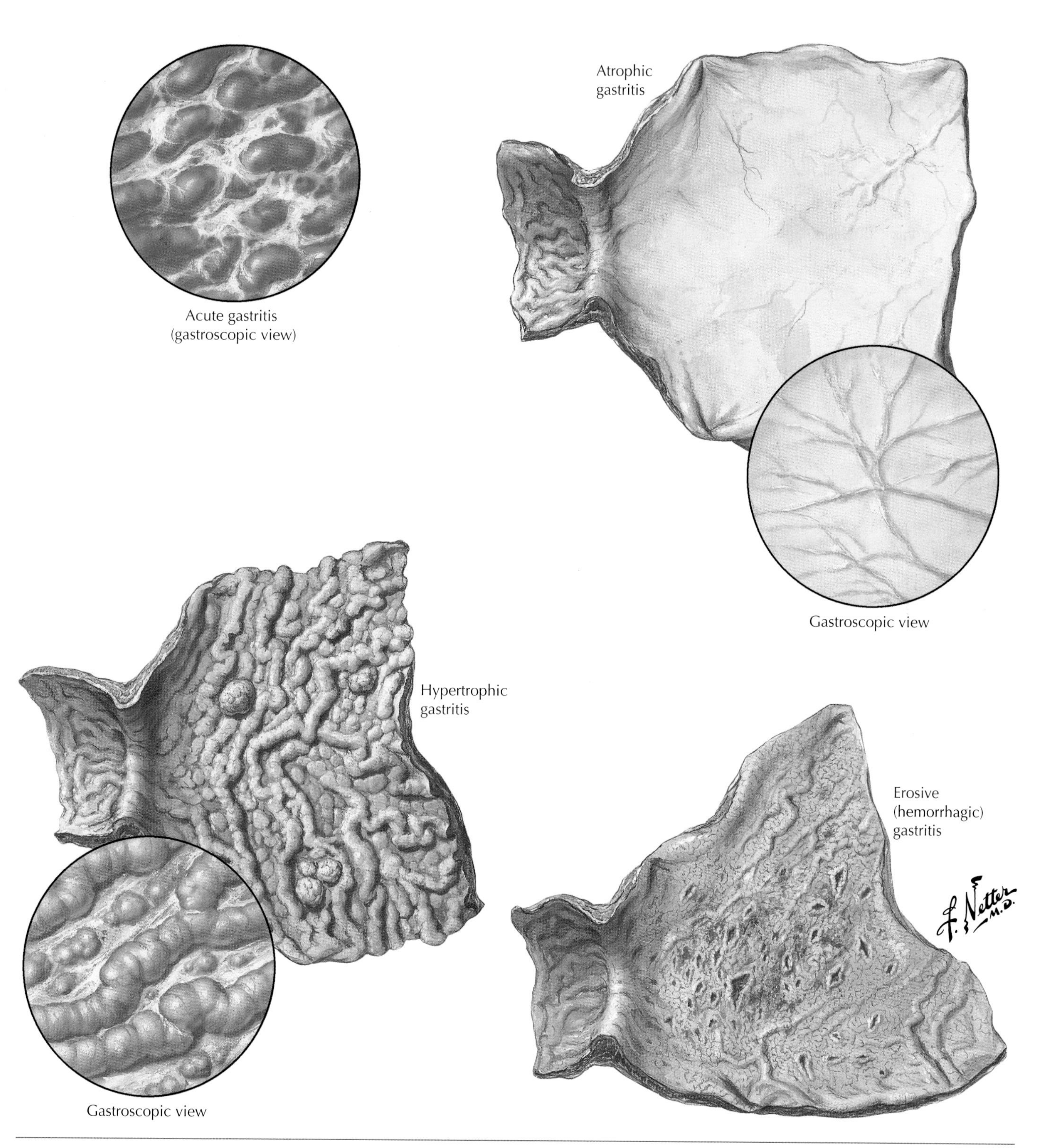

Figure 55-1 *Gastritis.*

Erosive Gastritis: Acute Gastric Ulcers

56

Martin H. Floch

Erosive gastritis consists of small, acute gastric ulcers that occur in the body or antrum of the stomach (**Fig. 56-1**). The cause may be an infectious organism (e.g., *Helicobacter pylori*, streptococci) or damage to the mucosa from numerous possible agents (e.g., nonsteroidal antiinflammatory drugs [NSAIDs]). Aspirin is a noted culprit. The ulcers may range in size from a few millimeters to a centimeter. They often appear as multiple lesions.

CLINICAL PICTURE

Clinical findings in patients with acute gastric ulcers vary greatly. If associated with a simple streptococcal infection, the ulcer may last for a few days and cause nausea, vomiting, and mild abdominal pain. The ulcer also may be asymptomatic or may cause severe bleeding and hematemesis. If the symptoms are short term, endoscopy is not necessary; if bleeding occurs or symptoms persist, endoscopy confirms the diagnosis.

DIAGNOSIS

Erosive ulcers are small. *H. pylori* must be ruled out as a cause (see Chapter 54). If serologic, breath test, stool antigen, or histologic findings are negative, the cause must be determined by history. Microscopically, the erosions reveal acute and perhaps mild chronic inflammation in the superficial mucosa, which can extend down to the muscularis. Endoscopy may reveal very small ulcers. A brown or a black spot indicates recent bleeding. The ulcers rarely may be chronic. The greatest danger from these ulcers is massive bleeding.

It is surprising that these small ulcers can cause great pain or bleed excessively. If they are duodenal and caused by *H. pylori*, therapy for the infection relieves the symptoms. If they result from NSAID therapy, the ulcers can become chronic in patients who continue using NSAIDs to relieve osteoarthritic pain (see Chapters 42 and 55).

TREATMENT AND MANAGEMENT

For effective long-term management of erosive gastritis, it is best to determine the cause of the ulcers. Any patient in the acute state should receive a proton pump inhibitor (PPI). If the ulcer is bleeding, the PPI can be administered intravenously. If the erosions are caused by a drug or by an infectious agent, or if they are of peptic origin, they invariably heal. Follow-up endoscopy is usually not needed, but if symptoms persist or there is any sign of continued disease, such as prolonged nausea, occasional vomiting, or mild pain, repeat endoscopy is essential to rule out an underlying malignancy. If a drug is the cause, eliminating the drug is the only therapy required. If an infection is the cause, eliminating the infection will cure the small ulcers. If the cause is unknown, the patient may require long-term antacid therapy.

COURSE AND PROGNOSIS

The course of erosive gastritis is usually short, with an excellent prognosis. Bleeding stops with treatment, and symptoms subside. Rarely, the bleeding may be massive, and the diagnosis is usually associated with another gastropathy or ischemia; emergency surgery is needed.

ADDITIONAL RESOURCES

Graham DH, Genta RM, Dixon MF: *Gastritis*, Philadelphia, 1999, Lippincott Williams & Wilkins.

Lee EL, Feldman M: Gastritis and gastropathies. In Feldman M, Friedman LS, Brandt LJ, editors: *Gastrointestinal and liver disease*, ed 8, Philadelphia, 2006, Saunders-Elsevier, pp 1067-1088.

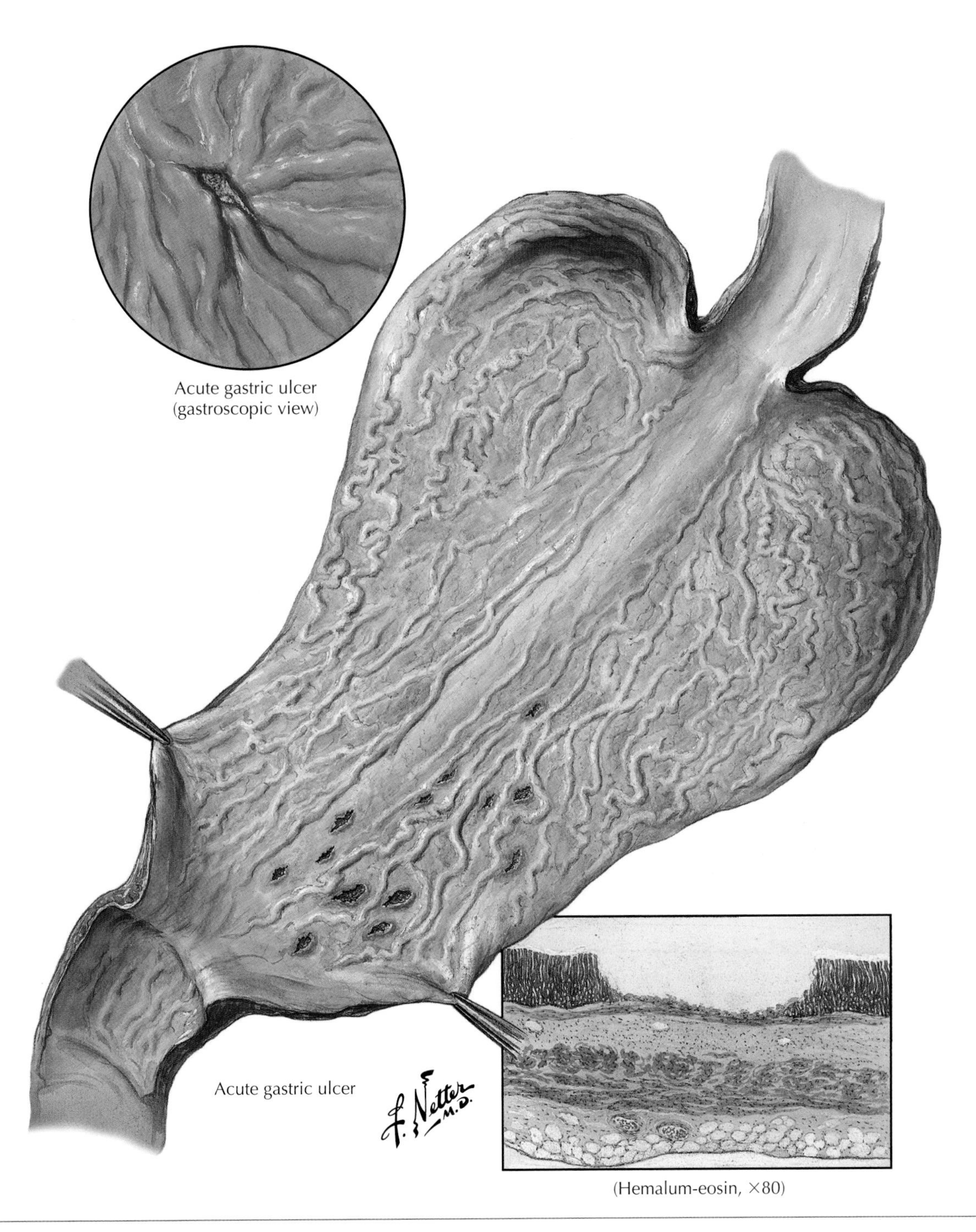

Figure 56-1 *Erosive Gastritis with Acute Gastric Ulcers.*

Peptic Ulcer Disease: Definition and Pathophysiology

57

Martin H. Floch

Peptic ulcer disease is a term used to refer to ulceration of the gastric or duodenal mucosa aggravated by penetration of the mucosal barrier by acid and pepsin (**Fig. 57-1**). The natural history of peptic ulcer disease was dramatically revised with the discovery of H_2 inhibition of acid secretion and then proton pump acid inhibition. The discovery that *Helicobacter pylori* is a major factor in all ulcer disease led to its treatment and to the cure of peptic ulcer when associated with *H. pylori* infection. The identification of iatrogenic causes (e.g., nonsteroidal anti-inflammatory drugs [NSAIDs]) further defined the nature and causes of peptic ulceration.

Before the advent of these findings, peptic ulcer had been considered an acute and chronic disease that required long-term diet, psychotherapy, and surgical therapy. These approaches have changed greatly, however, as have descriptions of the causes, ulcerations, and complications. Surgical removal has become a rarity, and the diagnosis of peptic ulcer disease is most often made endoscopically. Therefore, referring to "superficial" ulcers that do not penetrate is not clinically practical. Acute and chronic ulcers are currently difficult to define and no longer fit their historical definitions.

PATHOPHYSIOLOGY

The gastric and duodenal epithelium is protected by a mucous coat, which in turn is usually covered by an unstirred, bicarbonate-rich layer of water. Mucus and bicarbonate are secreted by gastric epithelial cells, as well as duodenal Brunner glands. Whenever acid and pepsin break these layers, cells may be injured. Minor injuries from irritants are usually rapidly healed. However, when the injury is prolonged by any of the causes elaborated here, ulceration may occur. Acid and pepsin overrun the defensive and regenerative processes to break down the mucosa.

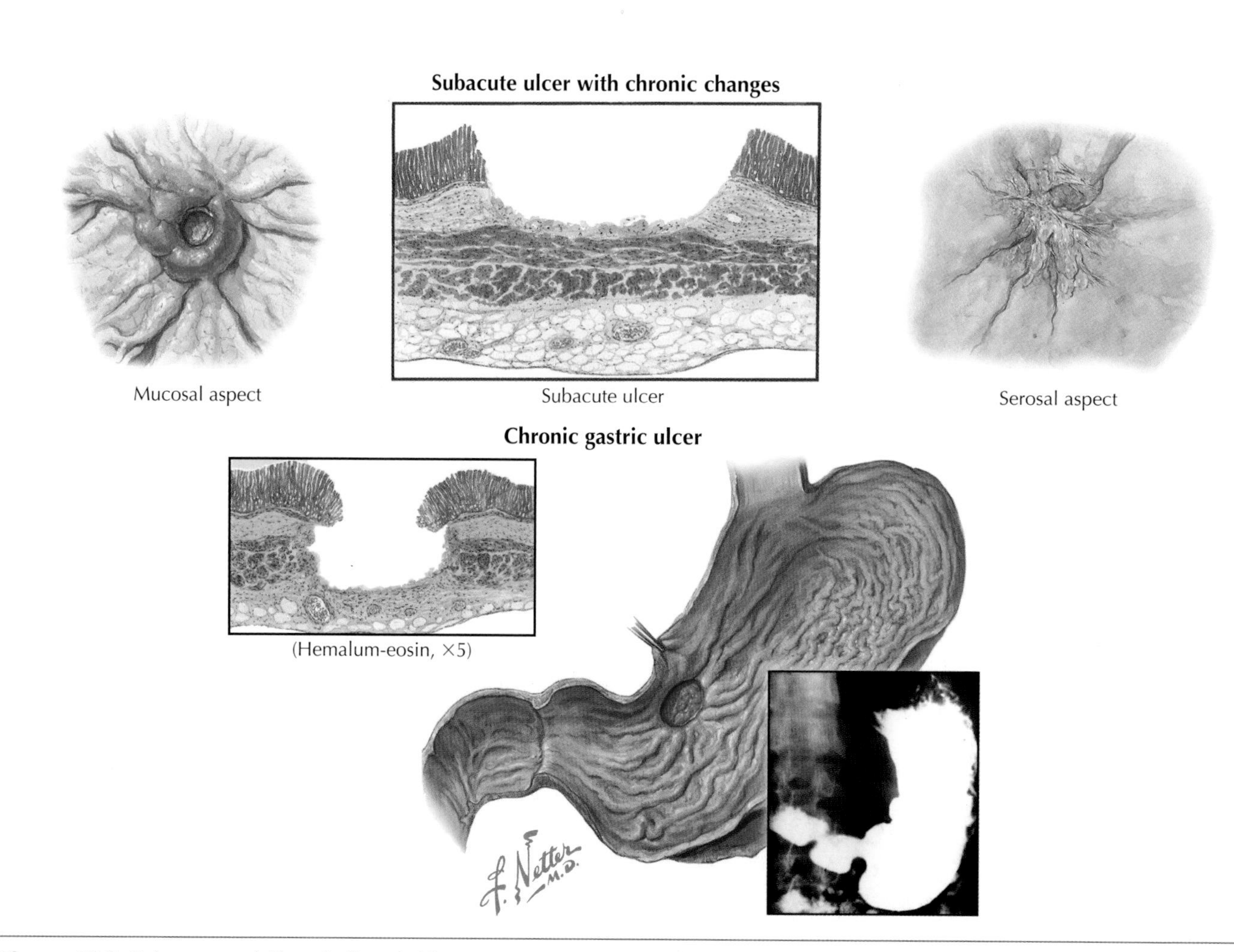

Figure 57-1 *Subacute and Chronic Gastric Ulcers.*

CAUSES AND ASSOCIATIONS

Causes and associations of peptic ulceration can be classified in four categories: infectious *(H. pylori)*, drug related (NSAIDs), hypersecretory, and miscellaneous.

Infectious Causes

Helicobacter pylori is probably the most common cause of peptic ulceration, depending on the prevalence in a particular area. As discussed in Chapter 54, *H. pylori* infection penetrates the mucous layer. It requires acid for its survival but can protect itself from acid destruction by an alkaline material that it secretes.

Because *H. pylori* affects approximately half the world's population, people with the bacterium are susceptible to peptic ulceration. Once the bacterium is present, its effects range from mild degrees of inflammation to ulceration. In the gastric mucosa, *H. pylori* causes inflammation, and in the duodenal mucosa, it causes metaplasia to gastric epithelium and then the resultant damage. Approximately 60% of patients with gastric ulcer and 80% of those with duodenal ulcer have chronic *H. pylori* infection. It is estimated that only 20% of those infected ever acquire peptic ulcer. Whether ulceration develops depends on several factors, including the strain of *H. pylori* and other risk factors in the host. Regardless, treating the *H. pylori* infection dramatically decreases the occurrence of ulcer.

Drug-Related Causes

Worldwide use of NSAIDs for the relief of pain and neurologic disorders has made these drugs the first or second most common cause of peptic ulceration, depending on the extent of their use. NSAIDs injure the gastrointestinal (GI) mucosa topically and systemically. Again, when the mucosal barrier is broken in the stomach or the duodenum, peptic ulceration develops. Attempts to decrease the harmful topical effects of NSAIDs have included applying enteric coating, but the systemic effects persist, resulting in simple, superficial petechiae becoming deep ulcerations. Careful endoscopic studies have revealed that mucosal petechiae or small erosions develop in 15% to 30% of patients who use NSAIDs. However, serious pain or bleeding is still relatively rare and is estimated to affect less than 1% of patients. Certain risk factors, including smoking, old age, and concomitant *H. pylori* infection, increase the chances of peptic ulceration or bleeding. It also appears that selective cyclooxygenase-2 (COX-2) inhibitors produce less damage than the standard COX-1 NSAIDs. Given all these facts, the physician still must consider NSAIDs as one of the most common causes of peptic ulceration.

Two other ulcerogenic drugs, alendronate and risedronate, are frequently used to treat osteoporosis. As newer and more potent drugs become available, clinicians must be aware that they are potential ulcerogenic agents.

Hypersecretory Causes

With regard to the third category, hypersecretory states, it is well known that patients with duodenal ulcer appear to produce higher amounts of gastric acid, but this fact is less important now that *H. pylori* is known to be a major cause of ulcer. *H. pylori* itself may stimulate acid production and may increase gastrin levels. Studies in acid production since the contribution of *H. pylori* became known are not readily available.

In *Zollinger-Ellison syndrome* (ZES), the gastrinoma secretes gastrin, and there is an associated proliferation of enterochromaffin-like (ECL) cells in the stomach, stimulating hypersecretion of acid (see Chapters 41 and 199). With this high level of acid secretion, the mucosal barrier becomes overwhelmed, and breaks occur in the gastric and duodenal mucosa to cause ulceration. Treating gastrinomas may simply require proton pump inhibitors (PPIs) in high doses, to combat hypersecretion, or chemotherapy, embolization, or surgical resection, depending on the patient and the extent of the lesion (see Section VII).

Another cause of acid hypersecretion is *systemic mastocytosis*, in which proliferating numbers of mast cells produce large amounts of histamine that affect gastric secretion and have systemic effects on the skin, liver, and bone marrow. In patients with systemic mastocytosis, treatment with H_1-receptor and H_2-receptor antagonists, anticholinergics, oral disodium chromoglycate, and even corticosteroids, with or without cyclophosphamide, may be helpful.

Massive resection of the small bowel in patients with short-bowel syndrome is often associated with hypergastrinemia and hypersecretion. These patients require selective therapy because of absorption problems. Also, antral G-cell hyperfunction syndrome may be confused with ZES and is usually treated medically.

Miscellaneous Causes

All clinicians encounter patients whose ulcers do not fit any of the categories just described. Since the discovery of *H. pylori*'s role, there is always suspicion that other infectious agents may cause chronic ulceration. Stress is no longer considered a major factor. However, results from Pavlov's experiments on stress ulceration still hold true. There is no question that psychologic stress can stimulate hormonal release, and peptic ulceration certainly may be caused by severe environmental or psychologic stress. However, the stress must be severe to be a factor in ulceration. Similarly, as indicated, cigarette smoking, alcohol, and consumption of hot spices or high amounts of caffeine (coffee, tea, colas) may be factors in the production of peptic ulceration.

ADDITIONAL RESOURCES

Cryer B: Mucosal defense and repair: role of prostaglandins in the stomach and duodenum, *Gastroenterol Clin North Am* 30:877-894, 2001.

Cryer B, Spechler SJ: Peptic ulcer disease. In Feldman M, Friedman LS, Brandt LJ, editors: *Gastrointestinal and liver disease*, ed 8, Philadelphia, 2006, Saunders-Elsevier, pp 1089-1110.

Hopkins RJ, Jirardi LS, Turney EA: Relationship between *Helicobacter pylori* eradication and reduced duodenal and gastric ulcer recurrence, *Gastroenterology* 110:1244-1252, 1996.

Kurata JH, Nogawa AN: Meta-analysis of risk factors for peptic ulcers: nonsteroidal anti-inflammatory drugs, *Helicobacter pylori*, and smoking, *J Clin Gastroenterol* 24:2-17, 1997.

Marshall BJ: *Helicobacter pylori* in peptic ulcer: have Koch's postulates been fulfilled? *Ann Med* 27:565-568, 1995.

Wolfe NN, Lichtenstein GH, Singh G: Gastrointestinal toxicity of nonsteroidal antiinflammatory drugs, *N Engl J Med* 340:1888-1899, 1999.

Peptic Ulcer Disease: Duodenitis and Ulcer of the Duodenal Bulb

58

Martin H. Floch

Peptic ulcer occurs when the duodenal bulb or the first or second part of the duodenum becomes ulcerated because of severe focal inflammation. The inflammation may be in areas of the bulb or proximal duodenum, referred to as *duodenitis* (**Fig. 58-1**). As discussed in Chapters 55 and 56, frequently the causes of this phenomenon are *Helicobacter pylori* infection or nonsteroidal antiinflammatory drugs (NSAIDs). This discussion is limited to duodenitis and disease of the bulb and the first and second parts of the duodenum. Duodenitis from other origins and affecting the entire duodenum is discussed in Section IV.

CLINICAL PICTURE

The most common symptom of duodenal ulcer or duodenitis is *epigastric pain.* However, nausea, recurrent vomiting, and occult or gross bleeding may be the presenting symptoms and the reason patients seek treatment. It is surprising how often (almost 50% of patients) duodenal ulcers and duodenitis manifest as bleeding. The bleeding may be in the form of chronic anemia or massive upper gastrointestinal (GI) hemorrhage (see Chapter 60).

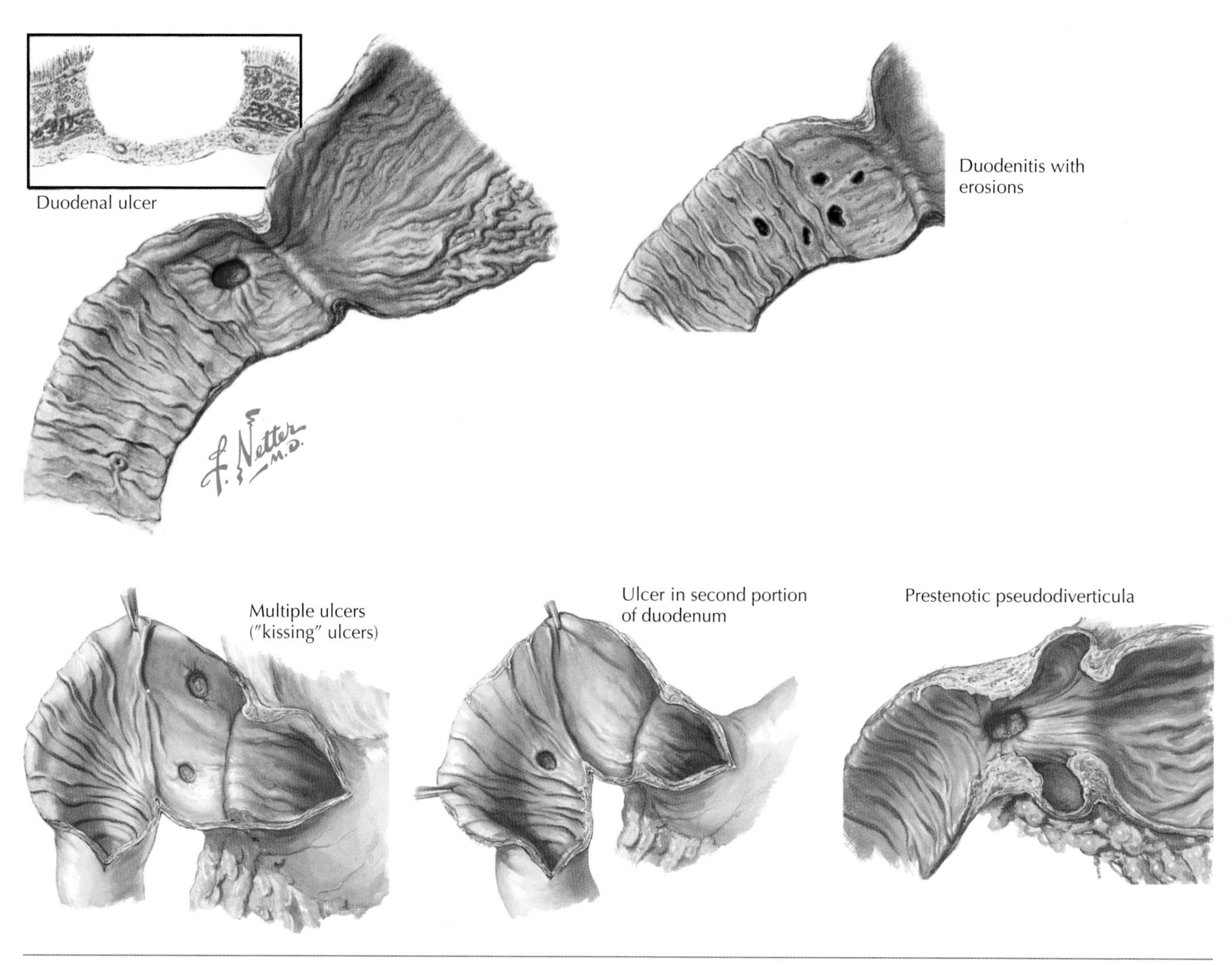

Figure 58-1 *Duodenal Ulcers.*

DIAGNOSIS

The most common cause of duodenal ulcer and duodenitis is *H. pylori* infection. Therefore, some clinicians diagnose *H. pylori* infection noninvasively and treat the patient. Symptoms may completely resolve without endoscopic evaluation. However, if the patient has anemia or acute bleeding, endoscopy is essential even with a noninvasive diagnosis of *H. pylori* infection.

During endoscopy, the duodenal bulb is often swollen and difficult to distend. Consequently, the endoscopist will observe the duodenitis but may miss the ulcer bed.

More common, and clinically important, is *chronic duodenal ulcer*. With few exceptions, this lesion is seated within the duodenal bulb. It develops with essentially the same frequency on the anterior or posterior wall. The average size of a duodenal ulcer is 0.5 cm, but ulcers on the posterior wall are usually larger than those on the anterior wall, primarily because the posterior wall ulcers, separated by the pancreas lying below the ulcer, can enlarge without free perforation.

Duodenal peptic ulcer is usually round and has a punched-out appearance. When small, the ulcer may be slitlike, crescent shaped, or triangular. Unlike acute ulcers, which stop at the submucosa, chronic ulcers involve all layers, penetrating to the muscular coat and beyond. An ulcer on the anterior wall may show a moderate amount of proliferation, but an ulcer on the posterior wall shows evidence of considerable edema and fibrosis. Healing may proceed as with a gastric ulcer—with the disappearance of the crater and bridging of the gap through the formation of fibrous tissue covered by new mucous membrane—but healing becomes more difficult once the destruction of the muscular layer has gone too far.

Symptoms of chronic ulcer are typical and are characterized by periodic episodes of gnawing pain, usually located in the epigastrium. The pain occurs 1 to 2 hours after meals and may be relieved by food.

Peptic ulcers in a region distal to the duodenal bulb occur infrequently (<5% of all duodenal ulcers), decreasing in frequency with their distance from the pylorus. Ulcers in the second portion of the duodenum cause the same symptoms and complications as ulcers of the duodenal bulb. The clinical picture and later the clinical significance of acute ulcers, however, may be much more complex because of the functional and anatomic implications for the adjoining structures. Because of the edema of their margins and surroundings and because of penetration or shrinkage, acute ulcers may cause obstruction and eventually stenosis of the papilla of Vater or the lower part of the common bile duct, along with one or both of the pancreatic ducts, so that chronic pancreatitis or biliary obstruction with jaundice, or even both, may result. Deep penetration may give rise to choledochoduodenal fistula.

Multiple chronic ulcers of the duodenum are common, occurring in 11% to 45% of autopsy cases. Rarely have more than two been found. Ulcers developing on both the anterior and the posterior wall are referred to as *kissing ulcers*.

Only a small percentage of patients with active duodenal ulcer also have active gastric ulcer. One of the most typical duodenal deformities occurring with the ulcerative process is the *prestenotic pseudodiverticulum*. Seen from the lumen, it represents a relatively flat, sinuslike indentation, usually between the pylorus and the site of the ulcer or proximal to a duodenal stricture resulting from a cicatricial remnant of an ulcer. Although all layers of the duodenal wall participate in the formation of such a pouch, this differs from a true duodenal diverticulum (see Chapter 52) in that the mucosa has not evaginated through a small muscular gap.

TREATMENT AND PROGNOSIS

Treatment for duodenitis and duodenal ulcer is similar to (or the same as) treatment for acute gastric ulcers (see Chapter 56) and depends on the cause. Once again, if infectious *(H. pylori)*, eradication of the organism is curative. If the cause is an irritating agent or a metabolic disease, treatment varies and is discussed according to conditions in Chapter 57, for which the cause must be identified and then the appropriate therapy used. In the acute phase, H_2 or proton pump inhibitors will usually relieve pain until specific therapy is available, and certainly until the ulcer heals, which may take 2 to 4 weeks. Antacids and sucralfate can also be helpful.

ADDITIONAL RESOURCES

Chan FKL, Lau YW: Treatment of peptic ulcer disease. In Feldman M, Friedman LS, Brandt LJ, editors: *Gastrointestinal and liver disease*, ed 8, Philadelphia, 2006, Saunders-Elsevier, pp 1111-1136.

Cryer B: Mucosal defense and repair: role of prostaglandins in the stomach and duodenum, *Gastroenterol Clin North Am* 30:877-894, 2001.

Cryer B, Spechler SJ: Peptic ulcer disease. In Feldman M, Friedman LS, Brandt LJ, editors: *Gastrointestinal and liver disease*, ed 8, Philadelphia, 2006, Saunders-Elsevier, pp 1089-1110.

Hopkins RJ, Jirard LS, Turney RA: Relationship between *Helicobacter pylori* eradication and reduced duodenal and gastric ulcer recurrence, *Gastroenterology* 110:1244-1252, 1996.

Kurata JH, Nogawa AN: Meta-analysis of risk factors for peptic ulcers: nonsteroidal anti-inflammatory drugs, *Helicobacter pylori*, and smoking, *J Clin Gastroenterol* 24:2-17, 1997.

Marshall BJ: *Helicobacter pylori* in peptic ulcer: have Koch's postulates been fulfilled? *Ann Med* 27:565-568, 1995.

Megraud F, Marshall BJ: How to treat *Helicobacter pylori:* first-line, second-line, and future therapies, *Gastroenterol Clin North Am* 29:759-777, 2000.

Wolfe NN, Lichtenstein GH, Singh G: Gastrointestinal toxicity of nonsteroidal anti-inflammatory drugs, *N Engl J Med* 340:1888-1899, 1999.

Peptic Ulcer Disease: Complications

59

Martin H. Floch

Numerous complications can occur from acute or chronic ulceration of the stomach or duodenum. These include perforation, bleeding, scarring stenosis and obstruction, and various forms of malabsorption (**Fig. 59-1**).

ACUTE PERFORATION

Clinical Picture and Pathophysiology

Perforation of an ulcer may be free and may extend into the peritoneal cavity or into an adjacent organ. Free perforation is an acute, life-threatening complication. It appears to be more common in smokers and elderly persons.

The duration of an ulcer (stomach or duodenum) does not seem to affect how quickly the ulcerative and inflammatory processes penetrate the muscular and serous layers. Acute peptic ulcer may penetrate the gastric or intestinal wall so rapidly that 10% to 25% of patients may have no history of previous symptoms. Chronic ulcers, on the other hand, may exist for years without progressing so deep as to implicate the serosa, although chronic ulcers that cause severe, persistent symptoms and recurrent or calloused ulcers can always cause perforation. The rapidity with which the digestive effect of the strongly acid gastric juice destroys the wall layers and approaches the serosa cannot be anticipated.

Once perforation has taken place, the location of the ulcer plays a dominant role in subsequent events. Ulcers of the anterior wall of the stomach and the duodenum have a greater access to the so-called free peritoneal cavity than those on the posterior wall. Ulcers of the posterior wall may penetrate the underlying organs, such as the left lobe of the liver, the pancreas, or the gastrohepatic ligament. These may block the ulcer and prevent the entry of gastric or duodenal contents into the peritoneal cavity. This blocked perforation, in which a new floor for the ulcer has been organized outside the visceral wall, is called *chronic perforation* or *penetration*, whereas the term *subacute perforation* is reserved for certain tiny ruptures in the serosa that occur only with the relatively slowly advancing penetration of a chronic gastric ulcer. In such cases, fibrous adhesions to contiguous parenchymal organs or peritoneal attachments result from periinflammatory tissue reactions, long before the ulcer has permeated the serosal layer. Adhesions intercept the small amount of gastric content that might escape through the usually tiny apertures, thus enveloping the fluid, which may lead to the development of localized abscesses.

Free perforation occurs most frequently with ulcers of the anterior wall of the duodenal bulb. The hole resulting from an acute perforation is usually round, varying in diameter from 2 to 4 mm. One of the characteristic features of these holes is their sharp edge, which makes them appear to have been punched out. Surrounding tissue may fail to show any signs of chronic induration, edema, or inflammation.

Acute and free perforations, whether in the stomach or duodenum, are dramatic episodes. Perforation causes sudden, excruciating, explosive pain throughout the abdomen that may radiate to the chest and shoulder. The patient becomes pale and often breaks into a cold sweat. To reduce the abdominal pain, the patient may flex the thighs toward the abdomen, which is extremely rigid and tender. This early phase may last 10 minutes to a few hours, and depending in part on the amount and type of gastrointestinal content released into the peritoneal cavity, the patient's body temperature becomes subnormal, but the pulse rate and blood pressure remain within normal ranges (pulse rate may even be rather slow). Breathing may become superficial and panting. Within a short time, sometimes introduced by a period of apparent subjective improvement, all the typical signs (nausea, vomiting, dry tongue, rapid pulse, fever, leukocytosis) of severe, acute, diffuse peritonitis appear. The tenderness, in the early phase confined mostly to the upper part of the abdomen, has spread over the total abdominal area. It may be excessive in the lower-right quadrant if, with a perforated duodenal ulcer, the intestinal material dissipates into the right lumbar gutter along the ascending colon.

Diagnosis

The differential diagnosis between a perforated gastric or duodenal ulcer and pancreatitis or mesenteric thrombosis may be difficult, but these signs are seldom encountered with a ruptured appendix. Other conditions, such as ectopic pregnancy, ruptured diverticulum, renal colic, acute episodes of biliary tract disease, acute intestinal obstruction or volvulus, and sometimes coronary thrombosis, must be considered.

The most helpful sign to confirm a suspected diagnosis of ulcer perforation is the presence of free air in the peritoneal cavity, particularly in the subphrenic space, demonstrable by upright x-ray examination. If the patient can sit or stand, the air will accumulate under the diaphragm. Escaped air is rarely present only under the left diaphragm and may be detected under both diaphragmatic leaves, but usually, air is found only under the right diaphragm. Computed tomography (CT) assists in making the correct diagnosis when there is clinical uncertainly or no free air is present.

Treatment and Prognosis

If escaped air is found, emergency surgery is indicated. The prognosis for the patient with a perforated gastric or duodenal ulcer is better when surgery is performed early. At present, the procedure of choice is closure of the perforation. If the facilities are suboptimal and the patient is in poor general condition, efforts to treat conservatively with suction through an indwell-

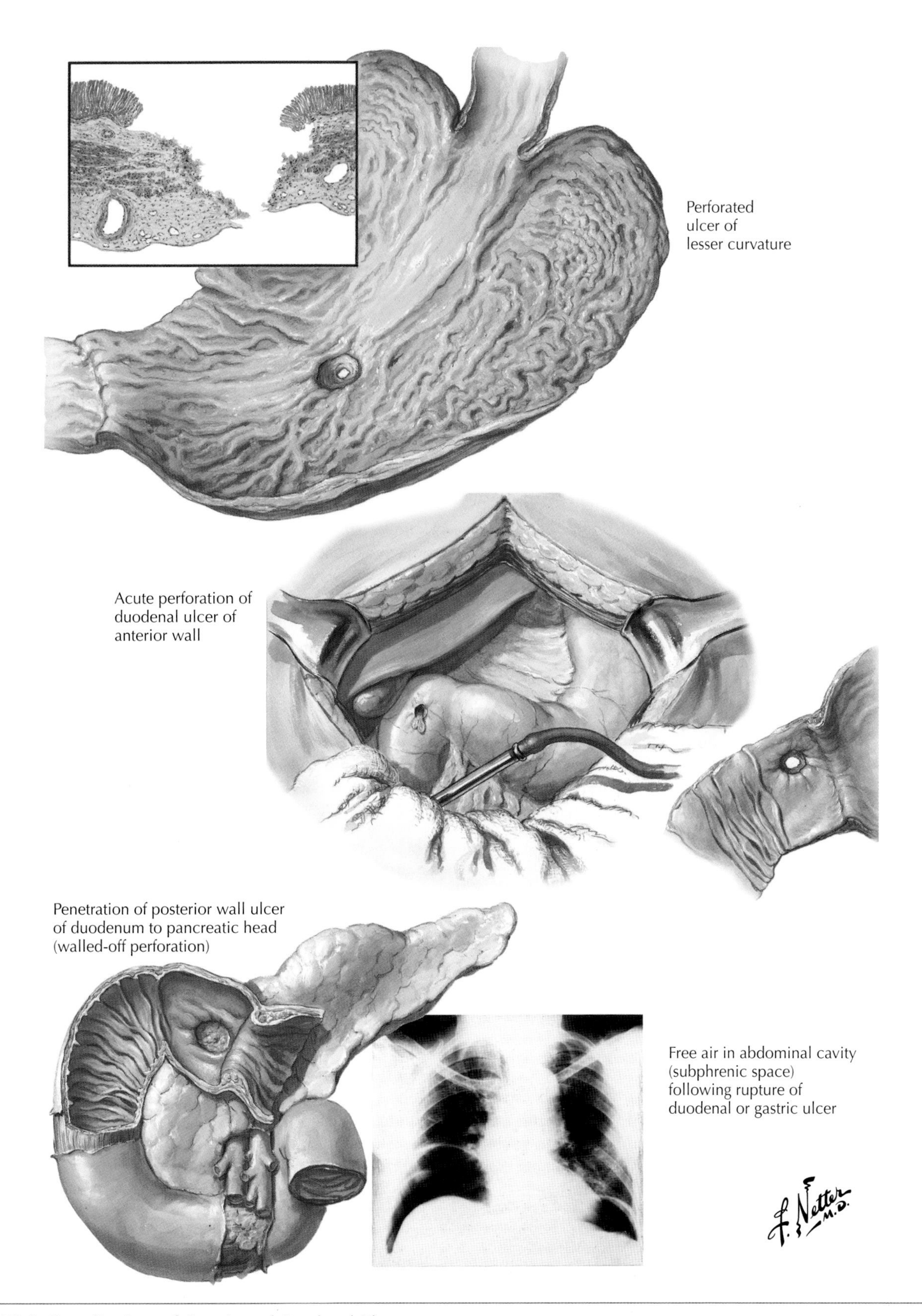

Figure 59-1 *Complications of Gastric and Duodenal Ulcers.*

ing catheter in the stomach, massive antibiotics, and supportive therapy entail greater risk and are less successful than surgery. Closing the perforation, irrigating the peritoneum, and administering antibiotics currently represent the usual treatment. In some patients, more definitive ulcer surgery, such as vagotomy with or without resection, may be considered.

CHRONIC PERFORATION

Clinical Picture and Pathophysiology

Erosion of the serosal layer by a chronic peptic ulcer on the posterior walls of the stomach and duodenum and its penetration into a contiguous organ is a slow process; the patient may not even feel the actual perforation. Typical ulcer pains, associated with and relieved by eating, gradually become continuous, gnawing, boring pain that no longer responds to the ingestion of food. The pain may radiate to the back, shoulder, clavicular areas, or umbilicus or down to the lumbar vertebrae and the pubic or inguinal regions. A classic example of chronic perforation is the ulcer of the posterior wall of the duodenal bulb, penetrating into and walled off by the pancreas. During surgery for this condition, while attempting to remove the entire ulcer with its floor in the pancreatic tissue, the surgeon risks producing a pancreatic lesion that may open accessory pancreatic ducts. Therefore, it may be advisable to leave the ulcer floor untouched after careful dissection of the ulcer from the duodenal wall.

Ulcers located in the upper parts of the posterior duodenal wall tend to penetrate the hepatoduodenal ligament. This process is usually accompanied by the development of extensive, fibrous, and thickened adhesions, to which the greater omentum may contribute. The supraduodenal and retroduodenal portions of the common bile duct, coursing within the leaves of the ligament, may become compromised in these adhesions. As a result of a constriction or distortion of the common duct, mild obstructive icterus may confuse the clinical picture. Fortunately, perforation into the duct with subsequent cholangitis is rare. Acute, perforated ulcers of the posterior gastric wall seldom release chyme into the bursa omentalis, only producing signs of localized peritonitis without free air in the abdominal cavity.

Diagnosis

Endoscopy is invariably performed when the symptoms are chronic. The presence or absence of *Helicobacter pylori* is established and the degree of damage assessed. CT is essential to further determine the damage caused by scarring and to assess whether surgery is needed. If bile or pancreatic ducts are involved, magnetic resonance imaging (MRI) or endoscopic retrograde cholangiopancreatography (ERCP) may be needed.

Treatment and Prognosis

Each patient must be evaluated carefully to decide whether medical treatment with eradication of any infection and proton pump inhibitor (PPI) therapy will enable healing, or whether surgical intervention is necessary. The location of the perforated area will influence the choice of surgical approach. Again, the course will depend on these factors. If the pancreas or bile ducts require revision, the course will be prolonged, and morbidity and mortality will be higher.

OBSTRUCTION

Clinical Picture and Pathophysiology

Another typical complication of chronic relapsing duodenal or juxtapyloric ulcer is *stenosis of the pylorus*, which develops as the result of the gradual thickening of the duodenal wall and the progressive fibrotic narrowing of the lumen. The incidence of complete pyloric stenosis as a sequela to ulcer has decreased in recent decades because of improved medical management of this type of ulcer and prompt recognition of its initial phases. When the pyloric lumen begins to narrow, the stomach tries to overcome the impediment by increased peristalsis, and its muscular wall becomes hypertrophic. This stage is called *compensated pyloric stenosis* because, with these adaptation phenomena, the stomach succeeds in expelling its contents with only mild degrees of gastric retention. Later, when the lumen is appreciably narrowed, the expulsive efforts of the stomach fail, and the clinical picture is dominated by incessant vomiting and distress resulting from progressive dilation of the stomach, which at times may be massive (see Chapter 48).

Treatment

Decompensated pyloric stenosis, resulting in the retention of ingested material and the products of gastric secretion, is irreversible and is an unequivocal indication for surgical intervention.

HEMORRHAGE

Minor bleeding occurs in most patients with acute or chronic peptic ulcer. Occult blood can be found in the stools or gastric juice of most ulcer patients, the result of the oozing characteristic of ulcerative lesions. Along with perforation, massive hemorrhage is the most dangerous ulcer complication, but fortunately, it occurs much less often than minor bleeding. It has been estimated that of all massive hemorrhages of the gastrointestinal tract, 50% stem from peptic ulcer.

The diagnosis, course, and treatment of acute and chronic bleeding are discussed in Chapter 60.

ADDITIONAL RESOURCES

Cryer B, Spechler SJ: Peptic ulcer disease. In Feldman M, Friedman LS, Brandt LJ, editors: *Gastrointestinal and liver disease*, Philadelphia, 2006, Saunders-Elsevier, pp 1089-1110.

Kelly KA, Sarr MG, Hinder RA: *Mayo Clinic gastrointestinal surgery*, Philadelphia, 2004, Saunders.

Shaffer HJ: Perforation of structure in the gastrointestinal tract: assessment by conventional radiology, *Radiol Clin North Am* 30:405-426, 1992.

Gastrointestinal Bleeding

60

Martin H. Floch

Gastrointestinal (GI) bleeding may indicate acute or chronic loss of blood from the GI tract. It may originate in the upper or the lower part of the tract. Acute bleeding may be life threatening; chronic bleeding is slow or even occult (**Fig. 60-1**). The GI tract can lose as much as 50 mL of blood per day, which is replaced, without anemia, but it is a signal of a GI lesion. This chapter discusses upper GI bleeding; Chapter 132 describes lower intestinal bleeding.

The incidence of bleeding from the upper GI tract is approximately 100 cases per 100,000 population. The most common cause of upper GI hemorrhage is peptic ulceration of the stomach or duodenum, with or without aspirin or nonsteroidal antiinflammatory drug (NSAID) use. Approximately 50% of duodenal ulcers are silent. The next most common cause is variceal bleeding, then Mallory-Weiss syndrome with lacerations from excessive vomiting with trauma, and then simple erosive disease of the stomach or duodenum. Less common causes are vascular ectasia, Dieulafoy vascular malformation, upper GI neoplasm, severe esophagitis, and rare causes of severe bleeding such fistula, hemobilia, and esophageal ulceration or lesion. Age appears to be a major factor; elderly persons are at increased risk and make up 30% to 45% of the population with upper GI bleeding. Portal hypotension and esophagogastric variceal bleeding are discussed in Section IX.

CLINICAL PICTURE

Patients with massive upper GI bleeding may also have hematochezia (blood in feces), causing them to quickly seek treatment. Other patients bleed slowly and seek treatment after significant anemia and paleness have developed. Often, patients with slow bleeding report weakness, but the paleness is apparent on first examination. In patients with acute bleeding, the massive loss of blood may manifest as sudden syncope or, in emergency situations, as hypovolemia with significant hypotension and occasionally shock.

The patient with slow GI bleeding may seek treatment for anemia, with findings of paleness and weakness. In other patients, the bleeding may truly be occult, and stool samples will be positive for occult blood at routine examination.

Syncope, hematochezia, melena, and hypotension are emergencies requiring hospitalization and circulatory stabilization.

If the bleeding is occult and the presentation is iron-deficiency anemia, the presentation may merely be of a patient seeking a cause for the iron loss. The workup and differential diagnosis may follow on an outpatient basis.

DIAGNOSIS

It is rare that a patient bleeds so heavily that he or she is brought to the operating room without a diagnostic evaluation; endoscopy is readily available in most emergency situations. Endoscopy is performed with or without prior cleansing lavage of the stomach, depending on the patient's condition and the endoscopist's preference. Identifying gastric or duodenal peptic disease may be simple but is often obscured by blood coating or clots in the stomach. Varices usually are easily identified, as are vascular lesions, which can include vascular ectasia, arteriovenous malformation, hereditary hemorrhagic telangiectasia, and angioma. Vascular lesions are difficult to identify and may not become apparent until the bleeding slows and the stomach is emptied. Upper endoscopy is the best procedure for diagnosing GI bleeding. Occasionally, however, it is difficult to establish the exact source, and angiography may be helpful. When direct endoscopy is nondiagnostic, wireless capsule endoscopy should be performed as soon as possible to identify the bleeding site or lesion.

For patients with occult bleeding, or when it is unclear from endoscopy whether a small lesion is causing the bleeding, the diagnostic tools available are angiography and capsule endoscopy. Enteroscopy with a longer upper endoscope may be helpful, but full-length, small-bowel enteroscopy of the ileum is difficult to perform and is used less often than wireless capsule endoscopy, which is simple and has become the procedure of choice when the diagnosis is uncertain. Double-balloon enteroscopy to visualize the small bowel is another choice when a lesion is suspected on capsule evaluation, but it is a difficult technique and is not used in many institutions. Whenever a patient has occult bleeding, the small bowel must be considered; thus, capsule endoscopy, enteroscopy, or computed enterography is used. Vascular malformations account for almost 70% to 80% of bleeding in the small bowel and can be diagnosed using capsule endoscopy. Tumors of the small bowel are visualized with capsule endoscopy or other imaging techniques (see Section IV).

TREATMENT AND MANAGEMENT

Approximately 85% of massive upper GI bleeding is controlled by transfusion and by proton pump inhibitor (PPI) or significant antacid therapy. When the endoscopist identifies a bleeding vessel, endoscopic therapy has proved extremely successful. Epinephrine injections, cauterization, and heater-probe cautery have been successful in controlling most lesions. Nevertheless, approximately 10% to 15% of patients with severe bleeding require surgical intervention. The criterion is usually lack of control of the bleeding, as evidenced by the need for a transfusion of more than 4 to 6 units of blood. When surgery is required, the options (simple sewing vs. resection) depend on the patient's age, anatomy, and clinical status.

In patients with occult bleeding and rare lesions, the final therapy usually involves surgical intervention. However, ectasias are a major problem. When ectasias are numerous or cannot be cauterized, therapy is frustrating; if localized, surgical resection is possible; if diffuse, various modalities are attempted. Estrogen therapy has been used in different trials but has not been universally successful. For the 5% to 10% of patients with unidentified GI bleeding, therapy becomes supportive, with intermittent blood replacement and continuing diagnostic efforts.

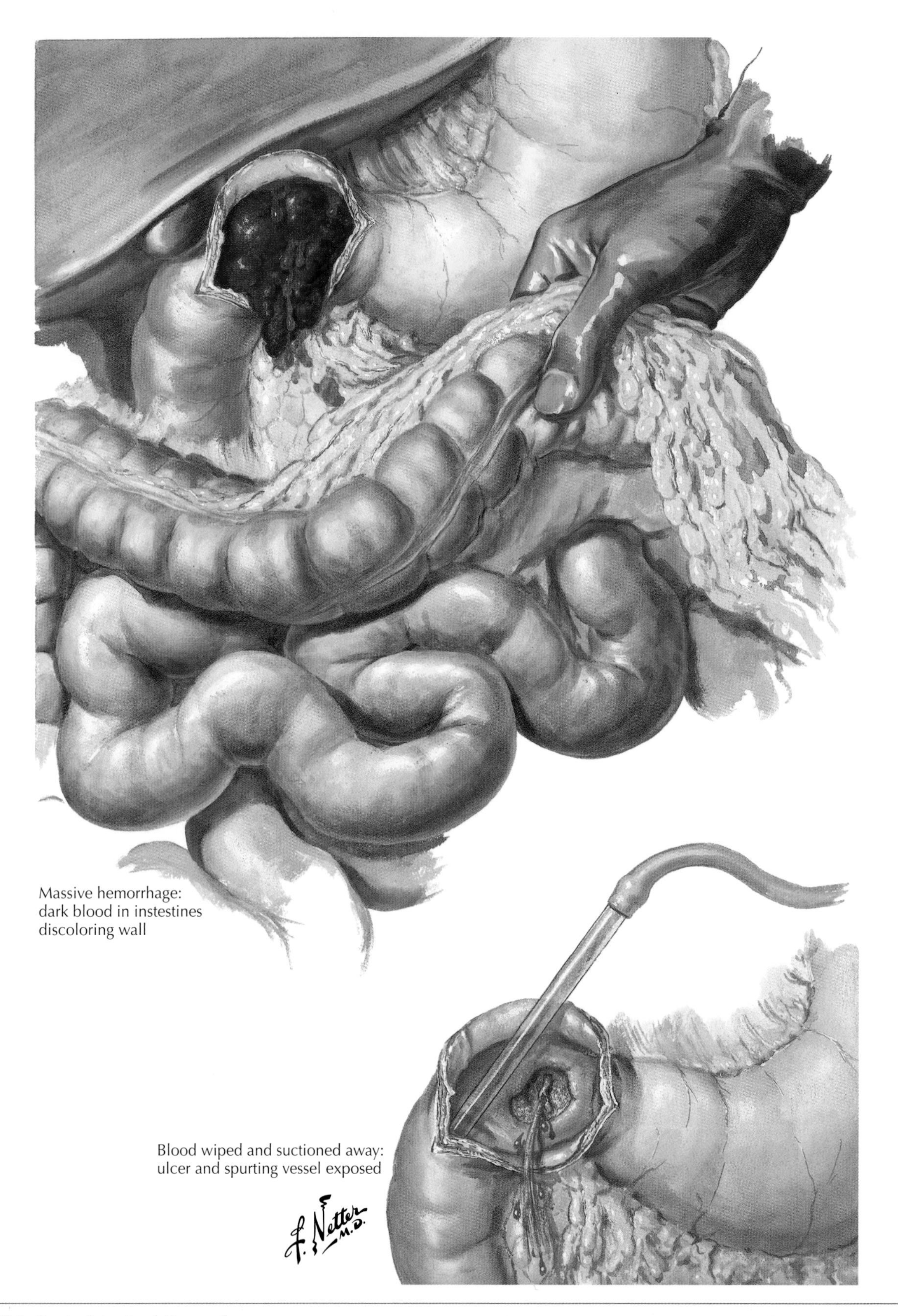

Figure 60-1 *Gastrointestinal Hemorrhage.*

COURSE AND PROGNOSIS

Most GI bleeding is successfully controlled by medical therapy, and the difficult cases are controlled by surgery. The prognosis in these patients is excellent. However, mortality from acute GI bleeding ranges from 5% to 12%. When the bleeding is stopped, the prognosis for the patient is good. Inability to stop the bleeding leads to a failed outcome. It is estimated that more than 20,000 deaths occur annually in the United States because of GI bleeding. Even during surgery, it may be difficult to identify the bleeding site. Intraoperative endoscopy has been helpful, but if the cause of the bleeding cannot be found, the prognosis is guarded, especially for the elderly patient.

If the cause of occult bleeding cannot be determined, patient and physician are similarly frustrated. Slow bleeds are easily handled with transfusion and iron replacement, but prognoses vary. Prognoses for specific lesions are discussed under the appropriate disease chapters (e.g., neoplasms, varices, lower GI bleeding).

ADDITIONAL RESOURCES

Barkun A, Bardou M, Marshall JK: Consensus recommendations for managing patients with nonvariceal upper gastrointestinal bleeding, *Ann Intern Med* 139:843-847, 2003.

Enestvedt BK, Gralneck IM, Mattck N, et al: An evaluation of endoscopic indications and findings related to nonvariceal upper-GI hemorrhage in a large multicenter consortium, *Gastrointest Endosc* 67:422-429, 2008.

Kim KE: *Acute gastrointestinal bleeding: diagnosis and treatment*, Totowa, NJ, 2003, Humana Press.

Lema LV, Ruano-Ravina A: Effectiveness and safety of capsule endoscopy in the diagnosis of small bowel diseases, *J Clin Gastroenterol* 42:466-471, 2008.

Lewis BS, Wengner JS, Waye JD. Small bowel enteroscopy and intraoperative enteroscopy for obscure gastrointestinal bleeding, *Am J Gastroenterol* 86:171-172, 1991.

Pennazio M, Santucci R, Rondomotti E, et al: Outcome of patients with obscure gastrointestinal bleeding after capsule endoscopy: report of 100 consecutive cases, *Gastroenterology* 126:643-653, 2004.

Pitcher JL: Therapeutic endoscopy and bleeding ulcers: historical overview, *Gastrointest Endosc* 36:2-7, 1990.

Rockey DC: Gastrointestinal bleeding. In Feldman M, Friedman LS, Brandt LJ, editors: *Gastrointestinal and liver disease*, ed 8, Philadelphia, 2006, Saunders-Elsevier, pp 255-299.

Therapeutic Gastrointestinal Endoscopy

61

Martin H. Floch

Visualization of the proximal gastrointestinal (GI) tract began in 1868 with Adolfe Kussmaul, who was able to visualize the stomach using a rigid tube. Later, Rudolf Schindler used lenses to make the lower half of the "gastroscope" flexible. By 1960, Basel Hirskowitz helped develop and popularize the use of fiberoptics with a flexible "fiberscope" that permitted intubation of the duodenum. Videoendoscopes were introduced in 1983, using a chip, a charge-coupled device, of a photosensitive silicon grid to produce current by means of photoactivation. Many revisions and technologic advances followed, and videoscopes are now standard equipment for all endoscopists. Additional developments enabled use of therapeutic techniques in the esophagus, stomach, and duodenum. Attaching an ultrasound receiver to the tip of the scope has further developed endoscopy as a diagnostic tool by combining endoscopy with ultrasound (EUS).

STANDARD PROCEDURE

Upper endoscopy, or esophagogastroduodenoscopy (EGD), is now performed as an ambulatory procedure. After appropriate consent is obtained, the patient is sedated. The type of sedation used varies greatly across countries and institutions. Some still use simple topical anesthesia, whereas others choose sedation or general anesthesia. The most common sedation is often referred to as "conscious" sedation and consists of pain relief and some form of sleep induction that may have an amnesic effect. Meperidine or fentanyl plus diazepam or midazolam is frequently used.

Standard endoscopy is used for diagnostic purposes and has channels that permit biopsy specimens to be obtained for histologic evaluation. The development of flexible videoscopes with larger channels and various angulations created a revolution in instrumentation and enhanced the development of therapeutic procedures (**Fig. 61-1**).

THERAPEUTIC PROCEDURES

In *pneumatic dilatation* for achalasia, guidewires are placed through an endoscope into the stomach. A pneumatic, dilating hydrostatic balloon is passed over the guidewire and positioned across the lower esophageal sphincter (LES). The balloon is then inflated, and the increased hydrostatic pressure tears the sphincter. Although most gastroenterologists recommend pneumatic dilatation, many prefer *surgical splitting* of the sphincter. *Clostridium botulinum* toxin (botulinum toxin A, Botox) administration is usually reserved for patients unable to undergo surgery or vigorous pneumatic dilatation.

Monopolar and *bipolar electrocoagulation* and *heater probe* are used to coagulate mucosa. They are especially helpful in treating bleeding ulcers. The tip of the probe is placed directly on the bleeding side or on the four quadrants surrounding it. Coagulation occurs, and the bleeding usually stops.

In *laser treatment*, the flexible quartz probe is passed through the endoscopic channel and can conduct either an argon or an Nd:YAG laser so that the light beam "hits" tissue. This may be used to coagulate or ablate bleeding sites, ectasias, or tumors. It has been a useful procedure but has not gained wide acceptance. Laser treatment for esophageal cancer is a palliative procedure, but the principle is the same as that used for bleeding.

Injection therapy to control bleeding requires a thin flexible catheter with a retractable needle at the tip. An agent is injected into the mucosa to occlude vessels around a bleeding site, including ethanol, diluted epinephrine, and other sclerosing agents. This is often used to stop bleeding vessels. Saline injection is also used to raise a small lesion from the wall so that it can be more easily snared. (This technique is used in the colon to remove flat polyps.)

The use of *injection and banding* for sclerosis of esophageal varices is described in Section IX. Both techniques are important and have been successful in controlling and treating variceal bleeding.

Once a stricture is identified in the esophagus, it is customary to dilate. *Esophageal dilatation* techniques include (1) "blind" passage with weighted catheters; (2) *savory dilatation*, in which a guidewire is passed into the stomach and successively larger catheters are passed over the guidewire through the stricture; and (3) *balloon dilatation*, in which a collapsed balloon is passed through the endoscope and, once placed at the stricture site, is inflated to stretch or tear the stricture. Metal dilators (Eder-Puestow) are available. Savory dilators are hard plastic, and hydrostatic balloon dilators consist of plastic polymers. Weighted bougies can be passed with no sedation or local sedation, whereas savory dilators and hydrostatic balloons require endoscopy.

Plastic or metal *stents* are passed through a large port channel of an endoscope and used to bridge malignant lesions in the esophagus. The stents are temporary but provide effective palliation for short periods.

With *polypectomy*, polyps in the esophagus, stomach, and colon can be removed by passing a snare through the endoscope. The endoscopist loops the snare around the base of the polyp, then applies cautery to cut the base. The polyp is removed through suction or, if too large, is retrieved by removing the scope with the polyp held by the snare's loop or by suction. The same technique can be used in any of the hollow organs. At times, the snare is flat. When the base can be lifted by saline injection, the mucosa can be snared and the tissue retrieved.

Endoscopic retrograde cholangiopancreatography is described in detail in Section VII. Using a lateral-viewing scope, the endoscopist can cut the sphincter in the ampulla with electric cautery, permitting removal of stones, passage of stents to dilate strictures in the common duct, and passage of special scopes to visualize the common duct or pancreatic ducts.

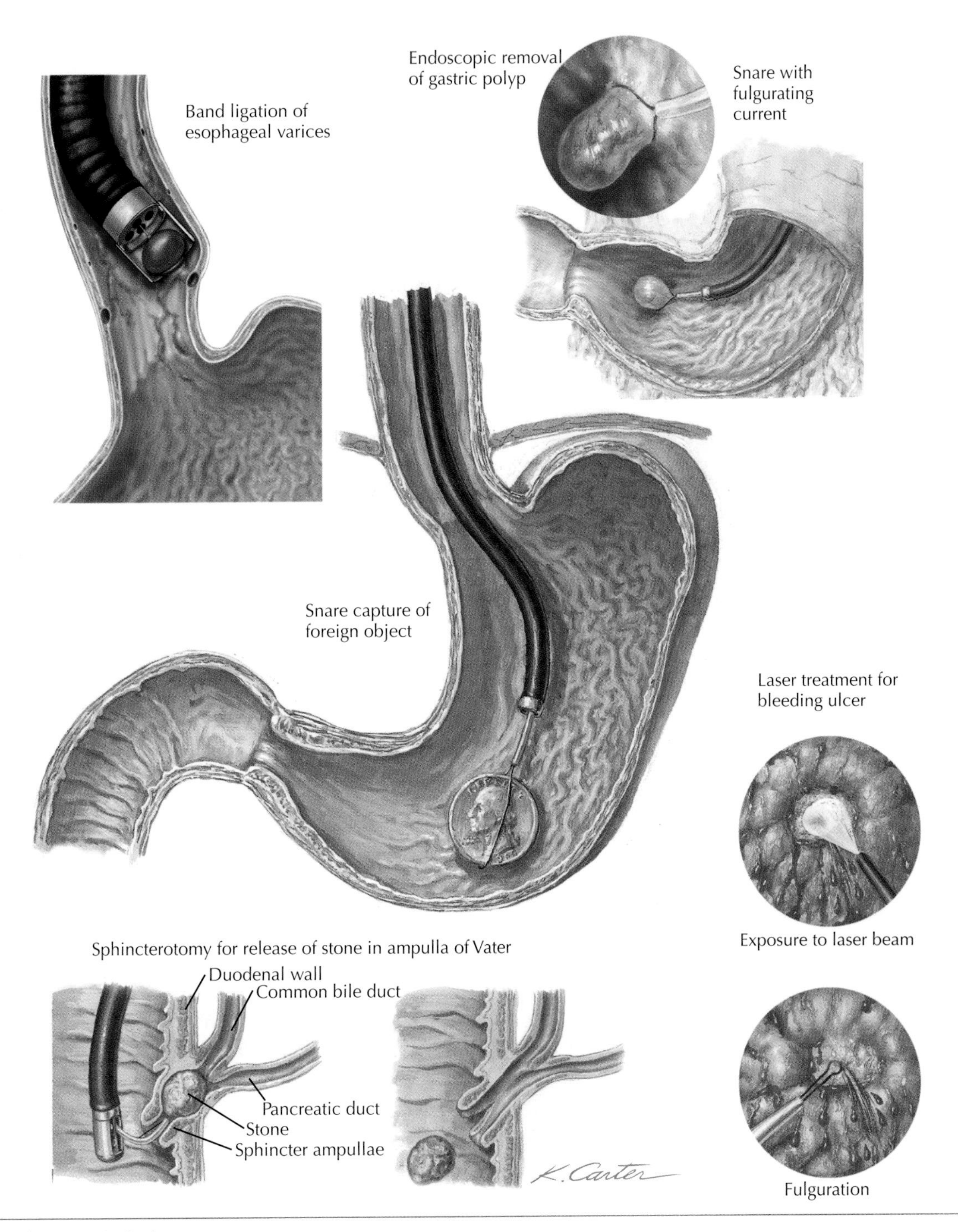

Figure 61-1 *Therapeutic Endoscopy.*

Esophageal and gastric foreign bodies (boluses) can be removed through an endoscope using loops, snares, and baskets. If the bolus is soft, it can be broken up and pushed into the stomach. If the bolus is firm or hard, it can be snared for retrieval. The technique used varies with the size and type of bolus and the patient's condition. Some endoscopists use an *overtube* to remove parts of the obstructing bolus with the scope and then pass the scope again to continue to remove the bolus. Boluses in the stomach can be large, and surgery may be needed if they cannot be broken up or removed. After a bolus is removed from the esophagus, follow-up endoscopy is usually performed when the traumatized area has healed, to ensure that the patient does not have a chronic disease and to establish the cause of the obstructing event.

ADDITIONAL RESOURCES

Dam JV: *Gastrointestinal endoscopy,* Boston, 2004, Landes Bioscience.

Ginsberg GG, Kochman ML, Norton I, Gustout CJ: *Clinical gastrointestinal endoscopy,* Philadelphia, 2005, Saunders-Elsevier.

Policy and procedure manual for gastrointestinal endoscopy: guidelines for training and practice, Manchester, Mass, 2002, American Society for Gastrointestinal Endoscopy.

Benign Tumors of the Stomach

62

Martin H. Floch

Benign tumors of the stomach are relatively rare. Numerous autopsy studies reveal that gastric polyps are present in approximately 0.1% of gastric specimens. However, with the advent of endoscopy, small tumors have been more readily identified (**Fig. 62-1**). These tumors can be classified as epithelial, submucosal, and ectopic.

Benign tumors of epithelial origin include hyperplastic, fundic gland, and adenomatous polyps. Submucosal lesions include leiomyomas, lipomas, fibromas, hamartomas, hemangiomas, neurofibromas, gastrointestinal stromal tumors (GISTs), eosinophilic granulomas, and inflammatory polyps. Ectopic tissue, such as pancreatic rests or Brunner gland hyperplasia, may result in apparently benign tumors.

CLINICAL PICTURE

In general, benign tumors are asymptomatic and are identified on radiography or endoscopy. Although benign, these tumors may be associated with bleeding or obstructive phenomena. Severe, acute bleeding may occur with a lipoma that has surface erosion and an active bleeding vessel on the surface. Chronic bleeding and anemia may develop from an intermittent or slow leak from the polyp. A rare presentation is gastric outlet obstruction caused by prolapse of a large polyp into the duodenum.

DIAGNOSIS

If the lesion is identified using radiography, barium contrast, or computed tomography (CT), endoscopy is required for biopsy and further evaluation. Endoscopic ultrasound (EUS) is helpful to evaluate the depth of the lesion, and characteristic findings of various benign lesions confirm diagnosis. Initial endoscopy with biopsy is often preferred. It may be possible to remove the entire lesion endoscopically; therefore, the procedure is therapeutic.

TREATMENT AND COURSE

Small lesions may be hyperplastic polyps, fundic gland polyps, or adenomas. All are benign and of epithelial origin. Once histologic diagnosis is made, these require no further therapy. Dilated, distorted fundic glands are usually very small polyps, smaller than 1 cm, but they may develop in large numbers. *Fundic gland polyps* are usually asymptomatic and are not associated with bleeding. Frequently, the polyps are hyperplastic and consist of a proliferation of epithelial elements without atypia. *Hyperplastic polyps* also are very small and are not premalignant or associated with bleeding. *Adenomatous polyps* (adenomas) may be larger than fundic gland or hyperplastic polyps. Adenomas have the potential for malignancy and may be associated with adenocarcinoma. The risk for carcinoma, once adenomatous polyps are identified, may be as high at 10%. Once the lesion is larger than 2 cm, it is associated with malignancy. These polyps should be removed, and patients with adenomas should be enrolled in a surveillance program once the adenomatous histology is identified.

Gastrointestinal stromal tumors, also referred to as *mesenchymal stromal tumors*, are derived from smooth muscle. GISTs make up to 70% of all GI mesenchymal tumors. They tend to be slow growing and often are not diagnosed until the fifth or sixth decade of life. However, they may be associated with bleeding and upper GI distress. Unfortunately, a certain percentage is malignant and does metastasize. When GISTs are identified at endoscopy but are too deep for histologic examination, EUS can help define the depth and identify possible lymph node involvement. The tumors must be surgically removed. At surgery, the specimen can be graded, depending on the number of mitoses. If the mitotic rate is greater than 2 per 10 high-power fields, the lesions are more likely to metastasize and spread (see Chapter 64).

Very small lesions (<1 cm), identified as stromal on EUS, may be monitored. However, if they become larger than 2 to 3 cm, they should be removed so that the histology and aggressiveness of the tumor can be determined and metastasis can be prevented.

Gastric tumors may develop as part of hereditary *gastrointestinal polyposis syndromes.* Gastric lesions are seen in conjunction with familial adenomatous polyposis. Although they are invariably benign, severe dysplasia has been reported in some patients. Given that these patients often undergo colectomy, their stomachs and their duodenums have to be monitored carefully.

Patients with *Peutz-Jeghers syndrome* have formation of *hamartomatous polyps* (hamartomas) in the small intestine and may have stomach hamartomas (24% in one series). These polyps are invariably benign, but malignancy has been reported; therefore, these patients must have their stomachs and their small bowels monitored. Patients with *juvenile polyposis* also have multiple hamartomas. Although the syndrome can occur rarely in the stomach, it is more often associated with colonic lesions. As in Peutz-Jeghers syndrome, potential malignancy is a concern; therefore, patients with juvenile polyposis should also be monitored.

PROGNOSIS

Simple polyps of the stomach that reveal no other risk factors or familial syndromes usually imply a benign prognosis. However, if a GIST is suspected, or if the polyps are associated with familial syndromes, patients should be monitored carefully for potential malignancy (see Chapter 64).

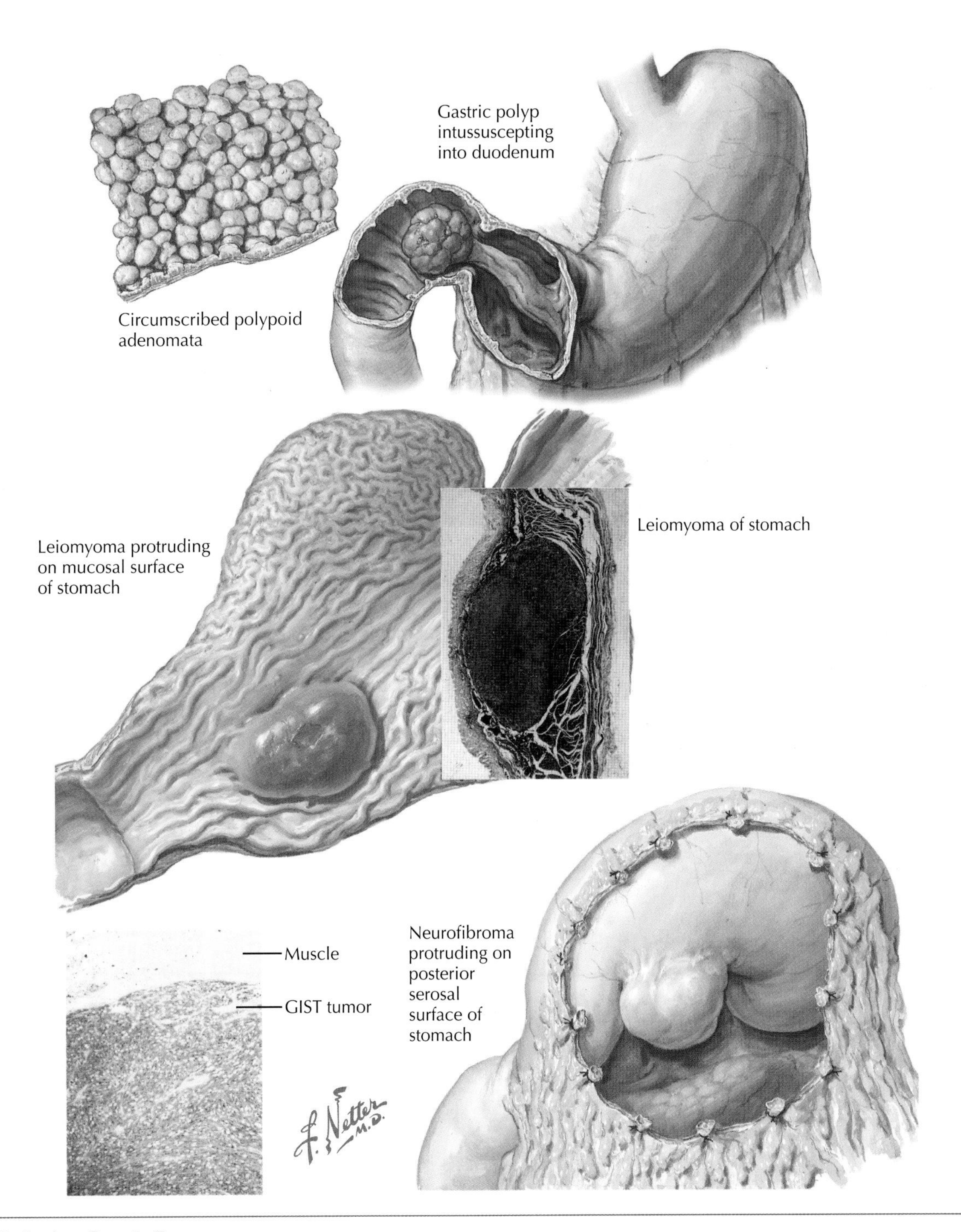

Figure 62-1 *Benign Gastric Tumors.*

ADDITIONAL RESOURCES

Burt RW: Gastric fundic gland polyps, *Gastroenterology* 125:1462-1469, 2003.

Davila RF, Faigel DO: GI stromal tumors, *Gastrointest Endosc* 58:80-88, 2003.

Demetri GD: Gastrointestinal stromal tumors (GISTs). In Feldman M, Friedman LS, Brandt LJ, editors: *Gastrointestinal and liver disease*, ed 8, Philadelphia, 2006, Saunders-Elsevier, pp 589-604.

Loo RJ, Ginsberg GG: Gastric polyps. In DiMarino AJ Jr, Benjamin SB, editors: *Gastrointestinal disease and endoscopic approach*, Thorofare, NJ, 2002, Slack, pp 467-480.

Gastric Lymphoma and Mucosa-Associated Lymphoid Tissue

Martin H. Floch

63

Lymphomas of the stomach are of two types, both of B-lymphocyte origin: marginal-zone B-cell lymphoma of the mucosa-associated lymphoid tissue (MALT) type and diffuse, large B-cell lymphoma (**Fig. 63-1**).

MUCOSA-ASSOCIATED LYMPHOID TISSUE TUMOR

The stomach is the most common site of lymphoma in Western societies. The MALT type arises from malignant transformation of B cells that exist normally in the gut but that proliferate in the inflammatory process. Gastric tissue acquires MALT in the pathologic response to *Helicobacter pylori* infection. The MALT type represents 40% of gastric lymphomas.

Because MALT tumor is related to *H. pylori* infection, the incidence is higher where there is a greater incidence of chronic infection. Evidence has established that *H. pylori* has a key role in the development of MALT lymphoma. It is believed that the disease starts with *H. pylori* infection, which causes gastritis and initiates the immune response of T and B cells. In this tissue, MALT is formed. *H. pylori*–reactive T cells drive B-cell proliferation, which eventually leads to genetic abnormalities that result in aggressive tumor activity.

Tumors are often located in the antrum, but they can be multifocal in as many as 33% of patients. MALT lymphomas may appear as erosions, erythema, or ulcers. Histologically, there is invasion and partial destruction of gastric glands by tumor cells. Cells are usually small and infiltrate the lamina propria. Often, it is difficult to make the diagnosis of lymphoma when the lesions are small. However, invasion and distortion of the tissue clarify the diagnosis. MALT lymphomas are usually staged as follows:

- Stage I: tumors are confined to the mucosa.
- Stage II: tumors extend into the abdomen.
- Stage III (II_E): tumors penetrate the serosa to involve adjacent organs.
- Stage IV: tumors are disseminated to nodal tissue and beyond, or they are supradiaphragmatic.

Clinical Picture

Patients with MALT tumors usually have epigastric pain or dyspepsia. However, some may present with nausea, bleeding, or weight loss.

Diagnosis

Diagnosis of MALT tumors is made by endoscopic visualization and tissue biopsy. Thorough evaluation must include computed tomography (CT) of the abdomen, pelvis, and chest, evaluation of bone marrow, and measurement of serum lactate dehydrogenase (LDH) concentration, which is usually normal in MALT and elevated in other lymphomas. Endoscopic ultrasound (EUS) is important to assess the extent through the wall and to assess nodal involvement. After the histologic diagnosis, staging is essential.

Treatment, Course, and Prognosis

During stage I, MALT is treated with antibiotic eradication of *H. pylori;* the recommended regimens are listed in Chapter 54. Monitoring and follow-up are essential. At 1 to 2 months, eradication must be established, with tumor regression recorded at endoscopy. When the tumor is eradicated, follow-up endoscopy must be performed at 6-month intervals for 2 years. Complete remission is reported in 70% of patients. Remissions occur rapidly but may take as long as 18 months.

Reinfection and relapse have been reported; incomplete remission or removal is always a concern. Controversy surrounds the use of surgery, which some believe is indicated for all MALT tumors.

In stage II or II_E (III), antibiotic eradication of *H. pylori* has been attempted, with reports of complete remission. However, most oncologists agree that chemotherapy should be added for these patients, and some recommend surgery with or without radiation. This vigorous regimen has resulted in an 82% survival rate.

Stage IV disease requires chemotherapy and evaluation for local radiation with or without surgery; prognosis is guarded.

DIFFUSE LARGE B-CELL LYMPHOMA

Diffuse, large B-cell lymphoma tumors make up 45% to 50% of gastric lymphomas, but their cause is not clear. Although associated, *H. pylori* infection has not been implicated as causing large B-cell lymphoma of the stomach. Therefore, eradicating *H. pylori* is not considered adequate therapy when diffuse, large B-cell lymphoma is identified. Lesions occur in the body and in the antrum of the stomach and tend to be multifocal. They typically invade the tunica muscularis, and histology reveals clusters or sheets of large cells. As many as 40% of these patients show evidence of MALT association, but once it evolves, a large B-cell mass is the primary diagnosis and concern.

Clinical Picture

These lesions are ulcerating and therefore may present with bleeding. Some lesions are large enough to cause symptoms of obstruction. Patients may have pain and associated nausea or anorexia. An elevated LDH concentration sometimes occurs.

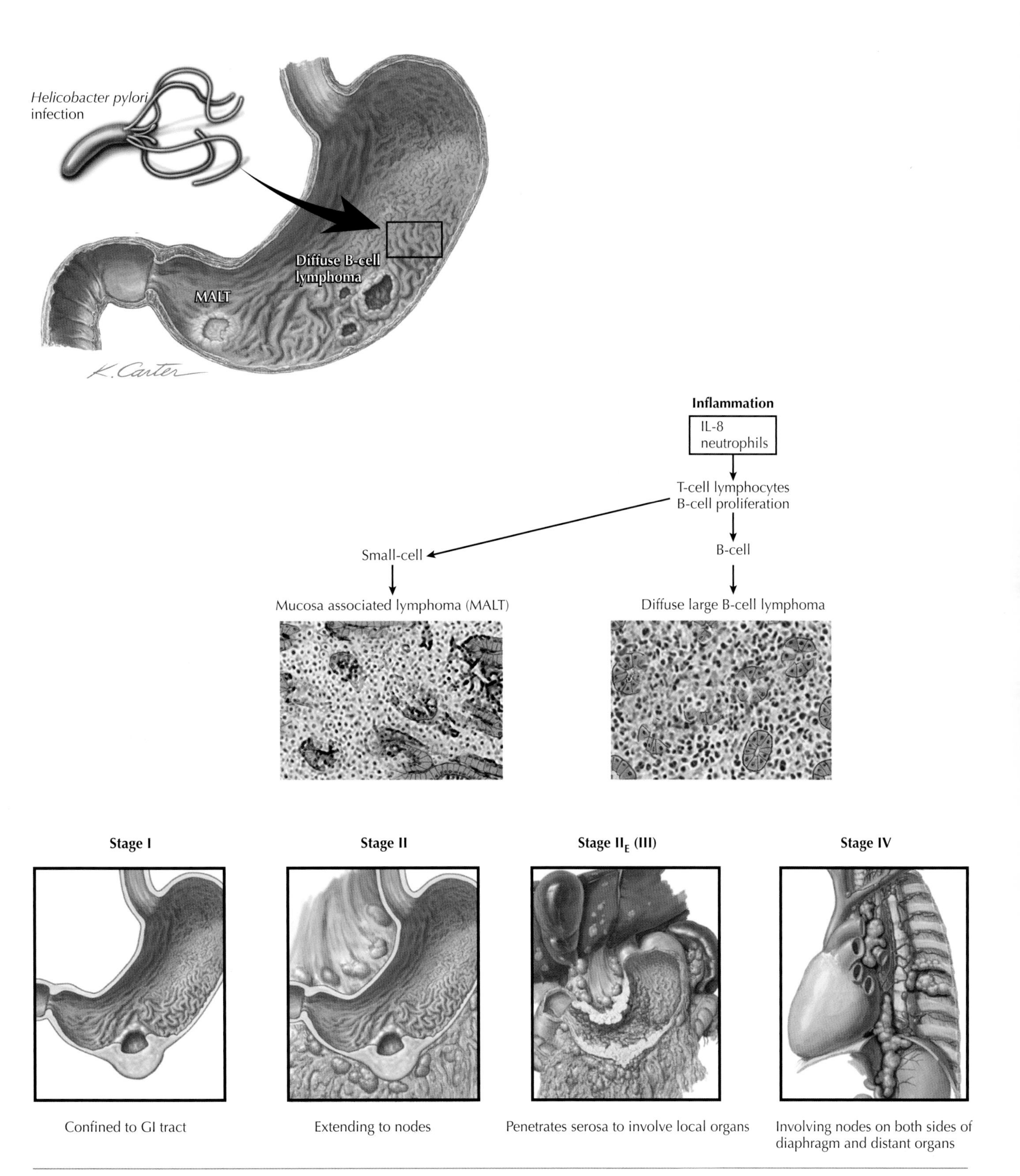

Figure 63-1 *Gastric Lymphoma and Mucosa-Associated Lymphoid Tissue (MALT) Tumor.*

Diagnosis

Evaluation must include upper endoscopy and CT evaluation of the chest, abdomen, and pelvis. EUS is helpful for assessing the full depth of the lesion in the wall of the stomach and the surrounding nodes.

Treatment, Course, and Prognosis

Approximately 70% of patients with stage I disease have no recurrence for 5 years with surgical therapy. However, some investigators believe that with adequate endoscopic diagnosis, patients can be treated with multiple chemotherapy regimens and radiation without surgery, providing there is thorough EUS evaluation. Large B-cell lymphoma appears to be responsive to radiation and to chemotherapy. Therefore, controversy surrounds the role of surgery and the relative roles of chemotherapy and radiation. Many oncologists think that radiation results in relapses, and that chemotherapy thus should be combined with radiation, which has become the standard of care when nodes are involved. Surgery should be considered in all these patients in combination with radiation and chemotherapy. One series reported survival at 40%.

When *H. pylori* infection is identified and treated, diffuse large B-cell lymphoma has responded to antibiotic therapy in rare cases, but such unimodal treatment is not recommended.

ADDITIONAL RESOURCES

Collins RH Jr: Gastrointestinal lymphomas. In Feldman M, Friedman LS, Brandt LJ, editors: *Gastrointestinal and liver disease*, ed 8, Philadelphia, 2006, Saunders-Elsevier, pp 565-587.

Dejong D, Boot H, Van Heerde P, Hart GAM: Histologic rating in gastric lymphoma: pretreatment criteria and clinical relevance, *Gastroenterology* 112:1466-1474, 1997.

Kuo SH, Chen LT, Wu MS, et al: Long-term follow-up of gastrointestinal patients with mucosa-associated lymphoid tissue lymphoma: need for a revisit of surgical treatment, *Ann Surg* 247:265-269, 2008.

Parsonnet J, Hansen S, Rodriguez L, et al: *Helicobacter pylori* infection in gastric lymphoma, *N Engl J Med* 330:1267-1271, 1994.

Wotherspoon AC, Doglioni C, Diss TC, et al: Regression of primary low-grade B-cell gastric lymphomas of mucosal-associated lymphoid tissue to *Helicobacter pylori*, *Lancet* (342):575-577, 1993.

Cancers of the Stomach

64

Martin H. Floch

Cancer involving the stomach is the second most common cancer in the world. Although it is decreasing in North America, stomach cancer continues to increase throughout the rest of the world. The incidence of adenocarcinoma has increased only in lesions at the cardioesophageal junction (**Fig. 64-1**).

The etiology of stomach cancer remains complex, and multiple factors are involved. Tobacco, alcohol, dietary nitrates, nitrites, and nitrosamines have all been implicated. High intake of salt has also been implicated in certain parts of the world, whereas increased refrigeration has been associated with a decrease in cancer. Epidemiologic studies show that *Helicobacter pylori* plays a role in gastric carcinogenesis. Chronic inflammation associated with *H. pylori* gastritis is the presumed mechanism. Atrophy of the gastric mucosa (as in pernicious anemia) and intestinal metaplasia are predisposing factors (see Chapter 55).

Most stomach cancers conform to one or two types: the *intestinal form*, which contains glandlike tubular structures, and the *diffuse form*, which contains poorly differentiated cells. The intestinal form is thought to involve steplike development, with the predisposing atrophic gastritis and intestinal metaplasia progressing to dysplasia and finally to cancer.

Genetic changes include loss of heterogenicity and loss of the effectiveness of P53G suppression and, in some patients, mutation of the APC/β-catenin pathway; expression of P16 and P27 is also decreased. Decreased expression of the epithelial cadherin gene in patients with diffuse gastric cancer may account for the morphologic differences between intestinal and diffuse gastric lesions. Families with hereditary diffuse gastric cancer have mutations in epithelial cadherin. Geneticists are beginning to understand the effect of environmental factors on the development of stomach cancer.

CLINICAL PICTURE

The clinical picture of stomach cancer varies greatly, depending on the site of the cancer. Lesions at the cardioesophageal junction, which are increasing in incidence in North America, tend to manifest early; they cause dysphagia or early dyspepsia and are diagnosed at endoscopy in the workup. Larger lesions in the fundus and body or ulcerating lesions may cause dyspeptic symptoms, anemia, or frank bleeding. Many are associated with anorexia and weight loss and manifest at a later stage. Of particular interest is *linitis plastica*, which diffusely involves the stomach with a fibrotic-type histology that results in anorexia and weight loss.

DIAGNOSIS

As endoscopy becomes more effective, the diagnosis of stomach cancer is frequently made on visualization and biopsy of the lesion. Radiographic studies, still used in many parts of the world, show the classic lesions, from a bottleneck linitis plastica to an ulceration that requires endoscopic evaluation and biopsy. It is still difficult to differentiate between a benign and a malignant ulcer. Differentiating gastric ulcer requires biopsy, as well as healing. Reevaluation is done in 3 to 6 weeks to ensure the ulcer has healed; if not, vigorous repeat biopsy evaluation is necessary.

Once the lesion has been identified on histologic evaluation, endoscopic ultrasound (EUS) can be helpful in determining the extent of the lesion through the stomach wall and whether lymph nodes are involved. Small nodes can still contain malignant cells and therefore can be diagnosed only during surgical exploration. Computed tomography (CT) can assist in identifying nodes missed on EUS and should be used for full evaluation. Endoscopy, EUS, and CT enable classification and staging of the disease.

TREATMENT AND MANAGEMENT

The only curative treatment for stomach cancer is surgery to remove the lesion. Surgery also provides important palliation for patients with large or obstructing lesions. Surgery is *not* used when there is extensive linitis plastica, or when the lesions have metastasized and carcinomatosis is present and resection or palliation would be useless.

The extent of surgery and lymph node resection varies, as decided by the patient and surgeon. Use of adjuvant chemotherapy or chemotherapy plus radiation is becoming more common. Depending on the center and experience of the staff, adjuvant therapy may be used before or after surgical resection. The benefits of vigorous chemotherapy and radiation after surgical resection also depend on the patient's clinical status.

Endoscopic mucosal resection of very small lesions is used in Japan. Endoscopic resection has been reserved for the intestinal histology in which EUS has demonstrated no lymph node involvement. The maximum size of the tumors has been less than 2 cm.

COURSE AND PROGNOSIS

Five-year survival rates of patients with stomach cancer range from 18% to 25% in the United States to as high as 50% in Japan. Survival may be higher in patients with very small lesions that are identified early. The intensive screening programs in Japan seem to be effective. Survival rates vary with the extent of the surgery and with the extent of the disease. Naturally, carcinoma in situ has the best prognosis. T1 lesions that involve the mucosa or submucosa have a better prognosis than T2 lesions that invade the tunica muscularis, T3 lesions that go beyond the serosa, or T4 lesions that invade adjacent organs. The stratification of lesions is further delineated by nodal involvement, which has a poor prognosis, as does any sign of metastasis.

Early gastric carcinoma, which involves T1 lesions, has up to an 85% to 90% cure rate. Because of early detection techniques, more lesions of early gastric cancer are discovered, and the prognosis has improved. Recent advances in chemotherapy have

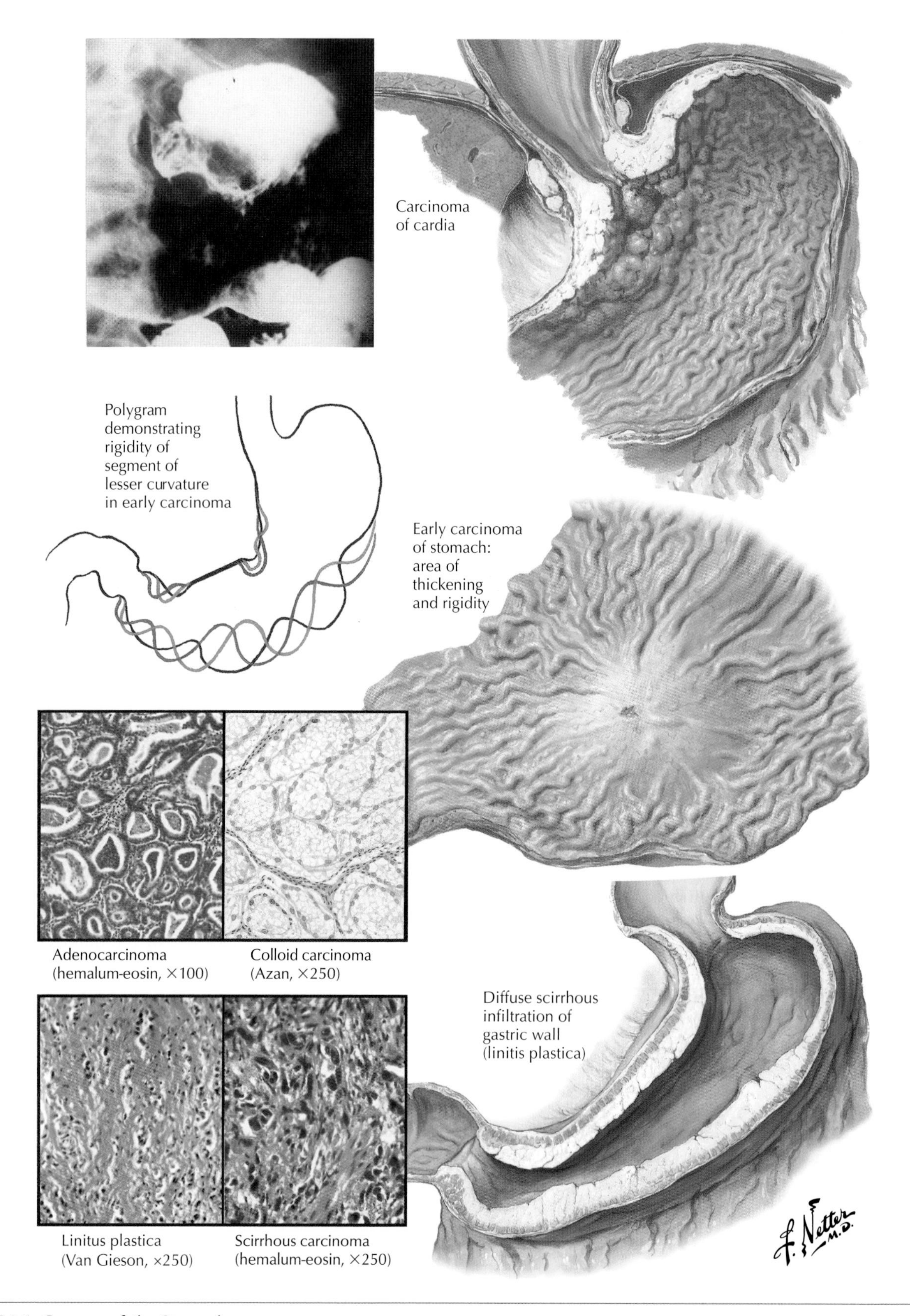

Figure 64-1 *Cancers of the Stomach.*

improved outcomes. Oral fluoropyrimidine S-1 has improved 3-year survival from 70% to 80% in East Asian patients who had a D2 dissection for locally advanced lesions.

GASTROINTESTINAL STROMAL TUMORS

Gastrointestinal stromal tumors (GISTs), also called *leiomyomas* or *leiomyosarcomas*, are rare (see also Chapter 62). Approximately 70% occur in the stomach, but they may occur in other parts of the GI tract as well. Their phenotypic feature is that of tumors originating from smooth muscle, neural elements, or both. GISTs appear to be relatively benign lesions, but when the pathologist can count more than five mitoses per 50 high-power fields, they are considered aggressive and can be of higher risk.

Clinically, GISTs often present with bleeding or as endoscopic findings on evaluation for dyspepsia. Endoscopically, GISTs may appear thimble shaped, and on EUS, they clearly arise from the stromal cell element. When larger than 4 cm, they are considered malignant.

Treatment is surgical removal. However, if the tumors are malignant and have metastasized, GISTs may respond to the tyrosine kinase inhibitor imatinib mesylate (Gleevec), used widely as adjuvant therapy. There are imatinib-resistant GISTs, and newer agents are being developed that appear effective. Therefore, an aggressive diagnostic and therapeutic approach is indicated to provide the appropriate therapy.

CARCINOID TUMORS

Approximately 2% to 3% of all carcinoid tumors arise in the stomach. They may be poorly differentiated or well differentiated, but carcinoids are thought to arise from neuroendocrine cells. They rarely have metabolic activity, unlike small-bowel carcinoids (see Chapter 121).

Carcinoid tumors are usually small and produce incidental findings when they originate in the stomach. However, there are rare reports of malignancy and spread, and these patients can have all the symptoms described for adenocarcinoma of the stomach. If the tumors are metabolically active, patients will present with the clinical picture of the small-bowel variety.

During endoscopy, carcinoid tumors may be seen as small ulcers, polyps, or tumors. They develop more frequently in patients with atrophic gastritis. The treatment of choice is removal of the lesion, but true *carcinoid syndrome* often metastasizes, and then the treatment is chemotherapy. Somatostatin has been used to control symptoms, although rarely in gastric carcinoid and primarily in the active, small-bowel lesions.

ADDITIONAL RESOURCES

Davila RE, Faigel DO: GI stromal tumors, *Gastrointest Endosc* 58:80-88, 2003.

Gupta P, Tewari M, Shukia HS: Gastrointestinal stromal tumor, *Surg Oncol* 17:129-138, 2008.

Haughton JM, Wang TC: Tumors of the stomach. In Feldman M, Friedman LS, Brandt LJ, editors: *Gastrointestinal and liver disease*, ed 8, Philadelphia, 2006, Saunders-Elsevier, pp 1139-1170.

Katz SC, Dematteo RP: Gastrointestinal stromal tumors and leiomyosarcomas, *Surg Oncol* 97:350-359, 2008.

Kaurah P, MacMillan A, Boyd N, et al: Founder and recurrent CDH1 mutations in families with hereditary diffuse gastric cancer, *JAMA* 297: 2360-2372, 2007.

Sakuramoto S, Sasako M, Yamaguchi T, et al: Adjuvant chemotherapy for gastric cancer with S-1, an oral fluoropyrimidine, *N Engl J Med* 357: 1810-1820, 2007.

Wang C, Yuan Y, Hunt RH: The association between *Helicobacter pylori* infection and early gastric cancer: a meta-analysis, *Am J Gastroenterol* 102:1789-1798, 2007.

Tumors of the Duodenum

Martin H. Floch

65

Tumors of the duodenum are rare. The *benign* neoplasms that may be encountered are Brunner gland hyperplasia, polypoid adenomas, lipomas, leiomyomas, neurofibromas, hemangiomas, and aberrant pancreatic tissue. All benign tumors are rare, and often only slightly elevated. A polyp may be on a pedicle and then appear mobile, shifting back and forth by peristaltic motion, and occasionally prolapsing into the pylorus (**Fig. 65-1**).

Carcinoma of the duodenum is also rare, but it is the most common site of primary small-bowel adenocarcinoma. The incidence of 0.35%, based on autopsy studies, is postulated to be a result of the rapid transit of material through the duodenum, lower bacterial load, neutralizing pH, and benzopyrene hydrolase, which is present in much higher concentrations and appears to detoxify benzopyrene, a carcinogen found in various foods. High-risk situations for adenocarcinoma are celiac sprue and familial adenomatous polyposis. Adenomas occur infrequently in patients with duodenal tumors, and the transition to malignancy does occur (see Chapter 120). Tumors of the ampulla are described in Chapter 207.

CLINICAL PICTURE

In sporadic reports, obstruction symptoms have occurred when a large polyp on a pedicle prolapsed into the pylorus, acting as a ball valve. However, the usual presentation of duodenal tumors is bleeding, anemia, or jaundice. Unfortunately, by the time of diagnosis, the lesions have frequently spread: In one series, as many as 70% were beyond the local site at presentation.

DIAGNOSIS

Endoscopy is the main tool for diagnosing duodenal tumors. Computed tomography may be helpful, and endoscopic ultrasound can assist in staging the disease. In some cases, barium contrast studies can help determine the extent of the lesion and evaluate the anatomy at surgery.

TREATMENT, COURSE, AND PROGNOSIS

Treatment is primarily surgical resection and removal of regional lymph nodes. Radical resection, including pancreaticoduodenectomy, may be of little help once node metastasis occurs. Chemotherapy is usually adjunctive or palliative.

Five-year survival rate for adenocarcinoma ranges from 10% to 20%. The presence of positive nodes reduces 5-year survival of 55% to 12%.

ADDITIONAL RESOURCES

Brophy C, Cahow CE: Primary small bowel malignant tumors, *Am J Surg* 55:408-416, 1989.

Gill SS, Heuman DM, Mihs AA: Small intestinal neoplasms, *J Clin Gastroenterol* 33:267-282, 2001.

Howe JR, Karnell LH, Menck HR, et al: The American College of Surgeons Commission on Cancer and the American Cancer Society: adenocarcinoma of the small bowel—review of the National Cancer Database, 1985-1999, *Cancer J* 1986:2693-2706, 1999.

Howe JR, Karnell LH, Scott-Conner C: Adenocarcinoma of the small bowel, *Cancer* 86:2693-2706, 1999.

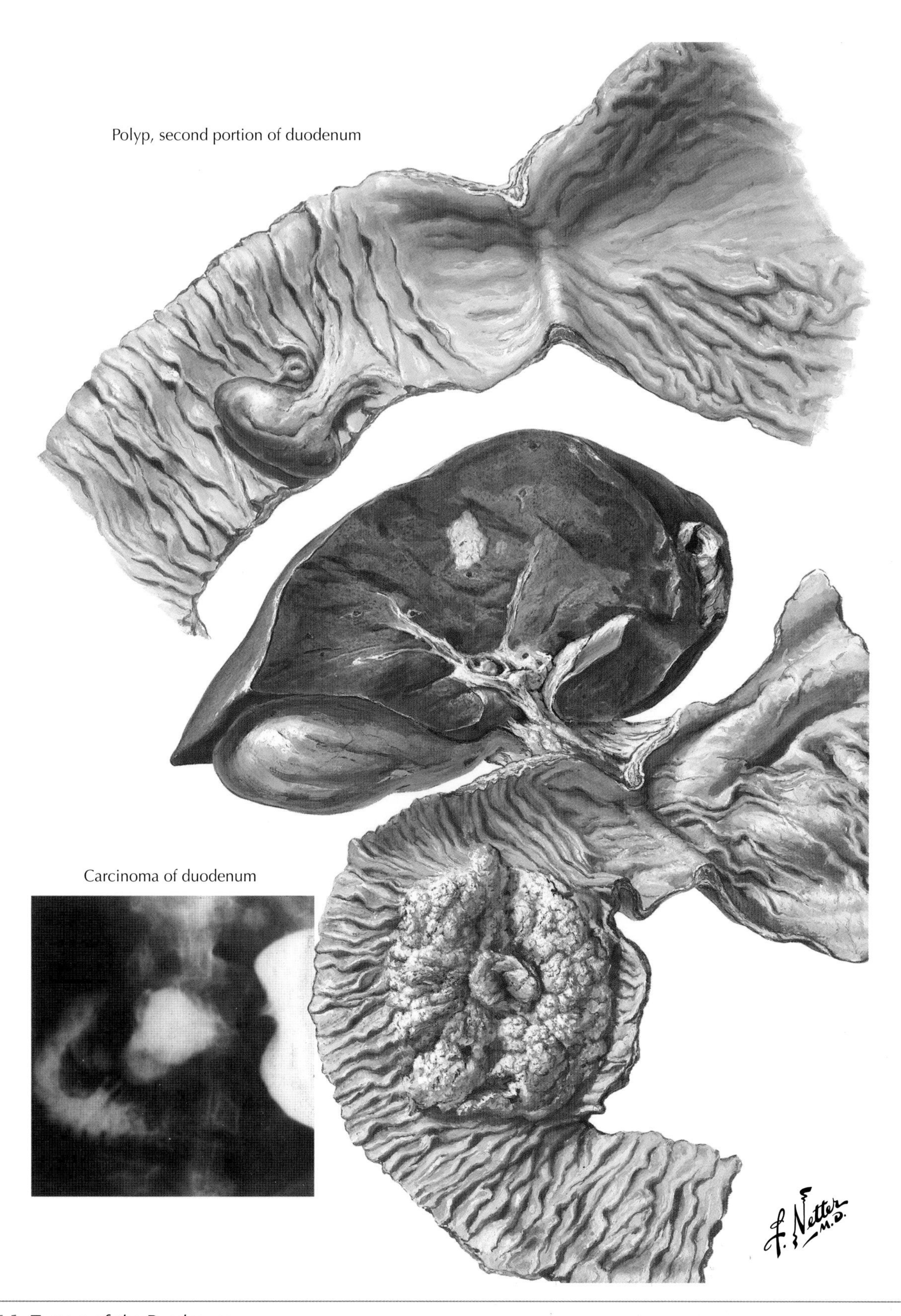

Figure 65-1 *Tumors of the Duodenum.*

Principles of Gastric Surgery

66

Martin H. Floch

With the advent of H_2 inhibitors and proton pump inhibitors (PPIs) and treatment of *Helicobacter pylori*, gastric surgery has declined significantly. Previously, the primary indication for gastric surgery was control of peptic ulcer disease, but the success of medical therapy greatly decreased the need for gastric surgery.

Indications for gastric surgery have not changed: uncontrollable gastrointestinal (GI) bleeding, perforation, or obstruction. A recent indication for gastric surgery is control of morbid obesity (see Chapters 67 and 68).

PEPTIC ULCER SURGERY AND VAGOTOMY

The procedure of choice for the surgical treatment of gastroduodenal ulcer is *subtotal gastrectomy* (**Fig. 66-1**). Two thirds to three quarters of the distal portion of the stomach is removed, to reduce the acid-secreting mucosa to such a degree that the gastric juice becomes *anacidic* (achlorhydric), or at least hypoacidic. Complete removal of the entire antrum is necessary. Several procedures have been developed, but only a few have stood the test of time.

The Viennese surgeon Billroth was the first to perform partial gastrectomy, which included the pylorus and connected the distal end of the remaining stomach with the open end of the duodenum (Billroth I). In some cases, however, because of technical difficulties, a sufficiently wide duodenal cuff is not available, or fibrosis in the area or anatomic restrictions may make the procedure difficult. Therefore, Billroth developed another type of gastrectomy (Billroth II), in which the duodenal stump is closed, and the stump of the stomach is connected to a loop of jejunum. Such a gastrojejunostomy can be constructed in front of the transverse colon or in retrocolic fashion. In the antecolic procedure, surgeons are careful to ensure that the afferent loop is free from the colon, and a side-to-side anastomosis of the afferent and efferent loops is created.

Vagotomies were performed before the development of PPIs and the excellent drugs that reduce acid secretion. Although now rarely used, vagotomy may be performed during a procedure for bleeding or may be necessary during radical surgery for cancer.

Surgery for peptic ulcer disease is rarely performed. Occasionally, however, it is performed after intractable bleeding or when a patient has difficulty using medications. For these patients, truncal vagotomy with drainage, highly selective vagotomy, or truncal vagotomy and antrectomy may be selected, depending on the surgical experience and the patient. Truncal vagotomy requires identification and destruction of the anterior and posterior vagi at the level of the distal esophagus. Highly selective vagotomy attempts to preserve other functions of the vagus but eliminates vagal innervation to the acid-producing stomach by dissecting the vagal distribution along the stomach. This procedure is difficult, and risk of ulcer recurrence depends on the surgical experience. When vagotomy is performed, complementary surgery—either gastrojejunostomy, to increase drainage through the pylorus, or antrectomy—is also usually performed to ensure the stomach can empty.

ROUX-EN-Y GASTROJEJUNOSTOMY

Previously reserved for complications of gastric surgery or as part of pancreatic resection for carcinoma, Roux-en-Y gastrojejunostomy is now used for obesity surgery (see Chapter 68).

TOTAL GASTRECTOMY

Total gastrectomy is usually performed with a resultant esophagojejunostomy or colon segment interposition to preserve food transit. These radical procedures are reserved for carcinoma of the upper part of the stomach or for unusual trauma to the upper abdomen.

ADDITIONAL RESOURCES

Kelly KA, Sarr MG, Hinder RA: *Mayo Clinic gastroenterology surgery*, Philadelphia, 2004, Saunders.

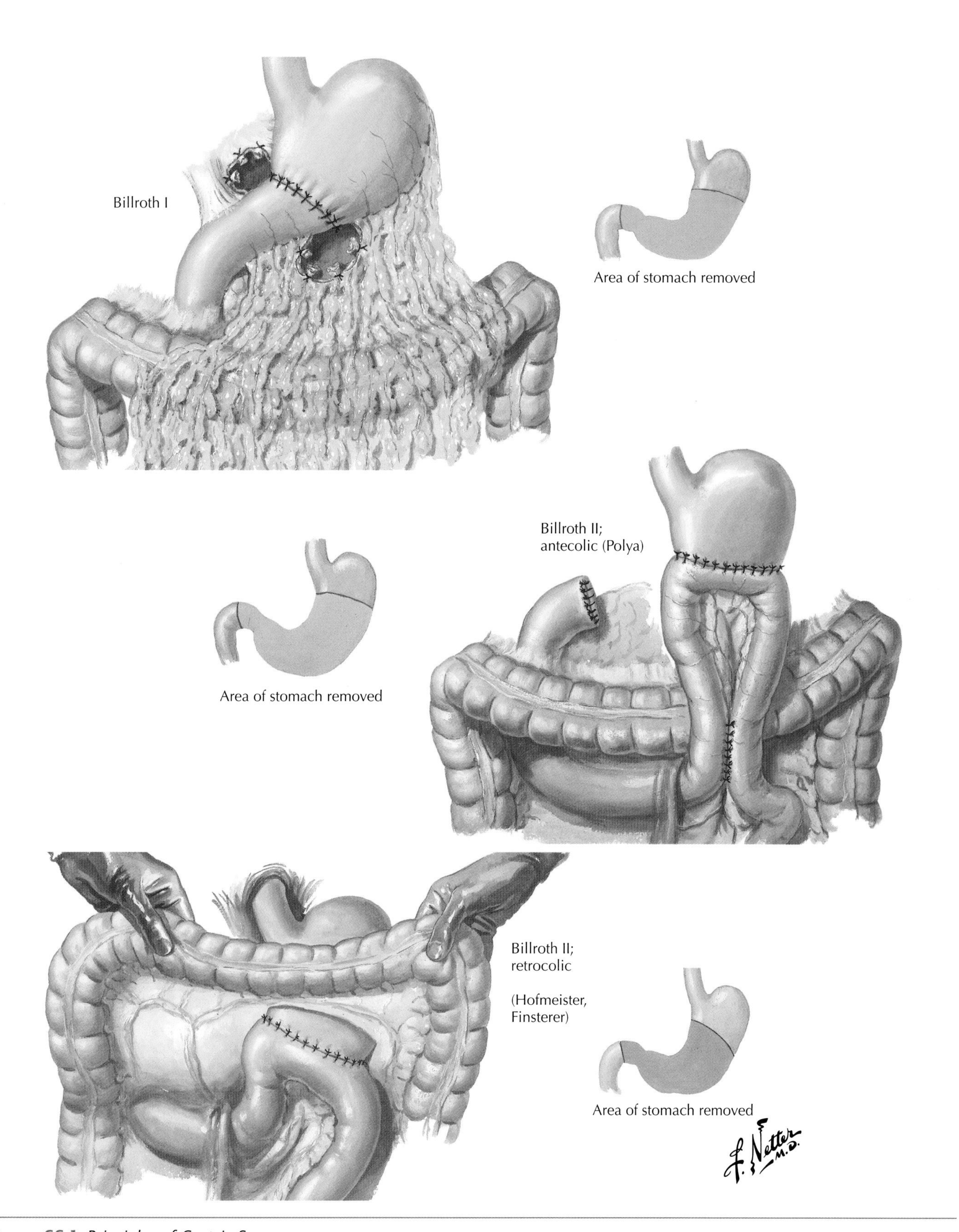

Figure 66-1 *Principles of Gastric Surgery.*

Treatment of Morbid Obesity

67

Neil R. Floch and Raul J. Rosenthal

Over the past 20 years, obesity has increased dramatically in the United States; currently, the total percentage of the U.S. population who are *overweight* is 64%. In 2006, 72 million U.S. adults were obese, or 33.3% of men and 35.3% of women. Adults age 40 to 59 had the highest obesity rates. Blacks and Mexican-American women had obesity rates of 53% and 51%, respectively, significantly higher than the Caucasian rate of 39%. Severely obese individuals now exceed 14 million, or 6.9% of U.S. women (10 million) and 2.8% of U.S. men (4 million). These patients are in the greatest need of assistance.

Obesity is responsible for at least 30 other diseases. Type 2 diabetes (T2DM) is associated with severe obesity in 20% of patients. Hypertension, hyperlipidemia, obstructive sleep apnea (OSA), hypoventilation, asthma, gastroesophageal reflux disease (GERD), coronary artery disease (CAD), chronic heart failure (CHF), cerebrovascular accident (CVA, stroke), nonalcoholic steatohepatitis (NASH), low back pain, degenerative joint disease (DJD), pseudomotor cerebri, urinary stress incontinence, and polycystic ovary syndrome (PCOS) are all associated with obesity. These patients have an increased risk of developing cancers of the esophagus, uterus, breast, prostate, liver, and kidney. Patients are also prone to ventral and incisional hernias. Severely obese people are at risk for mental disorders, including depression and anxiety, as well as eating disorders.

The result of increased morbidity in the obese population is a life span that shortens as body mass index (BMI) increases. Severely obese women are twice as likely to die as those of normal weight. The increase in prevalence of obesity worldwide and its growing effect on related medical diseases and mortality have empowered both medical and surgical efforts to combat this growing epidemic.

INDICATIONS

There are many behavioral, medical, and surgical options for treating obesity (**Fig. 67-1**). Nonsurgical options are open to all patients. Surgical therapy focuses on those with a BMI of 35 kg/m^2 or greater, although evidence now indicates that obese individuals with BMI lower than 35 may benefit from surgery as well, especially if they have T2DM.

Body mass index is the most practical and widely used measure of an individual's size. It is calculated by dividing the patient's weight in kilograms by the height in meters squared (kg/m^2). In adults, BMI is categorized as follows:

Underweight	<18.5 kg/m^2
Normal weight	18.5-24.9 kg/m^2
Overweight	25-29.9 kg/m^2
Class I obesity	30-34.9 kg/m^2
Class II obesity	35-39.9 kg/m^2
Class III obesity	40-49.9 kg/m^2 (severely, extremely, or morbidly obese)
Class IV obesity	>50 kg/m^2 ("superobese")

Bariatric surgery should be considered for patients who have failed medical management, such as dieting, exercise, and drug therapy. In 2004, the American Society for Metabolic and Bariatric Surgery (ASMBS) updated the 1991 National Institutes of Health (NIH) guidelines on obesity. Currently, candidates for surgery should have a BMI greater than 40 or BMI greater than 35 and major comorbidity, such as T2DM, OSA, obesity-related cardiomyopathy, or DJD. Although evidence indicates that patients with lower BMI, especially those with T2DM, would benefit from surgical intervention, this indication is not yet approved. Patients must have tried to lose weight by medical methods, must be motivated, and must be informed about the procedure and potential consequences. They must also be an acceptable surgical risk.

Patients who may not be considered for surgery include those with unstable CAD, severe pulmonary disease, portal hypertension, or active substance abuse, as well as those unable to carry out the necessary postsurgical lifestyle changes. Contraindications also include untreated major depression or psychosis, active binge eating, or severe coagulopathy. Bariatric surgery is controversial in those older than 65 or younger than 18, but these limits are now being relaxed as long-term positive outcome data in both populations are reported.

PREPARATION

Before surgical intervention, patients must attend an educational seminar and interact with former surgical patients. They receive an Internet tutorial on the procedure, are evaluated by a psychiatric therapist, and meet with a nutritionist. Many patients will need to have sleep apnea ruled out for if they have high BMI or severe symptoms. Cardiac and pulmonary evaluations are performed to assess the risk of anesthesia. Patients undergoing gastric bypass are evaluated by upper endoscopy, because after the gastric pouch is created and the stomach is divided, access to the stomach is extremely difficult. Patients also undergo blood testing for thyroid disease, liver disease, and T2DM. Postoperatively, patients are encouraged to follow up with nutrition and psychiatric support groups.

MANAGEMENT AND TREATMENT

The six strategies to obtain significant weight loss and maintain a healthy weight are (1) dieting, (2) physical activity, (3) behavior modification, (4) combination of three previous strategies, (5) pharmacotherapy, and (6) weight loss surgery. Evidence indicates that a low-calorie diet (LCD), when followed, will help reduce weight in overweight and obese individuals. When fol-

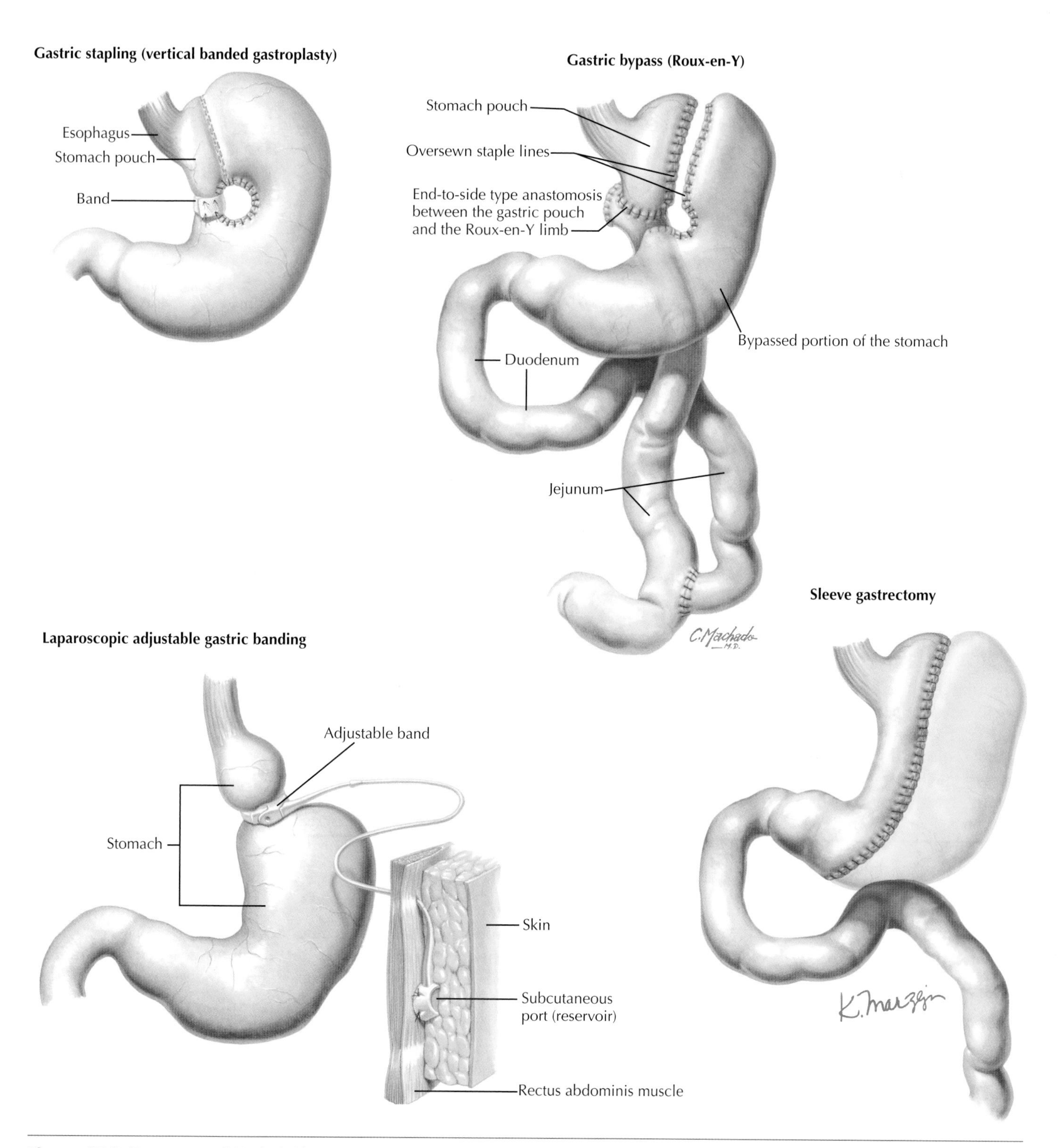

Figure 67-1 *Treatment of Morbid Obesity.*

lowed for 6 months, a diet of 800 to 1500 kcal/day will achieve an 8% weight loss. Fat reduction without caloric reduction does not work. A very-low-calorie diet (VLCD) of 400 to 500 kcal/day reduces weight more rapidly than an LCD of 1000 to 1500 kcal/day, but there is no difference in long-term weight loss between these two types of diets.

Physical activity can result in a caloric deficit and reduce BMI by 2% to 3%. It may also decrease abdominal fat and increase cardiovascular fitness. Physical activity is extremely important in preventing weight gain. Behavior modification to achieve dieting and physical activity is helpful; however, combining all these therapies has been found to be most successful.

Only two medications, orlistat and sibutramine, are approved by the U.S. Food and Drug Administration (FDA) for long-term weight loss. The side effects are fat malabsorption with orlistat and increased heart rate and blood pressure with sibutramine.

TYPES OF BARIATRIC PROCEDURES

Bariatric surgical procedures are classified according to their mechanism of action: malabsorptive or restrictive. The *restrictive* procedures limit caloric intake by creating a small stomach pouch, ranging from virtual in the Lap-Band, to a larger pouch in the vertical banded gastroplasty, to a long gastric tube in the gastric sleeve. Weight loss depends on a decrease in caloric intake and therefore is more gradual.

The primary mechanism of *malabsorptive* procedures is to create diverting pathways for food and digestive substances so that they meet distally within the small-bowel lumen, and therefore have a smaller length of absorptive surface area in which to interact. The biliopancreatic diversion and duodenal switch are examples of malabsorptive procedures. The Roux-en-Y gastric bypass combines features of both restriction and malabsorption with the creation of a small stomach pouch and a 25% to 30% functional small-bowel bypass.

The trend now is toward minimally invasive approaches to bariatric surgery; studies show better cost-effectiveness and safety than with open procedures.

RESTRICTIVE SURGERY

The *vertical banded gastroplasty* (VBG) is a purely restrictive procedure in which the upper cardia of the stomach is separated by a vertical staple line from the remainder of the stomach. The outlet is then wrapped with a piece of mesh or a band. The outlet aperture is not adjustable. When the stomach pouch is full, patients are satiated. Further eating may result in vomiting if the pouch is not allowed to empty. Weight loss occurs because of decreased caloric intake of solid food. *Excess weight loss* (EWL) is as much as 66% at 2 years and 55% at 9 years. The ability to consume high-calorie liquid meals and sweets and gradually increased pouch capacity caused by overeating are major disadvantages. The VBG has become antiquated because it combines the disadvantages of a higher complication rate and the inability to adjust the band. The rate of revision ranges from 20% to 56% and mainly results from staple line disruption, stomal stenosis, band erosion, band disruption, pouch dilatation, vomiting, and GERD.

Laparoscopic adjustable gastric banding (LAGB) is a purely restrictive procedure that separates a micropouch from the remainder of the stomach. American surgeon Lubomyr Kuzmak developed the open adjustable band. In 1991, Inamed Corp. developed the LAGB with Belgian surgeons M. Belachew and M. Legrand. The first human LAGB was placed in 1993 in Belgium. Starting in 1994, LAGB placement was performed extensively in Europe and Australia. The FDA approved its use in the United States in 2001. Since then, more than 400,000 Lap-Bands (Allergan) have been placed worldwide. Currently, a second band, Realize (Ethicon), has been introduced in the United States. Both bands are composed of (1) a silicone band with a balloon inner tube that wraps around the stomach, (2) a portacath (Port-A-Cath) that lies under the skin on the rectus muscle for access, and (3) tubing to connect the two. The band is accessed 4 to 6 weeks after surgery by inserting a needle and syringe into the port and injecting or withdrawing fluid. In this manner, the balloon increases in diameter, and the aperture between the two stomach compartments becomes smaller as the patient undergoes more restriction.

The LAGB is now widely used because of easy placement, quick recovery and same-day discharge or 1-day hospital stay, and lower complication rate. The adjustable band has almost replaced VBG as the main purely restrictive procedure. Advantages of the LAGB include no stapling of bowel, zero to 0.5% mortality with adjustability, and minimal nutritional complications. The band is reversible because it can be completely removed. Initial reports from Europe and Australia report EWL of 15% to 20% at 3 months, 40% to 53% at 12 months, and 45% to 58% at 24 months. The first American experience had poor results, but after improved treatment protocols, subsequent data showed similar experiences to worldwide data of EWL of 45% to 75% at 24 months. Weight loss after LAGB occurs gradually compared with malabsorptive procedures. Successful weight loss is more likely when patients are committed to close follow-up. LAGB is associated with improvement and resolution of comorbidities, such as diabetes, asthma, sleep apnea, and hypertension that depends on weight loss. Quality of life also greatly improves.

The *sleeve gastrectomy* (SG) is the first part of the biliopancreatic diversion (BPD) procedure, and in recent years, the procedure has been separated into two stages to lower the mortality rates seen in higher-BMI patients undergoing BPD. Since then, SG has gained support as a purely restrictive, stand-alone procedure. It entails the creation of a stomach tube from cardia to antrum that involves removal of the fundus and body of the stomach along the greater curvature. The antrum is left intact. Weight loss occurs from a smaller stomach from which the ghrelin-producing cells have been removed. Patients experience approximately 33% EWL in 1 year. Patients who reach a plateau or who regain weight may opt for completion to a laparoscopic Roux-en-Y gastric bypass (RYGB) or BPD.

The *intragastric balloon* (Allergan) is an endoscopically placed temporary solution for weight loss in obese patients. The soft balloon is inserted in the stomach and inflated with saline. The distended device fills the stomach and induces satiety while causing restriction. It is currently in trials in the United States, Europe, and Brazil. EWL is 38% and 48% for 500-mL and 600-mL balloons, respectively. Although appealing to individuals who desire to lose a moderate amount of weight, its greatest benefit may be as an initial procedure in superobese, high-risk patients to help them lose enough weight to be considered lower-risk candidates for definitive laparoscopic procedures. Intragastric balloon placement has the disadvantages of nausea, vomiting, abdominal pain, ulceration, and balloon migration, as well as its temporary effect.

MALABSORPTIVE SURGERY

Biliopancreatic diversion was developed because of poor results with jejunoileal bypass; many patients developed kidney problems and liver failure. The BPD involves a partial gastrectomy that is anastomosed distally to the ileum. There is a long segment of Roux limb and a short common channel where the food and biliopancreatic juices meet to allow for absorption. The process results in significant malabsorption and 72% EWL at up to 18 years postoperatively. The procedure is now performed laparoscopically with similar results. Disadvantages of BPD include mortality of 1% and high incidence of protein malnutrition, anemia, diarrhea, and stomal ulceration.

The BPD with duodenal switch (BPD/DS) is a BPD that differs by creating a partial SG with preservation of the pylorus. A Roux limb with a short common channel is also created. The BPD/DS has been recommended for patients with supermorbid obesity (BMI >50). Unfortunately, the high mortality rate has led to the development of a staged procedure, with the gastric sleeve done first, then BPD later if more weight loss is necessary. BPD/DS results in less stomal ulceration and diarrhea than BPD and can be performed laparoscopically. It is not performed routinely because of high morbidity and mortality.

MIXED PROCEDURES

The *Roux-en-Y gastric bypass* is the most common surgical procedure for weight loss in the United States. Developed in 1967 by Mason and Ito, it included a partial gastrectomy and loop gastrojejunostomy. It was originally performed after the observation that patients lost weight after ulcer surgery. The procedure was modified by Griffin in 1977 to include a Roux-en-Y. The first laparoscopic bypass was performed by Wittgrove in 1994. Laparoscopy has a lower incidence of incisional hernia and wound infection, a faster recovery, and more rapid return to work than open surgery. It is safely performed by well-trained and experienced surgeons.

Weight loss after gastric bypass is mostly attributed to restriction, but also to malabsorption. It has better weight loss averages than the purely restrictive procedures. The bypass is composed of a 30-mL gastric pouch separated from the gastric remnant and restricting food intake and emptying. The pouch is connected by a stoma to the small bowel, creating a Roux-en-Y segment. The remnant stomach and duodenum empty gastric acid, pancreatic enzymes, and bile through a 30-cm to 50-cm biliopancreatic limb, which is connected to the "food" or Roux-en-Y limb at 75 to 150 cm distally. At this location, significant digestion begins. A short Roux limb will not add the benefit of malabsorption, although limbs longer than 200 cm in long-limb bypass may lead to decreased absorption of vitamins and nutrients and complications of malnutrition.

The gastrojejunostomy may cause "dumping," which results in rapid emptying of concentrated contents into the small bowel and symptoms of lightheadedness, nausea, diaphoresis, abdominal pain, and diarrhea. Some patients develop a negative conditioning response when eating concentrated foods to prevent this response. Ghrelin is a peptide hormone secreted in the stomach and duodenum that stimulates people to eat. Its response to eating is inhibited by the gastric bypass and the SG. Recent evidence suggests that an exaggerated response of peptide-yy may also contribute to loss of appetite.

Early weight loss is rapid but reaches a plateau after 18 months, after which patients tend to gain weight if behavior modification is not instituted. Long-term studies include open techniques with a larger gastric pouch and range from 50% to 60% EWL for up to 16 years postoperatively. Most recent EWL is now 70% to 80% with the laparoscopic technique and a 30-mL gastric pouch. After bypass, overall comorbidity is reduced as much as 96%. T2DM is resolved or improved in 83% to 98%, hypertension in 52% to 92%, GERD in 88% to 98%, OSA in 86% to 93%, dyslipidemia in 70% to 96%, and osteoarthritis in 93% of patients.

COURSE AND PROGNOSIS

Surgery is the last option after medical failure in attempting to reverse the effects of comorbidities and improve quality of life and extend life expectancy. Evidence shows that bariatric surgery is effective in improving and resolving medical problems. Additional benefits include reduced medication costs and fewer lost workdays. However, bariatric surgery is also associated with significant perioperative complications and mortality.

It is now believed that bariatric surgery may be the best treatment for T2DM. Recently, the previously named "American Society for Bariatric Surgery" included the word "Metabolic" in its title (now ASMBS) to account for the emerging field of metabolic surgery, which includes treatment of diabetes, hypertension, and hyperlipidemia. Evidence suggests that T2DM improvement and resolution may occur in restrictive procedures directly related to the amount of EWL. A recent study comparing 30 obese T2DM patients who underwent 2 years of dieting and lifestyle modification and 30 who underwent adjustable LAGB reported diabetes remission in 73% of the surgical group versus 13% in the lifestyle modification group. Malabsorptive procedures may involve another mechanism that changes the body's response to gastrointestinal hormones, such as incretins, glucagon-like peptide, and glucose-dependent insulinotrophic polypeptide.

Two meta-analyses reported that mean overall percent of EWL was 61% to 64%, which varied according to procedure. Mortality at 30 days was 0.1% for purely restrictive procedures, 0.5% for gastric bypass, and 1.1% for BPD or DS. Diabetes completely resolved in 77% and resolved or improved in 86% of patients. Hyperlipidemia improved in 70% or more of patients. Hypertension resolved in 62% and resolved or improved in 79% of patients. OSA resolved in 86% and resolved or improved in 84% of patients. Greater weight loss occurred with gastric bypass compared with gastroplasty. Laparoscopic surgery resulted in less wound complications than open surgery.

Two recent studies report a 29% reduction in mortality after bariatric surgery and a 40% reduction in mortality from all causes. Deaths from diabetes decreased by 92%, from CAD by 56%, and from cancer by 60%. However, accidental death and suicide increased in incidence. A population study reported that obese surgical patients had a lower overall mortality rate of 0.7% versus 6.2% in the nonsurgical group.

Centers of Excellence (COE) have been established for bariatric facilities. Guidelines for the two current accreditation programs (ASMBS-affiliated Surgical Review Corporation and American College of Surgeons) include integrated preoperative/postoperative care with nutritional, behavioral, and medical programs and a commitment to follow up with 75% of postsurgical patients for 5 years.

The obesity epidemic must be reversed with a global effort to alter eating behavior and promote exercise. Until then, surgery will remain the most viable option to treat morbid obesity and related comorbidities.

ADDITIONAL RESOURCES

Adams TD, Gress RE, Smith SC, et al: Long-term mortality after gastric bypass surgery, *N Engl J Med* 357:753, 2007.

Belachew M, Legrand M, Vincenti VV, et al: Laparoscopic placement of adjustable silicone gastric band in the treatment of morbid obesity: how to do it, *Obes Surg* 5:66, 1995.

Buchwald H, Avidor Y, Braunwald E, et al: Bariatric surgery: a systematic review and meta-analysis, *JAMA* 292:1724, 2004.

Christou NV, Sampalis JS, Liberman M, et al: Surgery decreases long-term mortality, morbidity, and health care use in morbidly obese patients, *Ann Surg* 240:416, 2004.

Cummings DE, Weigle DS, Frayo RS, et al: Plasma ghrelin levels after diet-induced weight loss or gastric bypass surgery, *N Engl J Med* 346:1623, 2002.

Dixon JB, O'Brien PE, Playfair J, et al: Adjustable gastric banding and conventional therapy for type 2 diabetes, *JAMA* 299:316, 2008.

Hess DS, Hess DW: Biliopancreatic diversion with a duodenal switch, *Obes Surg* 8:267, 1998.

Jones SB, Jones DB: *Obesity surgery: patient safety and best practices*, Woodbury, Conn, 2009, Cine-Med, pp 33-34.

National Institutes of Health: Gastrointestinal surgery for severe obesity: Consensus Development Conference Panel, *Ann Intern Med* 115:956, 1991.

Pories WJ, Swanson MS, MacDonald KG, et al: Who would have thought it? An operation proves to be the most effective therapy for adult-onset diabetes mellitus, *Ann Surg* 222:339, 1995.

Regan JP, Inabnet WB, Gagner M, Pomp A: Early experience with two-stage laparoscopic Roux-en-Y gastric bypass as an alternative in the super-super obese patient, *Obes Surg* 13:861, 2003.

Roman S, Napoleon B, Mion F, et al: Intragastric balloon for "non-morbid" obesity: a retrospective evaluation of tolerance and efficacy, *Obes Surg* 14:539, 2004.

Schauer P, Ikramuddin S, Hamad G, Gourash W: The learning curve for laparoscopic Roux-en-Y gastric bypass is 100 cases, *Surg Endosc* 17:212, 2003.

Scopinaro N, Gianetta E, Adami GF, et al: Biliopancreatic diversion for obesity at eighteen years, *Surgery* 119:261, 1996.

Sjostrom L, Narbro K, Sjostrom CD, et al: Effects of bariatric surgery on mortality in Swedish obese subjects, *N Engl J Med* 357:741, 2007.

Suter M, Jayet C, Jayet A: Vertical banded gastroplasty: long-term results comparing three different techniques, *Obes Surg* 10:41, 2000.

US Department of Health and Human Services: Statistics related to overweight and obesity, Bethesda, Md, 2008, Weight Control Information Network (WIN). http://www.win.niddk.nih.gov/publications/PDFs/stat904z.pdf.

Complications of Bariatric Surgery

68

Raul Rosenthal, Neil R. Floch, and Jeremy Eckstein

Bariatric surgery is an elective procedure for the high-risk obese population. Despite great improvements in surgical technique and postoperative care, the morbidity is still about 30% with bariatric surgery. These patients have an *increased inflammatory response*, making them prone to overwhelming shock in the presence of complications. Patient monitoring is essential in the first postoperative days because small changes in vital signs can be the earliest indication of complications. Often, when the patient develops clear signs of sepsis, the complication has already developed.

Another clear sign that must be monitored in the postoperative patient is persistent *tachycardia*. A heart rate greater than 120 beats per minute in the bariatric patient is a concern, even if the patient otherwise appears stable. Tachycardia is the most sensitive sign of an anastomotic leak, present in up to 72% of patients with complications, and is often the first sign to appear. Tachycardia is not specific for leaks, however, and other causes of tachycardia (e.g., hypoxia, hypertension, hemorrhage, pain) must be excluded. All patients with postoperative tachycardia should be monitored until the cause is elucidated and resolved.

Roux-en-Y gastric bypass (RYGB) is the most common bariatric procedure performed in the United States and has the highest complication rate because of its complexity. *Laparoscopic adjustable gastric banding* (LAGB) and *laparoscopic sleeve gastrectomy* (LSG) are increasingly used.

This chapter discusses complications of bariatric surgery that include those specific to RYGB as well as LAGB and LSG. Early complications include bleeding, anastomotic leak, deep venous thrombosis (DVT), and pulmonary embolism (PE). Late complications include anastomotic stricture, marginal ulcer, gastrogastric fistula, bowel obstruction, metabolic disorders (**Fig. 68-1**).

EARLY COMPLICATIONS

Bleeding

Bleeding is a complication inherent to any surgical procedure; its incidence may be slightly increased in the bariatric population because of the rigorous DVT prophylactic regimen. In our experience, postoperative hemorrhage is seen in 4% of patients. The bleeding is intraluminal in 60% of bleeding cases and intraperitoneal in 40%. Postoperative bleeding can be diagnosed by clinical signs of tachycardia, hypotension, increased drain output, and decreased hemoglobin level or collection on computed tomography (CT) scan. The most common site of intraluminal bleeding is at the anastomotic level. Bleeding can be successfully managed with observation supported by blood transfusion in 75% of patients; 25% of our bleeding patients required reexploration.

Anastomotic Leak

Anastomotic leak is the second leading cause of death (after PE) following gastric bypass (bariatric) surgery. Leak occurs in 2% of patients regardless of the technique used to create the anastomosis, with most leaks occurring on day 3. As with bleeding, the most common site of leak is at the anastomotic level, with a higher prevalence at the gastrojejunostomy, followed by the jejunojejunostomy, the gastric pouch, and finally the gastric remnant.

Early detection of anastomotic leaks is vital to successful management. The most sensitive sign of leak is persistent tachycardia (72% of patients); left shoulder pain, abdominal pain, and fever are not as sensitive. Gastrointestinal (GI) studies and CT scan can detect leaks in only 30% and 56% of cases, respectively. Because of the lack of specificity in clinical presentation and imaging studies, surgical exploration should be part of the diagnostic algorithm. About 60% of patients will require surgical intervention. Patients who achieve good drainage through the original drains placed and show no signs of sepsis may be treated conservatively with strict nothing-by-mouth (*nil per os*, NPO) status, total parenteral nutrition, and intravenous (IV) antibiotics. Patients with signs of sepsis should undergo diagnostic laparoscopy, abdominal washout, and drain placement. Any attempt to redo the anastomosis is strongly discouraged because of the increased risk of anastomotic breakdown in the presence of inflammation.

Leaks are also a potential complication after LSG, seen in 0.8% of cases. In our experience, most leaks are located at the most proximal part of the staple line by the gastroesophageal (GE) junction. Patients presenting with anastomotic leak will ultimately need surgical management.

Pulmonary Embolism and Deep Venous Thrombosis

Pulmonary embolism is the leading cause of death after bariatric surgery. The incidence of PE after RYGB is 0.41%. Risk factors for PE and DVT development are older than 50 years of age, postoperative anastomotic leak, smoking, and previous DVT/PE. All patients undergoing RYGB should receive subcutaneous heparin before and after surgery, as well as compression stockings and early ambulation. Low-molecular-weight heparin should be avoided the first 48 hours postoperatively because of an increased incidence of bleeding. Patients with a previous history of PE/DVT, pulmonary hypertension or lymphedema or with severely impaired ambulation, should be considered for prophylactic placement of an inferior vena cava (IVC) filter.

LATE COMPLICATIONS

Anastomotic Stricture

Anastomotic stricture, mainly at the level of the gastrojejunostomy, is a relatively common complication, seen in 1.6% to 20% of bariatric patients. Symptoms usually start 3 weeks after surgery. Patients present with nausea, vomiting, and inability to progress to solid foods. Diagnostic studies should include upper GI series and upper endoscopy.

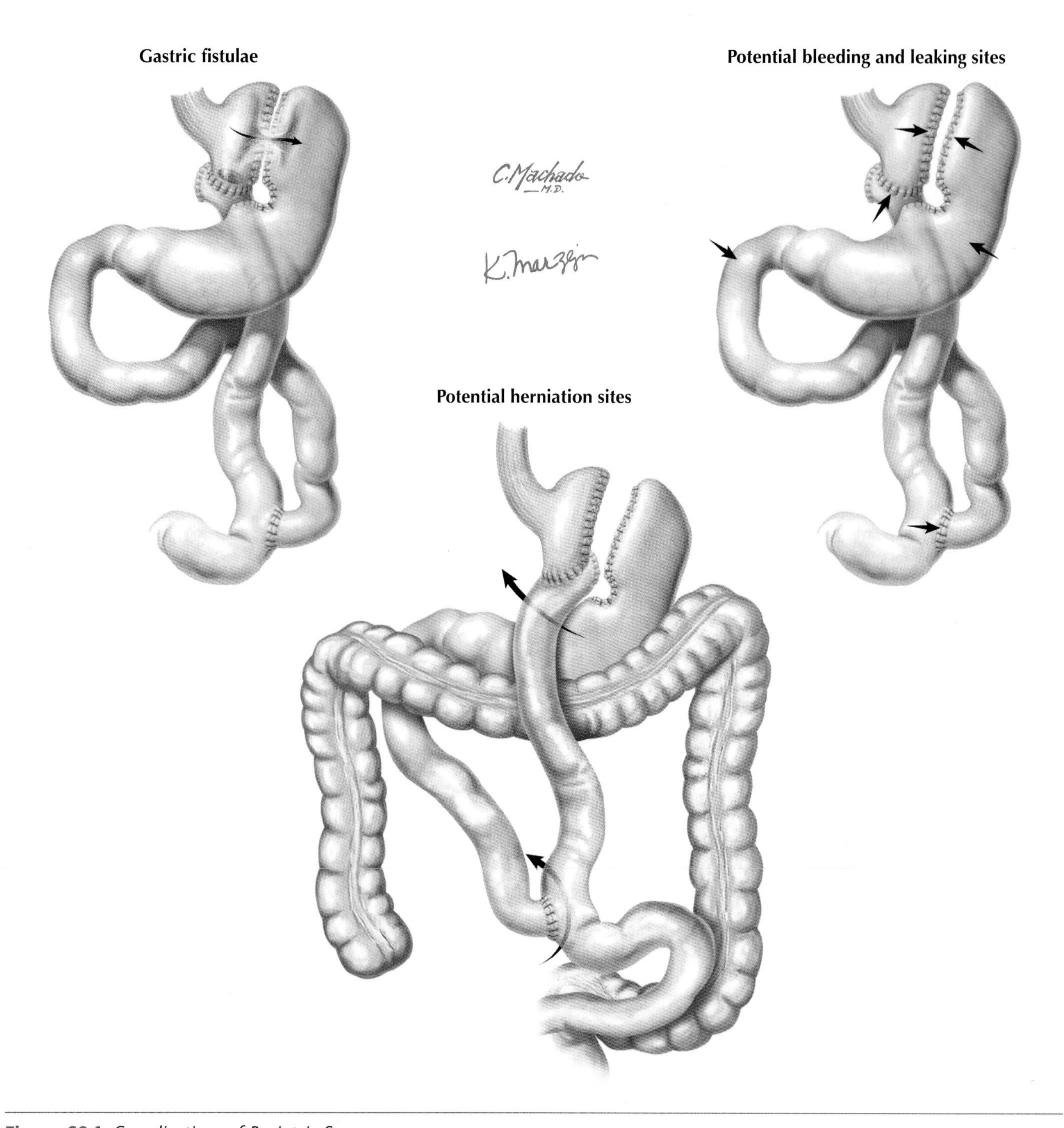

Figure 68-1 *Complications of Bariatric Surgery.*

Management of stricture consists of serial endoscopic balloon dilatations (usually up to three), separated by 2-week intervals. Initial dilatations should not exceed 8 mm because of the risk of perforation (2.2%); subsequent dilatations can be increased up to 18 mm. The patient with perforation will complain of abdominal pain and distention, and a plain abdominal film will reveal free air under the diaphragm. These perforations can be treated conservatively with IV antibiotics and NPO status if the patient is asymptomatic. Alternatively, if the patient shows any sign of sepsis, or if a large extravasation of contrast is seen on radiographic studies, diagnostic laparoscopy and drain placement are essential. Failed weight loss after dilatation is not seen in patients on long-term follow-up.

The incidence of stricture is much lower after LSG (0.7%), but it is still a potential complication. Patients will present with excessive nausea, vomiting, and dehydration. An upper GI study is diagnostic. Most strictures will resolve after endoscopic balloon dilatation, but conversion to RYGB may be necessary with a long stricture not amenable to dilatation.

Marginal Ulcer

Marginal ulceration after gastric bypass is diagnosed in 1% to 16% of patients. Predisposing factors are active smoking and preoperative *Helicobacter pylori* infection. Patients usually present with midepigastric pain and/or upper GI bleeding. Upper endoscopy is diagnostic. In most cases, ulceration is seen at the jejunal mucosa of the gastrojejunostomy and mucosal erythema of the gastric pouch. Most bleeding ulcers respond to endoscopic therapy. Patients should take proton pump inhibitors (PPIs) and sucralfate for 2 months. Smoking cessation is mandatory.

Gastrogastric Fistula

Gastrogastric fistula (GGF) is a challenging late complication, with an incidence of approximately 1%. The vast majority of GGFs are symptomatic at diagnosis. Symptoms consist of moderate to severe epigastric pain, nausea, and vomiting, with GI bleeding in fewer cases. CT may show oral contrast in the gastric remnant, but this finding is of low specificity because contrast may reflux from the biliopancreatic limb. The most sensitive test is a barium swallow with position changes (erect, supine, left lateral, and right lateral decubitus); esophagogastroduodenoscopy (EGD) may be diagnostic in some cases.

The GGF may resolve after medical treatment, so the presence of GGF is not an indication for prompt surgery. The surgical management of symptomatic GGF after medical failure consists of laparoscopic partial remnant gastrectomy with or without trimming of the gastric pouch, fistulous tract, or both, while leaving the gastrojejunostomy intact. Because remnant gastrectomy may be associated with a high complication rate, surgical treatment should be offered to patients with intractable disabling symptoms, evidence of anastomotic leak, or peritonitis.

Bowel Obstruction

Severe bowel obstruction (SBO) after laparoscopic RYGB has many causes, usually iatrogenic etiologies resulting from narrow anastomoses, tight closure of mesenteric defects, mesenteric or intramural hematomas, anastomotic leaks, incarcerated ventral hernias, internal hernias, and adhesions. Patients may develop symptoms in the immediate postoperative period up to years after surgery. The incidence of SBO varies from 0.4% to 7.45%. With the adoption of the laparoscopic approach, postoperative SBO secondary to adhesions and incisional hernias has been reduced. However, a higher incidence of SBO caused by internal hernias is seen, compared with the open procedure, if the retrocolic retrogastric route is chosen to advance the alimentary limb. Internal herniation can occur at the jejunojejunostomy, Petersen's space, or the transverse mesocolonic defect after a retrocolic approach.

If internal herniation is suspected, the patient must be taken to the operating room as soon as possible to reduce the herniation and prevent intestinal necrosis. Most SBO cases can be approached laparoscopically, but the conversion rate is higher due to loss of working space because of distended bowel. Procedures include lysis of adhesions, closure of mesenteric defects, and reconstruction of the jejunojejunostomy.

Metabolic Complications

Nutritional deficiencies are rare in purely restrictive procedures such as LAGB or LSG, unless the patient has significant changes in eating habits or complications. Macronutrient deficiency or protein-calorie malnutrition can be found in up to 5% of gastric bypass patients, and its prevalence is directly proportional to the length of the alimentary limb. Most of these patients can be managed by nutritional consult and guidance. Micronutrient deficiencies include vitamins, minerals, trace metals, and electrolytes, all of which are absorbed at specific sites in the small intestine. The bypassing of these sites and the reduction in the small intestine's absorptive capacity can lead to deficiencies. The most common deficiencies associated with RYGB are iron, vitamin B_{12}, calcium, and vitamin D. Complications of these deficiencies range from anemia to irreversible encephalopathy. All patients undergoing gastric bypass should be instructed to take nutritional supplements for the rest of their lives. Follow-up with annual testing of vitamin levels is mandatory.

When nutritional consult and vitamin supplements do not correct the patient's nutritional deficiencies, parenteral nutrition should be started, with consideration of revision or reversal of the bypass.

COMPLICATIONS WITH ADJUSTABLE GASTRIC BAND

Regarded as the least invasive of all bariatric procedures, LAGB has a much lower complication rate (13%) than gastric bypass (31%). However, LAGB has complications not seen with other bariatric procedures, including megaesophagus (2.6%), band prolapse (1.3%), and band erosion (1.3%). These complications are believed to be caused by overinflation and poor fixation at band placement.

Regardless of the LAGB complication, most patients will present with GERD-like symptoms and intolerance to food. The first test is a barium swallow, usually performed by the surgeon, and the band may be adjusted simultaneously. If any abnormality is seen on the swallow study, the band is deflated completely and the patient scheduled for endoscopy. In mega-esophagus or band prolapse, a trial of 8 weeks with an empty band is offered to the patient, followed by a repeat barium study. If there is no improvement, the patient is scheduled for surgical removal or repositioning of the band. With band erosion, the device should be removed in all cases.

ADDITIONAL RESOURCES

Ballesta C, Berindoague R, Cabrera M, et al: Management of anastomotic leaks after laparoscopic Roux-en-Y gastric bypass, *Obes Surg* 18(6):623-630, 2008.

Carrodeguas L, Szomstein S, Soto F, et al: Management of gastrogastric fistulas after divided Roux-en-Y gastric bypass surgery for morbid obesity: analysis of 1,292 consecutive patients and review of literature, *Surg Obes Relat Dis* 1(5):467-474, 2005.

Chevallier JM, Zinzindohoué F, Douard R, et al: Complications after laparoscopic adjustable gastric banding for morbid obesity: experience with 1,000 patients over 7 years, *Obes Surg* 14(3):407-414, 2004.

Cho M, Kaidar-Person O, Szomstein S, Rosenthal RJ: Laparoscopic remnant gastrectomy: a novel approach to gastrogastric fistula after Roux-en-Y gastric bypass for morbid obesity, *J Am Coll Surg* 204(4):617-624, 2007.

Filho AJ, Kondo W, Nassif LS, et al: Gastrogastric fistula: a possible complication of Roux-en-Y gastric bypass, *J Soc Laparoendosc Surg* 10(3):326-333, 2006.

Frezza EE, Reddy S, Gee LL, Wachtel MS: Complications after sleeve gastrectomy for morbid obesity, *Obes Surg* 19(6):684-687, 2009.

Gonzalez R, Haines K, Nelson LG, et al: Predictive factors of thromboembolic events in patients undergoing Roux-en-Y gastric bypass, *Surg Obes Relat Dis* 2(1):30-35, 2006.

Gonzalez R, Sarr MG, Smith CD, et al: Diagnosis and contemporary management of anastomotic leaks after gastric bypass for obesity, *J Am Coll Surg* 204(1):47-55, 2007.

Gumbs AA, Duffy AJ, Bell RL: Incidence and management of marginal ulceration after laparoscopic Roux-Y gastric bypass, *Surg Obes Relat Dis* 2(4):460-463, 2006.

Kothari SN, Lambert PJ, Mathiason MA: A comparison of thromboembolic and bleeding events following laparoscopic gastric bypass in patients treated with prophylactic regimens of unfractionated heparin or enoxaparin, *Am J Surg* 194(6):709-711, 2007.

Lalor PF, Tucker ON, Szomstein S, Rosenthal RJ: Complications after laparoscopic sleeve gastrectomy, *Surg Obes Relat Dis* 4(1):33-38, 2008.

Mehran A, Szomstein S, Zundel N, Rosenthal R: Management of acute bleeding after laparoscopic Roux-en-Y gastric bypass, *Obes Surg* 13(6):842-847, 2003.

Podnos YD, Jimenez JC, Wilson SE, et al: Complications after laparoscopic gastric bypass: a review of 3464 cases, *Arch Surg* 138(9):957-961, 2003.

Poitou Bernert C, Ciangura C, Coupaye M, et al: Nutritional deficiency after gastric bypass: diagnosis, prevention and treatment, *Diabetes Metab* 33(1):13-24, 2007.

Rasmussen JJ, Fuller W, Ali MR: Marginal ulceration after laparoscopic gastric bypass: an analysis of predisposing factors in 260 patients, *Surg Endosc* 21(7):1090-1094, 2007.

Rogula T, Yenumula PR, Schauer PR: A complication of Roux-en-Y gastric bypass: intestinal obstruction, *Surg Endosc* 21(11):1914-1918, 2007.

Rosenthal RJ, Szomstein S, Kennedy CI, et al: Laparoscopic surgery for morbid obesity: 1,001 consecutive bariatric operations performed at The Bariatric Institute, Cleveland Clinic Florida, *Obes Surg* 16(2):119-124, 2006.

Tucker ON, Escalante-Tattersfield T, Szomstein S, Rosenthal RJ: The ABC system: a simplified classification system for small bowel obstruction after laparoscopic Roux-en-Y gastric bypass, *Obes Surg* 17(12):1549-1554, 2007.

Ukleja A, Afonso BB, Pimentel R, et al: Outcome of endoscopic balloon dilation of strictures after laparoscopic gastric bypass, *Surg Endosc* 22(8):1746-1750, 2008.

Postgastrectomy Complications: Partial Gastrectomy

69

Martin H. Floch

Complications in the postgastrectomy period occur with both open and laparoscopic surgical techniques and may occur after complete healing. These include recurrent ulceration, gastroparesis (delayed gastric emptying), afferent loop syndrome, dumping syndrome and postvagotomy diarrhea, bile reflux gastritis, and gastric adenocarcinoma. Symptoms vary depending on the complication and are briefly described here.

RECURRENT ULCERATION

Marginal ulcers, or *jejunal ulcers*, are rare and may occur in less than 1% of postgastrectomy patients. These ulcers usually are caused by inadequate acid suppression, which is highly unlikely since the advent of proton pump inhibitors (PPIs), and Zollinger-Ellison syndrome is always suspected. Taking nonsteroidal antiinflammatory drugs (NSAIDs) can cause ulceration. The presenting symptom is most frequently pain, although occult bleeding and anemia have been reported. Endoscopic evaluation provides the diagnosis. Endoscopy and biopsy are essential to rule out early malignant carcinoma. These lesions can be controlled by removal of irritating drugs and PPI acid suppression. Although rarely required as treatment, if carcinoma is present, surgical resection is paramount.

GASTROPARESIS (DELAYED GASTRIC EMPTYING)

Gastroparesis is most often associated with truncal vagotomy or basic motility disturbance, as occurs in diabetes. The symptom usually is nausea or inability to eat and occasionally, vomiting of feedings. Attempts at medical treatment with prokinetic agents, such as metoclopramide or erythromycin, may be of help. Some patients have gastroparesis before their surgery, and treatment is complicated (see Chapter 47).

AFFERENT LOOP SYNDROME

Afferent loop syndrome occurs only in patients who have undergone gastrojejunostomy. The loop from the duodenum running to the gastrojejunostomy may become obstructed. The obstruction is partial and intermittent. It often creates the symptom of epigastric fullness or upper abdominal pain, which is relieved only by vomiting. The vomitus is most often bile colored to relieve the obstruction. The obstruction can be caused by scarring or adhesions or by a twisting of the intestinal loop. Diagnosis is made when the dilated loop is demonstrated on radiographic study. Endoscopy may or may not be helpful but is needed to evaluate for the presence of strictures or malignancy. However, demonstration of the dilated loop is essential, often achieved through imaging. The treatment for afferent loop syndrome is invariably surgery, with conversion of the gastrojejunostomy to either a Billroth I or a Roux-en-Y configuration. At times, the surgeon is able to reconform the gastrojejunostomy.

DUMPING SYNDROME AND POSTVAGOTOMY DIARRHEA

Dumping syndrome and postvagotomy diarrhea may be the most common complications of gastrojejunostomy, occurring in as many as 5% to 10% of procedures. Symptoms occur immediately after eating, and even during eating in some patients. Symptoms vary from epigastric discomfort and a vague feeling of oppression to sudden episodes of profuse sweating, tachycardia, tremor, and a tendency to syncope. Often, patients may deny their symptoms, but they feel relieved by eating or lying in a supine position. Some patients can adjust to the symptoms, but others cannot overcome the lightheadedness and tachycardia that may occur immediately on eating to 30 minutes after meals.

Dumping is usually triggered by jejunal distention as hypertonic gastric contents rapidly enter the small bowel (**Fig. 69-1**). Symptoms are usually caused by the shift in volume of fluid into the jejunum. Symptoms occurring later may be caused by hyperglycemia, which can occur 1 to 2 hours after meals. Intestinal and vasomotor symptoms develop after rapid passage of a hyperosmolar meal from the stomach pouch to the intestine. A concomitant drop in plasma volume occurs with symptoms, although this can also be demonstrated in healthy persons. The type of gastric resection is not related to the syndrome. Large quantities of hypertonic solution produce the syndrome in all patients after gastrectomy.

In the typical sequence of events, a susceptible patient eats a meal of hyperosmolar food (usually concentrated, simple carbohydrates), which rapidly enters the jejunum, where it causes a sudden shift in fluid so that a measurable fall occurs in plasma volume. Bloating, diarrhea, and varying degrees of weakness, dizziness, sweating, pallor, and tachycardia may develop. At this stage, the patient is usually hyperglycemic, but rebound hypoglycemia and hypokalemia have been recorded. Hypoglycemia is often referred to as the "late aspect" of the dumping syndrome and results from a rapid increase in blood sugar, followed by a rapid decrease that may cause symptoms.

Treatment is careful dietary management, which includes the elimination of simple-carbohydrate foods and fluid. It is usually successful in 80% of patients. Consuming high-fiber foods may also be helpful. It has been shown that guar (10 g) can prevent the syndrome. Five-gram sachets of pectin twice daily are usually effective. If all measures fail, surgical correction is needed, but this is rare.

After gastrectomy, all patients are advised to consume smaller meals, ideally six small feedings daily, and to restrict their intake

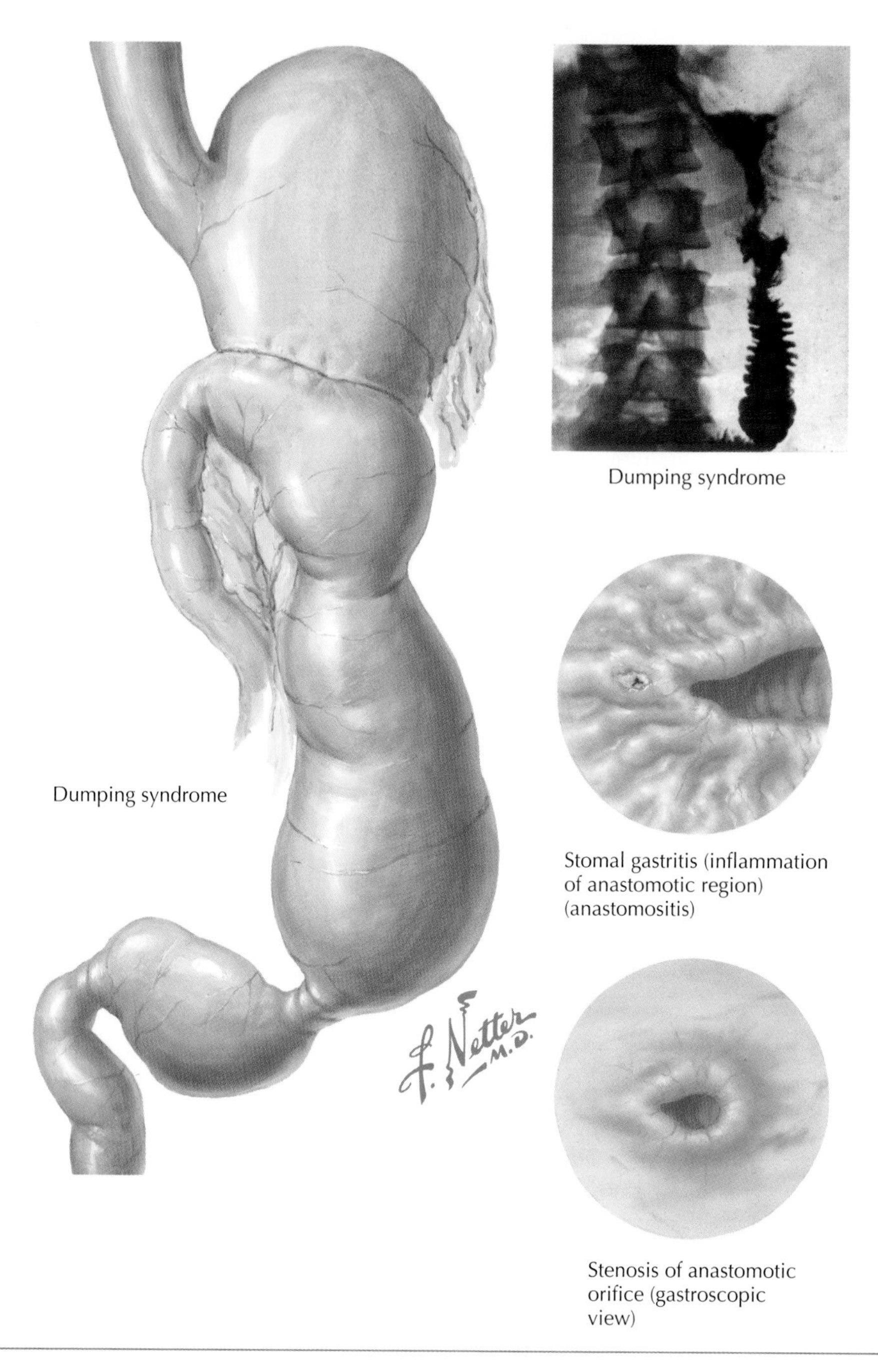

Figure 69-1 *Postgastrectomy Dumping Syndrome.*

of simple carbohydrates. It is best to increase the protein and fat content of the food. Most patients will learn to accommodate. Less than 1% acquire debilitating symptoms and may require conversion to a Roux-en-Y procedure. Before conversion surgery, a trial of somatostatin may be helpful now that daily injections can be decreased by depot injections. Somatostatin slows emptying and delays the onset of glucose and insulin release. However, if long-term somatostatin treatment if needed, conversion to Roux-en-Y may be preferred.

Because vagotomy is rarely performed, the complication rate is low. However, many patients experience diarrhea after vagotomy. Often, patients adjust, and treatment with antidiarrheal agents, such as codeine or loperamide, helps them during the period of adjustment. Again, somatostatin may be tried if symptoms are severe.

BILE REFLUX GASTRITIS

When bile easily refluxes from the gastrojejunostomy into the gastric stump, it can cause significant inflammation of the remaining gastric lining. Patients may become symptomatic and experience pain or occasional vomiting. Although an infrequent complication, bile reflux gastritis may rarely require revision of the procedure to a Roux-en-Y. Trials of therapy with coating agents, such as sucralfate, may be helpful in selected patients. Trials to increase gastric emptying that may be delayed by the use of prokinetic agents are also suggested. To make this diagnosis, gastritis must be demonstrated by endoscopy, and it must be clearly demonstrated that the bile is causing the gastritis. This is difficult to prove; thus, trials of therapy and finally surgical recorrection may be needed.

GASTRIC ADENOCARCINOMA

Patients who have undergone subtotal gastrectomy have been reported to have a twofold increase in cancer risk after 15 years. However, these epidemiologic data are controversial, and it is unclear how subtotal gastrectomy is related to *Helicobacter pylori* infection or reflux of duodenal content. Nevertheless, many clinicians believe that after gastrectomy patients should be surveyed by endoscopy every 2 to 3 years. Again, this is controversial. At this point, the surveillance recommendations are not clearly defined and depend on physician and patient in each case (see Chapter 68).

ADDITIONAL RESOURCES

Kelly KA, Sarr MC, Hinder RA: *Mayo Clinic gastroenterology surgery*, Philadelphia, 2004, Saunders.

Tanimura S, Higashino M, Fukunaga Y, et al: Laparoscopic gastrectomy for gastric cancer: experience with more than 600 cases, *Surg Endosc* 22:1161-1164, 2008.

Effects of Total Gastrectomy

70

Martin H. Floch

Total gastrectomy almost always results in nutritional problems (**Fig. 70-1**). Usually performed to attempt a curative procedure for carcinoma, total gastrectomy also may be done to treat trauma.

CLINICAL PICTURE

After total gastrectomy, patients have great difficulty gaining weight, and most never regain their preoperative weight. Others may develop selected deficiencies. The clinical picture is a patient having difficulty gaining weight after surgery who may report postprandial fullness, anorexia, and nausea (rarely vomiting, with or without diarrhea).

DIAGNOSIS

The diagnosis is clear from the patient's history, but when the presentation is of a selected deficiency, a more detailed workup is necessary. After barium study to demonstrate the anatomy, a full biochemical workup establishes immunologic status, electrolyte findings, and mineral levels.

Loss of the reservoir function of the stomach deprives the patient of the capacity to hold a normal meal, which necessitates the ingestion of frequent, small feedings and careful deglutition and chewing. The triturating (grinding) action of the stomach is lost; therefore, mastication is essential. Intestinal transit is frequently accelerated. Resultant problems include dumping syndrome and weight loss.

Selected deficiencies are caused by absorptive defects such as steatorrhea, protein malnutrition, osteoporosis and osteomalacia, anemia secondary to vitamin B_{12} deficiency or iron deficiency, hypoglycemia, and vitamin A deficiency.

Postoperative complications such as anastomotic ulcers, afferent loop syndrome, and intussusception may occur. Loss of the bacteriostatic function of the stomach results in altered microecology or bacterial overgrowth in the small intestine.

A diagnostic workup may also include a study of transit time to help in the therapeutic management. Selective absorptive studies may also be helpful to determine the function of the small bowel, such as a xylose absorption test for carbohydrates, quantitative stool studies to determine the degree of steatorrhea, and vitamin blood levels.

TREATMENT AND MANAGEMENT

Treatment is clear. Any complication such as the dumping syndrome must be treated, and the patient must have adequate caloric and nutrient intake. Weight should be monitored carefully. Patients must be encouraged to eat a high-protein, high-fat diet and to avoid simple carbohydrates. Usually, frequent feedings with snacks between meals and at bedtime can help maintain the body weight at a stable level. Patients should be reassured that they are not going to gain large amounts of weight and that maintaining a normal but low body mass index is adequate. In patients with steatorrhea, consuming too much fat compounds the problem; more readily absorbed fats, such as medium-chain triglycerides, may be helpful.

These patients need vitamin B_{12} injections for life to maintain adequate B_{12} stores. With any suggestion of osteoporosis, adequate calcium and vitamin D are essential for maintaining bone health. In one study, 50% of patients developed iron deficiency after subtotal gastrectomy; therefore, supplements are frequently needed. Vitamin A stores may become depleted after total gastrectomy and are absorbed poorly after subtotal gastrectomy. These patients frequently require vitamin A supplementation.

Nutrition supplementation may be necessary with liquid formulas when patients cannot tolerate or consume enough food. However, tube feeding through a jejunostomy or total parenteral nutrition should not be required and is rarely indicated.

COURSE AND PROGNOSIS

If gastrectomy is performed to remove a malignant tumor, the prognosis depends on the status of the cancer. If the cancer is curable, or if the gastrectomy was performed for trauma, patients can do fairly well and maintain normal life status. Weight will always be a problem, and patients will require frequent feedings with nutrition supplements, as previously described. All postgastrectomy patients should take at least one vitamin tablet daily and select nutrition replacement as indicated.

ADDITIONAL RESOURCES

Floch MH: *Nutrition and diet therapy in gastrointestinal disease*, New York, 1981, Plenum.

Hoffman WA, Spiro H: Afferent loop syndromes, *Gastroenterology* 40:201-209, 1961.

Jenkins DJA, Gassull MA, Leeds AR, et al: Effect of dietary fiber on complications of gastric surgery: prevention of postprandial hypoglycemia by pectin, *Gastroenterology* 72:215-220, 1977.

Kiefer ED: Life with a subtotal gastrectomy: a follow-up study 10 or more years after operation, *Gastroenterology* 37:434-440, 1959.

Lundh G: Intestinal digestion absorption after gastrectomy, *Acta Chir Scand* 114(suppl 231):1-83, 1958.

Metz G, Gassull MA, Drasar BS, et al: Breath hydrogen test for small intestinal bacterial colonization, *Lancet* 1:668-670, 1976.

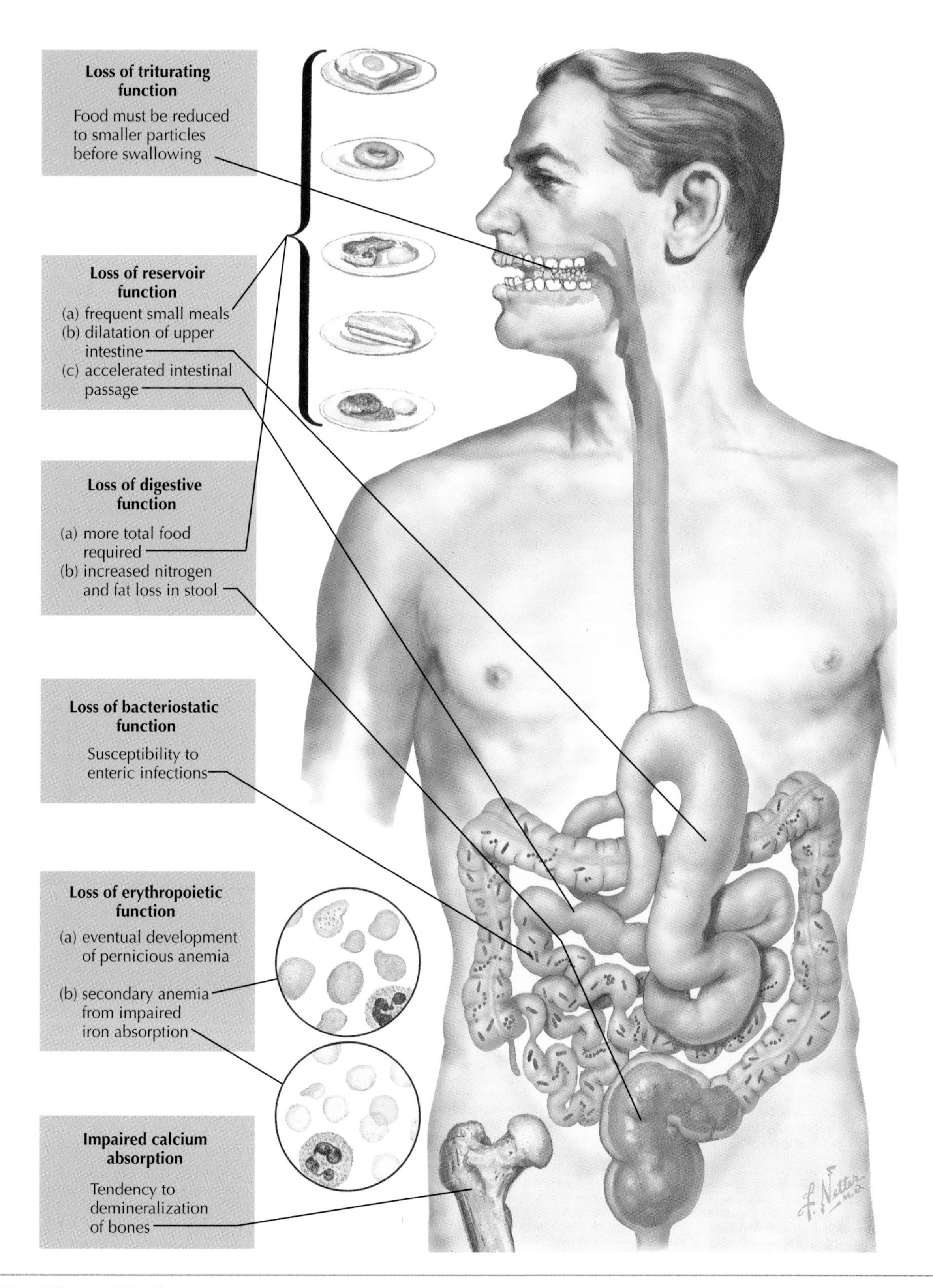

Figure 70-1 *Effects of Total Gastrectomy.*

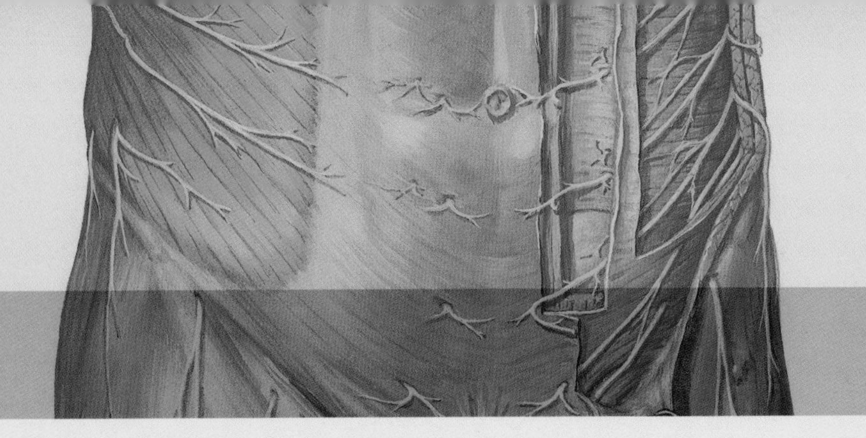

SECTION III

Abdominal Wall

Anterolateral Abdominal Wall

71

David Podkameni and Raul J. Rosenthal

Before describing the abdominal *parietes* (walls, or limits), the different ways in which the word *abdomen* is used should be discussed. For some, abdomen is synonymous with *abdominopelvic cavity*. For others, abdomen literally means the area of the body between the diaphragm and the pelvis minor (true pelvis). "Abdomen" is also used loosely to refer to a general area of the body.

For the purpose of a specific description, it is appropriate to label the portion of the body below the diaphragm as the abdominopelvic cavity and then to divide this into the *abdominal cavity proper* and the *pelvic cavity*, or pelvis minor, which is separated by the plane of the pelvic inlet (plane traversing sacral promontory and pubic crests). Of note, however, certain structures normally referred to as "abdominal" structures, such as loops of small bowel, are usually suspended into the pelvic cavity. Inferior and posteroinferior (dorsocaudal) support of the abdominal viscera is provided by the parietes of the pelvic cavity, not the theoretical plane at the pelvic inlet. It is practical to divide the parietes of the abdominopelvic cavity into four general parts: anterolateral abdominal wall; posterior wall of the abdominal cavity; diaphragm (superior wall or roof of abdominal and abdominopelvic cavities); and parietes of the pelvic cavity, which can be loosely termed the "floor" of the abdominopelvic cavity. However, the limits of these parietal sections are not all sharply defined because of the curved contours involved; therefore, certain arbitrary limits must be defined for descriptive purposes.

The *anterolateral abdominal wall* (or simply *abdominal wall*) fills in the bony cartilaginous frame between the costal margin above and the hipbones below (**Fig. 71-1**). "Anterolateral" is used here to mean that this part of the wall is anterior and lateral and follows the curve of the wall for a distance onto the posterior aspect. In this description, the quadratus lumborum muscle and the structures medial to it are included with the posterior wall of the abdominal cavity.

The anterolateral abdominal wall can contract and relax, which helps accommodate the size of the abdominopelvic cavity to any changes in the volume of contained viscera, and which controls the intraabdominal pressure. A surgical approach to the abdominopelvic cavity is usually made through this wall.

General layers of the anterolateral abdominal wall from the outside in are skin, subcutaneous tissue, superficial fascia (or membranous portion of subcutaneous layer), fascia, muscles with their related fascia, transversalis fascia, extraperitoneal fascia, and parietal peritoneum. The layers of the abdominal wall can be depicted on computed tomography and magnetic resonance imaging.

The abdominal skin is of average thickness, thicker dorsally than ventrally and laterally, and loosely attached to the underlying layers except in the umbilical area. The *superficial fascia* (tela subcutanea) is soft and mobile and contains a variable amount and distribution of fat, depending on the patient's nutritional status. The thickness of this layer can be estimated as the thickness of a fold, minus double the thickness of the skin, which would be approximately twice the thickness of this layer. At the area inferior to the level of the umbilicus, the superficial fascia is classically described as having a superficial fatty layer (Camper fascia) and a deep membranous layer (Scarpa fascia). This description is simplified because the layering is not always as clearly delineated as indicated here, but this adequately serves a descriptive purpose. *Camper fascia* is continuous with the fatty layer of surrounding areas, as seen in the fatty layer of the thigh. *Scarpa fascia* merges with the fascia lata in a parallel line to and just below the inguinal ligament. Medial to the public tubercle, both layers continue into the urogenital region, which is significant in relation to the path that extravasated urine takes, such as after rupture of the bladder neck. When entering the proper layer of the urogenital region, the urine may escape upward into the anterolateral abdominal wall. In males, the two layers merge into the scrotum and blend into a single smooth muscle–containing layer; at this point, the fat is abruptly lost as the layers begin to form the scrotum. Cephalad to the symphysis pubis, additional closely set strong bands of Scarpa fascia form the fundiform ligament of the penis, extending down into the dorsum and the sides of the penis.

The outer investing layer of the *deep fascia*, which is not readily distinguishable from the muscular fascia on the external surface of the external abdominal oblique muscle and its aponeurosis, is easily demonstrated over the fleshy portion of the muscle. Separating this from the aponeurotic portion of the muscle, however, is difficult. This layer is attached to the inguinal ligament and merges with the fascia emerging from the ligament to form the *fascia lata*. In addition, it joins the fascia on the inner surface of the external oblique at the subcutaneous inguinal ring to form the external spermatic fascia. External to the lower end of the linea alba, the outer investing layer thickens into the suspensory ligament of the penis, anchoring the penis to the symphysis and the arcuate ligament of the pubis. It is also continuous with the deep fascia investing the penis.

The nerve supply of the external abdominal oblique muscle is derived from the anterior primary division of the sixth to twelfth thoracic (T6-T12) spinal nerves. T7 to T11 are intercostal nerves that continue from the intercostal spaces into the anterolateral abdominal wall, to lie in the plane between the internal oblique and transversus muscles. T12 is the subcostal nerve and follows a similar course. The iliohypogastric nerve, from T12 and first lumbar (L1), also contributes to the nerve supply. The nerves probably have a segmental distribution corresponding to the primitive segmental condition of the muscle, with T10 extending toward the umbilicus and T12 toward a point halfway between the umbilicus and symphysis pubis.

The *cremaster muscle*, which is variably well developed only in males, represents an extension of the lower border of the internal oblique, and possibly the transversus, over the testis and the spermatic cord. Laterally, the cremaster is thicker and

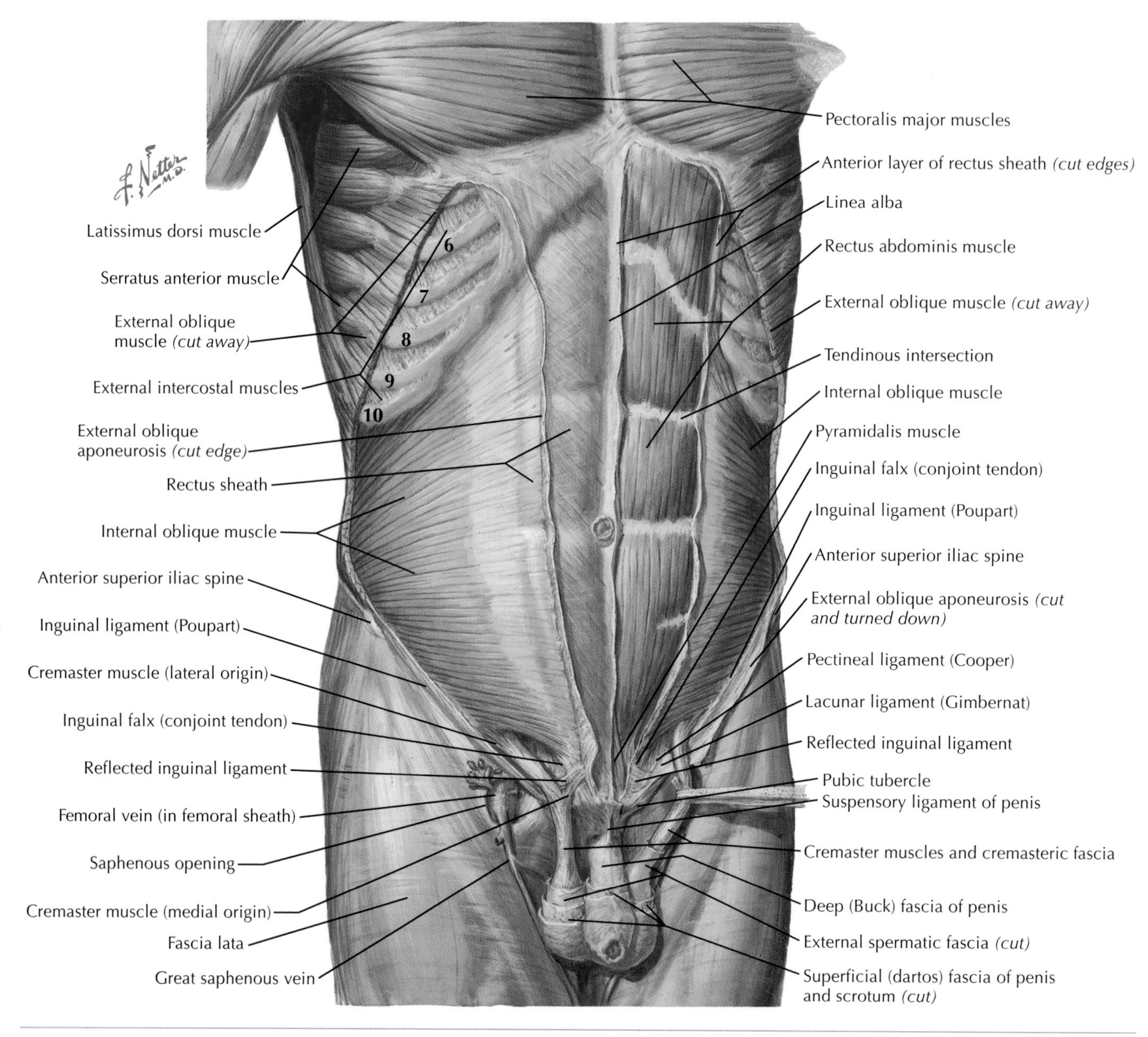

Figure 71-1 *Anterior Abdominal Wall: Intermediate Dissection.*

fleshier and attaches to the middle of the turned-under edge of the external oblique aponeurosis and to the inferior edge of the internal oblique. From here, the scattered fibers, interspersed with connective tissue (cremasteric fascia), spread in loops over the spermatic cord and testis to end at the pubic tubercle and the anterior layer of the rectus sheath. The cremaster's nerve supply is from the genital (external spermatic) branch of the genitofemoral nerve and, generally, a branch from the ileo-inguinal nerve. The action of the cremaster muscle is to lift the testis toward the subcutaneous inguinal ring.

The rectus abdominis muscle generally acts in conjunction with the external oblique, internal oblique, and transversus abdominis muscles, but it is specifically involved in producing forced expiration and flexion of the vertebral column.

Peritoneum

72

David Podkameni and Raul J. Rosenthal

The peritoneum is the extensive serous membrane that, in general, lines the parietes (walls) of the abdominopelvic cavity and reflects from these parietes to cover, to a differing extent, the various viscera within the abdominopelvic cavity (**Figs. 72-1** and **72-2**). The serous membrane lining the parietes is continuous with that on the surface of the viscera contained in the portions of the body cavity involved, referred to as the *parietal* and "virtual" space between the respective serous membranes. Only a film of fluid forming a virtual space separates the respective organs and parietes. These cavities are closed in males but are open in females, where the lumen of each uterine tube is continuous with the peritoneal cavity.

The peritoneum is more complex in its arrangement than the pericardium or the pleura, essentially because the organs or viscera that developed during the embryonic phase rotate between two distinct cavities, the pleuroperitoneal channels, and have an anterior and posterior fixation, or *mesons*. One example is the stomach, which develops between the pleuroperitoneal cavities with an anterior and posterior *mesogastrium*. After rotation of the stomach during fetal development, the right side becomes the posterior aspect of the stomach, and the left side becomes the anterior portion. Concomitantly, the liver and spleen develop in the anterior and the posterior meson, respectively. Because of the rotation, these organs are situated on the right and left sides of the abdominal cavity, allowing the formation of a small cavity behind the stomach, called the *lesser sac*. This communicates with the abdominal cavity through an opening or foramen (Winslow), limited anteriorly by the hepatic pedicle, posteriorly by the portal vein, inferiorly by the duodenum (second portion), and superiorly by the caudate lobe of the liver.

The most accurate way to conceptualize the arrangement of the peritoneum is to trace it in three planes: a midsagittal plane and two horizontal planes. This is done at the levels of the epiploic foramen and the umbilicus, preferably in a fresh specimen at autopsy. Lacking this opportunity, the use of these three planes remains one of the most informative approaches for studying the peritoneal continuity and its relationship to the abdominopelvic viscera. Currently, it has become almost mandatory to evaluate the anatomy from another standpoint, in which computed tomography and magnetic resonance imaging are the cornerstones of technology in the field of anatomic study.

The degree to which the abdominal viscera are covered by peritoneum (visceral peritoneum) varies, depending on whether the peritoneum covers only part of one surface of the viscus or covers the viscus entirely, except for the area of attachment of a suspending double-layered fold of peritoneum. Several terms are used to designate degrees of peritoneal covering. Given that most of these terms do not have a generally accepted connotation, however, a more specific description of the peritoneal covering of a certain viscus is to state *which parts* of *which surfaces* of the viscus are covered by peritoneum. *Retroperitoneal* is a common descriptive term that means "behind the peritoneum." Referring to certain organs as retroperitoneal is not a universally accepted designation.

The *greater omentum* (gastrocolic omentum, or ligament) is the largest peritoneal fold. It may drape like a large apron from the greater curvature of the stomach in front of the other viscera as far as the brim of the pelvis, or even into the pelvis or perhaps into an inguinal hernia, frequently on the left side. It may be shorter, just a fringe on the greater curvature of the stomach, or longer and folded between loops of small intestine, tucked into the left hypochondriac area or turned upward, in front of the stomach. The upper end of the left border is continuous with the gastrolienal ligament, and the upper end of the right border extends as far as the beginning of the duodenum. The greater omentum is usually thin, has a delicate layer of fibroelastic tissue as its framework, and is cribriform in appearance, although it usually contains some adipose tissue and may accumulate a large amount of fat in an obese person.

The *lesser sac* peritoneum on the posteroinferior surface of the stomach and the *greater sac* peritoneum on the anterosuperior surface meet at the greater curvature of the stomach and course inferiorly to the free border of the greater omentum, where they turn superiorly to the transverse colon. Early in development, these two layers (which are the elongated dorsal mesogastrium) course superiorly in front of the transverse colon and transverse mesocolon to the anterior surface of the pancreas. Because they fuse to each other and to the peritoneum on the transverse colon and anterior surface of the primitive transverse mesocolon, these two layers of peritoneum, running superiorly as the posterior layer of the greater omentum, apparently separate from each other to surround the quite-mobile greater omentum. These layers can shift to fill otherwise temporary gaps between viscera or build a barrier against bacterial invasion of the peritoneal cavity by adhering at potential danger spots.

The *lesser omentum* (gastrohepatic omentum, or ligament), hepatogastric ligaments, and hepatoduodenal ligaments extend from the posteroinferior surface of the liver to the lesser curvature of the stomach and the beginning of the duodenum. The lesser omentum is extremely thin, particularly the portion to the left, which is sometimes fenestrated. The portion to the right is thicker and ends in a free, rounded margin that contains the common bile duct to the right, the hepatic artery to the left, and the portal vein posterior to these; it then forms the anterior border of the epiploic foramen. In addition to these structures, the lesser omentum contains the right and left gastric arteries close to the lesser curvature of the stomach and the accompanying veins, lymphatics, and autonomic nerve plexuses. The peritoneum forming the anterior layer of this omentum and continuing onto the anterosuperior surface of the stomach is the *greater sac* peritoneum, and that forming the posterior layer and continuing onto the posteroinferior surface of the stomach is

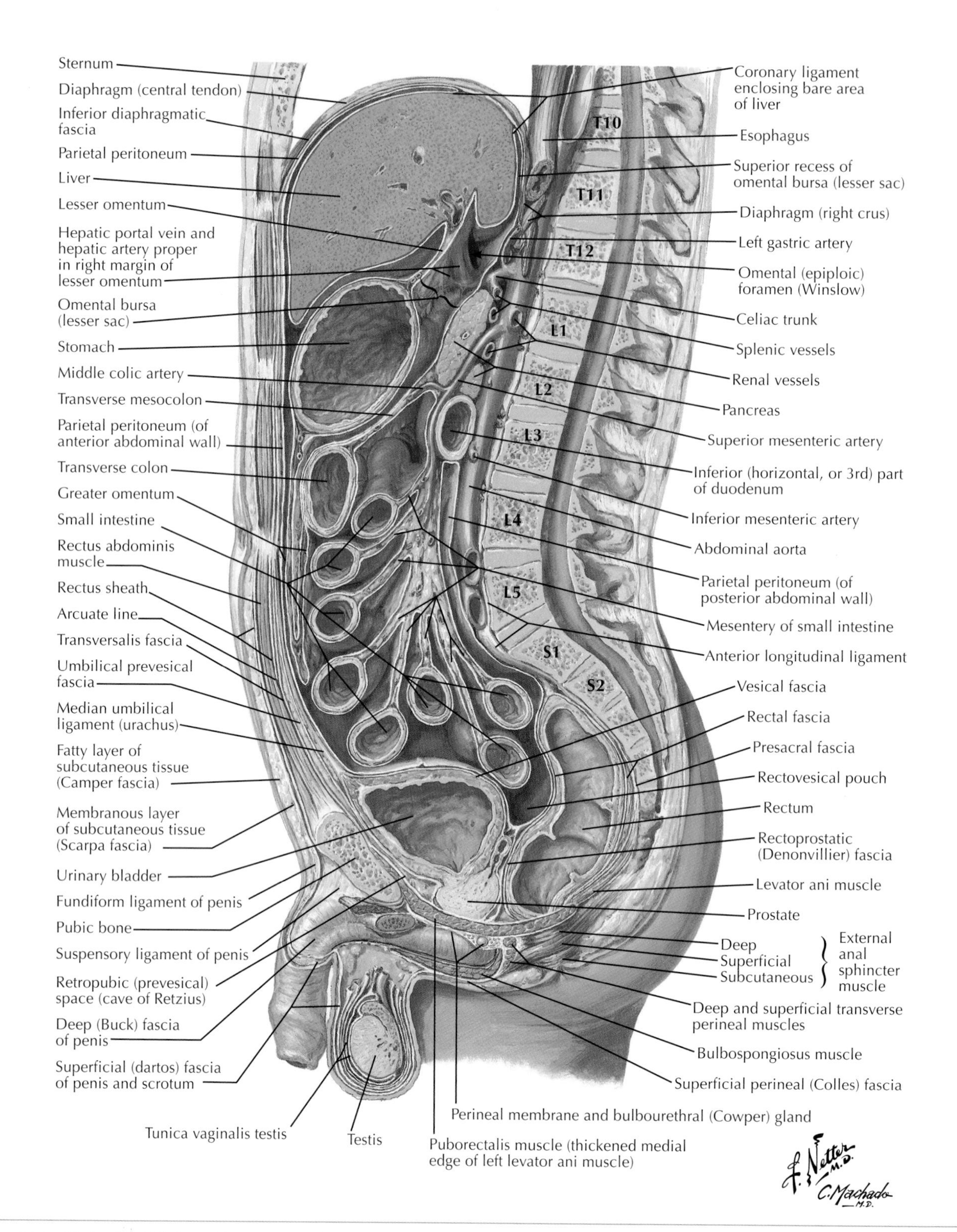

Figure 72-1 *Abdominal Wall and Viscera: Median (Sagittal) Section.*

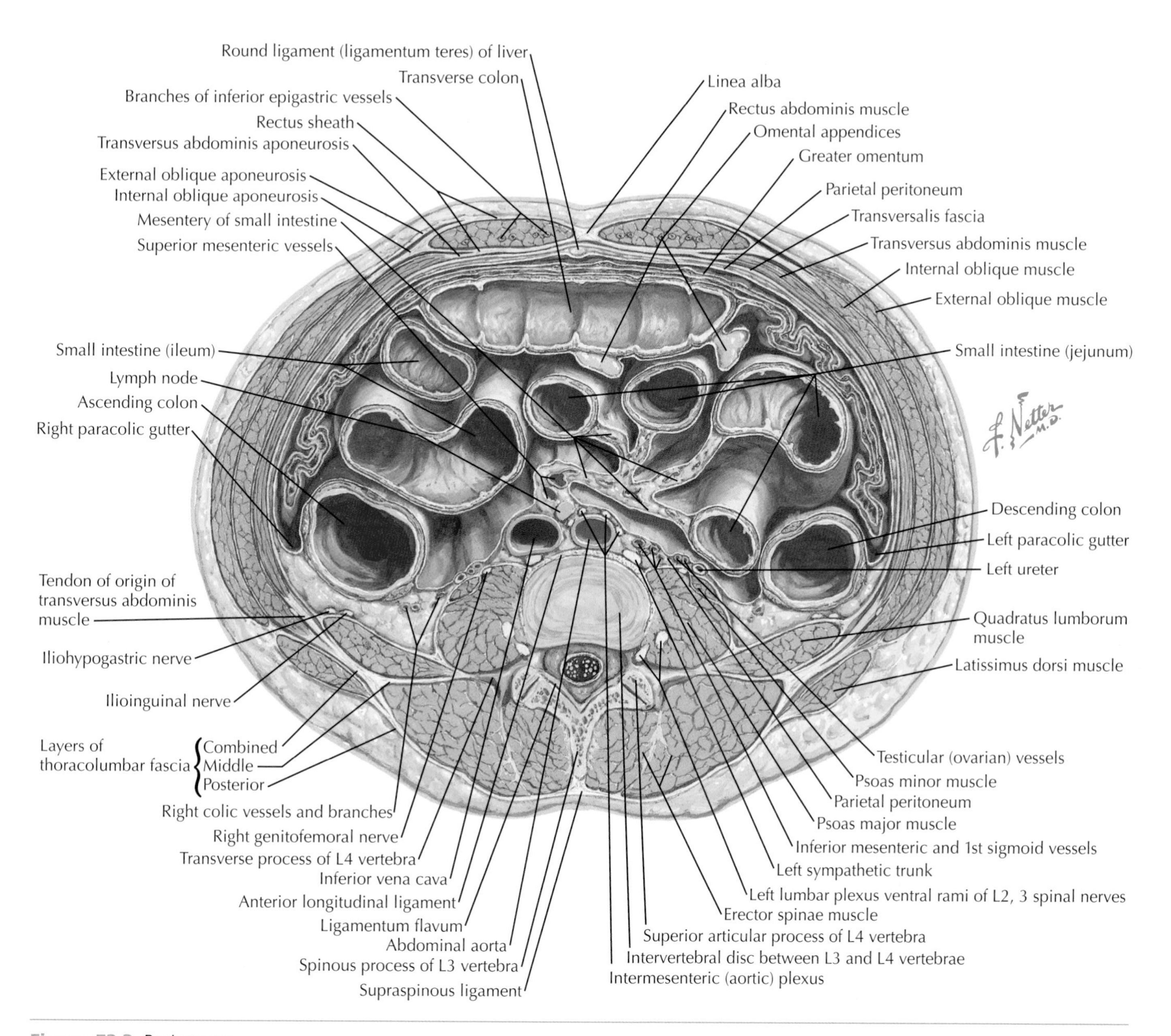

Figure 72-2 *Peritoneum.*

the *lesser sac* peritoneum. The lesser omentum reaches the liver at the porta, and to the left of the porta, it extends to the bottom of the fossa for the ligamentum venosum.

The *parietal peritoneum* is supplied by the nerves to the adjacent body wall and is thus sensitive to pain. The *visceral peritoneum* is insensitive to ordinary pain stimuli. The roots of the mesenteries contain receptors that give rise to pain in response to stretching of the mesentery. When moist surfaces of peritoneum are in contact and become irritated, adhesions tend to form and often become permanent.

Pelvic Fascia and Perineopelvic Spaces

73

David Podkameni and Raul J. Rosenthal

Steadily changing pressure and filling conditions in the pelvis require a unique adaptability of those structures that essentially support the viscera within the funnel-like frame of the pelvis. Part of such support derives from the anorectal musculature and the levator ani. However, because these muscles are greatly involved in the sphincteric and emptying functions of the anorectal canal, their supporting tasks must be assisted by connective tissue structures with adequate tensile strength, such as from the pelvic fascia. Thus, the pelvic fascia is removed from the traditionally passive role of an "undifferentiated subserous tissue." Such oversimplified descriptions mask the true physiologic and surgical significance of the pelvic fascia. Although its anatomic relation is complex, it is generally recognized that the pelvic fascia is best divided into a visceral and a parietal portion. The visceral part lies entirely above the pelvic diaphragm and forms the fascial investments of the pelvic viscera, the perivascular sheaths, and the intervisceral and pelvovisceral ligaments.

The *perineal fascia* consists of a superficial subcutaneous layer and a deep membranous layer (**Fig. 73-1**). The superficial layer is unnamed but is considered to correspond to Camper fascia of the abdominal wall. The deep layer is *Colles fascia*, corresponding to Scarpa fascia of the abdomen. The superficial layer varies considerably throughout the perineum. Over the anal triangle, it forms the fatty layer of the perianal space, whereas laterally over the ischial tuberosities, it consists of fibrous fascicles that connect to the underlying bone and form, directly over the ischial tuberosities, fibrous bursal sacs. The main part of the deep layer of Colles fascia has a firm attachment to the pubic rami and to the posterior margin of the urogenital diaphragm. It spreads medially across the urogenital triangle, constituting the floor of the superficial perineal compartment, which lies between it and the interior layer of the urogenital diaphragm and contains the superficial perineal musculature.

Presacral wings extend medially from the hypogastric sheath in front of the sacrum and the presacral fascia, lying in an almost vertical plane, in contrast to the superior and interior wings, which unfold in an almost horizontal plane. On reaching the sides of the rectum, the presacral wing splits into two leaves that encircle the rectum as the *rectal* (visceral) *fascia*. This wing contains the superior and middle hemorrhoidal (rectal) vessels, the inferior hypogastric or pelvic nerve plexus, and many lymphatics.

The origin, course, and insertion of the pelvic musculature and the configuration of the anorectal musculature, along with the supralevator and infralevator fasciae, give rise to several perineopelvic spaces. These are important for understanding the pathogenesis of infectious and malignant processes of the pelvis and perineum. As with the fasciae, these spaces are separated into supralevator and infralevator groups.

At the supralevator level in males, four main spaces can be identified: prevesical space (space of Retzius), rectovesical space, bilateral pararectal spaces, and retrorectal space. In males and females, the *prevesical space* is a potentially large cavity surrounding the front and lateral walls of the bladder. The main cavity in front of the bladder comprises two superimposed anteromedial recesses and two lateral compartments. The *upper recess* lies behind the anterior abdominal wall and is roofed by the peritoneal reflection from the dome of the bladder, supported by the umbilicovesical fascia (urachus) and the umbilical prevesical fascia. Its lateral borders are demarcated by the obliterated umbilical arteries (lateral umbilical ligaments). The *lower recess*, continuous with the upper recess, lies behind the symphysis and pubic bones and in front of the bladder. Its floor is formed by the pubovesical (pubourethral) ligaments in females and the puboprostatic ligaments in males (true ligaments of bladder). The *lateral compartments* of the prevesical space are bound by a lateral wall formed by the obturator and the supra-anal fasciae and a medial wall formed by the bladder and the inferior hypogastric fascial sheath. The lateral recesses contain the ureter and the main neurovascular supply to the bladder and, in males, to the prostate. The floor of the lateral recess is the supra-anal fascia, which affords attachment to the true lateral ligaments of the bladder. Dorsally, the lateral recess of the prevesical space extends as far as the root of the hypogastric sheath in the region of the ischial spine. The root is formed by the superior hypogastric fascial wing covered by the peritoneum, where these tissues are reflected from the lateral pelvic wall.

The *retrovesical space* (compartment) in males, divisible into three subspaces, lies between the bladder and the prostate, covered by the vesical and prostatic fasciae anteriorly, and the rectum, covered by the rectal fascia posteriorly. Its roof is formed by the rectovesical recess, or pouch of the peritoneum, which is formed by the continuity of the peritoneal reflection from the rectum to the bladder. Its floor is the posterior part of the urogenital diaphragm. Denonvillier fascia (rectogenital septum), originating from the undersurface of the rectovesical peritoneal pouch and extending caudally in a coronal plane, divides into two areas; an anterior leaf blends with the prostatic fascia or capsule, and a posterior leaf attaches below to the urogenital diaphragm medially and to the inferior hypogastric wing laterally. Thus, the rectovesical compartment becomes partitioned into the rectovesical and retroprostatic spaces anteriorly and the prerectal space posteriorly. The inferior wing of the hypogastric fascial sheath, with its contents, marks the lateral boundary of the two anterior spaces and is also the separation from the lateral recess of the space of Retzius. Caudally, the *prerectal space* terminates where the rectourethralis muscle, covered by a fibrous extension of the rectal fascia, attaches itself to the urogenital diaphragm or its superior fascia. The *retropros-*

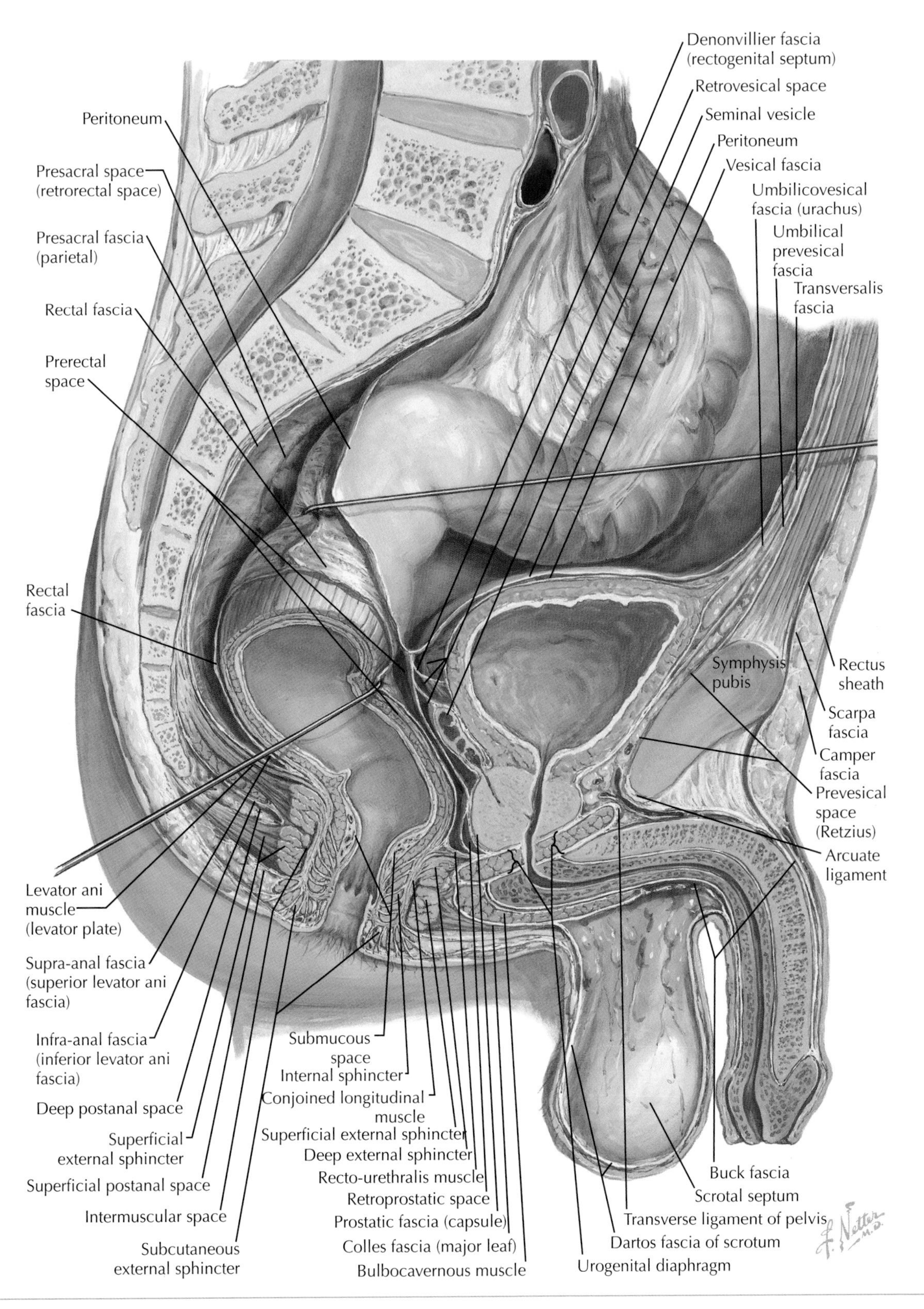

Figure 73-1 *Pelvic Fascia and Perineopelvic Spaces.*

tatic space (Proust space) terminates caudally in the same region but varies, depending on the variable caudal limits of the rectoprostatic fascia (Denonvillier fascia) and its attachments to the prostatic capsule.

In females, as with males, the area between the bladder and the rectum is divided into three spaces. The dominant dividing structure is *not* the rectogenital septum (Denonvillier fascia), but rather the more bulky vagina and cervix uteri. Anterior to these structures, two spaces are formed, the vesicocervical space above and the vesicovaginal space below. They are separated by a fascial septum, the supravaginal septum or vesicocervical ligament that forms the floor of the vesicocervical space, and the roof of the vesicovaginal space. The *vesicocervical space* is roofed by the uterovesical fold of the peritoneum and extends caudally to the point where the urethra and vagina are in apposition above the superior layer of the urogenital diaphragm. In the floor of this space, the medial and lateral pubourethral ligaments surround the urethra, which lies in a fused musculofascial sheath. Laterally, the *vesicovaginal space* is limited by the strong fascial connections between the bladder and the cervix, the uterovesical ligaments, or pillars of the bladder. In females, the posterior component of the *rectovesical space* is farther from the anterior compartments than in males, because the substantial mass of the cervix uteri and the vagina provide more separation. Separation into two spaces occurs through the rectovaginal septum (corresponding to Denonvillier septum in males), the existence of which, in females, remains unproved. Of more practical importance is that the *rectovaginal space* is roofed by the peritoneal fold, which forms the rectouterine pouch of Douglas. The boundaries of this space are the vaginal fascia anteriorly and the rectal fascia posteriorly. Laterally, the space extends to the fusion of the vaginal and rectal fascial collars, which in this region form the wings of the vagina. The space terminates caudally at the line of fusion between the posterior vaginal wall and the anal canal. In this region, numerous fascial and muscular elements apparently fuse, prompting the term *perineal body*, or central point of the perineum.

The *pararectal space* extends on each side from the rectogenital septum in males and the cardinal ligament in females to the presacral fascial wing. It lies on the supra-anal fascia covering the superior surface of the pubococcygeal muscle, along the inferolateral parts of the rectum or its fascial enclosure. In both males and females, its roof is composed of the peritoneum reflected from the lateral aspects of the rectum to the pelvic parietes, forming the floor of the pararectal peritoneal fossa.

The *presacral space*, similar in both genders, constitutes the interval between the parietal pelvic fascia, covering the sacrum and the piriformis, coccygeus, and pubococcygeus muscles, and the presacral fascial wing of the hypogastric sheath, which envelops the rectum as the rectal fascia. Where the posterior rectal wall lies almost horizontally, the rectal fascial collar produces the ventral lining of the presacral space. Cranially, the space becomes continuous with the prevertebral-retroperitoneal areolar tissue. A strong lateral barrier for this space is provided by the attachment of the hypogastric sheath to the parietal fascia, which explains why retrorectal abscesses are more apt to rupture into the rectum than to penetrate into the other supralevator spaces.

The largest and most important of the infralevator spaces are the paired *ischiorectal spaces*, averaging 6 cm anteroposteriorly, 2 to 4 cm wide, and 6 to 8 cm deep. Each ischiorectal space is shaped like an irregular wedge, with the apex at the pubic angle and the base at the gluteus maximus muscle. Circumanal and infra-anal fasciae covering the superficial and deep portions of the external sphincter and the superimposed puborectalis and pubococcygeus portions of the levator ani muscle form the wall. Attachments of this muscle and the infra-anal fascia to the urogenital diaphragm mark the medial wall of the anterior extension (Waldeyer fascia), which extends above the urogenital diaphragm. At the most cranial point of the ischiorectal fossa, the inner wall joins its outer lunate fasciae, overlying the obturator internus muscle and farther down the ischial tuberosity. The infra-anal fascia covering the iliococcygeus muscle roofs the ischiorectal space. The coccyx and the sacrospinous and sacrotuberal ligaments, overlapped by the gluteus maximus muscle, constitute the base or posterior wall of the fossa. These structures thus confine the posterior extension of the ischiorectal space, which has, posteriorly to the anal canal, no medial walls. The fossae of each side communicate with each other by what is known as the *deep postanal space*, which lies above the anococcygeal ligament or posterior extension of the external anal sphincter and below the levator plate. This deep postanal space is also known as the *posterior communicating space* because the right and left ischiorectal spaces communicate through it. The deep postanal space is thus the usual pathway for purulent infections to spread from one ischiorectal space to the other, resulting in the semicircular or "horseshoe" posterior anal fistula. The floor of the ischiorectal space behind the urogenital diaphragm is the transverse septum of the ischiorectal fossa. In the anterior recess, the urogenital diaphragm forms the floor. Large fat globules lying in a matrix of thin collagenous fibrils fill the ischiorectal space. Interior hemorrhoidal (rectal) vessels and nerves cross each space obliquely from its posterior-lateral angle en route from the pudendal vessels and nerves in Alcock (pudendal) canal to the anal canal. Superficial and deep compartments of the urogenital diaphragm occupy the space within the pubic arch and contain the urogenital musculature, which is in close functional relationship to the pelvic diaphragm and the anorectal sphincters.

Inguinal Canal

74

Colleen Kennedy and Raul J. Rosenthal

The inguinal canal is an oblique tunnel, 3 to 5 cm long, through the muscular and deep fascial layers of the anterior abdominal wall, parallel to and superior to the inguinal ligament (**Fig. 74-1**). The canal extends between the internal inguinal ring, located in the transversalis fascia approximately halfway between the anterior superior spine of the ilium and the pubic symphysis, and the external inguinal ring, located in the aponeurosis of the external abdominal oblique muscle just superior and lateral to the pubic tubercle. In the male, the canal conveys the spermatic cord, comprising the vas deferens and the vessels and nerves of the testes (**Fig. 74-2**). The anatomy of the inguinal canal is similar in the female but is somewhat less well developed. The canal in the female contains the round ligament of the uterus as it travels toward its termination in the labia majora.

The *internal inguinal ring*, a funnel-shaped opening in the transversalis fascia, is the site at which the transversalis fascia becomes the innermost covering of the spermatic cord, the internal spermatic fascia. Inferior epigastric vessels are just medial to the internal inguinal ring, and the most lateral point of the inferior border of the transverses muscle is just lateral to this ring. The *external inguinal ring* is formed by a division of the fibers of the external abdominal oblique aponeurosis with fibers that pass superomedial to the ring attaching to the pubic symphysis. This portion of the external oblique aponeurosis is called the *superior crus* of the external (superficial) ring. Fibers of the external oblique aponeurosis that pass inferolateral to the superficial inguinal ring are called the *inferior crura* of the ring.

The lower border of the external oblique *aponeurosis* is folded under itself, with the edge of the fold forming the inguinal ligament. The fascia lata on the anterior aspect of the thigh is closely blended to the full length of the ligament, and its lateral half is fused with the iliac fascia as the iliacus muscle passes into the thigh. Fibers of the aponeurosis form the medial half of the inguinal ligament, rolling under in such a way that the fibers forming the inferolateral margin of the external inguinal ring become the most inferior fibers at the attachment to the pubic bone, thus attaching most interiorly on the pubic tubercle, whereas fibers originally more inferior attach higher up on the tubercle, in sequence along the medial part of the pecten pubis for varying distances. The lowest fibers in the aponeurosis attach the farthest laterally on the pecten. The portion of the aponeurosis that runs posteriorly and superiorly from the folded edge of the ligament to the pecten pubis is called the pectineal part of the inguinal ligament, or the lacunar ligament.

The inguinal canal and associated structures can be viewed as a tubular tunnel having a roof, a floor, and anterior and posterior walls. The two openings are the deep inguinal ring in the transversalis fascia at the internal (lateral) end of the canal and the superficial inguinal ring in the aponeurosis of the external oblique muscle at the external (medial) end of the canal. The external oblique aponeurosis, strengthened by the intercrural fibers, is present in the entire length of the anterior wall of the canal. For approximately the lateral one fourth to one third of the canal, fibers of the internal oblique muscle, which arise from the inguinal ligament and the related iliac fascia, form the anterior wall of the canal deep to the external oblique aponeurosis. The floor of the canal is formed in its medial two thirds to three fourths by the rolled-under portion of the external oblique aponeurosis, together with the lacunar ligament, forming a shelf on which the spermatic cord rests.

The *transversalis fascia* is present for the entire length of the posterior wall of the inguinal canal. Toward the medial end of the canal, reinforcing the part of this wall posterior to the superficial inguinal ring, is the reflected inguinal ligament. The tendon of the rectus abdominis muscle fuses with the posterior aspect of the conjoined tendon. All the reinforcing structures are anterior to the transversalis fascia. The subserous fascia and the peritoneum are posterior or deep to the transversalis fascia and continue behind the deep inguinal ring. At the lateral end of the canal, the inferior epigastric artery and vein are posterior to the canal in the subperitoneal fascia. Overlying these vessels, thickening in the transversalis fascia may appear, called the interfoveolar ligament. A slight depression in the parietal peritoneum, as seen from within, may be present at the site of the deep inguinal ring. The roof of the inguinal canal is formed by the most inferior fasciculi of the internal oblique muscle as they gradually pass in a slightly arched fashion, from a position at their origin anterior to the canal to a position at their insertion posterior to the canal. At the lateral end of the canal, the lower fasciculi of the transversus abdominis arch similarly cover the canal.

The weakest area in the anterolateral wall in relation to the inguinal canal is the area of the superficial inguinal ring, which is reinforced by the reflected inguinal ligament, the conjoined tendon, and the expansion laterally and inferiorly from the tendon of the rectus muscle to the pecten pubis. This generally weakened area, through which direct inguinal hernias pass, is often described as a triangle bounded superolaterally by the inferior epigastric vessels, superomedially by the lateral margin of the rectus, and inferiorly by the inguinal ligament, known as the *inguinal* (Hesselbach) *triangle*.

Preperitoneal hernia repair has allowed a new view of the anatomy of the inguinal canal. The *preperitoneal space* is bordered posteriorly by the peritoneum and anteriorly by the transversalis fascia. This space contains connective tissue along with vascular structures and nerves. Vascular structures of the preperitoneal space include the iliac vessels. External iliac vessels run on the medial aspect of the psoas muscle before passing under the iliopubic tract and the inguinal ligament and becoming the femoral vessels. Five major nerves in the preperitoneal space are responsible for innervation of the lower abdominal wall, inguinal, and genital regions: the iliohypogastric, hypogastric, ilioinguinal, genitofemoral, and lateral femoral cutaneous nerves.

Anterior view

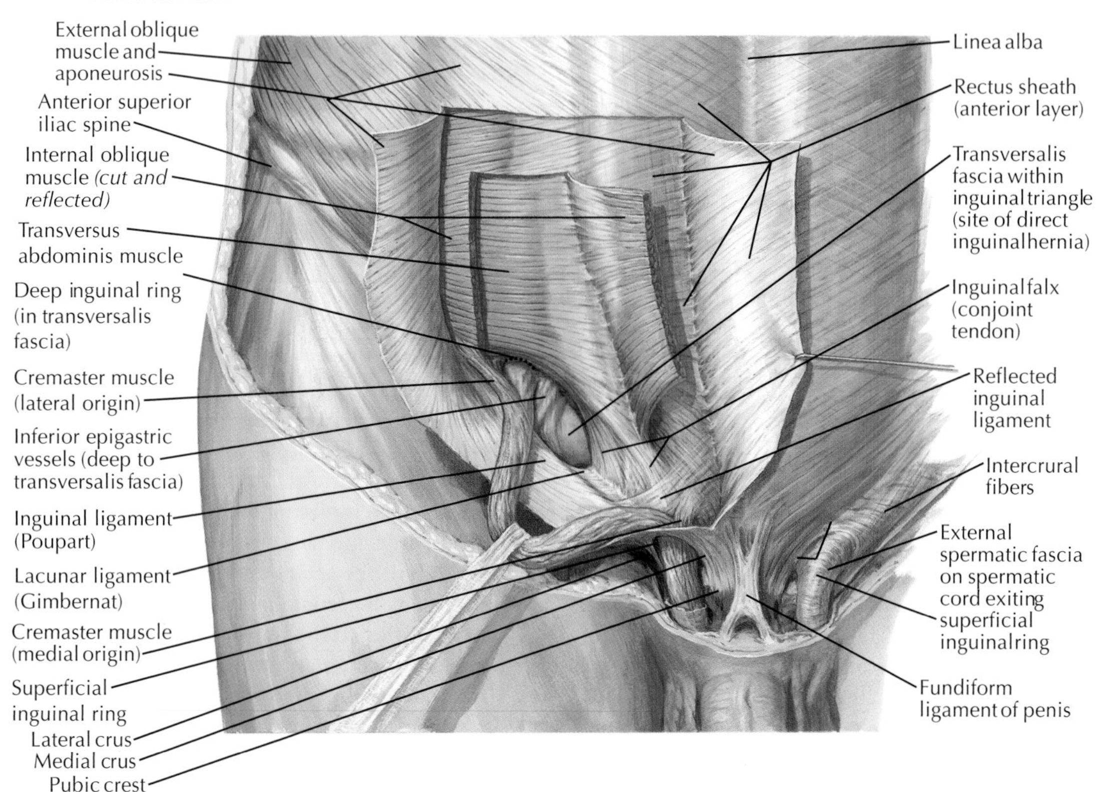

Posterior (internal) view

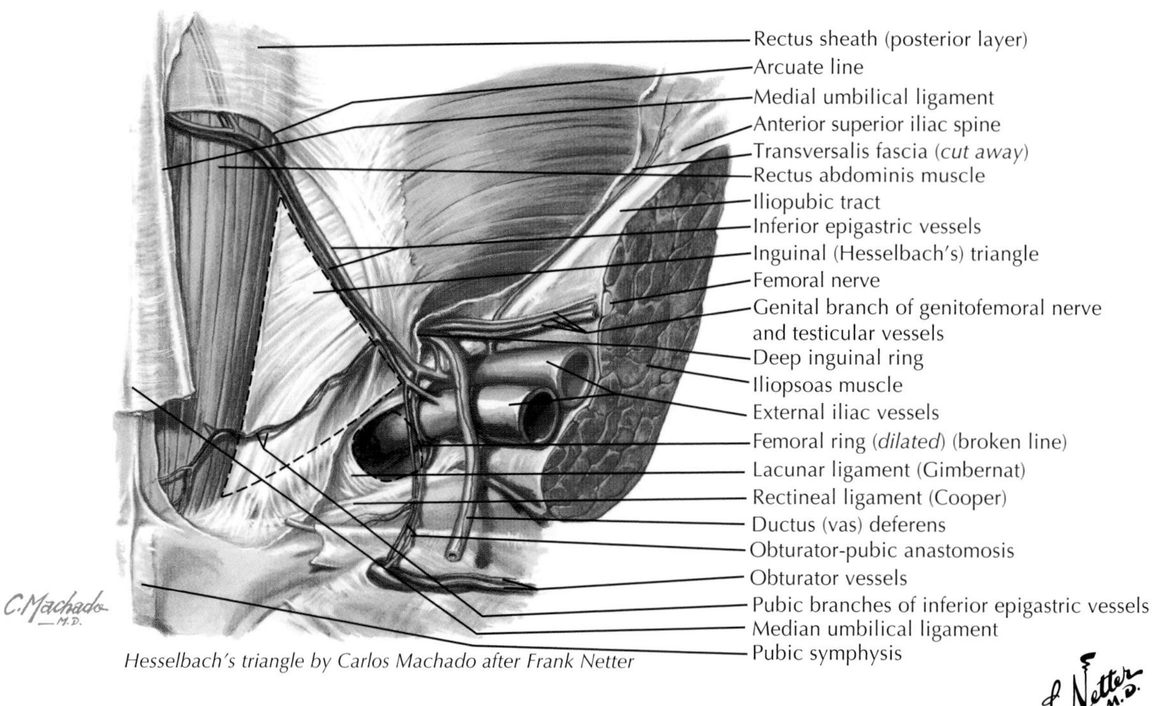

Hesselbach's triangle by Carlos Machado after Frank Netter

Figure 74-1 *Inguinal Region: Dissections.*

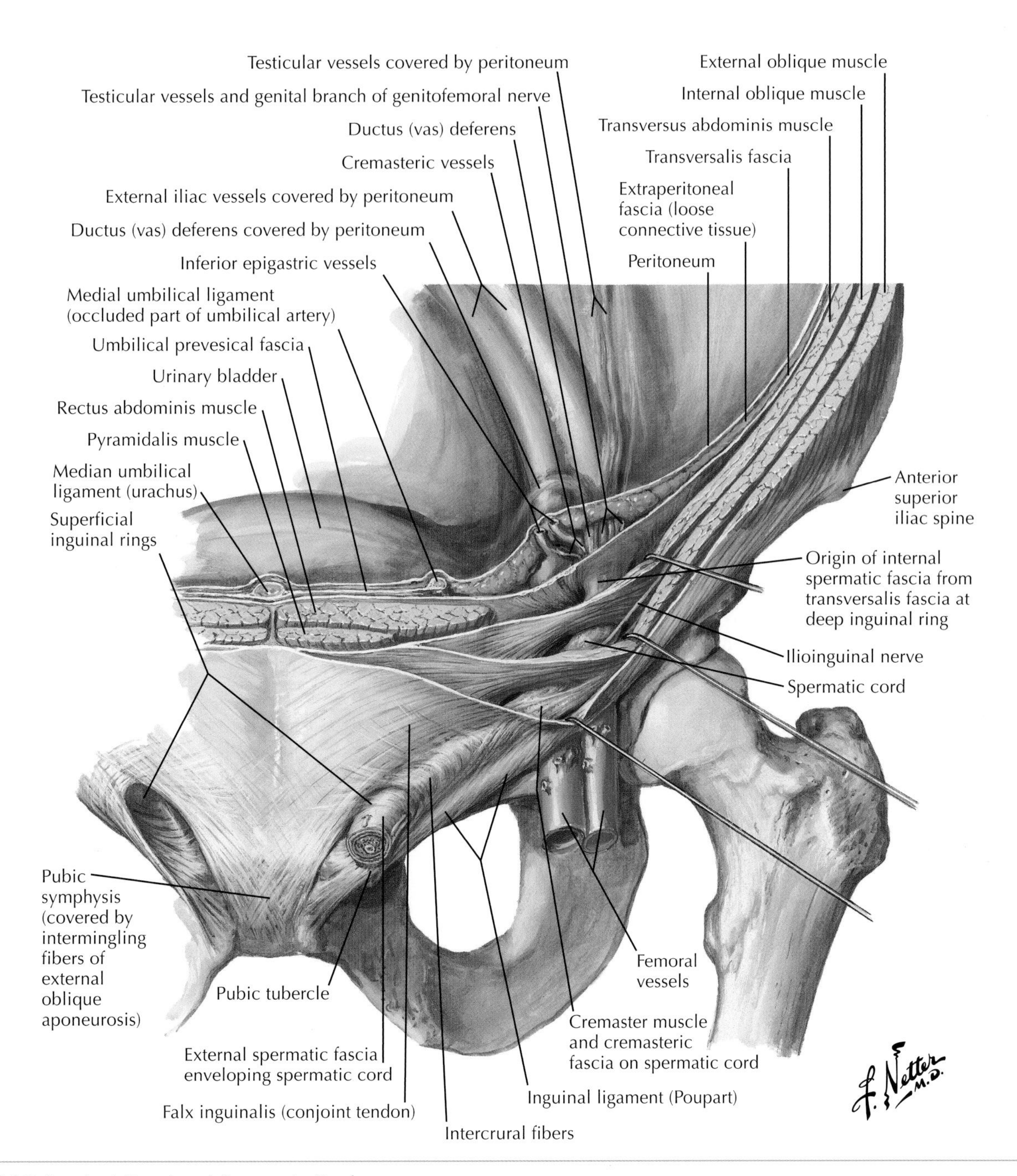

Figure 74-2 *Inguinal Canal and Spermatic Cord.*

ADDITIONAL RESOURCES

Annibali R, Fitzgibbons RJ: Laparoscopic anatomy of the abdominal wall. In Phillips EH, Rosenthal RJ, editors: *Operative strategies in laparoscopic surgery*, Berlin, 1995, Springer, pp 75-82.

Annibali R, Quinn TH, Fitzgibbons RJ: Surgical anatomy of the inguinal region and the lower abdominal wall: the laparoscopic perspective. In Bendavid R, editor: *Prostheses and abdominal wall hernias*, Austin, Texas, 1994, Landes Medical, pp 82-103.

Abdominal Regions and Planes

75

Flavia Soto and Raul J. Rosenthal

The abdomen is divided into what are called *areas* or *regions* (**Fig. 75-1**). A simple view shows two imaginary planes passing through the umbilicus, one vertically and the other horizontally. The abdomen is thus divided into four quadrants: a right and a left upper quadrant and a right and a left lower quadrant.

A division of the abdomen for descriptive purposes uses two vertical and two horizontal planes that divide the abdomen into nine regions (tic-tac-toe board). The zone above the upper of the two horizontal planes is divided by the two vertical planes into a centrally placed *epigastric region*. The epigastrium, with a *hypochondriac region* (hypochondrium) on either side, is designated as the right and left hypochondriac regions. The diaphragm is in the upper limit of the abdomen, so most of the hypochondriac regions and part of the epigastrium are beneath the ribs. Also, because these three regions make up much of the right and left upper quadrants, these quadrants also extend well into the ribs. The zone between the two horizontal planes is divided into a centrally placed *umbilical region*, with a specified left and right *lumbar* or *lateral abdominal* region on either side. The zone below the lower of the two horizontal planes has a centrally placed *hypogastric* or *pubic* (suprapubic) *region*, with a specified left and right *inguinal* or *iliac* region on either side.

Although localization using these nine regions is more specific than with the four quadrants, localization is still general. Much disparity exists regarding the ideal site of each of the four lines, or planes, used in this scheme. The upper horizontal (superior transverse) line, or plane, may be drawn midway between the upper border of the sternum and the upper border of the symphysis pubis. This plane passes through the pylorus and consequently is named the *transpyloric plane*. It has also been described as midway between the xiphisternal junction and the umbilicus, passing through the tip of the ninth costal cartilage, the fundus of the gallbladder, and the lower part of the body of the first lumbar vertebra. Another identifying landmark for locating the upper horizontal plane is the most caudal part of the costal margin (usually, most caudal part of tenth costal cartilage). This plane is called the *subcostal plane*.

Two vertical planes, or lines, one on either side, may be located halfway between the median plane and the anterior superior spine of the ilium (or halfway between pubic tubercle and anterior superior iliac spine, or midpoint of inguinal ligament, right and left midinguinal planes).

Vertical planes on each side may also be identified by using the lateral border of the rectus abdominis muscle or the semilunar line, which, if followed inferiorly and medially toward the pubic tubercle, brings the entire inguinal canal into the inguinal region.

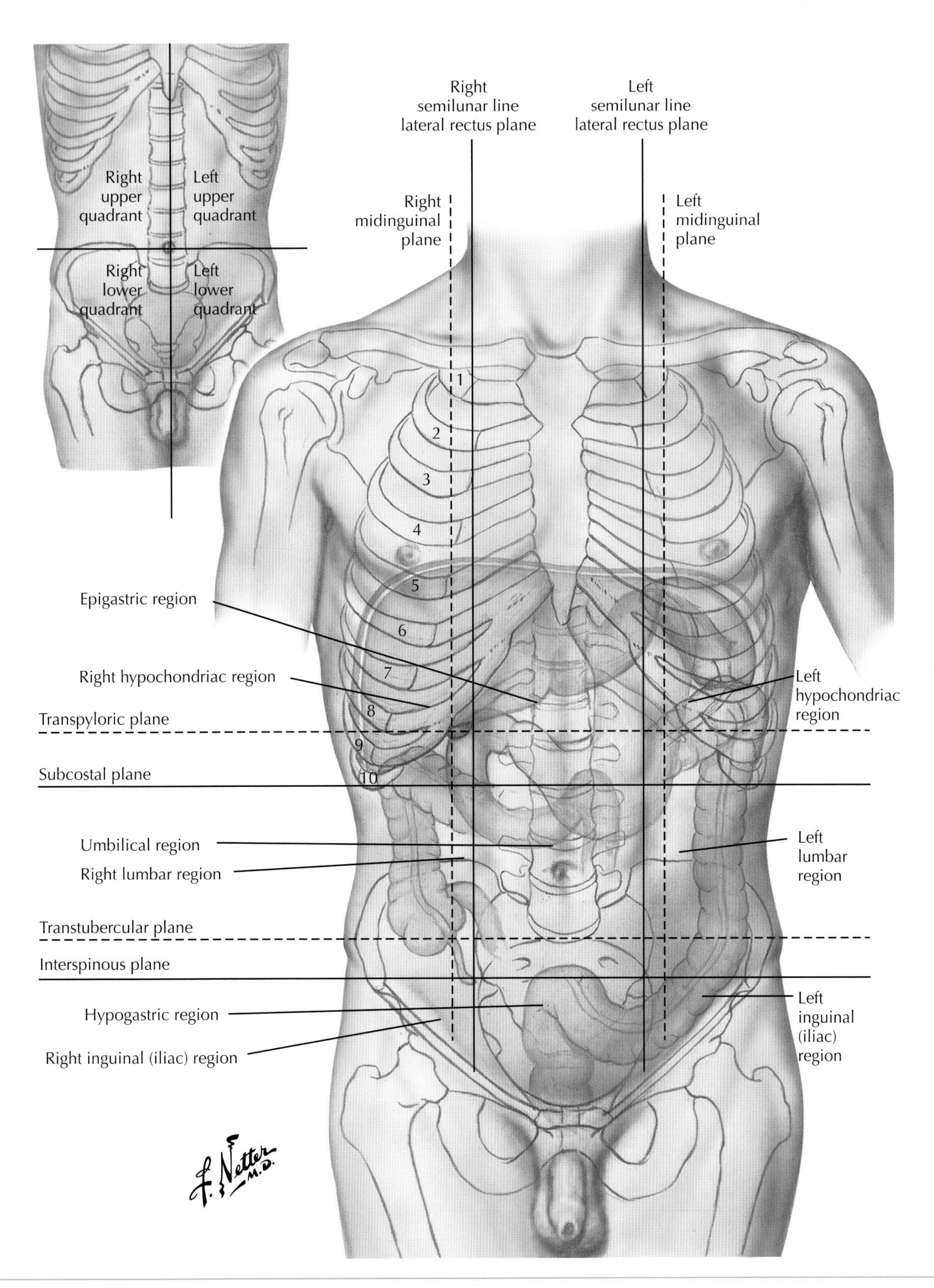

Figure 75-1 *Regions and Planes of the Abdomen.*

Abdominal Wall and Cavity: Congenital Abnormalities

76

Flavia Soto and Raul J. Rosenthal

The most common variant of the abdominal wall and congenital abnormality is the *diastasis recti*. This consists of an upper midline protrusion of the abdominal wall between the right and left rectus abdominis muscles. This is a weakness of the linea alba and does not require treatment unless an epigastric hernia develops in association with the diastasis recti.

Omphalocele may be seen in neonates and represents a defect in the closure of the umbilical ring (**Fig. 76-1**). The herniated viscera are usually covered with an amniotic sac.

Gastroschisis, a defect in the abdominal wall lateral to the umbilicus, is caused by failure of the body wall to close. The abdominal viscera protrude through the defect, and no sac is present to cover the herniated intestine.

Omphalomesenteric vitelline duct remnants may present as abnormalities related to the abdominal wall. In the fetus, the omphalomesenteric duct connects the fetal midgut to the yolk sac. This normally obliterates and disappears completely. However, any part of or the entire fetal duct may persist and give rise to symptoms.

Meckel diverticulum results when the intestinal end of the omphalomesenteric duct persists (see Fig. 76-1). This is a true diverticulum of the intestine, with all layers of the intestinal wall represented.

Anomalies of the urachus, a fetal structure that connects the developing bladder to the umbilicus, can occur. The urachus is normally obliterated at birth. It may persist in toto, resulting in a vesicoumbilical fistula manifested by the drainage of urine from the umbilicus. Proper treatment is excision of the fistula after distal urinary obstruction has been excluded.

Urachal sinus results when the umbilical end of the urachus does not obliterate normally. Such sinuses present as the chronic drainage of small amounts of urine from the umbilicus. They may become infected and should be totally excised.

ADDITIONAL RESOURCES

Sabiston A: Abdominal wall, umbilicus, peritoneum, mesenteries, omentum and retroperitoneum. In Sabiston A, editor: *Textbook of surgery: the biological basis of the modern surgical practice*, ed 15, Philadelphia, 1997, Saunders, pp 809-823.

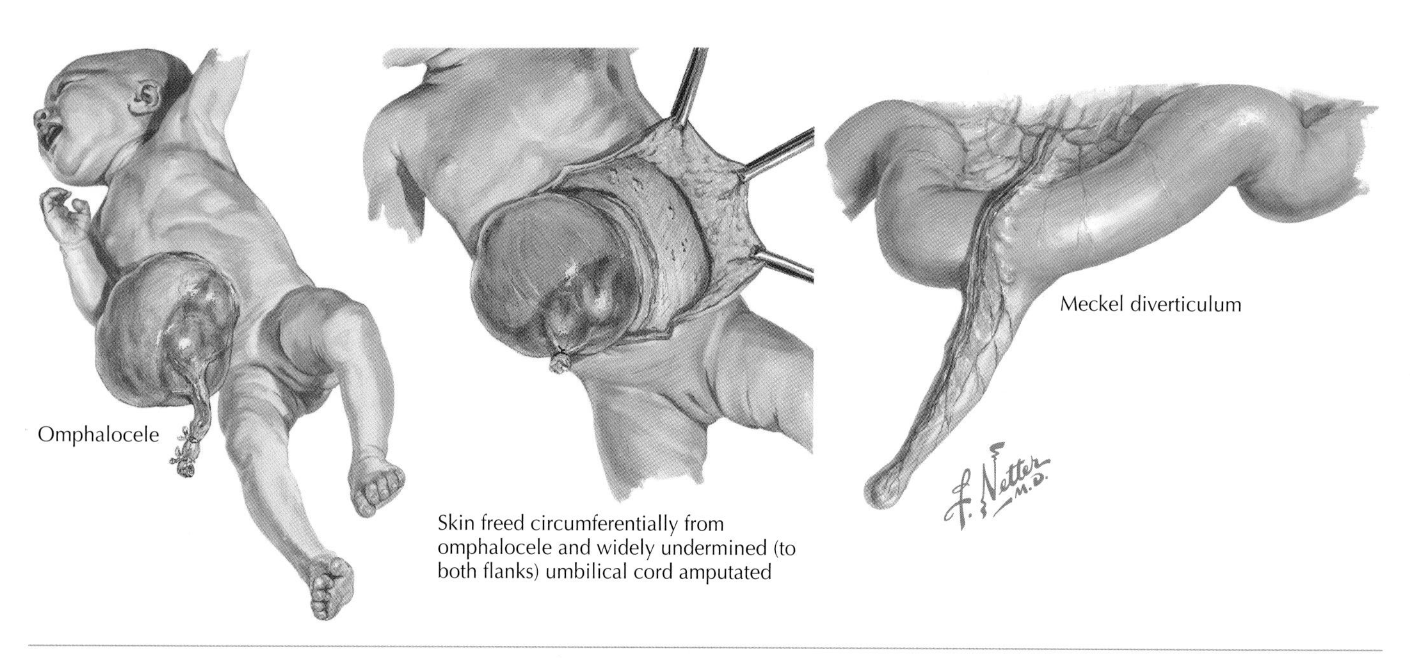

Figure 76-1 *Abdominal Wall and Cavity: Congenital Abnormalities.*

Acute Abdomen

77

Flavia Soto and Raul J. Rosenthal

The term *acute abdomen* theoretically identifies a patient with abdominal pain persisting for more than 6 hours and associated with rebound tenderness or other evidence of inflammatory reaction or visceral dysfunction that, if left untreated, could have a deleterious effect on the patient's health (**Fig. 77-1**).

CLINICAL PICTURE

The etiology of an acute abdominal condition remains one of the most difficult problems to identify. Many pathologic processes, both intraabdominal and extraabdominal, may account for these symptoms. Accurate history taking and thorough physical examination help to differentiate the variety of causes. Localization, nature, and progression of pain are essential for an accurate diagnosis.

Pain in the right upper abdominal quadrant may originate from cardiac, pulmonary, gastrointestinal, and renal conditions. Evidence of cardiac failure may implicate the heart, whereas pleuritic pain, cough, sputum, and auscultatory findings over the right lower lobe may indicate disease above the diaphragm. A prodromal period of nausea and anorexia, followed by pain, jaundice, and enlargement of the liver, suggests hepatitis; this must be differentiated from acute cholecystitis, which presents as colicky pain in the same quadrant. Urinalysis showing red or white blood cells suggests pyelonephritis or renal stones, whereas glycosuria and ketonuria may confirm initial clinical evidence of diabetic acidosis.

The most difficult area to diagnose in females is the right lower quadrant. Although persistent pain in this region could indicate appendicitis, the identical symptoms may result from a twisted, ruptured, or bleeding ovarian cyst; pelvic inflammatory process; or twisted pedunculated fibroid. However, all these conditions require surgical intervention, provided systemic and renal diseases have been excluded. If the patient has pain and tenderness on the left side, tumor or diverticulitis must be included in the differential diagnosis. If the patient has an abdominal scar from previous surgery and is reporting cramps and vomiting, intestinal obstruction should also be considered as a possible cause. Hernia and its complications can be ruled out if the pain is located in either lower quadrant.

Although the location of the pain usually identifies the site of the disease process, appendicitis frequently begins with epigastric or periumbilical pain before localizing in the right lower quadrant. Perforated peptic ulcer, acute cholecystitis, and pancreatitis may manifest as lower abdominal pain because of extravasation of inflammatory exudates to the lumbar gutter. *Rebound tenderness*, the most significant sign of peritoneal inflammation, indicates the need for surgical intervention in all patients, except those with systemic diseases such as porphyria and sickle cell crisis.

Expanding and ruptured abdominal aortic aneurysm often manifests with abdominal pain. Patients report upper abdominal tenderness and back pain and are often hypovolemic and in shock. Physical examination reveals a pulsatile mass. It is important to recognize the urgency of this situation and the need for rapid surgical intervention. Although ultrasound, computed tomography (CT), and angiography can confirm the diagnosis of abdominal aortic aneurysm, time usually does not permit their use (see Chapter 80).

DIAGNOSIS

Abdominal and chest radiographs should be used to confirm a diagnosis. The chest film will exclude or include pneumonia, pulmonary infarction, congestive heart failure, pericardial effusion, and fractured ribs, all of which can mimic acute abdomen. Free air under the diaphragm is pathognomonic of a perforated viscus. Opaque calculi may be visible and can lead to a diagnosis of cholecystitis, nephrolithiasis, or even gallstone ileus. In patients with injury manifesting as paralytic ileus, x-ray examination may disclose vertebral or pelvic fracture. In localized ileus, the sentinel loop may be seen in pancreatitis, appendicitis, or mesenteric infarction.

Ultrasound and CT should be reserved for cases requiring further confirmation to obtain an accurate diagnosis. Ultrasound provides a safe, painless method of evaluating acute abdomen and rapidly assessing multiple organs, including the liver, spleen, biliary tract, pancreas, appendix, kidneys, and ovaries. Also, pulsed Doppler ultrasound allows assessment of many vascular abnormalities, including aortic and visceral arterial aneurysm, arteriovenous fistula, and venous thrombus.

Computed tomography is also a safe, noninvasive, and efficient method of investigating the acute abdomen by providing detailed information on a variety of structures, with views of the bowel wall, mesentery, and retroperitoneum. In particular, the kidneys, pancreas, duodenum, and aorta are better delineated than with other diagnostic modalities. In addition, CT provides sensitive detection of free air, abscesses, calcifications, and collections of intraperitoneal fluid.

Magnetic resonance imaging requires more time and is generally less useful than CT.

Differential diagnoses generated after taking a clinical history then become a working diagnosis after the physical examination and laboratory and radiologic results are obtained. Subsequent management depends on the accepted treatment for the suspected condition.

It is strongly recommended that the patient with acute abdomen be continually reevaluated, preferably by the same examiner, even after a working diagnosis has been established. If the patient is unresponsive to appropriate treatment, the working diagnosis must be reassessed, returning to the initial differential diagnosis list.

ADDITIONAL RESOURCES

Delcore R, Cheung L: Acute abdominal pain. VIII. Common clinical problems. In *Care of the surgical patient: perioperative management and techniques*, Philadelphia, 1995, American College of Surgeons.

Ferzoco S, Becker J: The acute abdomen. In Sabiston A, editor: *Essentials of surgery*, ed 2, Philadelphia, 1994, Saunders, pp 274-280.

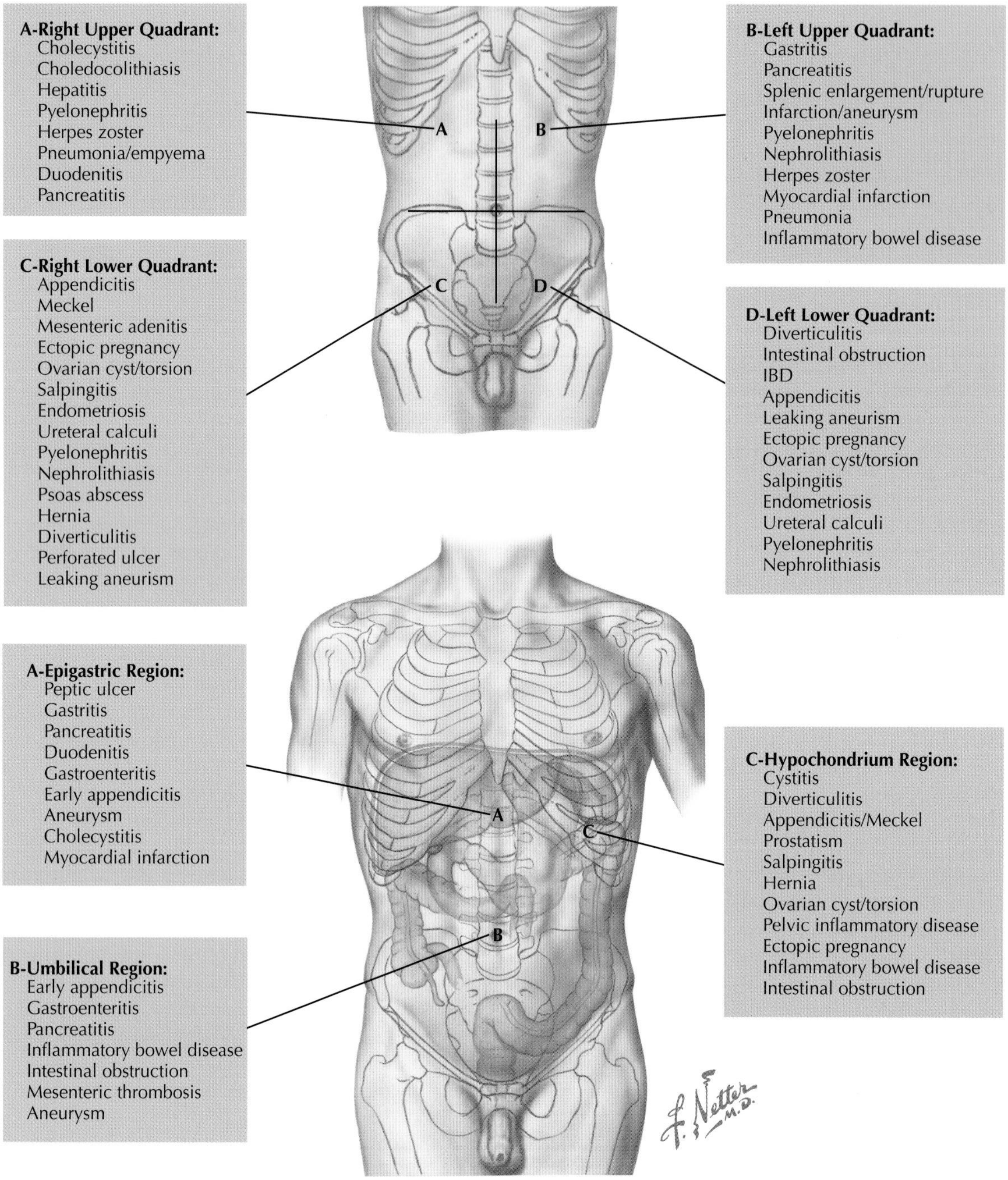

A-Right Upper Quadrant:
Cholecystitis
Choledocolithiasis
Hepatitis
Pyelonephritis
Herpes zoster
Pneumonia/empyema
Duodenitis
Pancreatitis

B-Left Upper Quadrant:
Gastritis
Pancreatitis
Splenic enlargement/rupture
Infarction/aneurysm
Pyelonephritis
Nephrolithiasis
Herpes zoster
Myocardial infarction
Pneumonia
Inflammatory bowel disease

C-Right Lower Quadrant:
Appendicitis
Meckel
Mesenteric adenitis
Ectopic pregnancy
Ovarian cyst/torsion
Salpingitis
Endometriosis
Ureteral calculi
Pyelonephritis
Nephrolithiasis
Psoas abscess
Hernia
Diverticulitis
Perforated ulcer
Leaking aneurism

D-Left Lower Quadrant:
Diverticulitis
Intestinal obstruction
IBD
Appendicitis
Leaking aneurism
Ectopic pregnancy
Ovarian cyst/torsion
Salpingitis
Endometriosis
Ureteral calculi
Pyelonephritis
Nephrolithiasis

A-Epigastric Region:
Peptic ulcer
Gastritis
Pancreatitis
Duodenitis
Gastroenteritis
Early appendicitis
Aneurysm
Cholecystitis
Myocardial infarction

C-Hypochondrium Region:
Cystitis
Diverticulitis
Appendicitis/Meckel
Prostatism
Salpingitis
Hernia
Ovarian cyst/torsion
Pelvic inflammatory disease
Ectopic pregnancy
Inflammatory bowel disease
Intestinal obstruction

B-Umbilical Region:
Early appendicitis
Gastroenteritis
Pancreatitis
Inflammatory bowel disease
Intestinal obstruction
Mesenteric thrombosis
Aneurysm

F. Netter M.D.

Figure 77-1 *Acute Abdomen.*

Alimentary Tract Obstruction

78

Guillermo Higa and Raul J. Rosenthal

Any organic or functional condition that primarily or indirectly impedes the normal propulsion of luminal contents from the esophageal inlet to the anus should be considered an obstruction of the alimentary tract (**Fig. 78-1**).

Although valid generalizations can be made about the alimentary tract, the spectrum of diseases affecting this system and their clinical manifestations are significantly related to the constituent organ(s) involved. Thus, esophageal disorders manifest mainly through their relationship to swallowing. Gastric disorders are dominated by features related to acid secretion, and diseases of the small and large intestine manifest primarily through alterations in nutrition and elimination.

CLINICAL PICTURE

The most common symptoms resulting from disorders involving the alimentary tract include pain and alterations in bowel habit, especially diarrhea and constipation. Of these symptoms, abdominal pain is the most frequent and variable and may reflect a broad spectrum of problems, from the least threatening to the most urgent.

Abdominal pain of abrupt onset is often encountered in serious illness requiring urgent intervention, whereas a history of chronic discomfort is frequently related to an indolent disorder. A change in the pattern or character of pain may be equally important because it may signify progression to a more critical stage of a problem (recent or chronic) that was mild in onset. Ascertaining the location of the pain (upper or lower, localized or diffuse), its character (sharp, burning, cramping), and its relationship to meals often provides significant insight into the most important diagnostic considerations. If eating produces the symptom, the physician should determine whether the discomfort occurs while the patient is eating (as in esophageal disorders), shortly after the meal (as often occurs in biliary tract disease and abdominal angina), or 30 to 90 minutes later (typical of peptic disease). Pain that is not affected by eating suggests a process outside the bowel lumen, such as abscess, peritonitis, and some malignancies. Relationship of the discomfort to bowel movement, especially in association with an altered bowel habit, should focus attention on a disorder of the small or large bowel, such as inflammatory bowel disease.

Alterations in bowel habit can result from either disruption of normal intestinal motility or significant structural abnormality. The onset of worsening constipation in an adult with previously regular habits, especially when accompanied by systemic symptoms such as weight loss, suggests an underlying obstructing process, particularly malignancy.

PATHOPHYSIOLOGY

In the newborn, various *congenital anomalies*—esophageal or intestinal abnormalities, anal atresias, colonic malrotation, volvulus of the midgut, meconium ileus, and aganglionic megacolon—result in obstruction. Early symptoms that suggest obstruction include increased salivation, feeding intolerance with regurgitation or vomiting, abdominal distention, and failure to pass meconium. Other causes of mechanical interference of intestinal function in early infancy include intestinal duplications, volvulus caused by mesenteric cysts and annular pancreas, incarceration in an internal or external (inguinal) hernia, and congenital peritoneal bands; the latter may not become clinically manifest until adulthood or older.

Duodenal obstruction may manifest early in the newborn period or within the first year of life, but sometimes it does not manifest until years into childhood. The common area of blockage is just beyond the ampulla of Vater.

Esophageal diseases can interfere with the normal passage of fluids and solids through the gullet (uppermost row in Fig. 78-1). Fibrotic narrowing has also been observed after anastomotic or plastic procedures at the lower end of the esophagus. An overly tight closure of the hiatus in the repair of a sliding esophageal hernia results in compression by obstruction at this point. The same picture results from extraluminal pressure on the esophagus by a tumor mass or a visceral abnormality in the neck or mediastinum.

Gastric obstruction may be caused by accumulation in the stomach of ingested material, such as hair (trichobezoar) or fruit/vegetable fibers (phytobezoar), or development of hypertrophic pyloric stenosis, spastic or cicatricial occlusion related to prepyloric or postpyloric peptic ulcer, or malignant neoplasm.

Small Intestine

Obstruction of the small bowel has four characteristics: abdominal pain, vomiting, abdominal distention, and obstipation. Symptoms are variable, depending on the anatomic level (proximal or distal), degree of obstruction (partial or complete), and presence or absence of strangulation.

Obstruction of the duodenum by extraluminal compression may occur in the superior mesenteric small intestine, most often as incarceration of a loop of small bowel in an inguinal or a femoral hernia. On occasion, incarceration occurs in an internal hernia, or a bowel segment may become caught in the ring of a congenital, traumatic, or surgical defect in the diaphragm.

A foreign body, either ingested by mouth or consisting of a large biliary calculus, may become impacted and result in obstruction; an accumulation of parasitic worms has been reported as a cause. Mechanical obstruction of the small intestine can also be caused by intussusception or by compression, torsion, or angulation of one or more bowel loops. Etiologic mechanisms include postoperative adhesions, congenital peritoneal bands, metastatic tumor implants, Meckel diverticulum, and plastic or adhesive peritonitis (tuberculosis, talc granuloma). On occasion, primary neoplasms of the small intestine (carcinoma, lymphosarcoma, Hodgkin granuloma) may cause obstructive manifestations.

Varying degrees of small-bowel obstruction may be caused by segmental fibrotic stricture formation (as in regional ileitis

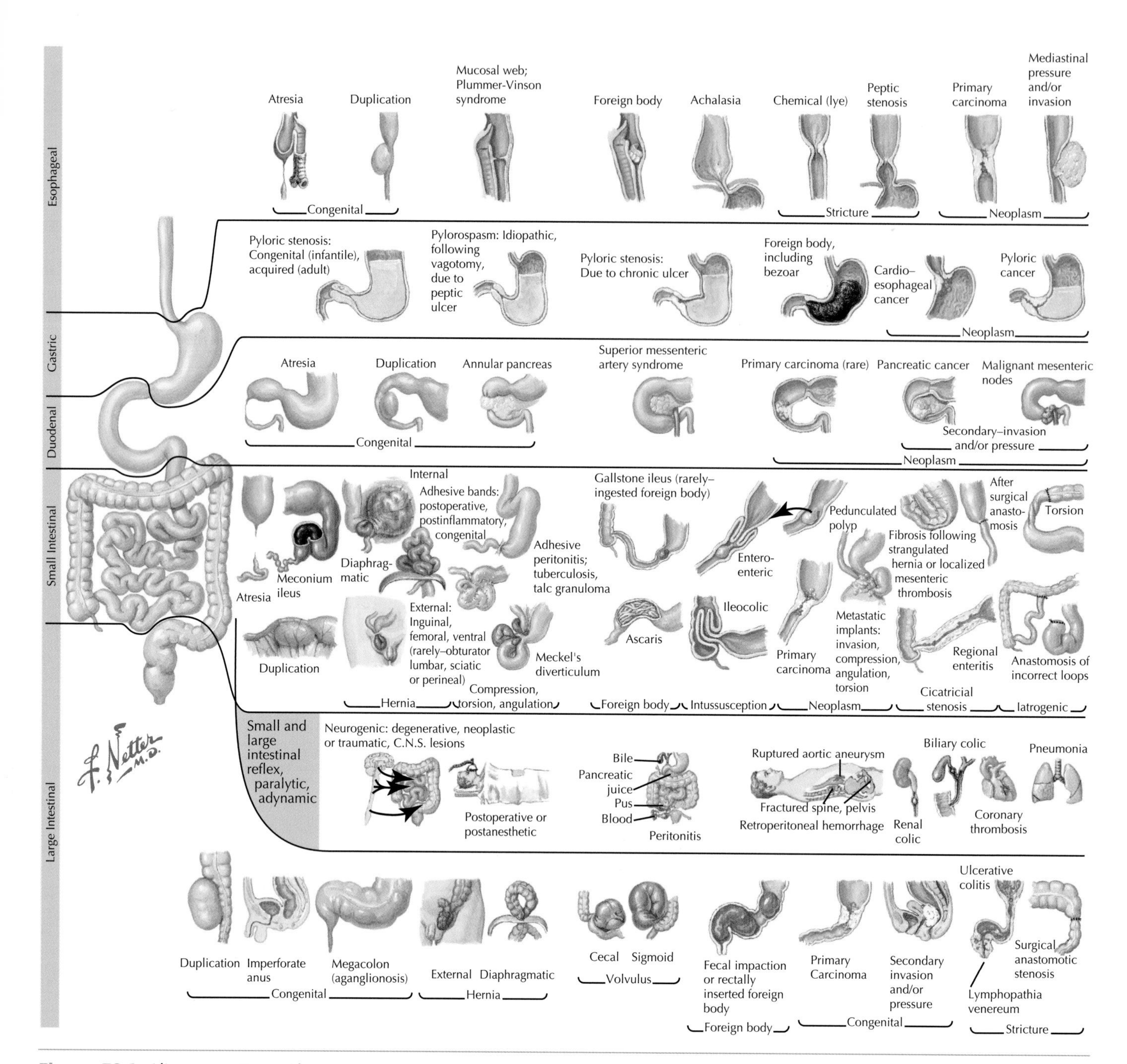

Figure 78-1 *Alimentary Tract Obstruction.*

or jejunitis) or stenosed bowel loop resulting from healing after localized infarction sustained in hernial incarceration or mesenteric vascular occlusion.

Iatrogenic intestinal obstruction (anastomotic stenosis, torsion or angulation, anastomosis of incorrect loops) may result from faulty surgical technique.

Large Intestine

Cancer of the large intestine constitutes the most frequent mechanism of obstruction of this viscus. Occlusion occurs more often where the bowel lumen is of smaller caliber. Occasionally, volvulus of the right colon can be seen if it has a long mesentery. Fecal impaction causing obstruction and sigmoid colon volvulus may develop in elderly patients. A large foreign body inserted rectally (perversion, psychosis) may also obstruct the lumen.

The most common symptoms of high-grade colonic obstruction are abdominal pain, distention, vomiting, and constipation. These symptoms may be superimposed on vague symptoms related to the developing obstruction, such as flatulence, diarrhea, and variable degrees of abdominal distention.

Strictures of the colon may result from cicatricial fibrosis (granulomatous ileocolitis or colitis, nonspecific ulcerative colitis, diverticulitis, venereal lymphopathy) or may occur as postoperative complications (anastomotic stricture, posthemorrhoidectomy). Extraluminal compression, usually caused by primary or metastatic pelvic tumors and rarely by pelvic inflammatory exudates or abscess formation, may obstruct the lower bowel at the level of the rectosigmoid or rectum.

Intestinal Ileus

Nonmechanical impairment of intestinal motor function has been descriptively termed reflex, adynamic, or paralytic ileus. As a complication of various causative clinical conditions, the patient has the syndrome of gastric retention, constipation and failure to pass flatus, abdominal distention, "silent abdomen," and radiographic findings of dilatation of the small and large intestine with gas and accumulated fluid.

Reflux ileus may be encountered in patients with various lesions of the central nervous system. Intestinal atony may follow surgical anesthesia or the trauma of intraabdominal surgical manipulation, extensive rib fractures, or blunt abdominal trauma (e.g., immersion blast injury). Other clinical conditions in which ileus has reportedly occurred as a so-called reflex phenomenon include renal or biliary colic, pneumonia, torsion infarction of an ovarian cyst, coronary thrombosis, and retroperitoneal hemorrhage (fracture of spine or pelvis, dissecting aortic aneurysm, urinary extravasation, rupture of kidney).

Paralytic or *adynamic ileus* occurs most often with purulent peritonitis caused by perforated appendicitis, perforation of a hollow viscus, pelvic inflammatory disease, leakage or dehiscence of an intestinal suture line, or wound evisceration. Ileus may follow the intraperitoneal extravasation of gastric or duodenal contents (perforated peptic ulcer), pancreatic juice (acute hemorrhagic pancreatitis), bile (perforated gallbladder, bile leakage from liver or bile ducts), and blood (postoperative hemorrhage; rupture of liver, spleen, or graafian follicle; ectopic gestation; chocolate cyst of ovary).

ADDITIONAL RESOURCES

Filston H: Pediatric surgery. In Sabiston A, editor: *Textbook of surgery*, vol 2, Philadelphia, 1986, Saunders, pp 1253-1298.

Isselbacber K, Podolsky D: Approach to the patient with gastrointestinal disease. In Fauci AS, editor: *Harrison's principles of internal medicine*, ed 14, vol 2, New York, 1998, McGraw-Hill, pp 1579-1583.

Welch J: Mechanical obstruction of the small and large intestines. In Moody FG, *Surgical treatment of digestive disease*, ed 2, Chicago, 1990, Mosby-Year Book, pp 624-639.

Mesenteric Vascular Occlusion

79

Guillermo Higa and Raul J. Rosenthal

The mesenteric circulation receives approximately 25% of cardiac output under resting conditions. Thus, it is one of the largest regional circulatory areas in the body. Mesenteric circulation is functionally complex, and its location within the peritoneal cavity makes objective assessment difficult.

Mesenteric ischemic disorders result from chronic or acute insufficiency of blood flow to all or part of the intestine (**Fig. 79-1**). Varied etiologies and degrees of ischemia produce different clinical manifestations and pathophysiologic changes requiring prompt diagnosis and aggressive management. Ischemic intestinal injury may be acute or chronic, extensive or segmental, recurrent or transient, and completely reversible.

Mesenteric ischemia is a function of the following:

- State of the systemic circulation
- Degree of functional or anatomic vascular compromise
- Number and caliber of vessels affected
- Response of the vascular bed to diminished perfusion

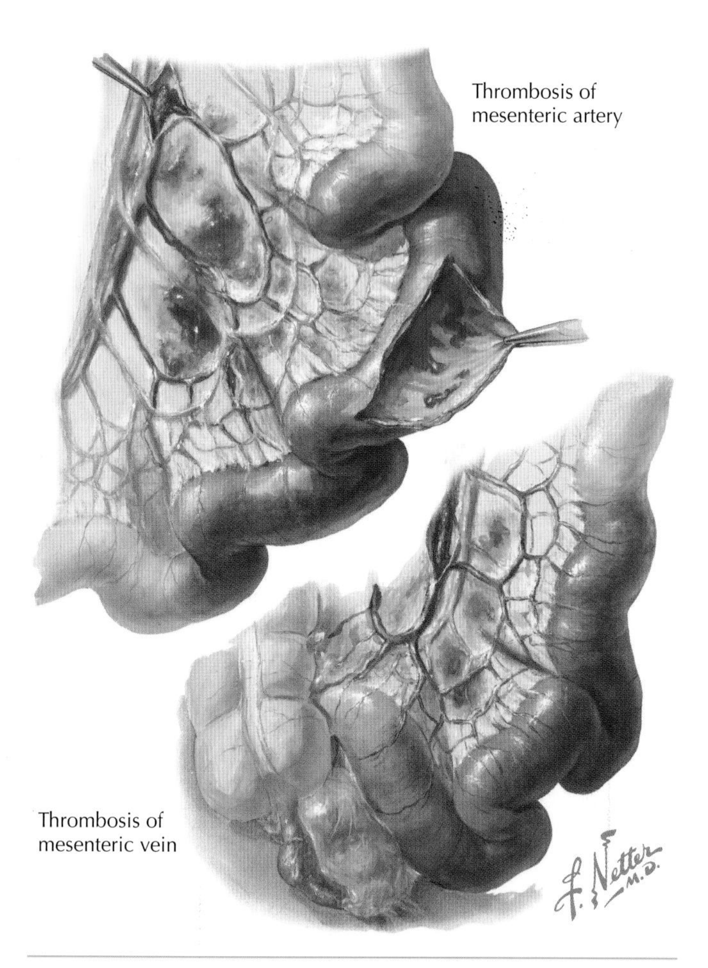

Figure 79-1 *Mesenteric Vascular Occlusion.*

- Nature and capacity of the collateral circulation to supply the needs of the dependent segment of bowel
- Duration of the insult
- Metabolic needs of the dependent segment, as dictated by its function and bacterial population

PATHOPHYSIOLOGY

Ischemic injury to the intestine occurs when it is deprived of oxygen and other nutrients necessary to maintain cellular metabolism and integrity. Reduced blood flow may reflect poor systemic perfusion or may result from local changes in the splanchnic vasculature. Narrowing of the major mesenteric vessels, focal atheromatous emboli, vasculitis, and mesenteric vasoconstriction can all lead to inadequate circulation at a cellular level. However, regardless of the cause, the end results include a spectrum of injury ranging from reversible functional alterations to transmural hemorrhagic necrosis throughout the bowel.

Factors influencing ischemic injury of the bowel include the state of the general circulation, the extent of collateral blood flow, the response of the mesenteric vasculature to autonomic stimuli, circulating vasoactive substances, local humoral factors, and the products of cellular metabolism before and after reperfusion of the ischemic segment of bowel.

Mesenteric Ischemia and Infarction

Acute intestinal ischemia may be classified as *occlusive* or *nonocclusive*. Occlusion may result from an arterial thrombus or an embolus of the celiac or superior mesenteric arteries or from venous occlusion in the same distribution. Arterial embolus most often occurs in patients with chronic or recurrent atrial fibrillation, artificial heart valves, or valvular heart disease; arterial thrombosis is usually associated with extensive atherosclerosis or low cardiac output.

Superior mesenteric artery embolus tends to lodge at points of normal anatomic narrowing, usually just distal to the origin of a major branch. An embolus may completely occlude the artery, but usually, partial obstruction of the vessel lumen occurs.

CLINICAL PICTURE

The major clinical feature of acute mesenteric ischemia is *severe abdominal pain*, often colicky and periumbilical at the onset and then becoming diffuse and constant. Vomiting, anorexia, diarrhea, and constipation also occur frequently but are of little diagnostic help. Examination of the abdomen may reveal tenderness and distention; bowel sounds are often normal, even in patients with severe infarction. Some patients have surprisingly normal findings on abdominal examination despite the severe pain. Mild gastrointestinal bleeding is often detected by examining the stool for occult blood; gross hemorrhage is unusual except in ischemic colitis. A typical laboratory finding is pronounced polymorphonuclear leukocytosis. Later in the disease course, gangrene of the bowel occurs, with diffuse peritonitis, sepsis, and shock.

Abdominal plain films in patients with mesenteric ischemia may reveal air-fluid levels and distention. Barium study of the small intestine reveals nonspecific dilatation, poor motility, and evidence of thick mucosal folds (thumbprinting).

DIAGNOSIS

Acute mesenteric ischemia is a serious condition associated with high morbidity and mortality. Patients thought to have acute arterial embolus should undergo immediate celiac and mesenteric angiography to localize the embolus, followed by embolectomy.

TREATMENT AND MANAGEMENT

Restoration of normal circulation may allow complete recovery if it can be accomplished before irreversible necrosis or gangrene occurs. Unfortunately, infarction and transmural necrosis are frequently found at surgery, necessitating resection. Arterial or venous thrombosis is not generally amenable to surgical removal of the thrombus; therefore, resection of the affected bowel is required. Similarly, patients with nonocclusive ischemia are not candidates for corrective vascular surgery because major vessels are patent.

Patients with nonocclusive ischemia often have extensive necrosis of the small or large intestine, resulting from the widespread nature of the ischemic event. The decision to operate when mesenteric ischemia is suspected is often difficult, given that the typical patient is a poor surgical risk because of advanced age, dehydration, sepsis, and other serious comorbid conditions.

COURSE AND PROGNOSIS

Superior mesenteric artery thrombosis occurs at areas of severe atherosclerotic narrowing, usually at the origin of the superior mesenteric artery. The acute ischemic episode is typically superimposed on chronic mesenteric ischemia. Approximately 20% to 50% of patients have a history of postprandial abdominal pain with or without malabsorption and weight loss during the weeks or months preceding the acute episode. Most patients with superior mesenteric artery thrombosis have a history of coronary, cerebrovascular, or peripheral arterial insufficiency.

Mesenteric venous thrombosis has long been known as the "great imitator" of other abdominal disorders. Symptoms are nonspecific except for abdominal pain, which is present in more than 90% of patients. The onset of abdominal pain, nausea, and vomiting may be insidious over 2 to 3 weeks before development of abdominal signs. No symptoms clearly indicate the diagnosis of mesenteric thrombosis.

Mesenteric venous thrombosis may follow intraabdominal infections or may develop in association with thrombotic hematologic disorders such as polycythemia vera or postsplenectomy thrombocythemia. Because the thrombotic process in the mesenteric veins is more slowly progressive, the symptoms of pain, nausea, and vomiting may be present for several days before the involved bowel becomes frankly gangrenous and signs and symptoms manifest.

The prognosis in mesenteric venous thrombosis is more favorable than in arterial occlusion but still serious because further thrombosis and infarction may occur. Anticoagulant therapy may be helpful after surgery to prevent secondary venous thrombosis. Reports of mesenteric venous thrombosis causing fatal intestinal bleeding without hemorrhagic infarction of the bowel have been verified at autopsy.

ADDITIONAL RESOURCES

Bergan JJ, Yao JST: Acute intestinal ischemia. In Rutherford RB, editor: *Vascular surgery*, ed 2, Philadelphia, 1984, Saunders, pp 948-963.

Brandt LJ, Boley SJ: Ischemic and vascular lesions of the bowel. In Sleisenger MH, Fordtran JS, editors: *Gastrointestinal disease*, Vol 2, Philadelphia, 1993, WB Saunders, pp 1927-1962.

Disorders of the mesenteric circulation. In Harrison TR, Fauci AS, editors: *Harrison's principles of internal medicine*, ed 14, Vol 2, New York, 1998, McGraw-Hill, pp 1651-1653.

Kaleya RN, Boley SJ: Mesenteric ischemic disorders. In Maingot R, Schwartz S, Ellis F, Husser WC, editors: *Maingot's abdominal operations*, Vol 1, Norwalk, Conn, 1990, Appleton & Lange, pp 655-689.

O'Mara CS, Ernst CB: Physiology of the mesenteric circulation. In Zuidema GD, Shackelford RT, editors: *Shackelford's surgery of the alimentary tract*, ed 3, Vol 5, Philadelphia, 1991, Saunders.

Other Vascular Lesions

80

Guillermo Higa and Raul J. Rosenthal

Aneurysm is defined as a pathologic dilatation of a segment of blood vessel. *True aneurysm* involves all three layers of the vessel wall and is distinguished from a *pseudoaneurysm*, in which the intimal and medial layers are disrupted and the dilatation is lined only by adventitia, and sometimes by perivascular clot. **Figure 80-1** shows arteries of the posterior abdominal wall.

DISEASES OF THE AORTA

Aortic Aneurysm

Abdominal aortic aneurysm (AAA) is a common condition, with an estimated incidence of 30 to 66 per 1000 persons, having tripled in the past three decades. AAAs tend to rupture suddenly, resulting in approximately 15,000 deaths per year, making AAA the thirteenth leading cause of death in the United States.

Aneurysms of the infrarenal aorta are by far the most common arterial aneurysms. Men are affected more than women by a ratio of 4:1. Other aneurysms, such as common or internal iliac and femoropopliteal aneurysms, are also frequently present in patients with AAA.

ETIOLOGY

The most common pathologic condition associated with AAA is atherosclerosis. Additional causes of AAA include cystic medial necrosis, syphilis, tuberculosis, other bacterial infections, Takayasu arteritis, giant cell arteritis, seronegative spondyloarthropathy, rheumatoid arthritis, and trauma.

CLINICAL PICTURE

From 70% to 75% of all infrarenal AAAs are asymptomatic when first discovered, often during routine physical examination or radiographic study for another reason (upper GI series, barium enema, intravenous pyelography, lumbosacral radiography, abdominal CT, ultrasound).

An AAA may cause symptoms as a result of rupture or expansion, pressure on adjacent structures, embolization, dissection, or thrombosis. Compression of the adjacent bowel can cause early nausea and vomiting. A large aneurysm can actually erode the spine and cause severe back pain in the absence of rupture.

Abrupt onset of severe back, flank, or abdominal pain is characteristic of aneurysmal rupture or expansion. Ruptured aneurysms constitute 20% to 25% in most series. Symptoms and time course vary, depending on the nature of the rupture. Small tears of the aneurysmal sac may result in a small leak that, at least temporarily, resolves with minimal blood loss. This is usually followed within a few hours by frank rupture, producing a catastrophic medical emergency.

The classic clinical manifestations of ruptured AAA are midabdominal or diffuse abdominal pain, shock, and a pulsatile abdominal mass. The pain may be more prominent in the back or flank or may radiate to the groin or thigh.

TREATMENT AND MANAGEMENT

Surgical excision and replacement with a graft is indicated for AAAs of any size that are expanding or associated with symptoms. Surgery is indicated for asymptomatic aneurysms with diameters greater than 6.5 cm.

Chronic Arteriosclerotic Occlusive Disease

Chronic occlusive disease involves the distal abdominal aorta below the renal arteries. Because of the slowly progressive nature of the atherosclerotic process, the natural history of aortic occlusion is usually chronic and insidious. *Claudication* characteristically involves the lower back, buttocks, and thighs and may be associated with impotence in males (Leriche syndrome). The severity of symptoms depends on the adequacy of collaterals. With sufficient collateral blood flow, complete occlusion of the abdominal aorta may occur without the development of ischemic symptoms.

Physical findings include the absence of femoral and other distal pulses bilaterally and the detection of an audible bruit over the abdomen (usually at or below the umbilicus) and the common femoral arteries. Atrophic skin, hair loss, and coolness of the lower extremities are usually observed. In advanced ischemia, rubor on dependency and pallor on elevation can be seen.

Acute Aortic Occlusion

Acute occlusion in the distal abdominal aorta represents a medical emergency because it threatens the viability of the lower extremities. This usually results from an occlusive embolus that almost always originates from the heart. Rarely, acute occlusion may result from in situ thrombosis in a preexisting, severely narrowed segment of the aorta or from plaque rupture and hemorrhage into such an area. Severe chest pain, coolness, and pallor of the lower extremities and the absence of distal pulses bilaterally are the usual manifestations. Diagnosis should be established rapidly by aortography. Emergency thrombectomy or revascularization is indicated.

Iliac Artery Aneurysm

Common iliac artery aneurysms frequently occur in continuity or in association with AAAs. Iliac aneurysms are often atherosclerotic (degenerative) and thus occur mainly in the population with, or at risk for, atherosclerosis. The natural history of iliac aneurysm is unfavorable, with a high rate of rupture within a few months of diagnosis, possibly because of their large size when diagnosed.

Vascular Kidney Injury

Involvement of the renal vessels by atherosclerotic, hypertensive, embolic, inflammatory, and hematologic disorders is usually a manifestation of generalized vascular disease.

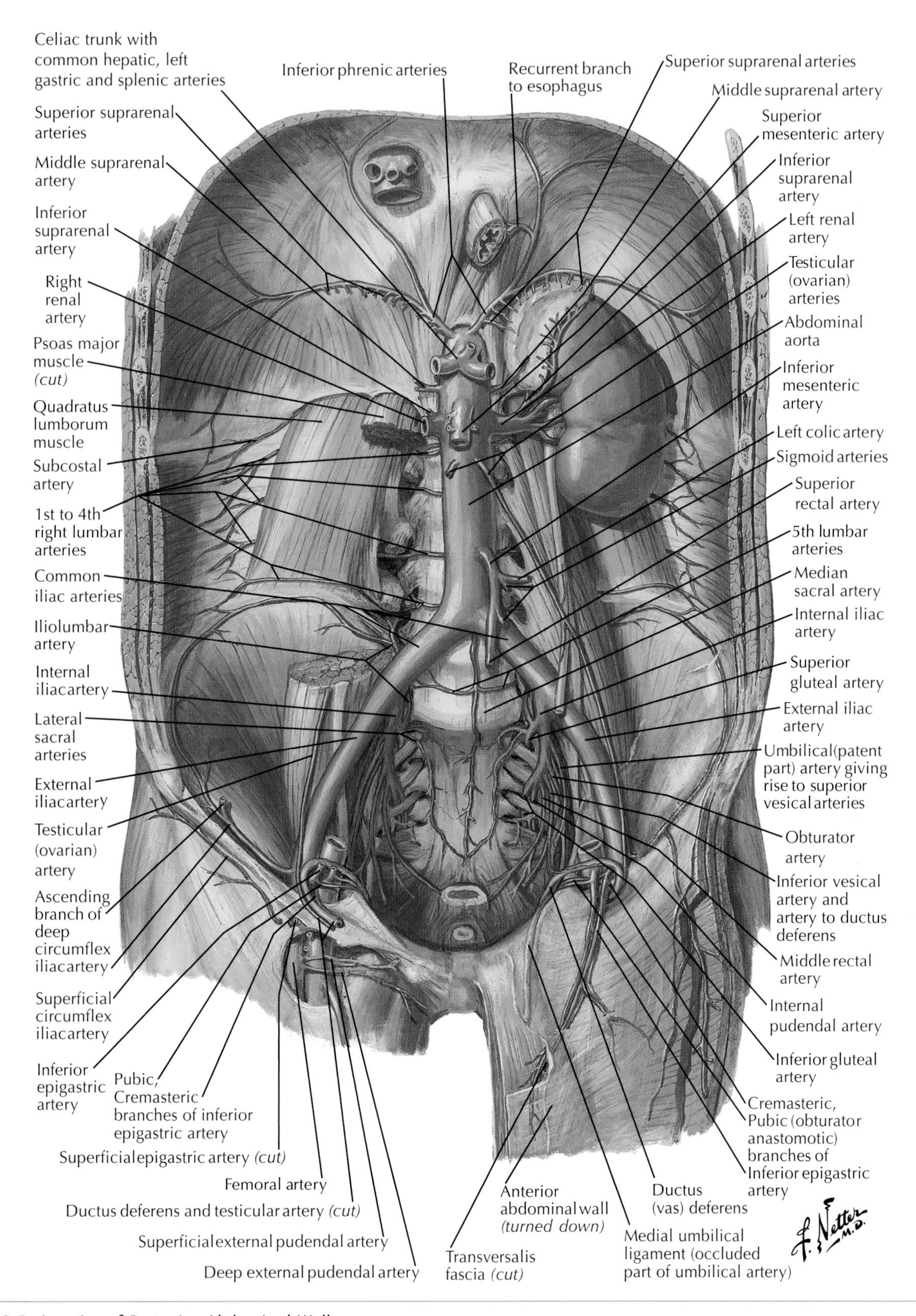

Figure 80-1 *Arteries of Posterior Abdominal Wall.*

THROMBOEMBOLIC DISEASES OF THE RENAL ARTERIES

Thrombosis may result from an intrinsic abnormality in the renal vessels (posttraumatic, atherosclerotic, or inflammatory) or from emboli originating in distant vessels, usually fat emboli, emboli originating in the left side of the heart (mural thrombi after myocardial infarction, bacterial endocarditis, or aseptic vegetations), or paradoxical emboli passing from the right side of the circulation through a patent foramen ovale or atrial septal defect.

The clinical presentation is variable, depending on the time and extent of the occlusive event. Acute thrombosis and infarction, such as after embolization, may result in the sudden onset of flank pain and tenderness, fever, hematuria, leukocytosis, nausea, and vomiting.

Renal Artery Thrombosis

Thrombosis of one or both main veins occurs with various other conditions: trauma, extrinsic compression (lymph nodes, aortic aneurysm, tumor), renal cell carcinoma, dehydration (infants), nephrotic syndrome, pregnancy, and oral contraceptive use.

Clinical manifestations depend on the severity and abruptness of onset of thrombosis. Acute thrombosis is characterized by sudden loss of renal function, often accompanied by fever, chills, lumbar tenderness (with kidney enlargement), leukocytosis, and hematuria. Hemorrhagic infarction and renal rupture may lead to hypovolemic shock. Definitive diagnosis can be established only through selective venography with visualization of the occluding thrombus.

Renal Artery Stenosis or Ischemic Renal Disease

The common cause of renal artery stenosis in middle-aged and elderly persons is an atheromatous plaque at the origin of the renal artery. In younger women, stenosis is caused by intrinsic structural abnormalities of the arterial wall caused by a heterogenous group of lesions termed *fibromuscular dysplasia*.

Renal artery stenosis should be suspected when hypertension develops in a previously normotensive person older than 50 or in persons younger than 30 who have suggestive symptoms, such as vascular insufficiency to other organs, high-pitched epigastric bruit on physical examination, symptoms of hypokalemia secondary to hyperaldosteronism (muscle weakness, tetany, polyuria), and metabolic alkalosis.

Renal Artery Macroaneurysms

Renal artery macroaneurysm represents an uncommon vascular disease. Most renal artery aneurysms are saccular. About 75% are located at primary or secondary renal artery bifurcations and appear to be caused by a medial degenerative process. Most are asymptomatic.

SPLANCHNIC ARTERIAL ANEURYSMS

Splanchnic arterial aneurysms constitute an uncommon but serious vascular disease. Major splanchnic vessels involved with these macroaneurysms, in decreasing order of frequency, are the splenic, hepatic, superior mesenteric, celiac, gastroepiploic, jejunal-ileocolic, pancreaticoduodenal-pancreatic, and gastroduodenal arteries, as follows:

- *Splenic arterial aneurysms* account for 60% of all splanchnic arterial aneurysms. Symptoms of left upper quadrant or epigastric pain occur in some patients with splenic aneurysms, and abdominal discomfort has been described in as many as 20%.
- *Hepatic arterial aneurysms* account for 20% of all splanchnic aneurysms, are usually solitary, and are extrahepatic in 80% and intrahepatic in 20% of patients. Few hepatic aneurysms are symptomatic, but characteristically manifest as right upper quadrant and epigastric pain.
- Proximal *superior mesenteric arterial aneurysms* are the third most common splanchnic aneurysm, accounting for 5.5% of these lesions. In these patients, abdominal discomfort varies from mild to severe pain and is often suggestive of intestinal angina.
- *Celiac arterial aneurysms* account for 4% of all splanchnic aneurysms. Celiac aneurysms can be asymptomatic or can be associated with vague abdominal discomfort.
- *Gastric and gastroepiploic arterial aneurysms* account for 4% of splanchnic aneurysms. Patients usually seek emergency treatment without preceding symptoms.
- *Jejunal, ileal, and colic arterial aneurysms* account for 3% of splanchnic aneurysms. Many of these aneurysms are undoubtedly asymptomatic and are recognized as incidental findings during arteriography for gastrointestinal bleeding.
- *Pancreatic and pancreaticoduodenal arterial aneurysms* account for 2% and *gastroduodenal arterial aneurysms* account for an additional 1.5% of all splanchnic arterial aneurysms. Most patients with these aneurysms have epigastric pain and discomfort.

ADDITIONAL RESOURCES

Badr K, Brenner B: Vascular injury to the kidney. In Harrison TR, Fanci AS, editors: *Harrison's principles of internal medicine*, ed 14, Vol 2, New York, 1998, McGraw Hill, pp 1558-1560.

Dzau V, Creager M: Diseases of the aorta. In *Harrison's principles of internal medicine*, ed 14, vol 1, Philadelphia, 1998, Saunders, pp 1394-1398.

Goldstone J: Aneurysms of the aorta and iliac arteries. In Moore W, editor: *Vascular surgery*, ed 4, Philadelphia, 1993, Saunders, pp 401-434.

Stanley J, Messina L: Splanchnic and renal artery aneurysms. In Moore W, editor: *Vascular surgery*, ed 4, Philadelphia, 1993, Saunders, pp 435-450.

Acute Peritonitis

81

Dario Berkowski and Raul J. Rosenthal

Peritonitis can result from any local trigger of inflammation, usually infection (**Fig. 81-1**). However, infection may not be present at the earliest stage of peritonitis. Inflammation of the peritoneum may be caused by a number of agents, including bacteria, fungi, viruses, chemical irritants, and foreign bodies. Therefore, management depends on identifying the etiology of the infectious process.

Primary peritonitis is defined as a monomicrobial infection without visceral perforation. *Secondary peritonitis* is related to a peritoneal infection arising from an intraabdominal source, with visceral perforation. *Tertiary peritonitis* develops after the treatment of secondary peritonitis resulting from failure of the inflammatory response.

The peritoneal cavity responds to infection in three ways. First, bacteria are rapidly absorbed through the diaphragmatic stomata. Second, bacteria are destroyed by mechanisms generated by the complement cascade and phagocytes. Third, the infection is localized as an abscess.

CLINICAL PICTURE

Patients with established peritonitis may report a history of onset that indicates acute appendicitis or salpingitis as the source of origin. The sudden onset of peritonitis should be suspected by an acute perforation of a hollow viscus.

Early features depend on the severity and extent of the peritonitis. Extensive peritonitis that involves the abdominal aspect of the diaphragm may be accompanied by shoulder tip pain. Vomiting often occurs early in the disease course; the patient has malaise, temperature is frequently elevated, and purulent exudates appear.

Peritonitis caused by appendicitis is usually accompanied by a temperature higher than 38° C (100° F), whereas in peritonitis caused by perforated peptic ulcer, the temperature seldom reaches this level, and the pulse tends to fluctuate.

Examination of the abdomen demonstrates tenderness localized to the affected area or generalized if the peritoneal cavity is extensively involved. The abdomen is silent on auscultation, and rectal examination reveals tenderness of the pelvic peritoneum. As the disease progresses, the abdomen becomes distended, signs of free fluid may be detected, and vomiting is effortless and feculent. The patient is reactive and has sunken eyes and pale, clammy skin, and the extremities are cyanotic.

DIAGNOSIS

Abdominal radiography reveals free subdiaphragmatic gas from hollow viscus perforation. Differential diagnosis includes intestinal obstruction, intraperitoneal hemorrhage, acute pancreatitis, dissection or leakage of an aortic aneurysm, or basal pneumonia.

ADDITIONAL RESOURCES

Caroline CJ, Baldessarre J, Levison ME: Peritonitis: update on pathophysiology, clinical manifestation and management, *Clin Infect Dis* 24:1035-1047, 1997.

Bouchier IA, Ellis H, Fleming P, editors: *French's index of differential diagnosis*, ed 13, Boston, 1996, Butterworth Heinemann, pp 3-5.

Hall JC, Heel KA, Papadimitrou JM, Platell C: The pathobiology of peritonitis, *Gastroenterology* 114:185-196, 1998.

Hines OJ, Ashley SW: Lesions of the mesentery, omentum and retroperitoneum. In Zinner MJ, Schwartz SI, Ellis H, editors: *Maingot's abdominal operations*, ed 10, Vol 1, Norwalk, Conn, 1997, Appleton & Lange, pp 709-722.

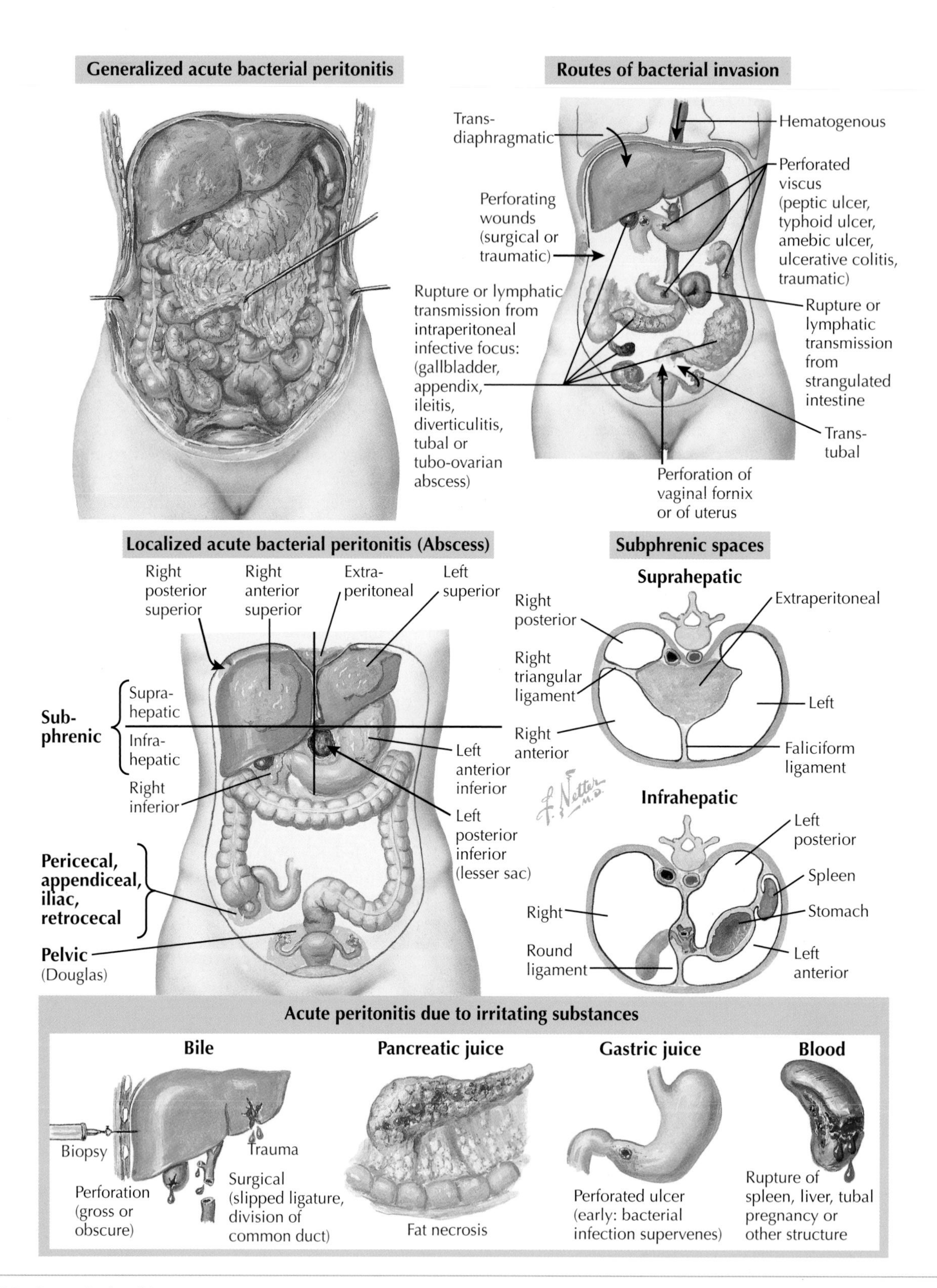

Figure 81-1 *Acute Peritonitis.*

Chronic Peritonitis

82

Dario Berkowski and Raul J. Rosenthal

Although abdominal tuberculosis is now considered a rare disease in industrialized countries, it is still a concern in many developing nations. *Tuberculous peritonitis* (chronic peritonitis) is the sixth most common site of extrapulmonary tuberculosis in the United States, following lymphatic, genitourinary, bone and joint, miliary, and meningeal peritonitis (tuberculosis). Recently, cases of tuberculosis have increased in Western countries, with outbreaks in inner cities, homeless shelters, nursing homes, and prisons and among immigrants from high-risk areas and patients with human immunodeficiency virus (HIV) infection.

CLINICAL PICTURE

Chronic peritonitis (tuberculous peritonitis) must be considered in the differential diagnosis of patients with fever and nonspecific abdominal symptoms. Abdominal tuberculosis may manifest either as an intestinal form or as tuberculous peritonitis (**Fig. 82-1**).

The intestinal form is frequently encountered in the differential diagnosis of an acute abdomen caused by gastrointestinal obstruction, perforation, or pseudo–acute appendicitis or accompanied by a palpable mass. Diagnosis is often made at laparotomy.

Tuberculous peritonitis is more insidious and is accompanied by vague symptoms of fever, abdominal pain, weight loss, anorexia, lethargy, or change in bowel habit. Abdominal distention, fluid in the peritoneal cavity, and another puckered and thickened omentum forms a tumor that traverses the midabdomen. Because the diagnosis can be elusive, it is essential to obtain confirmatory evidence before initiating therapy.

TREATMENT AND MANAGEMENT

Gram stain is frequently negative in primary bacterial peritonitis, so the initial choice of antimicrobial therapy is often empirical, based on the most likely pathogens. Some third-generation cephalosporin antibiotics have proved as efficacious as a combination of ampicillin and an aminoglycoside. If peritonitis develops during hospitalization, a therapeutic regimen of an aminoglycoside antibiotic plus antipseudomonal penicillin or cephalosporin should be active against *Pseudomonas aeruginosa*.

Medical management of secondary peritonitis includes antimicrobial therapy and supportive measures to maintain vital function and to improve circulation, nutrition, and oxygenation of vital organs. Antimicrobial agents must penetrate to the infection site in concentrations sufficient to overcome the effects of high bacterial density, metabolic inactivity, and slow growth rate of the bacterial inoculum, low pH, necrotic tissue of low oxidation-reduction (redox) potential, and bacterial products that may lower the drug's efficacy. For example, aminoglycosides and clindamycin are less active at acid pH values; aminoglycosides are less active at low redox potentials; and β-lactam drugs are less active against high bacterial densities.

Secondary peritonitis, or abdominal sepsis, is still associated with a high mortality rate of approximately 30%, despite improvements in antibiotic treatment and intensive care facilities. Surgical treatment of secondary peritonitis is usually threefold: laparotomy to eliminate the source of infection, perioperative peritoneal lavage to reduce bacterial load, and prevention of persistent or recurrent infection.

ADDITIONAL RESOURCES

Ahmad M, Ahmed A: Tuberculous peritonitis: fatality associated with delayed diagnosis, *South Med J* 92:406-408, 1999.

Johnson CC, Baldessarre J, Levison ME: Peritonitis: update on pathophysiology, clinical manifestation and management, *Clin Infect Dis* 24:1035-1047, 1997.

Lam KN, Rajasoorya C, Mah PK, Tan D: Diagnosis of tuberculosis peritonitis, *Singapore Med J* 40:601-604, 1999.

Lamme B, Boermeester MA, Reitsma JB, et al: Meta-analysis of relaparotomy for secondary peritonitis, *Br J Surg* 89:1516-1524, 2002.

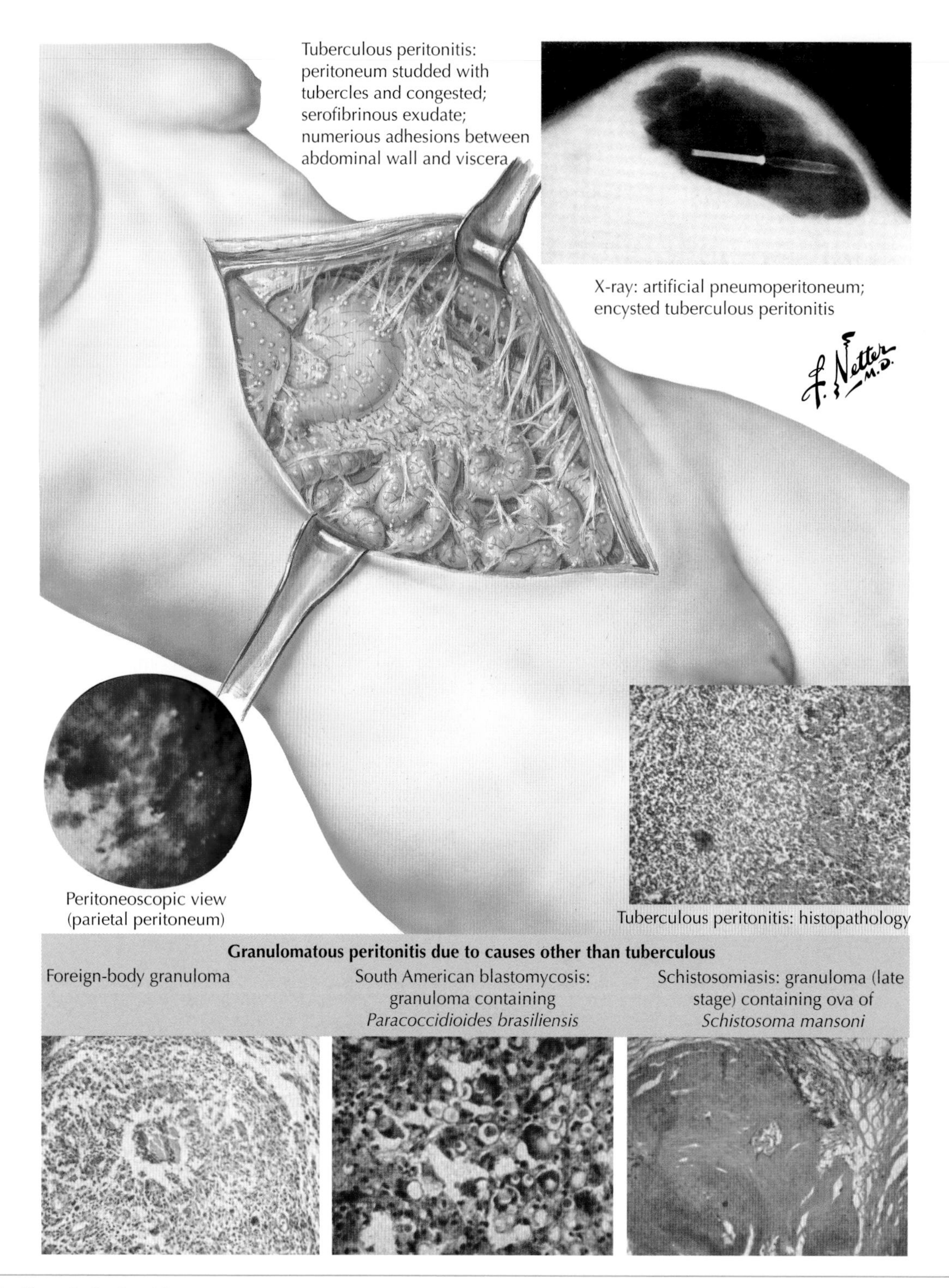

Figure 82-1 *Chronic Peritonitis.*

Cancer of the Peritoneum

83

Dario Berkowski and Raul J. Rosenthal

The peritoneum is a mesothelial membrane that lines the inner surface of the abdominal cavity (parietal peritoneum) and extends to cover various solid organs and most of the bowel (visceral peritoneum). Primary peritoneal tumors are infrequent, with most manifesting as secondary metastases from abdominal organs, such as the intestine, breast, and lung (**Fig. 83-1**). A variety of primary mesenteric tumors originate from the connective tissue, smooth muscle, adipose tissue, germ cells, epithelium, mesothelium, nerve, and vascular tissue. The two most common types are mesotheliomas and desmoid tumors of the mesentery.

Malignant peritoneal *mesothelioma* is rare. It was first described in 1908 and accounts for 20% of mesotheliomas. In the United States, there are approximately 2500 new cases annually, usually in men during the fifth and sixth decades of life. Approximately one third involve the peritoneum. Peritoneal mesotheliomas originate from the lining of the pleura, peritoneum, and pericardium. This tumor is classed into three types: malignant tubulopapillary mesothelioma, cystic mesothelioma, and benign adenomatoid mesothelioma. The etiology is unclear, and presentation is often atypical. An association has been made between asbestos exposure and malignant mesothelioma.

Desmoid tumors were first described in 1832 and are the most frequent primary tumor of the mesentery. They have four peaks of presentation, at age 5 years (juvenile), 27 years (fertile), 44 years (middle age), and 68 years (old age).

Peritoneal surface malignancy can result from seeding of a gastrointestinal cancer or an abdominopelvic sarcoma. An aggressive approach to peritoneal surface malignancy involves peritonectomy procedures and perioperative intraperitoneal chemotherapy. Peritoneal surface dissemination of appendiceal malignancy arises from a perforated appendiceal tumor. This primary tumor invades the appendiceal wall or produces a mucocele that eventually ruptures. In both situations, mucous-producing adenomatous epithelial cells are disseminated throughout the abdomen and pelvis. This results in extensive accumulation of mucinous tumor at characteristic sites within the peritoneal cavity.

The mortality rate in these patients is 100% because of progressive bowel obstruction and terminal starvation. Some studies have reported 5% survival 5 years after surgery.

CLINICAL PICTURE

Desmoid tumor manifests as a nontender, painful mass if the tumor is growing into surrounding structures. A tumor in the omentum may be palpable as an epigastric mass. Patients usually seek treatment with signs and symptoms of advanced disease, including nonspecific abdominal pain and signs of bowel obstruction. Most of these patients have anemia, weight loss, fever, melena, lymphadenopathy, and ascites.

DIAGNOSIS

No acceptable universal staging system has been proposed for peritoneal mesotheliomas. These tumors generally remain confined to the abdomen until late in the course, when they spread to one or both pleural cavities. Chest radiography reveals pleural plaques in approximately 50% of patients with peritoneal primary tumors, compared with 20% in patients with pleural mesothelioma, reflecting the higher level of asbestos exposure in patients with peritoneal disease.

Malignant mesotheliomas may be mistaken for carcinomatosis. Cytologic examination of peritoneal aspirate obtained by needle biopsy, by either laparotomy or laparoscopy, has been used for diagnosis and classification. This method is associated with low sensitivity and specificity and establishes the diagnosis in only 5% to 10% of patients. The specific antibody for cytoplasm of mesotheliomas can be used for immunohistochemistry.

Peritoneal mesotheliomas may be the source of ectopic hormone secretion, such as antidiuretic hormone, growth hormone, adrenocorticotropic hormone, and insulin-like hormone.

Although histology may reveal epithelial, mesenchymal, or mixed patterns, electron microscopy may be necessary to establish a mesothelial origin. Mesothelial cells demonstrate microvilli, cytoplasmic tonofilaments, and glycogen-like granules.

Desmoid tumors are variable in size and can grow as large as 22 kg. Although these tumors tend to invade surrounding tissues, they are not malignant. The differential diagnosis includes keloids, neurofibromas, lipomas, dermatofibromas, rhabdomyomas, subacute myositis, and hemangiomas. Surgical trauma may contribute to the formation of desmoid tumors because they have been discovered in scars from previous surgery. Furthermore, estrogen levels have a close correlation with tumor growth rate, which peaks at menopause.

Computed tomography (CT) and magnetic resonance imaging (MRI) assist in determining the extent of invasion of local structures. Classic findings on CT include mesenteric thickening, peritoneal studding, hemorrhage within the tumor mass, and ascites. However, patients may have advanced disease despite relatively normal CT findings. MRI offers potentially higher resolution and improved delineation than CT.

Peritoneal fluid from malignant ascites may be a watery transudate or a viscous fluid rich in mucopolysaccharides. Viscous ascites with a high fluid level of hyaluronidase may suggest the diagnosis. Massive ascites may result in incorrectly diagnosing severe cirrhosis rather than mesothelioma.

Thrombocytosis is common and associated with poor prognosis. Other associated abnormalities include phlebitis, emboli, hemolytic anemia, and disseminated intravascular coagulation.

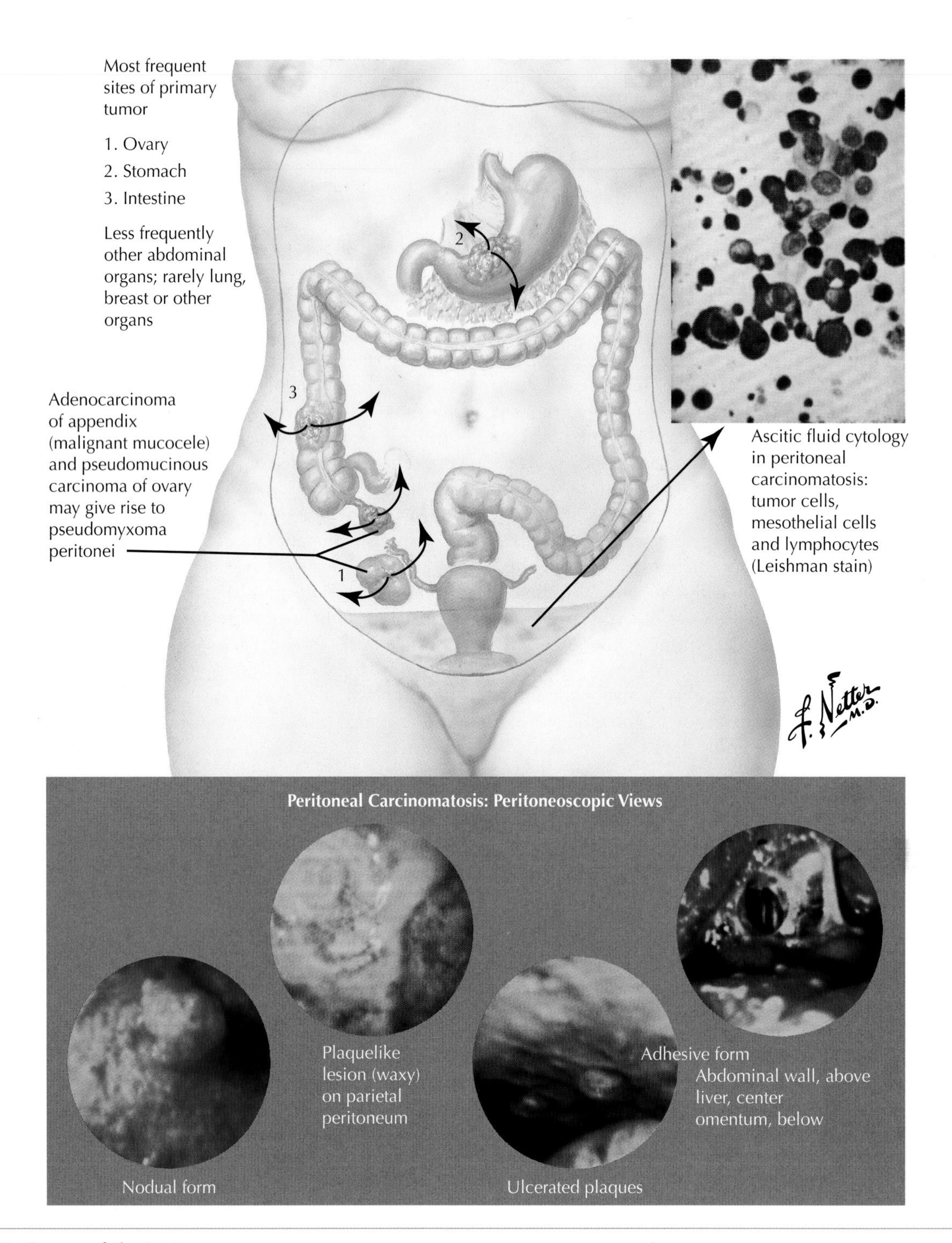

Figure 83-1 *Cancer of the Peritoneum.*

TREATMENT AND PROGNOSIS

Most patients die without metastases or involvement of the chest. Esophageal achalasia, secondary amyloidosis, and dermatomyositis have been reported. Survival of untreated patients in most series is short, 4 to 12 months.

Various surgical and autopsy series have shown that peritoneal mesothelioma involves all the peritoneal surfaces, often with masses of 5 cm or more. Sites of local invasion include the liver, abdominal wall, diaphragm, retroperitoneum, gastrointestinal tract, and bladder. Seeding of laparotomy scars and biopsy has also been observed.

Effective local therapy may have a substantial impact on the survival of these patients because the tumor is confined to the peritoneal cavity at initial diagnosis. Surgical intervention can provide palliation for small-bowel obstruction or relief of massive ascites by perivenous shunting.

Treating mesotheliomas involves a multimodality approach. Surgical intervention has been used for debulking and bypassing an intestinal obstruction. Chemotherapy with doxorubicin seems to be the most effective treatment. Prognosis is poor, and median survival from the time of diagnosis is approximately 18 months.

Desmoid tumors are primarily treated with surgery. Radiotherapy can successfully reduce these tumors, although postsurgical recurrence is high (40%).

ADDITIONAL RESOURCES

De Vita VT, Hellman S, Rosenberg SA: *Cancer: principles and practice of oncology*, ed 5, Philadelphia, 1997, Lippincott Williams & Wilkins, pp 1870-1871.

Gough DB, Donohue JH, Schutt AJ, et al: Pseudomyxoma peritonei: long-term patient survival with an aggressive regional approach, *Ann Surg* 219:112-119, 1994.

Hines OJ, Ashley SW: Lesions of the mesentery, omentum and retroperitoneum. In Zinner MJ, Schwartz SI, Ellis H, eds: *Maingot's abdominal operations*, ed 10, Vol 1, Norwalk, Conn, 1990, Appleton & Lange, pp 709-722.

Kerrigan SAJ, Cagle P, Churg A: Malignant mesothelioma of the peritoneum presenting as an inflammatory lesion, *Am J Surg Pathol* 27:248-253, 2003.

Mohamed F, Changand D, Sugarbaker P: Third look surgery and beyond for appendiceal malignancy with peritoneal dissemination, *J Surg Oncol* 83:5-13, 2003.

Raptopoulos R, Gourtsoyiannis N: Peritoneal carcinomatosis, *Eur Radiol* 11:2195-2206, 2001.

Sugarbaker PH: Management of peritoneal surface malignancy: the surgeon's role, *Arch Surg* 384:576-587, 1999.

Tandar A, Abraham G, Gurka J, et al: Recurrent peritoneal mesothelioma with long-delayed recurrence, *J Clin Gastroenterol* 33:247-250, 2001.

Benign Paroxysmal Peritonitis (Familial Mediterranean Fever)

84

Colleen Kennedy and Raul J. Rosenthal

Familial Mediterranean fever (FMF) is a disorder characterized by sporadic attacks of serosal inflammation (**Fig. 84-1**). It is inherited as a single-gene autosomal-recessive trait. FNF is associated primarily with ethnic groups originating in the Mediterranean region. Initial onset of disease occurs early in life, with 90% of patients experiencing the first attack before age 20 years.

The *FMF* gene is located on the short arm of chromosome 16. The gene codes for a protein, *pyrin*, found in the cytoplasm of cells of myeloid lineage, colonocytes, and prostate cells. Pyrin acts as a regulator of proinflammatory cytokine transcription and may also interfere with neutrophil activation. Patients with FMF have *missense* mutations that cause alterations in pyrin formation.

Incomplete penetrance and varying expression of FMF suggest a multifactorial influence on disease expression. Heterogeneity among the disease-modifying proteins may contribute to the variable phenotypes seen in patients with FMF.

The triggering mechanism for acute attacks of FMF is unknown. The neutrophil is thought to be the primary cell in the inflammatory response. Pyrin is expressed by monocytes and neutrophils and is associated with the interleukin-1 (IL-1) inflammatory pathway.

CLINICAL PICTURE

Familial Mediterranean fever, or benign paroxysmal peritonitis, is an antiinflammatory disease characterized by periodic episodes of fever and serositis. The patient's temperature may rise to 103° F (39.4° C), and the disease is associated with severe abdominal pain and peritonitis in 95% of patients. Marked tenderness is variable; peritoneal signs such as rebound tenderness occur frequently; and guarding is occasionally seen. Patients also have leukocytosis (left shift on differential), nausea, vomiting, and constipation. The clinical picture is indistinguishable from that of acute surgical abdomen. Abdominal pain typically precedes fever by a few hours and continues for 24 to 48 hours after the fever resolves.

Pleuritic chest pain is present in approximately 30% of patients with FMF, arthritis in 50%, and skin manifestations in 15%. Other manifestations include aseptic meningitis, pericarditis, febrile myalgia, and vasculitis. One of the main complications of FMF is amyloidosis, with deposition of amyloid in the kidney contributing the greatest morbidity. Secondary amyloidosis is characterized by deposition of amyloid A fibrils, derived from serum amyloid A protein in the tissue produced in response to IL-1 stimulation.

DIAGNOSIS

A diagnosis of FMF should be considered for patients who have recurrent and self-limited attacks of fever associated with serositis. Genetic analysis for the *FMF* gene is indicated for patients with periodic fever syndrome and a family history of FMF or amyloidosis. Levels of acute-phase reactants, such as erythrocyte sedimentation rate, C-reactive protein, and fibrinogen, are almost always elevated during these attacks. The diagnosis of FMF is made by clinical criteria.

Livneh and colleagues established a set of diagnostic criteria in 1997. Major criteria involve peritonitis (generalized), pleuritis (unilateral) or pericarditis, monoarthritis (hip, knee, ankle), or fever alone. Typical attacks may also involve incomplete abdominal attack. Minor criteria (incomplete attacks) involve the chest, joints, exertional leg pain, or favorable response to colchicine. Requirements for a diagnosis of FMF are one or more major criteria or two or more minor criteria.

TREATMENT AND MANAGEMENT

First-line treatment for FMF is *colchicine*, which prevents febrile attacks in 60% of patients and reduces the number of attacks in another 20% to 30%. The standard dose is 1 mg daily, which can be increased to 3 mg if needed and tolerated. Colchicine is effective only for preventive therapy and is not effective in stopping acute attacks. Treatment of acute attacks is supportive and includes management of the associated pain.

PROGNOSIS

The prognosis for patients with FMF (benign paroxysmal peritonitis) is determined by the presence of amyloidosis. In its absence, life expectancy is normal. The advent of colchicine treatment for FMF has greatly improved survival because adequate treatment has been shown to prevent amyloid deposition.

ADDITIONAL RESOURCES

Drenth JP, Van der Meer JW: Hereditary periodic fever, *N Engl J Med* 345:1748-1757, 2001.

Gertz MA, Petit RM, Perrault J: Autosomal dominant familial Mediterranean fever–like syndrome with amyloidosis, *Mayo Clinic Proc* 62:1095-1100, 1987.

Livneh A, Langevitz P, Zemer D, et al: Criteria for the diagnosis of familial Mediterranean fever, *Arthritis Rheum* 40:1879-1885, 1997.

Odabas AR, Cetinkaya R, Selcuk Y, Bilen H: Familial Mediterranean fever, *South Med J* 95:1400-1403, 2002.

Ozel AM, Demirturk L, Yazgan Y, et al: Familial Mediterranean fever: a review of the disease and clinical and laboratory findings in 105 patients, *Dig Liver Dis* 32:504-509, 2000.

Ozen S: Familial Mediterranean fever: revisiting an ancient disease, *Eur J Pediatr* 162:449-454, 2003.

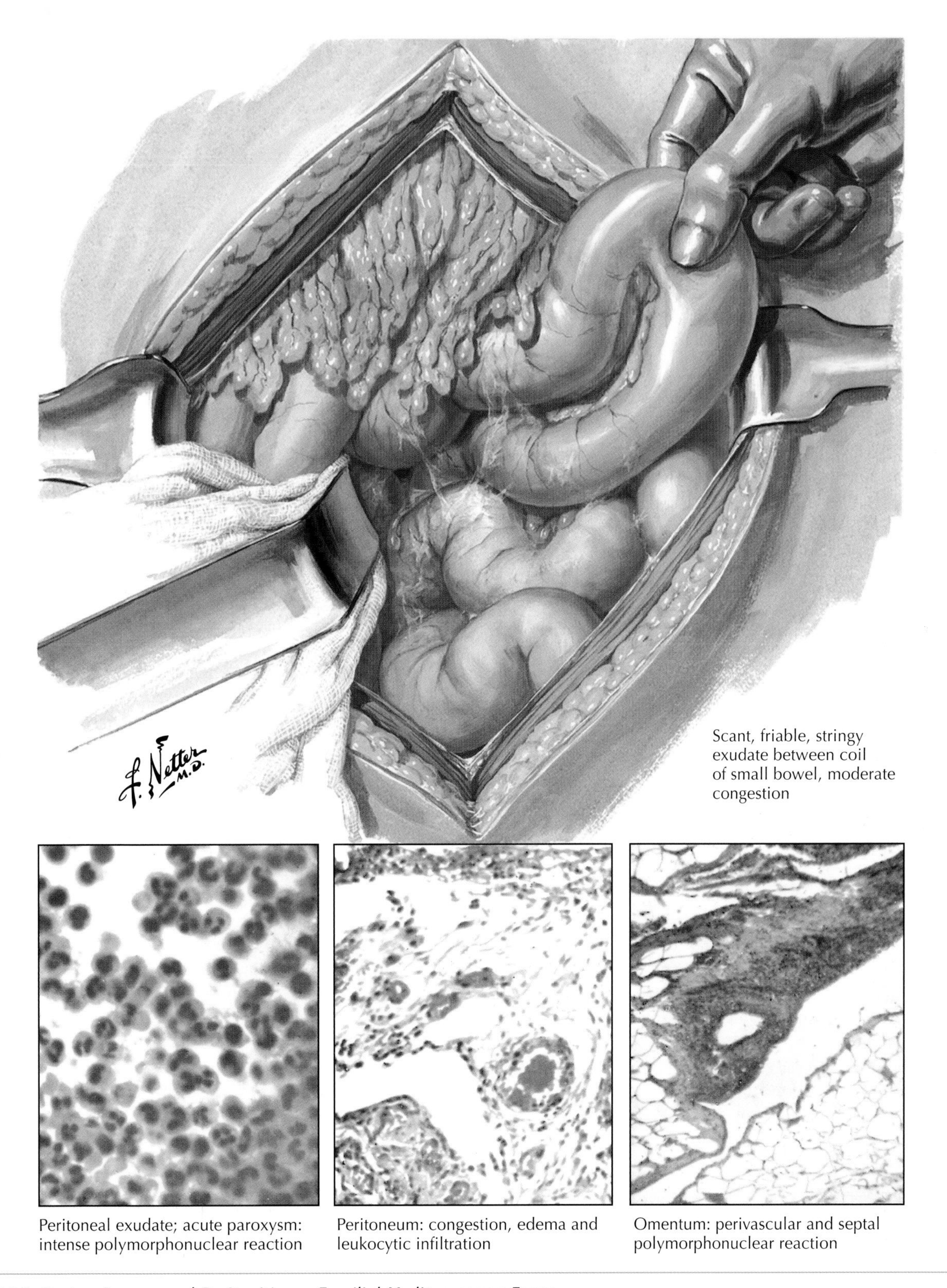

Figure 84-1 *Benign Paroxysmal Peritonitis, or Familial Mediterranean Fever.*

Abdominal Wounds of the Small Intestine

85

Colleen Kennedy and Raul J. Rosenthal

Injuries to the small intestine are common in patients who have experienced penetrating trauma (**Fig. 85-1**). Diagnosis and management of these types of injuries are relatively simple and direct. Injury to the small bowel secondary to blunt trauma, however, can present a much greater diagnostic dilemma. The importance of early diagnosis stems from the high morbidity and mortality rates seen in patients with delayed diagnosis of these injuries. Associated morbidity is often secondary to the identification and prompt treatment of small-bowel injury.

The small intestine is approximately 6.5 meters (21.5 feet) long; the first 40% includes the jejunum, and the remainder is ileum. This distinction is important in resection secondary to the different digestive capabilities of the ileum and jejunum. The ileum is responsible for the absorption of vitamin B_{12} and the reabsorption of bile salts into the enterohepatic circulation. Location of the wound is also important in that the distal small bowel has a higher anaerobic bacterial load. Therefore, distal small bowel injuries pose greater risk for postoperative infection.

CLINICAL PICTURE

Prompt identification of small bowel injuries depends on a high index of suspicion for injury. This is particularly true with blunt abdominal trauma, which often has multiple associated injuries that distract from subtle abdominal findings. Physical examination of awake and alert patients reveals progressive abdominal tenderness that is generalized. In patients with head injuries or who are intoxicated or have other reasons for altered mental status, this finding is unreliable.

Injury to the small bowel must be ruled out in patients with penetrating abdominal injury. The small intestine is the most frequently injured intraabdominal viscus in patients with penetrating trauma because it occupies most of the abdominal cavity and is highly mobile. Therefore, injury with penetration may occur in almost any area of the peritoneal cavity.

DIAGNOSIS AND TREATMENT

Traditionally, the diagnosis of penetrating abdominal trauma includes exploratory laparotomy for all penetrating injuries. Evaluation of the entire length of the small bowel begins with the ligament of Treitz and continues to the ileocecal valve. Inspection includes examination of the entire mesentery and identification of hematoma or possible injury to the blood supply of the bowel.

Any injuries encountered should be immediately repaired. This method eliminates further trauma when replacing the bowel in the peritoneal cavity, and it permits the surgeon to examine surrounding tissue to determine whether simple repair or resection is necessary. Small perforations may be repaired with a simple suture. For small lacerations, closure should be completed in a transverse fashion to the long axis of the bowel to prevent narrowing of the lumen. Resection should be performed for multiple injuries in a short segment or for a short segment with massive tissue destruction. In these circumstances, primary anastomosis can be performed. When managing mesenteric injuries, the viability of the bowel must be determined. For injuries involving bowel ischemia in a short segment, simple resection with reanastomosis can be performed. In the absence of ischemia, mesenteric injury should be primarily repaired. If a large segment of the bowel is ischemic or if associated injuries are severe, resection should be limited and should be followed by a planned second-look procedure and delayed anastomosis.

The diagnosis of small-bowel trauma secondary to blunt injury is a controversial area. Comparisons of ultrasonography and computed tomography (CT) with diagnostic peritoneal lavage for blunt trauma have shown a high rate of sensitivity for lavage, although this invasive test has its own complications. Ultrasound and CT for small intestinal trauma lacks sensitivity in the absence of associated injuries. The presence of free fluid in the peritoneal cavity in the absence of solid-organ injury is highly suggestive of small-bowel injury. Small intestinal injury may be missed on radiologic studies, particularly if the study is performed in the early postinjury period. Diagnostic laparoscopy has been used for evaluation of peritoneal penetration after stab wounds and diaphragmatic penetration after thoracoabdominal penetrating injuries. Recent studies have offered laparoscopy as an option for evaluating the small intestine and mesentery for blunt trauma injury. Management of small-bowel injuries secondary to blunt trauma is similar to the management of penetrating injuries discussed earlier.

Postoperative management of patients with small-bowel injuries is often dictated by the associated injuries. Early enteral nutrition is preferred in the postoperative period, if possible. However, this may be difficult in the patient who has sustained prolonged shock or a large mesenteric injury.

ADDITIONAL RESOURCES

Al-Salamah SM, Mirza SM, Ahmad SN, Khalid K: Role of ultrasonography, computed tomography and diagnostic peritoneal lavage in abdominal blunt trauma, *Saudi Med J* 23:1350-1355, 2002.

Fakhry SM, Brownstein M, Watts DD, et al: Relatively short diagnostic delays (<8 hours) produce morbidity and mortality in blunt small bowel injury: an analysis of time to operative intervention in 198 patients from a multicenter experience, *J Trauma* 48:408-414, 2000.

Lannelli A, Fabiani P, Karmdjee BS, et al: Therapeutic laparoscopy for blunt abdominal trauma with bowel injuries, *J Laparoendosc Adv Surg Tech A* 13:189-191, 2003.

McQuay N, Britt LD: Laparoscopy in the evaluation of penetrating thoracoabdominal trauma, *Am Surg* 69:788-791, 2003.

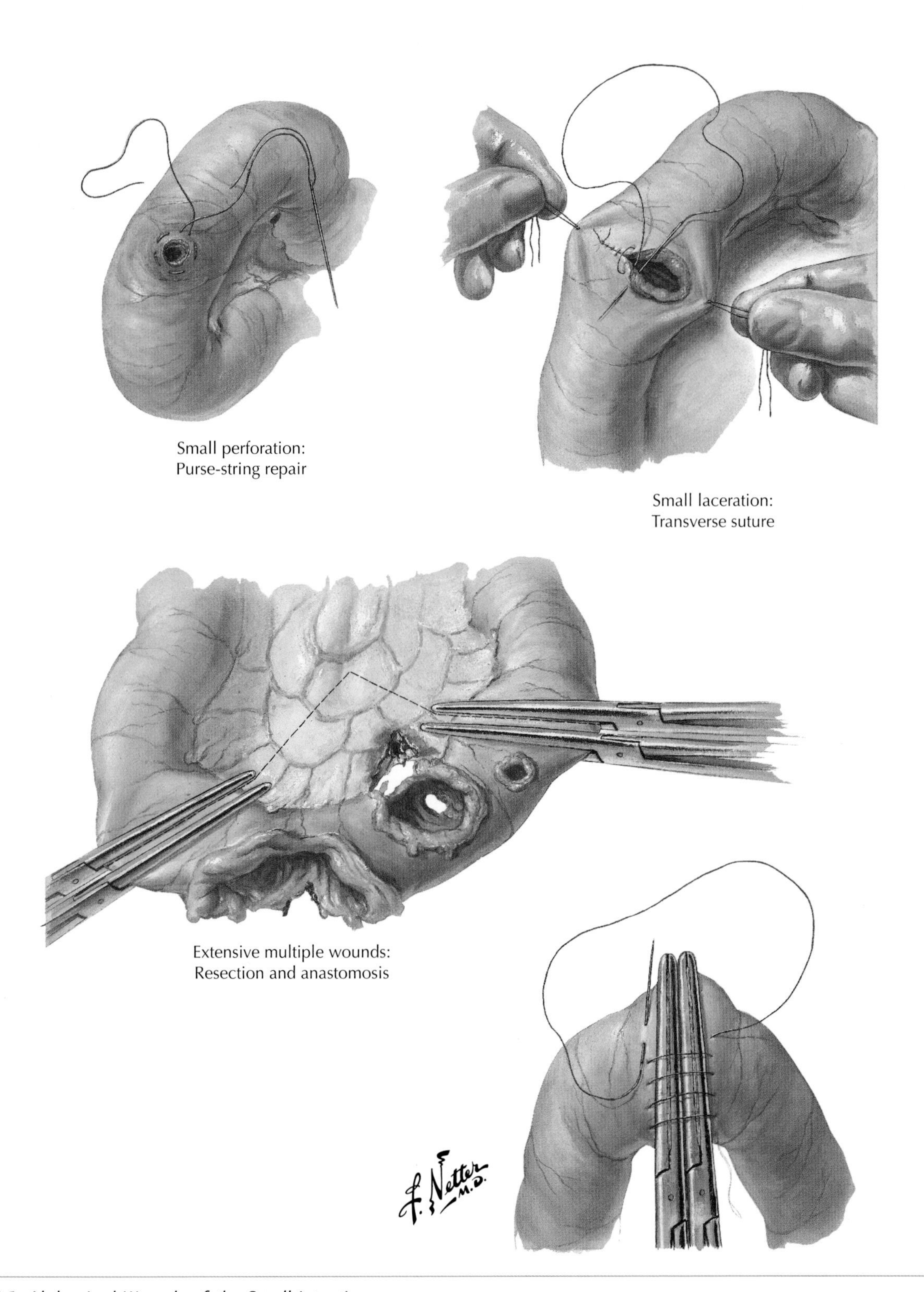

Figure 85-1 *Abdominal Wounds of the Small Intestine.*

Pikoulis E, Delis S, Psalides N, et al: Presentation of blunt small intestinal and mesenteric injuries, *Ann R Coll Surg Engl* 82:103-106, 2000.

Rossi P, Mullins D, Thal E: Role of laparoscopy in the evaluation of abdominal trauma, *Am J Surg* 166:707-710, 1993.

Stafford RE, McGonigal MD, Weigelt JA, Johnson TJ: Oral contrast solution and computed tomography for blunt abdominal trauma: a randomized study, *Arch Surg* 134:622-626, 1999.

Abdominal Wounds of the Colon

86

T. Cristina Sardinha and Raul J. Rosenthal

Trauma remains a major cause of death and disability. In North America, trauma is the leading cause of death in persons younger than 44 years. This leads to significant economic and social consequences.

The colon is the second most common abdominal organ injured in penetrating trauma. However, injury to the colon is rare in blunt trauma (2%-5%). Nonetheless, colonic injury after blunt abdominal trauma is associated with a higher risk for complications and increased hospitalization. Furthermore, no diagnostic modality or combination of findings can reliably exclude blunt injury to the colon. Therefore, a surgical approach is recommended early in the evaluation of abdominal trauma suspected to involve the colon (**Fig. 86-1**).

TREATMENT AND PROGNOSIS

Evolution in the management of colon injury to civilians resulting in significantly reduced deaths is a benefit of lessons learned from warfare. From the Civil War to the Vietnam War, the mortality rate from colonic injury declined from more than 90% to less than 10%. Several factors have contributed to the significant improvement in survival in this patient population, including the use of a diverting colostomy, fluid resuscitation, availability of blood products, and broad-spectrum antibiotics.

Preoperative antibiotics against aerobic and anaerobic flora should be initiated early after the injury. Infection rates are significantly reduced when prophylactic antibiotics are given before surgery rather than during surgery (7% vs. 33%, respectively). Moreover, advances in anesthesia and intensive care management continue to contribute to the overall decrease in morbidity and mortality.

The *mechanism of injury* is often one of the few factors distinguishing military from civilian injury. Combat lesions frequently result from high-velocity weapons and explosive devices, whereas civilian injuries often result from handguns, stab wounds, and blunt trauma. The degree of tissue damage is proportional to the kinetic energy delivered by high-velocity weapons. Therefore, the overall prognosis for civilian trauma is better than for war-related injuries because civilian injuries are often a result of low-velocity weapons. Penetrating wounds involving the intraperitoneal portions of the colon frequently occur in multiples, in contrast to lesions involving the ascending or descending colon. However, lesions in the retroperitoneal portions of the colon are often overlooked, leading to severe anaerobic infection because of the vulnerability of the retroperitoneal space.

Accurately assessing death from colon injury is difficult because of the frequent involvement of other abdominal organs in victims of trauma, particularly severe blunt trauma. The cause of death in victims of colon trauma in the early postoperative period may be associated injuries rather than injury to the colon itself. The reported mortality rate related to civilian colon injury alone is 1% to 3%.

The most frequently used scales to classify the severity of colon injury are the Penetrating Abdominal Trauma Index and the Colon Injury Scale (CIS). The Penetrating Abdominal Trauma Index score estimates organ injury on a scale of 1 to 5: 1 = minimal, 2 = minor, 3 = moderate, 4 = major, and 5 = maximal. The CIS grades the injury as follows: 1 = serosal injury, 2 = single wall injury, 3 = less than 25% wall involvement, 4 = 25% or more wall involvement, and 5 = entire colonic wall disruption and blood supply involvement. Some surgeons plan surgical management based on the degree of colonic injury. Patients with CIS grades 1 to 3 are eligible for primary repair, whereas those with CIS grades 4 and 5 are usually treated with fecal diversion.

Despite the impressive improvements in survival for victims of colon trauma over the years, controversy still surrounds the optimal management of colon injury. Diverse factors may influence the decision whether to choose fecal diversion or primary closure of the injury with or without resection. The 1979 criteria considered mandatory for the performance of colostomy in patients with colon injury included preoperative shock, delay in surgery longer than 8 hours from injury, significant peritoneal fecal contamination, intraabdominal blood loss greater than 1000 mL, and colon injury requiring resection. These criteria were challenged in 1996 by a prospective, randomized trial of 109 patients with penetrating colon wounds undergoing primary repair or fecal diversion (average age and trauma index similar in both groups). The incidence of sepsis-related complications was 20% in the primary repair group and 25% in the diversion group. The authors concluded that all colon injuries in the civilian population should be managed by primary repair. Moreover, a 2002 Cochrane Review favors primary repair over fecal diversion for the management of penetrating colon wounds. The most recent statement of the Eastern Association for the Surgery of Trauma regarding the management of penetrating intraperitoneal colon injuries recommends primary repair of nondestructive penetrating colon lesions. For destructive colon injuries, resection and anastomosis are recommended for stable patients without major associated trauma. Patients with significant comorbid conditions or severe associated injury have better outcomes when resection and colostomy constitute the treatment of choice.

Treatment of colon trauma must consider the surgeon's experience, a disciplined approach, the mechanism and extent of injury, the patient's status, and any associated injuries. Additionally, all patients should receive prophylactic antibiotics before surgery.

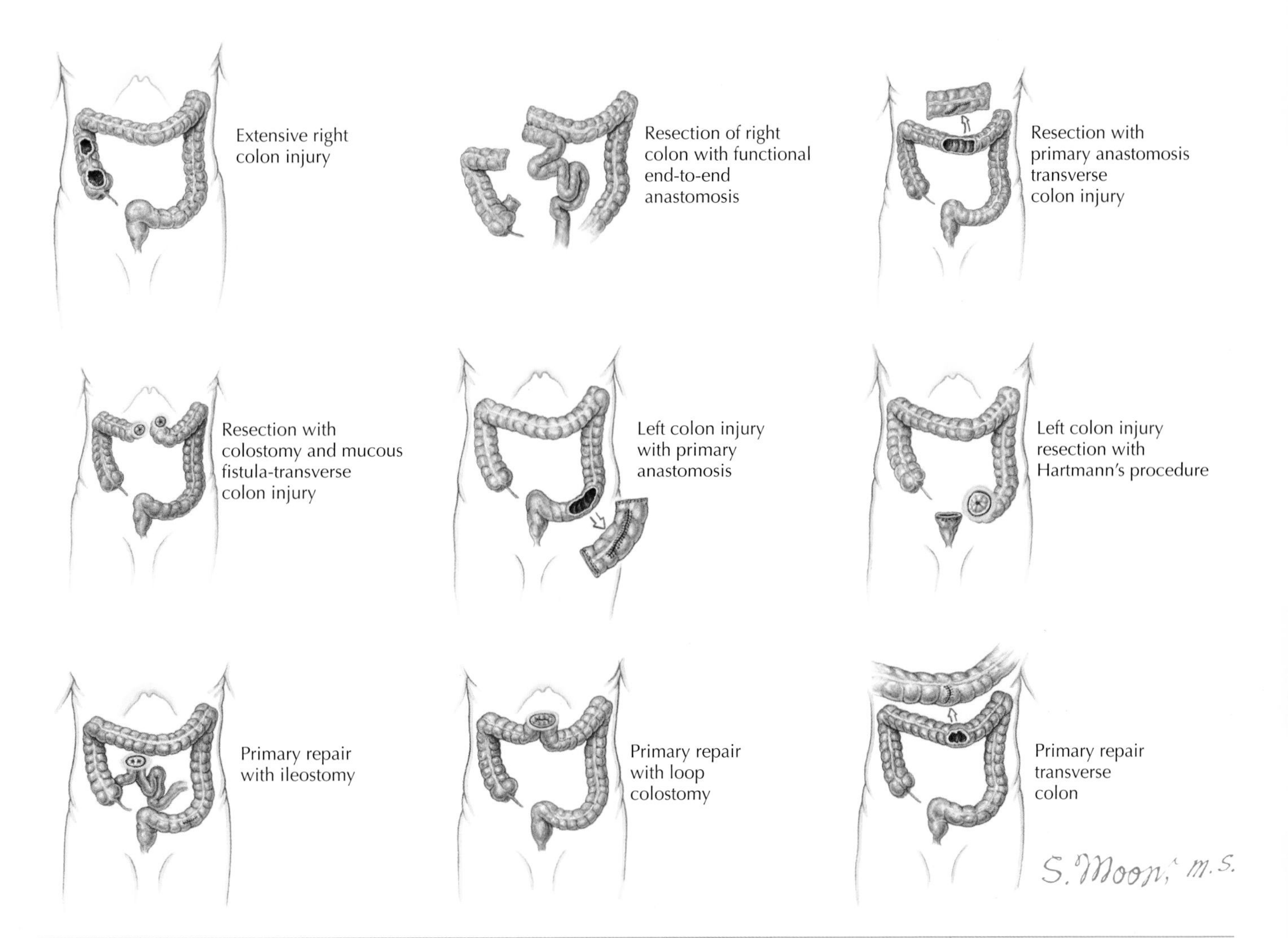

Figure 86-1 *Abdominal Wounds of the Colon.*

ADDITIONAL RESOURCES

Brohi K: Injury to the colon and rectum. http://www.trauma.org/abdo/COLONguidelines.html. Accessed July 2003.

Cayten CG, Fabian TC, Garcia VF, et al: Patient management guidelines for penetrating intraperitoneal colon injury. http:www.east.org. Accessed 1998.

Chappuis CW, Dietzen CD, Panetta TP, et al: Management of penetrating colon injuries: a prospective randomized trial, *Ann Surg* 213:492-497 (discussion 498), 1991.

Gonzalez RP, Merlotti GJ, Holevar MR: Colostomy in penetrating colon injury: is it necessary? *J Trauma* 41:271-275, 1996.

Karulf RE, Fitzharris G: Colon trauma. http:www.fascrs.org/coresubjects/2002/karulf.html.

Nelson R, Singer M: Primary repair for penetrating colon injuries, *Cochrane Database Syst Rev* CD002247, 2002.

Smith LE: Traumatic injuries. In Gordon PH, Nivatvongs S, editors: *Principle and practice for the surgery of the colon, rectum and anus*, ed 2, St Louis, 1999, Quality Medical Publishing, pp 1235-1284.

Williams MD, Watts D, Fakhry S: Colon injury after blunt abdominal trauma: results of the EAST Multi-Institutional Hollow Viscus Injury Study. *J Trauma* 55:906-912, 2003.

Indirect and Direct Inguinal Hernias

87

Emanuele Lo Menjo and Raul J. Rosenthal

Hernia derives from the Greek word *hernios*, which means "bud." It indicates a protrusion of organs through an abnormal defect in a natural cavity. The key part of the definition is the actual *opening* and not the protrusion of the organ itself. In fact, the protrusion may not be recognized in some patients, especially early in the process. The neck of the hernia sac corresponds to the hernial orifice. The dimension of the neck and the volume of the distended sac determine the size of the hernia. Its type depends on its location and cause, the mobility of the herniated organ, and the status of the blood supply (**Fig. 87-1**).

Hernias of the abdominal wall occur only where aponeurosis and fascia are devoid of the protecting support of striated muscle. Some may be acquired through muscular atrophy or surgery. Common sites of herniation are the groin, umbilicus, linea alba, semilunar line of Spieghel, diaphragm, and surgical incisions. Other similar but rare sites of herniation are the peritoneum, superior lumbar triangle of Grynfeltt, inferior lumbar triangle of Petit, and obturator and sciatic foramina of the pelvis.

Generally, symptoms of hernias result from the discomfort caused by the dimension of the hernia and its contents. The diagnosis is easily made on physical examination; a hernia sac transmits a palpable impulse when the patient strains or coughs, whether the patient is supine or upright. About 20% of males and 0.2% of females acquire hernias during their lifetime. Inguinal hernias are by far the most common. The highest incidence is during the first year of life, with a second, lower peak between ages 16 and 20. In women, most inguinal hernias are indirect. Some attribute this phenomenon to the stronger transversalis fascia from the stress of childbearing, making the floor of the inguinal canal more solid.

INDIRECT INGUINAL HERNIAS

To understand the pathophysiology of inguinal hernias, it is necessary to review the developmental stages of the groin. The testes, originating in the abdominal cavity, migrate downward and push a fold of peritoneum, the funicular process, or processus vaginalis. This eventually obliterates during the last trimester of pregnancy. Failure of this involution leaves a channel open for the formation of an indirect inguinal hernia. However, a patent funicular process does not necessarily imply the development of a hernia. Other factors play a role in permitting intraabdominal structures to enter this sac.

Another theory of indirect inguinal hernia formation involves a disparity between intraabdominal pressure and resistance of the muscular and fascial structures forming the deep inguinal ring. Increased abdominal pressure can be constant and prolonged (pregnancy, ascites), repeated but intermittent (prolonged coughing in patients with pulmonary emphysema), or sudden (trauma).

The hernial *sac* is generally composed of peritoneum and attenuated layers of the abdominal wall. The *ring* is the actual defect and sometimes is the only abnormality palpable on physical examination. The *contents* may vary in different parts of the large and small intestines, bladder, ovaries, and omentum. The proximal part of the sac is the narrowest and is therefore called the *neck*. The distal part is the *fundus*. The sac is anterosuperior to the spermatic cord in indirect hernias. The protrusion of the sac and the widening of the deep ring alter the relationship between the two inguinal rings, which begin to lie perpendicularly. In addition, the deep inferior epigastric vessels are displaced medially.

Clinical Picture

The spectrum of symptoms ranges from none to bowel strangulation. In states of rest or recumbency, most hernias are asymptomatic. Physical activity, especially increased intraabdominal pressure, elicits symptoms of fullness, pain, or simply a bulge. Occasionally, symptoms may be attributed to the specific organ involved (dysuria in bladder, constipation for sigmoid colon).

Diagnosis

History taking and physical examination confirm the diagnosis of hernia. The examination should be conducted with the patient in the supine and upright positions. With the examiner's finger gently invaginating the scrotal skin and covering the superficial ring, the patient is encouraged to abruptly increase abdominal pressure (Valsalva maneuver, cough).

For smaller hernias, especially in women and children, inspection is often more valuable than palpation.

When the scrotum is swollen ipsilaterally, the clinician must consider that another abnormality has developed, such as hydrocele, varicocele, or testicular mass.

Physical examination is more difficult in infants, but a thickened cord at the superficial ring is a reliable sign of hernia, especially if it is unilateral. When a left-sided inguinal hernia develops in a child, there is a 50% chance of bilaterality; however, this does not seem to be true for a right-sided inguinal hernia.

Treatment and Management

Surgery is required for inguinal hernia. Mechanical alternatives to surgery, such as a truss, are reserved for medically unfit patients, but the results have been disappointing. Yarn trusses had been used in infants to promote spontaneous closure of the funicular process, but this methodology has been abandoned, and surgery has become the preferred approach for patients of all ages.

Surgery for inguinal hernia repair is simpler in infants than in adults. In fact, the proximity of the deep and superficial

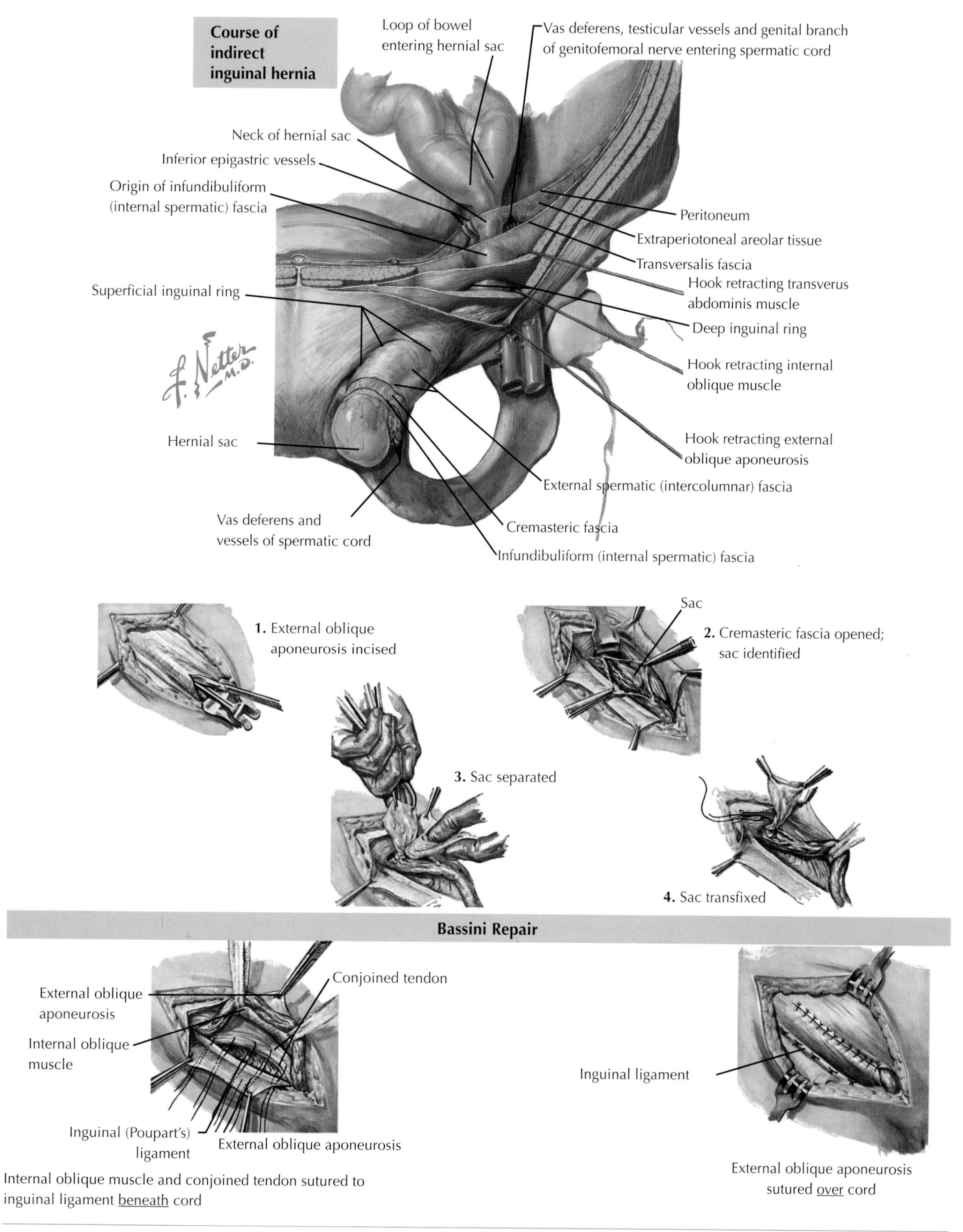

Figure 87-1 *Indirect and Direct Inguinal Hernias.*

inguinal rings makes the opening of the aponeurosis of the external oblique muscle unnecessary and dangerous for the cord structures. The sac is carefully opened, twisted onto itself, and tied as high as possible using a transfixion stitch; the excess is removed. The skin is then approximated, and the incision is sealed using collodion. Isolating the sac is more difficult in female infants because of the smaller size of the sac, in turn making it more difficult to identify anterior to the round ligament.

Numerous surgical techniques have been described. Bassini in Italy and Halsted in United States established the fundamental principles at the end of the 1800s.

Bassini repair is one of the cornerstones in the evolution of inguinal hernia repair. It involves approximation of the internal oblique, transversalis fascia, and transversus abdominis muscle complex to the inguinal ligament and the ileopubic tract to reinforce the posterior wall of the inguinal canal. Care must be taken to avoid injury to the iliohypogastric and ilioinguinal nerves. The sac is opened between clamps, and its contents are reduced in the abdominal cavity. The sac is twisted onto itself, and a transfixion suture is applied at its base and tied. The ligated sac retracts into the abdominal cavity if adequate dissection has been performed. In a patient with a sliding hernia, the herniated organ (colon, bladder) constitutes the wall and the contents of the hernia. The sac should not be opened, but only reduced into the abdominal cavity. The spermatic cord is displaced inferiorly, and the layer formed by the internal oblique muscle, transversus abdominis muscle, and transversalis fascia is sutured to the iliopubic tract superiorly and the inguinal ligament inferiorly. The most lateral part of the repair forms the new deep inguinal ring. To avoid weakening of the femoral canal caused by the upward pull of the inguinal ligament, the Cooper ligament may be included in the inferior part of the repair.

Halsted repair is almost identical to the Bassini procedure, except for the subcutaneous position of the cord structures and the addition of an extra layer to the repair. Halsted added the external oblique aponeurosis to the triple layer, believing that skeletonization of the cord reduces recurrence. This resulted in a high incidence of testicular ischemia and hydrocele, so Halsted modified the original technique, leading to the Halsted II or Ferguson-Andrews procedure. In the modified procedure, the internal oblique and the conjoint tendon are sutured together with the superior flap of the external oblique aponeurosis to the inguinal ligament of the cremaster muscle, which is reapproximated once the sac is removed. The external oblique aponeurosis is then imbricated to form the anterior portion of the repair.

In 1898, Lotheissen described using the iliopectineal, or Cooper, ligament rather than the inguinal ligament to anchor the triple layer (internal oblique, transversalis fascia, and transversus abdominis muscle) of the Bassini repair. This technique was applied after two failed Bassini repairs. McVay and Anson subsequently popularized this procedure 40 years later by adding a relaxing incision in the rectus sheath to avert tension. The external oblique was then closed over the cord. This variation has since become more widely used than the Bassini repair, especially for repair of direct hernias.

DIRECT INGUINAL HERNIAS

Direct inguinal hernias have a different pathophysiologic mechanism than indirect hernias. Direct hernia is considered an acquired condition and is prevalent in men age 40 to 50. A congenital developmental abnormality has been identified, and the fibers of the lower internal oblique muscle seem to be arranged in a transverse rather than an oblique configuration. Consequently, the conjoint tendon is attached to the rectus muscle at a more superior level. The protrusion is characteristically medial to the deep epigastric vessels, through the posterior wall of the inguinal canal.

Clinical Picture

Predisposing factors to direct hernia include chronic increase in intraabdominal pressure (obesity, ascites, chronic cough, constipation, occupational or recreational weight lifting) and atrophy of the abdominal wall musculature (malnutrition, aging). Symptoms are usually more subtle than with indirect inguinal hernias.

Diagnosis

The diagnosis of direct inguinal hernia is made by physical examination with the patient in the supine and upright positions.

Treatment and Management

Treatment of direct hernia is primarily surgical, and multiple techniques have been described. With lason repair, the incision is made over the protrusion and is extended to the superficial inguinal ring; the external oblique is then incised. After isolation and retraction of the cord structures, a circular incision is made through the attenuated conjoint tendon, facilitated by gently retracting the hernia sac outward. A similar incision is made on the underlying transversalis fascia. The peritoneal sac is opened and inspected. At this point, the sac is twisted onto itself, as previously described, and closed at its base with a transfixion suture. The transversalis is then approximated with interrupted sutures. The actual repair is accomplished by suturing the conjoint tendon to the inguinal or pectineal ligaments. The external oblique aponeurosis is imbricated above or below the cord.

Since the introduction of synthetic meshes in the 1950s, interest in the application of *tension-free repair* (e.g., Lichtenstein) in inguinal hernias increased exponentially. Initially, used to reinforce conventional repairs, the Marlex mesh was then extended to patients with more challenging hernias, such as large direct or recurrent hernias. The infection rate was reportedly less than 2%, and the recurrence rate was approximately 6%. Lichtenstein advocated the prosthetic mesh only to reinforce conventional repairs by decreasing tension at the suture line. Later, Lichtenstein reported a series of 300 consecutive inguinal hernias repaired primarily with a tension-free onlay prosthetic mesh. According to the original description, after conventional hernia incision, the external oblique muscle is incised to access the inguinal canal. The hernia is then reduced, and the sac is excised as previously described. At this point, a 5 × 10–cm mesh is anchored by a continuous nonabsorbable suture at the lower edge to the lacunar ligament medially and the inguinal ligament laterally. A slit in the mesh is fashioned to allow passage of the spermatic cord. The superior edge of the mesh is attached to the conjoint tendon and the rectus sheath. Advantages of this technique include decreased postoperative pain, lower recurrence rates, and faster return to work and normal activity. In addition, the learning curve is shorter for this procedure than for conventional repairs. A major concern of using this technique was the possibility of infection. Subsequent trials have failed to show a higher incidence of infection using mesh; rather, use of synthetic mesh has been shown to reduce the rate of recurrence and the likelihood of long-term pain.

Since the introduction of laparoscopic techniques in the treatment of inguinal hernias in the early 1990s, the popularity of *laparoscopic repair* has increased. Proponents of this technique claim that, although the recurrence rates are similar, postoperative pain and ability to return to work were significantly superior with the laparoscopic method. However, this approach was associated with a steep learning curve, the need for general anesthesia, and increased hospital costs. Indications for the laparoscopic procedure include bilateral and recurrent hernias.

Among the various techniques described, only transabdominal preperitoneal repair and total extraperitoneal polypropylene repair remain popular. The main difference between them is the approach to the preperitoneal space. With *transabdominal preperitoneal repair*, the transversalis fascia is exposed by way of the intraperitoneal route, whereas with *total extraperitoneal polypropylene repair*, dissection is entirely preperitoneal.

The key to success is to dissect the transversalis fascia fully into the space of Retzius. This is especially important if the mesh is not anchored, which then requires more extensive dissection and a larger mesh (10 × 15 cm) so that it can cross the midline and extend beyond the hernia defect by at least 4 cm. Advocates of the nonfixing technique claim no difference in recurrence than with fixing, but a higher incidence of postoperative neuralgia.

If the fixing technique is chosen, care must be taken to avoid injury to the ilioinguinal and lateral cutaneous nerves by placing staples, anchors, or tacks above the iliopubic tract.

ADDITIONAL RESOURCES

Arregui ME, Nagan RF, editors: *Inguinal hernia: advances or controversies?* New York, 1994, Radcliffe Medical Press, pp 435-436.

Bendavid R: The Shouldice method of inguinal herniorrhaphy. In Nyhus LM, Baker RJ, Fischer JE, editors: *Mastery of surgery*, ed 3, Boston, 1987, Little, Brown, pp 1826-1838.

Camps J, Nguyen N, Cornet DA, Fitzgibbons RJ: Laparoscopic transabdominal preperitoneal hernia repair. In Phillips EH, Rosenthal RJ, editors: *Operative strategies in laparoscopic surgery*, Berlin, 1995, Springer, pp 83-87.

Fallas MJ, Phillips EH: Laparoscopic near-total preperitoneal hernia repair. In Phillips EH, Rosenthal RJ, editors: *Operative strategies in laparoscopic surgery*, Berlin, 1995, Springer, pp 88-94.

Nyhus L: Recurrent groin hernia, *World J Surg* 13:541-544, 1989.

Read RC: Inguinofemoral herniation: evolution of repair by the posterior approach to the groin. In Nyhus LM, editor: *Shackelford's surgery of the alimentary tract*, ed 4, Philadelphia, 1996, Saunders, pp 129-137.

Rutledge RH: Cooper ligament repair of groin hernias. In Nyhus LM, Baker RJ, Fischer JE, editors: *Mastery of surgery*, ed 3, Boston, 1987, Little, Brown, pp 1817-1825.

Schwartz SI: Abdominal wall hernias. In Schwartz SI, editor: *Principles of surgery*, ed 7, Vol 2, New York, 1999, McGraw-Hill, pp 1437-1611.

Terranova O, Battocchio F: The Bassini operation. In Nyhus LM, Baker RJ, Fischer JE, editors: *Mastery of surgery*, ed 3, Boston, 1987, Little, Brown, pp 1807-1816.

Welsh DRJ: Inguinal hernia repair: a contemporary approach to a common procedure, *Mod Med* 2:49-54, 1974.

Femoral Hernias

88

Emanuele Lo Menzo and Raul J. Rosenthal

Femoral hernias are those in which the abdominal viscera protrude through the femoral ring (**Fig. 88-1**). In most patients, a sac is present. These types of hernias usually occur unilaterally, with a right-sided predominance. Femoral hernias are three times more common in females, and the overall incidence is much lower than for inguinal hernias.

Although the true etiology of femoral hernia is unknown, two theories have been postulated. The first is that these hernias are *congenital*, derived from a preformed sac. The finding of femoral hernias in fetuses supports this theory. Conversely, proponents of the *acquired* theory hypothesize that the increased intraabdominal pressure plays a key role. This would explain the higher incidence of femoral hernias in older, multiparous women. Also, the embryonic peritoneal protrusion is different from the hernial sac.

Pathophysiologically, the femoral ring represents the superior border of the femoral canal. The posterior border consists of the iliac fascia, which continues as the pectineal fascia. Anteriorly, the femoral canal is bound by the downward extension of the transversalis fascia. The two fascial layers fuse inferiorly with the adventitia of the femoral vessels to assume a funnel-like configuration. The femoral sheath below the inguinal ligament is covered by the fascia lata, except at the region of the fossa ovalis, where it is covered by a weaker layer (cribriform fascia). Thus, the pathophysiologic boundaries of the femoral canal are the cribriform fascia and the inguinal ligament anteriorly, the fascia lata over the pectineus muscle and Cooper ligament posteriorly, and the femoral vein and medial lacunar (Gimbernat) ligament laterally.

The herniating viscus pushes a peritoneal sac and some preperitoneal fat tissue behind the inguinal ligament and through the inguinal ring. It then emerges in the subcutaneous tissue at the level of the fossa ovalis. At this point, the herniated viscus can extend upward to the level of the inguinal hernia. The neck, however, is always below the inguinal ligament and lateral to the pubic tubercle. For this reason, a femoral hernia can be mistaken for an inguinal hernia, making reduction difficult if pressure is applied directly upward and toward the superficial ring. The sac may descend anterior to the femoral vessels (prevascular hernia) or behind the vessel (retrovascular hernia). When an aberrant obturator artery is present, the sac may be bisected (bilocular).

CLINICAL PICTURE

Femoral hernias are usually small and produce minimal symptoms until they become incarcerated or strangulated.

DIAGNOSIS

The diagnosis of femoral hernia is made by the presence of a mass at the medial aspect of the thigh, just beneath the inguinal ligament. The mass must be differentiated from an obturator hernia, saphenous vein varix, enlarged lymph node, lipoma, and psoas muscle abscess. Strangulation is the most common complication of femoral hernias, occurring 10 times more frequently than in inguinal hernias.

TREATMENT AND MANAGEMENT

The only successful treatment of femoral hernias is surgery. As with other types of hernia, various procedures have been described over the years, indicating the lack of a widely accepted approach. In general, the different approaches can be divided into inguinal and subinguinal, depending on the route chosen to access the hernia. The procedures may also be combined. In general, the *subinguinal* approach requires less dissection and is indicated for patients at high risk. The *inguinal* approach requires more dissection but offers better long-term results.

The lacunar ligament is frequently divided so that the sac can be reduced. Care must be taken during this step because an anomalous obturator artery originating from the deep epigastric artery can be injured. Once reduced, the sac is opened, ligated, and transected as high as possible. The inguinal canal is then repaired using one of the procedures previously described. Suturing the inguinal ligament or conjoint tendon to the iliopectineal ligament closes the opening into the femoral canal.

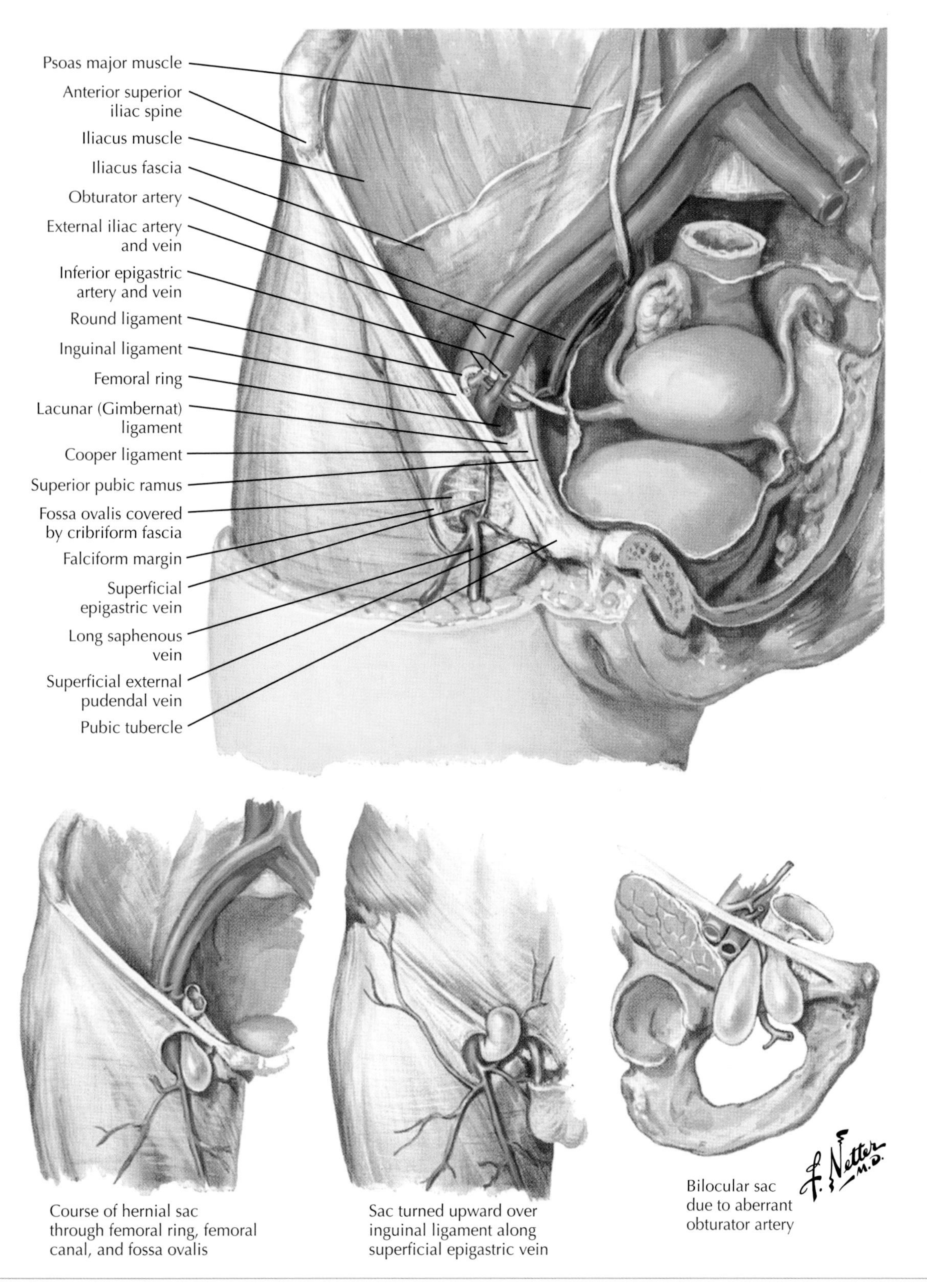

Figure 88-1 *Anatomy of Femoral Hernia.*

Ventral Hernias

89

Flavia Soto and Raul J. Rosenthal

Incisional (ventral) *hernias* usually result from inadequate healing of a previous surgical incision (**Fig. 89-1**). Many factors lead to the formation of incisional hernias; obesity is a leading cause. Bulk associated with a fatty omentum and excessive subcutaneous tissue increases the strain on the surgical wound in the early stages of healing. Many obese persons have an associated loss of muscle mass and tone that may result in weakened fascial layers to compensate for the strain.

Other factors that can contribute to ventral hernia formation are advanced age, malnutrition, ascites, postoperative hematoma or wound infection, peritoneal dialysis, pregnancy, and conditions that cause increased intraabdominal pressure and, therefore, increased strain in the abdominal wall. Certain medications that blunt the inflammatory response, such as steroids and chemotherapeutic agents, may also contribute to poor wound healing.

Incisional hernias are a problem because of high recurrence and complication rates. Hernias protruding through very large defects rarely become incarcerated or strangulated. Hernias with multiple components may compromise the integrity of their contents.

TREATMENT AND MANAGEMENT

Primary repair of an incisional hernia can occasionally be accomplished. However, most ventral hernias are best treated with prosthetic materials. Expandable polytetrafluoroethylene (PTFE) is the prosthetic material of choice. Timing of surgical repair must be individualized, and, if possible, patients should be in optimal medical condition. Preoperative preparation includes weight loss, smoking cessation, rigid control of diabetes, and avoidance of medications that impair wound healing. Previously, the presence of infection or possible contamination contraindicated the use of prosthetic materials. New bioprosthetic materials are now available for these types of situations; however, their use should be restricted to prospective clinical trials until more data are collected.

Surgery for ventral hernia repair may be conventional or laparoscopic. Several key factors determine which technique should be used.

All hernias are amenable to repair through laparoscopic techniques; however, some require extensive dissection and adhesiolysis. In patients who have undergone multiple previous abdominal surgeries, the laparoscopic approach may be compromised by the ability to access the abdominal cavity safely and establish the pneumoperitoneum. Obese patients and patients with significant comorbidities are generally benefited by laparoscopic repair.

UMBILICAL HERNIAS

Umbilical hernias are usually present at birth. Persons of African descent are predisposed to this condition. In the United States, umbilical hernias present eight times more often in black persons than in white persons. Most umbilical hernias close spontaneously, but those that persist in children older than 5 years may be surgically corrected.

Umbilical hernias that develop during adulthood are considered *acquired*. Conditions that result in increased intraabdominal pressure, such as pregnancy, ascites, and abdominal distention, may contribute to umbilical hernias in adults. It is rare for umbilical hernias to be complicated by incarceration or strangulation.

Simple transverse closure of the defect with proper reduction of the sac is generally adequate to treat umbilical hernias. Mesh placement is rarely required because recurrence is uncommon after primary closure.

SPIGELIAN HERNIAS

A Spigelian hernia is an abdominal wall defect located on the outer edge of the *crescent line;* the area between the anterior rectus muscle and the sheaths of the wide muscles of the abdomen in the area is called the *semilunar line*. Spigelian hernias usually occur inferiorly to the semicircular line of Douglas. The lack of posterior rectus fascia below the line of Douglas contributes to inherent weakness in this area. These hernias constitute 0.12% to 2% of all abdominal wall hernias. Classic etiologic factors associated with these defects include obesity, chronic obstructive pulmonary disease, previous surgery, and abdominal trauma.

Spigelian hernias may be found incidentally by ultrasonography and computed tomography; however, preoperative diagnosis has only a 50% accuracy rate. Spigelian hernias can be successfully repaired during the initial surgery by approximation of the tissues with interrupted sutures. If the defect is large, prosthetic reinforcement may be used.

EPIGASTRIC HERNIAS

Epigastric hernias are defects of the linea alba above the umbilicus; these are usually small and may be difficult to detect in obese patients. The most common symptom is a painful sensation of "pulling" at the midline on reclining. These hernias are usually repaired by simple closure. However, the surgeon must keep in mind that more than one defect may exist; therefore, proper exposure of the linea alba is mandatory.

RICHTER HERNIA

Any abdominal wall hernia in which the antimesenteric border of the intestine protrudes through the sac may be classified as a Richter hernia. The key characteristic of this type of hernia is that the entire circumference of the intestine is not involved. Symptoms and clinical course vary, depending on the degree of intestinal obstruction. Strangulation can occur as a painful mass, with abdominal distention, nausea, and vomiting.

In general, Richter hernia is found in the femoral canal. Interestingly, the incidence of Richter hernia has increased since

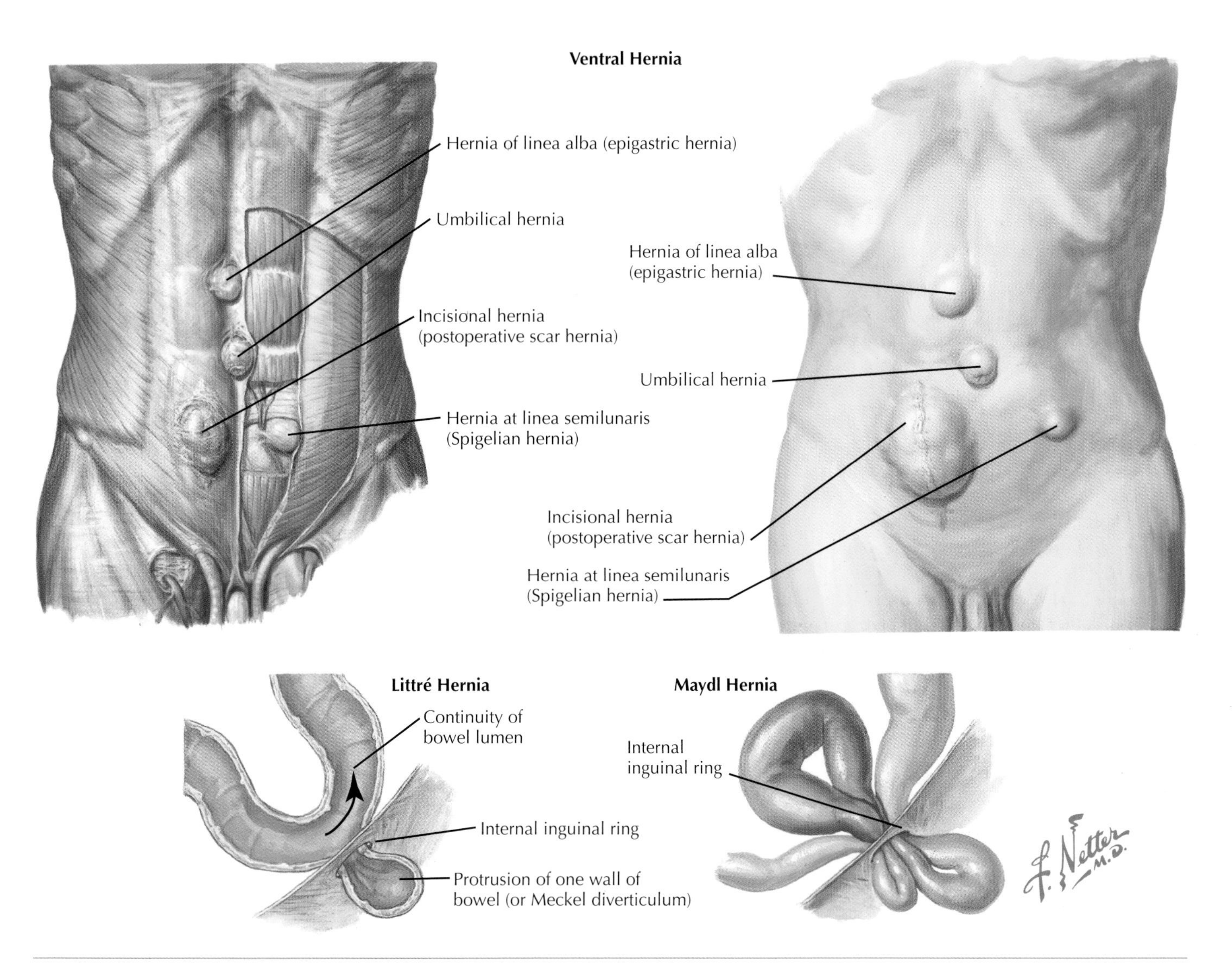

Figure 89-1 *Abdominal Wall: Ventral Hernias.*

the advent of laparoscopy because of unrepaired trocar sites. Proper evaluation of intestinal viability is critical to hernia repair; in patients with compromised viability, intestinal resection is required. The protrusion of more than one loop of intestine (usually two) through the sac is considered a "W" (retrograde) hernia, or Maydl hernia.

LITTRÉ HERNIA

In Littré hernia, the entire component of the sac must be a Meckel diverticulum. This type of hernia is infrequent and difficult to diagnose. Clinical manifestations of Littré hernia are localized pain over a preexisting hernia with intestinal obstruction, intestinal bleeding, perforation, fistula, and malignancy.

ADDITIONAL RESOURCES

Birgisson G, Park AE, Mastrangelo MJJ, et al: Obesity and laparoscopic repair of ventral hernias, *Surg Endosc* 14:1-5, 2001.

Dumanian GA, Denham W: Comparison of repair techniques for major incisional hernias, *Am J Surg* 185:61-65, 2003.

Kyzer S, Alis M, Aloni Y, Charuzi I: Laparoscopic repair of postoperative ventral hernia, *Surg Endosc* 13:928-931, 1999.

Sanchez Montes I, Deysine M: Spigelian and other uncommon hernia repairs, *Surg Clin North Am* 83:1235-1253, 2003.

Stigg KM, Rohr MS, McDonald JC: Abdominal wall, umbilicus, peritoneum mesenteries, omentum and retroperitoneum. In Townsend CM Jr, Sabiston A, editors: *Textbook of surgery*, ed 16, Philadelphia, 2001, Saunders, p 769.

Lumbar, Obturator, Sciatic, and Perineal Hernias

90

Samir Yebara and Raul J. Rosenthal

LUMBAR AND OBTURATOR HERNIAS

Lumbar hernia is an uncommon protrusion in the posterior abdominal wall that occurs in either the inferior (Petit) or the superior (Grynfeltt and Lesshaft) lumbar triangle (**Fig. 90-1**). Most hernias occur in the superior triangle, which is bordered by the twelfth rib superiorly, the internal oblique muscle anteriorly, and the quadratus abdominis muscle posteriorly. The latissimus dorsi forms the roof, and the transversus abdominis forms the floor. Hernias through the Grynfeltt triangle often develop after flank incisions for kidney surgery. Petit triangle hernias often occur in young, athletic women and usually contain fat, intestine, mesentery, or omentum. Bowel incarceration is the presenting sign in 24% of patients with lumbar hernia. The patient seeks treatment for a "lump in the flank" and reports dull, heavy, localized pain. Acquired hernias may be caused by direct trauma, penetrating wounds, poor healing of flank incisions, and abscesses. Diagnosis is confirmed on identification of a defect in the flank.

Computed tomography (CT) has been advocated as the diagnostic study of choice for suspected lumbar hernia in patients whose condition is stable. Lumbar hernia should be differentiated from soft tissue tumors, abscesses, hematomas, and renal tumors. Lumbar hernias tend to become larger and should be repaired when found. Simple suture repair of small hernias is practical, whereas the repair of larger hernias is more demanding. Overlapping and imbricating suture repairs can be performed; however, some large hernias may require mesh reinforcement, pedicle flaps, or free flaps.

Obturator hernia, a rare type of hernia, is a herniated viscus through the obturator canal. The obturator foramen is the largest foramen in the body and is formed by the rami of the ischium and the pubis. The obturator canal is 2 to 3 cm long and 1 cm wide; it contains the obturator nerve, artery, and vein surrounded by fat. A strong quadrilamellar musculoaponeurotic barrier, formed by internal and external obturator membranes and obturator internal and external muscles, closes the foramen. Deterioration of the obturator membrane and enlargement of the canal result in the formation of a hernial sac, causing significant intestinal obstruction and incarceration.

An obturator hernia develops in three stages. The first stage is the entry of periperitoneal tissue and fat into the pelvic orifice of the obturator canal. The second stage begins with the development of a dimple in the peritoneum over the internal opening and progresses to the invagination of a peritoneal sac. The third stage begins with the onset of symptoms produced by the entrance of an organ, usually the ileum, into the sac. Three specific signs indicate a strangulated hernia. *Obturator neuralgia* extends from the inguinal crease to the anteromedial aspect of the thigh. The *Howship-Romberg sign* is the classic sign, seen in 25% to 50% of diagnosed strangulated hernias, and is characterized by pain in the medial thigh and occasionally in the hip; flexion of the thigh usually relieves the pain. The *Hannington-Kiff sign* is an absent adductor reflex in the thigh, resulting from obturator nerve compression.

Sophisticated radiologic modalities such as CT, ultrasonography, and magnetic resonance imaging (MRI) can reliably aid in the diagnosis of obturator hernia. CT is recognized as the standard diagnostic modality. Differential diagnoses include psoas abscess, femoral and perineal hernias, intestinal obstructions, inguinal adenitis, and diseases of the hip joint.

Various approaches to obturator hernia repair have been suggested. The abdominal approach, open or laparoscopic, is preferable when compromised bowel is suspected. The retropubic (preperitoneal) approach is preferred when there are no signs of obstruction. More effective closure of the defect may be obtained by the use of polytetrafluoroethylene (PTFE) patch or other mesh. Regardless of the approach, reducing the contents and inverting the hernial sac are the first steps in the surgical treatment of obturator hernias.

SCIATIC AND PERINEAL HERNIAS

Sciatic hernia is also known as ischiatic hernia, gluteal hernia, hernia incisurae ischiadica, and sacrosciatic hernia. Sciatic hernias may emerge through the suprapiriform or infrapiriform spaces or through the lesser sciatic foramen. The sciatic notch on the inferior margin of the pelvis is transformed into the greater and lesser sciatic foramina by the sacrospinous and sacrotuberous ligaments. The greater sciatic foramen is subdivided by the piriform muscle, which traverses the space. Atrophy of the piriform muscle increases the risk for sciatic hernia.

Sciatic hernias are difficult to diagnose. Clinically, patients have pain patterns that originate mainly in the pelvis but that occasionally radiate to the buttocks and the posterior thigh. Patients rarely exhibit signs such as protrusion, bulge, or saccule because of the small size of the hernial sac. The sac may contain small bowel, ureter, or ovary. Other documented sac contents include Meckel diverticulum, colon, and bladder. Differential diagnoses include abscess, lipoma, and gluteal aneurysm. Sciatic hernias are usually diagnosed and treated during surgery, with most repairs performed through a transperitoneal or transgluteal approach.

Perineal hernia is another infrequent but well-recognized complication of major pelvic surgery. There are two basic types of perineal hernias: primary and secondary. *Primary*, or anterior, perineal hernias develop through the urogenital diaphragm with the bulbocavernosus muscle medially, the ischiocavernosus laterally, and the transverse perineal muscle posteriorly. Primary hernias result from acquired weakness of the pelvic floor structures, common in middle-aged women with a history of childbirth. *Secondary*, or posterior, perineal hernias involve a defect between the levator ani and the coccygeus muscle. Secondary hernias are rare, reported in only 1% of patients, and are related to previous surgery, particularly pelvic exenteration, abdomino-

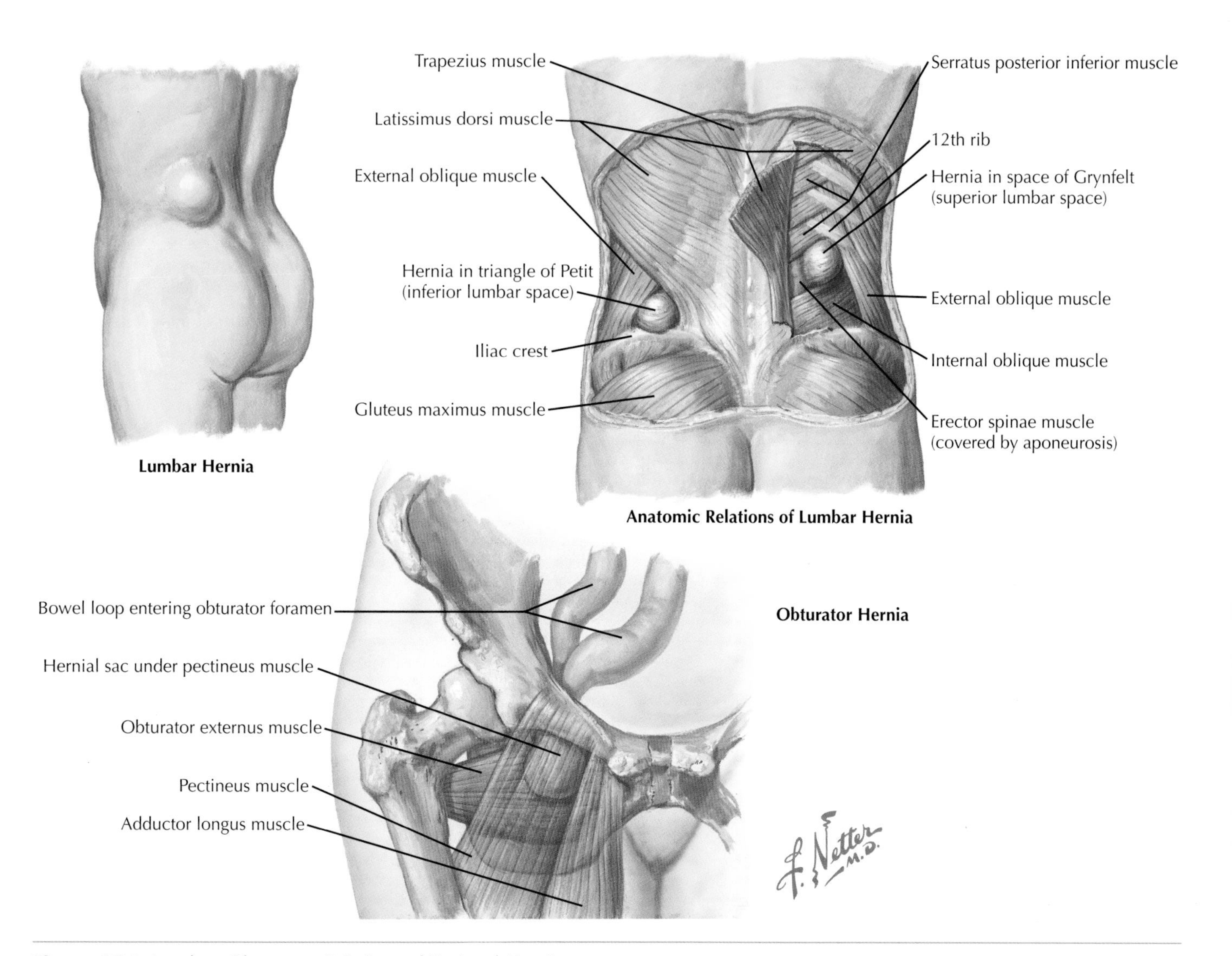

Figure 90-1 *Lumbar, Obturator, Sciatic, and Perineal Hernias.*

perineal resection, and perineal prostatectomy. The hernial sac may contain bowel, omentum, or bladder.

Common symptoms of perineal hernia are pain and a reducible bulge, but difficult urination may also be a symptom. Differential diagnoses include soft tissue tumors, cysts, abscesses, and hematomas.

Various approaches for surgical repair of perineal hernia have been proposed but not standardized. Although the abdominal or perineal approach can be used, myocutaneous flap or mesh reinforcement is required for repair.

ADDITIONAL RESOURCES

Brian EB, Kimball M: Traumatic lumbar hernia, *South Med J* 93:1067-1069, 2000.

Eubanks S: Hernias. In Townsend CM Jr, Sabiston A, editors: *Textbook of surgery*, ed 16, Philadelphia, 2001, Saunders, pp 783-800.

Franklin ME Jr, Abrego D, Parra E: Laparoscopic repair of postoperative perineal hernia, *Hernia* 6:42-44, 2002.

John RM, Michael JO, William BS: Sciatic hernia as a cause of chronic pelvic pain in women, *Obstet Gynecol* 91:998-1001, 1998.

Julian EL, Bruce WR, James WJ: Obturator hernia, *J Am Coll Surg* 194: 657-663, 2002.

Julian EL, Kirien TK: Diagnosis and treatment of primary incarcerated lumbar hernia, *Eur J Surg* 168:193-195, 2002.

Lee JS, John A, Gene LC, John ES: Obturator hernia, *Surg Clin North Am* 8:71-84, 2000.

Internal Hernias: Congenital Intraperitoneal Hernias

91

Elias Chousleb and Raul J. Rosenthal

Most intraperitoneal hernias result from anatomic variants that are usually present at birth (**Fig. 91-1**). Hernias that develop secondarily to alterations in normal intestinal rotation during embryologic development have sacs. These types of hernias are generally *paraduodenal* or *mesocolic*. Hernias that develop through defects in the mesentery or peritoneum do not have sacs. These types of hernia include those through the epiploic foramen, congenital defects in the mesentery of the small and large intestine, and less commonly, defects in the broad ligaments of the uterus.

Trauma and previous abdominal surgery also cause intraperitoneal hernias. Internal hernias are rarely diagnosed before surgery, regardless of their location. Suspected signs of intestinal obstruction and a palpable mass in the corresponding region may indicate an internal hernia. One of the most remarkable features of internal hernias is that they may remain asymptomatic for the person's lifetime, with diagnosis made only on postmortem study.

During surgical treatment of internal hernias, the surgeon must be aware that major blood vessels traverse the neck of the hernial sac; therefore, it may be prudent to open the sac beyond the neck to decompress the gut.

ADDITIONAL RESOURCES

Stigg KM, Rohr MS, McDonald JC: Abdominal wall, umbilicus, peritoneum mesenteries, omentum and retroperitoneum. In Townsend CM Jr, Sabiston A, editors: *Textbook of surgery*, ed 16, Philadelphia, 2001, Saunders, p 779.

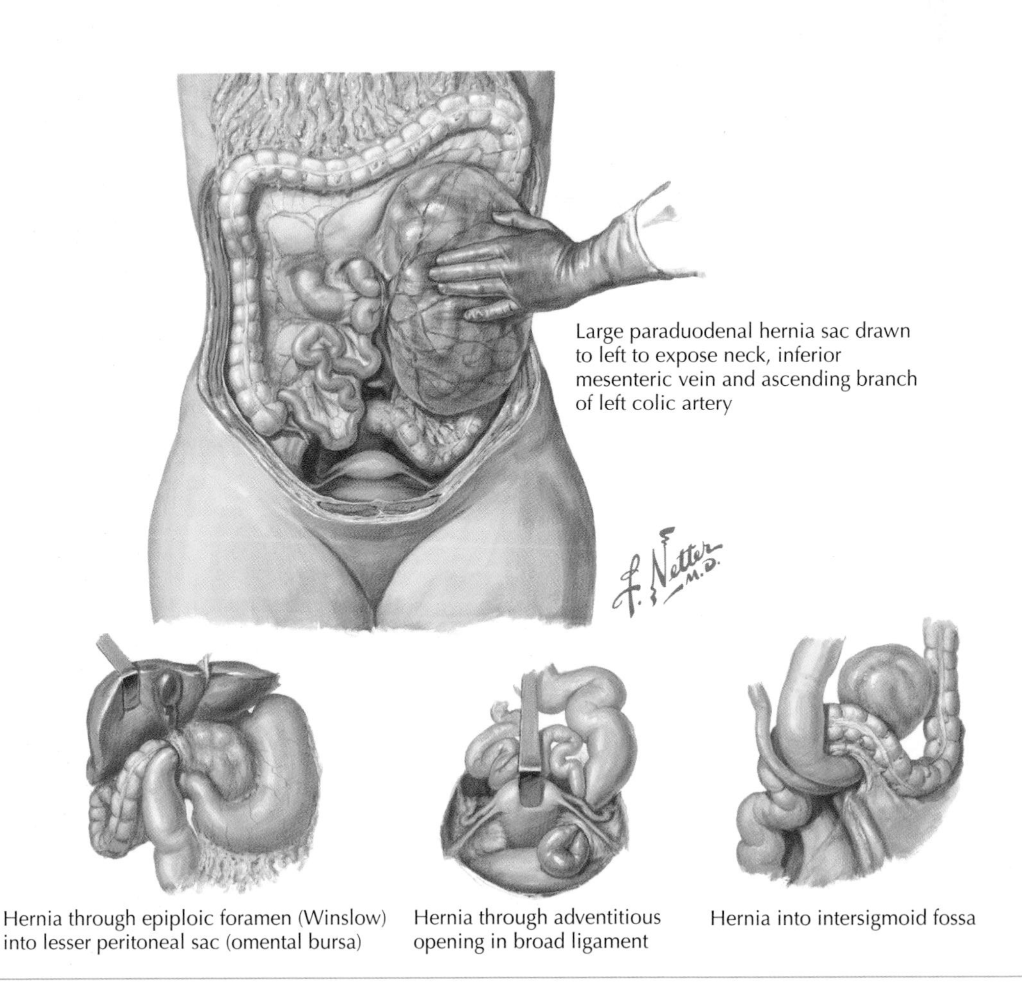

Figure 91-1 *Internal Hernias: Congenital Intraperitoneal Hernias.*

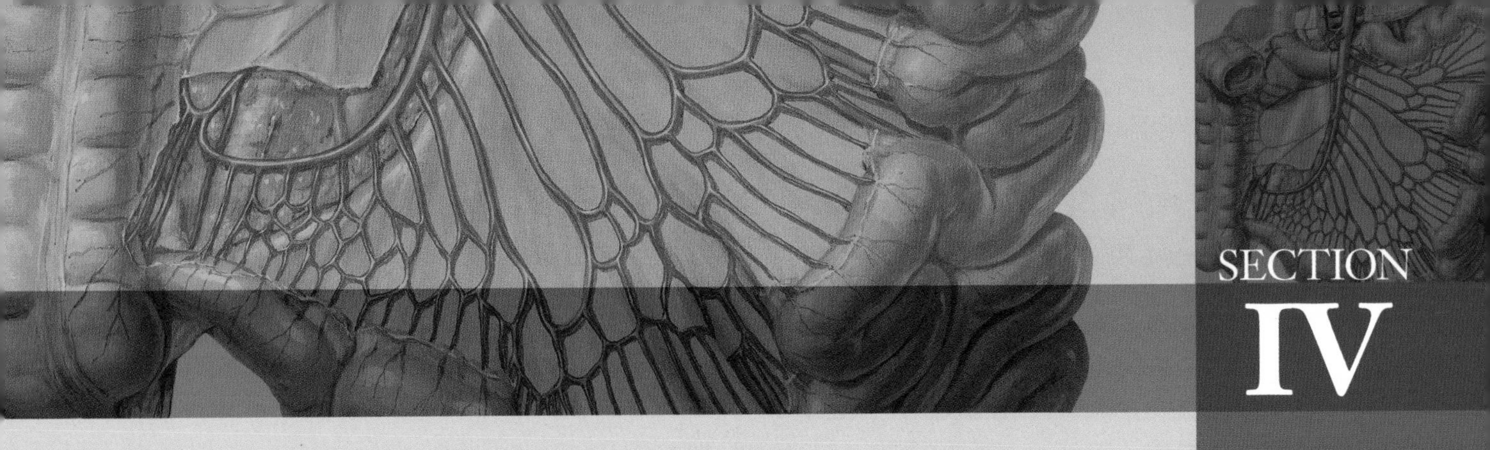

SECTION IV

Small Intestine

Topography of the Small Intestine

92

Martin H. Floch

The small intestine consists of a retroperitoneal portion, the duodenum, and a mesenteric portion comprising the coils of the jejunum and the ileum (**Fig. 92-1**). Given that the mesenteric portion of the small intestine is subject to considerable individual and functional variations, its total length varies considerably. The average length for adults is approximately 5 meters (15-20 feet), 40% of which is accounted for by the upper part, the jejunum, and 60% by the lower part, the ileum.

The *jejunum* begins at the duodenojejunal flexure on the left side of the second lumbar vertebra or, occasionally, somewhat more cranially (see Section II). The *ileum* joins the large intestine in the region of the right iliac fossa. The *duodenojejunal flexure* is situated high up in the inframesocolic zone of the peritoneal cavity and may be partially concealed by the attachment of the transverse mesocolon. Between the duodenojejunal flexure and the ileocolic junction, the parietal line of attachment of the small intestine mesentery runs obliquely from above on the left to below on the right, passing across the lumbar spine, large prevertebral blood vessels (aorta, inferior vena cava), right psoas major, and right ureter.

Because the *mesentery* is only about 15 to 20 cm (6-8 inches) long at its parietal line of attachment, rather than the several meters (corresponding to length of intestine) along its intestinal attachment, it splays fanlike toward the intestine. Mesentery, consisting of two layers of peritoneum, affords the intestinal coils a wide range of movement. The space between the two layers of peritoneum is filled with connective tissue and fat tissue, the latter varying greatly from one person to another. Embedded in this tissue are blood and lymph vessels running between the intestine and the dorsal wall of the abdomen, along with nerves and mesenteric lymph nodes.

The various portions of the large intestine appear as a horseshoe-shaped arch and form a frame enclosing the convolutions of the small intestine (see Section V). This frame, however, may be overlapped ventrally by the coils of the small intestine, particularly on the side of the descending colon. Similarly, depending on their filling and on their relationship to the pelvic organs, the coils of the small intestine may bulge downward into the true pelvis or, if the pelvic organs are greatly distended (e.g., in pregnancy), may be displaced in a cranial direction.

With a greatly variable shape and highly mobile position, the greater omentum hangs like an apron from the greater curvature of the stomach and spreads between the anterior abdominal wall and the coils of the small intestine.

The greater part of the coils of the jejunum lies upward to the left, whereas those forming the ileum are situated lower and to the right side. Because it is attached only to its mesentery, the small intestine is capable of considerable movement. Its coils vary greatly in position even in the same person, depending on the state of intestinal filling and peristalsis and on the position of the body as observed under x-ray examination after oral introduction of a rubber tube. In accordance with its progressively shortened mesentery, the only position that has a more or less "constant" position is the terminal ileum, which passes from the left across the right psoas major to the site of the ileocolic junction (see Chapter 95).

ADDITIONAL RESOURCES

Kahn E, Daum F: Anatomy, histology, embryology, and developmental anomalies of the small and large intestine. In Feldman M, Friedman LS, Brandt LJ, editors: *Gastrointestinal and liver disease*, ed 8, Philadelphia, 2006, Saunders-Elsevier, pp 2061-2091.

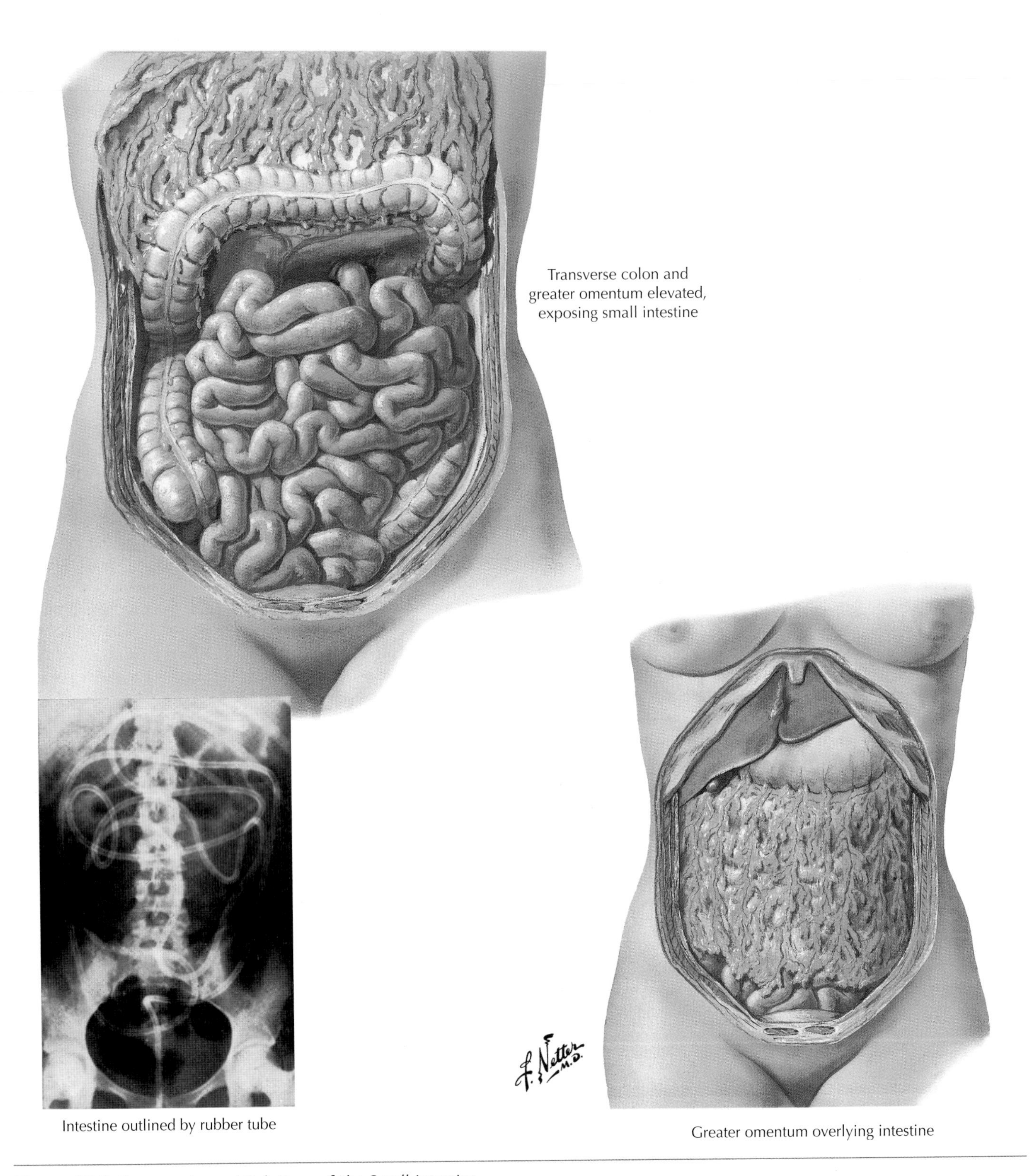

Figure 92-1 *Topography and Relations of the Small Intestine.*

Gross Structure of the Small Intestine

Martin H. Floch

93

The freely mobile portion of the small intestine extends from the duodenojejunal flexure to the ileocolic orifice. This portion of the small intestine consists of the jejunum and the ileum, which run imperceptibly into each other; the transition is marked by a gradual change in the diameter of the lumen and by various structural alterations. As with the entire gastrointestinal (GI) tract, the virtually identical walls of the jejunum and ileum consist of five coats: mucosa, submucosa, circular muscularis, longitudinal muscularis, and serosa (**Fig. 93-1**).

The innermost layer, the mucous membrane, is thickly plicated by macroscopically visible circular or convoluted folds, or *plicae*, known as *circular* or *Kerckring folds* (valves), or valvulae conniventes. These folds vary in height, projecting 3 to 10 mm into the lumen, and run in a transverse direction to the lumen's longitudinal axis. Some plicae extend all the way around the internal circumference, others go only halfway or two thirds the way around the circumference, and still others spiral around two or even more times. These do not act as a true valve; projecting into the lumen, Kerckring valves will slow down, to a certain extent, the progression of the luminal contents, but their essential function is to increase the absorptive surface area. This principle is all the more obvious because the fold's surface is further equipped with tiny, fingerlike projections, or *villi*.

Below the epithelial surface of the mucosa, but participating in the formation of Kerckring folds and the villi, is the tunica propria, or *lamina propria*, a loose coat of predominantly reticular connective tissue, assuming in some parts a lymphatic character. The lamina propria also contains thin fibers of smooth muscle radiating from the muscularis mucosae and extending upward to the tips of the villi, which have an even surface when these fibers are relaxed but become jagged or indented when the fibers contract (see Chapter 94). The muscular fibrils act as motors maintaining the pumping function of the villi. Situated in the lamina propria, and especially in the stroma of the villi, are the terminal ramifications of the blood vessels, the central lacteal or lymph vessels of the villi, and nerve fibers. Many solitary lymph nodes are embedded in the lamina propria, which may reach far into the submucosal layer.

The muscularis mucosae separates the mucous membrane from the *submucosal coat* and is composed of two thin, nonstriated muscle layers that keep the movable muscle layer in place. The outer longitudinal layer is thinner than the inner circular layer from which the muscle fibers in the core of the villi emanate. Tunica submucosa consists of collagen connective tissue, the fibers of which form a network of meshes. By altering the angles of its meshes, the submucosal network is able to adapt to changes in the diameter and length of the intestinal lumen. The submucosa contains a rich network of capillaries and larger vessels, numerous lymphatics, and the submucous nerve plexus of Meissner. The muscle layer is made of smooth muscle cells. The thick inner circular layer and the thinner outer longitudinal layer are connected by convoluted transitional fascicles where the layers border on each other. Between the two layers is spread a network of nonmyelinated nerve fibers and ganglion cells, the myenteric plexus of Auerbach (see Chapter 100).

Serosa is composed of a layer of flat, polygonal epithelia and a subserosa of loose connective tissue. It covers the entire circumference of the intestinal tube, except for a narrow strip at the posterior wall, where the visceral peritoneum connects with the two serous layers of the fan-shaped mesentery.

The jejunum and ileum differ in size and appearance. The ileal lumen is narrower and the diameter of the total ileal wall is thinner than in the jejunum. The average diameter of the jejunum is 3 to 3.5 cm, whereas the ileum is 2.5 cm or less. Because of this difference, the intestinal contents show up more clearly through the ileum than through the jejunum. When the abdomen is opened, the jejunum has a whitish red hue, whereas the ileum, during life and after death, takes on a darker appearance. The folds and the villi become smaller and decrease in number as the small intestine continues. In the lower reaches of the ileum, the folds appear only sporadically.

In the jejunum, lymphatic tissue is encountered only in the form of solitary nodules, which appear as pinhead-sized elevations on the surface of the mucosa. They become more numerous and more pronounced as they near the large intestine. In addition, aggregate nodules (Peyer patches) appear, confined to the ileum. Averaging 1 to 1.5 cm wide, Peyer patches are 2 to 10 cm long and vary greatly in number, usually 20 to 30. The ileal mesentery contains more fatty tissue and appears to be thicker than the jejunum.

ADDITIONAL RESOURCES

Kahn E, Daum F: Anatomy, histology, embryology, and developmental anomalies of the small and large intestine. In Feldman M, Friedman LS, Brandt LJ, editors: *Gastrointestinal and liver disease*, ed 8, Philadelphia, 2006, Saunders-Elsevier, pp 2061-2091.

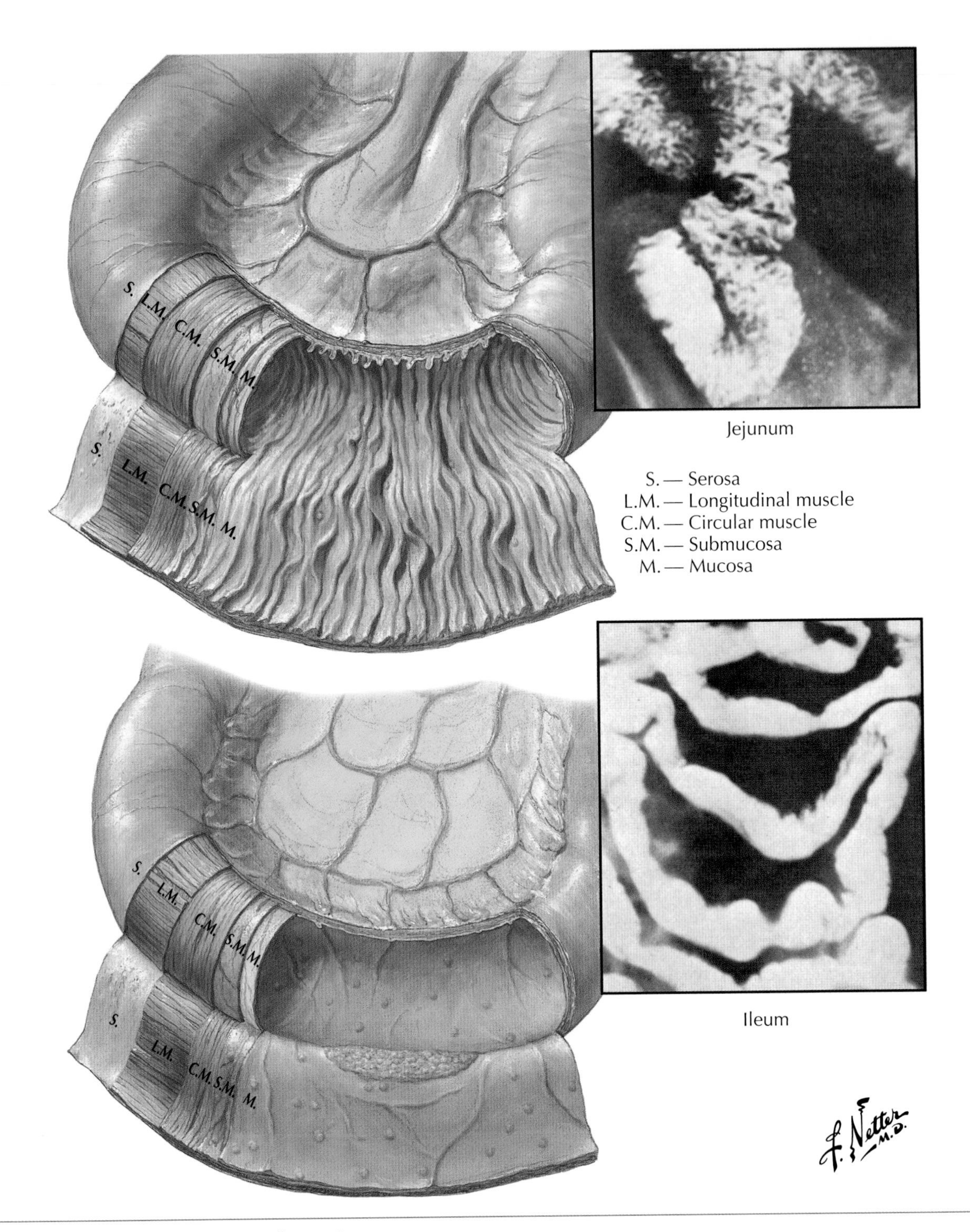

Figure 93-1 *Gross Structure of the Small Intestine.*

Microscopic Structure of the Small Intestine

94

Martin H. Floch

The entire mucosal surface of the small intestine is covered with projections 0.5 to 1.5 mm long, the intestinal *villi* (**Fig. 94-1**). The mass of these villi (estimated at 4 million for the jejunum and the ileum) accounts for the velvetlike appearance of the intestinal mucosa. In the jejunum, villi are longer and broader than in the ileum. Valleys, or indentations, between the villi result in nonramified pits, each of which harbors one or two tubular structures, the intestinal glands, or *crypts of Lieberkühn*.

The entire inner surface of the small intestine is covered by a single line of epithelial cells, most of which are cylindrical, highly prismatic *columnar cells* with a well-developed cuticular border on the surface. Between these columnar cells are interspersed three other types of cells: goblet cells, Paneth (oxyphilic granular) cells, and enterochromaffin cells. *Goblet cells* secrete an alkaline mucous fluid that coats the entire mucosa. As the small intestine moves closer to the large intestine, healthy anaerobic bacterial organisms tend to live in this mucus and function as probiotics. Most goblet cells are found in the crypts of Lieberkühn or along the lower parts of the villi, but some are located in the upper parts of the villi. The characteristic elements of the floor of the crypts are *Paneth cells*, also called *oxyphilic granular* cells because of the staining qualities of their granules. They secrete antimicrobial and growth protein substances. The third cell type is *enterochromaffin cells* (argentaffin or argyrophilic), which contain basal staining granules with a high affinity for silver and chromium. Their habitat is the crypts of Lieberkühn, where it is now believed stem cells exist and give rise to all intestinal cells. These cells have a definite neuroendocrine function.

Within the tunica (lamina) propria is a great variety of cells, most of which originate from reticular cells. In addition to the usual connective tissue cells, lymphocytes and plasma cells are present. *Lymphocytes* show a marked tendency to migrate through the epithelium toward the lumen. These cells make up the largest mass of immunoprotective tissue in the body. *Mast cells* are also present in the lamina propria and react to antigens. The *interstitial cells of Cajal* are present in the wall (see Chapter 100).

The principal task of the gastrointestinal tract is to serve as an organ of nutrition to satisfy caloric and nutritional requirements. Key steps in digestion occur within the lumen of the small bowel, and then absorption occurs through these epithelial cells. Villi, covered by the epithelial cells, function as the *organelles* of absorption.

The luminal surface of the epithelial cell is covered with fine, projecting rods called *microvilli*. Each epithelial cell contains approximately 1000 microvilli, which increases the cellular surface approximately 24 times. The average length of a microvillus is 1 μm, and the width is 0.07 μm. Covered by a continuation of the cell membrane, microvilli contain, in the core, fine fibrils connected by a network of fibrils called the *terminal web*. Microvilli form a sheet that can be seen under the microscope and that is often lost when the epithelium is damaged, as in celiac disease.

Shortly after the ingestion of a fatty meal, fine lipid droplets are observed in the intermicrovillus spaces, which then are seen in the terminal web; pinocytotic activity subsequently occurs. The droplets seem to proceed and can be found in the main body of the epithelial cell, where they coalesce into large units in vesicles connected to each other by intracellular tubules. The system is referred to as the *endoplasmic reticulum*. Through this reticulum, fat droplets pass toward the lateral cell surfaces, and from the intercellular spaces the droplets traverse the basement membrane to enter the central lacteals of the villi. In the region below the microvilli, the profile of the lateral surface is irregular because of end plates. Toward the base of each cell, the membrane is *plicated*, or underplayed, which means the adjacent cells become interdigitated.

ADDITIONAL RESOURCES

Kahn E, Daum F: Anatomy, histology, embryology, and developmental anomalies of the small and large intestine. In Feldman M, Friedman LS, Brandt LJ, editors: *Gastrointestinal and liver disease*, ed 8, Philadelphia, 2006, Saunders-Elsevier, pp 2061-2091.

Scoville DH, Sato T, He XC, Li L: Current view: intestinal stem cells and signaling, *Gastroenterology* 134:849-864, 2008.

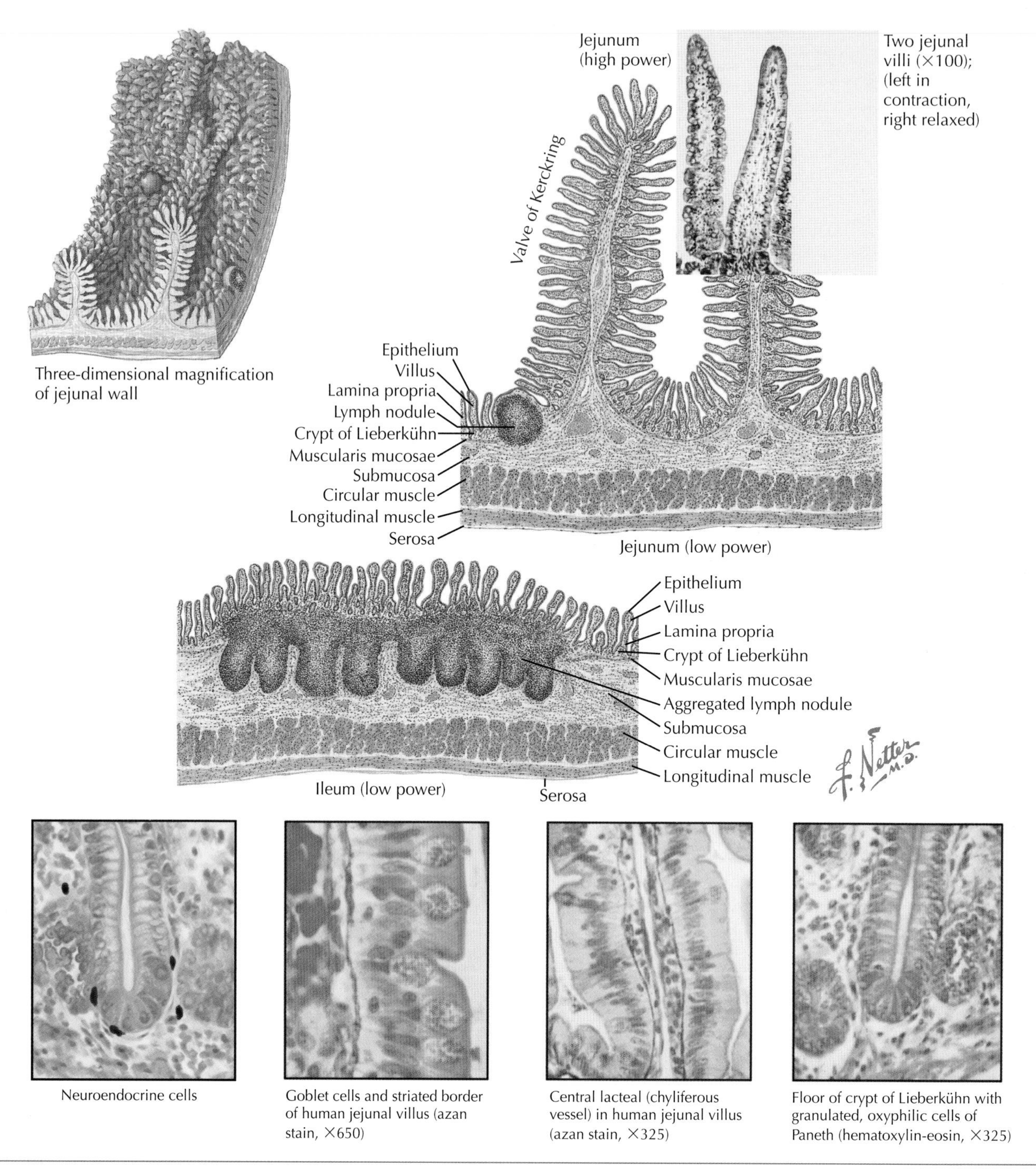

Figure 94-1 *Microscopic Structure of the Small Intestine.*

Terminal Ileum

95

Martin H. Floch

The terminal ileum is the most caudal part of the small intestine and usually lies in the pelvis over the right iliac fossa. It opens sideways from the left into the medial wall of the large intestine (**Fig. 95-1**). The section of the large intestine caudally or below this junction is a "blind" sac and thus is termed the *cecum*.

In most people, where the ileum joins the large intestine, the peritoneal fold extends from the terminal part of the ileomesentery, across the front of the ileum, to the cecum and lowest part of the ascending colon. This fold is known as the *ileocolic fold*, or *superior ileocecal fold*. It contains the anterior cecal artery and forms the anterior wall of the fossa, correspondingly termed the *ileocolic fossa*, or *superior ileocolic fossa*. The posterior wall of this fossa is made up of the terminal ileum and its mesentery. Its mouth opens downward and somewhat to the left. Another fold, known as the *ileocecal fold*, or *inferior ileocecal fold*, is often encountered in front of the mesoappendix, extending from the lower or right side of the terminal ileum to the cecum. Together with the mesoappendix as the posterior wall, the fold again forms a fossa, the *ileocecal fossa*, or *inferior ileocecal fossa*, of which the fold represents the interior wall. The ileocecal fold contains no important vessel and therefore has been named the "bloodless" *fold of Treves*. The third peritoneal extension, the mesoappendix, serves as a mesentery of the appendix.

At the ileocecal junction, the terminal ileum is thrust with all its coverings into the wall, invaginates the large intestine, and creates within the lumen of the latter what has been known as the *ileocecal valve*. On exposure of this sphincter at autopsy, the ileal aperture is seen as bounded by two almost horizontal folds, referred to as the upper and lower "lips" of the "valve," in approximately 60% of cases. At both ends of the lips, where they seem to coalesce, two mucosal ridges extend horizontally in the lumen of the large intestine, resembling the crescent-shaped bulbs of the colon. These ridges, known as the *frenulum* of the ileal orifice (ileocecal valve), form the dividing line between the cecum and the ascending colon. In vivo, the ileum may protrude into the large intestine in the form of a rounded *papilla*, the lumen of which assumes a starlike appearance when closed, often compared with the appearance of the cervix protruding into the vagina. When visualized endoscopically, the ileocecal valve may appear closed or open, or at times with motility, it may change shape.

It is thought that the ileocecal valve actually acts as a true sphincter, a sphincter that may be under neural and hormonal control. Dissection of the musculature of the area reveals that some fibers from the mesocolic taenia, ascending from the colon and cecum to the appendix, turn inward and pass into the ileocolic papilla, whereas others turn outward to become continuous with the longitudinal muscle of the ileum. Similarly, the longitudinal muscle of the ileum takes divergent courses, with some fibers passing into the papilla and others joining fibers of the taenia. It is postulated that the circular muscle layer, which is much stronger, closes the sphincter, whereas the longitudinal muscle layer opens it.

Functionally, the ileocecal valve prevents reflux of colonic contents into the small bowel. Motility studies confirm that the terminal ileum and its muscular anatomic structure behave as a valve, allowing ileal contents to empty in a pulsatile manner.

ADDITIONAL RESOURCES

Kahn E, Daum F: Anatomy, histology, embryology, and developmental anomalies of the small and large intestine. In Feldman M, Friedman LS, Brandt LJ, editors: *Gastrointestinal and liver disease*, ed 8, Philadelphia, 2006, Saunders-Elsevier, pp 2061-2091.

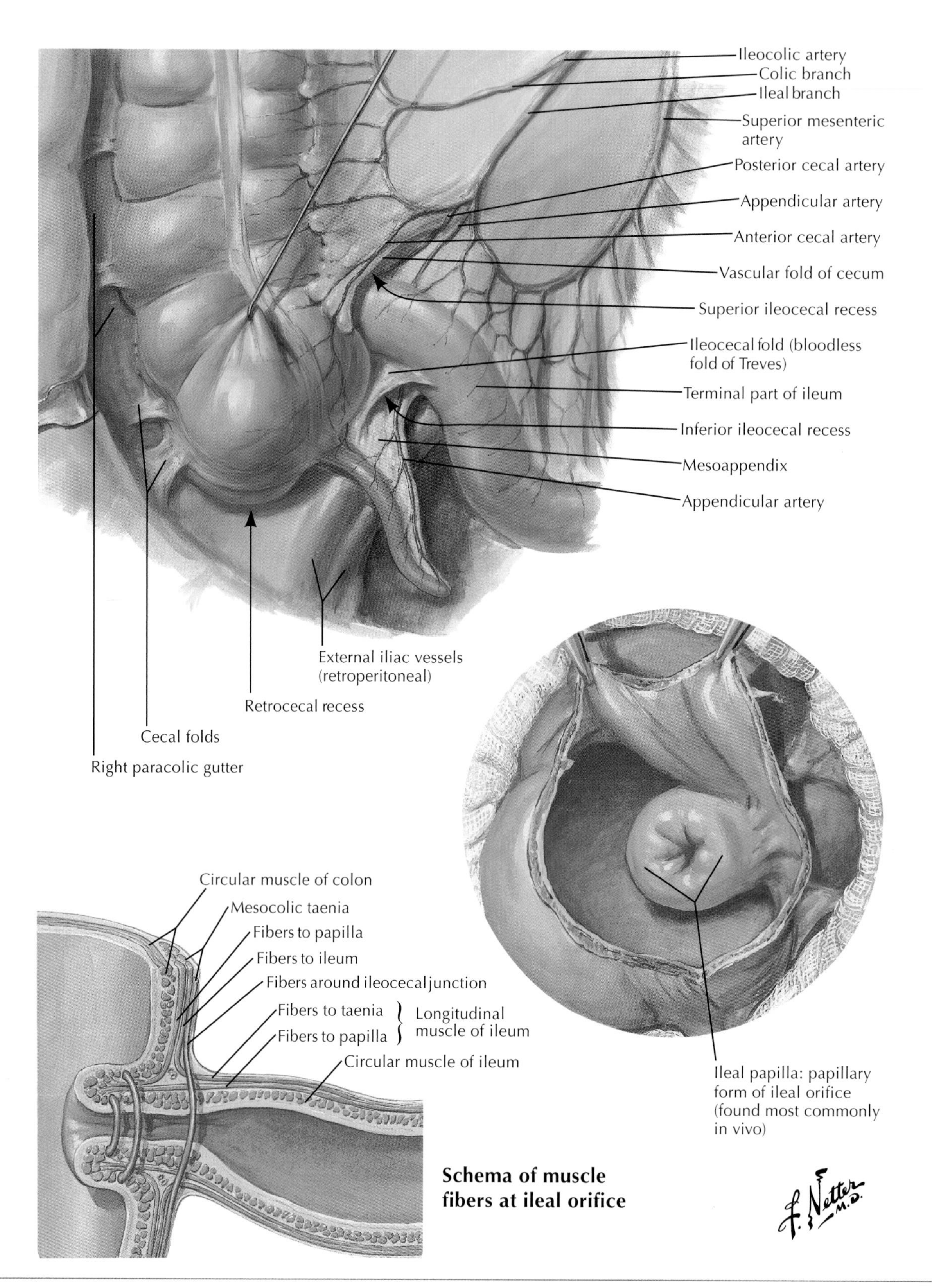

Figure 95-1 *Ileocecal Region.*

Secretory, Digestive, and Absorptive Functions of the Small Intestine

96

Martin H. Floch

The mucosa of the gut, throughout its entire length, is equipped with secretory cells. The secretory product of the duodenal glands is an alkaline, pale-yellow, viscous fluid consisting essentially of mucus, a primary function of which is to protect the proximal duodenum against the corrosive action of gastric chyme. The glandular apparatus of the jejunum and ileum produces the succus entericus (intestinal juice). The epithelial secretions also contain enzymes, including peptidases, nucleases, nucleosidases, phosphatase, lipase, maltase, sucrase, lactase, and the coenzyme enterokinase, which activates trypsinogen and chymotrypsinogen of pancreatic origin to form active trypsin and chymotrypsin, respectively. The flow of the succus entericus is stimulated by acid secretion in the upper intestine; by local mechanical and chemical stimuli; by the administration of secretin, enterokinin, and pilocarpine; and by sympathectomy.

DIGESTION

In *protein digestion*, the breakdown of food protein begins in the stomach through the action of pepsins, the effectiveness of which depends on the rate of emptying from the stomach and the pH of both the stomach and the duodenum (**Fig. 96-1**). However, it is apparent from patients who have achlorhydria or who have undergone surgical bypass that gastric proteolysis is not necessary to break down and absorb most proteins. Pancreatic proteolytic enzymes are secreted in the *proenzyme phase*. Through the action of enterokinase, which is secreted in the succus entericus, these enzymes are activated to trypsin, chymotrypsin, elastase, and carboxypeptidases A and B. The final product of the intraluminal enzyme activity yields peptide chains of two to six amino acids, which make up approximately two thirds of the content, with the other third in the form of simple amino acids. Digestion then occurs further in the *brush border* of the enterocytes as the amino acids and oligopeptides are absorbed. The brush border contains several peptidases, and several within the cytoplasm of the enterocytes complete digestion and some transformation of amino acids for metabolic activity.

Dipeptides are more effectively and actively absorbed than simple amino acids and tripeptides. *Protein absorption* occurs primarily in the duodenum and jejunum and requires a complex transporter system in the brush border with separate sodium-dependent, acid, and basic amino acid systems. Congenital disorders of amino acid transport result in serious growth and developmental disorders and nutritional disease. Epidermal growth factor, neurotensin, cholecystokinin, and secretin enhance transport, whereas somatostatin and vasoactive intestinal polypeptides decrease transport.

The specific method of vitamin B_{12} absorption transport is discussed in Section II.

Digestive and absorptive processes involving the nutrients of *carbohydrates* generally consist of enzymatic cleavage of polysaccharides and oligosaccharides into disaccharides and monosaccharides (**Fig. 96-2**). The process is relatively simple compared with the digestive and absorptive process that proteins and fats require. *Starches* are the main energy-producing nutrients of all plant foods and consist primarily of amylose and amylopectin. Dietary carbohydrate also includes lactose from milk; fructose, glucose, and sucrose from vegetables and fruits; and sugars as additives in all drinks.

Nonstarch polysaccharides are poorly digested by human enzymes and make up the major component of dietary fiber. Other unavailable carbohydrates that are poorly digested by human enzymes are pectins, gums, lignins, and alginates. These are readily metabolized by the enteric flora, and their products can be absorbed through the small and large intestinal mucosa (see Section V).

Starch and sugar digestion occurs with the release of salivary and pancreatic amylases that cleave the α-1,4 link of simple disaccharides and a long-chain starch. The degree of activity of salivary amylase depends on the duration of chewing and the proximity of the enzymes in the chyme, as well as the churning of the stomach. However, the major enzyme breakdown of starch occurs through pancreatic amylase activity in the duodenum, and hydrolysis mainly occurs intraluminally in the proximal small intestine. Monosaccharides and disaccharides are presented to the brush border of the enterocytes, where membrane hydrolysis occurs. Lactase, maltase, sucrase-isomaltase, isomaltase, and trehalase—enzymes in the brush border—are most active in the duodenum and jejunum. These enzymes cleave the disaccharides into glucose, galactose, and fructose, which are transported across the mucosa. Transport is active and passive. Transport mechanisms are controversial, but it is agreed that an active sodium-glucose co-transporter exists and that much of the process is sodium dependent.

The problems of disaccharidase deficiencies result in significant diarrhea and symptomatic syndromes. Disaccharidases are synthesized within the endoplasmic reticulum. The classic deficiency of lactase results in lactose malabsorption. Less common are sucrose- and fructose-absorptive deficiencies, and less common and less well known are the trehalose deficiencies that result from absorption after mushroom ingestion.

During *fat digestion and absorption*, fats are passed into the stomach, where gastric lipase may be active and prefers action at one-ester bonds. The chyme passes the fats into the duodenum, where pancreatic lipase has its greatest activity in both

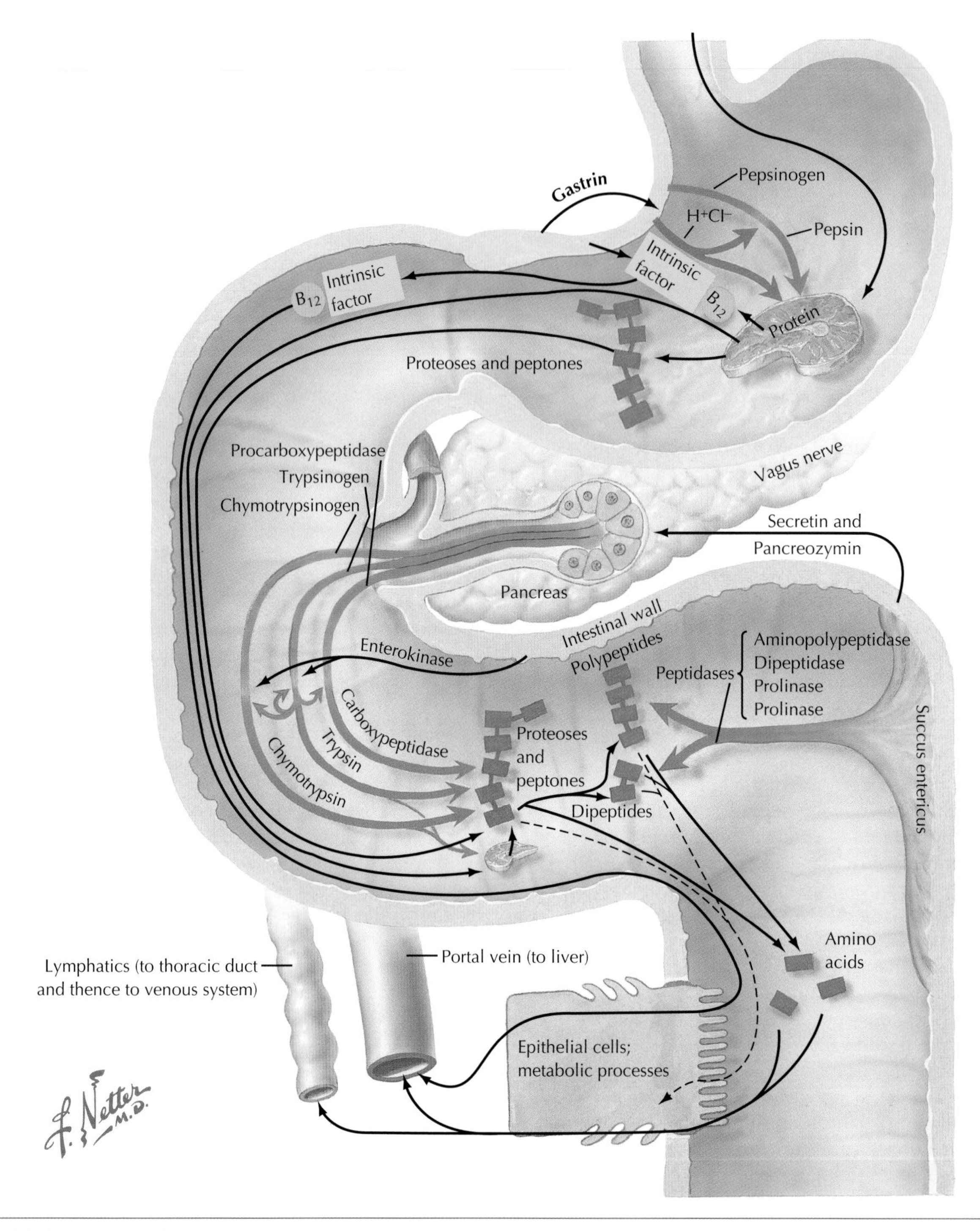

Figure 96-1 *Digestion of Protein.*

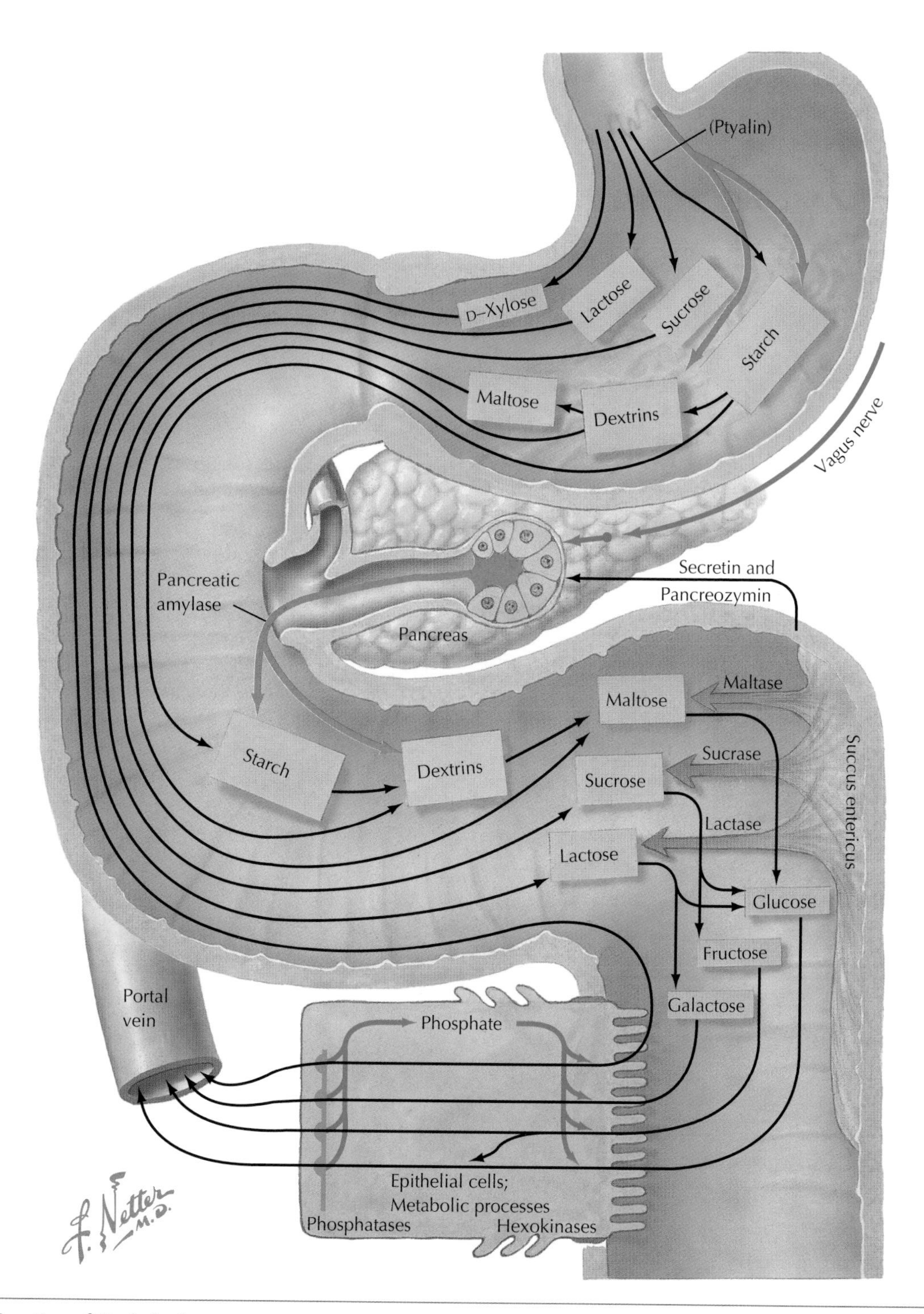

Figure 96-2 *Digestion of Carbohydrates.*

one-ester and three-ester bonds. Approximately 95% of ingested fat is absorbed, undergoing complicated mechanisms to pass into the lymphatics and bloodstream. Intraluminally, the fat is broken down into emulsion droplets, which requires droplets to be coated with a phospholipid. *Lipolysis* begins in the stomach, but pancreatic lipase exerts greater effect. Colipase, lipase, phospholipids, and bile salts are all involved in a complex mechanism of forming *micelles*, and a lipid phase transformed into an aqueous phase permits transport across the brush border. During this process, triglycerides are broken down into diglycerides and monoglycerides, in addition to the monoglycerides and diglycerides present in foods and absorbed. Bile salts are capable of forming micelles because they are ideal emulsifying agents. An unstirred water layer at the surface of the brush border readily permits short-chain or medium-chain fatty acid absorption but limits long-chain fatty acid absorption. Cholesterol, as well as monoglycerides, diglycerides, and triglycerides, is readily absorbed through these mechanisms. Once within the enterocytes, triglycerides resynthesize, lipoproteins form, and then chylomicrons form. The material is passed into the lymphatic and portal vein systems to the circulation.

The process of fat digestion and absorption is complex; Figure 96-1 helps clarify the intraluminal process. The interested reader is referred to important descriptions in the literature.

It is also now clear that the small intestine has a great ability to adapt to changes through both hypertrophy and development of new transport systems.

ADDITIONAL RESOURCES

Carey MC, Small DN, Bliss CM: Lipid digestion and absorption, *Annu Rev Physiol* 45:651-665, 1983.

Dahlquist A, Semenza G: Disaccharidases of small intestine mucosa, *J Pediatr Gastroeneterol Nutr* 4:857-868, 1985.

Farrow JJ: Digestion and absorption of nutrients and vitamins. In Feldman M, Friedman LS, Brandt LJ, editors: *Gastrointestinal and liver disease*, ed 8, Philadelphia, 2006, Saunders-Elsevier, pp 2147-2197.

Gastrointestinal Hormones

97

Martin H. Floch

Secretin became the first gastrointestinal (GI) hormone identified in humans in 1902. Since then, a myriad of GI peptides have been identified as hormones. Some have been verified, others have not; regardless, their functions are important. GI hormones act in one of the following four ways:

1. *Endocrine function.* Epithelial cells secrete a substance into the circulation that acts at a distance.
2. *Autocrine function.* The substance secreted by the epithelial cell affects processes in the cell itself.
3. *Paracrine signaling.* The peptide secreted by the cell affects processes in adjacent cells.
4. *Neurocrine function.* Neurons secrete chemical transmitters with peptides into synapses or onto other cell types that signal neurotransmission.

Box 97-1 lists the peptides and hormonal actions identified in the GI tract. Although controversy surrounds many, the list includes and categorizes GI hormones confirmed by evidence from human and animal experimentation. **Figure 97-1** demonstrates the diffuse and integrating effects of the peptide hormone substances gastrin, cholecystokinin, and serotonin.

Some hormones have reached clinical significance. Produced by specialized G cells, *gastrin* regulates gastric secretion and has two major forms, G_{34} and G_{17}, formed primarily in the gastric

Box 97-1 Peptides with Action in the Gastrointestinal Tract

Gut Peptides That Function Mainly as Hormones
Gastrin
Ghrelin
Glucose-dependent insulinotropic peptide
Glucagon and related gene products (GLP-1, GLP-2, glicentin, oxyntomodulin)
Insulin
Leptin
Motilin
Pancreatic polypeptide
Peptide tyrosine tyrosine
Secretin

Gut Peptides That May Function as Hormones, Neuropeptides, or Paracrine Agents
Cholecystokinin (CCK)
Corticotropin-releasing factor
Endothelin
Neurotensin
Somatostatin

Gut Peptides That Act Mainly as Neuropeptides
Calcitonin gene-related peptide
Dynorphin and related gene products
Enkephalin and related gene products
Galanin
Gastrin-releasing peptide
Neuromedin U
Neuropeptide Y
Peptide histidine isoleucine or peptide histidine methionine
Pituitary adenylate cyclase–activating peptide
Substance P and other tachykinins (neurokinin A, neurokinin B)
Thyrotropin-releasing hormone
Vasoactive intestinal polypeptide

Peptides That Act as Growth Factors
Epidermal growth factor
Fibroblast growth factor
Insulin-like factors
Nerve growth factor
Platelet-derived growth factor
Transforming growth factor-β
Vascular endothelial growth factor

Peptides That Act as Inflammatory Mediators
Interferons
Interleukins
Lymphokines
Monokines
Tumor necrosis factor-α

Gut Peptides That Act on Neurons
CCK
Gastrin
Motilin

Modified from Liddle RA: Gastrointestinal hormones and neurotransmitters. In Feldman M, Friedman LS, Brandt LJ, editors: Gastrointestinal and liver disease, *ed 8, Philadelphia, 2006, Saunders-Elsevier, pp 3-25.*

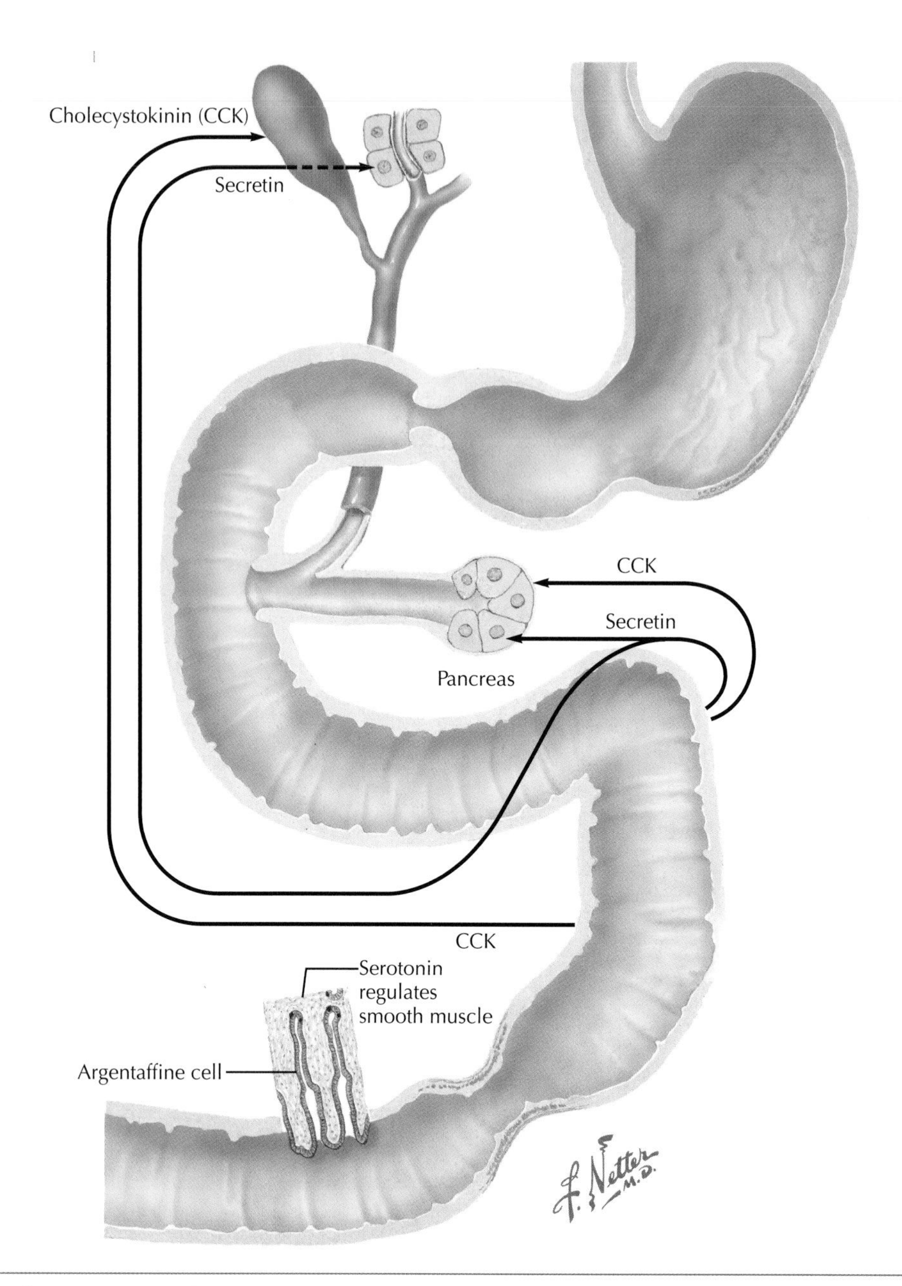

Figure 97-1 *Example of Gastrointestinal Hormone Physiology.*

antrum and secreted by enterochromaffin cells (see Chapters 38-41). Gastrin secretion into the circulation and production by G cells are stimulated by food, and gastrin acts directly on parietal cells to stimulate acid production. Proteins and high-protein foods have a greater influence than other nutrients on the production of gastrin and on the pH of the stomach. High acid production inhibits gastrin release, and a high gastric pH is a good stimulus for its secretion. Hypergastrinemia occurs in several pathologic states, including Zollinger-Ellison syndrome (see Chapters 41 and 57).

Secretin is a 27–amino acid peptide that stimulates pancreatic fluid and bicarbonate secretion (see Section VII). When the pH is raised in the duodenum, secretin release is inhibited. When gastric acid and chyme pass into the duodenum, secretin secretion is stimulated. Enteroendocrine cells, called *S cells*, produce secretin in the small intestine.

Cholecystokinin (CCK) is produced by I cells of the small intestine and is secreted into the blood when food passes into the small bowel. CCK has many actions, including gallbladder stimulation and stimulation of pancreatic secretion, and it helps regulate gastric and intestinal motility. Evidence also indicates that CCK induces satiety, so this hormone has many actions that help regulate feeding. The many forms of CCK include 33–, 58–, and 8–amino acid peptides, with all having biologically similar activities. CCK-A receptors reside primarily in the GI tract. The CCK-B receptor resides in the brain, however, and has been used clinically to stimulate pancreatic secretion in function tests. No disease is known to be associated with CCK.

Vasoactive intestinal polypeptide (VIP; also vasoactive intestinal peptide) has broad activity in the intestine. VIP acts as a vasodilator that increases blood flow, relaxes smooth muscle, and stimulates epithelial cell secretion. It also acts as a chemical messenger that is released from nerve terminals. VIP is chemically related to secretin and glucagons and is an important neurotransmitter. It is not produced by endocrine cells or the GI tract, but rather is produced and released from neurons. VIP has great clinical significance in certain watery diarrhea syndromes (e.g., Verner-Morrison), which demonstrate greatly increased VIP activity.

Glucagon is produced by pancreatic alpha cells and in the ileum and colon by L cells. It has several receptors, and it is known to participate in glucose homeostasis.

Epidermal growth factor (EGF) hormones are numerous throughout the GI tract and help regulate cell growth and activity. EGF hormones are complex, but their action appears to occur primarily through paracrine effects, although some growth factors may have autocrine action. EGF, the first growth factor discovered, is secreted from submaxillary glands and Brunner glands of the duodenum. EGF is believed to act with luminal cells of the GI tract to regulate cell proliferation and thus has an important trophic effect.

Recently recognized GI hormones are amylin, ghrelin, leptin, and guanylin/uroguanylin.

Other substances, as well as VIP, are now often associated with significant neural regulation of the GI tract. Particularly, serotonin and somatostatin and their agonists are used clinically. *Somatostatin* is found in interneurons and has an inhibitory effect by causing muscle relaxation. *Serotonin* (5-hydroxytryptamine, 5-HT) is found within the myenteric plexus and acts as a transmitter. Recently, 5-HT_4 and 5-HT_3 receptor agonists have been used in the treatment of irritable bowel syndrome (see Chapter 108).

ADDITIONAL RESOURCES

Bray GA: Afferent signals regulating food intake, *Proc Nutr Soc* 59:373-384, 2000.

Geoghen J, Pappas TN: Clinical use of gut peptides, *Ann Surg* 225:145-154, 1997.

Holst JJ, Fahrenkreig J, Stadile F, Rehfeld JF: Gastrointestinal endocrinology, *Scand J Gastroenterol* 216:S27-S38, 1996.

Liddle RA: Gastrointestinal hormones and neurotransmitters. In Feldman M, Friedman LS, Brandt LJ, editors: *Gastrointestinal and liver disease*, ed 8, Philadelphia, 2006, Saunders-Elsevier, pp 3-25.

Tulassay Z: Somatostatin and the gastrointestinal tract, *Scand J Gastroenterol* 228:S115-S121, 1998.

Imaging of the Small Intestine

98

Martin H. Floch

Although the small bowel is long and convoluted, imaging modalities allow its visualization (**Figs. 98-1** and **98-2**). Barium contrast studies, computed tomography (CT) with enterography, and several forms of endoscopy (direct, wireless capsule, double balloon) are used to visualize select areas.

BARIUM CONTRAST STUDIES

Approximately 16 oz of liquid with very fine, pulverized barium is given to the patient, with serial x-ray films taken before and immediately after the drink, then followed until the terminal

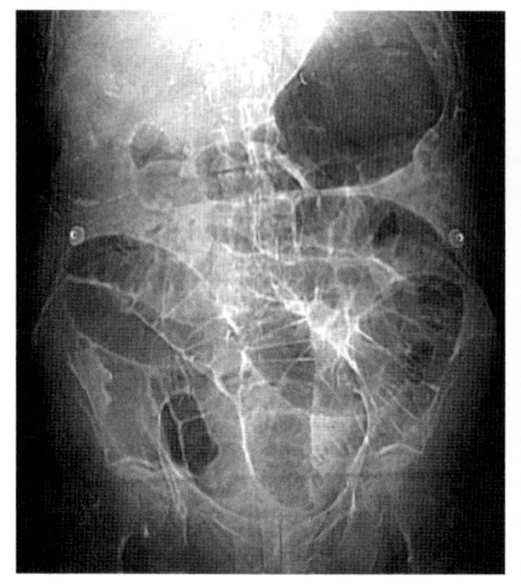

Radiograph demonstrating air-filled loops of distended small bowel.

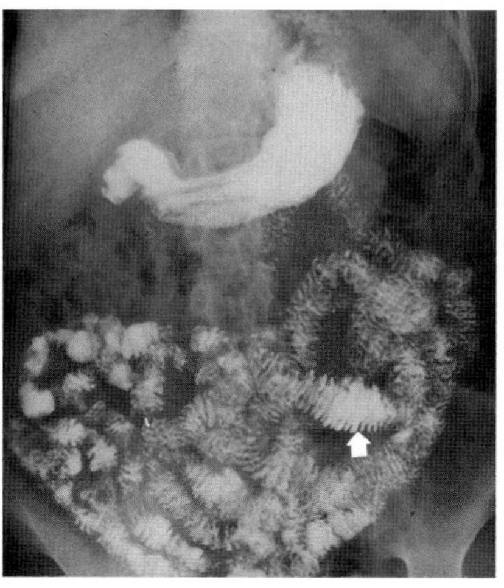

Barium contrast study demonstrating the normal feathery pattern of the jejunum. *Arrow* points to a filled loop with normal valvulae conniventes.

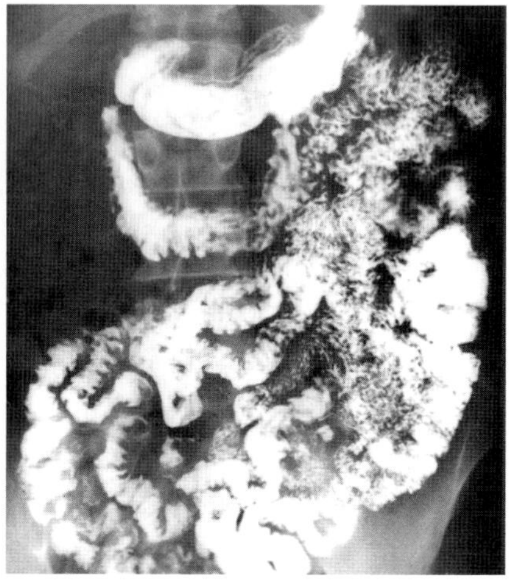

Barium contrast study of a patient with malabsorption syndrome demonstrating the loss of normal folds, clumping of barium, and separation of the meal.

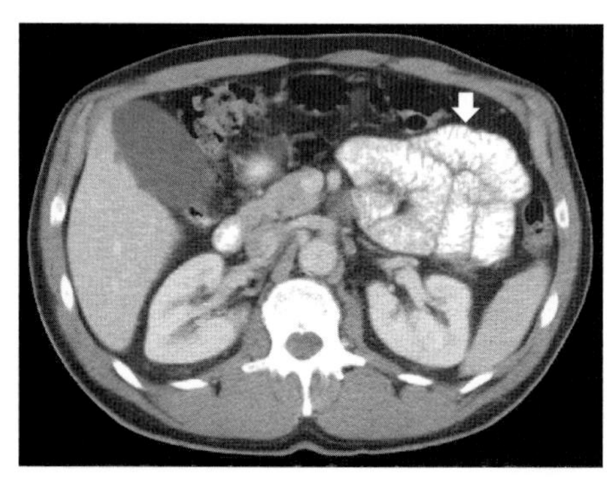

Contrast CT image demonstrating the normal appearance of filled loops of jejunum.

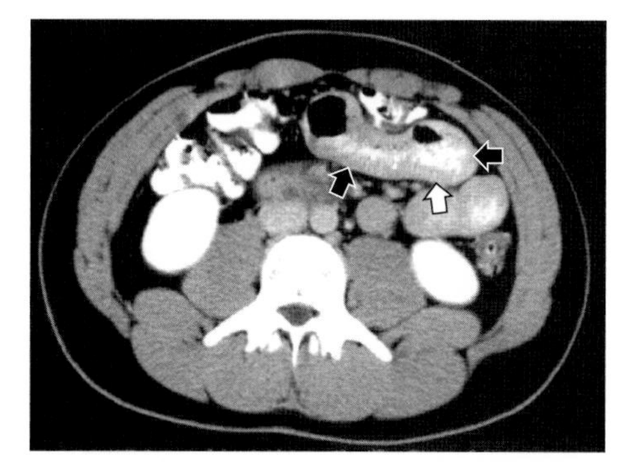

CT image of the abdomen of a patient with Henoch-Schönlein syndrome. The *arrows* point to an abnormal hemorrhagic loop of small bowel.

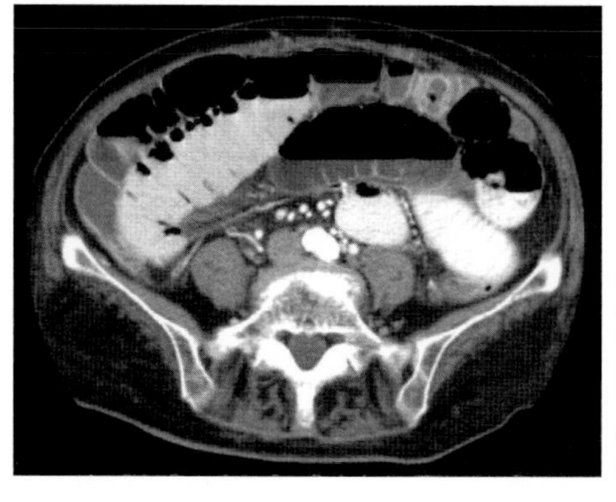

CT image of the abdomen of a patient with small bowel obstruction. Note the large dilated loops of small bowel with fluid levels.

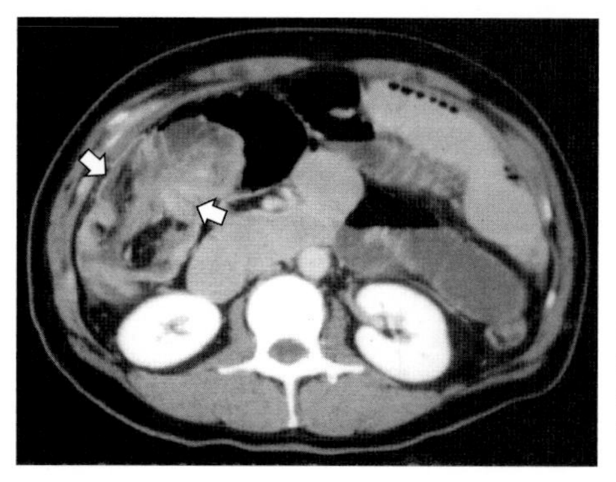

CT image of the abdomen demonstrating small bowel obstruction caused by intussusception *(arrows)*.

Figure 98-1 *Imaging Studies of the Small Intestine.*

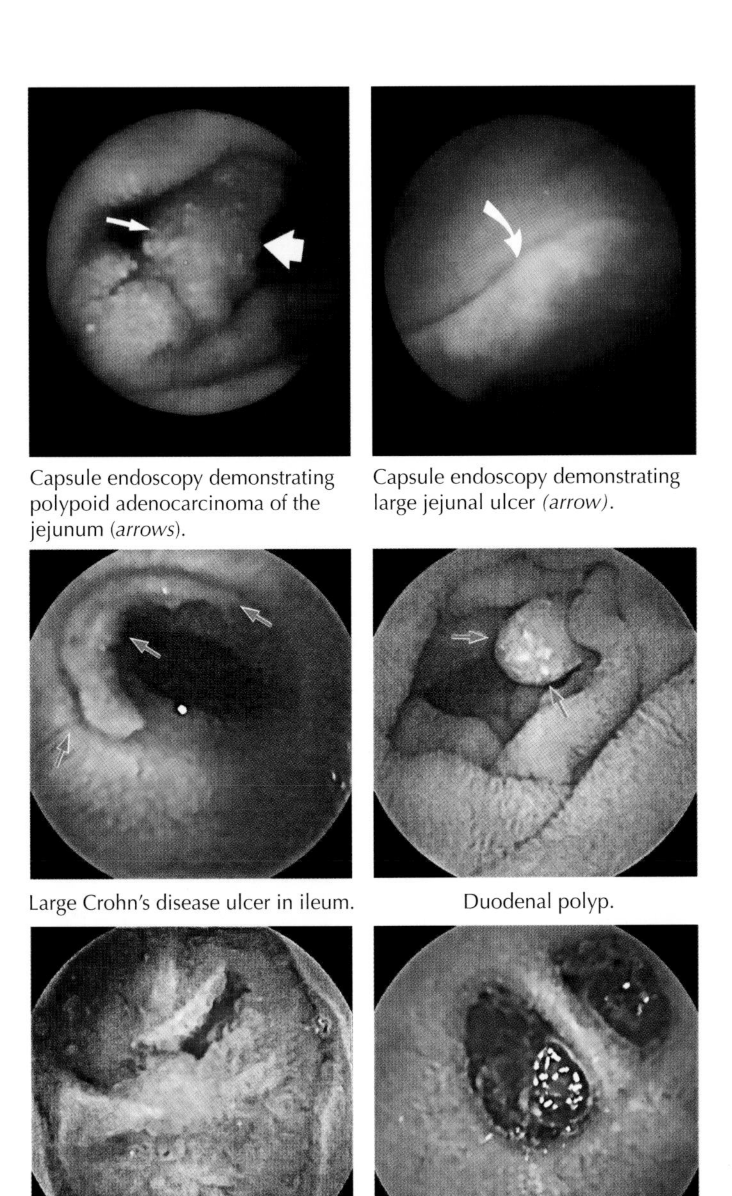
Capsule endoscopy demonstrating polypoid adenocarcinoma of the jejunum (*arrows*).

Capsule endoscopy demonstrating large jejunal ulcer (*arrow*).

Large Crohn's disease ulcer in ileum.

Duodenal polyp.

Small bleeding duodenal ulcer.

Opening to ileal diverticula.

Figure 98-2 *Wireless Capsule Endoscopy.*

ileum is visualized. The test usually is completed in 1 to 2 hours, but some patients have slow transit, and it may take several hours to visualize the terminal ileum. Under fluoroscopy, the radiologist can obtain the many views of the terminal ileal area. The jejunum has characteristic folds, and the ileum is flatter. The jejunum lies primarily in the left side and the ileum in the right side of the abdominal cavity. The terminal ileum has a normal appearance, at times likened to a bird's beak. The jejunum should measure no more than 3 to 3.5 cm in width and the ileum no more than 2.5 to 3 cm in its maximum width. Obstructive lesions, filling defects, irregular mucosa, and the so-called malabsorption pattern (scattered, dilated loops of bowel) can be diagnosed with barium studies. These are described under disease topics in this section.

Enteroclysis, or a small bowel enema, is performed at some centers by inserting a tube into the duodenum and then flooding it with barium contrast solution. This enables the radiologist to better control the timing of images and fluoroscopy so as to identify difficult lesions.

COMPUTED TOMOGRAPHY

CT is used extensively in many institutions. When a contrast dye is given before the procedure, the progress of the dye through the small bowel can be observed on CT scan, similarly discerning lesions seen on contrast barium study. However, CT also allows visualization of the thickness of the bowel wall and any inflammatory response surrounding the wall. Therefore, many radiologists prefer CT to barium contrast study for obtaining initial diagnosis. However, barium contrast can be more helpful if a specific intraluminal lesion must be delineated carefully. CT can also be extremely helpful in diagnosing appendicitis and evaluating the colon and possible inflammatory lesions.

ENDOSCOPY

Endoscopy is used to evaluate the most proximal part of the small bowel and the terminal small bowel through a *colonoscope*. Adding enteroscopes and capsule endoscopy has made visualization possible. The *push enteroscope* is a short scope that can be passed into the proximal or middle jejunum. If it is necessary to visualize the proximal small bowel, this instrument offers direct visualization. The *long enteroscope*, which can pass to the terminal ileum, has been available but has not gained wide use or been successful because it takes an inordinate amount of time to pass the instrument.

Wireless capsule endoscopy (WCE) is now reaching its potential. With this procedure, a small capsule containing a camera captures continuous images in a recorder. The patient swallows the capsule, and the images are recorded. The endoscopist or interpreter must then carefully review all the images. This is time-consuming but allows the identification of small lesions, such as bleeding arteriovenous malformations, as well as strictures and ulcerations. WCE has become the procedure of choice to identify obscure bleeding, early Crohn's disease, and small lesions of the bowel. It also complements other imaging studies, often identifying a lesion that is then carefully confirmed by either CT enterography or direct endoscopy (see Fig. 98-2).

Double-balloon endoscopy has not gained wide acceptance because it is a difficult, time-consuming technique. It is done either from above or through the terminal ileum and requires the use of balloons to tease the scope either caudad or aborally. Balloon endoscopy is used to confirm WCE findings or to identify occult gastrointestinal bleeding. Its advantage is that biopsy material can be obtained through the endoscope.

ADDITIONAL RESOURCES

Feigel DO, Cave D: *Capsule endoscopy*, Philadelphia, 2008, Saunders-Elsevier.

Gore RM, Levine MS: *Textbook of gastrointestinal radiology*, ed 3, Philadelphia, 2008, Saunders-Elsevier.

Gross SA, Stark ME: Initial experience with double-balloon enteroscopy at a U.S. center, *Gastrointest Endosc*, 67(6):898-901, 2008.

Gustout CJ: Clinical update: capsule endoscopy, *Gastrointest Endosc* 10:1-4, 2002.

Lee JKT, Sagel SS, Stanley RJ, Heiken JP: *Computed tomography with MRI correlation*, ed 3, Philadelphia, 1998, Lippincott-Raven.

Tanaka S, Mitsui K, Tatsuguchi A, et al: Current status of double-balloon endoscopy: indications, insertion route, sedation, complications, technical matters, *Gastrointest Endosc* 66(3 Suppl):S30-S33, 2007.

Vascular Supply and Drainage in the Small Intestine

99

Martin H. Floch

The blood supply to the small and large intestines is extremely variable and, in many cases, uncertain and unpredictable. Variations in the origin, course, anastomoses, and distribution of the intestinal vessels are so common and significant that the conventional textbook descriptions are inadequate and, in many respects, even misleading, similar to descriptions regarding the blood supply of the upper abdominal organs. It is important for surgeons working in this area and for radiologists interpreting angiograms to understand these variations, as detailed in comprehensive anatomy and radiology texts.

ARTERIAL CIRCULATION

The *superior mesenteric artery* arises from the front of the aorta, typically at the level of mid-L1 (first lumbar vertebra), but as far down as the upper third of L2. The distance between the origin of the celiac and superior mesenteric arteries is usually 1 to 6 mm but varies from 1 to 23 mm. Thus, contiguous origins of the two vessels are often found, but a common origin from a celiac-mesenteric trunk is rare.

The superior mesenteric artery, passing downward and forward and swinging to the left, particularly in its lower third, gives off a variable number (13-21) of intestinal arteries from its convex (left) side, ranging from three to seven (average five) above and 8 to 17 (average 11) below the origin of the ileocolic artery. The first group supplies the jejunum, and the second supplies part of the jejunum and the entire ileum. Intestinal arteries for the jejunum and ileum, running between the layers of the mesentery, follow the pattern shown in **Figure 99-1**.

Each vessel courses fairly straight, for a variable distance, before it divides into branches that unite with branches from the adjacent primary stem vessels to form a series of anastomosing arches, the *arterial arcades*. From these primary arcades arise the secondary and shorter intestinal arteries, which in turn form secondary arcades. Further arcades, although smaller, are formed similarly, essentially by the more distal arteries. In the terminal arcades, small, straight vessels (arteriae rectae) arise. Except for the blood supply of the first part of the duodenum, where the first arcade is small with short arteriae rectae, the jejunal arteries are long, have a large caliber, and establish primary and secondary arcades, from which arise multiple long arteriae rectae. Stem arteries for the ileum become progressively shorter, the arcades become smaller, and the arteriae become less elongated.

The vascularization pattern of the jejunum is so characteristically different from the ileum that, through simple inspection of the gut, the examiner can usually distinguish between jejunum and ileum. The jejunum has a thicker wall and a greater digestive surface than the ileum and receives the larger intestinal branches.

The first jejunal branch of the superior mesenteric origin may be large (6 mm in diameter) and may have four large arcades forming branches 6 to 8 cm long and 3 to 4 mm in diameter. In many cases, however, the first jejunal branch is small (1-2 mm) and is anastomosed with the inferior pancreaticoduodenal artery or shares a common origin. A large primary jejunal artery may be followed by a slender second jejunal artery. The distribution and caliber of the intestinal branches of the superior mesenteric artery vary in the same person; large and small branches alternate without rule or order. Although the first and second jejunal arteries are thought to communicate through an arcade, such an arcade is missing in many patients, in whom the first jejunal artery is found to have no connection with the second.

Anatomic studies show great variation in the arteriae rectae as they pass from the arcades to the walls of the small intestine, entering directly, overlapping, or forming small arcades (see Additional Resources). However, the result is a rich blood supply to the small intestine, where the biologic need for absorption is served.

VENOUS DRAINAGE

In number, point of origin, and mode of distribution, the veins involved in the drainage of the small intestine follow the same design as the corresponding arteries (**Fig. 99-2**). Accordingly, the veins have been given the same terminology. An exception is the *superior mesenteric vein*, in that it reaches the right gastroepiploic vein just before entering the portal vein. Other tributaries of the superior mesenteric vein are concordant with the arteries of the same name, which leave the superior mesenteric artery.

In the region where the left colic and upper sigmoid arteries originate from the inferior mesenteric artery, the corresponding vein follows a course of its own, separating from the respective artery. The inferior mesenteric vein takes a straight-upward course, ascending behind the peritoneum, over the psoas muscles, and to the left of the fourth portion of the duodenum. The vein continues behind the body of the pancreas to enter most frequently (38% of observed cases) the splenic vein from the latter's union with the superior mesenteric vein (i.e., origin of portal vein). In 29% of persons, the inferior mesenteric vein enters the superior mesenteric vein, and in 32%, it joins the latter and the splenic vein at the junction. In a few persons, a second inferior mesenteric vein has been found. The portal vein, especially variations of its tributaries, is extremely important (see Section IX).

LYMPH DRAINAGE

The intramural lymph vessels of the small intestine begin with the central lacteals of the villi. At the base of the villi, the central lacteals join with the lymph capillaries, draining the region of the crypts of Lieberkühn, thus forming a fine network within the tunica propria, in which the first lymphatic valves are

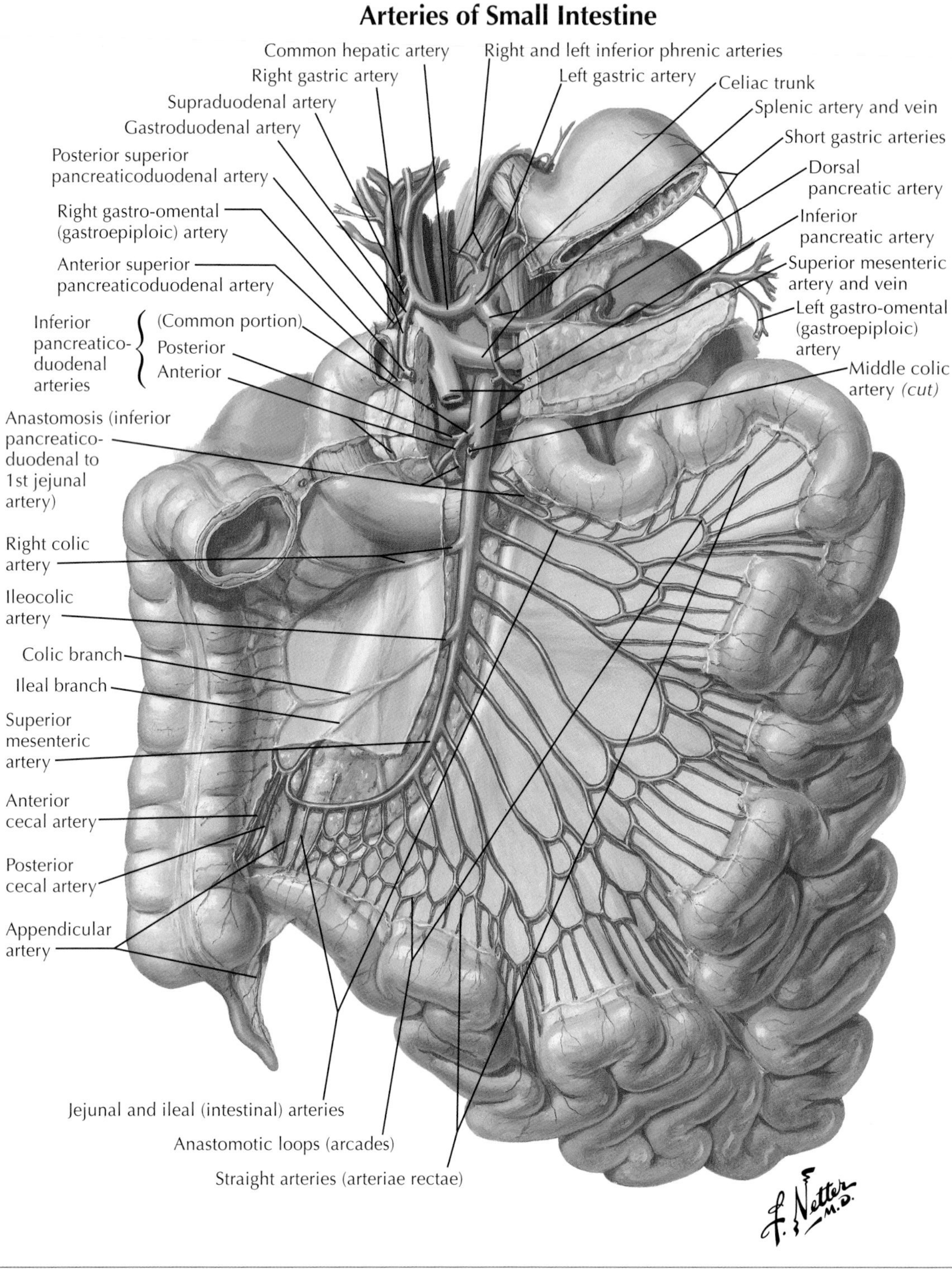

Arteries of Small Intestine

Common hepatic artery
Right and left inferior phrenic arteries
Right gastric artery
Left gastric artery
Celiac trunk
Supraduodenal artery
Splenic artery and vein
Gastroduodenal artery
Short gastric arteries
Posterior superior pancreaticoduodenal artery
Dorsal pancreatic artery
Right gastro-omental (gastroepiploic) artery
Inferior pancreatic artery
Anterior superior pancreaticoduodenal artery
Superior mesenteric artery and vein
Inferior pancreatico-duodenal arteries (Common portion) Posterior Anterior
Left gastro-omental (gastroepiploic) artery
Middle colic artery *(cut)*
Anastomosis (inferior pancreatico-duodenal to 1st jejunal artery)
Right colic artery
Ileocolic artery
Colic branch
Ileal branch
Superior mesenteric artery
Anterior cecal artery
Posterior cecal artery
Appendicular artery
Jejunal and ileal (intestinal) arteries
Anastomotic loops (arcades)
Straight arteries (arteriae rectae)
F. Netter M.D.

Figure 99-1 *Arterial Circulation of the Small Intestine.*

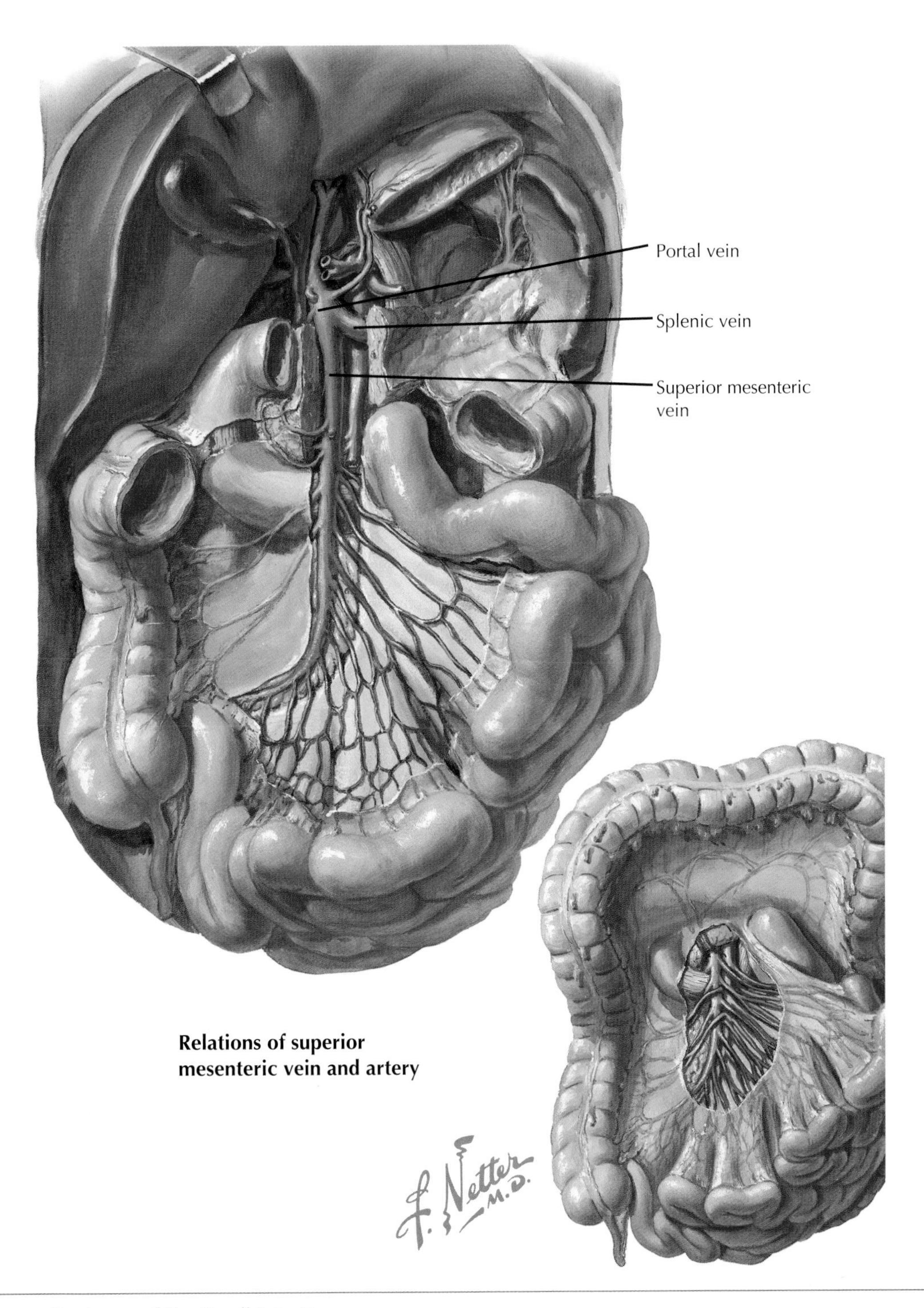

Figure 99-2 *Venous Drainage of the Small Intestine.*

already encountered. Many minute branches emerge from this network, penetrating through the muscularis mucosae into the submucosa, where a further network of lymphatic vessels spreads. From this network, in which valves are a conspicuous feature, large lymph vessels receive lymph from the muscle layers and from the serosa and subserosa and pass to the line of attachment of the mesentery, where, together with the arteries and veins, the lymph vessels leave the intestinal wall to enter the mesentery. Lymph vessels of the small intestine have long been referred to as *lacteals* or *chyliferous vessels* because they transport absorbed fat in emulsified form and therefore appear as milky-white threads after the ingestion of fat-containing food.

Lymph vessels of the mesentery drain through masses of *mesenteric lymph nodes*, which number approximately 100 to 200, and constitute the largest aggregate of lymph nodes in the body. They increase in number and size toward the root of the mesentery. In the root of the mesentery, larger lymphatic branches are situated that lead into the superior mesenteric nodes where the superior mesenteric artery arises from the aorta. From the duodenum, the lymph vessels—some of which run through the pancreatic tissue (see Section VII)—pass the lymph nodes lying cranial, caudal, and dorsal to the head of the pancreas. Of these, the upper are known as the *subpyloric* and right *suprapancreatic* nodes, the lower as the *mesenteric root* nodes, and the dorsal as the *retropancreatic* nodes. Lymph flows from these various nodes into the group of celiac lymph nodes.

From the superior mesenteric nodes and the celiac nodes, the lymph passes through the short intestinal or gastrointestinal lymph trunk, which is sometimes divided, like a river delta, into several smaller, parallel trunks. Lymph then enters the *cisterna chyli*, a saclike expansion of the beginning of the thoracic duct. The intestinal trunk drains not only the entire small intestine but also all organs whose lymph is collected in the celiac and superior mesenteric lymph nodes, especially the stomach, liver, pancreas, and extensive portions of the large intestine.

ADDITIONAL RESOURCES

Kornblith PL, Boley SJ, Whitehouse BS: Anatomy of the splanchnic circulation, *Surg Clin North Am* 72:1-30, 1992.

Rosenblum JD, Boytle CM, Schwartz IB: The mesenteric circulation: anatomy and physiology, *Surg Clin North Am* 77:289-306, 1997.

Innervation of the Small and Large Intestines

Martin H. Floch

The nerves supplying the small and large intestines contain sympathetic and parasympathetic efferent and afferent fibers (**Figs. 100-1** and **100-2**). These nerves have branches of the celiac plexus, the superior and inferior mesenteric plexuses, and the superior and inferior hypogastric plexuses. The *hypothalamus* is a source and terminus of pathways involved in visceral activities, with extensive cortical connections to the premotor areas of the frontal cortex, the cingulate gyrus, and the orbital surfaces of the frontal lobes. Descending fibers important in parasympathetic functioning arise mainly from the anterior region of the hypothalamus and form synapses with cells in the dorsal vagal nuclei and in the second to fourth sacral segments of the spinal cord (S2-S4). Axons of these cells constitute the *preganglionic* (efferent) fibers in the vagal and pelvic splanchnic nerves, which are distributed to many viscera. Vagus nerves supply those parts derived from the foregut and midgut, and the pelvic splanchnic nerves innervate the parts derived from the hindgut. Intestinal preganglionic fibers carried to the vagal and pelvic splanchnic nerves terminate by relaying around the ganglia cells in the enteric plexuses, and the axons of these

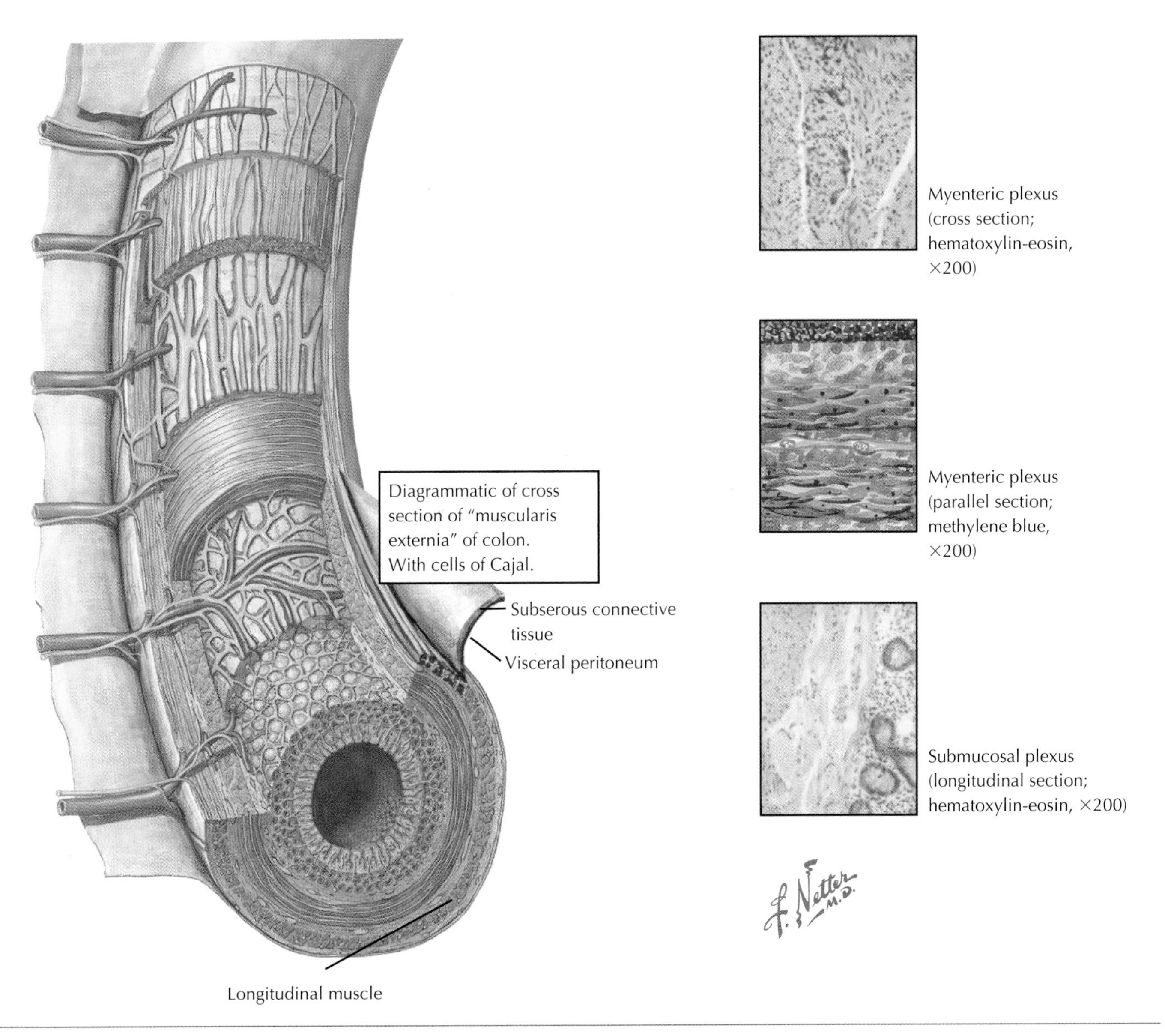

Figure 100-1 *Innervation of the Intestines.*

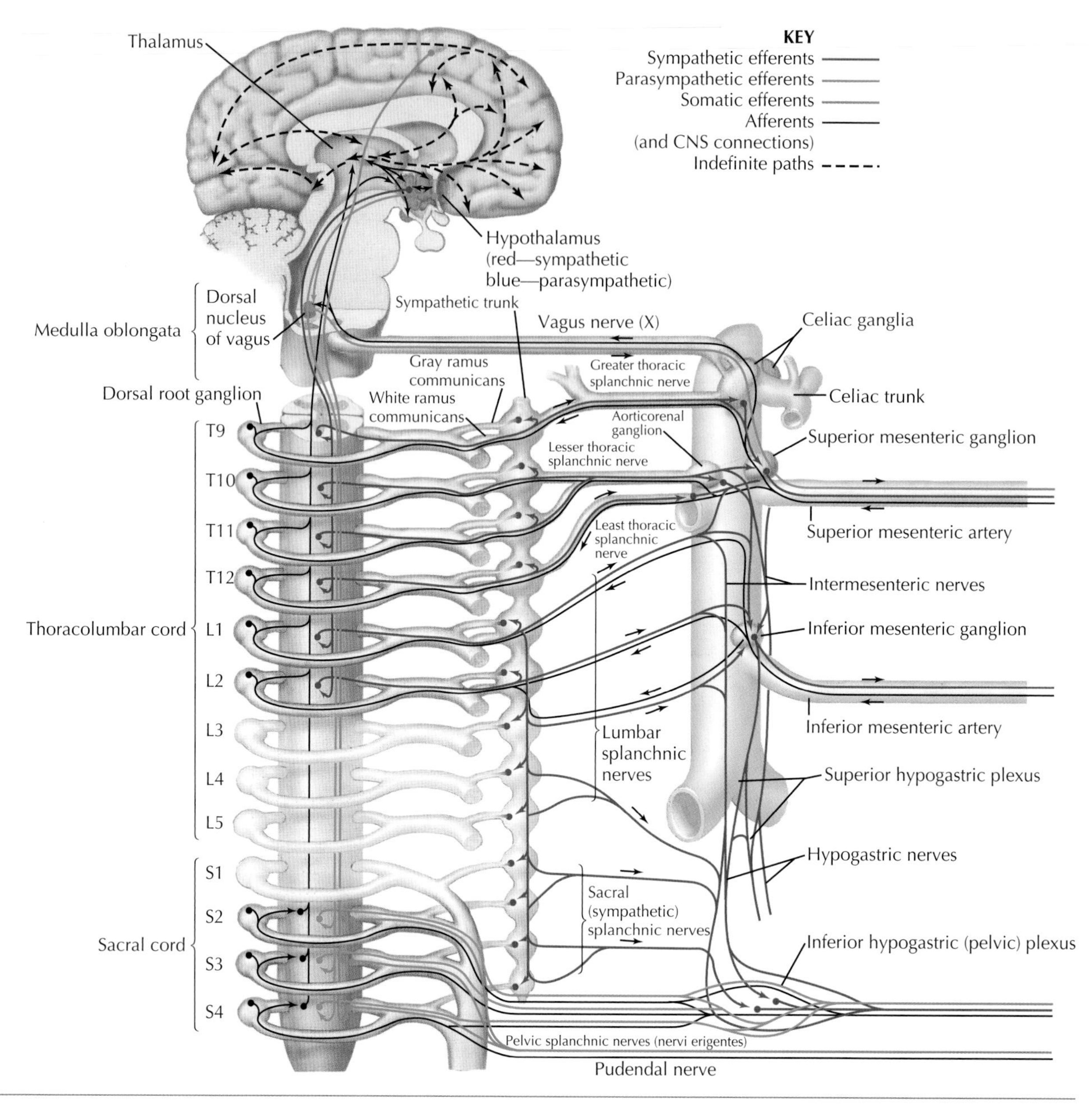

Figure 100-2 *Intestinal Innervation: Efferent and Afferent Pathways and Plexuses.*

ganglionic cells become the *postganglionic* parasympathetic fibers, which, together with the corresponding sympathetic fibers, serve the smooth muscle of the intestinal wall, the intramural vessels, and the intestinal glands.

Fibers descending from the central nervous system (CNS), carrying sympathetic impulses on intestinal activities, relay around lateral cornual cells in the four or five lowest thoracic and the two or three upper lumbar segments of the spinal cord. Axons of these cells, representing the preganglionic sympathetic fibers, emerge from the ventral nerve roots of the corresponding segments and pass in white rami communicantes to the adjacent ganglia of the sympathetic trunks. Some fibers relay within these ganglia, whereas others traverse the trunk uninterruptedly, leaving it in medially directed branches as thoracic, lumbar, or sacral splanchnic nerves, which end in the plexuses previously cited to enter synapses with ganglionic cells. Axons of these cells, the postganglionic fibers, accompany the branches of the various arteries supplying the intestine.

The chief segmental sources of the sympathetic fibers innervating different regions of the intestinal tract are indicated in Figure 100-2, but because of overlap, minor contributions may derive from adjacent segments.

Certain alimentary functions are probably controlled by simple reflex arcs located in the intestinal wall, but other reactions are mediated through more elaborate reflex arcs involving the CNS and consisting of the usual afferent, internuncial, and efferent neurons. Numerous *afferent* fibers of relatively large caliber traverse the enteric plexus without relaying and are carried centripetally through approximately the same sympathetic splanchnic and parasympathetic nerves that transmit the preganglionic, or *efferent*, fibers. Afferent fibers are the peripheral processes of pseudo-unipolar cells in the inferior vagal

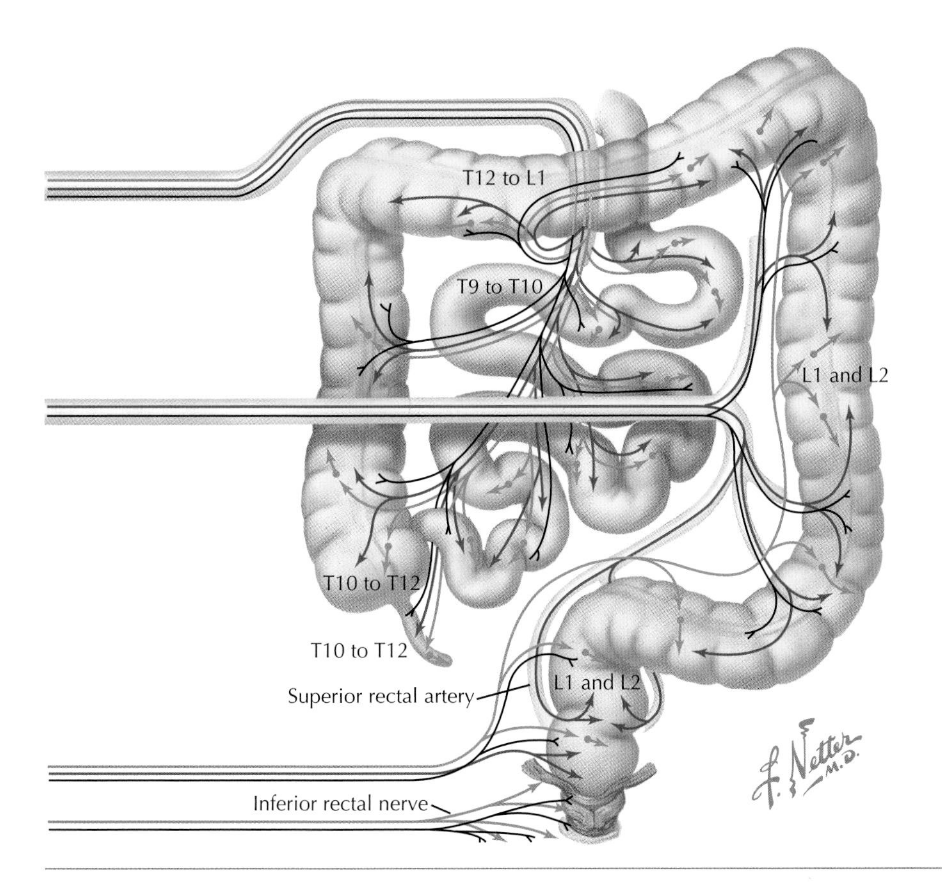

ganglia or in the dorsal root ganglia of those spinal ganglia that carry preganglionic intestinal fibers. Central processes enter the brainstem or the spinal cord.

Although insensitive to ordinary tactile, painful, or thermal stimuli, the intestines respond to tension, anoxia, chemicals, and other stimuli. Specialized cutaneous nerve endings in the intestine are absent, except for the Vater-Pacini (pacinian) corpuscles in the adjacent mesentery. As with the efferent fibers, the exact mode of termination of the visceral afferent fibers remains controversial, but *whorl*, *skein*, *great*, *looplike*, and *free* endings have been described in the mucosal, muscular, and serosal coats.

Intrinsic innervation is affected through the enteric plexus in the alimentary tract from the esophagus to the rectum. This plexus consists of small groups of nerve cells interconnected by networks of fibers, and it is subdivided into the *myenteric* (Auerbach) *plexus* and the *submucosal* (Meissner) *plexus*. The Auerbach plexus is relatively coarse, with thicker meshes and larger ganglia at the intersections than the Meissner plexus, which consists of fine meshes with small ganglia. The myenteric plexus lies in the interval between the circular and longitudinal muscular coats and the main (primary) meshes and gives off fascicles of fibers that form finer secondary, and even finer tertiary, plexuses and that ramify within and between the adjacent layers of muscle. Some fibers from the longitudinal intramuscular plexus enter the subserous plexus and constitute a rarified subserous plexus. The submucosal plexus is also subdivided into more superficial and deeper fibers. *Interstitial cells of Cajal* are a network of non-neuronally derived cells from smooth muscles; those in the intramuscular, myenteric plexus and submucosal layers are interconnected. Interstitial cells of Cajal are proving important in understanding motility disturbances.

Nerve bundles contain postganglionic sympathetic, preganglionic and postganglionic parasympathetic, and afferent fibers in addition to elongated dendrites. There is a rich network of dendrites.

The *superior mesenteric plexus* is a continuation of the lowest part of the celiac plexus and surrounds the origin of the superior mesenteric artery. It is interconnected by stout filaments to the celiac and aorticorenal ganglia. The large superior mesenteric ganglia is located usually just above the root of the artery and is incorporated in the commencement of the superior mesenteric plexus. The main plexus divides into subsidiary plexuses corresponding to all the branches of the artery (inferior pancreatic duodenal, jejunal, ileal, ileocolic, right and middle colic), and it innervates those parts of the intestine indicated by their names. Nerves and arteries follow the same route, except for the patterns by which they approach the gut wall. Vessels advance toward the wall and form characteristic arcades, but nerves pass straight outward without arcade formation. This rich network of ganglia and plexuses is distributed throughout the small and large bowel and corresponds to the arteries.

The *superior hypogastric plexus* (presacral nerves), situated in front of the dichotomized aorta and between the divergent common iliac arteries, is a flattened band of intercommunicating nerves extending from the level of the lower border of the third lumbar vertebra (L3) to the upper border of the sacrum, where it ends by dividing into the right and left groups of hypogastric nerves. These nerves are then distributed through the inferior hypogastric plexuses.

Specialized sensory endings exist in the part of the anal canal that develops from the proctodeum and that is supplied by the inferior hemorrhoidal nerve. Sensory endings are absent in the region above the Hilton white line, whereas the afferent fibers end by breaking up to form fibrils or delicate plexuses between the epithelial cells. Thus, below the pectin, this innervation resembles that of the skin, whereas above the pectin, the mucosa is supplied by sympathetic nerves derived from the inferior mesenteric and inferior hypogastric plexuses, following the paths of the hemorrhoidal arteries and the parasympathetic fibers from the pelvic splanchnic nerves. All these nerves convey efferent and afferent fibers to and from the terminal part of the gut. In accord with this difference in nerve supply of the anoderm are the differing sensory responses. The lower part, supplied by somatic nerves, is sensitive to tactile, painful, and thermal stimuli, whereas the upper part of the anal canal is almost insensitive to such stimuli but responds readily to alternations in tension. From a practical point of view, this neuroanatomic situation explains why an anal fissure is so painful and why, with an injection for hemorrhoids, the puncture is scarcely felt if the needle is inserted through the mucosa.

ADDITIONAL RESOURCES

Cook IJ, Brooks SJ: Colonic motor and sensory function and dysfunction. In Feldman M, Friedman LS, Brandt LJ, editors: *Gastrointestinal and liver disease*, ed 8, Philadelphia, 2006, Saunders-Elsevier, pp 2111-2126.

Visceral Reflexes

Martin H. Floch

101

Visceral reflexes explain a number of clinical signs and symptoms. Afferent impulses from the hypertonic sigmoid initiate reflexes to cranial structures, to the bronchial tree, to the stomach, and to the abdominal skin.

The afferent limb of *viscerosomatic reflexes*, originating from viscera and affecting somatic structures, may be by way of sympathetic or parasympathetic nerves. The efferent limb is usually through somatic nerves or autonomic paths. In *viscerovisceral reflexes*, the afferent and efferent limbs may contain both sympathetic and parasympathetic nerves, but they may be mediated by the intrinsic nerve plexuses only. *Somatovisceral reflexes* involve somatic afferents and sympathetic or parasympathetic efferents to the viscus.

Because they lack a true efferent limb to the arc, *viscerosensory reflexes* are not true reflexes and are believed to result from a shunt or transfer of sensory impulses from autonomic afferents to somatic afferents. Exactly where the shunt or transfer takes place is conjectural. Viscerosensory reflexes explain the phenomena of *referred pain* and *skin hyperalgesia.* In the case of sympathetic reflexes, hyperalgesia occurs in skin areas innervated by the same spinal segment from which the nerve supply of the diseased viscus derives, and in the case of parasympathetic reflexes, it may manifest in more remote areas.

Figure 101-1 illustrates some of the major visceral reflexes. These are essential in clinical medicine because they explain why somatic or sensory stimuli can cause gastrointestinal symptoms. Pain from muscle or bone may cause vomiting; psychologic or sensory stimuli can cause diarrhea; and abdominal pain can cause headache. The clinician must remember that the symptom may come from a distant stimulus, as listed in **Table 101-1**.

Table 101-1 Origin, Effect, and Clinical Significance of Visceral Reflexes

Reflex	Origin	Effect	Clinical Significance
Viscerosomatic			
Visceromuscular	Diseased abdominal organ	Contraction of voluntary muscles and erectores pili muscles innervated by corresponding spinal segment; also neck and laryngeal muscles	Involuntary guarding suggests underlying visceral irritative process.
Visceroglandular	Diseased abdominal organ	Sweating in area of corresponding dermatomes	Aids in identifying level of visceral involvement.
Viscerovascular	Diseased abdominal organ	Dilatation of blood vessels; dermographia; sense of warmth in corresponding dermatomes	Aids in identifying level of visceral involvement.
Viscerosensory	Diseased abdominal organ	Hyperalgesia in corresponding dermatomes	In absence of distention, explains tenderness and intolerance of tight garments.
Viscerovisceral			
Gastroileocolic and duodenoileocolic	Food entering stomach and duodenum	Stimulation of ileac and colic motility	Accounts for postcoffee defecation reflex; postprandial distress in irritable colon syndrome.
Esophagosalivary and gastrosalivary	Esophagus and stomach	Paroxysmal sialorrhea	Clue to neoplasm of esophagus.
Enterogastric	Distention or irritation of enteric canal	Inhibition of stomach; antral spasm	One of the mechanisms of indigestion; biliousness; nausea.
Cologastric	Distention or irritation of colon	Inhibition of stomach; antral spasm	Instigating epigastric distress in irritable colon syndrome; vomiting in appendicitis.
Urinary tract–gut	Disease of urinary tract	Inhibition and distention of gut	Acute abdominal symptoms may be of genitourinary origin.
Viscerocardiac	Disease of GI organs	Diminution of coronary flow; changes in heart rhythm and rate	Myocardial disturbances (tachycardia, bradycardia, arrhythmia) may occur in GI disorders.
Visceropulmonary	Disease of GI organs	Spasm of bronchioles	Accounts for sense of difficult breathing in irritable colon syndrome.

GI, *gastrointestinal.*

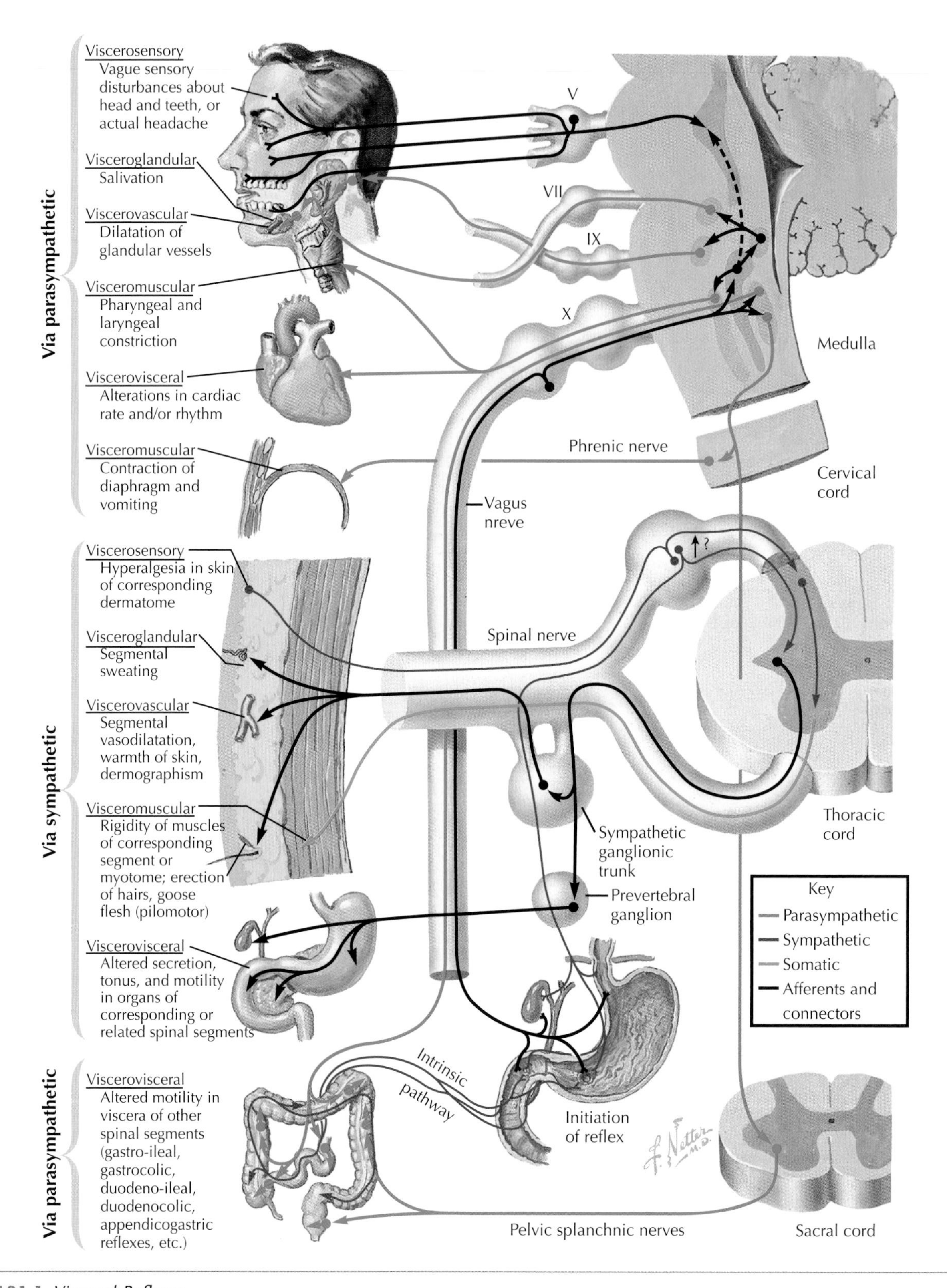

Figure 101-1 *Visceral Reflexes.*

ADDITIONAL RESOURCES

Glasgow RE, Mulvhill SJ: Acute abdominal pain. In Feldman M, Friedman LS, Brandt LJ, editors: *Gastrointestinal and liver disease*, ed 8, Philadelphia, 2006, Saunders-Elsevier, pp 87-98.

Kuo B: Chronic abdominal pain. In Feldman M, Friedman LS, Brandt LJ, editors: *Gastrointestinal and liver disease*, ed 8, Philadelphia, 2006, Saunders-Elsevier, pp 99-108.

Congenital Abnormalities of the Small Intestine

102

Martin H. Floch

Congenital lesions develop in the gastrointestinal (GI) tract and may cause intestinal obstruction (**Fig. 102-1**). Almost all presentations are in newborns and necessitate immediate surgery. The most common site of complete obstruction or *atresia* (congenital absence or closure) is in the ileum. The duodenum is the second most common site. Obstruction may result from atresia, malrotation of the colon, volvulus, meconium ileus, or imperforate anus.

CLINICAL PICTURE

Vomiting, absence of stool, and abdominal distention are the clinical triad indicating a significant problem in an infant. Certain atresias are evident within the first 24 hours. Malrotation, volvulus, and meconium ileus may manifest immediately. An infant with an imperforate anus produces no stool; diagnosis should be made on initial examination of the newborn.

DIAGNOSIS

Careful examination of the infant and, if necessary, radiographic examination reveal the obvious diagnosis. At times, a barium contrast study (upper or lower) may be necessary to prove the diagnosis.

TREATMENT AND MANAGEMENT

Intestinal obstruction caused by a congenital lesion is life threatening, and surgical intervention is rapidly needed. Atresia requires end-to-end anastomosis to maintain the continuity of the bowel. A malrotation or volvulus requires cutting of the mesentery, and an imperforate anus requires surgery to create an outlet.

Meconium ileus may present differently than the other forms of congenital obstruction in that the mass of meconium can create irregular loops of distended bowel rather than obstruction at the site of the occlusion, such as in atresia. Meconium ileus also may develop in infants born with fibrocystic disease of the pancreas. Less than 10% of infants with cystic fibrosis develop meconium ileus. However, it may occur within the first few months, and the presentation may not be as immediate as with other forms of congenital intestinal obstruction.

PROGNOSIS

If the surgeon is able to perform a corrective procedure, the prognosis is good. However, the problems of infancy and childhood created by a congenital defect may affect growth and development. Early surgery can cause great psychologic stress in the family and can result in long-term psychosocial problems.

Other anomalies of the GI tract, such as abdominal wall defects, intestinal duplication, mesenteric cysts, and omphalomesenteric cysts, are rare. Hirschsprung disease is discussed separately in Section V.

ADDITIONAL RESOURCES

Kahn E, Daum F: Anatomy, histology, embryology, and developmental anomalies of the small and large intestine. In Feldman M, Friedman LS, Brandt LJ, editors: *Gastrointestinal and liver disease*, ed 8, Philadelphia, 2006, Saunders-Elsevier, pp 2061-2091.

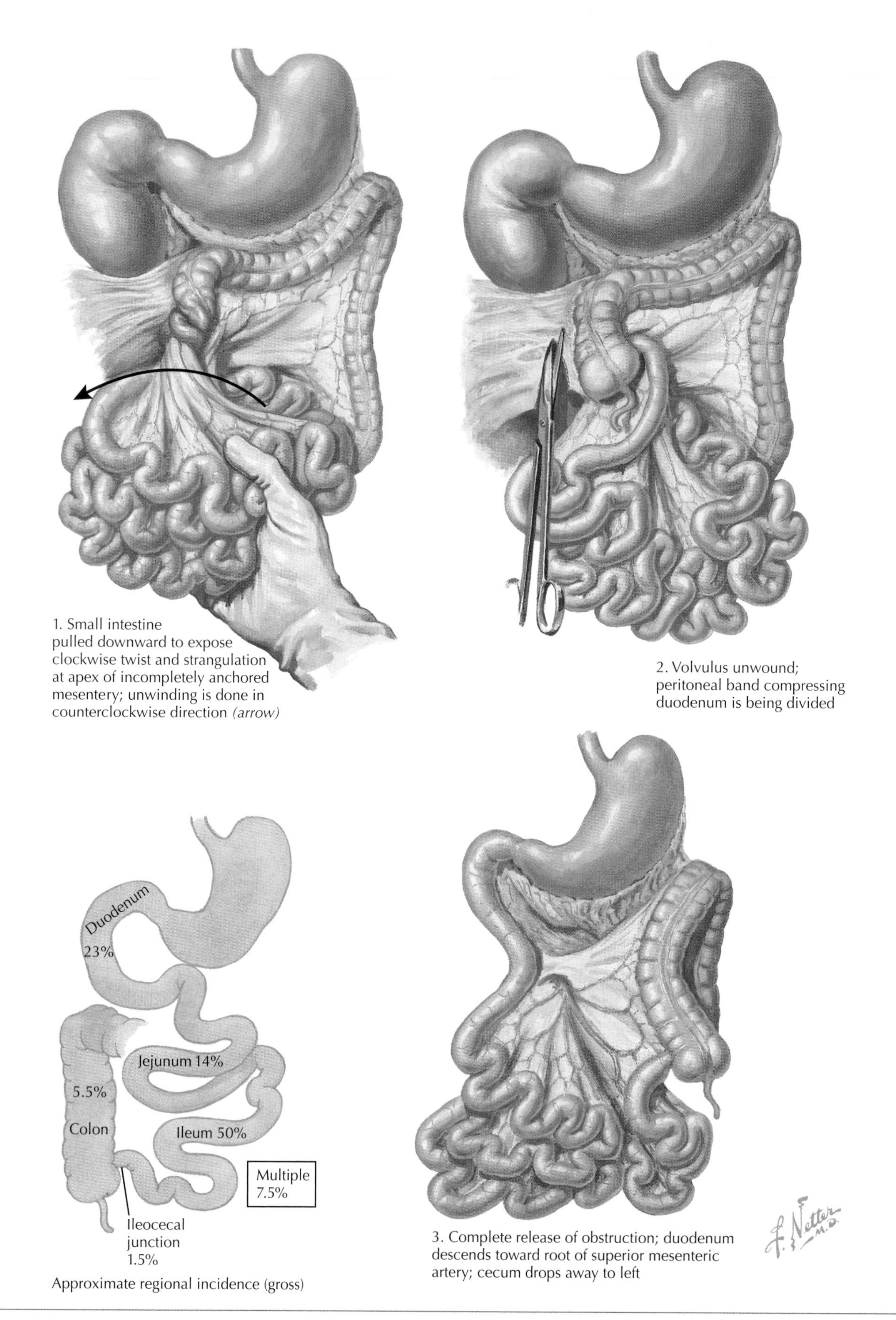

Figure 102-1 *Congenital Intestinal Abnormalities, Including Malrotation of the Colon with Volvulus of the Midgut.*

Meckel Diverticulum

103

Martin H. Floch

The yolk sac is connected to the primitive tubular gut by the *vitelline* (omphalomesenteric) *duct* in early embryonic stages and is normally obliterated about the seventh week of fetal life. Failure of the duct to disappear results in a variety of remnants; the most common presentation is a *sacculation*, or pouch, attached to the ileum and best known as Meckel diverticulum (**Fig. 103-1**).

Meckel diverticulum is the most frequent congenital anomaly of the gastrointestinal tract and occurs in 1% to 3% of the population. It is located 30 to 90 cm proximal to the ileocecal junction and is always attached to the antimesenteric side of the ileal wall. The diverticulum varies from 1 to 10 cm long and 1 to 3 cm wide. In contrast to the *acquired* intestinal diverticulum, the wall of Meckel diverticulum is composed of all the layers and is thus a true diverticulum. Mucosal lining usually corresponds to that of the ileum, but it may contain ectopic gastric mucosa, or nodules of pancreatic tissue, which can cause serious complications. The rest of the vitelline duct is obliterated in most patients, but it might remain as a fibrous band. The persistence of the entire vitelline duct as a permanent tube leads to an umbilical-intestinal fistula that is usually discovered in infancy.

CLINICAL PICTURE

The clinical picture varies greatly, depending on the complications. Most patients with Meckel diverticulum have no symptoms and no complications. When patients become symptomatic, however, their symptoms vary with the condition. A typical presentation is intestinal bleeding, which is more common in children and often manifests with maroon stools. If the bleeding is slow, melena may be present. Peptic ulceration has been reported when ectopic gastric mucosa is present in the diverticulum. Strangulation, intussusception, torsion, incarceration of Meckel diverticulum into a hernia, and adhesions with obstruction caused by the hernia have all been reported. These findings may present a clinical picture of acute abdomen or chronic pain. Also, neoplasm has been reported to develop in the diverticulum, and all the complications of an intraabdominal neoplasm may be present.

DIAGNOSIS

Diagnosis can be made by technetium-99m pertechnetate imaging. However, if Meckel diverticulum is suspected, computed tomography (CT) is often performed, and the diagnosis is made at canning. The differential diagnosis includes appendicitis, cholecystitis, diverticulitis, salpingitis, and any other inflammatory condition leading to a colonic or gastroduodenal lesion or pathologic condition of the small bowel. Although the diagnosis is made most frequently through CT, the surgeon is often surprised at laparotomy.

TREATMENT AND MANAGEMENT

The treatment of any symptomatic Meckel diverticulum is surgical resection. It is now possible to reach the lesion easily by laparoscopy, and simple resection is possible.

COURSE AND PROGNOSIS

The prognosis for Meckel diverticulum is excellent for all the abnormalities except neoplasia, which depends on the type of lesion.

ADDITIONAL RESOURCES

Sanders LE: Laparoscopic treatment for Meckel's diverticulum: obstruction and bleeding is managed with minimal morbidity, *Surg Endosc* 9:724-727, 1995.

St Vil D, Brandy ML, Panic S, et al: Meckel's diverticulum in children: a 20-year review, *J Pediatr Surg* 26:1289-1292, 1991.

Turgeon DK, Barnett JL: Meckel's diverticulum, *Am J Gastroenterol* 85: 777-781, 1990.

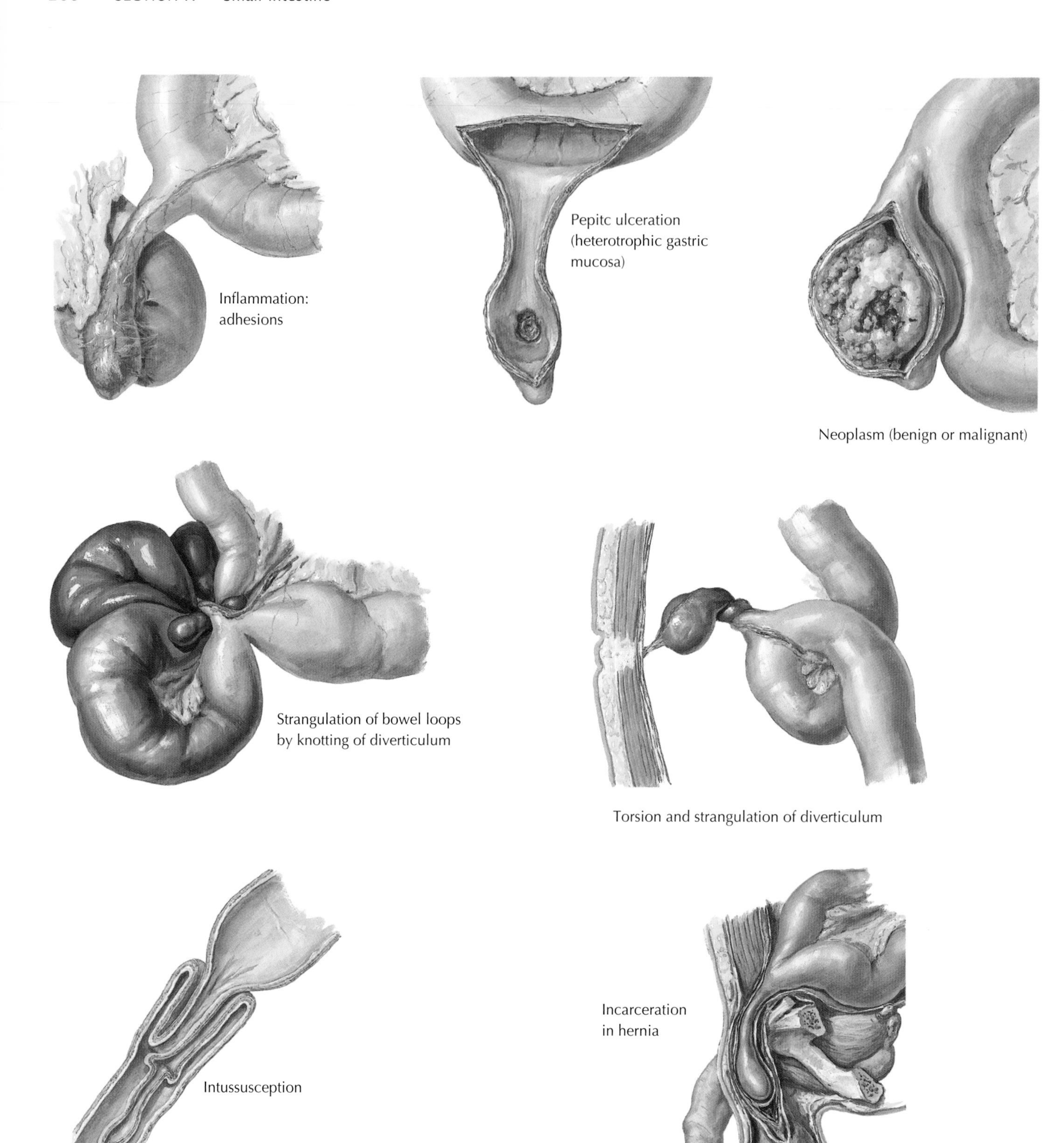

Figure 103-1 *Meckel Diverticulum (Vitelline Duct Remnants).*

Diverticula of the Small Intestine

104

Martin H. Floch

A diverticulum of the small bowel is a "blind" outpocket from the hollow viscus that consists of one or more layers (**Fig. 104-1**). Incidence at autopsy ranges from 0.2% to 0.6%. Diverticula are less common in distal areas of the small bowel. Their etiology is unknown, but most appear to be acquired and to consist of mucosal and submucosal layers only. Colonic diverticulosis is associated with small bowel diverticulosis in 35% to 44% of patients. Complications of inflammation and diverticulitis, obstruction associated with enteroliths, bleeding, perforation, volvulus, bacterial overgrowth, and multiple diverticula have all been reported.

CLINICAL PICTURE

The clinical picture may be an incidental finding and *asymptomatic*, an *acute* complication, or *chronic* symptoms. Symptoms of acute abdomen caused by free perforation, volvulus, or obstruction associated with an enterolith or gastrointestinal (GI) bleeding may cause a patient to seek treatment. Chronic presenting symptoms, including dyspepsia, nausea, occasional vomiting, mild pain, flatulence, and diarrhea, may be caused by bacterial overgrowth or mild inflammation associated with one of the diverticula. Therefore, a broad presenting clinical picture is possible.

Once diagnosed, symptoms should be fully evaluated. Clinicians tend to view diverticula as benign, but they can be a significant finding. Once inflammation occurs in a diverticulum, an acute abdominal infection can result that mimics appendicitis or inflammation, as in Meckel diverticulum. When inflammation develops, erosion of a blood vessel can cause a slow bleed or a massive hemorrhage.

The clinicopathologic correlation of jejunal diverticula is similar to that seen in colonic diverticulosis. The association of jejunal diverticula with steatorrhea and macrocytic anemia has been reported since 1954. There are many cases in the literature of proven bacterial overgrowth and malabsorption. Many patients with jejunal diverticula also have symptoms similar to those of scleroderma, such as esophageal motility disturbances and Raynaud phenomenon.

DIAGNOSIS

The diagnosis is established with small bowel barium contrast study or computed tomography scan with contrast. When chronic anemia or bacterial overgrowth is suspected, studies to evaluate for malabsorption are helpful (see Chapter 109). Anemia and steatorrhea can be demonstrated in patients with bacterial overgrowth.

TREATMENT AND MANAGEMENT

Patients with acute diverticula require surgery, usually by open laparotomy or by laparoscopic surgery when necessary. The treatment for bacterial overgrowth is long-term antibiotic therapy. Ampicillin, tetracycline, or a second- or third-generation drug such as ciprofloxacin can be used daily. It is recommended to alternate antibiotic therapy and "to give the bowel a rest." Some clinicians administer antibiotics every other month to treat this type of bacterial overgrowth. Malabsorption and anemia can be corrected by antibiotic therapy.

COURSE AND PROGNOSIS

Depending on the age of the patient and the acute nature of the clinical situation, the prognosis for diverticulum is similar to that for any condition with a perforated bowel or massive GI hemorrhage. The prognosis for patients with bacterial overgrowth is excellent once therapy is instituted. The mere presence of jejunal or small bowel diverticula without symptoms is benign. However, the clinician should be aware of the risks and should check the patient intermittently for potential insidious complications.

ADDITIONAL RESOURCES

Badnoch J, Bedford PD: Massive diverticula of the upper intestine presenting with steatorrhea and megaloblastic anemia, *Q J Med* 23:462-470, 1954.

DeBree E, Grammatikakis J, Christodoulakis M, Tsiftsis D: The clinical significance of acquired jejunal diverticula, *Am J Gastroenterol* 93:2523-2528, 1998.

Krishnamurthy S, Kelly MM, Rohrmann CA, Schuffler MD: Jejunal diverticulosis: a heterogeneous disorder caused by a variety of abnormalities of smooth muscle or myenteric plexus, *Gastroenterology* 85:538-547, 1988.

Lobo GN, Braithwaite BD, Fairbrother BJ: Enterolithiasis complicating jejunal diverticulosis, *J Clin Gastroenterol* 29:192-193, 1999.

Michinzi F, Pelliccia O: Diverticulosis and acute diverticulitis of the jejunum and ileum, *N Engl J Med* 261:1015-1019, 1959.

Rodriguez HE, Ziauddin MF, Quiros ED, et al: Jejunal diverticulosis in gastrointestinal bleeding, *J Clin Gastroenterol* 33:412-414, 2001.

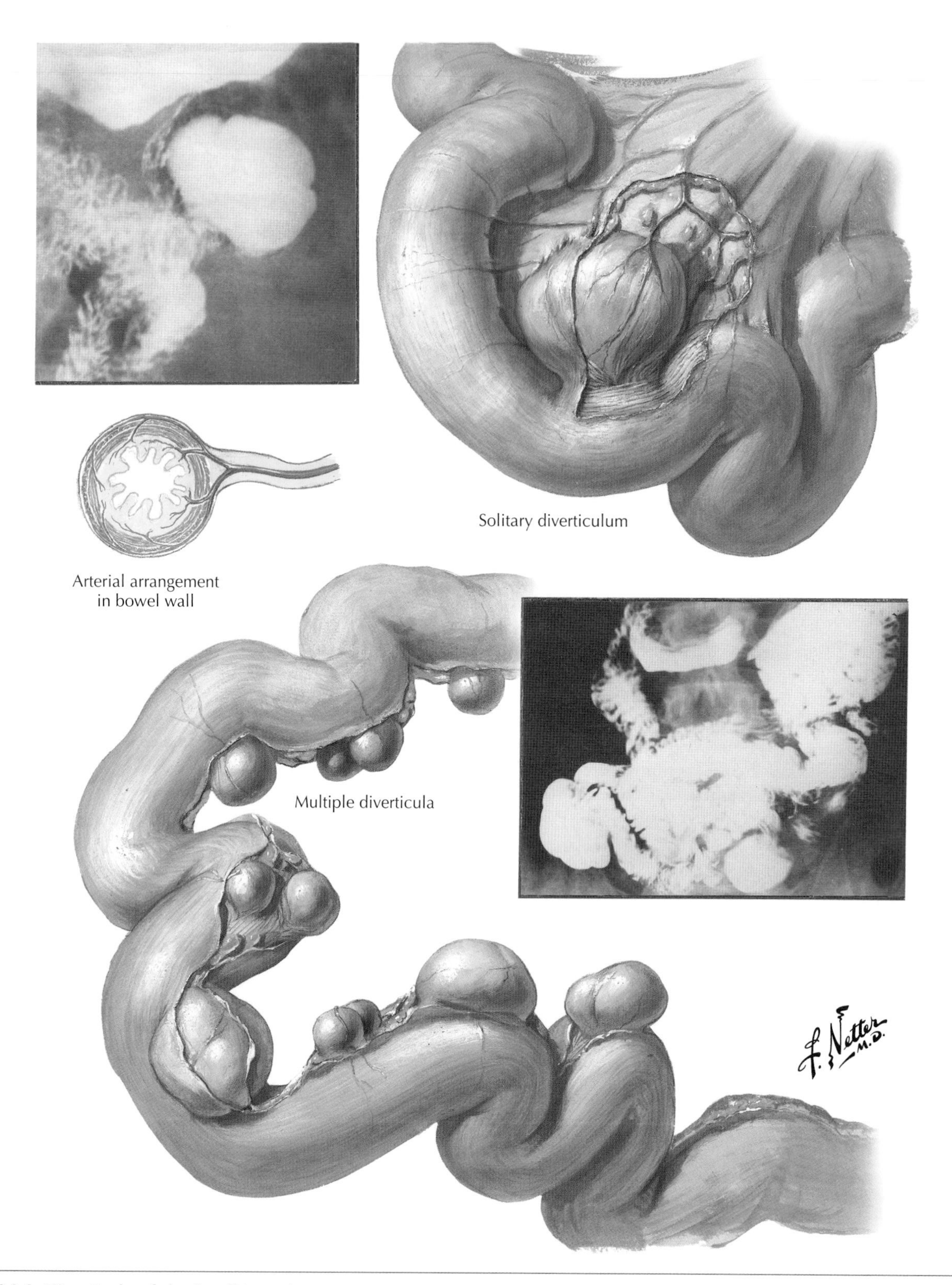

Figure 104-1 *Diverticula of the Small Intestine.*

Motility and Dysmotility of the Small Intestine

105

Martin H. Floch

Many patients undergo full evaluation for nausea, vomiting, abdominal pain, or dyspepsia, but routine evaluations reveal no abnormalities (see Section II for details of these symptoms and their evaluation). Occasionally, it is readily apparent that a patient may have a true gastric-duodenal motility disturbance (see Chapter 47). Progressive motility from the antrum into the duodenum, then from the duodenum into the small bowel, is important for maintaining normal motor function for absorption and digestion and for cleansing the bowel to prevent bacterial overgrowth (**Fig. 105-1**).

Any indication of disturbed absorption of nutrients or bacterial overgrowth should lead to further small bowel motility evaluation. True small bowel transit disturbances and dysmotility are rare and difficult to prove. These evaluations are made primarily at selective research centers. However, motility disturbances do exist, and patients can be helped if indications of small bowel dysmotility can be identified.

The small bowel has both a vigorous working longitudinal muscle layer and a circular muscle layer that are governed by an intrinsic nervous system and an extrinsic nervous system modified by the function of the interstitial cells of Cajal. Intrinsic neurons have their cell bodies within the wall and are divided into sensory (afferent) neurons, motor (efferent) neurons, and interneurons, whereas extrinsic neurons are derived from the vagal and spinal pathways of the parasympathetic and sympathetic divisions of the autonomic nervous system.

MEASUREMENTS OF TRANSIT AND MOTILITY

The *lactulose breath test* is used primarily to measure the small intestine's ability to absorb hydrogen or labeled carbon that is freed after the lactulose reaches the distal small intestine; thus, it really is a measure of *orocecal transit time*. These tests are used primarily in research centers and are difficult to interpret in a clinical setting. Some institutions evaluate for scintigraphic orocecal transit time to measure the transit time, but these also are primarily research institution techniques. Barium contrast studies and fluoroscopy have been used, but these are imprecise for determining true transit time and are most effective when there is some obstructive lesion. At times, without any explanation, transit time using barium contrast may be as short as 30 to 45 minutes.

Small bowel *manometry* is used with great skill in research institutions. Gastric and duodenal probes are placed, and metric measurements can be made of antral, pyloric, and duodenal contractions. The normal patterns of motility, a regular and continuous series of contractions (phase 3), occur within the gut during fasting, approximately three cycles per minute in the stomach and 11 cycles per minute in the duodenum. The regular string of contractions lasts several minutes and propagates downward to the bowel from the antrum as a *migrating motor complex* (MMC). Passage of the MMC is followed by a quiescent period (phase 1), and then the contractile pattern becomes irregular (phase 2), which usually occurs during fasting. Immediately upon eating, the fasting pattern of motility is replaced with a fed pattern, both in the stomach and in the small bowel. Contractile activity increases, and there are no migrating bursts of contractions. The increase in contractions in the small intestine mixes intestinal contents and enhances absorption. Although performed mainly at research institutions, these studies are indicated in the diagnosis of subtle intestinal obstruction and in the explanation for small intestinal bacterial overgrowth (Chapter 114), differentiation of intestinal myopathy from neuropathy, gastroparesis refractory to prokinetic agents, and differentiation of visceral pain syndrome from other motility disorders.

Clinicians most often must rely on radiologic transit with barium or computed tomography with contrast to observe small bowel motility. These techniques provide only broad evaluation but can suggest disturbances requiring more sophisticated study and may help guide the clinical treatment course.

DIAGNOSIS

Diseases associated with disordered intestinal motility are nonulcerative dyspepsia (see Chapter 53), irritable bowel syndrome (Chapter 108), chronic intestinal pseudo-obstruction (Chapter 107), partial small bowel obstruction (Chapter 106), gastric resection, acute illness, pregnancy, diabetes, metabolic disturbances, drugs, scleroderma and other connective tissue diseases, neurologic syndromes, rare myopathies, and biliary dyskinesia. Diabetes mellitus often is associated with altered neurotransmission, as can occur with numerous drugs that decrease or alter contractions.

It is important to note that some patients may require evaluation at a research institution that performs the necessary tests. These patients usually have chronic symptoms and have undergone evaluation but now require detailed research studies and analysis.

Colonoscopy offers repeat visualization of the distal ileum. The end of the ileum appears to be wedged into the wall of the colon and at times can be seen to flutter. However, observations on living patients indicate that the ileocecal junction functions as a *true* sphincter, which means it regulates the flow of material from the ileum to the cecum and prevents retrograde passage. However, there is controversy as to whether this is merely a function of the angulation of the anatomy or a true muscular function based on distention of the ileum with a final peristaltic contraction that empties ileal contents into the cecum. At present, no clinical disease entity is known to be related to this physiology.

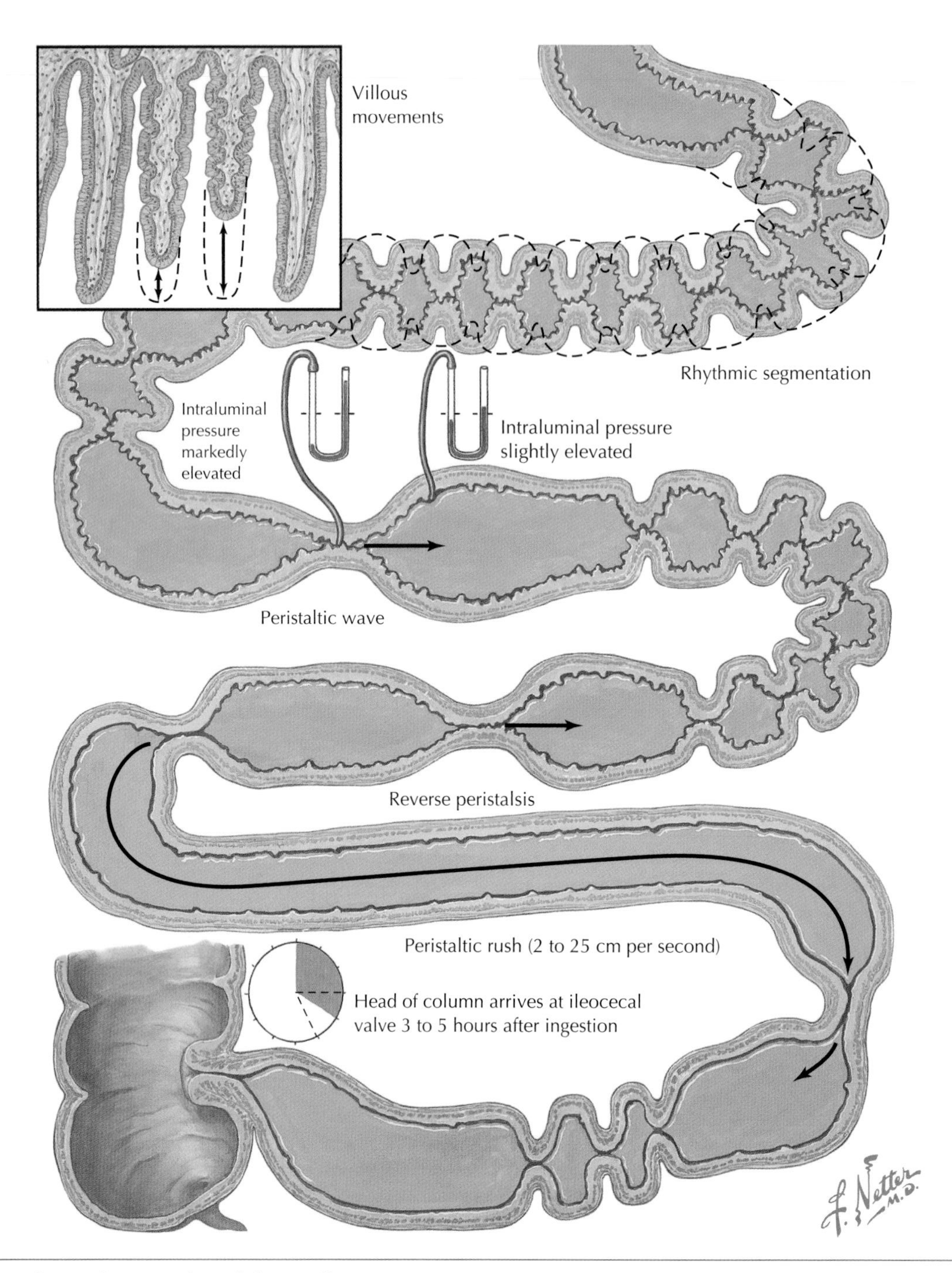

Figure 105-1 *Motility and Dysmotility of the Small Intestine.*

ADDITIONAL RESOURCES

Andrews JM, Blackshaw LA: Small intestinal motor and sensory function and dysfunction. In Feldman M, Friedman LS, Brandt LJ, editors: *Gastrointestinal and liver disease*, ed 8, Philadelphia, 2006, Saunders-Elsevier, pp 2093-2110.

Bratton JR, Jones MP: Small intestinal motility, *Curr Opin Gastroenterol* 23:127-133, 2007.

Snape WJ Jr: Role of motility measurements in managing upper gastrointestinal dysfunction, *Gastroenterologist* 6:44-59, 1998.

Yamamoto T, Watabe K, Nakahara M, et al: Disturbed gastrointestinal motility and decreased interstitial cells of Cajal in diabetic db/db mice, *J Gastroenterol Hepatol*, 23:660-667, 2008.

Obstruction and Ileus of the Small Intestine

106

Martin H. Floch

Intestinal obstruction occurs when the onward passage of intestinal contents is limited by mechanical abnormalities or a functional disturbance (**Fig. 106-1**). Peristaltic activity may be abolished by reflexes originating from diseased structures within, or remote from, the abdominal cavity. When obstructive lesions occur, some motility may persist. The obstruction may be partial or complete, and when it occurs with vascular compromise, it may indicate simple ischemia or may be attributed to strangulation.

The most common cause of small intestinal obstruction is *adhesions* (>50%). *Herniation* accounts for approximately 25% of obstructions. Adhesions are *extrinsic*, as are hernias, and other causes of extrinsic lesions may be congenital bands, volvulus, or carcinomatosis outside the bowel wall. The remaining 25% of small bowel obstructions are caused by inflammatory lesions, intussusceptions, neoplasms, foreign bodies, or atresias and stenosis. These lesions are primarily *intrinsic* to the bowel.

Ileus occurs when the bowel ceases to pass its contents, becoming *adynamic*. Ileus usually occurs after surgery but also results from inflammatory, metabolic, and neurogenic lesions. It also may be associated with electrolyte imbalances or drug administration. Acute inflammation in the bowel, such as appendicitis, diverticulitis, or peritonitis, can cause loss of motility and ileus. Ileus is also associated with acute pancreatitis, ischemic lesions of the bowel, and occasionally, chest lesions causing systemic sepsis. Any of the hernias described in Section III can cause intestinal obstruction and manifest acutely.

CLINICAL PICTURE

Depending on the cause, acute obstruction of the bowel may present with abdominal pain, nausea, and vomiting of reflux origin. Abdominal examination reveals increased peristalsis, which gives way to alternating periods of hyperactivity and quiescence until the latter predominates.

In adynamic ileus, peristalsis ceases from the start; therefore, the bowel is extremely quiet and emits poor bowel sounds. Invariably, however, there is increased fluid secretion into the gut and accumulation of gas, with the bowel becoming more and more distended. In patients with *mechanical ileus*, the distention is proximal to the point of obstruction, whereas in patients with *reflex ileus*, the distention is more generalized.

As peristalsis fails, the absorptive functions also fail, permeability is altered, and intestinal bacteria and toxic substances can translocate. Water and electrolytes enter the bowel lumen, aggravating the distention and invariably increasing the vomiting. The overall effect on the patient is dehydration of body tissue and circulatory failure. Patients may become hypotensive and may experience shock and sepsis.

In mild intermittent obstruction, the symptoms are less severe, and the only clinical presentation may be recurrent abdominal pain. However, anorexia always correlates with the pain and recurrent bloating of abdominal distention. When obstruction is complete, bowel movements cease. With partial obstruction, the patient may have some bowel movement or even mild diarrhea.

DIAGNOSIS

The clinical picture varies from a mild, partial obstruction to an acute, life-threatening situation. Abdominal radiographs are essential to define the type and degree of obstruction, with films of the patient in the supine and upright positions. Radiographs usually help determine whether the obstruction involves the small intestine, with or without the colon. In patients with ileus, the colon is invariably involved, whereas in patients with mechanical obstruction, the area of the dilated loops of bowel with possible fluid levels can identify the site of the obstruction.

Computed tomography (CT) usually clearly delineates the point of obstruction and invariably makes the diagnosis. CT can define closed-loop obstruction, strangulations, characteristic pictures of volvulus, extrinsic or intrinsic neoplasia, and it can clearly indicate the need for medical or surgical therapy.

When the diagnosis is unclear, contrast studies may be helpful. CT scan is performed with contrast, and the lesions at the point of the obstruction can be identified. Occasionally, however, CT findings are unclear, and barium contrast study may be indicated. Barium studies are more feasible in the small intestine than in the large intestine, because barium hardens when water is absorbed in the large bowel. Barium should not be used to diagnose a suspected colonic lesion. Absorbable contrast should be used; in the small bowel, lesions can be clearly defined before therapy is indicated.

It is important to confirm the diagnosis of strangulation or impending gangrenous bowel so that the bowel can be saved. This is particularly important in elderly patients, who may have few symptoms before the bowel becomes strangulated and gangrenous.

TREATMENT AND MANAGEMENT

Immediate therapy involves nasogastric suction to treat distention, electrolyte and fluid replacement, and antibiotics to clear infection and prevent sepsis. In adynamic ileus, particularly after surgery, patience and fluid replacement frequently correct the problem.

If the patient is thought to have mechanical obstruction, surgical intervention is indicated for partial and complete obstruction. For complete obstruction, the surgery is performed on an emergency basis to prevent gangrenous bowel.

COURSE AND PROGNOSIS

Ileus may develop after surgery and can last several days. The patient will need reassurance, nasogastric suction, careful elec-

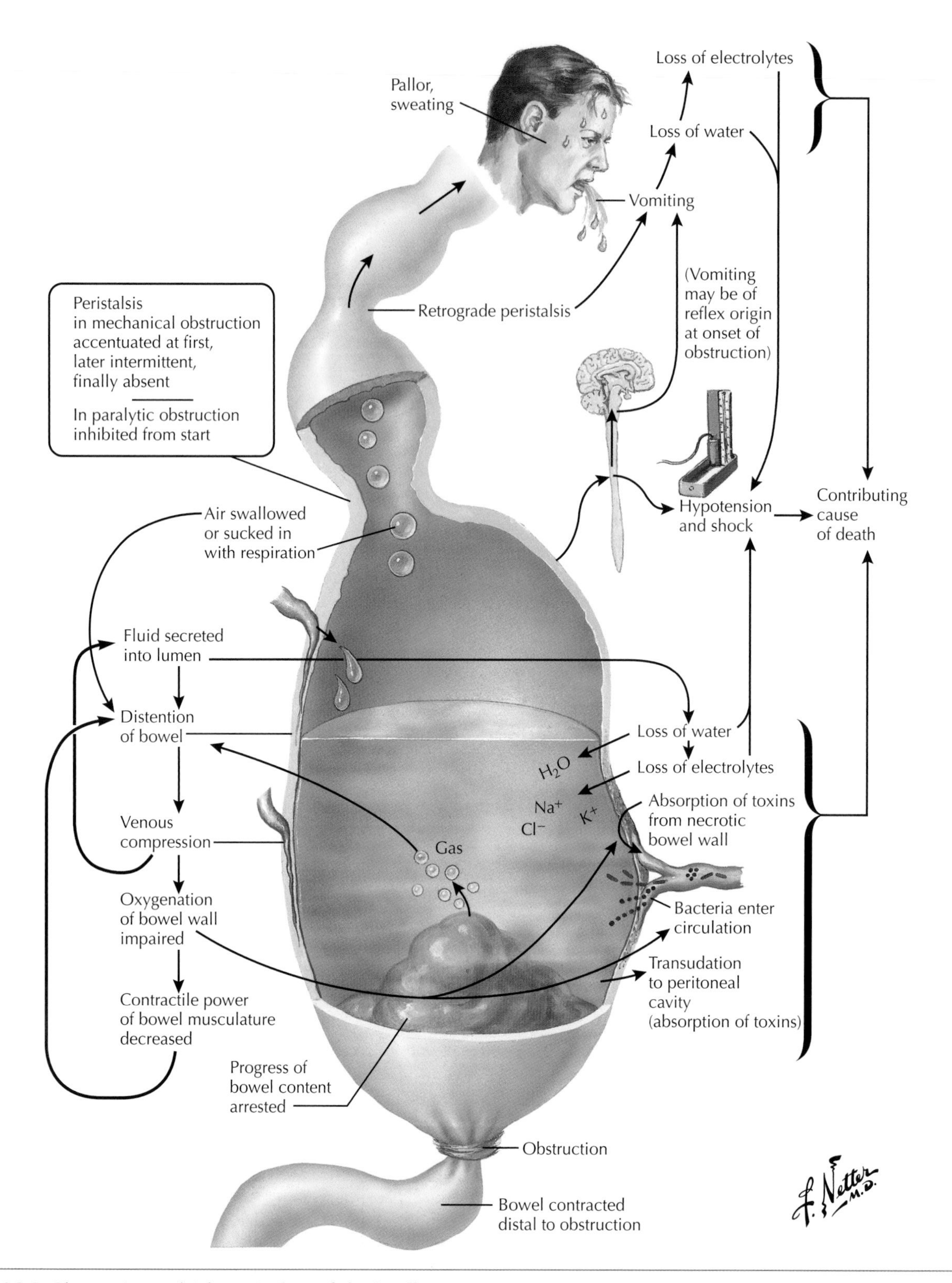

Figure 106-1 *Obstruction and Adynamic Ileus of the Small Intestine.*

trolyte replacement, and nutritional support. Prognosis is usually good when diagnosis shows the initial disease to be benign. If the diagnosis indicates malignancy or severe infection and abscess, or conditions such as severe pancreatitis, the prognosis is guarded, and mortality is significant.

If surgery is performed quickly and intervenes in mechanical obstruction, the prognosis is excellent. However, once gangrenous bowel develops, particularly in the elderly patient, the prognosis is guarded. Again, if a malignant lesion causes the obstruction, the prognosis is that for the particular malignancy.

The rapid advent of laparoscopic surgery has facilitated diagnosis, corrected mechanical obstruction, and resulted in shorter postoperative periods. Adynamic ileus rarely requires surgery, but when the bowel remains distended and the patient obtains minimal relief through suction and electrolyte replacement, decompression and drainage may be necessary.

ADDITIONAL RESOURCES

Kelly KA, Sarr MG, Hinder RA: *Mayo Clinic gastrointestinal surgery,* Philadelphia, 2004, Saunders.

Turnage RH, Heldman M, Cole P: Intestinal obstruction and ileus. In Feldman M, Friedman LS, Brandt LJ, editors: *Gastrointestinal and liver disease,* ed 8, Philadelphia, 2006, Saunders-Elsevier, pp 2653-2678.

Chronic Intestinal Pseudo-Obstruction

107

Martin H. Floch

Chronic intestinal pseudo-obstruction (CIPO) is a clinical syndrome that mimics the signs and symptoms of intestinal obstruction (**Fig. 107-1**). In patients with CIPO, no mechanical obstructive lesion is found on evaluation. CIPO is now thought to be a type of dysmotility and may be related to a functional disturbance in the interstitial cells of Cajal. CIPO may be associated with many disorders of the intestinal or extraintestinal nervous system, including neurologic diseases, small intestinal visceral myopathy, and endocrine and metabolic disorders, and may be precipitated by drugs (**Box 107-1**). Visceral myopathies and neuropathies are associated with numerous rare congenital abnormalities and are often referred to as *primary* CIPO. Acute colonic pseudo-obstruction (Ogilvie syndrome, false colonic obstruction) is different and presents in critically ill and surgical patients.

CLINICAL PICTURE

CIPO may manifest at any age. Developmental and congenital conditions result in presentation in infancy or early childhood. Associated systemic diseases, such as scleroderma, or paraneoplastic causes may occur later in life. The presenting symptoms are usually obstruction with abdominal distention and pain, vomiting, and inability to have a bowel movement. When the disease is associated with visceral myopathy or neuropathy, pain may become predominant. Rarely, it may be associated with a segmental disease such as megaduodenum or megacolon. An esophageal motility disorder may coexist. Symptoms of those localized lesions, in association with CIPO, may predominate or may obfuscate the diagnosis. In addition, some patients have bacterial overgrowth and can develop malabsorption and diarrhea, with a confusing picture of alternating diarrhea and constipation. Invariably, the physical findings are intestinal obstruction with a distended abdomen, no bowel sounds, and a tympanitic bowel.

DIAGNOSIS

History and physical examination are essential to establish the presence of any neurologic or associated diseases. Often, physical examination reveals weight loss and possibly malnutrition, depending on the degree and chronicity of disease. Radiographs of the abdomen reveal the intestinal obstruction. Patients usually have large, dilated loops of bowel with fluid levels. Barium contrast study or computed tomography with contrast media reveals the classic obstructive lesion. Gastric emptying studies or esophageal manometry may further elucidate the associated condition, such as gastroparesis or esophageal symptoms. Complete thyroid evaluation and tests for collagen and vascular diseases should be performed. Finally, it may be necessary to

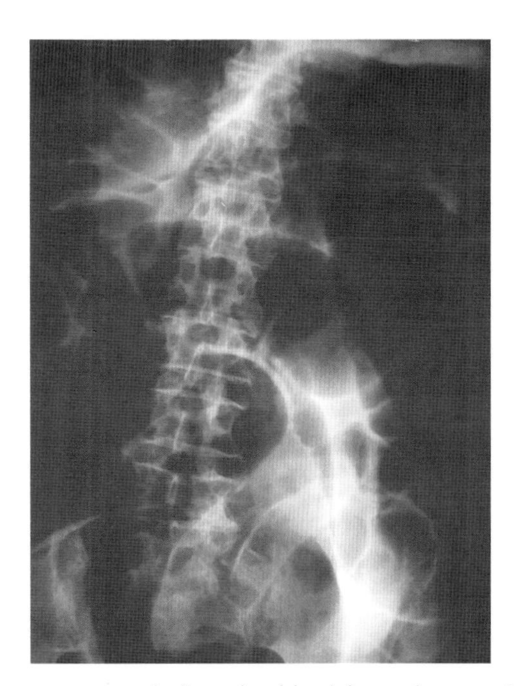

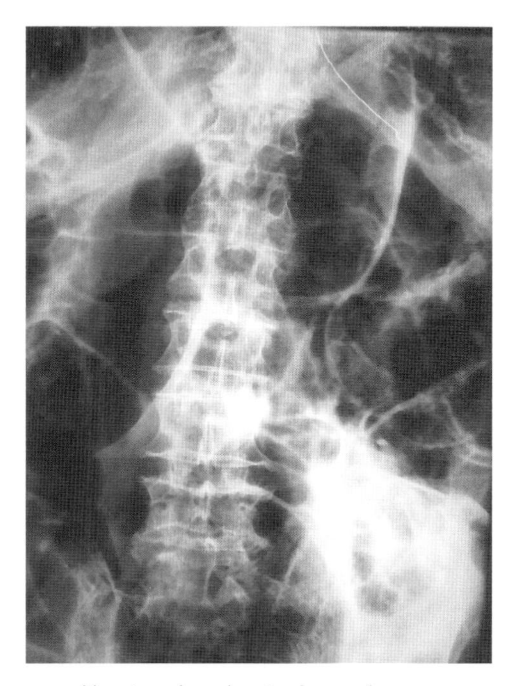

Radiographs of the abdomen demonstrating severe dilatation of overlapping loops of small and large bowel, which is characteristic of CIPO.

Figure 107-1 *Chronic Intestinal Pseudo-Obstruction.*

Box 107-1 Clinicopathologic Classification of Chronic Intestinal Pseudo-Obstruction

I. Disorders of the smooth muscle
   A. Primary
      1. Familial visceral myopathies
         a. Type 1 (autosomal dominant)
         b. Type 2 (autosomal recessive, with ptosis and external ophthalmoplegia)
         c. Type 3 (autosomal recessive, with total gastrointestinal tract dilatation)
      2. Sporadic visceral myopathy
      3. Congenital, in infants
   B. Secondary
      1. Progressive systemic sclerosis/polymyositis
      2. Muscular dystrophy syndromes
      3. Systemic lupus erythematosus
      4. Amyloidosis
      5. Radiation injury
      6. Ehlers-Danlos syndrome
      7. Mitochondrial myopathy
   C. Diffuse lymphoid infiltration
   D. Others (muscle cell inclusions; absence of actin)
II. Disorders of the myenteric plexus
   A. Familial visceral neuropathies
      1. Recessive, with intranuclear inclusions (neuronal intranuclear inclusion disease)
      2. Recessive (familial steatorrhea with calcification of the basal ganglia and mental retardation)
      3. Dominant, with neither of the above
      4. POLIP syndrome (*p*olyneuropathy, *o*phthalmoplegia, *l*eukoencephalopathy, *i*ntestinal *p*seudo-obstruction)
      5. Infantile short bowel, malrotation, and pyloric hypertrophy
      6. With progressive neurologic disease at young age
   B. Sporadic visceral neuropathies
      1. Degenerative, noninflammatory (at least two types)
      2. Degenerative, inflammatory (with lymphocytes, plasma cells, or both in myenteric, and sometimes submucosal, plexus)
         a. Paraneoplastic
         b. Infectious (Chagas disease, cytomegalovirus)
         c. Idiopathic
         d. Isolated axonopathy
   C. Developmental abnormalities
      1. Total colonic aganglionosis (sometimes with small intestinal aganglionosis)
      2. Maturational arrest
         a. Isolated to myenteric plexus
         b. With mental retardation
         c. With other neurologic abnormalities
      3. Neuronal intestinal dysplasia
         a. Isolated to intestine
         b. With neurofibromatosis
         c. With multiple endocrine neoplasia, type 2b
   D. Myotonic dystrophy
III. Neurologic disorders
   A. Parkinson disease
   B. Autonomic dysfunction, familial and sporadic
   C. Total autonomic or selective cholinergic dysfunction after Epstein-Barr virus infection
   D. Brainstem tumor
IV. Small intestinal diverticulosis
   A. With muscle resembling visceral myopathy
   B. With muscle resembling progressive systemic sclerosis
   C. With visceral neuropathy and neuronal intranuclear inclusions
   D. Secondary to Fabry disease
V. Endocrine and metabolic disorders
   A. Myxedema
   B. Pheochromocytoma
   C. Hypoparathyroidism
   D. Acute intermittent porphyria
VI. Drugs
   A. Opiates (narcotic bowel syndrome)
   B. Anticholinergics
   C. Phenothiazines
   D. Clonidine
   E. Tricyclic antidepressants
   F. Vinca alkaloids (e.g., vincristine)
   G. Calcium channel blockers
   H. Fetal alcohol syndrome
VII. Miscellaneous
   A. Jejunoileal bypass
   B. Sclerosing mesenteritis
   C. Celiac sprue
   D. Ceroidosis?

From Camilleri M: Acute and chronic pseudo-obstruction. In Feldman M, Friedman LS, Brandt LJ, editors: Gastrointestinal and liver disease, *ed 8, Philadelphia, 2006, Saunders-Elsevier, pp 2679-2702.*

perform a deep biopsy of intestinal muscle; laparoscopy may be helpful.

TREATMENT AND MANAGEMENT

Treatment and management of patients with CIPO depend on whether they have an associated disease. In patients with idiopathic disease and in most with associated disease, standard therapies include prokinetic agents, diet control, and palliative surgery. When obtaining proper nutrition becomes a severe problem, total parenteral nutrition (TPN) on a home basis may be necessary.

Drugs include metoclopramide, 10 to 20 mg three times daily (tid) or four times daily (qid); domperidone, 10 to 20 mg tid or qid (when available); octreotide, 50 to 100 μg daily subcutaneously; and erythromycin, 250 mg tid. These prokinetic agents can be given by mouth or by intravenous injection, depending on the drug and the availability.

Diet therapy varies with the particular condition. During the acute phase, patients are supported intravenously. During the chronic phase, they are usually on low-fat, low-fiber diets with frequent small feedings, gradually progressing to normal feeding as symptoms slowly subside.

Palliative surgery is a drastic procedure that may become necessary to decompress the bowel or to remove or bypass a nonfunctional area. Colectomy may be used, depending on the degree and type of intestinal involvement. In severely affected patients and those with rare conditions, small intestine transplantation has recently become a treatment consideration, although the procedure is life threatening and offers only limited life expectancy.

COURSE AND PROGNOSIS

When CIPO is associated with another disease, the prognosis of that disease modifies the outcome. Unfortunately, when associated with any significant disease, the finding of CIPO usually heralds the beginning of a fatal prognosis; these patients may die in a few months to a few years. Patients with the idiopathic chronic form may live for many years once a regimen is designed that relieves the obstruction and adequate nutrition can be maintained. Some of these patients receive a combination of prokinetics and nutritional support; life can be maintained, but quality of life decreases. CIPO has a marked effect on the development of children.

ADDITIONAL RESOURCES

Camilleri M: Acute and chronic pseudo-obstruction. In Feldman M, Friedman LS, Brandt LJ, editors: *Gastrointestinal and liver disease*, ed 8, Philadelphia, 2006, Saunders-Elsevier, pp 2679-2702.

Connor FL, DiLorenzo C: Chronic intestinal pseudo-obstruction: assessment and management, *Gastroenterology* 130:S29-S36, 2006.

Gambarara M, Knafelz D, Diamanti A, et al: Indication for small bowel transplant in patients affected by chronic intestinal pseudo-obstruction, *Transplant Proc* 34:866-867, 2002.

Heneyke S, Smith VB, Spitz L, Milla PJ: Chronic intestinal pseudo-obstruction: treatment and long-term follow-up of 44 patients, *Arch Dis Child* 81:21-27, 1999.

Quigley EM: Chronic intestinal pseudo-obstruction, *Curr Treatment Options Gastroenterol* 2:239-250, 1999.

Schwankovsky L, Mousa H, Rowhani A, et al: Quality of life outcomes in congenital chronic intestinal pseudo-obstruction, *Dig Dis Sci* 47:1965-1968, 2002.

Scolapio J, Ukleja A, Bouras E, et al: Nutritional management of chronic intestinal pseudo-obstruction, *J Clin Gastroenterol* 28:306-312, 1999.

Sutton DH, Harrell SP, Wo JM: Diagnosis and management of adult patients with chronic intestinal pseudo-obstruction, *Nutr Clin Pract* 21: 16-22, 2006.

Irritable Bowel Syndrome and Functional Gastrointestinal Disorders

108

Martin H. Floch

Irritable bowel syndrome (IBS) is one of a group of functional gastrointestinal (GI) disorders characterized by abdominal discomfort or pain and frequently associated with a change in bowel movements (**Fig. 108-1**). Worldwide, IBS is the most frequent symptom complex in patients seeking GI consultation. In the United States, it is estimated that more than 3 million office and hospital visits are attributed to this symptom complex.

In disease classification, IBS is thought to be the intestinal or bowel disorder of a group of functional GI disorders that include (1) esophageal disorders such as globus hystericus and functional chest pain; (2) GI disorders such as functional dyspepsia (Chapter 53), aerophagia, functional vomiting (Chapter 45), and functional abdominal pain; (3) functional biliary disorders such as sphincter of Oddi dysfunction (Chapter 206); and (4) anorectal disorders such as anorectal pain and pelvic floor dyssynergia (see Section V).

As understanding of GI physiology broadens, the cause of IBS is thought to be a disturbance in the autonomic and enteric nervous systems of the gut and the gut-brain axis. This disturbance results in abnormal motility and visceral hypersensitivity. Early inflammation of the gut may be a trigger for the onset of the syndrome. The discovery of lactose intolerance has removed many patients from the category of IBS, and, as science evolves, it is certain that other causes may explain the symptoms in select groups of patients. The wide range of food sensitivities and food allergies that exist, as well as slow development of knowledge in this area, may progress and prove helpful in the future.

CLINICAL PICTURE

Three classic symptoms are associated with IBS: abdominal pain, diarrhea, and constipation. Alternating diarrhea and constipation, or diarrhea only, or constipation only, may prevail. In the classic model of the female patient, a woman seeks treatment for recurrent low abdominal pain associated with an inability to have a bowel movement. Constipation is persistent and becomes a lifelong problem. The patient frequently will present with many evaluations that have revealed no significant abnormality. Often, the abdominal pain is relieved by a bowel movement. Patients may or may not notice whether the character of the stool has changed, and often they describe pebblelike stools. However, the pain is persistent and may be nagging or severe, and patients may require pain relief. Characteristic is the history of frequent visits to physician offices with a workup that has revealed no disease.

The other common symptom complex is severe diarrhea, noted more often in male patients and consisting of loose or explosive bowel movements that may almost herald incontinence. The diarrhea is associated with severe abdominal cramps, and frequently, patients describe a formed stool on first motion, then liquid stools. Once the diarrhea has stopped, the cramps frequently cease. However, the cramps can be severe and have caused sweating during the bowel movement.

The third pattern is constant, recurrent lower abdominal pain associated with diarrheal bowel movements and, once they have subsided, days of no bowel movements or constipation with a feeling of an inability to evacuate. It is estimated that many patients never report these symptoms to physicians, but, as stated, IBS is one of the major symptom complexes treated by physicians and GI consultants.

DIAGNOSIS

The latest consensus, as expressed in Rome Criteria II and III, is that there should be 12 weeks or more, in the past 12 months, of abdominal discomfort or pain that has two or three features: it must be relieved with defecation, the onset must be associated with a change in frequency of stool, and/or the onset must be associated with a change in the caliber or appearance of the stool. This complex may be associated with increased stool frequency, abnormal forms of stool, marked increased straining or urgency, or a feeling of incomplete evacuation, the appearance of mucus in the stool, or an associated feeling of bloating and abdominal discomfort. The Rome Criteria requires this symptom complex for the diagnosis of IBS.

Some clinicians insist they can make the diagnosis of IBS without a detailed workup to rule out organic disease. However, most will insist that the diagnosis should not be made without laboratory and imaging evaluation. Certainly, the stool must be free of blood, and a simple microscopic evaluation for leukocytes and overt parasites should be conducted. Screening should be performed to rule out thyroid disease and lactose intolerance. Proponents cite fructose intolerance as a cause in selected patients, and if suspected, this should be ruled out. Most clinicians will insist that colonoscopy or barium enema evaluation of the colon be performed, depending on the resources available. Furthermore, many will also insist that the small bowel should undergo screening barium study, or again, depending on the resources, capsule endoscopy or computed tomography of the bowel.

Because the symptoms are recurrent and persistent, conducting a full evaluation is recommended to be able to treat the IBS patient long term. Also, negative tests reassure the patient that there is no organic disease. Differential diagnosis does include malignancy of the colon, inflammatory bowel disorders, chronic infections, endocrine disorders, and psychiatric disorders.

A theory of bacterial overgrowth or *dysbiosis* has emerged in which some patients have a treatable increase in the small bowel bacterial flora or disturbance in the flora. A lactulose tolerance

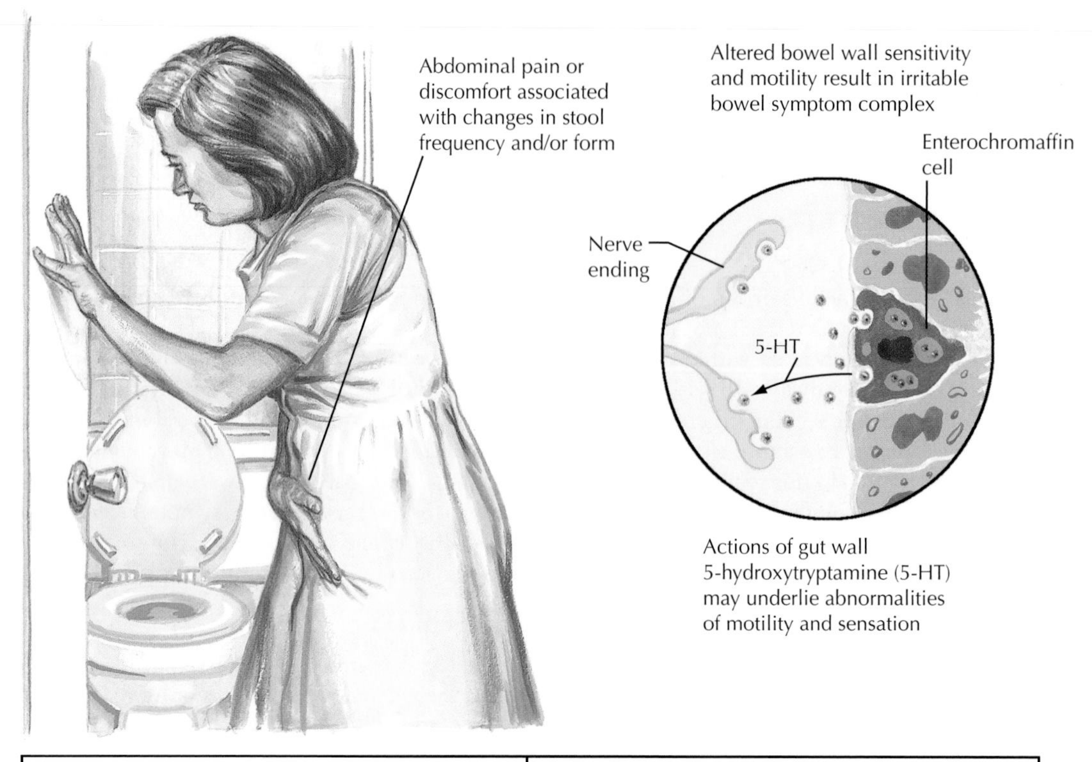

Rome II diagnostic criteria* for IBS	**Symptoms not essential for the diagnosis, but if present increase the confidence in the diagnosis and help to identify subgroups of IBS:**
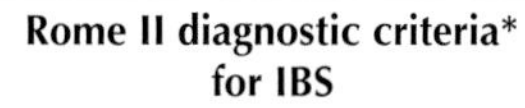 Twelve weeks** or more in the past 12 months of abdominal discomfort or pain that has two out of three features: a. Relieved with defecation b. Onset associated with change in frequency of stool c. Onset associated with change in form (appearance) of stool * In the absence of structural or metabolic abnormalities to explain the symptoms. ** The 12 weeks need not be consecutive.	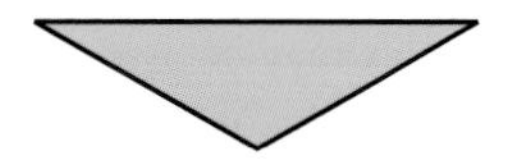 • Abnormal stool frequency (>3 daily or <3 weekly • Abnormal stool form (lumpy/hard or loose/watery stool) >1/4 of defecations • Abnormal stool passage (straining, urgency, or feeling of incomplete evacuation) >1/4 of defecations • Passage of mucus >1/4 of defecations • Bloating or feeling of abdominal distention >1/4 of days

Figure 108-1 *Irritable Bowel Syndrome.*

test is usually positive and is needed to make this diagnosis. Direct bacterial aspiration studies are rarely available, but when performed in research centers, these can prove the small bowel overgrowth. Another theory claims that the normal flora is disturbed, and therefore the addition of probiotic organisms will be helpful. No consensus exists on these theories, but clinicians do treat based on theories, with varied results.

TREATMENT AND MANAGEMENT

There is no single treatment for IBS. Treatment varies with the symptom complex. Because IBS is considered a biopsychologic disorder with abnormal motility and abnormal visceral perceptions, treatments vary. To individualize the treatment, symptoms must be addressed, usually in the categories of constipation, diarrhea, or pain associated with bloating.

The constipated patient is usually first put on a high-fiber diet that requires increased intake of dietary fiber from the usual low amount in Western diets to the normal recommended amounts of 25 to 35 g daily (see Section X). If the constipation is intractable and a workup reveals no organic cause, laxatives or stimulants must be used (see Chapter 136). The new approach has been to use serotonin ($5\text{-}HT_4$) receptor agonists. Tegaserod was a helpful drug but has been removed from the market in the United States.

If diarrhea is the problem, the approach is different. These patients must be carefully evaluated for lactose intolerance, but once that is ruled out, anticholinergics may be tried. If they are not successful, diphenoxylate or loperamide may be used to decrease the number of bowel movements; certain agents are helpful when used before social engagements. Dosage is usually as described with the drug product. Patients may want to take one or two tablets before leaving home.

The $5\text{-}HT_3$ antagonists have now proved to be effective. Alosetron is available and effectively decreases colonic motility. Type 3 alosetron and type 4 tegaserod are first-line serotonin products. We can expect that their future development as antagonists will be helpful in the treatment of IBS.

Clinicians claiming bacterial overgrowth will use rifaximin and other antibiotics in cyclic treatment. This has had varied results, but when correlated with good evidence of overgrowth, it can be helpful.

Proponents of probiotic use have had good results using *Bifidobacterium infantis* in a daily dose, reportedly effective in a double-blind controlled study.

Abdominal pain and bloating can be the predominant factor over the uncontrollable symptom. When treatment of the constipation or diarrhea is unsuccessful and the pain predominates, psychotherapeutic agents may be helpful. Most clinicians use tricyclic antidepressants such as desipramine, 50 mg three times daily, or amitriptyline, 10 to 25 mg twice daily or at bedtime.

COURSE AND PROGNOSIS

Irritable bowel syndrome can become a lifetime problem. These patients may have abdominal pain with altered bowel movements as children and progress into adulthood with the same symptoms, or they may seek treatment for postinfectious problems that persist for years. Symptoms may resolve with treatment, and then recur years later. However, with understanding, reassurance, and careful treatment, these patients frequently adjust well to their gut abnormality, although their symptoms persist for life. Importantly, many patients with these symptoms never seek medical attention because they learn to control the symptoms. Altered motility and visceral hypersensitivity appear to exist for long periods, even for a lifetime, with exacerbation of severity, but most patients learn to control and live with the symptoms.

ADDITIONAL RESOURCES

Camilleri N: Management of the irritable bowel syndrome, *Gastroenterology* 120:652-668, 2001.

Choi YK, Kraft N, Zimmerman B, et al: Fructose intolerance in IBS and utility of fructose-restricted diet, *J Clin Gastroenterol* 42:233-238, 2008.

Drossman DA, Corazziari E, Talley NJ, et al: *The functional gastrointestinal disorders: diagnosis, pathophysiology, and treatment—a multinational consensus*, ed 2, McLean, Va, 2000, Degnon Associates.

Floch MH, Narayan R: Diet and the irritable bowel syndrome, *J Clin Gastroenterol* 35:S45-S54, 2002.

Halvorson HA, Schlett CD, Riddle MS: Postinfectious irritable bowel syndrome: a meta-analysis, *Am J Gastroenterol* 101:1894-1899, 2006.

Hunt RH: Evolving concepts in the pathophysiology of functional gastrointestinal disorder, *J Clin Gastroenterol* 35:S2-S6, 2002.

Posserud I, Stotser PQ, Bjornsson ES, et al: Small intestinal bacterial overgrowth in patients with irritable bowel syndrome, *Gut* 56:802-808, 2007.

Ringel Y, Drossman DA: Irritable bowel syndrome: classification and conceptualization, *J Clin Gastroenterol* 35:S7-S11, 2002.

Rosemore JG, Lacy BE: Irritable bowel syndrome: basis of clinical management strategies, *J Clin Gastroenterol* 35:S37-S44, 2002.

Wald A: Psychotropic agents in irritable bowel syndrome, *J Clin Gastroenterol* 35:S58-S67, 2002.

Whorwell PJ, Altringer L, Morel J, et al: Efficacy of an encapsulated probiotic *Bifidobacterium infantis* 35624 in woman with irritable bowel syndrome, *Am J Gastroenterol* 102:1581-1590, 2006.

Wood JD: Neuropathophysiology of irritable bowel syndrome, *J Clin Gastroenterol* 35:S11-S22, 2002.

Evaluation of the Small Bowel

Martin H. Floch

109

The rapid development of imaging and histopathologic techniques, in conjunction with an altered pattern of disease and treatments late in the twentieth century, has changed the "standard" of testing for small bowel function and anatomy. Testing has now become a sophisticated evaluation performed for outpatients in most clinical practices and in detail at university research centers. Evaluation of the patient thought to have small bowel disease should include the following:

1. Detailed history and physical examination for clues to etiology and physical findings of malabsorption
2. Careful imaging technique
3. Biopsy of the small intestine, if needed
4. Biochemical function evaluation
5. Stool analysis for parasitic and infectious etiologies

Key findings in the history and physical examination and analysis are described under each disease. However, small bowel disease must be suspected whenever a patient experiences weight loss, diarrhea, anemia, or any sign or symptom of a selective malabsorption deficiency.

IMAGING TECHNIQUES

In the twentieth century, the barium contrast study was used to demonstrate the classic malabsorption pattern, including loss of normal small bowel folds, dilatation of the bowel, and segmentation of the meal. Subtle and dramatic presentations were often recorded. In addition, other defects of the small bowel, including neoplasia, strictures, and diverticula, were frequently demonstrated.

The development of computed tomography (CT) and enterography now allows sophisticated imaging of select areas of the bowel and demonstrates the thickness of the bowel wall and any reaction outside the bowel wall. CT has enhanced the clinician's ability to evaluate for many diseases. At times, however, it is still necessary to do both barium contrast and CT studies.

Endoscopy

The small bowel can be visualized using direct endoscopes, but visualization is usually limited to the duodenum with upper endoscopy, to the proximal jejunum with push enteroscopes, and to the terminal ileum with colonoscopes. Long enteroscopes traversing the entire small bowel are used only in select institutions and have not gained wide acceptance.

Capsule endoscopy is the best procedure for imaging the mucosa (but cannot be used for biopsy) and has proved successful in identifying lesions missed by standard radiographic imaging. It is the preferred technique for occult gastrointestinal (GI) bleeding, to identify such lesions as angiodysplasia, and has found early ulcerations of Crohn's disease, as well as the lesions of nonsteroidal antiinflammatory drug (NSAID) ulcerations. These techniques are discussed in Chapter 61 and relevant disease chapters.

HISTOPATHOLOGY

Small bowel biopsy is essential to confirm the diagnosis of many diseases. Histopathologic evaluation is discussed as it applies to each disease in the corresponding chapter.

BIOCHEMICAL EVALUATION

Sophisticated tests have evolved to define selective malabsorption. However, the clinician in practice can still rely on a battery of simple tests to determine whether bowel function is causing malabsorption. It is helpful in differentiating among biliary, pancreatic, and small bowel malabsorption.

A test for carbohydrate malabsorption and small bowel function is the *D-xylose absorption* test (**Fig. 109-1**). A dose of 5 or 25 g can be given orally, and then the urine is collected for 5 hours to determine the amount excreted in the urine. Provided the patient is well hydrated and has no evidence of bacterial overgrowth, this test reliably assesses small bowel mucosal function. It cannot be used in the patient with renal disease, however, and adequate urinary collections must be ensured. However, if less than 20% of the ingested dose is excreted, and if a false-positive result is unlikely, D-xylose absorption is a reliable test for determining small bowel malabsorption.

Fat malabsorption is determined best by *72-hour stool fat collection* while the patient is receiving at least 60 to 80 g of fat. Stool must be collected, and this remains a social problem. Nevertheless, there are many simple containers available to help make the collection easier for the patient. The stool should be stored overnight under refrigeration. When this test is conducted carefully, it sets the standard for malabsorption. At least 94% of ingested fat should be absorbed; thus the test is reliable if a 6-g fat excretion in feces is used as the lower limit of normal. From 6 to 8 g should arouse suspicion, and excretion greater than 8 to 10 g indicates malabsorption of fat.

Negative D-xylose and positive stool fat findings usually indicate that the cause is not the small bowel but rather is pancreatic or biliary. Quantitative fecal analysis is the standard; qualitative and semiqualitative tests and breath tests have been too inconsistent to diagnose fat malabsorption. Some clinicians screen for serum *β-carotene*, but this is merely a screening test. Low β-carotene levels do correlate excellently with malabsorption, however, and very high levels can be seen in certain thyroid diseases.

The *lactose absorption* test is extremely important in determining whether lactose intolerance or malabsorption exists.

The *breath hydrogen* test is now the standard for lactose testing and is easily done in most laboratories. At baseline, high hydrogen may indicate bacterial overgrowth or a very high intake of dietary fiber substances, but increasing breath hydrogen, after the ingestion of a standard dose of lactose, is sensitive and specific for lactose intolerance and malabsorption.

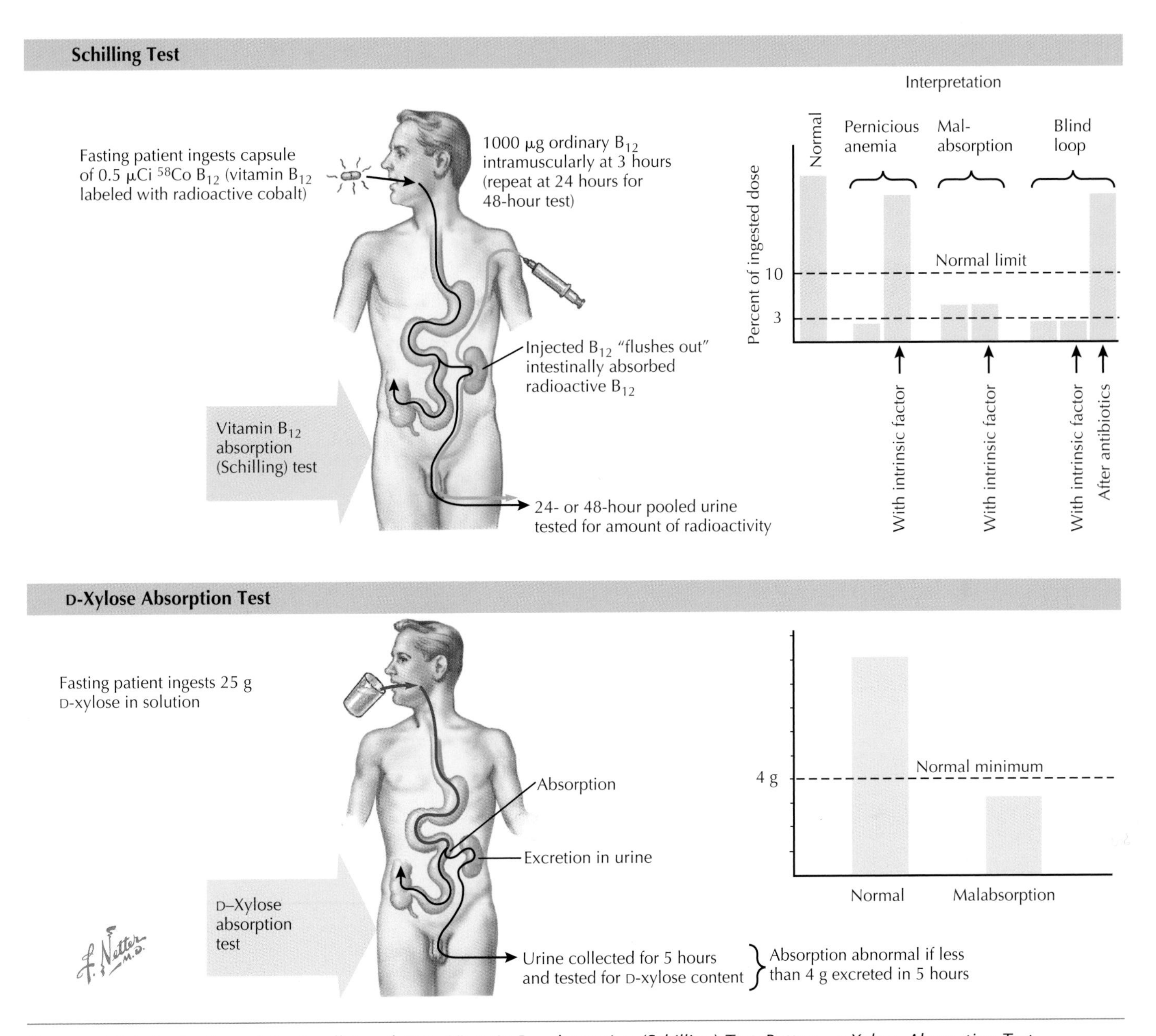

Figure 109-1 *Evaluation of the Small Bowel:* Top, *Vitamin B_{12} absorption (Schilling) Test;* Bottom, *D-Xylose Absorption Test.*

Tests for *protein malabsorption* remain difficult to perform. Quantitative *fecal nitrogen* is excellent but is rarely used because it requires stool collection and laboratory analysis. It is actually simple to conduct but is not used by most clinicians. Sophisticated *isotope studies* are used in research laboratories but are rarely used in clinical medicine.

The *Schilling test* is used regularly to determine vitamin B_{12} absorption (see **Fig. 109-1**). It is based on the observation that small amounts of intestinally absorbed B_{12} are flushed out with the urine when larger quantities administered parenterally are present. Thus, B_{12} absorption can be evaluated by measuring radioactivity in a 24-hour urine specimen after the patient has ingested a capsule containing 0.5 μCi of radioactive cobalt with which B_{12} has been labeled. Three hours after administration of the test dose, the patient receives a parenteral injection of 1 mg of ordinary vitamin B_{12}. In healthy persons, at least 10% of the ingested radioactivity appears in the urine within 24 hours. In the absence of *intrinsic factor*, radioactivity excretion is minimal. When intrinsic factor is administered simultaneously with the oral B_{12}, radioactivity excretion increases to normal levels. This is the second part of the Schilling test, differentiating pernicious anemia from malabsorption syndrome. A third part is often performed to distinguish bacterial overgrowth, requiring the patient to be treated with antibiotics after the previous parts are done. Excretion improves after administration of antibiotics, and bacterial overgrowth presumably has caused the malabsorption. Similarly, Schilling results may be abnormal because of pancreatic insufficiency and a lack of freeing R factor. Adding pancreatic enzymes modifies the Schilling test, and results may be normal if the enzymes are added to the oral dose of radioactive vitamin B_{12}.

Other breath tests and permeability tests to evaluate the small intestine are performed at research centers. The *lactulose breath test* is described in Chapter 114.

ADDITIONAL RESOURCES

Benson JA, Culver PJ, Ragland S, et al: The D-xylose test in malabsorption syndromes, *N Engl J Med* 256:335-338, 1957.

Butterworth CE, Perez-Santiago E, Montinez de Jesos J, Santini R: Studies on the oral and parenteral administration of d(+) xylose, *N Engl J Med* 261:157-162, 1959.

Faigel DO, Cave DR: *Capsule endoscopy*, Philadelphia, 2008, Saunders-Elsevier.

Gore RM, Levine MS: *Textbook of gastrointestinal radiology*, ed 2, Philadelphia, 2000, Saunders.

Högenauer C, Hammer HF: Maldigestion and malabsorption. In Feldman M, Friedman LS, Brandt LJ, editors: *Gastrointestinal and liver disease*, ed 8, Philadelphia, 2006, Saunders-Elsevier, pp 2199-2142.

Lee JKT, Sagel SS, Stanley RJ, Heiken JP: *Computed tomography with MRI correlation*, ed 3, Philadelphia, 1998, Lippincott-Raven.

Romagnuolo J, Schiller D, Bailey, RJ: Using breath test wisely in a gastroenterology practice: an evidence-based review of indications and pitfalls in interpretation, *Am J Gastroenterol* 97:1113-1126, 2002.

Van de Kamer JH, ten Bokbel Huinik H, Weijens HH: Rapid method for determination of fat in feces, *J Biol Chem* 177:547-552, 1949.

Lactose Intolerance

110

Martin H. Floch

Lactose intolerance occurs when there is a deficiency of lactase in the brush border of the small intestine. When a person with lactase deficiency ingests lactose, the poorly digested lactose is fermented in the small and large intestines, resulting in abdominal bloating, discomfort, or diarrhea (**Fig. 110-1**).

Lactose, a disaccharide sugar found in milk, is a structural combination of *glucose* and *galactose*. Holzel described defective lactose absorption that resulted in infant disease in 1959. Later, the clinical picture was also noted in adults.

The incidence of lactose intolerance varies throughout the world. Rates are 85% to 100% in persons of African and Asian descent, 40% to 90% in those of Mediterranean descent, and 5% to 20% in persons of English and Nordic descent. Initially, the great ethnic and racial differences were thought to occur because of adaptation, but now genetic differences are believed the probable cause.

When lactose is digested, it is normally hydrolyzed in the small intestine. The lactase level appears to fall below 3 U/g tissue for intestinal protein in symptomatic lactase-deficient patients. When the intestinal microflora ferment the lactose, they produce tannic acids, lactic acid, carbon dioxide, and hydrogen. In addition, lactose itself has an osmotic effect in the colon.

CLINICAL PICTURE

The clinical picture varies with the amount of lactose ingested and malabsorbed and the degree of the lactase deficiency. Some patients develop clinical symptoms when they consume very small amounts of lactose, whereas others are less sensitive. Relating this to actual intake of food, some subjects can tolerate a glass of milk, whereas others cannot. The basis for the *lactose tolerance* test is 50 g of lactose, which would be the equivalent of four glasses of milk. Relating this to symptoms, some subjects may never have diarrhea, whereas others will have a violent episode. Those without diarrhea may have mild discomfort, bloating, or cramps after ingesting lactose.

Many persons are lactose intolerant but are unaware of the degree of deficiency. In addition, many are thought to have irritable bowel syndrome (IBS; see Chapter 108). Any patient with postprandial symptoms should undergo lactose tolerance testing.

DIAGNOSIS

Lactose tolerance tests are simple to perform and based on malabsorption resulting in bacterial fermentation in the intestine with hydrogen production. Hydrogen absorption then results in its recovery in expired air. The patient is given a standard dose of 50 g lactose in water. After a baseline hydrogen measurement, expired air is collected every 30 minutes for 3 hours. A 10-ppm increase in hydrogen is classified as a positive finding. There is usually no difficulty making a positive diagnosis, with no false-positive results. At times, patients may be thought to have a high baseline hydrogen level because of a diet very high in fiber or because of bacterial overgrowth in the small intestine. This is taken into consideration during the differential diagnosis.

If the small intestine is damaged because of acute infectious processes, lactose deficiency may be transient, and the test result may be positive. However, in an otherwise healthy patient with no acute infectious disease, a positive finding is indicative of lactose intolerance. Although it is a primary deficiency that occurs in infancy, as infants mature into childhood, lactase levels decrease in the small bowel mucosa, and the adult, or acquired, deficiency becomes apparent. If the small bowel is damaged, the enzyme level becomes deficient because of the damage to the small bowel mucosa that occurs in diseases such as celiac disease. Many clinicians believe that the deficiency is increased in inflammatory bowel disease, but this apparently is not true and is merely related to the small patient population. Whenever there is confusion concerning intolerance, a lactose tolerance test should be performed.

Measuring breath hydrogen level is the most accurate test. Arriving at the diagnosis by comparing glucose tolerance results with lactose tolerance results (i.e., measuring blood glucose level after patient ingested lactose) was the initial method used. Blood tests are now unnecessary. The breath hydrogen test is readily available and should be used.

TREATMENT AND MANAGEMENT

There are two alternatives for treatment and management. First, removing all lactose-containing foods from the diet may be necessary for extremely sensitive patients. For those who are only mildly sensitive, small amounts of lactose can be ingested without symptoms.

The second alternative is to use lactase substitutes. The enzyme is now readily available worldwide in many over-the-counter products. In addition, in many industrialized countries, milk and milk products are now produced with predigested lactose and are labeled "lactose free."

It is important to review the patient's diet so that the extremely sensitive patient can be fully educated regarding which foods contain lactose. Most are dairy products, and any product containing milk or cheese or made from milk or cheese contains lactose. In modern food production, lactose is frequently used as filler; therefore, all labels must be read carefully, and all commercial products must be carefully analyzed to ensure that there is no lactose. Also, some pharmaceutical companies add lactose as filler to capsules and pills, which can activate symptoms in patients with extreme sensitivity to lactose. Therefore, the content of capsules or pills should be checked for these sensitive patients.

COURSE AND PROGNOSIS

Removing lactose from the diet should relieve all symptoms. If not, the diagnosis should be questioned. The availability of lactase substitutes and lactose-free dairy products has been

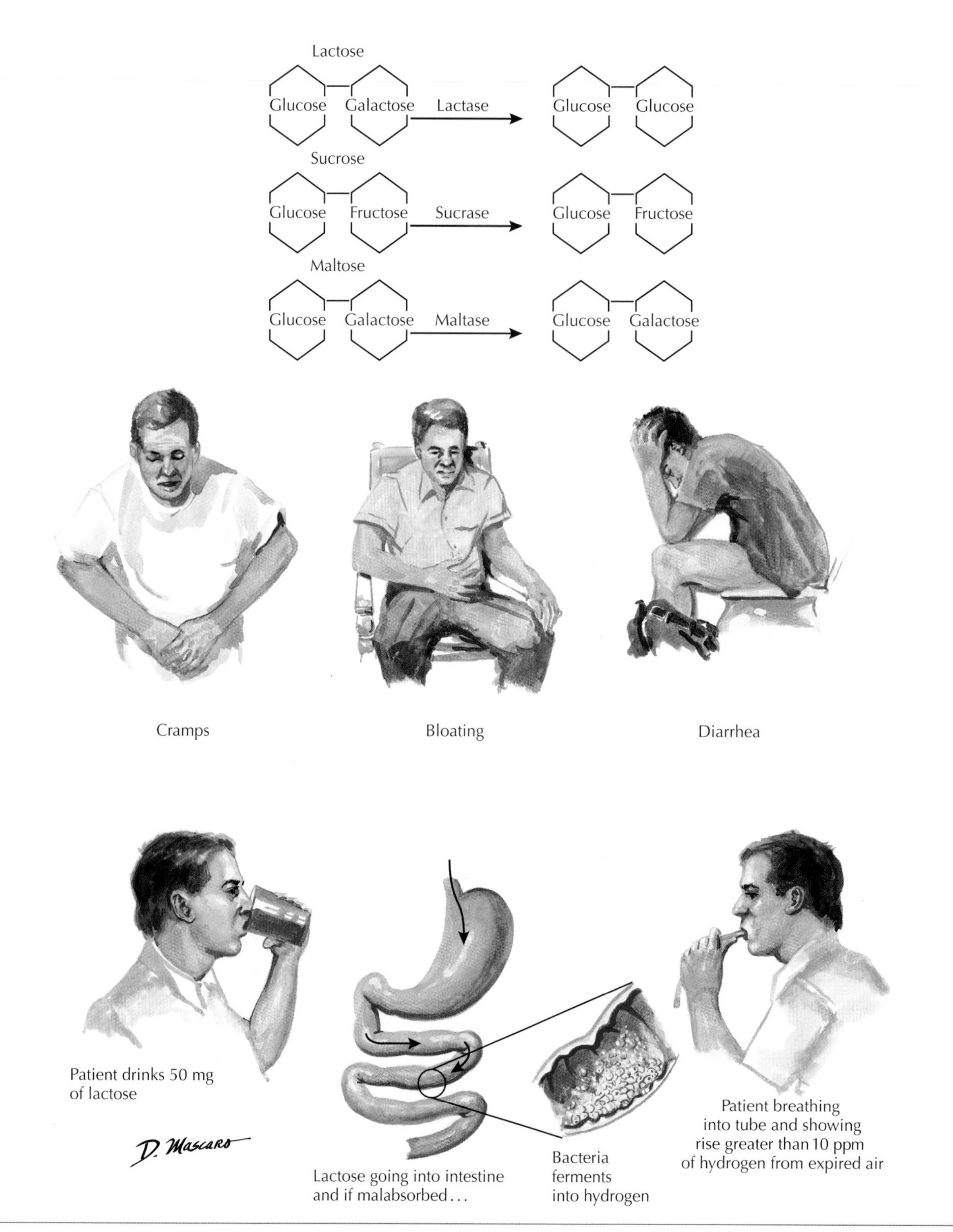

Figure 110-1 *Lactose Intolerance.*

helpful to patients with lactose intolerance. Educating the patient and controlling the diet usually result in relief of symptoms. Prognosis is excellent because this is merely an intolerance that can be corrected.

Patients who remove lactose from their diets should be cautioned about adequate calcium intake because many relied on milk and milk products for their calcium supply.

FRUCTOSE INTOLERANCE

Fructose intolerance results from inborn errors of metabolism or is acquired in adulthood. Because of its rarity, fructose intolerance is less understood than lactose intolerance. Some persons definitely have *sucrose intolerance*, which manifests when disaccharide sucrose is digested to free glucose and fructose. Fructose is present in some fruits and notably present in honey. It is also added to some candies.

Sucrase deficiency in the small bowel, or malabsorption of fructose through the brush border of the enterocytes, results in symptoms. Diarrhea, chronic abdominal pain, and bloating have been reported. Testing for this symptom complex is similar to the lactose breath test except that fructose is used. An increase in hydrogen level indicates malabsorption and usually correlates with the symptoms.

The best treatment is to remove fructose from the diet. In patients with many symptoms, sucrose is also removed. This fructose-intolerant population is more difficult to manage than lactose-intolerant patients.

Physician and patient awareness of possible fructose intolerance has resulted in reports of more cases. As much as 30 g fructose may be present in some sodas and some artificially flavored drinks, so it is important to be aware of the possibility of fructose malabsorption and intolerance. A careful study in which patients were advised to reduce sucrose and fructose intake resulted in a significant reduction in symptoms of irritable bowel syndrome. Reducing intake is difficult but can be rewarding.

ADDITIONAL RESOURCES

Bayless TM, Rosenzweig MS: Racial difference in the incidence of lactase deficiency, *JAMA* 197:968-972, 1966.

Bond JH, Levitt MD: Fate of soluble carbohydrate in the colon of rats and man, *J Clin Invest* 57:1158-1164, 1976.

Gibson PR, Newnham E, Barrett JS, et al: Review article: fructose malabsorption and the bigger picture, *Aliment Pharmacol Ther* 25:349-363, 2007.

Holzel A, Schwarz V, Sutcliff KW: Defective lactose absorption causing malnutrition in infancy, *Lancet* 1:1126-1129, 1959.

Hommes FA: Inborn errors of fructose metabolism, *Am J Clin Nutr* 58:788-795, 1993.

Kuokkanen M, Enattah NS, Oksanen A, et al: Transcriptional regulation of the lactase-phlorizin hydrolase gene by polymorphisms associated with adult-type hypolactasia, *Gut* 52:647-652, 2003.

Mann NS, Cheung EC: Fructose-induced breath hydrogen in patient with fruit intolerance, *J Clin Gatroenterol* 42:157-159, 2008.

Quinn RJ, Montgomery RA, Chitkra TK, Hirschorn JN: Changing genes: losing lactase, *Gut* 52:617-619, 2003.

Riby JE, Fujisawa T, Kretchmer M: Fructose absorption, *Am J Clin Nutr* 58(suppl):748-753, 1993.

Sahi T: Genetics and epidemiology of adult-type hypolactasia, *Scand J Gastroenterol* 29(suppl):7-20, 1994.

Shepherd SJ, Gibson PR: Fructose malabsorption and symptoms of irritable bowel syndrome: guidelines for effective dietary management, *J Am Diet Assoc* 106:1631-1639, 2006.

Diarrhea

111

Martin H. Floch

Since the time of Hippocrates, the term *diarrhea* has been used to designate abnormally frequent passage of loose stools. However, it is a subjective symptom. Patients describe any increased frequency or fluidity to mean diarrhea. It is generally accepted that more than three bowel movements per day represent diarrhea, or a consistency of watery, large-volume liquid, or both. The clinician must obtain an accurate history and ensure that the patient is not reporting incontinence or staining.

The volume of stool in diarrhea is important but is usually difficult to determine. A constipated stool amounts to less than 100 mL per day. In Western societies, with diets of moderate fiber, stool volumes amount to approximately 200 mL daily. When stools are measured in vegetarians or those with a high intake of fiber, the volume may be 400 mL daily. Watery, voluminous stools exceed 500 mL and may reach 1 to 2 L per day in diarrhea.

CLINICAL PICTURE

The presentations of acute and chronic diarrhea are different. The patient with *acute diarrhea* is frequently ill and may have a fever, dehydration, severe abdominal cramps, or uncontrollable passage of watery stools. If the cause is severe gastroenteritis, diarrhea is frequently associated with vomiting.

The patient with *chronic diarrhea* frequently has abdominal cramps or bloating and discomfort associated with an increased number of bowel movements. It is important to ascertain whether movements are nocturnal because this usually heralds organic disease rather than irritable bowel syndrome (IBS). If the diarrhea has lasted more than 10 to 30 days, it is considered chronic. Chronic diarrhea frequently entails the loss of important nutrients and the beginning of malabsorption syndromes.

In either acute or chronic diarrhea, the patient may appear toxic or dehydrated. When dehydration is a major component, the patient has lost a great amount of water and electrolytes.

DIAGNOSIS

Worldwide, the major cause of diarrhea is infection, which may be acute or chronic. Acute viral or bacterial infections can have a violent onset with severe, watery diarrhea; cholera is a classic form. The chronic diarrheas are caused by (1) lactose and food intolerance, (2) endocrinopathy such as hyperthyroidism, (3) carcinoid-secreting tumors, and (4) IBS. **Figure 111-1** categorizes causes of diarrhea.

With acute infections, gastroenteritis usually is limited, and differential diagnosis is not pursued. However, in epidemics of the common Norwalk virus, rotavirus, or bacterial food contamination or in extensive cholera epidemics, the agent is identified for public health purposes (see Section VI).

There is definite overlap among the inflammatory bowel diseases, such as chronic ulcerative colitis, Crohn's disease, and the other colitides (see Section V). It is important to note for the differential diagnosis that these diseases frequently manifest with abdominal cramps; the course may be prolonged; blood may or may not be present in the stool; and testing for an infectious agent in the stool is negative. Although the onset of inflammatory bowel disease may be acute, it usually becomes chronic and then falls into the differential diagnosis of the other metabolic causes listed in Figure 111-1. IBS associated with diarrhea rather than constipation is common, and with a negative workup, the final diagnosis becomes apparent.

Whenever the cause of diarrhea is uncertain, it is helpful to determine whether the patient has so-called osmotic or secretory diarrhea. Classic *osmotic* presentations are lactose intolerance and increased intake of magnesium salts or other agents that are poorly absorbed. Classic *secretory* diarrhea is demonstrated by enterotoxins, such as cholera and neuroendocrine tumors, and is seen with surgical and severely damaged bowel. To make this differential, fecal electrolytes and osmolality are determined. The osmolality of colonic fecal water should be approximately that of body fluids; therefore the composite should be no more than 290 mOsm/kg. Osmotic and secretory diarrhea can be differentiated by the "osmotic gap." Secretory diarrhea stool fluid electrolytes reach a level of body fluid electrolytes, whereas the osmolality in the osmotic diarrheas is low and usually caused by ingestion of nonelectrolyte substances. Electrolyte studies of stool are rarely used now because of the difficulty in collection and the confusing analysis attributed to ingested salts, making them impractical. As discussed for malabsorption disorders, chronic diarrhea may be large and muddy, and may contain fat, resulting in greasy and foul-smelling stool, which helps in the differential diagnosis (see Chapters 109 and 112).

TREATMENT AND MANAGEMENT

Treatment of acute diarrhea is largely symptomatic and supportive. With extreme diarrhea, or in a patient who has an ileostomy, dehydration may be rapid and may necessitate intravenous fluid and electrolyte replacement. Most patients with gastroenteritis do not require such vigorous therapy. However, fluid replacement is frequently necessary, and in patients with cholera, it is mandatory. Commonly used in Western and Eastern societies, *oral rehydration solutions* are usually based on a high-sodium/potassium electrolyte content associated with simple glucose that enhances absorption in the small intestine. However, other carbohydrates have recently been added to the solutions to enhance short-chain fatty acid absorption from the colon. These solutions aid in the positive absorption of water and electrolytes to treat the excretion caused by diarrheal toxins. Home oral rehydration formulas usually contain 90 to 120 mEq sodium, 25 to 35 mEq potassium, and 25 to 50 g carbohydrate per liter.

Because infectious causes are common, physicians and paramedics often empirically use antibiotic therapy. Using antibiotics without a confirmed infectious cause is not recommended for the short course. When the course is prolonged over days

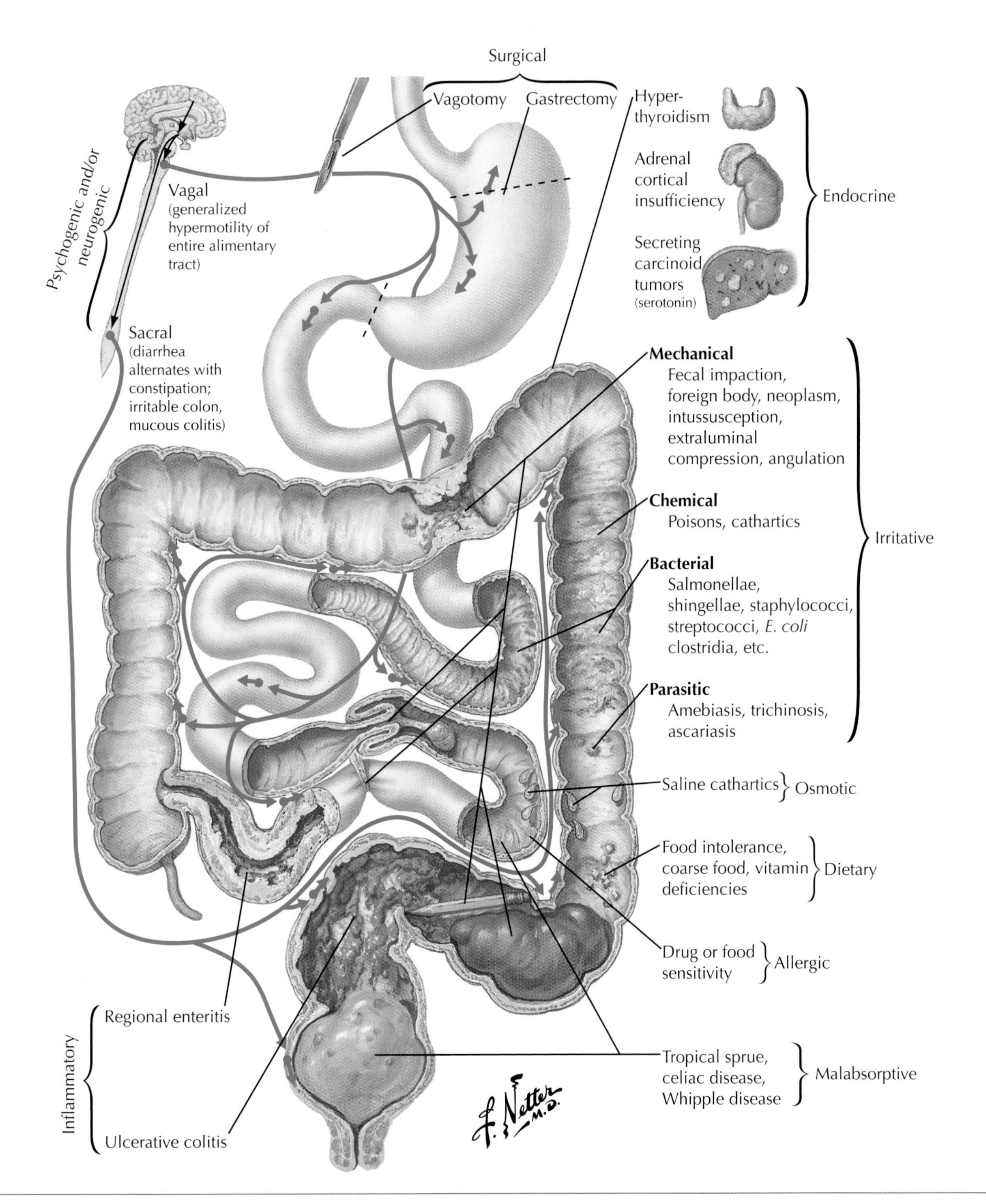

Figure 111-1 *Categories of Diarrhea and Causes.*

into a week, fluoroquinolone is often given and may be effective. As discussed in Section VI, whether to treat salmonellosis is always controversial.

Nonspecific antidiarrheal agents can reduce symptoms. The initial concern about reducing the clearance of pathogens by using such agents has been unsubstantiated. The agents most frequently available are Diphenoxylate, 2.5 to 5 mg four times a day (qid); loperamide, 2 to 4 mg qid; codeine, 15 to 60 mg qid; and tincture of opium, 2 to 20 drops qid. Recently, probiotics have been shown to be effective in shortening the course of acute diarrhea. *Lactobacillus* GG and other organisms are recommended.

Treatment of chronic diarrhea and the use of α-adrenergic agonists (e.g., clonidine), octreotide (somatostatin analog), bile salt–binding resins (e.g., cholestyramine), and psyllium seed (to solidify stool) are discussed under particular disease chapters in this section as well as Sections V and VI.

COURSE AND PROGNOSIS

Although most episodes resolve in a few days, acute infectious diarrhea may result in death in some children and immunocompromised or frail patients. Therefore, although rare, severe dehydration and vascular instability must be treated aggressively. The prognosis for chronic diarrhea is discussed under each particular disease.

ADDITIONAL RESOURCES

Floch MH, Walker WA, Guandalini S, et al: Recommendations for probiotic use—2008, *J Clin Gastroenterol* 42(3 Suppl):S104-S108, 2008.

Ramakrishna BS, Venkataraman S, Srinivasan P, et al: Amylase-resistant starch plus oral rehydration solution for cholera, *N Engl J Med* 342:308-314, 2000.

Schiller LR, Sellin DH: Diarrhea. In Feldman M, Friedman LS, Brandt LJ, editors: *Gastroenterology and liver disease*, ed 8, Philadelphia, 2006, Saunders-Elsevier, pp 159-186.

Szajewska H, Skorka A, Ruszczynski M, Gieruszczak-Bialek D: Meta-analysis: *Lactobacilllus* GG for treating acute diarrhea in children, *Aliment Pharmacol Ther* 25:871-881, 2007.

Tesjeux HL, Briend A, Butzner JD: Oral rehydration solution in the year 2000: pathophysiology, efficacy, and effectiveness, *Baillieres Clin Gastroenterol* 11:509-515, 1997.

Celiac Disease and Malabsorption

112

Martin H. Floch

Any disease that affects nutrient digestion or small bowel function, or that compromises bowel circulation or motility, may result in a malabsorption syndrome, which includes systemic vascular, infectious, and neoplastic diseases (**Fig. 112-1**). This chapter and others in this section discuss common intestinal diseases that cause malabsorption disorders.

Gluten enteropathy, or *celiac disease*, is a malabsorption syndrome that results from gluten-sensitive damage to the intestinal microvilli and villi, producing an abnormal villous architecture and resulting in malabsorption. This disease process exemplifies the classic signs and symptoms of malabsorption disorders. When a person with gluten enteropathy ingests gluten, the epithelium becomes damaged, the cellular maturation of epithelial cells of the villus becomes disturbed, the small bowel mucosa becomes inflamed, and mild villous atrophy to total loss of villi results in atrophic-looking mucosa.

Understanding this classic disease entity leads to an understanding of small bowel function and all its possible diseases. Almost all manifestations of abnormal digestion and absorption and all systemic manifestations, from skin disorders to malignancy, are associated with celiac disease.

Samuel Gee first described celiac disease in 1888 as "celiac affliction," later publicized by Herter in 1908. Finally, in 1950, Dicke and colleagues found that removing wheat from the diet caused the signs and symptoms to disappear. "Celiac disease" should be used only when gluten is demonstrated to be the cause. Some persons have "refractory sprue," in which gluten removal does not reverse the disease. Other names associated with this disorder include "idiopathic steatorrhea" and "celiac sprue."

Celiac disease occurs worldwide, affecting an estimated 1% of the world's population. Ireland has the highest incidence, but recently, a high incidence has been reported in Italy. Gluten enteropathy appears primarily in Europe, North and South America, East Asia, and Australia.

When glutens from wheat, rye, and barley are presented to the intestinal mucosa, they react to form *gliadins*, which can cause the damage. Genetic studies show that a person must have alleles that encode for HLA-DQ2 or HLA-DQ8 proteins. However, many persons have DQ2 or DQ8 human lymphocyte antigen (HLA) gene expression and do not have celiac disease. Patients with celiac disease appear to be able to tolerate oats, which contain some of the biochemical products present in the other grains. Along with oats, patients tolerate rice, corn, sorghum, and millet without intestinal villous damage.

CLINICAL PICTURE

Figure 112-2 demonstrates all possible presentations and signs/symptoms associated with celiac disease and severe malabsorption. The cardinal presentation of weight loss associated with steatorrhea is seen only occasionally. The clinician should check for celiac disease in the current environment of plentiful food when a patient experiences anemia, osteoporosis, unexplained diarrhea, or any vitamin deficiency, even if weight loss is not apparent. With the availability of numerous serologic tests, latent celiac disease has become more apparent in patients with such conditions as occult anemia, osteoporosis, and some associated malignancies.

DIAGNOSIS

Serum testing has improved the ability to make an early diagnosis of celiac disease and malabsorption syndrome. Immunoglobulin A (IgA) endomysial antibodies and IgA tissue transglutaminase antibodies have reached almost 98% sensitivity and specificity. IgA and IgG antigliadin antibodies are less sensitive and less specific but more helpful.

The standard for diagnosing celiac disease is to demonstrate the histologic lesions on small bowel biopsy. Although initial understanding of this disease resulted from biopsy specimens obtained by suction through tubes or capsules, current biopsy specimens are obtained by punch biopsy at endoscopy. To make the diagnosis, the clinician must be certain that the specimens are obtained from the second or third part of the duodenum, to avoid distortion by glands of the proximal duodenum. Furthermore, adequately sized specimens must be obtained. However, a lesion is not specific for the diagnosis, and significant inflammation or atrophy of villi may be seen in many other disorders, such as tropical sprue and infectious enteropathies. Marked inflammation may also be noted in lymphoproliferative disorders.

Therefore, the diagnosis of celiac disease can be elusive, although capsule endoscopy has become an additional tool to help in difficult cases.

Other tests of small bowel absorption, such as the 72-hour stool collection for fecal fat and the D-xylose absorption test (see Chapter 109), are often helpful when there is some confusion about serologic findings. Although imaging was important in the early stages of understanding celiac disease, now it is less helpful, although it remains useful for ruling out lesions that might be secondary to gluten enteropathy (e.g., malignancy) or for diagnosing other causative disorders (e.g., collagen disease). A diagnosis of gluten enteropathy must be established by serology and histology and demonstration of malabsorption. Finally, many clinicians insist that reversing the primary or secondary findings must be demonstrated on the gluten-free diet.

Dermatitis herpetiformis is a skin disease that produces papulovesicular lesions symmetrically on the extensor surfaces of the extremities, trunk, buttocks, neck, and scalp. It usually manifests in adulthood and is associated with celiac disease in 80% of patients. However, less than 10% of patients with celiac disease have this skin disorder.

Other diseases associated with celiac disease include microscopic colitis, although these associations are sporadic and reported without evidence of celiac disease. The most disturbing association is the increased incidence of lymphoma of the bowel and other malignancies. Diligence is required when following up each patient.

TREATMENT AND MANAGEMENT

The treatment for celiac disease is a gluten-free diet. There is no other treatment. Removing all gluten from the diet is essential. All diet manuals carry clear recommendations. A celiac society has formed and helps patients to obtain appropriate grains and cereals for cooking and to follow the diet. It has been shown repeatedly that once wheat, barley, or rye is instilled in the small bowel, a lesion develops within hours. Therefore, patients may have attacks when they are "fooled" by foods that contain any of the incriminating grains. Dietitians are well versed on the disorder and have the adequate educational tools to treat patients.

For patients whose disease is refractory, some clinicians have added glucocorticoids to the therapy.

When a patient is seriously depleted, supplemental therapy is necessary, including vitamin replacement and specific iron or B_{12} (or folic acid) replacement for anemia. Because these deficiencies have been reported in many patients, replacement may be necessary for all nutrients.

COURSE AND PROGNOSIS

Patients who follow a gluten-free diet can have a normal life expectancy. However, the complications of malignancy or the inability to comply with the diet may result in serious morbidity and mortality. Approximately 3% of patients with gluten enteropathy acquire some malignancy, which is an extremely high number. Potential malignancy should be sufficient incentive for patients to comply with a gluten-free diet, but poor compliance is prevalent. Many patients do not comply because they do not have symptoms. However, the high incidence of related malignancy and occult disease indicates that the gluten-free diet is essential to decrease morbidity and mortality in all patients with celiac disease. It is also important to remember that many patients are asymptomatic, and symptoms may not develop until a complication presents.

ADDITIONAL RESOURCES

Benson JA, Culver PJ, Ragland S, et al: The D-xylose test in malabsorption syndromes, *N Engl J Med* 256:335-338, 1957.

Collin P, Reunala T, Pukkla E, et al: Celiac disease–associated disorders and survival, *Gut* 35:1215-1220, 1994.

DuBois RN, Lazenby AJ, Yardley JH, et al: Lymphocytic enterocolitis in patients with "refractory sprue", *JAMA* 262:935-938, 1989.

Green PH, Cellier C: Celiac disease, *N Engl J Med* 357:1731-1743, 2007.

Holmes GK, Stokes PL, Sorahan TM, et al: Celiac disease, gluten-free diet, and malignancy, *Gut* 17:612-618, 1976.

London KE, Scott H, Hansen T, et al: Gliadin-specific HLA-DQ restricted T-cells isolated from the small intestinal mucosa of celiac disease patients, *J Exp Med* 178:187-192, 1993.

Nelson JK, Moxness KE, Jensen MD, Gastineau CF: *Mayo Clinic diet manual: a handbook of nutrition practices*, St Louis, 1994, Mosby–Year Book.

Pemberton CM, Moxness KE, German MJ, et al: *Mayo Clin diet manual*, ed 6, Philadelphia, 1988, Decker.

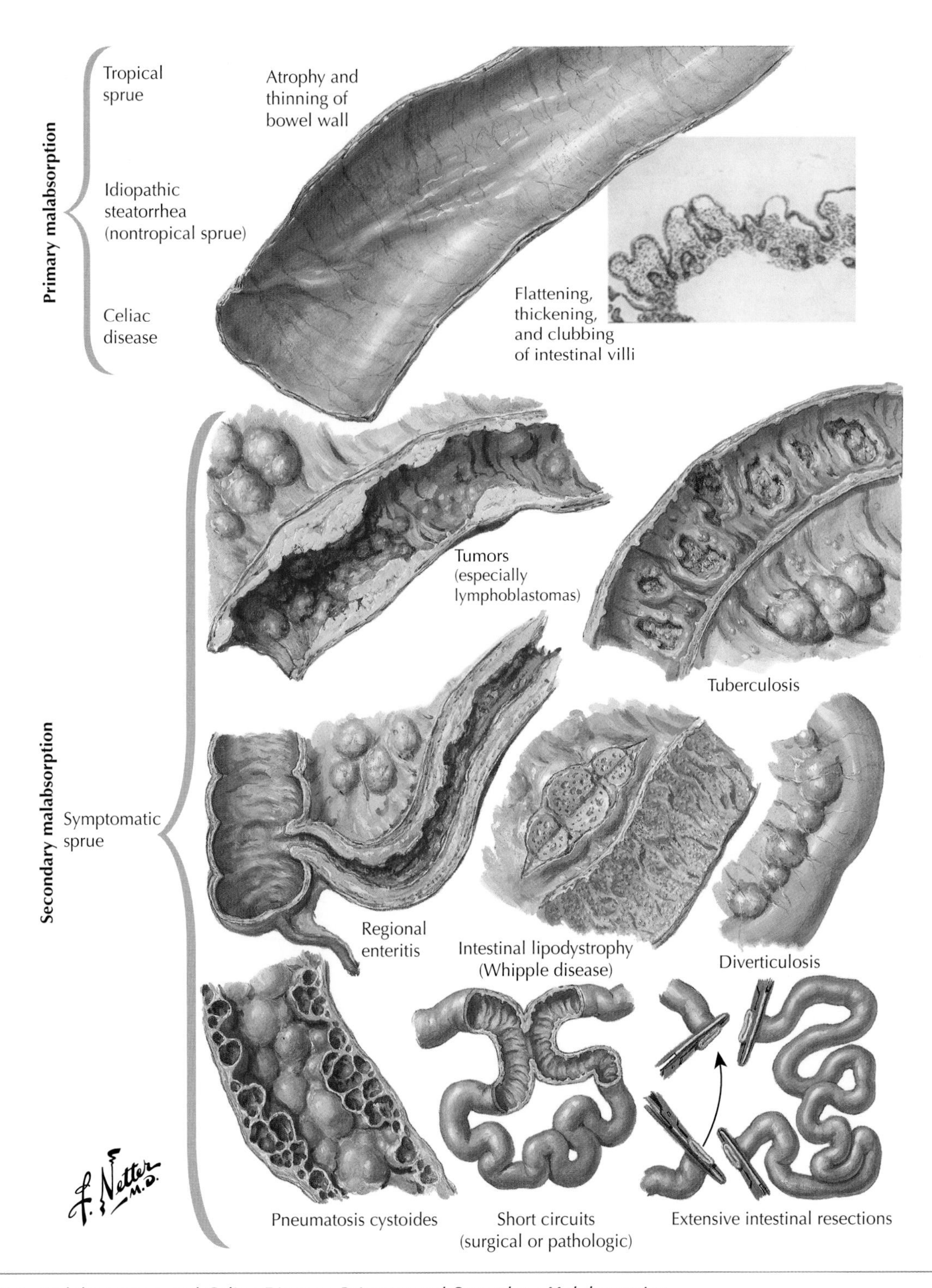

Figure 112-1 *Malabsorption and Celiac Disease: Primary and Secondary Malabsorption.*

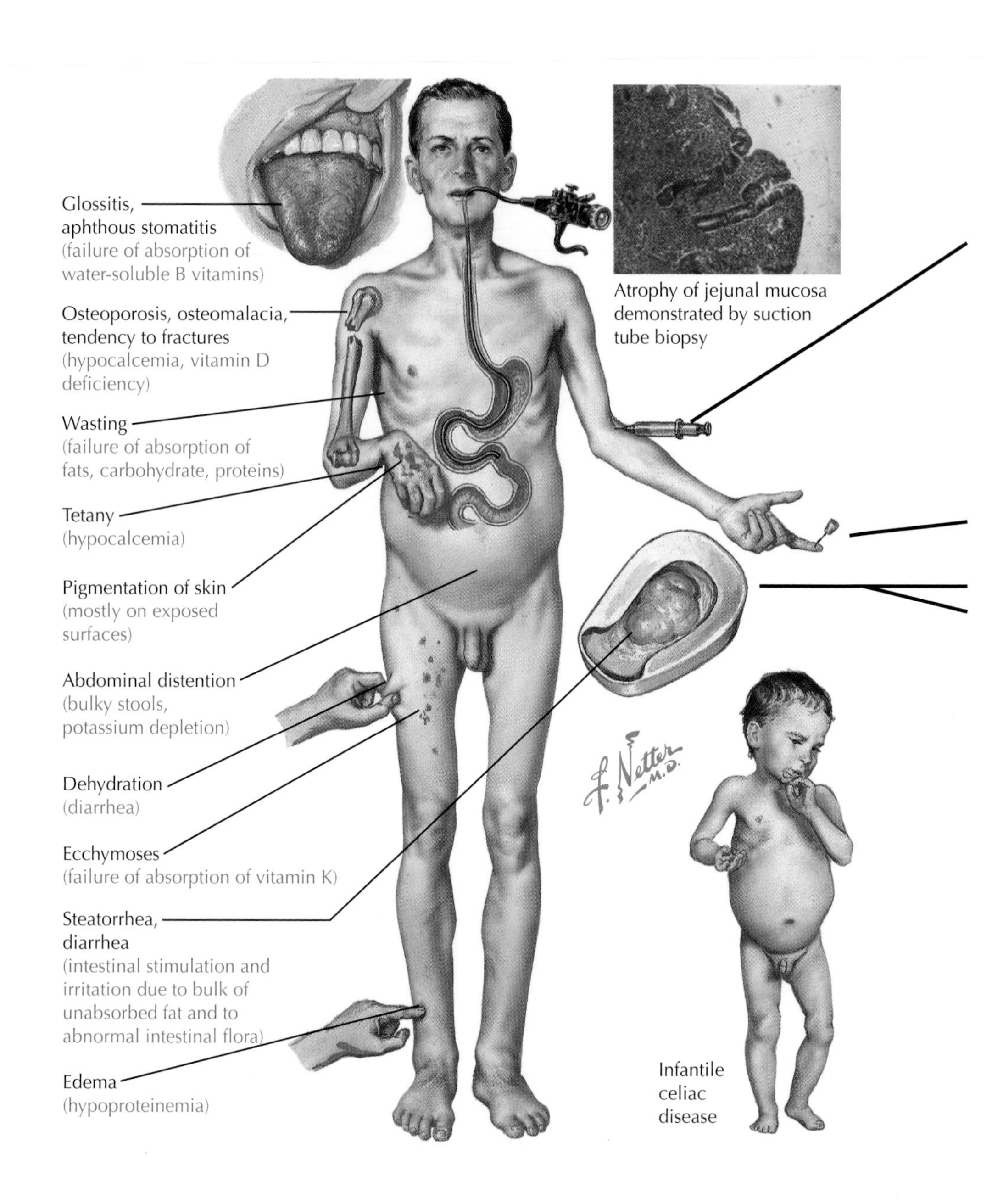

Figure 112-2 *Signs and Symptoms of Malabsorption.*

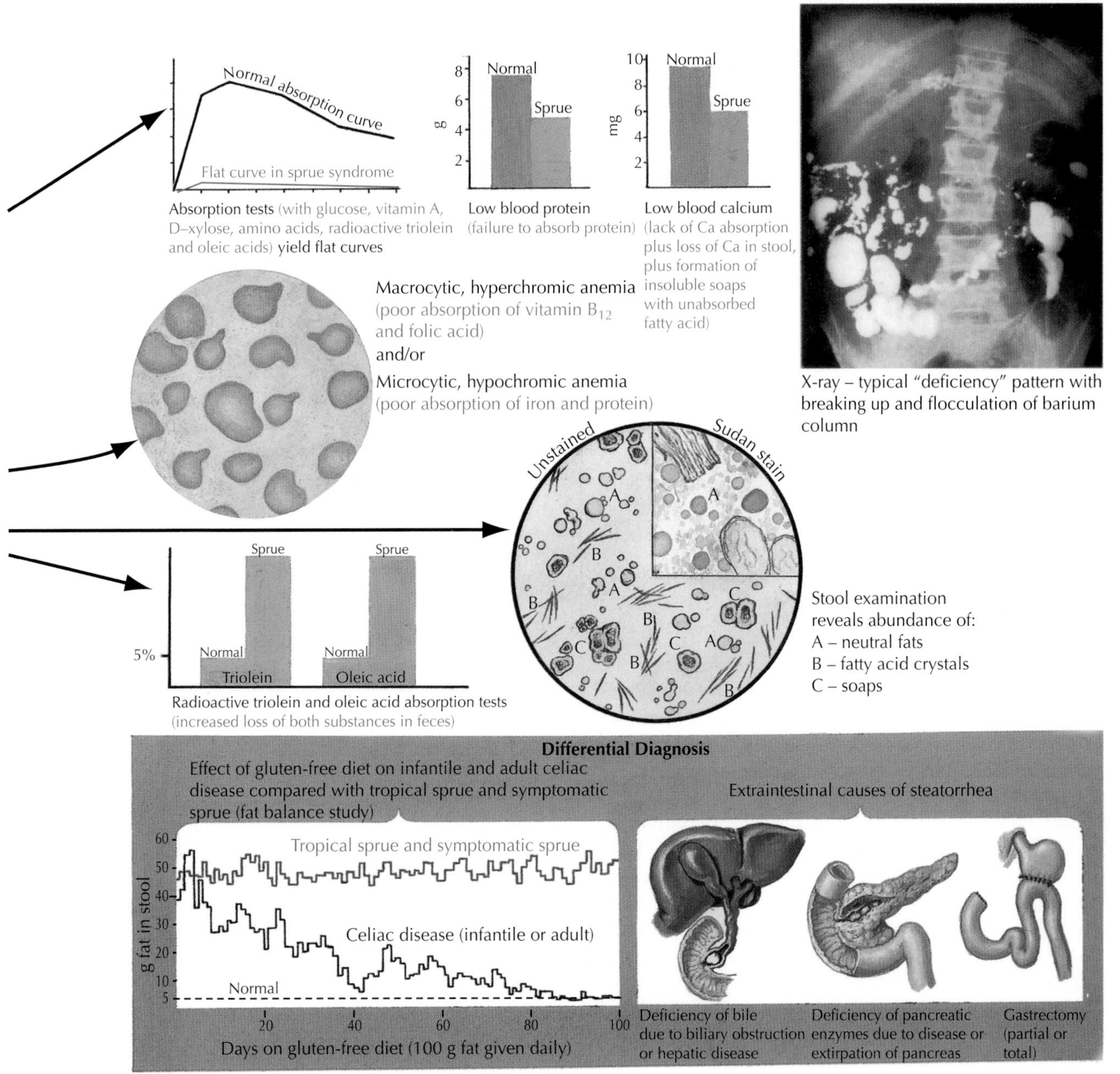
Normal absorption curve
Flat curve in sprue syndrome
Absorption tests (with glucose, vitamin A, D–xylose, amino acids, radioactive triolein and oleic acids) yield flat curves
Normal
Sprue
g
Low blood protein (failure to absorb protein)
Normal
Sprue
mg
Low blood calcium (lack of Ca absorption plus loss of Ca in stool, plus formation of insoluble soaps with unabsorbed fatty acid)
Macrocytic, hyperchromic anemia (poor absorption of vitamin B12 and folic acid)
and/or
Microcytic, hypochromic anemia (poor absorption of iron and protein)
X-ray – typical "deficiency" pattern with breaking up and flocculation of barium column
Unstained
Sudan stain
Stool examination reveals abundance of:
A – neutral fats
B – fatty acid crystals
C – soaps
Sprue
Sprue
5%
Normal
Normal
Triolein
Oleic acid
Radioactive triolein and oleic acid absorption tests (increased loss of both substances in feces)
Differential Diagnosis
Effect of gluten-free diet on infantile and adult celiac disease compared with tropical sprue and symptomatic sprue (fat balance study)
Extraintestinal causes of steatorrhea
Tropical sprue and symptomatic sprue
Celiac disease (infantile or adult)
Normal
g fat in stool
Days on gluten-free diet (100 g fat given daily)
Deficiency of bile due to biliary obstruction or hepatic disease
Deficiency of pancreatic enzymes due to disease or extirpation of pancreas
Gastrectomy (partial or total)
F. Netter M.D.

Whipple Disease

Martin H. Floch

113

Whipple disease is a systemic, infectious disease that primarily affects the small intestine and its lymphatic drainage, although it also has many extraintestinal manifestations, comparable to classic celiac disease (**Fig. 113-1**). George Whipple first described the clinical syndrome in 1907, but it was not until the bacterium that causes the disease, *Tropheryma whippelii*, was isolated and identified in 1991 and 1992 that full understanding of the entity and its response to therapy became clear. Ribosomal RNA sequences reveal that *T. whippelii* is a novel actinomycete not closely related to other characteristic organisms. The exact sequence of infectivity is not clearly understood, but *T. whippelii* is probably transmitted by the oral route.

CLINICAL PICTURE

Only approximately 700 cases of Whipple disease have been described in the literature, many from the state of Connecticut. The syndrome is so classic and the treatment so readily available, however, that it must enter into any differential diagnosis when a male patient has malabsorption syndrome. For unknown reasons, Whipple disease occurs more often in men.

A frequent symptom is the extraintestinal manifestation of *arthritis*. Patients may seek treatment for swollen joints that may be associated with any part of a malabsorption syndrome, usually steatorrhea, but they may also have anemia or weight loss. Early in the disease process, the patient may have arthritis with abdominal bloating or pain. The patient may not be able to describe the steatorrhea clearly but will report diarrhea. About 95% of patients with Whipple disease lose weight, 78% have diarrhea, 65% have arthralgias, and 60% have abdominal pain. Often, the arthralgias predate the other clinical manifestations by several years. Because the organism causes lipodystrophy and lymph nodes are involved, lymph node enlargement is common.

On physical examination, the patient may have abdominal tenderness, hyperpigmentation of the skin, and fever. When the disease is progressive, the patient may have full-blown malnutrition, often because the diagnosis was obscure and attributed to the aging process associated with arthritis. Many reports cite central nervous system involvement with neurologic manifestations, as well as signs of aging, including dementia, ophthalmoplegia, and myoclonus. There have been case reports of cardiac involvement, and fibrinous pericarditis and endocarditis are frequently seen postmortem.

Arthralgias are the primary extraintestinal manifestation and frequently precede full-blown Whipple disease by many years. Arthralgias tend to be migratory and show little evidence of inflammation. However, aspiration of the joints may reveal the organism and lipid-filled macrophages.

DIAGNOSIS

It is important to consider a diagnosis of Whipple disease because it can affect men with arthralgias, which may precede the signs and symptoms of malabsorption. All diseases that may cause malabsorption enter the differential diagnosis once there is evidence of steatorrhea and weight loss. Similarly, all differential diagnoses for arthralgias are present after the intestinal disease develops.

Once the clinician suspects Whipple disease, the diagnosis is confirmed through small bowel biopsy. All test findings that are positive in other malabsorption disorders may be positive in Whipple disease, and the biopsy specimen becomes pathognomonic. Lipid-filled macrophages that stain positively with periodic acid–Schiff (PAS) stain are dramatic and diagnostic. The same lipid deposition occurs in lymph nodes. The bacterium can be demonstrated on electron microscopy. Because it can be fatal if not treated, it is essential that the physician consider Whipple disease.

TREATMENT AND MANAGEMENT

After the discovery that Whipple disease was caused by a bacterium, single-antibiotic therapy proved successful in most patients, but there was a high degree of relapse. *T. whippelii* apparently behaves similarly to mycobacteria and chronic infections. Therefore, the current recommendation (and most successful treatment) is a combination of initial treatment and long-term therapy. The current recommendation is penicillin G plus streptomycin, 6 to 24 million U intravenously daily, plus 1 g intramuscularly daily for 10 to 14 days. An alternative for patients sensitive to penicillin is ceftriaxone, 2 g intravenously daily for a similar period. The recommendation for long-term therapy to prevent relapse is maintenance with trimethoprim-sulfamethoxazole, 160 to 800 mg orally two or three times daily for at least 1 year. Tetracycline was initially the drug of choice, but patients often relapsed, and it is no longer a first-line agent. Because of antibiotic allergies and sensitivities, many other regimens are used and described in the literature.

COURSE AND PROGNOSIS

When first described and before antibiotic treatment, Whipple disease was a progressive and unrelenting fatal disease. Antibiotics now make this a curable, treatable disease. Therapy is rewarding, although some neurologic signs, including dementia, may not reverse. The PAS-positive macrophages disappear slowly and may persist up to 1 year (or in rare cases, 8 years) before being completely resolved. However, most extraintestinal manifestations and positive signs disappear within 1 year. It is important to remember that long-term antibiotic therapy is usually needed.

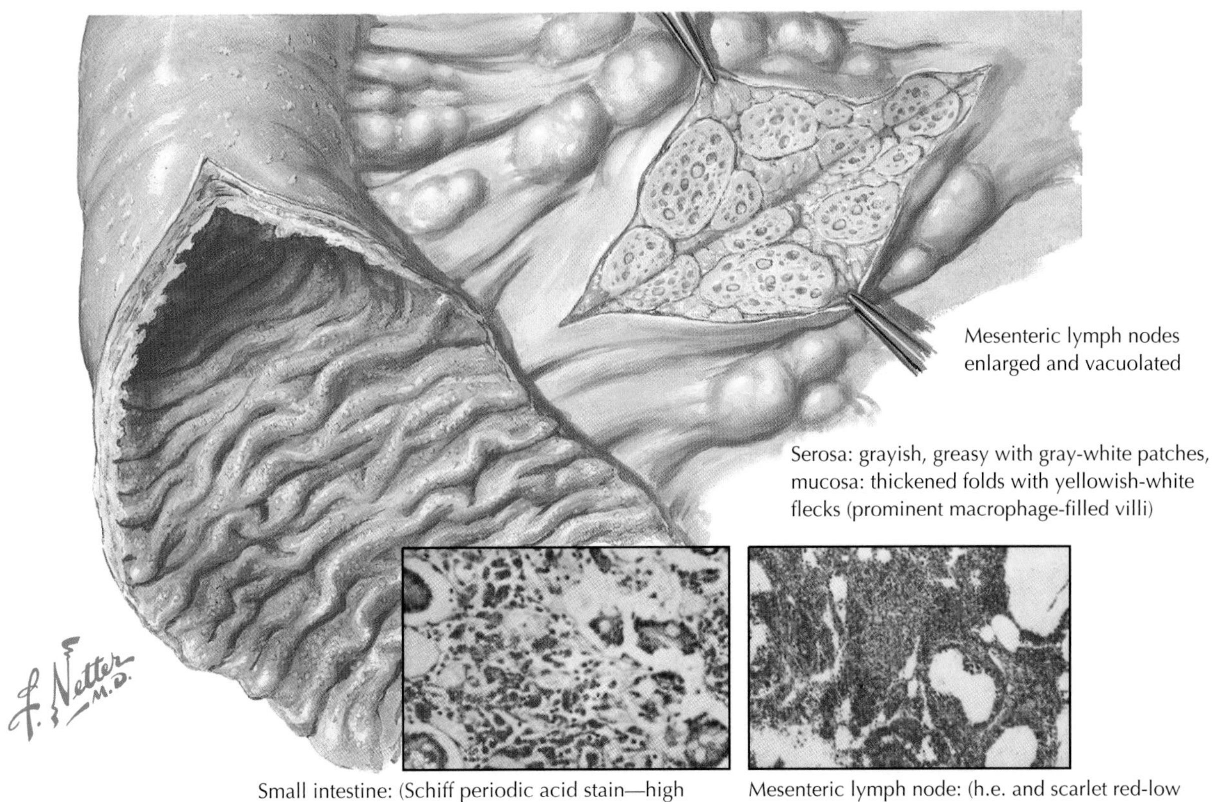

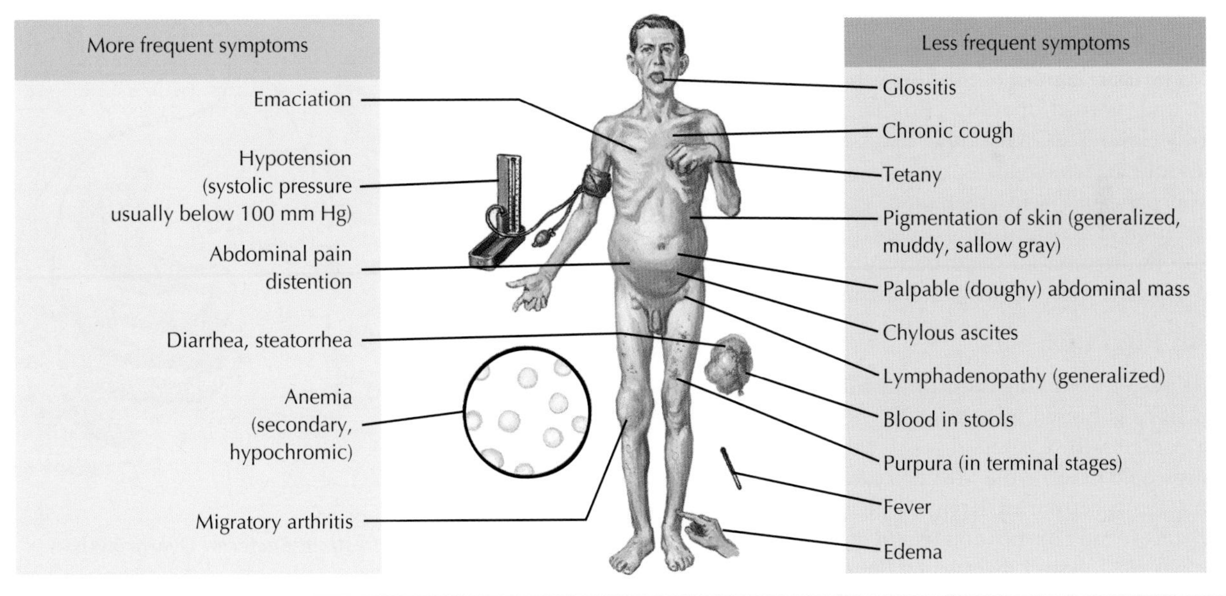

Figure 113-1 *Whipple Disease.*

ADDITIONAL RESOURCES

Fleming JL, Wiesner RH, Shorter RG: Whipple's disease: clinical, biochemical, and histopathologic features and assessment of treatment in 29 patients, *Mayo Clin Proc* 63:539-551, 1988.

Maiwald M, von Herbay A, Relman DA: Whipple's disease. In Feldman M, Friedman LS, Brandt LJ, editors: *Gastrointestinal and liver disease*, ed 8, Philadelphia, 2006, Saunders-Elsevier, pp 2319-2331.

Maizel H, Ruffin JM, Dobbins WO: Whipple's disease: a review of nineteen patients from one hospital, and a review of the literature since 1950, *Medicine* 49:175-205, 1970.

Marth T: Whipple's disease. In Mandell GL, Bennett JE, Folin R, editors: *Principles and practices of infectious diseases*, ed 6, Philadelphia, 2005, Churchill Livingstone–Elsevier, pp 1306-1310.

Ralmen DA, Schmidt TM, MacDermott RP, et al: Identification of the uncultured bacillus of Whipple's disease, *N Engl J Med* 327:293-302, 1992.

Wilson KH, Blitchington R, Frothingham R, et al: Phylogeny of the Whipple's disease–associated bacteria, *Lancet* 338:474-475, 1991.

Small Intestinal Bacterial Overgrowth

114

Martin H. Floch

The small intestinal bacterial overgrowth (SIBO) syndrome may be caused by a disturbance in gastrointestinal (GI) motility, by an alteration in the anatomy of the intestine, or less likely, by a loss of gastric acid secretion. The disturbance usually causes malnutrition and weight loss from malabsorption. In the healthy small bowel, indigenous bacterial flora growth in the jejunum ranges from 10^2 to 10^5 colony-forming units per milliliter (CFU/mL). Bacteria consist primarily of gram-positive and gram-negative aerobic organisms. Farther down in the small bowel, toward the midileum and into the terminal ileum, the bacterial flora proliferate, and anaerobic organisms increase dramatically, becoming as numerous as the aerobe organisms, and the total counts in the distal ileum can reach 10^8 or 10^9 CFU/mL. Intestinal flora and the increase in anaerobes are discussed in Section VI.

Aerobic and anaerobic organisms proliferate when stasis occurs in the small bowel (**Fig. 114-1**). *Stasis* results from prolonged strictures or delayed emptying of the small bowel, and organisms are also seen in jejunal diverticula. When the aerobic and anaerobic flora proliferate, they can interfere with absorption, compete for nutrients, and produce substances that cause symptoms. The usual control of the flora is based on gastric acid production and normal motility. Proliferating bacteria can compete for vitamin B_{12}. On the other hand, they can synthesize and produce folic acid. Fat malabsorption caused by interference with enterohepatic bile acid circulation, carbohydrate use by bacteria, and protein malabsorption caused by decreased absorption and use by bacteria all contribute to the malabsorption. Because of bacterial toxic substances, diarrhea might increase.

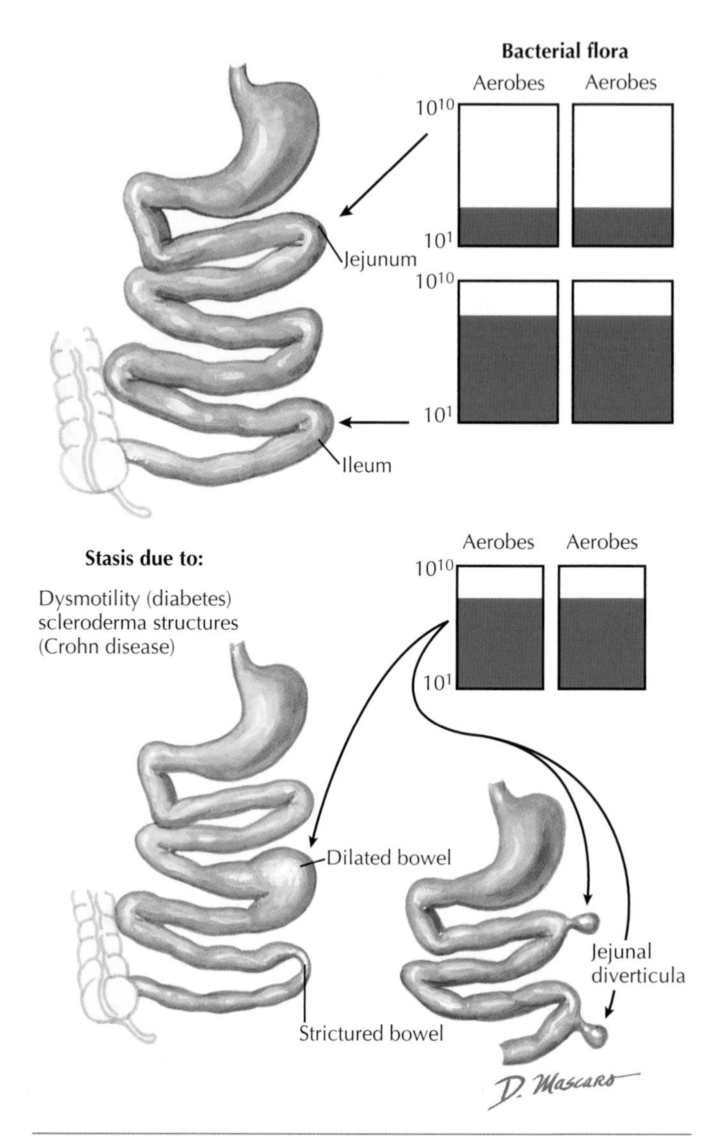

Figure 114-1 *Small Intestinal Bacterial Overgrowth.*

CLINICAL PICTURE

Clinical features vary with the cause of the SIBO syndrome. For example, if stasis is caused by a stricture with dilatation of the bowel in a patient with Crohn's disease, pain and a change in bowel habits would precede the SIBO picture. If the patient has scleroderma, the spectrum of Raynaud phenomenon, esophageal disturbances, and constipation would accompany SIBO. If the patient has small bowel diverticula, the presentation might be malabsorption. In rare cases, the presentation might simply be anemia or malnutrition with no other symptoms before presentation.

The usual causes of SIBO are disturbed anatomy (e.g., blind loops, diverticulosis, strictures), dysmotility (e.g., diabetic enteropathy, scleroderma, pseudointestinal obstruction), and various other conditions in which the small bowel becomes diseased and stasis occurs.

DIAGNOSIS

Diagnosis depends on demonstrating a cause for the bacterial overgrowth and the resulting malnutrition and malabsorption. Radiographic techniques are essential for demonstrating dilated loops of small bowel, diverticula, and strictures. In some patients, bacterial overgrowth can be documented. Most institutions do not perform colony counts on small bowel aspirates; when available, it is helpful. Currently, however, malabsorption is often documented with breath tests, including the carbon 14 (^{14}C)–bile acid breath test, ^{14}C-xylose breath test, fasting hydrogen breath test, lactulose-hydrogen breath test, and glucose-hydrogen breath test. Use of these tests varies greatly among institutions.

All these tests are relatively simple to perform, but the ^{14}C-xylose breath test appears to be the most sensitive and the most specific. After a 1-g oral dose of ^{14}C-xylose, elevated levels of $^{14}CO_2$ are detected in the breath within 1 hour in more than 85% of patients with SIBO syndrome. Findings on the lactulose-hydrogen breath test and baseline H_2 levels can both be positive, but these measurements appear to be less sensitive than the xylose test. However, they are preferred for pregnant women and children because they do not require significant use of isotopes. In addition, Schilling test findings are classically positive in bacterial overgrowth.

Many clinicians avoid detailed workups. Once they have demonstrated an anatomic reason for bacterial overgrowth, they institute a trial of antibiotic therapy to determine whether the syndrome can be corrected. This is a valid approach when isotope and absorption studies are difficult to conduct.

TREATMENT AND MANAGEMENT

All attempts should be made to correct the cause of the stasis or obstruction. However, this is often difficult surgery that is not feasible or warranted in patients with scleroderma, diabetes, or small bowel dysmotility. Therefore, it becomes necessary to use antibiotic therapy for extended periods of time. Most clinicians believe in alternating antibiotic treatment with periods of no treatment. This cyclic therapy has been used in all types of combinations, such as 1 week off and 3 weeks on per month or 1 month on and 1 month off. This type of cyclic therapy has been individualized.

Antibiotics used to treat SIBO include amoxicillin–clavulanic acid, twice daily (bid); cephalexin, 250 mg four times daily (qid), plus metronidazole, 250 mg three times daily; trimethoprim-sulfamethoxazole, bid; or chloramphenicol, 250 mg qid. In addition, ciprofloxacin, norfloxacin, and rifaximin have recently been used successfully in a variety of protocols. Somatostatin and prokinetic agents have also been used to help in dysmotility syndromes. Therapy must be individualized for each patient depending on the disease and disorder causing the bacterial overgrowth.

It is important that the antibiotic selected is well tolerated and can be used cyclically for the individual patient, and that symptoms and malabsorption are relieved by the particular regimen. Some patients are successfully treated for short periods with long intervals between treatment. Each patient should be monitored closely.

Nutritional deficiencies should be corrected. Correcting the bacterial overgrowth will address future deficiencies if therapy is successful. However, any vitamin or mineral deficiencies should be corrected by increased supplements. When bacterial growth decreases, absorption of macronutrients and vitamins corrects any weight loss, as monitored with absorption testing. Outcome evaluation (e.g., use of serum prealbumin) and repeat evaluation for anemia and breath analysis should be performed.

COURSE AND PROGNOSIS

The prognosis for SIBO depends on the disease process causing the overgrowth. If caused by scleroderma, the prognosis is that of scleroderma's natural history. If the cause is Crohn's disease or diabetes, all their comorbidities are in effect. Once a regimen is found for the overgrowth, SIBO can be managed, and the morbidity of the primary disease can be decreased with the hope of decreasing mortality risk.

ADDITIONAL RESOURCES

Gregg CR: Enteric bacterial flora and bacterial overgrowth syndrome, *Semin Gastrointest Dis* 13:200-209, 2002.

O'Mahoney S, Shanahan F: Enteric bacterial flora and bacterial overgrowth. In Feldman M, Friedman LS, Brandt LJ, editors: *Gastrointestinal and liver disease*, ed 8, Philadelphia, 2006, Saunders-Elsevier, pp 2243-2256.

Quigley E, Quera R: Small intestinal bacterial overgrowth: roles of antibiotics, prebiotics, and probiotics, *Gastroenterology* 130:S78-S90, 2006.

Saltzman JR, Russell RM: Nutritional consequences of intestinal bacterial overgrowth, *Comp Ther* 20:523-530, 1994.

Schiller LR: Evaluation of small bowel overgrowth, *Curr Gastroenterol Rep* 9:373-377, 2007.

Short Bowel Syndrome

115

Martin H. Floch

Short bowel syndrome (SBS) usually occurs when less than 200 cm of small intestine remains after intestinal surgery. Normally, the bowel measures 450 to 500 cm (18-20 feet). When an insult occurs to the bowel resulting in surgical removal of major parts, the patient is left with SBS (**Fig. 115-1**). Although resection of less than 75% of the bowel, leaving larger amounts of the small bowel, may not result in the full SBS, it still can precipitate some aspects of the syndrome. For example, vitamin B_{12} deficiency occurs when there is total ileal resection but an intact jejunum.

Surgery-related causes of SBS that require treatment are vascular compromise, Crohn's disease, volvulus, radiation enteritis, trauma to the bowel, and neoplasm. In infants, congenital atresias and stricture are noted, but volvulus with malrotation may be a cause, as well as necrotizing enterocolitis. Fortunately, SBS is not common, although the incidence and prevalence of survivors is increasing as therapy improves. At least 40,000 to 50,000 patients with SBS require total parenteral nutrition (TPN) each year, and at least as many are managed without TPN.

CLINICAL PICTURE

The clinical syndrome is usually divided into three phases: acute, adaptation, and maintenance. The acute phase occurs immediately after surgery, when patients require significant amounts of electrolytes and fluids. Features of SBS may include hypovolemia, dehydration, metabolic acidosis, hypoalbuminemia, and deficiencies of potassium, calcium, zinc, magnesium, copper, body acids, fat-soluble vitamins, folic acid, and B_{12}.

Timing, onset, and severity of the clinical manifestations are compounded by numerous variables. One of the major factors is whether other organs are involved. The most significant is whether the colon remains after surgical resection and, if so, how much of the colon is available. The preservation of the ileocecal valve is another major variable.

Phase 1 lasts weeks to months. After that, the patient moves into phase 2, with a variable clinical picture (see Treatment and Management). If major electrolyte and mineral deficiencies develop, clinical evidence may include chronic dehydration. Diarrhea and electrolyte loss may develop when attempts are made to compensate through greatly increased intake. Subtle skin and blood abnormalities may develop because of slowly developing micronutrient and vitamin deficiencies. Weakness, dizziness, unsteady gait, and lack of mental acuity may all be present.

DIAGNOSIS

The diagnosis of SBS clearly can be made at surgery. If SBS is not apparent, the clinician obtains appropriate history and performs simple barium contrast studies to determine bowel length. Less than 200 cm, or less than 120 cm, of remaining bowel is a clinically arbitrary figure; symptoms and deficiencies are the important factors. Certainly, if more than 75% of the bowel remains, SBS is highly unlikely. Nevertheless, selective deficiencies can occur because of ileal resection. Once it is established that less than 200 cm of bowel remains, a correlation must be made with symptoms and deficiencies. To perform these evaluations, a full analysis for anemia, iron and B_{12} absorption, and lactose absorption and tolerance will provide the clinician with a clear therapeutic guideline.

TREATMENT AND MANAGEMENT

During the acute phase, lasting up to 2 months, intravenous electrolyte and nutrient replacement may be essential. It is important to add antisecretory therapy, which includes a proton pump inhibitor, and antimotility therapy, which is usually best administered with the somatostatin analog octreotide. During the acute phase, the patient is evaluated and TPN initiated.

During the adaptation phase, lasting from 2 to 24 months, the bowel begins to adapt by slightly lengthening and increasing individual villous height, thereby increasing the absorptive surface. A major factor in phase 2 is whether the colon is still present and whether the ileocecal valve remains intact. It appears that jejunum has greater ability to adapt than ileum.

After the adaptation period, the patient moves into the long-term management (maintenance) period. In phase 3, little changes, and determinations are made as to how much can be accomplished by oral feeding and how much TPN will be needed. Some clinicians do use an enteral feeding period during the TPN rest period. Some place a percutaneous gastrostomy or enterostomy tube so that the bowel can be fed slowly during the sleeping hours. However, this creates more discomfort and is socially problematic.

Regulation of oral intake requires trial and error. The goal is to obtain enough energy and protein intake to maintain body homeostasis. Proteins are the key nutrient. An intake of 0.8 to 1.2 mg/kg body weight must be maintained. To metabolize the protein, an adequate amount of energy is essential, usually from a combination of carbohydrates and fats. If the colon is present, medium-chain triglycerides can be absorbed through colonic mucosa and may be helpful if added to the diet. Otherwise, long-chain triglycerides are used. Fat intake may result in more fat excretion in the stool, with the problem of steatorrhea if the colon is present. However, therapy to bind bile salts is usually not warranted, considering how easily the short bowel can become deficient in bile salts. Consequently, fat absorption is essential to maintain energy requirements. Carbohydrates in the form of simple and complex polysaccharides are a major energy source. Some patients who cannot tolerate simple sugars because of bacterial breakdown develop D-lactic acidosis, but this is rare. Most patients can tolerate sugars and complex polysaccharides, which can make up 60% and 70% of the energy requirement. Absorption of fat-soluble vitamins may become a problem. However, all vitamins and trace minerals should be replaced in adequate amounts (see Section X).

Short bowel types

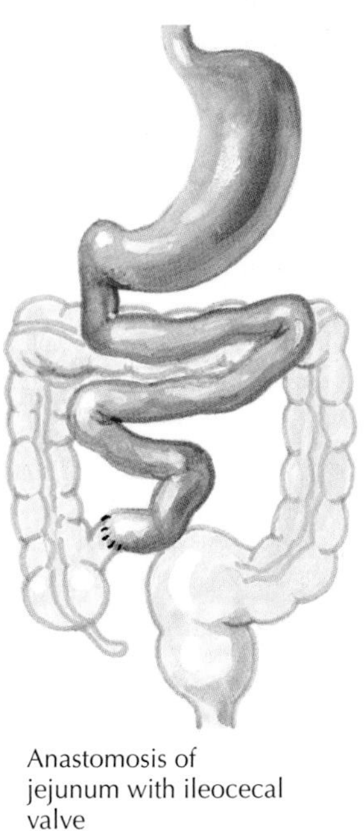

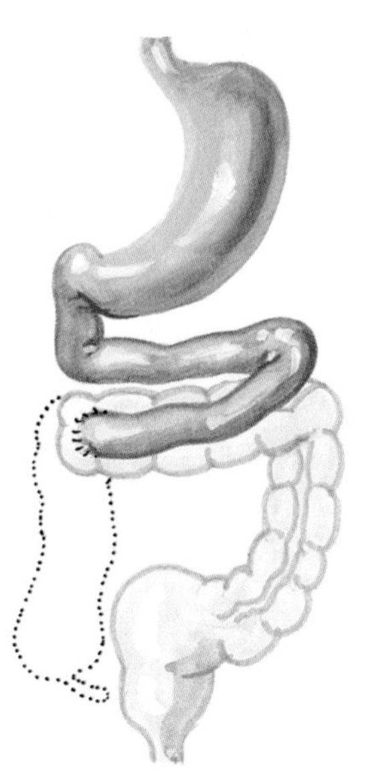

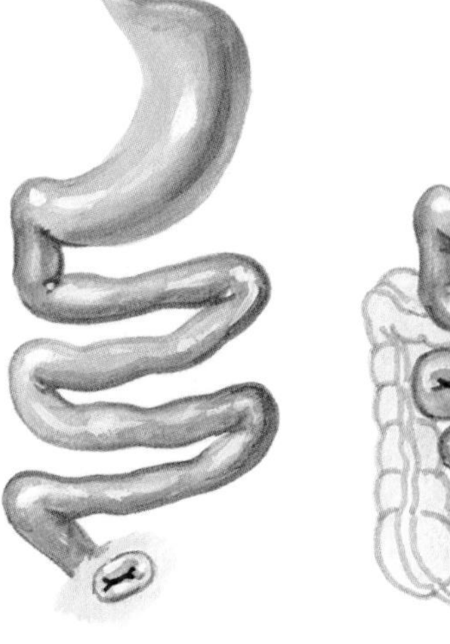

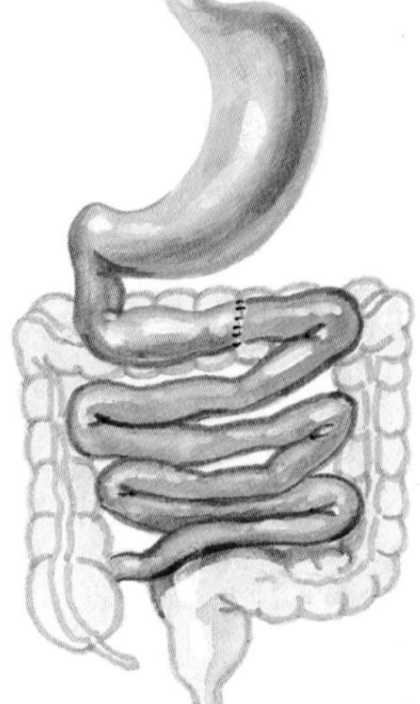

Surgery because of:
- Vascular compromise
- IBD
- Volvulus
- Atresia
- Trauma
- Neoplasm

Symptoms
- Dehydration
- Electrolyte imbalance
- Mineral imbalance
- Vitamin deficiency
- Oxalate stones

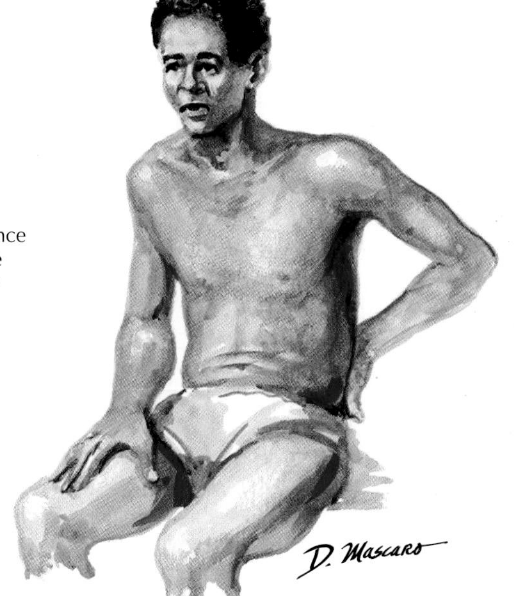

Figure 115-1 *Short Bowel Syndrome.*

The degree and amount of nutrient material that can be taken orally varies greatly in patients with SBS. However, after the 2-year period of adaptation, most will require some TPN. For severe cases, the requirement may be total. Most patients may take some of their nutrients orally, and many will require at least half their nutrition through TPN, thereby requiring a permanent catheter.

When there is an intact colon, increased oxalate absorption can occur because of decreased bile salts, resulting in oxalate nephrolithiasis. Therefore, these patients should receive a low-oxalate diet.

Antimotility agents may be helpful during any stage of SBS and include diphenoxylate, loperamide, codeine, and tincture of opium (see Chapter 111).

Oral hydration solutions might be helpful for select patients. Patients can try 250 mL to 1 L during the adaptive phase and for longer periods if they have selective electrolyte absorption difficulties. However, these are rare situations, and TPN usually corrects these problems.

Whenever bacterial overgrowth is suspected in SBS, such as occurs with D-lactic acidosis, intermittent antibiotic therapy may be necessary. Antibiotics used for alternating 2-week or 1-month periods are often tried (see Chapter 114).

It has been suggested that growth hormone increases bowel adaptation, but this is controversial. Nevertheless, some patients may request or attempt this therapy. At present, growth hormone is not generally recommended, although it may be tried in low-dose form.

COURSE AND PROGNOSIS

The prognosis is usually favorable if the duodenum and at least 60 cm of jejunum or ileum remains intact in SBS. The presence of an ileocecal valve greatly enhances the ability of the remaining jejunum or ileum to work effectively, by slowing down transit and helping water and electrolyte absorption. Furthermore, the availability of a colon to absorb electrolytes and some minerals is helpful, and colonic microflora become important. If the colon is present, fermentation of soluble fibers and the usual undigested polysaccharides occurs, producing short-chain fatty acids, primarily butyric, propionic, and acetic. The colon then becomes a digestive organ, and as much as 10% to 20% of caloric requirements can be supplied through the absorption of short-chain fatty acids. Their formation is enhanced by healthy bacterial flora; some clinicians use probiotics to enhance the effectiveness of the microflora in the colon. Consequently, the availability of the colon as a digestive organ improves the course and prognosis of SBS.

Comorbidities significantly affect morbidity and mortality. In an elderly patient with other organ illness, the prognosis for long life becomes guarded. Associated diseases or malignancies, severe radiation enteritis, and unrelenting inflammatory bowel disease all increase morbidity. Possible complications of associated chronic liver disease further increases the morbidity and mortality risk.

Finally, the expanding role of transplantation surgery is now offering new insight into the future for patients with SBS. Although still new, small bowel transplantation has been effective in select cases. Its future role remains to be determined, but it is available for patients with intractable disease who cannot be sustained on TPN.

ADDITIONAL RESOURCES

Buchman AL: Short bowel syndrome. In Feldman M, Friedman LS, Brandt LJ, editors: *Gastrointestinal and liver disease*, ed 8, Philadelphia, 2006, Saunders-Elsevier, pp 2257-2276.

Buchman AL, Scolapio J, Fryer J: AGA technical review on short bowel syndrome and intestinal transplantation, *Gastroenterology* 124:1111-1134, 2003.

Nordgaard I, Hansen BS, Mortensen PB: Importance of colonic support for energy absorption as small-bowel failure proceeds, *Am J Clin Nutr* 64:222-231, 1996.

Scolapio JS, Camilleri M, Fleming CR, et al: Effect of growth hormone, glutamine, and diet on adaptation in short bowel syndrome: a randomized, controlled study, *Gastroenterology* 113:1074-1081, 1997.

Seguy D, Vahedi K, Kapel N, et al: Low-dose growth hormone in adult home parenteral nutrition–dependent short bowel syndrome patients: a positive study, *Gastroenterology* 124:293-302, 2003.

Steiger E: Guidelines for pharmacotherapy in short bowel syndrome, *J Clin Gastroenterol* 40:S73-S106, 2006.

Sundaram A, Koutkia P, Apovin CM: Nutritional management of the short bowel syndrome in adults, *J Clin Gastroenterol* 34:207-220, 2002.

Food Allergy

Martin H. Floch

The terms *food allergy* and food *hypersensitivity* are synonymous. Food "allergy" is distinguished from food "intolerance" in that it includes a true allergic, immunologic response (**Fig. 116-1**). *Food intolerances* are not immunologic responses. Allergic responses may be acute and may result in anaphylaxis (a rare occurrence) or may be chronic.

Although its incidence is difficult to determine, food hypersensitivity occurs in approximately 2% of the general population and in as many as 6% of young children. However, anecdotal experience indicates a much higher incidence, with many patients not seeking assistance or medical advice because they eliminate foods they identify as causing symptoms. Common

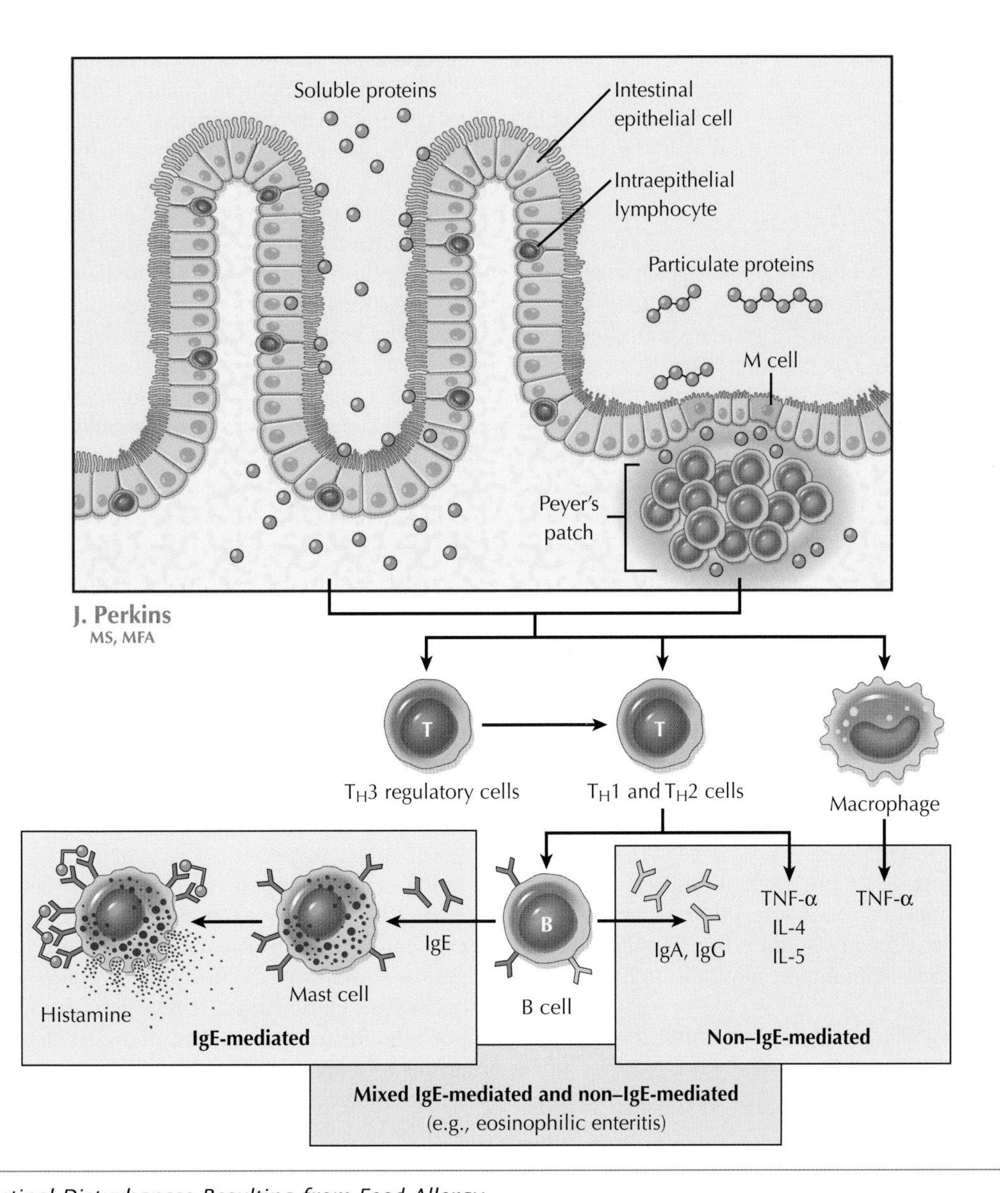

Figure 116-1 *Intestinal Disturbances Resulting from Food Allergy.*
This diagram demonstrates the entry of proteins through the intestinal epithelium that become attached or react with lymphocytes within Peyer's patches or with macrophages. The reaction is different if it is a T cell or a B cell. The T cells produce non–IgE-mediated reactions through the mechanisms listed in the box on the right. The B cells can stimulate this process or stimulate the IgE-mediated process and activate mast cells. There are also mixed IgE and non–IgE-mediated conditions such as eosinophilic enteritis.

foods associated with symptoms are milk, eggs, fish, tree nuts, shellfish, soybeans, fruits, and wheat. Children tend to outgrow the symptoms, and in adulthood, the foods most often associated with food hypersensitivity are peanuts, tree nuts, fruits, fish, and shellfish.

The pathophysiology of food allergy is becoming clearer. True hypersensitivity reactions usually are mediated through immunoglobulin E (IgE) and associated with atopy. However, these findings are not essential when an alleged exposure can be demonstrated by an exclusion diet. Although IgG and IgA antibodies are demonstrated in celiac disease, the demonstration of these antibodies in other food allergies has been controversial. Immediate-phase reactions are IgE mediated, with inflammation mediators released from mast cells. However, some hypersensitivity reactions are non–IgE mediated and involve histamine release from mast cells. Therefore, both IgE hypersensitivity and non-IgE hypersensitivity exist.

Non-IgE food allergy is T-cell mediated, but mixed IgE-mediated and non–IgE-mediated conditions appear to involve the gut, including eosinophilic esophagitis, gastritis, or gastroenteritis.

CLINICAL PICTURE

The allergic response may manifest as an emergency anaphylactic reaction (angioneurotic edema, urticaria, asthmatic attacks, allergic rhinitis) or by less dramatic responses (rashes, focal edema. migraine headache). The gastrointestinal tract can certainly react by producing acute symptoms of gastroenteritis or, more chronically, of epigastric distress or diarrhea. The entire spectrum of acute and chronic symptoms may be caused by any of the food allergens. A clear relationship to foods is considered probable in acute, severe, and dramatic reactions; in oral allergy syndromes that produce pruritus; in celiac disease; and in dietary protein-induced enteropathy or enterocolitis in infancy.

Syndromes less dramatic in presentation that require workup include gastroesophageal reflux in infants, eosinophilic esophagitis or gastroenteritis, and enteropathies. Food allergy is often suspected but is extremely difficult to prove or diagnose. The symptoms may be minimal but are persistent and annoying.

DIAGNOSIS

Acute reactions and those temporally associated with symptoms (e.g., pruritus, erythema) are more easily diagnosed. Reactions causing vague symptoms (e.g., migraine, diarrhea) can be difficult. Most allergists and scientists agree that the diagnosis of food allergy can be made from one or more of the following:

1. Clear history and an allergic-like reaction after the ingestion of food.
2. Exclusion of any anatomic, functional, metabolic, or infectious cause.
3. Finding of certain pathologic features consistent with allergy, such as eosinophilia.
4. Confirmation of a relationship between ingestion of a dietary protein and symptoms (e.g., clinical challenges, repeated inadvertent exposures).
5. Evidence of food-specific IgE antibody.
6. Failure to respond to conventional therapies, and improvement in symptoms with elimination diets.

Some clinicians rely on a response to treatment of allergic inflammation with corticosteroids or other drugs. However, the final accepted diagnostic step is elimination of the food to relieve symptoms and food challenge that recreates the symptomatic picture.

Primary tests for specific IgE antibody or particular foods, radioallergosorbent test (RAST), and skin-prick test are frequently used, but their high false-positive rates create confusion. Adjunctive tests such as endoscopic biopsies, absorption studies, and stool analysis for eosinophils are all used but are not pathognomonic. Specific IgE antibody may be used to rule out the potential for severe, acute reactions before conducting oral challenges in patients with atopic disease or possible history of severe, acute reactions. In patients with chronic, symptomatic disorders without atopic disease, food-specific immunoglobulin test results are usually negative. Negative skin-prick tests are the most diagnostic, because the negative predictive value of the skin test is usually greater than 95% and excellent. Unfortunately, the positive predictive value of a positive finding is approximately 50%. In vitro tests for a specific IgE with RAST are helpful in evaluating IgE-mediated food allergy. Problems in interpretation are compounded by immunologic cross-reactivity between botanical families and animal species.

Elimination diets are an essential tool in determining whether a food is causing the allergy. There are three types of *elimination diets:* (1) elimination of one or several foods suspected of provoking symptoms; (2) elimination of all but a defined group of allowed foods; and (3) use of an amino acid–based formula or the elemental diet. These types of elimination evaluation should be performed by experienced allergists. Failure to have symptoms resolved on the elimination diet would rule out food allergy.

Physician-supervised oral *food challenges* are required for the diagnosis of food allergy. In general, when several foods are under consideration as the cause of symptoms, when test results for food-specific IgE are positive, and when elimination has resulted in the resolution of symptoms, oral challenge testing for each food eliminated is used to diagnose specific sensitivities and to allow expansion of the diet. In the patient with acute anaphylactoid-type reaction but with no evidence of a food-specific IgE suspected of provoking the reaction, physician-supervised challenge is used to reintroduce the food safely in case of false-negative findings on the skin/RAST test. If suspicion concerning a particular food remains high despite its elimination without a resolution of symptoms, challenges may be needed for clarification. If tests for specific IgE antibodies are not relevant to the disorder, oral challenges are often the only means of diagnosis. This is the case for most of the GI hypersensitivity reactions. Oral challenges are also required to determine when clinical tolerance has developed. However, oral challenges may be optional or contraindicated in some patients, and a physician with experience in the field of food allergy must be consulted.

Research with probiotics indicates that some children have a deficient immune response. Probiotic organisms can stimulate the immune response to correct an allergic response caused by certain foods in infancy and by milk in children.

TREATMENT AND MANAGEMENT

Eliminating the allergic agent is the treatment for food allergy. As discussed, oral challenges may be required to clarify which foods are causing the symptoms. However, judicious use of elimination diets and challenges by experienced physicians can result in adequate control of the problem.

When the responsible foods cannot be identified, some physicians use antihistamines and, in extreme circumstances, corticosteroids to prevent symptoms. Double-blind placebo-controlled food challenges are frequently used, considered by some as the standard by which to diagnose food allergy and thus determine elimination.

Some clinicians use mast cell stabilizers, such as cromolyn sodium and ketotifen fumarate. The key to good management is patient education, with careful instruction on how to eliminate all protein-inciting agents in foods.

Evidence indicates that probiotics may be helpful in the treatment of milk allergy or atopy in children. *Lactobacillus rhamnosus*, also called *Lactobacillus* GG, can be used.

COURSE AND PROGNOSIS

The prognosis is usually excellent in patients who know the food responsible for the allergy and who have eliminated it from the diet. The patient who has experienced an anaphylactoid or severe reaction makes sure that epinephrine injections are available whenever there is potential exposure to an allergen. Patients experiencing chronic conditions may be plagued by the concept of food allergy, especially when specific allergens are not proved. The field is open and prone to questionable diagnostic and therapeutic procedures. However, most credible allergists can set up a program so that their patients can tolerate the diet and receive adequate nutrition despite elimination of inciting foods. Experience with documented testing indicates that some food allergy reactions can be transient. Therefore, patients should be tested after a reasonable time (1 year) to determine whether the allergy persists.

The field of infant allergy reports that most infants requiring hypoallergenic formulas usually become tolerant within 1 to 2 years, and the prognosis is then excellent. These patients are treated by a pediatric allergist.

ADDITIONAL RESOURCES

Atkins D: Food allergy: diagnosis and management, *Prim Care* 35:119-140, 2008.

Bischoff S, Crowe SE: Gastrointestinal food allergy: new insights into pathophysiology and clinical perspectives, *Gastroenterology* 128:1089-1113, 2005.

Isolauri E, Salminen S: Probiotics: use in allergic disorders, *J Clin Gastroenterol* 42:S91-S96, 2008.

Kirjavainen PV, Salminen SJ, Isolauri E: Probiotic bacteria in the management of atopic diseases: underscoring the importance of viability, *J Pediatr Gastroentrol Nutr* 36:223-227, 2003.

Nowak-Wegrzyn A, Sampson HA: Adverse reactions to foods, *Med Clin North Am* 90:97-127, 2006.

O'Leary PF, Shanahan F: Food allergies, *Curr Gastroenterol Rep* 4:373-382, 2002.

Sampson HA: Food allergies. In Feldman M, Friedman LS, Brandt LJ, editors: *Gastrointestinal and liver disease*, ed 8, Philadelphia, 2006, Saunders-Elsevier, pp 427-442.

Shah U, Walker WA: Pathophysiology of intestinal food allergy, *Adv Pediatr* 49:299-316, 2002.

Sicherer SH: Advances in anaphylaxis and hypersensitivity reaction to foods, drugs, and insects, *J Allergy Clin Immunol* 119:1462-1469, 2007.

Eosinophilic Gastroenteritis

117

Martin H. Floch

Eosinophilic gastroenteritis is a disease in which tissue eosinophilia occurs in any segment of the gastrointestinal (GI) tract and is associated with GI symptoms (**Fig. 117-1**). Although first described in 1937, with frequent case reports corroborating the syndrome, the following criteria to establish eosinophilic gastroenteritis were not defined until 1990:

1. The patient must have GI symptoms.
2. Eosinophilic infiltration must be demonstrated on biopsy in one or more areas of the GI tract.
3. The patient should have no eosinophilic involvement of other organs outside the GI tract.
4. The patient should have no parasitic infestation.

Eosinophilic gastroenteritis is a relatively rare disease. The actual incidence is unknown, but most gastroenterologists believe that many cases are unreported. The typical presentation is in the third to fifth decade of life. It may affect any age group, with no apparent gender predilection. The cause of the infiltrate is unknown. It is often associated with a hyperallergic state. Eosinophils may actually damage the GI tract by releasing protein and toxic substances.

Although eosinophils are present in many bowel disorders (e.g., parasitic disease, inflammatory bowel disease), the only evidence in eosinophilic gastroenteritis is the infiltrate in the small bowel. Because it is frequently associated with multiple allergies, the assumption is that patients with eosinophilic gastroenteritis are allergic to some food or substance in the environment. However, this has not been proved. Eosinophilic gastroenteritis is now thought to be a mixture of IgE-mediated and non-IgE-mediated allergic reactions.

CLINICAL PICTURE

Most often, the stomach and the small bowel are involved with the mucosal and submucosal eosinophilic infiltrate. These

Symptoms
- Dyspepsia
- Malnutrition and malabsorption
- Diarrhea
- Weight loss
- Allergy symptoms (asthma)

Biopsy appearance

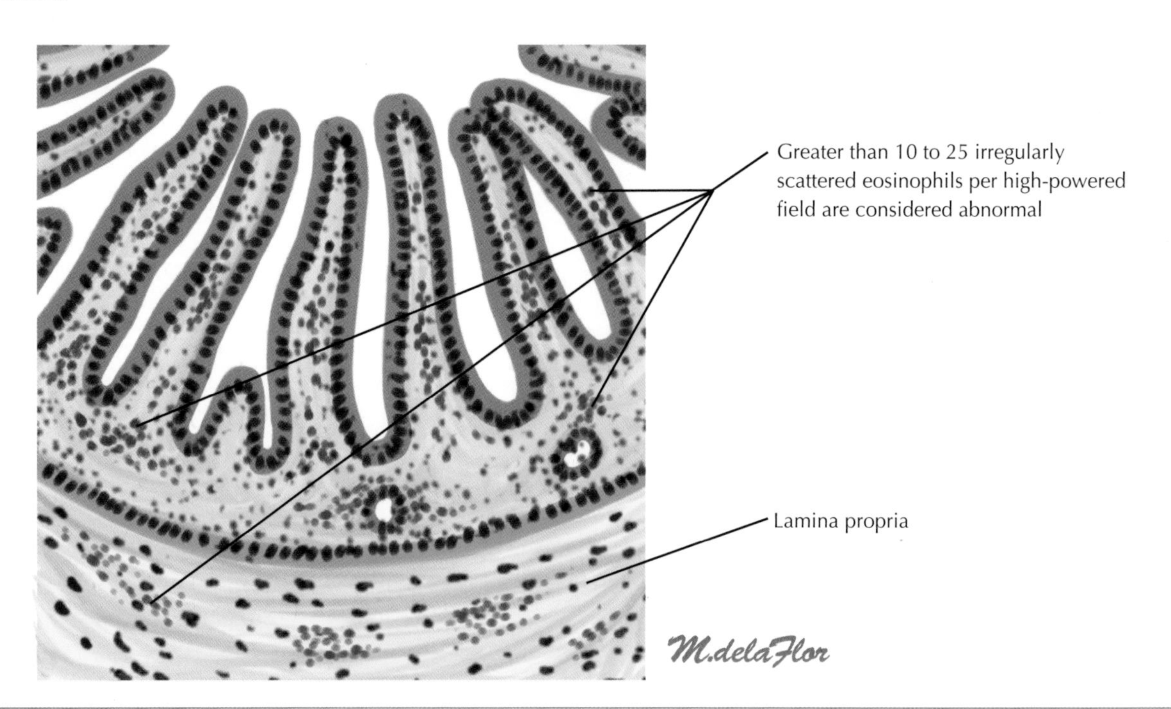

Figure 117-1 *Eosinophilic Gastroenteritis.*

patients seek treatment for dyspepsia or diarrhea or for mild, intermittent abdominal pain. The syndrome may be intermittent or may be longstanding and accompanied by weight loss and associated malnutrition and malabsorption. Depending on the extent and degree of eosinophil concentration and damage to the stomach and small bowel, erosions and weeping lesions can develop, but these are rarely reported. In young children, an allergy frequently manifests with mild asthma, atopy, severe hay fever, or multiple food intolerances. Almost every part of the GI tract has been involved. Such rare entities as eosinophilic ascites and eosinophilic polyps have been reported.

Eosinophilic esophagitis is usually reported as a separate entity when only the esophagus is involved. It behaves differently from generalized eosinophilic gastroenteritis. Rings form in the esophagus, and symptoms can be chronic. It is treated as severe reflux disease but also requires topical therapy (see Section I).

DIAGNOSIS

To fulfill the criteria for eosinophilic gastroenteritis, there must be a tissue diagnosis. Once suspicion is raised, upper endoscopy is essential to obtain esophagus, stomach, and small bowel samples for biopsy. The clinical picture usually raises the index of suspicion. Peripheral eosinophilia is present in almost 80% of patients, although it may be low grade and is often overlooked. It is important to rule out parasitic disease. Evaluation for ankylostomiasis, or intestinal parasites, must always be performed. It is also important to rule out other systemic diseases that can cause eosinophilic infiltrate, such as lymphoma, vasculitis, and Addison disease. Another concomitant finding may be a low serum albumin.

It is important to rule out all other diseases; thus, imaging of the upper GI and lower intestinal tracts becomes important. If diarrhea is part of the picture, the usual diarrhea workup for pathogens and colonoscopy are important to obtain biopsy specimens and to rule out inflammatory bowel disease.

Biopsy evaluations are critical. Eosinophilic infiltrate may be part of any inflammatory process. Therefore, the pathologist must determine whether there is an abnormal increase in eosinophils. Usually, more than 10 eosinophils per high-power field (hpf) is considered abnormal, but in classic eosinophilic gastroenteritis, there are more than 25 eosinophils/hpf. The infiltrate may be scattered, however, so several areas should undergo biopsy. The appearance of intraepithelial eosinophils is a strong indicator of the syndrome, even if the numbers of eosinophils in the lamina propria and the submucosa are not high.

Eosinophilic gastroenteritis should not be confused with *hypereosinophilic syndrome* or *diffuse vasculitis*, both of which involve multiple organs.

TREATMENT AND MANAGEMENT

Corticosteroids are the most successful agents once the diagnosis of eosinophilic gastroenteritis is firmly established. Prednisone is begun in relatively high doses, then tapered over 1 to 2 weeks. Most patients respond dramatically, but as many as 15% have relapses and require increasing doses. Many patients are carried with low-dose maintenance prednisone of 5 to 10 mg daily (similar to the treatment of chronic asthma).

Cromolyn sodium has been used with varying success. Because of its safety, cromolyn is often administered before corticosteroids.

If testing reveals a food allergy, it is helpful to eliminate any inciting foods. However, the test for food allergy usually fails in patients with eosinophilic gastroenteritis.

COURSE AND PROGNOSIS

The prognosis for patients with eosinophilic gastroenteritis is excellent. Most patients respond extremely well, and after initial therapy, or in months, they can discontinue prednisone treatment. Relapses occur but are usually easily controlled.

ADDITIONAL RESOURCES

Bischoff SC, Mayer J, Nguyen OT, et al: Immunohistochemical assessment of intestinal eosinophil activation in patients with eosinophilic gastroenteritis and inflammatory bowel disease, *Am J Gastroenterol* 94:3521-3529, 1999.

Kalantar SJ, Marks R, Lambert JR, et al: Dyspepsia due to eosinophilic gastroenteritis, *Dig Dis Sci* 42:2327-2332, 1997.

Khan S, Orenstein SR: Eosinophilic disorders of the gastrointestinal tract. In Feldman M, Friedman LS, Brandt LJ, editors: *Gastrointestinal and liver disease*, ed 8, Philadelphia, 2006, Saunders-Elsevier, pp 543-556.

Klein NC, Hargrove R, Sleisinger MH, et al: Eosinophilic gastroenteritis, *Medicine* 49:299-304, 1970.

Sampson HA: Food allergies. In Feldman M, Friedman LS, Brandt LJ, editors: *Gastrointestinal and liver disease*, ed 8, Philadelphia, 2006, Saunders-Elsevier, pp 427-442.

Talley NJ, Shorter RG, Phillips SF, et al: Eosinophilic gastroenteritis: a clinical pathologic study of patients with disease of the mucosae, musculae, and subserosal tissue, *Gut* 31:54-61, 1990.

Yang CU, West AB: What do eosinophils tell us in biopsies of patients with inflammatory bowel disease? *J Clin Gastroenterol* 36:93-98, 2003.

Intussusception of the Small Intestine

118

Martin H. Floch

Intussusception is the invagination of a portion of the intestine into the contiguous distal segment of the enteric tube (**Fig. 118-1**). It usually occurs in infants at 4 to 10 months of age and is associated with acute enteritis, allergic reactions, and conditions that cause hypermotility. In older persons, intussusception is associated with a polyp or malignancy, an enlarged Peyer's patch or diverticulum (e.g., Meckel), and a large number of rare entities that cause intrusion into the bowel.

Intussusceptions are classified according to the part of the digestive tube that telescopes into the *intussuscipiens*, the receiving part, including ileo-ileal, jejunoileal, and ileocolic invaginations. The most common is the ileocolic intussusception. A double invagination, or an intussusception within an intussusception, may also occur, called ileo-ileocolic. How far the *intussusceptum*, the part that becomes ensheathed by the more distal portion, enters the intussuscipiens depends on the length and motility of the mesentery. The intussuscipiens can be compressed, after which edema, peritoneal exudation, vascular strangulation, and finally, intestinal gangrene can develop.

In children, the most common causes of intussusception are associated infections, and in adults, the most common causes are neoplasms, although these primary causes affect the other group as well. Approximately 30% to 50% of small bowel intussusceptions and 50% to 65% of colonic intussusceptions are associated with malignant neoplasms.

CLINICAL PICTURE

Clinical presentation of intussusception may be alarming, with the sudden development of abdominal pain and cramplike sensations occurring every 10 to 20 minutes. Children may appear to be in shock. In approximately 85% of children, a movable mass may be palpated. If the symptoms progress, blood may be found in the stool. In adults, the presentation may be acute, but often it is intermittent and accompanied by cramplike abdominal pain with nausea and vomiting. With chronic presentation lasting more than 1 week, the patient may lose weight. At times, the chronicity of symptoms can fool the clinician.

DIAGNOSIS

Diagnosis is made and confirmed by radiographic imaging. Ultrasonography is helpful, but barium contrast or computed tomography with contrast media often demonstrates the intussusception. Different radiologic patterns are observed, with classic findings of target lesions, sausage-shaped masses, and associated obstructive phenomena.

TREATMENT AND MANAGEMENT

For most infants and children, intussusception is reduced by pressure from the contrast enema when it is ileocolic. However, the high incidence of neoplasia in adults requires surgical intervention. At times, surgery is performed on an emergency basis to prevent ischemic and gangrenous bowel formation. Once the diagnosis is made, surgical evaluation and intervention are essential. The surgeon may reduce the intussusception when the bowel is entered, then use an end-to-end or end-to-side anastomosis to remove any suspicious lesion. Laparoscopic surgery is now used to perform these resections, although this requires a skilled surgeon.

PROGNOSIS

The immediate prognosis is usually excellent. However, if a neoplasm is identified, the prognosis depends on the type of lesion. A high incidence of polyps in association with intussusception in children and adults may require follow-up monitoring and possibly chemotherapy. When intussusception is associated with hypermotility of viral cause in a child, the prognosis is excellent after the acute infection is controlled.

ADDITIONAL RESOURCES

DeFiore JW: Intussusception, *Semin Pediatr Surg* 8:214-220, 1999.

Dennehy PH, Bresee JS: Rotavirus vaccine and intussusception: where do we go from here? *Infect Dis Clin North Am* 15:189-207, 2001.

Eisen LK, Cunningham JD, Aufses AH Jr: Intussusception in adults: institutional review, *J Am Coll Surg* 188:390-395, 1999.

Maconi G, Radice E, Greco S, et al: Transient small-bowel intussusceptions in adults: significance of ultrasonographic detection, *Clin Radiol* 62:792-797, 2007.

Merine D, Fishman EK, Jones B: Enteroenteric intussusception: CT findings in nine patients, *Am J Roentgenol* 148:1129-1136, 1987.

Nagurney TM, Sarr MG, McIlrath DC: Surgical management of intussusception in the adult, *Ann Surg* 193:230-238, 1981.

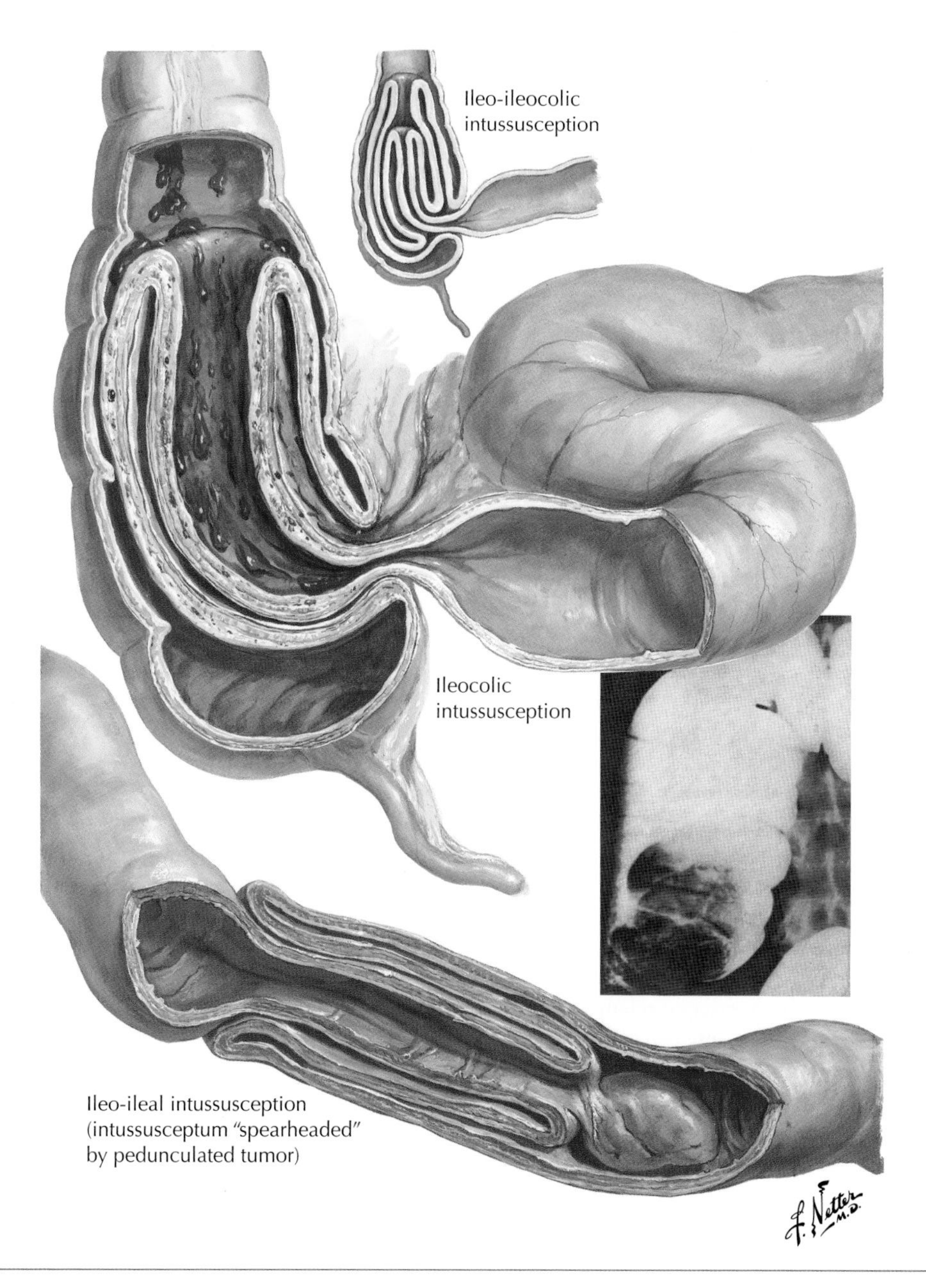

Figure 118-1 *Intussusception of the Small Intestine.*

Benign Tumors of the Small Intestine

119

Martin H. Floch

Benign tumors of the small intestine are rare. In a series of 22,810 autopsies, the incidence was 0.16%. The neoplasms may be located anywhere in the small bowel. *Stromal tumors*, or *leiomyomas*, are more common in the jejunum, and *adenomas* are more common in the ileum. Both are benign but have significant malignant counterparts or potential. Benign tumors may be single or multiple. Adenomas, neurofibromas, and angiomas have a familial occurrence. Leiomyomas and adenomas are most common and, along with lipomas, myomas, some angiomas, and neurogenic tumors, make up more than two thirds of lesions (**Fig. 119-1**).

Rarer benign tumors include fibromas, lymphangiomas, myxomas, and osteomas. Tumors may be intraluminal, extraluminal, or intramural. They may vary in size from millimeters to "very large" at 3 cm. Neurogenic tumors tend to appear in multiples, whereas adenomas, lipomas, and leiomyomas tend to appear singly. Neurogenic tumors in the small intestine may be part of a generalized neurofibromatosis (von Recklinghausen disease).

Benign vascular tumors of the small intestine comprise true tumors of blood vessels, or *angiomas*, and congenital vascular malformations, or *hamartomas*. These are difficult to differentiate and may be part of a generalized vascular dysplasia. *Rendu-Osler-Weber syndrome* (hereditary hemorrhagic telangiectasia), a mendelian-dominant disease, includes angiomas of the skin, mucous membranes, and viscera. Another mendelian-dominant inherited syndrome is *Peutz-Jeghers syndrome*, characterized by the association of gastrointestinal GI adenomas, polyposis, and a distinct type of mucocutaneous pigmentation (**Fig. 119-2**). These polyps may be adenomatous or hamartomatous.

Brunner gland hamartomas and nodular lymphoid hyperplasia of the small intestine are benign tumors or polyps. Both are rare and frequently asymptomatic, but they require histologic definition.

CLINICAL PICTURE

In general, many of these benign lesions never cause symptoms, or cause such vague symptoms that a clinical diagnosis cannot be established. Serious clinical symptoms occur usually as a complication, such as intestinal obstruction, necrotic change in the tumor with resultant hemorrhage, infection, rupture, or malignant degeneration (see **Fig. 119-2**). Intraluminal tumors are likely to cause symptoms earlier than extraluminal tumors. Extraluminal lesions may grow to large dimensions before they become symptomatic. Intraluminal polypoid tumors may lead to obstruction by intussusception. This will manifest with an acute obstructive phenomenon or insidiously, with mild intermittent small bowel obstruction and symptoms of abdominal pain, vomiting, and either diarrhea or constipation. Severity of the symptom depends on the degree of obstruction and the site of the tumor.

Bleeding from the tumor is attributed to necrotic erosion of a vessel. It may be slow and insidious, or it may be massive and result in severe GI bleeding. Extraluminal tumors may rupture into the peritoneal cavity, or they may become necrotic after torsion of a pedicle and thus may lead to intraabdominal hemorrhage or an acute abdomen. In rare cases, fistulae may form through intramural or extraluminal tumors connecting the intestinal lumen with the abdominal cavity, resulting in peritonitis.

Malignant degeneration of benign tumors of the small bowel may occur. Adenomatous lesions are reported to develop into carcinoma in as many as 40% of patients. Similarly, leiomyomas (stromal tumors) are developing into leiomyosarcomas at an increasingly alarming rate.

DIAGNOSIS

Large tumors can be diagnosed easily through barium contrast study or contrast computed tomography. However, small bleeding lesions may often be missed. Small bowel barium studies are often helpful. However, direct-view small bowel endoscopy, double-balloon endoscopy, and capsule endoscopy now make these lesions visible to the endoscopist. Capsule endoscopy has become the procedure of choice to identify lesions early. Depending on the symptom presentation, however, a contrast study of the small bowel, enteroscopy, and if necessary, double-balloon endoscopy and/or capsule endoscopy are indicated. With these tools, a diagnosis should be made.

TREATMENT AND MANAGEMENT

Most benign tumors are asymptomatic and therefore may be incidental findings during workup. However, once the tumor is identified or symptoms have developed, it is often mandatory to obtain a histologic diagnosis. Laparoscopic surgery has facilitated diagnosis of these lesions, but when small, they may require open laparotomy for identification. If a complication such as intussusception, bleeding, or perforation develops, emergency surgery is essential to remove the lesion.

COURSE AND PROGNOSIS

The prognosis for patients with benign tumors of the small intestine is excellent. Occasionally, however, a stromal tumor is found to be a leiomyosarcoma, and the prognosis for what appeared to be a benign adenoma becomes one for an adenocarcinoma (see Chapter 120).

ADDITIONAL RESOURCES

Feigel DO, Cave DR: *Capsule endoscopy*, Philadelphia, 2008, Saunders-Elsevier.

Gill SS, Heuman DM, Mihs AA: Small intestinal neoplasms, *J Clin Gastroenterol* 33:267-282, 2001.

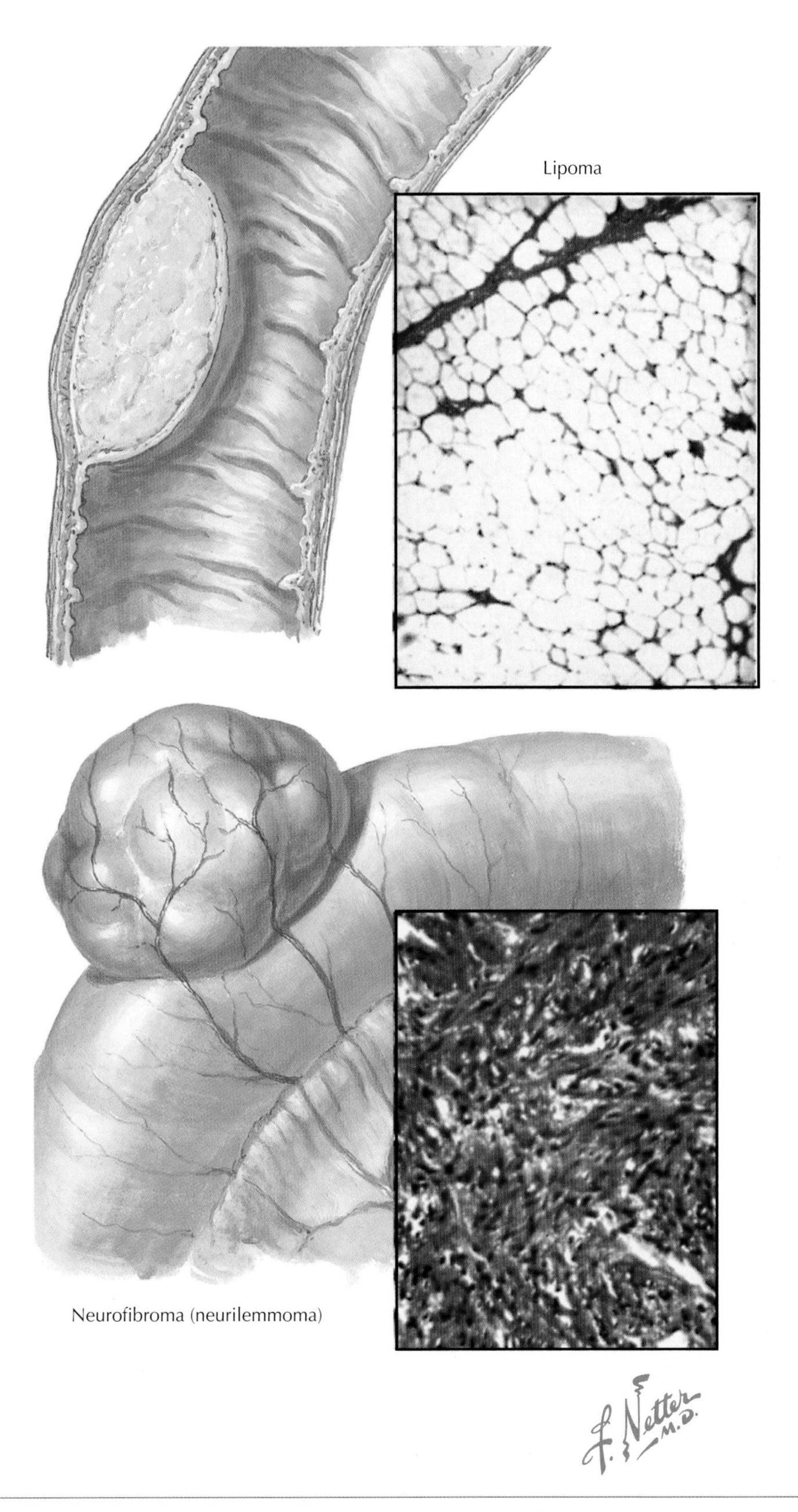

Figure 119-1 *Benign Tumors of the Small Intestine:* Top, *Lipoma;* Bottom, *Neurofibroma.*

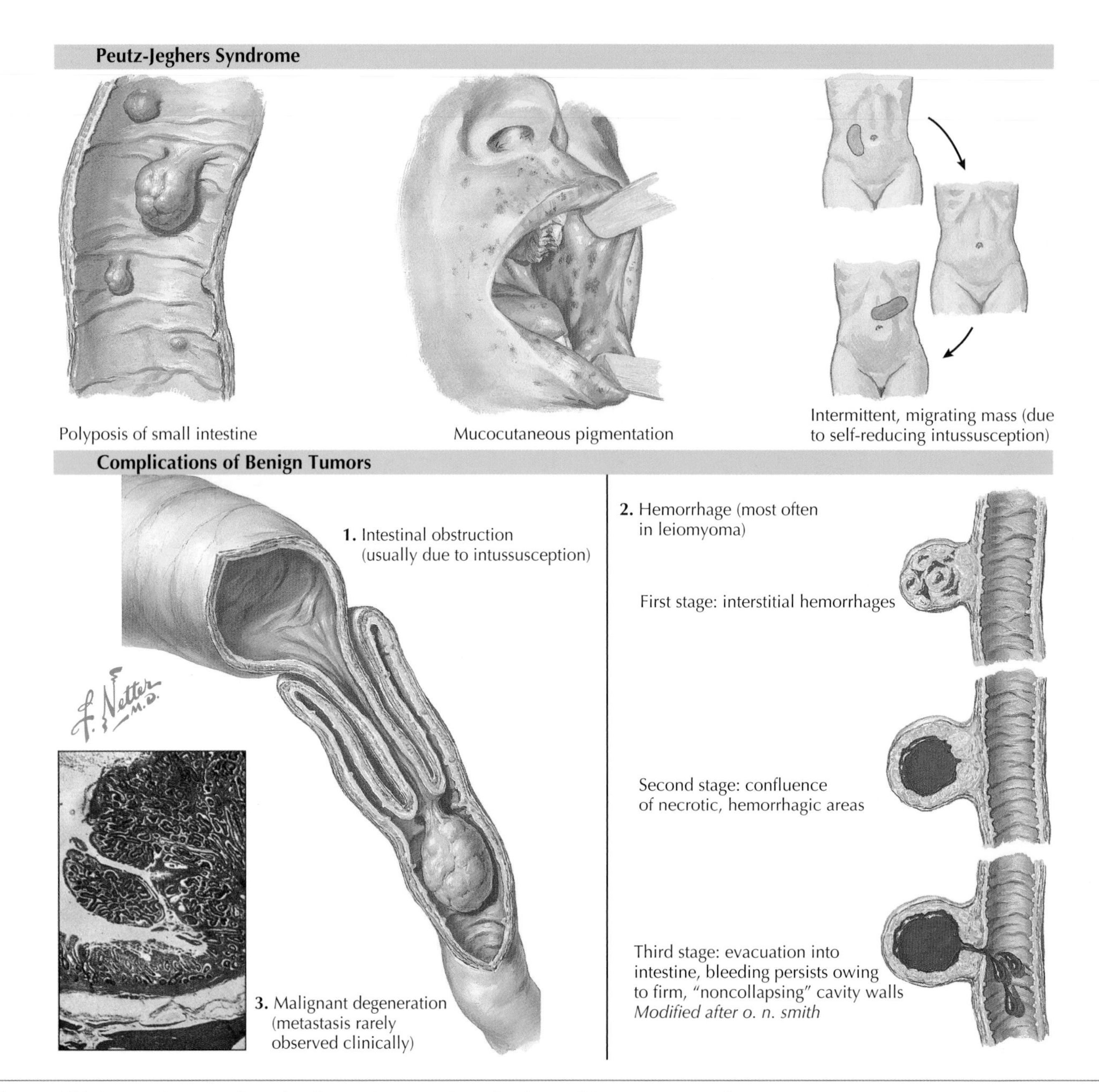

Figure 119-2 Top, *Peutz-Jeghers Syndrome;* Bottom, *Complications of Benign Intestinal Tumors.*

Lappas JC, Maglinte DDT, Sandresagaran K: Benign tumors of the small bowel. In Gore RM, Levine MS, editors: *Textbook of gastrointestinal radiology*, ed 3, Philadelphia, 2008, Saunders-Elsevier, pp 845-851.

Levine JA, Burgart LS, Batts KP, et al: Brunner's gland hamartomas: clinical presentation and pathologic feature of 27 cases, *Am J Gastroenterol* 90:290-294, 1995.

Maglinte DDT, Lappas JC, Sandresegaran K: Contrast imaging. In Gore RM, Levine MS, editors: *Textbook of gastrointestinal radiology*, ed 3, Philadelphia, 2008, Saunders-Elsevier, pp 755-764.

Rustgi AK: Small intestinal neoplasms. In Feldman M, Friedman LS, Brandt LJ, editors: *Gastrointestinal and liver disease*, ed 8, Philadelphia, 2006, Saunders-Elsevier, pp 2703-2712.

Tanaka S, Mitsui K, Tatsaguchi A, et al: Current status of double-balloon endoscopy: indications, insertion routes, sedation, complications, technical matters, *Gastrointest Endosc* 66:S30-S33, 2007.

Wittemann BJ, Janssen AR, Griffioen G, Lamers CB: Villous tumors of the duodenum: an analysis of the literature with emphasis on malignant transformation, *Neth J Med* 42:5-11, 1993.

Malignant Tumors of the Small Intestine

120

Martin H. Floch

Malignant tumors of the small intestine are rare. Their frequency in a large autopsy study was lower than 0.1%. Although the small bowel is the largest gastrointestinal organ, less than 5% of malignant tumors arise in the small intestine. The reason for this remains unclear. *Adenocarcinoma* (nonampullary) is the most common malignancy and accounts for 30% to 50% of malignant tumors. Most develop in the duodenum or the jejunum. Predisposing factors appear to be alcohol intake, Crohn's disease, celiac disease, and neurofibromatosis. A predisposing factor appears to be a preexisting adenoma, and more than 40% of patients with familial adenomatous polyposis (FAP) have polyps in the proximal small bowel, and more than 5% develop adenocarcinoma. **Figure 120-1** shows the morphologic types and local consequences of malignant tumors of the small intestine.

Carcinoma of the ampulla of Vater, together with the other adenocarcinomas, makes up most of the malignancies in the small bowel. Again, these tumors are rare, but they are the most

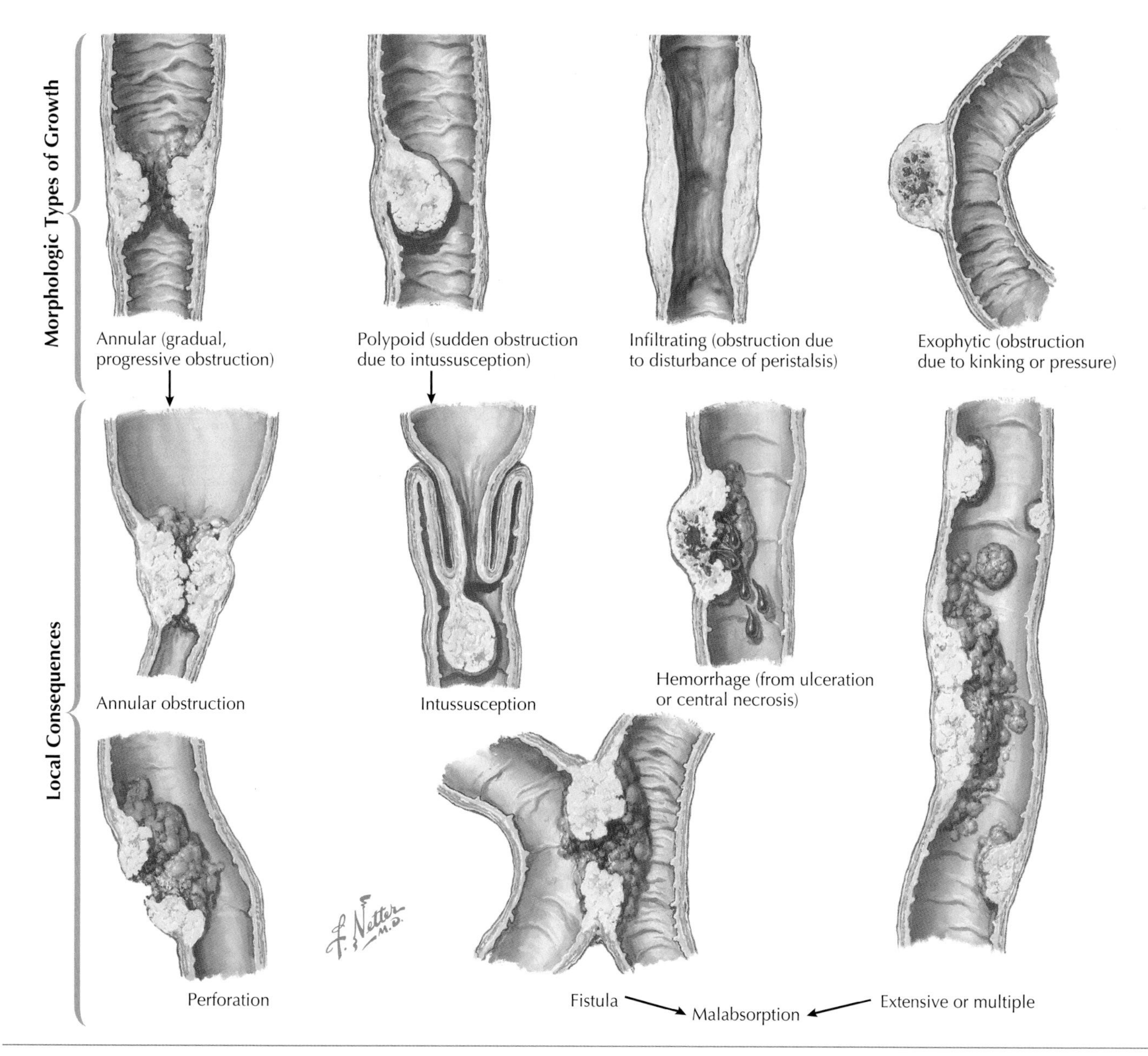

Figure 120-1 *Malignant Tumors of the Small Intestine.*

common sites of extracolonic malignancy in FAP. Lymphomas are the third and leiomyosarcomas the fourth most common small intestinal malignancies. Lymphomas make up 15% to 20% of all malignant small bowel tumors, of which non-Hodgkin lymphomas are most common.

Leiomyosarcomas are now classified as *gastrointestinal stromal tumors* (GISTs). It is often difficult for pathologists to define whether these are benign or malignant. Differentiation is usually based on a mitotic index. If there is less than 1 mitosis per 30 high-power fields, metastases are less than 1%. If there are more than 10 mitoses per high-power field, metastases are 100% and 5-year survival rates decrease to 5%, with no 10-year survival reported. A mitotic index between 0 and 10 has grades of 5-year survival and metastases. The GIST tumors are of great interest (see Chapter 62), and more are being reported. It is not yet clear whether the increase results from better diagnostic studies or whether there is a true increased incidence.

CLINICAL PICTURE

As with benign tumors of the small bowel, the presentation of malignant lesions may be slow and insidious. Patients may have low-grade anemia, slow bleeding, mild abdominal pain, weight loss, and if there is an infiltrating large lesion, slowly developing malabsorption or intermittent cramps. Patients may have acute symptoms of obstructing intussusception or massive bleed. Patients with malignant lesions usually do not seek emergency treatment but have persistent symptoms over several months.

DIAGNOSIS

As with benign tumors, imaging is paramount for malignant tumors, which are usually easily identified using barium contrast or computed tomography contrast evaluation. Because of their nature, these tumors are usually larger than benign tumors, but enteroscopy, double-balloon enteroscopy, and capsule endoscopy now allow earlier diagnosis if symptoms are also present. Low-grade anemia resulting from slow blood loss from a small lesion may prompt an earlier presentation and permit earlier diagnosis for a better prognosis.

Ampullary lesions may involve bile duct obstruction and jaundice. Because of the position of these lesions, patients may seek treatment early, and the lesions may be resected early. Current endoscopic and endoscopic ultrasound (EUS) techniques may even allow endoscopic resection. EUS now permits determination of tumor depth and is an important technique for evaluation at ampullary sites.

TREATMENT AND MANAGEMENT

It is essential that all tumors be resected. Benign or malignant, the histology must be established; thus, resection is imperative. When tumors are very small, attempts to resect by endoscopic techniques may be used. EUS is used in some of these cases.

Once the lesion is resected, histologic analysis will indicate what type of chemotherapy is needed. Chemotherapy is changing rapidly, and the preferred drugs must be determined by the latest oncologic evaluations. This holds true for adenocarcinomas, as well as for lymphomas and leiomyosarcomas.

Therapy for lymphoma is usually more complex. Primary lymphomas should be classified so that management can be determined. Once again, the oncologist provides guidance on the latest therapy.

COURSE AND PROGNOSIS

Findings from a detailed study reveal that the 5-year survival rate for adenocarcinoma of the small bowel is 30% and that the median survival time is less than 20 months. Ampullary adenocarcinoma has a better prognosis, with 5-year survival of 36% and an even better prognosis for patients who undergo early resection.

The overall prognosis for lymphoma varies greatly with the disease stage. In advanced stages, 5-year survival generally ranges from 25% to 30%. However, with radical surgery and rapidly improving drug therapy, 5-year survival has increased, reported as high as 60% to 70%. The prognosis for patients with these tumors is still not good, but advances in chemotherapy may improve prognostic indicators in the future.

ADDITIONAL RESOURCES

Cao J, Zuo Y, Chen Z, Li J: Primary small intestinal malignant tumors: survival analysis of 48 postoperative patients, *J Clin Gastroenterol* 42:167-173, 2008.

Gill SS, Heuman DM, Mihs AA: Small intestinal neoplasms, *J Clin Gastroenterol* 33:267-282, 2001.

Gross SA, Stark ME: Initial experience with double-balloon enteroscopy in a US center, *Gastrointest Endosc* 67(6):890-897, 2008.

Howe JR, Karnell LH, Menck HR, et al: The American College of Surgeons Commission on Cancer and the American Cancer Society: adenocarcinoma of the small bowel—review of the National Cancer Data Base, *Cancer* 86:2693-2706, 1999.

Ito K, Fujita N, Noda Y, et al: Preoperative evaluation of ampullary neoplasm with EUS and transpapillary intraductal US: a prospective and histopathologically controlled study, *Gastrointest Endosc* 66:S740-S747, 2007.

Rustgi AK: Small intestinal neoplasms. In Feldman M, Friedman LS, Brandt LJ, editors: *Gastrointestinal and liver disease*, Philadelphia, 2008, Saunders-Elsevier, pp 2703-2712.

Shalow TA: Primary malignant disease of the small intestine, *Am J Surg* 69:372-380, 1945.

Tran TC, Vitale GC: Ampullary tumors: endoscopic versus operative management, *Surg Innov* 11:255-263, 2004.

Carcinoid Syndrome and Neuroendocrine Tumors

121

Martin H. Floch

Eighty percent of gastrointestinal (GI) neuroendocrine tumors are made up of enterochromaffin-like cell carcinoids, duodenal gastrin G-cell tumors, and rectal trabecular L-cell carcinoids. Less common neuroendocrine tumors are gangliocytic paragangliomas, somatostatinomas, lipomas, and schwannomas. Pathologic study has revealed that the term *carcinoid* represents a wide spectrum of neoplasms that originate from different neuroendocrine cells.

Carcinoid tumors of the small intestine that secrete serotonin (5-hydroxytryptamine, 5-HT) are associated with the *carcinoid syndrome*. These neuroendocrine tumors may originate from the foregut, midgut, or hindgut, and the enterochromaffin cells making up the neoplasm may secrete serotonin, gastrin, or adrenocorticotropic hormone. Consequently, they can produce different syndromes. However, the most frequent site for carcinoid tumors is in the GI tract, and the second most common site is the bronchopulmonary system. The most common site within the GI tract is the small bowel, followed by the appendix, with a significant number occurring in the rectum. Neoplasms are usually small and yellowish. Those found in the appendix rarely metastasize, whereas those originating in the small bowel can be more virulent. They can develop anywhere in the GI tract and have been reported in the esophagus, stomach, pancreas, and large bowel.

Carcinoid syndrome occurs when the tumor secretes a large amount of an active substance such as serotonin. Symptoms occur from the carcinoid when the tumors grow larger or when they metastasize. Carcinoid tumors smaller than 1 cm rarely metastasize. Lesions larger than 2 cm should be treated aggressively.

Rectal carcinoids do not result in the carcinoid syndrome, and most are asymptomatic.

CLINICAL PICTURE

Frequently, carcinoids are discovered inadvertently during surgery in the appendix or during a workup. Often, the carcinoids are asymptomatic. In other cases, the carcinoid can cause mild abdominal pain, bleeding, or intussusception, which would then manifest as intermittent abdominal pain or acute obstruction (**Fig. 121-1**). Rarely, carcinoids manifest as a palpable mass.

When the tumor secretes an active substance, as in approximately 10% of patients, the resulting symptoms are referred to as carcinoid syndrome. The patient typically experiences intermittent abdominal cramps associated with diarrhea, flushing of the face and entire body, and extragastrointestinal symptoms of bronchospasm or even cyanosis.

On physical examination, again rarely or occasionally, a palpable mass heralds the diagnosis. Most often, however, there are no findings except when acute intussusception and obstruction occur. In patients with carcinoid syndrome, the physical examination may reveal murmurs of tricuspid valve disease. Carcinoid is associated with fixation of the tricuspid valve leaflets, resulting in typical murmurs. Left-sided heart disease occurs in 10% to 15% of patients. The presentation of the patient with flushing, diarrhea, and tricuspid murmur is classic and almost pathognomonic of the carcinoid syndrome.

DIAGNOSIS

As indicated, often the neuroendocrine tumor is asymptomatic and identified during a workup for other diseases. However, once symptoms develop, it is essential to perform radiographic contrast studies. These may reveal the site of the tumor. Computed tomography and magnetic resonance imaging are also helpful. Somatostatin-labeled isotope scanning is highly effective for identifying the primary hematostatic sites.

Biochemically, carcinoid syndrome is clearly diagnosed with the finding of more than 10 mg 5-hydroxyindolacetic acid in a 24-hour urine collection.

TREATMENT AND MANAGEMENT

Removing very small lesions is usually curative. Lesions that grow as large as 2 cm have a significant chance to metastasize and require vigorous therapy. The primary lesion must be removed and any metastases treated. Surgical resection of metastatic lesions has resulted in some cures. Chemotherapy has had marginal effect. Chemoembolization and hepatic artery embolization have been effective when performed by experienced practitioners.

Once the carcinoid syndrome evolves, pharmacologic therapy is important, particularly if the lesions cannot be totally removed surgically. Somatostatin receptors are present in more than 80% of carcinoid tumors. The use of *somatostatin* or *octreotide* has proved to be highly effective in relieving the carcinoid syndrome symptoms. Long-acting octreotide injections are now available, making this therapy more feasible and effective. Other inhibitors of serotonin synthesis, such as parachlorophenylalanine and methyldopa, have been used to block the conversion of tryptophan to serotonin. However, the most effective treatment is with somatostatin.

COURSE AND PROGNOSIS

Resecting tumors smaller than 1 cm results in an excellent prognosis. However, if tumors have grown at other sites in the GI tract, metastases occur and the prognosis varies.

The 5-year survival rate for patients with gastric carcinoids is 49% if localized. For pancreatic lesions, which tend to grow large and are discovered late, 5-year survival is 34%. The presentation is variable in the small bowel. When the lesions are larger than 2 cm, metastases have been reported in various series at rates of 33% to 80%. Colon lesions tend to be discovered

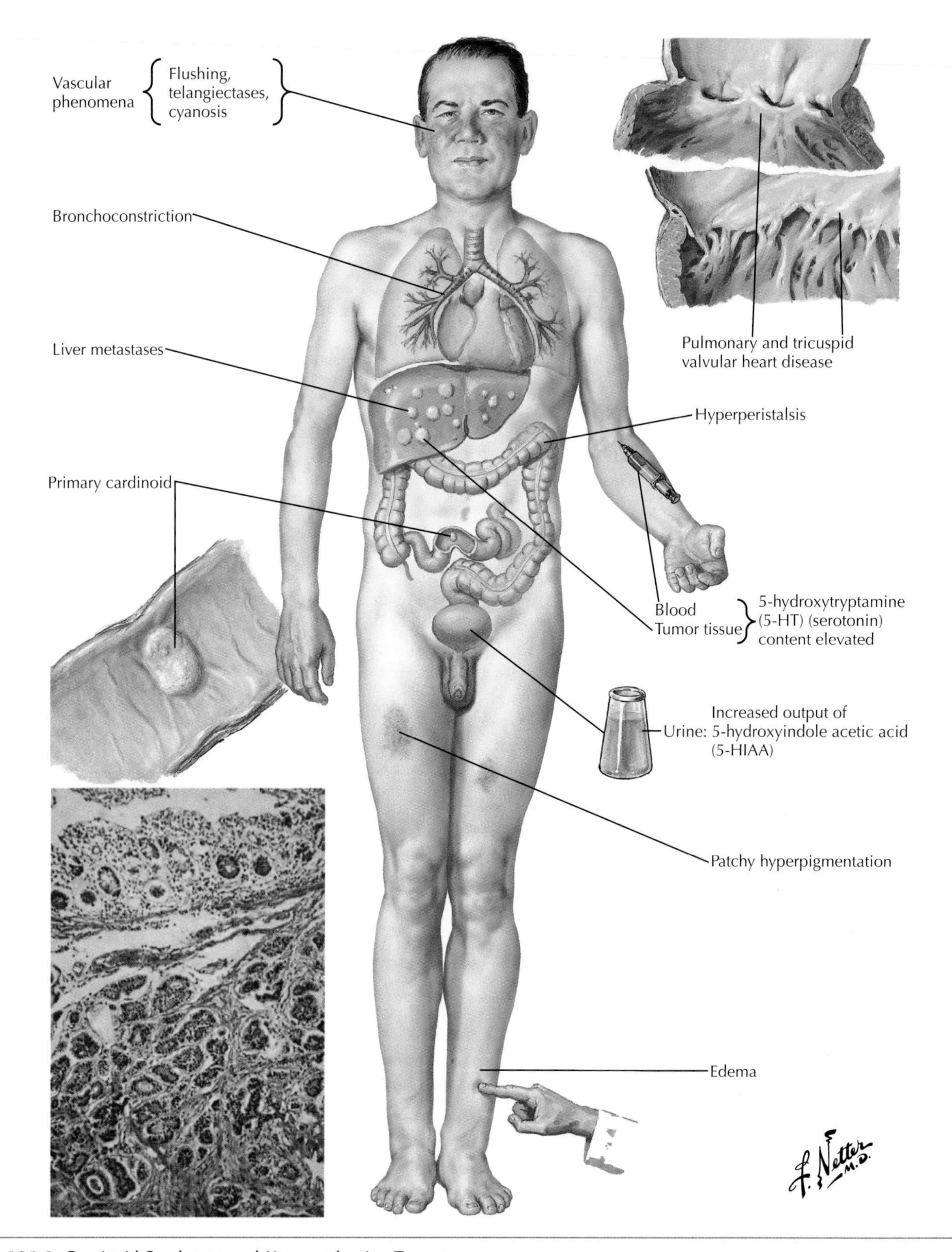

Figure 121-1 *Carcinoid Syndrome and Neuroendocrine Tumors.*

later, and 5-year survival with these is 42%. Local treatment of rectal carcinoids yields good results; however, if the lesions are larger than 2 cm, the probability of metastasis is 60% to 80%. When rectal lesions are smaller than 1 cm, the cure rate is 98%.

Overall prognosis for carcinoid tumors varies. Once carcinoids evolve into carcinoid syndrome, the prognosis is guarded.

ADDITIONAL RESOURCES

Anthony T, Kim L: Gastrointestinal carcinoid tumors and the carcinoid syndrome. In Feldman M, Friedman LS, Brandt LJ, editors: *Gastrointestinal and liver disease*, ed 8, Philadelphia, 2006, Saunders-Elsevier, pp 605-624.

Kulke MH, Meyer RJ: Carcinoid tumors, *N Engl J Med* 340:858-868, 1999.

Modlin IM, Sandor A: An analysis of 8,305 cases of carcinoid tumors, *Cancer* 79:813-829, 1997.

Que FG, Nagorney DM, Batts KP, et al: Hepatic resection for metastatic neuroendocrine carcinomas, *Am J Surg* 169:36-42, 1995.

Ruszniewski P, Ducreux M, Chayvialle JA, et al: Treatment of the carcinoid syndrome with a long-acting somatostatin analog lanreolide: a prospective study in 39 patients, *Gut* 39:279-283, 1996.

Ileostomy, Colostomy, and Gastroenteric Stomas

122

Martin H. Floch

This chapter discusses ileostomies and colostomies. Gastroenteric anastomoses are discussed in Chapters 66 to 70 and in Chapter 271. Ileal pouch anal anastomoses are discussed in Chapter 149.

Total or partial *colectomy* with resultant *ileostomy* or *colostomy* is performed primarily to treat inflammatory bowel disease (IBD), familial adenomatous polyposis (FAP), cancer involving pelvic organs, and abdominal trauma. Various types of ileostomies were performed in the past, such as the Kock pouch, but they are no longer popular and rarely used. Temporary colostomies are frequently performed for patients with acute diverticulitis and perforation. These patients often undergo repeat anastomosis and usually have a temporary colostomy for 3 to 6 months. **Figure 122-1** demonstrates the varied function of gastroenteric stomas, depending on the degree of resection.

Since the advent of the Brooke ileostomy, so-called ileostomy malfunction has rarely been seen. The major problem with ileostomy is electrolyte imbalance in the patient with gastroenteritis or food intolerance. After a colostomy has been established and the patient is properly educated to maintain it, colostomy function is compatible with a normal life pattern.

CLINICAL PICTURE

The problems of ileostomy occur when the output is greatly increased, as occurs in any form of gastroenteritis. The patient can rapidly become dehydrated as the symptoms of abdominal pain and effluence increase.

Colostomy function, once established and regular, rarely represents any problems other than the care of the stoma and peristomal areas. In elderly patients who have undergone colostomy because of incontinence or severe constipation, managing the colostomy is a geriatric care issue. The clinician must always be aware of possible complications, such as intestinal obstruction, from other causes, such as adhesions or recurrent malignancy. When obstruction occurs, it is followed by the classic decrease in output through the ileostomy or colostomy, with resultant distention and all the signs of intestinal obstruction.

DIAGNOSIS

Normal ileostomy output varies from 300 to 800 mL/day. Depending on what the patient eats and drinks, output may increase. However, the ileostomy effluent should be no more than 1 L/day. When it increases dramatically, fluids and electrolytes must be replaced. Ordinarily, sodium chloride and potassium intake is maintained through a regular diet. Salt loss is a major problem; therefore, these patients should be encouraged to increase their salt intake.

When the terminal ileum is resected, vitamin B_{12} malabsorption may occur. Vitamin B_{12} deficiency may be subtle and present as neurologic findings or may be a full-scale megaloblastic anemia. The clinician must be aware of these deficiencies. Vitamin B_{12} should be intramuscularly replaced intermittently in these patients.

TREATMENT AND MANAGEMENT

As mentioned, salt replacement is essential whenever ileostomy effluent increases. Patients with severe output may have to be admitted to an emergency service to receive supplementary intravenous (IV) fluid and electrolyte replacement. After clinical evaluation, most patients can be maintained on adequate oral intake without IV treatment.

The classic World Health Organization liquid formulas that depend on rapid absorption of sodium and glucose can be effective in helping the patient maintain fluid and electrolyte balance. These standard formulas are available in commercial products (e.g., Pedialyte), or they can be made at home (e.g., add salt amount plus table sugar amount to 1 L of fluid, in addition to desired flavoring) (see Chapter 111). Careful monitoring of the patient's blood pressure might be necessary to ensure that vascular compromise has not occurred.

At times, it may be necessary to perform x-ray studies to determine that no obstruction has developed, particularly for patients with a history of adhesions or malignancy. Proper irrigation of the sigmoid colostomy may be necessary to ensure evacuation. This becomes part of the routine colostomy care for patients with low colostomy, but not for those with transverse colostomy.

Care of the peristomal skin areas is now routine. All communities have centers from which patients can receive advice for the care of any skin sensitivity or lesions that may occur.

A stoma rarely needs to be revised. In certain situations, however, the stoma malfunctions or becomes stenotic, and repair is necessary. This is a surgical decision, and examination is heralded by complaints of pain or symptoms of partial obstruction.

COURSE AND PROGNOSIS

If the ileostomy is performed to manage IBD, trauma, or any benign disease, the prognosis is excellent, and these patients are able to lead a normal life. Similarly, if colostomy is performed for benign disease, patients can live normally. However, if the ostomy is for malignancy or for FAP, monitoring must include checking for recurrence of the neoplasia. Elderly patients often require concomitant geriatric care.

ADDITIONAL RESOURCES

Brooke BN: Management of ileostomy, including its complications, *Lancet* 2:102-104, 1952.

Cima RR, Pemberton JH: Ileostomy, colostomy, and pouches. In Feldman M, Friedman LS, Brandt LJ, editors: *Gastrointestinal and liver disease*, ed 8, Philadelphia, 2006, Saunders-Elsevier, pp 2549-2561.

Kelly KA, Sarr MG, Hinder RA: *Mayo Clinic gastrointestinal surgery*, Philadelphia, 2004, Saunders.

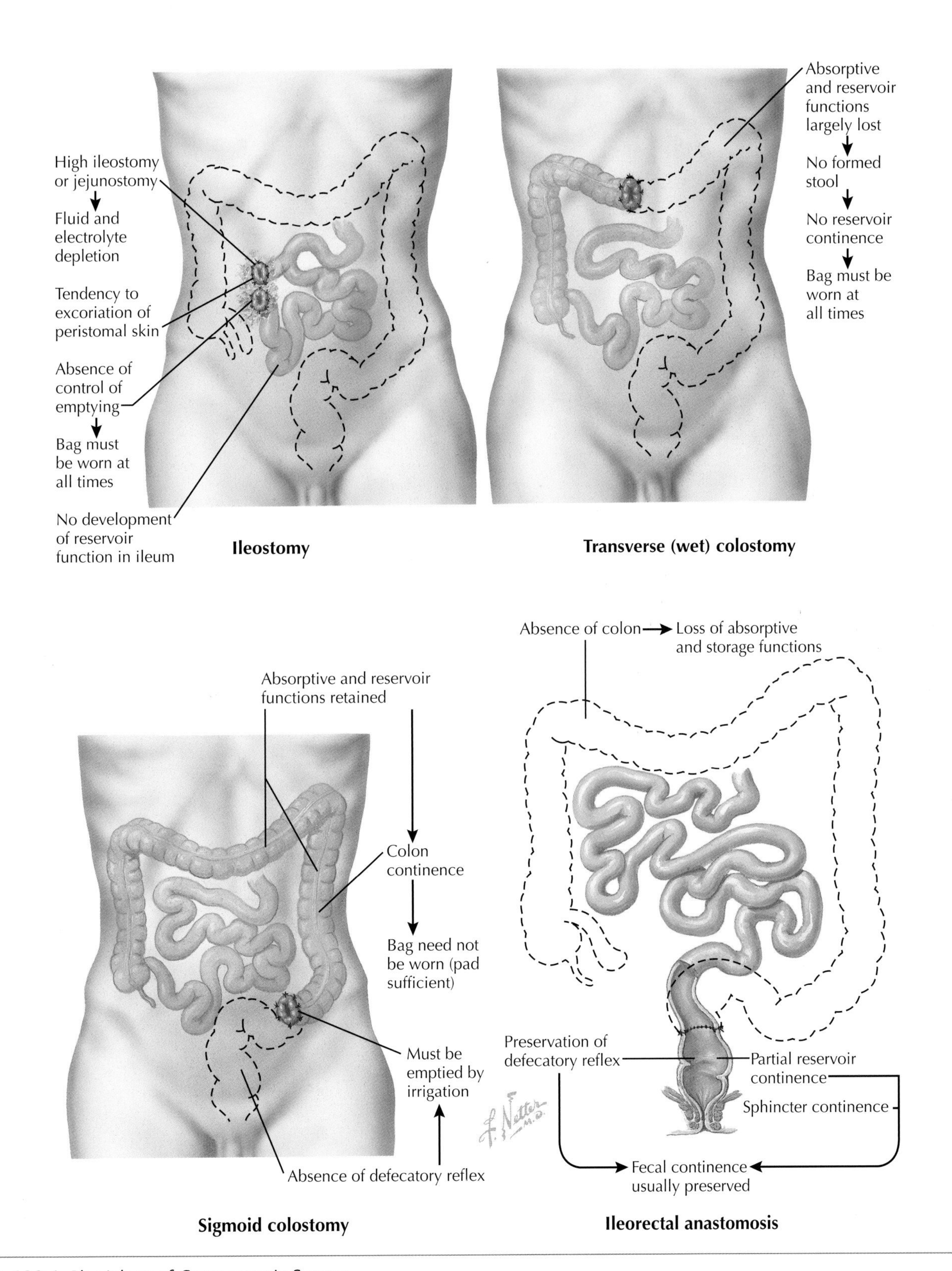

Figure 122-1 *Physiology of Gastroenteric Stomas.*

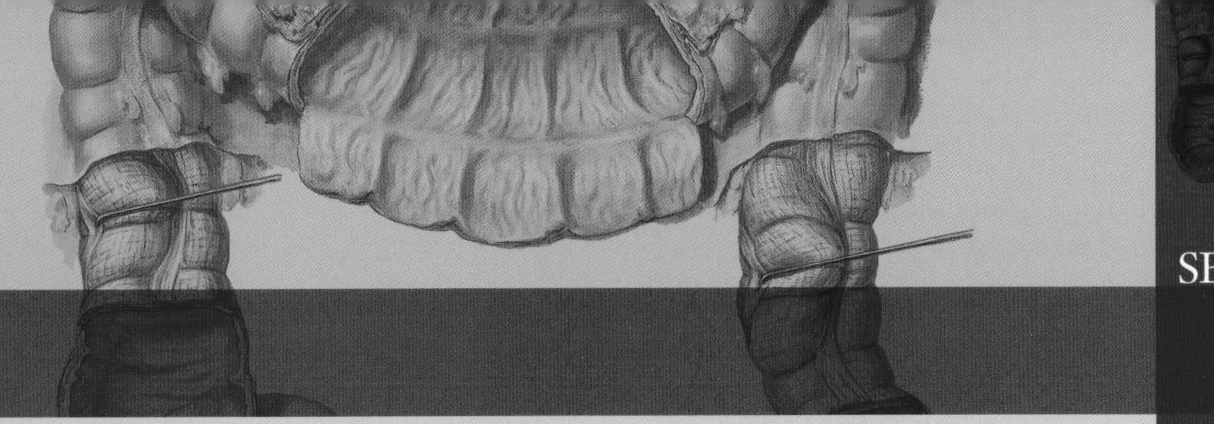

SECTION
V

Colon, Rectum, and Anus

Structure and Histology of the Colon

123

Martin H. Floch

The large intestine varies in caliber, depending on its functional state. Haustra form sacculations that are separated by constricting furrows, so that the lumen bulges and contracts alternately. The caliber is greatest at the commencement of the large intestine (cecum) and narrows toward the rectum. Viewed as a whole, the various parts of the large intestine describe a horseshoe-shaped arc (**Fig. 123-1**). The total length of the large intestine is approximately 120 to 150 cm (4-5 ft). The four segments of the colon are known as the ascending, transverse, descending, and sigmoid colon. The ascending and descending colon are situated retroperitoneally, and the transverse and sigmoid colon are situated intraperitoneally.

The *ascending colon* averages approximately 15 to 20 cm (6-8 inches) in length and runs in a more or less straight course from the upper lip of the ileocecal valve to the right colic or hepatic flexure, where it passes into the transverse colon. The right colic flexure is usually on the undersurface of the right lobe of the liver. The *transverse colon,* varying from 30 to 60 cm in length, extends from the hepatic flexure to the left colic or splenic flexure, situated slightly more cranially. It lies intraperitoneally and thus is attached to the posterior abdominal wall by a peritoneal fold (mesentery), the transverse mesocolon, which is very short in the region of the flexures and is longest in the middle of the transverse colon. The retroperitoneal *descending colon,* approximately 20 to 25 cm in length, extends downward from the left colic flexure to the iliac crest or beyond it into the left iliac fossa. After running from the angle between the lateral edge of the kidney and the quadratus lumborum muscle and then over the iliac muscle, the colon finally passes in front of the psoas major, crossing the femoral and genitofemoral nerves, and continues with no sharp dividing line into the pelvic colon, or *sigmoid colon,* at which point the colon becomes intraperitoneal again. On its anterior surface, the descending colon is overlapped by the greater omentum and generally by coils of the small intestine.

Corresponding to the structure of the entire intestinal tract, the wall of the colon and cecum consists of a mucosa, a submucosa, a double-layered muscularis, and depending on its relation to the peritoneum, a serosa and subserosa or an adventitia. The external aspects of the colon, however, differ from those of the small intestine not only because of its greater caliber but also because of the appearance of three typical formations: (1) the three taeniae, (2) the haustra, and (3) the appendices epiploicae. The three *taeniae* are longitudinal bands, approximately 8 mm in width, running along the total length of the colon; they exist because the outer muscle layer (i.e., longitudinal muscle) does not constitute a uniform coat. In the region of these three bands, the longitudinal musculature is conspicuous by its thickness, whereas in the spaces between them, it consists merely of a very thin coating. Each taenia is named by reference to its topographic situation in relation to the transverse colon. The *taenia mesocolica* is situated dorsal to the transverse colon at the line of attachment of the transverse mesocolon and comes to lie dorsomedially on the ascending and descending colon. The *taenia omentalis* is related to the line of attachment of the greater omentum on the ventrocranial surface of the transverse colon and runs along the dorsolateral aspect of the ascending and descending portions. The *taenia libera* is free (not related to any mesenteric or omental attachment) and generally found on the caudal (inferior) surface of the transverse and on the interior aspect of the ascending and descending colon. Where the appendix joins the cecum, and where the sigmoid passes into the rectum, the three taeniae merge into one uniform muscle coat, which in the proximal rectum is more strongly developed at its anterior and posterior parts than laterally. Generally, the posterior, lateral, and anterior taeniae coalesce into a broad longitudinal band in the region of the middle and lower sigmoid.

The *haustra* are more or less prominent sacculations formed in the spaces between the taeniae. They are separated from each other by constricting, circular furrows of varying lengths. The degree of their prominence depends on contraction of the taeniae; the more the taeniae contract, the more marked the haustra intestine becomes, whereas it is almost completely absent when the taeniae are totally relaxed.

The third structural characteristic, the *appendices epiploicae,* consists of grape-shaped subserosal pockets filled with fat and varying in size according to the patient's nutritional state. On the ascending and descending colon, the epiploic appendices are generally distributed in two rows, whereas on the transverse colon, they form only one row along the line of the taenia libera. These fat pads can become extremely large in obese patients.

Corresponding to the furrows between the haustra, visible on the outer surface, the mucous membrane of the large intestine forms crescent-shaped transverse folds, known as the *plicae semilunares.* As a rule, the lengths of these semilunar folds correspond to the distance between two taeniae, although they may be longer. Whereas Kerckring folds in the small intestine consist merely of mucosa and submucosa, the plicae semilunares also include the circular muscle layer.

In contrast to the small intestine, the mucosa of the large intestine is not covered with villous projections but contains deep, tubular pits that increase in depth toward the rectum and extend as far as the muscularis mucosae. In the submucosa, in addition to the usual structures (blood vessels, lymphatics, Meissner submucosal plexus), numerous solitary lymphatic nodules are present, originating in the reticular tissue with the tunica (lamina) propria and penetrating through the muscularis mucosae into the submucosa. The mucosal epithelium of the large intestine comprises one layer built of tall, prismatic cells that, when fixed in a fresh state, display a cuticular border on

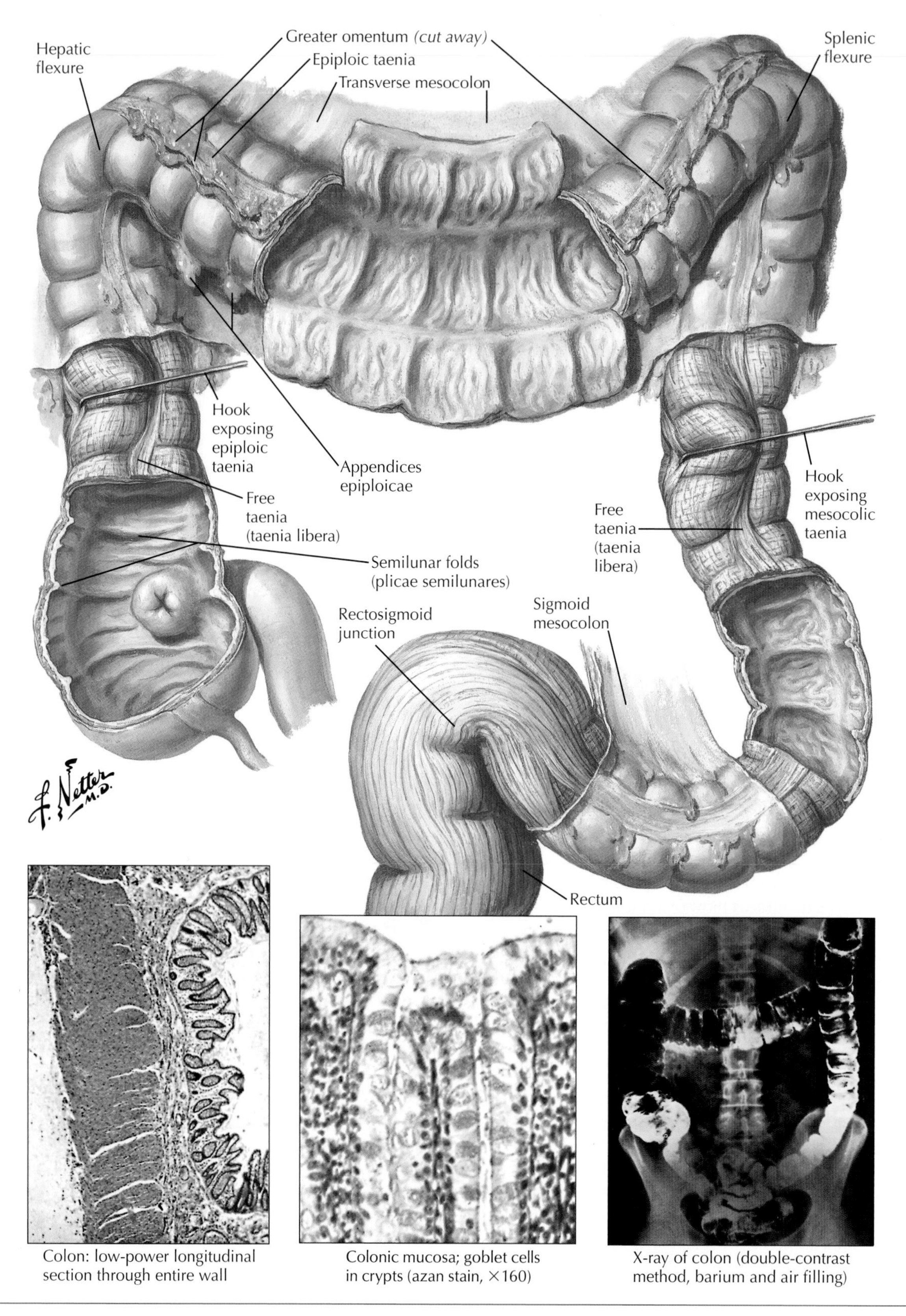

Colon: low-power longitudinal section through entire wall

Colonic mucosa; goblet cells in crypts (azan stain, ×160)

X-ray of colon (double-contrast method, barium and air filling)

Figure 123-1 *Topography and Structure of the Colon.*

their surface. Goblet cells are numerous, especially at the base of the pits. In contrast to the small intestine, the colonic epithelium is relatively simple, but enterochromaffin cells are present, producing serotonin, as are L cells, producing GLI/PYY, and rare D cells, producing somatostatin.

Details of the sigmoid colon are discussed in Chapter 124; unique features of the appendix are described in Chapter 141.

ADDITIONAL RESOURCES

Kahn E, Daum F: Anatomy, histology, embryology, and developmental anomalies of the small and large intestine. In Feldman M, Friedman LS, Brandt LJ, editors: *Gastrointestinal and liver disease*, ed 8, Philadelphia, 2006, Saunders-Elsevier, pp 2061-2091.

Sigmoid Colon

Martin H. Floch

124

The sigmoid colon is a specialized section of the colon that has specific motility function, as described in Chapter 133. In populations of the Westernized world, the sigmoid typically develops diverticula and then the pathology of diverticulitis (see Chapter 142). Colonoscopists are frequently challenged by passage of the instrument through this area because of chronic diverticulosis. Therefore, the sigmoid's anatomic structure is of particular importance (**Fig. 124-1**).

The exact point of commencement of the sigmoid colon—in other words, the transition of descending colon to sigmoid colon—is indefinite. The sigmoid is generally considered to be the part of the large bowel between the descending colon and the rectum that, as a result of its attachment to the mesentery, is freely movable. Because this mesentery is subject to great variations, the extent of the sigmoid also becomes variable. It has been described as beginning at the left iliac crest and the margin of the left psoas muscle or brim of the pelvis minor. Other authorities regard the sigmoid colon as comprising the iliac colon (an iliac portion that has no mesentery) and the pelvic portion (pelvic colon), with a mesentery beginning at the brim of the pelvis minor. The mesenteriolized sigmoid colon generally assumes an omega-shaped flexure arching over the pelvic inlet toward the first or second sacral vertebra (S1 or S2) or toward the right side of the pelvis. It finally joins the rectum at an acute angle at about the S3 level. This typical shape of the sigmoid is not a constant finding.

The sigmoid colon may be short, running straight and obliquely into the pelvis, or it may be so long that the loop extends far to the right or, in extreme cases, high into the abdomen. Its average length is approximately 40 cm (16 inches) in adults and 18 cm in children. With the variations mentioned, it may reach 84 cm (and even longer).

The root of the mesentery (i.e., of the mesosigmoid) is variable but, characteristically, starts in the upper left iliac fossa, proceeds downward a few inches, and proceeds mesially and again upward to a point on the psoas muscle, slightly to the left of the fourth lumbar vertebra (L4), where it turns downward into the pelvis. The line of mesenteric attachment takes the shape of an irregular and blunted inverted V. Turning caudally after having reached its highest point, the attachment line of the mesosigmoid courses over the left common iliac artery and vein, just above the division of the artery. The length of the mesosigmoid (i.e., distance from root of bowel wall) is extremely variable. A small peritoneal fossa, the *intersigmoid fossa* or recess, is formed by the mesosigmoid while twisting around the vascular pedicle. Rarely, this causes retroperitoneal hernia. Nevertheless, the fossa is a valuable guide to the vascular stalk. The left ureter passes retroperitoneally behind the intersigmoid recess. These relationships are variable but still important when disease occurs in this area.

The looping, arching variations in the sigmoid colon complicate the passage of instruments and make diagnosis difficult when disease occurs in the sigmoid. Diverticulitis can be confused with appendicitis, and diverticulitis with perforation can affect the ureter and pelvic organs.

The mucosa and submucosa of the sigmoid colon are almost identical with the corresponding structures of other parts of the colon. The same holds true for the arrangement of the circular and longitudinal muscle layers, except for the most distal parts of the sigmoid colon, where the three flat, longitudinal muscle bands (taeniae), typical of the large intestine, spread out to form a completely encircling longitudinal muscle layer of the rectosigmoid junction. In the same region, the circular layer thickens, in some cases to such an extent that its prominence is alluded to as the "sphincter muscle" of the junction. It is questionable, however, whether this thickening has a true sphincteric function. Throughout the course of the sigmoid colon, the appendices epiploicae of the serous coat diminish gradually in number and size.

The importance of the anatomic variations in the sigmoid colon is discussed under colonoscopy (see Chapter 129).

ADDITIONAL RESOURCES

Kahn E, Daum F: Anatomy, histology, embryology, and developmental anomalies of the small and large intestine. In Feldman M, Friedman LS, Brandt LJ, editors: *Gastrointestinal and liver disease*, ed 8, Philadelphia, 2006, Saunders-Elsevier, pp 2061-2091.

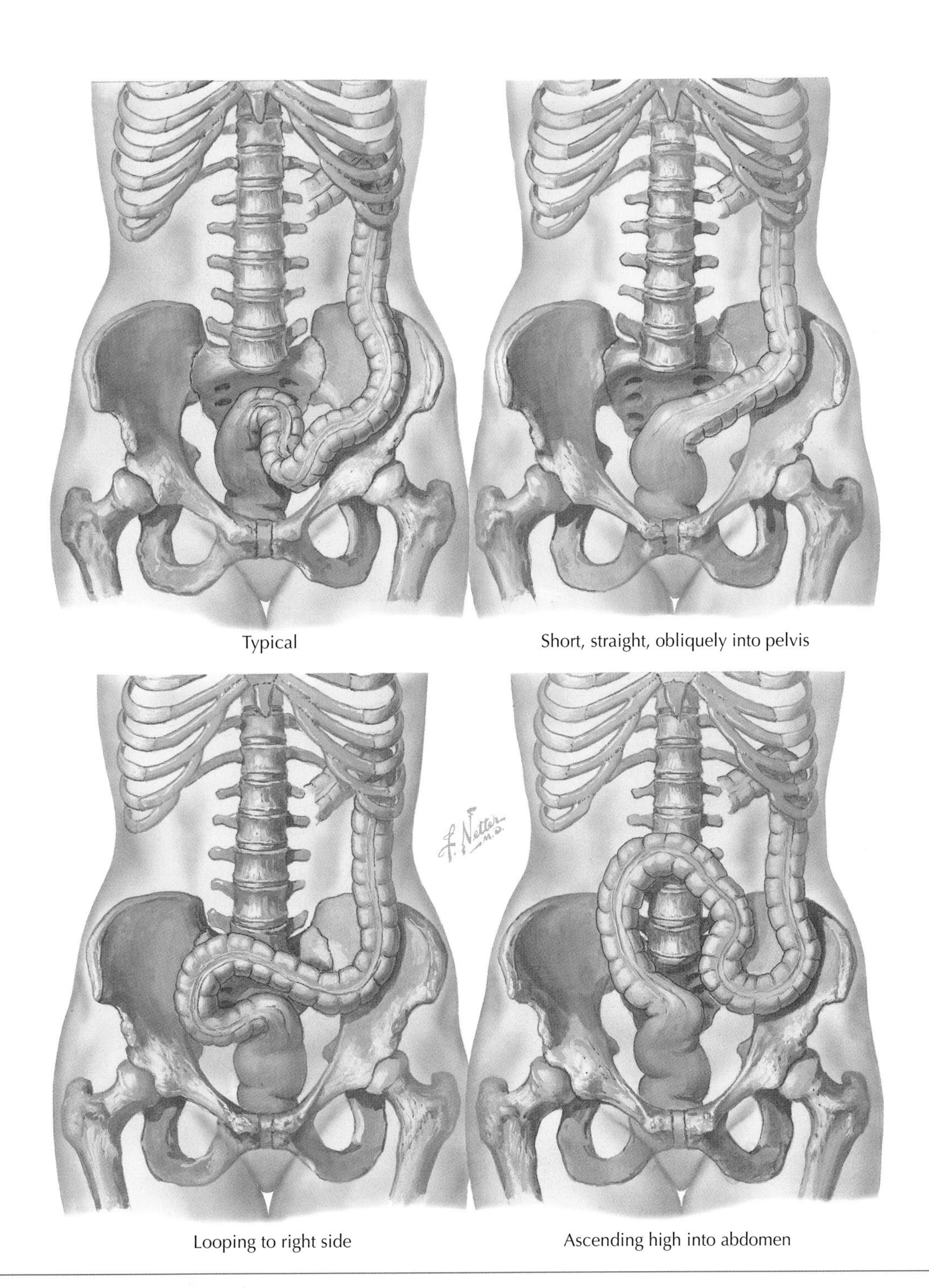

Figure 124-1 *Typical Sigmoid Colon and Variations.*

Rectum and Anal Canal

Martin H. Floch

To understand the numerous diseases and pathologic conditions that occur in the rectal and anal areas, it is essential to understand the details of the anatomy in this area. Functional problems of defecation, vascular disorders of hemorrhoids, and secondary problems of inflammatory bowel disease are all related to this anatomy.

The terminal part of the intestine consists of the rectum and the anal canal, which extends from the rectosigmoid junction, at the level of the third sacral vertebra (S3), 10 to 15 cm (4-6 inches) downward to the anorectal line (**Figs. 125-1** and **125-2**). The peritoneal coat continues down from the sigmoid, but only over the anterior and lateral rectal walls, for 1 to 2 cm. A very small mesorectum may occasionally be present, but only close to the rectosigmoid junction. The rectum is thus generally a truly *retroperitoneal* organ.

From the upper anterior rectal surface, the peritoneum is reflected into the interval between the rectum and the bladder in the male or the rectum and uterus in the female, forming the *retrovesical* or *retrouterine* recess or pouch, respectively. The depth to which these reflections extend varies an average of

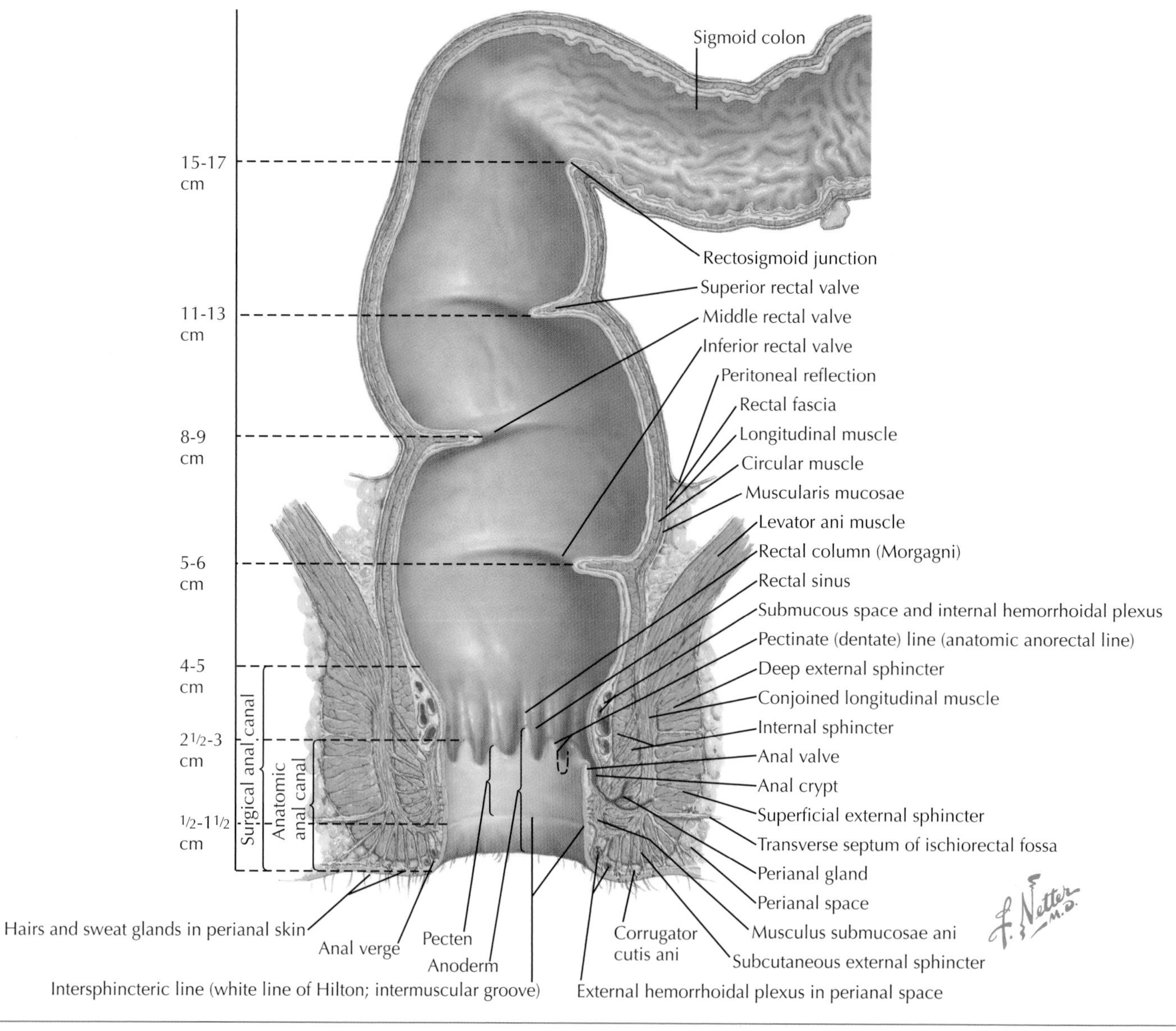

Figure 125-1 *Rectum from the Rectosigmoid Junction to the Anal Verge.*

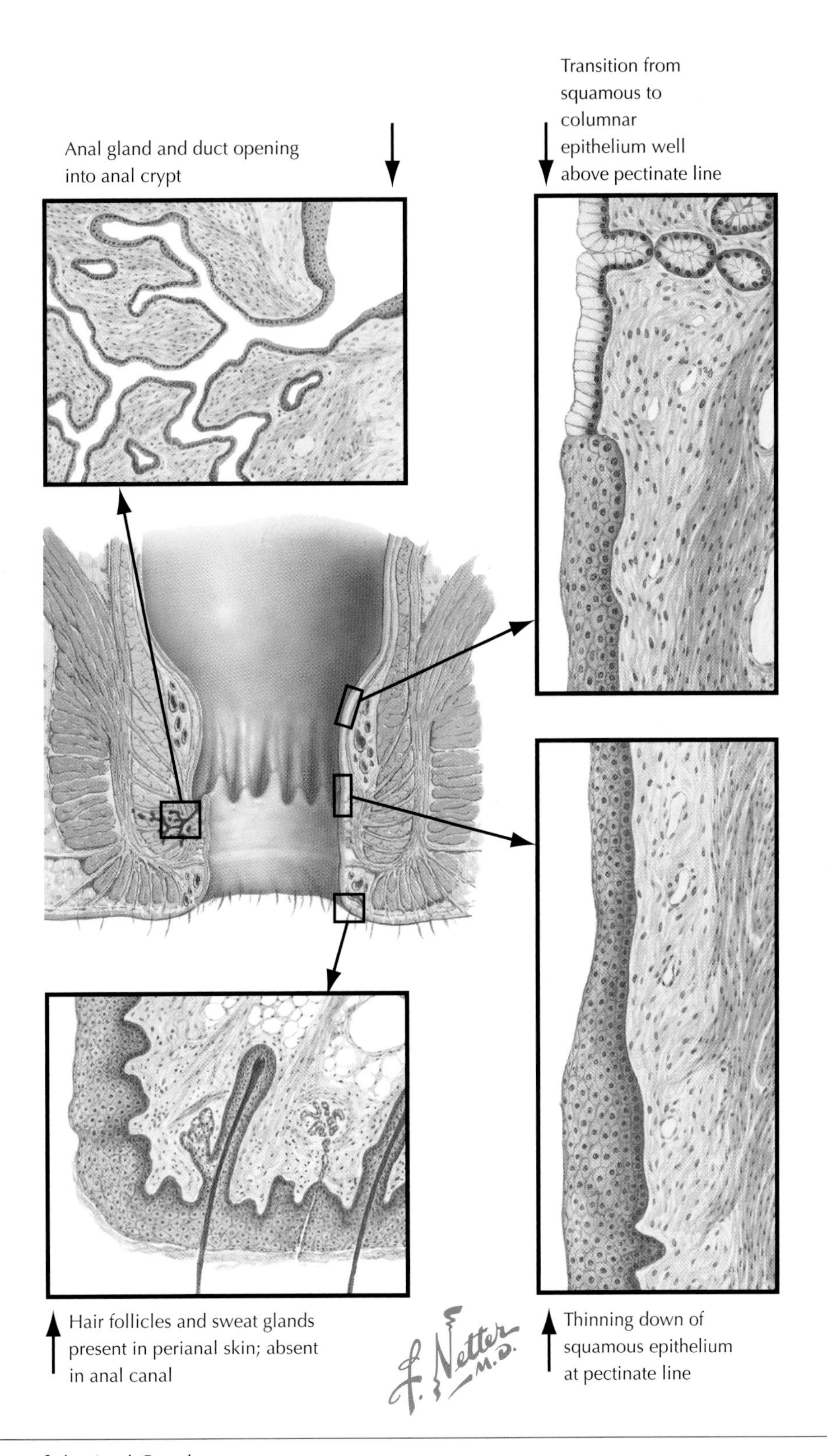

Figure 125-2 *Histology of the Anal Canal.*

7 cm in males and 4 cm in females. The length and diameter of the rectum vary greatly. The rectum in women is typically much smaller than in men.

The posterior rectal wall hugs the anterior aspect from the sacrum to the sacrococcygeal articulation, where it comes to lie more or less horizontally over the levator shelf. The anterior wall is comparatively straight and follows, closely aligned, parallel to the posterior cephalic axis of the vagina in the female or the rectogenital septum in the male. The rectal ampullary portion usually has three frequently prominent lateral curvatures, or *flexures*. All three bends correspond to the indentations on the opposite side of the internal rectal wall, which are produced by crescentlike plications of the mucosa and the submucosa, including the circular musculature, but not the longitudinal musculature. These more or less marked folds, known as *rectal valves*, encircle approximately one third to one half of the rectal circumference. The superior and inferior rectal valves are located on the left side, the superior valves approximately 4 cm below the rectosigmoid junction and the inferior valves approximately 2 to 3 cm above the dentate line. The middle rectal valve lies usually on the right side at or slightly above the level of the peritoneal reflection, approximately 6 to 7 cm above the dentate line. Digital examination may reach the distance of the middle valve.

The sphincteric portion of the rectum, often considered to be the upper third of the surgical anal canal, begins at the clinically palpable upper edge of the anorectal muscle ring, usually about 4 cm above the *anal verge*, where the rectum narrows considerably. It extends down to the anatomic *anorectal line* (dentate line), an irregular or undulating demarcation in the rectal mucosa, about 2 to 3 cm above the anal verge. It has been assumed that this line marks the junction of the endodermal primitive gut with the ectodermal proctodeum. However, histologic evidence now shows that this transition of the fetal structures is not abrupt but spreads gradually over several centimeters. Nevertheless, the dentate line is visually recognizable, circling the bowel, presenting from 6 to 12 cranial extensions and an equal number of intervening caudal sinuosities. The dentate line projects cranially a variable distance, correlative to the rectal columns. These projections may be 1 to 1.5 cm long, and the peninsulas of anoderm they enclose are designated *anal columns*. Interdigitating rectal columns, with their rectal sinuses and anal columns, make a rich network at the junction of the mucosa and the ectodermal tissue. Occasionally, the anal columns form papillary projections into the rectal lumen, and the name *anal papillae* has been applied to these teatlike processes. In most cases, the anal papillae are absent, but when present and exposed to chronic infection, they may hypertrophy and become so prominent as to appear like fibrous polyps, which may even prolapse through the anal canal.

The anatomic anal canal starts at the dentate line and extends to the anal verge, or the margin where the anal tube opens outwardly, or to the circumference. It is difficult to define but is roughly identical with the margin of the anal skin, where hair stops growing.

Within the submucosa of the sphincteric rectal portion lies the internal hemorrhoidal venous plexus in the submucosal space. The submucosa is also particularly rich in lymphatics and in terminal nerve fibers. Compared with the sigmoid colon, the mucous membrane in the rectum is thicker. It also becomes increasingly redder and more vascularized as it reaches the surgical anal canal, until its lowermost portion assumes an almost plum color. The extreme vascularity predisposes it to hemorrhagic disorders.

The cuboidal or columnar epithelium of the rectum extends downward into the upper third of the surgical anal canal, where it changes irregularly above the dentate line into a stratified squamocuboidal type, which directly covers the internal hemorrhoid plexus, the rectal columns, and the sinuses of Morgagni. This epithelial transition zone, which is sometimes referred to as the *anal mucosa*, does not start abruptly and is arranged haphazardly.

In the subepithelial muscular start of the anal canal and in the lower rectum, simple tubular and racemose (grapelike) glands may be found, described as perianal glands, intramuscular glands, and anal ducts. Usually, the mouths of the ducts open into the bottom of the anal crypts, but the ducts and glands may extend for a variable distance into the adjacent tissues, even into or through the sphincteric muscles ("intramuscular" glands). They are significant as possible sites of anorectal infection and fistula.

The anorectal musculature is important in understanding many problems with defecation and sphincter control. The significance of the *conjoined longitudinal muscle* for the physiology, pathology, and surgery of the anorectum cannot be overemphasized. Together with the *levator ani muscle*, it exercises its levator and sphincteric action on the anal canal by fibers taking their course through the entire length of the conjoined longitudinal muscle. By its extension at the level of the intermuscular groove, and by its fascial frame in the upper third of the surgical anal canal, the muscle influences the spread of anorectal infections and the sites of the openings and main tracts of fistulas. **Figures 125-3** and **125-4** help clarify some of the interrelationships.

The outermost and also most caudad musculature elements of the anal canal belong to the *external anal sphincter*, which is a trilaminar striated muscle. Its three parts—subcutaneous, superficial, and deep—are easily recognized. The *subcutaneous* portion, approximately 3 to 5 mm in diameter, surrounds the anal orifice directly above the anal margin and is rarely palpable and often discernible as a distinct angular ridge. The male anatomy may show anterior muscular extensions to the median raphe, and posterior extensions of the subcutaneous external sphincter may connect with the coccyx. In the female, the subcutaneous portion is more strongly developed, particularly anteriorly, where it forms a prominent angular band that is frequently incised at episiotomy. The subcutaneous portion is functionally integrated with the levator ani muscle through extensions of the conjoined longitudinal muscle, which pass fanlike through it, to terminate as fibers of the corrugator cutis ani. The elliptically shaped *superficial portion*, the next deep layer, is the largest and strongest of the three lamina of the external anal sphincter. It is complex and interdigitating, arises from the tip of the coccyx, and helps form the right and left muscular components of the anococcygeal ligament. The *deep portion* of the external anosphincter is mainly an annular muscle bundle, usually not attached to the coccyx. It is intimately blended with the puborectalis muscle as the fibers of that muscle pass, sleevelike, around the terminal rectum.

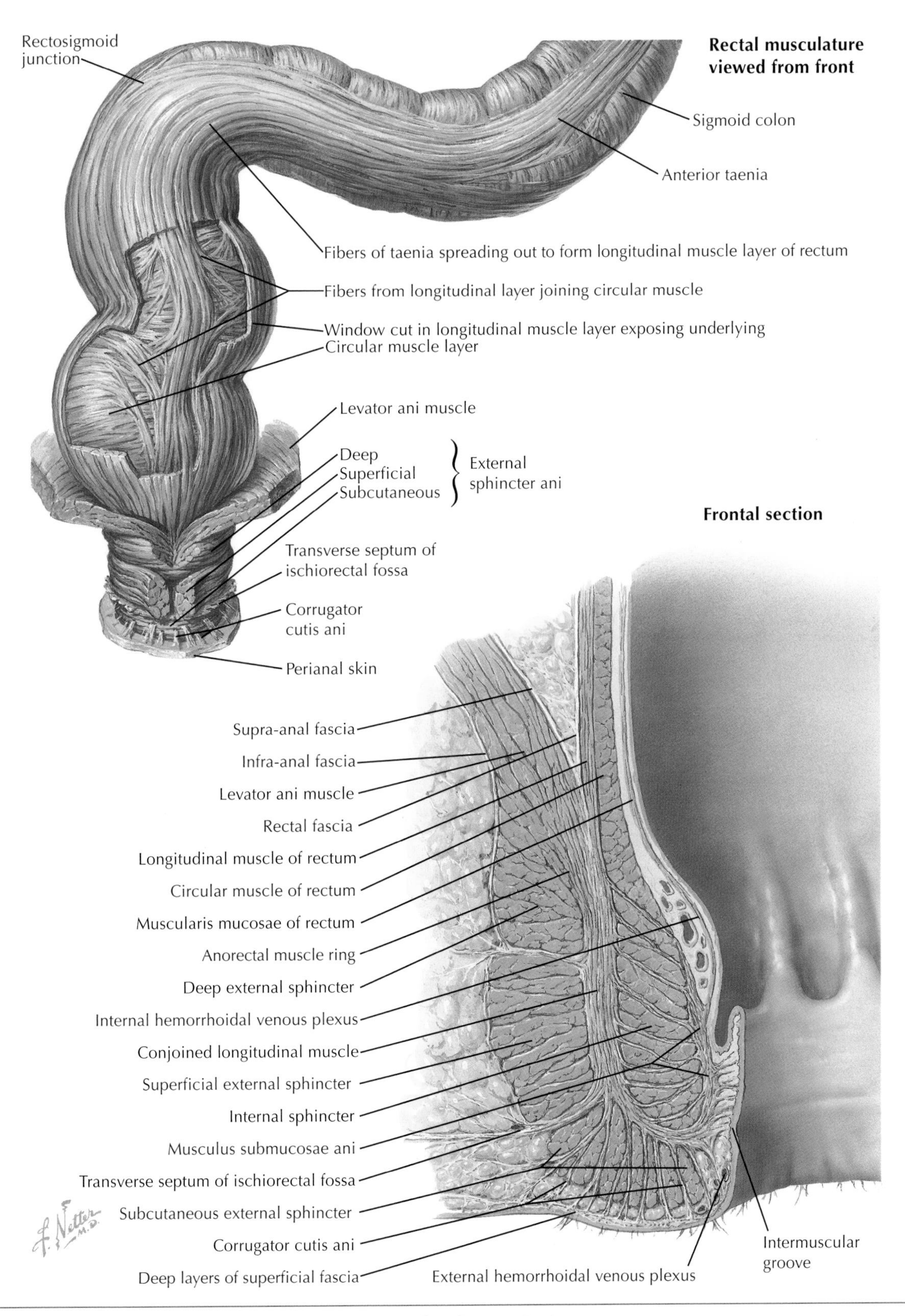

Figure 125-3 *Anorectal Musculature: Continuity with Sigmoid and Cross Section.*

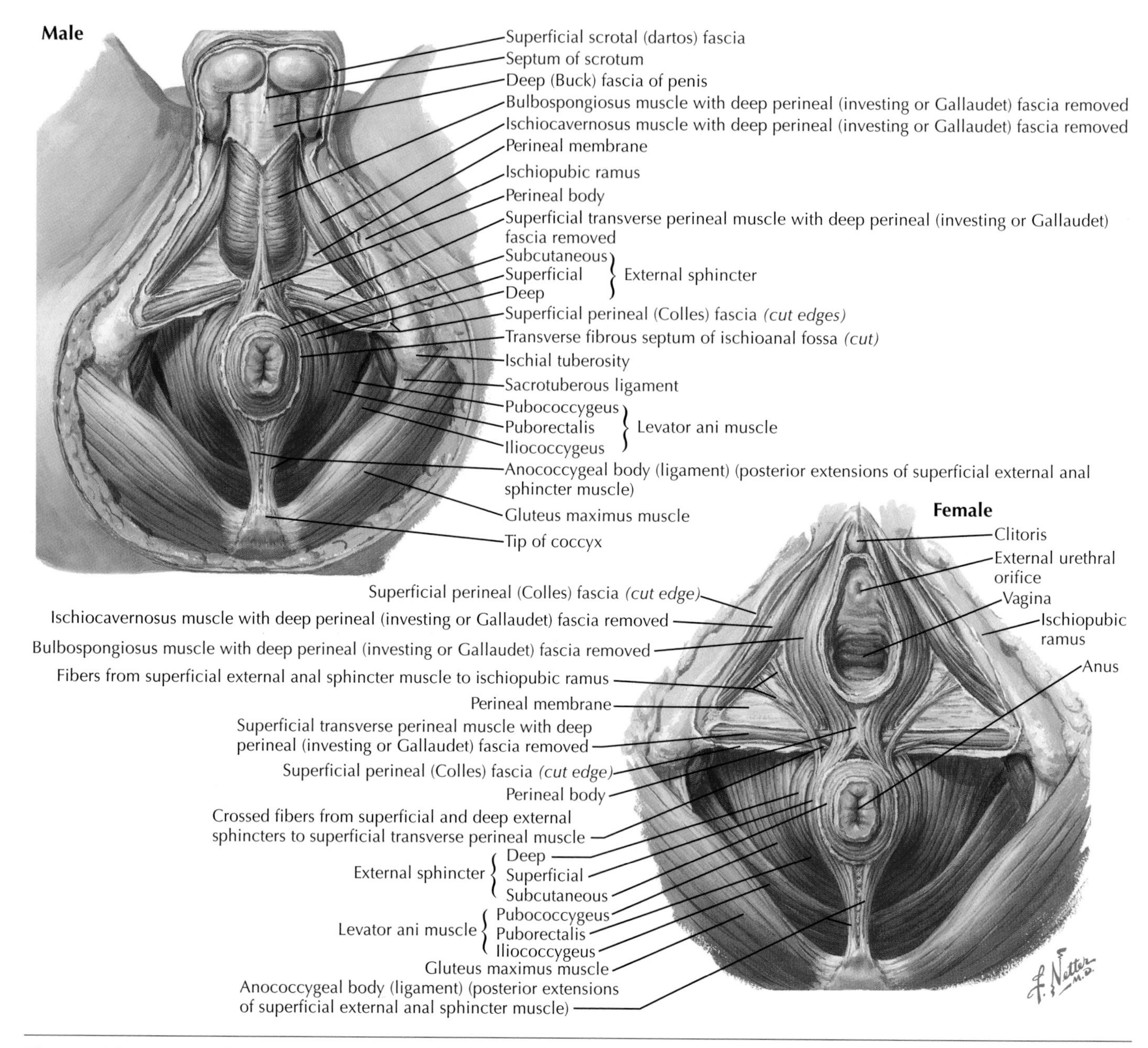

Figure 125-4 *Anorectal Musculature: Pelvic Floor (Male and Female).*

The essential forces that keep the rectoanal canal in position derive from muscles forming the pelvic floor, the levator ani muscle, which is assumed to be composed of three individual components: pubococcygeus, puborectalis, and ileococcygeus muscles. These muscles create a diaphragmatic plane. The action of all these muscles, as demonstrated in Figures 125-3 and 125-4, are integral for normal defecation. They are also pivotal in retraining when sphincters become incompetent. The levator ani fixes the pelvic floor and acts as a fulcrum, against which increased abdominal pressure, as occurs in lifting, coughing, and defecation, may be exerted. Levator ani integration with the other muscles is essential in maintaining the integrity of rectal and urogenital function.

ADDITIONAL RESOURCES

Kahn E, Daum F: Anatomy, histology, embryology, and developmental anomalies of the small and large intestine. In Feldman M, Friedman LS, Brandt LJ, editors: *Gastrointestinal and liver disease*, ed 8, Philadelphia, 2006, Saunders-Elsevier, pp 2061-2091.

Vascular, Lymphatic, and Nerve Supply of the Large Intestine

126

Martin H. Floch

As in the small intestine, major arterial vessels that supply the colon are usually paralleled by similar venous vessels, except for the rectum, where the venous drainage is rich (**Fig. 126-1**).

The *middle colic artery*, according to conventional descriptions, arises at the lower border of the pancreas and passes into the right half of the transverse mesocolon, where, at a variable distance from the colonic wall, it typically divides into two branches. One branch courses to the right to anastomose with the ascending branch of the right colic artery; the other branch turns to the left to anastomose with the ascending branch of the left colic artery, derived from the inferior mesenteric artery (see Fig. 126-1). Both divisions of the middle colic undergo subsequent branchings, forming primary and secondary arcades that direct arteriae rectae to the transverse colon. This description, however, does not apply to all cases. As a separate branch of the superior mesenteric artery, the middle colic artery is frequently absent. In such cases, the artery is usually replaced by a common right middle colic trunk and occasionally by a branch of the left colic; at times, the latter reaches the hepatic flexure. An accessory middle colic artery may be present, generally arising from the aorta above the chief middle colic artery, and it usually anastomoses with branches from the left colic artery, forming a secondary arc in the left transverse mesocolon. There may be many variations of these formations.

Conventionally, the *right colic artery* is described as rising from the superior mesenteric artery, dividing, halfway between its origin and the ascending colon, into a descending branch that unites with the ileocolic artery and an ascending branch that unites with the left branch of the middle colic artery (see **Fig. 126-1**). Again, variations are seen in these formations. In as many as 18% of studies, the right colic artery is entirely absent. Many variations occur in the integration of the flow to the right column from the superior mesenteric artery and to the left colon from the inferior mesenteric artery. However, the ileocolic artery, the last branch of the superior mesenteric on its right side, is always present. Therefore, as a rule, the two chief arterial branches, the ascending colic and the descending ileocolic, branch. Ten different sites of origin and courses are described for the appendicular artery.

The *inferior mesenteric artery* arises typically from the anterior aspect or left side of the aorta, 3 to 5 cm above its bifurcation, which is situated on the level of the lower third of the fourth lumbar vertebra (L4). The inferior mesenteric proceeds down and forks into the last of the sigmoid arteries and the superior rectal artery. This creates a rich blood supply in the rectum (see **Fig. 126-1**). The sigmoid arteries, in combination with the right pudendal arteries, the rectal arteries, and some of the blood vessels that supply the muscles of the pelvis, create a rich arterial blood flow to the rectum and anal area.

The venous drainage of major arteries largely parallels the arterial drainage, except for the rectum and anus, where the same veins and arteries are present but rich internal and external *rectal hemorrhoidal plexuses* serve essentially the mucosal, submucosal, and perianal tissue. The plexuses encompass the rectal circumference completely, but the greatest aggregation of small and large veins takes place in the rectal columns. Generally, the vessels returning the blood from the plexuses course 10 cm upward in the submucosa. Branches derived from these veins pierce the muscular layer of the rectum and communicate with the perimuscular plexus and directly with the superior rectal vein. Because these piercings of the musculature occur mostly above the level at which the perimuscular plexus connects with the middle rectal veins, it is apparent that the latter assist in draining the internal rectal plexus. The blood of this plexus returns mostly through the superior rectal vein.

Dilatation of the internal rectal plexus results in internal hemorrhoids (see Chapter 162). Dilatation of the external rectal plexus or thrombosis of its vessels constitutes the external hemorrhoids. The two plexuses (internal and external rectal) are separated by the muscularis submucosa ani and by the dense tissue of the rectum, but they communicate with each other through these tissues by slender vessels that increase in size and number with age and that are voluminous in the presence of hemorrhoids. Further, the inferior and middle rectal veins and their collecting vessels, the internal pudendal veins, have valves, whereas the superior rectal veins are devoid of such valves. Increased pressure in the portal vein, as in cirrhosis or other causes of portal hypertension, may reverse the circulation in the superior rectal veins. Portal blood flows through the superior rectal veins, traversing the rectal plexus and then carried away by the inferior rectal vein and shunted through the internal iliac vein of the cable system.

LYMPH DRAINAGE

Lymph drainage of the large intestine is complex. The first important regional lymph nodes are the ileocolic, right colic, middle colic, and left colic, pertaining to the respective regions of the large intestine. They start with a chain of nodes, collectively called *pericolic nodes*, which lie along the medial margin of the ascending, transverse, and descending colon and lie dorsal to these portions of the gut in the retroperitoneal tissue and, to a lesser extent, in the mesosigmoid. Each of these groups of lymph nodes pours its lymph into lymph ducts that run side by side with the respective blood vessels in a median direction toward the large prevertebral vessels. Lymphatics emanating from the rectum and anal canal run in two main directions. In the lower part of the anal canal, they pass over the peritoneum, alongside the scrotum or labia majora and the inner margin of the thigh, to the superior inguinal nodes. The upper part of the

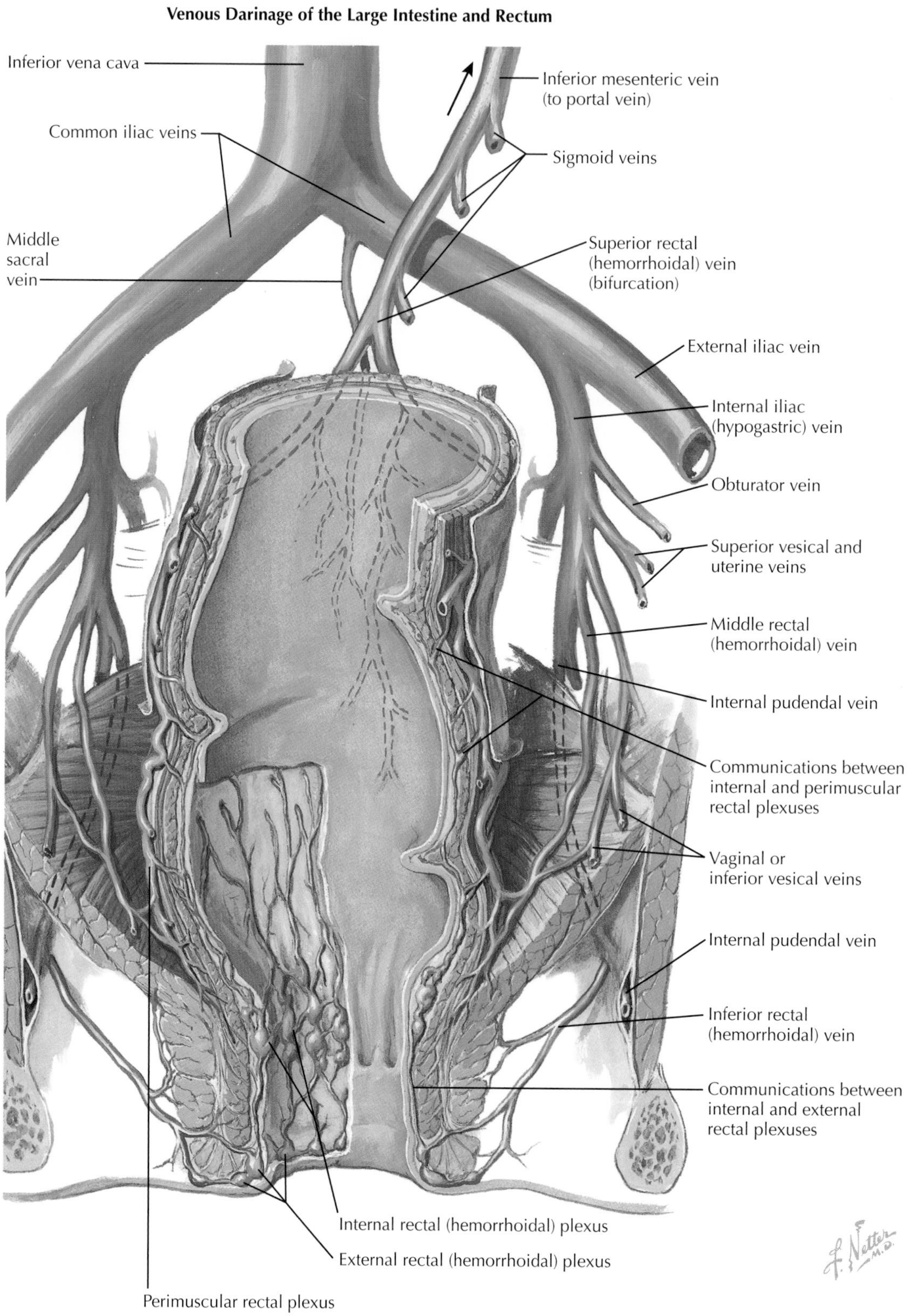

Venous Darinage of the Large Intestine and Rectum

Inferior vena cava
Inferior mesenteric vein (to portal vein)
Common iliac veins
Sigmoid veins
Middle sacral vein
Superior rectal (hemorrhoidal) vein (bifurcation)
External iliac vein
Internal iliac (hypogastric) vein
Obturator vein
Superior vesical and uterine veins
Middle rectal (hemorrhoidal) vein
Internal pudendal vein
Communications between internal and perimuscular rectal plexuses
Vaginal or inferior vesical veins
Internal pudendal vein
Inferior rectal (hemorrhoidal) vein
Communications between internal and external rectal plexuses
Internal rectal (hemorrhoidal) plexus
External rectal (hemorrhoidal) plexus
Perimuscular rectal plexus

Figure 126-1 *Vascular, Lymphatic, and Nerve Supply of the Large Intestine.*

Arterial Circulation of the Large Intestine

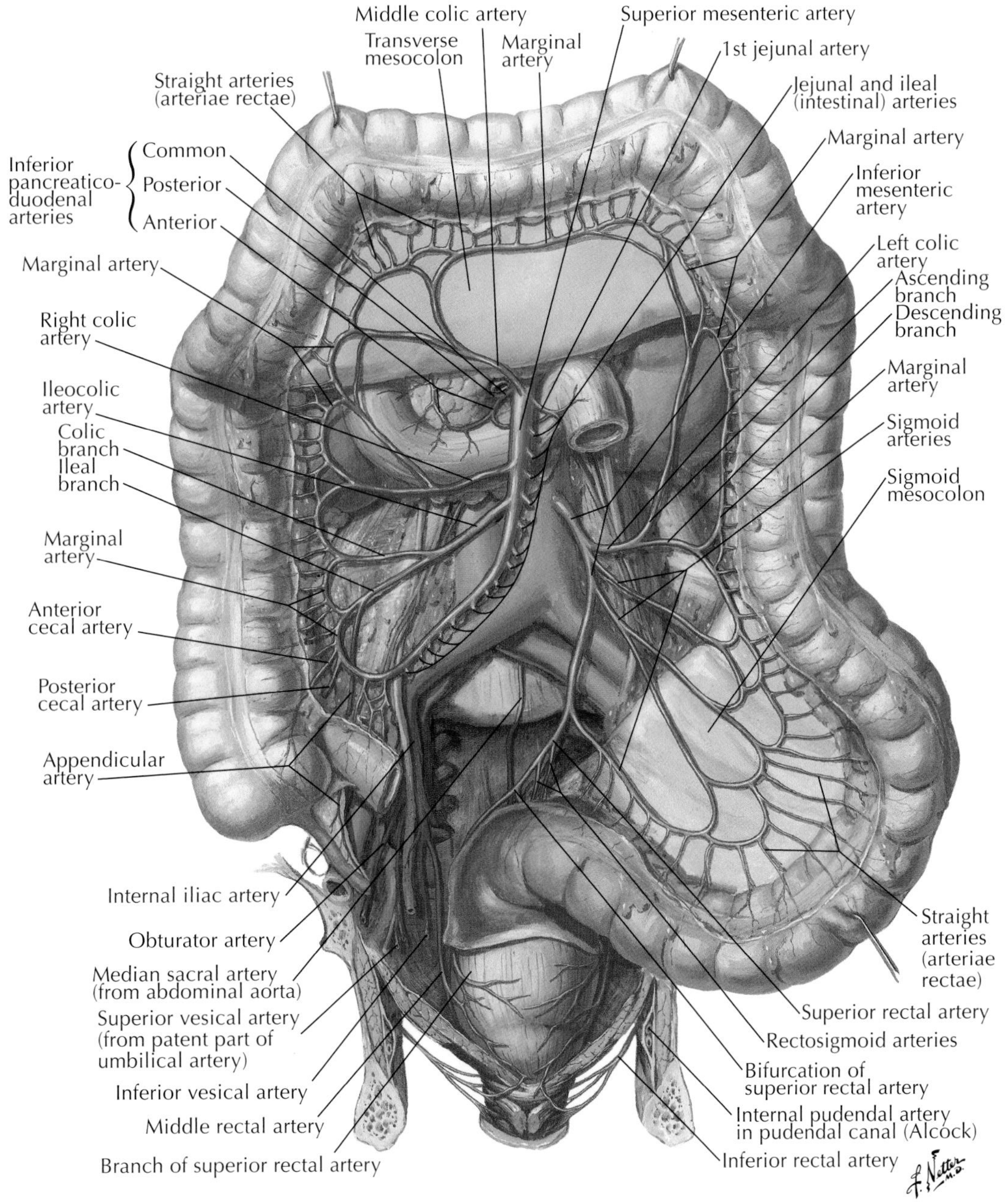

Middle colic artery
Superior mesenteric artery
Transverse mesocolon
Marginal artery
1st jejunal artery
Straight arteries (arteriae rectae)
Jejunal and ileal (intestinal) arteries
Marginal artery
Inferior pancreatico-duodenal arteries { Common, Posterior, Anterior
Inferior mesenteric artery
Left colic artery
Ascending branch
Descending branch
Marginal artery
Marginal artery
Right colic artery
Sigmoid arteries
Ileocolic artery
Colic branch
Ileal branch
Sigmoid mesocolon
Marginal artery
Anterior cecal artery
Posterior cecal artery
Appendicular artery
Internal iliac artery
Straight arteries (arteriae rectae)
Obturator artery
Median sacral artery (from abdominal aorta)
Superior rectal artery
Superior vesical artery (from patent part of umbilical artery)
Rectosigmoid arteries
Bifurcation of superior rectal artery
Inferior vesical artery
Internal pudendal artery in pudendal canal (Alcock)
Middle rectal artery
Branch of superior rectal artery
Inferior rectal artery
F. Netter M.D.

anal canal is drained cranially into preaortic and inferior mesenteric nodes.

INNERVATION

Innervation of the large intestine is similar to that of the small intestine, the central nervous system, and the intrinsic nerves of the colon, as described in Section IV.

ADDITIONAL RESOURCES

Kahn E, Daum F: Anatomy, histology, embryology, and developmental anomalies of the small and large intestine. In Feldman M, Friedman LS, Brandt LJ, editors: *Gastrointestinal and liver disease*, ed 8, Philadelphia, 2006, Saunders-Elsevier, pp 2061-2091.

Secretory, Digestive, and Absorptive Functions of the Colon and Colonic Flora

127

Martin H. Floch

The mucous membrane of the large intestine secretes an opalescent, mucoid, alkaline fluid composed essentially of water, mucus, and electrolytes (**Fig. 127-1**). Chemical and mechanical irritation enhances the secretion of this fluid. The colonic epithelium also has an excretory function, in that it is used as an elimination route for metals (lead, mercury, bismuth; possibly silver and calcium). The mucous gel consists of large mucin glycoproteins, trefoil factors, defensins, secretory immunoglobulins, electrolytes, phospholipids, bacteria, sloughing epithelial cells, and numerous other components still to be identified. This mucoid gel layer serves as a protective coat that is secreted primarily by the goblet cells.

The digestive function of the colon is significant. Varying but usually small amounts of fat and proteins, proteases, peptones, and peptides escape digestion in the jejunum and ileum and may be digested in the colon by bacterial enzymes capable of breaking down these substances. Putrefactive action of bacteria can produce fatty acids. Certain amino acids, primarily tryptophan but also tyrosine, phenylalanine, and histidine, can be digested to form such compounds as skatole, indole, phenol, creosol, and histamine. These products may be absorbed in relatively small amounts by the mucosa and can be transported to the liver, where they are detoxified. They are usually excreted by the kidney in the form of sulfates and glucuronides. The bulk of this material remains in the colonic lumen and leaves the intestine with the feces. Indole, skatole, mercaptan, hydrogen sulfide, and breakdown products of cystine give the feces its unpleasant odor. The color of the feces derives chiefly from stercobilin, the bacterial reduction product of bile pigment.

The greatest activity of fermentation in the colon is the breakdown of starch and nonstarch polysaccharides, such as cellulose. These compounds are not broken down by human enzymes but by bacterial enzymes, producing the important *short-chain fatty acids* (SCFAs). The molecular ratio of *butyric* acid, *acetic* acid, and *propionic* acid is approximately 20:60:20. People whose diets include a large proportion of fiber, which contains many nonstarch polysaccharides, nurture the bacterial flora, and more SCFAs are produced. Depending on dietary intake, the amount of polysaccharide that reaches the colonic flora can range from 5% to 20% of oral intake. This leads to the production of 300 to 800 μmol of SCFA, which makes up as much as 5% to 10% of the host's energy. Most of the bran and insoluble types of fiber are not broken down and remain in the feces to form bulk and to hold onto water. Butyric acid is the preferred fuel of colonocytes, and acetic acid and propionic acid are involved in cholesterol synthesis and control when they are absorbed into the hepatic circulation.

The main absorptive function of the colon is to balance the excretion of electrolytes and water by absorbing large amounts of fluid that reach the cecum with sodium. Elaborate mechanisms of epithelial chloride-bicarbonate and sodium-potassium exchange permit homeostatic balancing for the host. A series of sodium-hydrogen exchanges enables sodium absorption. Intercellular cyclic adenosine monophosphate (cAMP) and calcium regulate sodium-hydrogen exchanges. In addition, substances such as nitric oxide, protein kinases, and cytoskeletal proteins all affect sodium-hydrogen exchange. The final effect of water excretion is that the average 1500 to 2000 mL of liquid that reaches the cecum is absorbed, and, depending on the amount of dietary fiber and stool bulk from fiber (bacterial or insoluble bran), leaves a stool of 100 to 400 mL. SCFAs are the principal luminal products of the colon, are heavily involved in this balance of water and electrolyte absorption, and enhance sodium absorption. SCFAs appear to be able to diffuse readily across membranes and are absorbed by several mechanisms. Because other water-soluble substances can also be absorbed through these mechanisms, some drugs (e.g., chloral hydrate, anticholinergics, xanthenes, digitalis glycosides) are administered rectally.

The *enteropathic circulation* includes the secretion of bile acids and bile into the upper small intestine, with the major reabsorption occurring in the ileum. It does continue somewhat in the ascending colon, where primary and secondary bile acids are reabsorbed.

In summary, a major function of the large intestine is to absorb fluid and balance electrolytes. This occurs primarily in the ascending colon. Colonic flora has a major role in fermenting starch and nonstarch polysaccharides to produce SCFAs. SCFAs are the primary fuel for the colonocyte and are absorbed into the enterohepatic circulation, affecting cholesterol metabolism. Dietary fiber acts to nurture the bacterial flora and to increase the bulk of fecal material so that stool bulk and content are maintained for normal colonic motility and function.

COLONIC MICROFLORA

The gastrointestinal tract and the organisms within its lumen constitute an *ecologic unit*, often referred to as *intestinal microecology*. The metabolism and function of this ecologic unit affect the host. The four interacting components of the unit are the wall of the gut, the fluid secreted into the lumen, the food that enters the gut, and the intestinal microbiota.

Anaerobic culture and RNA/DNA techniques enable the identification of the most common aerobic and anaerobic organisms in humans. An estimated 500 species reside in the large intestine, and the healthy bowel has approximately 100 trillion organisms. Although the flora in the distal ileum is rich, it is greatest in the colon. Anaerobes are approximately 100 to 1000

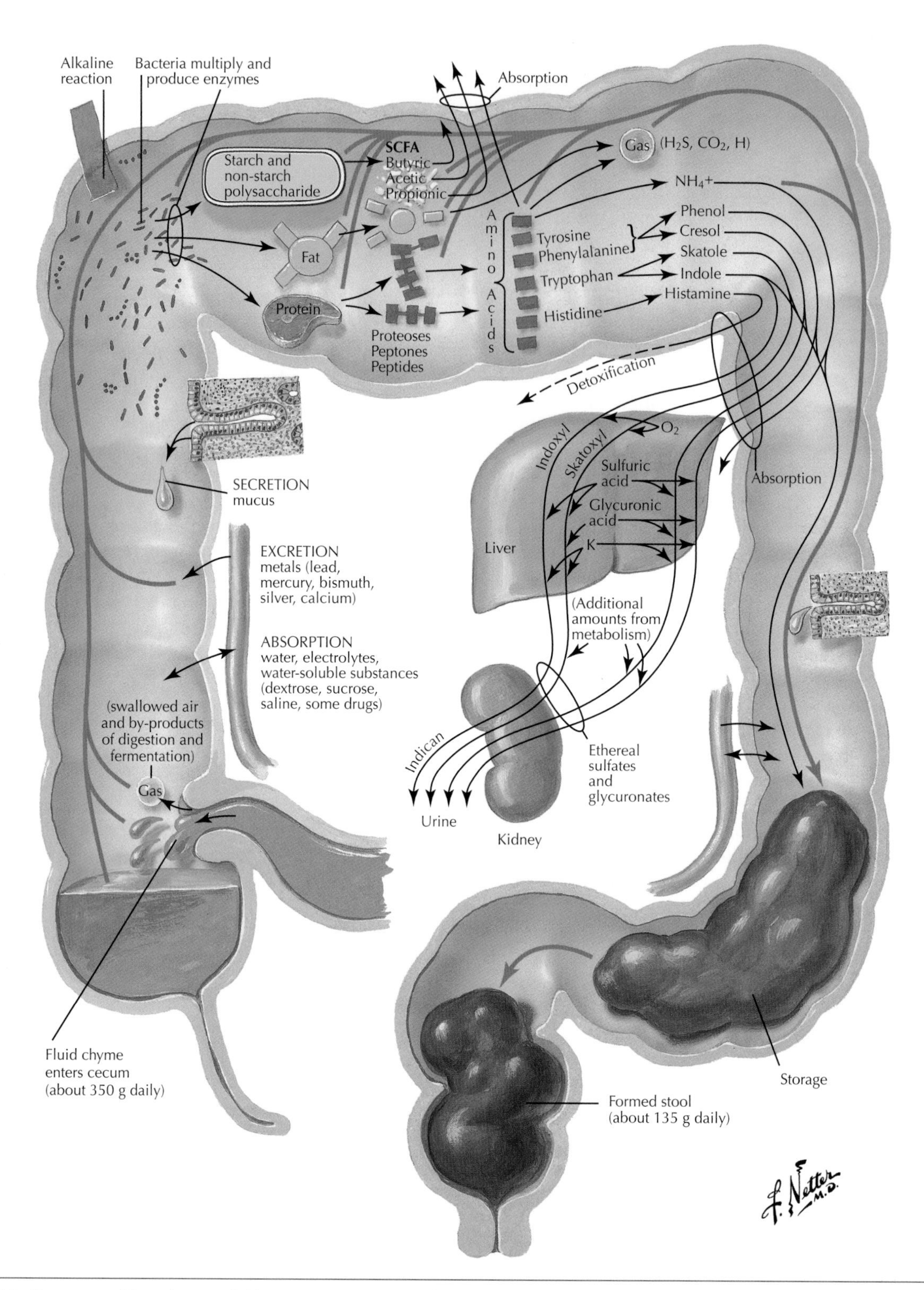

Figure 127-1 *Secretory, Digestive, and Absorptive Functions of the Colon and Colonic Flora.*

times more abundant than aerobes. Major anaerobic species are anaerobic cocci, *Bacteroides*, *Eubacterium*, *Bifidobacterium*, *Lactobacillus*, Veillonellae, and *Fusobacterium*. Major aerobic organisms belong to species of *Escherichia*, *Enterococcus*, *Streptococcus*, *Bacillus*, *Citrobacter*, and *Klebsiella*.

The major role identified for the bacterial flora is fermenting undigested carbohydrates and converting some fats and proteins into waste products. It is now appreciated, however, that carbohydrate fermentation products are absorbed and can make up as much as 5% to 10% of absorbed energy. Also, a major role for the bacteria is to deconjugate and transfer bile acids to the enterohepatic circulation.

ADDITIONAL RESOURCES

Cummings J: Quantitating short-chain fatty acid production in humans. In Binder HJ, Cummings J, Soergel KH, editors: *Short-chain fatty acids*, London, 1994, Kluwer Academic, pp 11-19.

Dudeja PK, Gill R, Ramaswamy K: Absorption-secretion and epithelial cell function. In Koch TR, editor: *Colonic diseases*, Totowa, NJ, 2003, Humana Press, pp 3-22.

Feingold SM, Atteberg HR, Sutter DL: Effect of diet on human fecal flora: comparison of Japanese and American diets, *Am J Clin Nutr* 27:1456-1459, 1974.

Floch MH, Gershengoren W, Freedman LR: Methods for the quantitative study of the aerobic and anaerobic intestinal bacterial flora in man, *Yale J Biol Med* 41:50-59, 1968.

Ho SD, Shekels LL: Mucin and goblet cell function. In Koch TR, editor: *Colonic diseases*, Totowa, NJ, 2003, Humana Press, pp 53-71.

Moote WEC, Holdeman LB: Human fecal flora: the normal flora of 20 Japanese-Hawaiians, *Appl Microbiol* 27:961-979, 1974.

Ramaswamy K, Harig JM, Soergel KH: Short-chain fatty acid transport by human intestinal apical membranes. In Binder HJ, Cummings J, Soergel KH, editors: *Short-chain fatty acids*, London, 1994, Kluwer Academic, pp 93-103.

Wells AL, Sauliner DMA, Gibson GR: Gastrointestinal microflora and interactions with gut mucosa. In Gibson GR, Roberfroid MB, editors: *Handbook of prebiotics*, Boca Raton, Fla, 2008, CRC Press/Taylor & Francis Group, pp 13-38.

Probiotics

Martin H. Floch

Probiotics are live-microbial food supplements that benefit the person by improving microbial balance (**Fig. 128-1**). They are usually strains of lactobacilli or bifidobacteria, but yeasts such as *Saccharomyces* have also been used. Probiotics are typically administered in yogurts, capsules, or powders and have the following properties:

1. Consist of bacteria of human origin.
2. Survive passage through the gut.
3. Resist stomach secretions, hydrochloric acid, and liver bile acids.
4. Produce substances called *adhesins*, which help probiotics adhere to human intestinal cells and help the epithelium prevent invasion by pathogens.
5. Have the ability to colonize the human intestinal tract, particularly within the mucous layer.
6. Produce antimicrobial substances, and antagonize carcinogenic and pathogenic flora.
7. Are safe to use in large amounts.

Box 128-1 Probiotic Microorganisms

***Lactobacillus* Species**	***Bifidobacterium* Species**
L. acidophilus	*B. adolescentis*
L. amylovorus	*B. animalis*
L. casei	*B. lactis*
L. crispatus	*B. bifidum*
L. gasseri	*B. breve*
L. johnsonii	*B. infantis*
L. paracasei	*B. longum*
L. plantarum	
L. reuteri	
L. rhamnosus	
Other Lactic Acid Bacteria	
Enterococcus faecium	
Leuconostoc mesenteroides	
Streptococcus thermophilus	
Non-Lactic Acid Bacteria	
Bacillus cereus var. *toyoi*	
Escherichia coli strain Nissle	
Propionibacterium freudenreichii	
Saccharomyces cerevisiae	
Saccharomyces boulardii	

Modified from Floch MH, Hong-Curtiss J: Probiotics in functional foods and gastrointestinal disorders, Curr Treat Opt Gastroenterol *5:311-321, 2002.*

The concept of probiotics was first developed by Metchnikoff, who won the Nobel Prize in 1907 for his theory that aging was related to the products of putrefactive bacteria in the intestine. He proposed administering good bacteria to prevent putrefaction and to enhance fermentation. Kiploff promulgated the importance of lactobacilli and Roetger stressed their therapeutic application in the last half of the twentieth century. Stillwell finally coined the term "probiotic," which was coined and finally used by Parker. In 1989, Fuller defined probiotics as "microbial supplements that benefit the host animal by improving its intestinal microbial balance." Studies reveal that the benefits of probiotics may derive from local intestinal effects as well as their resultant systemic action.

Numerous organisms are used as probiotics, alone or in combination (**Box 128-1**). Anecdotal evidence and findings from human and animal experiments propose that all the listed probiotics may be helpful. More studies are becoming available; those usually accepted as clinically important are described here.

Probiotics improve immune status. Their growth clearly increases immunoglobulin A (IgA) production, and they have been used to help treat and prevent childhood infectious diarrhea. They are also reportedly helpful in treating atopy and cow's milk allergy. They are used to prevent and treat antibiotic-associated diarrhea and *Clostridium difficile* colitis, which occurs after antibiotic treatment in a significant percentage of hospital patients. *Lactobacillus* GG and *Saccharomyces* have been used for this purpose.

Genitourinary infections have been prevented and treated successfully with yogurts and *Lactobacillus acidophilus*. Results have been mixed, although studies show excellent findings with vaginal and oral use of different organisms. A combination of eight probiotic organisms, VSL#3 has been used successfully to treat pouchitis (see Chapter 149). Some have treated inflammatory bowel disease with *Escherichia coli* strain Nissle, and others have used *Lactobacillus* in combination with bifidobacteria. Probiotics have also been used for irritable bowel syndrome; *Bifidobacterium infantis* is reported as helpful, as are *B. animalis* and *Lactobacillus plantarum* in a yogurt product (**Table 128-1**).

The gastrointestinal tract is an ecologic unit that is significantly affected by the use of probiotics and foods. Alteration of the bacterial flora can result in disease. The use of prebiotics and probiotics can correct that disease process in select patients. The field of prebiotics has just emerged, and use of prebiotic products to nurture organisms is beginning to prove effective. A combination of a probiotic with a prebiotic is referred to as a *symbiotic*. The effectiveness of probiotics has been documented (see Additional Resources), but it is clear that larger studies are needed to understand specific effects in specific diseases.

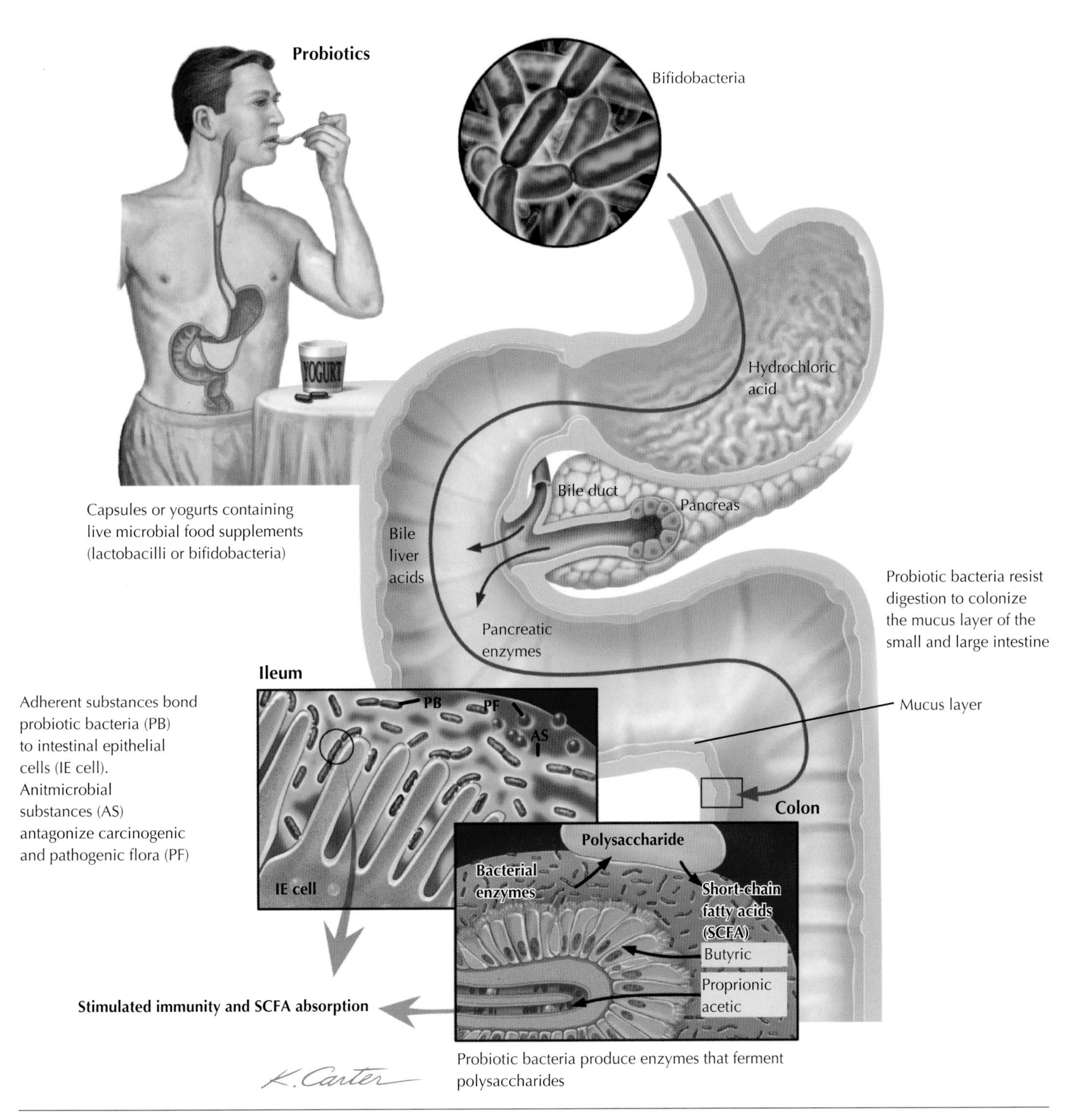

Figure 128-1 *How Probiotics Assist in Digestion and Affect the Large Intestine* (SCFA, *Short-Chain Fatty Acid*).

Table 128-1 Probiotics with Reported Clinical Relevance

Probiotic	Clinical Relevance
Lactobacillus acidophilus	Prevention of recurrent candidal vaginitis
L. acidophilus LCI	Adherence to human intestinal cells, balancing of intestinal microflora, and immune enhancing
L. acidophilus NCFO1748	Treatment of constipation and decreased food enzymes Prevention of radiotherapy-related diarrhea
L. acidophilus NFCM	Treatment of lactose intolerance, production of bacteriocins, decreased fecal enzyme activity, high lactase activity
L. casei Shirota	Balancing of intestinal bacteria, decreased fecal enzymes, superficial bladder cancer control
L. rhamnosus GG	Treatment and prevention of rotavirus diarrhea and relapsing *Clostridium difficile* colitis Prevention of acute diarrhea and antibiotic-associated diarrhea Treatment of atopy and milk allergy eczema
Bifidobacterium bifidum	Treatment of rotavirus and viral diarrhea Balancing of intestinal microflora
Lactobacillus reuteri	Colonizing of intestinal tract, shortening of rotavirus diarrhea
Saccharomyces boulardi	Prevention and treatment of *Clostridium difficile* colitis and antibiotic-associated diarrhea
Escherichia coli strain Nissle	Used in inflammatory bowel disease to maintain remission
Eight strains VSL 3 (see text and reference)	Decreased recurrence of pouchitis
Lactobacillus plantarum *Bifidobacterium infantis*	Decreased symptoms of irritable bowel syndrome

Modified from Floch MH, Hong-Curtiss J: Probiotics in functional foods and gastrointestinal disorders, Curr Treat Opt Gastroenterol *5:311-321, 2002.*

ADDITIONAL RESOURCES

Floch MH, Binder HJ, Filborn B, Gershengoren W: The effect of bile acids on intestinal microflora, *Am J Clin Nutr* 25:1418-1426, 1972.

Floch MH, Montrose DC: Use of probiotics in humans: an analysis of the literature, *Gastroenterol Clin North Am* 34:547-570, 2005.

Floch MH, Walker WA, Guandalini S, et al: Recommendations for probiotic use—2008, *J Clin Gastroentrol* 42:S104-S108, 2008.

Gibson GR, Roberfroid MB: *Handbook of prebiotics*, Boca Raton, Fla, 2008, CRC Press/Taylor & Francis Group.

Ghosh S, van Heel D, Playford RJ: Probiotics in inflammatory bowel disease: is it all gut flora modulation? *Gut* 53:620-622, 2004.

Gobach SL: Probiotics and gastrointestinal health, *Am J Gastroenterol* 95:S2-S4, 2000.

MacFarlane GT, Cummings JH: Probiotics and prebiotics: can regulating the activities of intestinal bacteria benefit health? *BMJ* 318:999-1003, 1999.

Reid G: Probiotics, agents to protect the urogenital tract against infection, *Am J Nutr* 73:437S-443S, 2001.

Tannock GW: *Probiotics: a critical review*, Wymondham, UK, 1999, Horizon Scientific Press.

Anoscopy, Sigmoidoscopy, and Colonoscopy

129

Martin H. Floch

Clinicians can now examine the anal canal, rectum, and colon with ease and with little discomfort to the patient. The indication for each procedure depends on the clinical presentation. For example, red blood on toilet paper or blood dripping into the toilet bowl frequently requires anoscopy and sigmoidoscopy, but not colonoscopy. Chronic diarrhea or abdominal pain and signs of intermittent obstruction necessitate colonoscopy (**Fig. 129-1**).

Use of these techniques for cancer prevention and screening varies throughout the world. Most gastroenterologists agree that colonoscopy is the preferred procedure, whereas occult blood testing and sigmoidoscopy are practiced in many countries as part of colon cancer prevention programs. However, after risk factors (e.g., family history of colon cancer) are identified, colonoscopy is recommended.

ANOSCOPY

Anatomy and abnormalities of the anal canal are examined through anoscopy. The anoscope is a rigid instrument that may be short (proctoscope) or as long as 10 cm (anoscope). Examination requires little preparation, with the patient in the left lateral decubitus position and the buttocks spread by the left hand. The instrument is gently inserted into the rectum after digital examination. If digital examination is too difficult to perform because of marked obesity or severe pain, the procedure may have to be limited and deferred to later, when the patient can be sedated and full sigmoidoscopy or colonoscopy performed. After the anoscope or proctoscope is inserted, the anal ring and distal rectum can be carefully inspected. It may be necessary to place the patient in the knee-chest position to spread the buttocks wide. This is especially true in some obese patients, who may feel discomfort in the rectum.

SIGMOIDOSCOPY

Rigid sigmoidoscopes are no longer used. The advent of the flexible sigmoidoscope now permits the prepared patient to be examined in the left lateral decubitus position without sedation, although some clinicians prefer using mild sedation. After appropriate cleansing, which may include a cathartic or simply low enemas, depending on patient cooperation, the instrument is inserted into the rectum after digital examination.

The valves of Houston can be observed, and with appropriate technique, the distal sigmoid loops can be traversed, and the flexible instrument can be passed into the descending colon. The instrument is usually passed 60 cm (24 inches), unless marked spasm in the sigmoid or an abnormality such as severe diverticular disease prevents passage. Most patients experience some pain as the instrument passes through the sigmoid loop. In approximately 25% of patients, the sigmoid loop is tortuous, creating difficulty for less experienced endoscopists. In these patients, if it is essential to examine the entire sigmoid, sedation and full colonoscopy are indicated. Passage through the difficult sigmoid loops requires training and endoscopic skills in rotating the instrument and the patient (see Additional Resources).

Aspiration specimens may be obtained for bacteriologic and parasitic study, and mucosal biopsy specimens may be obtained to evaluate for chronic diarrhea, parasitic diseases, and certain systemic diseases.

COLONOSCOPY

Advances in technology and skills now allow examination of the entire colon and the terminal ileum in a single examination. Indications for colonoscopy range from a screening procedure to rule out polyps or early malignancy to evaluation of all symptoms referable to the intestine (**Fig. 129-2**).

The patient must be well prepared for successful completion of colonoscopy. This includes adequate education and then bowel preparation, in which the entire colon is cleansed. Usually, there is a 2-day limitation on intake of fiber-containing foods, with a careful purging of the bowel the day before the examination. Purging agents include magnesium salts, nonabsorbable carbohydrates, and balanced electrolyte solutions in large volume. The type of purge depends on the patient's ability to tolerate substances and the staff's instruction skills.

For routine colonoscopy, sedation is usually required. Rarely, some patients prefer no sedation, but the average patient requires mild sedation with a narcotic plus a sedative. Some patients are so apprehensive or uncooperative that anesthesia is required.

Bacteremia, which is usually harmless, may develop during the procedure, but for patients who have prosthetic heart valves, previous endocarditis, or recent vascular surgery, antibiotics are used for prophylaxis during and immediately after colonoscopy. Although some controversy exists, antibiotics are not usually recommended for patients with prosthetic joints, pacemakers, or simple mitral valve prolapse. Usual doses are 1 g of ampicillin and 80 mg of gentamicin administered intravenously 10 to 30 minutes before the procedure. Other patients may simply use an oral antibiotic before and several hours after the procedure.

Colonoscopy requires significant training. However, once an endoscopist is skilled, the colon and terminal ileum can be examined and biopsy material obtained for study. Therapeutic colonoscopy is gaining wide acceptance for the removal of polyps, dilatation of certain strictures, reduction of sigmoid volvulus, and intraoperative evaluation to assist in surgery. These procedures are described under the various diseases in later chapters of Section V.

The endoscopic techniques of passage through a difficult sigmoid or redundant splenic, transverse, or hepatic flexure can

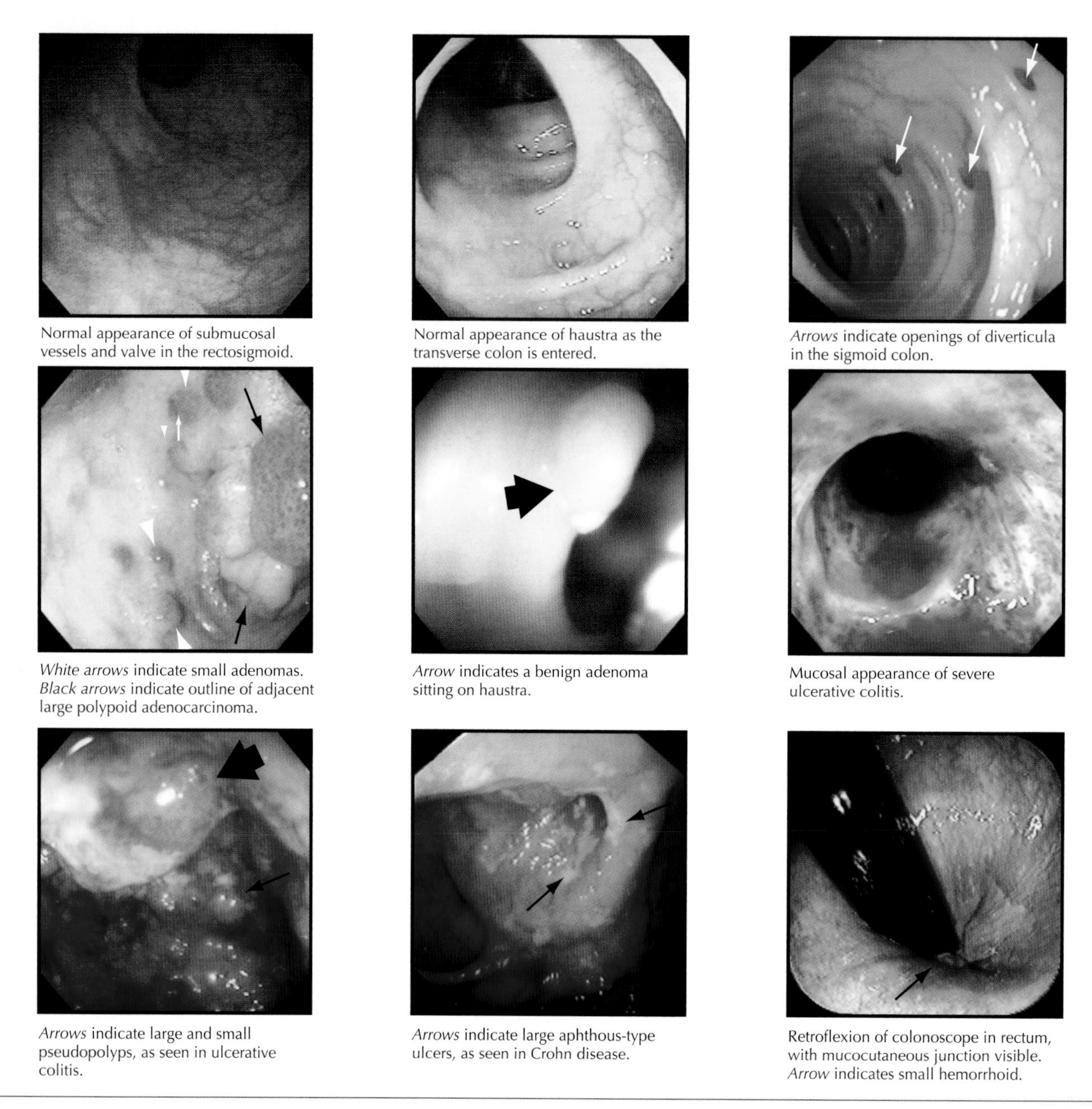

Normal appearance of submucosal vessels and valve in the rectosigmoid.

Normal appearance of haustra as the transverse colon is entered.

Arrows indicate openings of diverticula in the sigmoid colon.

White arrows indicate small adenomas. *Black arrows* indicate outline of adjacent large polypoid adenocarcinoma.

Arrow indicates a benign adenoma sitting on haustra.

Mucosal appearance of severe ulcerative colitis.

Arrows indicate large and small pseudopolyps, as seen in ulcerative colitis.

Arrows indicate large aphthous-type ulcers, as seen in Crohn disease.

Retroflexion of colonoscope in rectum, with mucocutaneous junction visible. *Arrow* indicates small hemorrhoid.

Figure 129-1 *Anoscopic, Sigmoidoscopic, and Colonoscopic Views of Normal Intestinal Structures and Various Lesions.*

be challenging to perform. Most skilled endoscopists are able to evaluate and intubate the terminal ileum, and specimens can be obtained for evaluating diseases in those areas. The complication rate of bleeding or perforation was only 0.35% in a study involving 25,000 diagnostic procedures.

To evaluate the anal ring thoroughly, it is important for the endoscopist to *retroflex* the instrument so that the entire rectal vault can be seen. Because the rectum and anal ring contract as the instrument is pulled through the rectum, small lesions may be missed. These lesions are exposed when the rectal vault is distended and examined from the inside out. It is also easy to obtain biopsy specimens through the retroflexed instrument.

Newer techniques, such as high resolution with chromoscopy and narrow-band imaging, are being used at research and university centers, but not widely. Capsule endoscopy is rapidly becoming a useful tool.

ADDITIONAL IMAGING METHODS

Radiographic imaging techniques and virtual colonoscopy are helpful but are usually reserved for patients who cannot, or choose not to, undergo direct endoscopic viewing procedures.

ADDITIONAL RESOURCES

American Society for Gastrointestinal Endoscopy: Complications of colonoscopy, *Gastrointest Endosc* 57:441-444, 2003.

Barkun A, Chiba N, Enns R, et al: Commonly used preparations for colonoscopy: efficacy, tolerability, and safety—a Canadian Association of Gastroenterology position paper, *Can J Gastroenterol* 20(11):699-710, 2006.

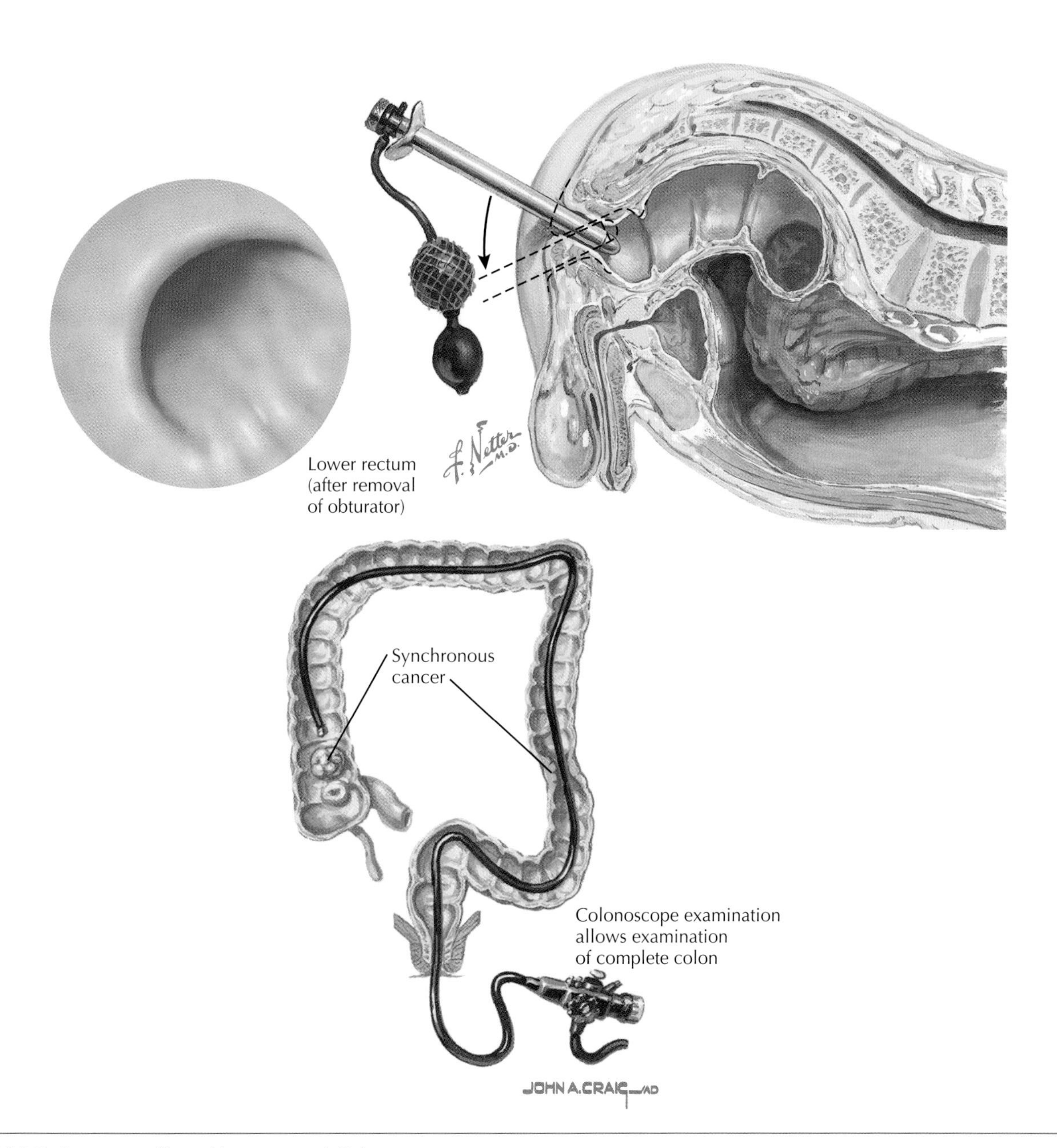

Figure 129-2 *Anoscopy, Sigmoidoscopy, and Colonoscopy.*

Cotton PB: *Advanced digestive endoscopy,* Malden, Mass, 2008, Wiley-Blackwell.

Hassan C, Zullo A, Winn S, Morini S: Cost-effectiveness of capsule endoscopy in screening for colorectal cancer, *Endoscopy,* February 2008 (Epub).

Rex DK: Is virtual endoscopy ready for widespread application? *Gastroenterology* 125:608-614, 2003.

Rex DK, Lieberman D: ACG colorectal cancer prevention plan: update on CT-colonography, *Am J Gastroenterol* 101:1410-1413, 2006.

Waye JD, Rex D, Williams CB: *Colonoscopy: principles and practice,* Malden, Mass, 2008, Wiley-Blackwell.

Williams CB: *Practical gastrointestinal endoscopy: the fundamentals,* ed 5, Oxford, England, 2003, Blackwell.

Laparoscopy

Martin H. Floch

130

Laparoscopy, or peritoneoscopy, is the direct inspection of the peritoneal cavity and its contents by means of an endoscopic instrument introduced through the abdominal wall (**Fig. 130-1**). The procedure is used in gastroenterology and gynecology to diagnose conditions that cannot be determined through simpler methods. Its major value is that it can frequently obviate the need for exploratory laparotomy.

Modern laparoscopy was preceded by rigid peritoneoscopy. A trocar was entered in the abdominal cavity after insufflation with air into the peritoneal cavity, and a peritoneoscope was introduced through the trocar to visualize the liver and intraabdominal organs. The laparoscope has replaced the peritoneoscope. It is a flexible instrument that transmits an image from the tip of the laparoscope to a monitor. Video-laparoscopy techniques have greatly advanced and now include laparoscopic therapy and surgery. The broad, ever-expanding indications for diagnostic laparoscopy include assessment of the liver, undiagnosed acute and chronic abdominal pain, and unexplained ascites. Laparoscopy is also used to stage malignancy and to obtain guided liver biopsy specimens or specimens of any suspected lesion through radiographic imaging.

Indications for laparoscopy in trauma and therapeutic procedures are also rapidly increasing. Laparoscopic cholecystectomy has been well accepted as the preferred procedure, except when the abdomen must be entered in complicated cases. At present, technology and skilled laparoscopists are also able to explore the common bile duct. Resecting major organs such as the spleen and colon is now common practice and will become even more prevalent with greater clinical skills. Furthermore, laparoscopic inguinal hernia repair is now standard in many institutions.

With the advent of the need for fundoplication in esophageal reflux disease, laparoscopic fundoplication has become a standard procedure in many institutions, but skilled laparoscopic surgeons are required. Similarly bariatric surgeons use laparoscopy to do most of their surgery (see Chapters 66 and 67).

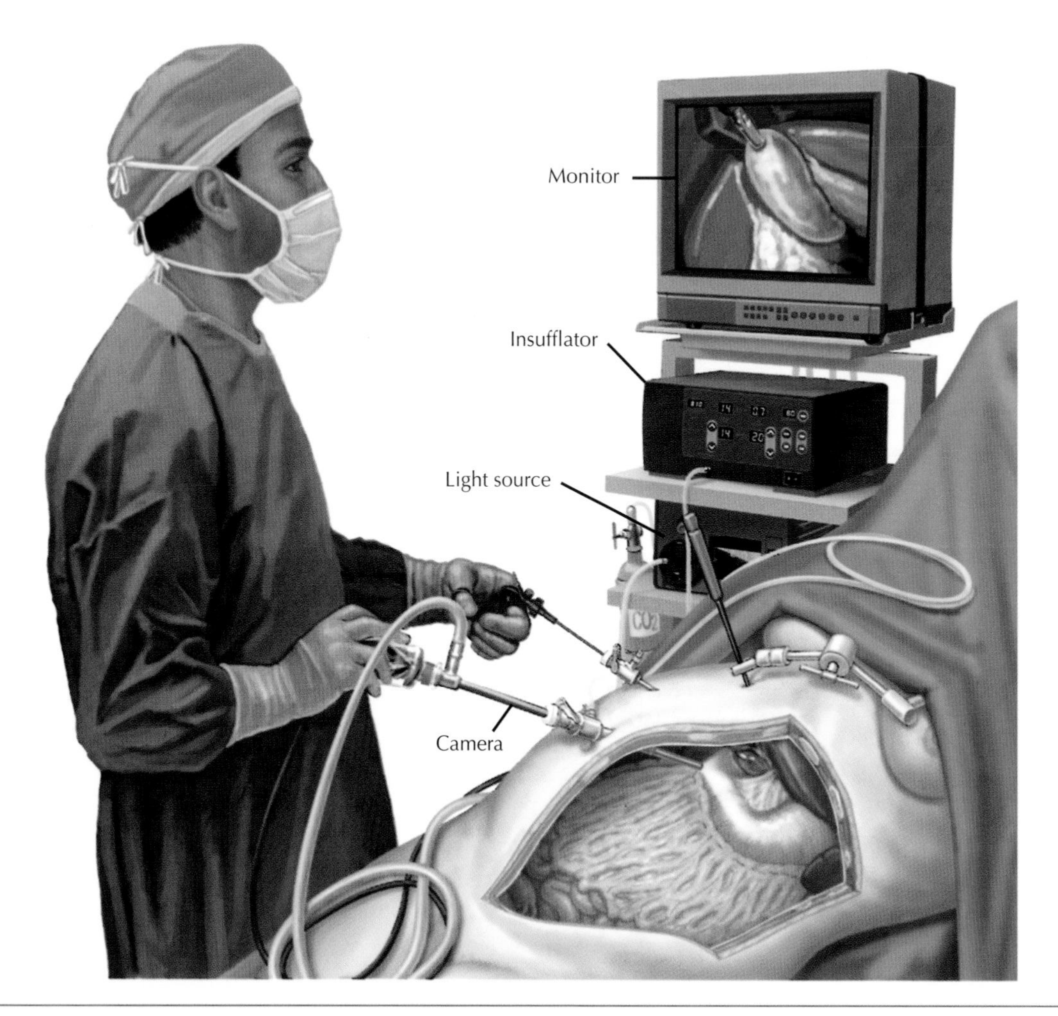

Figure 130-1 *Laparoscopy (Peritoneoscopy).*

Laparoscopy and therapeutic laparoscopy procedures are becoming simpler, and advancing technologic and surgical skills may make them the major standard for most abdominal procedures.

The main advantage of laparoscopy is prevention of invasive surgery. Small lesions measuring 1 mm may be identified and correlated with computed tomography (CT) and magnetic resonance imaging (MRI) findings. Furthermore, direct biopsy is relatively safe and diminishes any possible bleeding. There are some disadvantages because an instrument is used. In patients who have undergone previous surgery with multiple adhesions, it is difficult to visualize many of the areas intraabdominally. Organs in the retroperitoneal area and some abdominal areas may not be visualized well. The experience of the laparoscopist and the laparoscopic surgeon is a major factor; the more experience each has, the more that can be accomplished.

Although laparoscopy can be performed without anesthesia, a complication is always possible. Therefore, the patient must be a candidate for general anesthesia. Local anesthesia with intravenous sedation or inhaled nitrous oxide may be used. Most clinicians prefer using nitrous oxide to create the pneumoperitoneum, but some still prefer carbon dioxide. Laparoscopes with forward viewing range from 2 to 10 mm in diameter. Some have oblique-angled lenses. Other standard instruments used by the laparoscopist include scissors, grasping devices, probes, clamps, forceps, retractors, and biopsy forceps.

The patient is invariably in a lithotomy position and a 10- to 15-degree Trendelenburg position when a small skin incision is made for injection. Incisions made to permit entrance of the laparoscope vary, depending on the patient's status and any previous surgery.

ADDITIONAL RESOURCES

Katada N, Hinder RA, Raiser F, et al: Laparoscopic Nissen fundoplication, *Gastroenterologist* 3:95-104, 1995.

Kelly NA, Sarr MG, Hinder RA: *Mayo Clinic gastrointestinal surgery*, St Louis, 2003, Elsevier.

Parra JL, Reddy KR: Diagnostic laparoscopy, *Endoscopy* 36:289-293, 2004.

Scheidbach H, Schneider C, Huegel O, et al: Laparoscopic sigmoid resection for cancer, *Dis Colon Rect* 45:1641-1647, 2002.

Soper NJ: Laparoscopy and laparotomy. In Yamada T, editor: *Textbook of gastroenterology*, ed 4, Philadelphia, 2003, Lippincott–Williams & Wilkins, pp 3273-3295.

Stool Examination

Martin H. Floch

131

Examination of the stool directly and through microscopy and chemical tests provides much useful information, although patients, physicians, and laboratories tend to avoid stool evaluation (**Fig. 131-1**).

GROSS AND MICROSCOPIC INSPECTION

Acholic stool suggests biliary obstruction; *tarry* stool indicates gastrointestinal (GI) bleeding; and *red* stool signifies bleeding from the lower GI tract. Usually, the shape of the stool is reliably reported by the patient, and the appropriate deductions may be made without the physician observing the stool. Diarrheal stool is loose and watery, but stool associated with malabsorption can be large and greasy, often staining the toilet bowl. The Bristol Stool Scale clearly describes the foam, form, and character of the bowel movement material and has been useful in clinical and research settings (**Fig. 131-2**). The character of stools is also discussed thoroughly in Chapters 111 (diarrhea) and 136 (constipation).

Microscopic examination of stool can be helpful when inflammatory cells (polymorphonuclear or eosinophilic) are identified. Furthermore, ova and parasites are identified as eggs or larvae.

CHEMICAL ANALYSIS

Chemical analysis of stool is important for assessing *fecal occult blood test* findings. Many different tests are available at this time, all using a technique similar to the standard *guaiac test*. False-positive results can occur. The patient must not have ingested large amounts of red meat or peroxidase-containing vegetables and fruits (e.g., broccoli, cantaloupe, cauliflower, radishes, turnips). It is also reported that heavy intake of iron supplements, vitamin C, aspirin, and nonsteroidal antiinflammatory drugs (NSAIDs) give false-positive results. Guaiac-impregnated slide tests are readily available and used most often. False-negative results can occur from using outdated reagents and old slides.

Stool weight is important in that it is greatly increased in most malabsorption syndromes and in diarrheas of metabolic origin. Stool weight in the person on a low-fiber diet may be as low as 100 g, whereas that in the person on a high-fiber diet can be as high as 300 to 400 g. Stool electrolyte studies revealing stool osmolality can aid in evaluating various forms of diarrhea. This information is used only to classify diarrhea as secretory. In *secretory diarrhea*, an osmotic gap exists in stool osmolality, approximately 290 mOsm, and is calculated as twice the concentration of sodium plus potassium. Normally, the gap is less than 50 mOsm, and certainly less than 125 mOsm. Lack of a gap indicates secretory diarrhea.

The chemical analysis of stool can be important for evaluating malabsorption of small bowel, pancreatic, or metabolic disease. This requires a 72-hour stool collection. Several kits are available for fat analysis (see Chapter 109). Routine stool staining for fat has been unreliable and is no longer recommended.

BACTERIOLOGIC AND PARASITIC TESTING

Stool culture is important. Stool should be as fresh as possible so that the laboratory can inoculate it on appropriate media. This is the preferred method of identifying bacterial pathogens, and it is reliable in acute states. *Clostridium difficile* is primarily identified using available toxins. Laboratories should be equipped to identify A and B toxins because the latter can often be missed. Identification of parasites is discussed in Section VI.

ANTIGEN AND GENETIC TESTING

The stool antigen test for *Helicobacter pylori* is as sensitive as other tests to determine the presence of *H. pylori*, and it can be useful in diagnosing and treating gastric and peptic ulcer disease. Furthermore, the stool antigen test for *Giardia lamblia* is also reliable and helpful to identify this protozoan, which tends to live in the duodenum and is difficult to identify in stools.

Many laboratories are developing *fecal water tests* to evaluate stool for colon cancer cells and for colon cancer screening. Although these are still too new to be used in clinical practice, they are developing rapidly and should become important soon. Similarly, stool tests are being developed for genetic analysis of cells, although these are not yet clinically effective tools.

ADDITIONAL RESOURCES

Fine KB, Schuler LR: AGA technical review on the evaluation and management of chronic diarrhea, *Gastroenterology* 116:1464, 1999.

Lewis SJ, Heaton KW: Stool scale as a useful guide to intestinal transit time, *Scand J Gastroenterol* 32:920-924, 1997.

Mandel JS, Bond JH, Curch TR, et al: Reducing mortality from colorectal cancer by screening for fecal occult blood, *N Engl J Med* 328:1365, 1993.

Nair P, Langcrhoirn S, Dutta S, et al: Coprocytobiology: on the nature of cellular elements from stools in the pathophysiology of colonic disease, *J Clin Gastroenterol* 36:584-593, 2003.

O'Donnell LJD, Virjee J, Heaton KW: Detection of pseudodiarrhea by simple clinical assessment of intestinal transit rate, *BMJ* 300:439-440, 1990.

Thompson WG: Functional bowel disorders. Functional abdominal pain. In Drossman DA, editor: *The functional gastrointestinal disorders*, ed 2, McLean, Va, 2000, Degnon Associates, pp 351-432.

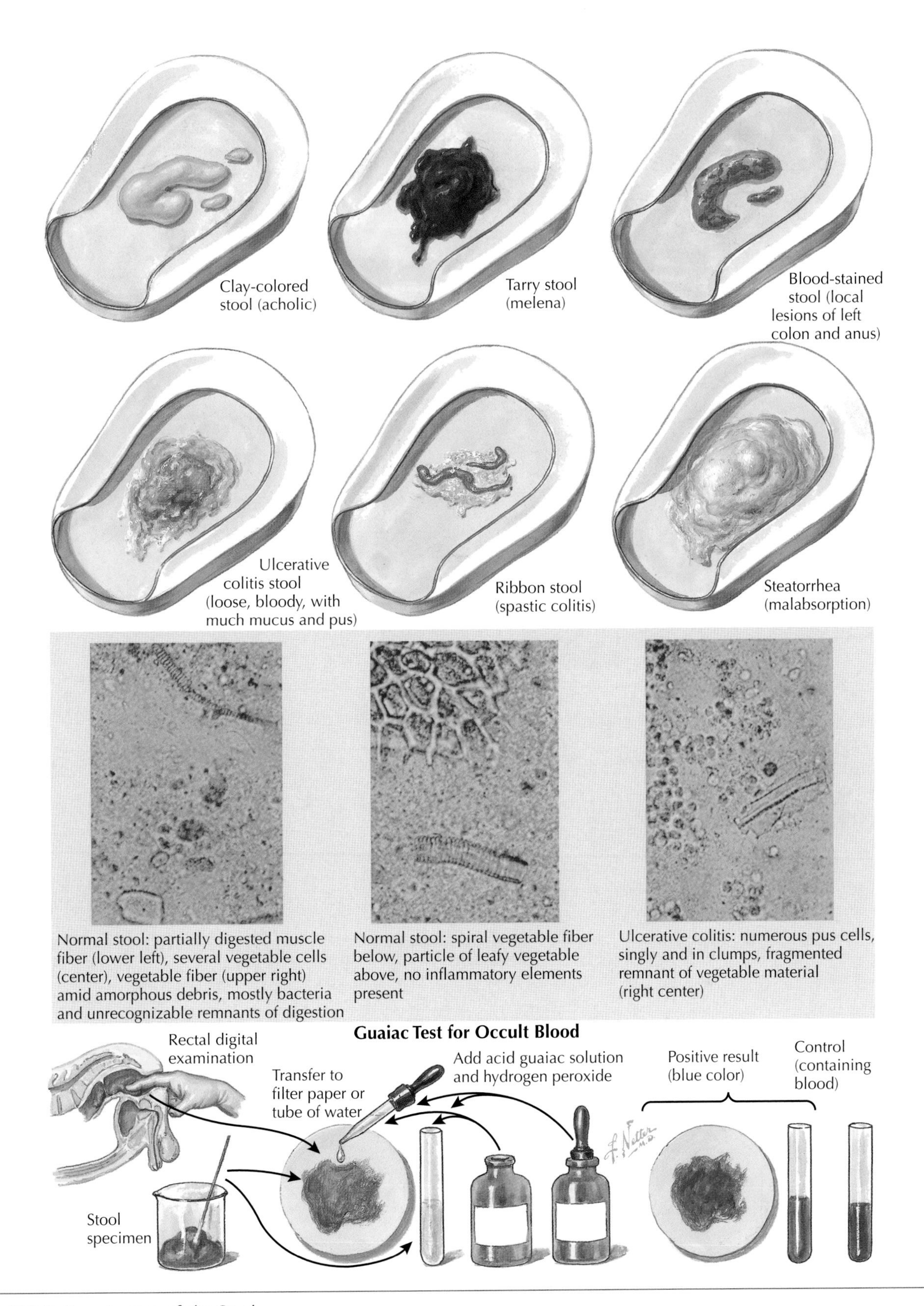

Figure 131-1 *Examination of the Stool.*

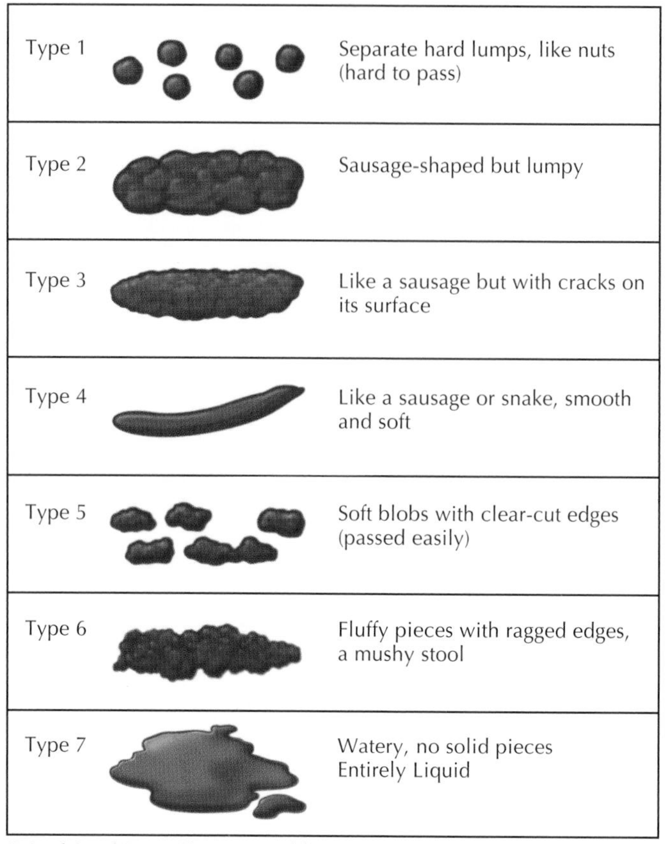

Bristol Stool Form Chart created by Heaton and Lewis at the University of Bristol. Originally published in Scand J Gastroenterol, *32(9):920-924,1997.*

Figure 131-2 *Bristol Stool Form Scale.*

Intestinal Bleeding

132

Martin H. Floch

Gastrointestinal (GI) bleeding is acute or chronic loss of blood from the GI tract (**Fig. 132-1**). Blood may come from the upper or lower part of the GI tract. Acute bleeding may be life threatening, whereas chronic bleeding may be slow or even occult. The GI tract can lose as much as 50 mL of blood daily; it is replaced without anemia, but indicates a GI lesion. This chapter discusses lower GI bleeding; Chapter 60 discusses upper GI hemorrhage (see also Section IV).

Lower GI bleeding can be defined as blood loss from below the ligament of Treitz. It accounts for approximately one fourth to one third of all bleeding events. The incidence appears to increase with age, with more than a 200-fold increase from the third to the ninth decade of life, and it is more common in men than in women. The mortality rate is less than 5%. Upper GI bleeding has a higher mortality rate than lower GI bleeding.

CLINICAL PICTURE

If patient has anemia, it is important to examine stool for occult blood. If findings are positive, the GI tract must be ruled out as the probable cause of the anemia. After the upper GI tract is ruled out, the intestine must be investigated. Severe hemorrhages reveal themselves by the appearance of visible blood in the stool. Blood may be bright red, or it may cause *melena*, a black discoloration indicating that blood has been exposed to digestive activity and usually that the bleeding is from above the ligament of Treitz. Often, however, melena may be from the small bowel or from as far down as the cecum. Red blood in the stool results from severe bleeding. It is important not to be fooled by black stool that is caused by the ingestion of iron supplements, bismuth preparations, or foods such as blackberries. Furthermore, red beets, when consumed in large amounts, can produce red stool.

Patients with occult bleeding usually have few GI symptoms. However, when the symptoms point to the upper GI tract, that workup precedes the lower GI workup. Some patients who are severely anemic may be weak, which usually signifies that occult bleeding has been long term. Acute rectal bleeding that presents as *hematochezia* (blood in feces) and bright-red blood may start with a bout of syncope or as a blood-filled toilet bowl. Patients with acute rectal bleeding must be monitored immediately.

DIAGNOSIS

Once it is established that bleeding is from the intestine, full evaluation of the lower GI tract is indicated. When the patient has anemia or occult bleeding, the diagnostic workup can be done in an orderly fashion. However, if the bleeding is massive, the patient must be immediately stabilized, upper GI bleeding ruled out, and a rapid workup performed. History and physical examination are essential. It is important to note whether abdominal pain or diarrhea is associated with the bleeding and whether the patient has any other GI symptoms. The anal ring must be examined, including anoscopy to rule out severe internal or hemorrhoidal bleeding.

Once the patient is stabilized, some physicians prefer radionucleotide imaging to determine whether the bleeding is from the left or the right side of the colon. This usually requires a bleeding rate of more than 1 mL/min. Institutions report varying degrees of success with radionucleotide imaging, and some prefer sigmoidoscopy or colonoscopy. Endoscopy has replaced barium enema examination.

An attempt is made to purge the bowel, and then careful colonoscopy is performed. Emergency colonoscopy is reported to make a final diagnosis for colonic lesions in 74% to 90% of patients. Some clinicians prefer angiography before the colonoscopic examination if the bleeding is massive and if surgery is being contemplated.

Once the site is identified through angiography, *embolization* can be attempted with polyvinyl alcohol or intraarterial vasopressin. Some studies indicate effectiveness in 79% of patients. Angiographers are now able to place coaxial catheters into the vessel, and microcoils can be used to deliver an embolic agent. This technique appears to lower the risk for bowel ischemia.

The most common cause of intestinal bleeding is *diverticular disease* of the colon. The next most common causes are angiodysplasias, cancer or polyps, and inflammatory bowel disease (IBD). Anorectal, hemorrhoidal, and fissure bleeding are relatively uncommon for massive bleeding or anemia, although they are reported in as many as 9% of patients in some series. Other causes are rare and include Meckel diverticulum and intussusception, colitis secondary to ischemia, infections or radiation, and rare entities such as Dieulafoy lesion of the rectum and solitary rectal ulcer. Varices of the colon and rectum are also rarely reported.

Bleeding is usually on the left side but does occur from right-sided lesions; therefore, it is important to identify the site if surgery is contemplated.

TREATMENT AND MANAGEMENT

Treatment depends on the lesion. Because diverticular disease is the most common cause of intestinal bleeding, and because bleeding stops with minimal transfusion in more than 80% of patients, the prognosis is excellent. However, bleeding may be massive and persistent; therefore, imaging or angiography must identify whether it is the right or the left colon for the surgeon to perform the correct partial colectomy. Angiodysplasias are often embolized or are seen on colonoscopy, and attempts to cauterize or ablate them are possible. Certainly, neoplastic lesions can be removed endoscopically. If they are not, surgical resection of the area is required. IBD is treated as described in Chapters 147, 152, and 154. Anorectal disease is treated locally with appropriate therapy (see Chapters 162 and 165).

The prognosis in hematochezia is excellent because 80% stop spontaneously, and mortality is well below 5%.

When the colon is not the cause, and bleeding comes from the small bowel, it may be caused by neoplasia identified on

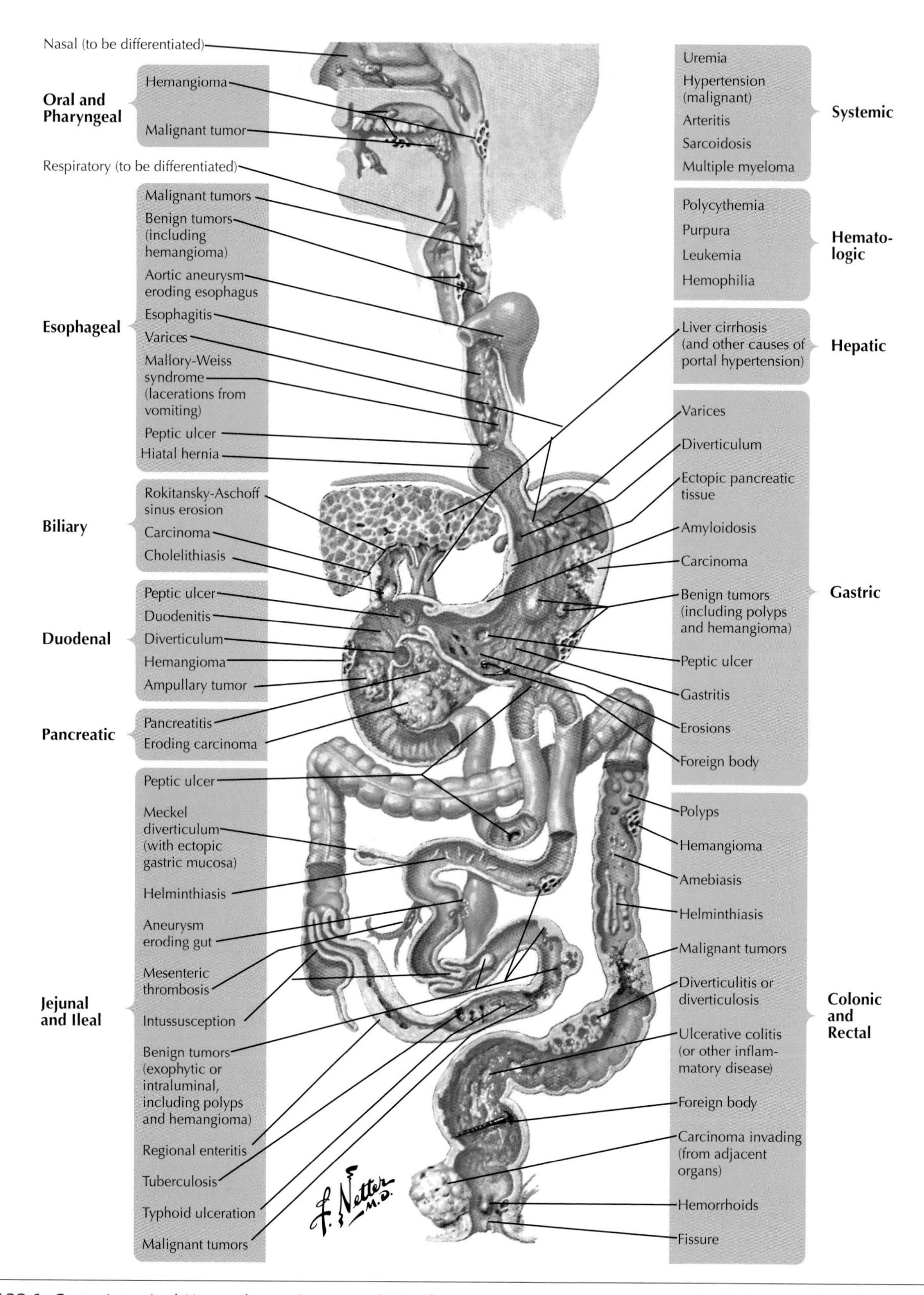

Figure 132-1 *Gastrointestinal Hemorrhage: Causes and Manifestations.*

radiographic imaging or enteroscopy. The advent of capsule endoscopy has identified ectasias and small lesions caused by Crohn's disease that were not seen radiographically. It is now the preferred technique to evaluate occult, and at times active, small bowel bleeding. The closer it is done to the bleeding event, the better the capsule diagnostic findings. These tests require treatment of the IBD. If bleeding persists, surgical resection is needed. When Meckel diverticulum causes bleeding, surgical correction is required; the prognosis is excellent. If neoplasias of the small bowel are identified, they are usually resected, and the prognosis depends on the neoplasm (see Section IV).

COURSE AND PROGNOSIS

The course and prognosis vary for patients with intestinal bleeding, depending on the primary disease. With a benign disorder, the prognosis is excellent. If it is neoplastic, prognosis depends on the type of tumor. A small group of patients with chronic iron-deficiency anemia have recurrent positive stool test results for occult blood and are defined as "occult bleeders." When no lesion is identified, blood replacement and maintenance therapy are performed, and the long-term prognosis is guarded.

ADDITIONAL RESOURCES

Jensen DM, Machicado GA, Jutabha R, Kovacs TO: Urgent colonoscopy for the diagnosis and treatment of severe diverticular hemorrhage, *N Engl J Med* 342:78-82, 2000.

Kim ST, Nemcek AA, Vogelzang RL: Angiography and interventional radiology. In Gore RM, Levine MS, editors: *Gastrointestinal radiology*, Philadelphia, 2008, Saunders-Elsevier, pp 117-140.

Rockey DC: Gastrointestinal bleeding. In Feldman M, Friedman LS, Brandt LJ, editors: *Gastrointestinal and liver disease*, ed 8, Philadelphia, 2006, Saunders-Elsevier, pp 255-291.

Zuccaro G Jr: Management of the adult patient with acute lower gastrointestinal bleeding. American College of Gastroenterology Practice Parameters Committee, *Am J Gastroenterol* 93:1202-1208, 1998.

Zuckerman GR, Prakash C: Acute lower intestinal bleeding. II. Etiology, therapy, and outcomes, *Gastrointest Endosc* 49:228, 1999.

Motility and Dysmotility of the Large Intestine

133

Martin H. Floch

Normal colonic motility involves the integrated function of many components: the central nervous system (CNS); the colonic nervous system; the circular and longitudinal involuntary muscle of the colon, sigmoid, and rectum; and the voluntary muscle of the pelvis and rectum. This complex system of neuromuscular activity is further integrated with impulses arising in the upper gastrointestinal (GI) and small bowel tracts and is influenced by GI hormone activity (see Chapter 105).

Various types of colonic movements can be differentiated (**Fig. 133-1**). There appears to be a receptive relaxation of the cecal musculature as the terminal ileum evacuates its contents and permits the accommodation of adequate quantities of intestinal chyme before the activation of stretch receptors. Adaptive relaxation in other parts of the colon provides similar accommodation of the fecal contents without distress and without premature propulsion, which does occur in the rectum when the rectum becomes filled. Such a *reservoir of continence* function is a property particularly of the descending colon, and it is this feature that renders colostomy a fairly tolerable and practical condition.

Contraction of the longitudinal muscular bands (taeniae) shortens the bowel and forms pleats or sacculations (haustra) in which the residues of the chyme are retained to allow time for the absorption of water and a number of digestive products. This function is abetted by contractions of the circular muscle, which may create small indentations within the haustra. These contractions of the longitudinal and circular musculature must be considered analogous to the rhythmic segments of the small bowel. There are *nonpropulsive* movements of segmental contractions and *propulsive* movements that arise in proximal segments and pass in a caudad direction, obliterating some haustra. *Mass peristalsis*, analogous to the peristaltic rush of the jejunum or ileum, occurs only two or three times in a 24-hour period and is initiated by (or is related to) the gastrocolic reflex and propels colonic contents toward the rectosigmoid colon.

The innervation of the colon is similar to that of the small bowel (see Chapter 100). Extrinsic nerves exert an effect on the intrinsic nerve network, which appears to function autonomously and is capable of coordinating movements of adjacent segments necessary for peristaltic progression. These effects may be altered in many pathologic conditions. The broad concept that the parasympathetic nerves generally augment, and the sympathetic nerves inhibit, muscular contraction is acceptable as a working hypothesis, but it becomes more complex when the intrinsic nervous system of the colon is considered.

Understanding the *interstitial cells of Cajal* has helped clarify colonic physiology. They serve two important functions as smooth muscle pacemakers: controlling myogenic activity and mediating or amplifying the effects of the enteric neurons. The cells of Cajal are spread throughout the muscle and are nonneuronal, derived from smooth muscle.

Dietary factors affect regional transit; a solid-food diet hinders transit through the cecum and ascending colon, whereas a mixed diet, particularly liquids, is stored in the ascending and the transverse colon. The volume and consistency of the contents affect the rate of emptying and correlate with stool frequency and weight. Chemical substances and distention can stimulate propulsive activity. *Antiperistaltic waves* assist in retaining contents within the colon.

Transit of material through the colon is normally slow, and contents move from the cecum to the rectum over 24 to 48 hours. Transit appears to be, in general, faster in men than in women and faster in younger women than in middle-aged or older women. *Propulsion* varies tremendously among persons. The material does tend to dwell in the rectosigmoid colon, indicating storage in this area. In some patients, radiographic observations reveal a marked delay in the ascending or the transverse colon. As the volume increases in the ascending colon, retroperistaltic patterns decrease and give way to aboral propulsive contractions.

The functions of the rectum, anus, and pelvic floor are discussed in Chapters 134 to 136.

Manometry of the colon is largely restricted to research centers and has not proved to be adaptable or helpful in clinical situations. These centers can demonstrate abnormalities in the interdigestive migratory motor complex and relate them to functional abnormalities. Using motility-measuring material and electronic barostat is helpful in understanding the physiology but is not readily adaptable to clinical situations.

DISORDERS OF COLONIC MOTILITY

Three major areas of colon motility disorders concern clinicians: constipation, diarrhea, and irritable bowel syndrome, as discussed in Chapters 108, 111, and 136. Disturbances in anorectal motility are discussed in Chapters 134 and 135.

Colonic motility disturbances can be secondary to nonmotor disorders, such as occur with endocrinopathy and neurologic disease. Furthermore, once the intestine is diseased, motility is also easily disturbed, and either digestion or absorption becomes altered. This occurs when inflammation in the intestinal tract caused by acute or chronic disease alters the balance of the neuromuscular activity and severely alters the motility of the colon.

ADDITIONAL RESOURCES

Cook IJ, Brookes SJ: Colonic motor and sensory function and dysfunction. In Feldman M, Friedman LS, Brandt LJ, editors: *Gastrointestinal and liver disease*, ed 8, Philadelphia, 2006, Saunders-Elsevier, pp 2111-2123.

Wiley JW, Nostrant TB, Owang C: Evaluation of gastrointestinal motility. In Yamada T, editor: *Atlas of gastroenterology*, Philadelphia, 2003, Lippincott–Williams & Wilkins, pp 987-1014.

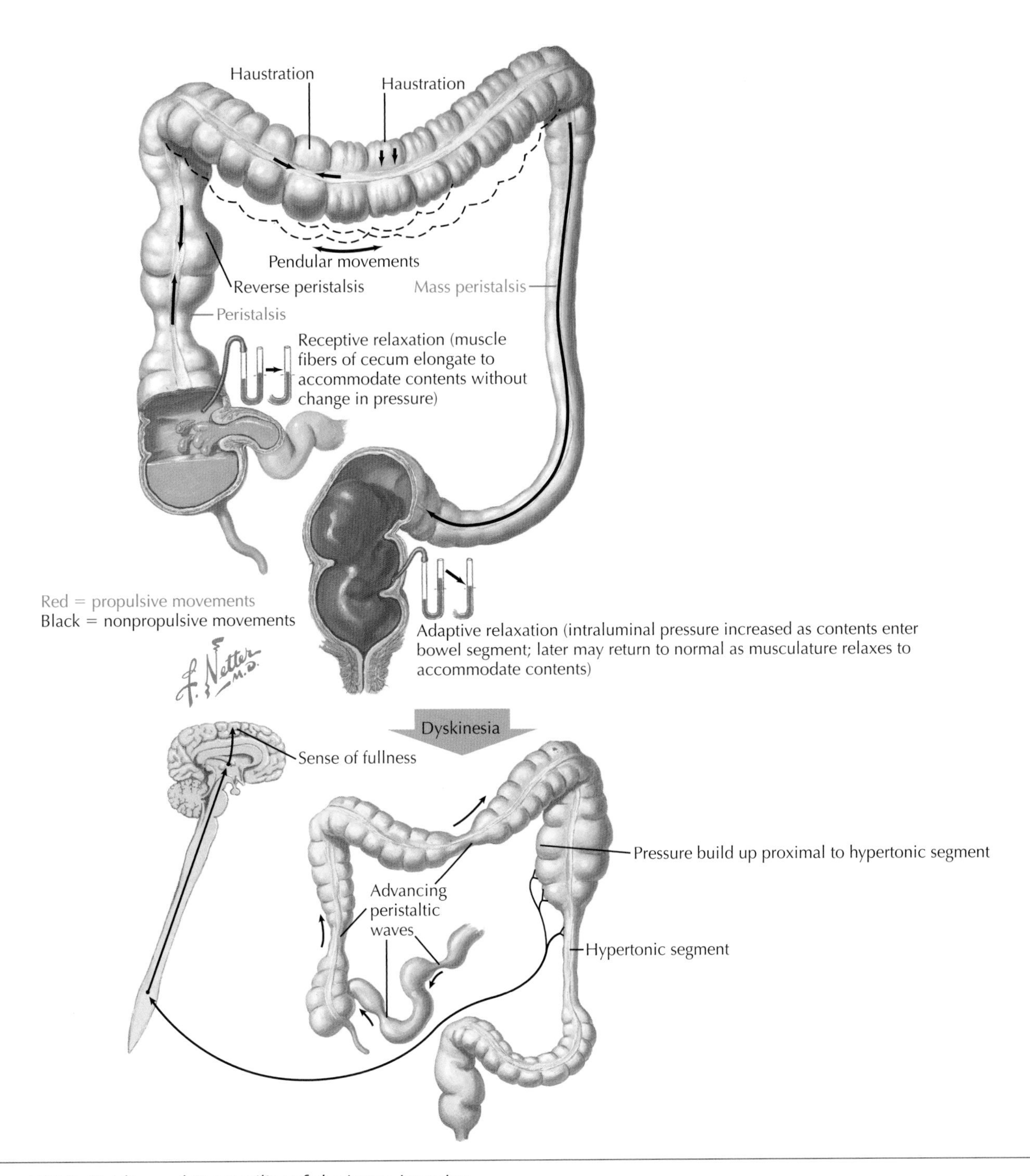

Figure 133-1 *Motility and Dysmotility of the Large Intestine.*

Defecation

Martin H. Floch

134

A peristaltic wave moving the contents of the left colon into the rectum is usually considered the initial phenomenon in the sequence of events in defecation (**Fig. 134-1**).

The urge to defecate normally occurs when residue accumulates in the rectum, at intervals varying from several times daily to every fourth or fifth day. When the rectum fills with approximately 400 mL in a healthy person with intact nerves and reflexes, the urge to defecate is usually uncontrollable. Most people feel an urge daily, usually in the morning after breakfast, after awakening from sleep, assuming the erect position, and moving about. Ingesting food and liquids favors the initiation of *mass peristalsis* (gastrocolic reflex). Increased intrarectal pressure brings about a reciprocal relaxation of the anal sphincters, which may be counteracted by voluntary contraction of the external sphincter, permitting delay of defection. Prolonged delay may result in a temporary reduction in the intensity of the urge. The entire act of defecation is a series of contractions and relaxations of muscles in the rectum and the pelvis.

When the urge to defecate occurs, the person usually assumes a squatting position facilitated by a reflex contraction of the hamstrings. The squatting position supports an increase in intraabdominal pressure, which is accomplished by contraction and fixation of the diaphragm, closure of the glottis, and contraction of the muscles of the abdominal wall. Voluntary control of contraction of the external sphincter is released, and the fecal mass is expelled by the increasing rectal contraction, which leads to intrarectal pressure of 100 to 200 mm Hg. Simultaneously, the muscles of the pelvic floor contract and contribute to the forces that increase the intraabdominal pressure. The contents of the left colon, or part of it, may be emptied in a single continuous peristaltic progression, or the anorectal structures may return to the resting state after the first bolus has been evacuated, until another contraction of the colon delivers more fecal material into the rectum.

This integrated function involves neurons of the motor cortex, sympathetic and parasympathetic pathways, and numerous reflex mechanisms (**Fig. 134-2**). Electromyographic studies indicate that the pelvic floor musculature behaves as a single muscle during defecation, and that the anorectal angle created by the puborectalis muscle produces a functional obstruction to prevent accidental loss of stool.

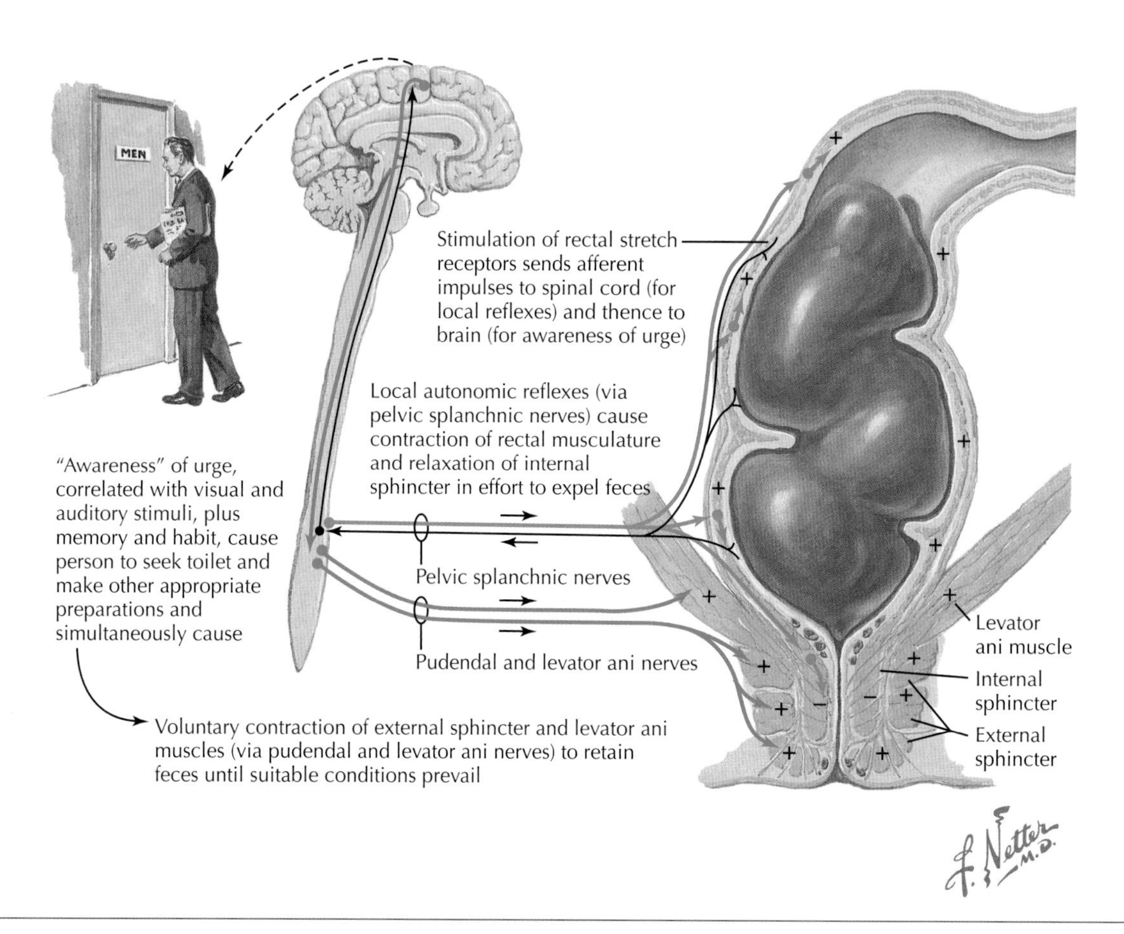

Figure 134-1 *Sequence of Events Leading to the Act of Defecation.*

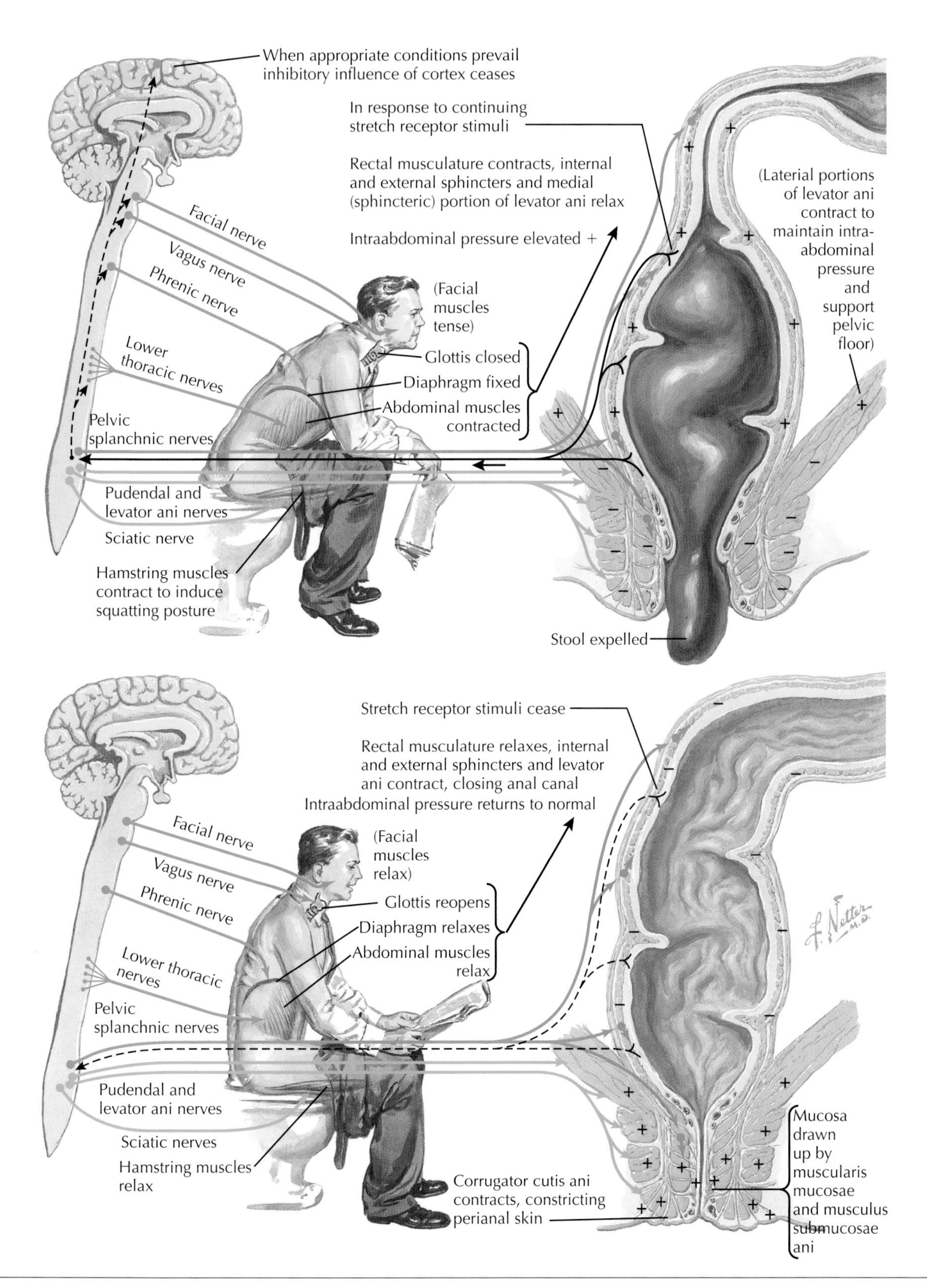

Figure 134-2 *Neuronal Functions and Anatomic Correlations in Defecation.*

Disease and malfunction of the anal sphincter and rectum and anal manometry are discussed Chapter 164.

ADDITIONAL RESOURCES

Cook IJ, Brookes SJ: Colonic motor and sensory function and dysfunction. In Feldman M, Friedman LS, Brandt LJ, editors: *Gastrointestinal and liver disease*, ed 8, Philadelphia, 2006, Saunders-Elsevier, pp 2111-2126.

Hasler WL: Motility of the small intestine and colon. In Yamada T, editor: *Textbook of gastroenterology*, ed 4, Philadelphia, 2003, Lippincott–Williams & Wilkins, pp 220-247.

Wiley JW, Nostrant TB, Owang C: Evaluation of gastrointestinal motility. In Yamada T, editor: *Atlas of gastroenterology*, Philadelphia, 2003, Lippincott–Williams & Wilkins, pp 987-1014.

Pathophysiology of Defecation and Fecal Incontinence

135

Martin H. Floch

During the period of spinal shock, which supervenes for some weeks immediately after transection of the spinal cord above the origin of the lumbar sympathetic nerves, the rectum and sphincters are completely paralyzed, and the patient is incontinent (**Fig. 135-1**). Thereafter, the tonus of the sphincters returns, and defecation occurs reflexively by way of the lumbosacral center. Because voluntary contraction of the external sphincter is no longer possible and distention of the rectum no longer perceived, the patient has no control over the act of defecation. In paraplegic patients, this poses a difficult problem, usually managed by the regular use of enemas and digital evacuation of the rectum.

When the cord lesion involves the cauda equina, with destruction of the sacral innervation, the reflexes are abolished and defecation becomes automatic, or dependent entirely on intrinsic nervous mechanisms. In these patients, the rectum still responds to distention, although with limited force, and the reciprocal relaxation of the already patulous sphincters enables feces to be extruded. Some awareness of rectal distention may be present if the transection is below the lumbar sympathetic outflow, and the persistence of sympathetic connections in the absence of a sacral outflow may contribute to the sluggishness of rectal contractions.

The presence of excretory material in the rectum is not, in itself, sufficient to excite the urge to defecate. The content must be sufficiently large to exceed the threshold of the distention stimulus characteristic of the person. In many patients with regular bowel movements, digital examination reveals a considerable mass of varying consistency in the rectum. However, the accumulation of a large mass in a greatly dilated rectum, especially in older persons, suggests loss of tonicity of the rectal musculature that could be attributed to the longstanding habit of ignoring or suppressing the urge to defecate, or to degeneration of nerve and muscle pathways involved in defecation reflexes. Painful lesions of the anal canal (e.g., ulcers, fissures, thrombosed hemorrhoidal veins) impede defecation by exciting a spasm of the sphincters and by voluntary suppression to avoid the resultant pain.

Dietary factors greatly affect defecation. Persons who eat diets high in fiber (30-50 g daily) will have loose stools and easy defecation, whereas subjects eating low-fiber diets will have small, hard, and infrequent stools. The type of fiber also affects the character of the stool and defecation. Diets high in insoluble fiber, such as the African maize diet, will produce soft and watery stool, whereas diets high in soluble fiber will produce increased gases because of fermentation, with a softer, more gel-like stool.

Distention of the rectum often provokes a repeated, almost continuous urge to defecate (tenesmus), but the rocklike character of the feces prevents molding for passage through the sphincters. If the condition cannot be dealt with by rectal infusions of oil or by surface-acting agents such as dioctyl sodium sulfosuccinate (docusate sodium), digital evacuation is often necessary.

The constant urge to defecate in the absence of appreciable content in the rectum may be caused by external compression of the rectum, by intrinsic neoplasms, and particularly by inflammation of the rectal mucosa.

FECAL INCONTINENCE

Fecal incontinence occurs when voluntary control of the external sphincter fails because of a large bolus of feces (see Chapter 167). This occurs most often when there is a rapid, sudden transit. Rapid transit occurs in infectious diarrheas and in irritable bowel syndrome. Incontinence may also be caused by damage of the anal sphincter or the pelvic musculature (see Chapters 164 and 165). Fecal incontinence, the involuntary passage of stool to the anus, is estimated to occur in approximately 7% of the general population in the form of soiling, but in less than 1% in the form of gross incontinence, although the incidence in nursing homes is as high as 50%. Soiling usually results from some abnormality in the anal sphincter, but large amounts of feces are a result of damage to the anal sphincter or severe neuropathy.

A detailed workup includes physical examination of the anal ring. If the gross pathology is not defined, pelvic magnetic resonance imaging (MRI), barium defecography, anal sonography, and anal manometry may be helpful, as well as electromyography or pudendal nerve terminal motor latency testing if surgery is needed. These tests may not be readily available in clinical offices and are used primarily in university or research centers (see Additional Resources).

The treatment of incontinence depends on the associated cause. If there is anal disease, the treatments discussed in the appropriate chapters on anal disorders are used. If caused by other conditions, treatment is provided as discussed under the specific condition, but typically, a general pattern develops in which a trial of either diphenoxylate or loperamide is indicated. Patients may rely on rectal colon plugs or may use diapers, depending on the degree of incontinence. As often used in nursing homes, studies have shown diapers to be helpful to decrease the incidence in patients who experience incontinence where no definite cause can be established.

Attempts at using biofeedback training have shown varied success. However, it has been extremely helpful, especially when pelvic musculature needs to be strengthened.

Finally, there are surgical alternatives to strengthen the musculature or to bypass the rectum with an ostomy.

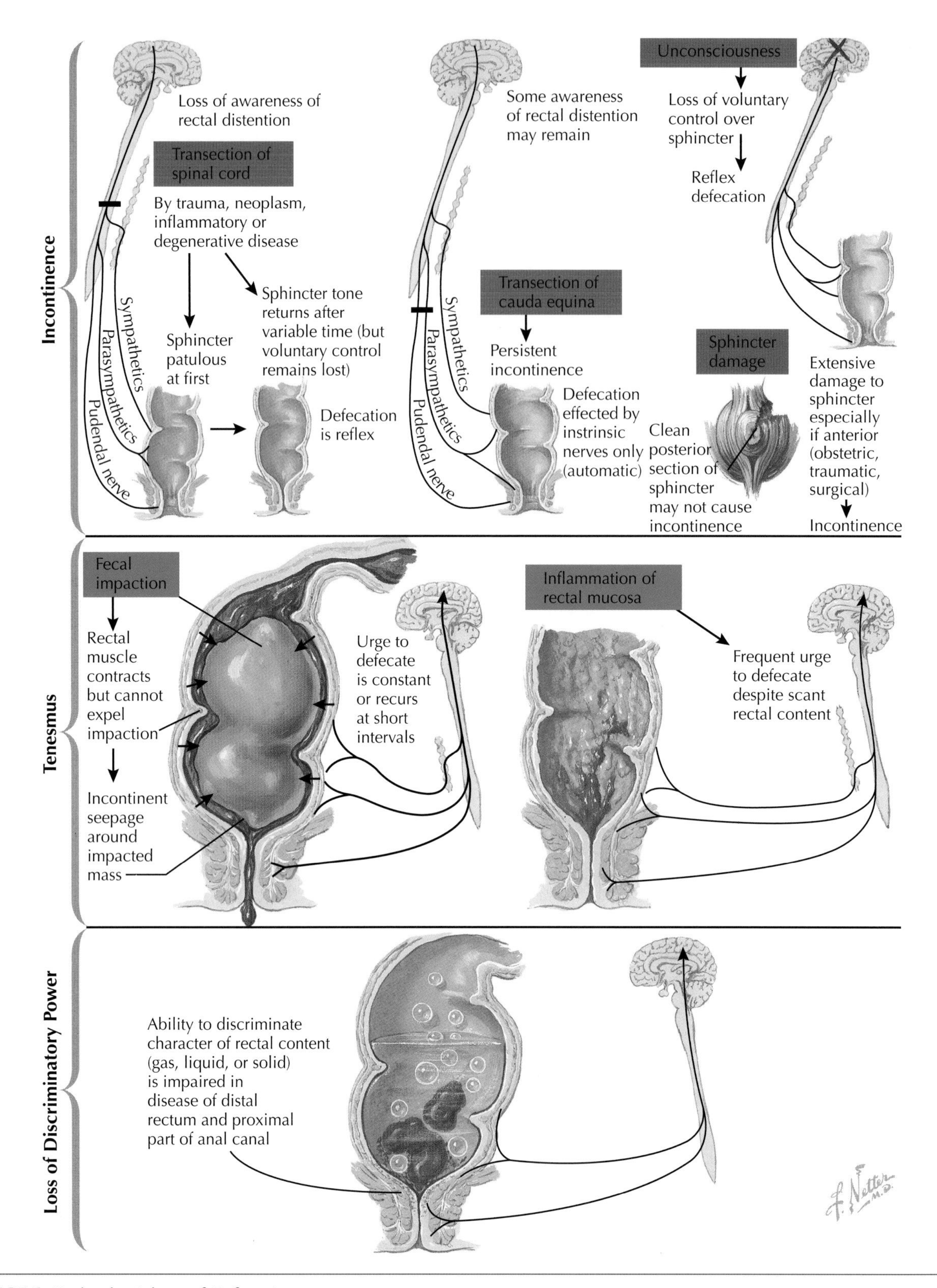

Figure 135-1 *Pathophysiology of Defecation.*

ADDITIONAL RESOURCES

DiLorenzo C, Youssef N: Childhood constipation: evaluation and treatment, *J Clin Gastroenterol* 33:199-205, 2001.

Drossman DA, editor: *The functional gastrointestinal disorders*, ed 2, McLean, Va, 2000, Degnon Associates.

Patel SM, Lembo AJ: Constipation. In Feldman N, Friedman LS, Brandt LJ, editors: *Gastrointestinal and liver disease*, ed 8, Philadelphia, 2006, Saunders-Elsevier, pp 221-253.

Rao SSC: Diagnosis and management of fecal incontinence, *Am J Gastroenterol* 99:1585-1604, 2004.

Schiller LR: Fecal incontinence. In Feldman N, Friedman LS, Brandt LS, editors: *Gastrointestinal and liver diseases*, ed 8, Philadelphia, 2006, Saunders-Elsevier, pp 199-219.

Wald A: Fecal incontinence in adults, *N Engl J Med* 356:1648-1655, 2007.

Constipation

Martin H. Floch

136

Constipation can generally be defined as fewer than three bowel movements per week. However, if it is included as a "functional disorder" in accordance with the Rome II criteria, the definition would state that the diagnosis of constipation requires at least 12 weeks of symptoms during the preceding 12 months, and that two of the following criteria should be present: straining, lumpy or hard stool, sensation of incomplete evacuation, sensation of anorectal obstruction, and fewer than three bowel movements per week.

Constipation is a symptom. Therefore, interpretation is subjective regarding sensations of bowel movement and stool size.

In the United States, studies indicate there are approximately 2.5 million patient visits to physician offices per year for constipation. Most people living in industrialized countries have one bowel movement per day, and it varies in weight from 120 to 130 g. However, there is great individual variation. In general, stool weight is related to the transit time, and colonic transit is usually delayed in most patients with constipation.

Etiologies or pathophysiologic mechanisms that cause constipation can be classified into six groups: (1) inadequate intake of dietary fiber, (2) drug-induced constipation, (3) metabolic and endocrine problems, (4) neurologic problems, (5) local or systemic disease involving the intestine, and (6) functional disorder or irritable bowel syndrome (IBS; **Fig. 136-1**).

CLINICAL PICTURE

Because constipation is a subjective interpretation, patient reports vary greatly. However, as defined, the stool must be small or infrequent. The symptom must be correlated with the cause. If the patient is otherwise healthy, the constipation is usually part of a functional gastrointestinal disorder, and the patient has many symptoms, but none of organic cause. If the constipation has an organic cause, the symptoms will be associated with metabolic, neurologic, or local disease. For example, if the patient has Parkinson disease, the associated tremors will accompany constipation. If it is associated with severe rectal pain, the problems of anal ring disease will be evident. If the constipation is associated with severe hypothyroidism, the patient would also report weakness and other symptoms of the endocrine disorder.

DIAGNOSIS

Diagnosis depends on the initial history and physical findings. If evident on physical examination, an associated disease usually explains the constipation. If no disease is evident, the workup includes history, colonoscopy or barium study of intestine, and transit time using opaque markers or colonic scintigraphy. Thorough evaluation of anorectal function requires inspection of the area and, if necessary, anorectal manometry and defecography.

The importance of *transit time* to evaluate the function versus the symptom is essential in making the diagnosis of any organic obstructive disease versus a functional disorder. The transit time is simple to measure and requires swallowing Sitz markers, which are then traced by simple abdominal x-ray films for 5 days. The clinician can determine whether there is slow, even transit or whether an obstruction exists and should be evaluated.

TREATMENT AND MANAGEMENT

If the constipation has an organic cause, that disease must be treated to treat the constipation. Withdrawing drugs that might be necessary therapy often is difficult; thus the constipation may require treatment even if the patient is taking a narcotic and pain-relieving therapy. The treatment of constipation is dietary, behavioral, pharmacologic, or surgical. Drugs used to treat constipation are described in Chapter 137 (see Box 137-1).

When constipation is caused by fiber deficiency, results of *dietary treatment* are often successful. Dietary fiber intake should be increased to 20 to 25 g, if tolerated by the patient without abdominal discomfort. Fiber increase should be gradual. Foods that include fiber substances are listed in Chapter 268, but often it is important to recommend a high-fiber cereal in the morning, depending on the patient's work and eating habits. Dietary recommendations then can be made to increase the intake of fruits, vegetables, and grains. A high-fiber breakfast cereal plus 4 to 5 portions of fruits or vegetables can easily reach the 25 g average of daily dietary fiber needed to maintain normal colon function.

When fiber does not solve the problem, *pharmacologic therapy* may or must be instituted. Pharmacologic agents available are listed in all drug formularies and derive from seven categories: bulk formers, emollients, lubricants, saline products, stimulants, hyperosmolar agents, and prokinetic agents (see Chapter 137). It is a challenge for a physician to select the correct combination of therapies. Most recently, work with 5-HT_4 antagonists has shown some promise, although tegaserod (Zelnorm), the latest drug in this category, was taken off the U.S. market but did help some patients. A chloride channel absorption-blocking agent, lubiprostone (Amitiza), has proved helpful in some patients.

In children with constipation, *behavioral change and retraining* may be helpful in as many as 50% to 75% of patients. Enemas and laxation are used to empty the bowel frequently and are gradually withdrawn as the child learns the behavior of regular defecation.

In patients with functional disorders and certain psychiatric disorders, successful treatment can be accomplished if the psychologic state of the child or adult is treated.

Indications for *surgical treatment* are relatively few, but it may be successful in some patients. Certainly, for organic causes such as Hirschsprung disease, surgery is needed. Some patients with severe colonic inertia are refractory to all therapies, and in rare, severely intractable cases, ileorectal anastomosis or cecostomy is used.

Treating constipation in a patient with a functional disorder or a patient taking drugs is often challenging. Some patients require simple laxation, whereas others require different

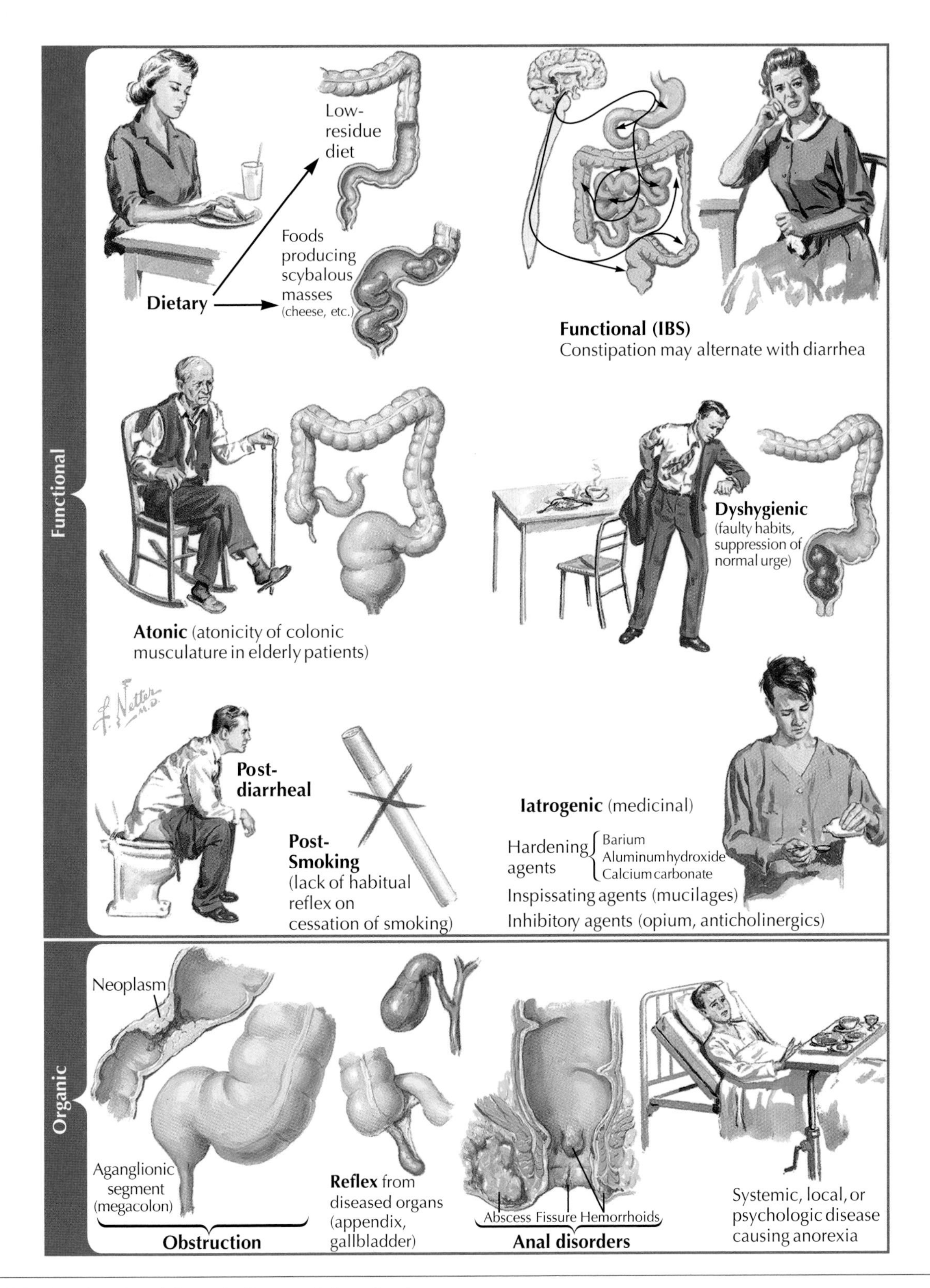

Figure 136-1 *Functional and Organic Causes of Constipation.*

combinations of laxation. The clinician should start simply with one of the agents described in the seven categories, then add a wetting or osmotic agent. If unsuccessful, stimulants can be added. An example would be to begin with magnesium hydroxide (Milk of Magnesia) or polyethylene glycol (MiraLax) on a daily or an every-other-day basis. If this fails, a wetting agent or an osmotic agent, such as lactulose, could be tried. If that does not succeed, a stimulant such as senna extract, or a trial of a new agent such as lubiprostone, could be used.

COURSE AND PROGNOSIS

Most cases of simple constipation can be treated by dietary means with increased soluble or insoluble fibers, or both. When this approach fails and a cause cannot be determined, therapy that incorporates behavioral approaches or pharmacologic agents is necessary. Surgery is rarely indicated.

ADDITIONAL RESOURCES

Brand LJ, Schoenfeld P, Prather CM, et al: An evidence-based approach to the management of chronic constipation in North America, *Am J Gastroenterol* 100(suppl):1-22, 2005.

DiLorenzo C, Youssef N: Childhood constipation: evaluation and treatment, *J Clin Gastroenterol* 33:199-205, 2001.

Drossman DA, editor: *The functional gastrointestinal disorders*, ed 2, McLean, Va, 2000, Degnon Associates.

Patel SM, Lembo AJ: Constipation. In Feldman N, Friedman LS, Brandt LJ, editors: *Gastrointestinal and liver disease*, ed 8, Philadelphia, 2006, Saunders-Elsevier, pp 221-252.

Effects of Drugs on the Colon

137

Martin H. Floch

Folklore and society have maintained the philosophy that regular bowel movements are essential to health. Consequently, people have attempted to maintain bowel regularity and experience a "good" bowel movement by using laxatives and drugs that affect the gastrointestinal (GI) tract, particularly the colon (**Fig. 137-1** and **Box 137-1**).

Drugs may cause colonic injury. Enemas can actually damage the mucosa of the colon, especially when substances are added to the enema that can be injurious, such as hydrogen peroxide or caustic soaps. Routine laxatives such as those containing senna may color the bowel and cause melanosis coli. Oral contraceptives, vasopressin, ergotamine, cocaine, dextroamphetamine, digitalis, neuroleptics, and alosetron have all been shown to cause an ischemic condition and ischemic colitis. Nonsteroidal antiinflammatory drugs (NSAIDs), cyclooxygenase (COX-2) inhibitors, certain antibiotics, and chemotherapy can cause a single or numerous ulcerations in the bowel. So-called cathartic colon may result from frequent and abusive use of laxatives. The colon could become shortened or dilated from laxative abuse.

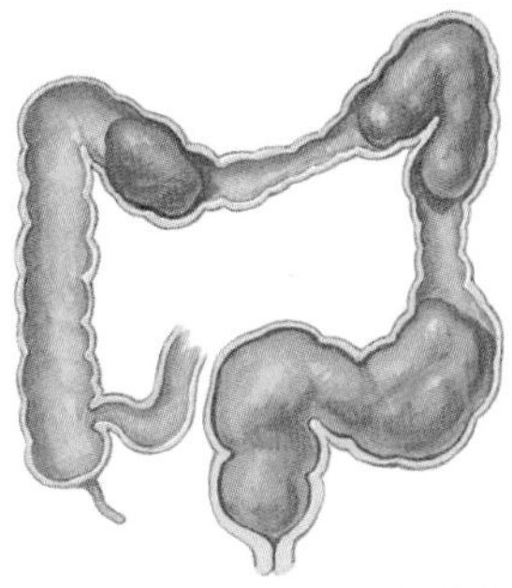

Bulk agents (bran, psyllium, methylcellulose) provide increased size, promote peristalsis by distention

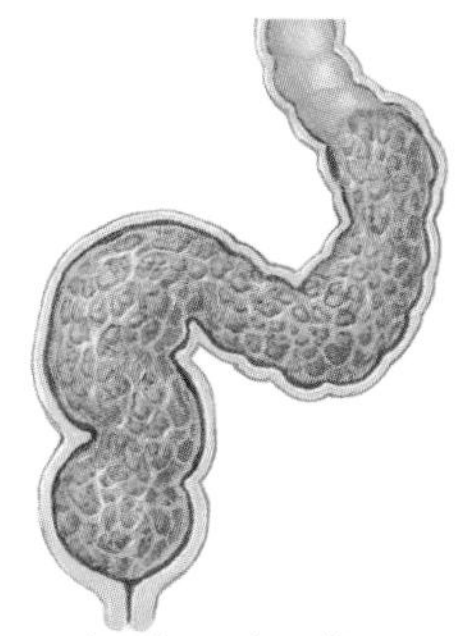

Wetting agents (dioctyl sodium sulfosuccinate) soften stool by coating and dispersion of component particles

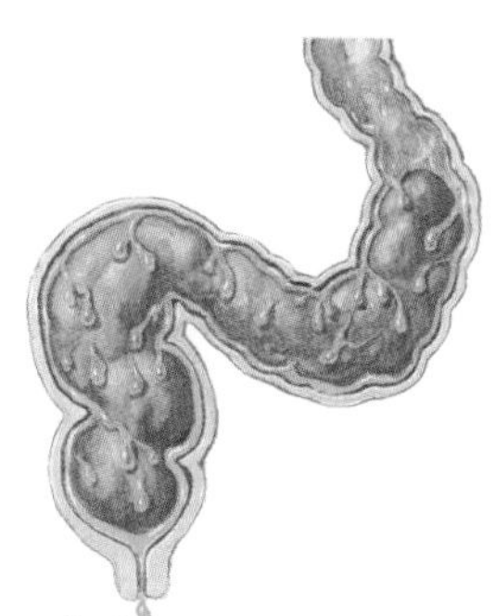

Mineral oil lubricates and mixes with stool to soften it

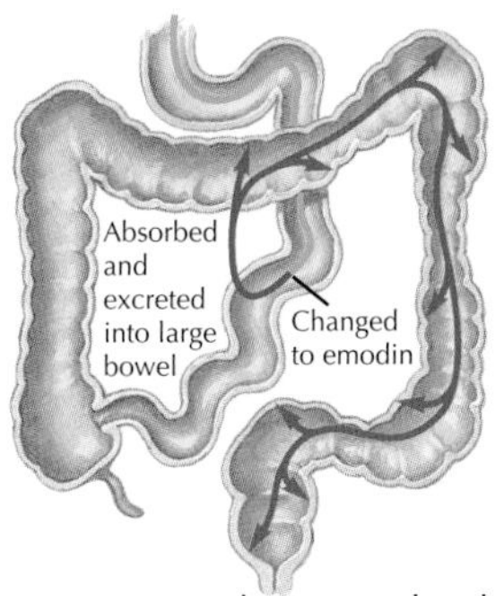

Emodins (cascara, senna, aloes) stimulate large bowel peristalsis and secretion by irritation

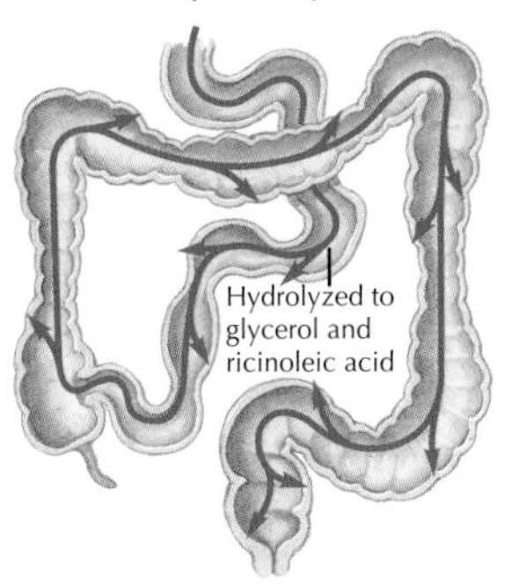

Castor oil and derivatives stimulate activity of small and large bowel by irritation

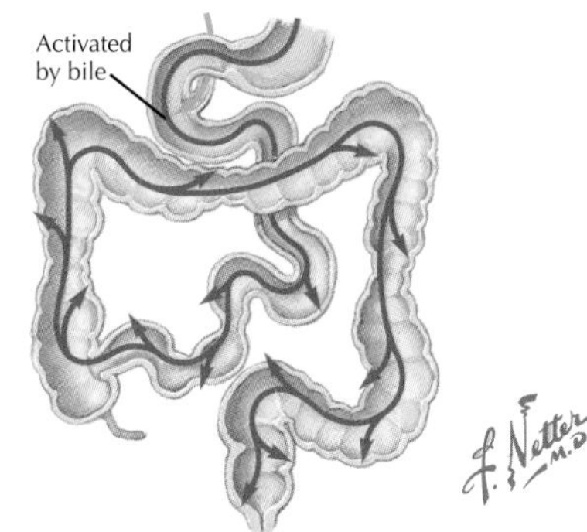

Phenolphthalein stimulates peristalsis and secretion by irritation; site of major action undetermined, probably widespread

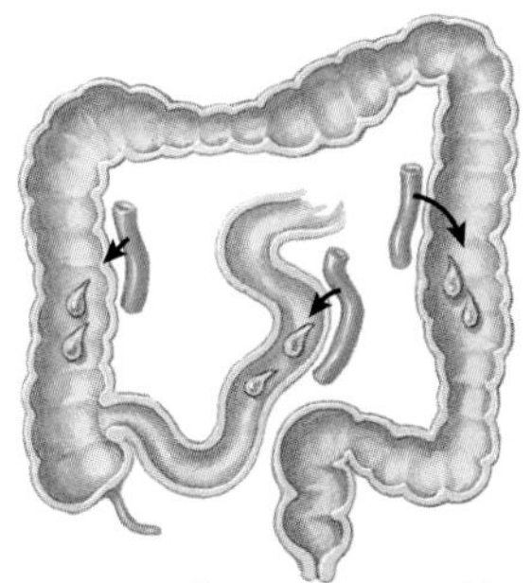

Salines (magnesium sulfate, citrate, and hydroxide; sodium phosphate) draw and hold fluid in lumen osmotically, also have some irritant action

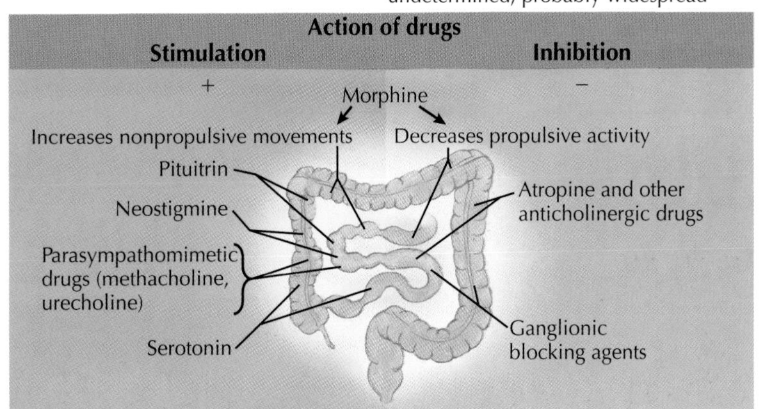

Figure 137-1 *Effects of Drugs on the Large Intestine.*

Box 137-1 Cathartic/Stimulant/Laxative Drug Categories

Bulk Agents
Bran (wheat fiber): retains water
Psyllium (ispaghula): nurtures and increases size of microflora
Guarana
Sterculia
Methylcellulose (inert chemical)
Nonabsorbed Sugars
Lactulose (osmotic; nurtures flora)
Lactitol (osmotic)
Salts (extract water)
Magnesium citrate
Magnesium hydroxide ($Mg[OH]_2$)
Sulfate (SO_4)
Sodium sulfate ($NaSO_4$)
Polyethylene Glycol (retains water)
Cleansing: 1 L
Daily: 250 mL
Anthranoid Compounds
Aloe
Cascara
Senna
Castor oil
Polyphenolic Compounds
Phenolphthalein
Bisacodyl
Sodium picosulfate
Detergents
Docusate sodium
Liquid Paraffin (lubricant)
Mineral oil
Prokinetic Agents (stimulate motility)
Tegaserod (Zelnorm)* (5-HT_4 agonist)
Chloride Channel Blocker
Lubiprostone (Amitiza)

**Pulled from U.S. market.*

Drugs may affect the motor and secretory activities in the intestine directly or indirectly. The classic parasympathomimetic drug *methacholine* and the agents that inhibit hydrolysis of acetylcholine by blocking cholinesterase action (*physostigmine* and *neostigmine*) stimulate intestinal contractions. *Atropine* and a legion of synthetic anticholinergic drugs block the transmission of parasympathetic stimuli to the effector organs and thus inhibit intestinal contractions. Drugs that stimulate or inhibit the effects of sympathetic nerves are less active in the GI tract than other systems. Therefore, sympathomimetic drugs would have to be used in very large doses to affect intestinal motility. *Ganglionic blocking agents*, interfering with the transmission of nerve impulses at the sympathetic and the parasympathetic ganglionic synapses, inhibit intestinal contractions.

Morphine and all related opiates, used for centuries as antidiarrheal agents, generally decrease the propulsive motility and increase tonus, particularly of the large intestine, sometimes to the point of spasms, which may explain the abdominal discomfort associated with opiates.

The endless number of drugs and preparations that promote defecation are called *cathartics* and are frequently listed as "laxatives" or "purgatives." Classification according to different mechanisms or actions is usually the basis for their selection in a given clinical situation (see Box 137-1 and Chapters 133-136).

ADDITIONAL RESOURCES

Drossman DA, editor: *The functional gastrointestinal disorders*, ed 2, McLean, Va, 2000, Degnon Associates.

Patel SM, Lembo AJ: Constipation. In Feldman M, Friedman LS, Brandt LJ, editors: *Gastrointestinal and liver disease*, ed 8, Philadelphia, 2006, Saunders-Elsevier, pp 221-253.

Schiller LR, Sellin JH: Diarrhea. In Feldman M, Friedman LS, Brandt LJ, editors: *Gastrointestinal and liver disease*, ed 8, Philadelphia, 2006, Saunders-Elsevier, pp 159-185.

Megacolon

Martin H. Floch

138

Megacolon is divided into the congenital and acquired types. The congenital form is Hirschsprung disease and includes classic, short-segment, and ultrashort-segment types, as well as total colonic aganglionosis (**Fig. 138-1**).

Acquired megacolon includes many disorders. However, *idiopathic* acquired megacolon has no cause and is often associated with an acute form of Ogilvie syndrome (see Chapter 107). Acquired forms are also associated with a variety of neurologic diseases, intestinal smooth muscle disease, and metabolic disorders (see Chapter 106).

Congenital megacolon usually presents during infancy (although it is now well documented in adolescents and adults). It is more

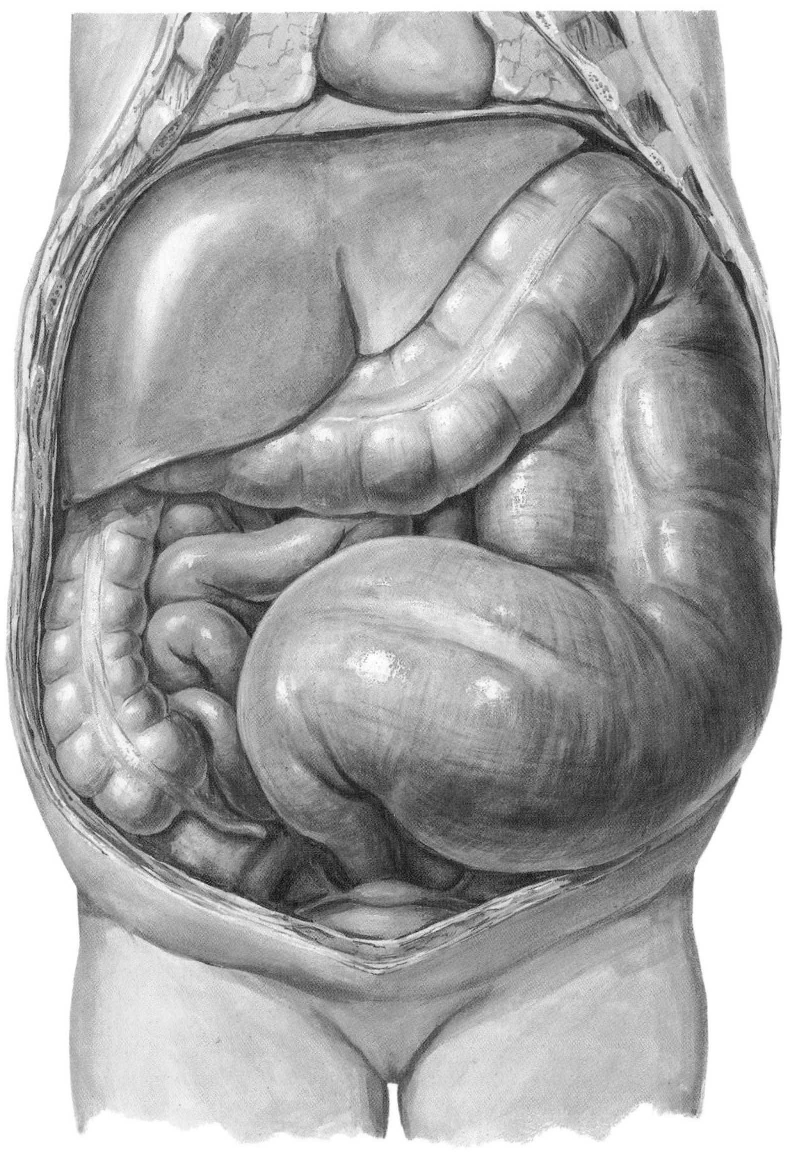

Tremendous distention and hypertrophy of sigmoid and descending colon; moderate involvement of transverse colon; distal constricted segment

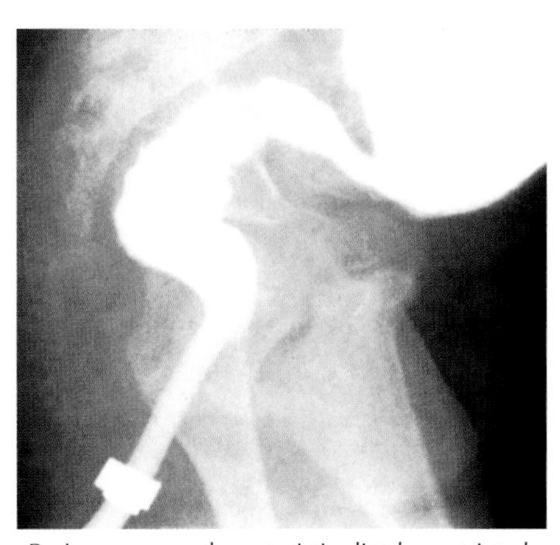

Barium enema; characteristic distal constricted segment

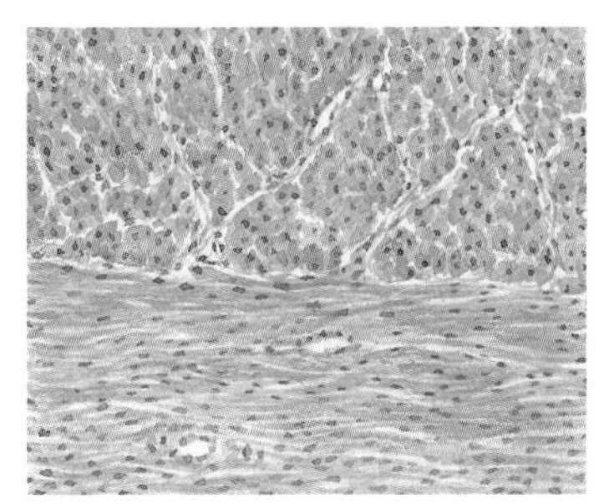

Ganglion cells absent

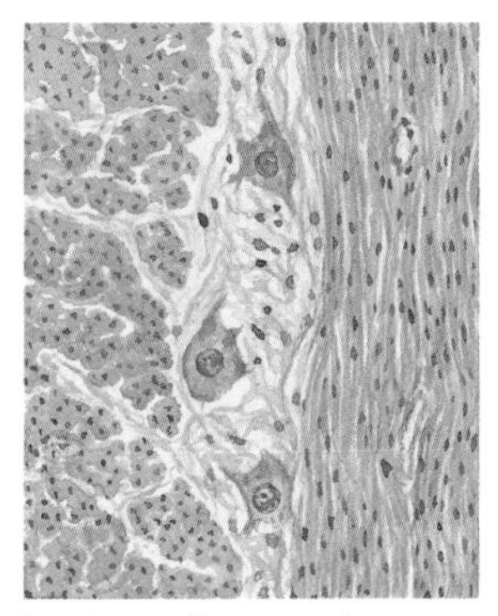

Ganglion cells present between longitudinal and circular muscle layers

Figure 138-1 *Megacolon (Hirschsprung Disease).*

common in males than in females and classically has an *aganglionic* segment in the rectum or sigmoid with a dilated segment above. Findings from anal manometry are classically abnormal. The pathophysiology is attributed to a loss of ganglionic cells in the segment of bowel. It is reported as dominant and recessive inheritance. In addition, patients have classic cases with the idiopathic form.

CLINICAL PICTURE

Constipation is the symptom that drives the parent or patient to seek medical advice. Functional constipation brings more children to medical attention than does aganglionic megacolon, but the differential diagnosis includes these two conditions. Great variations in clinical signs and symptoms are expected, but typical situations are readily characterized. The child with chronic functional constipation is a healthy-looking youngster of normal body appearance, whereas a child with aganglionic megacolon appears to be chronically ill, has a protuberant abdomen, and bears the stigma of malnutrition in growth and development. However, well-documented cases of normal growth and development show the disease progressing into adolescence, with rare reports in adults, up to age 40.

Typically, the child or adolescent has never had a normal bowel movement and requires laxatives and enemas to evacuate. Occasionally, the presentation is diarrhea characterized by liquid stool moving around an impaction. On rectal examination, the sphincter may be normal or relaxed, and stool or a dilated rectum is palpated; fecal impaction is often severe, and the rectal sphincter may be extremely tight.

DIAGNOSIS

The diagnosis is usually straightforward in the classic infant presentation. In early childhood or adolescence or in adulthood, however, making the diagnosis becomes more difficult. Barium enema shows the classic presentation of a narrowed segment with dilated bowel above. It is now known that in the atypical presentations, the aganglionic or narrowed segment may be extremely short and may involve only the internal anal sphincter.

Proctosigmoidoscopy and colonoscopy reveal a dilated sigmoid. At times, if the dilatation is longstanding, it may be difficult to examine the entire bowel endoscopically. However, no obstructive lesions are demonstrated. If the fecal impaction is chronic, traumatic ulceration of the mucosa may be noted. Barium contrast enema may reveal the narrow segment and confirm the suspected diagnosis, but further evaluation is needed before surgery can be performed.

In the classic presentation, physiologic testing reveals that the anal sphincter fails to relax after distention of the rectum. This may be helpful when the narrowed segment is not easily demonstrated. However, manometry has not been universally reproducible.

The diagnosis of Hirschsprung disease is definitively made by examining biopsy specimens that reveal *aganglionosis*, or a patchy area, in the segment. Often, suction or punch biopsy is attempted first, but deeper biopsy is required to confirm the diagnosis. However, the observation of ganglion cells on colon biopsy tends to rule out the diagnosis of Hirschsprung disease.

Acquired megacolon is often confused with congenital megacolon (Hirschsprung disease). Certainly, when the rectum is greatly dilated and short-segment Hirschsprung disease is suspected, the diagnosis depends on biopsy and histologic interpretation. When the two conditions are difficult to distinguish, full-thickness biopsy is essential.

TREATMENT AND MANAGEMENT

Classic megacolon requires surgical correction, the treatment of choice. The goal of surgery is to establish regular defecation. The aganglionic segment must be removed.

COURSE AND PROGNOSIS

When the presentation and findings are not classic and are seen in early childhood or adolescence, the patient has had numerous trials of laxatives. The adult patient will have undergone repeated laxative and probably enema therapy. Some patients can be managed by laxative and enema evacuation to provide comfort and to ensure the bowel is not dilated and causing obstruction to normal digestion and absorption.

Because of the great variation in presentations and laxative programs used, course and prognosis vary. In the classic case of Hirschsprung disease, the prognosis is good after successful surgery. Long-term follow-up studies reveal approximately a 90% cure rate, with residual fecal soiling in some cases. Even though course and prognosis can vary greatly, especially in older age groups, the prognosis is good if the aganglionic segment can be removed.

ADDITIONAL RESOURCES

Barnes PRH, Lennard-Jones JE, Howley PR, Todd IP: Hirschsprung's disease and idiopathic megacolon in adults and adolescents, *Gut* 27:534-541, 1996.

Kahn E, Daum F: Anatomy, histology, embryology, and developmental anomalies of the small and large intestine. In Feldman M, Friedman LS, Brandt LJ, editors: *Gastrointestinal and liver disease*, ed 8, Philadelphia, 2006, Saunders-Elsevier, pp 2061-2091.

Kim HJ, Kim AY, Lee CW, et al: Hirschsprung disease and hypoganglionosis in adults: radiologic findings and differentiation, *Radiology* 247:428-434, 2008.

Marty TL, Leo T, Matlak ME, et al: Gastrointestinal function after surgical correction of Hirschsprung's disease: long-term follow-up in 135 patients, *J Pediatr Surg* 40:655-663, 1995.

McConnell EJ, Pemberton JH: Megacolon: congenital and acquired. In Feldman N, Friedman LS, Schleisenger MH, editors: *Gastrointestinal and liver disease*, ed 7, Philadelphia, 2002, Saunders, pp 2129-2139.

Todd IP: Adult Hirschsprung's disease, *Br J Surg* 64:311-312, 1977.

Sigmoid Volvulus

139

Martin H. Floch

Intestinal volvulus is the rotation of a segment of bowel around its mesentery (**Fig. 139-1**). This usually causes a closed-loop obstruction and may cause vascular compromise. Primary volvulus of the colon usually occurs in the sigmoid colon and in the cecum; the other parts of the large bowel are well fixed in the posterior abdominal wall. This chapter discusses volvulus of the sigmoid, and Chapter 140 discusses volvulus of the cecum.

Sigmoid volvulus is a comparatively rare form of intestinal obstruction in the Western world, occurring more frequently in middle-aged and elderly patients. It is more common in Eastern Europe and Asia, presumably reflecting different dietary habits. Cultures with a diet high in bulky vegetables have a higher incidence of sigmoid volvulus. The high-fiber bulk causes larger fecal residue and results in a more distended and elongated bowel prone to rotation.

CLINICAL PICTURE

In the United States, patients with sigmoid volvulus are usually constipated and frequently use laxatives. Sigmoid volvulus develops more frequently in patients with neurologic or psychiatric disorders such as Alzheimer or Parkinson disease. Chronic symptoms may be difficult to differentiate from the patient's usual constipation. Onset of symptoms is usually sudden, however, with lower abdominal pain, obstipation, and abdominal distention. When vascular compromise occurs, ischemia and peritoneal signs may evolve rapidly. Possible ischemia and perforation, as well as acute volvulus, mandate rapid evaluation.

DIAGNOSIS

Once abdominal distention raises the index of suspicion, immediate radiographic evaluation of the abdomen is necessary (see radiograph in Fig. 140-1). The classic dilated sigmoid loop may be seen on a simple obstructive series. Usually, no feces are present in the rectum, but there may be dilatation of other loops in the colon. Plain x-ray films of the abdomen confirm the diagnosis in more than 60% of patients. If there is any doubt, an absorbable-dye enema can assist the diagnosis. *Barium should not be used*, because it may result in perforation. Computed tomography can also make the diagnosis promptly.

TREATMENT AND MANAGEMENT

Management consists of attempted enema, use of a rectal tube, and flexible sigmoidoscope or colonoscope to decompress and detorse the sigmoid bowel. The success rate using a flexible sigmoidoscope is 95% to 98%, depending on the endoscopist's experience. The colonoscope is usually not necessary unless the shorter sigmoidoscope cannot reach the suspected area. If bowel is ischemic, immediate surgery should be seriously considered to relieve any compromise in the vasculature.

In rare cases, the symptoms might be mild and recurrent, and a thorough evaluation is necessary. A long loop of sigmoid may twist, necessitating surgical correction. Twisting of this loop may be discovered only by chance.

COURSE AND PROGNOSIS

Recurrence rate for sigmoid volvulus is approximately 40% and is high even after endoscopic decompression. Therefore, if symptoms recur, elective resection is recommended. It is important to consider whether the patient is ambulatory or institutionalized. If the patient experiences vascular compromise and gangrene has set in, mortality is very high (>50%). Surgical evaluation and timing of surgery are extremely important in predicting the course and progress for each patient.

ADDITIONAL RESOURCES

Ballentyne G: Review of sigmoid volvulus: history and results of treatment, *Dis Colon Rect* 25:494-501, 1982.

Ballantyne G, Brandner M, Beart R, Ilstrup D: Volvulus of the colon, *Ann Surg* 202:83-92, 1985.

Brothers TE, Strodel WE, Eckhauser FE: Endoscopy in colonic volvulus, *Ann Surg* 206:1-7, 1987.

Burrell H, Baker V, Wardrop B, Evans A: Significant plain film findings in sigmoid volvulus, *Clin Radiol* 49:317-319, 1994.

Turnage RH, Heldmann M, Cole P: Intestinal obstruction. In Feldman M, Friedman LS, Brandt LJ, editors: *Gastrointestinal and liver disease*, ed 8, Philadelphia, 2006, Saunders-Elsevier, pp 2653-2667.

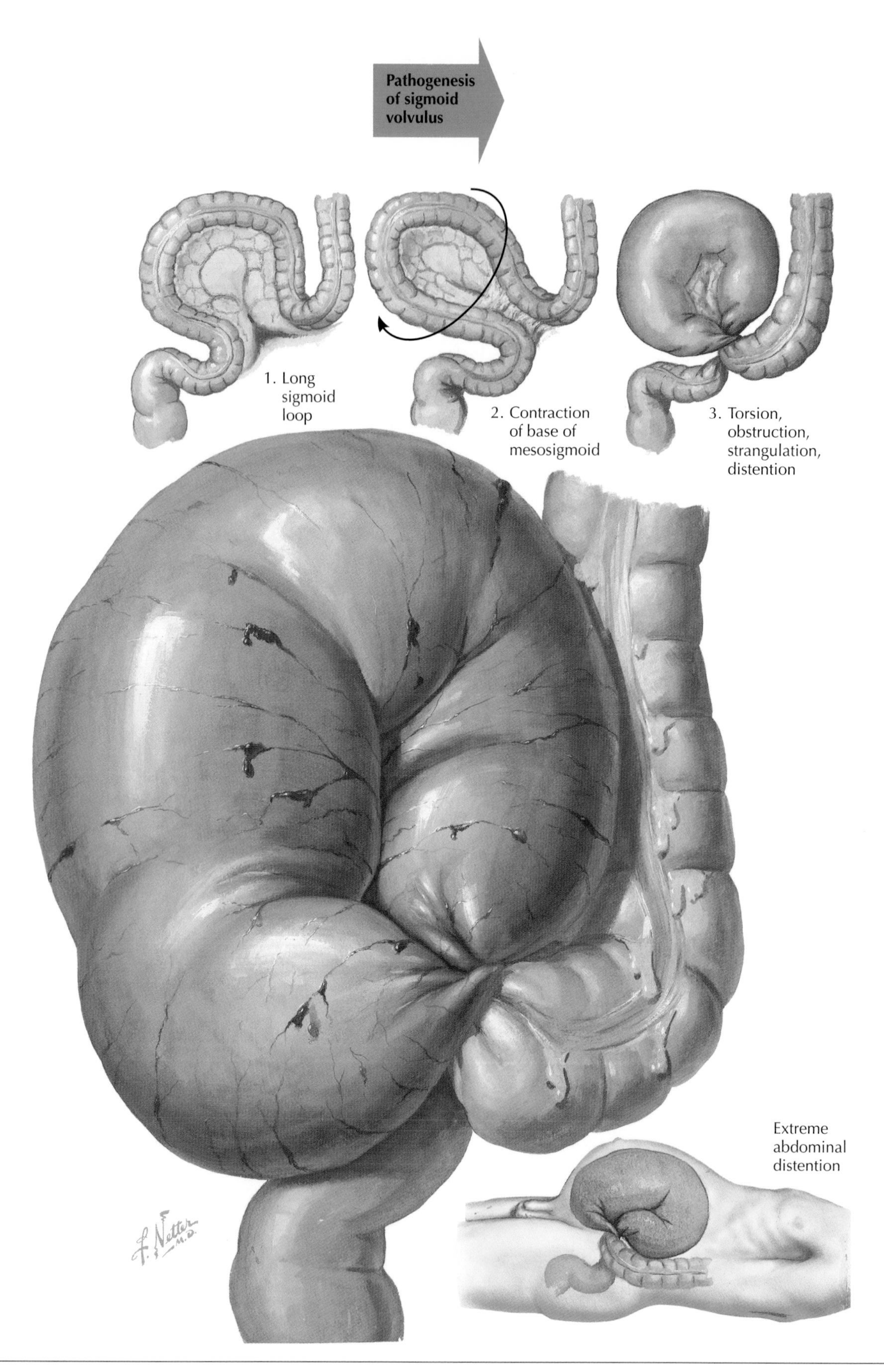

Figure 139-1 *Volvulus of the Sigmoid Colon.*

Volvulus of the Cecum

Martin H. Floch

140

Volvulus of the cecum is an abnormal rotation of the cecum from an apparent anomalous attachment to the mesentery, resulting in increased mobility (**Fig. 140-1**). It occurs infrequently in Western countries, where it accounts for approximately 1% of cases of intestinal obstruction. As with volvulus of the sigmoid, volvulus of the cecum appears to be more common in populations who have a diet high in vegetables and fibrous foods. Persistent loading of the bowel with a large fecal mass may play a role in the etiology.

The predisposing factor is inadequate fixation of the cecum and the ascending colon to the posterior abdominal wall, a result of an incomplete embryologic third stage of intestinal rotation. Twisting of the mesentery may be loose and intermittent without becoming complete, but when it is complete, the vessels may strangulate, and the bowel may become gangrenous.

A variant of cecal volvulus called *cecal bascule* can occur when the ascending colon fills up without twisting. It accounts for approximately 10% of cases of cecal volvulus. Although this type is not a true volvulus, cecal bascule can present with the same symptoms.

CLINICAL PICTURE

Patients with volvulus of the cecum are usually younger than those with sigmoid volvulus; most are 20 to 40 years of age. Onset with severe central abdominal pain is usually sudden, and vomiting soon follows. The pain is constant, but intermittently severe. Symptoms are rarely sporadic. If the pain ceases, it is presumed the volvulus resolves, then recurs spontaneously. On examination, the abdomen is usually distended. In most patients, the distended cecum may be distinguished as a palpable tympanic swelling in the midabdomen.

DIAGNOSIS

The best way to diagnose volvulus of the cecum is by viewing plain x-ray films of the abdomen, which reveal a greatly distended central coil, possibly with a fluid level. Loops of ileum behind it may be obstructed. Plain abdominal radiographs and clinical status confirm the diagnosis in 50% of patients. In some patients, a Gastrografin barium enema or computed tomography is necessary to delineate the site of the destruction. Using all the studies still reveals a diagnosis in only 60% to 90% of patients. However, with the picture of abdominal pain, abdominal distention, vomiting, constipation, and the dilated loop, there is enough suspicion to warrant immediate laparotomy. One study reported that as many as 25% of patients have gangrenous bowel at laparotomy. Therefore, there should be no delay when the suspicion is significant.

TREATMENT AND MANAGEMENT

After the diagnosis is made, especially if gangrene is suspected, laparotomy and resection are urgent. When there is no suspicion of peritonitis or gangrene, attempts at reducing the

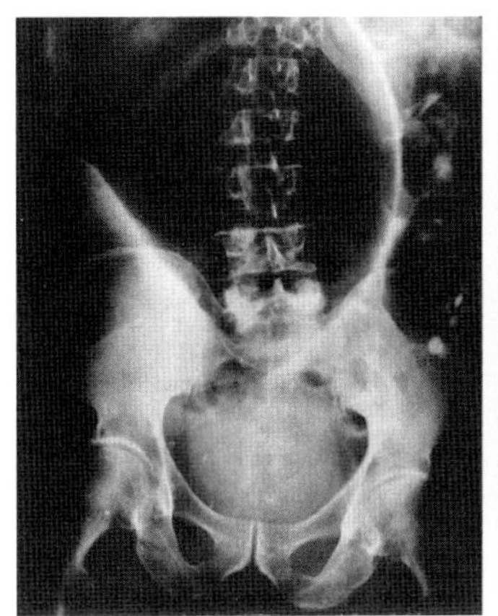

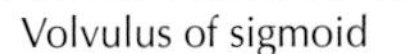

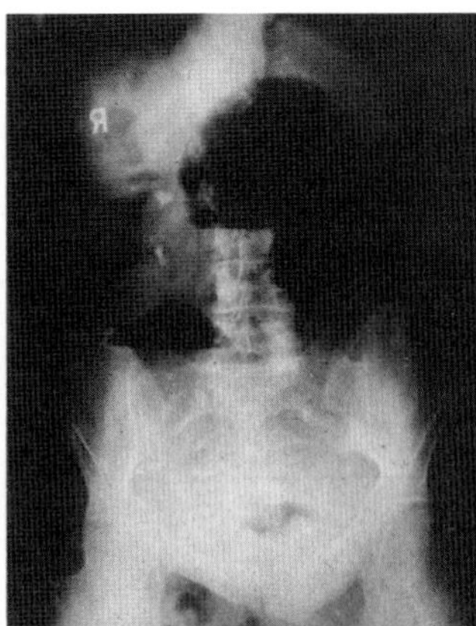

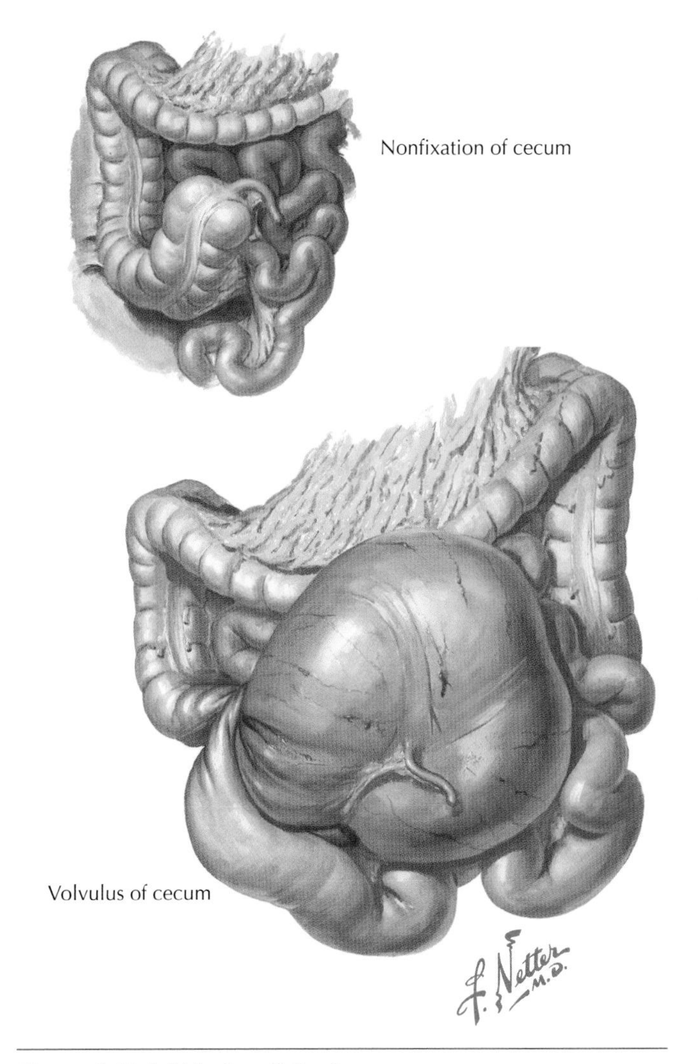

Figure 140-1 *Volvulus of the Cecum.*

volvulus with colonoscopy are indicated. However, this is less successful than in sigmoid volvulus. Consequently, most surgeons believe surgery is indicated without delay when cecal volvulus is suspected. Surgical detorsion is performed, followed by cecal fixation to prevent recurrence. Depending on the patient and on the condition, right hemicolectomy with cecal resection may be indicated to eliminate possible recurrence.

COURSE AND PROGNOSIS

The mortality rate may be as high as 10%, and if gangrene develops, it may approach 40%. These statistics emphasize the need for early surgery to treat volvulus of the cecum.

ADDITIONAL RESOURCES

Ballantyne G, Brandner M, Beart R, Ilstrup D: Volvulus of the colon, *Ann Surg* 202:83-92, 1985.

Chiao GZ, Rex DK: Motor disorders of the colon. In DeMarino AJ, Benjamin SB, editors: *Gastrointestinal disease*, Thorofare, NJ, 2002, Slack, pp 881-913.

Delabrousse E, Sarlieve P, Sailliey N, et al: Cecal volvulus: CT findings and correlation with pathophysiology, *Emerg Radiol* 14:411-415, 2007.

Rabinovici R, Simansky DA, Kaplan O, et al: Cecal volvulus, *Dis Col Rect* 55:765-771, 1990.

Tejler G, Jiborn H: Volvulus of the cecum: report of 26 cases and review of the literature, *Dis Col Rect* 31:445-449, 1988.

Diseases of the Appendix: Inflammation, Mucocele, and Tumors

141

Martin H. Floch

Inflammatory changes of the vermiform appendix are the most frequent cause of laparotomy in Western countries, where the incidence of appendicitis is approximately 10 times that in Eastern countries.

The *appendix* is part of the cecum, and its abdominal marking usually lies in the transition of the outer to the middle third of an imaginary line drawn from the anterior superior iliac spine to the umbilicus (McBurney point). However, because of embryologic development and mobility of the cecum, the appendix may lie anywhere in the abdominal cavity (**Fig. 141-1**). Furthermore, it is a rudimentary organ, narrow and thin, and may be positioned retrocecally or may fall into the pelvis. The layers of the appendix are the same as those in other parts of the intestinal tract, but it is not enwrapped by visceral peritoneum, and the longitudinal muscle envelops the entire circumference.

It is proposed that obstruction of the lumen of the appendix is followed by infection to cause acute appendicitis. In the initial stage, acute appendicitis is confined to the mucous membranes, which become edematous and hyperemic, and are invaded by white blood cells (WBC). If the inflammatory process subsides, which rarely occurs, but involves other coats of the appendix in an acute suppurative process, the whole organ becomes enlarged and a fibrinous or fibropurulent exudate covers the appendix. *Abscesses* may form in the appendiceal wall and lead to gangrenous appendicitis. Necrosis and putrefaction of the entire appendiceal tissues may lead to perforation. Once infected material spills into the abdominal cavity, peritonitis may form a periappendiceal abscess or diffuse peritonitis. Depending on the position of the appendix, it may spill into the pelvis or any location in the abdominal cavity.

Rarely, the suppurative process may heal without medical attention. The resultant fibrosis of the wall could lead to a future cyst or *mucocele* formation when mucus is secreted into the fibrosed area. This type of cystic tumor may be benign but must be differentiated from adenocarcinoma. Malignant tumors can

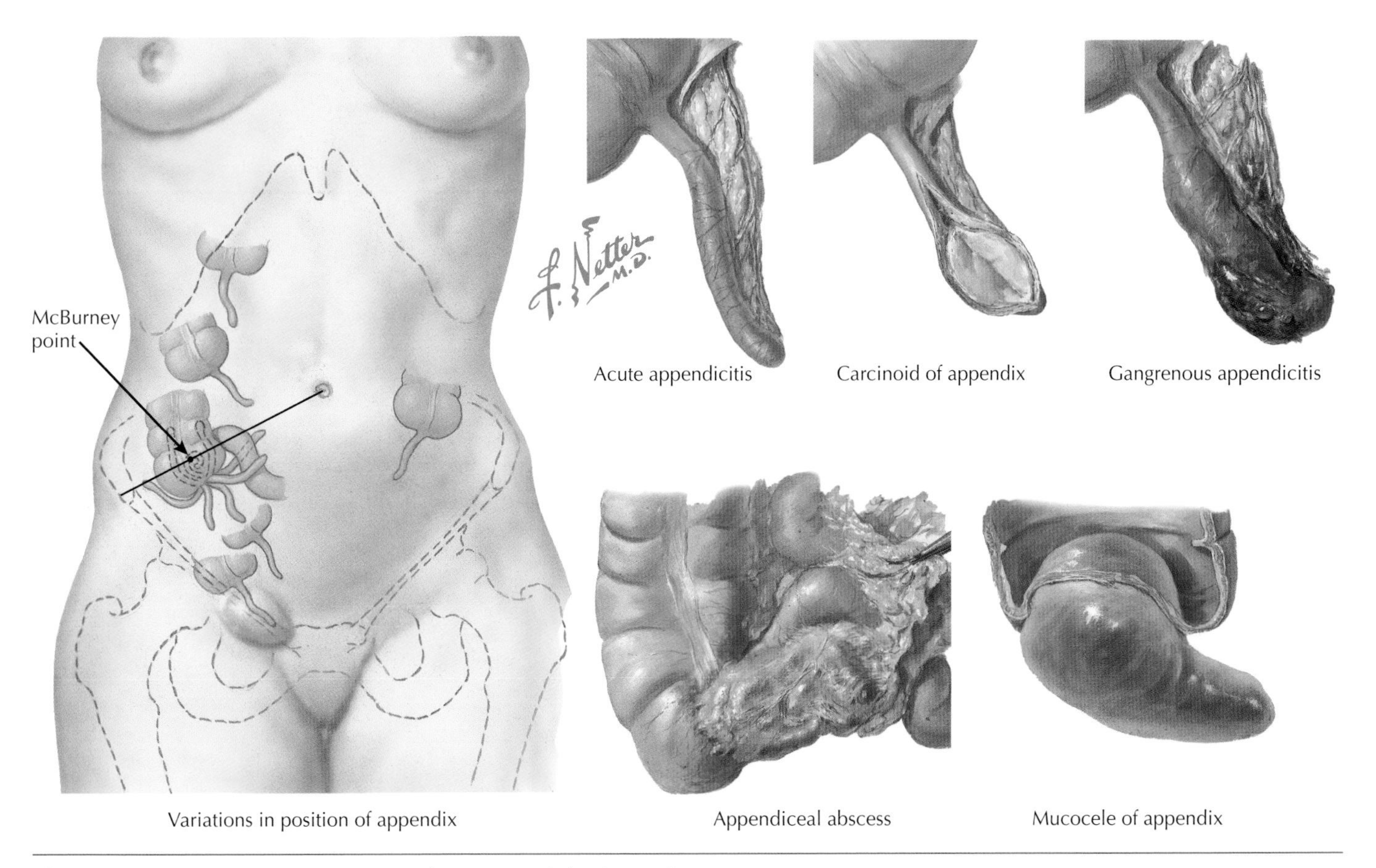

Figure 141-1 *Variations in Position and Diseases of the Appendix.*

spill into the entire abdominal cavity, causing pseudomyxoma peritonei.

CLINICAL PICTURE

The classic clinical picture of appendicitis is periumbilical pain that radiates to the right lower quadrant and that may be associated with nausea or vomiting. The picture usually evolves over 24 hours and may be associated with fever, but the pain usually is persistent, grows in intensity, and is associated with significant nausea or vomiting. Unfortunately, most cases are not typical, and symptoms are often vague. In elderly persons, the picture may be masked, and the acute signs may not be present until massive perforation appears. Also, the appendix may be positioned in different parts of the abdominal cavity, and symptoms may be atypical, such as left lower quadrant pain, right upper quadrant pain, or even left upper quadrant pain (see **Fig. 141-1**).

All diagnosticians should consider appendicitis in any presentation of acute abdominal pain or mild nausea and vomiting. Appendicitis does not cause violent vomiting or extensive diarrhea and thus should not be confused with severe, acute gastroenteritis.

DIAGNOSIS

When clinical findings point to appendicitis, surgical intervention is indicated. Studies show that a higher index of complexity in diagnosis leads to more perforations. Atypical presentations, as previously described, occur in approximately 20% of patients. Good clinical practice indicates that simple laparotomy or laparoscopy should be performed rather than waiting for perforation.

The differential diagnosis includes Meckel diverticulitis, cecal diverticulitis, mesenteric lymphadenitis, and ileitis. Other causes, such as renal, sigmoid, and gynecologic disease, can be diagnosed by careful examination and imaging techniques.

Computed tomography (CT), especially with helical imaging, has become the most accurate diagnostic test for appendicitis, greater than 90%, and can be extremely helpful when diagnosis is uncertain. Ultrasound examination may also yield high accuracy, but the procedure is operator dependent and varies from institution to institution. However, when appendicitis is suspected during pregnancy and x-ray imaging should be avoided, ultrasound is used. Helical CT will make the diagnosis and help rule out any complications, such as perforation or abscess, or any renal or gynecologic disease.

It is also useful to conduct a simple WBC count. The WBC count is elevated in approximately 80% of patients with appendicitis. A normal WBC count usually alerts the diagnostician to look for other causes of the pain.

TREATMENT AND MANAGEMENT

Treatment is surgical removal of the appendix. Unfortunately, if perforation has occurred, local drainage of the abscess may be required. In rare, fascinating cases, obstruction of the appendix occurs because of parasites, such as pinworm. The parasitic infestation should be treated along with the appendicitis (see Section VI).

Simple excision of the appendix can be accomplished with a small incision. Also, surgeons adept in laparoscopic surgery now perform appendectomy with ease. The choice depends on the patient and the surgeon. Both procedures produce equally excellent results, provided no complications have arisen.

COURSE AND PROGNOSIS

Once the correct diagnosis is made and there is no perforation, patients usually leave the hospital within 24 to 48 hours and can return to full activity in 2 weeks. However, if perforation occurs, mortality rates increase and can become significant in elderly patients. Complication rates may be as high as 12% to 20%. Mortality as high as 30% has been reported in the elderly population. Therefore, physicians need to make an early diagnosis and intervene to prevent complications.

Tumors of the appendix other than mucocele and cystadenocarcinoma do occur at significant rates (see Chapter 156 and 157). Carcinoid of the appendix is the most common appendiceal tumor (see Chapter 121).

ADDITIONAL RESOURCES

Addiss DG, Shaffer N, Fowler BS, Tauxe RV: The epidemiology of appendicitis and appendectomy in the United States, *Am J Epidemiol* 132:910-919, 1990.

Birnbaum BA, Jeffrey RB: CT and sonographic evaluation of acute right lower quadrant abdominal pain, *Am J Roentgenol* 170:361-370, 1998.

Franz NG, Norman J, Fabri PJ: Increased morbidity of appendicitis with advancing age, *Am Surg* 61:40-46, 1995.

Horton MD, Counter SF, Florence NG, Hart MJ: A prospective trial of computed tomography and ultrasonography for diagnosing appendicitis in the atypical patient, *Am J Surg* 179:379-388, 2000.

Kelly K, Sarr M, Hinder R: *Mayo Clinic gastrointestinal surgery*, Philadelphia, 2004, Elsevier.

Rao PM, Rhea JT, Novelline RA: Helical CT of appendicitis and diverticulitis, *Radiol Clin North Am* 37:895-906, 1999.

Walker AR, Segal I: Appendicitis: an African perspective, *J R Soc Med* 88:616-620, 1995.

Diverticulosis: Diverticular Disease of the Colon

142

Martin H. Floch

Diverticulosis of the colon is an acquired condition that results from herniation of the mucosa through defects in the muscle coats (**Fig. 142-1**). Defects are usually located where the blood vessels pierce the muscular wall to gain access to the submucosal plane. These vessels enter at a constant position, just on the mesenteric side of the two lateral taeniae coli, so diverticula typically occur in two parallel rows along the bowel. Appendices epiploicae (omentales) are also situated in this part of the circumference.

Diverticula probably arise from pulsion as a result of increased intraluminal pressure from uncoordinated peristalsis or inadequate luminal contents, possibly resulting from a low-fiber diet. Diverticula do not occur in the rectum but may be found throughout the entire colon. They are more common in the left side and usually affect the sigmoid colon. Diverticula are relatively rare in persons younger than 40 but are common (60%) in persons older than 60 living in Western countries. Because the incidence is dramatically higher in Western countries, where the diet is much lower in fiber, the decreased fiber intake results in decreased colonic luminal content, and pressure from wall contraction is transmitted to the wall rather than to the luminal content. Therefore, diverticula formation is related to fiber deficiency. Because contractions and their force are greatest in the sigmoid; this supports the theory that diverticula are more prominent in the sigmoid than in the rest of the bowel. Associated with chronic diverticula formation is the gradually increased deposition of connective tissue by *elastin*, resulting in the thickening of the sigmoid, with some bowel rigidity where large amounts of elastin are deposited.

Although diverticula may form in the second or third decade of life, they are usually asymptomatic at that time.

CLINICAL PICTURE

If discovered during barium enema, computed tomography (CT), or colonoscopy performed for different clinical reasons (e.g., gastrointestinal bleeding, irritable bowel syndrome [IBS]), diverticula are assumed to be asymptomatic. Approximately 60% of the Western population older than 80 has significant but asymptomatic diverticula formation. Furthermore, the high percentage of patients with IBS often includes those who have some diverticula formation, but not diverticulitis (see Chapter 143).

Many physicians assume the diverticula are not causing symptoms when in reality they may cause mild symptoms. In addition, many patients with diverticula have symptoms but do not develop acute or chronic diverticulitis. Some may report a low-grade, dull ache in the left lower quadrant or a change in bowel habit. Some thought to have IBS may have diarrhea or constipation. However, diverticula are rarely thought to cause the symptoms unless temperature and white blood cell (WBC) count are elevated, or imaging findings demonstrate spasm in the area of the diverticula.

On the basis of its natural history, diverticular disease is now classified into *asymptomatic*, *symptomatic uncomplicated*, and *complicated* types.

DIAGNOSIS

The diagnosis of diverticula formation is made through barium enema, CT, or colonoscopy. When diverticula are discovered incidentally, no further diagnostic evaluation is needed. However, if they are accompanied by symptoms, colonoscopy may be necessary if the discovery was made during barium enema or CT. It may also be necessary to perform a WBC count and to monitor the patient carefully to identify any associated inflammation.

TREATMENT AND MANAGEMENT

Most clinicians do not treat diverticula when they are discovered incidentally. Diverticula can result from a low-fiber diet, so all patients with diverticula should be advised to increase their fiber intake to 25 to 35 g daily (see Chapter 268). There is no evidence that increased fiber intake will reverse the diverticula formation, but increased intake should prevent further diverticula formation, and increased fiber in the bowel should decrease the force of pressure on the bowel wall.

COURSE AND PROGNOSIS

In Western societies, there is a concomitant increase in diverticula with age, and 60% of the population has diverticula by age 85. Studies also reveal that 5% to 20% of patients develop true inflammation associated with diverticula. Asymptomatic diverticula are benign and have an excellent course and prognosis.

In patients who do have complications, the course and prognosis vary and are described in Chapter 143.

ADDITIONAL RESOURCES

Alny TP, Howell DA: Diverticular disease of the colon, *N Engl J Med* 302:324-329, 1980.

Hughes LE: Postmortem survey of diverticular disease of the colon, *Gut* 10:326-335, 1969.

Panter NS, Burkett BP: Diverticular disease of the colon: a 20th century problem, *Clin Gastroenterol* 4:3-21, 1975.

Sheth AA, Longo W, Floch MH: Diverticular disease and diverticulitis, *Am J Gastroenterol* 103(6):1550-1556, 2008.

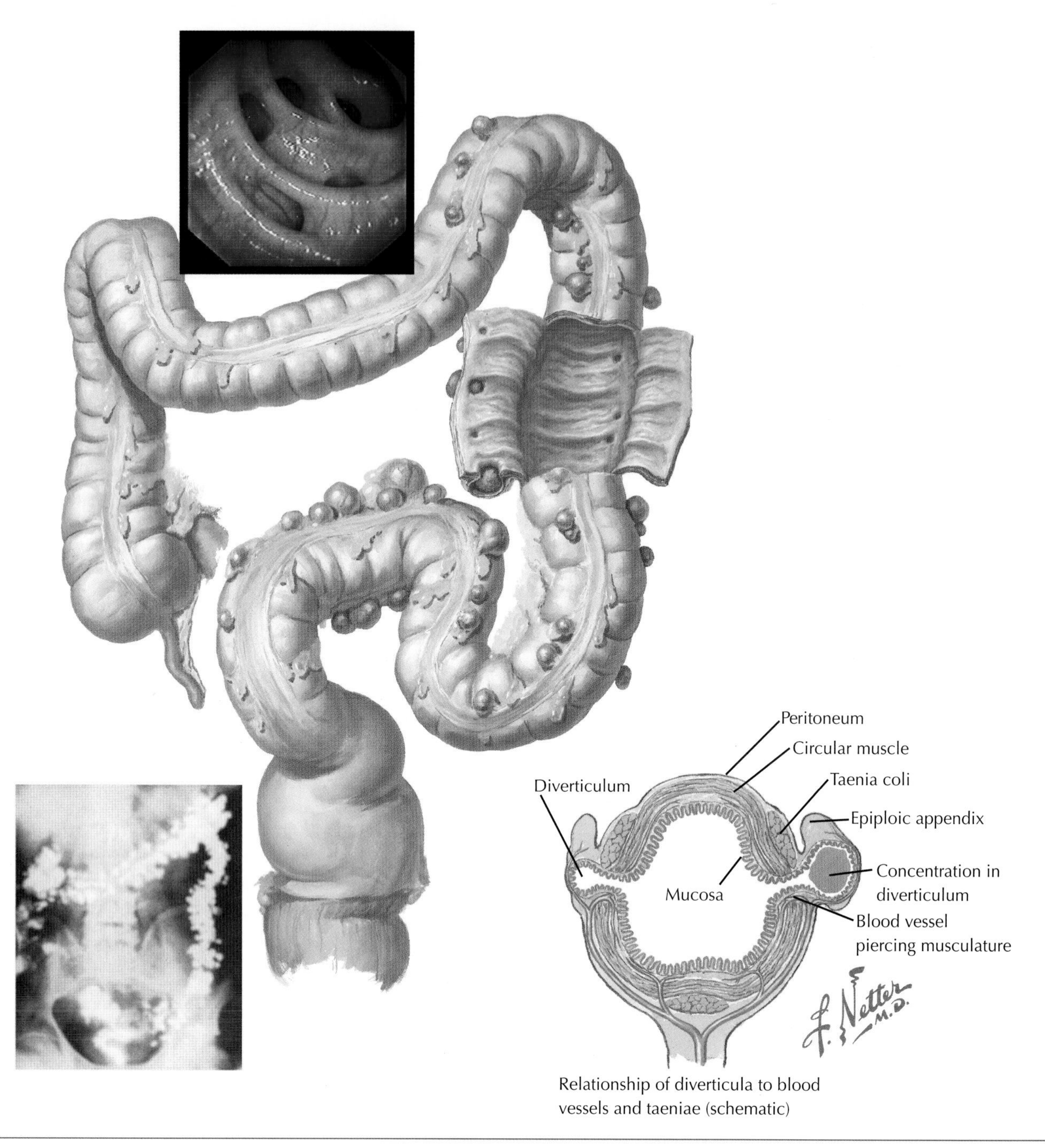

Figure 142-1 *Colonoscopic View of Diverticula.*

Diverticulosis: Diverticulitis and Its Complications and Diverticular Bleeding

143

Martin H. Floch

DIVERTICULITIS

Diverticula formation may be considered the first stage of the disease process. *Diverticular disease* represents the entire spectrum of, first, diverticula formation (see Chapter 142), then diverticulitis with its complications (**Fig. 143-1**). Inflammation in and around the diverticula occurs in only a small percentage of patients, for unclear reasons. Obstruction at the mouth of the diverticulum was thought to result in diverticulitis, similar to appendicitis, but this is no longer the accepted theory, and some believe that chronic inflammation precedes clinical diverticulitis.

Once diverticulitis develops, it may manifest as symptomatic disease or may progress to cause complications, with pericolitis (microabscess), pericolic phlegmon, pericolic abscess, pelvic or intraabdominal abscess, and free perforation leading to bowel obstruction, fistulization, or bacteremia and septicemia. Any of these complications may occur after inflammation begins in the diverticula.

Clinical Picture

The clinical presentation depends largely on the degree and extent of the inflammation. The most common presentation is left lower quadrant pain associated with low-grade fever, mild leukocytosis, and obstipation. Symptoms vary, however, depending on the severity of the inflammation.

If the pathologic process is localized in the left lower quadrant, the pain will present as left lower quadrant pain. The pain might be mild or severe and is usually associated with decreased number of bowel movements or intestinal obstruction. Occasionally, patients have mild diarrhea. If the obstructive symptoms persist, nausea or vomiting may occur. Depending on the site of the diverticula, the pain may be distributed across the midabdomen or to the suprapubic and right side of the colon. Any fever is usually low grade, but temperature greater than 102° F (38.8° C) indicates that the process is extensive or that bacteremia or septicemia might be present.

The classic picture is tenderness in the left lower quadrant with or without a palpable mass. However, if *rebound* occurs and if the mass is very tender, an abscess might have formed. Diffuse rebound could indicate peritonitis and perforation. Examination of the rectum usually reveals no significant findings.

Diagnosis

Laboratory results often reveal an elevated white blood cell (WBC) count with a predominance of polymorphonuclear leukocytes. In elderly or immunocompromised patients, leukocytosis may not be present. Depending on the symptoms (obstructive more than inflammatory), the differential diagnosis includes carcinoma of the colon, pelvic mass of gynecologic origin, or atypical appendicitis. At this point, radiologic evaluation is critical. Barium enema examination is not indicated and may even be contraindicated because of potential perforation and leakage from the pressure. Computed tomography (CT) currently is the best technique; it reveals the location of any abscess formation, any sign of perforation, and any thickening of the bowel associated with diverticula. Furthermore, CT clearly reveals the extent of the disease by demonstrating the gray areas in the mesentery adjacent to the diverticula. Colorectal examination findings are often negative, and pelvic examination might be necessary to rule out gynecologic disease; again, CT can help in defining the disease.

In rare cases, it is impossible to differentiate between diverticulitis and Crohn's disease or carcinoma. The diagnosis of the so-called overlap syndrome, in which both diverticula and Crohn's disease are present in the sigmoid, can be made only by pathologists. Because perforation may result from colonoscopy, as from barium enema, medical treatment is pursued until diagnostic colonoscopy can be performed. At times, especially when toxicity or perforation is suspected, the clinical situation indicates surgical intervention, and then a definitive diagnosis is made at laparoscopy.

Treatment and Management

Uncomplicated symptomatic diverticulitis is usually treated with bowel rest, antibiotic therapy, and slow progression from clear liquids to a low-fiber diet and then to a high-fiber diet. Once complications are present, however, vigorous treatment is indicated. The patient may require hospital admission, with intravenous antibiotics the preferred therapy and broad enough to cover the spectrum of aerobic and anaerobic microflora; this is essential. At present, most institutions use gentamicin plus another broad-spectrum antibiotic.

If an abscess has formed, percutaneous drainage should be considered. When drainage is unsuccessful or is used only as a temporizing condition, surgical intervention is necessary. Indications for surgery include uncontrollable sepsis, perforation, obstruction, fistula, and uncontrolled hemorrhage, as well as persistent symptoms of diverticulitis in the patient receiving antibiotic therapy. Two-stage procedures are most common at present. A colostomy is created, and after an extended period (usually months), the continuity of the bowel is recreated. However, one-stage resections are used when the segment is small with minimal inflammation or when the patient has symptoms without signs of abscess inflammation.

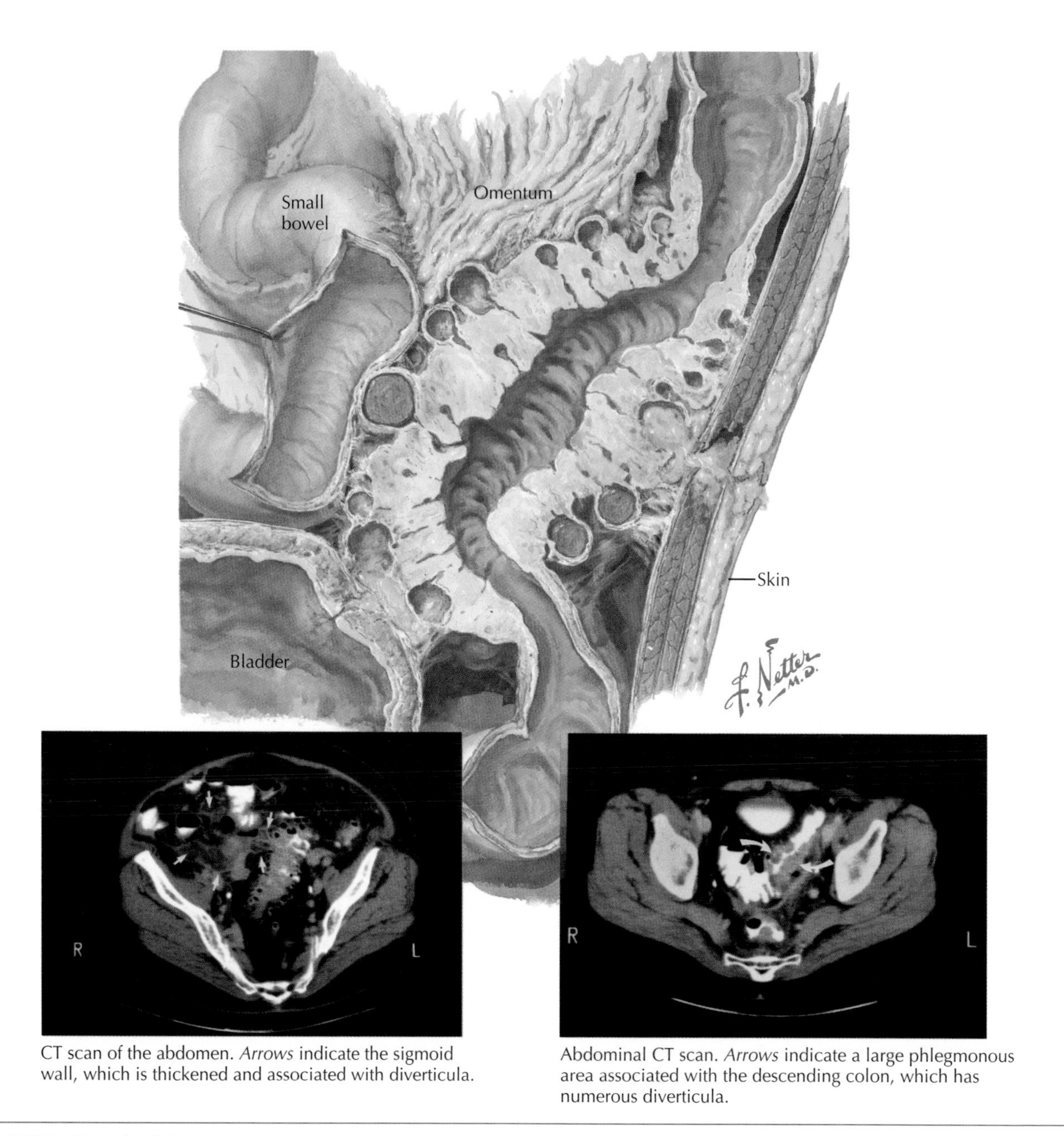

CT scan of the abdomen. *Arrows* indicate the sigmoid wall, which is thickened and associated with diverticula.

Abdominal CT scan. *Arrows* indicate a large phlegmonous area associated with the descending colon, which has numerous diverticula.

Figure 143-1 *Diverticulitis.*

Course and Prognosis

The risk for recurrent symptoms of diverticulitis or attacks ranges from 7% to 45%. It is thought that a high-fiber diet will reduce the risk, with a 70% chance of response to medical therapy after a first attack but only a 6% chance after the third attack. Thus, some recommend resection of the sigmoid after two attacks of uncomplicated diverticulitis. However, treatment of recurrent attacks is changing, and some now believe these cases can be followed medically if uncomplicated. Based on early reports, some are using mesalamine (5-aminosalicylic acid) to decrease attacks, although this recommendation is still under research. Mortality rates range from 3% to 35%, depending on the patient population. Early diagnosis, appropriate therapy, and surgical intervention (when indicated) are essential.

DIVERTICULAR HEMORRHAGE

The most common cause of massive colonic bleeding in Western societies is diverticular bleeding. An estimated 3% to 5% of patients with diverticular disease will have severe blood loss. In most patients, bleeding occurs from a single diverticulum. Surprisingly, the diverticula that bleed are often on the right side but are not inflamed. The most persistent finding is thinning of the wall of the diverticulum, with erosion of a vessel. Patients typically have massive lower gastrointestinal (GI) bleeding.

There are standard therapeutic protocols for GI hemorrhage, as well as a standard workup, but the clinician should determine whether the bleeding is from the right or the left side of the colon, because surgical intervention may be necessary. Therefore, radionucleotide imaging, arteriography, and colonoscopy should all be used to determine the location of the bleeding. Barium contrast has no place in this evaluation.

Therapeutically, arteriography can be used to inject sclerosing agents to occlude the bleeding vessel. If this is not successful, surgical intervention is necessary if the bleeding is not controlled. Usually, angiography identifies the site, and vasopressin infusions or embolization of a vessel can be used to stop the bleeding. If these do not succeed, surgical intervention is necessary. If the segment is identified, rebleeding is relatively rare. Segmental resection, when possible, is associated with lower morbidity than hemicolectomy. Unfortunately, surgical mortality is 9% to 11% and depends on comorbid conditions and patient age. When identifying the site is difficult, morbidity and mortality rates increase to approximately 35%, and a total colectomy is considered. Fortunately, most cases stop spontaneously, and massive bleeding is relatively rare.

There are some caveats with therapy. Diverticulitis in the immunocompromised patient may manifest with few signs and symptoms; therefore, treatment may be delayed, and morbidity and mortality may increase dramatically. Also, the index of suspicion should be high when the patient has abdominal pain, fever, or obstipation.

Diverticulitis is unusual in younger people, especially those under age 40. When it does occur in this age group, it may have a virulent course, with as many as two thirds of these patients requiring surgical intervention in the past. However, recent studies indicate that the course is becoming less severe in younger patients and that a conservative approach may be followed.

ADDITIONAL RESOURCES

Browder W, Cerise EJ, Litwin MS: Impact of emergency angiography in massive lower gastrointestinal bleeding, *Ann Surg* 204:530-538, 1986.

Floch CL: Diagnosis and management of acute diverticulitis, *J Clin Gastroentol* 40:S136-S144, 2006.

Floch MH, White JA: Management of diverticular disease is changing, *World J Gastroenterol* 12:3225-3228, 2006.

Fox JM, Stollman NH: Diverticular disease of the colon. In Feldman M, Friedman LS, Brandt LJ, editors: *Gastrointestinal and liver disease*, ed 8, Philadelphia, 2006, Saunders-Elsevier, pp 2613-2632.

Ghorai S, Ulbright TM, Rex DK: Endoscopic findings of diverticular inflammation in colonoscopy patients without clinical acute diverticulitis: prevalence and endoscopic spectrum, *Am J Gastroenterol* 98:802-806, 2003.

Jensen DN, Machicado AN: Diagnosis and treatment of severe hematochezia: the role of urgent colonoscopy after purge, *Gastroenterology* 95:1569-1574, 1988.

Parks TG, Connell AM: The outcome of 455 patients admitted for treatment of diverticular disease of the colon, *Br J Surg* 57:775-778, 1970.

Pohlman P: Diverticulitis, *Gastroenterol Clin* 17:357-358, 1988.

Sheth AA, Longo W, Floch MH: Diverticular disease and diverticulitis, *Am J Gastroenterol* 103(6):1550-1556, 2008.

Tursi A: New physiopathological and therapeutic approaches to diverticular disease of the colon, *Expert Opin Pharmacother* 8:299-307, 2007.

Wexner SD, Rosen L, Lowry A, et al: Practice parameters for the treatment of mucosal ulcerative colitis—supporting documentation: the Standards Practice Task Force, American Society of Colon and Rectal Surgeons, *Dis Col Rect* 40:1277-1285, 1997.

Ulcerative Colitis: Definition and General Description

144

Martin H. Floch

Ulcerative colitis (UC) is a disease of unknown etiology characterized by diffuse mucosal inflammation of the large bowel (**Fig. 144-1**). The disease is variable in extent, severity, and clinical course, and knowledge of many aspects remains incomplete. The incidence and prevalence of UC vary with geography and race. Most studies have been performed in North America and England; incidence varies from 0 to 10 per 100,000, with prevalence as high as 200 per 100,000 population.

Environmental and genetic factors undoubtedly create an altered immune response that results in inflammation of the mucosa of the bowel. Although infectious etiologies have been suspect and do cause an acute colitis, no organism has been identified as the cause of the chronic idiopathic form. It is also clear that genetic factors have a greater influence in Crohn's disease than in UC. Nevertheless, alterations in the T-cell and cytokine responses unquestionably occur in UC patients. The lesion may be limited to the rectum *(proctitis)* or may extend from the rectum into the left side and then into the transverse colon (*universal colitis* or *pancolitis*).

CLINICAL PICTURE

The cardinal sign of UC is usually *diarrhea*, often associated with bleeding. Onset may be sudden, or bowel habits may change gradually. Sudden onset is often confused with disease of infectious etiology. In reality, many of the early descriptions of chronic UC followed infections with dysentery, but after the infectious organisms were clear, a chronic disease evolved. Gradual onset or progression of disease may be associated with elevated temperature, pain and cramps in the lower abdomen, and gradual weight loss. If the bleeding is severe, the patient will have anemia. If the symptoms are persistent and severe, anorexia, nausea, and occasional vomiting may follow, leading to weight loss and possible malnutrition, depending on when the patient seeks treatment. Rarely, the diseased bowel may distend and present as toxic megacolon.

Atypical presentations of UC can involve the skin, joints, eyes, or liver and may precede full-blown colonic involvement (see Chapter 145). In children, the presentation may simply be growth failure.

DIAGNOSIS

The clinical presentation of diarrhea and blood loss should result in a thorough evaluation of diarrhea, including stool cultures, stool for ova and parasites, and blood workup for anemia and infectious agents. Once this initial screening is accomplished, endoscopy is essential. The appropriate endoscopic and biopsy evaluations in UC can be obtained through sigmoidoscopy, but many gastroenterologists perform the examination with a colonoscope and obtain a full evaluation of the entire colon. However, this may not be necessary to make the diagnosis of UC, in which the lesion begins in the rectum and progresses into the left side of the colon.

Biopsy material reveals the classic lesions of nonspecific UC and can differentiate among infectious, ischemic, and possibly Crohn's colitis. The latter may be difficult to differentiate on mucosal biopsy. The advent of serologic markers as part of the clinical diagnostic armamentarium has been helpful. Finding an elevation of perinuclear antineutrophil cytoplasmic antibody (pANCA) is helpful diagnostically. Classic patterns have been described; the finding of anti–*Saccharomyces cerevisiae* antibody (ASCA) is helpful, and the combination of positive pANCA and negative ASCA findings has a positive predictive value for UC of 88% to 92%. New markers are being evaluated, but interpretations continue to be controversial.

Radiographic imaging can be helpful but often does not make the diagnosis of UC. Computed tomography reveals thickening of the bowel, and barium enema examination reveals classic spicules of the mucosa, strictures, or loss of haustral markings. Histologic diagnosis, however, is still necessary. Therefore, most clinicians choose colonoscopy over barium enema, although the latter is helpful in select cases. Barium enema and colonoscopy may be traumatic for patients with acute or toxic UC, so caution is warranted, and initial evaluation should be performed with gentle sigmoidoscopy.

TREATMENT AND MANAGEMENT

Numerous drugs are available to treat UC. Initial treatment usually starts with a variety of antiinflammatory agents, such as sulfasalazine (the former top choice) or mesalamine. When control is not obtained in a reasonable time, clinicians turn to corticosteroids or budesonide and then immunosuppressive agents, which include 6-mercaptopurine or azathioprine, methotrexate, cyclophosphamide, and most recently, anti–tissue necrosis factor. Some experts are being aggressive and treating with *biologicals* (e.g., infliximab) early in an attempt to induce longer-lasting remission. Newer biological drugs are also becoming available.

If all medical treatment fails, or the course is fulminant and unrelenting (e.g., severe bleeding, toxic megacolon), surgery is appropriate. Numerous subtleties are associated with UC, including its varied complications, occurrence during pregnancy, and malignant potential (see Chapters 146 and 147).

COURSE AND PROGNOSIS

The usual course of UC is benign in most patients, who are treated on an ambulatory basis with antiinflammatory or immunosuppressive agents. Research and university centers may use sophisticated indices to evaluate the course and end points of treatments. Most clinicians do not use these indices in daily practice. However, if a patient must be admitted for hospital

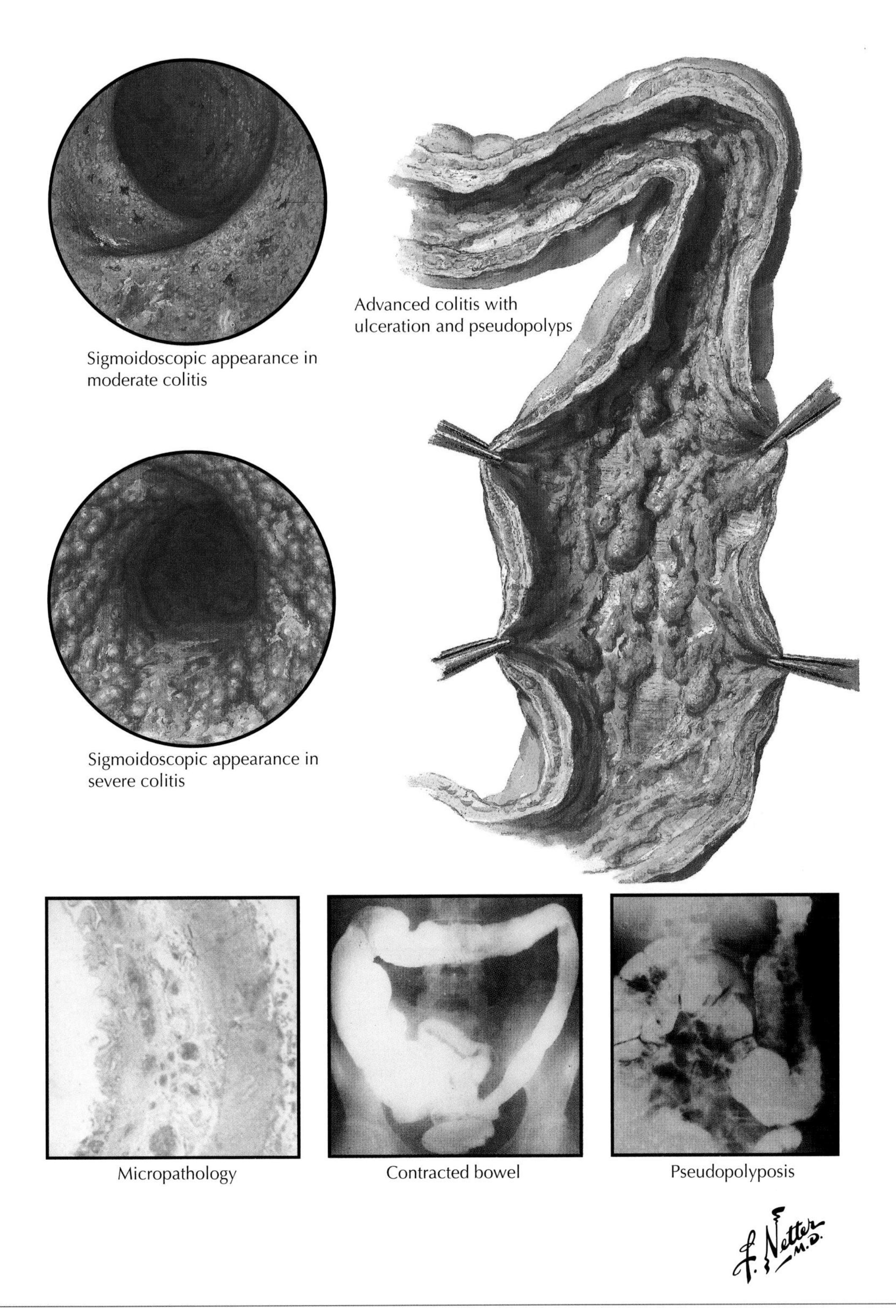

Figure 144-1 *Ulcerative Colitis: Sigmoidoscopic and Radiographic Appearance and Advanced Disease.*

care during the first year of the disease, the probability of further admissions and surgery increases. An estimated 30% of UC patients undergo surgery. Their lifetime incidence of colorectal cancer approaches 6%, and the cancer-related mortality rate is approximately 3%. Considering the incidence and prevalence of UC, these percentages are significant, and patients should be aware of all the possibilities in this chronic disease.

ADDITIONAL RESOURCES

Bonen DK, Cho JD: The genetics of inflammatory bowel disease, *Gastroenterology* 124:521-536, 2003.

Cohn RD: *Inflammatory bowel disease: diagnosis and therapeutics*, Totowa, NJ, 2003, Humana Press.

D'Haens G, Sanborn WJ, Faegan BG, et al: A review of activity indices and efficacy end points for clinical trials of medical therapy in adults with ulcerative colitis, *Gastroenterology* 132:763-786, 2007.

Ekbom A, Helmick C, Zack M, et al: Ulcerative colitis in colorectal cancer: a population-based study, *N Engl J Med* 323:1228-1232, 1990.

Ferrante M, Henckaerts L, Joossens M, et al: New serological markers in inflammatory bowel disease are associated with complicated disease behavior, *Gut* 56:1394-1403, 2007.

Peters M, Joossens S, Dermeire S, et al: Diagnostic value of anti–*Saccharomyces cerevisiae* and antineutrophilic cytoplasmic autoantibodies in inflammatory bowel disease, *Am J Gastroenterol* 96:730-734, 2001.

Podolsky BK: Inflammatory bowel disease, *N Engl J Med* 347:417-429, 2002.

Rutgeerts P, Sanborn WJ, Feagin BG, et al: Infliximab for the induction and maintenance therapy for ulcerative colitis, *N Engl J Med* 353:2462-2476, 2005.

Tanaka M, Riddell RH, Salto H, et al: Morphologic criteria applicable to biopsy specimens for effective distinction of inflammatory bowel disease from other forms of colitis and of Crohn's disease from ulcerative colitis, *Scand J Gastroenterol* 34:55-67, 1999.

Vermeire S, Assche GV, Rutgeerts P: Laboratory markers in IBD: useful, magic, or unnecessary toys? *Gut* 55:426-431, 2006.

Ulcerative Colitis: Extraintestinal Manifestations and Complications

145

Martin H. Floch

Extraintestinal manifestations of ulcerative colitis (UC) are numerous, and the complications can be severe, with significant morbidity and mortality (**Fig. 145-1**). This chapter lists all the complications, but detailed discussions of each may be found throughout this text and in the literature.

EXTRAINTESTINAL MANIFESTATIONS

An epidemiologic study of more than 1000 patients with UC reveals that the overall prevalence of extraintestinal manifestations is 21%, and that incidence increases with increased severity of illness. More than one extracolonic manifestation occurred in approximately 25% of patients. The cause of the extracolonic manifestations is unknown; **Box 145-1** lists the sites most often involved. However, the literature includes reports of scattered cases in the bronchopulmonary, renal and genitourinary, cardiac, endocrine, and neurologic organs.

The clinical presentation of each of these extracolonic manifestations is typical of the disease without UC. Manifestation may occur without active disease, but it occurs more frequently when disease activity is increased. Therefore, it is imperative that the disease be controlled and treated to ameliorate the extraintestinal manifestations.

Pyoderma gangrenosum is a classic manifestation for which intensive therapy may be necessary. Treatment of the pyoderma may require corticosteroids, which would not be used for the disease itself. Once again, each manifestation must be treated individually, in accordance with the experience readily available in the literature. Biological therapy is now being used more frequently in an attempt to control the symptoms rapidly when steroids are not immediately effective.

Importantly, disease manifestation may change. The best example is *sclerosing cholangitis*, which may occur before UC develops. It is hypothesized that this is part of an autoimmune process relating to the disturbance of the immunologic response in UC patients. Sclerosing cholangitis may progress to liver failure and liver complications, resulting in liver transplantation. Patients with UC are able to tolerate liver transplantation. UC appears to come under control with the combined treatments and therapies involved during transplantation, and patients may do well with their inflammatory bowel disease (IBD) and are able to sustain transplantation and recover. Reports indicate that as many as 4% of IBD patients develop sclerosing cholangitis.

An increased incidence of *osteoporosis* is also reported in IBD patients. Treatment with steroids and some immunosuppressive drugs aggravates osteoporosis. Therefore, patients must be diligently placed on calcium and vitamin D intake and any necessary bone-stimulatory drugs.

MAJOR COMPLICATIONS

Major complications of chronic UC are massive bleeding, perforation with or without dilatation, carcinoma, and perianal

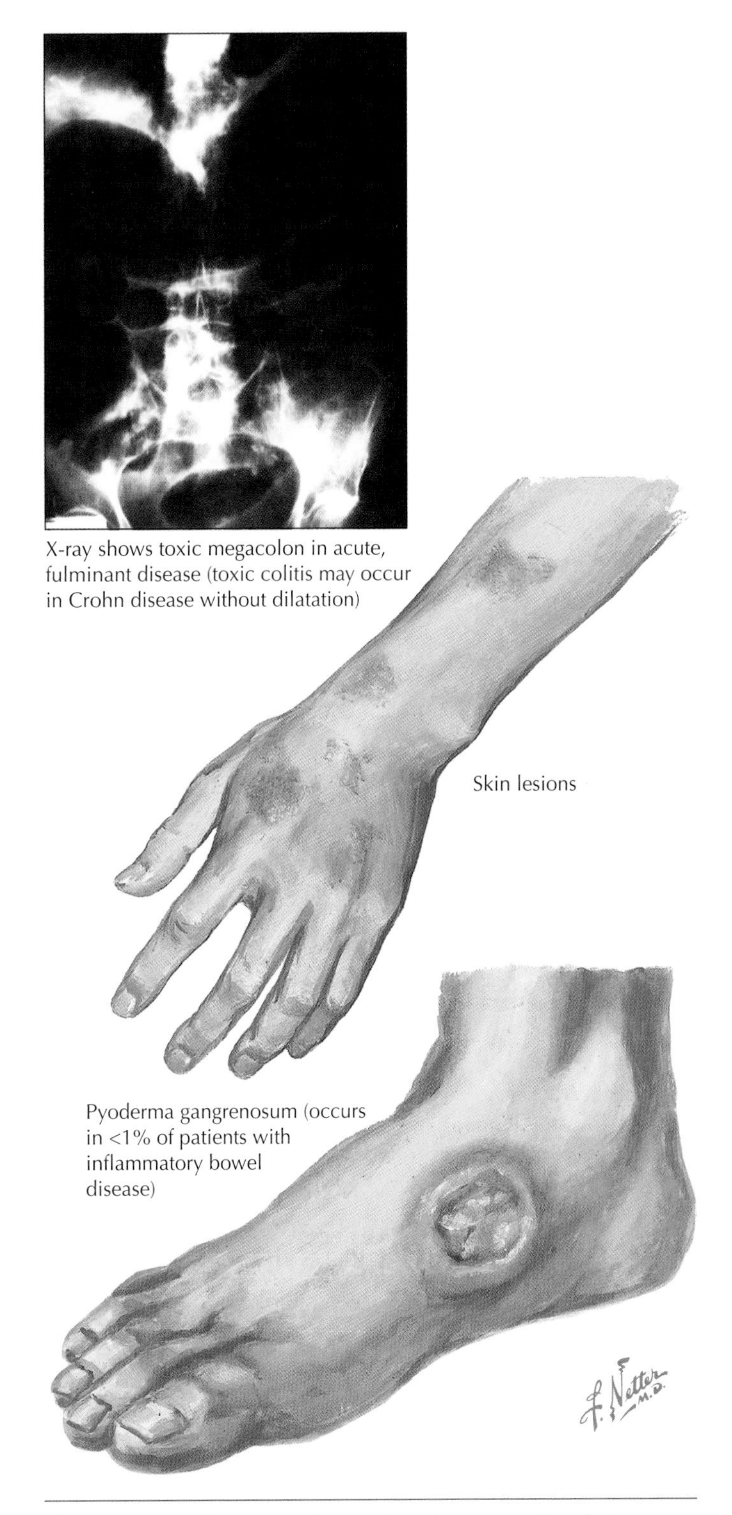

Figure 145-1 *Ulcerative Colitis: Extraintestinal Manifestations and Complications.*

Box 145-1 Extraintestinal Sites of Ulcerative Colitis Involvement

Joints (7%)*
Arthritis
Sacroiliitis
Ankylosing spondylitis
Hypertrophic osteoarthropathy
Osteoporosis/osteomalacia
Granulomatous synovitis
Rheumatoid arthritis
Osteonecrosis
Steroid-induced myopathy

Skin (2.6%)
Erythema nodosum
Pyoderma gangrenosum
Psoriasis

Eyes (1.6%)
Uveitis/iritis
Episcleritis
Chorioretinitis
Retinal vascular disease

Hepatobiliary (11%)
Primary sclerosing cholangitis
Steatosis
Cholelithiasis
Cholangiocarcinoma
Autoimmune hepatitis
Pericholangitis

**Percentage of UC patients affected.*

disease. Fortunately, *massive hemorrhaging* from the bowel is not as frequent as once believed. However, when a massive hemorrhage does occur and bleeding does not stop with standard intravenous therapy, colectomy may be necessary. Slow bleeding with its attendant anemia is more prevalent. Iron-deficiency anemia is most common and is treated with iron replacement. Deficiencies of other vitamins or of folic acid that occur with sulfasalazine therapy are treated with replacement of those vitamins.

Free perforation usually complicates toxic megacolon; therefore, toxic dilated bowel requires immediate surgical intervention when medical therapy fails. A colonic diameter greater than 6 to 7 cm may indicate toxic colon. When the patient is febrile, has abdominal distention, and has a high pulse rate or a greatly elevated white blood cell (WBC) count, emergency surgery may be required. Actually, the patient's condition should be stabilized and medical treatment attempted, but careful observation is necessary to ensure that perforation does not occur, because mortality rates can be as high as 40% to 50% with perforation. Surprisingly, toxic megacolon, which was thought to result from drug therapy (e.g., opiates, anticholinergics), seems to occur most often in first episodes of colitis. Free perforation may result from small lesions, which require good surgical support with appropriate surgical intervention as each case indicates. Again, morbidity and mortality are high.

Strictures occur in chronic UC, and most are short (<2-5 cm). Once a stricture is identified, the clinician should investigate whether a lesion is local (e.g., from scarring) or is caused by malignant infiltration. Because stent placement may be attempted, obtaining a specimen through biopsy or resection of the area is recommended, so as not to overlook a malignancy.

Chapter 146 discusses cancer and the role of surveillance and surgical resection. Importantly, however, the incidence of cancer increases dramatically after patients have had UC for more than 10 years. Careful surveillance and surgical intervention are required whenever a severely dysplastic lesion is suspected.

Perianal disease occurs frequently in UC patients. The main problem related to UC is irritation of the anal ring with resultant cryptitis, fissures, and perianal irritation. It is essential to control the diarrhea to control the irritation of the anal ring. After the disease is controlled, perianal disease usually subsides. Treatment of perianal disease is discussed in Chapters 162, 164, and 165.

Numerous extraintestinal manifestations and disease complications affect the quality of life of patients with UC. Although many are not life threatening, the burden is great on these patients. Reassurance and frequently psychotherapy are needed for them to cope with the recurrent manifestations of the disease. When the quality of life is seriously affected, colectomy should be considered. However, complications or extraintestinal manifestations such as sclerosing cholangitis can occur even after total colectomy. Fortunately, the rate of extraintestinal manifestations and complications is less than 50%, and those unaffected can lead normal lives.

Abscesses and fistulae are much more common in Crohn's disease (20%-40%) than in nonspecific UC (see Chapters 151-153).

ADDITIONAL RESOURCES

Barrie A, Regueiro M: Biologic therapy in the management of extraintestinal manifestations of inflammatory bowel disease, *Inflamm Bowel Dis* 13:1424-1429, 2007.

Greenstein AJ, Janowitz HD, Sachar DB: The extraintestinal complications of Crohn's disease and ulcerative colitis: a study of 700 patients, *Medicine* 55:401-411, 1976.

Monsen U, Sorstad J, Hellers G, Johansson C: Extracolonic diagnosis in ulcerative colitis: an epidemiologic study, *Am J Gastroenterol* 85:711-716, 1990.

Olsson R, Danielsson A, Jarnerot G, et al: Prevalence of primary sclerosing cholangitis in patients with ulcerative colitis, *Gastroenterology* 100:1319-1323, 1991.

Su C, Lichtenstein GR: Ulcerative colitis. In Feldman M, Friedman LS, Brandt LJ, editors: *Gastrointestinal and liver disease*, ed 8, Philadelphia, 2006, Saunders-Elsevier, pp 2499-2548.

Ulcerative Colitis: Histologic Diagnosis and Dysplasia

146

Martin H. Floch

Histologic diagnosis of ulcerative colitis (UC) or inflammatory bowel disease (IBD) is imprecise and often difficult to make (**Fig. 146-1**). In addition, the factors that differentiate UC from Crohn's disease are often not evident through simple mucosal biopsy. However, when accepted criteria are present, a diagnosis can be made. The following factors define UC:

- *Distorted mucosal architecture*, as evidenced by crypt distortion with or without crypt atrophy. In severe cases, mucosal erosions and ulcerations can be seen.
- *Inflammatory cell infiltrate in the lamina propria*, which is usually lymphoplasmacytic in what is described as the "chronic" form and neutrophilic or eosinophilic in the "acute" form. So-called crypt abscesses are usually described in the acute form and contain neutrophils (see **Fig. 146-1**). The intensity of the white blood cell (WBC) infiltrate ranges from mild to severe. An often-described, ambiguous feature is an excess of intraepithelial neutrophils or lymphocytes, in which the infiltrate in the lamina propria is not severe. *Basal plasmocytosis*, when at least three plasma cells are found below the crypt area, may be a significant finding.
- *Mucin depletion, distal Paneth cell metaplasia*, and areas of *intense eosinophilia* have been described in UC. Detailed analysis of clinical status with histologic findings has revealed that crypt atrophy, crypt distortion, basal plasmocytosis, and severe mononuclear cell infiltration, with Paneth cell metaplasia distal to the hepatic flexure, were significant findings in UC, with sensitivity greater than 99% and specificity greater than 97%. In addition, the presence of epithelioid granulomas revealed a sensitivity of 86% to 94% and specificity of 97% to 100% for Crohn's disease.

Other colitides that can be diagnosed by mucosal biopsy may confuse the differential diagnosis of UC, but usually these are distinguished by unique characteristics. *Infectious colitis* causes an acute inflammatory response, but rarely are crypt areas distorted. *Chronic ischemic colitis* can cause ulceration similar to UC, but also usually shows lamina propria fibrosis, and the classic inflammatory response is minimal or absent. *Chronic radiation disease* and *chronic graft-versus-host disease* may elicit similar responses. However, the history in these patients is significant. Nonsteroidal antiinflammatory drugs (NSAIDs) have been associated with acute or chronic colitis, along with microscopic, lymphocytic, and collagenous colitis, but rarely with crypt architectural disturbance (see Chapter 155).

Numerous eosinophils in the clinical and endoscopic findings of UC may further confuse the diagnosis. *Eosinophilic gastroenteritis* may be included in the differential diagnosis (see Chapter 117). In many UC patients, the biopsy material shows many eosinophils; degree and distribution may be confusing. Classically, eosinophils represent a response to allergic or parasitic disease. However, their presence in UC is of uncertain significance, and no clinical distinguishing features exist to necessitate different treatment.

DYSPLASIA

The presence of dysplasia in mucosal biopsy material from patients with UC is classified as high, low, or indeterminate. Studies reveal that pathologists' interpretations vary greatly, as do reports among institutions. Although it is difficult to differentiate the three categories, certain criteria can establish a category. The clinician must emphasize the importance of a correct diagnosis to the pathologist interpreting the specimens.

Low-grade dysplasia can be defined as "nuclear crowding," with some stratification of nuclear pleomorphism and hyperchromasia. The glandular architecture is minimally abnormal. *High-grade dysplasia* has more marked nuclear pleomorphism, hypochromasia, and stratification, and there may or may not be architectural abnormalities such as villi-formed surfaces. As seen in Figure 146-1, it may be difficult and completely arbitrary when uncertainty exists in defining low-grade and high-grade dysplasia. Whenever there is marked inflammation and apparently significant reparatory changes, it may be difficult or impossible to distinguish reactive changes from true dysplasia. In these patients, a diagnosis of *indefinite dysplasia* is made.

Dysplasia often resembles adenoma; it may be impossible to distinguish them. So-called sporadic adenomas may develop in areas of UC and thus may enter into the clinical decision process. If a sporadic adenoma develops in an area distant from active UC, such as in the right colon, and if the colitis is present only in the left colon, the sporadic adenoma can be considered as truly benign. Patients with dysplastic lesions appear to have UC longer (>10 years) and are more likely to have active disease.

DYSPLASIA AND MONITORING IN INFLAMMATORY BOWEL DISEASE

It is now apparent that there are molecular factors for colitis-associated colon cancer (see **Fig. 146-1**). The bowel with no dysplasia presumably goes through a formation process of indefinite dysplasia, low-grade dysplasia, high-grade dysplasia, and then carcinoma. *Aneuploidy* (altered DNA content) and microsatellite instability develop, and tumor-suppressor genes such as *APC* and *p53* become altered. Other changes in genetic regulation, such as induction of the K-*ras* oncogene, and loss of other functioning tumor-suppressor genes, in addition to the *p53* gene function, appear to permit adenoma to progress to carcinoma.

Therefore, the question is whether careful surveillance can prevent cancer in IBD patients. Most pathologists and clinical gastroenterologists conclude that the increased incidence of colorectal cancer in UC patients warrants careful monitoring

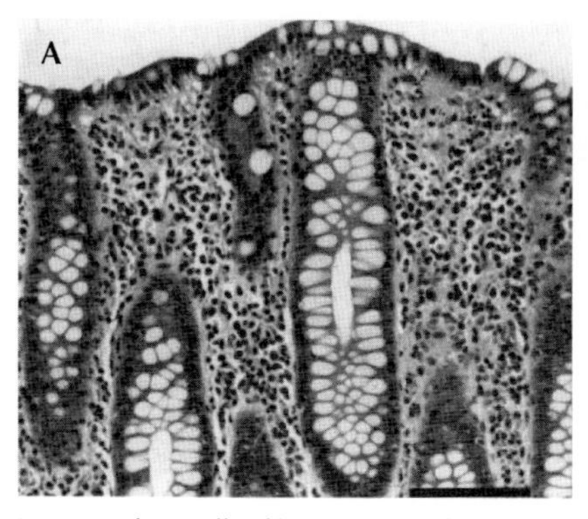

Mononuclear cell infiltration in the lamina propria. (A) Upper limit of "mild" mononuclear cell infiltration.

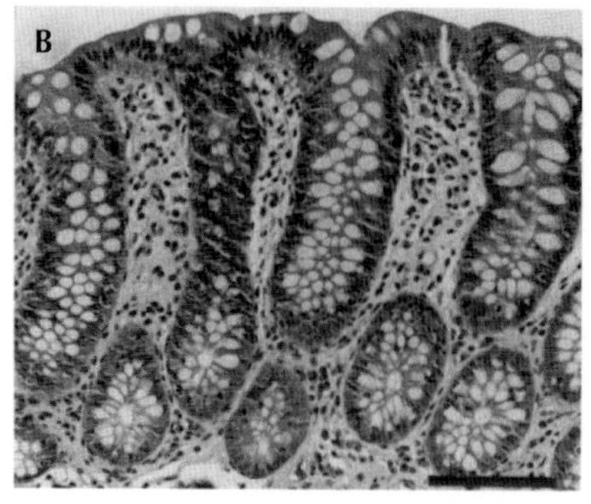

(B) Upper limit of "minimal" mononuclear cell infiltration.

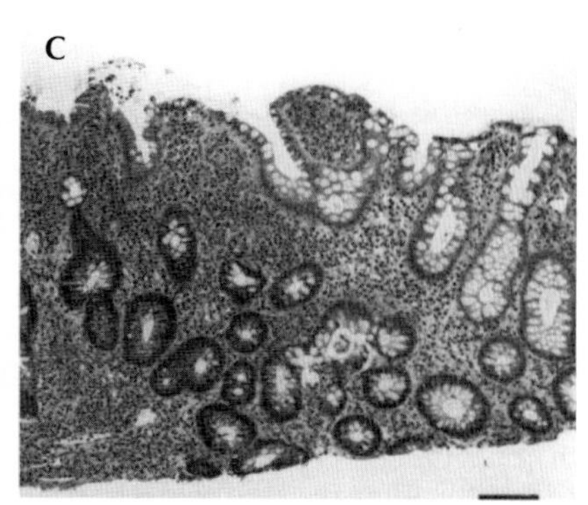

(C) Borderline figure narrowly judged as "focal" mononuclear cell infiltration. (Hematoxylin and eosin; bar = 100 μm.)

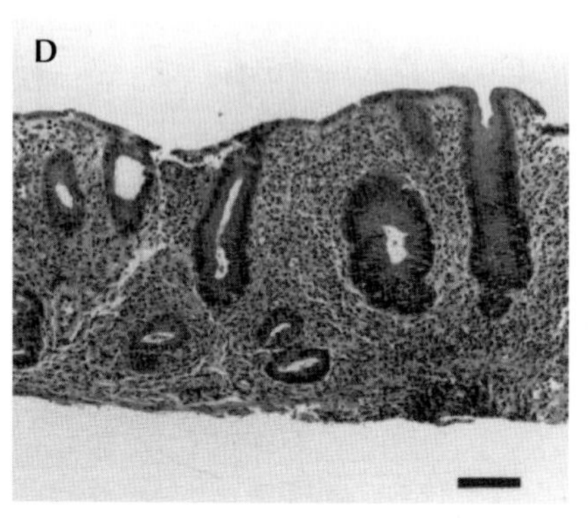

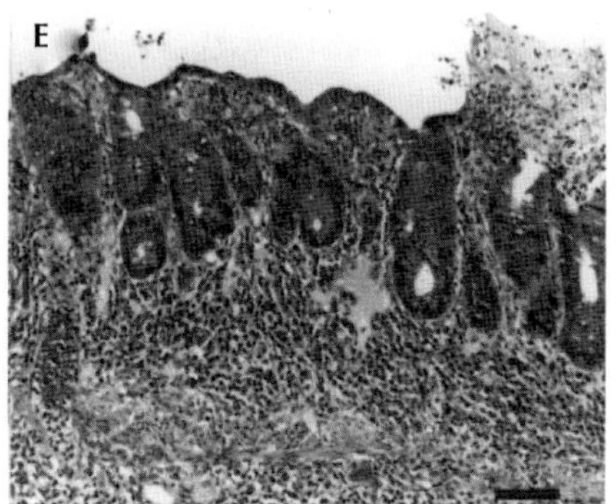

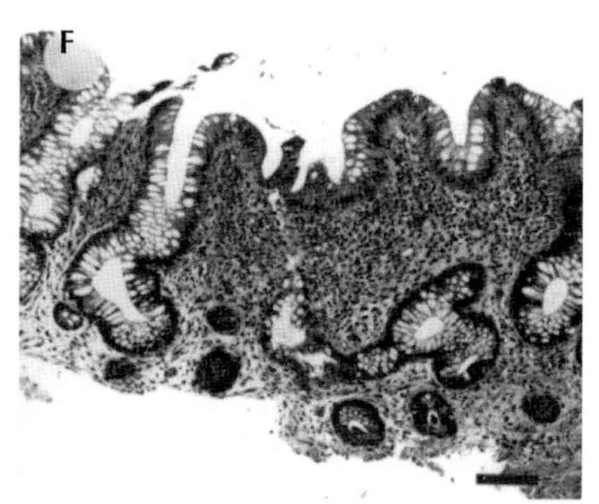

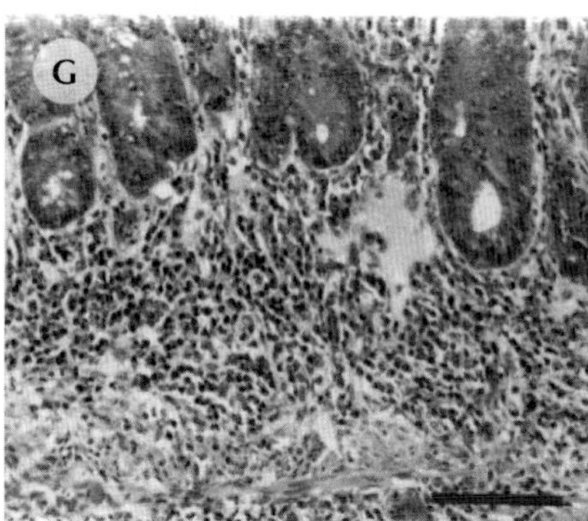

Crypt architectural abnormalities and basal plasmacytosis (**D** and **E**). Typical figures judged as "presence" of crypt atrophy, which was recognized by generally increased distance of more than one crypt diameter between crypts (**D**), or a general increase in the distance between crypts and the muscularis mucosae (**E**). (**F**) Typical figure judged as "presence" of crypt distortion, which was recognized by branched crypts with nonparallelism. (**G**) Typical basal plasmacytosis. (Hematoxylin and eosin; bar = 100 μm.)

Figure 146-1 *Ulcerative Colitis: Histologic Diagnosis and Dysplasia.* Schematic modified from Itzkowitz S: *J Clin Gastroenterol* 36:S70-S74, 2003; histologic images from Tanaka M, Riddell RH, Salto H, et al: *Scand J Gastroenterol* 34:55-67, 1999.

and surveillance. If no dysplasia or adenoma formation is seen on first colonoscopy, it is safe to monitor the patient every 2 years. Most accept that because the incidence of carcinoma increases after UC is present for 10 years, surveillance should be considered annually. Although no random clinical trials have been conducted, several cohort and case series indicate that monitoring does have a role in preventing cancer. Surveillance colonoscopy includes thorough evaluation of the colon, with specimens for mucosal biopsy taken every 5 to 10 cm. Physicians disagree on the number of biopsy specimens needed, but most agree at least 10 areas should be sampled. Controversy surrounds whether low-grade dysplasia is an indication for surgery, but most agree that the presence of high-grade dysplasia, or adenoma within active disease, warrants total colectomy.

ADDITIONAL RESOURCES

Collins PD, Mpofu C, Watson AJ, Rhodes JM: Strategies for detecting colon cancer or dysplasia in patients with inflammatory bowel disease, *Cochrane Database Syst Rev* 19:CD000279, 2006.

Goldblum JR: The histologic diagnosis of dysplasia, dysplasia-associated lesion or mass, and adenoma, *J Clin Gastroenterol* 36:S63-S69, 2003.

Hart J: Pathologic features of inflammatory bowel disease. In Cohen RD, editor: *Inflammatory bowel disease,* Totowa, NJ, 2003, Humana Press, pp 327-350.

Itzkowitz S: Colon carcinogenesis in inflammatory bowel disease: applying molecular genetics to clinical practice, *J Clin Gastroenterol* 36:S70-S74, 2003.

Loftus ED Jr: Does monitoring prevent cancer in inflammatory bowel disease? *J Clin Gastroenterol* 36:S79–S83, 2003.

McKenna BJ, Appelman HD: Dysplasia of the gut: the diagnosis is harder than it seems, *J Clin Gastroenterol* 34:111-116, 2002.

Tanaka M, Riddell RH, Saito H, et al: Morphologic criteria applicable to biopsy specimens for effective distinction of inflammatory bowel disease from other forms of colitis, and of Crohn's disease from ulcerative colitis, *Scand J Gastroenterol* 34:55-67, 1999.

Vleggaar FP, Lutgens MW, Claessen MM: The relevance of surveillance endoscopy in long-lasting inflammatory bowel disease (review), *Aliment Pharmacol Ther* 26(suppl 2):47-52, 2007.

Yang GU, West AB: What do eosinophils tell us in biopsies of patients with inflammatory bowel disease? *J Clin Gastroenterol* 36:93-102, 2003.

Ulcerative Colitis: Surgical Treatment

147

Martin H. Floch

The American College of Gastroenterology Practice Guidelines on treatment of ulcerative colitis (UC) states, "Absolute indications for surgery are exsanguinating hemorrhage, perforation, and documented or strongly suspected carcinoma. Other indications for surgery are (1) severe colitis with or without toxic megacolon, unresponsive to conventional maximal medical therapy, and (2) the patient with less severe, but medically intractable symptoms or intolerable steroid side effects."

Although indications seem clear in clinical practice, one unclear area is the *suspicion of cancer*. Some clinicians believe that low-grade dysplasia is an indication for colectomy, and most believe that high-grade dysplasia is a clear indication for total colectomy in UC patients.

CLINICAL PICTURE

Eventually, the patient who is undergoing extensive medical therapy for UC and who requires large doses of steroids or medication, whether continuous or intermittent, must consider surgery. The choice is frequently up to the patient. Some patients do not want to deal with the lifestyle-limiting effects of a chronic disorder and choose surgery, whereas others fear surgery. However, surgery can relieve the symptoms of UC and cure the disease.

DIAGNOSIS

The clinician must be certain of the diagnosis when referring a patient with UC for surgery. The extent of the disease is clearly evaluated on colonoscopy, and histologic and biopsy specimens can make a definitive diagnosis (see Chapter 146). Unfortunately, as many as 20% to 50% of patients have indeterminate diagnoses. The surgeon cannot be sure whether these patients have Crohn's colitis or UC; the type of surgery to perform then becomes debatable. Serologic examination may be 97% to 100% specific for Crohn's colitis and 80% to 90% specific for UC. Therefore, serologic findings that demonstrate high anti–*Saccharomyces cerevisiae* antibody (ASCA) level raise the index of suspicion that the patient has Crohn's colitis (see Chapter 144). The small bowel should be evaluated if time permits. If the patient has acute hemorrhage or acute perforation, a two-stage procedure should be performed to give the patient an ileostomy, with an ileal pouch inserted later. All physicians involved in patient evaluation should determine whether they are dealing with nonspecific UC or Crohn's colitis.

TREATMENT AND MANAGEMENT

It is estimated that approximately 30% of UC patients will require surgery. During the past decade, ileal pouch anastomosis has become the treatment of choice. However, some patients do not have the appropriate body build for a pouch, or they may have a relative contraindication. For these patients, total colectomy with ileostomy is the treatment of choice (see Chapters 148 and 149). With either procedure, the patient will be free of UC unless it is a variant form or is indeterminate, or unless Crohn's ileitis develops after surgery. Ileostomy is a safe procedure, and long-term analysis now reveals that ileoanal pouches are safe, although some patients do experience pouchitis. Surgery may be performed even in patients with sclerosing cholangitis.

Figure 147-1 demonstrates the ileostomy site and the removal of the rectum with closure of the perianal area. This is usually a two-stage procedure. After the ileostomy is created, the patient is given several months to recover, and then the rectum is removed and the perineum closed. This procedure is associated with less sexual dysfunction than other procedures and does not hinder female reproduction. Details of ileal pouch anal anastomosis are described in Chapter 149.

ADDITIONAL RESOURCES

Cima RR, Pemberton JH: Ileostomy, colostomy, and pouches. In Feldman M, Friedman LS, Brandt LJ, editors: *Gastrointestinal and liver disease*, ed 8, Philadelphia, 2006, Saunders-Elsevier, pp 2549-2562.

Fazio WV, Ziv Y, Church JM, et al: Ileal pouch anal anastomoses: complications and function in 1005 patients, *Ann Surg* 222:120-126, 1995.

Ferrante M, Declerck S, De Hertogh G, et al: Outcome after proctocolectomy with ileal pouch–anal anastomosis for ulcerative colitis, *Inflamm Bowel Dis* 14:20-28, 2008.

Kornbluth A, Sachar DB: Ulcerative colitis practice guidelines in adults (update). American College of Gastroenterology, Practice Parameters Committee, *Am J Gastroenterol* 99:1371-1385, 2004.

Olsen KO, Juul S, Berndtsson I, et al: Ulcerative colitis: female fecundity before diagnosis, during disease, and after surgery compared with a population sample, *Gastroenterology* 122:15-19, 2002.

Poritz LS, Koltun WA: Surgical management of ulcerative colitis in the presence of primary sclerosing cholangitis, *Dis Col Rect* 46:173-178, 2003.

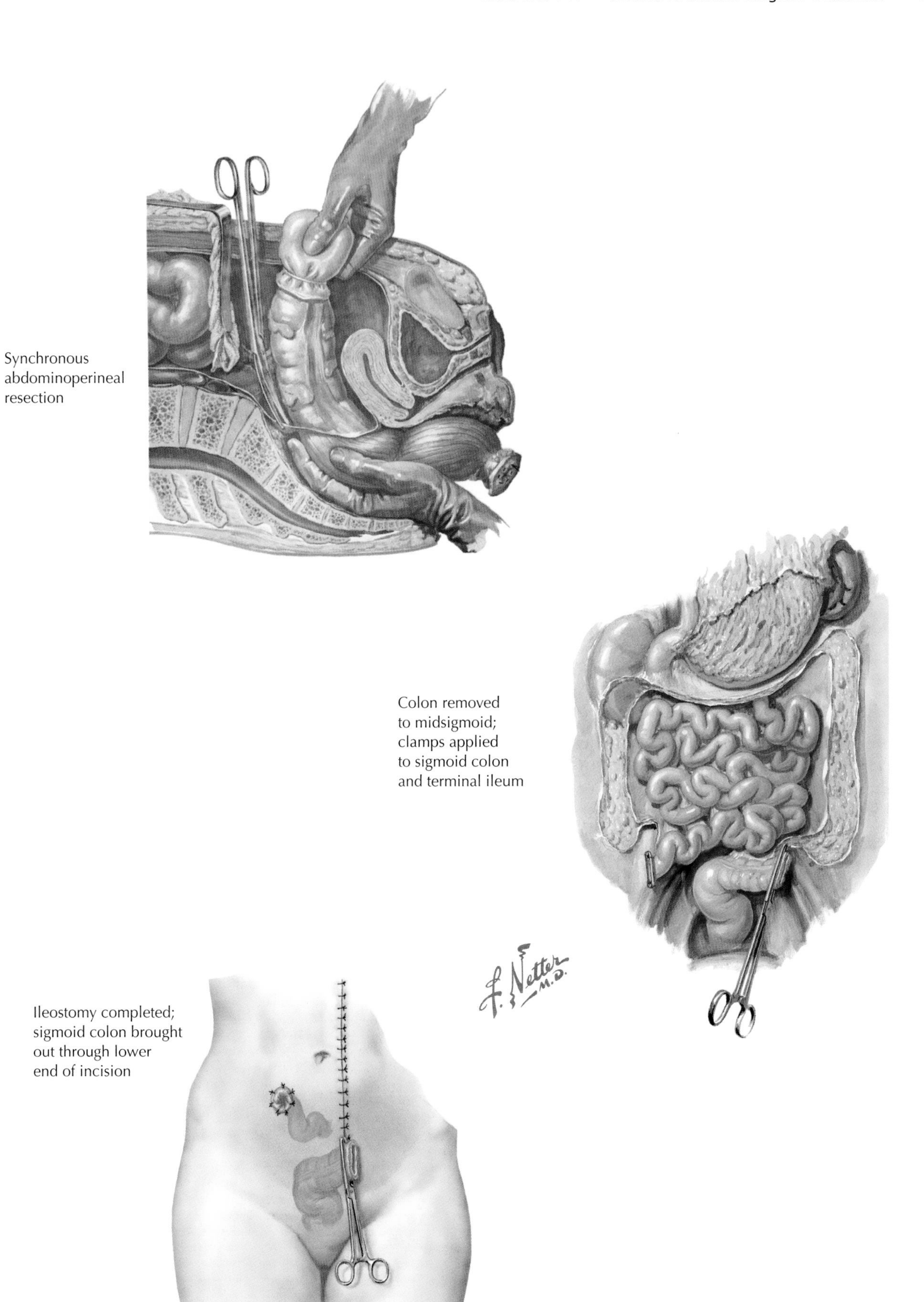

Figure 147-1 *Ulcerative Colitis: Surgical Treatment.*

Ileostomy and Colostomy

148

Martin H. Floch

Colostomies are performed for numerous reasons, including obstructive lesions caused by cancer, severe diverticulitis, severe intractable constipation, and trauma. Ileostomies are performed for the same reasons, but permanent ileostomy is reserved almost exclusively for patients with ulcerative colitis and diffuse polyposis of the colon. Although ileal pouch anal anastomosis (IPAA) is now the preferred procedure (see Chapter 149), some patients may not be surgical candidates for IPAA or may not be able to tolerate the pouch and must be converted to an ileostomy. Both ileostomies and colostomies may be permanent or temporary, depending on the clinical indication.

Because it involves the removal of the colon, *ileostomy* often results in more complications than colostomy. Colostomy excreta usually have a free or an intermittent flow that may be facilitated by simple irrigation. Ileostomy effluent has a free flow that should range from 800 to 1000 mL/day, depending on the diet. Ileostomy effluent may be dramatically increased in patients with gastroenteritis. Complications of ileostomies include malfunctioning, prestomal ileitis, irritation of the peristomal skin, and obstruction.

CLINICAL PICTURE

The carefully placed ileostomy is usually situated in the right lower quadrant. Most patients become well versed in taking care of the site. **Figure 148-1** demonstrates the method of placing an appliance over the fistula. Most patients become comfortable with it and empty the appliance several times daily.

DIAGNOSIS

The diagnosis of an obstructive lesion is the same as for any intestinal obstruction. After ileostomy, however, a patient may have an early obstruction, and the ileal discharge first may increase because of bowel dilatation and increased intestinal secretions. Examination of the stoma with a small finger or an endoscope can reveal the dilated ileum, and radiographic studies can confirm the findings. Surgery is indicated to remove any obstruction.

Prestomal ileitis fortunately occurs rarely, but patients with this condition may also exhibit features of mechanical obstruction. They may become dehydrated or at times anemic. Ileal mucosa may demonstrate ulcerations. It is thought that prestomal ileitis is often a recurrence of Crohn's disease. However, it may be difficult to determine whether there is an obstruction at the stoma site or whether Crohn's disease has recurred.

TREATMENT AND MANAGEMENT

When an obstruction or prestomal ileitis is suspected, endoscopy and medicosurgical intervention are indicated, depending on the cause. If the condition appears to be inflammatory, medical treatment should help resolve the problem. However, if it is mechanical, surgical intervention is necessary, and resection of the stoma must be considered.

Care of the peristomal area is a routine part of postoperative nursing. Most institutions have a specialized nurse to educate patients. Emollient salves and skin care are administered for lesions.

Most people who undergo ileostomy can lead normal lives. Occasionally, it is necessary to revise a stoma. Most surveys indicate that approximately 90% of patients are satisfied after ileostomy, but some require a revision to the procedure. Patients learn to change their diets and to limit their intake of foods not digested by human enzymes, such as high-fiber foods of insoluble fiber, including fruits and grains. Most patients learn to chew their food thoroughly so that it does not obstruct the stoma. However, every clinician has occasionally treated a patient who has swallowed unchewed nuts that caused obstruction. Patients learn what to do and what not to do.

COURSE AND PROGNOSIS

The major disease for which the ileostomy or colostomy is performed dictates the prognosis. If the disease is malignant, it may result in a poor outcome. If the disease is benign or is caused by trauma, the patient can lead a normal life. Colostomy presents less risk for dehydration if gastroenteritis occurs. Ileostomy patients must be monitored carefully. If severe enteritis occurs and the ileostomy effluent is not controlled by simple medication, patients frequently need intravenous replacement of fluid and electrolytes.

ADDITIONAL RESOURCES

Cataldo P, MacKeigen JM: *Intestinal stomas: principles, techniques, and management*, ed 2, New York, 2004, Informa Healthcare.

Cima RR, Pemberton JH: Ileostomy, colostomy, and pouches. In Feldman M, Friedman LS, Brandt LJ, editors: *Gastrointestinal and liver disease*, ed 8, Philadelphia, 2006, Saunders-Elsevier, pp 2549-2559.

Evans JP, Brown MH, Wilkes GH, et al: Revising the troublesome stoma: combined abdominal wall recontouring and revision of stomas, *Dis Col Rect* 46:122-126, 2003.

Fulham J: Providing dietary advice for the individual with a stoma, *Br J Nurs* 17: S22-S27, 2008.

Gordon P, Nivatvonas S: *Principles and practices of surgery: colon, rectum and anus*, 2007, New York, Informa Healthcare.

Roy PH, Sauer WG, Beahrs OH, Farrow GN: Experience with ileostomies: evaluation of long-term rehabilitation in 497 patients, *Am J Surg* 119:77-84, 1970.

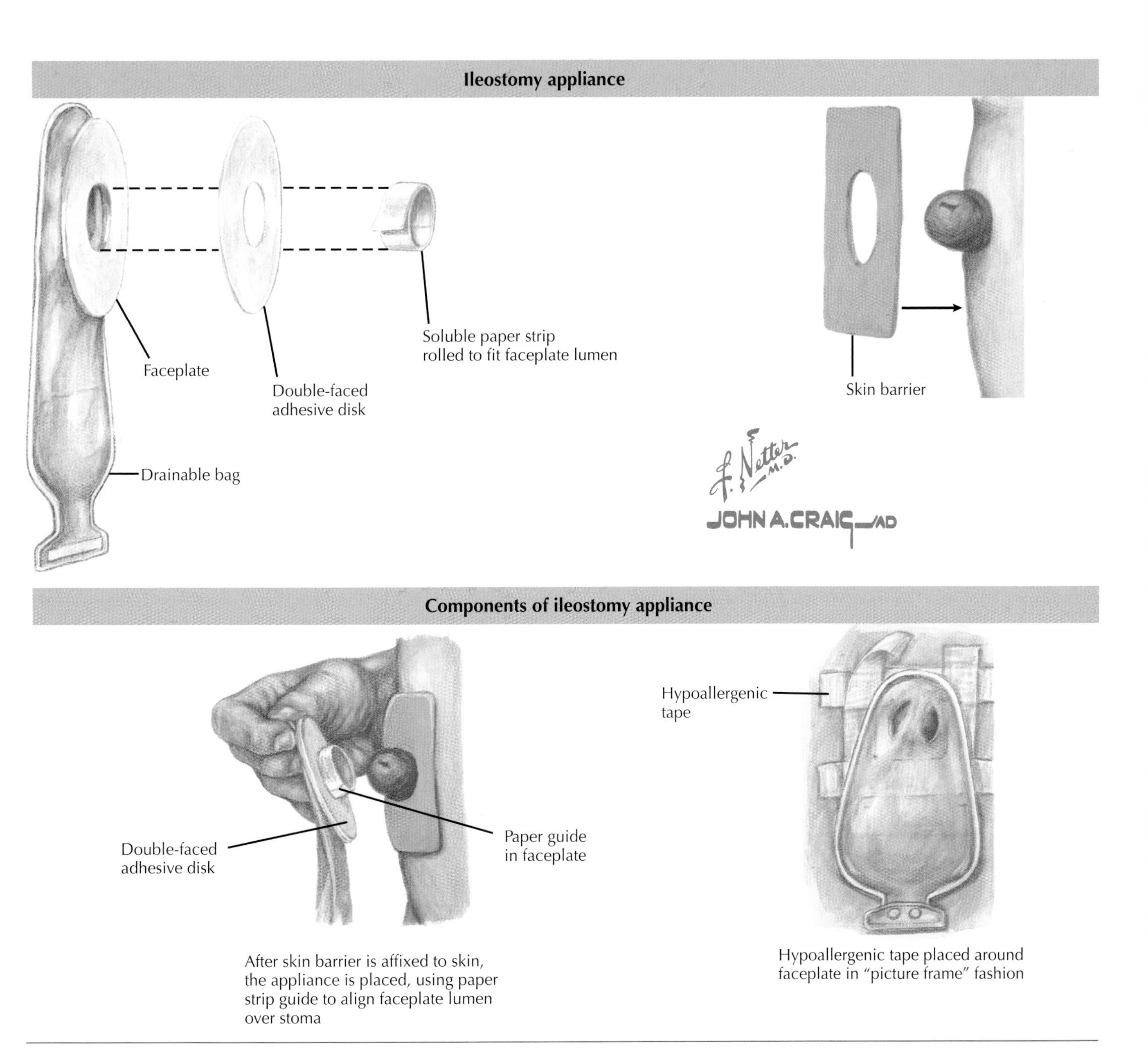

Figure 148-1 *Ileostomy and Colostomy.*

Ileal Pouch Anal Anastomosis and Pouchitis

149

Martin H. Floch

The surgical treatment of choice for patients requiring colectomy for ulcerative colitis or familial adenomatous polyposis (FAP) is *restorative proctocolectomy* with ileal pouch anal anastomosis (IPAA). The "pouch" is an ileal reservoir that acts as a neorectum to store fecal material and to avoid constant output from the ileostomy. Anastomosis of the ileal-to-rectal mucosa is controversial, but most surgeons prefer *mucosectomy*, bringing the ileal mucosa to the anal skin line. This is best accomplished with suturing, but some prefer stapling.

There are three types of pouches—J, S, and W—but the J pouch has become the most popular (**Fig. 149-1**). Unfortunately, as many as 40% of patients who undergo this procedure develop at least one episode of *pouchitis*, and 60% of these patients may have recurrent disease. Only 15% of patients with pouchitis have chronic disease that requires maintenance therapy.

Patients with IPAA to manage FAP rarely experience pouchitis. Furthermore, pouchitis appears to occur more frequently in patients with extraintestinal manifestations of colitis. Patients with primary *sclerosing cholangitis* have the highest prevalence of pouchitis.

CLINICAL PICTURE

Pouchitis causes diarrhea, lower abdominal pain, and bloody discharge. Patients may have fever, associated malaise, and nausea. Some lose weight or have night sweats or extraintestinal arthritis. When patients are routinely observed through sigmoidoscopy, some are found to have active pouchitis but few or no symptoms.

DIAGNOSIS

Because pouchitis consists of a triad of variable clinical (history), endoscopic, and histologic criteria, many clinicians use a *pouchitis disease activity index* consisting of a series of points assigned to the criteria (**Table 149-1**). Scores higher than 7 points indicate pouchitis. Clinicians also use this index to monitor the progress of therapy.

Positive histologic findings are necessary for a diagnosis of pouchitis, but this is controversial among pathologists, and most patients with pouches have low-grade inflammation. Therefore, the pouchitis activity index is helpful. Severe cases, determined by histologic findings, involve acute inflammatory infiltrates with crypt abscesses and ulceration. Most clinicians consider a diagnosis of pouchitis unwarranted if the inflammation and ulceration are found only through routine sigmoidoscopy, or if the patient does not have clinical symptoms or an increased number of stools or rectal bleeding. Therefore, the history (clinical) component is important for making the diagnosis. The Cleveland Clinic group defines *irritable pouch syndrome* as disease activity index less than 7 and little evidence of pouchitis other than its symptoms.

TREATMENT AND MANAGEMENT

Again, 60% of patients with pouchitis have recurrent attacks, and as many as 15% have significant long-term problems.

The primary treatment for pouchitis has been *metronidazole*. Initial studies demonstrated a change in the bacterial flora, with an increase in aerobic organisms and a decrease in anaerobes, as well as a decrease in the normal probiotic organisms. Metronidazole was excellent in inducing remissions, but symptoms recurred. Many clinicians used broad-spectrum coverage of ciprofloxacin and ampicillin if unable to use metronidazole, or if its use might have caused complications. Some clinicians have used other antiinflammatory agents, such as mesalamine, and some have had to use steroid enemas or steroids systemically to control symptoms.

The use of *probiotic agents* has been promising; one study reported almost 100% induction of remission and maintenance of the asymptomatic state. However, therapy requires large doses of VSL#3 daily, consisting of eight probiotic organisms. Studies over several years have proved the effectiveness of this therapy, but long-term evaluations are still necessary.

Previous data analysis showed that metronidazole was effective therapy for active pouchitis but that VSL#3 was effective for maintaining remission in chronic pouchitis. With judicious use of antibiotics and probiotics, pouchitis symptoms can be controlled.

It is important to monitor the pouch because dysplasia may develop. The incidence of dysplasia is unknown because IPAA has been broadly used for less than 10 years. Nevertheless, possible dysplasia in some patients indicates that pouches must be serially monitored.

Removal of the pouch and ileostomy rarely becomes necessary. Some studies report revision in as many as 2% to 3% of patients.

COURSE AND PROGNOSIS

Prognosis after IPAA is excellent. The Mayo Clinic reports that 94% of patients have successful outcomes. Nocturnal incontinence occurs in approximately 10% of patients, at least one episode of pouchitis occurs in 48% of patients, and cumulative pouch failure at 1 and 10 years occurs at rates of 2% and 9%, respectively. The most common early postoperative problems are bowel obstruction, urinary tract infection, and wound sepsis.

Postoperative impotence and retrograde ejaculation are noted in up to 3% of men with pouches. Pregnancy and delivery are well tolerated by women with pouches.

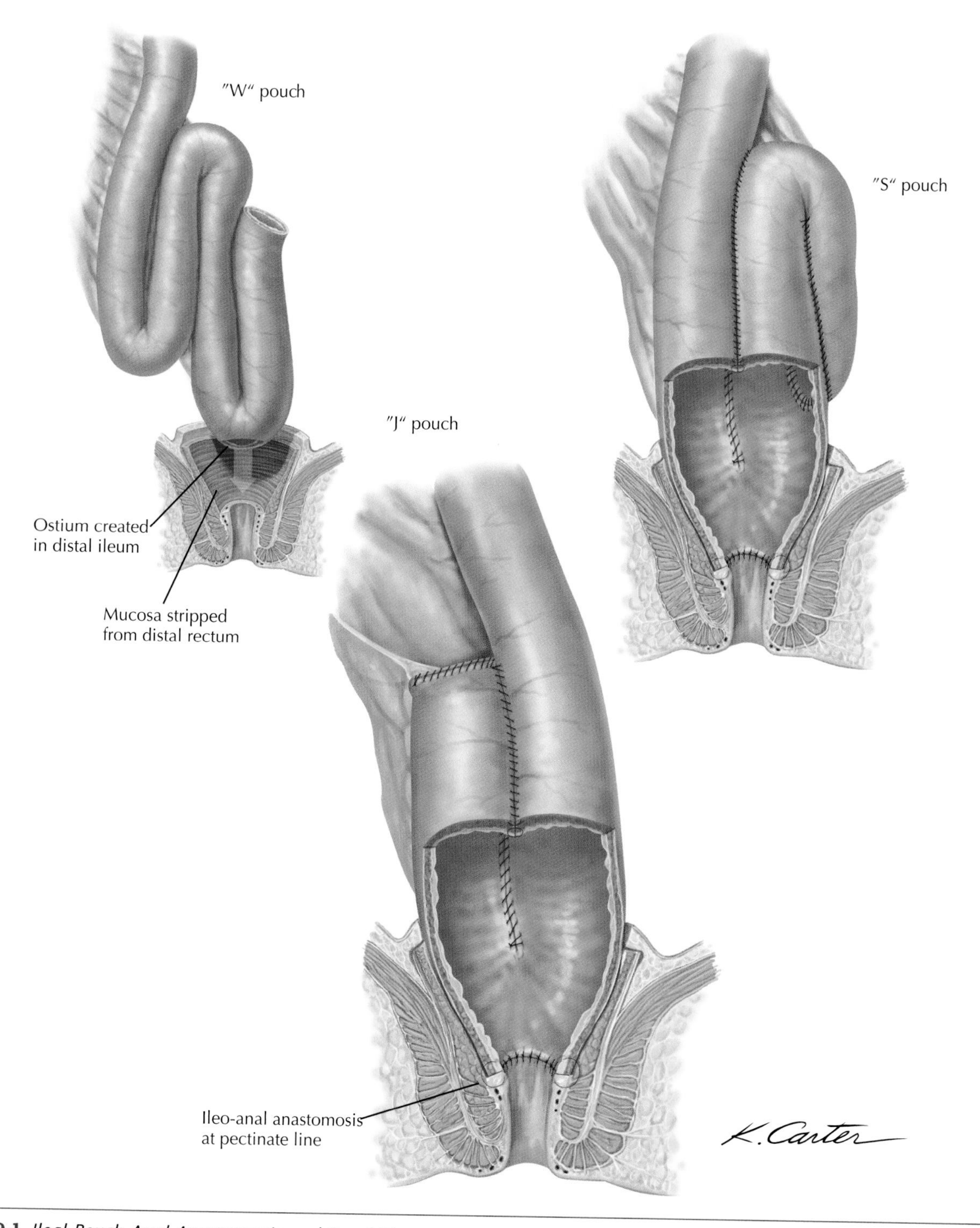

Figure 149-1 *Ileal Pouch Anal Anastomosis and Pouchitis.*

Table 149-1 Pouchitis Disease Activity Index*

Criterion	No. Points
History (Clinical)	
One or two more stools than usual	1
More than three stools	2
Daily bleeding	1
Occasional fecal urgency or cramps	1
Usual fecal urgency or cramps	2
Fever	1
Endoscopy	
Edema	1
Granularity	1
Friability	1
Loss of vascularity	1
Mucus	1
Ulceration	1
Histology	
Mild PMN infiltrate	1
Moderate PMN infiltrate and crypt abscesses	2
Severe PMN infiltrate and crypt abscesses	3
Less than 25% ulceration/low-power field	1
25% to <50%/low-power field	2
More than 50% ulceration/low-power field	3

**Score greater than 7 points indicates pouchitis.*
PMN, *polymorphonuclear leukocyte.*

ADDITIONAL RESOURCES

Bauer JJ, Gorfine SR, Gelernt IM, et al: Restorative proctocolectomy in patients older than 50 years, *Dis Col Rect* 40:562-568, 1997.

Floch MH: Probiotics, irritable bowel syndrome, and inflammatory bowel disease, *Curr Treat Opt Gastroenterol* 6:283-288, 2003.

Gionchetti P, Rizzello F, Venturi A, et al: Oral bacterial therapy as maintenance treatment in patients with chronic pouchitis: a double-blind, procedure-controlled trial, *Gastroenterology* 119:305-309, 2000.

Meagher AP, Farouk R, Dozois RR, et al: J ileal pouch and anastomosis for chronic ulcerative colitis: complications and long-term outcome in 1310 patients, *Br J Surg* 85:800-805, 1998.

Rubenstein MC, Fisher RL: Pouchitis: pathogenesis, diagnosis, and management, *Gastroenterologist* 2:129-133, 1996.

Sandborn WJ, McLeod R, Jewel DP: Medical therapy for induction and maintenance of remission in pouchitis: a systemic review, *Inflamm Bowel Dis* 5:33-39, 1999.

Sandborn WJ, Tremaine WJ, Batts KP, et al: Pouchitis after ileal pouch anastomosis: a pouchitis disease activity index, *Mayo Clin Proc* 69:409-413, 1994.

Shen B, Achker JP, Lashner BA, et al: Irritable pouch syndrome: a new category of diagnosis for symptomatic patients with ileal pouch–anastomosis, *Am J Gastroenterol* 97:972-977, 2002.

Wexner SD, Rosen L, Lowry A, et al: Practice parameters for the treatment of mucosal ulcerative colitis: supporting documentation. Standards Practice Task Force, American Society of Colon and Rectal Surgeons, *Dis Col Rect* 40:1277-1285, 1997.

Differentiating Features of Ulcerative Colitis and Crohn's Disease

150

Martin H. Floch

A classification of *inflammatory bowel disease* (IBD) based on scientific data has not gained acceptance. An early attempt was made to classify IBD according to two categories: (1) ulcerative colitis and Crohn's disease and (2) collagenous colitis, eosinophilic enteritis, Behçet disease, transient colitis, microscopic colitis, prestomal ileitis, pouchitis, and solitary rectal ulcer. None of these entities had a definitive etiology, resulting in overlap and confusion.

Most authorities group these IBD entities into *ulcerative colitis* (UC), *Crohn's disease*, and *indeterminate colitis*. However, IBD study still has no classification, with difficult clinical situations in which patients seem to have overlapping diseases, such as diverticulitis and Crohn's disease. Nevertheless, because the etiology of each is unknown and because the most common disease entities are UC and Crohn's disease, this chapter discusses the differentiating factors because these are important for clinical management and determining the course and prognosis of each. Many of the same therapeutic regimens are used for UC and Crohn's disease, with subtle variations, and the clinician should understand the differences (**Fig. 150-1**).

The differential diagnosis of UC and Crohn's disease greatly overlaps, and for indeterminate colitis, a differentiation cannot be made. Furthermore, the IBD pattern appears to be changing with the changing environment and more information. The following differentiating features and comparisons, however, have held over the past few decades.

PRESENTING SYMPTOMS

Patients with UC often have diarrhea that contains mucus or blood. At times, the patient may appear to have infectious gastroenteritis. However, the diarrhea persists even after therapeutic trials. Rarely, an extraintestinal manifestation, such as migratory joint pain, pyoderma gangrenosum, or acute toxic megacolon, may be the presenting symptom.

Patients with Crohn's disease usually have abdominal pain and a change in bowel habits. They may have diarrhea. The presentation of Crohn's colitis may be similar to that of UC. However, the diffuseness of Crohn's disease may result in extracolonic symptoms in many patients. If Crohn's disease has developed in the small bowel, symptoms of pain, fever, mass, or anemia may be present. If extracolonic symptoms include joint pain or liver disease, these may be the presenting symptoms.

GENETICS AND HEREDITY

UC and Crohn's disease are complex genetic disorders with multiple contributing genes. Linkage studies implicated some unique genomic regions containing IBD susceptibility genes, with some unique to Crohn's disease, some unique to UC, and some unique to both diseases. Inflammatory bowel disease II, on chromosome 12q, is observed more in UC. Inflammatory bowel disease I, on chromosome 16q, contains the Crohn's disease–susceptible gene *NOD2/CARD15*. Patients of European descent show three major coding-region polymorphisms within *NOD2/CARD15*. One copy of the risk allele means the patient has a twofold to fourfold risk for Crohn's disease, but with a double-dose carriage, the risk increases to 20-fold to 40-fold. Patients with ileal disease, earlier age of onset, and stricturing carry the *NOD2/CARD15* allele.

Ethnic, racial, family, and twin studies reveal a leveling off in incidence and prevalence of UC and an increasing rate of Crohn's disease. In addition, nonwhite persons appear to have a consistently lower rate of Crohn's disease. The prevalence of Crohn's disease per 100,000 population is 43.6 among white persons, 29.8 among black persons, 4.1 among Hispanic persons, and 5.6 among Asian persons. Furthermore, Jewish persons in the United States seem to be at greatest risk for twofold to fourfold higher incidence and twofold to ninefold higher prevalence. Family studies show IBD clusters within families, suggesting that genetics plays a major role. First-degree relatives have about a 10-fold to 15-fold increased risk. The risk overlaps; one family member may acquire Crohn's disease and another, UC. In addition, asymptomatic family members appear to have an increased rate of positive serologic findings. Monozygotic twin concordance for Crohn's disease is reported to be as high as 42% to 58%, a key factor supporting a major genetic influence.

GROSS AND MICROSCOPIC PATHOLOGY

UC is a diffuse mucosal inflammatory disease limited to the colon. UC almost always affects the rectum and appears to spread from the rectum aborally. There are rarely "skipped" areas, and the disease is contiguous. Studies of large numbers of UC patients reveal that approximately 45% have gross rectal and sigmoid involvement, and 37% have involvement of the entire colon (17% on left side only). UC is mucosal, so biopsy of the mucosa reveals the classic histology. Although some cases are confusing, the most typical findings can usually distinguish acute infectious-limiting disease. Patients with UC have distorted crypt architecture, atrophy of crypts, mixed-lymphocytic plasma cells, and neutrophilic infiltrates with varying degrees of eosinophilia. Although often present, crypt abscesses are not pathognomonic of UC.

Gross findings in Crohn's disease, when involving only the colon, may be similar to those of UC, but often manifest with aphthous-type ulcerations. Ulcerations are deeper in Crohn's disease than in UC. The disease is *transmural* in that inflamma-

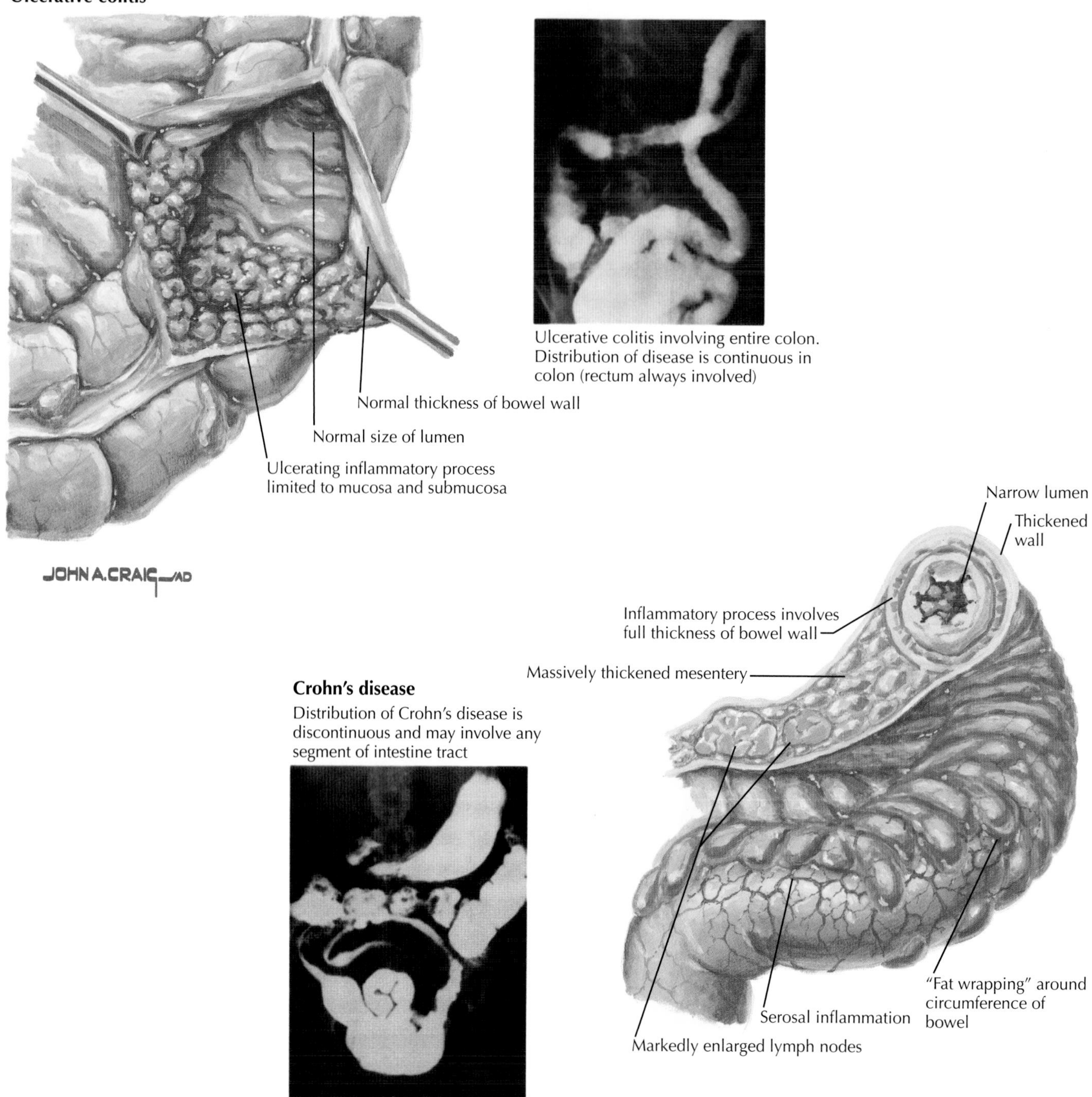

Figure 150-1 *Differentiating Features of Ulcerative Colitis and Crohn's Disease.*

tion penetrates the entire wall of the colon. Classic histologic findings reveal granulomas and a variety of histologic changes, including some similar to those in UC. However, the defining feature is that Crohn's disease is transmural. This may not be appreciated on simple mucosal biopsy, but it certainly is present on larger specimens. Furthermore, because of Crohn's transmural nature, gross findings include fibrosis, stricturing, and classic "skipped" areas. The rectum may or may not be involved. Defining gross features also include lesions with deeper ulceration than in UC. Once the small bowel is involved, the patient unquestionably has Crohn's disease.

In biopsy material, changes in the M-cell population in the epithelium overlying Peyer patches, prominent lymphoid aggregates, and dilated submucosal lymphatics and granulomas are pathognomonic of Crohn's colitis and are not seen in UC.

ENDOSCOPIC FINDINGS

The classic endoscopic finding in UC is a friable, finely ulcerated mucosa that may be limited to the rectum *(proctitis)* or that may extend aborally to any point in the large bowel.

In Crohn's disease, the endoscopic finding may be similar to that of UC. Often in Crohn's disease, however, ulcerations are deeper, there are skipped areas of normal and abnormal bowel in the colon, and the rectum appears normal. Also, areas of erythema and ulceration are scattered around the colon.

SEROLOGIC FINDINGS

Although serologic findings are not always positive, the serologic pattern can be helpful. Approximately 80% of patients with UC and as many as 45% of patients with Crohn's disease have perinuclear antineutrophil cytoplasmic antibody (pANCA). Anti–*Saccharomyces cerevisiae* antibody (ASCA) occurs in 60% to 70% of patients with Crohn's disease and is specific in that it rarely occurs in UC patients.

Positive ASCA and negative pANCA help determine that the disease is Crohn's colitis if simple mucosal biopsy does not elucidate the diagnosis and the classic skipped areas are absent. Strong positive pANCA and negative ASCA help determine that the disease is UC. This combination might have a predictive value in 95% to 96% of patients. Newer antibody tests are being developed and may become helpful in some cases.

Although the clinician cannot easily observe the serologic differences, UC is more of an antibody-mediated hypersensitivity, whereas Crohn's disease shows an ongoing T-helper cell type 1 response with excessive interleukin-12, interferon-γ, and tissue necrosis factor-α.

INDETERMINATE COLITIS

Even with clear identification of clinical and histologic factors distinguishing UC from Crohn's colitis, some patients develop Crohn's disease in the remaining small bowel after ileal pouch anal anastomosis (IPAA) for UC. These patients are then categorized as having indeterminate colitis. The clinician cannot determine which form of colitis is present, and variations of both Crohn's disease and UC may be observed in the same patient over the course of the disease.

Another indeterminate area is the overlap of *diverticular disease* with colitis. When these diseases cannot be differentiated pathologically and seem to occur together, they are cured when the area is resected. Therefore, the pathology may be confusing, but the course indicates the disease is neither the classic ulcerative type nor Crohn's type of colitis (see Chapter 143).

ADDITIONAL RESOURCES

Bonen DK, Cho JH: The genetics of inflammatory bowel disease, *Gastroenterology* 124:521-536, 2003.

Farmer RG, Easley KA, Rankin GB: Clinical patterns, natural history, and progression of ulcerative colitis: a long-term follow-up of 1116 patients, *Dig Dis Sci* 38:1137-1146, 1993.

Ferrante M, Henckaerts L, Joossens M, et al: New serological markers in inflammatory bowel disease are associated with complicated disease behavior, *Gut* 56:1394-1403, 2007.

Lennard-Jones JE: Classification of inflammatory bowel disease, *Scand J Gastroenterol* 24(suppl 170):2-6, 1989.

Panaccione R, Sandborn WJ: Is antibody testing for inflammatory bowel disease clinically useful? *Gastroenterology* 116:1001-1002, 1999.

Sands BE: Crohn's disease. In Feldman M, Friedman LS, Brandt LJ, editors: *Gastrointestinal and liver disease*, ed 8, Philadelphia, 2006, Saunders-Elsevier, pp 2459-2498.

Sartor RB: Mechanisms of disease: pathogenesis of Crohn's disease and ulcerative colitis, *Nat Clin Pract Gastroenterol Hepatol* 3:390-407, 2006.

Su C, Lichtenstein GR: Ulcerative colitis. In Feldman M, Friedman LS, Brandt LJ, editors: *Gastrointestinal and liver disease*, ed 8, Philadelphia, 2006, Saunders-Elsevier, pp 2499-2548.

Surawicz CM, Belie L: Rectal biopsy helps to distinguish acute self-limiting colitis from idiopathic inflammatory bowel disease, *Gastroenterology* 96:104-113, 1984.

Crohn's Disease

Martin H. Floch

151

Crohn's disease is a transmural inflammation of the gastrointestinal (GI) tract characterized by granulomas. It involves primarily the colon and the ileum *(regional enteritis)* but may involve any part of the GI tract (**Fig. 151-1**). Approximately 40% of patients have a pattern that involves the small and large intestines, 30% have only small bowel involvement, and 25% have large bowel involvement.

The incidence of Crohn's disease appears to be slightly higher in females. In Western countries, the incidence now ranges from 6 to 10 per 100,000 population, with a prevalence rate of 130 per 100,000, and reports indicate the incidence is rising. However, the incidence is much lower in Japan, South America, and Africa.

The cause is of Crohn's disease unknown. However, most authorities believe an infectious, environmental, or at times drug (NSAID) trigger results in an altered immune inflammatory response in the correct genetic setting. Most cases are diagnosed in patients younger than age 40. Often, however, the disease goes undiagnosed or is mild until a complication develops later in life.

Because Crohn's disease involves transmural inflammation, it appears to "skip" areas in the GI tract. A healthy area of small or large bowel can be adjacent to a diseased area (see Fig. 151-1). Transmural inflammation may lead to internal or external fistulization. Genetic susceptibility is apparent; as many as 25% of patients have a positive family history of Crohn's disease. The *NOD2/CARD15* gene of chromosome 16, found in monocytes involved in the immune response to pathogenic organisms, appears responsible for the increased susceptibility.

CLINICAL PICTURE

The major presenting symptoms in Crohn's disease are abdominal pain, diarrhea or change in bowel habits, and weight loss. Symptoms vary greatly. In ulcerative colitis (UC), diarrhea is the overriding factor. The bowel pattern varies with the area of the intestine involved. When the terminal ileum is involved, the major presenting symptom is usually right lower quadrant pain, which can be confused with appendicitis. At times, the abdominal pain may be diffuse, but it varies with the area of the bowel involved.

Because of the chronicity of the symptoms, most patients lose weight. Depending on the extent of Crohn's disease, the symptoms can be accompanied by obstruction, and thus obstipation and abdominal distention; by perianal rectal drainage resulting from fistulization; or in rare cases, by anemia from blood loss and malabsorption. Symptoms vary depending on the site of GI disease.

Physical examination may reveal no findings or a mass in the right lower quadrant. The patient with Crohn's disease may have diffuse tenderness or obstruction or may be anemic, pale, and febrile. Once again, physical findings depend on the extent of disease. Atypical presentations from extraintestinal manifestations may be seen.

DIAGNOSIS

Initial laboratory evaluation may reveal anemia, elevated erythrocyte sedimentation rate and C-reactive protein, leukocytosis, and thrombocytosis, again depending on the extent of disease. If the patient is losing weight and has diarrhea, hypoalbuminemia may be present.

The next step in the diagnosis involves examining the GI tract. Symptoms will determine whether to begin with endoscopy. If the patient has diarrhea, colonoscopy will likely reveal the classic aphthous ulcerations of the bowel and "skipped" areas. Histologic examination may confirm the diagnosis. Most colonoscopists can examine the terminal ileum. Biopsy can be performed on that area to confirm or rule out a diagnosis of terminal ileitis.

Barium contrast studies or computed tomography (CT) can be used to image the GI tract. Either technique may make the diagnosis of Crohn's disease, or either may be needed to complement the other's findings. Barium studies may reveal a classic stricture or an inflamed loop of bowel. Most often, Crohn's disease is in the terminal ileum, but skipped areas may be present. Barium studies may also reveal strictures and enteroenteric fistulae. Because of the symptoms, or possibly a masked presentation, the clinician may prefer CT or may need CT to complement findings of the barium study. CT classically reveals evidence of the increasing fat pad deposition noted in most pathology specimens and the "smudgy" area outside the bowel. The bowel wall shows areas of thickness, depending on the site of involvement.

If the patient has perianal or anal disease, anoscopy should be performed in addition to colonoscopy to determine the sites of fistulization and fissures.

Serologic markers are now available and have been helpful in confirming, and at times making, the diagnosis. About 60% to 70% of patients with Crohn's disease have elevated anti–*Saccharomyces cerevisiae* antibody (ASCA) levels. However, ASCA has specificity greater than 95%. If the perinuclear antineutrophil cytoplasmic antibody (pANCA) is negative, the patient almost certainly has Crohn's disease.

Capsule endoscopy is gaining wide acceptance and can demonstrate ulceration and stricture of the small bowel. Because some are subtle, small bowel changes can be missed on barium contrast studies or CT scans. Push enteroscopy, with scopes of distance, often do not cover the entire small bowel. Double-balloon enteroscopy may be helpful in uncertain cases, and biopsy material can be obtained. Positive findings on capsule endoscopy in conjunction with positive serologic markers strongly suggest that the patient has Crohn's disease.

Extraintestinal manifestations develop in approximately 25% of the patients. These may complicate Crohn's disease, may occur at any time during the disease, or may be a presenting finding, to which the clinician must be alert. Areas involved are the skin, eyes, liver, and musculoskeletal system. In addition,

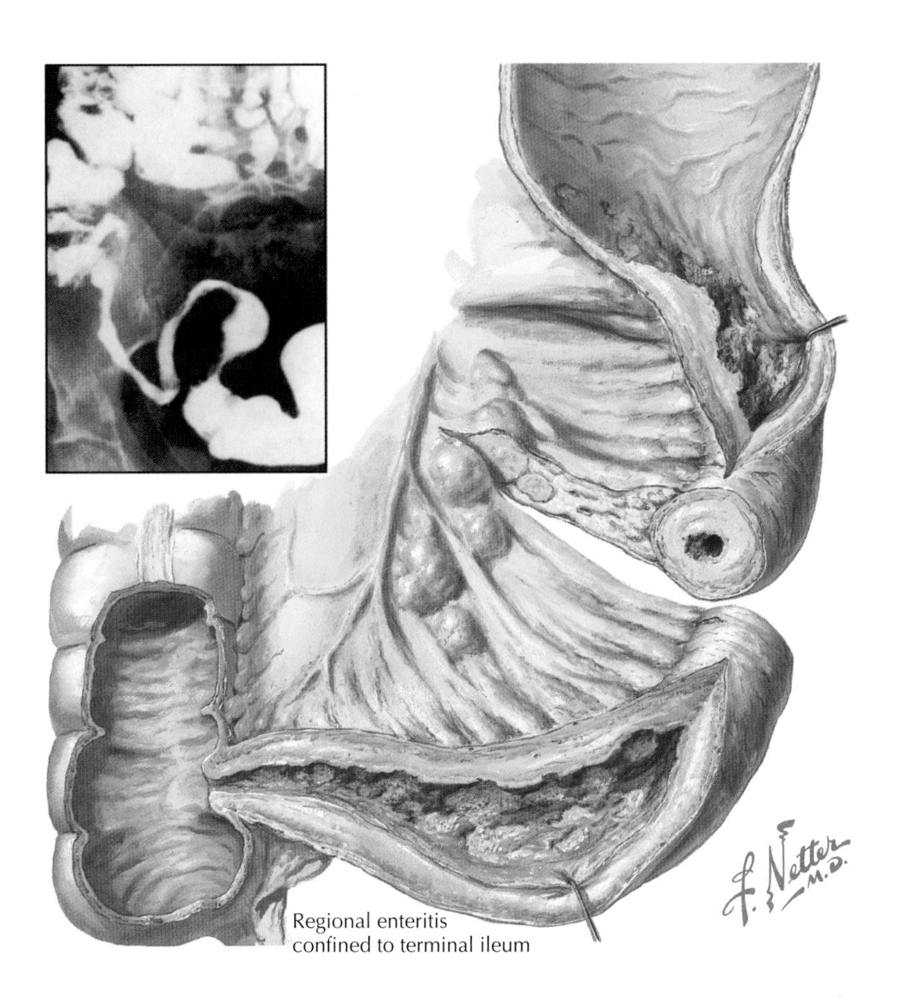

Regional variations

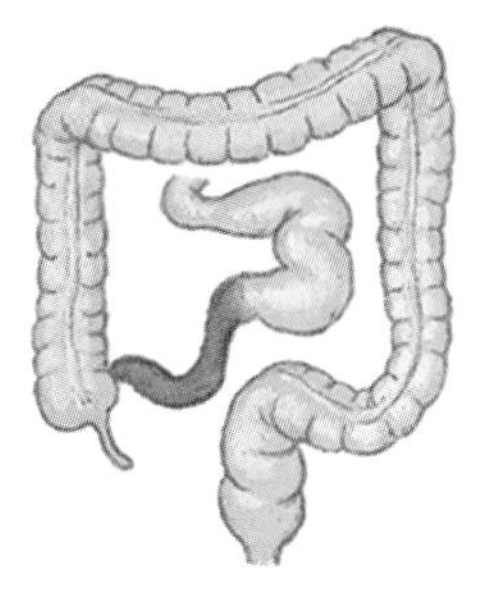

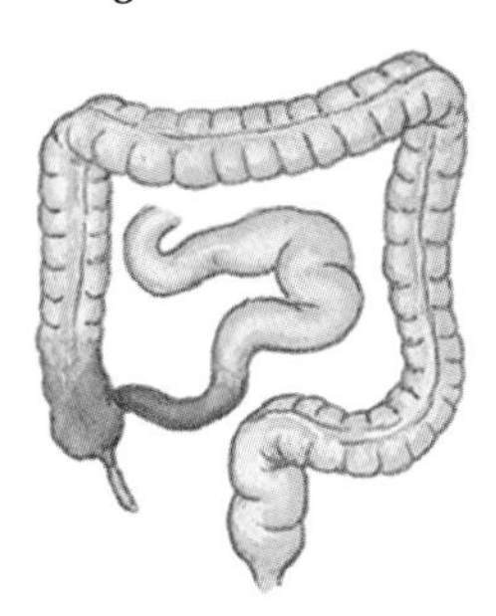

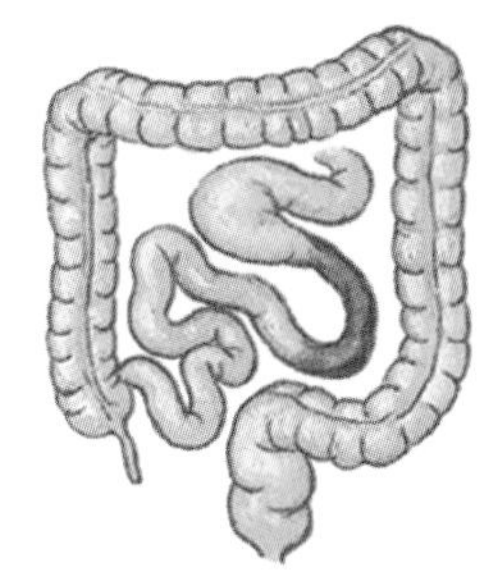

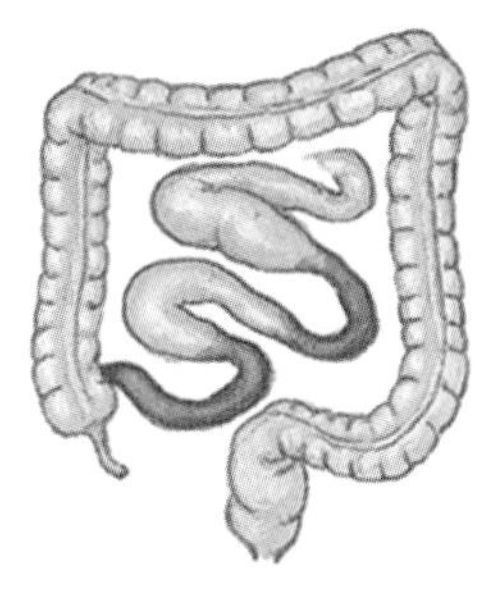

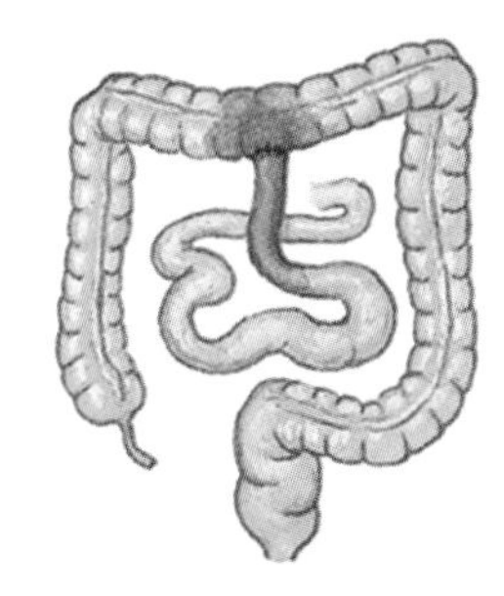

Figure 151-1 *Crohn's Disease (Regional Enteritis).*

other areas of the gut are reportedly involved, including the mouth, esophagus, stomach, and duodenum. The most dramatic extraintestinal manifestations are pyoderma gangrenosum and perianal tags (skin), uveitis, iritis, and conjunctivitis (eye), and peripheral arthritides, sacroiliitis, and osteoporosis (bones). Primary sclerosing cholangitis may be the presenting symptom if Crohn's disease develops later in life.

Diseases that can mimic Crohn's disease include bowel lymphomas, other malignancies, tuberculosis of the bowel, and at times, a chronic *Yersinia* infection. However, laboratory test results and endoscopy can differentiate these diseases readily.

TREATMENT AND MANAGEMENT

Pharmacologic treatment for Crohn's disease includes a host of immunosuppressive agents (see Chapter 154). Surgical management is discussed in Chapter 152 and nutritional management in Section X.

Nutritional management for patients with Crohn's disease is different than that for patients with UC. In Crohn's disease, patients may achieve remission on elemental diets; UC patients do not. Although it is slow and medical therapy is faster, nutritional management is effective. The decision to stay with medical therapy when the disease is insidious and causes changes in the activities of daily living is difficult. Surgery can be effective and can cause remission of symptoms, but surgical intervention has limits.

COURSE AND PROGNOSIS

Crohn's disease results in high morbidity and diminished quality of life, but it does not greatly shorten a patient's life span. It is estimated that 70% of patients will require some type of surgical procedure during their lifetime. However, most patients can achieve remission with one of the therapeutic medical regimens. When surgery is necessary, patients frequently achieve remission (see Chapter 152). When repeated surgeries and resections are necessary, short bowel syndrome may develop.

Failure to thrive in children, malnutrition, and anemia are problems associated with Crohn's disease. Diligent medical and nutritional therapy can maintain good quality of life in most patients.

ADDITIONAL RESOURCES

Borelli O, Cordischi L, Cirulli M, et al: Polymeric diet alone versus corticosteroids in the treatment of active pediatric Crohn's disease: a randomized controlled open-label trial, *Clin Gastroenterol Hepatol* 4:744-753, 2006.

Glayan TO, Kandil HM: Nutritional factors in inflammatory bowel disease, *Gastroenterol Clin North Am* 31:203-218, 2002.

Griffiths AM, Ohlsson A, Sherman PN, Sutherland LR: Meta-analysis of enteral nutrition as a primary treatment of active Crohn's disease, *Gastroenterology* 108:1056-1067, 1995.

Johnson T, Macdonald, Hill SM, et al: Treatment of active Crohn's disease in children using partial enteral nutrition with liquid formula: a randomized controlled trial, *Gut* 55:356-361, 2006.

Murphy SJ, Ullman TA, Abreu MD: Gut microbes in Crohn's disease: getting to know you better? *Am J Gastroenterol* 102:397-398, 2008.

O'Sullivan NA, O'Morain CA: Nutritional therapy in Crohn's disease, *Inflamm Bowel Dis* 4:45-53, 1998.

Sands B: Crohn's disease. In Feldman M, Friedman LS, Brandt LJ, editors: Gastrointestinal and liver disease, ed 8, Philadelphia, 2006, Saunders-Elsevier, pp 2459-2498.

Siebold F: ASCA: genetic marker, predictor of disease, or marker of a response to an environmental agent? *Gut* 54:1212-1213, 2005.

Crohn's Disease: Complications and Surgical Therapy

152

Martin H. Floch

The most common complications of Crohn's disease are abscesses and fistulae, obstruction, and perianal disease (**Fig. 152-1**). Abscesses develop in 15% to 20% of patients and arise from an infected area. Fistulae are more common and develop in 20% to 40% of patients, and most are enteroenteric or enterocutaneous.

Antibiotic and medical treatment may be helpful and in some studies have proved effective, but surgical therapy is often necessary. Obstruction is a common complication, and although stents and dilatation are attempted, surgical intervention is necessary to relieve progressive obstruction, which is most common in the small bowel. Perianal disease is probably the most difficult complication to treat. Antibiotics, bowel rest, and intensive medical therapy, including anti–tissue necrosis factor, vary in their ability to heal fistulae. Hemorrhage and free perforation are rare but do occur. At times, bleeding may be occult and may require treatment.

An estimated 60% of patients with Crohn's disease will require some form of surgery within 10 years of diagnosis. These data are several decades old, however, and new, more dramatic medical therapies may be able to avert the need for surgery in some patients. Nevertheless, the list of indications for surgery is long, and its necessity becomes apparent in individual patients (**Box 152-1**).

CLINICAL PICTURE

The clinical presentation varies with each complication. Symptoms and physical findings of pain, elevated temperature,

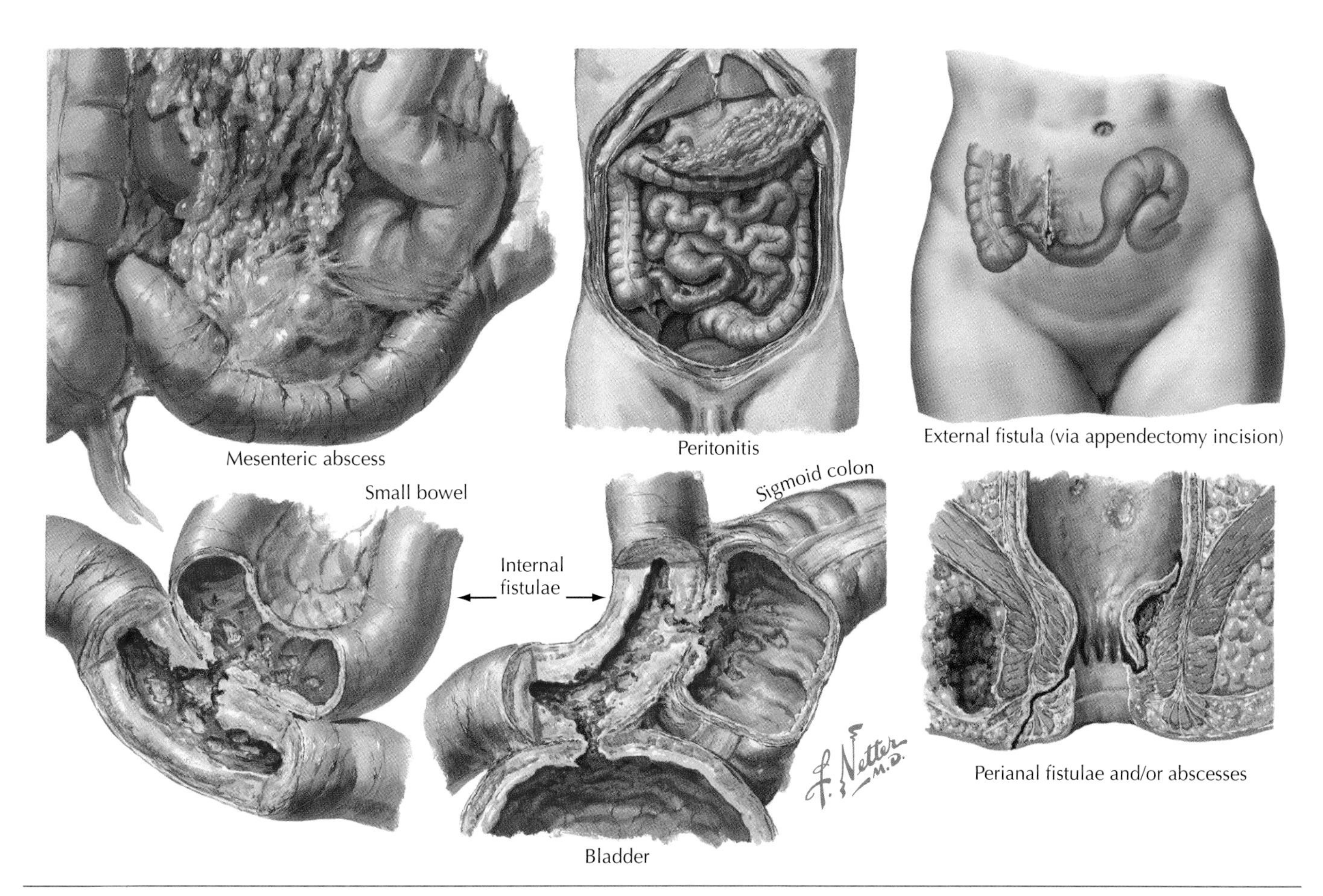

Figure 152-1 *Crohn's Disease: Complications.*

Box 152-1 Indications for Surgery for Crohn's Disease

Medical management failure
Intestinal obstruction
- Partial or complete

Intestinal fistulae
- Symptomatic enteroenteric fistula
- Enterocutaneous fistula
- Enterovesical fistula
- Enterovaginal fistula

Intraabdominal abscess
Inflammatory mass
Hemorrhage
Perforation
Perineal disease
- Perianal abscess
- Superficial fistula in ano unresponsive to medical therapy
- Complex fistula in ano unresponsive to medical therapy

Modified from Cohen RV, editor: Inflammatory bowel disease, *Totowa, NJ, 2003, Humana Press.*

abdominal mass, and rectal abnormality combined with computed tomography findings confirm the diagnostic impressions and indicate which medical or surgical therapy is needed.

TREATMENT AND MANAGEMENT

When bowel resection is needed to remove obstruction, there should always be an attempt to limit the amount of bowel resected. At times, however, there are "skipped" areas, and several segments have to be resected. Severe recurrent obstruction leads to significant resection and short bowel syndrome (see Chapter 115).

Enteroenteric fistulae and internal fistulae are not an indication for surgery unless symptomatic. Fistulization of the skin, bladder, or vagina may require a more aggressive or a surgical approach because other organ systems are involved.

If an abscess forms intraabdominally, medical management may still be effective. However, if the abscess can no longer be controlled, surgical drainage and resection are necessary.

Treatment of perianal disease is discussed in Chapter 153.

Finally, it must be emphasized that patients with Crohn's disease are at increased risk for adenocarcinoma of the colon and the small bowel. If screening demonstrates dysplasia, careful consideration must be given to resection. This is easier to perform in colon disease, but suspicious lesions in the small bowel must be investigated. Resection is indicated if the lesions may be neoplastic.

COURSE AND PROGNOSIS

Surgery for Crohn's disease is not curative. Endoscopic evidence for recurrence ranges from 28% to 73% at 1 year and 77% to 85% at 3 years after ileal resection. Studies from various institutions give different rates, but most clinicians believe that Crohn's disease will recur within 5 years. Unfortunately, if repeat surgery is necessary, short bowel syndrome and malabsorption may develop. Recurrences are most likely to occur close to the area of resection. However, long-term studies have not been conducted to evaluate the effectiveness of the newer biological therapies postoperatively to prevent recurrent resection.

Surgical treatment for Crohn's colitis includes segmental resection, subtotal colectomy with ileoproctostomy, or if performed because of an uncertain primary diagnosis, ileorectal anal anastomosis and total colectomy with ileostomy. Older studies report recurrence in approximately 75% of patients at 10 years; again, long-term studies of the newer therapies are needed to determine whether repeat surgical treatment is necessary. Nevertheless, total colectomy with ileoanal anastomosis is not recommended for patients with Crohn's colitis because of the high recurrence rate in the area of the new pouch. Unfortunately, some patients with indeterminate colitis inadvertently undergo total colectomy with an ileoanal procedure only to experience a recurrence of ileitis. Laparoscopic surgery can be performed for Crohn's disease and ileocolic resection. Early studies reveal that laparoscopic surgery results in lower incidence of minor complications and shorter periods of postoperative recovery. Indications for the use of laparoscopic surgery do not differ from those for open surgery.

ADDITIONAL RESOURCES

Fazio VW, Ziv Y, Church JM, et al: Ileal pouch anal anastomoses: complications and function in 1005 patients, *Ann* Surg 222:120-126, 1995.

Froehlich F, Juillerat P, Pittet V, et al: Maintenance of surgically induced remission of Crohn's disease, *Digestion* 76:130-135, 2007.

Hurst RB: Surgical management of inflammatory bowel disease. In Cohen RV, editor: *Inflammatory bowel disease*, Totowa, NJ, 2003, Humana Press, pp 157-200.

O'Sullivan NA, O'Morain CA: Nutritional therapy in Crohn's disease, *Inflamm Bowel Dis* 4:45-53, 1998.

Sands B: Crohn's disease. In Feldman M, Friedman LS, Brandt LJ, editors: *Gastrointestinal and liver disease*, ed 8, Philadelphia, 2006, Saunders-Elsevier, pp 2459-2498.

Schraut WH: The surgical management of Crohn's disease, *Gastroenterol Clin North Am* 31:255-263, 2002.

Perianal Disease in Crohn's Disease

153

Martin H. Floch

Anorectal Crohn's disease may consist of fissures, ulcers, abscesses, fistulae, strictures, edematous skin tags, and benign skin tags (**Fig. 153-1**). The overall prevalence is 36%, but it is 46% when the colon is involved and only 25% when the small bowel is involved. Often, one of these lesions may appear before there is evidence of intestinal disease.

CLINICAL PICTURE

The patient may have only mild diarrhea or may have severe rectal pain or bulging abscesses. A simple skin tag may be enlarged and painless but can be rectally disfiguring ("elephant ears"). Such pain is severe and necessitates emergency treatment.

Physical examination on the exterior can reveal the lesion. It is important to test for local tenderness and look for discharge from a fistula. At times, it may be too difficult to perform a rectal examination, but the anal ring can be teased, and often a small fissure can be demonstrated without entering the anal ring. Bulging of a large abscess is obvious.

DIAGNOSIS

To diagnose a fissure, the fissure must be seen, but its presence can be suspected when there is great tenderness on digital examination. Anoscopy may have to be delayed or performed only with sedation. It is important to rule out any malignant lesion during early presentation in a patient who has not undergone endoscopic examination of the anal ring, and anesthesia may be necessary to evaluate the rectum.

TREATMENT AND MANAGEMENT

Therapy for fissures, edematous skin tags, and associated hemorrhoidal problems is local, as described Chapters 162 to 167. Topical treatment may be successful, but if it does not result in improvement, surgical intervention may be necessary. Ischiorectal abscesses require prompt incision and drainage. Most heal, but as many as 35% develop a fistula in ano. Although many Crohn's lesions do not heal, most do resolve.

The goal of therapy in perianal disease is relief of local pain and preservation of the sphincter. Although abscesses must be drained, the presence of a fistula does not necessarily require more surgery. Aggressive medical therapy helps healing. Local cleansing, Sitz baths, suppositories, and cleansing ointments to the area all are important.

Fissures and fistulae have healed with antibiotic therapy. Metronidazole has been effective in as many as 75% of patients, but several months of therapy are required, and disease does tend to recur. Lesions also heal with adequate disease control; therefore, vigorous immunosuppression or biological therapy is indicated.

If medical therapy fails, surgical drainage and placement of setons and mushroom catheters may be necessary. This therapy usually requires months, but healing occurs slowly. Occasionally, more aggressive surgery is necessary and includes fistulotomy. If perianal disease is severe and incontinence develops, proctectomy is necessary. Removal of the fecal flow eventually leads to healing.

ASSOCIATED RESOURCES

Ingle SB, Loftus EV Jr: The natural history of perianal Crohn's disease, *Dig Liver Dis* 39:963-969, 2007.

Rankin GB, Watts D, Melnyk CS, Kelley ML: National Cooperative Crohn's Disease Study: extraintestinal manifestations and perianal complications, *Gastroenterology* 77:914-920, 1979.

Steele SR: Operative management of Crohn's disease of the colon including anorectal disease, *Surg Clin North Am* 87:611-631, 2007.

Vermeire S, Van Assche G, Rutgeerts P: Perianal Crohn's disease: classification and clinical evaluation, *Dig Liver Dis* 39:959-962, 2007.

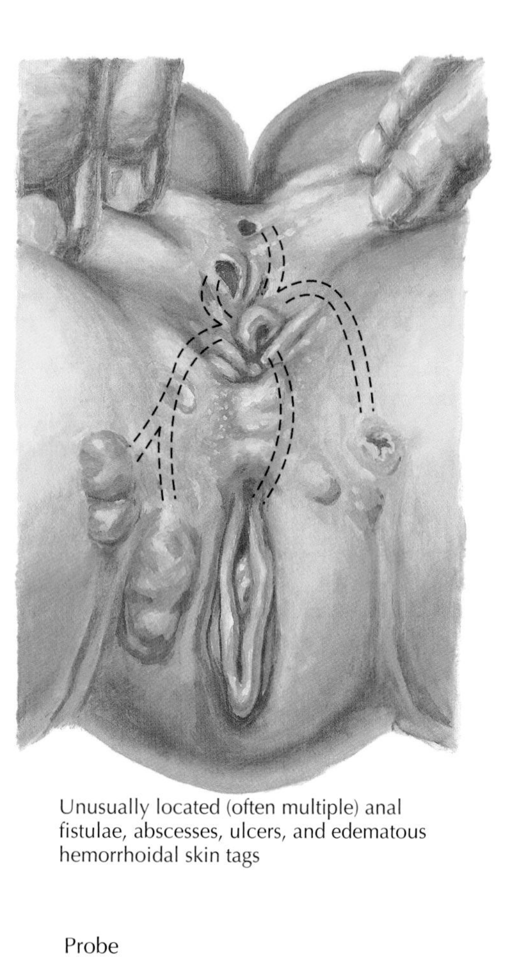

Unusually located (often multiple) anal fistulae, abscesses, ulcers, and edematous hemorrhoidal skin tags

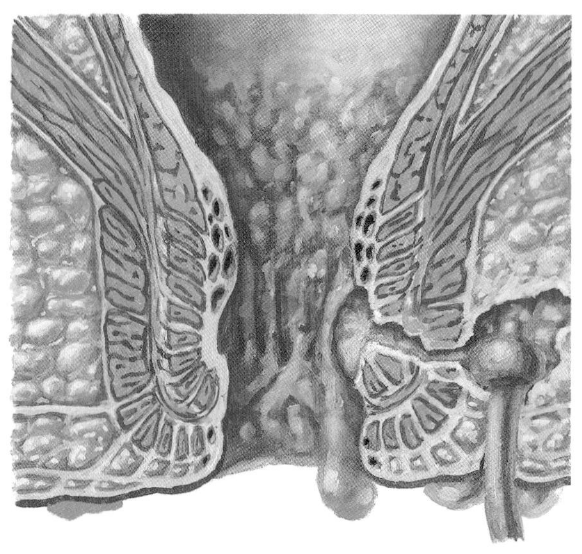

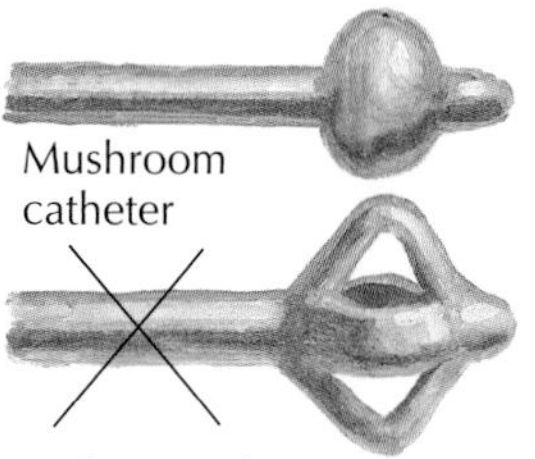

Abscess drained by placing small mushroom catheter as close to anus as possible to avoid subsequent long fistula tract

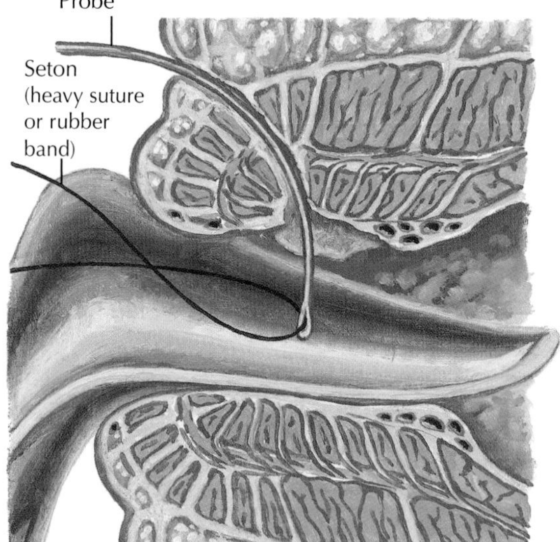

Sepsis of fistula tract controlled by placing seton (avoids fistulotomy wounds, which heal poorly)

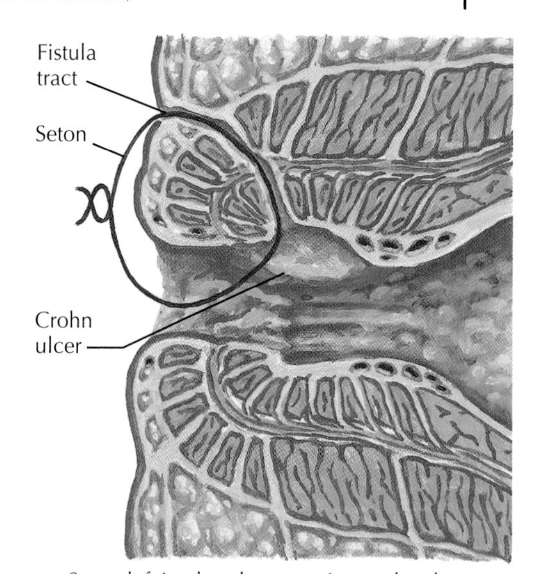

Seton left in place between internal and external openings to prevent abscess formation and further destruction of sphincter mechanism

Figure 153-1 *Crohn's Disease: Perianal Disease.*

Drug Therapy for Inflammatory Bowel Disease

154

Martin H. Floch

Pharmacologic agents used for the treatment of inflammatory bowel disease (IBD) are similar for both ulcerative colitis (UC) and Crohn's disease, with two major differences. Rectally instilled topical agents are effective in UC, but usually not in Crohn's disease, and certainly not in Crohn's ileitis. Nutritional therapy, such as the use of elemental and polymeric diets for the induction of remission, has been effective in Crohn's disease, but not in UC. With these differences in mind, most drugs can be used for patients with UC or Crohn's disease. The clinician's experience in the use of medications facilitates successful management. Importantly, U.S. Food and Drug Administration (FDA) approval may apply only to some uses of these drugs, but most practitioners use the drugs based on the literature.

Management of UC tends to vary with the severity and extent of illness; active disease is treated more vigorously than inactive disease (**Box 154-1**). Inactive clinical symptomatic disease may reflect *complete remission*, with no evidence of a pathologic condition or with a low-grade, chronic condition. In either case, maintenance therapy is recommended.

Depending on patient tolerance and physician preference, topical therapy in the form of enemas or suppositories is often effective for acute and maintenance therapy. Some patients cannot tolerate topical therapy, and oral medications are necessary. When severe or fulminant disease progresses, intravenous (IV) therapy is necessary. Nutrition therapy has no role in patients with UC other than to maintain adequate energy and nutrient intakes.

The same medications in UC therapy are used for patients with Crohn's disease, with some variations (**Box 154-2**). The increased incidence of perianal disease requires different approaches, and the effectiveness of infliximab has been more dramatic and is used earlier in Crohn's disease. In addition, complete bowel rest and elemental or polymeric liquid diets have been effective in inducing remission when patients or clinicians do not want to use corticosteroid therapy. Although nutritional therapy is slower, significant studies have shown that it works.

Table 154-1 outlines proven drug therapies for IBD.

Box 154-1 Pharmacologic Treatment of Active Ulcerative Colitis to Control Symptoms or Induce Clinical Remission

Mild to Moderate Disease

Distal Colitis

Sulfasalazine or 5-ASA (oral or topical)

Topical corticosteroid (or in combination with oral therapy)

Pancolitis

Sulfasalazine or oral 5-ASA

Moderate to Severe Disease

Distal Colitis

Topical or oral 5-ASA

Topical corticosteroid (or in combination with oral therapy)

Prednisone

Pancolitis

Prednisone

Azathioprine or 6-MP

Severe to Fulminant Disease

Distal Colitis or Pancolitis

Infliximab or other biological*

Intravenous corticosteroids

Intravenous cyclosporine

Maintenance Therapy

Distal Colitis

Sulfasalazine or 5-ASA (oral or topical)

Azathioprine or 6-MP

Pancolitis

Sulfasalazine or oral 5-ASA

Azathioprine or 6-MP

Modified from Stein RB, Hanauer SB: Gastroenterol Clin North Am *28:297, 1999.*
**Adalimumab, certolizumab, or natalizumab.*
5-ASA, *5-Aminosalicylic acid;* 6-MP, *6-mercaptopurine.*

DRUG THERAPIES

Aminosalicylates

The broad group of medications known as aminosalicylates relies on *5-aminosalicylic acid* (5-ASA) as its active compound. In the past, *sulfasalazine* was the preferred choice and dependent on the bacterial release of 5-ASA from sulfapyridine. Because the intolerance rate was 10% to 15%, the need to remove sulfa from the compound led to the development of other forms of 5-ASA therapy.

Therefore, conjugated, unconjugated, and rectal forms are now available, as well as a formulation that can be given once daily. It appears that delayed-release (DR) *mesalamine* in several compounds can be used in the small bowel, whereas all the others, plus the DR form, are effective in colon disease. Other-

Box 154-2 Pharmacologic Treatment of Active Crohn's Disease to Induce Clinical Remission or Control Symptoms

Mild to Moderate Disease
- Sulfasalazine or 5-ASA
- Metronidazole
- Prednisone
- Azathioprine or 6-MP
- Infliximab or other biological*

Severe Disease
- Prednisone
- Intravenous corticosteroids
- TPN or elemental diet
- Infliximab or other biological*
- Intravenous cyclosporine

Perianal or Fistulizing Disease
- Metronidazole or alternative antibiotic
- Azathioprine or 6-MP
- Infliximab or other biological*
- TPN on temporary basis

Maintenance Therapy
- Sulfasalazine or 5-ASA
- Metronidazole or ciprofloxacin
- Azathioprine or 6-MP
- Infliximab or other biological*

Modified from Stein RB. Hanauer SB: Gastroenterol Clin North Am *28:297, 1999.*
**Adalimumab, certolizumab, or natalizumab.*
5-ASA, *5-Aminosalicylic acid;* 6-MP, *6-mercaptopurine;* TPN, *total parenteral nutrition.*

wise, the effectiveness of these medications is comparable, and remission and maintenance rates of 35% to 95% have been reported.

It is important to remember that the 5-ASA compounds are associated with complications. They may cause diarrhea, and some clinicians prefer not to use them in patients with severe, acute UC. Many reports cite impaired renal function from 5-ASA use; therefore, renal function should be monitored in patients receiving long-term therapy.

Corticosteroids

Corticosteroids are the most effective drugs to induce remission of IBD, but they are not preferred for maintenance therapy. Many patients with chronic or severe disease require long-term corticosteroid therapy and may develop the associated complications. Great effort should be made to use an immunomodulator to replace corticosteroids if patients become dependent on their use.

There are increasing reports that a single course of IV corticosteroid therapy will induce remissions as effectively as other maintenance therapies. More research is needed on this form of therapy. Corticosteroids are also rapidly effective in topical form, when disease is limited to the rectum or the distal colon. Topical corticosteroid or mesalamine is the drug of choice for proctitis.

Immunomodulators

Azathioprine and *6-mercaptopurine* (6-MP, actively produced from metabolized azathioprine) are effectively used for maintenance therapy in UC and Crohn's disease. Doses are variable, and patients must be checked carefully for any hematologic complications. However, long-term therapy with these drugs has proved to be relatively safe. *Cyclosporine* has been used for severe disease when patients are hospitalized and are not responsive to IV corticosteroids. Cyclosporine can induce a remission of symptoms, and patients may be placed on oral cyclosporine or other immunosuppressive agents. However, this is drastic therapy, and these patients are close to undergoing total colectomy or surgery.

Methotrexate has been used less frequently; studies show mixed benefit. Large doses are necessary to induce remission of symptoms. Some clinicians use methotrexate when other therapies are failing.

Biologicals

Infliximab (anti–tumor necrosis factor–α antibody) has been shown to be effective in the treatment of Crohn's disease and UC. It must be administered intravenously. Usually, the patient receives three treatments of 5 or 10 mg/kg body weight at 2-week intervals; then it is decided whether to administer long-term maintenance therapy, with doses every 8 weeks. Because allergic and possible hematologic adverse effects may develop, infliximab must be used with caution. However, its wide use in rheumatoid arthritis has given gastroenterologists confidence about its safety. Infliximab has been effective for many patients who could not tolerate other maintenance or corticosteroid therapy and is now accepted as corticosteroid-sparing therapy.

Natalizumab, an α_4-integrin, was shown in early trials to be effective in inducing remission and improving quality of life. It was taken off the market because of cases of progressive multifocal leukoencephalopathy (PMI). Natalizumab has now been placed back on the market and joins infliximab, *adalimumab*, and *certolizumab* as other biological therapy choices. Extensive research is now being done to decide the best biological drugs and select clinical situations in which biologicals may be effective against IBD.

An increased incidence of lymphoma has been suspected with the use of infliximab, but recent evaluations indicate the immunomodulator may be incriminated. Intensive evaluation is forthcoming,

Placebo

All the acute and chronic trials with biologicals have been matched against placebo. Placebos have worked in 5% to 50% of the trials. Therefore, all the previously mentioned drugs have

Table 154-1 Oral Drug Therapy of Proven Value in Inflammatory Bowel Disease

Drug	Site	Ulcerative Colitis	Crohn's Disease
Aminosalicylates			
Topical (enemas and suppositories)			
Mesalamine	Rectum to distal colon	++	
Azo-Bond Compounds			
Sulfasalazine (sulfapyridine + 5-ASA)	Colon	++	++
Olsalazine (5-ASA dimer)	Colon	++	++
Balsalazide (5-ASA + 4-ABBA)	Colon	++	++
Mesalamine			
Delayed release (Asacol)	Ileum, colon	++	++
Multimatrix delivery system (Lialda)	Ileum, colon	++	++
Sustained release (Pentasa)	Stomach to colon	++	++
Steroids			
Topical			
Hydrocortisone suppositories	Rectum, distal colon	++	++
Hydrocortisone, prednisone, and betamethasone enemas	Rectum, distal colon		
Systemic (PO or IV)			
Prednisone, methylprednisolone, hydrocortisone	Stomach to colon	++	++
Immunomodulators			
Azathioprine	Stomach to colon	++	++
6-MP	Stomach to colon	++	++
Cyclosporine	Stomach to colon	+	+
Methotrexate	Stomach to colon	+	+
Antibiotics			
Metronidazole	Stomach to colon	+	++
Ciprofloxacin	Stomach to colon	+	++
Biologicals			
Infliximab (IV) (anti-TNF-α)	Stomach to colon	++	++
Natalizumab (α_4-integrin)	Stomach to colon	++?	++
Adalimumab	Stomach to colon	++?	
Certolizumab	Stomach to colon	++?	

ABBA, *4-Aminobenzoyl-beta-alanine;* 5-ASA, *5-Aminosalicylic acid;* 6-MP, *6-mercaptopurine;* PO, *oral;* IV, *intravenous;* TNF, *tumor necrosis factor.* +, *Effective;* ++, *very effective.*

statistical significance, but patients with IBDs have been shown to enter remission, which is maintained without therapy, through placebo medication.

NONDRUG THERAPIES

Forms of therapy other than pharmaceutical agents have included *probiotics*, *fish oils*, and *nutrient supplements* and *antioxidants*. These are mentioned in the chapters on the particular diseases for which they are used. It must be stressed, however, that these therapies are not *pharmacologic* agents and have not been exposed to the rigorous controls applied to the pharmaceutical industry.

PREGNANCY AND FERTILITY

Because IBD is a disease of the young, the concern of an effect on pregnancy or fertility is always present.

Most drugs are categorized into pregnancy classes A, B, C, and D. The actual disease process is associated with only a slight increase in spontaneous abortion. Statistical data reveal that sulfasalazine, mesalamine, and corticosteroids appear safe during pregnancy. Furthermore, a recent study reports statistically significant safety with the use of 6-MP. Sulfasalazine has been shown to affect sperm and thus is contraindicated in men who hope to father children. Other information on drugs in spermiogenesis is minimal. The long-term effect of biologicals is not yet known.

Classifications change with the most recent information available; therefore, the pregnancy class of a drug should be checked with the physician prescribing and the pharmacy delivering the drug to a pregnant patient.

ADDITIONAL RESOURCES

Behm BW, Bickston SJ: Tumor necrosis factor-alpha antibody for the maintenance of remission in Crohn's disease, *Cochrane Database Syst Rev* 1:CD006893, 2008.

Francella A, Dyan A, Bodian C, et al: The safety of 6-mercaptopurine for childbearing patients with inflammatory bowel disease: a retrospective cohort study, *Gastroenterology* 124:9-17, 2003.

Hanauer SB, Feagan BV, Lichtenstein GR, et al: Maintenance infliximab for Crohn's disease: the Accent I randomized trial, *Lancet* 359:1541-1549, 2002.

Ho GT, Smith L, Aitken S, et al: The use of adalimumab in the management of refractory Crohn's disease, *Aliment Pharmacol Ther* 27:308-315, 2008.

Jani N, Regueiro MD: Medical therapy for ulcerative colitis, *Gastroenterol Clin North Am* 31:147-166, 2002.

Lichtenstein GR, Abreu MT, Cohen R, Tremaine W: American Gastroenterological Association Institute medical position statement on corticosteroids, immunomodulators, and infliximab in inflammatory bowel disease, *Gastroenterology* 130:935-939, 2006.

MacDonald JB, McDonald JW: Natalizumab for induction of remission in Crohn's disease, *Cochrane Database Syst Rev* 1:CD006097, 2007.

O'Sullivan NA, O'Morain CA: Nutritional therapy in Crohn's disease, *Inflamm Bowel Dis* 4:45-53, 1998.

Stein RB, Hanauer SB: Medical therapy for inflammatory bowel disease, *Gastroenterol Clin North Am* 28:297, 1999.

Targan SR, Feagan BG, Fedorak RN, et al: Natalizumab for the treatment of active Crohn's disease: results of the ENCORE trial, *Gastroenterology* 132:1672-1683, 2007.

Yousry TA, Major EO, Ryschkewitsch C, et al: Evaluation of patients treated with natalizumab for progressive multifocal leukoencephalopathy, *N Engl J Med* 353:924-933, 2006.

Microscopic Colitis (Lymphocytic or Collagenous Colitis)

155

Martin H. Floch

Microscopic colitis is a syndrome in which patients have (1) chronic diarrhea, (2) normal findings on colonoscopy and a normal-appearing mucosa, (3) histologic evidence of increased cellular infiltrate in the lamina propria, and (4) either (a) full-blown *lymphocytic colitis* with intraepithelial lymphocytes and increased infiltrate in the lamina propria or (b) *collagenous colitis* with a collagen band below the epithelium larger than 10 μm (**Fig. 155-1**).

Microscopic colitis has been difficult to define, but numerous reports have clarified the syndrome. Initially described as "collagenous" colitis, with watery diarrhea and deposition of collagen, it was later noted that many patients with chronic watery diarrhea have a lymphocytic infiltrate and a definite intraepithelial lymphocytic infiltrate. During the past decade, many patients have been described with watery diarrhea responsive to therapy and with increased infiltrate only in the lamina propria. It is

Lymphocytic colitis

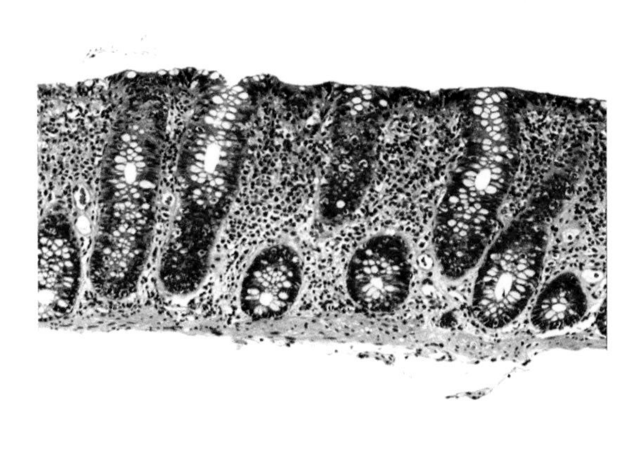

Low-power microphotograph of lymphocytic colitis that shows increased lymphocytic and round cell infiltration in the lamina propria. The crypts appear normal.

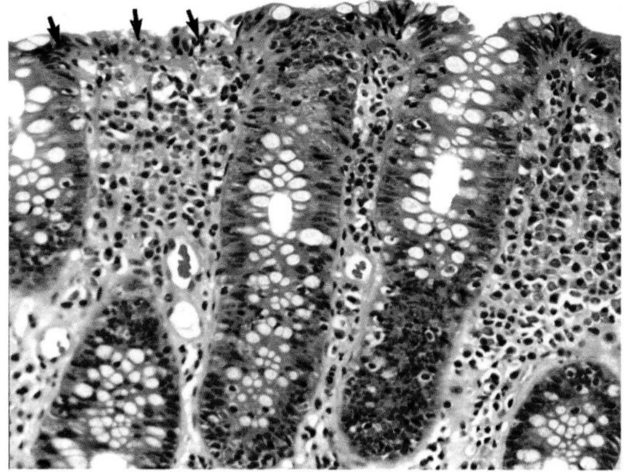

High-power microphotograph of lymphocytic colitis (same patient and biopsy as in figure at left). *Arrows* indicate the classic infiltrate of lymphocytes in the epithelium.

Collagenous colitis

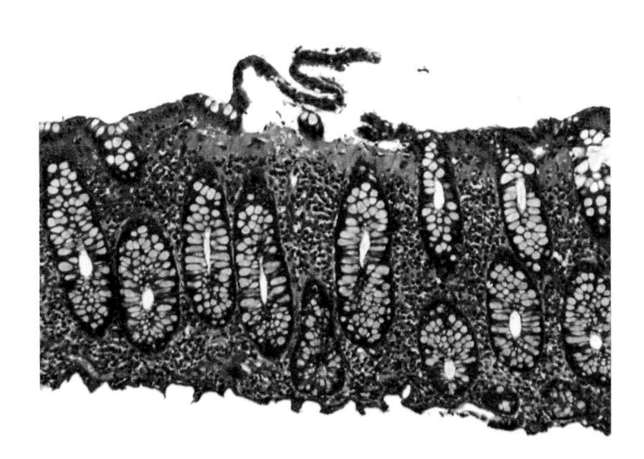

Colon biopsy specimen showing collagenous colitis (trichrome stain [blue]). Note the enlarged subepithelial collagen layer.

We are grateful to Dr. Marie Robert, who supplied these photomicrographs.

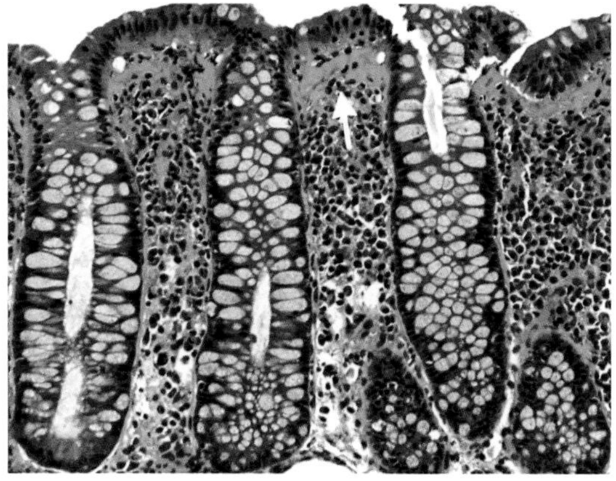

High-power microphotograph of collagenous colitis that shows an enlarged (pink) collagen layer and an increased lymphocytic infiltrate in the lamina propria. The *arrow* indicates a cellular element (fibroblast nucleus) entrapped in the enlarged collagen layer.

Figure 155-1 *Microscopic Colitis: Lymphocytic and Collagenous.*

important to note that watery diarrhea and absence of gross findings on colonoscopy are components of the syndrome. Histologic examination of the biopsy specimen shows no distortion of the crypts, as seen in ulcerative colitis.

One study found microscopic colitis in 9.5% of patients with watery diarrhea. Incidence is reported as 4.2 per 100,000 population, with the lymphocytic type three times more common than the collagenous form. The cause and etiology remain unknown, but it is thought that microscopic colitis is associated with toxins, drugs, or latent autoimmune enteropathy, as well as various diseases (e.g., celiac). Almost 30% of patients with celiac disease have been reported to have some form of microscopic colitis; thus a similar cause is suspected.

CLINICAL PICTURE

The cardinal feature of microscopic colitis is watery diarrhea, generally manifesting during the sixth decade of life. Less than 50% of patients have abdominal pain along with diarrhea, and less than 33% experience weight loss. Reports indicate an association with lansoprazole and cholestyramine, omeprazole, nonsteroidal antiinflammatory drugs (NSAIDs), and celiac disease.

The effects of watery diarrhea can be severe, and many patients become hypokalemic. Systemic inflammation is rare. Microscopic colitis appears to be twice as prevalent in women as in men.

DIAGNOSIS

The diagnosis is made through histopathology. Often, a patient with low-grade watery diarrhea receives a diagnosis of irritable bowel syndrome on the basis of normal gross colonoscopy findings, but no biopsy specimen was taken because the mucosa looked normal. Mucosal biopsy of the colon must be performed to determine whether a patient with watery diarrhea has microscopic colitis. More detailed research will allow a determination of the incidence and prevalence of the disease.

The histopathology of microscopic colitis has been controversial. An initial criterion was "increased collagen deposition," and then the "presence of intraepithelial lymphocytes" was added as a criterion. It is now accepted that increased levels of lymphocytes and plasma cells, with the possibility of eosinophils in the lamina propria, correlating with the clinical picture and resolving with treatment, make the diagnosis. The pathologist must determine whether crypt distortion has occurred. If so, the diagnosis would be nonspecific ulcerative colitis.

TREATMENT AND MANAGEMENT

Eliminating any agent thought to cause microscopic colitis is paramount. The medications most often used are *mesalamine* and *budesonide*, with the latter favored in a recent study. Trials have been conducted using bismuth subsalicylate, antibiotics, and other agents, but 5-aminosalicylic acid (5-ASA) compounds can induce remission. Some experts prefer a 4- to 8-week trial of bismuth subsalicylate before administering mesalamine. It may be necessary to use prednisone to induce remission if bismuth subsalicylate and mesalamine fail to do so. Some reports show that calcium channel blockers can be helpful.

It is important to document microscopically that the patient's condition has returned to normal with therapy or with withdrawal of a suspected etiologic agent.

COURSE AND PROGNOSIS

Recent large studies indicate the difficulty of determining the relevance of histopathologic findings to the clinical course. All microscopic colitides seem to behave similarly. The course of microscopic colitis is benign, and most patients respond to mesalamine, budesonide, bismuth subsalicylate, and if needed, corticosteroids or an experimental therapeutic agent.

More than 70% of patients experience long-term cessation of diarrhea. Other patients, however, experience relapse and require repeat therapy after being in remission. In either case, the prognosis is excellent.

ADDITIONAL RESOURCES

Bonderup OK, Hansen JB, Birket-Smith L, et al: Budesonide treatment of collagenous colitis: a randomized, double-blind, placebo-controlled trial with morphometric analysis, *Gut* 52:248-251, 2003.

Fernandez-Banares F, Salas A, Esteve N, et al: Collagenous and lymphocytic colitis: evaluation of clinical and histologic features, response to treatment, and long-term follow-up, *Am J Gastroenterol* 908:340-347, 2003.

Fraser AG, Warren BF, Chandrapala R, Jewell PP: Microscopic colitis: a clinical and pathologic review, *Scand J Gastroenterol* 37:1241-1245, 2002.

Lindstrom CG: Collagenous colitis with watery diarrhea: a new entity, *Pathol Urol* 11:87-91, 1976.

Pardi DS, Loftus EV Jr, Smyrk TC, et al: The epidemiology of microscopic colitis: a population-based study in Olmstead County, Minnesota, *Gut* 56:504-508, 2007.

Read NW, Kregs GR, Reid MG, et al: Chronic diarrhea of unknown origin, *Gastroenterology* 68:264-270, 1980.

Robert M: Pathology of microscopic, lymphocytic, and collagenous colitis, *J Clin Gastroenterol* 38:S17-S26, 2004.

Schiller L: The clinical spectrum of microscopic colitis, *J Clin Gastroenterol* 38:S27-S30, 2004.

Stroehlein JR: Microscopic colitis, *Curr Treat Options Gastroenterol* 3:231-236, 2007.

Townsend RD, Lestina LS, Bensen SP, et al: Lansoprazole-associated microscopic colitis: a case series, *Am J Gastroenterol* 97:2908-2913, 2002.

Wald A: Other diseases of the colon and rectum. In Feldman M, Friedman LS, Brandt LJ, editors: *Gastrointestinal and liver disease*, ed 8, Philadelphia, 2006, Saunders-Elsevier, pp 2811-2832.

Neoplasms of the Large Bowel: Colon Polyps

156

Martin H. Floch

A colon *polyp* is any elevation of the colon mucosal surface (**Fig. 156-1**). It may be of any size, sessile or pedunculated, and benign or malignant. *Benign polyps* are categorized as neoplastic, nonneoplastic, and submucosal. *Neoplastic polyps* are usually considered premalignant, and most are adenomas. *Adenomas* may be classified histologically as tubular, tubulovillous, or villous. The polyp may demonstrate low-grade or high-grade dysplasia. High-grade dysplasia in a polyp is often referred to as "intermucosal carcinoma (malignant)," or *carcinoma in situ*. Carcinomatous polyps are discussed in Chapters 157 to 159.

Nonneoplastic polyps include mucosal, hyperplastic, and inflammatory (pseudopolyps) types, hamartomas; and other rare types. *Submucosal polyps* are lipomas, lymphoid collections, leiomyomas, hemangiomas, fibromas, and rare presentations of endometriosis, pneumatosis cystoides intestinalis, colitis cystica profunda, or metastatic lesions.

The prevalence of colorectal neoplasia varies worldwide: 30 to 40 per 100,000 population in the United States; 15 to 30 per 100,000 in Europe; and less than 5 to 10 in South America and Asia. The prevalence varies with the population being studied.

Hereditary polyposis syndromes show that colorectal neoplasia has a *genetic* component (see Chapter 158). The progression of normal mucosa to neoplasia is associated with a loss of the *APC* gene in the cell, and the progression to carcinoma is associated with K-*ras*, *DCC*, and *p*53 activity. The genetic component is complex; the interested reader is referred to the Additional Resources. It is estimated that up to 20% of neoplastic polyps result from genetic effects. As genetic studies evolve, this picture may become clearer.

Strong epidemiologic evidence shows that *dietary* factors play a major role in polyp formation through the microflora and intestinal microbiologic relationships. Diets with high levels of fat and red meat and low levels of fiber are associated with a higher incidence of neoplastic polyp formation. Australian patients who combined a low-fat diet with increased bran intake had a lower incidence of recurrent polyps; increased bran or low-fat diet alone did not result in a decrease. Not all studies are in agreement, but the very low incidence of neoplastic polyps in societies whose diets are high in fiber and low in saturated fats is incriminating for these nutrient factors. Dietary carcinogens and micronutrient deficiencies are also thought to play a role, but no proof of a cause-and-effect sequence is yet available.

Polyps usually develop in the rectum and the sigmoid and descending colon. Incidence seems to be greater on the right side of the colon. Most polyps can progress from an adenoma to a carcinoma, *flat adenomas* have a potential for malignancy, and the amount of severely dysplastic tissue in a polyp is related to its size. Flat polyps are of great interest to endoscopists because of their malignant potential and because they are more challenging to identify and remove completely. Polyps shown on histologic examination to contain villous elements are associated with a higher incidence of malignancy. Therefore, patients with flat adenomas and polyps with significant villous elements are at higher risk for malignancy.

Patients with hyperplastic or inflammatory polyps are also at risk for carcinoma. *Serrated polyps*, a combination of hyperplastic and adenomatous elements, are being found with increasing frequency and have malignant tendencies. Most submucosal polyps are benign. Many patients with carcinoids, metastatic lesions, melanomas, lymphomas, and Kaposi sarcoma have malignant polypoid formation in the colon. Except for its association with some malabsorption syndromes, lymphoid hyperplasia has no malignant significance.

CLINICAL PICTURE

Polypoid lesions are often accompanied by occult or gross bleeding. Depending on their position, they may cause intussusception or obstruction of the bowel; therefore, they rarely cause pain. Usually, polyps are detected during colonoscopic or barium enema screening for other symptoms. If the polyp is large, which is now unusual, the patient experiences a change in bowel habits and obstruction. Large lesions are rarely benign. Unlike benign polyps, malignant formation is life threatening.

Three findings with a polyp are important in risk for malignancy: size (>2 cm), histologic type (villus formation), and degree of dysplasia (severe).

Malignant polyps are discussed in Chapters 159 and 160. Screening and treatment are discussed in Chapter 157.

ADDITIONAL RESOURCES

Alberts DS, Ritenbaugh C, Story JA, et al: Randomized, double-blinded, placebo-controlled study of wheat bran fiber and calcium on fecal bile acids in patients with resected adenomatous colon polyps, *J Natl Cancer Inst* 88:81-92, 1996.

East JE, Saunders BP, Jass JR: Sporadic and syndromic hyperplastic polyps and serrated adenomas of the colon: classification, molecular genetics, natural history, and clinical management, *Gastroenterol Clin North Am* 37: 25-46, 2008.

Giovannucci E, Stampfer MJ, Colditz G, et al: Relationship of diet to risk of colorectal adenoma in men, *J Natl Cancer Inst* 84:91-98, 1992.

Hurstone DP, Cross SS, Adam I, et al: A prospective clinicopathological and endoscopic evaluation of flat and depressed colorectal lesion in the United Kingdom, *Am J Gastroenterol* 98:2543-2549, 2003.

Itzkowitz SH, Rochester J: Colonic polyps and polyposis syndromes. In Feldman N, Friedman LS, Brandt LJ, editors: *Gastrointestinal and liver disease*, ed 8, Philadelphia, 2006, Saunders-Elsevier, pp 2713-2757.

Jacobs ET, Thompson PA, Martinez ME: Diet, gender, and colorectal neoplasia, *J Clin Gastroenterol* 41:731-746, 2007.

Koushik A, Hunter DJ, Spiegelman D, et al: Fruits, vegetables and colon cancer risk in a pooled analysis of 14 cohort studies, *J Natl Cancer Inst* 99:1471-1483, 2007.

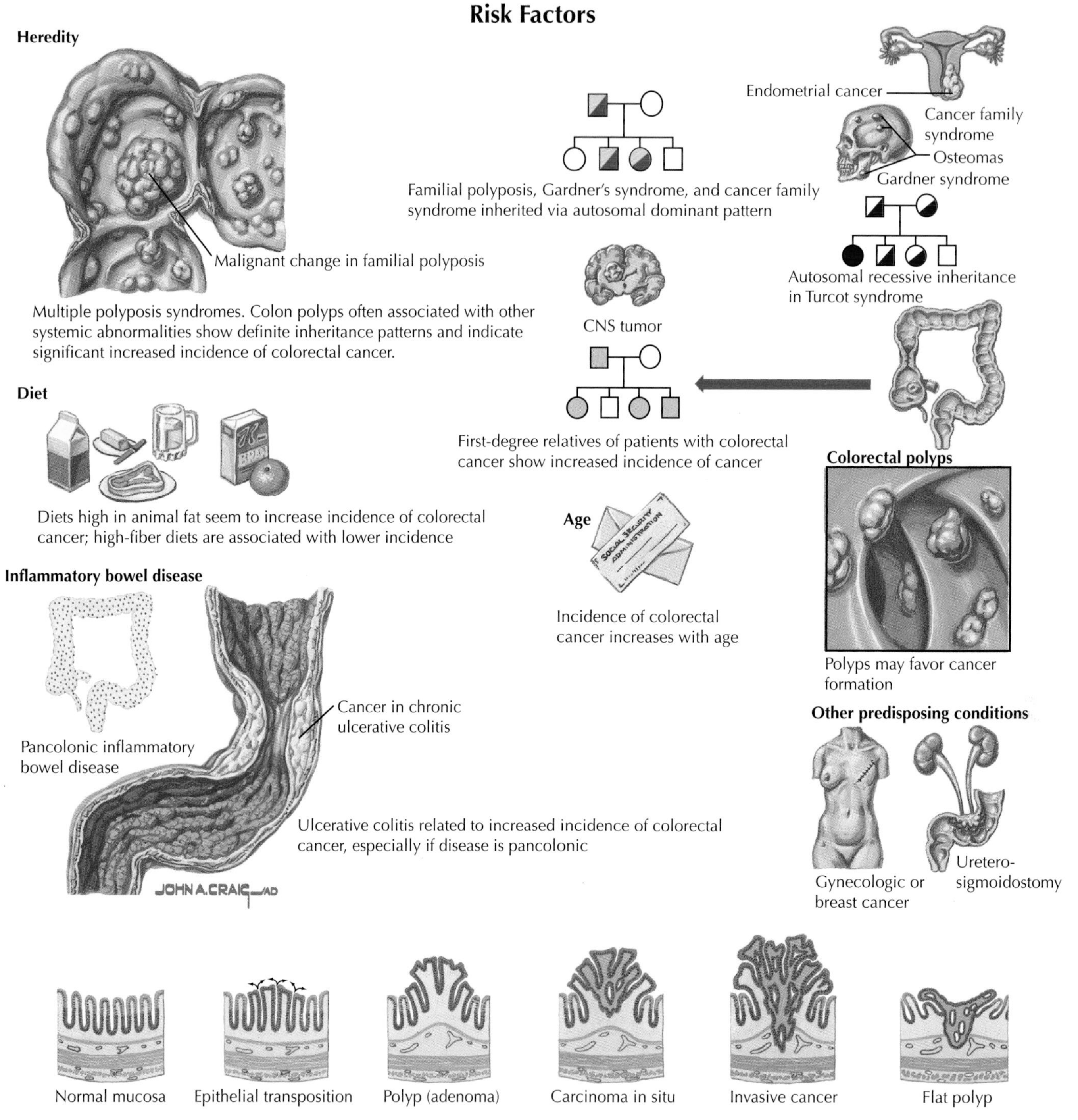

Risk Factors

Heredity

Malignant change in familial polyposis

Multiple polyposis syndromes. Colon polyps often associated with other systemic abnormalities show definite inheritance patterns and indicate significant increased incidence of colorectal cancer.

Familial polyposis, Gardner's syndrome, and cancer family syndrome inherited via autosomal dominant pattern

Endometrial cancer

Cancer family syndrome

Osteomas

Gardner syndrome

Autosomal recessive inheritance in Turcot syndrome

CNS tumor

First-degree relatives of patients with colorectal cancer show increased incidence of cancer

Colorectal polyps

Polyps may favor cancer formation

Diet

Diets high in animal fat seem to increase incidence of colorectal cancer; high-fiber diets are associated with lower incidence

Age

Incidence of colorectal cancer increases with age

Inflammatory bowel disease

Pancolonic inflammatory bowel disease

Cancer in chronic ulcerative colitis

Ulcerative colitis related to increased incidence of colorectal cancer, especially if disease is pancolonic

JOHN A.CRAIG_AD

Other predisposing conditions

Gynecologic or breast cancer

Uretero-sigmoidostomy

Normal mucosa | Epithelial transposition | Polyp (adenoma) | Carcinoma in situ | Invasive cancer | Flat polyp

Figure 156-1 *Neoplasms of the Large Bowel: Colon Polyps.*

Makinen MJ: Colorectal serrated adenocarcinoma, *Histopathology* 50:131-150, 2007.

McGarr SE, Ridlon JM, Hylemon PB: Diet, anaerobic bacterial metabolism and colon cancer: a review of the literature, *J Clin Gastroenterol* 39:98-109, 2005.

Pasche B: Familial colorectal cancer: a genetics treasure trove for medical discovery, *JAMA* 299:2564-2565, 2008.

Potter JD: Fiber and colorectal cancer: where to now? *N Engl J Med* 340:223-224, 1999.

Soetikno RM, Kalternbach T, Rouse RV, et al: Prevalence of nonpolypoid (flat and depressed) colorectal neoplasms in asymptomatic and symptomatic adults, *JAMA* 299:1027-1035, 2008.

Neoplasms of the Large Bowel: Screening and Treatment of Colon Polyps

157

Martin H. Floch

The diagnosis of colon polyps can be made through sigmoidoscopy, colonoscopy, barium enema, or virtual colonoscopy (**Fig. 157-1**). The diagnosis is made when symptoms indicate the need a polyp search or a screening procedure to prevent colon cancer. Screening allows small lesions to be detected and removed before they can advance to carcinoma.

Screening procedures include fecal occult blood testing, digital rectal examination, sigmoidoscopy, colonoscopy, and virtual colonoscopy. All may detect a lesion, but *colonoscopy* results in the greatest yield and enables biopsy and possible removal and identification of the polyp. Therefore, colonoscopy is the screening procedure of choice for most gastroenterologists. Statistical evidence shows that a combination of fecal occult blood testing and sigmoidoscopy can be as effective as colonoscopy or double-contrast barium enema in preventing mortality from colon cancer, if cost is an issue and colonoscopy is unavailable. Endoscopy and stool screening procedures are described in Chapters 129 and 131.

TREATMENT AND MANAGEMENT

Once identified, a polypoid lesion should be removed. Most lesions smaller than 3 cm can be removed during endoscopy. Biopsy, *snaring*, elevating the polyp with water injection and removing it piecemeal, and cauterization are all used effectively to ablate polyps. It is hoped that the entire polyp has been removed; the removed tissue is evaluated histopathologically for possible carcinoma in situ. The histopathology is important to ensure that all neoplastic tissue and highly dysplastic or serrated lesions are removed with no possibility of spread. The histology and correlation with genetic and growth factors (e.g., microsatellite instability, hereditary nonpolyposis colorectal cancer) are important (see Additional Resources). Complication rates of screening procedures are low.

After a polyp is removed, the question is how often colonoscopy should be repeated. If there is any question about total removal, or if the colonoscopy was difficult, a repeat procedure is performed within 1 year. However, if there is only a single polyp, the question is whether repeat colonoscopy should be performed in 3 to 5 years. Recent studies reveal that small tumors can grow rapidly. A physician or patient may want a repeat examination in 3 years, but if examination findings are excellent, screening can be delayed until 5 years, with fecal occult blood testing in the "off" years. The guidelines are clear, but variations occur depending on each patient's health and circumstances.

The patient identified with polyps should be treated for risk prevention and possible polyp chemoprevention. The patient should also be advised to maintain adequate weight, decrease the intake of foods high in polyunsaturated fat, and increase the intake of foods high in dietary fiber. An 81-mg dose of aspirin daily seems effective in decreasing the risk for cancer. Other chemopreventive agents, such as nonsteroidal antiinflammatory drugs (NSAIDs; e.g., sulindac, COX-2 inhibitors), and increased calcium intake show some statistical effectiveness, but low-dose aspirin is the most widely accepted chemopreventive regimen.

COURSE AND PROGNOSIS

In the United States, vigorous screening programs in conjunction with colonoscopic removal of polyps and small lesions, combined with chemopreventive agents, can decrease the incidence of colon cancer and related deaths. In other countries, incidence and mortality appear to be stabilizing but vary with environmental and economic conditions.

The chemopreventive agents include aspirin, celecoxib, and now statins. These drugs can be associated with cardiovascular complications and gastrointestinal bleeding, and their use and selection depend on the physician's recommendations for the individual patient.

ADDITIONAL RESOURCES

Baron JA, Cole BF, Sandler RS, et al: A randomized trial of aspirin to prevent colorectal adenomas, *N Engl J Med* 348:891-898, 2003.

Fletcher RH: Screening sigmoidoscopy: how often and how good? *JAMA* 290:106-108, 2003.

Pinol V, Castells A, Andreu M, et al: Accuracy of revised Bethesda guidelines, microsatellite instability, and immunohistochemistry for the identification of patients with hereditary nonpolyposis colorectal cancer, *JAMA* 293(16):1986-1994, 2005.

Poynter MPH, Gruber SB, Higgins PDR, et al: Statins and the risk of colorectal cancer, *N Engl J Med* 352:2184-2192, 2005.

Psaty BM, Potter JD: Risks and benefits of celecoxib to prevent recurrent adenomas, *N Engl J Med* 355:950-952, 2006.

Spring KJ, Zhao ZZ, Karamatic R, et al: High prevalence of sessile serrated adenomas with *BRAF* mutations: a prospective study of patients undergoing colonoscopy, *Gastroenterology* 131:1400-1407, 2006.

Rex DK, Johnson DA, Lieberman DA, et al: Colorectal cancer prevention 2000: screening recommendations of the American College of Gastroenterology, *Am J Gastroenterol* 95:868-877, 2000.

Riegert-Johnson DL, Johnson RA, Rabe KG, et al: The value of MUTYH testing in patients with early-onset microsatellite stable colorectal cancer referred for hereditary nonpolyposis colon cancer syndrome testing, *Genet Test* 11:361-365, 2007.

Rostom A, Dube C, Lewin G, et al: Nonsteroidal and anti-inflammatory drugs and cyclooxygenase-2 inhibitors for primary prevention of colorectal cancer: a systematic review prepared for the U.S. Prevention Services Task Force, *Ann Intern Med* 146:376-389, 2007.

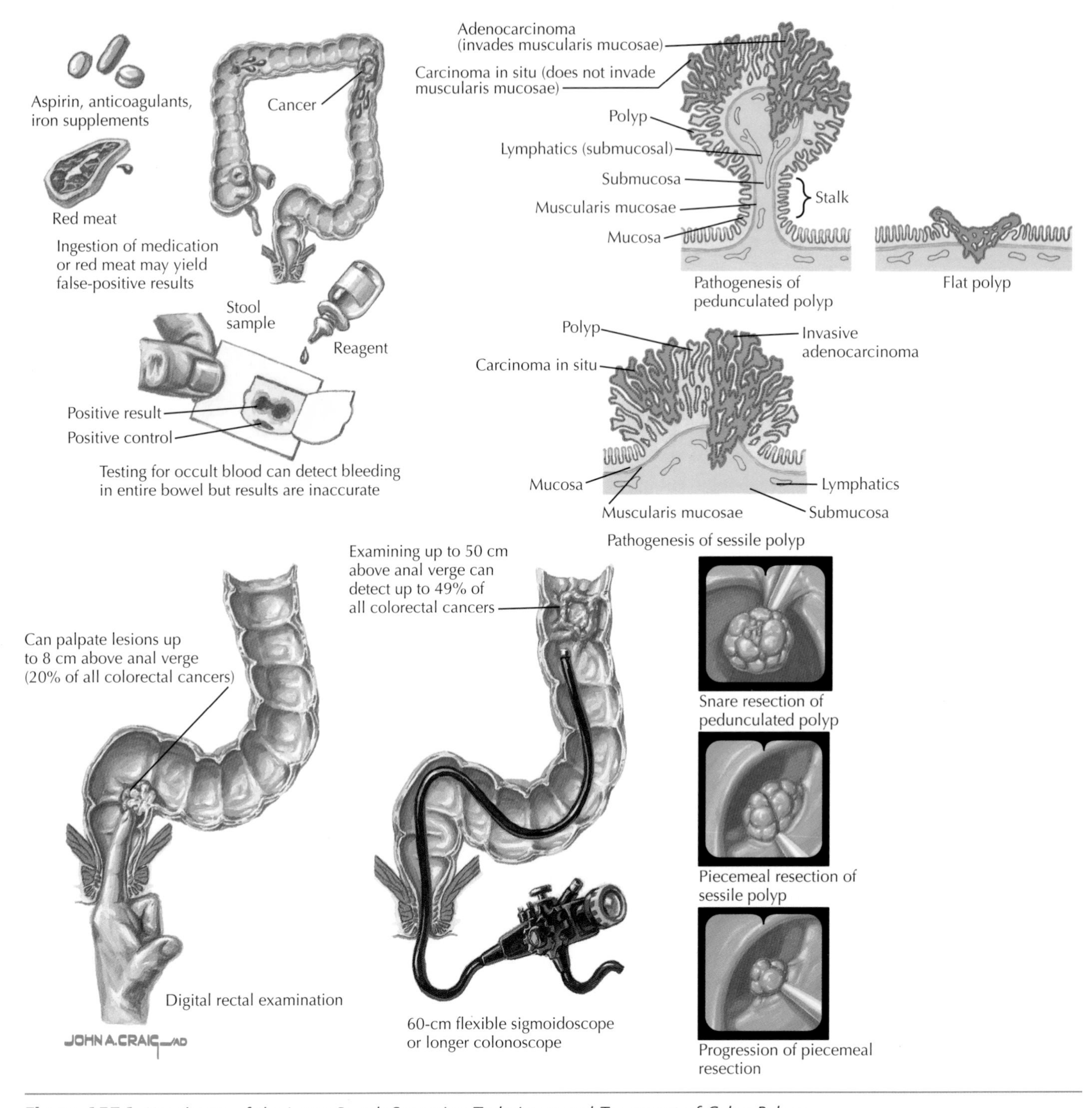

Figure 157-1 *Neoplasms of the Large Bowel: Screening Techniques and Treatment of Colon Polyps.*

Sandler RS, Halabi S, Baron JA, et al: A randomized trial of aspirin to prevent colorectal adenomas in patients with previous colorectal cancer, *N Engl J Med* 348:883-890, 2003.

Schoen RE, Pinsky PF, Weissfeld JL, et al: Results of repeat sigmoidoscopy three years after a negative examination, *JAMA* 290:41-48, 2003.

Walsh JNE, Terdiman JP: Colorectal cancer screening: scientific review, *JAMA* 289:1296, 2003.

Walsh JNE, Terdiman JP: Colorectal cancer screening: clinical applications, *JAMA* 289:1297-1302, 2003.

Familial Adenomatous Polyposis and Polyposis Syndromes

158

Martin H. Floch

Gastrointestinal (GI) polyposis is the presence of multiple polypoid lesions in the GI tract (**Fig. 158-1**). Numerous syndromes have now been classified (**Box 158-1**).

Familial adenomatous polyposis (FAP) is the most common and best known of the polyposis syndromes. It has an autosomal dominant inheritance pattern, and it occurs from germline mutations of the *APC* gene (see Additional Resources for details on the genetics of these disorders). The prevalence is approximately 3 cases per 100,000 population. There appears to be no geographic or significant ethnic variation.

Recently, a germline *MYH* mutation has been described with a syndrome similar to that in patients who do not have the *APC* mutation. These patients were identified in the United Kingdom with a clinical syndrome similar to FAP. The significance of this genetic finding requires further evaluation.

CLINICAL PICTURE

Patients with GI polyposis often have a family history of the syndrome or rectal bleeding, diarrhea, or abdominal pain. Most patients seek treatment in the fourth decade of life if the family history is unknown. If a patient has a variant syndrome, disease manifestation may be that of the syndrome, such as a bone lesion in Gardner syndrome, a neurologic lesion in Turcot syndrome, or an intussusception in Peutz-Jeghers syndrome. The details of the syndromes listed in Box 158-1 are not described in this chapter; they are unique to those particular syndromes, particularly when associated polyps and lesions develop outside the colon.

DIAGNOSIS

Multiple polyps in the colon can be identified during endoscopy. Colonoscopy is essential for evaluating the degree and extent of the polyps and for ruling out malignant transformation. Polyposis may be detected through barium air contrast study, but colonoscopy is still essential to rule out malignant changes in any of the polyps. Once polyposis syndrome is identified, upper endoscopy and upper barium contrast studies should be performed to rule out gastric and duodenal polyps. Polyps can occur in any location. The rate of occurrence in the duodenal area is 3% to 5% in FAP. Up to 2% of patients may have pancreatic or thyroid carcinoma. Hepatoblastoma has similarly been reported in up to 1.6% of patients.

Genetic testing has become available worldwide, but methods change rapidly. Laboratories can be identified through the Internet (www.gtest.org). Although testing is sophisticated and is not performed regularly in most clinical settings, it is recommended when the clinical diagnosis is uncertain, when polyposis syndrome appears certain but the patient has no family history, and when relatives are tested in a family with FAP.

Box 158-1 Gastrointestinal Polyposis Syndromes

Inherited Polyposis Syndromes

Adenomatous Polyposis Syndromes

Familial adenomatous polyposis (FAP)

Variants of FAP

- Gardner syndrome
- Turcot syndrome
- Attenuated adenomatous polyposis coli

Hamartomatous Polyposis Syndromes

Peutz-Jeghers syndrome

Juvenile polyposis

Syndromes related to juvenile polyposis

- Cowden disease
- Bannayan-Ruvalcaba-Riley syndrome

Rare hamartomatous polyposis syndromes

- Hereditary mixed polyposis syndrome
- Intestinal ganglioneuromatosis and neurofibromatosis
- Devon family syndrome
- Basal cell nevus syndrome

Noninherited Polyposis Syndromes

Cronkhite-Canada syndrome

Hyperplastic polyposis syndrome

Lymphomatous polyposis

Nodular lymphoid hyperplasia

Modified from Feldman N, Friedman LS, Brandt LJ, editors: Gastrointestinal and liver disease, *ed 8, Philadelphia, 2006, Saunders-Elsevier, p 2737.*

TREATMENT AND MANAGEMENT

Once polyposis syndrome is diagnosed, treatment depends on the particular syndrome. In FAP, the patient may have few polyps that gradually increase in number. It is essential that the patient undergo colon screening if genetic testing indicates the potential for FAP or if a first-degree relative has FAP. Screening should occur early, by age 10 to 12 years. If polyps are not found, screening can be performed at longer intervals. If polyps are found, the timing of colectomy varies. Clinicians prefer to wait until adulthood, but given the high risk for colon cancer in these patients, ileal pouch anal anastomosis (IPAA) is now recommended early and in place of colostomy, particularly because there is a lower incidence of pouchitis in these patients than in

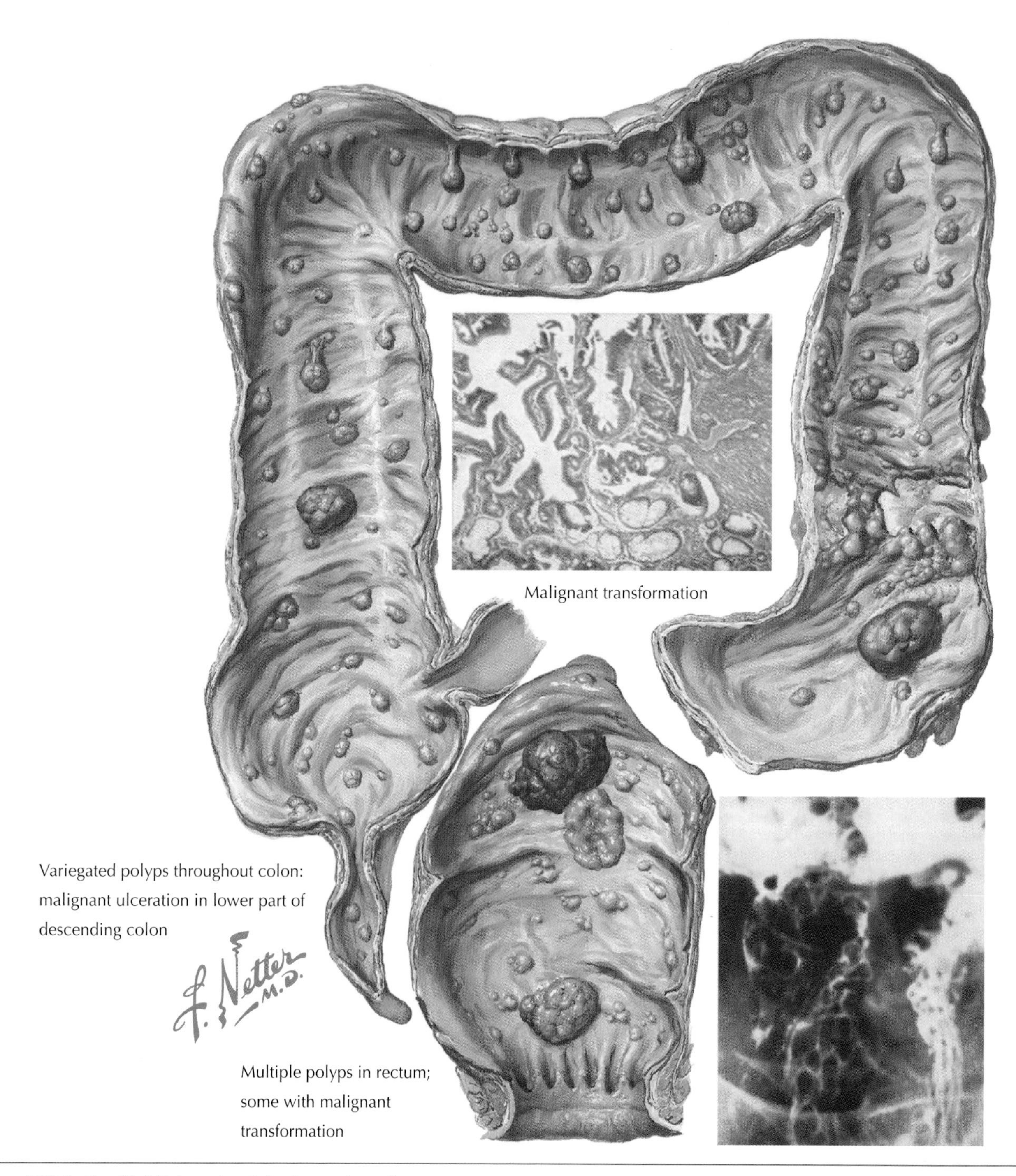

Figure 158-1 *Familial Polyposis of the Large Intestine.*

those who undergo the procedure for ulcerative colitis. Some clinicians still prefer total colectomy with ileorectal anastomosis. However, the risk for rectal carcinoma is always present, thus the IPAA recommendation (see Chapter 149).

In all other polyposis syndromes, colon screening is recommended at different intervals. Given the high rates of duodenal lesions and duodenal adenocarcinoma in patients with FAP, however, upper endoscopy evaluation should be performed regularly.

COURSE AND PROGNOSIS

With early screening and surgical correction, the prognosis for patients with polyposis syndromes should improve. Life expectancy for a person with untreated FAP is not much longer than 40 years, but it improves dramatically if the patient undergoes total colectomy. Duodenal cancer and desmoid tumors are the major complications and mortality risks after colectomy. Other cancers are reported, but the rate of carcinomatosis does not increase in these patients. Rectal cancer occurs in a high percentage of patients with polyposis syndrome, so they must be placed under careful surveillance or, even better, should undergo IPAA after colectomy.

ADDITIONAL RESOURCES

Bulow S, Bulow C, Vassen H, et al: Colectomy and ileorectal anastomosis is still an option for selected patients with familial adenomatous polyposis, *Dis Col Rect*, June 4, 2008 (Epub).

Desai TK, Barkel D: Syndromic colon cancer: Lynch syndrome and familial adenomatous polyposis, *Gastroenterol Clin North Am* 37:47-72, 2008.

Itzkowitz SH, Rochester J: Colonic polyps and polyposis syndromes. In Feldman N, Friedman LS, Brandt LJ, editors: *Gastrointestinal and liver disease*, ed 8, Philadelphia, 2006, Saunders-Elsevier, pp 2713-2757.

Levin B, Lieberman DA, McFarland B, et al: Screening and surveillance for early detection of colorectal cancer and adenomatous polyps, 2008: a joint guideline from the American Cancer Society, the US Multi-Society Task Force on Colorectal Cancer, and the American College of Radiology, *Gastroenterology* 134:1570-1595, 2008.

Sieber OM, Lipton L, Crabtree M, et al: Multiple colonic adenomas, classic adenomatous polyposis, and germ-line mutations in *MYH*, *N Engl J Med* 348:791-799, 2003.

Vassen HF, Moslein G, Alonso A, et al: Guidelines for the clinical management of familial adenomatous polyposis (FAP), *Gut* 57:704-713, 2008.

Colon Cancer: Clinical Picture and Diagnosis

159

Martin H. Floch

Cancer of the colon is a major cause of morbidity and mortality in Western countries with similar dietary and lifestyle habits. Colon cancer is believed to be an acquired genetic disease caused by exposure to environmental carcinogens. A genetic instability occurs, followed by mutational changes that produce neoplastic clones. Free from homeostatic growth controls, the cancer increases with time and total carcinogenic exposure. The clinical result is that cancer incidence rises as an exponential function of age. Deaths from colorectal cancer typically begin to increase slowly in the fifth decade of life, rising steeply with advancing age. In the United States, 56,700 persons died of colorectal cancer in 2001; the incidence varies greatly with the population, from 30% to 35% per 100,000.

The prevalence in whites is highest in the ascending colon and cecum (22% in men, 27% in women) and in the sigmoid colon (25% in men, 23% in women). The incidence is approximately 3 to 4 per 100,000 population in undeveloped countries of Asia and 35 per 100,000 in some areas in the United States. The epidemiologic association of high-fat, high-meat, low-fiber diets correlates with worldwide incidence discrepancies. Nutrients such as calcium, selenium, and antioxidant vitamins may decrease this incidence, but chemoprevention programs based on these substances have not yet been effective.

Colorectal cancer develops throughout the large bowel. Approximately 45% of cases are observed in the rectum and sigmoid, but the distribution seems to be increasing on the right side, with as many as 25% in the cecum and ascending colon. Alterations in K-*ras*, *APC*, *DCC*, and *p*53 genes have all been demonstrated in association with malignant transformation. Other significant genetic findings are unfolding. Syndromes such as *hereditary nonpolyposis colorectal cancer* (HNPCC), or Lynch syndrome, are being discovered and understood. Families with this syndrome must be followed closely. Patients at particular risk for colorectal cancer are those who have polyposis syndromes (see Chapter 158).

CLINICAL PICTURE

Family history of cancer of the colon, polyps, or inflammatory bowel disease (IBD) raises the index of suspicion for colorectal cancer in patient with classic signs, such as rectal bleeding, change in bowel habits, signs of obstruction, or anemia in the presence of occult blood (**Fig. 159-1**).

DIAGNOSIS

Any patient with clinical suspicion of colon cancer must undergo colonoscopic evaluation. Findings from the physical examination, which must include digital rectal examination, may indicate the need for endoscopic evaluation. If the patient has an anorectal lesion (see Chapter 161), diagnosis can be made through proctoscopy or sigmoidoscopy. Even if the patient has a low lesion, however, full colonoscopic examination should be performed because synchronous lesions develop in as many as 30% of patients; treatment depends on understanding the distribution of a polyp or a malignant lesion anywhere in the colon. If full examination is not possible, contrast air enemas can be used with absorbable media or with barium if there is no chance of obstruction. Large biopsy specimens should be obtained for diagnosis. Evaluation should be followed by computed tomography (CT) of the abdomen and by chest evaluation to stage the cancer for the appropriate treatment (see Chapter 160). Endoscopic ultrasound can be helpful, particularly in the rectum, in staging the disease to determine whether it has extended through the wall and whether nodes are present.

Many lesions can mimic those of colon cancer, including benign tumors, areas of diverticulitis, IBD, masses caused by parasites, and other tumors of the colon, such as lymphoma, carcinoid, Kaposi sarcoma, and extrinsic lesions involving the peritoneum. When direct biopsy material cannot be obtained, biopsy through laparoscopy or laparotomy is necessary to determine appropriate treatment.

Full serum chemistry profiles are needed to determine whether the patient has anemia. Carcinoembryonic antigen (CEA) level may be elevated, and CEA level does have some relationship to outcome.

Colorectal cancer is a major clinical problem, but preventive and control measures can be effective. Recent data indicate that its incidence is stabilizing, probably associated with intensive screening programs in the United States. Any patient with a suspicious lesion or with a high-risk background should undergo colonoscopic screening (see Chapters 156 and 157). Fecal occult blood testing, sigmoidoscopic examination, and barium contrast enemas have been shown to be effective in large screening programs. However, most gastroenterologists prefer colonoscopy. If it is unavailable, fecal occult blood testing and sigmoidoscopy can be used for screening, and barium air contrast can be used for diagnostic evaluation. Contrast studies are more effective at detecting larger cancerous lesions than small, benign polyps. In fact, before the advent of colonoscopy, the barium air contrast procedure was used for preoperative evaluation.

Chapter 160 discusses the treatment, management, course, and prognosis of colon cancer.

ADDITIONAL RESOURCES

Bresalier RS: Malignant neoplasms of the large intestine. In Feldman M, Friedman LS, Brandt LJ, edtors: *Gastrointestinal and liver disease*, ed 8, Philadelphia, 2006, Saunders-Elsevier, pp 2759-2810.

Lagerstedt Robinson K, Liu T, Vandrovcova J, et al: Lynch syndrome (hereditary nonpolyposis colorectal cancer) diagnostics, *J Natl Cancer Inst* 99:291-299, 2007.

Vasen HF: The Lynch syndrome (hereditary nonpolyposis colorectal cancer) (review), *Aliment Pharmacol Ther* 26(suppl 2):113-126, 2007.

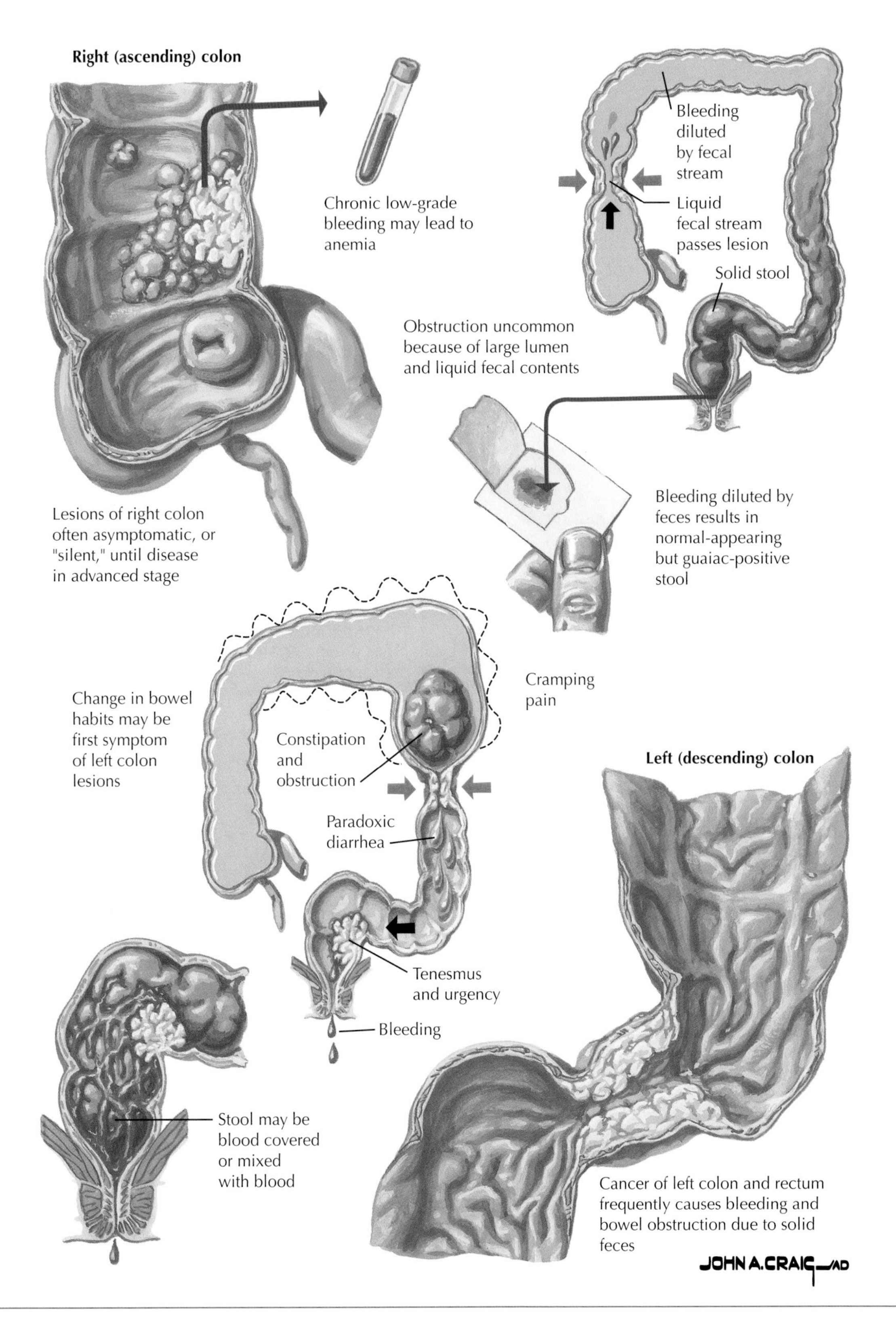

Figure 159-1 *Clinical Manifestations of Colorectal Cancer.*

Colon Cancer: Staging, Treatment, and Outcome

160

Martin H. Floch

STAGING

To determine the correct treatment for a patient with colon cancer, the disease must be staged. The malignancy may be limited to an area within a polyp, often referred to as *carcinoma in situ*. Lesions may be limited to the muscularis mucosae, or they may extend into the muscularis propria or into the subserosa. Lesions also may invade and perforate the peritoneum and other organs. Lymph nodes may or may not be involved. The classification commonly used is the tumor-node-metastasis (TNM) classification (**Box 160-1**). However, the Dukes stages are still popular and categorize lesions as limited to the bowel wall (Dukes A), extending through the wall (B), extending to nodal or regional metastases (C), and extending to distant metastases (D).

Approximately 20% of colon cancer tumors are poorly differentiated or undifferentiated, and 10% to 20% may have a strong mucinous cellular element. Each factor usually indicates poor outcome.

TREATMENT

Surgical resection is the primary therapy for any neoplastic lesion of the colon. **Figure 160-1** demonstrates the segmental and hemicolectomy resections usually performed.

Before surgery, blood screening, computed tomography (CT), and colonoscopic or barium contrast study should be performed. Carcinoembryonic antigen (CEA) is not used for diagnostic purposes but is helpful in monitoring postoperative care. Anemia must be treated. Any evidence of another tumor in the colon will govern the surgical decision on how much of the colon to resect.

Adjuvant therapy is not used for carcinoma in situ or for stage I disease. Some clinicians do not want or advise adjuvant therapy in stage II, but most do recommend it when the mucosa is invaded. Numerous therapeutic regimens are available, but levamisole plus 5-fluorouracil (5-FU) is effective in increasing survival for patients with Dukes stages B and C tumors. Five-year survival rates according to Dukes classification are 99% for stage A, 85% for stage B, 67% for stage C, and 14% for stage D. Of all patients with colon cancer, approximately 70% have resectable disease that can be cured, but 45% of these surgical patients experience recurrent disease. In patients with stage C tumors, adjuvant therapy has improved recurrence rates from 63% to 47% and has reduced mortality in approximately 33% of treated patients compared with controls. The addition of leucovorin to the available drugs has resulted in continued success in treatment. A 5-FU/leucovorin combination for 6 months for all patients with stage III is now suggested (see Bresalier in Additional Resources).

Again, CEA is useful only as a preoperative staging and a postoperative follow-up tool. It is not useful for screening, but a sharp increase in CEA level after surgery or a decrease as a result of treatment does have statistical clinical significance.

Radiation treatment appears to have little role in colon cancer, but it has a definite role in rectal cancer, as discussed in Chapter 161.

Box 160-1 Colorectal Cancer Staging (American Joint Committee on Cancer TNM Classification)

Stage 0

Carcinoma in situ, intraepithelial, or invasion of lamina propria* (Tis N0 M0).

Stage I

Tumor invades submucosa (T1 N0 M0); Dukes A.

Tumor invades muscularis propria (T2 N0 M0).

Stage II

Tumor invades muscularis propria into submucosa or into nonperitonealized pericolic or perirectal tissues (T3 N0 M0); Dukes B.†

Tumor perforates the visceral peritoneum or directly invades other organs or structures and/or perforates visceral peritoneum.‡

Stage III

Any degree of bowel wall perforation with regional lymph node metastasis.

N1 metastasis in 1 to 3 regional lymph nodes.

N2 metastasis in 4 or more regional lymph nodes.

Any T N1 M0; Dukes C.†

Any T N2 M0.

Stage IV

Any invasion of the bowel wall with or without lymph node metastasis, but with evidence of distant metastasis.

Any T, any N M1; Dukes D.

Modified from Feldman M, Friedman LS, Brandt LJ, editors: Gastrointestinal and liver disease, *ed 8, Philadelphia, 2006, Saunders-Elsevier, p 2784.*
NX, *Regional lymph nodes cannot be assessed;* N0, *no regional lymph node metastasis;* MX, *distant metastasis cannot be assessed;* M0, *no distant metastasis;* M1, *distant metastasis.*
*Tis *includes cancer cells confined within the glandular basement membrane (intraepithelial) or lamina propria (intraepithelial) with no extension through the muscularis mucosae into the submucosa.*
†Dukes stage B (corresponds to stage II) is a composite of better (T0 N0 M0) and worse (T4 N0 M0) prognosis groups, as is Dukes stage C (corresponds to stage III) (any T N1 M0 and any T N2 M0).
‡Direct invasion in T4 includes invasion of other segments of the colorectum by way of the serosa, such as invasion of the sigmoid colon by carcinoma of the cecum.

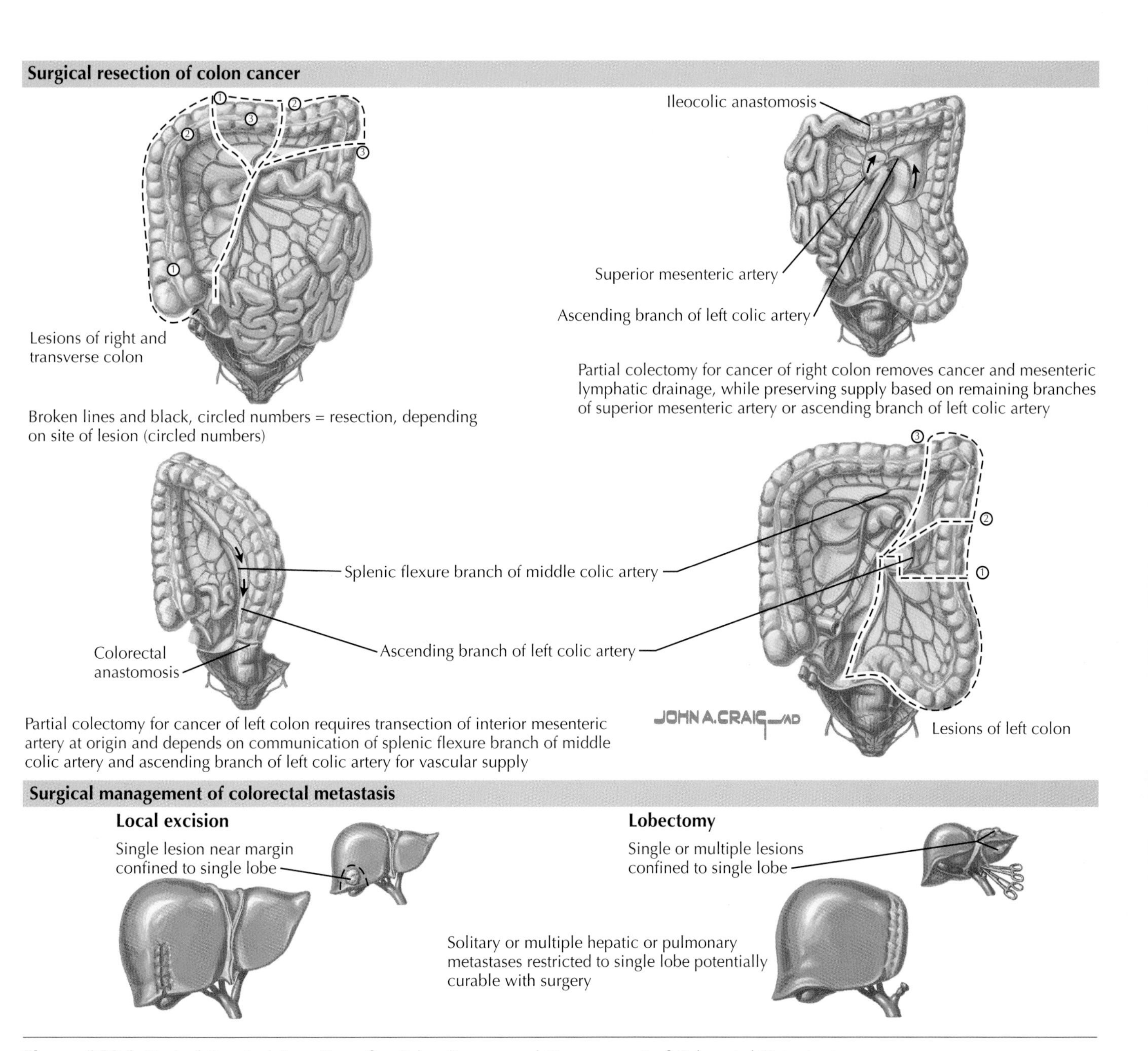

Figure 160-1 *Typical Surgical Resections for Colon Cancer and Management of Colorectal Metastasis.*

COURSE AND PROGNOSIS

Postsurgical follow-up should include CT, colonoscopy, and CEA determinations. The incidence of metachronous colorectal cancer ranges from 1% to 5%. Therefore, colonoscopy is essential for patient follow-up. The first follow-up should occur 1 year after surgery, and then, depending on the patient's status, serial follow-up should occur at 1, 2, or 3 years. The incidence of recurrent cancer is high whenever the tumor has penetrated the wall and the nodes have become involved (stage III or above). Therefore, follow-up is essential to rule out metastasis and to determine whether further surgery or chemotherapy is necessary.

Metastatic disease holds a poor prognosis. The major site of metastasis is the liver; more than 80% of patients who die of metastatic colorectal cancer have liver involvement. Aggressive approaches to resect a single metastasis and even multiple metastases have resulted in some success. Patients with untreated hepatic metastases have a median survival of approximately 10 to 11 months. Mortality from surgery for resection of liver metastases is less than 5%, increasing the 5-year survival rate for patients to 25% to 35%, depending on the series evaluated.

Unfortunately, chemotherapy has not proved successful for these patients; with a wide range of reported responses, the maximum improvement rate is 20%. New chemotherapeutic agents have been developed and others are emerging, and although therapy now includes aggressive surgery, results are still uncertain.

Initial dietary studies reveal an increased rate of recurrence in patients who continue to eat a Western diet. This work is preliminary but should encourage patients with stage III colon cancer to eat a high-fiber, non-Western type of diet.

ADDITIONAL RESOURCES

Bresalier RS: Malignant neoplasms of the large intestine. In Feldman M, Friedman LS, Brandt LJ, editors: *Gastrointestinal and liver disease*, ed 8, Philadelphia, 2006, Saunders-Elsevier, pp 2759-2811.

Fry RD, Fleshman JW, Kodner IJ: *Cancer of the colon and rectum*, Ciba-Geigy Clinical Symposia, vol 41, Summit, NJ, 1989, Ciba Pharmaceutical.

Myerhardt JA, Niedzwiecki D, Hollis D, et al: Association of dietary patterns with cancer recurrence and survival in patients with stage III colon cancer, *JAMA* 298:754-764, 2007.

Young GP, Macrae FA: Neoplastic and non-neoplastic polyps of the colon and rectum. In Yamada T, editor: *Textbook of gastroenterology*, ed 4, Philadelphia, 2003, Lippincott–Williams & Wilkins, pp 1882-1913.

Rectal Cancer

Martin H. Floch

161

Rectal cancer appears to be a different disease than colon cancer. Mortality from rectal cancer in the United States is slowly decreasing, even though the incidence is slowly increasing. This indicates that early detection and treatment of rectal polyps may be preventing the formation of malignant lesions. Furthermore, there appears to be an epidemiologic difference between rectal and colon cancer, because the incidence of rectal cancer is similar for Japan and the United States but different for colon cancer. Treatment and response to treatment differ in the rectum and the colon.

CLINICAL PICTURE

Patients with rectal cancer seek treatment for bleeding or difficulty with defecation. Rectal examination usually reveals a lesion. The lesion may be soft or polypoid and therefore may be missed on rectal examination, but it is clearly seen on proctoscopy (**Fig. 161-1**). Once rectal cancer has developed, it is invariably symptomatic. Rectal polyps can be asymptomatic but, fortunately, are noted during cancer screening and are removed through sigmoidoscopy or colonoscopy. Again, this may be a reason for the declining incidence of rectal cancer.

DIAGNOSIS

The diagnosis of rectal cancer is made through visualization of the lesion by sigmoidoscopy or colonoscopy and biopsy of the lesion. Colonoscopy is recommended to rule out other colon lesions before initiating therapy. Synchronous lesions are present at significant rates. Ultrasound is now recommended to evaluate the extent of the lesion, stage the lesion, and determine whether nodes are present. Once the lesion is identified, a full workup should be performed to evaluate for metastasis, including computed tomography (CT) of the abdomen and pelvis.

TREATMENT AND MANAGEMENT

It is generally well accepted that surgery is still the best therapy for rectal cancer. Smaller lesions that show no evidence of metastasis can be excised, but most authorities believe that any lesion larger than Dukes stage A (stage I) requires wide excision. Also, preoperative combined-modality therapy is now preferred and accepted as beneficial. Past studies revealed that patients undergoing resection for cure had 2- and 5-year survival rates of 56% and 43%, respectively, if they underwent preoperative radiation therapy. Recent studies have confirmed the importance of preoperative chemoradiation for patients with advanced disease. Chemoradiation is effective in downstaging most lesions and permits a better cure rate. Downstaging after chemoradiation can be accomplished in 60% to 90% of patients, with a complete resectability rate of approximately 60% and a pathologic response rate of 10% to 20%.

COURSE AND PROGNOSIS

The course and prognosis of rectal cancer depend completely on the stage at which the disease is recognized. Again, early removal of simple rectal polyps seems to have decreased the incidence of rectal carcinoma. Dukes stage A lesions can be resected locally if there is no evidence of metastasis. Patients with lesions categorized as Dukes stage B (stage II) and greater should undergo preoperative chemoradiation to downstage the disease. Cure depends largely on the extent of the lesion and whether metastasis has occurred. Prognosis is poor if there is preoperative nodal involvement. Prognosis is better if downstaging results in no evidence of local metastases.

ADDITIONAL RESOURCES

Daniels IR, Fisher SE, Heald RJ, Moran BJ: Accurate staging, selective preoperative therapy and optimal surgery improves outcome in rectal cancer: a review of the recent evidence, *Colorectal Dis* 9:290-301, 2007.

Dosoretz DE, Gunderson LR, Hedberg S, et al: Preoperative irradiation for unresectable rectal and rectosigmoid carcinomas, *Cancer* 52:814-820, 1983.

Fry RD, Fleshman JW, Kodner IJ: *Cancer of the colon and rectum*, Ciba-Geigy Clinical Symposia, vol 41, Summit, NJ, 1989, Ciba Pharmaceutical.

Garcia-Aguilar J, Anda E, Sirvongs P, et al: Pathologic complete response to preoperative chemoradiation is associated with lower local recurrence and improved survival in rectal cancer patients treated by mesorectal excision, *Dis Col Rect* 46:298-304, 2003.

O'Connell MJ: Current status of adjuvant therapy for colorectal cancer, *Oncology* 18:751-755, 2004.

Ratto C, Valentini D, Organti AG, et al: Combined-modality therapy in locally advanced primary rectal cancer, *Dis Col Rect* 46:59-67, 2003.

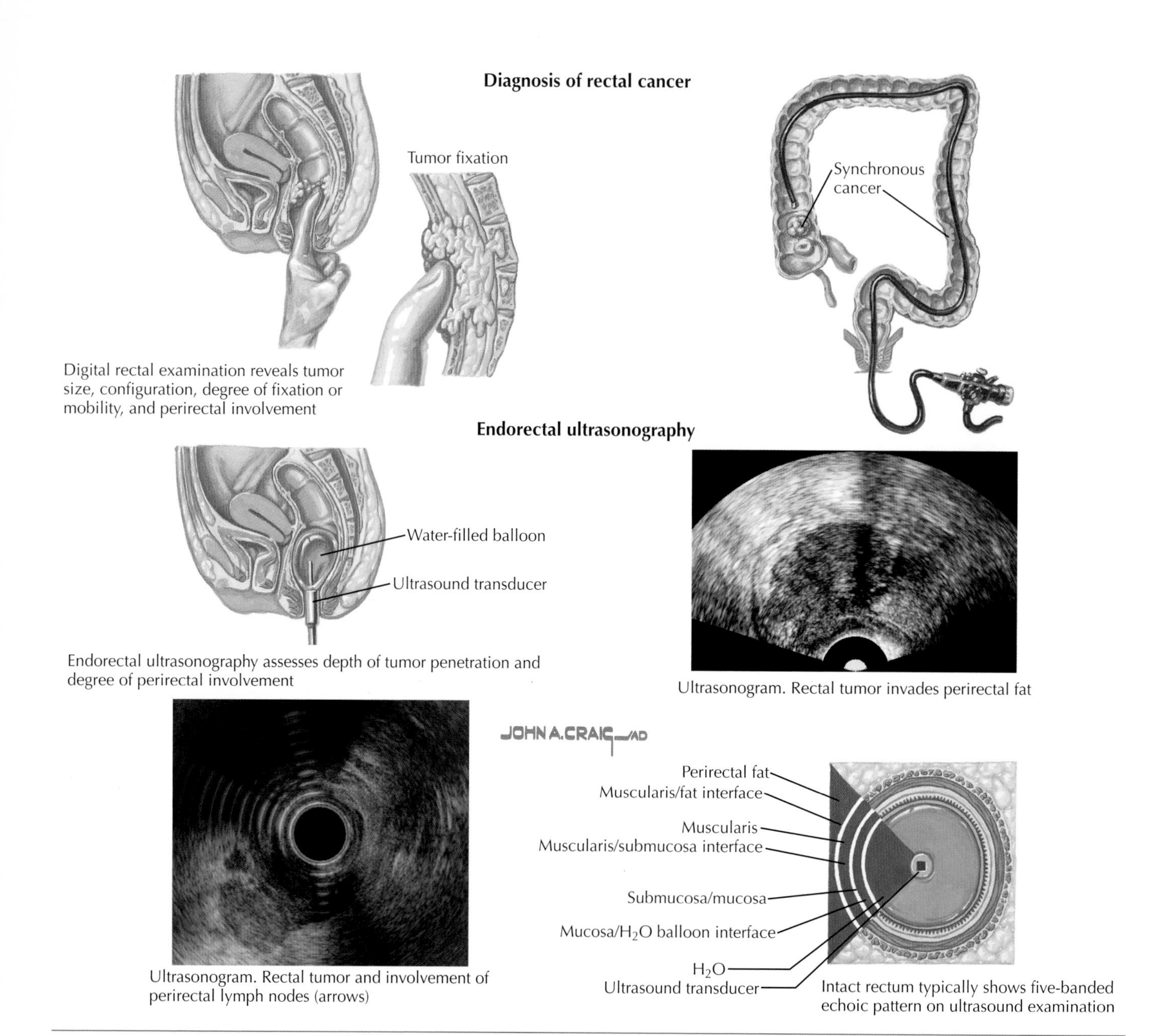

Figure 161-1 *Rectal Cancer.*

Hemorrhoids

162

Martin H. Floch

Hemorrhoids are varicose dilatations of the radicals of the superior or inferior plexus of the hemorrhoidal veins. Varicosities are accompanied, in varying degrees, by hypertrophy and round cell infiltration of the perivascular connective tissue. Approximately 50% of the population is affected. Hemorrhoids usually develop in persons between 25 and 55 years of age, and they seldom develop in children.

Factors in hemorrhoid formation include (1) genetic predisposition, (2) absence of valves in the portal venous system attributed to the erect human posture, and (3) conditions that cause transient or constant increased pressure or stasis within the rectal venous plexuses, such as straining from constipation, frequency of bowel movements from diarrhea, rectal tumors or strictures, pregnancy, and pelvic tumors that increase pressure. Varicosities of the inferior hemorrhoidal plexus cause *external hemorrhoids* that are situated below the pectinate line and are covered by the modified skin of the anus. Formation of a thrombus within a vein, or from rupture of a vein with extravasation of blood into the cellular tissue, causes a thrombotic external hemorrhoid, which is an acute variety of the external type. This usually results from straining and appears as a sudden, painful lump that can be seen as a round, bluish, tender swelling (**Fig. 162-1**). When these resolve, they remain as skin tags.

Internal hemorrhoids are enlargements of the veins of the superior hemorrhoidal plexus. In early stages, they do not protrude through the anal ring, and they are seen only on endoscopic evaluation. In later stages, they may protrude through the anal canal, and if they protrude through the anal ring and persist, they can become ulcerated.

CLINICAL PICTURE

The clinical picture of a thrombosed external hemorrhoid is one of a painful, tender mass. Its onset is usually sudden. However, most hemorrhoids manifest as simple rectal bleeding, with blood seen on the toilet tissue or in the toilet bowl. At times, the bleeding might be brisk enough to result in bright-red blood in the toilet bowl. It is rarely severe enough to cause anemia. Internal hemorrhoids are usually reducible, but when they cannot be reduced, they may become strangulated, and vessels may become thrombosed, resulting in a very tender, painful presentation.

DIAGNOSIS

Simple external thrombosed hemorrhoids can be readily observed on inspection. Often, the anal ring is too tender to permit full rectal examination, but when full examination is possible, no lesions are felt inside the anal ring. Inspection and digital examination are essential in all evaluations of external hemorrhoids. If anemia or significant bleeding develops, endoscopic examination is essential. Proctologic or sigmoidoscopic examination is required when internal hemorrhoids are suspected: these are visualized in the rectal vault.

Hemorrhoids are graded from 1 to 4 depending on severity. When hemorrhoids are recurrent and prolonged, a full diagnostic evaluation is warranted. Complete blood count should be performed to check for anemia, with full colonic examination to rule out any associated malignancy. Computed tomography of the pelvis may also be important if an associated tumor may be causing intraabdominal pressure and formation of the hemorrhoid.

TREATMENT AND MANAGEMENT

Depending on the symptoms, initial treatment is invariably medical for grades 1 and 2 hemorrhoids. A tender, severe, thrombosed hemorrhoid that is not reduced within 24 to 48 hours might require lancing for relief of the pain. Medical therapy usually includes stool softeners, increased dietary fiber to soften stools, Sitz baths with or without salts in warm water, astringents such as witch hazel, and topical analgesics. If patients cannot change their diet, supplements such as psyllium seed are encouraged.

Many clinicians add hydrocortisone suppositories or creams or mesalamine suppositories to reduce the inflammatory reaction surrounding the internal or external hemorrhoids. These are effective but need to be used for at least 1 to 2 weeks. Temporary relief of pain from external hemorrhoids can be obtained by using sprays (e.g., benzocaine 20%) or ointments (e.g., dibucaine 1%).

When conservative medical therapy fails, surgery is usually necessary. Surgical procedures include rubber band ligation, injection sclerotherapy, cryosurgery, electrocoagulation, photocoagulation, and hemorrhoidectomy. Meta-analysis of controlled, randomized trials reveals that rubber band ligation is superior to sclerotherapy, and that patients undergoing ligation are unlikely to require therapy later. Furthermore, although it results in better responses than rubber band ligation, hemorrhoidectomy also causes more pain and complications. Most clinicians and proctologists recommend rubber band ligation as first-line surgical treatment.

COURSE AND PROGNOSIS

Most hemorrhoids resolve with medical therapy, but many patients require surgery. After surgery, the acute situation is under control, but some patients may have mild, recurrent bleeding that is usually controlled by medical therapy. Occasionally, repeat surgery is needed.

ADDITIONAL RESOURCES

Chand M, Nash GF, Dabbas N: The management of haemorrhoids, *Br J Hosp Med* 69:35-40, 2008.

Hull TL: Diseases of the anorectum. In Feldman M, Friedman LS, Brandt LJ, editors: *Gastrointestinal and liver disease*, ed 8, Philadelphia, 2006, Saunders-Elsevier, pp 2833-2854.

MacRae HN, McLeod RS: Comparison of hemorrhoidal treatments: a meta-analysis, *Can J Surg* 40:14-22, 1997.

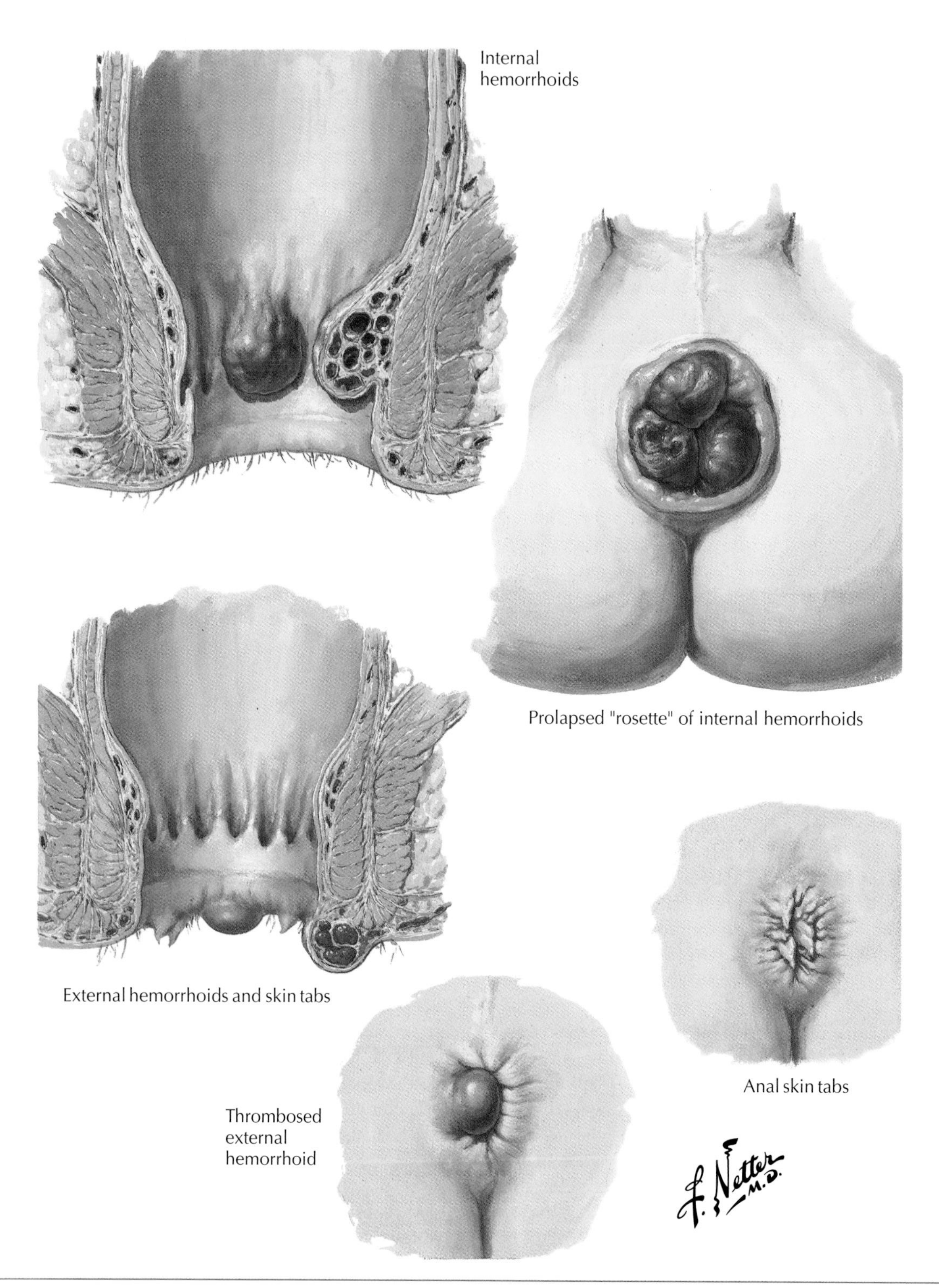

Figure 162-1 *Hemorrhoids.*

Rectal Prolapse and Procidentia

163

Martin H. Floch

Prolapse of the rectum is a condition in which one or more layers of the rectum or anal canal protrudes through the anal orifice (**Fig. 163-1**). It can be partial or complete. *Partial prolapse* involves only the mucosa, which usually extends no more than ½ to 1 inch (2.5 cm) outside the anal canal. *Procidentia* is total prolapse, involving all the layers of the rectum. The mass is larger and bulbous, and it may eventually contain a hernial sac of peritoneum with a segment of bowel in the interior. Rectal prolapse is uncommon in children, but it may occur during infancy. Although usually idiopathic, it may be associated with congenital defects. Prolapse occurs with defecation, usually reduces spontaneously with conservative treatment, and is self-limited.

Rectal prolapse in adults occurs more often in women than in men. It is associated with poor pelvic musculature tone, chronic straining, fecal incontinence, and often neurologic or traumatic damage associated with the pelvis. Its etiology remains unknown, but a defect in the supporting structures may permit increased intraabdominal pressure to produce the prolapse. In elderly or debilitated persons, prolapse is usually caused by a loss of sphincteric tone.

CLINICAL PICTURE

Patients usually seek treatment for rectal staining or incontinence. Careful history reveals they can feel there is prolapse of tissue with defecation. Degree of prolapse varies by patient. When it is significant, it is troublesome. Prolapse is often associated with straining and with a sensation of incomplete evacuation and of the mass.

Complete rectal prolapse (procidentia) is large, and patients seek treatment fearing that the mass they can sense is malignant. The mass can cause pain and bleeding.

DIAGNOSIS

Examination with the patient in the left lateral position and straining slightly often reveals the prolapse and a weak anal sphincter. At times, the diagnosis cannot be made in this position, and the patient must be placed in the upright position, or sitting, so that straining will produce the prolapse. A detailed workup is needed to rule out malignancy and should include colonoscopic examination and computed tomography of the pelvis to ensure that no perirectal lesion increases intraabdominal pressure on the rectum.

Procidentia necessitates digital rectal examination, and an attempt should be made to reduce the mass. It may be tender, and the patient may need sedation to undergo the examination. For these patients, a colorectal surgeon should be involved in evaluating the procidentia.

TREATMENT AND MANAGEMENT

Very small prolapses can be handled medically by making sure that straining is reduced to a minimum. If the prolapse is significant and recurrent, it must be removed surgically.

Numerous surgical procedures are available for rectal prolapse. Some are simple and others are complex, involving a combined procedure in which the rectum is fixed to the sacral hollow and the redundant sigmoid colon is removed. The type of procedure and the outcome vary and depend on the degree of the prolapse and the experience of the colorectal surgeon.

In very elderly patients, simple extraabdominal approaches through perineal rectosigmoidectomy are possible. Some studies of these approaches in this select patient group have shown great success.

COURSE AND PROGNOSIS

Many persons live with a small degree of prolapse and are able to control their bowel movements. When incontinence becomes a problem, or when the prolapse is complete, various surgical interventions are possible. Studies report 80% good to excellent results with no mortality. A 90% success rate is reported in very elderly patients who require an extraabdominal approach. If the patient has uncontrolled prolapse and colostomy is necessary, prognosis depends on the debilitating disease, but colostomy is usually successful.

ADDITIONAL RESOURCES

Bachoo P, Brazzelli M, Grant A: Surgery for complete rectal prolapse in adults, *Cochrane Database Syst Rev* 1:CD001758, 2000.

Gourgiotis S, Baratsis S: Rectal prolapse, *Int J Colorectal Dis* 22:231-243, 2007.

Madiba TE, Baig MK, Wexner SD: Surgical management of rectal prolapse, *Arch Surg* 140:63-73, 2005.

Williams JG, Rothenberger DA, Madoff RD, et al: Treatment of rectal prolapse in the elderly by perineal rectosigmoidectomy, *Dis Col Rect* 35:830-839, 1992.

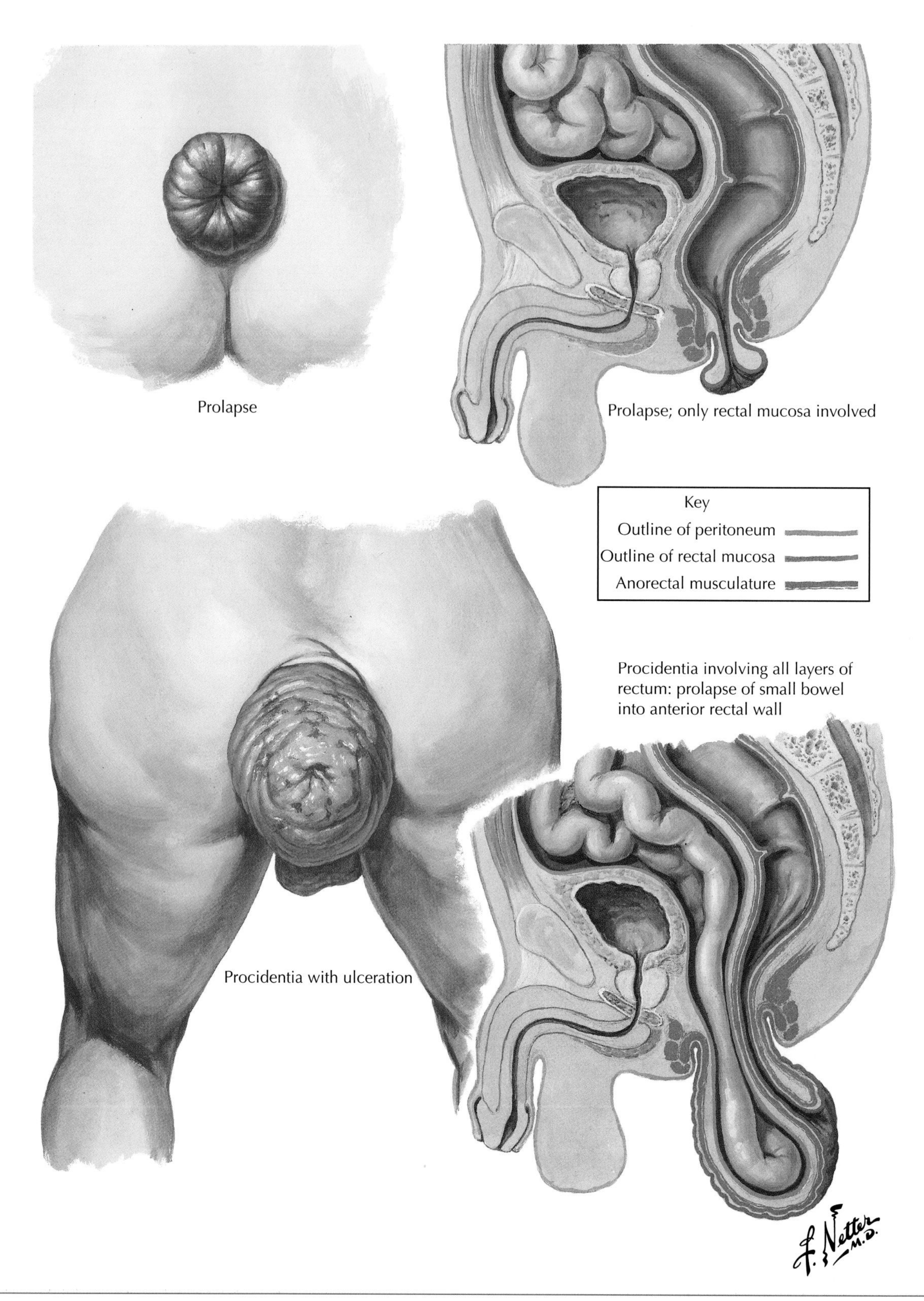

Figure 163-1 *Rectal Prolapse and Procidentia.*

Anal Fissure, Pruritus Ani, Papillitis, and Cryptitis

164

Martin H. Floch

ANAL FISSURE

Anal fissure is a tear of the skin in the distal anal canal, usually in the posterior midline (**Fig. 164-1**). Occasionally, it is in the anterior midline. When not in the midline, anal fissure is often associated with an abnormality such as Crohn's disease, human immunodeficiency virus (HIV) infection, tuberculosis, syphilis, or anal malignancy. Fissures usually are acute but may become chronic. The cause of an anal fissure typically is unknown, but it is clearly associated with increased resting anal pressure. Identifying this physiology has led to some of the most recent treatments. The exact incidence is not known, but anal fissure is relatively common.

Clinical Picture

The classic presentation of anal fissure is acute, severe pain on defecation that may persist for hours after passage of the fecal bolus. Bleeding may be associated with the fissure, with blood on the toilet paper or in the bowl. However, the pain is significant and is the hallmark of the clinical presentation.

Diagnosis

The diagnosis of anal fissure is best made by spreading apart the patient's buttocks with gloved fingers. Digital rectal examination is extremely painful and should be avoided. If the fissure is seen and is in the midline, it is not secondary to other diseases, and treatment can be used as a therapeutic trial. At times, a sentinel skin tag lies distal to the fissure. Endoscopy should be deferred until the patient is asymptomatic. If there is any uncertainty about the diagnosis, sedation and anesthesia may be necessary to enable appropriate examination of the patient.

Treatment and Management

Many options exist for treating anal fissure. The therapy chosen often depends on the experience of local physicians. Medical therapy should be tried first, including a course of Sitz baths, psyllium fiber supplements to soften the stool, or emollient suppositories. Treatment may be successful in 27% to 44% of patients. Some clinicians have used Sitz baths plus bran supplementation, at least 15 g daily, which has also been successful. Use of 2% lidocaine ointment plus 2% hydrocortisone cream has proved to be successful in approximately 60% of patients.

When simple medical therapy does not succeed, other options include topical therapy. Topical nitrates (0.2% glyceryl trinitrate) and calcium channel blockers have proved effective, with more data on the nitrates. Healing has been reported in as many as 77% of patients with the use of four suppositories for 8 weeks. These agents significantly lower anal sphincter pressure.

Botulin toxin (Botox) has yielded excellent results and healing rates of 82% and 79% at 3 and 6 months, respectively. Other studies have resulted in slightly lower percentages, and recurrence has been reported with botulin toxin.

If medical therapy for anal fissure does not succeed, surgery is warranted. Most patients heal after lateral internal sphincterotomy. Some may experience incontinence, although the rate of this surgical complication is extremely low.

Course and Prognosis

As indicated, outcomes are excellent for most patients with anal fissure, and cure rates range from 50% to 70% after medical therapy and 70% to 90% after surgical therapy. Some fecal incontinence may occur after surgical treatment, but recurrence is low.

Simple posterior or anterior fissure is benign, but clinicians should always be on guard that perianal disease may be the first manifestation of Crohn's disease or a venereal disease (see Chapter 166).

PRURITUS ANI

Pruritus ani is a symptom that may accompany any anorectal disease, but frequently no evidence of primary disease is found. Perianal itching without any apparent cause is thought to be neurodermatitis. When a cause is identified, it should be treated. Causes of pruritus ani include parasitic infections, local irritants (e.g., food allergies), and dermatologic conditions (e.g., psoriasis, atopic dermatitis, moisture collecting in obese person).

Treatment for pruritus ani is whichever therapy is necessary for the primary disease and symptoms. For troublesome pruritus, the patient should discontinue irritating soap and use hypoallergenic soap instead, apply a hydrocortisone cream regularly, keep the anal area clean, and use a protective ointment such as zinc oxide. Severe chronic pruritus can be treated.

PAPILLITIS AND CRYPTITIS

Inflammatory processes of the papillae usually start in one of the crypts and cause pain disproportionate to the size and severity of the lesion. In acute *papillitis*, the area is swollen, edematous, and congested. In the chronic stage of papillitis, the area becomes fibrosed and hypertrophied. Gradually, the hypertrophic papillae may develop a stalk and then change to a so-called fibrous polyp, which may produce the sensation of a foreign body in the anal canal.

Cryptitis may remain restricted to circumscribed reactions in and around the crypts, or it may spread to the surrounding tissues, including the formation of abscesses and fistulae (see Chapter 165). Symptoms of cryptitis occasionally resemble those of a fissure and often include itching and radiation of pain, which is aggravated by defecation. Visualization of the cryptitis and any purulent discharge or granulation tissue, or of the hypertrophied papillae, is necessary. Treatment is usually

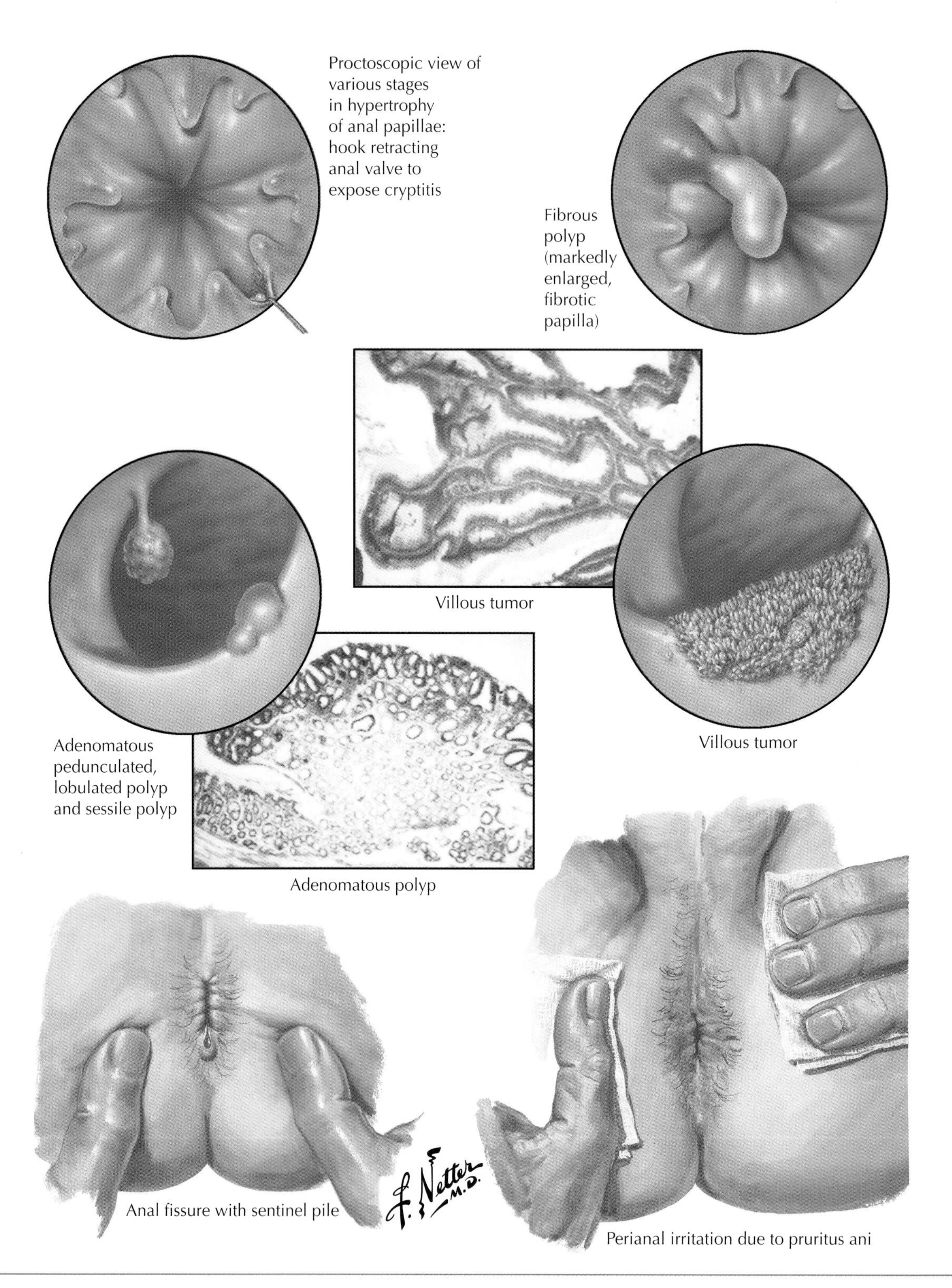

Figure 164-1 *Anal Fissure and Pruritus Ani.*

successful with medical regimens to decrease local irritation by using hydrocortisone or mesalamine suppositories, softening the stool through psyllium or bran intake, and using relaxing agents, if necessary.

ADDITIONAL RESOURCES

Brisinda G, Cadeddu F, Brandara F, et al: Randomized clinical trial comparing botulinum toxin injections with 0.2% nitroglycerin ointment for chronic anal fissure, *Br J Surg* 94:162-167, 2007.

Freuhauf H, Fried M, Wegmueller B, et al: Efficacy and safety of botulinum toxin A injection compared with topical nitroglycerine ointment for the treatment of chronic anal fissure: a prospective randomized study, *Am J Gastroenterol* 101:2107-2112, 2005.

Hull TL: Diseases of the anorectum. In Feldman M, Friedman LS, Brandt LJ, editors: *Gastrointestinal and liver disease*, ed 8, Philadelphia, 2006, Saunders-Elsevier, pp 2833-2854.

Madoff RD, Fleshman JW: AGA technical review on the diagnosis and care of patients with anal fissure, *Gastroenterology* 124:235-245, 2003.

Anorectal Abscess and Fistula

Martin H. Floch

165

Localized infection with a collection of pus in the anorectal area is designated an *anorectal abscess*. Usually, it results from the invasion of the normal rectal flora into the perirectal or perianal tissues. The pathologic process seems to start with inflammation of one or more of the crypts (see Chapter 164), spreads to the anal ducts and anal glands, and then spreads submucosally, subcutaneously, or transsphincterally to the surrounding tissue. This sequence of events closes with the spontaneous rupture of the abscess, either into the anorectal canal or through the perianal skin, if the abscess has not been drained surgically. After the abscess has perforated, the cavity and its outlet shrink, leaving a tubelike structure, an *anorectal fistula*, which invariably is the result of the abscess. Therefore, the abscess is the acute phase, and the fistula is the chronic phase.

The *levator ani plane*, demarcating the various perineal pelvic spaces, is used to classify anorectal abscesses according to localization. Retrorectal, pelvirectal, and submucosal abscesses belong to the *supralevator* abscesses and have a somatic sensory nerve supply; therefore, these cause a sensation of discomfort from pressure rather than from pain in the anorectal region. *Infralevator* abscesses may produce signs of toxemia and prostration. *Retrorectal* and *pelvirectal* abscesses originate from infectious processes in other pelvic organs and thus are not anorectal lesions in the strict sense, although they usually rupture into the rectum or the anal canal. Infralevator abscesses are also divided according to site into subcutaneous, intramuscular, fistulorectal, and cutaneous abscesses. The fistula is called *complete* when both openings, the primary and the secondary, can be detected and are accessible. Such a complete variety usually connects the rectal lumen with the anal or perianal skin. If there is only one opening, it is called a *blind* fistula or a sinus.

Figure 165-1 depicts the various types of fistulas (or fistulae) and the *Goodsall-Salmon law*, in which an imaginary transverse line across the center of the anus can be used to predict the location of the tract and the primary opening.

Anorectal abscess and fistula are associated with specific diseases, such as Crohn's disease, malignancy, radiation proctitis, leukemia, lymphoma, tuberculosis, actinomycosis, and lymphogranuloma venereum. Other diseases may cause a similar picture, such as diverticulitis and Bartholin abscesses.

CLINICAL PICTURE

Swelling in the perianal area accompanied by acute pain are the most common symptoms. The patient reports that a change in sitting position, moving, or a bowel movement makes the pain worse. Onset is usually slower than in fistula formation, and the patient may experience fever and fatigue. Discharge from the abscess may occur. Chronic purulent discharge is a major problem. Depending on the site and the amount of drainage, the abscess may be minor or large, and the perianal area may be excoriated.

DIAGNOSIS

Endosonography and magnetic resonance imaging (MRI) can be extremely helpful in delineating the extent of the abscess and the fistulous tracts.

TREATMENT

Healthy persons with small, superficial abscesses can undergo outpatient drainage under local anesthesia. Past treatments used setons, but this form of therapy alone is now used less often and is combined with other therapy, such as infliximab. For large abscesses or those in patients with underlying disease, surgery is necessary. Standard surgical treatment is usually performed on an emergency basis. Mortality is as high as 50% if surgery is delayed and a necrotizing anorectal infection develops.

Although antibiotics may not be necessary for small lesions, most clinicians believe antibiotics are important, especially if the patient has an underlying associated disease. Antibiotic therapy must treat the aerobic and the anaerobic flora and therefore is usually a combination of drugs.

Postoperative management consists of routine surgical management and may include warm baths, prevention of any constipation, and appropriate nutrition. Novel therapies have been developed to close fistulae. Fibrin glue has been reported to be successful in up to 69% of patients. This therapy has worked after initial failures with other therapies.

Finally, special consideration must be given to patients with Crohn's disease. The primary disease must be treated, which may include 5-acetylsalicylic acid (5-ASA) products, immunosuppressive agents, or biologicals and antibiotic therapy. Infliximab (anti–tumor necrosis factor-α) has been used successfully. Attempts at healing fistulae in patients with Crohn's disease through bowel rest or enteral feedings have shown some success, but the fistulae tend to return when regular foods are added. Newer studies indicate a relationship between genetic variants of *NOD2/CARD15* and antibiotic response.

COURSE AND PROGNOSIS

Surprisingly, MRI has been shown to predict clinical outcome based on the initial severity of the fistulous tracts. As indicated, if there is delay in treatment, mortality may increase to as much as 50%. If treatment is introduced early in a relatively healthy patient, the prognosis is good. However, if there is associated significant morbidity, such as severe diabetes or autoimmune disease, the fistula and the abscess can constitute significant comorbidity, leading to death. Therefore, aggressive therapy is indicated in these patients.

ADDITIONAL RESOURCES

Angelberger S, Reinisch W, Dejaco C, et al: *NOD2/CARD15* gene variants are linked to failure of antibiotic treatment in perianal fistulating Crohn's disease, *Am J Gastroenterol* 103:1197-1202, 2008.

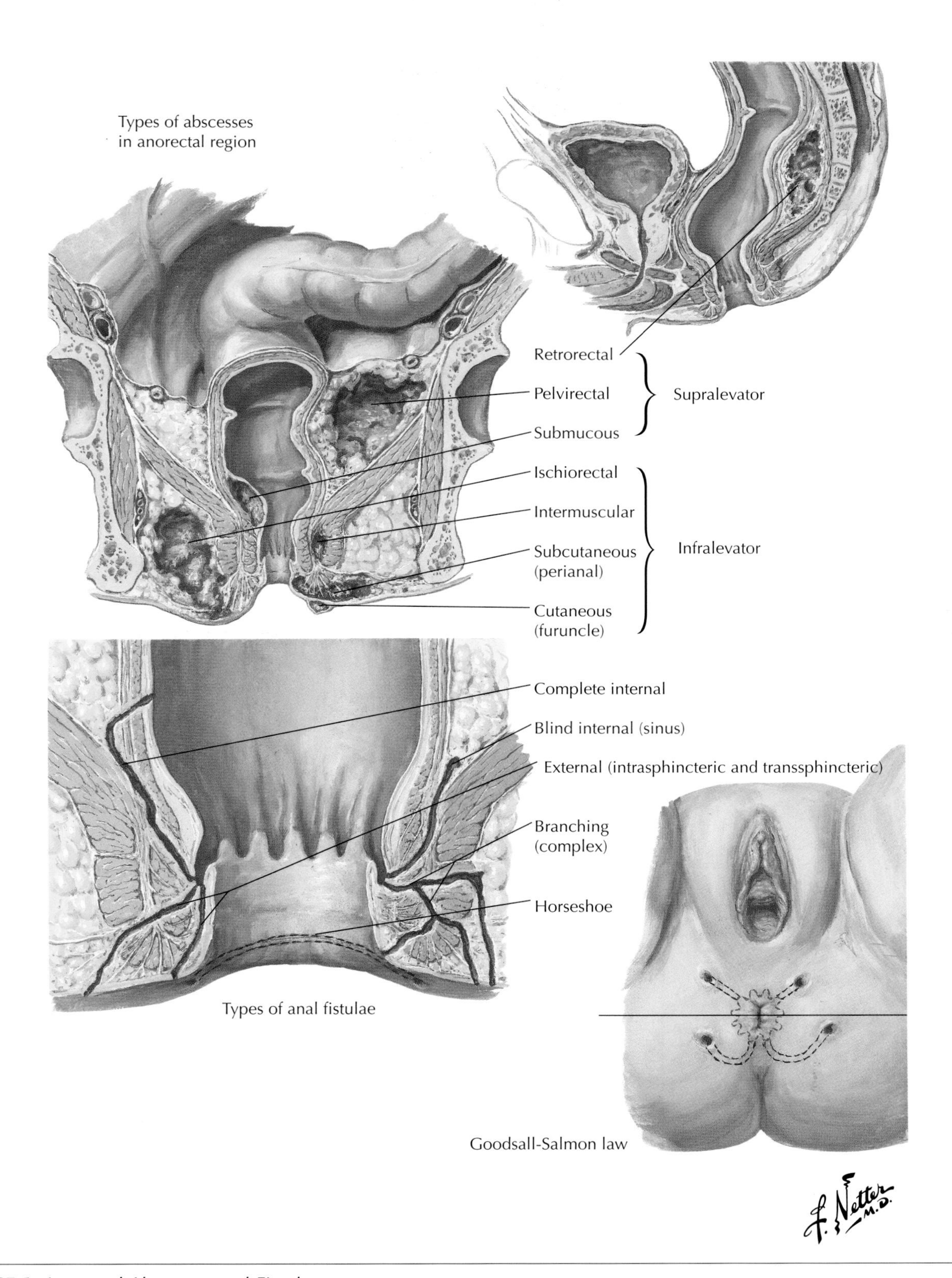

Figure 165-1 *Anorectal Abscesses and Fistulae.*

Bubrick NP, Hitchcock CR: Necrotizing anorectal and perineal infections, *Surgery* 86:655-661, 1979.

Hull TL: Diseases of the anorectum. In Feldman M, Friedman LS, Brandt LJ, editors: *Gastrointestinal and liver disease*, ed 8, Philadelphia, 2006, Saunders-Elsevier, pp 2833-2854.

Hyder SA, Travis SP, Jewell DP, et al: Fistulizing anal Crohn's disease: results of combined surgical and infliximab treatment, *Dis Colon Rect* 49:1837-1841, 2006.

Sandborn WT, Fazio VW, Feagan BG, Hanauer SB: AGA technical review on perianal Crohn's disease, *Gastroenterology* 125:1508-1530, 2003.

Schwartz DA, Herdman CR: The medical treatment of Crohn's perianal fistulas (review), *Aliment Pharmacol Ther* 19:953-967, 2004.

Schwartz DA, White CM, Wise PE, Herline AJ: Use of endoscopic ultrasound to guide combination medical and surgical therapy for patients with Crohn's perianal fistulas, *Inflamm Bowel Dis* 11:727-732, 2005.

Sentovich SM: Fibrin glue for anal fistulas: long-term results, *Dis Col Rect* 46:498-502, 2003.

Lymphogranuloma Venereum and Sexually Transmitted Proctitis

Martin H. Floch

166

Lymphogranuloma venereum (LGV) and sexually transmitted proctitis are relatively uncommon in the general population, occurring more often in male homosexuals and promiscuous heterosexuals. Certain infections are typically seen when proctitis is identified in a patient with a history of promiscuous sexual activity (**Table 166-1**). LGV is rare but leads to a pathologic condition that can be confused with other granulomatous diseases (**Fig. 166-1**).

LYMPHOGRANULOMA VENEREUM

The organism that causes LGV is *Chlamydia trachomatis*. Worldwide, *C. trachomatis* accounts for approximately 50 million new infections of LGV each year. It is the leading bacterial cause of sexually transmitted disease (STD) in the United States. *C. trachomatis* is divided into three biovars based partly on host susceptibility and DNA homology. The trichoma biovar and the LGV biovar cause human infections, whereas the third biovar does not. The trichoma biovar multiplies in columnar epithelial cells, and the LGV strains also are capable of multiplying in macrophages. The biovars have been subdivided into 15 serovars, which appear to cause specific infections. Anorectal LGV infections occur in homosexual men and heterosexual women engaging in anal intercourse. Infection also may be spread by infected vaginal secretions in women or through lymphatic spread from genital infection.

Clinical Picture

The clinical picture for patients with LGV can vary greatly, from absence of symptoms to severe granulomatous disease (see Fig. 166-1). Variation occurs with the different serovars. If LGV leads to proctocolitis, patients may have severe itching, discharge, diarrhea or constipation, rectal bleeding, fever, lymphadenopathy, and lower abdominal pain. If left untreated, LGV may progress to perirectal abscess with fibrosis stricture, stenosis, and Crohn's disease presentation. At times, obstruction of the lymphatic system can cause marked hemorrhoids or perianal condylomata.

DIAGNOSIS

In any patient with suspected sexually transmitted proctitis, the differential diagnosis should include all possible conditions. Certainly, in any patient with Crohn's disease with an appropriate history, LGV should be suspected. Diagnosis is made by culture of the rectal exudate and identification of infected cells with a fluorescein-labeled monoclonal antibody to chlamydial antigens.

A fourfold increase in the titer in acute and convalescent sera has been used as supportive evidence of infection. However, positive serology alone does not make the diagnosis of LGV because a high percentage of adults have antibodies to

Table 166-1 Sexually Transmitted Causes of Proctocolitis

Organism	Symptoms	Diagnosis	Preferred Treatment
Chlamydia trachomatis (non-LGV)	+ or −	Culture, DFA	Doxycycline Azithromycin
C. trachomatis (LGV)	+	Culture, DFA, serology	Doxycycline
Neisseria gonorrhoeae	+ or −	Culture, Gram stain	Ciprofloxacin Ofloxacin
Treponema pallidum	+ or −	Darkfield microscopy, immunohistochemical stain, serology	Penicillin
Herpes simplex virus	+	Clinical syndrome (sacral radiculomyelopathy), viral culture, Tzanck prep; serology (whole blood, IgG immunoassay)	Acyclovir
Human papillomavirus	+ or −	Clinical appearance, Papanicolaou smear, biopsy, PCR/Southern blot	Cryotherapy Trichloroacetic acid Surgical excision
Entamoeba histolytica	+	Serology (invasive disease), stool examination for O&P (3-6 samples), erythrophagocytosis zymodeme analysis	Metronidazole plus iodoquinol (invasive disease); or iodoquinol; or paromomycin; or diloxanide furoate (colonization)

Modified from Blaser MJ et al: Infections of the gastrointestinal tract, *ed 2, Philadelphia, 2002, Lippincott-Williams & Wilkins.*
DFA, *Direct fluorescence assay;* IgG, *immunoglobulin G;* LGV, *Lymphogranuloma venereum;* PCR, *polymerase chain reaction;* O&P, *ova and parasites.*

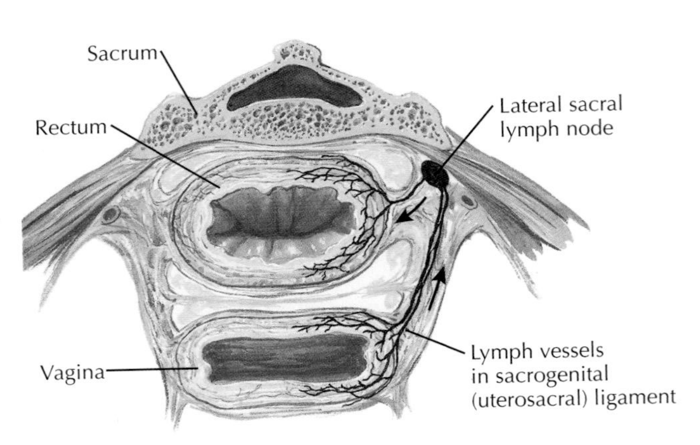

Pathway of spread of lymphogranuloma (lymphopathia) venereum from upper vagina and/or cervix uteri to rectum via lymph vessels

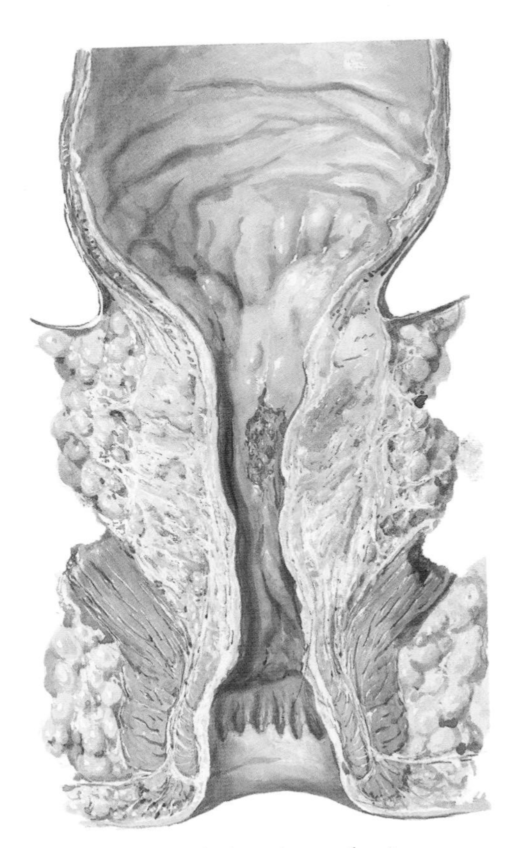

Long tubular stricture of rectum

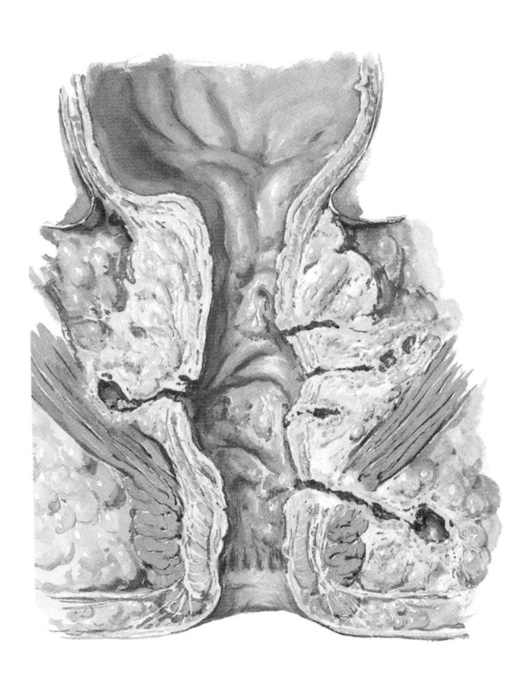

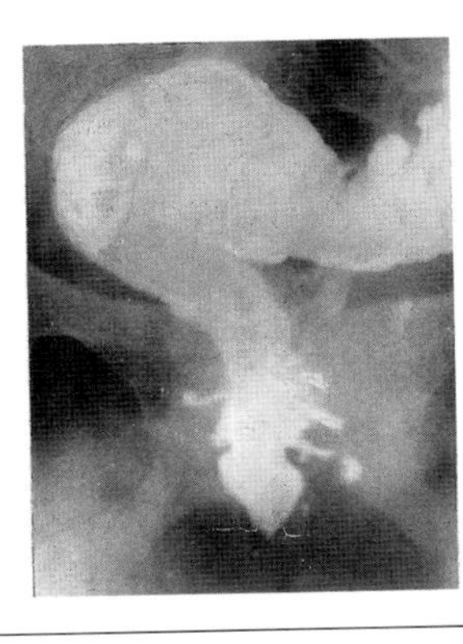

Stricture of rectum with multiple blind sinuses

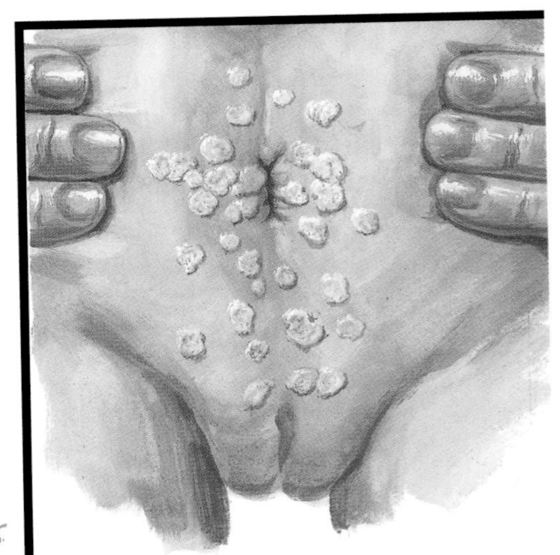

Condylomata lata (2° syphilis)

Figure 166-1 *Lymphogranuloma Venereum.*

C. trachomatis. Therefore, a single serum sample is usually of little diagnostic value.

Treatment and Management

Treatment is effective with tetracycline, doxycycline, and erythromycin. Resistant *C. trachomatis* strains have been reported. However, current treatment of non-LGV biovars is 100 mg of doxycycline twice daily for 7 to 10 days. A single 1-g dose of azithromycin appears to be as effective as a 7-day course of doxycycline for genital infections and uncomplicated rectal infections. Some recommend a 21-day course of doxycycline. If a patient has full-blown rectal disease with strictures, long-term antibiotic therapy plus surgery may be necessary.

SEXUALLY TRANSMITTED PROCTITIS

Table 166-1 describes the organisms that can cause sexually transmitted proctitis. Furthermore, in the patient with human immunodeficiency virus (HIV), a wide range of organisms may be involved in producing the same symptoms. The diagnosis must be made by identifying the causative organism. Treatment is then specific for the organism (see Section VI). Although antibiotics may cause acute and fulminating colitis, most cases are usually self-limiting. *C. trachomatis* infection can be confused with Crohn's granulomatous disease, and treatment and prognosis vary with each. Also, antibiosis can mimic fulminant colitis, and the other infections can cause persistent and recurrent proctitis.

ADDITIONAL RESOURCES

Collins L, White JA: Lymphogranuloma venereum, *BMJ* 332:66, 2006.

Smith PB, Janoff EN: Gastrointestinal complications of the acquired immunodeficiency syndrome. In Yamada T, editor: *Textbook of gastroenterology*, ed 4, Philadelphia, 2003, Lippincott–Williams & Wilkins, pp 2567-2589.

Stamm WE, Jones RB, Batteiger BE: *Chlamydia trachomatis* (trachoma, perinatal infections, lymphogranuloma venereum, and other genital infections). In Mandel GL, Bennett JE, Dolin R, editors: *Mandel, Douglas, and Bennett's principles and practice of infectious diseases*, ed 6, Philadelphia, 2005, Churchill Livingstone–Elsevier, pp 2239-2256.

van Hal SJ, Hillman R, Stark DJ, et al: Lymphogranuloma venereum: an emerging anorectal disease in Australia, *Med J Aust* 187:309-310, 2007.

Verley JR, Quinn TC: Sexually transmitted infections of the anus and rectum. In Blaser MJ, editor: *Infections of the gastrointestinal tract*, ed 2, Philadelphia, 2002, Lippincott–Williams & Wilkins, pp 357-382.

Fecal Incontinence

167

Martin H. Floch

Fecal incontinence is the involuntary passage of fecal material. It may be a single occurrence, but when recurrent, incontinence can be devastating. Several surveys indicate that fecal incontinence may occur in 7% to 15% of the population. It is more frequent in women than in men. In acutely ill hospital patients, the rate of fecal incontinence is as high as 33% to 43%. It is the leading cause of admission of patients to nursing homes, where episodes of incontinence occur in approximately 20% of residents. **Box 167-1** lists some of the many causes of fecal incontinence.

Incontinence associated with infectious diarrhea or a bout of irritable bowel syndrome is different from incontinence that results from traumatic injury of the anal sphincter or from neurologic disease in a nursing home patient. Normal continence depends on normal function of the internal anal sphincter and the external sphincter and puborectalis muscle, as well as normal neurologic innervation for rectal sensation and distention (see Chapters 134 and 135). When any of these functions is significantly disturbed, incontinence can occur.

CLINICAL PICTURE

The patient with fecal incontinence may seek treatment after having a large, uncontrolled, loose bowel movement or after regular fecal staining or recurrent episodes of significant incontinence, or the patient may be brought in by a family member and may be unaware of any rectal staining. It is important to be certain that the patient has fecal incontinence and not merely anal discharge of mucus or blood. Certainly, the latter infers an abnormality other than a true loss of stool.

Box 167-1 Causes of Fecal Incontinence

Normal Sphincters and Pelvic Floor

- Diarrhea
  - Infection
  - Inflammatory bowel disease
  - Intestinal resection

Anatomic Derangements and Rectal Disease

- Congenital abnormalities of anorectum
- Fistula
- Rectal prolapse
- Anorectal trauma
  - Injury
  - Childbirth injury
  - Surgery (including hemorrhoidectomy)
- Sequelae of anorectal infections, Crohn's disease

Neurologic Diseases

Central Nervous System Disease

- Dementia, sedation, mental disease
- Stroke, brain tumors
- Spinal cord lesions
- Multiple sclerosis
- Tabes dorsalis

Peripheral Nervous System Disease

- Cauda equina lesions
- Polyneuropathies
  - Diabetes mellitus
  - Shy-Drager syndrome
  - Toxic neuropathy
- Traumatic neuropathy
  - Idiopathic incontinence
  - Perineal descent
  - Postpartum
- Altered rectal sensation (site of lesion unknown)
  - Fecal impaction
  - Delayed-sensation syndrome

Skeletal Muscle Diseases

- Myasthenia gravis
- Myopathies, muscular dystrophy

Smooth Muscle Dysfunction

Abnormal Rectal Compliance

- Proctitis caused by inflammatory bowel disease
- Radiation proctitis
- Rectal ischemia
- Fecal impaction

Internal anal Sphincter Weakness

- Radiation proctitis
- Diabetes mellitus
- Childhood encopresis

Modified from Feldman N, Friedman LS, Schlesinger MH, editors: Gastrointestinal and liver disease, *ed 7, Philadelphia, 2002, Saunders.*

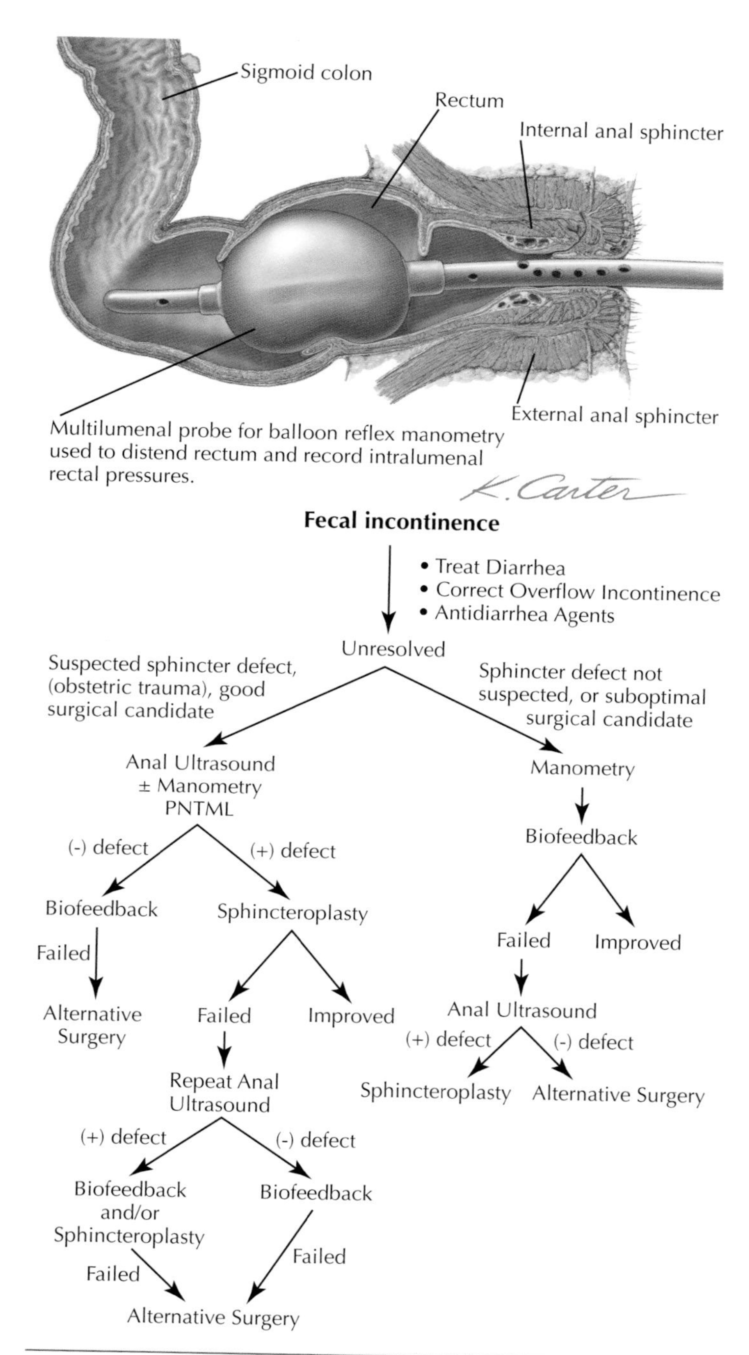

Figure 167-1 *Fecal Incontinence.* Algorithm from Soffer E: Practical approach to fecal incontinence, *Am J Gastroenterol* 95:1879, 2000. PNTML, *Pudental nerve terminal motor latency.*

DIAGNOSIS

The patient's history is essential to determine whether trauma to the rectum has occurred or whether the patient has a neurologic disorder. In addition, inspection and examination of the rectum are essential. The voluntary squeeze response on digital rectal examination can help localize the problem to the internal or external sphincter. Voluntary contraction suggests an internal sphincter problem, with a disturbance in the autonomic nervous system or with smooth muscle function, as occurs in patients with diabetes or scleroderma. Striated muscle may be affected because of neuropathy or longstanding damage.

If the diagnosis is not obvious, as in traumatic injury, a combination of anal ultrasound and anal manometry can be helpful (see algorithm in **Fig. 167-1**). Both procedures are usually performed in special centers, but these are limited in number. Magnetic resonance imaging (MRI) is the best way to determine whether the integrity of the internal anal sphincter has been compromised. Although endosonography appears to be more accurate and costs less, MRI is available in many areas and may be the only modality that clearly defines the anatomy. Defecography may be used to evaluate the rectoanal angle but adds little to the treatment algorithm.

Anal manometry is important for both diagnosis and treatment and is very provider dependent (see Fig. 167-1). After documenting weakness or incompetence of the sphincters, biofeedback therapy can be used effectively to correct many causes of incontinence.

TREATMENT AND MANAGEMENT

Medical therapy includes appropriate bowel management. Uncontrollable diarrhea, overflow incontinence from impaction, and other obvious causes should be treated medically. Biofeedback should be tried if the situation indicates that it can help the patient's condition (see algorithm).

If the sphincter has been significantly damaged, surgical therapy and repair are indicated. Many techniques are available, and successful repair depends on the surgeon's skill and experience. Colostomy is a last resort but may be necessary when other therapies fail and it is determined that this would be best for the patient.

COURSE AND PROGNOSIS

As indicated by its cause, if incontinence occurs in an elderly patient or in a young patient with a neurologic disorder, outcome is poor, and appropriate hygiene is necessary. If incontinence is caused by sphincter damage, biofeedback mechanisms and surgical repair may correct the situation. Biofeedback is labor intensive with variable success, however, and requires skilled therapists or physicians.

ADDITIONAL RESOURCES

Bellicini N, Malloy PJ, Caushaj P, Koslowski P: Fecal incontinence: a review, *Dig Dis Sci* 53:41-46, 2008.

Bharucha AE, Seide BM, Zinsmeister AR, Melton LJ III: Relation of bowel habits to fecal incontinence in woman, *Am J Gastroenterol* 103:1470-1475, 2008.

Coller JA: Clinical application of anorectal manometry, *Gastroenterol Clin North Am* 16:17-33, 1987.

Hannaway CD, Hull TL: Fecal incontinence, *Obstet Gynecol Clin North Am* 35:249-269, 2008.

Henry MM: Pathogenesis and management of fecal incontinence in the adult, *Gastroenterol Clin North Am* 16:35-46, 1987.

Heymen S, Scarlett Y, Jones K, et al: Randomized, controlled trial shows biofeedback to be superior to alternative treatments for patients with pelvic floor dyssynergia-type constipation, *Dis Colon Rect* 50:428-441, 2007.

Law PJ, Kamm MA, Bartram CR: A comparison between electromyography and anal endosonography in mapping external sphincter defects, *Dis Colon Rect* 33:370-373, 1990.

Rao SS: Diagnosis and management of fecal incontinence, *Am J Gastroenterol* 99:1585-1616, 2004.

Rao SS, Seaton K, Miller M, et al: Randomized controlled trial of biofeedback, sham feedback, and standard therapy for dyssynergic defecation, *Clin Gastroenterol Hepatol* 5:331-338, 2007.

Schiller LR: Fecal incontinence. In Feldman N, Friedman LS, Brandt LJ, editors: *Gastrointestinal and liver disease*, ed 8, Philadelphia, 2006, Saunders-Elsevier, pp 199-219.

Wald A: Fecal incontinence in adults, *N Engl J Med* 356:1648-1655, 2007.

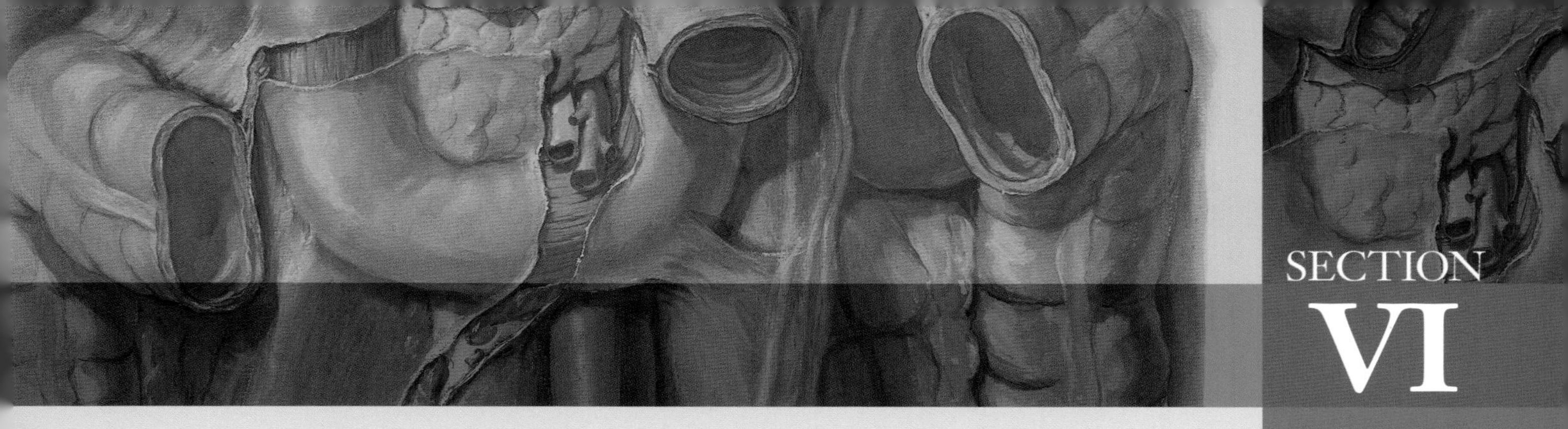

Infectious and Parasitic Diseases of the Alimentary Tract

Esophageal Infections

Martin H. Floch

Infections of the esophagus are relatively rare in patients with healthy immunity. However, after a chronic disease develops, such as diabetes, neoplasia, or immunosuppressive disease (e.g., acquired immunodeficiency syndrome [AIDS]), the esophagus becomes susceptible to infection. The most common invading organisms are fungi, followed by viruses (e.g., cytomegalovirus).

CLINICAL PICTURE

The presentation of esophageal infections varies greatly. Some patients may have only heartburn or chronic nausea, whereas others may have significant bleeding or severe dysphagia with odynophagia. Because of the variable symptoms, esophageal disease should be suspected whenever the presentation is unclear or is associated with any chronic or immunocompromising illness.

DIAGNOSIS

The diagnosis of esophageal infectious disease is made using endoscopy and biopsy. If the patient has suspected esophageal disease, upper endoscopy must be performed. The classic endoscopic appearance of esophageal disease—erosions, ulcerations, or plaquelike lesions—is easily seen. Biopsy specimens must be obtained from the margins of the lesions and sent for interpretation by a pathologist; brushings also must be obtained. Viral infections are often identified through serology and cultures. Because there may be multiple infectious etiologies, viral culture is recommended when the differential diagnosis includes a viral etiology.

The presence of *oral thrush* in a patient with immunosuppression may be sufficient evidence to treat for fungal infection. Typically, the oral lesions should be diagnosed by brush studies before any therapy is undertaken.

Controversy surrounds intervention without endoscopy and biopsy. Some clinicians do treat before an accurate diagnosis is obtained. It is also important to remember that as many as 20% of patients may have two pathogens.

The following infectious agents have been reported in the esophagus:

- **Fungal:** *Candida albicans*, *Torulopsis glabrata*, *Histoplasma capsulatum*
- **Viral:** Cytomegalovirus (CMV), herpes simplex virus (HSV), human immunodeficiency virus (HIV-1), Epstein-Barr virus (EBV)
- **Bacterial:** *Mycobacterium avium-intracellulare* complex, *Mycobacterium tuberculosis*, bacteriosis (unidentified)
- **Protozoal:** *Cryptosporidium* spp., *Pneumocystis carinii*, *Leishmania* spp.

Fungal Infections

Candida albicans is a yeast found in the normal human flora. Its prevalence is approximately 20%, but it is the most common organism causing esophageal infection in the patient with compromised immune response. When it becomes invasive, *C. albicans* usually forms plaques that have an erythematous base. At times, patients with thrush swallow some of the plaquelike material. Biopsy and endoscopy usually reveal no erosive base. In patients with AIDS or symptomatic diabetes, the finding of *C. albicans* in oral infection indicates a sensitivity of approximately 88% and a specificity of 81% that there will be esophagitis. *C. albicans* may grow throughout the esophagus, usually in the lower half, and frequently in HIV-1–infected patients. Because the neutrophils are usually intact in these patients, the bacterium rarely invades below the lamina propria. Nevertheless, the lower the lymphocyte count, the more prevalent is the infection.

Histoplasma capsulatum has been reported in immunocompromised patients but is a rare entity. Similarly, *Aspergillus*, *Cryptococcus*, and *Torulopsis* organisms are rare but have been reported.

Viral Infections

CMV is the second most common cause of esophagitis in the immunocompromised patient and the most common viral infection. The virus invades cells and then lies dormant. It becomes active during the severely immunocompromised phase of any disease, proliferating and causing severe inflammation and ulceration. CMV in these patients may involve any part of the gastrointestinal (GI) tract, including the liver and the colon, to cause hepatitis and colitis. It may cause large or small ulcers in the esophagus.

The second most common viral infection is HSV, which can develop in healthy persons. HSV is usually a mild infection, with occasional reactivation episodes, but because of normal immune processes, it is contained. In immunocompromised patients, however, the reactivation may occur frequently and may cause recurrent and severe symptoms. In addition, HSV can spread throughout the body into many other organs.

Herpes zoster (shingles) and varicella (chickenpox) can similarly infect the esophagus and cause severe infection in the immunocompromised patient. In the HIV patient, herpes zoster can cause small, aphthoid lesions in the esophagus without any other infection. These HIV patients receive specialized therapy.

EBV and human papillomavirus (HPV) can invade the esophageal mucosa. EBV is seen during acute infections, whereas HPV is asymptomatic.

Bacterial and Protozoal Infections

Bacterial infections of the esophagus are rare, but the major risk factor is *granulocytopenia*. These infections usually occur in patients with malignancy receiving chemotherapy and often are multibacterial.

Areas with a high incidence of tuberculosis and the HIV-positive population have an increased prevalence of active *Mycobacterium tuberculosis* infection. Often, extension into the esophagus is direct, but the clinical picture is rarely one of

Table 168-1 Recommended Treatments for Common Viral and Fungal Infections of Esophagus

Pathogen	Treatment	Alternative
Viral		
Cytomegalovirus*	Ganciclovir, 5 mg/kg IV bid for 2 to 3 weeks	Foscarnet, 60 or 90 mg/kg IV bid for 2 to 3 weeks
Herpes simplex virus*	Acyclovir, 400 mg PO five times daily for 7 to 14 days	Famciclovir, 500 mg PO bid for 7 days Valacyclovir, 1000 mg PO tid for 7 days
Fungal		
Candida albicans	Fluconazole, 100 mg PO qd for 14 days	Itraconazole, 200 mg qd for 14 days
Histoplasma capsulatum	Amphotericin B, 0.5 to 0.6 mg/kg IV qd for 4 to 8 weeks	Itraconazole, 200 mg bid

Modified from Blaser MJ, Smith PD, Ravdin JI, et al, editors: Infections of the gastrointestinal tract, *ed 2, Philadelphia, 2002, Lippincott–Williams & Wilkins.*
**Chronic suppression at reduced dosage may be necessary. Reduce dosage for decreased creatinine clearance. Resistance may develop; susceptibility testing is necessary.*
IV, *Intravenously;* PO, *orally;* qd, *every day;* bid, *twice daily;* tid, *three times daily.*

heartburn and odynophagia. Patients with tuberculosis usually have chest pain, weight loss, catastrophic bleeding, perforation, or fistula formation.

Protozoal infections are rare, as are other bacterial infections, and should be treated in accordance with published case reports of those diseases.

TREATMENT AND MANAGEMENT

Table 168-1 lists some of the recommended treatments for the common viral and fungal agents. Topical agents such as nystatin are effective in treating oral *Candida* infections and may be tried before diagnosis or therapy for esophageal disease. Topical agents are usually ineffective in treating the esophageal disease, and some clinicians administer clotrimazole troches, 10 mg five times daily for 7 to 14 days, as an inexpensive treatment that may be effective before using more potent systemic drugs, such as fluconazole or itraconazole. In patients receiving highly active antiretroviral therapy (HAART), the lesser therapy may be tried, but when the CD4+ count is less than 50, systemic oral therapy is usually needed.

Ganciclovir may be effective for CMV, but resistance occurs. Also, esophageal CMV infection is greatly reduced with HAART. Acyclovir is effective for HSV and well tolerated but may be required long term with recurrent HSV.

COURSE AND PROGNOSIS

The treatments outlined are usually effective in more than 50% and in as many as 100% of patients, but results depend largely on the associated disease. If the patient has mild diabetes, for example, treatment can be successful if the diabetes is controlled. On the other hand, if the patient has severe HIV for which HAART is ineffective, the prognosis is poor, and the esophageal disease may be a mixed infection that progresses.

ADDITIONAL RESOURCES

Graman PS: Esophagitis. In Mandell GL, Bennett JE, Dolin R, editors: *Mandell, Douglas and Bennett's principles and practice of infectious diseases*, ed 6, Philadelphia, 2005, Elsevier–Churchill Livingstone, pp 1231-1236.

Smith PD, Janoff EN: Gastrointestinal infections in HIV-1 disease. In Blaser MJ, Smith PD, Ravdin JI, et al, editors: *Infections of the gastrointestinal tract*, ed 2, Philadelphia, 2002, Lippincott–Williams & Wilkins, pp 415-443.

Wilcox CM: Esophageal infections and other human immunodeficiency virus–associated esophageal disorders. In DeMarino AJ Jr, Benjamin SB, editors: *Gastrointestinal disease*, ed 2, Thorofare, NJ, 2002, Slack, pp 213-230.

Wilcox CM: Gastrointestinal consequences of infection with human immunodeficiency virus. In Feldman M, Friedman LS, Brandt LJ, editors: *Gastrointestinal and liver disease*, ed 8, Philadelphia, 2006, Saunders-Elsevier, pp 667-682.

Typhoid Fever (Paratyphoid Fever, Enteric Fever)

Martin H. Floch

Classic typhoid fever is caused by *Salmonella typhi*. Less severe fever syndromes are caused by *Salmonella paratyphi A*, *S. paratyphi B* (*S. schottmüelleri*), and *S. paratyphi C* (*S. hirschfeldii*). The enteric fever of *S. typhi* is referred to as "typhoid fever" and the other three as "paratyphoid fever." There are no animal reservoirs. This is a disease of humans. Transmission of the organism occurs from human feces or urine, but flies or shellfish such as oysters and clams can transmit the organism (**Fig. 169-1**).

The disease still occurs in epidemic form in developing countries; mortality rates in Asia and Africa are as high as 30%. In developed countries such as the United States, no more than 400 to 500 cases occur each year, but approximately 0.4% can be fatal, especially if the strains are resistant. Classic small bowel and systemic findings are observed. The disease is easily understood, and it elucidates many pathophysiologic phenomena of infectious diseases in the small bowel.

CLINICAL PICTURE

Once the organisms enter by ingestion, they are filtered through Peyer patches and lymph nodes of the small bowel. If they become invasive, an incubation period of 7 to 14 days is usually followed by high fever, headache, and abdominal pain (**Fig. 169-2**). The pulse rate is usually slow, inconsistent with the high temperatures. Depending on where the intestinal Peyer patches are swollen and ulcerated, the abdominal pain may be periumbilical, in the right lower quadrant, or diffuse. Characteristically, the spleen is enlarged, swollen, and easily palpable. Rose spots may develop on the chest and abdomen. Surprisingly, only 50% of patients have diarrhea, and some may even be constipated.

In the second phase or second week of the disease, temperature remains consistent, and the patient looks debilitated. As the fever persists into the third week, patients may become delirious and possibly dehydrated and debilitated. The patient is severely anorexic and discharges diarrheic or classic "pea soup" stool and has a distended, tender abdomen. Untreated patients who survive start to improve gradually in the fourth week as temperatures decline. The natural history of the untreated severe disease is 4 weeks to 1 month.

DIAGNOSIS

Diagnosis is usually made by isolating *S. typhi* or *S. paratyphi* from blood cultures, stool, or urine. Bone marrow aspiration may be performed for diagnostic evaluation and to evaluate anemia or severe leukopenia. Blood cultures are positive in 50% to 70% of patients, and bone marrow cultures are positive in 90%. Organisms can be obtained from punch biopsies of the rose spots. When the diagnosis is difficult, the duodenal string test may be used; findings may be positive when the bone marrow findings are negative.

Serologic tests are used in some laboratories around the world, but culture techniques are more reliable. Agglutinin titers against a somatic antigen (zero-Widal test) usually rise in the second and third week of the illness. A titer of 1:320 is diagnostic, as is a fourfold rise in titer. The so-called H antigen is less significant, but high titers also suggest typhoid fever. If it is not possible to grow cultures, as occurs in some parts of the world, serologic tests must be performed.

TREATMENT AND MANAGEMENT

Classic enteric typhoid fever may be fatal in as many as 30% of patients. The standard since 1948 and treatment of choice is *chloramphenicol*, 500 mg orally four times daily; it is inexpensive and highly effective. Chloramphenicol reduces mortality to 1% and duration of fever to 3 to 5 days.

Because some organisms have developed resistance to chloramphenicol, ciprofloxacin and amoxicillin are alternative therapies. Oral quinolone and parenteral third-generation cephalosporins have also been substituted to treat resistant organisms.

COURSE AND PROGNOSIS

Although the natural history of typhoid fever ranges up to 1 month, it may be shortened to 1 week with adequate, successful antibiotic treatment. Most complications occur in the third or fourth week of infection. Intestinal perforation may occur at the site of ulceration from infected lymphoid tissue. Rare complications have included endocarditis, pericarditis, liver and splenic abscesses, and spontaneous rupture of the spleen. The leukocytosis that may occur in children resolves, as do the leukopenia and anemia in adults.

A chronic carrier state for *S. typhi* may occur in 1% to 4% of patients, more often affecting women and associated with biliary abnormalities. Careful treatment of biliary disease, antibiotic therapy, and thorough evaluation are necessary.

Vaccination is recommended only for persons at high risk for *Salmonella* infection, such as those traveling to the Indian subcontinent or laboratory personnel who work with the organisms. Vaccination is not recommended for travelers to other locations because of the difficulty involved, and because most vaccines produce significant adverse effects. The three vaccines available for *S. typhi* differ in their adverse effects and effectiveness, and their use should first be discussed and carefully planned with the patient.

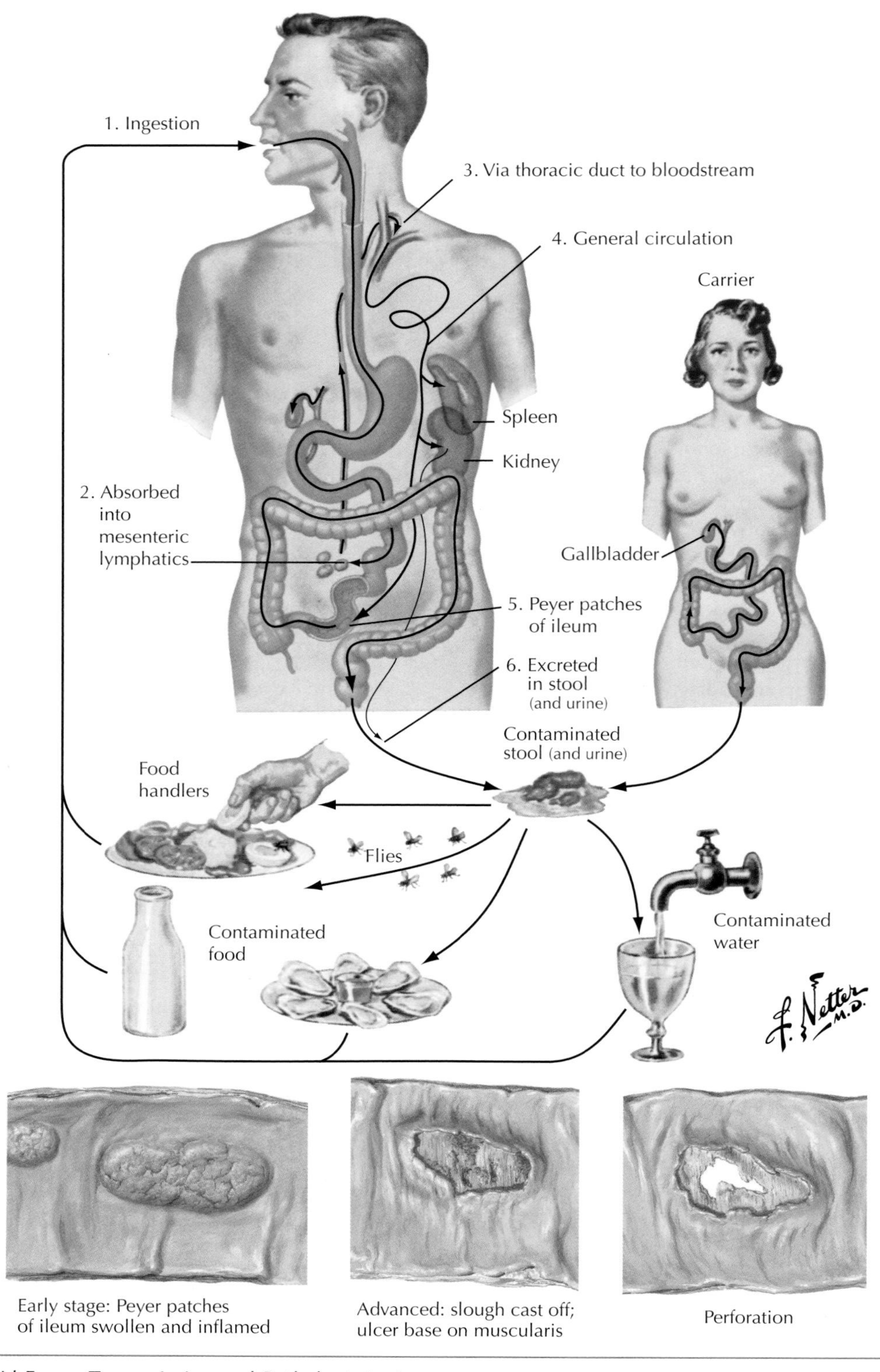

Figure 169-1 *Typhoid Fever: Transmission and Pathologic Lesions.*

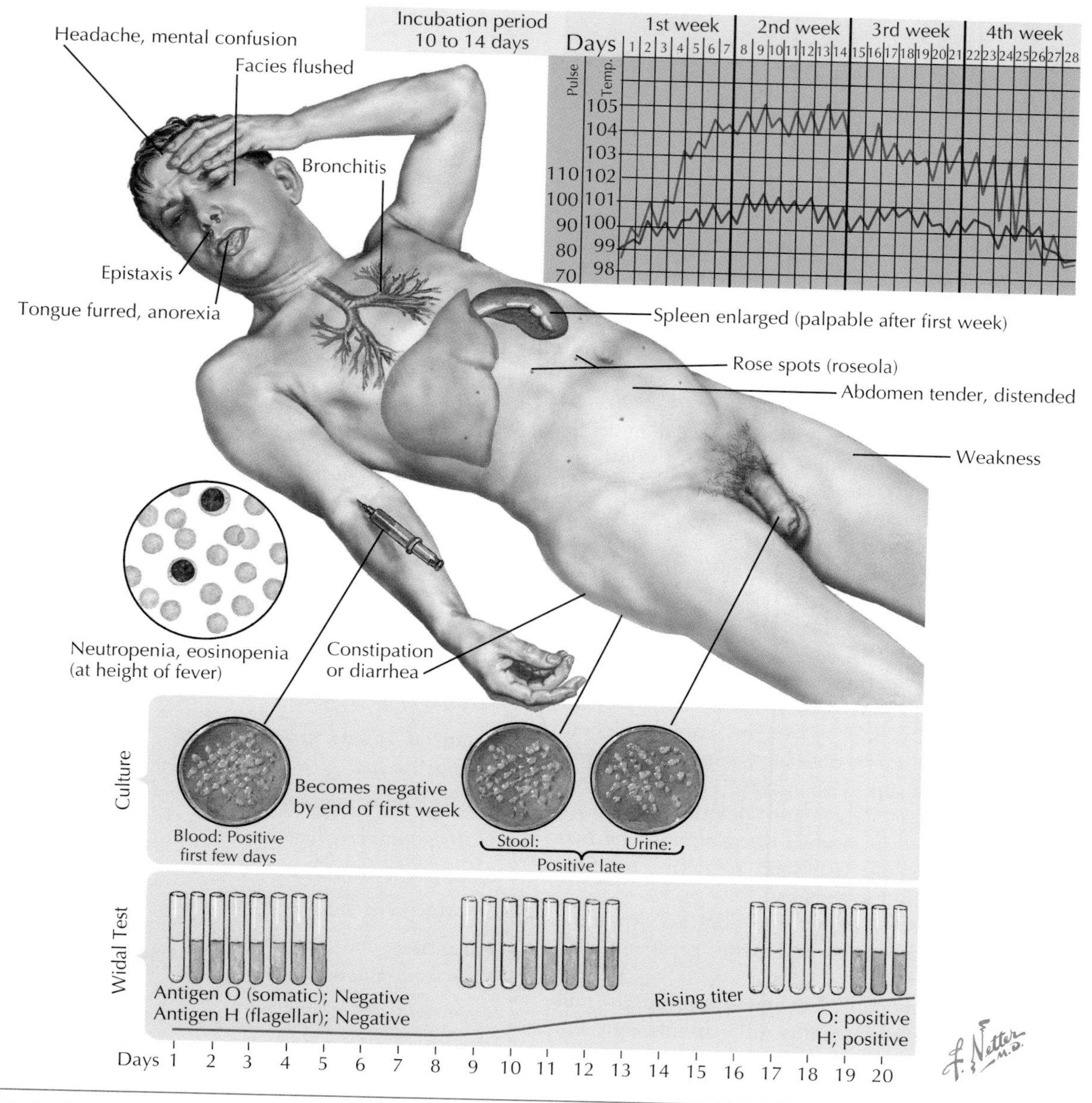

Figure 169-2 *Typhoid Fever: Clinical and Laboratory Diagnostic Features.*

ADDITIONAL RESOURCES

Fraser A, Goldberg E, Acosta CJ, et al: Vaccines for preventing typhoid fever, *Cochrane Database Syst Rev* 3:CD001261, 2007.

Gianella RA: Infectious enteritis and proctocolitis and bacterial food poisoning. In Feldman M, Freedman LS, Brandt LJ, editors: *Gastrointestinal and liver disease*, ed 8, Philadelphia, 2006, Saunders-Elsevier, pp 2333-2391.

Parry CM, Hien TT, Dougan G, et al: Typhoid fever, *N Engl J Med* 347:1770-1782, 2002.

Pegues DA, Ohl ME, Miller SI: *Salmonella* species, including *Salmonella typhi*. In Mandell GL, Bennett JE, Dolin R, editors: *Mandell, Douglas and Bennett's principles and practice of infectious diseases*, ed 6, Philadelphia, 2005, Elsevier–Churchill Livingstone, pp 2636-2654.

Sirinavin S, Garner P: Antibiotics for treating *Salmonella* gut infections, *Cochrane Database Syst Rev* 93:CD001167, 2000.

Food Poisoning and Enteric Pathogens

170

Martin H. Floch

Food poisoning can be defined as a clinical state characterized chiefly by acute gastroenteritis developing within hours or days of ingesting contaminated food. The food may contain organisms that grow within the host and can be designated *infectious*. Foods also may contain *toxins* produced by organisms growing in the food. In addition, foods such as mushrooms, fish, and mussels may contain poisonous chemicals.

An estimated 38 to 78 million food poisonings occur annually in the United States, resulting in approximately 325,000 hospitalizations and 2000 to 5000 deaths, depending on comorbidities. This chapter emphasizes gastroenteritis, but other food-borne illnesses exist. Depending on the area and the outbreak, about 50% of food-borne poisoning can be attributed to bacteria and 50% to viral agents (**Box 170-1**).

Box 170-1 Organisms That Cause Food Poisoning

Bacteria
- *Brucella* spp.
- *Campylobacter* spp.
- *Escherichia coli*, O157:H7
- *Escherichia coli*, non-O157:H7
- *Listeria monocytogenes*
- *Salmonella typhi*
- Nontyphoid *Salmonella* spp.
- *Shigella* spp.
- *Vibrio cholerae*
- Noncholera *Vibrio* spp.
- *Vibrio vulnificus*
- *Yersinia enterocolitica*

Bacterial Toxins Produced by:
- *Bacillus cereus*
- *Clostridium botulinum*
- *Clostridium perfringens*
- *Staphylococcus aureus*
- *Streptococcus* spp.

Parasites
- *Cryptosporidium parvum*
- *Cyclospora cayetanensis*
- *Giardia lamblia*
- *Toxoplasma gondii*
- *Trichinella spiralis*

Viruses
- Norwalk-like virus
- Rotavirus
- Astrovirus
- Hepatitis A virus

CLINICAL PICTURE

The causative organisms vary in their incubation period. Usually acute in onset, toxins are produced in foods by *Staphylococcus aureus*, *Bacillus cereus*, *Clostridium perfringens*, enterotoxigenic strains of *Escherichia coli* and *Vibrio cholerae*, *Campylobacter jejuni*, and *Salmonella* and *Shigella* spp. The invasive bacterial organisms are usually *Salmonella* or *Shigella* and sometimes *Campylobacter*, *Vibrio*, invasive *E. coli*, or *Yersinia enterocolitica*. The two common viral agents, Norwalk virus and rotavirus, produce symptoms 16 to 72 hours after ingestion.

Figures 170-1 and **170-2** depict the clinical symptoms and presentation in the infection and toxin types of food poisoning.

Symptoms vary depending on patient age, comorbidities, and the toxin. Diarrhea develops in almost all patients, but the amount of vomiting may vary; for example, *B. cereus* toxin can produce primarily a diarrhea-type or a vomiting-type food poisoning. In general, although confusing overlap can occur, if onset of disease is short, the organism is toxin producing; if onset is longer, the organism is infectious.

DIAGNOSIS AND TREATMENT

Diagnosis largely depends on obtaining stool for culture and, in select patients, other body secretions for culture. Because these infections are relatively short lived, serology is of little value, although it may be helpful for studying epidemics. Good cultures cannot be grown from viral agents, so serology is essential for evaluating the epidemiologic pattern.

Cholera

Cholera is usually caused by *V. cholerae*, a curved gram-negative rod capable of causing death within hours. Once these bacteria invade the intestine, their virulent toxin increases adenylate cyclase activity, which prevents water absorption and increases fluid and electrolyte secretion, rapidly dehydrating the patient. Daily fecal output can be as much as 20 L. Without fluid replacement, the patient can die. Other strains of *Vibrio* are rampant on the Indian subcontinent, and occasionally, cases are reported in the southern United States. There have been epidemics in South America.

Vibrio species grow in the lumen but are not invasive, and there is no bacteremia. Electrolyte and fluid replacement are essential. Patients with severe cholera require intravenous therapy, whereas those with milder cases can be treated with

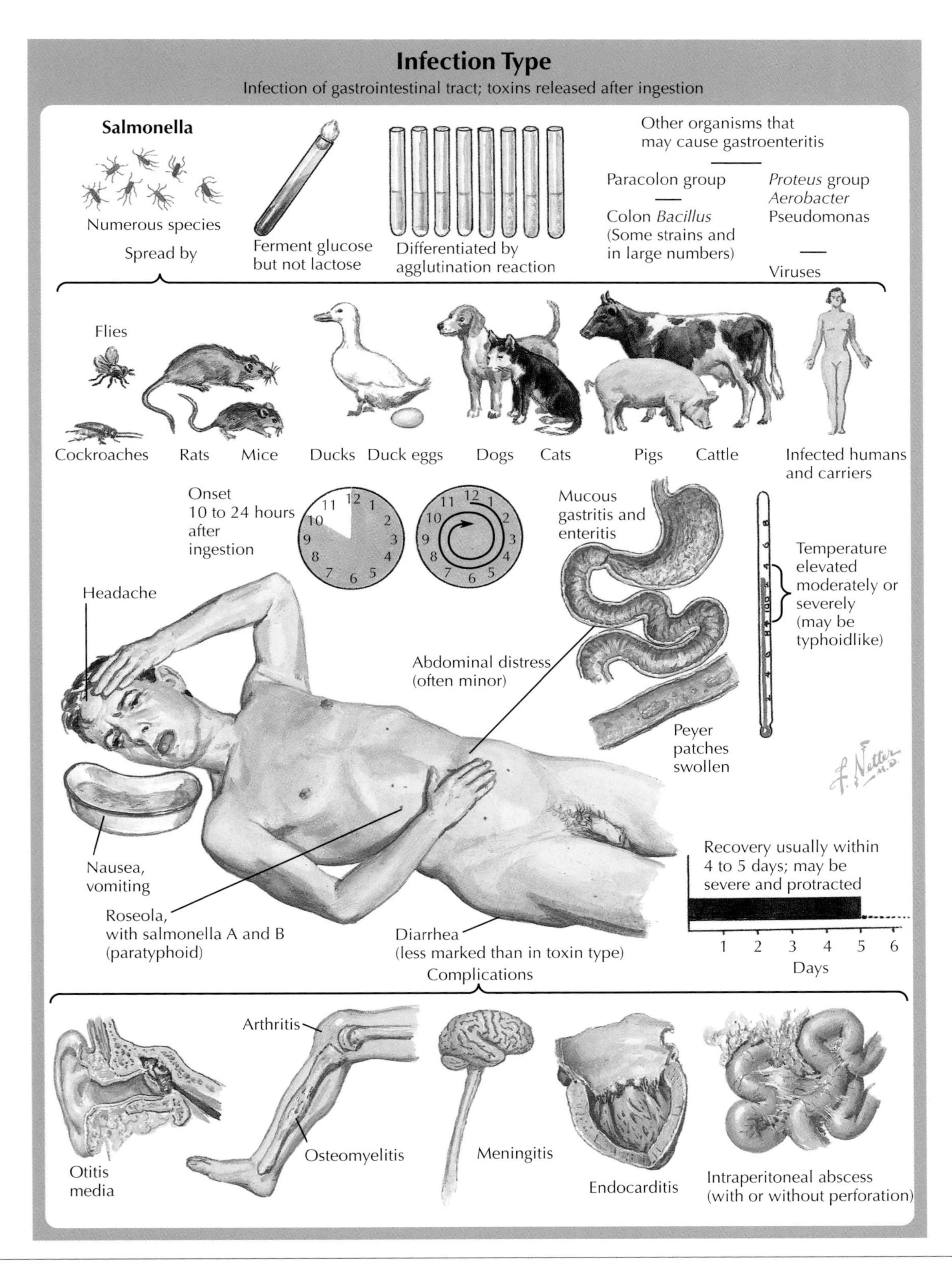

Figure 170-1 *Food Poisoning: Infection Type.*

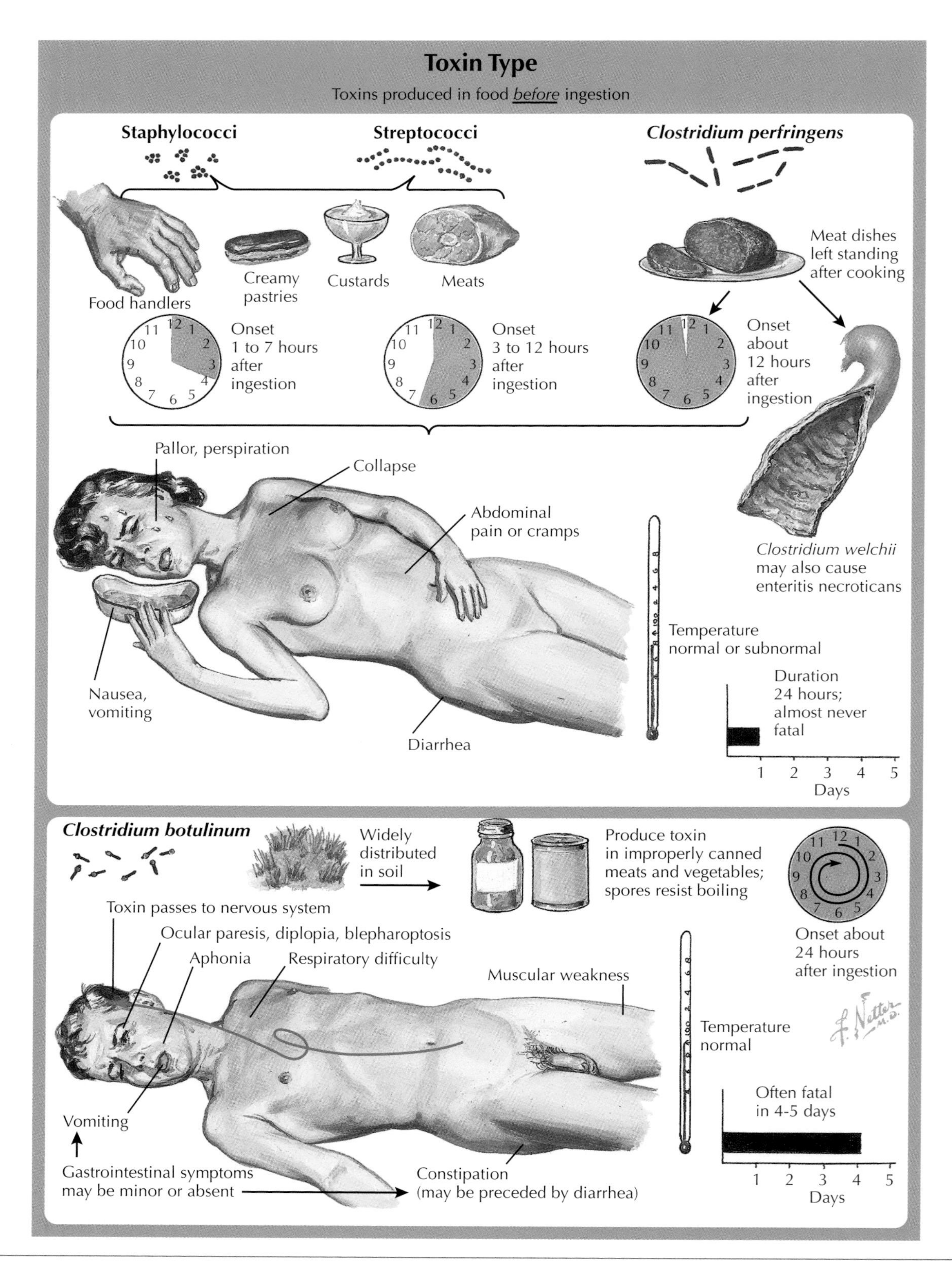

Figure 170-2 *Food Poisoning: Toxin Type.*

oral electrolyte solutions. A typical oral rehydration solution for adults contains 124 mmol/L of sodium, 16 mmol/L potassium, 90 mmol/L chloride, and 48 mmol/L bicarbonate, resulting in passive absorption of the electrolytes and fluid and preventing massive dehydration. Antibiotics may be added, usually tetracycline and doxycycline; if resistance develops, ciprofloxacin may be successful with a single dose. Similarly, trimethoprim-sulfamethoxazole can be used as an alternative.

Before the advent of fluid replacement and antibiotics, cholera mortality was as high as 50% to 75%, but is now less than 1% with proper therapy.

Salmonellosis

Nontyphoid salmonellosis is a common cause of food-borne enteric infections in the industrialized world, causing more than 1.5 million cases in the United States and accounting for 13% of all food poisoning outbreaks and 45% of deaths. It is caused by *Salmonella enteritidis*, *Salmonella typhimurium*, and other serologic types. Common reservoirs for these organisms include poultry, domestic livestock, and house pets.

Clinically, these bacteria can cause one of three syndromes—gastroenteritis, enteric fever, or bacteremia—or patients may be asymptomatic. This varied picture results because salmonellae do not seem to grow extensively in the intestine, but rather invade the lymph and phagocytic tissue. Any diarrhea is usually self-limiting, lasting 3 to 7 days, and the symptoms are largely resolved within 72 hours; however, organisms may exist in stool for 1 month.

Antibiotics are usually not recommended for routine *Salmonella* infection because of the increased risk for bacterial resistance and the usually short period of infection. However, it may be necessary to treat severely ill patients with comorbidities. If the patient has bacteremia, antibiotic therapy is indicated for 7 to 14 days, often with two antibiotics; choices include ampicillin, amoxicillin, trimethoprim-sulfamethoxazole, cefotaxime or ceftriaxone, chloramphenicol, and fluoroquinolones. If a patient has an intravascular infection, a 6-week course of antibiotics is indicated.

Other Bacterial Pathogens

The enteropathogenic, enteroinvasive, and enterohemorrhagic forms of *E. coli* produce different types of syndromes (see Additional Resources). *Shigella* produces a classic, bacillary form of dysentery.

The interested reader is referred to the appropriate resources for *C. perfringens*, *Campylobacter*, *Listeria*, *Yersinia*, and *B. cereus* infections. As a caveat in modern society, *B. cereus* can come from fried rice that is not fresh or that is kept at low-grade refrigeration for 24 to 48 hours.

Viral Pathogens

Viral pathogens causing food poisoning are as common as bacterial pathogens and include five groups: rotaviruses, caliciviruses (Norwalk viruses), enteric adenoviruses, astroviruses, and toroviruses (enveloped single-stranded RNA that causes 3% of diarrhea in children). Symptoms and syndromes are similar for all these viruses, with some different epidemiologic findings. Norwalk viruses are more common in adults, whereas rotaviruses, adenoviruses, astroviruses, and toroviruses are more common in infants and children. Adenoviruses also are associated with upper respiratory infection and may cause vomiting, diarrhea, and severe dehydration over 5 to 7 days.

Laboratory diagnosis is primarily made using immunoassay or electron microscopy. Successful growth of these viruses routinely in the laboratory has not been accomplished.

Treatment of viral food poisoning is largely supportive. When diarrhea is severe, replacing electrolytes and fluids is essential. Most cases resolve spontaneously. However, for patients with comorbidities and infants, the dehydration may be life threatening, especially when fluid replacement is not possible.

COURSE AND PROGNOSIS

Because transmission is through the oral and fecal routes, prevention is the best cure. When a pandemic or epidemic of food poisoning seems to be evolving, exposure to contaminated persons must be avoided. Oral rehydration solutions are effective and, if the diarrhea causes dehydration, should be used before intravenous therapy becomes necessary.

ADDITIONAL RESOURCES

Fry AM, Braden CR, Griffin PM, Hughes JM: Foodborne disease. In Mandell GL, Bennett JE, Dolin R, editors: *Mandell, Douglas, and Bennett's principles and practice of infectious diseases*, ed 6, Philadelphia, 2005, Elsevier–Churchill Livingstone, pp 1286-1301.

Gianella RA: Infectious enteritis and proctocolitis and bacterial food poisoning. In Friedman M, Feldman LS, Brandt LJ, editors: *Gastrointestinal and liver disease*, ed 8, Philadelphia, 2006, Saunders–Elsevier, pp 2333-2391.

Clostridium difficile, Pseudomembranous Enterocolitis, and Antibiotic-Associated Diarrhea

17

Martin H. Floch

Pseudomembranous enterocolitis (PMC) is believed to be caused by *Clostridium difficile*. Early in the evaluation of PMC, *Staphylococcus aureus* was also identified as a possible etiologic agent. Because some patients have involvement of the large and small bowels, and because only it is identified in rare cases, *S. aureus* is still considered a possible cause of the syndrome. However, the classic finding of PMC in the large bowel and the presence of *C. difficile* indicate that *C. difficile* is the primary cause of PMC.

C. difficile is a gram-negative, spore-forming anaerobe that produces two large, single-unit toxins: toxin A, an enterotoxin, and toxin B, a cytotoxin. The spores are subterminal and are not associated with toxin production. However, spore formation makes the organism clinically fastidious. *C. difficile* has been isolated from feces in as many as 20% of infants but as few as 3% of the general population. It does not always cause clinical symptoms. *C. difficile*–associated diarrhea is associated with two risk factors: a hospital stay and antimicrobial treatment. Antibiotics appear more likely to cause a *C. difficile* PMC. Clindamycin, erythromycin, tetracycline, and chloramphenicol were previously thought to be the precipitating agents, but it is now known that any disturbance in the microflora ecosystem is associated with an increased incidence of *C. difficile* diarrhea. The hospital setting is the significant risk factor, and it is assumed that hospital patients can become infested with the organism if they are not carriers. Community-acquired *C. difficile*–associated diarrhea is reported at a low incidence of 7 per 100,000 population, whereas in hospital patients, the incidence increases to more than 20 per 100,000. Because it is a spore-forming organism, *C. difficile* spores can be reactivated easily in a relatively asymptomatic patient.

Again, *C. difficile* is not recovered in many cases, and diarrhea may occur after use of antibiotics. The syndrome is then called *antibiotic-associated diarrhea* (AAD). It is not treated as *C. difficile* but as diarrhea until the cause of the diarrhea is discovered.

CLINICAL PICTURE

Patients have diarrhea or abdominal pain; the symptoms and findings are variable. Every hospital patient with diarrhea must be checked for *C. difficile*–associated diarrhea, as well as any hospital patient with evidence of colitis. Rarely, patients develop catastrophic toxic megacolon, including distended abdomen, high temperature, and progression to perforation.

DIAGNOSIS

Hospital patients with diarrhea must be checked for *C. difficile*, which includes testing for toxins A and B. These tests are now performed in all general hospitals and should also be performed when colitis is suspected. Endoscopy is often helpful. At times, a classic pseudomembrane can be seen, and biopsy reveals PMC. Other patients may not have a pseudomembrane but only low-grade colitis. *C. difficile* is rarely a superimposed infection in nonspecific ulcerative colitis or Crohn's disease, but it can be part of multiple infections complicating the course of a debilitated immunosuppressed patient.

TREATMENT AND MANAGEMENT

The two antibiotics routinely used for PMC are metronidazole and vancomycin. *Metronidazole* is usually given initially, 250 or 500 mg three times daily for 10 days, because of its lower cost. This therapy is reported to be effective in 95% of patients. Improvement usually occurs in a few days, and all symptoms are resolved within 2 weeks. However, some patients have intolerance to metronidazole, and not all patients respond rapidly. For those patients, oral *vancomycin* is given, 150 or 250 mg four times daily for 10 days. Vancomycin is poorly absorbed and is more expensive than metronidazole, but it is highly effective.

Treatment of AAD is described in Chapter 111. Some clinicians may try to treat AAD as *C. difficile*, even though the organism is not recovered; this is a clinical decision made by individual physicians and is not usually encouraged.

COURSE AND PROGNOSIS

The problem with *C. difficile* is that it recurs in 5% to 30% of patients 7 to 14 days after therapy is discontinued. This may be attributed to relapse or to the growth of spores, or it may represent reinfection in hospital patients. Clinicians turn to various courses to treat the second infection and to prevent recurrent infections. In the first relapse, usually a 2-week course of metronidazole or vancomycin is tried. If a patient has a second relapse, I prefer to use vancomycin in tapered doses starting with 125 mg four times a day, then twice a day after 1 week, then once every day after 1 more week, then once every other day, and then once every second or third day.

Probiotics are sometimes used to stimulate the normal bacterial flora and may prevent reinfection. *Saccharomyces boulardii* has been effective in some patients, and *Lactobacillus rhamnosus* (GG strain) is effective in others. Although most cases can be controlled by one of these regimens, vancomycin with rifampin has been necessary in some patients. In highly resistant cases, bacteriotherapy has been used.

It is important that adequate infectious disease control be instituted in hospitals to prevent reinfection and the spread of organisms among patients. Infection control should also occur in the home, with detailed patient education.

If PMC is severe and megacolon evolves, it must be treated similar to megacolon in ulcerative colitis. To avert catastrophe, surgery may be needed if the infection cannot be brought under control rapidly (see Chapter 138).

ADDITIONAL RESOURCES

Bartlett GJ, Chang TW, Gerwith M, et al: Antibiotic-associated pseudomembranous colitis due to toxin-producing *Clostridium*, *N Engl J Med* 298:531-534, 1978.

Doron SI, Hibbard PL, Gorbach SL: Probiotics for prevention of antibiotic-associated diarrhea, *J Clin Gastroenterol* 42:S58-S63, 2008.

Floch MH, Walker WA, Guandalini S, et al: Recommendations for probiotic use—2008, *J Clin Gastroenterol* 42:S104-S108, 2008.

Kelly CP, Lamont JT: Antibiotic-associated diarrhea, pseudomembranous enterocolitis, and *Clostridium difficile*–associated diarrhea and colitis. In Feldman M, Friedman LS, Brandt LJ, editors: *Gastrointestinal and liver disease*, ed 8, Philadelphia, 2006, Saunders-Elsevier, pp 2393-2412.

Surawicz CM: Role of probiotics in antibiotic-associated diarrhea, *Clostridium difficile*–associated diarrhea and recurrent *Clostridium difficile*–associated diarrhea, *J Clin Gastroenterol* 42:S64-S70, 2008.

Gastrointestinal Tuberculosis

172

Martin H. Floch

Tuberculosis (TB) is caused by *Mycobacterium tuberculosis*. Humans and cattle are the major reservoirs for this ancient disease, recognized for its prevalence in the pharaohs' Egypt. The two forms of the mycobacterium complex are *M. tuberculosis*, the common species transmitted orally through the diet, and *Mycobacterium bovis*, a related species in cattle transmitted through unpasteurized milk.

Found mainly in developing countries, TB infects almost one third of the world's population and causes more deaths per year than any other infectious disease. Before the development of biotics and antibiotics, TB was prevalent in developed countries, and 70% of patients with pulmonary TB progressed to gastrointestinal (GI) disease; now, less than 1% progress to GI involvement. Incidence of TB was recently at an all-time low of 7.4 per 100,000 population.

Organisms invade the intestinal tract through swallowing or by contaminated implements or foods (**Fig. 172-1**). Invasion of the gut occurs primarily in lymphoid tissue. Therefore, Peyer patches are susceptible, and the area of the terminal ileum, rich in lymphoid tissue, is most susceptible to intestinal TB. TB rarely develops in the esophagus through extension from the chest or through invasion from the stomach, duodenum, or other sites in the small bowel. The most common site is actually the distal ileum and cecum. Before antibiotics, TB was more common in children, but currently, in the Western world, the age range of patients with both types is 50 to 75 years, and in developing countries, 20 to 40 years. Patients with human immunodeficiency virus (HIV-1) infection are uniquely susceptible to TB. In Africa, TB occurs in 20% to 26% of the HIV-infected population.

Two types of TB occur: ulcerated and hypertrophic. In the *ulcerated* type, after invasion of the lymphoid follicles of a Peyer patch, ulceration slowly develops. A necrotic base forms in the ulcer, which rarely may perforate, or multiple nodules may occur around the ulcer and spread into the peritoneum. The less common *hypertrophic* type leads to extensive granular formation and fibrosis. It can form a "napkin ring" and mimic carcinoma.

Tuberculous peritonitis may occur from dissemination or direct extension. Caseating granulomas are characteristic of the disease. TB is reported in the appendix and in the rectum.

CLINICAL PICTURE

The most common symptom of intestinal TB is abdominal pain. Fever, anorexia, diarrhea, weight loss, constipation, bloating, and infrequently hemorrhage have been reported. When TB is isolated to the stomach or duodenum, symptoms referable to these organs may predominate. However, the intestinal phase is the most common, and abdominal pain is the most frequent symptom. When peritonitis or perforation occurs, symptoms of these complications are manifest.

DIAGNOSIS

Confirming the diagnosis of gastrointestinal TB depends on histologic demonstration of the organism. Therefore, biopsy through endoscopy or laparoscopy is essential. Caseating granulomas, growth of the organism from the specimen, and characteristic clinical findings usually make the diagnosis.

At times, the differential diagnosis can be difficult. It may include Crohn's disease, lymphoma, carcinoma, diverticular disease, appendicitis, and other infections of the GI tract, such as *Yersinia*. Rare, confusing pictures can be seen with histoplasmosis or *Mycobacterium avium*, or even with cryptosporidiosis. Cytomegalovirus (CMV) should also be considered in the differential diagnosis. Positive findings on purified protein derivative (PPD) testing may be helpful because it is unusual for TB to produce negative test results. However, one of the following criteria is necessary to make a firm diagnosis:

- Growth of the organism from infected tissue
- Histologic demonstration of *M. tuberculosis* in tissue
- Histologic demonstration of granulomas with caseating necrosis
- Typical gross pathologic findings in the bowel
- Histologic findings of granulomas with caseation necrosis in associated lymph nodes

TREATMENT AND MANAGEMENT

Gastrointestinal TB responds to antimycobacterial therapy. The treatment is the same as that used for pulmonary TB. Usually, patients are started on therapy with three or more agents, often isoniazid, rifampin, pyrazinamide, and either ethambutol or streptomycin. Susceptibility test results are often unavailable initially, but they become available approximately 8 weeks after culturing. Treatment is usually continued for at least 6 months. Other drugs are now becoming available, as are alternative drugs, including amikacin, levofloxacin, cycloserine, ethionamide, kanamycin, capreomycin, and *p*-aminosalicylic acid.

COURSE AND PROGNOSIS

Prognosis depends on the TB patient's comorbidities and basic immune status. Therapy with multiple antibiotics is usually successful when started early in the disease course. In severely affected HIV patients, TB may be fatal. If TB is suspected early, treatment should be started, pending histologic and microbiologic confirmation, because early treatment is critical for immunocompromised patients. Even with resistance to antibiotics, treatment can be successful because numerous drugs are effective.

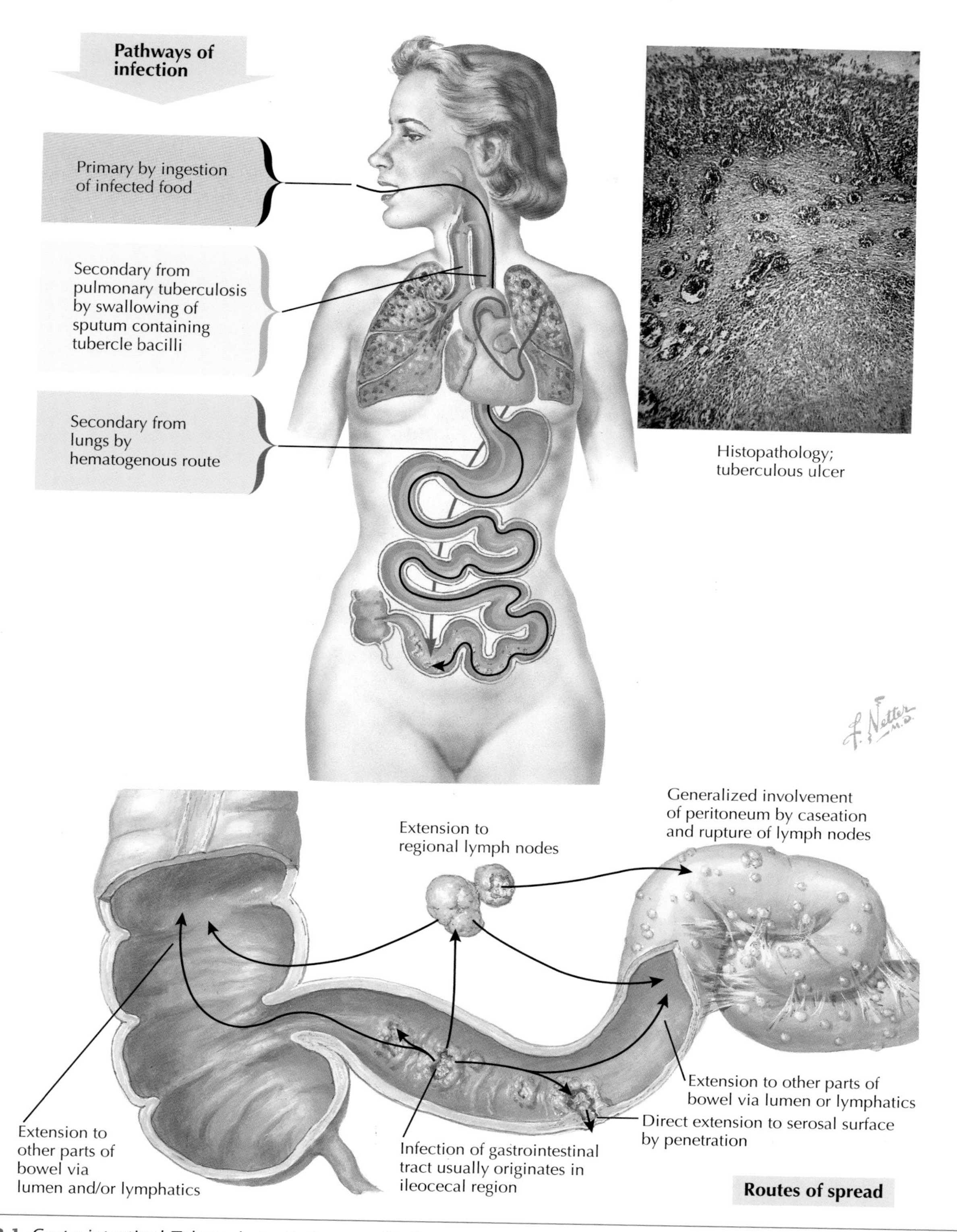

Figure 172-1 *Gastrointestinal Tuberculosis: Pathways of Infection and Routes of Spread.*

ADDITIONAL RESOURCES

Fitzgerald D, Haas DW: *Mycobacterium tuberculosis*. In Mandell GL, Bennett JE, Dolin R, editors: *Mandell, Douglas, and Bennett's principles and practice of infectious diseases*, ed 6, Philadelphia, 2005, Elsevier–Churchill Livingstone, pp 2852-2886.

Gianella RA: Infectious enteritis and proctocolitis and bacterial food poisoning. In Feldman M, Friedman LS, Brandt LJ, editors: *Gastrointestinal and liver disease*, ed 8, Philadelphia, 2006, Saunders-Elsevier, pp 2333-2391.

Horsburgh CR Jr, Nelson AN: Mycobacterial disease of the gastrointestinal tract. In Blaser MJ, Smith PD, Radvin JI, et al, editors: *Infections of the gastrointestinal tract*, ed 2, Philadelphia, 2002, Lippincott–Williams & Wilkins, pp 831-845.

Rasheed S, Zinicola R, Watson D, et al: Intra-abdominal and gastrointestinal tuberculosis, *Colorectal Dis* 9:773-783, 2007.

Abdominal Actinomycosis

173

Martin H. Floch

Actinomycosis of the abdomen is caused most often by a gram-positive anaerobic bacterium, *Actinomyces israelii*. However, many other species can cause the same syndrome. There are three major types of clinical infection syndromes in humans: cervicofacial, thoracic, and abdominal. *Cervicofacial* disease is the most common form and accounts for approximately 80% of infections. However, it may occur in intraabdominal disease and may be a problem in the differential diagnosis. The proportion of cases involving the abdomen averages 20%, with a range of 0 to 63%. More than 500 cases of actinomycosis were documented and reviewed from 1938 to 1998.

The most common site in the gastrointestinal tract is in the ileocecal valve area, but any part may be involved. It is assumed that once the disease is present in the abdominal cavity, drainage into the liver is the cause of liver abscess. Fortunately, when the disease manifests, it usually is in only one organ and has not disseminated. When infection occurs, other organs may be involved in the abscess formation.

Actinomyces is a part of the indigenous flora. Infections develop when the patient is in a susceptible state, such as during surgery, trauma, debilitating disease (e.g., malignancy, diabetes mellitus), or chronic corticosteroid therapy.

CLINICAL PICTURE

Most patients have pain, weight loss, and fever, and they may have anorexia and chills if there is visceral involvement. Hepatic *Actinomyces* is nonspecific and varied and thus must be considered in any patient with hepatic abscess, or it may accompany chronic fistula or manifest as an indurated mass or abscess.

DIAGNOSIS

Usually, the clinical picture is confusing, and the diagnosis is difficult and is only made after surgical exploration or drainage of an abscess. Because abdominal *Actinomyces* has a predilection for the right lower quadrant, it must be considered in any confusing presentation of Crohn's disease, tuberculosis of the ileum, or appendicitis. Classic symptoms include indurated mass, sinus tract and fistula, and abscesses (**Fig. 173-1**). Abdominal *Actinomyces* can mask carcinoma of the cecum or appendix.

The workup reveals leukocytosis, elevated erythrocyte sedimentation rate, and C-reactive protein. Once the alkaline phosphatase level is elevated, liver abscess must be considered. Cultures are positive in only 25% to 50% of patients, but Gram stain reveals the classic gram-positive branching filaments. The organisms also may be recovered from the blood. Definitive diagnosis depends on demonstrating the characteristic disease and *sulfa granules* on biopsy of the appropriate material. Sulfa granules are actually microcolonies of the organism, and classic eosinophilic material at the edges of the granules represents the host response.

TREATMENT AND MANAGEMENT

Once the diagnosis of actinomycosis is established, long-term antibiotic therapy is instituted, with possible debridement and drainage of any significant abscess or mass. Effective antibiotics include penicillin G, ampicillin, tetracycline, erythromycin, clindamycin, chloramphenicol, and imipenem. The drug of choice is *penicillin G*, which should be given in high doses. Depending on the species, antibiotics may need to be varied. The length of therapy depends on the extent of disease, but 6 to 9 months may be needed.

COURSE AND PROGNOSIS

Because the disease course is indolent, abdominal actinomycosis requires long-term therapy. Monitoring with computed tomography may be essential to follow therapeutic progress. If abdominal *Actinomyces* is associated with a malignancy, the course of the malignancy predicts the prognosis. However, if associated with chronic disease (e.g., diabetes mellitus), actinomycosis may be cured. In some persistent cases, long-term therapy resulted in relapse.

ADDITIONAL RESOURCES

Garner JP, MacDonald M, Kumar PK: Abdominal actinomycosis, *Int J Surg* 5:441-448, 2007.

Hecht DW, Feingold SM: Peritonitis and intra-abdominal abscess. In Blaser MJ, Smith PD, Ravdin JI, et al, editors: *Infections of the gastrointestinal tract*, ed 2, Philadelphia, 2002, Lippincott–Williams & Wilkins, pp 317-349.

Russo TA: Agents of actinomycosis. In Mandell GL, Bennett JE, Dorin R, editors: *Mandell, Douglas and Bennett's principles and practice of infectious diseases*, Philadelphia, 2005, Elsevier–Churchill Livingstone, pp 2924-2934.

Yang S, Li A, Lin J: Colonoscopy in abdominal actinomyces, *Gastrointest Endosc* 51:236-240, 2000.

Yeguez JF, Martinez SA, Sands LR, et al: Pelvic *Actinomyces* presenting as malignant large bowel obstruction: case report and review of the literature, *Am Surg* 66:85-90, 2000.

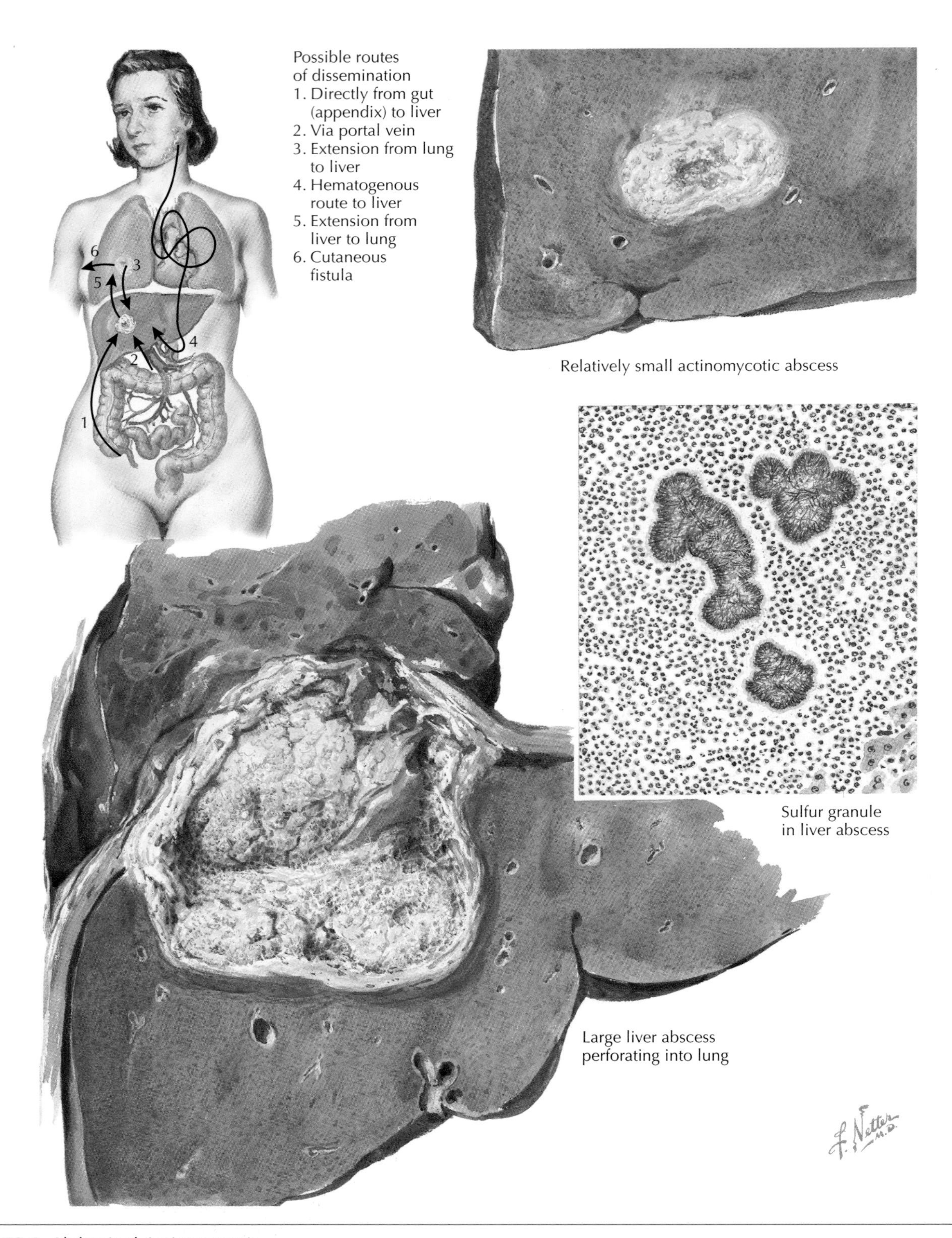

Figure 173-1 *Abdominal Actinomycosis.*

Amebiasis

174

Martin H. Floch

Human invasive amebiasis is caused by *Entamoeba histolytica*. Worldwide, amebiasis is the third most common parasitic disease. In the United States, *E. histolytica* represents the third most frequently identified protozoan from human specimens, following *Giardia lamblia* and *Blastocystis hominis*. **Figure 174-1** demonstrates how amebiasis is spread through the fecal-oral route by contaminated food and water. Societies in countries with poor sanitation have the highest infection rates, but minimal contamination is required, and some believe it takes only one cyst to cause infection in a susceptible host. An estimated 50,000 cysts are produced annually in a patient with invasive *E. histolytica*.

Entamoeba dispar is noninvasive and is limited to the intestine, but its infection rate is estimated to be 7-fold to 10-fold higher than that of *E. histolytica*.

The *E. histolytica* trophozoites are liberated from the cysts and may become invasive. They may spread through the colonic mucosa, entering the portal circulation to spread to the liver and

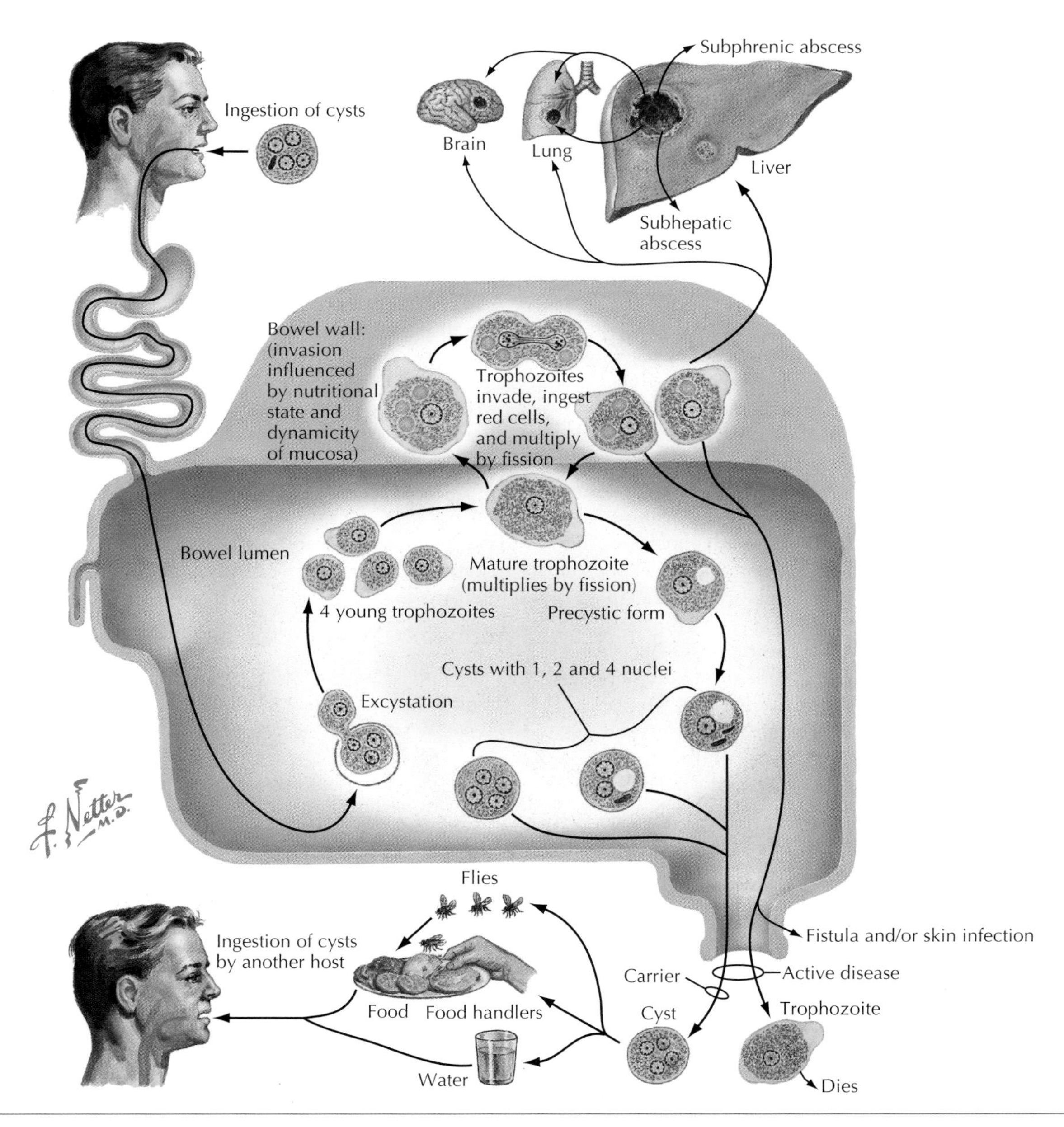

Figure 174-1 *Amebiasis: Fecal-Oral Spread of Disease.*

form liver abscess, or spread through the vascular tree into other sites. Rare cases have been reported of spread to almost all parts of the body. Amebic brain abscess is the most serious manifestation and has a high mortality rate. Intestinal disease is the most common form of *E. histolytica* infection.

CLINICAL PICTURE

Most cases of amebiasis manifest with diarrhea that may be mild or that may be full-blown colitis with bloody bowel movements, tenesmus, cramps, and in rare cases, fulminant, toxic colitis. Fortunately, only 10% of patients harboring *E. histolytica* or *E. dispar* have symptoms. Of the symptomatic group, diarrhea has been noted in all patients in most epidemiologic studies. Abdominal pain and tenderness are common, but fever occurs in only 40% of patients. When fulminant colitis develops, aggressive therapy is needed. Perforations of the colon are reported, with amebiasis diagnosed only after surgical exploration and treatment. Some symptoms are caused by liver abscess, the most common complication of invasive amebiasis. In patients with liver abscess, abdominal pain, fever, and hepatomegaly are evident. Some patients have wasting disease.

Pleuropulmonary complications are reported in up to 10% of patients, and rupture of the liver abscess into the peritoneum has been reported in 2% to 5%. Dramatic rupture of, and amebic liver abscess into, the pericardium has been reported as a fatal complication.

DIAGNOSIS

Early diagnosis is based on clear identification of the organisms in the patient's stool. Patients with active colitis have mobile trophozoites, whereas those with less active disease have the cystic form. When extraintestinal disease occurs, it may be difficult to identify *E. histolytica* in feces. Identifying the organisms from the intestine occurs in less than 30% of patients with liver abscess. It is also difficult for routine laboratory tests to distinguish between *E. dispar* and the more dangerous *E. histolytica*. A peripheral blood count is not helpful because leukocytosis may or may not be present, but there is no eosinophilia. When the liver is involved, alkaline phosphatase levels are elevated in 84% of patients.

Serologic tests may be helpful. Counterimmunoelectrophoresis, agar gel diffusion, direct hemagglutination, and enzyme-linked immunosorbent assay are positive in 85% to 95% of patients with amebic colitis or liver abscess. During the first 2 weeks, most titers are low, but they rise rapidly. The height of the titer usually correlates with the length of the disease, not its severity. Titers usually correlate with invasive disease, and carriers are usually negative. Antigen detection assays are available to help find *E. histolytica* in the stool.

The differential diagnosis for any patient with colitis should always include an evaluation for *E. histolytica*, especially in any area of the world in which infection is prominent and in select areas of developed countries with poor sanitation and immigrant populations.

During the workup of the patient with an extraintestinal manifestation, such as a liver abscess, aspirated material becomes important for the diagnosis. The lesion is initially identified through radiography, computed tomography, magnetic resonance imaging, or classic ultrasonography. However, the final diagnosis is based on examining aspirated material and identifying the classic pathology or trophozoites.

TREATMENT AND MANAGEMENT

Treatment varies with the type of amebiasis. Asymptomatic cyst passers can be treated with luminal agents such as iodoquinol, 650 mg three times daily (tid) for 20 days; paromomycin, 500 mg tid for 7 days; or diloxanide furoate, 500 mg for 10 days. If a patient has acute colitis, metronidazole (750 mg) plus one of the luminal agents is recommended for 7 to 10 days.*

For amebic liver abscess, metronidazole, 750 mg tid intravenously or orally, plus one of the luminal agents is recommended for 7 to 10 days. Tinidazole, 2 g orally daily for 5 days, is an alternative and is used frequently in other parts of the world.*

Metronidazole has been an effective agent for treating symptomatic disease. Some clinicians may not treat asymptomatic patients, but if any serologic test results are positive and disease is suspected, most recommend treatment. Pregnant women are usually administered paromomycin.*

With adequate treatment and early, diligent therapy, the mortality rate from liver abscess falls to less than 1%. The response is usually dramatic, within 3 days. Percutaneous drainage may be necessary, especially if an abscess might have ruptured. Opinions vary on the use of emetine, but if a laboratory cannot distinguish between *E. histolytica* and *E. dispar*, treatment with emetine should be instituted.

PROGNOSIS

The prognosis is guarded for patients with amebiasis and HIV infection or autoimmune disease, but it is good for patients with amebiasis and minimal comorbidity.

ADDITIONAL RESOURCES

Drugs for parasitic infections, New Rochelle, NY, 2007, Medical Letter.

Huston CD: Intestinal protozoa: In Feldman M, Friedman LS, Brandt LJ, editors: *Gastrointestinal and liver disease*, ed 8, Philadelphia, 2006, Saunders-Elsevier, pp 2413-2433.

Ravdin JI, Staufer WM: *Entamoeba histolytica* (amebiasis). In Mandell GL, Bennett JE, Dolin R, editors: *Mandell, Douglas and Bennett's principles and practice of infectious diseases*, ed 6, Philadelphia, 2005, Elsevier–Churchill Livingstone, pp 3097-3111.

*Pediatric doses are given according to weight (kg).

Giardia lamblia and Other Protozoan Infections

175

Martin H. Floch

Giardia lamblia, also called *Giardia intestinalis* (or *G. duodenalis*), is a flagellated intestinal protozoan (**Fig. 175-1**). Epidemiologists might debate whether *Entamoeba histolytica* or *G. lamblia* is the most common intestinal parasite worldwide, but *G. lamblia* may be present in as many as 20% to 30% of persons in the developing world and in as many as 2% to 5% of persons in the industrialized world. Certainly, *G. lamblia* is the most frequently identified intestinal parasite in the United States.

The life cycle of *G. lamblia* includes a trophozoite (active) phase in the intestine, which encysts to a cystic phase. The cysts are transmitted easily in water or through contamination from numerous hosts, including domestic animals and wild animals. Surface water and person-to-person contact are the most common modes of transmission. Infants, children, elderly persons, and immunocompromised patients are at particularly high risk for infection. It is estimated that as few as 10 cysts may result in infection. Once the cysts pass through the stomach, acid stimulates them to form trophozoites, which enter the duodenum and attach to the mucosa. In some persons, this action is harmless, but in others, it produces disease. Some persons in areas with high infection rates develop immunity. *Giardia* attaches to the intestinal cells by virtue of its ventral disc. Once attached, it can cause a pathologic response, resulting in the clinical disease spectrum. Trophozoites multiply by binary fission, and when exposed to a hostile environment in the intestine, they can encyst. A heavily infected host may pass thousands of cysts into the environment.

CLINICAL PICTURE

Once *G. lamblia* trophozoites are formed in the duodenum, they may cause symptoms. The incubation period lasts 7 to 14 days, but cysts do not appear in the stool until 1 week after symptoms develop. Hosts may be asymptomatic and become carriers, or they may acquire acute, self-limiting diarrhea or chronic diarrhea with complications. Not all patients have chronic diarrhea, and the diarrhea may be self-limiting and short lived so that the infection goes unnoticed or is ignored. From 25% to 30% of patients do have chronic diarrhea, and as many as 50% of infected patients may have malabsorption with loss of weight. Full-blown malabsorption can develop. The most severe presentations have been in persons with immunodeficiency or immunoglobulin A (IgA) deficiency. Malabsorption also has been associated with nodular lymphoid hyperplasia.

DIAGNOSIS

Whenever a patient has prolonged diarrhea and has visited (or lives in) an area with *Giardia* in its freshwater or with epidemic diarrhea, the diagnosis of *G. lamblia* infection should be pursued. Some clinicians, because of difficulty making the diagnosis, use a so-called therapeutic trial of treatment. Cysts or trophozoites can be identified in the stool or duodenal aspirates, but the yield is usually very low, and identification varies in laboratories from 25% to 75%, depending on the technician's skill. Serology tests are available, but the most reliable test now used is the stool antigen for *Giardia*. This is reported to be as effective as concentration methods for microscopic analysis, and in some cases, it is much more effective in making the diagnosis. When full-blown malabsorption develops, endoscopy and biopsy of the duodenum are indicated to rule out other diseases in the differential diagnosis of malabsorption. *Giardia* can be identified with a significantly high positive index in the biopsy and aspirate specimens.

TREATMENT AND MANAGEMENT

Metronidazole is available in industrialized and developing countries to treat *G. lamblia* infection. At 250 mg (5 mg/kg for children) three times daily (tid) for 7 days, metronidazole is more than 90% effective.

Other drugs that are effective but that are not available in all countries include tinidazole, nitazoxanide, quinacrine, furazolidone, and paromomycin. Each is effective in 50% to 90% of patients, depending on each patient's clinical status. Only paromomycin is recommended for pregnant women, even though it is less effective than the other antibiotics.

COURSE AND PROGNOSIS

Unfortunately, the diagnosis of *Giardia* is often missed because of the difficulty in identifying the organisms. However, using fecal antigens has improved the accuracy of the diagnosis. Once the diagnosis is made, some patients may require several courses of therapy because of resistant infection, but most patients are treated successfully, and the course of *G. lamblia* infection is benign.

OTHER INTESTINAL PROTOZOANS

Dientamoeba fragilis is a flagellated protozoan that does not appear to form cysts. It is transmitted from person to person. Although rare, it can cause diarrhea with abdominal pain, anorexia, fatigue, and fever. Its exact incidence and pathologic significance remain evasive. Once identified, *D. fragilis* should be treated with iodoquinol, 650 mg orally (PO) tid for 20 days, or tetracycline, 500 mg four times daily (qid) for 10 days. Metronidazole, 500 to 750 mg tid for 10 days (adults) and 35 to 50 mg/kg/day in three doses for 10 days (children), is usually the drug of choice for children. Paromomycin is another alternative.

Balantidium coli is a ciliated protozoan (see Fig. 175-1), and infections have been reported after transmission from pigs. It is uncommon cause of diarrhea, but on occasion the bacterium has

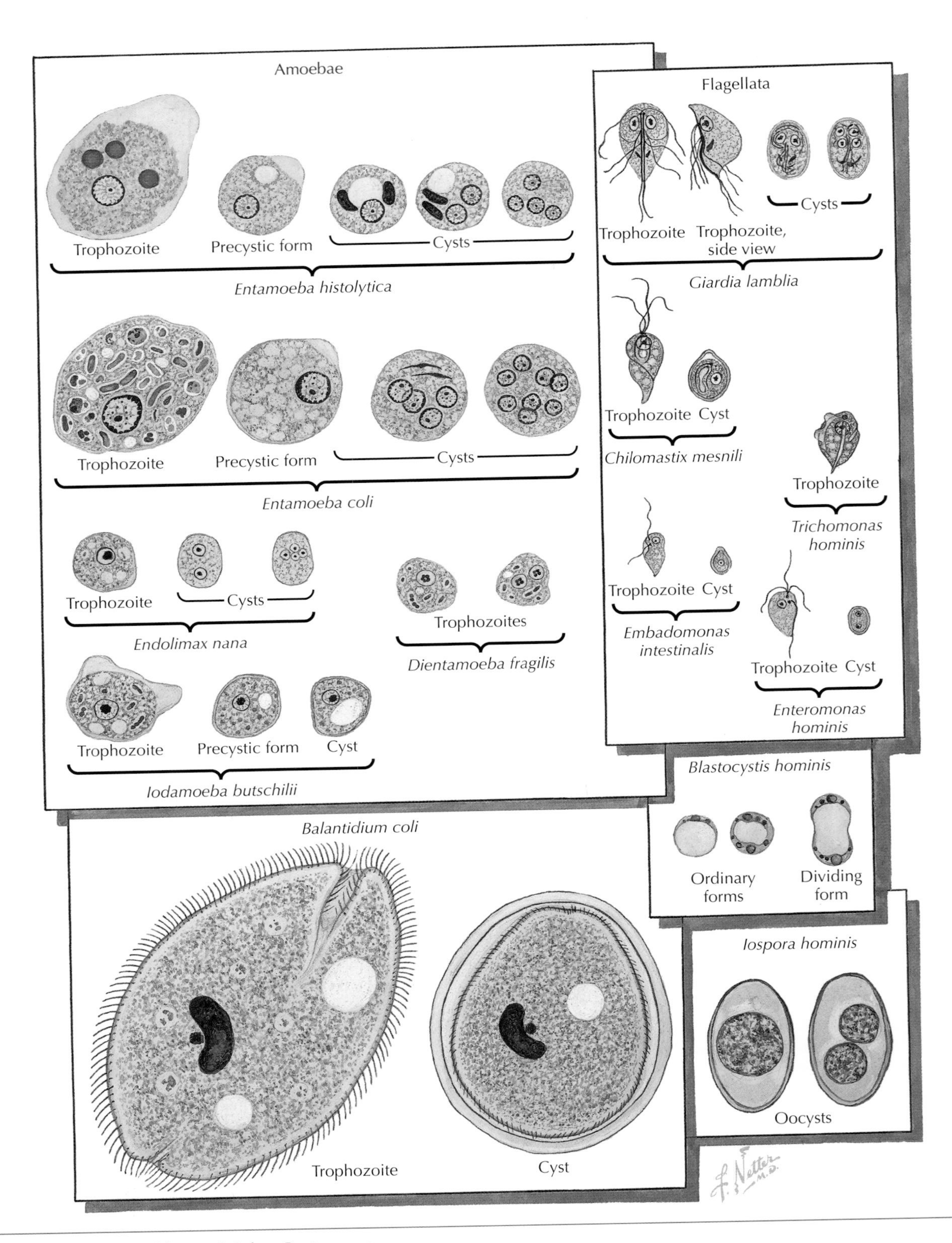

Figure 175-1 Giardia lamblia *and Other Protozoans.*

been reported to be invasive. If *B. coli* is found in the stool in a symptomatic patient, treatment is indicated. Tetracycline, 500 mg PO qid for 10 days, is the drug of choice. Alternatives are metronidazole and iodoquinol.

Blastocystis hominis is a parasite frequently found in stools; whether it can be pathogenic is controversial. Some clinicians think that after *B. hominis* is identified and treated, another organism is identified as the actual cause of the symptoms, and that cure results from treating the unknown organism. Regardless, when *B. hominis* is associated with diarrhea, it must be treated. The drug of choice is usually metronidazole (750 mg PO tid for 10 days) or iodoquinol (650 mg PO tid for 20 days). Trimethoprim-sulfamethoxazole and nitazoxanide are alternatives.

INTRACELLULAR PROTOZOAN PARASITES

Intracellular protozoan parasites have been reported less frequently than the other protozoans just discussed, but their incidence (although low) is significant, and they must be considered in the differential diagnosis of diarrheas. They are more prominent in elderly persons and immunocompromised patients. *Cryptosporidium parvum*, *Cyclospora cayetanensis*, and two species of microsporidia, *Enterocytozoon bieneusi* and *Encephalitozoon intestinalis*, are invasive organisms that can cause full-blown diarrhea along with fever, abdominal pain, and weight loss. The diagnosis is made by identifying the organisms in the feces or in biopsy specimens from the intestine. Treatments vary with each organism but are usually successful.

Isospora belli is related to *Cryptosporidium* and *Sarcocystis* and may cause protracted diarrhea, particularly in immunocompromised patients. Occasionally, *I. belli* has been identified in traveler's diarrhea. This obligate intracellular coccidian protozoan is found worldwide, but its infection rate is low and its epidemiology poorly understood. Its presentation may be similar to that of cryptosporidiosis or giardiasis, and the diagnosis of *I. belli* infection is made from examining stool samples or, when invasive, biopsy specimens of the small bowel.

Sarcocystosis, caused by protozoa of the *Sarcocystis* genus, is a rare infection in humans and is reported primarily in developing countries. It can cause necrotizing enteritis and, as with other protozoans, is diagnosed by the discovery of oocysts in the stool or the parasite in biopsy specimens.

ADDITIONAL RESOURCES

Behr MA, Koskin E, Guyorkos PW, et al: Laboratory diagnosis for *Giardia lamblia* infection: comparison of microscopy, coprodiagnosis, and serology, *Can J Infect Dis* 8:33-38, 1997.

Drugs for parasitic infections, New Rochelle, NY, 2001, Medical Letter.

Fisk TL, Keystone JS, Kozarskt P: *Cyclospora cayetanensis*, *Isospora belli*, *Sarcocystis* species, *Balantidium coli*, and *Blastocystis hominis*. In Mandell GL, Bennett JE, Dolin R, editors: *Mandell, Douglas and Bennett's principles and practice of infectious diseases*, ed 6, Philadelphia, 2005, Elsevier–Churchill Livingstone, pp 3228-3237.

Hill DL: *Giardia lamblia*. In Mandell GL, Bennett JE, Dolin R, editors: *Mandell, Douglas and Bennett's principles and practice of infectious diseases*, ed 6, Philadelphia, 2005, Elsevier–Churchill Livingstone, pp 3198-3205.

Huston CD: Intestinal protozoa. In Feldman M, Friedman LS, Brandt LJ, editors: *Gastrointestinal and liver disease*, ed 8, Philadelphia, 2006, Saunders-Elsevier, pp 2413-2433.

Moore GT, Cross WM, McGuire D, et al: Epidemic giardiasis at a ski resort, *N Engl J Med* 281:402-405, 1969.

Ward H, Jalen KN, Maitra TK, et al: Small intestinal nodular lymphoid hyperplasia in patients with giardiasis and normal serum immunoglobulins, *Gut* 24:120-124, 1983.

Weiss LM: Microsporidiosis. In Mandell GL, Bennett JE, Dolin R, editors: *Mandell, Douglas and Bennett's principles and practice of infectious diseases*, ed 6, Philadelphia, 2005, Elsevier–Churchill Livingstone, pp 3237-3254.

Intestinal Helminths: Trichuriasis

176

Martin H. Floch

Intestinal helminths are common worldwide and most infectious in areas of poor sanitation and warm climate. Intestinal helminths are divided into *roundworms*, or Nematoda, and *flatworms*, or Platyhelminthes. Platyhelminthes are subdivided into *cestodes* and *trematodes*.

Any nematode of the genus *Trichuris* is commonly known as the "whipworm" because of its morphology. Its life cycle is simpler than that of the other helminths (**Fig. 176-1**). *Trichuris* eggs are ingested in contaminated food and water (see Fig. 185-1). They mature in the distal small bowel and then pass into the colon. Adult worms migrate to the cecum and the appendix, where they live, copulate, and deposit eggs. The eggs pass through the feces to complete the life cycle.

Trichuriasis currently affects an estimated 1 billion persons worldwide, with most infections concentrated in the tropics or the semitropics; *Trichuris trichiura* most often infects humans, who are the only host of the species. It is identified in approximately 1% of stool specimens in the United States, most often in young children. Most humans harbor only a few worms, but the infection can be extremely heavy in some patients. The life span of the worm can range from 1 to 8 years, and each female may produce as many as 3000 to 20,000 eggs. The eggs may penetrate or attach to the mucosa and cause a significant pathologic response.

CLINICAL PICTURE

Mild *Trichuris* infections are asymptomatic. However, when the worm burden reaches more than 50 to 100, it may cause lower abdominal pain, diarrhea, distention, anorexia, and weight loss within a year. In children, it may cause dysentery. In developing countries, chronic infection can impair growth, and anemia may be severe and prolonged if trichuriasis is untreated.

DIAGNOSIS

Diagnosis of trichuriasis is usually made easily by the characteristic presence of eggs in stool specimens. The eggs are easy to identify because of their large number. It is surprising for an endoscopist to see the worms on sigmoidoscopy or colonoscopy, but they often can be seen hanging into the intestinal lumen. The accompanying anemia is iron deficient and microcytic and is usually associated with low-grade eosinophilia.

TREATMENT AND MANAGEMENT

The present drugs of choice for trichuriasis are mebendazole, 100 mg orally twice daily for 3 days or 500 mg for one dose, and albendazole, 400 mg orally for 3 days. Cure rates with these drugs are approximately 40%. The worm burden is decreased with single-dose therapy, but decreasing the worm burden is often difficult, and 3-day therapy is required for any attempt at a cure. Repeat stool analysis should be performed.

PROGNOSIS

The prognosis for patients with trichuriasis is excellent. However, clinicians must remember that clearing the worm burden can be difficult and that repeat therapy may be necessary.

ADDITIONAL RESOURCES

Drugs for parasitic infections, New Rochelle, NY, 2007, Medical Letter.

MacGuire JH: Intestinal nematodes (roundworms). In Mandell GL, Bennett JE, Dolin R, editors: *Mandell, Douglas and Bennett's principles and practice of infectious diseases*, ed 6, Philadelphia, 2005, Elsevier–Churchill Livingstone, pp 3260-3267.

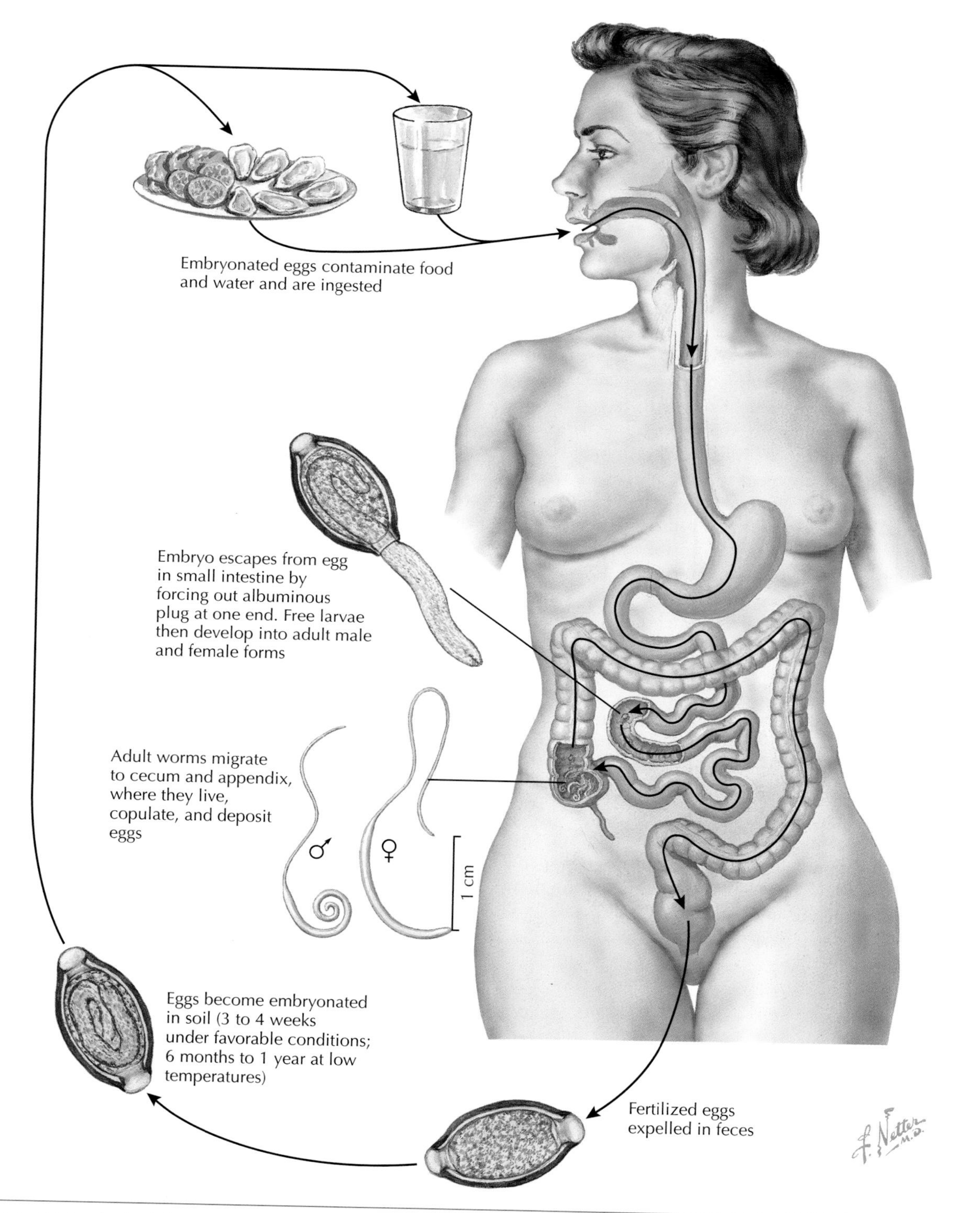

Figure 176-1 *Trichuriasis: Life Cycle of* Trichuris *Nematode Helminth (Whipworm).*

Enterobiasis

177

Martin H. Floch

Enterobiasis is caused by the pinworm *Enterobius vermicularis*. This nematode is probably the most common parasite to host on humans because it flourishes in temperate and tropical climates.

The small, spindle-shaped, round adult worms inhabit the cecum and appendix and adjacent parts of the large and small intestines; their heads attach to the intestinal mucosa. A male worm measures 2 to 5 mm in length, and the female, 9 to 11 mm. The female produces eggs in its ovary and releases them into a reservoir, or uterus, where fecundation takes place. When the reservoir is filled, the worm detaches itself from the bowel wall and migrates down the colon to the rectum. Some parasites are expelled with feces, but others migrate through the anal canal and, while crawling, deposit eggs in the perianal and the genitocrural folds. On average, one female deposits 11,000 eggs. Within hours of passage, the eggs enter an infective stage, and they may be passed to humans by hand contact, from sheets and pillowcases, or directly onto food and water (**Fig. 177-1**; see also Fig. 185-1). Once the eggs are ingested, the larvae escape from the eggs into the stomach and the duodenum, molt twice, and pass into the large intestine to complete the life cycle.

CLINICAL PICTURE

The most frequent clinical scenario for enterobiasis is a child brought to the physician's office because of severe anal itching. However, adults also may be infected and may seek treatment for the same symptom. Perianal reactions by large burdens of worms can be intense. When they migrate from the perianal area to the vagina, the worms can cause vaginitis; migration to the intestine can cause appendicitis. All the symptoms of appendicitis or vaginitis may develop. Reportedly, the worms may even reach the peritoneum and infect the ovary.

DIAGNOSIS

Demonstrating the *E. vermicularis* ova in feces or on a perianal specimen (using a strip of transparent tape) establishes the diagnosis. It is best to collect the specimen early in the morning, before the patient has bathed. Stool specimens are positive in only 10% to 15% of patients; the diagnosis is invariably made by collecting material from the perianal area.

TREATMENT AND MANAGEMENT

The drug of choice for treating enterobiasis is mebendazole (100 mg), pyrantel pamoate (11 mg/kg base; maximum 1 g) or albendazole (400 mg), all given orally in a single dose and repeated in 2 weeks. The worms live from 7 to 13 weeks, and treatment is usually successful if the patient is not reinfected. However, reinfection is a major problem. Sheets and pillowcases must be cleaned thoroughly, and all infected family members must be treated to prevent reinfection. At times, the entire household, including curtains and floors, must be cleaned to eradicate the eggs.

ADDITIONAL RESOURCES

Drugs for parasitic infections, New Rochelle, NY, 2007, Medical Letter.

Elliott DE: Intestinal worms. In Feldman M, Friedman LS, Brandt LJ, editors: *Gastrointestinal and liver disease*, ed 8, Philadelphia, 2006, Saunders-Elsevier, pp 2435-2457.

MacGuire JH: Intestinal nematodes (roundworms). In Mandell GL, Bennett JE, Dolin R, editors: *Mandell, Douglas and Bennett's principles and practice of infectious diseases*, ed 6, Philadelphia, 2005, Elsevier–Churchill Livingstone, pp 3260-3267.

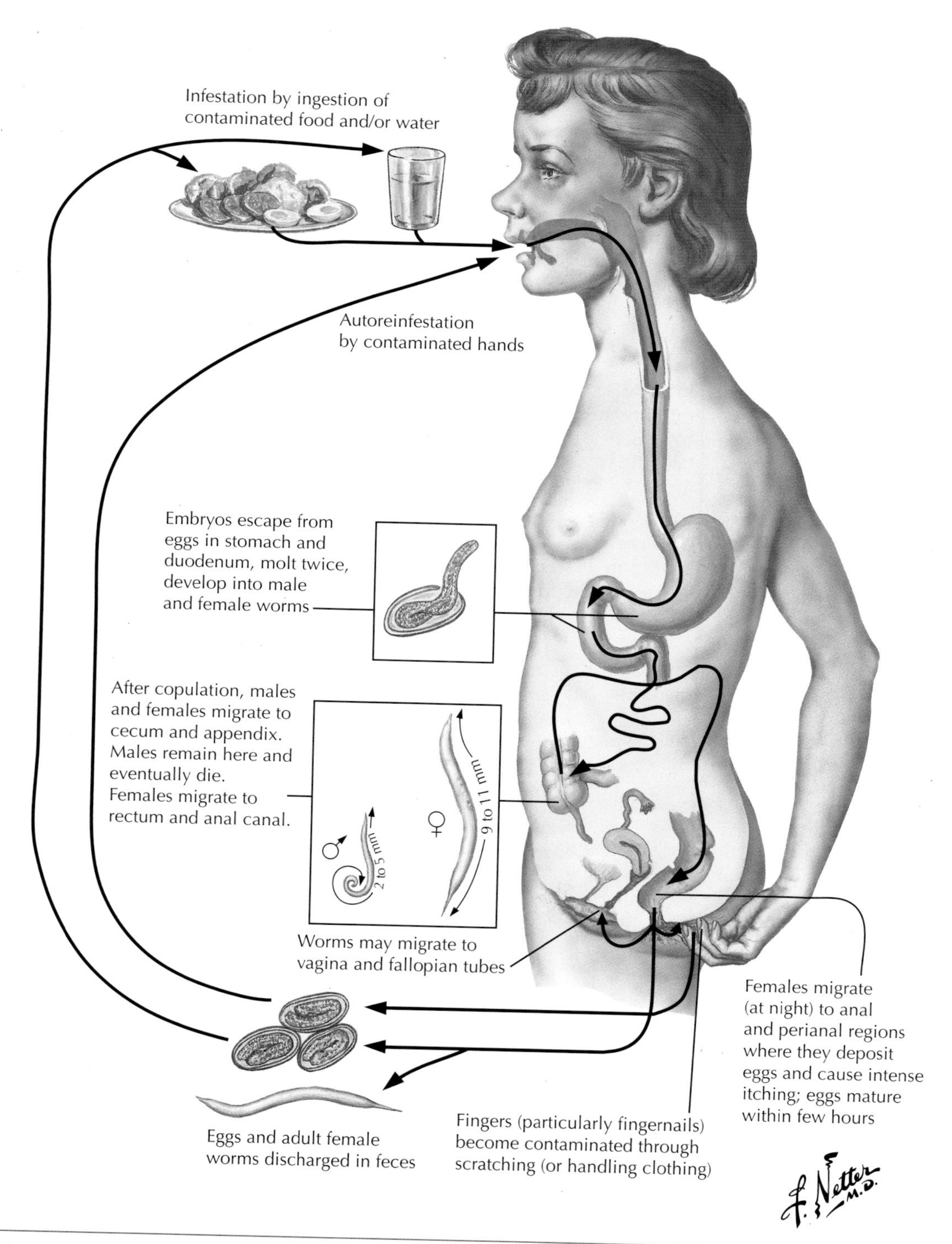

Figure 177-1 *Enterobiasis: Life Cycle of* Enterobius vermicularis *Nematode (Pinworm).*

Ascariasis

Martin H. Floch

178

Ascariasis is caused by *Ascaris lumbricoides*, which is the largest helminth to infect humans and one of the most prevalent. The adult male measures 15 to 25 cm and is smaller than the female, which may be as large as 35 cm. These helminths may live for 10 to 18 months and usually copulate in the lumen of the small intestine; the mature female may produce up to 200,000 eggs daily.

The life cycle begins with the eggs, which are passed into the soil and become fertilized and embryonated in 2 to 3 weeks. The eggs are ingested in contaminated food or water, then pass into the duodenum, where they liberate a larva that penetrates the small intestine and may reach the lungs (**Fig. 178-1**; see also Fig. 185-1).

A patient may have hypersensitivity reaction in the lung, causing the clinical manifestation of *Löffler syndrome*. For symptomatic patients, ascariasis causing Löffler syndrome is usually self-limiting, and this phase does not respond to antihelminth (anthelmintic) therapy.

The larvae pass into the bronchi and are swallowed. They mature in the small intestine, where they copulate, and their eggs are passed into feces to complete the life cycle.

CLINICAL PICTURE

As noted, Löffler syndrome may occur in the lungs, but it is a self-limiting disease. The patient may have pneumonia. Once the worms reach the lower intestine, symptoms may disappear, but heavier infestations can cause local disturbances, including true intestinal obstruction. The patient may have abdominal pain, loss of appetite, nausea, or diarrhea or constipation. When the worms cause obstruction, full intestinal obstruction can evolve. At times, the worms can cause obstruction in the appendix, or they may cause symptoms of biliary or pancreatic disease because of the migration to bile ducts and gallbladder or pancreatic duct. Eosinophilia is common. Obstruction is more common in children than adults. Varying degrees of extraintestinal manifestations are reported, with rates as high as 50%. Live worms have been noted in sputum and vomitus.

DIAGNOSIS

The diagnosis of ascariasis is made by demonstration of the larvae, ova, or worms. In the pulmonary phase, eosinophils and Charcot-Leyden crystals may be found in sputum, and larvae also have been recovered from sputum. Lung involvement precedes any intestinal phase by 8 to 10 weeks, but eggs do not appear in the feces in the early stage.

Given that these are the largest adult worms, they are easy to identify when they appear in feces. The eggs are characteristic of *A. lumbricoides* and are also easy to identify.

TREATMENT AND MANAGEMENT

As indicated, the pulmonary phase of ascariasis is self-limited but requires treatment to control all symptoms, sometimes necessitating the use of steroids. Anthelmintic treatment is ineffective for pneumonia. Once the diagnosis of *Ascaris* infection is made, the intestinal phase should be treated; the drug of choice is albendazole, 400 mg orally (PO) once; mebendazole, 100 mg PO twice daily for 3 days or 500 mg once; or ivermectin, 150 to 200 μg/kg PO once. Safety in children or pregnant women remains to be established.

PROGNOSIS

Ascariasis is successfully treated. Extraintestinal manifestations in themselves, however, may be serious problems, as in the rare case of pancreatitis.

The most important intervention is to prevent reinfection, especially in areas of the world that have poor sanitation. Mass chemotherapy in epidemic areas has been tried, but there is no substitute for improved sanitation.

ADDITIONAL RESOURCES

Drugs for parasitic infections, New Rochelle, NY, 2007, Medical Letter.

Elliott DE: Intestinal worms. In Feldman M, Friedman LS, Brandt LJ, editors: *Gastrointestinal and liver disease*, ed 8, Philadelphia, 2006, Saunders-Elsevier, pp 2435-2457.

MacGuire JH: Intestinal nematodes (roundworms). In Mandell GL, Bennett JE, Dolin R, editors: *Mandell, Douglas and Bennett's principles and practice of infectious diseases*, ed 6, Philadelphia, 2005, Elsevier–Churchill Livingstone, pp 3260-3267.

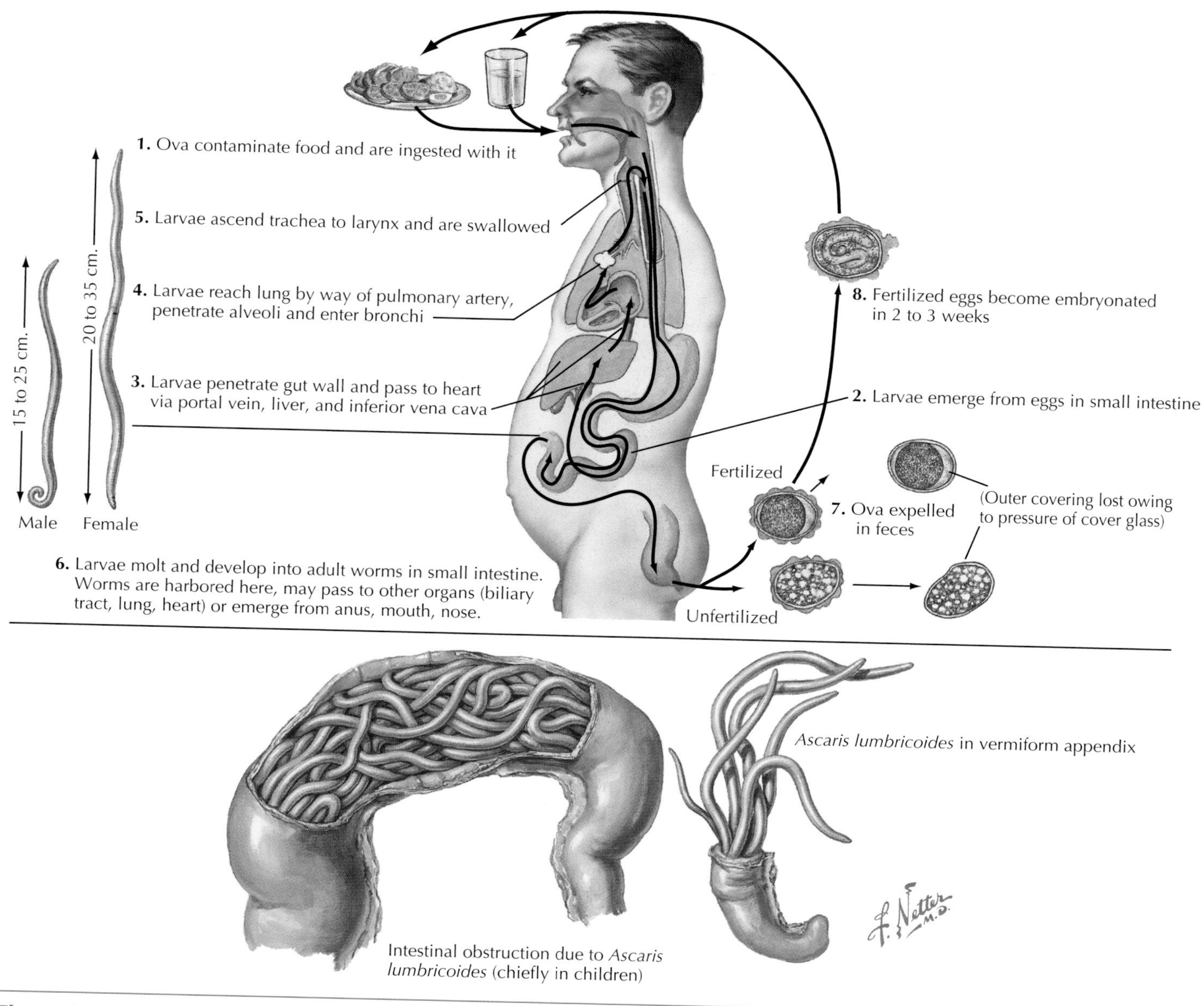

Figure 178-1 *Ascariasis: Life Cycle of* Ascaris lumbricoides.

Strongyloidiasis

179

Martin H. Floch

Strongyloidiasis is caused by the nematode *Strongyloides stercoralis*, referred to as the "threadworm." Its prevalence is unknown but ranges from 3 to 100 million on worldwide estimates. *S. stercoralis* infection occurs primarily in tropical and subtropical areas. Obtaining epidemiologic data on the infection has been difficult because stool analyses vary, infections can be low grade, and few larvae may be passed in the stool. However, strongyloidiasis is found in both poorly developed countries and industrialized nations. The most common presentation is low-grade infestation, but hyperinfection does occur and may result in severe gastroenteritis and dissemination of the organism.

The life cycle of *S. stercoralis* is complex because humans become infected by the filariform larvae, which penetrate the skin from the soil, then migrate through the bloodstream to the lungs (**Fig. 179-1**). The larvae are then swallowed and enter the duodenum, where the adults attach to or penetrate the wall. The female is a tiny worm no more than 2 mm in length. It enters the small bowel mucosa, where it can extrude eggs. The eggs hatch into rhabditiform larvae that migrate into the colon and are passed in the feces (see Fig. 185-1). A sexual cycle of development then occurs in which the rhabditiform larvae develop into males and females and pass eggs into the soil, which then form filariform larvae that can restart the cycle.

However, a short life cycle can occur in which the rhabditiform larvae mature and penetrate the skin in the perianal area and autoinfect the host. Some parasitologists still debate these varieties in the life cycle. Nevertheless, autoinfection is well documented, and rhabditiform larvae do develop adult sexual forms in soil. Larvae do migrate, and they have been found in the heart, liver, gallbladder, brain, genitourinary organs, nervous system, and pulmonary area.

Hyperinfection can occur if large numbers of organisms enter the host, usually an immunocompromised patient or a patient treated with corticosteroids, in whom autoinfection might become prolific. The worms are able to regulate their own populations, but when a host is treated with corticosteroids, the eggs produce increased amounts of ecdysteroid substances in host tissue, which allow proliferation of adult female worms and eggs and a massive number of larvae. Hyperinfection has been associated with millions of adult worms or filarial larvae in the mucosa of the small and large intestines.

CLINICAL PICTURE

Most patients with strongyloidiasis are asymptomatic, and larvae may be fortuitously encountered in the stool. However, if the infestation persists, symptoms may affect many organs. The patient may have a characteristic cutaneous lesion at the site in the perianal area by autoinfection or on the feet if the infection occurred through the soil. No other nematode has been associated with such a broad range of clinical symptoms. Patients with hyperinfection may have gastroenteritis with diarrhea and abdominal pain or massive diarrhea and malabsorption. Pulmonary manifestations of *Strongyloides* infection are well known, and pneumonia occurs. Rare cases of neurologic disease have been identified, as well as disease of almost every organ where the larvae have migrated. The clinical picture is varied, and the astute clinician must be alert to suspect the infection.

DIAGNOSIS

Eosinophilia is invariably present. If there is clinical suspicion that strongyloidiasis and eosinophilia are present, the stool should be screened carefully for larvae. However, the worms may be difficult to find. Reports indicate that three stools can increase the recovery rate to 50%, and seven stools, up to 100%. Many techniques are used to identify the larvae and depend on the parasitology laboratory. A negative test result means nothing. Several types of serologic tests can be performed, but vary in sensitivity; cross-reactivity with other parasites occurs. Eosinophilia may be absent in immunocompromised patients. The agar plate method now appears to be the most sensitive test for identification of the larvae.

TREATMENT AND MANAGEMENT

The drug of choice for strongyloidiasis is *ivermectin*, 200 μg/kg/day orally for 2 days. Safety in children less than 15 kg (33 lb) or pregnant woman remains to be determined. Ivermectin is well tolerated compared with the previously used thiabendazole, which is no longer available in many countries. An alternative drug, albendazole, 400 mg twice daily for 7 days, is reported to be less effective.

COURSE AND PROGNOSIS

Most *S. stercoralis* infections are asymptomatic and may go unnoticed. In immunocompromised patients or those receiving corticosteroids, however, the autoinfection rate may increase significantly to cause symptoms. Infections may become symptomatic and may even be fatal.

Hyperinfection can cause severe, debilitating gastroenteritis, and moderate infections can cause malabsorption with chronic, debilitating disease. Relapses are reported, and treatment with ivermectin may need to be followed with a second course of a second drug at a later date. Most infections are well controlled. When hyperinfection or a high-risk debilitating state develops, the outcome may be guarded.

ADDITIONAL RESOURCES

Drugs for parasitic infections, New Rochelle, NY, 2007, Medical Letter.

MacGuire JH: Intestinal nematodes (roundworms). In Mandell GL, Bennett JE, Dolin R, editors: *Mandell, Douglas and Bennett's principles and practice of infectious diseases*, ed 6, Philadelphia, 2005, Elsevier–Churchill Livingstone, pp 3260-3267.

Marathe A, Date V: *Strongyloides stercoralis* hyperinfection in an immunocompetent patient with extreme esosinophilia, *J Parasitol* 94:759-760, 2008.

Scowden EB, Schaffner W, Stone WJ: Overwhelming strongyloidiasis: an unappreciated opportunistic infection, *Medicine* 57:527-544, 1978.

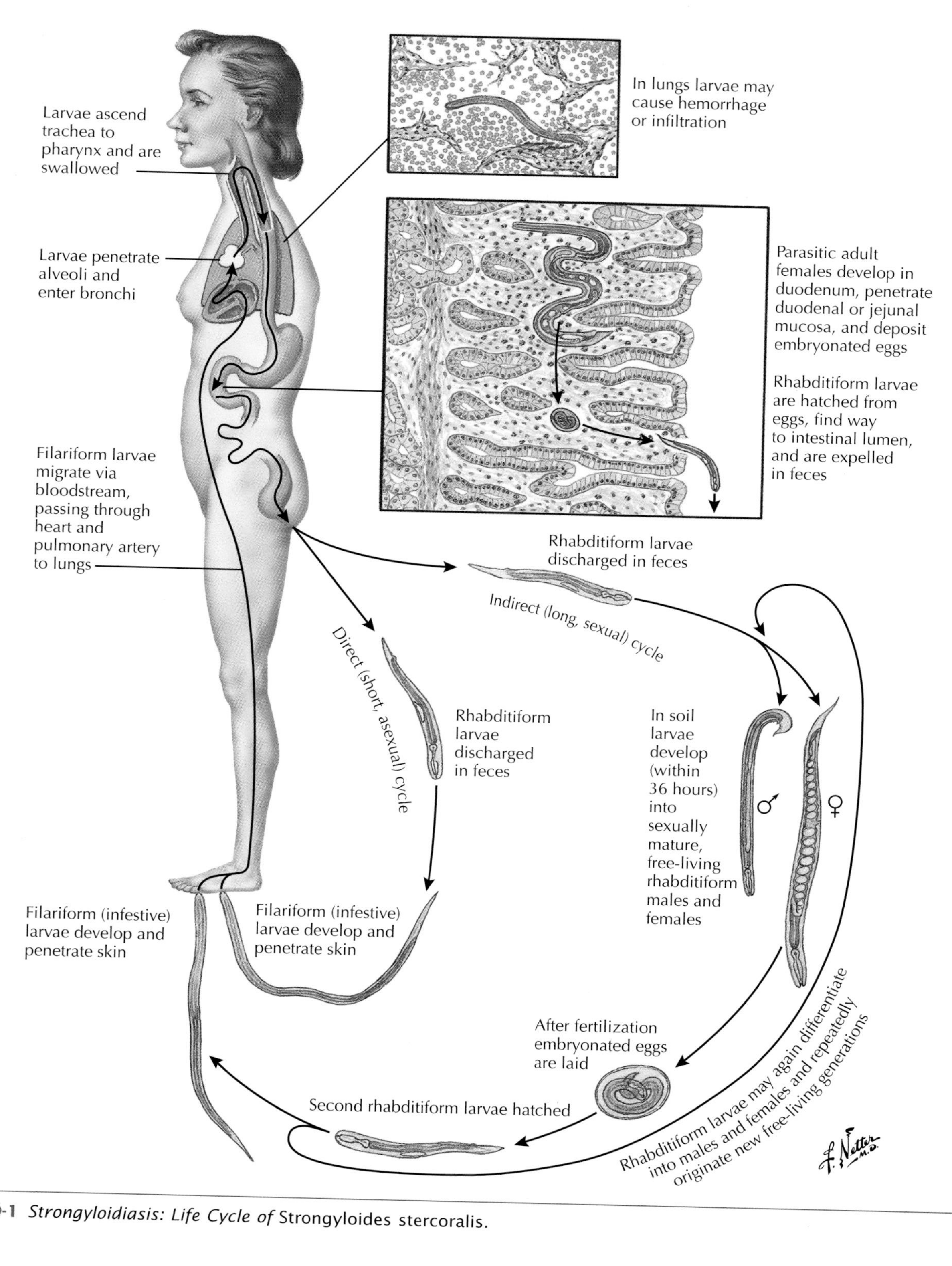

Figure 179-1 *Strongyloidiasis: Life Cycle of* Strongyloides stercoralis.

Hookworm Disease (Necatoriasis and Ancylostomiasis)

180

Martin H. Floch

Hookworm disease is caused by either of two nematodes, *Necator americanus* (New World hookworm) or *Ancylostoma duodenale* (Old World hookworm). Each of these helminths can cause the classic hookworm disease syndrome. *N. americanus* is found in the Western Hemisphere in tropical and subtropical areas and also in Africa and Asia. *A. duodenale* is found in the Mediterranean region and in parts of Europe and Asia; it is rarely, if ever, seen in the Western Hemisphere. The worms measure 7 to 9 mm, but *Ancylostoma* may be much larger than *Necator* and is more prolific, producing 10,000 to 30,000 eggs.

The life cycles of *A. duodenale* and *N. americanus* are essentially the same (**Fig. 180-1**). The worms attach to the small intestine, where they feed on the blood and lymph of the host, and the fertilized female lays eggs. Large numbers of eggs are passed in the feces. Rhabditiform larvae develop in warm moist soil, and the larvae penetrate human skin, pass through the circulation to the lungs, and are then swallowed to complete the life cycle in the duodenum (see Fig. 185-1). The worms may live for years in the host.

An estimated 1 billion persons harbor hookworms, which surprisingly were not identified as human parasites until the mid-19th century. They heavily infect children but can be unusually virulent and cause chronic infections and anemia in elderly persons. They probably are a major cause of iron deficiency anemia worldwide. Each worm can cause 0.03 to 0.26 mL of blood loss per day. *Ancylostoma* is larger and more aggressive in its drainage. Hookworms develop in tropical and subtropical areas, where larvae can grow in the soil.

CLINICAL PICTURE

Anemia is the distinguishing characteristic of hookworm infestation. The anemia relates to the burden of the infestation. Young children in underdeveloped countries may experience the impaired growth and development resulting from anemia, whereas elderly persons may experience the debilitating effects of anemia.

When infection occurs, skin lesions with eruptions may cause severe pruritus. As the larvae migrate through the lungs, they may cause hypersensitivity pneumonia and Löffler eosinophilic syndrome. When the worms inhabit the duodenum, they may cause abdominal pain and mild cramping. Eosinophilia is usually, but not always, present. When the worm burden is high, gastrointestinal bleeding can result. If the anemia is chronic, hypoalbuminemia and malabsorption syndrome may develop.

The migratory phase of the larvae may encompass creeping eruption or cutaneous larva migrans, typically caused by canine or feline hookworm larvae *(Ancylostoma braziliense)*. In some parts of the world, such as Australia, the dog hookworm *(Ancylostoma caninum)* has been reported to cause eosinophilic gastroenteritis.

Most patients have only mild infection; thus the infestation may go unnoticed or the patient might have only mild anemia. Large infestations can cause severe symptoms.

DIAGNOSIS

The diagnosis of hookworm disease is made on finding the characteristic eggs in the stools. Classic concentration techniques are used to find the eggs, and they are easily identified in most parasitology laboratories. Hookworm eggs of all types are similar, and the species is distinguished by identifying the larvae or the actual worm.

TREATMENT AND MANAGEMENT

Many drugs are available to rid the host of hookworms, but results may be variable. Mebendazole, 100 mg twice daily for 3 consecutive days or 500 mg once, and albendazole, 400 mg in a single dose,) are highly effective. However, these drugs are associated with rare toxicity and thus are discouraged in patients with blood dyscrasia, leukopenia, or liver disease. Therefore, pyrantel pamoate, 11 mg/kg (maximum 1 g), can be given daily for 3 days.

It may be necessary to institute iron replacement therapy if a patient has experienced continuous blood loss, and anemia should be monitored in any patient with chronic infestation. Children with impaired growth should be followed closely to ensure that hookworm was the cause and that it is corrected by iron and blood replacement.

COURSE AND PROGNOSIS

The prognosis is excellent once the host is rid of the hookworm burden. It is important that prevention be part of the therapy because reinfection can occur. Unfortunately, in parts of the world where sanitation and social conditions do not permit improvement to the environment, reinfection occurs. At present, there is no vaccine to prevent hookworm infection.

ADDITIONAL RESOURCES

Drugs for parasitic infections, New Rochelle, NY, 2007, Medical Letter.

Elliott DE: Intestinal worms. In Feldman M, Friedman LS, Brandt LJ, editors: *Gastrointestinal and liver disease*, ed 8, Philadelphia, 2006, Saunders-Elsevier, pp 2435-2457.

Keiser J, Utzinger J: Efficacy of current drugs against soil-transmitted helminth infections: systematic review and meta-analysis, *JAMA* 299: 1937-1948, 2008.

MacGuire JH: Intestinal nematodes (roundworms). In Mandell GL, Bennett JE, Dolin R, editors: *Mandell, Douglas and Bennett's principles and practice of infectious diseases*, ed 6, Philadelphia, 2005, Elsevier–Churchill Livingstone, pp 3260-3267.

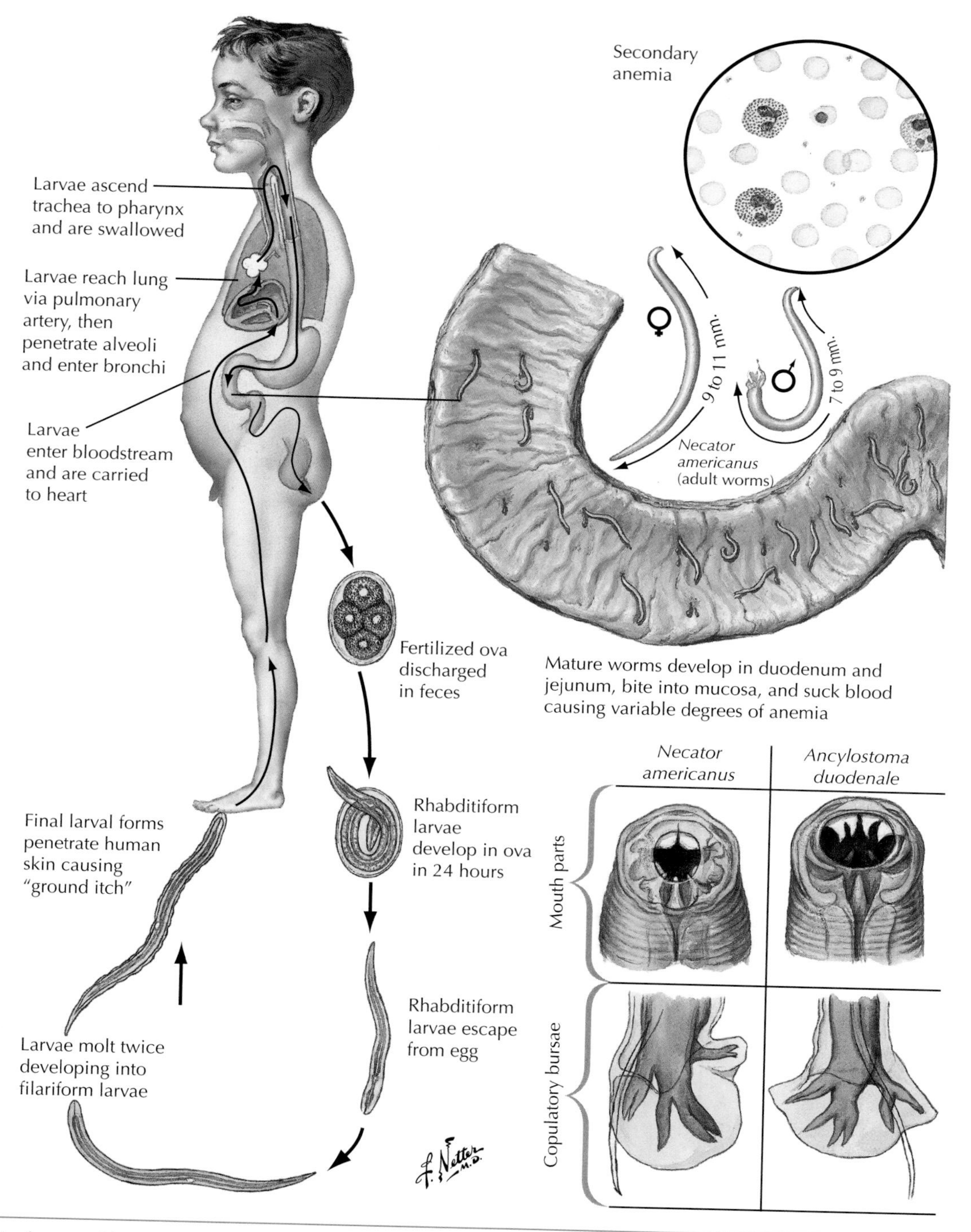

Figure 180-1 *Hookworm Disease: Life Cycle of* Necator americanus *(Necatoriasis) and* Ancylostoma duodenale *(Ancylostomiasis).*

Tapeworm (Cestode) Infection (Beef): *Taenia saginata*

18

Martin H. Floch

Essentially, four cestodes infect humans: *Taenia saginata*, *Taenia solium* (Chapter 182), *Hymenolepis nana* (Chapter 183), and *Diphyllobothrium latum* (Chapter 184). Humans are the only definite host for *T. saginata* and *T. solium*, but other cestodes become involved in the life cycle, usually by accident. This chapter describes *T. saginata*, more commonly known as the "beef tapeworm."

The beef tapeworm is relatively common in Latin America and Africa but also occurs in Europe and Asia; it is rarely seen in Australia, Canada, or the United States. In some areas of the world, its prevalence can be as high as 90%, usually where cattle are raised and where exposure to human excreta is common. Tapeworm bodies are flat and segmented and consist of a head, or scolex, and a series of segments known as *proglottides.* A tapeworm does not have a true body cavity. The body grows in a process called *strobilation*, or formation of new proglottides, in which the most posterior proglottides become mature and then gravid. Tapeworms range in length from 1 to 2 cm to 25 to 30 m. They are hermaphroditic, and proglottides contain ovaries and testes.

T. saginata can survive for up to 25 years in the human intestine (**Fig. 181-1**). Eggs are expelled with the proglottides and feces. Once they are ingested by the intermediate host, in this case, cattle, the eggs form oncospheres that invade the intestine and migrate to form cysticerci in beef. When a person eats undercooked beef, the cysticercus releases the scolex, which attaches to the small intestine, and proglottides begin to develop.

CLINICAL PICTURE

It may take 2 to 3 months after infestation before proglottid segments are passed in the stool. Patients are surprisingly asymptomatic or may have mild gastrointestinal symptoms, such as diarrhea, abdominal pain, nausea, or postprandial fullness. Although reported, violent vomiting and diarrhea are rare considering the prominence of this infection in certain areas. In some studies, more than half the patients were completely asymptomatic. The tapeworm can live within most human hosts and, amazingly, cause no malabsorption or weight loss. However, the host usually has significant eosinophilia.

Patients may exhibit some loss of appetite. Occasionally, intestinal obstruction results when a large worm with many proglottides becomes twisted and blocks the intestine.

DIAGNOSIS

Diagnosis of *T. saginata* infection depends on identifying the eggs in the proglottides. If expelled, proglottides may be studied on microscopy. Antigens can be detected in stool; but cross-reactivity occurs between *T. solium* and *T. saginata.*

TREATMENT AND MANAGEMENT

Praziquantel, 5 to 10 mg/kg in a single oral dose, is extremely effective against *T. saginata.* An alternative drug is paromomycin, but adverse effects are common. Paromomycin is preferred for pregnant women, however, because it is not absorbed. Niclosamide, 2 g orally once, is another alternative but is not approved by the U.S. Food and Drug Administration.

COURSE AND PROGNOSIS

Proglottides may continue to be expelled for many days after treatment. Therefore, follow-up stool examination should be done in 1 to 3 months to ensure success of therapy. However, reinfection often occurs in areas where *T. saginata* infections are common, and controlling beef tapeworm disease in individual patients depends on local sanitation and other societal factors.

ADDITIONAL RESOURCES

Drugs for parasitic infections, New Rochelle, NY, 2007, Medical Letter.

Elliott DE: Intestinal worms. In Feldman M, Friedman LS, Brandt LJ, editors: *Gastrointestinal and liver disease*, ed 8, Philadelphia, 2006, Saunders-Elsevier, pp 2435-2457.

King CH: Cestodes (tapeworms). In Mandell GL, Bennett JE, Dolin R, editors: *Mandell, Douglas and Bennett's principles and practice of infectious diseases*, ed 6, Philadelphia, 2005, Elsevier–Churchill Livingstone, pp 3285-3293.

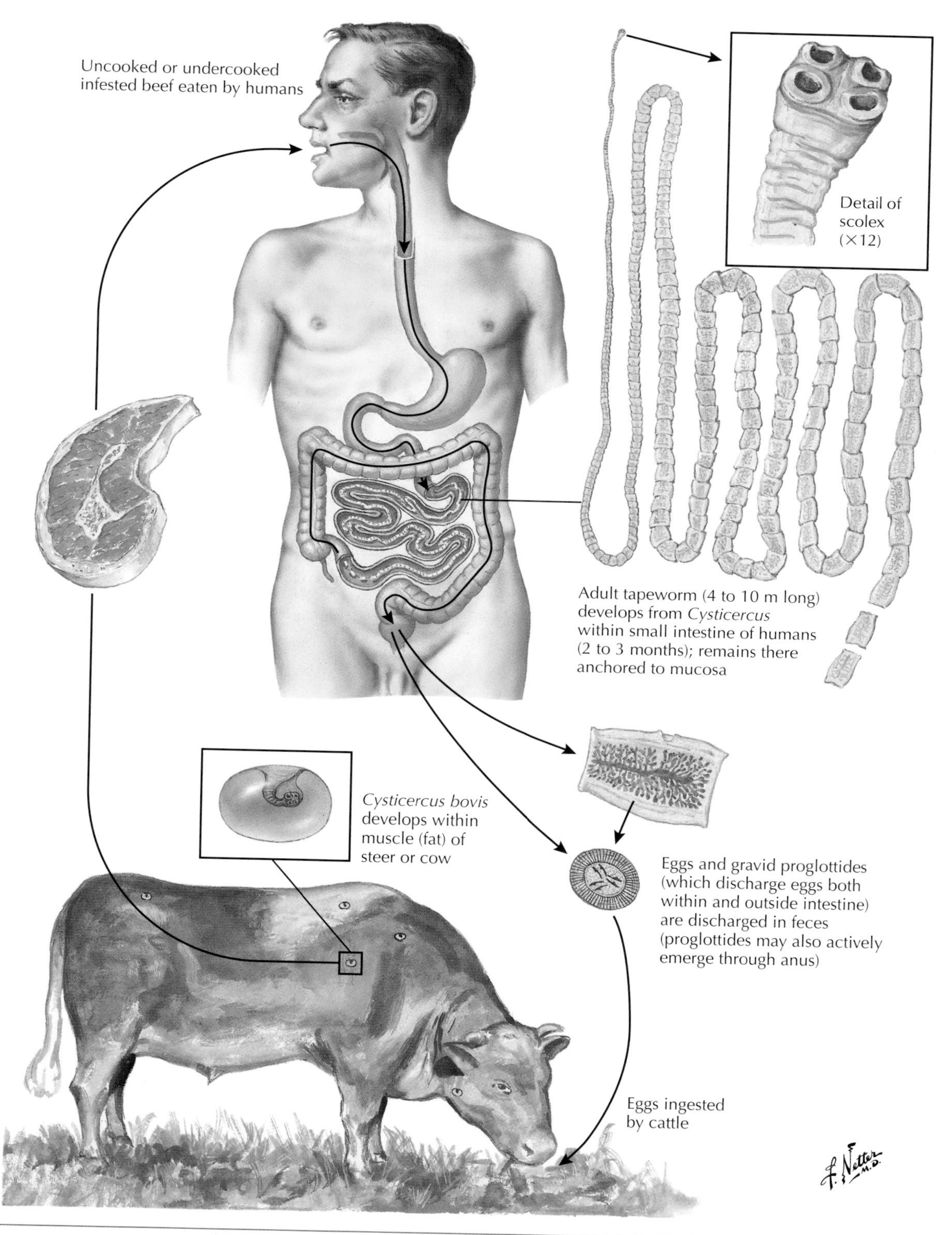

Figure 181-1 *Beef Tapeworm Infection: Life Cycle of* Taenia saginata *Cestode.*

Tapeworm Infection (Pork): *Taenia solium*

182

Martin H. Floch

The "pork tapeworm" *Taenia solium* causes significant illness in developing countries. Unlike the beef tapeworm *(Taenia saginata)*, both the adult and the larval stage cause significant disease. The life cycle of the pork tapeworm is similar to that of the beef tapeworm, but eating undercooked pork disseminates cysts, which may spread throughout the body (**Fig. 182-1**). *T. solium* has a worldwide distribution. It is relatively rare in the United States, and pork tapeworm disease usually involves the larval stages, *cysticercosis*, rather than the intestinal stage.

The adult worm is formed in approximately 1 to 2 months and then sheds 50,000 to 100,000 eggs daily in the proglottides. When pigs ingest the eggs, they hatch in the stomach, releasing larvae that penetrate the wall of the intestine and migrate to the muscle, where *Cysticercus cellulosae* larvae are formed. The life cycle is completed when humans eat undercooked pork. Another potential route of infection is ingestion of eggs from contaminated soil. The eggs then release the larvae, which can disseminate throughout the body, resulting in cysticercosis.

The most common manifestation of cysticercosis in areas of high prevalence is seizure. The term used for this phenomenon is *neurocysticercosis*.

CLINICAL PICTURE

It takes the pork tapeworm approximately 6 weeks to 2 months to begin to pass proglottides, which are smaller than those found in the beef tapeworm. *T. solium* proglottides are also less motile than those of *T. saginata*, and the intestinal phase may include vague abdominal pain, some nausea or anorexia, or occasionally, increased appetite and mild diarrhea. However, symptoms are usually mild in the intestinal phase. If a patient has a heavy infection and many proglottides are passed, anal pruritus can be a problem. Eosinophilia is usually present.

The more serious form of the disease is cysticercosis, because the brain and nervous tissue may be invaded. The beef tapeworm may form cysticercosis, but the cysticerci do not involve the brain and are not fatal. Full-blown symptoms of meningitis or encephalitis may be present from pork tapeworm cysticercosis. Symptoms can mimic those of brain tumor and have been reported to cause general paralysis. An estimated 50,000 deaths per year result from neurocysticercosis.

DIAGNOSIS

Diagnosis is made easily when a proglottid can be obtained and the eggs can be identified. Parasitology laboratories can easily differentiate between *T. saginata* and *T. solium*. Microscopic examination is most reliable. The scolex is rarely passed, and identification is usually made on the eggs. Coproantigens might be helpful for patients with mild infections. If neurocysticercosis develops, tissue must be obtained to confirm the diagnosis, which requires neurosurgical biopsy.

TREATMENT AND MANAGEMENT

Praziquantel, 5 to 10 mg/kg in a single oral dose, is the drug of choice for treating pork tapeworm. When the nervous system is involved, it is dangerous to give higher doses because this could trigger occult central nervous system involvement. As with *T. saginata*, niclosamide, 2 g orally once, is a non–FDA-approved alternative. A constant concern is that infection may occur by regurgitation of eggs from the intestine to the stomach, which would expose the patient to larval invasion. Purgatives have not been of benefit.

PREVENTION AND CONTROL

T. solium can be totally prevented by protecting pigs from eating contaminated human feces. Furthermore, infection can be prevented by cooking pork thoroughly or by freezing it at –20° C for 12 to 24 hours.

ADDITIONAL RESOURCES

Drugs for parasitic infections, New Rochelle, NY, 2007, Medical Letter.

Elliott DE: Intestinal worms. In Feldman M, Friedman LS, Brandt LJ, editors: *Gastrointestinal and liver disease*, ed 8, Philadelphia, 2006, Saunders-Elsevier, pp 2435-2457.

White AC Jr: Neurocysticercosis: updates on epidemiology, pathogenesis, diagnosis, and management, *Annu Rev Med* 51:186-206, 2000.

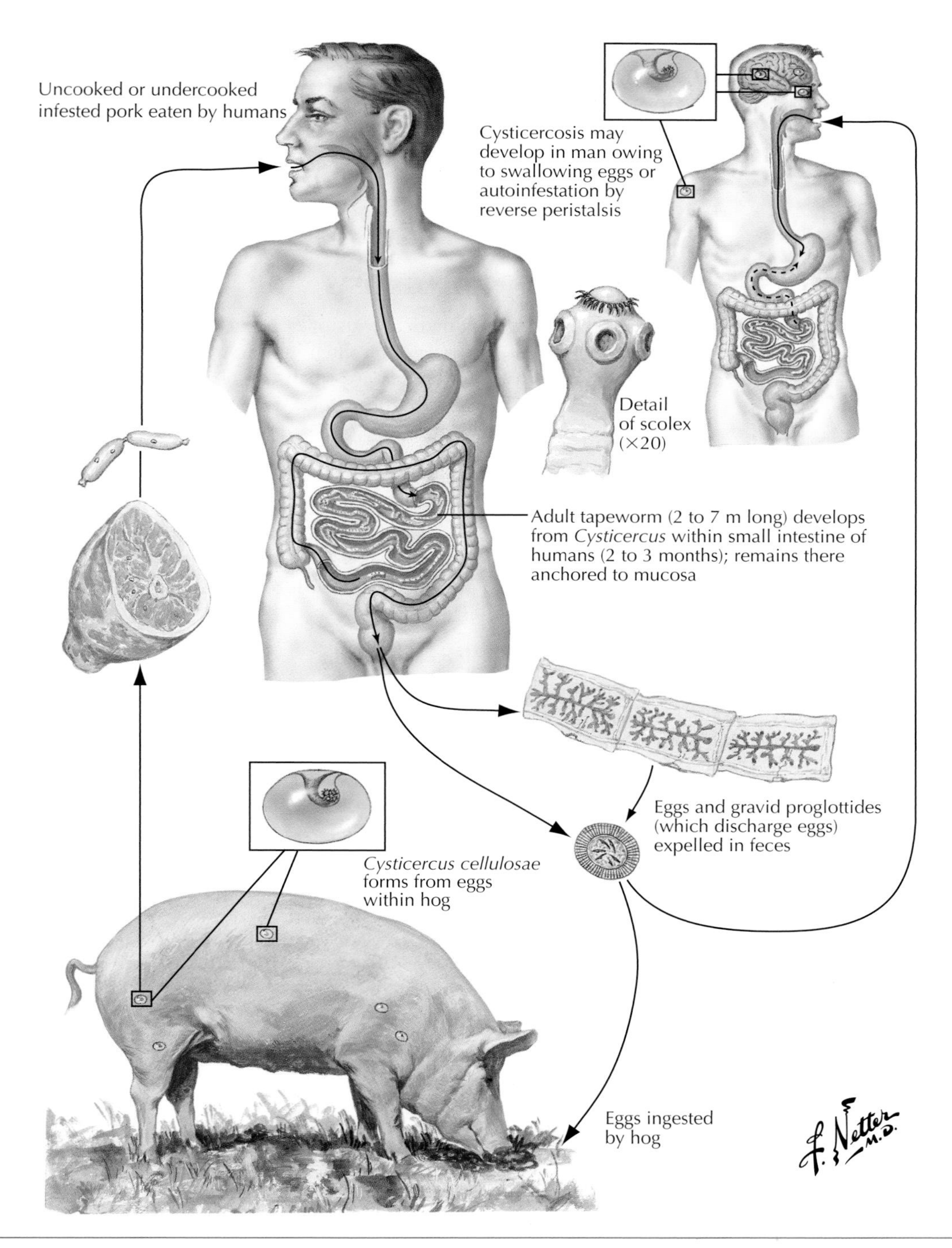

Figure 182-1 *Pork Tapeworm Infection: Life Cycle of* Taenia solium.

Tapeworm Infection (Dwarf): *Hymenolepis nana*

18

Martin H. Floch

Hymenolepis nana, also known as the "dwarf tapeworm," is the most common tapeworm of humans. It infects up to 50 million persons worldwide, including Europe, Africa, Asia, and South America. Transmission is primarily through the fecal-oral route, although autoinfection is common in children.

When infected eggs are ingested, they hatch in the duodenum, and the oncosphere invades the small bowel, where the larvae develop into a scolex that is passed into the lumen and attaches to the wall (**Fig. 183-1**). The scolex then matures, and the adult tapeworm, which is relatively short (2-4 cm), develops. Autoinfection regularly occurs; this type is referred to as *internal* autoinfection. Each of the small proglottides may produce as many as 200 eggs.

Hymenolepis diminuta is a tapeworm of rodents, but rarely, it may cause disease in humans. Human cases are usually associated with poor sanitation and living situations. The beetles in some oriental foods may transmit *H. diminuta* to humans. (See Fig. 185-1.)

CLINICAL PICTURE

Infections with *H. nana* are usually asymptomatic. Some patients may have mild diarrhea, abdominal pain, anorexia, and dyspepsia. Rarely, keratoconjunctivitis has been reported as an allergic inflammation of the eyes, thought to be caused by a toxin released by the parasite. Again, autoinfection is possible, and large burdens can occur in severely immunocompromised patients.

DIAGNOSIS

The diagnosis of *H. nana* is made by identifying the characteristic double-membrane eggs in the stool of infected persons. The eggs are small but characteristic. Low-grade eosinophilia may be noted. However, the disease must be suspected even when symptoms are minimal.

TREATMENT AND MANAGEMENT

As with beef and pork cestode infections, the drug of choice for dwarf tapeworm is praziquantel in a single oral dose of 25 mg/kg. Cure rates are usually greater than 98%. An alternative drug is nitazoxanide, 500 mg once or twice daily for 3 days.

PROGNOSIS

Treatment of *H. nana* infection is usually highly effective, but reinfection is possible. Sanitation conditions usually must be improved, and instructions for appropriate hygiene must be given to families when this type of infection occurs.

ADDITIONAL RESOURCES

Drugs for parasitic infections, New Rochelle, NY, 2007, Medical Letter.

Elliott DE: Intestinal worms. In Feldman M, Friedman LS, Brandt LJ, editors: *Gastrointestinal and liver disease*, ed 8, Philadelphia, 2006, Saunders-Elsevier, pp 2435-2457.

King CH: Cestodes (tapeworms). In Mandell GL, Bennett JE, Dolin R, editors: *Mandell, Douglas and Bennett's principles and practice of infectious diseases*, ed 6, Philadelphia, 2005, Elsevier–Churchill Livingstone, pp 3285-3293.

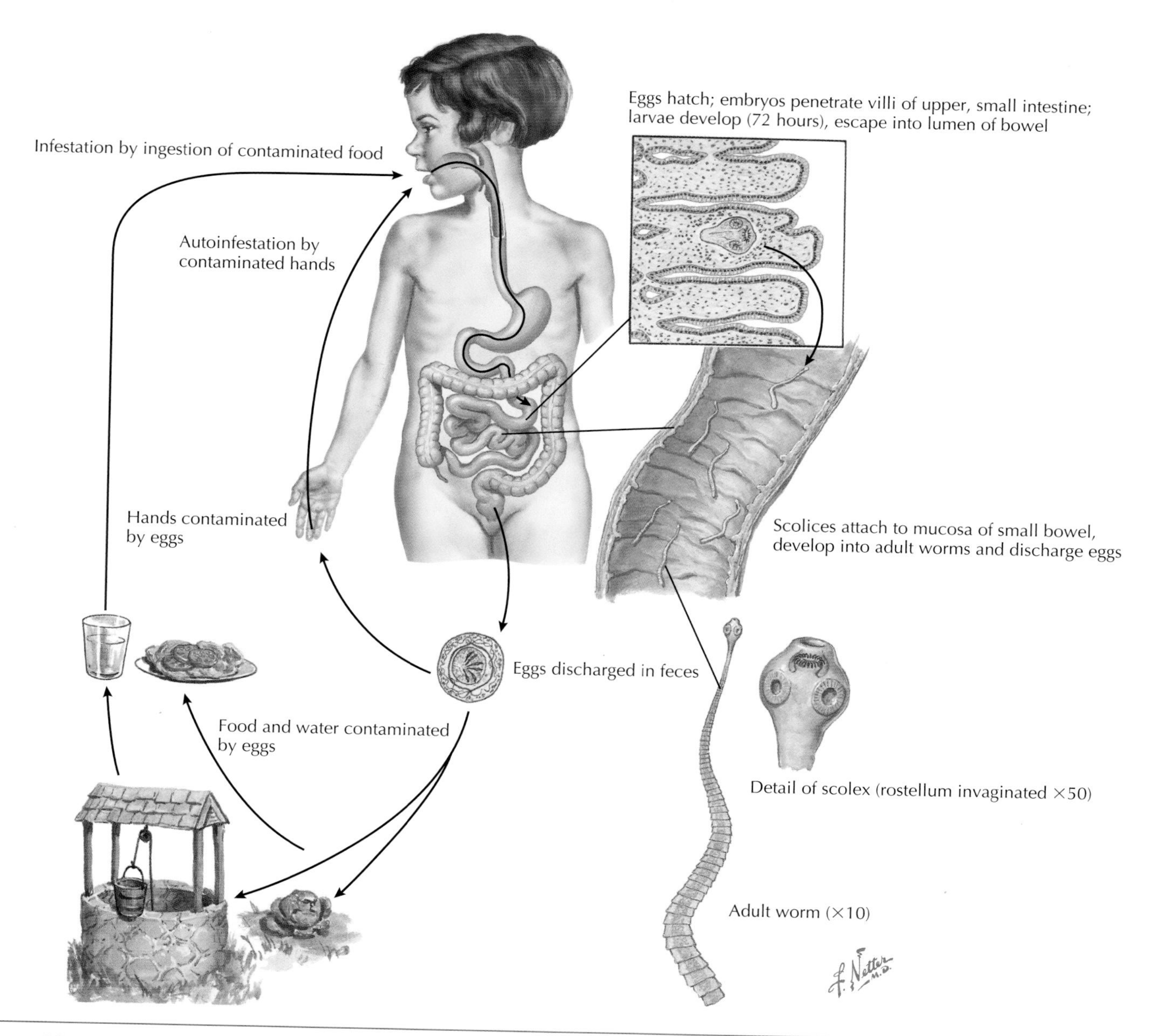

Figure 183-1 *Dwarf Tapeworm Infection: Life Cycle of* Hymenolepis nana.

Tapeworm Infection (Fish): *Diphyllobothrium latum*

184

Martin H. Floch

Diphyllobothriasis is caused by the "fish tapeworm" *Diphyllobothrium latum*. It occurs primarily in northern temperate regions, where freshwater fish constitute a major portion of the diet. Numerous cases are reported in the Baltic countries, northern parts of Europe, northern Asia and Siberia, northern Manchuria, and Japan. Several foci are known in North America in the Great Lakes area.

The life cycle of *D. latum* is relatively complex and requires three hosts (**Fig. 184-1**). Once the eggs are passed into freshwater, they hatch in approximately 2 weeks at favorable temperatures, and an embryo called the *coracidium* is passed into the water. The coracidium is usually eaten by a crustacean of the genus *Cyclops* or *Diaptomus*. Within the crustacean, the coracidium metamorphoses in 2 to 3 weeks into a *procercoid* larva. When the second host—a freshwater fish such as pike, eel, salmon, trout, perch, or pickerel—ingests the infested crustacean, the procercoid penetrates the viscera, muscles, and connective tissue and develops within 1 to 4 weeks into an elongated, wormlike *plerocercoid*. When a raw or a poorly cooked infected fish is eaten, the plerocercoid larva attaches to the mucosa of the small intestine of the third host and grows in approximately 3 weeks into an adult *D. latum*. The adult worm is small and may range from 2 to 10 mm, but a long chain of more than 3000 proglottides may develop, extruding as many as 1 million eggs daily. A single parasite may live for years within the host. (See Fig. 185-1.)

Other *Diphyllobothrium* tapeworms may infest humans. Also, other animals, including bears, wolves, foxes, and cats, can serve as hosts for many of these species, in addition to *D. latum*.

CLINICAL PICTURE

Patients with *D. latum* are usually asymptomatic, but diarrhea, fatigue, dizziness, and paresthesias are all reported. Again, worms can become extremely long. *D. latum* is unique in that it can compete for vitamin B_{12}, thereby causing megaloblastic anemia and possibly full-blown vitamin B_{12} deficiency, with glossitis and peripheral neuropathy. Fortunately, only 2% of persons develop vitamin B_{12} deficiency.

DIAGNOSIS

The diagnosis of diphyllobothriasis depends on finding the eggs in the stool. If the proglottides are passed, they are distinctive and can be identified. Patients may have mild eosinophilia, but megaloblastic anemia from a region thought to be at risk should warrant thorough evaluation of stool for the parasite. Eating sushi or raw fish from a high-risk area should also raise the index of suspicion of infestation.

TREATMENT AND MANAGEMENT

Praziquantel, 5 to 10 mg/kg in a single oral dose, is effective treatment. An alternative drug is niclosamide (2 g). Patients may pass a very long worm after treatment. Follow-up stool examination should be performed in 6 to 8 weeks because proglottides and eggs can be passed after successful treatment.

PROGNOSIS

The prognosis is excellent when appropriate treatment is given. However, reinfection must be controlled. In endemic areas where infection has been noted, all fish should be thoroughly cooked or frozen at −18° C for at least 24 hours.

ADDITIONAL RESOURCES

Drugs for parasitic infections, New Rochelle, NY, 2007, Medical Letter.

Elliott DE: Intestinal worms. In Feldman M, Friedman LS, Brandt LJ, editors: *Gastrointestinal and liver disease*, ed 8, Philadelphia, 2006, Saunders-Elsevier, pp 2435-2457.

King CH: Cestodes (tapeworms). In Mandell GL, Bennett JE, Dolin R, editors: *Mandell, Douglas and Bennett's principles and practice of infectious diseases*, ed 6, Philadelphia, 2005, Elsevier–Churchill Livingstone, pp 3285-3293.

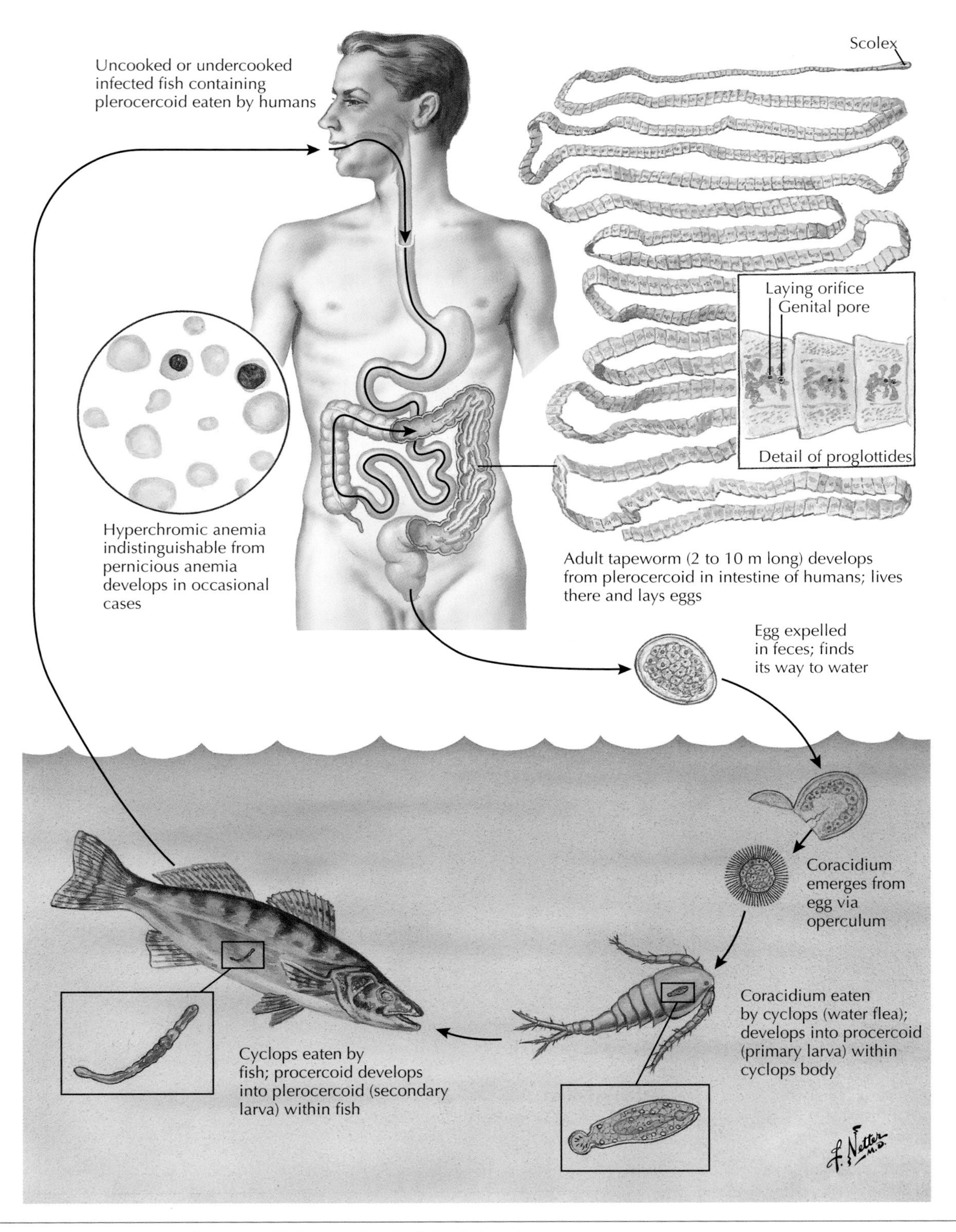

Figure 184-1 *Fish Tapeworm Infection: Life Cycle of* Diphyllobothrium latum.

Other Helminth Infections: *Trichinella spiralis* and Flukes

185

Martin H. Floch

TRICHINOSIS

The most common *Trichinella* species in the United States is *Trichinella spiralis*, which is almost worldwide in its distribution. However, other *Trichinella* species infect humans in Africa and in arctic regions. Humans become infected when they eat poorly cooked meat. The main reservoir is the pig.

The life cycle of *Trichinella* is complex. When the uncooked meat is eaten, larvae are released in the small intestine, where they go through four phases of development to adulthood. Afterward, male and female worms mate in the intestine, where they deposit larvae that invade the mucosa, enter the circulation, and encyst in the skeletal muscle. A single worm may produce as many as 1500 larvae. Humans who eat undercooked meat or meat from wild animals are particularly susceptible to infestation. In the United States, the number of trichinosis cases reported each year is usually less than 50.

Clinical Picture

During the initial stage of trichinosis, a patient may have mild gastroenteritis with nausea, vomiting, or abdominal pain and diarrhea. This phase may last up to 1 week. When the larvae enter the muscle, myalgia, fever, edema, and systemic signs of allergic response develop. The more larvae present, the more severe will be the disease. Generalized edema may develop, and proteinuria can follow. Central nervous system signs and symptoms, cardiomyopathy, and extraocular muscle involvement occur in patients with moderate to severe infection.

Muscle tenderness can be readily detected, along with an unusually high white blood cell count and significant eosinophilia.

Diagnosis

A diagnosis of trichinosis is made by obtaining muscle tissue and demonstrating the parasitic complexes. The diagnosis can also be made by detecting *Trichinella*-specific DNA using polymerase chain reaction. Helpful findings are elevated creatine kinase and lactic dehydrogenase levels. After approximately 2 weeks, enzyme-linked immunosorbent assay can detect the antibodies in some patients.

Treatment and Management

Albendazole or mebendazole is the treatment of choice. If symptoms are severe, corticosteroids may be helpful in decreasing inflammation.

Prevention and Control

Prevention is still the best method for avoiding trichinosis. All meat should be cooked thoroughly at 58.5° C for 10 minutes. Freezing also kills the worms, but the meat must be frozen at −20° C for at least 3 days. One species, *Trichinella nativa*, is resistant to freezing. Hunters and humans who eat wildlife should be cautious because they are particularly susceptible to *Trichinella* infestation if they eat meat that is not cooked properly.

FLUKE INFECTIONS

Flukes are trematodes with complex life cycles. The most common are the schistosomes (see Chapter 257). Rarer fluke infections involve liver flukes and other intestinal flukes.

Liver Flukes

Clonorchis sinensis and *Opisthorchis viverrini* are liver flukes found in Asia (Japan, China, Indochina, South Korea, Taiwan). Humans become infected because they eat raw freshwater fish. It is estimated that as many as 7 million people are infected in some parts of Asia. Liver flukes live in the biliary tract, and eggs pass into stool. There are two intermediate hosts: snails, then freshwater fish.

Fasciola hepatica can infect humans who eat watercress contaminated with excysted metacercariae. Snails are the intermediate host, and herbivorous mammals are easily infected. Larvae in the small intestine can penetrate the gut, migrate through the peritoneal cavity, and enter the liver. They then migrate through the liver and enter the bile ducts, where they spend the rest of their lives. Patients may have acute infections that can appear as acute hepatitis. Diagnosis is usually confirmed by finding the ova in the bile or the stool. However, imaging reveals defects in the liver.

Figure 185-1 shows ova of *C. sinensis* and *F. hepatica*, as well as those of intestinal helminths described in previous chapters.

Intestinal Fluke

Fasciolopsis buski is the intestinal fluke. It is seen throughout Asia and infects humans, pigs, and dogs. Snails are the intermediate host, and the cercariae subsist on water plants. Ingestion of the plants results in the encystment of larvae and the formation of mucosal abscesses in the small intestine. Diagnosis is made by finding the eggs in the stool, and the treatment of choice is praziquantel.

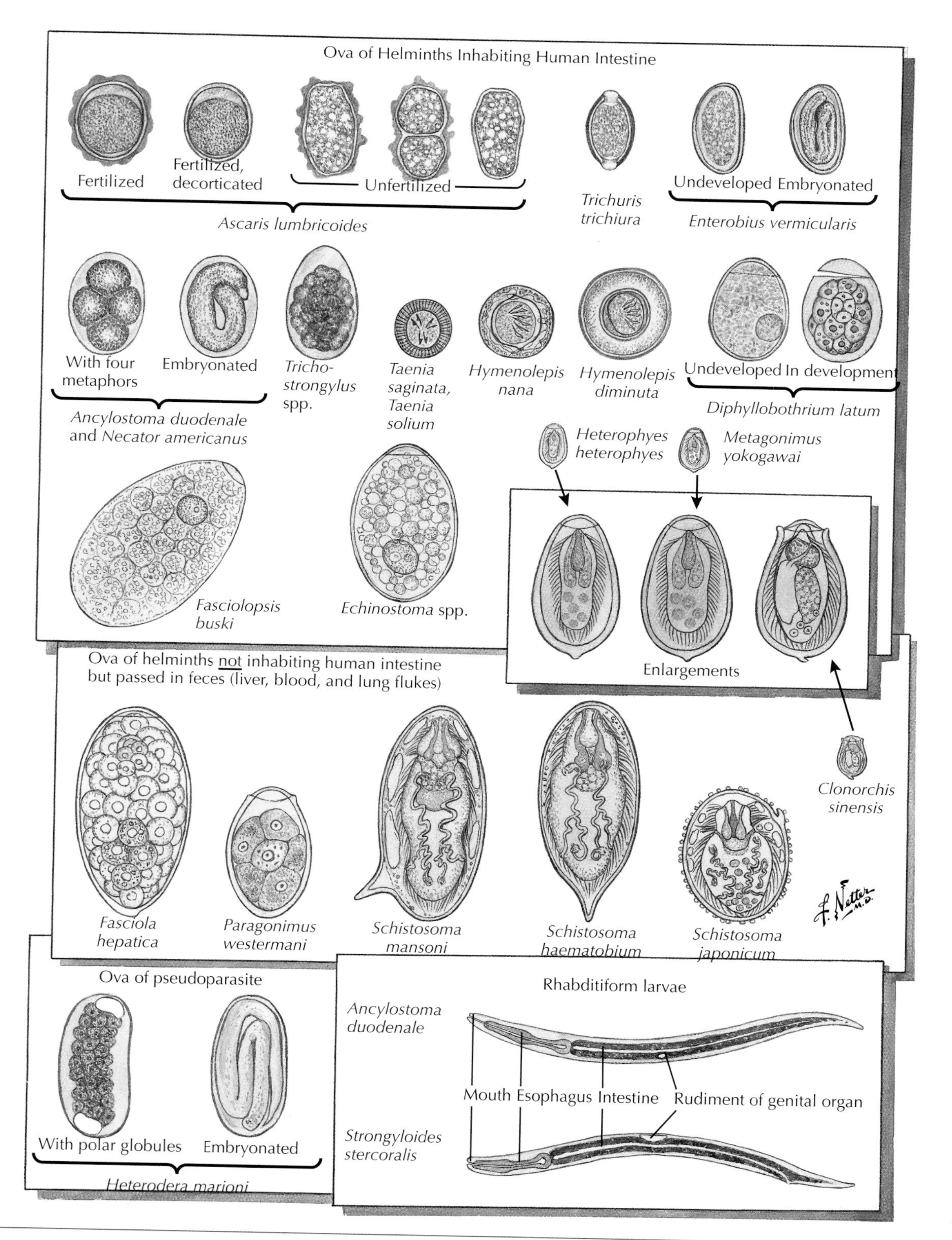

Figure 185-1 *Ova of Helminth Parasites and Pseudoparasites and Rhabditiform Larvae* (inset).

Clinical Picture

Flukes may live in the bile ducts for as long as 10 to 30 years. They may cause liver enlargement and every type of biliary condition, including adenomas, fibrosis, and stricture of the bile ducts. Bouts of cholangitis can lead to cholangiocarcinoma.

Diagnosis and Treatment

The diagnosis of fluke infection is made by identifying ova in the stool or in bile obtained by endoscopic retrograde cholangiopancreatography. The treatment of choice is praziquantel.

ADDITIONAL RESOURCES

Drugs for parasitic infections, New Rochelle, NY, 2007, Medical Letter.

Elliott DE: Intestinal worms. In Feldman M, Friedman LS, Brandt LJ, editors: *Gastrointestinal and liver disease*, ed 8, Philadelphia, 2006, Saunders-Elsevier, pp 2435-2457.

Grove DI: Tissue nematodes including trichinosis, dracunculiasis, and the filariases. In Mandell GL, Bennett JE, Dolin R, editors: *Mandell, Douglas and Bennett's principles and practice of infectious diseases*, ed 6, Philadelphia, 2005, Elsevier–Churchill Livingstone, pp 3267-3275.

Human Immunodeficiency Virus and the Gastrointestinal Tract

Martin H. Floch

186

Human immunodeficiency virus (HIV) causes acquired immunodeficiency syndrome (AIDS). First described in 1980 as a disease transmitted from chimpanzees, HIV infection is now a worldwide pandemic. It is estimated that 40 million people are infected with HIV-1. The initial focus was sub-Saharan Africa, where estimates indicate that as many as 25% of the population are infected. In some countries, as many as 30% of pregnant women have HIV-1 infection.

Initially, transmission had been by homosexual, orogenital, or anogenital contact, but the major mode of transmission is now heterosexual contact. The virus can enter as a result of mucosal trauma, but major cellular routes by which HIV-1 enters the nuclear cells of the lamina propria are through epithelial cells, M cells, and dendritic cells. The virus may be swallowed in amniotic fluid by the fetus. Once it enters the gastrointestinal (GI) mucosa, it is transmitted throughout the body. High viral loads during primary HIV-1 infection lead to the highest risk for infection. Other high-risk factors for transmission include high viral loads during end-stage HIV disease, mucosal trauma or inflammation, mucosal infection of recipient partner, increased or unprotected sexual contact, receptive anal intercourse, and lack of circumcision in a male index partner.

Gastrointestinal involvement may first be manifested by HIV-related diarrhea that can occur during the primary infection after 1 to 4 weeks of exposure. Therapies such as highly active antiretroviral therapy (HAART) have been effective at decreasing GI complications and involvement.

Sites of involvement in the GI tract are numerous and include the esophagus, stomach, biliary tract, gallbladder, pancreas, small bowel, colon, and rectum (**Fig. 186-1**).

CLINICAL PICTURE

The most common presenting symptoms in patients with AIDS are those related to esophageal disease and diarrhea. Symptoms of acute diarrhea occur in as many as 90% of patients 1 to 4 weeks after infection and may last for 1 to 3 weeks. Debilitating symptoms may be associated with fatigue, weight loss, and myalgias. However, most of the acute HIV-1 syndrome resolves. Chronic diarrhea may occur in as many as 50% of HIV-infected patients. Diarrhea may be mild or severe, and it may be associated with a full-blown picture of cramps, anorexia, fever, weight loss, and wasting. Patients with low CD4+ T-cell counts are most likely to have severe symptoms and malabsorption. Cachexia is often seen in AIDS patients in Africa. With the advent of HAART, diarrhea has been less prominent, but it still plagues severely infected AIDS patients.

Esophageal symptoms in HIV patients, primarily odynophagia and dysphagia, are caused by three main infectious agents, *Candida albicans*, cytomegalovirus (CMV), and herpes simplex virus (HSV) (see Chapter 168). These organisms cause symptoms in approximately 30% of patients and are second only to diarrhea as GI complications. In approximately 20% of patients, two organisms are the cause. The most common infectious agent is *Candida*, followed by CMV, and HSV has been reported in up to 30% of patients. Odynophagia can be severe and forces patients to limit their food intake, causing further debilitation.

The presence of abdominal pain necessitates the differential diagnosis of hepatobiliary and pancreatic disorders. Furthermore, any of the patient's lesions can bleed. Opportunistic infection can involve any part of the GI tract and may compound the clinical symptoms.

DIAGNOSIS

The workup for diarrhea is standard, with evaluation for all infectious agents (protozoa, bacteria, viruses, fungi) and for lymphoma or Kaposi sarcoma, as well as elimination of drug-induced causes. Protozoans are the most common infectious agents (see Chapter 175). *Cryptosporidium*, *Microsporidia*, *Isospora belli*, and *Cyclospora* are the most common parasitic infections. CMV, HSV, and adenovirus are the most common viral infections. Bacteria include *Mycobacterium avium-intracellulare* complex, *Salmonella* spp., *Shigella flexneri*, and *Campylobacter jejuni*. The fungal infections *Candida albicans* and *Histoplasma capsulatum* are also noted. Most of these infections occur late during the HIV disease course.

After the initial stage of infection, AIDS enteropathy may develop, which can be diagnosed only after completely negative findings for infectious agents. Enteropathy can cause wasting disease, but it may be controlled by HAART. The cause of the diarrhea must be determined from stool analysis and biopsy specimens.

Endoscopy and biopsy are used to diagnose esophageal disease. Lesions in the esophagus vary greatly but may appear as deep ulcerations and possibly as neoplasms. Severe odynophagia may be associated with recurrent chest pain. Careful endoscopic examination and histologic examination of biopsy specimens are needed to make the diagnosis.

Of special note is CMV infection, which is common in patients with AIDS. CMV may cause symptoms when it affects the esophagus, stomach, small bowel, and colon. It may cause full-blown enteritis or colitis. Less often, CMV has caused calculous cholecystitis, sclerosing cholangitis, pancreatitis, and appendicitis. CMV infection should be suspected in an HIV patient.

These patients require upper endoscopy and colonoscopy for evaluation. Lesions in the colon can be patchy and may involve only the right colon, underscoring the importance of colonoscopy. Anorectal symptoms require special differential diagnosis and are included in the evaluation of specific anal symptoms. Biopsy is essential, as for any intestinal disease. Because anorec-

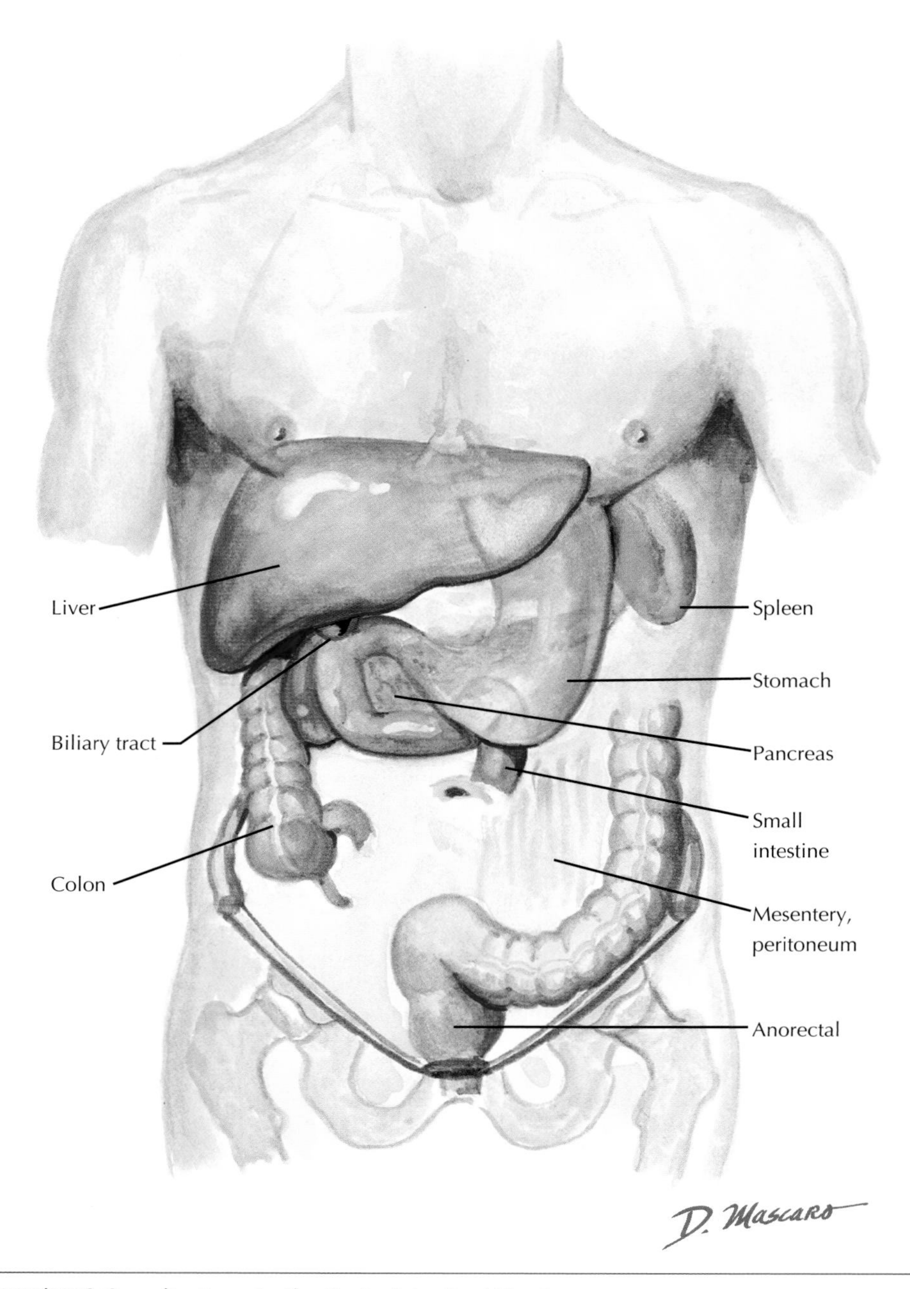

Figure 186-1 *Sites of HIV/AIDS Complications in the Gastrointestinal Tract.*

tal carcinoma is common in homosexual men and the risk increases with HIV infection, any suspicious lesion requires endoscopic workup and thorough biopsy.

TREATMENT AND MANAGEMENT

With HAART, mortality and morbidity rates improved in patients with HIV. From 1995 to 1997, mortality declined from 29.4 to 8.4 per 100 patients, with a concurrent decrease in opportunistic infections. The natural history of HIV-associated infections was affected in the following three ways:

1. Secondary infections and reactivation of latent infections decreased.
2. Prophylactic antimicrobials are now often unnecessary.
3. Ongoing infections appear to be resolved because of the repopulation of peripheral T cells.

Treatment of specific causes of diarrhea, either protozoan or infectious, is described in the chapters for those organisms.

Treatment of esophageal disease invariably involves treatment for *C. albicans*, which is fluconazole, 100 mg orally daily for 2 weeks. An alternative is itraconazole, 200 mg daily for 2 weeks.

Therapy for CMV is ganciclovir. Although effective in 63% to 100% of patients, ganciclovir is *virustatic* (inhibits replication), not virucidal, and most patients have recurrences. It also produces significant adverse effects. Foscarnet has shown some usefulness in ganciclovir-resistant CMV, but there is much less experience with its use. Fortunately, HAART has resulted in a decrease in the incidence of CMV infection.

Patients must be monitored for all GI symptoms because of the risk for recurrence.

COURSE AND PROGNOSIS

The course and prognosis for HIV infection have greatly improved with HAART. However, even though mortality and morbidity rates have significantly declined, HIV is still a lethal disease. Unfortunately, in parts of the world where therapy is less available, the disease runs its course and increases mortality. A major challenge is to bring improved therapy to patients in underdeveloped countries.

ADDITIONAL RESOURCES

Sulcowski MS, Chaisson RE: Gastrointestinal and hepatobiliary manifestations of human immunodeficiency virus infection. In Mandell GL, Bennett JE, Dolin R, editors: *Mandell, Douglas and Bennett's principles and practice of infectious diseases*, ed 6, Philadelphia, 2005, Elsevier–Churchill Livingstone, pp 1575-1582.

Wilcox CM: Gastrointestinal consequences of infection with human immunodeficiency virus. In Feldman M, Freedman LS, Brandt LJ, editors: *Gastrointestinal and liver disease*, ed 8, Philadelphia, 2006, Saunders-Elsevier, pp 667-682.

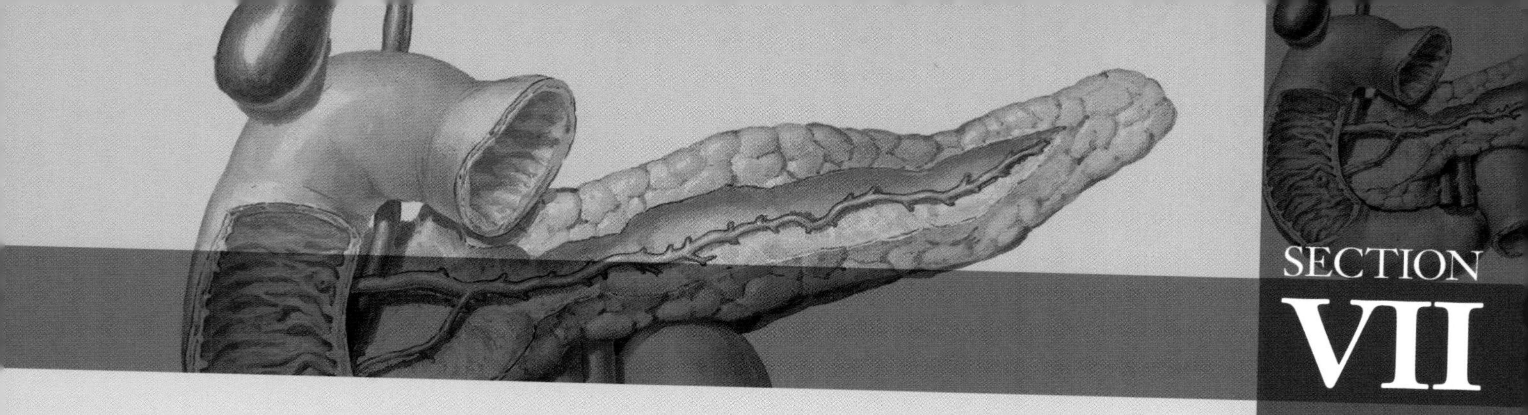

SECTION

VII

Pancreas

Development and Anatomy of the Pancreas

187

C. S. Pitchumoni

One of the earliest references to the pancreas as a distinct organ occurs in the Talmud, which refers to the organ as "the finger of the liver." Aristotle (384-322 BC) believed the organ had the sole function of protecting the neighboring vessels. Herophilus of Chaldikon, in the third century BC, made some initial anatomic descriptions. Almost 200 years later, Rufus of Ephesus used the term "pancreas" (pan, all; kreas, flesh). George Wirsung described the structure of the major pancreatic duct in humans in 1642. About 100 years later, Giovanni Santorini described the accessory pancreatic duct. Paul Langerhans discovered the endocrine pancreas in 1869.

The pancreas arises from two diverticula of the foregut, in a region that later becomes the duodenum (**Fig. 187-1**). Early in the fifth week of gestation, a larger dorsal bud develops proximally just above the level of the hepatic diverticulum. The ventral outpouch appears soon afterward. Growing fairly rapidly and extending into the dorsal mesentery of the duodenum near the developing omental bursa, the dorsal pancreatic bud passes in front of the developing portal vein. Because of the more rapid growth of the duodenum, the ventral bud, together with the developing *common bile duct* (CBD), rotates backward behind the duodenum. When the rotation is completed, the original ventral bud comes to lie close to, below, and somewhat behind the dorsal pancreas, and eventually its tip lies behind the superior mesenteric vein and the root of the portal vein.

From the larger dorsal bud originates the cephalad part of the head, as well as the neck, body, and tail of the pancreas, whereas the caudate part of the head and the uncinate process derive from the smaller ventral bud. Ducts develop in both buds but anastomose. During the seventh week, the dorsal and ventral buds fuse, enclosing the vena cava. The secretions of the neck, body, and tail are subsequently shunted into the duct of the smaller ventral pancreas, which thus becomes the principal pancreatic *duct of Wirsung*. Only the upper portion of the head is finally drained by the original duct of the dorsal pancreas, the accessory *duct of Santorini*.

The pancreas is 10 to 15 cm (4-6 inches) long and extends transversely across the abdomen from the concavity of the duodenum to the spleen. It has four parts: head, neck, body, and tail (**Fig. 187-2**). Located deep in the epigastrium and the left hypochondrium behind the lesser omental sac, approximately on the level of the first and second lumbar vertebrae (L1-L2), the gland escapes physical examination. The head is globular and has an inferior extension, and the *lingula* (uncinate process) projects hooklike to the left and is crossed anteriorly by the superior mesenteric vessels. Covered anteriorly by the pylorus and transverse colon, the head fits snugly into the loop of the duodenum, so that the CBD passes either through a groove or through the substance of the gland. The posterior surface of the head touches the inferior vena cava, left renal vein, and aorta. The splenic artery and vein extend along its upper border. Its anterior surface, covered by serosa, is separated by the omental bursa from the posterior wall of the stomach. The inferior surface, below the attachment of the transverse mesocolon, is related to the duodenojejunal junction and to the splenic flexure of the colon. The posterior surface is in contact with the aorta, splenic vein, and left kidney, where the body tapers off into a short tail.

The lymphatics of the pancreas arise as fine periacinar and perilobular capillary networks extending along the blood vessels to the surface of the gland. Direct lymphatic connections exist wherever the pancreas is attached to other organs.

The arterial blood supply to the head of the pancreas, together with the duodenum, is by an anterior and posterior pancreaticoduodenal arcade formed by the union of anterior and posterior branches of the superior and inferior pancreaticoduodenal arteries. Inferior and superior pancreatic arteries supply the neck and the body. The splenic artery supplies the tail and body of the pancreas through several branches.

The sympathetic nerves reach the pancreas through the greater and lesser splanchnic trunks arising from the fifth to the ninth thoracic ganglia. Major sympathetic innervation is through the greater splanchnic nerve. Parasympathetic fibers reach through vagi.

ADDITIONAL RESOURCES

Bockman DE: Anatomy of the pancreas. In Go VLW: *The pancreas: biology, pathobiology, and disease*, New York, 1993, Raven Press, pp 1-8.

Steer ML: Exocrine pancreas. In Townsend CM, Beauchamp RD, Evers BM, Mattox KL: *Sabiston textbook of surgery*, ed 18, Philadelphia, 2008, Saunders-Elsevier, pp 1589-1623.

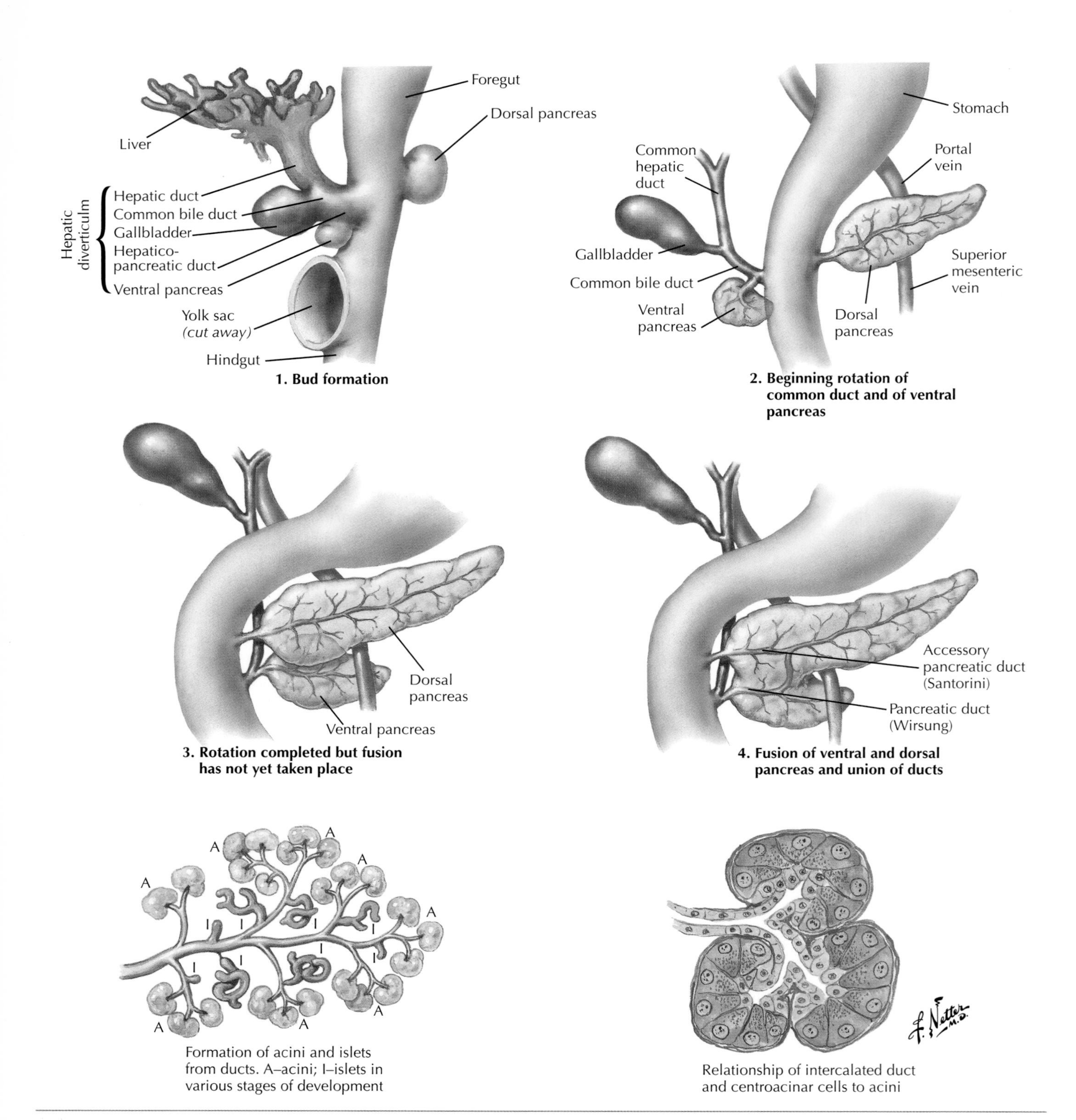

Figure 187-1 *Development and Anatomy of the Pancreas.*

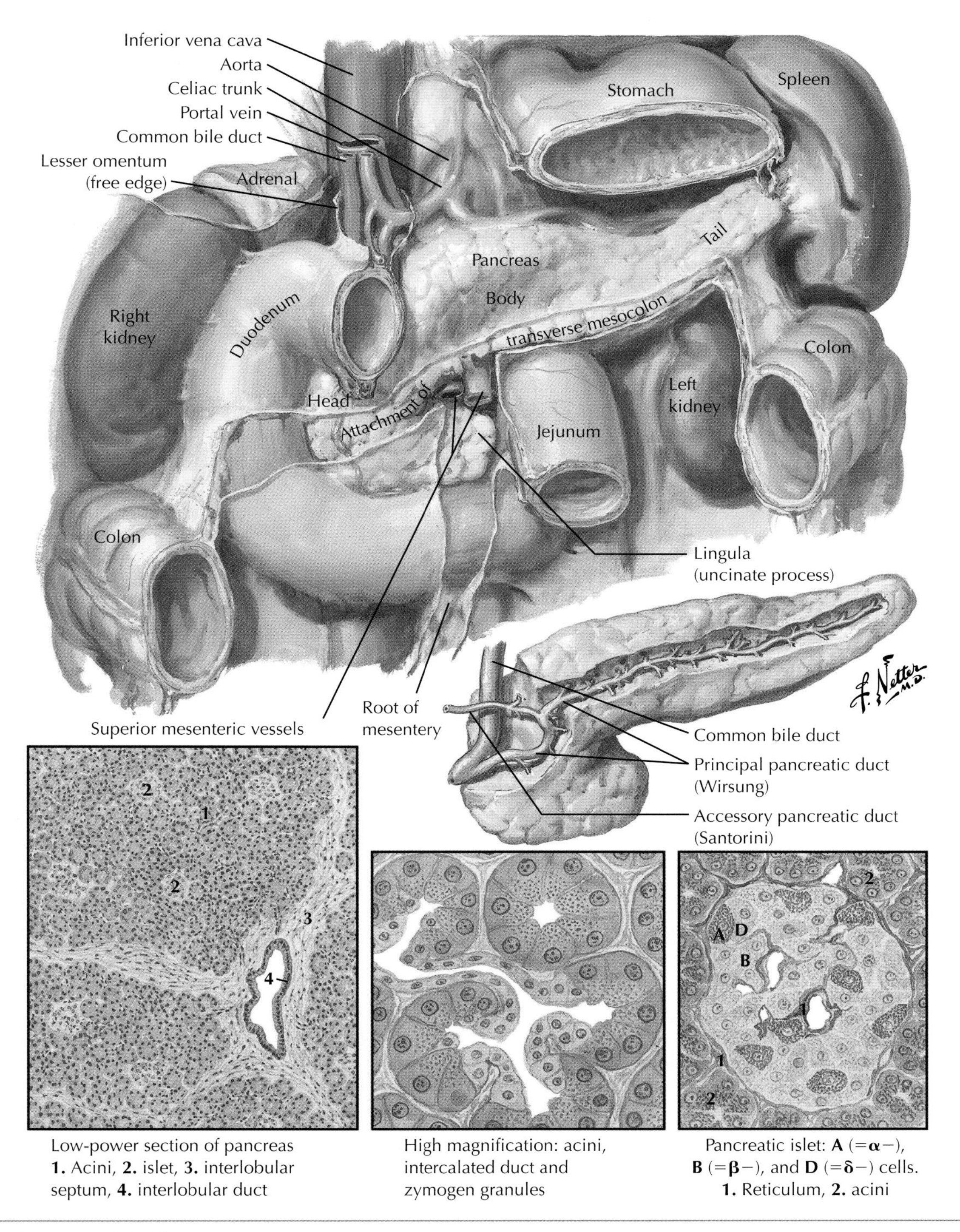

Low-power section of pancreas **1.** Acini, **2.** islet, **3.** interlobular septum, **4.** interlobular duct

High magnification: acini, intercalated duct and zymogen granules

Pancreatic islet: **A** (=α−), **B** (=β−), and **D** (=δ−) cells. **1.** Reticulum, **2.** acini

Figure 187-2 *Anatomy of the Pancreas.*

Pancreatic Ducts and Exocrine and Endocrine Organs

188

C. S. Pitchumoni

The pancreas is a complex organ with endocrine and exocrine functions. The acinar cells make up almost 90% of the gland's mass, ductal tissue about 5%, and 10^5 to 10^6 islets of Langerhans occupy about 2% of the pancreas. The arrangement of the pancreatic ducts within the gland varies considerably and even more so in relationship with the terminal common bile duct (CBD) (**Fig. 188-1**). Occasionally, the relative size of the two ducts (Santorini and Wirsung) is reversed so that the duct of Santorini remains the main duct. The accessory duct ordinarily inserts into the duodenum proximally, on a separate papilla (minor), but may enter through the papilla of Vater. The main pancreatic duct enters the duodenum either through a separate orifice or through a common channel, the ampulla of Vater.

The main pancreatic duct begins in the tail. Centrally located, it courses to the right through the body and neck and is joined by tributaries that usually enter at right angles, alternating from opposite sides. In the head, the main duct usually turns caudally and dorsally and comes close to or joins the CBD while coursing through the substance of the pancreas. The main duct usually drains the tail, neck, and body and the caudal and dorsal portions of the head. Ductal diameter is an important parameter for evaluation of pancreatograms from endoscopic retrograde cholangiopancreatography (ERCP). Diameters of 3.6 mm at the head, 2.7 mm in the body, and 1.6 mm in the tail and a length of 17.2 cm should be considered normal.

The *sphincter of Oddi* is formed by smooth muscles surrounding three sites: the lower end of the CBD, the major pancreatic duct, and the ampulla.

Pancreas divisum is a congenital variant of the ductal morphology in which the dorsal and ventral pancreatic ducts do not fuse. This abnormality is reported in up to 14% of autopsy studies and in 2% to 7% of patients undergoing ERCP. In pancreas divisum, the bulk of the enzyme flow occurs through the accessory papilla, which is narrower than the main duct. If the orifice is too small or is stenotic, the intraductal pressure is increased and may give rise to pancreatitis.

Annular pancreas is a rare malformation in which a band of pancreatic tissue surrounds the descending portion of the duodenum, with smooth continuation to the head of the pancreas. Partial obstruction of the duodenum may be a clinical presentation in childhood or later. *Heterotopia* of the pancreas is healthy pancreatic tissue developing in an abnormal location with no vascular, neuronal, or anatomic continuity with the main pancreas.

The *exocrine pancreas* is a compound acinar gland, similar in structure to the salivary gland, which lacks islets of Langerhans. The functional units consist of an acinus and its draining duct, and the units are separated by fine, connective tissue septa that have blood vessels, lymphatics, nerves, and secretory ducts. The nucleus of the acinar cell is located near the broad base. The cytoplasm is basophilic and contains numerous acidophilic, highly refractile zymogen granules, which contain the proenzymes. The *acinar cells* of the pancreas are among the richest cells of the body in RNA content and have the highest protein turnover. *Centroacinar cells* are smaller than acinar cells, are located at the junction of acini and ducts, and are devoid of zymogen granules. Carbonic anhydrase is associated with bicarbonate production and is present in the centroacinar cells and in the ductal epithelium. Intercalated ducts, which partially penetrate and drain the acini, are lined by centroacinar and clear cells. Intralobular ducts are also lined by cells with clear cytoplasm.

The *pancreatic islets (islets of Langerhans)*, scattered over the gland but especially in the body and tail, consist of cells structurally and functionally different from those of the exocrine parenchyma. Adult human islets consist of 10% alpha (α) cells, 70% beta (β) cells, 15% pancreatic polypeptide (PP) cells, and 5% delta (δ) cells. The α cells secrete glucagon; the β cells, insulin; the PP cells, pancreatic polypeptide; and the δ cells, somatostatin.

ADDITIONAL RESOURCES

Ballian N, Hu M, Liu S, Brunicardi C: Proliferation, hyperplasia, neogenesis and neoplasia in the islets of Langerhans: a review, *Pancreas* 35:199-206, 2007.

Johnson LR: Pancreatic secretion. In Johnson LR: *Gastrointestinal physiology,* ed 7, Philadelphia, 2007, Mosby-Elsevier, pp 85-95.

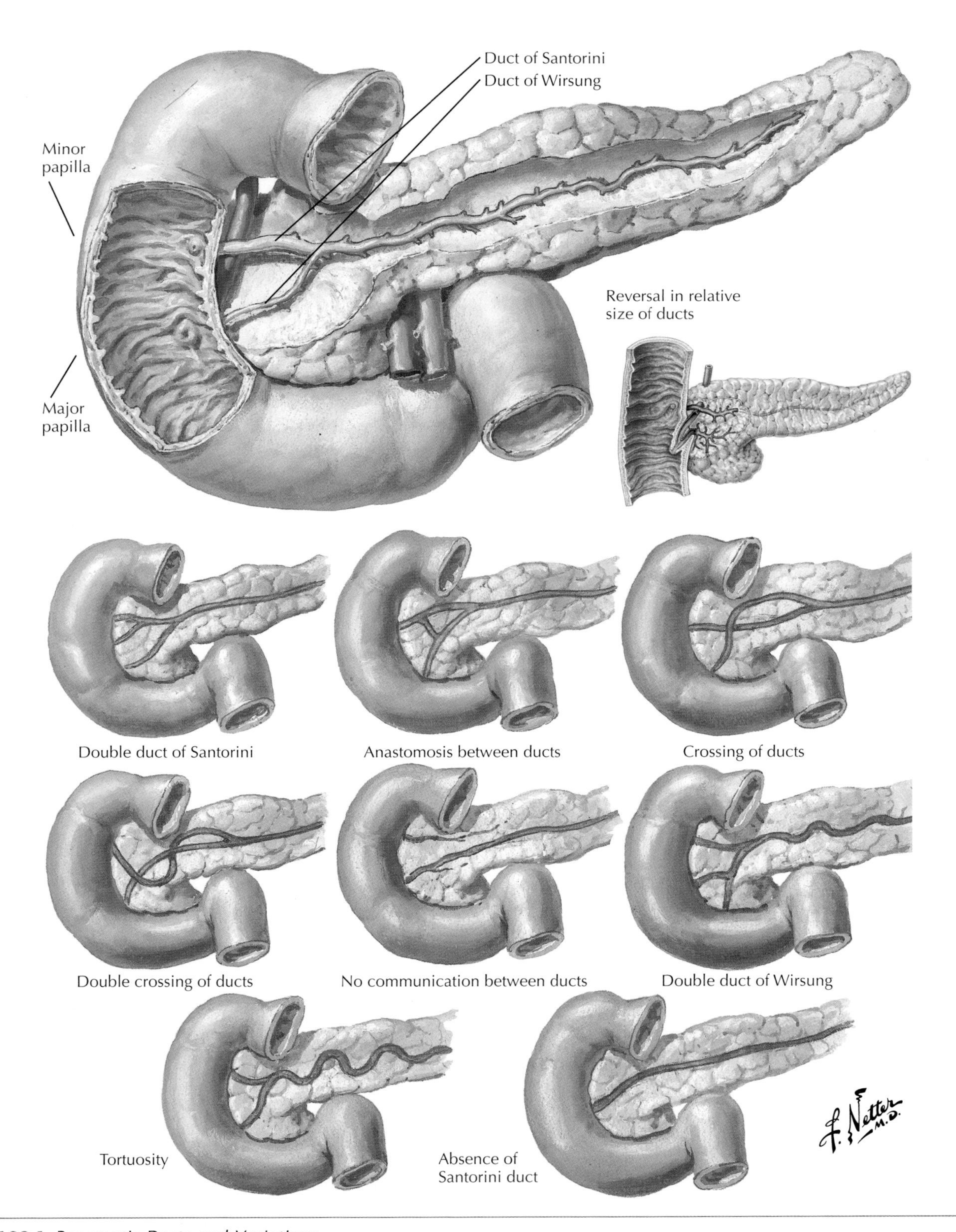

Figure 188-1 *Pancreatic Ducts and Variations.*

Pancreatic Physiology

C. S. Pitchumoni

Each day, 500 to 1000 mL of colorless, bicarbonate-rich fluid is secreted by the exocrine pancreas. The pancreas secretes approximately 0.2 to 0.3 mL of juice each minute in the interdigestive phase, which can increase to 3.15 mL/min in response to hormonal stimulation and a meal (**Fig. 189-1**). Digestive enzymes are secreted by the acinar cells, and a large volume of bicarbonate-rich solution is produced by the centroacinar and ductal cells. The role of the pancreatic secretion is to digest fat, proteins, and starch. Together with bile, it is responsible for fat absorption. The bicarbonate solution neutralizes the gastric hydrochloric acid in the duodenum, thus providing an ideal pH for enzymatic action. It also prevents inactivation of bile acids by gastric acid.

The basal bicarbonate concentration at low secretory rates is approximately 30 to 60 mmol/L, increasing to as high as 135 mmol/L after stimulation by gastric acid entering the duodenum. The juice is rich in proteins and has a concentration of 7 mg/mL after addition of cholecystokinin. The acinar cells secrete 19 different proteins. Only amylase and lipase are secreted in the active form; all the proteolytic enzymes are secreted as inactive proenzymes.

Pancreatic amylase hydrolyzes dietary starch, glycogen, and other carbohydrates (except cellulose) to produce disaccharides and a few trisaccharides. Digestion of starch by pancreatic amylase is a continuation of the process initiated by salivary amylase. *Pancreatic lipase* hydrolyzes neutral fat into fatty acids and monoglycerides. Lipase has a molecular weight of 48,000 daltons and isoelectric point of 6.5, a major difference from salivary or gastric lipase, which is stable at acid pH. Pancreatic lipase activity decreases from a maximal value at pH 7.5 to total inhibition at pH 4.5. *Colipase* is secreted as a proenzyme and activated by tryptic digestion, and it enhances lipase activity in the presence of bile salts. The breakdown products of fat digestion have low solubility in water, but the solubility is substantially enhanced after micelle formation by primary bile acids. Cholesterol esterase of pancreatic secretion hydrolyzes the ester bonds to yield free cholesterol. Phospholipases split lecithin to lysolecithins and free fatty acid.

Trypsinogen is converted to trypsin by the intestinal brush border enzyme enterokinase. Active trypsin then converts other zymogens into their active forms. Three forms of trypsinogen have been identified in human pancreatic juice—cationic, anionic, and intermediate—representing more than 20% of the total proteins of pancreatic juice. Proteolytic enzymes secreted as zymogens degrade dietary proteins into amino acids and oligopeptides composed of up to six amino acid residues. To protect the pancreas from autodigestion, the acinar cells also produce trypsin inhibitors. Proteases also play a part in clearing the complex dietary B_{12} with R-binding protein so that intrinsic factor can bind with vitamin B_{12} for further absorption in the terminal ileum.

Pancreatic secretion is regulated by interdependent hormonal and neuronal mechanisms. Secretin and cholecystokinin are the two important duodenal hormonal agents. Neuronal influence occurs through not only hormonal release but also direct cholinergic control of the exocrine pancreas. There is also a paracrine influence in the acinar cells. Pancreatic secretion has three phases: cephalic, gastric, and intestinal. In the gastric phase, vagovagal reflex effects are seen from the stomach to the pancreas. A *pyloropancreatic reflex* mechanism for pancreatic protein secretion is also reported. Food in the antrum of the stomach is a strong stimulant of pancreatic enzymes, in addition to its role in stimulating gastrin production. The most important phase of pancreatic secretion is the intestinal phase.

Secretin is released in response to acid chyme and, to a lesser extent, by sodium oleate and bile acids, entering the duodenum with pH less than 4.5 to 5. Secretin stimulates the ductal epithelium and stimulates the secretion of water and bicarbonate (up to 145 mEq/L concentration). Again, bicarbonate neutralizes gastric acid, protects the duodenal mucosa, and provides optimal pH for pancreatic enzymes to act on food.

Cholecystokinin (CCK) release is stimulated predominantly by products of partial protein digestion, proteases, and peptones and, to a small extent, by long-chain fatty acids and hydrochloric acid. The most important action of CCK through neural mechanisms by activation of cholinergic pathways is the stimulation of acinar cells to produce pancreatic enzymes. The two CCK receptors, CCK1 and CCK2, mediate the secretory responses of the acinar cells to CCK. There is a negative feedback inhibition of CCK release by pancreatic proteases in the duodenum. Administration of trypsin or other proteases suppresses duodenal CCK release, thereby inhibiting pancreatic enzyme secretion. A CCK-releasing factor is reported to be the mediator.

CCK primarily stimulates the acinar cells by acting on neuronal pathways. Acetylcholine is a major stimulant. Vagotomy or administration of atropine greatly reduces the pancreatic enzyme response to food in the intestine. Acinar cells do not contain CCK receptors.

Somatostatin is produced by the delta cells of the pancreas and is a potent inhibitor of pancreatic bicarbonate and enzyme secretion. Another inhibitory agent is *calcitonin gene–related peptide*, localized in endocrine or paracrine cells and nerve fibers of the pancreas. *Pancreatic polypeptide* produced by the pancreatic endocrine cells also inhibits pancreatic secretion, as does *peptide YY*, released from the ileum or colonic mucosa in response to free fatty acids and carbohydrates in the ileum. *Leptin* is a 167–amino acid protein secreted by adipose tissue, muscle, reproductive tract, and stomach that has an inhibitory role in CCK-stimulated secretion.

Evidence indicates that *insulin* and *glucagon* influence pancreatic enzyme synthesis and release. Distinct insulin receptors are identified in the acinar cells. Glucagon exerts an inhibitory effect on pancreatic secretion.

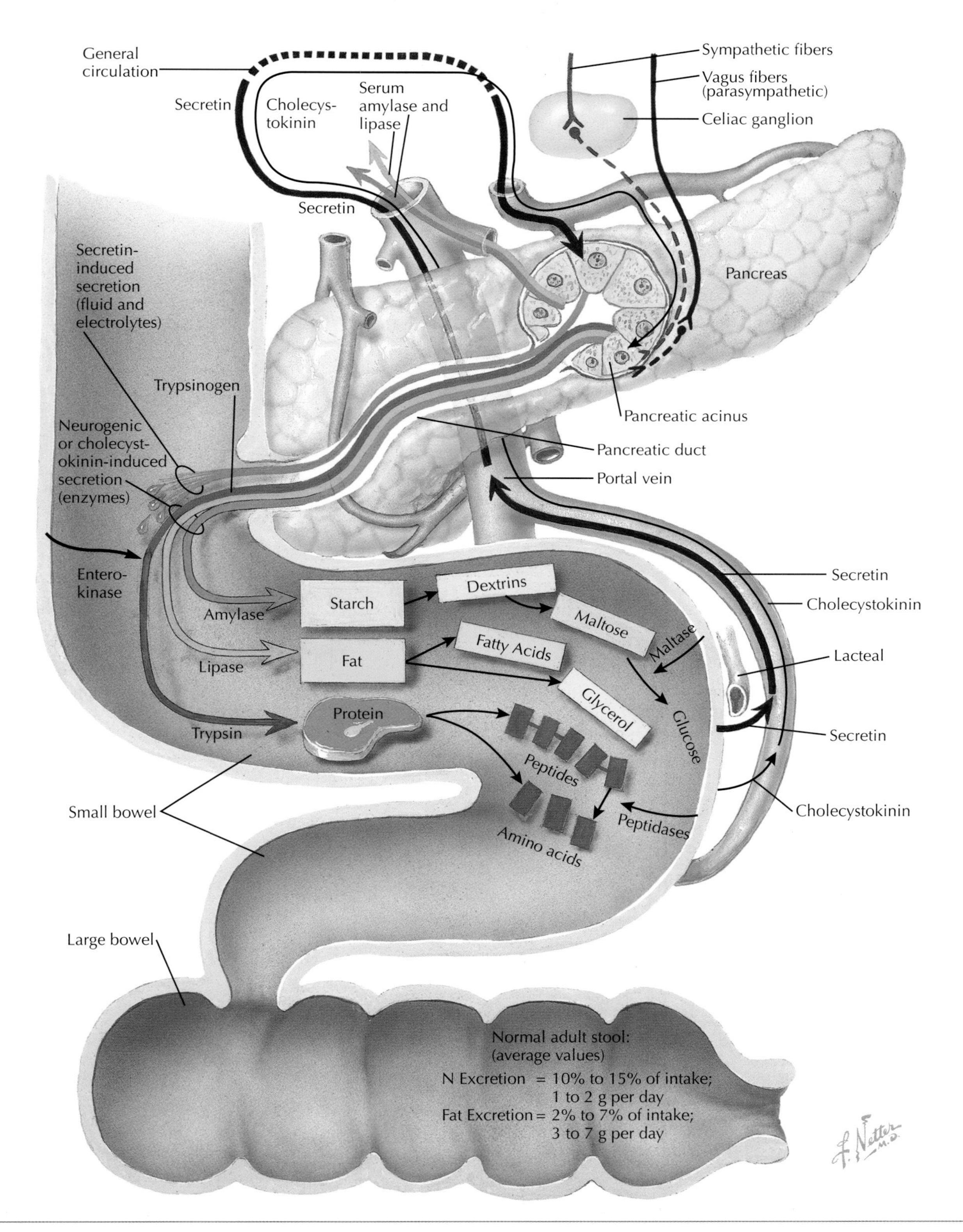

Figure 189-1 *Normal Secretory Functions of the Pancreas.*

ADDITIONAL RESOURCES

Asakawa A, Inui A, Yuzuriha, et al: Characterization of the effects of pancreatic polypeptide in the regulation of energy balance, *Gastroenterology* 124:1325-1329, 2003.

Pandol SJ: Neurohumoral control of exocrine pancreatic secretion, *Curr Opin Gastroenterol* 20:435-436, 2004.

Cystic Fibrosis

190

C. S. Pitchumoni

Cystic fibrosis (CF) is the most common autosomal recessive disease among white populations, with a frequency of 1 in 2000 to 3000 live births. The carrier rate in white persons is 1 in 25. Improved management of CF has led to the current survival of almost 37 years.

The classic form of CF is characterized clinically by sinopulmonary disease, *exocrine pancreatic insufficiency*, and male infertility and biochemically by abnormal sweat electrolytes (**Fig. 190-1**). Depending on the degree of genetic mutation, the disease may have the classic picture or features involving only the pancreas. Approximately 5% of individuals with CF are diagnosed after age 16 years because they lack classic symptoms.

Mutations are found in the *CFTR* (cystic fibrosis transmembrane conductance regulator) gene, located on the long arm of chromosome 7. The resulting abnormalities in epithelial ion and water transport are associated with derangements in airway mucociliary clearance and other cellular functions related to normal cell biology. The genetic defect causes increased sodium chloride content in sweat and increased electrical potential difference across the respiratory epithelium. The secretions become viscid, sticky, and dry and obstruct the ducts, leading to dysfunction at the organ level. In the pancreas, the secretions precipitate within the ducts, causing blockage and duct dilatation. Epithelial cells lacking *CFTR* fail to produce the normal level of the ubiquitous regulator nitric oxide, contributing to increased sodium reabsorption, enhanced inflammatory response, and inefficient destruction of bacteria.

CLINICAL PICTURE

During infancy, CF manifests as *meconium ileus* (obstruction of distal ileum or proximal colon with thickened viscid meconium), meconium peritonitis or prolonged jaundice caused by steatosis, and biliary obstruction caused by thick mucus. After infancy, failure to thrive associated with frequent, bulky, foul-smelling oily stools should alert to the possibility of CF secondary to exocrine pancreatic insufficiency. Deficiencies of fat-soluble vitamins A, D, E, and K and prolonged prothrombin time occur.

Recurrent bronchopulmonary infections in infancy and early childhood and colonization of the lungs with *Haemophilus influenzae, Staphylococcus aureus, Pseudomonas aeruginosa*, and *Burkholderia cepacia* are important features. Severe bronchiectasis, massive hemoptysis, and spontaneous pneumothorax are complications. A high incidence of nasal polyps is a classic association.

Intestinal obstruction near the ileocecal junction (distal intestinal obstruction syndrome) from accumulation of solid stool and intussusception may occur. Rectal prolapse is common. In children and adults, recurrent small bowel obstruction termed *meconium ileus equivalent* occurs. Other gastrointestinal problems include elevation of serum liver enzymes, cirrhosis with portal hypertension, and increased risk for cholelithiasis. Computed tomography may show "fatty replacement" of the pancreas.

Mutation of the *CFTR* gene is a risk factor for chronic pancreatitis, often considered "idiopathic." Although all CF patients are at risk for pancreatitis, the majority of patients with classic CF, the "pancreas insufficient" (PI) group, develop pancreatitis in utero, leading to pancreatic destruction by early childhood, and they are at low risk for subsequent attacks of acute pancreatitis. However, the "pancreas sufficient" (PS) group is particularly vulnerable to attacks of acute and chronic pancreatitis (appears to be idiopathic) later in life, explaining the frequent incidence of pancreatitis in adult CF patients.

Hepatic steatosis and gallstones, focal biliary stricture, and common bile duct (CBD) obstructions are the hepatobiliary manifestations. Portal hypertension, splenomegaly, esophageal varices and bleeding, and hepatic encephalopathy develop in advanced disease.

DIAGNOSIS

A clinical picture consistent with CF with laboratory evidence of *CFTR* dysfunction confirms the diagnosis. In many Western countries, newborns are screened for CF by measuring immunoreactive trypsinogen levels in blood, which are elevated in infants with CF but normal in adult PS patients with CF.

The sweat electrolyte test, the preferred test for the diagnosis of CF, should be performed only in a specialized laboratory by experienced personnel. An electrolyte concentration greater than 60 mEq/L is confirmatory.

Although there are hundreds of mutations of the *CFTR* gene, most U.S. laboratories screen for 20 to 30 of the common mutations.

TREATMENT AND MANAGEMENT

Management of CF consists of nutritional support, treatment of exocrine insufficiency, and appropriate therapy for bronchopulmonary infections and other complications. Proper planning for adulthood, education, and occupation are other important considerations.

Along with a high-calorie, balanced diet, supplementary oral pancreatic enzyme therapy in adequate dose is important to manage steatorrhea. To correct deficiencies of fat-soluble vitamins, a double-dose multivitamin preparation should be administered, along with vitamins E and K in water-soluble form. Medium-chain triglyceride supplementation is particularly needed in infants. The use of pancreatic preparations with a high concentration of proteolytic enzymes is reported to cause colonic stricture in children.

Aggressive management of bronchopulmonary problems is important. Lung or heart-lung transplantation is an option for a few patients. Ursodeoxycholic acid therapy improves bile flow and ameliorates liver function abnormalities.

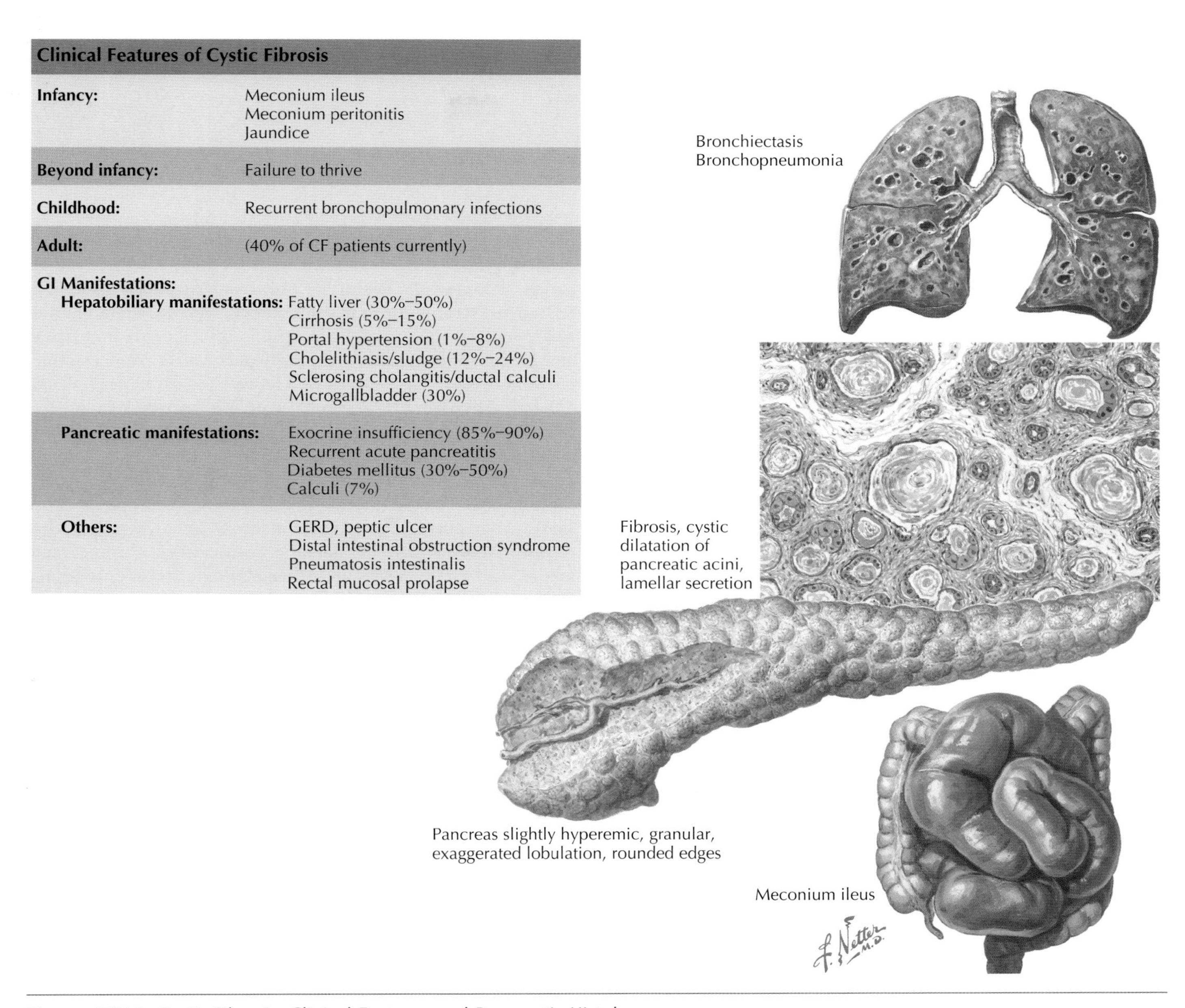

Clinical Features of Cystic Fibrosis	
Infancy:	Meconium ileus Meconium peritonitis Jaundice
Beyond infancy:	Failure to thrive
Childhood:	Recurrent bronchopulmonary infections
Adult:	(40% of CF patients currently)
GI Manifestations:	
Hepatobiliary manifestations:	Fatty liver (30%–50%) Cirrhosis (5%–15%) Portal hypertension (1%–8%) Cholelithiasis/sludge (12%–24%) Sclerosing cholangitis/ductal calculi Microgallbladder (30%)
Pancreatic manifestations:	Exocrine insufficiency (85%–90%) Recurrent acute pancreatitis Diabetes mellitus (30%–50%) Calculi (7%)
Others:	GERD, peptic ulcer Distal intestinal obstruction syndrome Pneumatosis intestinalis Rectal mucosal prolapse

Figure 190-1 *Cystic Fibrosis: Clinical Features and Pancreatic Histology.*

COURSE AND PROGNOSIS

Overall, with early diagnosis and aggressive management, the prognosis for CF has greatly improved, and many patients survive until the third or even fourth decade of life. All patients with at least one severe mutation of the *CFTR* gene should be longitudinally assessed for progressive pancreatic dysfunction.

Other congenital diseases of the pancreas of clinical importance include pancreas divisum (Chapter 188) and hereditary pancreatitis (Chapter 194).

ADDITIONAL RESOURCES

Boyle MP: Adult cystic fibrosis, *JAMA* 298:1787-1793, 2007.

Dray X, Bienvenu T, Desmazes-Dufeu D, et al: Distal intestinal obstruction syndrome in adults with cystic fibrosis, *Clin Gastroenterol Hepatol* 2:498-503, 2004.

Krysa J, Steger A: Pancreas and cystic fibrosis: the implications of increased survival in cystic fibrosis, *Pancreatology* 7:447-450, 2007.

Robertson MB, Choe K, Joseph PM: Review of the abdominal manifestations of cystic fibrosis in the adult patient, *Radiographics* 26:679-690, 2006.

Acute Pancreatitis: Etiology and Clinical Picture

191

C. S. Pitchumoni

Acute pancreatitis, an inflammatory disorder of the pancreas characterized clinically by abdominal pain and biochemically by elevated levels of serum amylase and lipase, may present as a mild, self-remitting disorder or as a fulminant disease with multiple–organ system failure (**Fig. 191-1**).

Common terms used to describe acute pancreatitis include those based on the Atlanta System of classification (**Box 191-1**). *Acute interstitial pancreatitis* (edematous pancreatitis) and *acute necrotizing pancreatitis* are the two histologic types of the disease based on severity. Hemorrhage is only a complication of severe acute pancreatitis; thus the term "hemorrhagic pancreatitis" is no longer used.

ETIOLOGY

Alcoholism and gallstone disease account for almost 80% of cases of acute pancreatitis (**Box 191-2**). Gallstone pancreatitis is caused by migration of stones, biliary sludge, or microlithiasis from the gallbladder through the common bile duct (CBD) to the ampulla of Vater and then into the duodenum. Pancreatitis results from the transitory presence of the stone near the duodenal outlet of the pancreatic duct. The pathogenesis of alcoholic pancreatitis is unclear and may involve direct toxic effects on the acinar cells and alterations in the composition of pancreatic secretion causing protein plugs (small duct hypothesis).

Traumatic acute pancreatitis may develop after penetrating or blunt injury to the abdomen or after abdominal surgery. Acute pancreatitis is the most common and feared complication of endoscopic retrograde cholangiopancreatography (ERCP), occurring in 2% to 8% of patients. Risk factors include female gender, younger age, history of previous post-ERCP pancreatitis, normal serum bilirubin, sphincter of Oddi dysfunction, multiple pancreatic duct contrast injections, difficulty in cannulation, and precut sphincterotomy (in particular, minor papilla sphincterotomy); these can be additive. Hyperlipidemia (types 1, 4, and 5; familial hyperlipoproteinemia) with triglyceride level greater than 1000 mg/dL is a rare cause of acute pancreatitis. Many drugs, including azathioprine, sulfonamides, sulindac, tetracycline, didanosine, pentamidine, estrogens, furosemide, acetylsalicylic acid (ASA) compounds, and valproic acid, rarely cause acute pancreatitis. Hypercalcemia secondary to hyperparathyroidism or other causes, diabetic ketoacidosis, chronic renal failure, and hemodialysis are other etiologic factors. Anatomic abnormalities of the pancreatic duct system, particularly *pancreas divisum*, are infrequent causes. Ampullary tumors and pancreatic

Box 191-1 Terminology for Acute Pancreatitis

Mild acute pancreatitis: Uneventful recovery in 3 to 5 days without organ dysfunction.

Severe acute pancreatitis: Associated with multiple–organ system failure and local complications such as necrosis, abscess, and pseudocyst.

Sterile pancreatic necrosis: Diffuse or focal areas of nonviable parenchyma.

Infected pancreatic necrosis: Nonviable parenchyma that is infected; diagnosed by needle aspiration.

Acute pancreatic fluid collections: Occur in severe acute pancreatitis and noted on CT; no well-defined wall seen.

Pancreatic pseudocyst: Collection of pancreatic juice and inflammatory exudate enclosed by a wall of fibrous or granulation tissue.

Pancreatic abscess: Circumscribed intraabdominal collection of pus in the pancreatic region.

Modified from Bradley EL III: Arch Surg *128:586-590, 1993.*

Box 191-2 Etiologic Associations for Acute Pancreatitis

- Chronic alcoholism
- Biliary tract disease
- Drug-induced causes
- Infectious agents (viral, bacterial, parasitic)
- Traumatic causes
- Hypertriglyceridemia
- Hypercalcemia
- Post-ERCP*
- Idiopathic causes
- Postsurgical causes
- Toxic causes: exposure to organophosphorus
- Anatomic abnormality: pancreas divisum
- Genetic causes
- Vascular diseases
- Scorpion bite
- Miscellaneous causes
  - Eating disorders: bulimia, anorexia nervosa, refeeding pancreatitis
  - Diabetic ketoacidosis (DKA)
  - Pancreatic cancer
  - Chronic renal failure
  - Hypothermia

**Endoscopic retrograde cholangiopancreatography.*

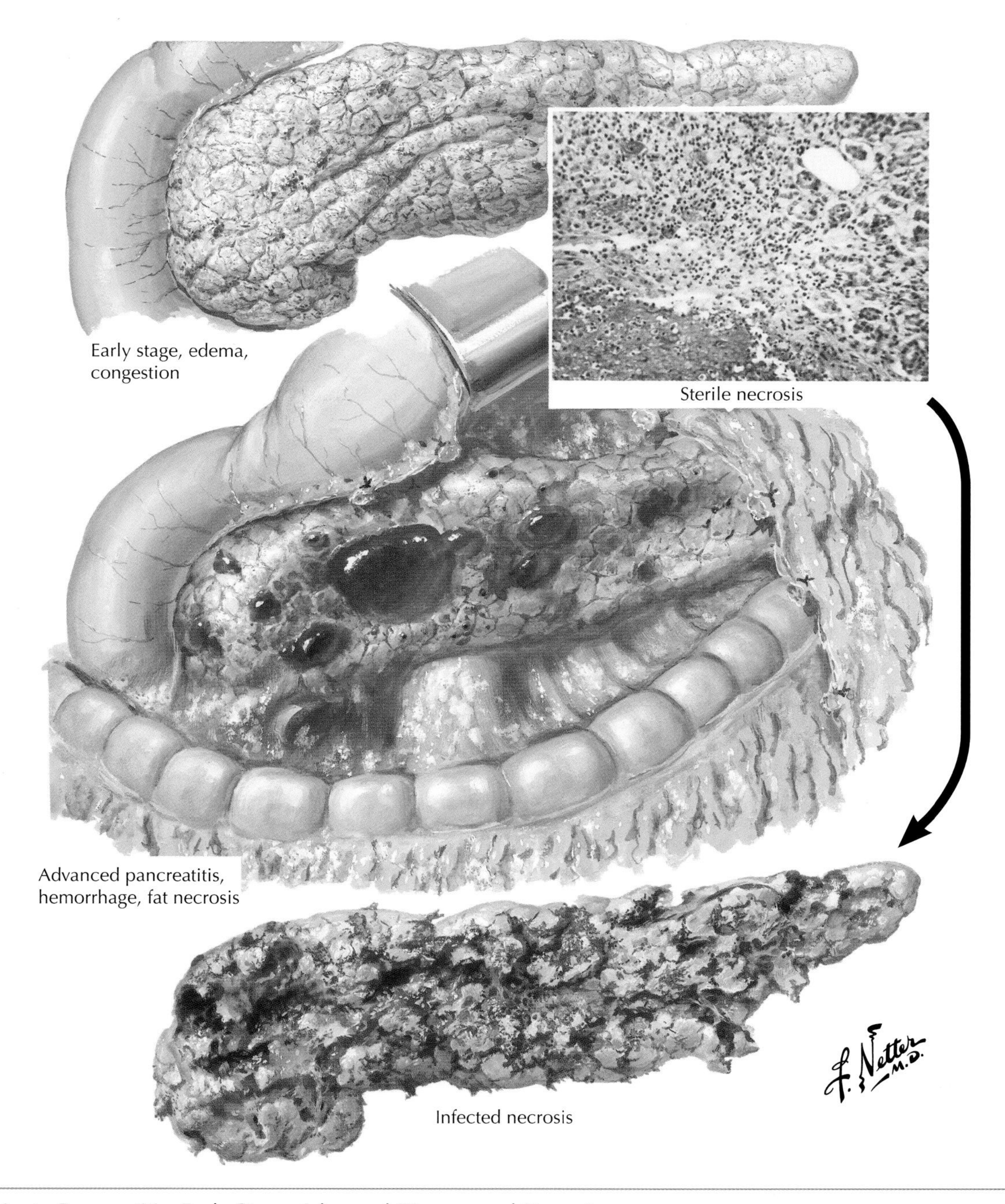

Figure 191-1 *Acute Pancreatitis: Early Stage, Advanced Disease, and Necrosis.*

cancer may present as acute pancreatitis. At least one third of the cases of idiopathic pancreatitis are secondary to microcrystals. Genetic abnormalities, particularly those associated with *CFTR*, *PRSS1*, and *SPINK1* genes, are currently proposed mechanisms to describe idiopathic pancreatitis.

Premature intracytoplasmic activation of trypsinogen to trypsin is considered the initial pathogenetic mechanism of acute pancreatitis. As a consequence, active phospholipase A_2, elastase, and lipase have been proposed to play major roles in the autodigestion of the pancreatic acinar cell, which is characteristic of the disease. A number of active enzymes, vasoactive peptides, proinflammatory cytokines, nitric oxide, and unopposed free radicals aggravate the disease and participate in the systemic complications of acute pancreatitis.

CLINICAL PICTURE

The cardinal manifestation of acute pancreatitis is sudden onset of epigastric abdominal pain that is moderate to severe and that lasts for many hours to days. The characteristic radiation of the pain straight to the back is moderately relieved by sitting forward or lying down on one side with knees flexed. Nausea, vomiting, low-grade fever, and tachycardia are associated features.

Physical examination of the abdomen in early stages may show mild to moderate tenderness on palpation with hypoactive or even absent bowel sounds, depending on the severity of illness. Abdominal rigidity is not a feature of acute pancreatitis. A palpable mass 2 or more weeks after onset may represent an inflammatory mass, infected necrosis, or early pseudocyst. Examination of the chest may reveal findings of pleural effusion, more often on the left side but occasionally on both sides. Other findings may include atelectasis, pneumonia, or congestive heart failure. Scleral icterus; subcutaneous fat necrosis over the buttocks, trunk, and extremities; lipemia retinalis; and eruptive xanthomas (indicative of preexisting hyperlipidemias) are rare. Ecchymotic discoloring of the flanks (Grey Turner sign) or periumbilical region (Cullen sign) are rare and nonspecific.

Physical findings of acute pancreatitis also vary depending on the etiologic factor (e.g., alcoholic vs. biliary or hyperlipidemic), degree of severity, stage of illness (at onset vs. 2 weeks later), and complications (cholangitis, pancreatic necrosis).

ADDITIONAL RESOURCES

Bradley EL III: A clinically based classification system for acute pancreatitis: summary of the International Symposium on Acute Pancreatitis, Atlanta, 1992, *Arch Surg* 128:586-590, 1993.

Cheng CL, Sherman S, Watkins J, et al: Risk factors for post-ERCP pancreatitis: a prospective multicenter study, *Am J Gastroenterol* 101:139-147, 2006.

DiMagno MJ, DiMagno EP: New advances in acute pancreatitis, *Curr Opin Gastroenterol* 23:494-501, 2007.

Acute Pancreatitis: Diagnosis, Treatment, and Prognosis

192

C. S. Pitchumoni

DIAGNOSIS

The initial evaluation of acute pancreatitis includes a number of conditions in the differential diagnosis. Uncomplicated cholecystitis, perforated peptic ulcer disease, splenic infarction, intestinal infarction, and ectopic pregnancy may mimic acute pancreatitis. Serum *amylase* or *lipase* levels at least three times above normal values on initial evaluation of a patient with recent onset of abdominal pain is almost diagnostic of acute pancreatitis and is unlikely to be secondary to other causes. **Box 192-1** lists other causes of hyperamylasemia.

It is important to note that in hypertriglyceridemic pancreatitis, serum amylase levels may be normal or only modestly elevated. Although high levels are often considered specific, falsely elevated lipase levels are as common as falsely elevated amylase levels. In macroamylasemia, the molecular weight of serum amylase is greater than 150,000 daltons (normal range, 50,000-55,000). Renal clearance is affected, and serum amylase level is disproportionately elevated above the near-normal or below-normal urinary amylase level. Familial benign hyperamylasemia and hyperlipasemia have been recently reported.

The initial evaluation includes hemoglobin/hematocrit (Hb/Hct) levels, leukocyte count, and aspartate transaminase, alanine transaminase, and lactate dehydrogenase levels. Within 1 day, it is appropriate to repeat Hb/Hct and leukocyte count as well as blood urea nitrogen, creatinine, calcium, and albumin levels, to help determine the prognosis.

Box 192-1 Selected Causes of Elevated Serum Amylase Levels (Hyperamylasemia)

- Pancreatitis
- Diabetic ketoacidosis (DKA)
- Mumps
- Perforated bowel
- Intestinal obstruction
- Cholecystitis
- Appendicitis
- Peritonitis
- Inflammatory bowel disease
- Renal failure
- Ruptured ectopic pregnancy
- Ovarian tumor, cyst
- Salpingitis
- Post-ERCP*
- Pancreatic cancer

**Endoscopic retrograde cholangiopancreatography.*

IMAGING PROCEDURES

It is important to perform initial chest (posteroanterior view) and flat-plate and upright abdominal radiography. Chest x-ray film may show elevation of the diaphragm, left-sided pleural effusion (occasionally bilateral), and atelectasis. Evidence of pericardial effusion, congestive heart failure, or acute respiratory distress syndrome may be seen later in the disease course. Abdominal radiography is not diagnostic of acute pancreatitis. Its value is in excluding other causes of acute abdominal pain associated with or without elevated serum amylase levels.

Abdominal ultrasound is a valuable initial diagnostic procedure in all patients with acute pancreatitis. Gallbladder stones, choledocholithiasis, dilated common bile duct (CBD), and sometimes enlargement of the pancreas can be identified. When the diagnosis of acute pancreatitis is clear and appears to be uncomplicated and mild, no indication exists for computed tomography (CT) within the first 72 hours of admission. The three indications for CT are (1) to clarify the diagnosis and exclude conditions such as mesenteric infarction, (2) to grade the severity of acute pancreatitis, and (3) to evaluate pancreatic and extrapancreatic intraabdominal complications (**Fig. 192-1**). Rapid-sequence, contrast-enhanced CT with 5-mm slicing is highly recommended as the best imaging study to visualize the inflammatory changes associated with acute pancreatitis. Magnetic resonance imaging and magnetic resonance cholangiopancreatography provide good evaluation of the biliary system to assess duct size and to rule out a stone in the CBD. The diagnosis of sterile or infected pancreatic necrosis is differentiated using CT, especially with fine-needle aspiration of the necrotic area.

TREATMENT AND MANAGEMENT

Principles of therapy include managing pain, providing *rest to the pancreas* functionally, and correcting fluid and electrolyte imbalances. Abdominal pain is relieved by meperidine, 25 to 50 mg intramuscularly every 4 to 6 hours as needed. Repeated doses of meperidine rarely causes seizures and many physicians prefer hydromorphone 0.2 to 0.6 mg (based on a 70-kg patient) every 2 to 3 hours. Patients with prior opiate exposure may tolerate higher doses. Only patients with severe acute pancreatitis need a nasogastric tube to keep their stomach empty. There is no benefit in administering an H_2-receptor antagonist or a proton pump inhibitor.

The role of prophylactic antibiotic therapy in patients with acute pancreatitis and sterile necrosis is controversial. Prophylactic antibiotics are considered for those with severe acute pancreatitis, dilatation of the CBD, and an impacted stone at the ampulla on sonography. Antibiotic therapy is given to treat complications such as pulmonary and biliary infections and infected pancreatic necrosis. Antibiotics with good diffusion capacity into the pancreas include imipenem, ofloxacin, metronidazole, and mezlocillin.

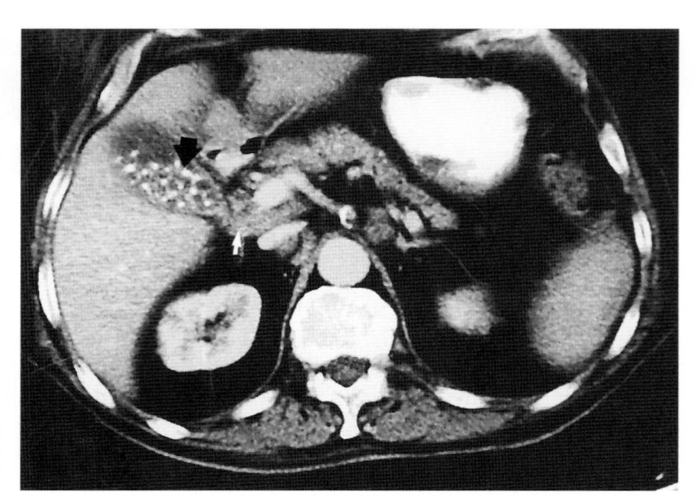

Mild biliary pancreatitis. Note stones in gallbladder *(black arrow)* and swollen pancreas *(white arrow)*.

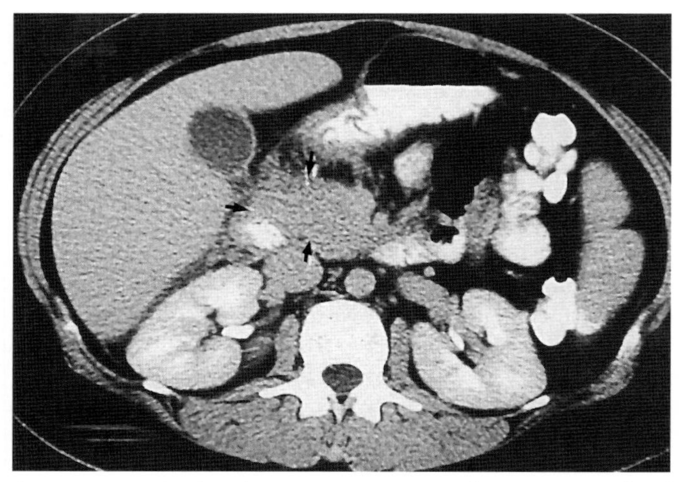

Acute pancreatitis of moderate severity. Note swollen pancreas outlined by *arrows*.

Figure 192-1 *Acute Pancreatitis: CT Images of Mild and Moderately Severe Disease.*

Oral intake of all food is discontinued early, until abdominal pain is completely resolved. After 3 or 4 days of intravenous fluids, peripheral parenteral nutrition (PPN), and tube feedings are options for providing continuous nutrition if pain in the abdomen continues. Current evidence indicates that intrajejunal tube feeding is preferred over total parenteral nutrition (TPN) because it is less expensive, avoids TPN complications, and bypasses the duodenum, thereby avoiding stimulation of the pancreas. In addition, intrajejunal tube feeding helps maintain the healthy bacterial flora of the gut and may help reduce translocation of colonic bacteria to the pancreas.

Patients with uncomplicated gallstone pancreatitis should undergo elective cholecystectomy in the same hospitalization soon after resolution of the acute episode of pancreatitis. Patients with severe acute pancreatitis should be managed in the intensive care unit from the beginning. Cholecystectomy may be electively performed later.

Peritoneal lavage and octreotide therapy are of no proven benefit. Urgent endoscopic retrograde cholangiopancreatography with papillotomy and stone extraction is indicated in those with demonstrated dilated CBD with evidence of an impacted stone and impending ascending cholangitis.

Box 192-2 Ranson Criteria for Determination of Poor Prognosis in Acute Pancreatitis

Admission

Age >55 years

White blood cell (WBC) count >16,000 cells/mm^3

Lactate dehydrogenase (LDH) >350 IU/L

Aspartate transaminase (AST) >250 IU/L

Glucose >200 mg/dL

Initial 48 Hours

Hematocrit (Hct) decrease >10%

Blood urea nitrogen (BUN) increase >5 mg/dL

Calcium <8 mg/dL

Oxygen partial pressure (Po_2) <60 mm Hg

Base deficit >4

Estimated fluid sequestration >6 L

COURSE AND PROGNOSIS

Although the majority of cases of acute pancreatitis take a mild course, almost 20% of patients have severe disease. Overall mortality in acute pancreatitis is 5%, but in severe cases may be 20% or higher. Death occurs either in 1 or 2 days of onset or 1 or 2 weeks later, from multiple–organ system failure. Telltale evidence of poor prognosis is a patient whose appearance shows toxicity and who has a distended and silent abdomen, acidosis, uremia, tachypnea, confusion, irritability, and evidence of shock. Age over 65 years, obesity, elevated Hb/Hct on admission, persistent organ failure, "early-onset" organ failure within the first 3 days, and pleural effusion are markers of poor prognosis.

There are many scoring systems to assess the prognosis. The criteria proposed by Ranson (**Box 192-2**) or the Acute Physiology and Chronic Health Evaluation II and other single markers of prognosis are helpful. Rapid decrease in serum albumin level, increase in C-reactive protein level (>150 mg/L), elevated urinary trypsinogen activation peptide (TAP), and elevated interleukin-6 serum levels are useful tests.

ADDITIONAL RESOURCES

Banks PA, Freeman ML: Practice guidelines in acute pancreatitis, *Am J Gastroenterol* 101:2379-2400, 2006.

Brown A: Prophylactic antibiotic use in severe acute pancreatitis: hemlock, help or hype? *Gastroenterology* 126:1195-1198, 2004.

Ranson JHC, Rifkind KM, Rosen OF: Objective early identification of severe acute pancreatitis, *Am J Gastroenterol* 61:443-451, 1974.

Rau BM, Bothe A, Kron M et al: Role of early multisystem organ failure as major risk factor for pancreatic infections and death in severe acute pancreatitis, *Clin Gastroenterol Hepatol* 4:1053 -1061, 2006.

Whitcomb D: Acute pancreatitis, *N Engl J Med* 354:2142-2150, 2006.

Working Party of the British Society of Gastroenterology, Association of Surgeons of Great Britain and Ireland, Pancreatic Society of Great Britain and Ireland: UK guidelines for the management of acute pancreatitis, *Gut* 54(suppl 3):iii1-iii9, 2005.

Acute Pancreatitis: Complications

193

C. S. Pitchumoni

SYSTEMIC COMPLICATIONS

Systemic complications in patients with acute pancreatitis occur within 2 to 7 days of admission and include hypocalcemia, hyperglycemia, hypoalbuminemia (three easily available markers of severity), hyperlipidemia, and coagulation abnormalities (**Box 193-1**). Systemic complications include shock (systolic blood pressure <80 mm Hg), respiratory distress (Po_2 <60 mm Hg), and renal distress (serum creatinine >1.4 mg/dL). Organ failure unresponsive to intensive care unit (ICU) management in the first week is closely related to pancreatic infections and death.

Hypocalcemia, with calcium levels less than 8 mg/dL, correlates strongly with poor prognosis. Calcium soap formation in areas of fat necrosis, decreased level secondary to hypoalbuminemia, sequestration of calcium in tissues as calcium/free fatty acid complexes, and influence of parathormone, calcitonin, and glucagon are probable factors in the pathogenesis of acute pancreatitis.

Hypoxemia is frequently seen in acute pancreatitis. Pleural effusion, particularly on the left side, is a marker of severity. Usually, pleural effusion resolves without any specific treatment. Pulmonary infiltrates, left lower lobe consolidation, and atelectasis are frequent radiologic findings. The pathogenesis of *acute respiratory distress syndrome* is multifactorial, and patients often require ICU support. It is important to avoid fluid overload and institute endotracheal intubation promptly and to provide positive end-expiratory pressure ventilation when necessary, as well as appropriate hemodynamic and nutritional support.

Box 193-1 Complications of Acute Pancreatitis

Systemic

Pulmonary: hypoxemia, pleural effusion, acute respiratory distress syndrome

Cardiac: shock, pericardial effusion, electrocardiograph changes, arrhythmias

Hematologic: disseminated intravascular coagulation, thrombotic thrombocytopenic purpura

Renal: azotemia, oliguria, myoglobinuria

Metabolic: hypocalcemia, hyperglycemia, acidosis, hypoalbuminemia

Central nervous system: psychosis, Purtscher retinopathy

Peripheral: rhabdomyolysis, fat necrosis, bone necrosis, arthritis

Intraabdominal

Pancreatic

Necrosis: (sterile vs. infected; nonviable parenchyma)

Fluid collections: peripancreatic or pseudocyst (infection, rupture, hemorrhage, abscess)

Local Nonpancreatic

Pancreatic ascites: high-protein, high-amylase ascites

Contiguous involvement: gastrointestinal bleeding, thrombosis of splenic vein, colonic infarction, lower GI bleeding

Obstructive jaundice

INTRAABDOMINAL COMPLICATIONS

Pancreatic and other intraabdominal complications occur 7 to 21 days after the onset of acute pancreatitis, while treatment for the acute attack is in progress (**Box 193-1**).

Pancreatic necrosis, both sterile and infected, may be characterized clinically by fever, leukocytosis, and persistent abdominal pain. Pancreatic necrosis may be diagnosed as nonenhanced parenchyma in contrast-enhanced computed tomography (CT) examination. CT-guided fine-needle aspiration, aspiration of pus, Gram stain, and culture are needed to diagnose infected necrosis (**Fig. 193-1**). In addition to antibiotics, therapy requires surgical exploration, removal of necrotic tissue, and retroperitoneal irrigation. The management of sterile pancreatic necrosis is purely supportive in most patients.

Pancreatic pseudocysts can complicate the course of both acute and chronic pancreatitis. Although fluid collections are common in an acute episode of pancreatitis, they usually disappear. In approximately 25% of patients, however, fluid collections develop into pseudocysts of the pancreas. Approximately 60% of the pseudocysts resolve spontaneously (see Fig. 193-1). Symptomatic pseudocysts, cysts that grow during observation, and cysts compressing adjacent viscera should undergo surgical, endoscopic, or percutaneous drainage. Complications of a pseudocyst include enlargement and pressure on adjacent organs, rupture, pancreatic ascites, pleural effusions, and aneurysm formation in blood vessels and hemorrhage. Communication with the pancreatic ductal system occurs in about 25% of pseudocysts. Endoscopic, radiologic, and surgical management approaches are available. Endoscopic transmural drainage with endoscopic ultrasound guidance is a major advancement. Transpapillary drainage with pancreatic sphincterotomy requires demonstration of a communication between the pseudocyst and the pancreatic duct, which may require an Endoscopic retrograde cholangiopancreatography (ERCP) study.

Pancreatic abscess is a loculated, walled-off collection of pus resulting from liquefaction of necrotic areas or secondary infection of a pseudocyst. Abdominal CT and fine-needle aspiration of the suspected area establish the diagnosis. Surgical drainage is required. Percutaneous drainage of pus may be helpful. To drain collections, multiple catheters or large-bore catheters up to 30-French are needed. After appropriate therapy, the prognosis is better than for infected pancreatic necrosis.

In *pancreatic ascites*, a cyst or a ruptured duct leaks into the peritoneal cavity. Ascitic fluid shows elevated amylase activity (usually >1000 U/dL; range, 200-100,000 U/dL) and protein (>2.5 g/dL; range, 2-5.7 g/dL). Medical management is to allow

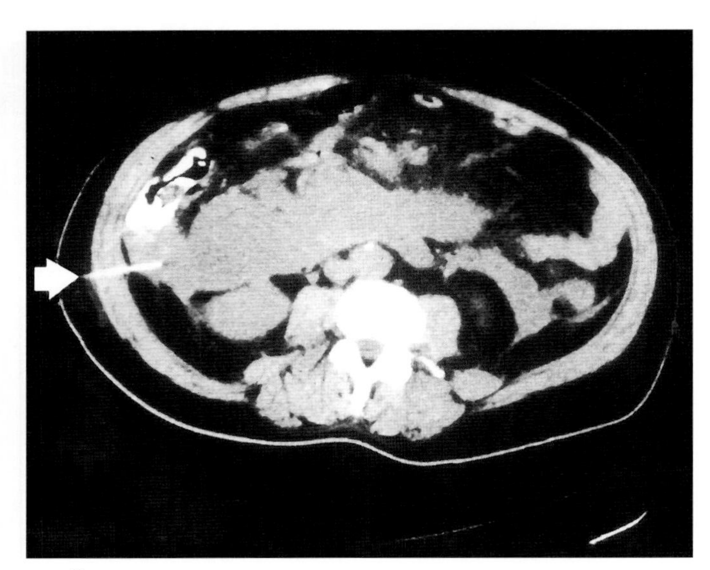

Needle *(arrow)* to aspirate necrotic pancreas.

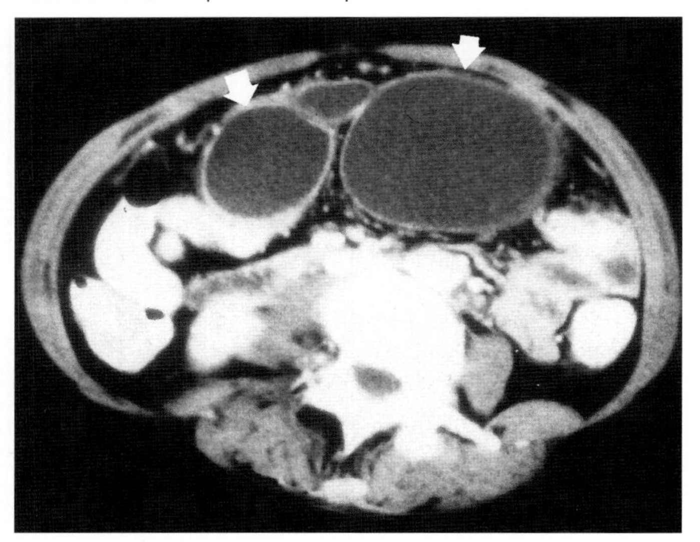

Large pseudocysts (see *arrows*).

Figure 193-1 *Complications of Acute Pancreatitis: CT Images of Necrosis and Pseudocysts.*

the patient nothing by mouth and to provide total parenteral nutrition (TPN) and intravenous octreotide therapy. Surgery may be needed to correct leakage from a cyst or a duct identified on ERCP.

ADDITIONAL RESOURCES

Bai Y, Gao J, Zou D, et al: Prophylactic antibiotic cannot reduce infected pancreatic necrosis and mortality in acute necrotizing pancreatitis: evidence from a meta-analysis of randomized controlled trials, *Am J Gastroenterol* 103:104-110, 2008.

Balthazar EJ: Complications of acute pancreatitis, *Radiol Clin North Am* 40:1211-1227, 2002.

Ho HS: Clinical usefulness of a treatment algorithm for pancreatic pseudocysts, *Gastrointest Endosc* 67:253-254, 2008.

Chronic Pancreatitis: Disease Forms and Clinical Picture

194

C. S. Pitchumoni

The cardinal manifestations of chronic pancreatitis are recurrent or persistent abdominal pain that lasts for months to years, accompanied by diabetes, steatorrhea, and pancreatic calculi. Morphologically, the disease is characterized by destruction and loss of exocrine parenchyma that may be focal, segmental, or diffuse, as well as by fibrosis of the pancreas (**Fig. 194-1**). These changes may be associated with strictures and dilatation of segments of the pancreatic duct and with intraductal protein plugs or calculi.

The epidemiology of chronic pancreatitis parallels the prevalence of alcohol abuse in the community. In the United States, almost 75% of cases of chronic pancreatitis are associated with chronic alcoholism, which is thus the most important etiologic factor.

Although the pathogenesis of pancreatic injury is still speculative, current data indicate that acinar cells are injured first, with subsequent development of secretory changes and morphologic alterations such as fibrosis, ductular abnormalities, and stone formation.

Toxic metabolites, unopposed free-radical injury, and genetic mutations that promote premature activation of trypsinogen to trypsin within the acinar cell are associated with the pathogenesis of chronic pancreatitis. Three genes are currently believed to play a major role: cationic trypsinogen gene (*PRSS1;* 7q35), *CFTR*, and *SPINK1*.

FORMS OF DISEASE

In *chronic alcoholism*, onset of chronic pancreatitis usually occurs after 15 or more years of 80 to 150 g of daily alcohol consumption. Individual susceptibility varies, and only about 10% of alcoholics who drink heavily develop clinical pancreatitis, but a larger number may have histologic changes in the pancreas. The susceptibility for pancreatic injury is increased with cigarette smoking and genetic abnormalities.

Tropical calculous pancreatitis (TCP) is a nonalcoholic form of chronic pancreatitis that occurs mostly in children and young adults of many developing nations. Cardinal clinical manifestations of TCP are recurrent abdominal pain in childhood, followed by onset of diabetes mellitus a few years later. Diabetes, almost an inevitable consequence of TCP (in contrast to alcoholic pancreatitis, in which diabetes develops in 30%-60% of patients), is characteristically brittle. The etiologic factors for TCP are unknown, but genetic factors (mainly *SPINK1* mutations) in association with other environmental factors are suspected.

Hereditary pancreatitis (HP) is one of the most common causes of recurrent pancreatitis in children, who may be as young as 7 years of age when HP is identified. Relatives of patients with HP may have painless disease, but more often they have pancreatic calculi, steatorrhea, or diabetes. A major milestone in the history of HP is the recent discovery of an abnormal gene and the feasibility of identifying the genetic abnormality. Mutations in the cationic trypsinogen gene *(R1224, N291)* cause the disease in 60% to 70% of kindreds. The most dreaded complication of HP is a 50- to 70-fold increased risk for pancreatic cancer.

Autoimmune pancreatitis (AIP) is a curable form of chronic pancreatitis that affects middle-aged men, in association with other autoimmune disorders (Sjögren syndrome, psoriasis, inflammatory bowel disease). AIP is characterized by elevated IgG_4 levels in serum. Extrapancreatic involvement is common and includes sclerosing cholangitis and lymphoplasmacytic forms of kidney, salivary gland, and lung disease. Typical computed tomography findings are diffusely enlarged pancreatic gland with rim enhancement and irregularly attenuated main pancreatic duct. Response to steroid therapy is characteristic. (See Fig. 195-1.)

Idiopathic pancreatitis has two subsets. The *juvenile* form is characterized by male preponderance, age of onset before 25 years, and a long history of recurrent attacks of abdominal pain. Hallmarks of chronic pancreatitis, such as calculi formation, pancreatic insufficiency, and diabetes, develop 25 to 28 years after onset. The prognosis is poor because of the absence of a removable cause. *Late-onset* (usually after age 60) idiopathic chronic pancreatitis *(senile pancreatitis)* may be painless, diagnosed by incidental discovery of calculi during routine abdominal radiography or during workup of a patient with steatorrhea of uncertain etiology.

Other rare causes of chronic pancreatitis include obstruction, hyperlipidemia, pancreas divisum, hyperparathyroidism, gastrectomy, and celiac disease.

CLINICAL PICTURE

Recurrent Abdominal Pain

Postprandial pain is the dominant symptom in about 85% of patients with chronic pancreatitis and the most common reason they seek medical attention. Pain can be debilitating and intractable, often leading to functional incapacity, drug and alcohol addiction, poor quality of life, and even suicidal tendencies. Steady, boring, and agonizing pain in the epigastrium or sometimes in the left upper quadrant with radiation directly to the back, between the twelfth thoracic (T12) and second lumbar (L2) vertebrae, or to left shoulder is the typical presentation. Patients sit up and lean forward to the so-called pancreatic position or lie in the knee-chest position on the side. The severity and frequency of painful attacks vary. The duration of pain-free intervals is unpredictable, from weeks to many months. Pain may decrease or disappear, become stable, or worsen with advancing disease.

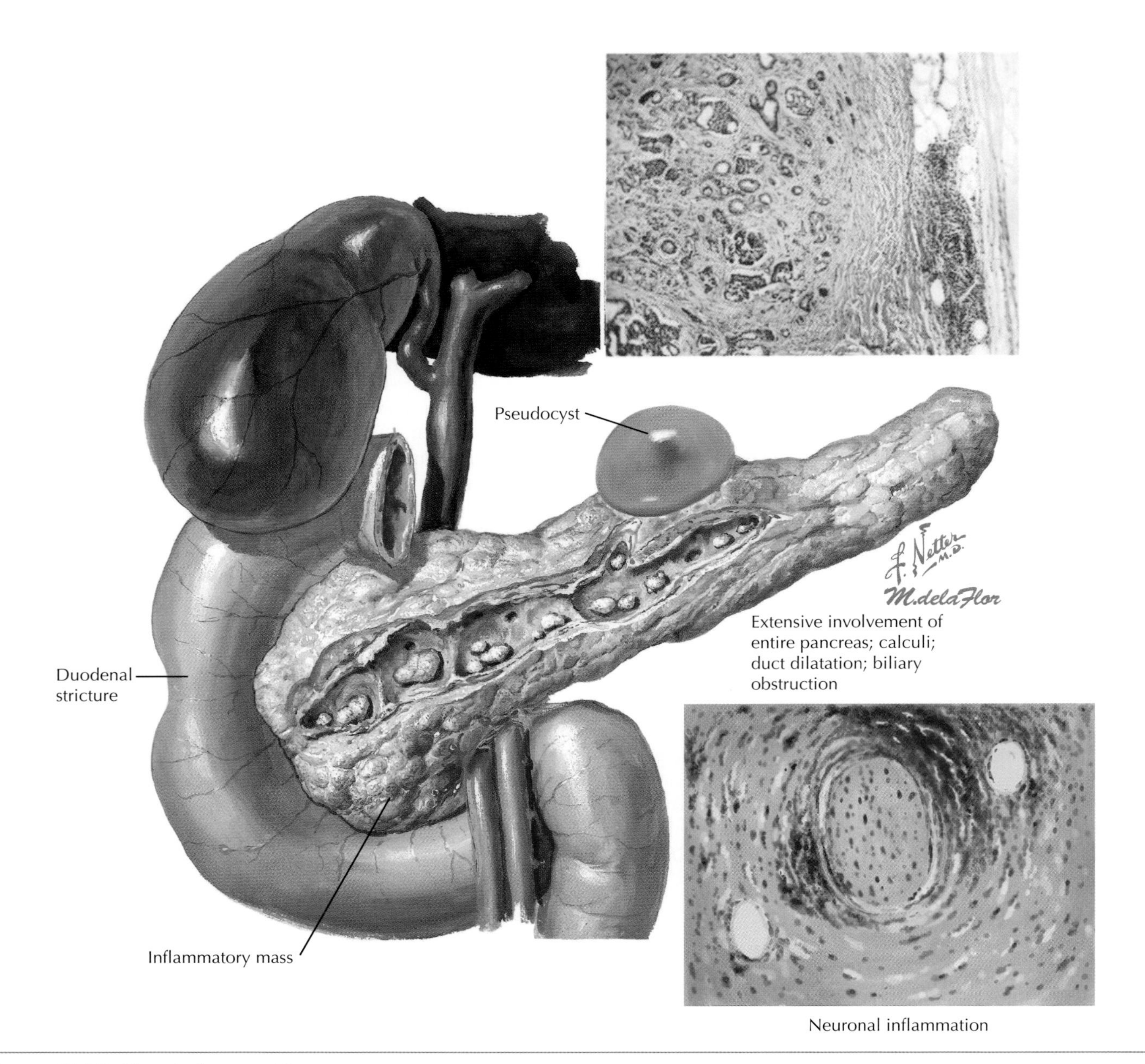

Figure 194-1 *Chronic Pancreatitis.*

Pancreatic pain appears to be multifactorial in pathogenesis, accounting for the difficulty encountered in pain management. Intraductal or interstitial hypertension, neuronal hypertrophy, perineural inflammation, and ongoing pancreatic injury are proposed mechanisms. Treatable complications such as pseudocyst and common bile duct (CBD) obstruction may contribute to the pain.

Malabsorption

Steatorrhea does not occur until enzyme secretion is reduced to less than 10% of normal. Lipolytic activity decreases more quickly than does tryptic activity, and it explains why steatorrhea occurs sooner and is more severe than protein malabsorption. In addition to reduced secretion of pancreatic enzymes, patients with severe chronic pancreatitis have decreased bicarbonate secretion. The stools may be bulky and formed, in contrast to the frank, watery diarrhea seen in malabsorptive disorders secondary to small intestinal causes. Oil droplets in the stool are often reported in chronic pancreatitis. Fecal weight tends to be lower and fat content of the stool greater (>20 g/24 hr) in patients with pancreatic insufficiency than in patients with steatorrhea from other causes.

DIABETES

Diabetes develops in almost 30% of patients with alcoholic pancreatitis 10 years after onset of disease. Diabetes in chronic pancreatitis is an example of acquired beta- and alpha-cell insufficiency associated with insulin resistance. Ketosis is rare, but other complications (e.g., nephropathy, retinopathy) are as common as in type 2 diabetes, and neuropathy may be more common in view of the addictive effect of alcoholism.

ADDITIONAL RESOURCES

Chari ST, Smyrck TC, Levy M et al: Diagnosis of autoimmune pancreatitis: the Mayo Clinic experience, *Clin Gastroenterol Hepatol* 4:1010-1016, 2006.

Kloppel G, Lutges J, Lohr M, et al: Autoimmune pancreatitis: pathological, clinical and immunological factors, *Pancreas* 27:14-19, 2003.

Lowenfels AB, Mainosnneuve P, Whitcomb DC: Risk factors for cancer in hereditary pancreatitis study group, *Med Clin North Am* 84:565-573, 2000.

Chronic Pancreatitis: Diagnosis, Complications, and Treatment

195

C. S. Pitchumoni

The diagnosis of chronic pancreatitis in its early stage remains a clinical challenge. Chronic pancreatitis should be considered in all patients with unexplained abdominal pain. Few patients have the classic triad of pancreatic calculi, diabetes mellitus, and steatorrhea. In most patients, the diagnosis is made based on a typical history of abdominal pain with a longstanding history of alcoholism. Physical examination findings in alcoholic pancreatitis are nonspecific.

DIAGNOSIS

Elevations of serum amylase or lipase levels are often not seen except in acute exacerbations and in early stages of chronic pancreatitis. The secretin-cholecystokinin (CCK) stimulation test, which involves placement of a tube into the duodenum, although considered the "gold standard," is performed by very few centers. A 72-hour fecal fat estimation or a random fecal elastase-1 test will help in establishing pancreatic insufficiency.

The imaging tests available are plain radiography, ultrasound or computed tomography (CT) of the abdomen, endoscopic ultrasound (EUS), magnetic resonance cholangiopancreatography (MRCP), and endoscopic retrograde cholangiopancreatography (ERCP). All imaging modalities have limited sensitivity or specificity. Plain film of the abdomen, including a chest film, is the initial radiologic study in all patients to detect pancreatic calculi and to exclude other intraabdominal causes of pain. Pancreatic calculi, although rare in early stages, are highly specific for the diagnosis of chronic pancreatitis and make other tests unnecessary.

Because it is noninvasive and widely available, abdominal CT is often performed to exclude other intraabdominal disorders and to identify complications such as pseudocyst, splenic artery aneurysms, enlarged common bile duct (CBD), intrahepatic ducts, and neoplastic masses. Small calculi missed on plain radiographs are identified on CT. Patchy atrophy of the pancreas, fatty replacement, and pancreatic ductal dilatation are visible in advanced cases. Helical CT using a pancreas-optimized protocol is recommended.

Diagnostic ERCP, until recently the preferred method for the diagnosis of chronic pancreatitis (almost replacing the secretin-CCK secretory test) has become unpopular. However, ERCP findings are highly sensitive (70%-90%) and specific (90%-100%). Limitations of ERCP are its high cost and its risk for acute pancreatitis (2%-3%) and even death (0.1%–0.2%). Ductal changes may be absent or minimal in many patients with mild alcoholic pancreatitis and in early idiopathic pancreatitis, where they are delayed (**Fig. 195-1**).

In the early diagnosis of chronic pancreatitis, EUS is a useful tool. An excellent correlation exists between EUS and histologic findings. MRCP has generally replaced ERCP for the diagnosis of ductal changes. Secretin-stimulated MRCP provides excellent visualization of the main duct and its side branches compared with non-stimulated MRCP. The diagnosis of autoimmune pancreatitis is discussed in Chapter 194.

COMPLICATIONS

Obstructive jaundice in chronic pancreatitis, as a result of CBD stenosis, is a frequent complication. The distal CBD that traverses the head of the pancreas is involved transiently in acute exacerbations of chronic pancreatitis or permanently as a result of fibrosis of the region or by compression from a pseudocyst. Clinical manifestations are similar to those of pancreatic carcinoma and include chronic pain, jaundice, and persistent elevation of serum alkaline phosphatase levels. Ultrasonography, percutaneous transhepatic cholangiography, ERCP, MRCP, and EUS delineate the stricture and proximal dilatation of CBD and intrahepatic biliary radicles. Surgical drainage by choledochojejunostomy relieves symptoms.

Fibrosis of the head of the pancreas may involve the adjacent duodenum, causing epigastric pain, postprandial fullness, nausea, and vomiting. Diagnosis is established by performing an upper gastrointestinal series and esophagogastroduodenoscopy. Surgical treatment may be needed to relieve obstruction.

Thrombosis of the splenic vein, splenic artery aneurysm, and pseudoaneurysm are complications of chronic pancreatitis. The splenic and portal veins may also be compressed by a pseudocyst or may be occluded by fibrosis from adjacent inflammation. A segmental form of portal hypertension, characterized by gastric and esophageal varices, may develop, and life-threatening variceal bleeding may occur. Splenectomy is curative.

Unilateral and bilateral pleural effusions are often the result of a leaking pseudocyst or a pleuropancreatic fistula, which can be demonstrated by ERCP. Surgical therapy is needed.

It has been reported that in men with chronic pancreatitis, cancer of the tongue, larynx, bronchus, colon, rectum, liver, skin/lip, bladder, and stomach occur more frequently than expected, whereas cancer of the breast, bone, and liver are more common in women with the disease. Cigarette smoking may be the etiologic factor for some of these cancers.

Any form of chronic pancreatitis (alcoholic, tropical, hereditary, or idiopathic) is associated with a high incidence of cancer of the pancreas.

TREATMENT AND MANAGEMENT

Managing chronic pancreatitis includes treating the associated pain, diabetes, and steatorrhea. A multidisciplinary team approach is needed, with an internist coordinating treatment and gastroenterologist, psychiatrist, pancreatic surgeon, social worker, and diabetologist as consultants.

Before therapy is initiated, treatable complications such as pseudocysts, CBD obstruction, and peptic ulcer disease should be ruled out. Pain relief is usually higher and deterioration of pancreatic function slower in alcohol-abstinent patients. Nono-

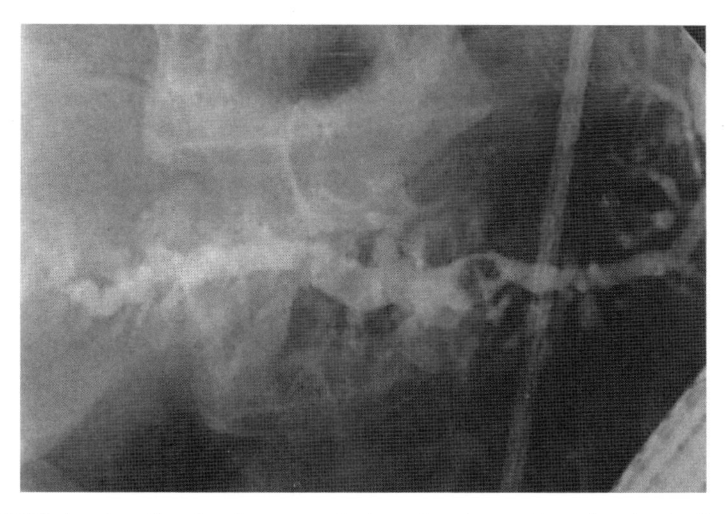

ERCP showing dilated main pancreatic duct, ductules, and intraductal calculi.

Autoimmune pancreatitis

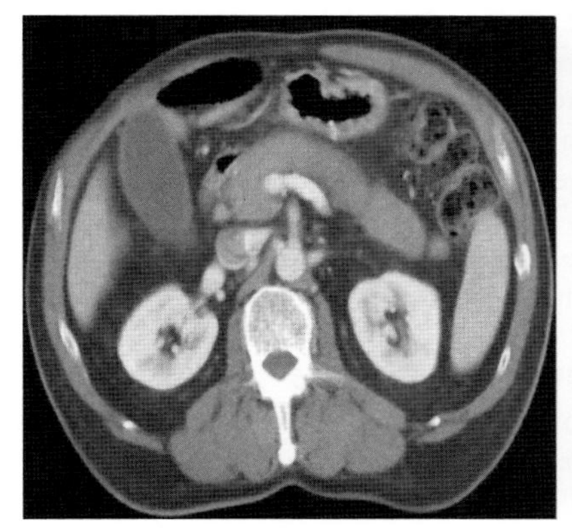

CT scan showing diffusely enlarged pancreas with delayed and thin enhancement.

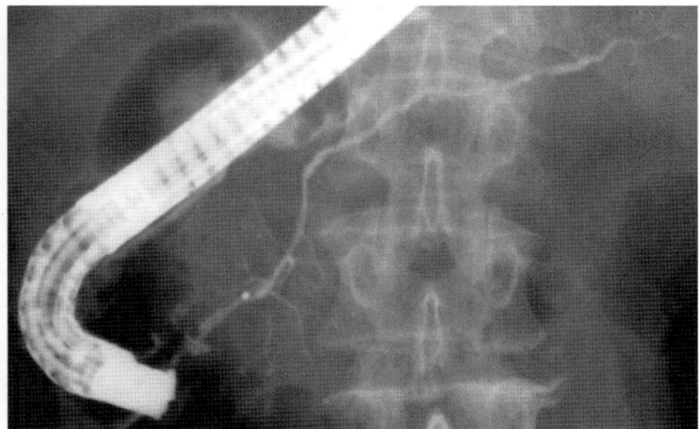

ERCP showing diffusely irregular narrowing of pancreatic duct.

Figure 195-1 *Chronic and Autoimmune Pancreatitis. From Chari S et al:* Clin Gastroenterol Hepatol *4:1010, 2006.*

pioid analgesics, such as acetaminophen, salicylates, and nonsteroidal antiinflammatory drugs (NSAIDs), with or without antidepressants, are preferred, but most patients need opioid analgesics for symptom relief.

The diet should be adequate in calories, high in protein (~24% of calories), moderate in carbohydrate (46%), and low in fat (30%). Small, frequent, low-fat meals are recommended to minimize pancreatic stimulation. Hospitalization, total cessation of oral intake of food, and short-term use of peripheral parenteral nutrition (PPN) or total parenteral nutrition (TPN) may be needed in acute exacerbations. Supplementation with antioxidant vitamins may reduce the recurrence of pain. Intrajejunal feeding, preferably with medium-chain triglyceride-rich food, stimulates CCK production less than oral feedings and is preferred over TPN.

The role of oral pancreatic enzyme therapy in the management of pain is controversial. A protease-dependent, negative-feedback mechanism contributes to the physiologic regulation of pancreatic secretion. It is speculated that extremely low levels of intraduodenal protease levels in chronic pancreatitis stimulate pancreatic enzyme secretion and increase intraductal pressure proximal to strictures. Conversely, adequate amounts of proteases in the duodenum decrease pancreatic secretion and intraductal pressure. Proponents of enzyme therapy recommend administering large doses of non-enteric-coated pancreatic enzyme preparations with high concentrations of proteases.

Octreotide acetate, a synthetic analog of somatostatin, when administered subcutaneously, inhibits pancreatic enzyme release and suppresses insulin and glucagon secretion. Pain relief is transient, and continuous treatment is expensive and impractical.

Types of endoscopic therapy for pain are (1) sphincterotomy, (2) internal drainage of pancreatic cysts, (3) extraction of stones from the pancreatic duct, with or without extracorporeal shock wave lithotripsy (ESWL), (4) guidewire catheter dilatation of strictures, and (5) pancreatic stents. A procedure is chosen based on the location of calculi or strictures.

The efficacy of *celiac plexus block* (CPB), either percutaneously or with surgical or EUS guidance, has not been established. CPB is associated with complications such as epidural or intraperitoneal hematomas, hypotension, gastroparesis, diarrhea, and sexual dysfunction.

Surgery is a consideration when pain is severe enough to interfere with daily life and when it cannot be managed medically. Surgery is tailored to ductal morphology, presence of an inflammatory mass in the head of the pancreas, and the expertise of the surgeon. Surgical candidates can be divided into two broad groups: those with dilated pancreatic ducts, who are more likely to benefit from ductal drainage, and those with normal-sized ducts, who may need pancreatic resection or denervation.

Longitudinal or lateral *pancreaticojejunostomy* (modified Puestow procedure) is preferred when the pancreatic ducts are large enough for anastomosis. Long-term pain relief is achieved in more than two thirds of patients with chronic pancreatitis and a dilated (>7-mm diameter) pancreatic duct. Recurrence of pain after initially successful longitudinal pancreaticojejunostomy suggests stricture formation and may indicate the need for reoperation.

In patients whose ducts are not dilated and in whom previous drainage failed, or in whom pathologic changes predominantly involve a particular area of the gland, resection offers good pain relief that tends to be more permanent than after pancreaticojejunostomy.

ADDITIONAL RESOURCES

Cahen DL, Gouma DJ, Nio Y, et al: Endoscopic versus surgical drainage of the pancreatic duct in chronic pancreatitis, *N Engl J Med* 356:676-684, 2007.

Etemad B, Whitcomb DC: Chronic pancreatitis: diagnosis, classification, and new genetic development, *Gastroenterology* 120:682-707, 2001.

Strate T, Bachmann K, Busch P, et al: Resection vs. drainage in treatment of chronic pancreatitis: long-term results of a randomized trial, *Gastroenterology* 134:1406-1411, 2008.

Pancreatic Cancer: Clinical Picture

196

C. S. Pitchumoni

Pancreatic cancer is the fifth leading cause of cancer mortality in the United States, with more than 37,000 deaths annually, and it is increasing in incidence. Risk increases sharply after age 50, with the peak incidence in the seventh and eighth decades of life. Mortality and incidence rates are often identical because of the delay in diagnosis and the lack of effective treatment. In the United States, the incidence of pancreatic cancer is slightly more common in men than women (10.6 vs. 8 per 100,000) and is higher in black and Jewish populations.

Almost 90% of the pancreatic cancers are moderately differentiated *adenocarcinomas* that arise from the ductular epithelium (**Fig. 196-1**). Most of these tumors (60%-70%) are localized in the head of the pancreas, and the rest (18%-20%) arise in the body and tail or are multifocal and infiltrate diffusely throughout the gland (20%). Tumors of the head of the pancreas are usually diagnosed early because of the early onset of obstructive jaundice. Tumors of the body and tail are often diagnosed at more advanced stages, carrying poor prognosis. Overall 5-year survival (5%-10%) reflects the late presentation of patients with advanced disease.

Acinar cell carcinoma, giant cell carcinoma, adenosquamous carcinoma, intraductal papillary mucinous tumors, cystadenocarcinoma, leiomyosarcoma, and lymphoma account for less than 10% of other pancreatic cancers.

Cigarette smoking is the greatest risk factor for pancreatic cancer. About 30% of the cases are attributed to heavy smoking in terms of number of cigarettes smoked and duration of smoking. Chronic pancreatitis of any cause (alcoholic, tropical, or hereditary) is associated with high risk.

Other disorders associated with a high risk for pancreatic cancer include hereditary nonpolyposis colorectal cancer, ataxia telangiectasia, Peutz-Jeghers syndrome, familial breast cancer, and familial atypical multiple-mole melanoma. High-fat diets are linked to pancreatic cancer. Diabetes mellitus has a controversial relationship, with some studies noting an increased risk and others finding no association. Occupational exposure to chemicals such as β-naphthylamine, benzidine dichlorodiphenyltrichloroethane, and metal dusts increases the risk. Partial gastrectomy and cholecystectomy are other noted risk factors. Genetic predisposition to pancreatic cancer is caused by inherited and acquired mutation of genes.

In its early stages, pancreatic cancer mimics a number of other conditions that cause epigastric or back pain. Other symptoms include weight loss, anorexia, nausea, jaundice, diarrhea, and depression. Diabetes mellitus or glucose intolerance may develop a few months to a year preceding the onset of other symptoms. Progressive obstructive jaundice develops in most patients with carcinoma of the head of the pancreas. Tumors of the body or tail may become large before symptoms develop. Small tumors of the head or ampulla cause early obstructive jaundice or symptoms of acute recurrent pancreatitis.

Pain is frequent, as is weight loss, particularly in patients with tumors of the body and tail of the pancreas. Hematemesis and melena may occur in late stages because of the development of gastric varices resulting from splenic vein occlusion. Ascending cholangitis characterized by fever and chills may occur. A palpable, distended gallbladder, usually the result of bile duct obstruction, is termed *Courvoisier sign* or *Courvoisier gallbladder* and is seen in 25% of patients. Supraclavicular lymph node enlargement (Virchow node), periumbilical mass (Sister Mary Joseph nodule), and palpable rectovaginal or rectovesical nodularity (Blumer shelf) are signs of advanced disease (**Fig. 196-2**). A nonspecific sign (Trousseau sign) is the migrating thrombophlebitis that occurs in all advanced cancers and is not specific for pancreatic cancer.

Cancer of the pancreas manifests with mild, tolerable, nonspecific symptoms mimicking many other gastrointestinal disorders. Even when appropriate diagnostic tests such as computed tomography are performed early, the findings may be nondiagnostic and may require follow-up studies weeks or months later for development of more specific radiologic signs.

ADDITIONAL RESOURCES

Li D, Xie K, Wolff R, et al: Pancreatic cancer, *Lancet* 363:1049-1057, 2004.

Lowenfels AB, Maisonneuve P: Environmental factors and risk of pancreatic cancer, *Pancreatology* 3:1-8, 2003.

Rebours V, Boutron-Ruault M, Schnee M, et al: Risk of pancreatic adenocarcinoma in patients with hereditary pancreatitis: a national exhaustive series, *Am J Gastroenterol* 103:111-119, 2008.

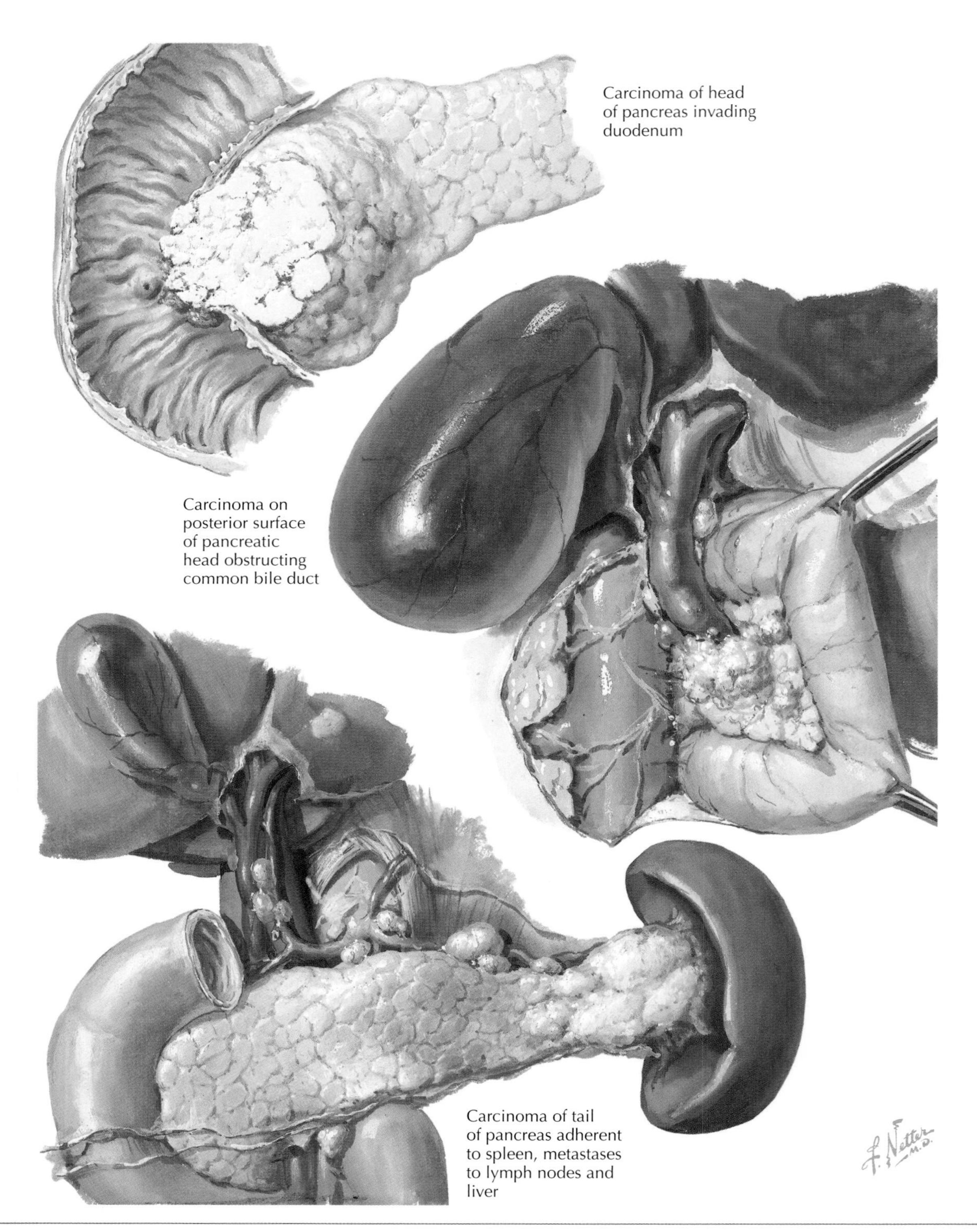

Figure 196-1 *Pancreatic Cancer: Clinical Features.*

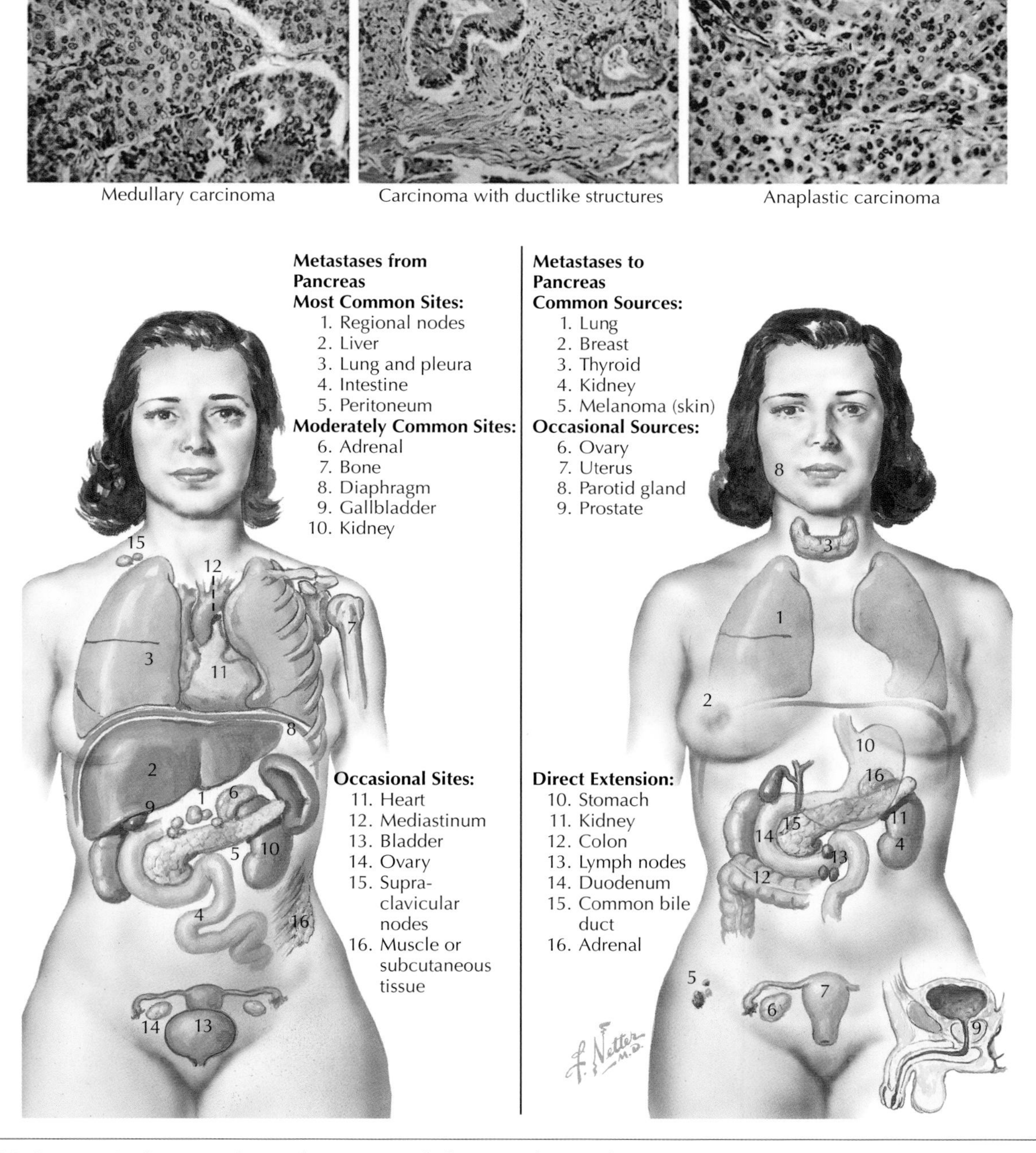

Figure 196-2 *Pancreatic Cancer: Advanced Disease, Including Histology and Metastases.*

Pancreatic Cancer: Diagnosis and Treatment

197

C. S. Pitchumoni

In any patient older than 50 who has unexplained jaundice, weight loss, upper abdominal or back pain, anorexia, or idiopathic pancreatitis, the index of suspicion for pancreatic cancer should be high. New-onset diabetes mellitus without obesity or family history, particularly combined with any of these symptoms, should alert the physician to the possibility of pancreatic cancer.

DIAGNOSIS

Depending on tumor size and degree of histologic differentiation, the level of cancer antigen (CA) 19-9, the most widely used serum marker, varies. In tumors less than 2 cm in diameter, the level may be immeasurable. Elevated CA 19-9 levels are highly correlated with the presence of pancreatic cancer. False elevations are seen in chronic pancreatitis and in benign obstructive jaundice. Tissue polypeptide–specific antigen is another recently described serum marker.

Improvements in imaging technology, including spiral computed tomography (CT), magnetic resonance imaging (MRI), positron emission tomography (PET), and endoscopic ultrasound (EUS) have helped us identify pancreatic cancer amenable to resection.

Abdominal ultrasonographic findings are suboptimal for diagnosing pancreatic cancer. Contrast-enhanced, thin-section helical CT of the pancreas helps diagnose, evaluate the extent of the disease (sensitivity and specificity more than 80% and 95%, respectively), and assess tumor resectability. Contrast-enhanced MRI using intravenous gadolinium-DTPA is useful for detecting small pancreatic tumors. Ductal size is best evaluated by magnetic resonance cholangiopancreatography (MRCP). Functional imaging modalities such as PET scanning are superior to conventional CT imaging. Demonstration of obstruction of the biliary and pancreatic ducts (double-duct sign) on endoscopic retrograde cholangiopancreatography (ERCP) is not always reliable. EUS is a minimally invasive technique in the evaluation of cancer resectability and the most accurate method for detecting vascular and lymph node enlargement. EUS-guided, fine-needle aspiration allows cytologic evaluation. Angiography has limited value.

TREATMENT AND MANAGEMENT

Depending on the stage, management of pancreatic cancer includes nutritional support, surgical resection, chemotherapy, radiation therapy, endoscopic and surgical palliation of pain, and chemoradiation for unresectable cancer. Approximately 85% to 90% of pancreatic cancers are unresectable. Optimal therapy requires a multidisciplinary team approach by a medical oncologist, interventional radiologist, gastroenterologist, radiotherapist, internist, and pain management specialist.

The Whipple procedure *(pancreaticoduodenectomy)* is the standard surgical procedure. The technique involves resection of the distal stomach, gallbladder, proximal jejunum, and regional lymph nodes. The mortality rate is less than 5% in expert hands. The new pylorus-preserving Whipple procedure reduces the incidence of postgastrectomy symptoms. Palliative surgery for the relief of biliary obstruction eliminates pruritus and probably has some benefit in promoting nutrition.

Current data suggest that the combination of pancreaticoduodenectomy with postoperative adjuvant *5-fluorouracil* (5-FU) and external beam radiation therapy improves the duration of survival. Chemoradiation has been suggested for patients with locally advanced unresectable pancreatic cancer to improve survival and quality of life and to downstage advanced locoregional disease to allow surgical resection. *Gemcitabine* (2′,2′-difluorodeoxycytidine; Gemzar) is a deoxycytidine analog capable of inhibiting DNA replication and repair. In addition to prolonging survival, gemcitabine improves quality of life by decreasing pain and the need for opioid analgesics. External beam radiation therapy with 5-FU chemotherapy, intraoperative radiation therapy (brachytherapy or electron beam), and external beam radiation therapy with novel (radiosensitizing) chemotherapeutic agents are options. Palliation of pain using percutaneous or EUS-guided celiac ganglion block, an intraoperative approach, or thoracoscopic splanchnicectomy improves quality of life.

Chapter 207 discusses ampullary carcinoma.

ADDITIONAL RESOURCES

Abbruzzese JL: Adjuvant therapy for surgically resected pancreatic adenocarcinoma, *JAMA* 299:1066-1067, 2008.

Pappas S, Federle MP, Lokshin AE, et al: Early detection and staging of adenocarcinoma of the pancreas, *Gastroenterol Clin North Am* 36:413-429, 2007.

Standards of Practice Committee: The role of endoscopy in the evaluation and treatment of patients with pancreaticobiliary malignancy, *Gastrointest Endosc* 58:643-649, 2003.

Cystic Tumors of the Pancreas

198

C.S. Pitchumoni

Cystic tumors of the pancreas are a heterogenous group of pancreatic neoplasms that include mucinous cystic neoplasms (50%), serous cystadenomas (30%), intraductal papillary mucinous neoplasms (12%), papillary cystic tumors (3%), and miscellaneous growths (5%) (**Figs. 198-1** and **198-2**). Often, these asymptomatic cysts are detected inadvertently during computed tomography (CT) of the abdomen. **Table 198-1** outlines the differentiating features of neoplastic pancreatic cysts.

Serous cystadenomas (SCAs) are generally benign, well-circumscribed, solitary microcystic lesions that arise and grow to large size in any part of the pancreas. Low viscosity of the aspirated fluid and low levels of amylase and carcinoembryonic antigen (CEA) are the findings. CT characteristically shows a lesion in the head or neck that has a honeycomb appearance with an area of central fibrosis (stellate star) of calcification. Small SCAs can be monitored. Surgery may be needed in large and symptomatic cysts and when malignancy cannot be ruled out.

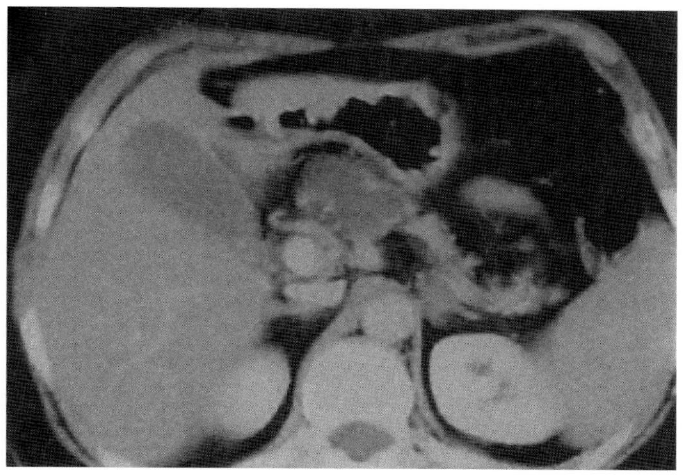

Serous cystadenoma of the pancreas.

Figure 198-1 *Cystic Tumors: Serous Cystadenoma of the Pancreas.*

Mucinous cystic neoplasms (MCNs; mucinous cystadenomas) are cystic lesions predominantly in the body or tail of the pancreas that occur mostly in women. The diagnosis of *cystadenocarcinoma* is suggested when there is a mass in the wall of the cyst. Fine-needle aspiration guided by CT or endoscopic ultrasound (EUS) may be necessary for a diagnosis. The aspirated fluid is viscous. Elevated CEA levels in the cyst fluid indicate malignancy. In contrast to IPMN, MCN does not communicate with the pancreatic duct. Surgical resection is the appropriate treatment.

Intraductal papillary mucinous neoplasms (IPMNs; mucinous ductal ectasia, intraductal mucinous cystadenocarcinoma, ductal ectasia, and duct-ectatic mucinous cystadenoma) affect men and women equally. Most patients have a long history of recurrent acute pancreatitis associated with steatorrhea and glucose intolerance. IPMN frequently affects the head of the pancreas (main duct IPMN), with dilatation and filling of the main pancreatic duct or its side branches with thick, viscous mucus. When localized to a branch (side branch IPMN), a "cluster of grapes" appearance is seen. The dramatic picture on endoscopic retrograde cholangiopancreatography is a patulous ampulla of Vater with extruding mucus ("fish mouth" appearance). Contrast-enhanced thin-cut CT and EUS show multiple dilated pancreatic duct side branches. The risk of malignancy is high in main duct IPMN. The appropriate treatment is surgical resection to relieve symptoms and to prevent invasive carcinoma.

Table 198-1 Differential Diagnosis of Pancreatic Neoplastic Cysts

	SCA	MCN	IPMN
Gender	F > M	Mainly F	F = M
Usual age	Middle age	Middle age	Elderly
History of pancreatitis	No	No	Possible
Morphology (imaging studies)	Microcystic and multicystic* Honeycomb appearance Sunburst calcification (20%)	Unilocular cyst, septations, and wall calcifications Solid component indicates malignancy	Main duct dilatation or limited to side branches
Location	Diffuse	Body and tail	Head
Fluid	Thin	Viscous	Thick
Cytology	Cuboidal cells Stains for glycogen	Positive for mucin Columnar cells with variable atypia Ovarian-like stroma present	Positive for mucin Columnar cells with variable atypia
Malignant potential	Rare	Yes	Yes

*More than six cysts.
F, Female; *IPMN*, intraductal papillary mucinous neoplasm; *M*, male; *MCN*, mucinous cystic neoplasm (cystadenoma); *SCA*, Serous cystadenoma.

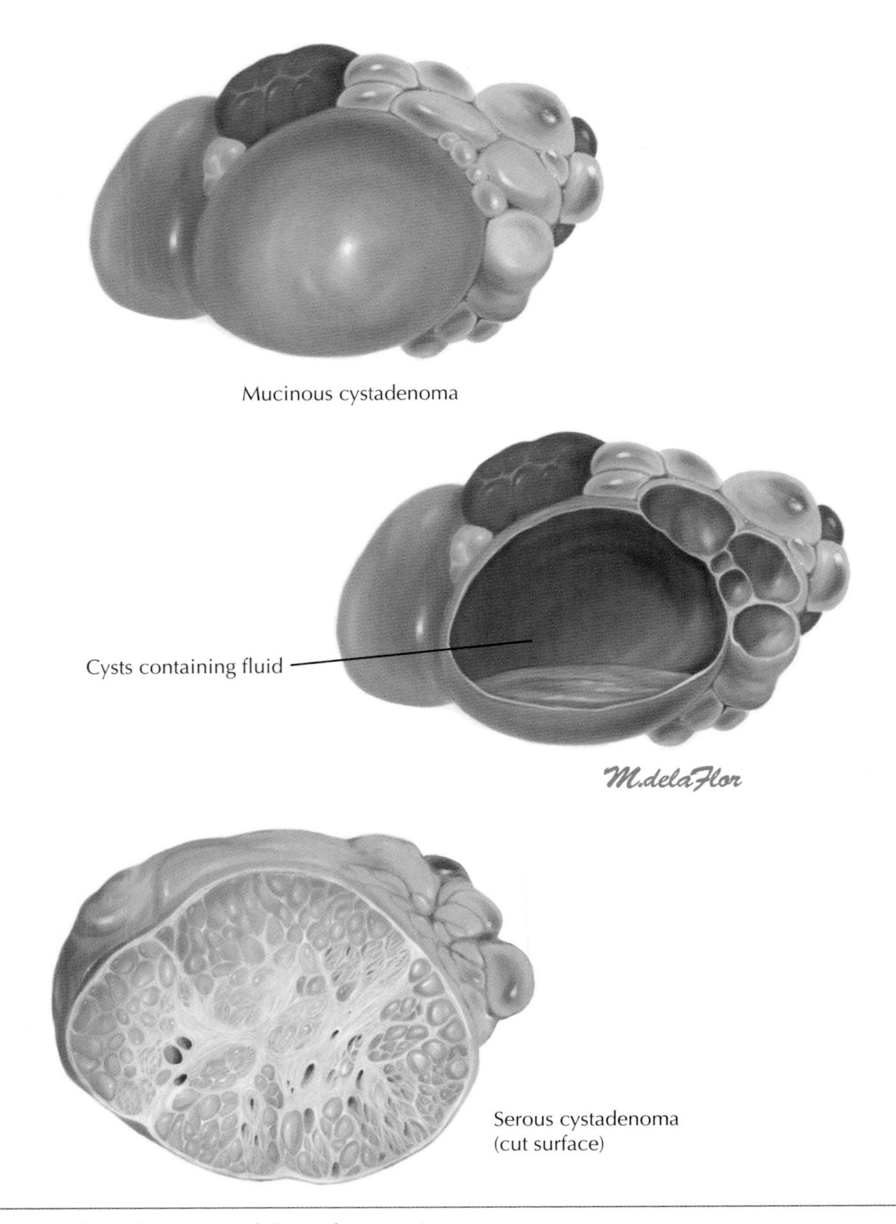

Figure 198-2 *Cystic Tumors: Cystadenoma and Cystadenocarcinoma.*

Papillary cystic neoplasm (solid and cystic tumor, solid and papillary neoplasm of pancreas) is an extremely rare tumor mostly seen in young women. It may be asymptomatic or may be accompanied by low-grade abdominal pain. An abdominal mass may be palpable.

ADDITIONAL RESOURCES

Khalid A, Brugge W: ACG practice guideline for the diagnosis and management of neoplastic pancreatic cysts, *Am J Gastroenterol* 102:2339-2349, 2007.

Oh HC, Kim MH, Hwang CY, et al: Cystic lesions of the pancreas: challenging issues in clinical practice, *Am J Gastroenterol* 103:229-239, 2008.

Tanaka M, Chari S, Adsay V, et al: International consensus guidelines for management of intraductal papillary mucinous neoplasms and mucinous cystic neoplasms of the pancreas, *Pancreatology* 6:17-32, 2006.

Pancreatic Neuroendocrine Tumors (Islet Cell Tumors)

199

C. S. Pitchumoni

Neuroendocrine tumors (NETs), also known as "islet cell tumors," represent a group of rare neoplasms of the neuroendocrine cells of the gastropancreatic system (**Fig. 199-1**). They may be functioning tumors (85%), based on the hormone produced, or nonfunctioning tumors (15%), diagnosed because of their mass effect or malignant behavior.

Polypeptide hormone–producing cells of the pancreatic islets (islets of Langerhans) have a common embryologic origin within the neural crest and subsequently migrate to the foregut enlargement of the pancreas. These tumors have similar histology but can be distinguished by immunohistochemistry.

Clinical picture, diagnosis, and treatment of NETs vary greatly depending on the syndrome and secretory production, as discussed under insulinoma, gastrinoma, and glucagonoma.

Pancreatic NETs may occur with an autosomal dominant inheritance as part of *multiple endocrine neoplasia* type 1 (MEN-1) syndrome. These tumors are benign or malignant. MEN-1 syndrome is a disorder of three glands: parathyroid, pancreatic islets, and pituitary. MEN-2 tumors include pheochromocytoma and adenoma or hyperplasia of parathyroid glands.

INSULINOMA

Insulinomas are the most common functional NETs. Almost 60% occur in middle-aged women. Insulinomas can be sporadic or familial, a component of MEN-1 syndrome. Most of the tumors are solitary, relatively benign, less than 2.5 cm in diameter, and evenly distributed throughout the pancreas. Approximately 10% of patients with insulinoma have the MEN-1 syndrome. A nonislet origin of the tumor has recently been noted.

Clinical Picture

Fasting hypoglycemia is a common clinical manifestation of insulinoma. Headache, visual disturbances, dizziness, lightheadedness, confusion, weakness, grand mal seizures, and coma are the neuroglycopenic symptoms of insulinoma. The catecholamine response to hypoglycemia causes diaphoresis, tremulousness, palpitations, irritability, and hunger. Hypoglycemic symptoms are precipitated by fasting or exercise and respond to carbohydrate ingestion. Frequent eating may lead to obesity.

Diagnosis

After 72 hours of supervised fasting, the patient displays the *Whipple's triad:* hypoglycemic symptoms (central nervous system, vasomotor), documented hypoglycemia, and relief of symptoms after glucose intake. Hypoglycemia usually develops within 24 hours of fasting. A serum insulin level of 5 μU/mL or more with concomitant plasma glucose level less than 45 mg/dL (2.5 mmol/L) indicates insulinoma. Factitious hypoglycemia is ruled out by fasting plasma levels of C-peptide greater than 1 to 7 ng/mL, by antibodies to insulin, and by sulfonylurea in the blood.

Abdominal computed tomography (CT), magnetic resonance imaging (MRI), and endoscopic ultrasound (EUS) are all useful. Small insulinomas are often missed by these techniques. Visceral angiography and Indium 111–labeled octreotide nuclear imaging are other diagnostic modalities.

Treatment and Management

The goal of management is to prevent hypoglycemia through frequent small meals. Diazoxide, 100 to 150 mg every 8 hours, prevents insulin release. Other therapies include calcium channel blockers, corticosteroids, and glucagon. Octreotide therapy is useful in some patients with insulinoma.

Surgical exploration and enucleation of the tumor is needed in most patients. More than 90% of patients can be cured because insulinomas are often single and benign.

In metastatic insulinomas, streptozotocin alone or in combination with 5-fluorouracil (5-FU), or doxorubicin, or both, has been recommended.

GASTRINOMA

Affecting young and middle-aged adults, gastrinoma, or *Zollinger-Ellison syndrome* (ZES), is a rare cause of peptic ulcer disease. Tumors of the pancreatic or duodenal wall G cells are responsible for the signs and symptoms. The most common NET found in MEN-1 patients is gastrinoma. Hyperparathyroidism is the most common associated endocrine abnormality.

Clinical Picture

Epigastric abdominal pain from recurrent peptic ulcer, associated with secretory diarrhea, raises suspicion for ZES. Pathophysiologic effects are secondary to hypergastrinemia and hypersecretion of gastric hydrochloric acid (HCl). Severe esophagitis is thus an accompanying lesion. Precipitation of bile acids by excessive HCl and inappropriate pH for pancreatic lipase activity cause steatorrhea.

Diagnosis

Elevated plasma gastrin level (normal <100 pg/mL; ZES >1000 pg/mL) and basal acid output of 15 mEq/hr or greater are typical of gastrinoma. Other hypergastrinemic conditions include pernicious anemia (achlorhydria), atrophic gastritis, chronic renal failure, and the rare postoperative condition of retained antral syndrome after Billroth II surgery. In patients with gastrinoma, intravenous injection of secretin causes a paradoxic increase in plasma/gastrin of 20 pg/mL or greater above basal level.

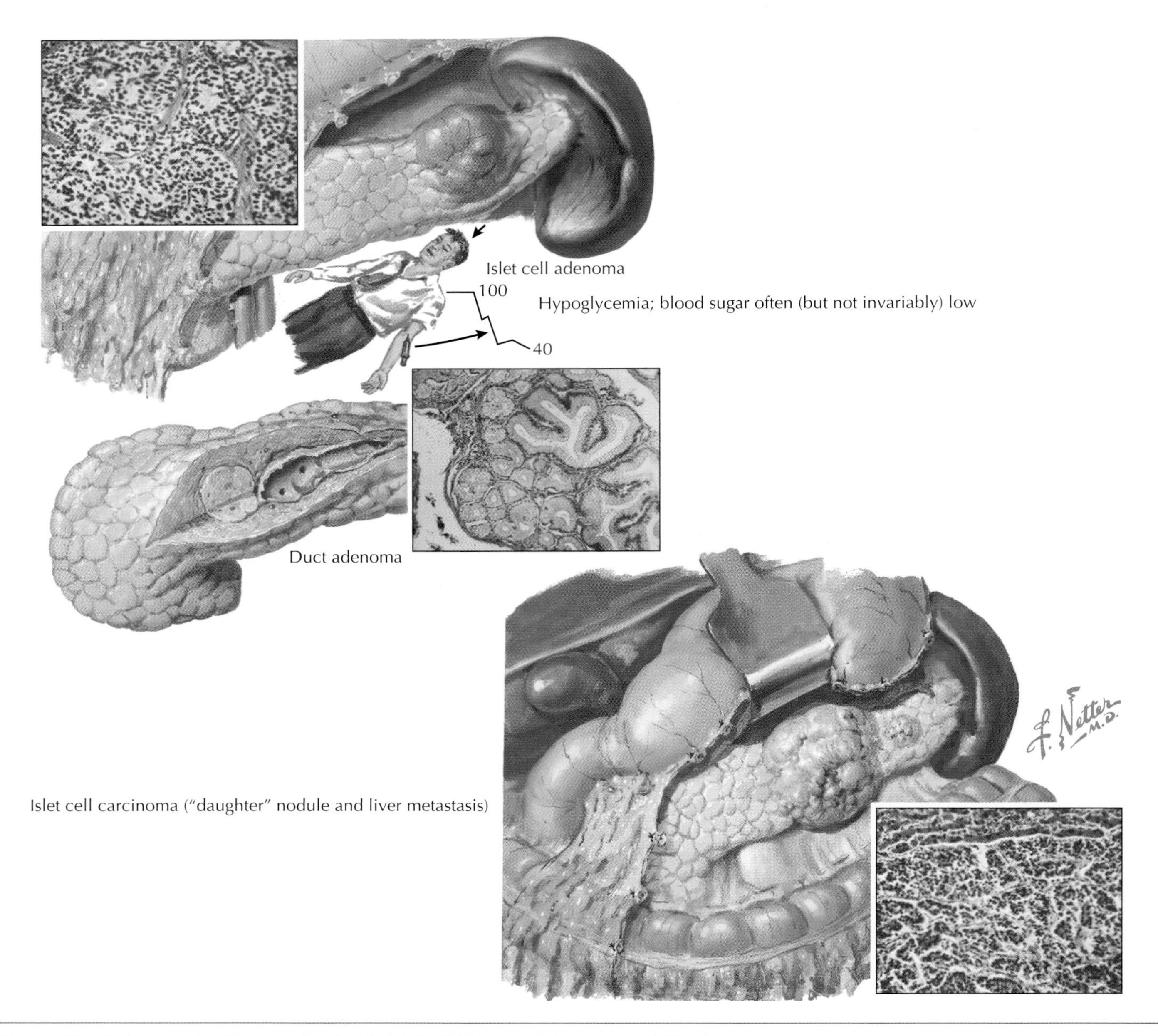

Figure 199-1 *Pancreatic Neuroendocrine (Islet Cell) Tumors: Adenoma and Carcinoma.*

Tumor localization is highly controversial. EUS is currently the most sensitive and specific localization procedure for NETs, superior to CT, MRI, and angiography. Selective angiography is a good imaging modality to identify primary and metastatic gastrinomas. Selective venous sampling for gastrin from portal tributaries has a sensitivity of 70% to 90% in identifying gastrinomas. Gastrinomas and other NETs are visualized after injection of isotope-labeled somatostatin analogs, such as ^{111}In-octreotide.

Treatment and Management

Effective control of gastric hypersecretion is possible with a proton pump inhibitor. Total gastrectomy is not recommended to deal with the complications of peptic ulcer disease. Octreotide is effective in suppressing gastrin release.

Most gastrinomas in the pancreas can be removed by enucleation, and large tumors can be removed by resective procedures. In patients with hepatic metastases, newer techniques include chemoembolization, cryotherapy, and alcohol ablation.

GLUCAGONOMA

Glucagonoma arising from pancreatic alpha cells is characterized by diabetes mellitus, severe dermatitis *(necrolytic migratory erythema)*, neuropsychiatric symptoms, glossitis or stomatitis, diarrhea, weight loss, anemia, and venous thromboses. Hypoaminoacidemia as a result of catabolic effects of glucagon is responsible for the skin rash.

Tumors are solitary and large (>6 cm). The plasma glucagon level is greater than 500 pg/mL (normal <100 pg/mL). CT, MRI, and EUS are helpful in the diagnosis.

Octreotide therapy reverses skin rash, reduces weight loss, and decreases diarrhea. Surgical therapy requires major pancreatic resection. Nutritional support is a major component of therapy.

VIPOMA

The characteristic features of VIPoma—vasoactive intestinal polypeptide (VIP)–secreting NET, also known as Verner-Morrison syndrome, pancreatic cholera, and WDHA syndrome (watery diarrhea, hypokalemia, and achlorhydria)—are voluminous secretory diarrhea, hypokalemia, and achlorhydria. About 50% of the tumors are malignant, and 75% also secrete pancreatic polypeptide. Fasting plasma VIP level is greater than 500 pg/mL when the patient has diarrhea. Medical therapy includes rehydration and correction of hypokalemia. Octreotide therapy gives prompt relief from diarrhea. Tumor enucleation and partial pancreatectomy are surgical options.

Other tumors of the pancreatic islets (other NETs) are rare.

ADDITIONAL RESOURCES

Hochwald SN, Zee S, Conlon-KC, et al: Prognostic factors in pancreatic endocrine neoplasms: an analysis of 136 cases with a proposal for low-grade and intermediate-grade groups, *J Clin Oncol* 20:2633, 2002.

Toumpanakis CG, Caplin ME: Molecular genetics of gastropancreatic neuroendocrine tumors, *Am J Gastroenterol* 103:729-732, 2008.

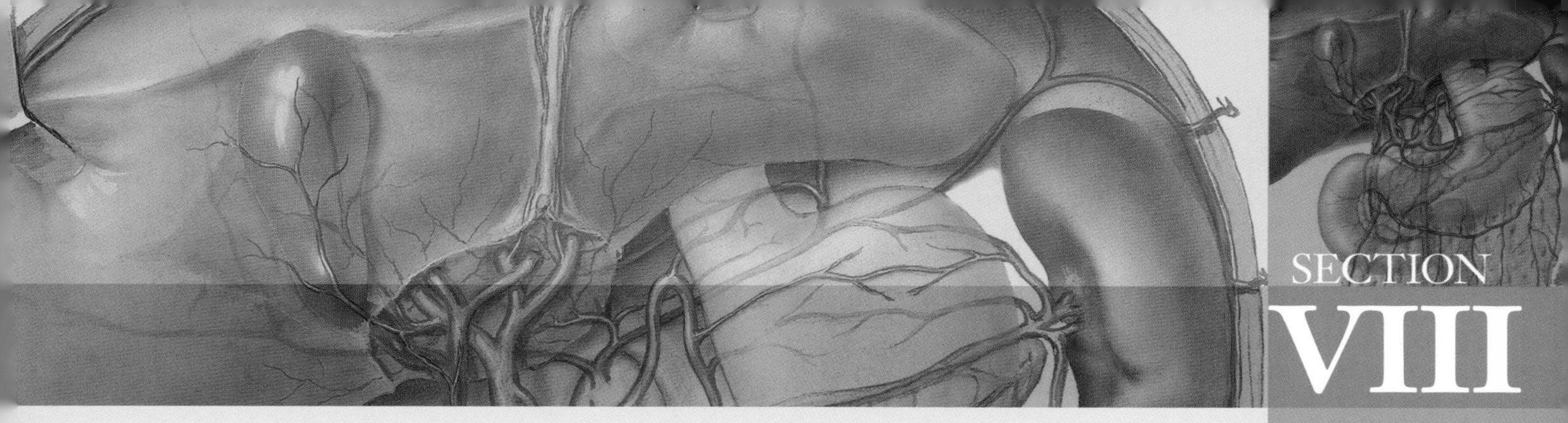

Gallbladder and Bile Ducts

Anatomy and Function of the Gallbladder

200

C. S. Pitchumoni

The gallbladder is a pear-shaped organ, usually 10 cm (4 inches) in length and 3 to 5 cm in diameter, that is attached to the inferior surface of the liver (**Fig. 200-1**). Two thirds of the gallbladder is covered by peritoneum. The fundus of the gallbladder projects beyond the liver; the body (or corpus) is in contact with the second portion of the duodenum and the colon; and the infundibulum (Hartmann pouch), located at the free edge of the lesser omentum, bulges forward toward the cystic duct. The neck is the part of the gallbladder between the body and the cystic duct.

The gallbladder has four layers: mucosa, muscularis, connective tissue, and in most parts, serosa. Right and left hepatic ducts unite to form the 2- to 3-cm-long hepatic duct, which in turn combines with the cystic duct to form the *common bile duct* (CBD), which is 10 to 15 cm long. In approximately 85% of people, the CBD is partially or totally covered by pancreatic tissue posteriorly as it approaches the duodenum. In more than 66%, the CBD and the major pancreatic duct share a common channel, 2 to 7 mm in length, before emptying into the duodenum. The *sphincter of Oddi* (SO) is an area of distal CBD, approximately 4 to 10 mm in length and mostly within the wall of the duodenum, that regulates the flow of bile and pancreatic juice.

Normally, the gallbladder has a volume capacity of approximately 30 to 60 mL. However, when it is actively reabsorbing water, sodium, chloride, and other electrolytes continuously, as much as 450 mL of secretion can be stored. Although the most potent stimulus for gallbladder contraction is the duodenal hormone *cholecystokinin* (CCK), it is also stimulated by acetylcholine from vagi and the enteric nervous system. The normal bile contains 70% bile salts, 22% phospholipids, 4% cholesterol, 3% proteins, and 0.3% bilirubin, along with the electrolytes of plasma. Water and electrolytes are reabsorbed, whereas cholesterol, lecithin, and the bile salts become concentrated during fasting.

When the gallbladder contracts, the SO relaxes. The SO has variable basal pressure and phasic contractile activity that are regulated by nerves, hormones, CCK, and secretin.

ADDITIONAL RESOURCES

Hernandez-Nazara A, Curiel-Lopez F, Martinez-Lopez E, et al: Genetic predisposition of cholesterol gallstone disease, *Ann Hepatol* 5:140-149, 2006.

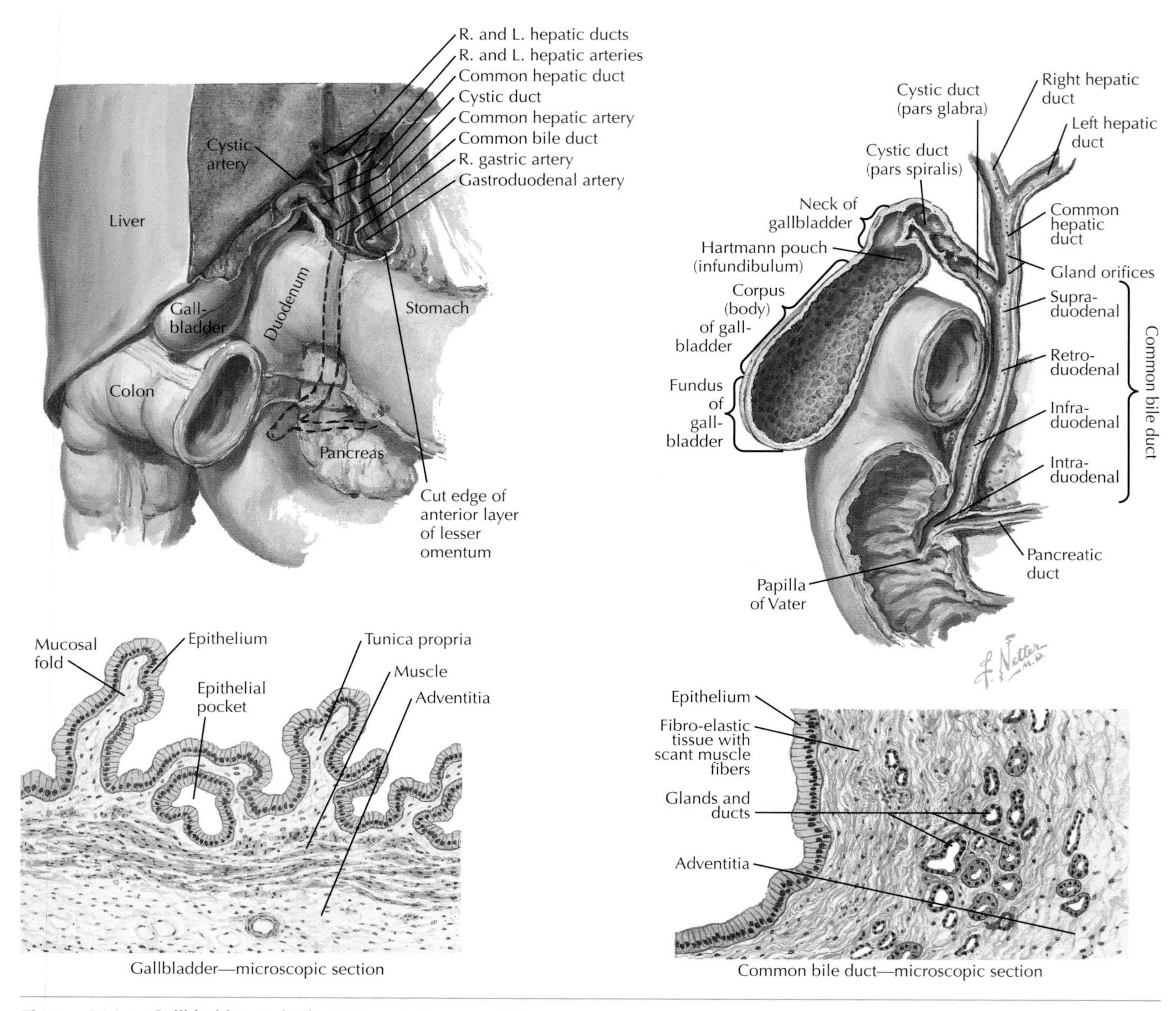

Figure 200-1 *Gallbladder and Bile Ducts: Anatomy and Histology.*

Gallstones

C. S. Pitchumoni

201

EPIDEMIOLOGY OF GALLSTONE DISEASE

An estimated 15% of the U.S. population has gallstones, and about 700,000 cholecystectomies are performed each year. The prevalence of gallstone disease is low in Asians and Africans. The highest prevalence is in Native Americans (Pima Indians) and in Latin American populations. In white women, the prevalence is 5% to 15% in those younger than 50 years and about 25% in those older than 50, compared with 4% to 10% and 10% to 15%, respectively, in white men.

More than 75% of the gallstones in the U.S. population are *cholesterol* stones, and most risk factors are attributable to cholesterol stones (**Table 201-1**). Several genes are associated with the principal metabolic pathways involved in the formation of gallstones.

TYPES OF GALLSTONES

Gallstones are divided into three major types—cholesterol, black pigment, and brown pigment—based on their composition and pathogenesis. In general, all gallstones form as a result of a change in bile composition, either an increase in the composition of a normal biliary component that exceeds its solubility or a decrease in a solubilizing component, or both (**Fig. 201-1**). Consequently, an insoluble substance called a *nidus* becomes supersaturated, and insoluble particles become sequestered and aggregate.

Cholesterol gallstone disease contributes to the occurrence of more than 80% of gallstones. These stones consist of pure cholesterol monohydrate crystals agglomerated by a mucin-glycoprotein matrix. Other constituents include unconjugated bilirubin and small amounts of calcium phosphate.

The pathogenesis of cholesterol gallstones is well studied (see Fig. 201-1). Major components of bile are bile salts, phospholipids, and cholesterol. *Bile salts*, synthesized from cholesterol, constitute the two primary *bile acids*, cholic acid and chenodeoxycholic acid. *Cholesterol* is only slightly soluble in aqueous media but is made soluble through formation of mixed micelles with bile salts and *phospholipids*, mainly lecithin. The enterohepatic circulation of bile acids resulting from bile reabsorption from the terminal ileum, along with hepatic synthesis, keeps the bile acid pool physiologically optimal to maintain the cholesterol in solution.

At least four mechanisms are involved in the formation of cholesterol gallstones: (1) supersaturation of bile with cholesterol, (2) nucleation of cholesterol monohydrate with subsequent crystallization and stone growth, (3) delayed emptying or gallbladder stasis, and (4) decreased enterohepatic circulation of bile acids. When the rate of bile acid secretion or the return of bile acid through enterohepatic circulation is decreased, as in patients with terminal ileal disease (e.g., Crohn's), the relative cholesterol content increases, and bile becomes supersaturated

Table 201-1 Risk Factors for Cholesterol Stones

Risk Factor	Comments
Age	Uncommon before age 20 years (exception: Mexican American girls).
Gender	Female/male ratio highest in youngest patients; narrows to 2:1 after age 50.
Nationality	Highest: Scandinavia, Northern Europe, Chile, and northern parts of India. Lowest: sub-Saharan Africa and Asia.
Race/ethnicity	Highest: Pima Indians of southern Arizona (70% of Pima women older than 25 years), other Native American tribes, and Alaskans. Lowest: American blacks.
Family history	Higher risk in first-degree relatives of gallstone patients.
Obesity	Relative risk rises sharply as degree of obesity increases; women more often affected. Metabolic syndrome. High body mass index (BMI) is a risk factor.
Rapid active weight loss	Bile is lithogenic because of reduced bile acid secretion.
Parity	Moderately elevated risk with increased parity.
Diabetes mellitus	Good association in Mexican Americans. Unclear association in other groups.
Ileal/Crohn's disease	Bile is lithogenic when ileal reabsorption is decreased.
Total parenteral nutrition (TPN)	Usually, sludge and pigment stones caused by bile stasis and gallbladder distention.
Medications	Estrogen therapy, oral contraceptive use, and octreotide therapy.
Diet	Association with high consumption of simple sugars. Low prevalence in vegetarians.
Spinal cord injury	Abnormal gallbladder motility may be a factor.
Miscellaneous	Celiac disease, vagotomy, and duodenal diverticula are rare associations. No significant relationship with hyperlipoproteinemia.

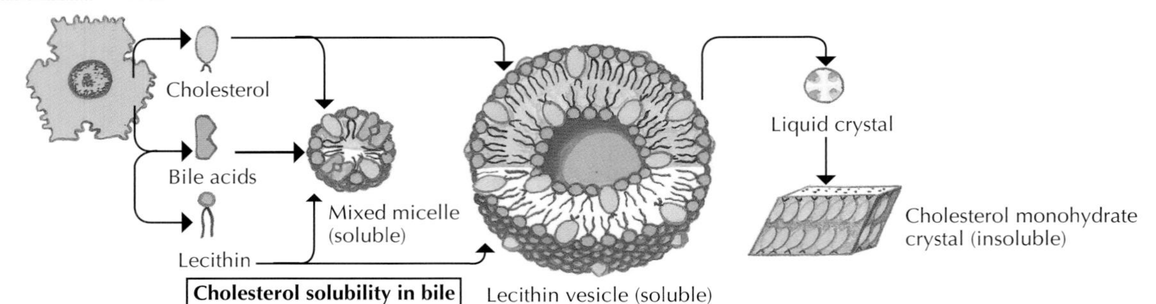

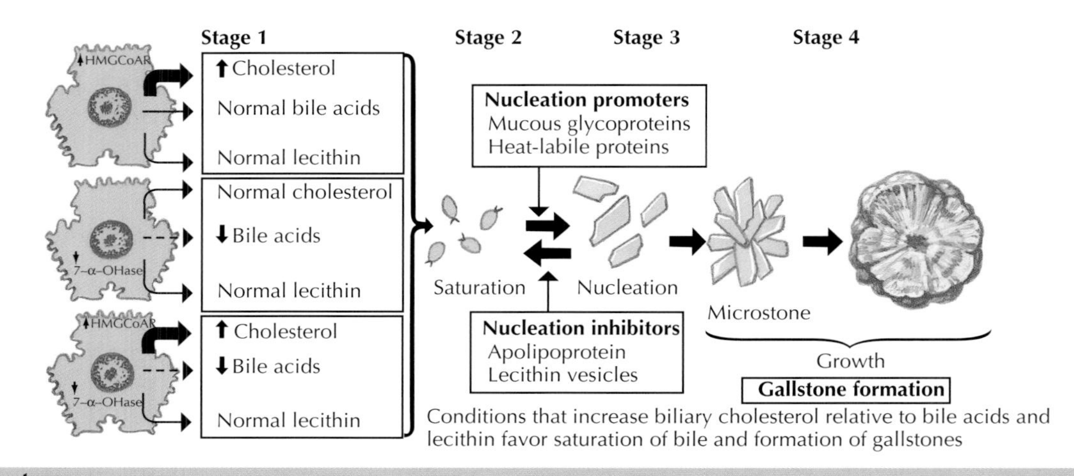

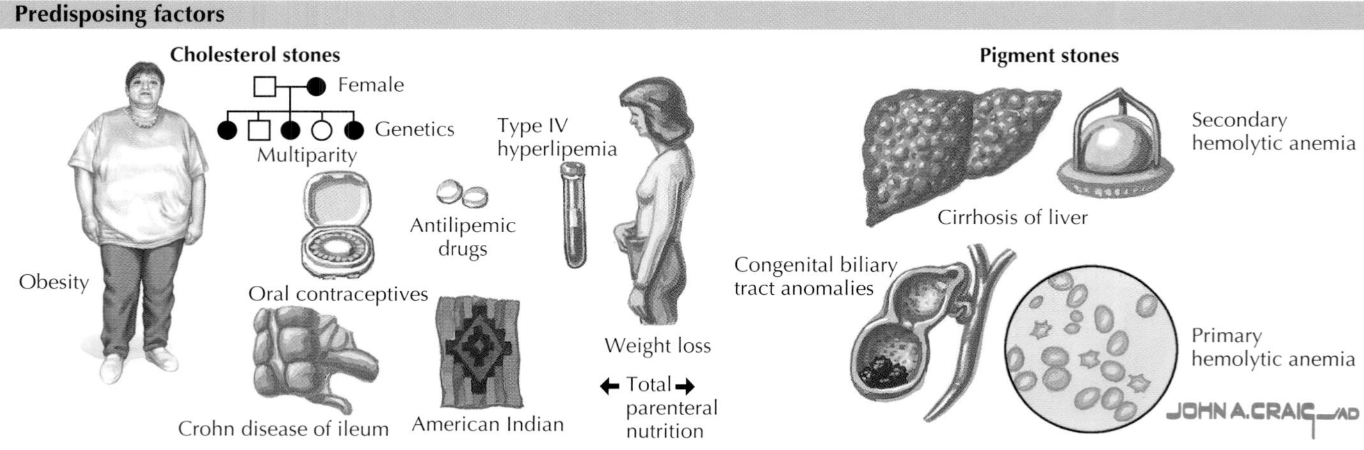

Figure 201-1 *Gallstones: Pathogenesis of Stone Formation.*

(lithogenic). Lithogenic bile that stays within the gallbladder alters gallbladder motility and stimulates mucous secretion by the gallbladder epithelium.

Gallbladder sludge, the reversible but early stage of gallstone formation, is the suspension of precipitated bile dispersed in a viscous, mucin-rich liquid phase. Its chemical composition is mostly cholesterol monohydrate crystals, calcium bilirubinate, calcium phosphate, and calcium carbonate. Sludge may disappear, or it may progress to develop gallstones. Sludge may cause cholecystitis, pancreatitis, or biliary pain. Sludge formation is promoted by prolonged fasting, total parenteral nutrition (TPN), and other causes of gallstone formation. Sludge formation can be prevented in patients receiving TPN and in those undergoing rapid weight loss by oral administration of ursodeoxycholic acid, 8 to 10 mg/kg daily.

Brown pigment stones are morphologically, chemically, and clinically distinct from black pigment stones. The brown pigment stone is laminated with alternating regions of brown and tan material and tends to cake when powdered. Brown pigment stones contain only small amounts of calcium phosphates and calcium carbonates. They develop in the gallbladder and in the intrahepatic and extrahepatic ducts and are associated with polymicrobial infection (e.g., *Escherichia coli*). Bacterial degradation by enzymes, primarily β-glucuronidase, deconjugates bilirubin and lecithin to free fatty acids. Brown pigment stones are mostly radiolucent. A decrease in effective bile salt micelles promotes cholesterol supersaturation. The predominant symptoms of brown pigment stone disease are jaundice, chills, fever, and abdominal pain. Cholangitis is common.

Black pigment stones are composed largely of calcium salts of unconjugated bilirubin, carbonate, and phosphate. The bilirubinate salts of black pigment stones are amorphous. Black pigment stones are clinically associated with hemolytic syndromes, cirrhosis of the liver, chronic alcoholism, malaria, TPN, and old age. Almost 50% of patients with sickle cell anemia and 15% to 40% of those with sickle cell disease have pigment stones by age 20 years.

More than 66% of black pigment stones but only 10% of cholesterol stones are radiopaque on abdominal plain films. The increased concentration of unconjugated bilirubin in the pathogenesis of black pigment stones is probably nonbacterial and nonenzymatic. Gallbladder stasis and defective acidification of gallbladder bile in an alkaline environment favor the formation of calcium phosphate and calcium carbonate. In hemolytic anemias, bilirubin levels increase 10-fold, with elevated gallbladder volume and stasis. Black pigment stones also develop in children and young adults.

ADDITIONAL RESOURCES

Tazuma S: Epidemiology, pathogenesis, and classification of biliary stones (common bile duct and intrahepatic), *Best Pract Res Clin Gastroenterol* 20:1075-1083, 2006.

van Erpecum KJ: Cholesterol-gallstone formation: more than a biliary lipid defect (editorial), *J Lab Clin Med* 144:121-123, 2004.

Acute Cholecystitis

C. S. Pitchumoni

Acute cholecystitis is inflammation of the gallbladder after persistent obstruction of the gallbladder outlet from an impacted stone, resulting in increased gallbladder pressure, rapid distention, decreased blood supply, and gallbladder ischemia, with subsequent bacterial invasion, inflammation, and possible perforation (**Fig. 202-1**). Approximately 10% to 20% of patients with symptomatic gallstones develop acute cholecystitis.

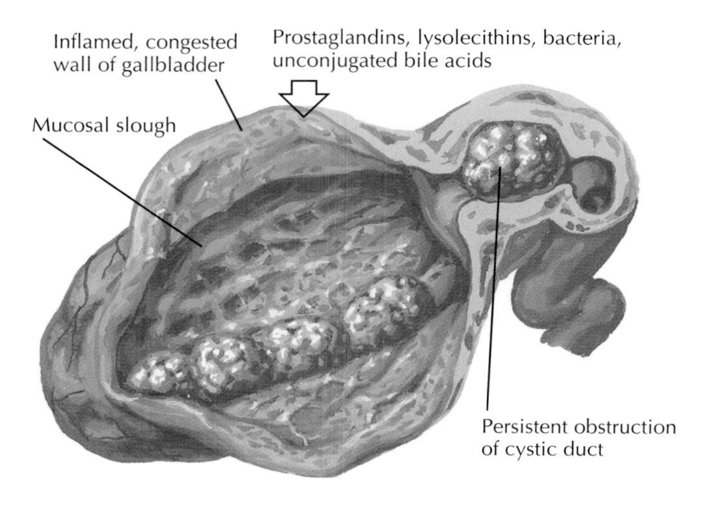

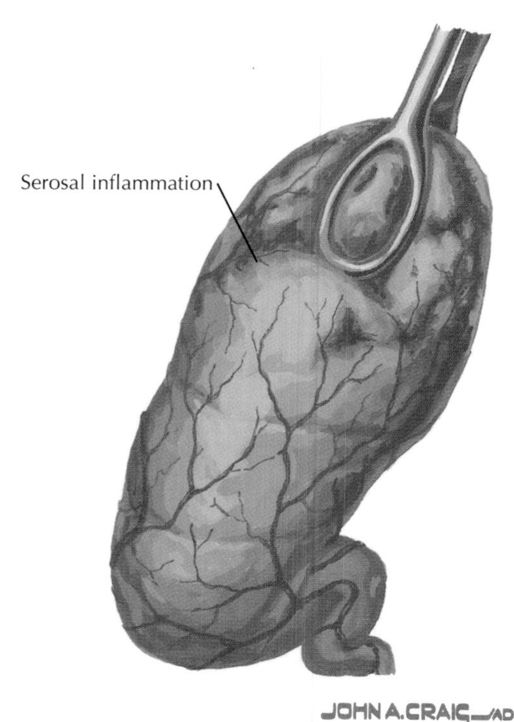

Figure 202-1 *Acute Calculous Cholecystitis.*

CLINICAL PICTURE

Steady and severe abdominal pain over the right upper quadrant radiating to the back, the right scapula, or the right clavicular area and associated with fever, nausea, anorexia, and vomiting is the main symptom of acute cholecystitis. Physical examination reveals tenderness over the gallbladder area. As the gallbladder area is palpated, the patient is asked to take a deep breath that brings the gallbladder down to the palpating hand. At the height of inspiration, as the gallbladder touches the palpating hand, the breath is arrested with a gasp *(Murphy sign)*. The sign is not found in chronic cholecystitis. Sensitivity to Murphy sign may be diminished in elderly patients. Complications of acute cholecystitis are empyema of the gallbladder, gangrene with perforation, intraabdominal abscess, and diffuse peritonitis.

Laboratory findings include leukocytosis with a shift to the left and mildly elevated bilirubin and alkaline phosphatase levels. Serum amylase and lipase levels are normal or only mildly elevated unless there is concomitant acute pancreatitis. The differential diagnosis includes acute pancreatitis, appendicitis, acute hepatitis, peptic ulcer disease, disease of the right kidney, right-sided pneumonia, Fitz-Hugh-Curtis syndrome (gonococcal perihepatitis), liver abscess, perforated viscus, and cardiac ischemia.

DIAGNOSIS

Imaging studies are needed to confirm the diagnosis and to exclude other causes of right upper quadrant pain. Imaging studies also help diagnose severe complications such as emphysematous cholecystitis and perforation, which require emergency surgery. Abdominal ultrasound findings include gallstones, sludge, lumen distention, mural thickening with a hypoechoic or anechoic zone within the thickened wall, increased flow on color Doppler sonography, and pericholecystic fluid; none of these, however, is pathognomonic of acute cholecystitis. Sonographic Murphy sign is defined as the presence of maximal tenderness elicited by direct pressure of the transducer over the gallbladder. Abdominal computed tomography (CT) is needed only when the diagnosis is vague or when abscess formation or gangrene is suspected. Gallstones, sludge, gallbladder distention, mural thickening, pericholecystic fluid, and subserosal edema are major findings.

Magnetic resonance cholangiopancreatography (MRCP), a noninvasive technique for evaluating the intrahepatic and extrahepatic bile ducts, is superior to ultrasound for detecting stones in the cystic duct. Magnetic resonance imaging (MRI) also helps in diagnosing complications of acute cholecystitis. Hepatobiliary iminodiacetic acid (HIDA) scintigraphy involves intravenous injection of a nuclear isotope to determine the patency of the cystic duct. It also demonstrates patency of the common bile duct (CBD) and ampulla. Positive HIDA testing is nonvisualization of the gallbladder with normal excretion of the isotope into the CBD and duodenum. HIDA scanning is highly sensitive

(95%) and specific (90%) for acute calculous cholecystitis. In acute *acalculous* cholecystitis, HIDA findings may be falsely negative because the cystic duct remains patent. False-positive results occur when the gallbladder cannot be visualized despite an unobstructed cystic duct, such as in severe liver disease and hyperbilirubinemia, after biliary sphincterotomy, and in fasting patients receiving total parenteral nutrition (TPN) who already have a maximally full gallbladder because of prolonged lack of stimulation.

TREATMENT AND MANAGEMENT

Management of acute cholecystitis includes bowel rest, parenteral fluids and nutrition, and intravenous antibiotics. Common organisms include *Escherichia coli*, *Enterococcus*, *Klebsiella*, and *Enterobacter*. The need for antibiotics in uncomplicated cholecystitis is debatable, although in clinical practice, most patients receive antibiotics. A combination of ampicillin (2 g intravenously every 4 hours) and gentamicin (dosed according to weight and renal function) is one of the many choices for empiric treatment. β-Lactam–based therapy and fluoroquinolones are other options.

The definitive therapy for acute cholecystitis is *cholecystectomy*. Early laparoscopic cholecystectomy, within 7 days of onset of symptoms, has become the preferred approach. Laparoscopic cholecystectomy eliminates the need to incise the rectus abdominis muscle, reduces postoperative pain, and shortens hospital stay and convalescence. The risk for CBD injury is 0.2% in both laparoscopic and open surgical approaches. Patients at high risk may be treated with percutaneous cholecystostomy in association with antibiotic therapy as a temporary measure. Other complications of laparoscopic cholecystectomy include bowel and liver lacerations, bile leak, gallstone spillage and abscess formation, and major bleeding. However, the risk for most of these complications is low with the increasing experience of the surgeon.

A new era in gastrointestinal surgery involves cholecystectomy with access to the peritoneal cavity through normal anatomy, known as "natural orifice" transluminal endoscopic surgery.

ADDITIONAL RESOURCES

Guruswamy KS, Samaj K: Early versus delayed laparoscopic cholecystectomy for acute cholecystitis, *Cochrane Database. Syst Rev* (4):CD00540, 2006.

Keus F, de Jong JA, Gooszen HG, et al: Laparoscopic versus small-incision cholecystectomy for patients with symptomatic cholecystolithiasis, *Cochrane Database Syst Rev* (4):CD06229, 2006.

Trowbridge RL, Rutkowski NK, Shojana KG: Does this patient have acute cholecystitis? *JAMA* 299:80-86, 2003.

Cholecystitis: Complications

203

C. S. Pitchumoni

Gangrenous cholecystitis is a term used to describe severe gallbladder inflammation with mural necrosis associated with an increased risk for perforation. Computed tomography of the abdomen may show mural necrosis, gas in the wall or lumen, intramural hemorrhage, pericholecystic abscess, or absent gallbladder wall enhancement (**Fig. 203-1**).

Emphysematous cholecystitis is severe acute cholecystitis caused by a gas-forming organism such as *Clostridium perfringens*, noted often in elderly persons and diabetic patients. Perforation is a complication. Abdominal ultrasound is helpful in the diagnosis.

Acalculous cholecystitis refers to inflammation of the gallbladder without gallstones, seen in 2% to 15% of patients undergoing cholecystectomy. It usually occurs in critically ill adults or after trauma, burns, or major surgery. Risk factors include diabetes mellitus, sepsis, prolonged fasting, acquired immunodeficiency syndrome (AIDS), and hepatic arterial chemotherapy. Its pathogenesis is attributed to the occlusion of the cystic duct by viscous bile. The prognosis is poor, and mortality is 60%. Complications include mural necrosis, gangrene, and perforation. Abdominal ultrasonography reveals gallbladder distention, mural thickening (>5 mm), pericholecystic fluid, positive sonographic Murphy sign, and emphysematous cholecystitis with gas bubbles arising in the fundus of the gallbladder (champagne sign). Percutaneous aspiration of the bile under sonographic guidance may help with the diagnosis. Percutaneous cholecystostomy is an emergency procedure required to temporize the critical illness.

Chronic cholecystitis may be secondary to repeated attacks of uncomplicated acute cholecystitis or without prior attacks (see Chapter 204). The symptoms are vague and may be only nonspecific epigastric or right upper quadrant pain. Histologically, patients have chronic inflammatory cell infiltration of the gallbladder associated with gallstones and thickening of the gallbladder wall.

Mirizzi syndrome is a rare condition in which a stone impacted in the cystic duct causes severe inflammation and may erode into the common bile duct (CBD), producing an inflammatory mass around the cholecystocholedochal fistula and obstructing the CBD. The clinical relevance is that the surgeon should not mistake the CBD for the cystic duct.

AIDS cholangiopathy manifesting late in the course of illness is most often caused by cytomegalovirus, *Cryptosporidium*, *Microsporum*, and *Mycobacterium avium*. The disease may manifest as papillary stenosis, sclerosing cholangitis, and extrahepatic duct strictures. The view on endoscopic retrograde cholangiopancreatography resembles primary sclerosing cholangitis. Treatment of AIDS cholangiopathy is unsatisfactory. Endoscopic sphincterotomy, stricture dilatations, and stenting are options.

ADDITIONAL RESOURCES

Elwood DR: Cholecystitis, *Surg Clin North Am,* 88:1241-1252, 2008.

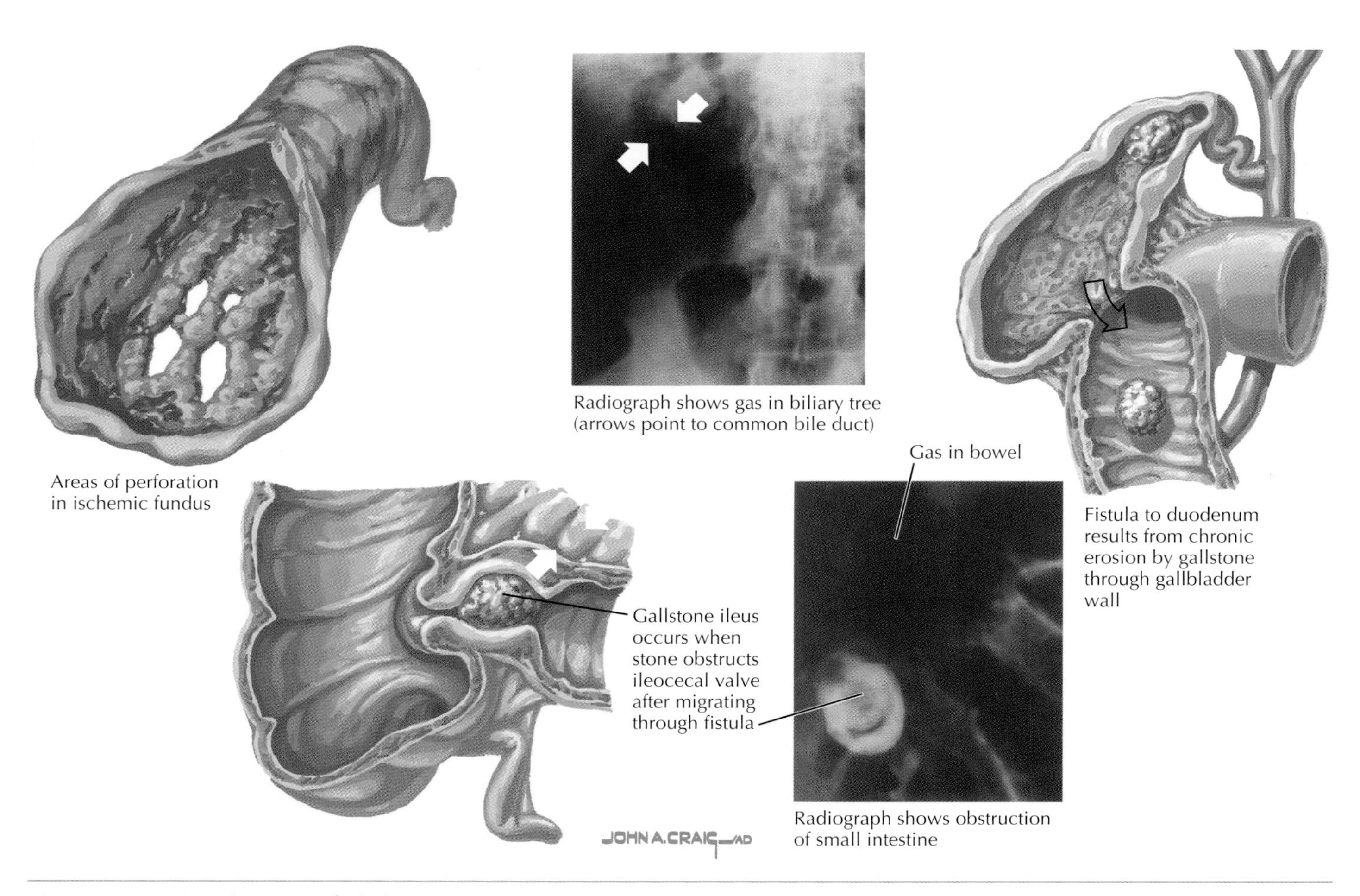

Figure 203-1 *Complications of Cholecystitis.*

Chronic Cholecystitis

204

C.S. Pitchumoni

Cholelithiasis and cholecystitis are classified into three types based on presentation: (1) silent gallstones, in which gallstones are detected accidentally and are truly asymptomatic; (2) symptomatic gallstone disease; and (3) abdominal symptoms caused by a comorbid condition such as peptic ulcer.

ASYMPTOMATIC GALLSTONE DISEASE

Almost 60% to 80% of the gallstones are incidentally found during routine abdominal sonography. Most patients do not have symptoms, even after follow-up periods as long as 20 years. Approximately 20% of persons with silent gallstones may develop symptoms by 15 years. Patients with asymptomatic gallstones do not require cholecystectomy except in countries where gallbladder carcinoma is prevalent. **Box 204-1** lists indications for prophylactic cholecystectomy.

Porcelain gallbladder is a rare, asymptomatic, chronic cholecystitis characterized by intramural calcification of the gallbladder wall. Plain abdominal radiographs reveal an incidental calcified gallbladder. The incidence of gallbladder carcinoma is as high as 33%, and prophylactic cholecystectomy is warranted.

SYMPTOMATIC GALLSTONE DISEASE

Clinical Picture

Patients with symptomatic gallstones have epigastric (or right upper quadrant) pain radiating to the back, right scapula, or right shoulder and lasting longer than 30 minutes (**Fig. 204-1**). Although frequently known as "biliary colic," the term is a misnomer because generally, the pain is constant and not colicky. The pain may be mild, moderate, or severe. Nonspecific symptoms such as bloating, flatulence, and heartburn are no more frequent in patients with gallstones than in the general population.

Box 204-1 Indications for Prophylactic Cholecystectomy in Patients with Silent Gallstones*

1. High risk for gallbladder cancer
   - Native American women with gallstones
   - Solitary stone or stone burden >3 cm
   - Porcelain gallbladder (calcification in wall)
   - Gallbladder polyps >12 mm
2. Carriers of *Salmonella typhosa*
3. Sickle cell disease (calcium bilirubinate stones)
4. Patients undergoing bariatric surgery
5. Lifestyle: working for long periods in remote parts of world with poor medical facilities

**Diabetes mellitus is not an indication for prophylactic cholecystectomy.*

Gallstones can cause biliary colic, acute cholecystitis, ascending cholangitis, and acute pancreatitis. Stones may fistulate into the duodenum from the gallbladder. *Bouveret's syndrome* refers to an impacted stone in the duodenum causing obstruction. A gallstone may become impacted near the terminal ileum, leading to small-bowel obstruction; in *gallstone ileus*, a supine abdominal radiograph may show air in the biliary tree. Physical examination will reveal right upper quadrant tenderness. Fever, abdominal rigidity, and rebound tenderness are signs of acute cholecystitis.

Diagnosis

Abdominal ultrasound, the "gold standard" for the diagnosis of gallbladder stones, has an excellent sensitivity and specificity. The test is conducted with the patient having fasted for at least 8 hours so that the stones can be seen in a distended gallbladder, surrounded by bile. Ultrasonographic criteria constitute an echogenic focus that casts an acoustic shadow and seeks gravitational dependency. Although echogenic, *sludge* (or multiple small gallstones) does not cast an acoustic shadow. Sludge is viscous and does not seek gravitational dependency as rapidly as "gravel."

Plain films of the abdomen and oral cholecystography are seldom used to diagnose gallstone disease. Only 15% to 20% of gallstones are seen on plain films. Computed tomography and magnetic resonance imaging are not indicated to diagnose gallstone disease but may be helpful in diagnosing its complications.

Treatment and Management

Therapy for gallstone disease is surgical, and laparoscopic surgery has made the procedure more acceptable than open cholecystectomy. Nonsurgical therapy is now seldom used. Ursodeoxycholic acid is advocated when lithogenic bile production is increased or microlithiasis is suspected, as occurs in patients with idiopathic pancreatitis, but has no role in symptomatic gallstone disease. Extracorporeal shock wave lithotripsy has also declined in use. A number of percutaneous radiologic interventions are available in emergency situations.

ADDITIONAL RESOURCES

Marshall HU, Einarsson C: Gallstone disease, *J Intern Med* 261:529-542, 2007.

Portincasa P, Moschetta A, Petruzzelli M, et al: Symptoms and signs of gallbladder stones, *Best Pract Res Clin Gastroenterol* 20:1017-1029, 2006.

Sanders G, Kingsworth AN: Gallstones, *BMJ* 335:295-299, 2007.

Sudden obstruction (biliary colic)

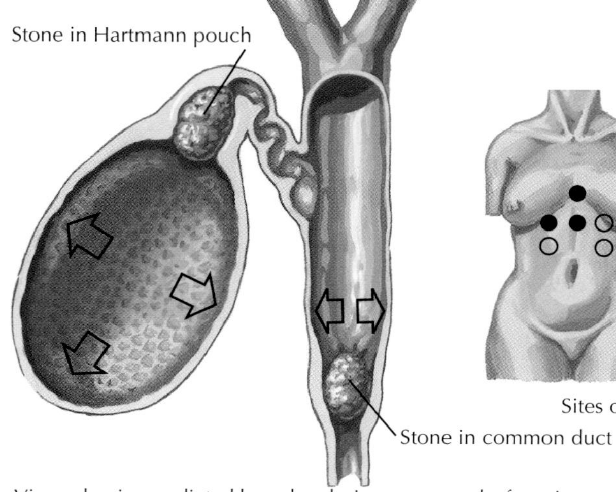

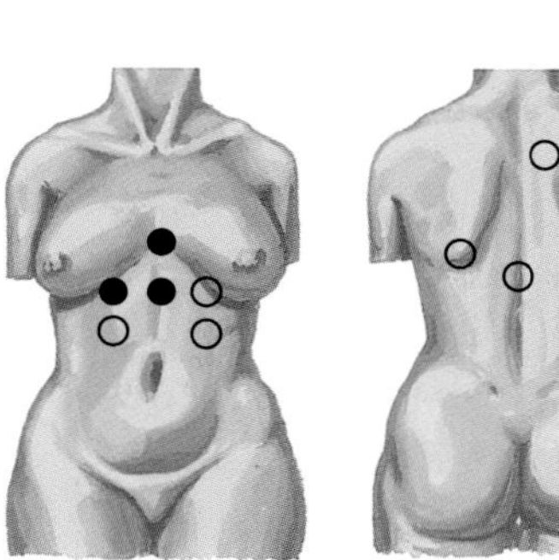

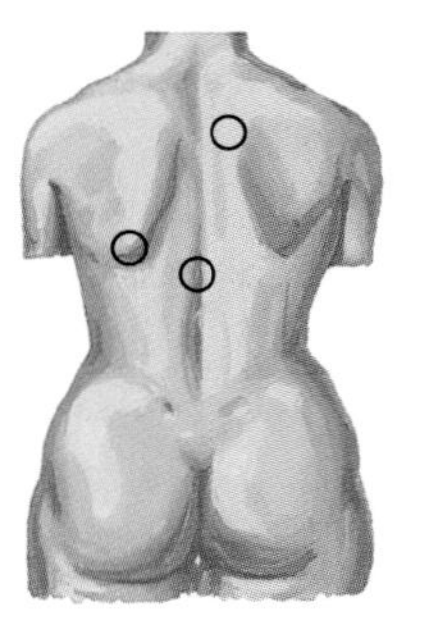

Visceral pain, mediated by splanchnic nerve, results from increased intraluminal pressure and distention caused by sudden calculous obstruction of cystic or common duct

Persistent obstruction (acute cholecystitis)

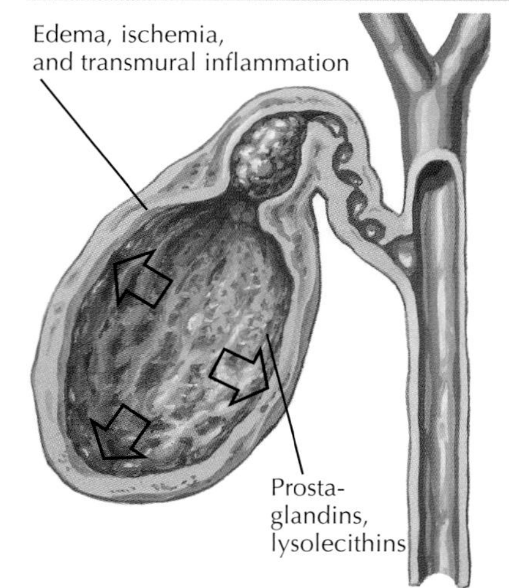

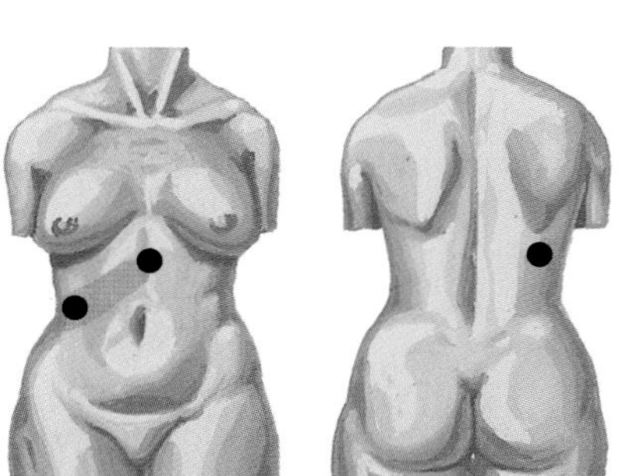

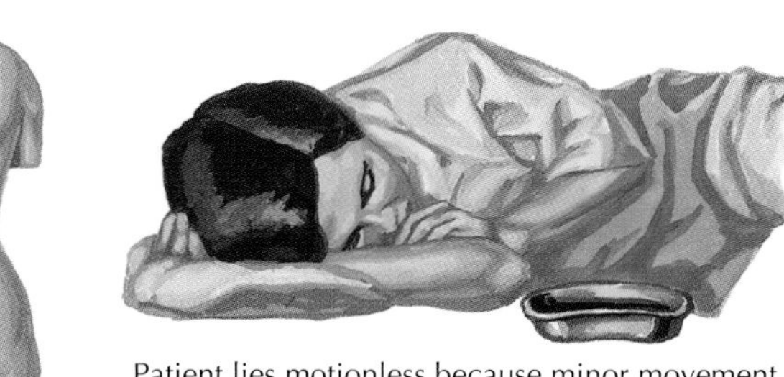

Parietal epigastric or right upper quadrant pain results from ischemia and inflammation of gallbladder wall caused by persistent calculous obstruction of cystic duct. Prostaglandins and lysolecithins released.

JOHN A.CRAIG—AD

Figure 204-1 *Mechanisms of Biliary Pain.*

Choledocholithiasis and Cholangitis

205

C. S. Pitchumoni

Stones in the common bile duct (CBD) are either primary or secondary. Primary stones are formed de novo in the CBD as a result of bacterial action on phospholipid and bilirubin (see Chapter 201). Secondary stones, including cholesterol and pigment stones, are formed in the gallbladder and are passed into the CBD.

CHOLEDOCHOLITHIASIS

The prevalence of CBD stones varies depending on several factors. Along with stones in the gallbladder, choledocholithiasis is noted in 12% to 15% of patients undergoing routine cholecystectomy. Also, in the immediate postoperative period, almost 1% of cholecystectomy patients have a retained stone. The prevalence of concomitant CBD stones increases with advancing age. Hemolysis predisposes patients to black pigment stones. Bacterial or parasitic infection of the biliary tract, a foreign body in the duct (surgical sutures and clips), and juxtapapillary duodenal diverticula increase the prevalence of brown pigment stones. *Ascaris* infection is a rare cause of CBD stones in endemic areas. Anatomic abnormalities such as low entry of the cystic duct (<3.5 cm from ampulla) and abnormal sphincter of Oddi motility predispose patients to CBD stones. Intrahepatic calcium bilirubinate stones (Asian cholangitis) are noted in Japanese and Korean patients. An iatrogenic form of choledocholithiasis is the development of pigment stones (regardless of original type of gallstones) after endoscopic sphincterotomy, which permits bacterial colonization of the CBD, deconjugation of bilirubin, and formation of pigment stones.

Clinical Picture

If the patient has an intact gallbladder, choledocholithiasis and cholecystolithiasis are clinically similar, and the diagnosis of CBD stones cannot be made based solely on symptomatology. CBD stones may be asymptomatic for many years or they may present with asymptomatic jaundice or with biliary colic, pancreatitis, or acute suppurative cholangitis (**Fig. 205-1**). "Biliary colic," a misnomer for the constant pain over the right upper quadrant, lasts for 30 minutes to several hours and is associated with nausea and vomiting but is not related to food intake. Prolonged obstruction for 4 to 5 years without therapy leads to biliary cirrhosis. Abnormal laboratory values indicate cholestasis with an elevated serum bilirubin concentration (2-14 mg/dL) and a greatly elevated alkaline phosphatase level.

Diagnosis

Abdominal ultrasound and computed tomography (CT) are initial diagnostic procedures. However small bile duct stones are missed by the procedures. Magnetic resonance cholangiopancreatography (MRCP) is an important alternative to exclude choledocholithiasis in patients at low or intermediate risk. Endoscopic ultrasound (EUS) is less invasive, entails no complications, and has a sensitivity and specificity of 90% to 100%. As a diagnostic modality for CBD stones, Endoscopic retrograde cholangiopancreatography (ERCP) has lost its popularity in view of complications. In some cases, CBD stones are diagnosed at surgery by palpation of the duct, intraoperative cholangiography, or choledochoscopy.

Treatment and Management

The predominant method of treating choledocholithiasis is by ERCP with sphincterotomy and stone extraction (**Fig. 205-2**). About 85% to 90% of CBD stones can be removed with Dormia basket or balloon catheter. Other options available for difficult stones include mechanical lithotripsy, intraductal shock wave lithotripsy, laser lithotripsy, extracorporeal shock wave lithotripsy, and biliary stenting.

CBD stones diagnosed at laparoscopic cholecystectomy may require conversion to open surgery and CBD exploration. Advances in laparoscopic biliary surgery now allow experienced surgeons to manage CBD stones at the same time. Postoperative ERCP is necessary if intraoperative removal of CBD stones is unsuccessful. Elective cholecystectomy is recommended in most patients after endoscopic clearance of CBD stones, if the gallbladder is intact. In elderly patients with multiple comorbid conditions or with cirrhosis, surgery may be risky, and endoscopic therapy alone may be acceptable.

CHOLANGITIS

In 1877, Charcot first described *pyogenic cholangitis* in patients with right upper quadrant pain, fever, and jaundice (Charcot triad). A severe form of cholangitis includes two additional features: hypotension and mental confusion (Reynolds pentad). In more than 80% of patients, the most important contributory cause for pyogenic cholangitis is CBD obstruction by choledocholithiasis, promoting bacterial overgrowth. Cholangitis may be secondary to malignant CBD obstruction or may be iatrogenic, caused by instrumentation (ERCP, stricture dilatation), postoperative biliary stricture, or papillary stenosis. Normally sterile bile becomes infected when local defenses are impaired. Elevated intraductal pressure decreases resistance to bacterial growth in bile, an otherwise excellent culture medium. Once cholangitis starts, infection may spread locally to the liver and to the systemic circulation, along with toxemia. Malignant obstruction, often total, is less likely to cause cholangitis, presumably because reflux of duodenal contents does not occur.

Clinical Picture

The classic picture of intermittent fever, pain, and jaundice is seen in 50% to 70% of patients with cholangitis. Shaking chills suggesting bacteremia occurs in two thirds of patients. Cholan-

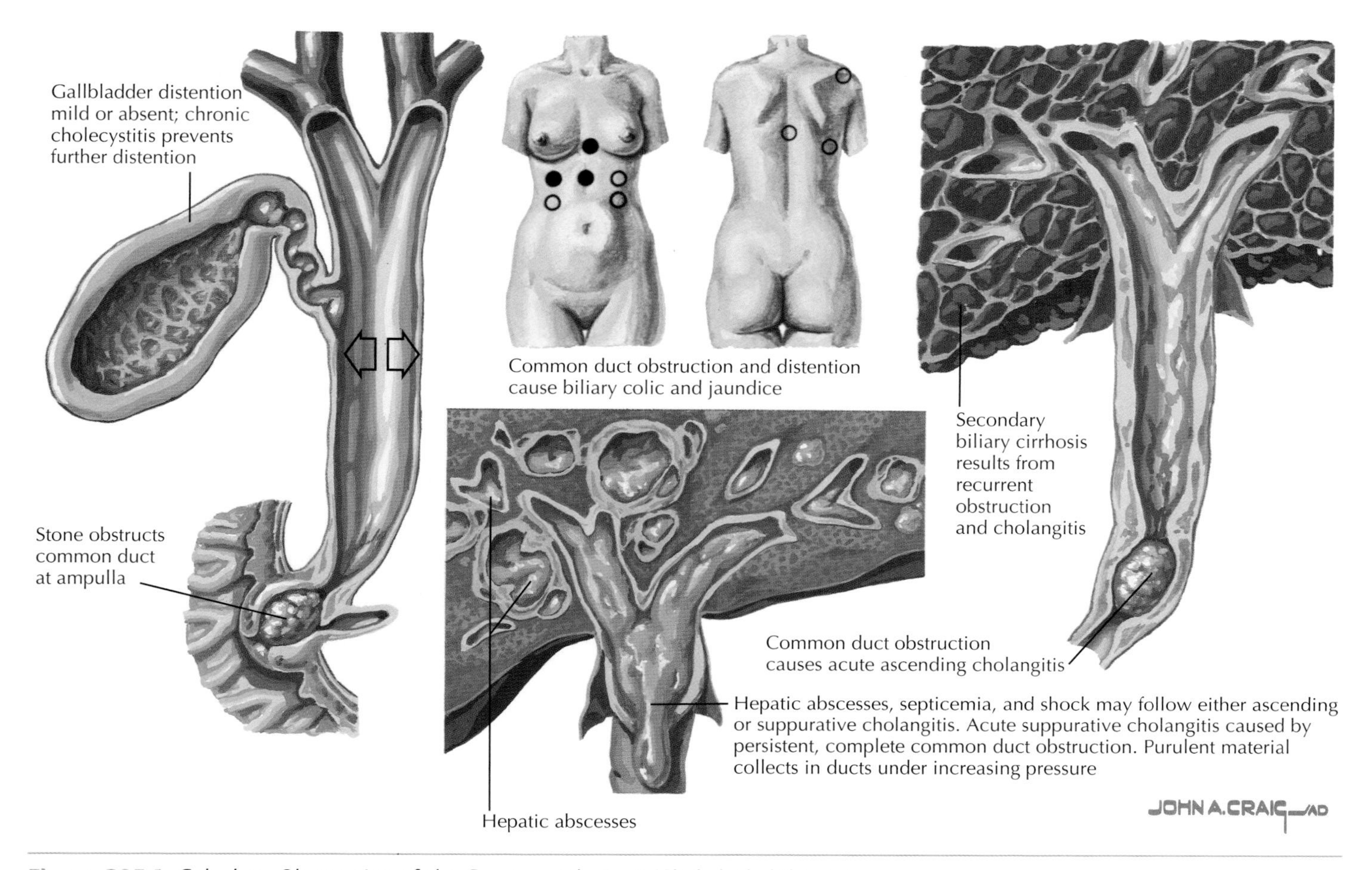

Figure 205-1 *Calculous Obstruction of the Common Bile Duct (Choledocholithiasis).*

gitis in the elderly patient should be suspected when there is sudden onset, mental confusion, lethargy, and delirium. Abdominal pain may be mild or even absent. Physical findings include fever, right upper quadrant tenderness, and jaundice. Rarely, hypotension and mental confusion occur, indicating a severe form of the disease. Untreated bacterial cholangitis has a poor prognosis, and even with treatment, mortality rates range from 5% to 30%.

Diagnosis

Abnormal laboratory findings include leukocytosis, mildly elevated bilirubin levels, and elevated alkaline phosphatase levels. Blood culture should be performed early in the evaluation. Rarely, liver function abnormalities mimic acute hepatitis, with greatly elevated serum levels of aspartate transaminase and alanine transaminase. However, normal liver enzyme levels do not exclude cholangitis. Hyperamylasemia, when noted, is mild and less than three times the upper limit of normal.

The diagnosis of cholangitis is supported by abnormal laboratory values and imaging studies. Abdominal ultrasonography helps in evaluating the size of the CBD and the presence of stones. Abdominal CT may show the same findings although less precisely, and MRCP better delineates the ductal morphology. Blood culture is usually positive for enteric organisms. Organisms that frequently cause cholangitis are *Escherichia coli, Klebsiella, Enterococcus, Enterobacter, Streptococcus*, and *Pseudomonas aeruginosa;* anaerobic bacteria are found in less than 10% of patients. Leukopenia, thrombocytopenia, coagulopathy, and renal failure suggest severe disease.

Treatment and Management

Cholangitis is managed with appropriate antibiotics and aggressive fluid resuscitation. In patients with impacted stones and evidence of cholangitis, ERCP with sphincterotomy and stone extraction is warranted after vital signs are stabilized. The goal is to relieve the obstruction to bile flow as early as possible. A nasobiliary drainage catheter helps in bile drainage and in subsequent cholangiographic studies. Percutaneous transhepatic cholangiography under local anesthesia is another option, although when used to guide biliary drainage, it may be complicated by bile leak, biliary vascular fistula, pneumothorax, bile peritonitis, and catheter-related sepsis. Surgical exploration for ductal decompression is performed after initial management. Failure of biliary decompression as a result of an inability to perform ERCP indicates the need for emergency surgery.

Asian (Oriental) *cholangitis*, or recurrent pyogenic cholangitis, refers to intrahepatic stone disease associated with episodes of pyogenic cholangitis reported from Hong Kong, China, Korea, and Japan. Intrahepatic stones are mostly calcium bilirubinate stones that develop in patients with chronic bacterial and parasitic infections, bile stasis, or environmental factors such as a low-protein diet.

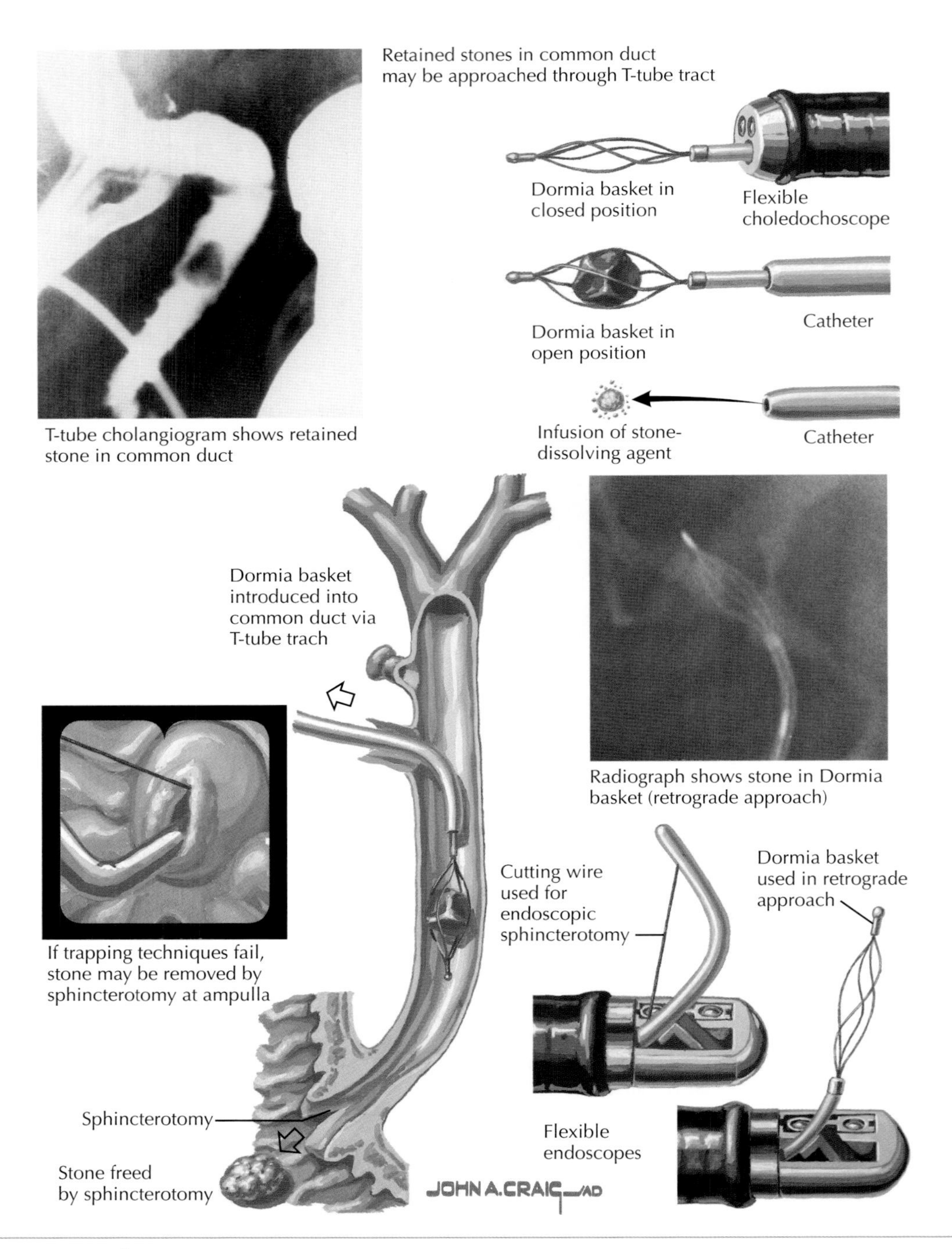

Figure 205-2 *Management of Stones.*

ADDITIONAL RESOURCES

Ayub K, Imada R, Slavin J: Endoscopic retrograde cholangiopancreatography in gallstone associated acute pancreatitis, *Cochrane Database Syst Rev* (4):CD 003630, 2004.

Caddy GR, Tham CKT: Symptoms, diagnosis, and endoscopic management of common bile duct stones, *Best Pract Res Clin Gastroenterol* 20:1085-1101, 2006.

Qureshi WA: Approach to the patient who has suspected acute bacterial cholangitis, *Gastroenterol Clin North Am* 35:409-423, 2006.

Weinberg BM, Shindy W, Lo S: Endoscopic balloon sphincter (sphincteroplasty) versus sphincterotomy for common bile duct stones, *Cochrane Database Syst Rev* (4):CD004890, 2006.

Sphincter of Oddi Dysfunction

206

C. S. Pitchumoni

The sphincter of Oddi (SO) is a fibromuscular sheath encircling the distal common bile duct (CBD), pancreatic duct, and common channel. Mechanical and physiologic dysfunction of the SO is a poorly understood disorder. The SO controls the flow of bile and pancreatic secretions into the duodenum, preventing reflux of duodenal juice into the pancreaticobiliary system. Papillary stenosis refers to the type of SO dysfunction (SOD) related to structural abnormality. Biliary dyskinesia is a functional blockage at the high-pressure zone, attributed to spasm, hypertrophy, or neuropathy of the sphincteric nerves. The advent of biliary manometry has revealed better data on this obscure disorder, but confusion remains with regard to many aspects of SOD.

ANATOMY AND PHYSIOLOGY

Although the SO is often considered a single sphincter, it is in fact composed of a biliary sphincter and a pancreatic sphincter. The common ductal segment passes obliquely through the duodenal wall and terminates at the papilla of Vater, a small nipple-like protrusion less than 1 cm in diameter. Numerous variations are observed as the common bile and pancreatic ducts join (**Fig. 206-1**).

The SO has a rich supply of nerves—intrinsic catecholamine-containing neurons, nonadrenergic noncholinergic (NANC) including substance P, vasoactive intestinal peptide (VIP), somatostatin, calcitonin gene–related peptide, met-enkephalin-like immunoreactivity, prominent galanin-like immunoreactivity, and nitric oxide—most of which may play a role in SO function. Spontaneous motor activity of the SO is primarily myogenic in nature and is presumed to be regulated by interstitial cells of Cajal. During the fasting stage, SO motility is integrated with migrating motor complex, which alters the flow of bile into the duodenum. During the fed state, myoelectrical potentials within the SO vary and may be influenced by endogenous hormones such as cholecystokinin.

CLINICAL PICTURE

A poorly understood functional disorder, SOD involves the biliary or pancreatic sphincter, and the presentation varies (see Fig. 206-1). SOD is more common in women in the fourth to sixth decades of life. Lack of clarity about the clinical picture of SOD is attributed to many factors. Confusion surrounds the terminology. Although they refer to SOD, the terms *papillary stenosis*, *sclerosing papillitis*, *biliary dyskinesia*, and *postcholecystectomy syndrome* may be etiologically different. Some differences are structural, and others are functional.

Diagnostic studies are imprecise. SOD can cause two clinical syndromes: recurrent biliary or pancreatic-type pain. Motility abnormalities of the biliary sphincter cause biliary pain with or without enzyme abnormalities (**Box 206-1**). Biliary symptoms starting or persisting after cholecystectomy may be attributed to unmasking preexisting SOD or precipitating SOD as a result of the severing of nerve fibers that pass between the gallbladder and the SO through the cystic duct. However, SOD after cholecystectomy is rare; the rate is less than 1%. SOD is also a reported cause for *idiopathic acute pancreatitis* or *chronic pancreatitis*. Increased basal pressure of SO has been reported in some patients with pancreatitis of undetermined cause. **Box 206-2** lists the modified classification of pancreatic-type SOD.

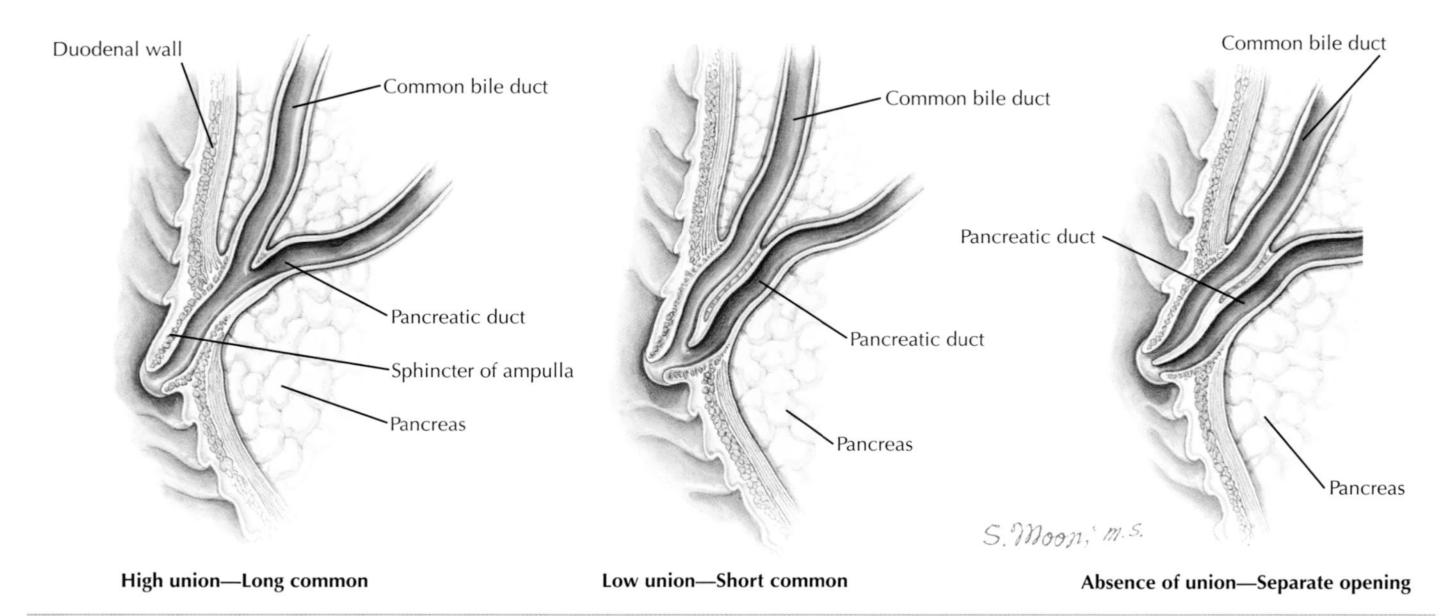

Figure 206-1 *Variations in Ductal Anatomy in Sphincter of Oddi Dysfunction.*

Box 206-1 Milwaukee Biliary Group Classification of Biliary Sphincter of Oddi Dysfunction

Type 1 (Biliary Stenosis)

Biliary-type pain

Abnormal liver function results (AST/ALP >2 × normal on two or more occasions)

Dilated common bile duct (>12 mm)

Delayed drainage of ERCP contrast (>45 minutes)

Type 2

Biliary-type pain and one or two of type 1 conditions

Type 3

Biliary-type pain with no other abnormalities

Data from Hogan WJ, Geenen JE: Endoscopy *20:179-183, 1988.*
ALP, Alkaline phosphatase; AST, aspartate transaminase, ECRP, endoscopic retrograde cholangiopancreatography.

Box 206-2 Modified Classification of Pancreatic-Type Sphincter of Oddi Dysfunction

Type 1

Pancreatic-type pain

Elevated amylase/lipase levels (>1½-2 × normal)

Dilated pancreatic duct diameter (>6 mm in head or >5 mm in body)

Type 2

Pancreatic-type pain with only one of type 1 criteria

Type 3

Pancreatic-type pain only

Data from Sherman S, Troiano FP, Hawes RH, et al: Am J Gastroenterol *86:586-590, 1991.*

DIAGNOSIS

Rome III criteria proposed for diagnosis of functional gallbladder and SOD specify three subsets: functional gallbladder disorder, functional biliary SO disorder, and functional pancreatic SO disorder. For a diagnosis of SOD, SO *manometry* must be performed by a competent professional at endoscopic retrograde cholangiopancreatography (ERCP). Basal SO pressure equal to or greater than 40 mm Hg is considered abnormal. SO pressure is not constant and may vary with the precipitation of symptoms. Separate evaluations of biliary sphincter and pancreatic sphincter may be necessary to diagnose the two sphincteric abnormalities. Frequently, when one sphincter pressure is normal, the other may be elevated.

Although the procedure of choice, SO manometry is invasive, technically difficult, and unavailable even in some major medical centers. Additionally, even when experts perform the procedure, results may be inconclusive, and abnormal pressures may be intermittent. The procedure is associated with complications, notably post-ERCP pancreatitis.

TREATMENT AND MANAGEMENT

The management of SOD is unsatisfactory, but it includes pharmacologic, endoscopic, and surgical approaches. The goal is to improve the flow of biliary and pancreatic secretions into the duodenum. Drugs causing smooth muscle relaxation, including calcium channel blockers and nitrates, have been used with limited success. Endoscopic sphincterotomy of the biliary or pancreatic segment, performed by an experienced endoscopist, relieves the symptoms. Biliary sphincterotomy has the greatest benefit in patients with elevated basal SO pressures (>40 mm Hg). Endoscopic injection of botulinum toxin for biliary SOD has been tried with some success.

The benefit of endoscopic sphincterotomy for patients with recurrent pancreatitis is controversial. Transduodenal surgical sphincterotomy is a substitute for pancreatic ductal obstruction.

ADDITIONAL RESOURCES

Geenen JE, Hogan WJ, Dodds WJ, et al: The efficacy of endoscopic sphincterotomy after cholecystectomy in patients with sphincter of Oddi dysfunction, *N Engl J Med* 320:82-88, 1989.

Hogan WJ, Geenen JE: Biliary dyskinesia, *Endoscopy* 20:179-183, 1988.

McLoughlin MT, Mitchell RMS: Sphincter of Oddi dysfunction and pancreatitis, *World J Gastroenterol* 47:6333-6343, 2007.

Prajapati D, Hogan WJ: Sphincter of Oddi dysfunction and other functional biliary disorders: evaluation and treatment, *Gastroenterol Clin North Am* 32:601-618, 2003.

Sherman S, Troiano FP, Hawes RH, et al: Frequency of abnormal sphincter of Oddi manometry compared with the clinical suspicion of sphincter of Oddi dysfunction, *Am J Gastroenterol* 86:586-590, 1991.

Woods CM, Saccone TP: Neurohormonal regulation of the sphincter of Oddi, *Curr Gastroenterol Rep* 9:165-170, 2007.

Ampullary and Gallbladder Carcinoma

207

C. S. Pitchumoni

A tumor arising as a periampullary mass from the area of interface of the pancreas, duodenum, and pancreaticobiliary ductal system may masquerade as an ampullary tumor (**Fig. 207-1**). Periampullary tumors not arising from the pancreatic duct have a much better outcome than carcinomas of the pancreas, common bile duct (CBD), and duodenum.

PERIAMPULLARY CARCINOMA

Clinical Picture

Patients typically manifest symptoms early in the course of the disease, usually in the seventh decade of life, with abdominal pain, obstructive jaundice, malaise, anorexia, and weight loss. This accounts for the early diagnosis of periampullary carcinoma and the relatively successful resection rate; reported 5-year survival is 37%. Jaundice is progressive but occasionally may be associated with cholangitis. Iron deficiency anemia as a result of chronic low-grade bleeding is a clinical association. The triad of intermittent painless jaundice, anemia, and enlarged palpable gallbladder (Courvoisier gallbladder) is seen in less than 10% of patients. The stool may be gray or silver as a result of acholic feces mixed with melena. Recurrent acute pancreatitis of no readily identifiable etiology may be the presenting feature.

Risk factors for periampullary tumors are similar to those for pancreatic cancer. Patients with familial adenomatous polyposis are predisposed to ampullary adenomas (see Chapter 158).

Diagnosis

Appropriate imaging studies are contrast-enhanced computed tomography (CT), magnetic resonance imaging (MRI) of the abdomen, and magnetic resonance cholangiopancreatography (MRCP). Gastroduodenoscopy allows visualization and biopsy of the tumor. Endoscopic ultrasound (EUS) helps in further evaluating tumor origin and in staging. Endoscopic retrograde cholangiopancreatography (ERCP) defines the extent, size, and gross appearance of the tumor and is useful in palliative stenting to relieve obstructive jaundice.

Treatment and Management

Pancreaticoduodenectomy is the most effective treatment for periampullary carcinoma. Small tumors at the ampulla have a good prognosis if they are resectable and there is no evidence of metastasis.

GALLBLADDER CANCER

Although highly lethal, gallbladder cancer is extremely rare. The annual incidence in the United States is less than 7000 cases, and it is the fifth most common malignancy of the gastrointestinal (GI) tract. When the diagnosis of carcinoma of the gallbladder is incidental on routine cholecystectomy for gallstone disease, the prognosis is excellent. Incidental carcinoma of the gallbladder is noted in 1% to 3% of cholecystectomy specimens and 0.5% to 7.4% of autopsies.

Risk factors for gallbladder carcinoma include gallstones, history of chronic cholecystitis, and porcelain gallbladder (see Chapter 204). Adenomas of the gallbladder may progress to cancer, but the frequency is unknown. The risk is related to the size of the polyp. Polyps smaller than 1 cm seldom undergo malignant changes. Other risk factors are anomalous drainage of the pancreatic duct into the CBD, congenital biliary cysts, and *Salmonella* infection (chronic gallbladder infection).

In contrast to the general U.S. population, gallbladder cancer is the most common GI malignancy in Native Americans who live in the southwest and in Mexican Americans. Worldwide, the incidence of carcinoma of the gallbladder is highest in Chile, Bolivia, and northern regions of India. The risk is higher in women and elderly populations.

Although gallstones are frequently associated with carcinoma, the incidence of carcinoma in patients with gallstone disease is extremely low. Symptomatic gallstone disease—large size of the stones (>2.5 cm) and long duration of gallstone disease (notably >40 years)—are other observed risk factors. Another risk factor for gallbladder carcinoma involves an anomalous pancreatobiliary duct junction. A strong association of *Salmonella* infection and its carrier state has been shown in many studies. Most gallbladder carcinomas are adenocarcinomas, but squamous cell tumors, mixed tumors, and adenoepidermoid tumors occasionally develop.

Clinical Picture

Most patients have nonspecific findings of right upper quadrant pain, malaise, weight loss, jaundice, anorexia, and vomiting mimicking symptomatic gallstone disease. Few patients have acute cholecystitis. At diagnosis, most patients have tumors that have invaded adjacent organs, local lymph node metastasis, or even distinct metastasis. The 5-year survival rate is less than 5%, except when the diagnosis is incidental on routine cholecystectomy.

Diagnosis

Diagnostic studies include abdominal ultrasound, CT, and MRI. The role of EUS is under evaluation. Major findings are focal or diffuse mural thickening, intraluminal polypoid mass usually larger than 2 cm originating in the gallbladder wall, and most often (45%-65% of patients), a subhepatic mass replacing or obscuring the gallbladder and often invading the adjacent liver. ERCP, MRCP, and percutaneous transhepatic cholangiography provide additional information for tumor staging and resectability. ERCP may also provide the opportunity for brush

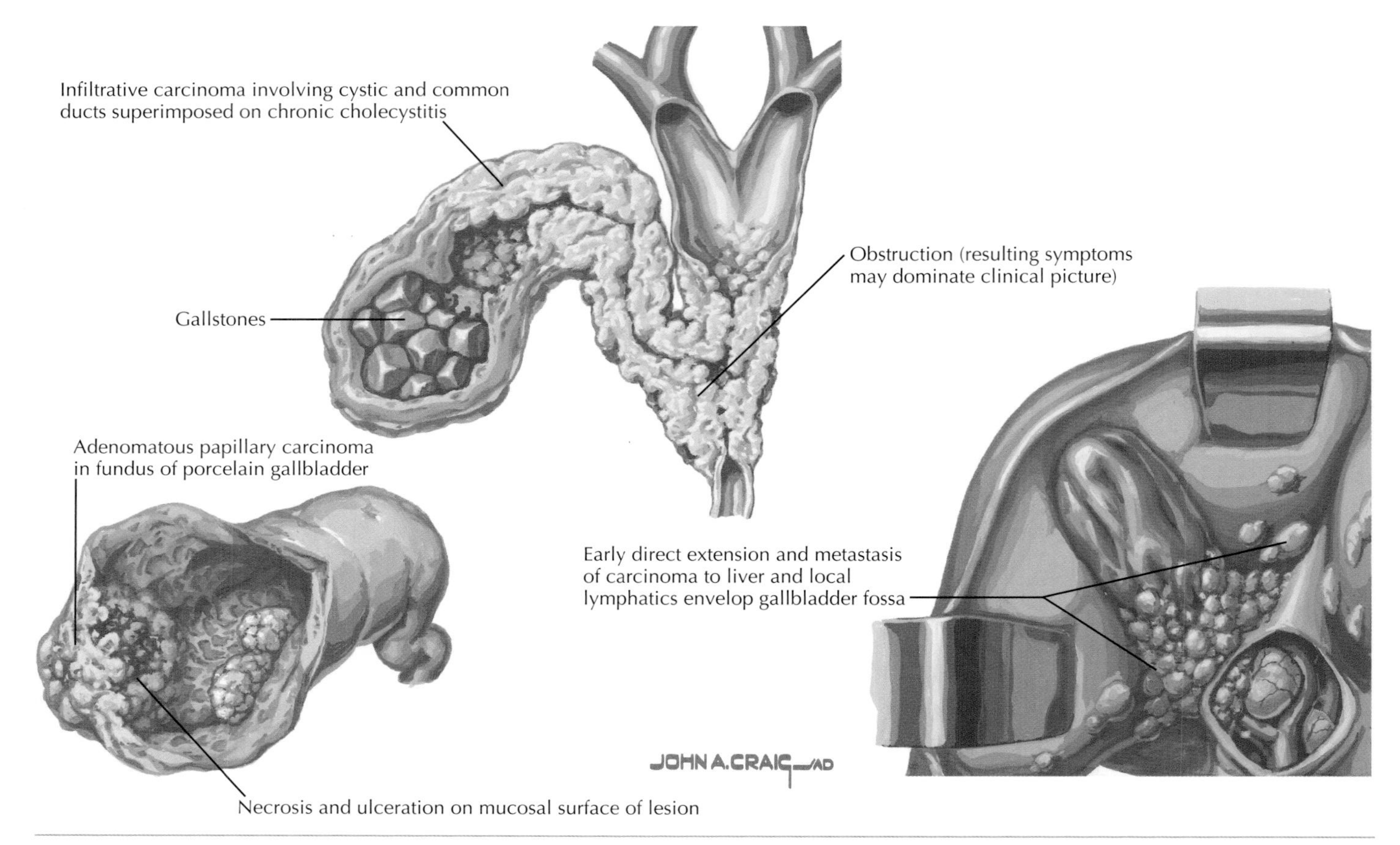

Figure 207-1 *Gallbladder Cancer.*

cytology and biopsy. Biochemical abnormalities indicate obstructive jaundice. Tumor markers (e.g., CEA, CA 19-9) are not helpful.

Treatment and Management

Management is surgical, and the prognosis depends on the stage of the cancer. Most tumors at diagnosis are unresectable. If the patient is thought to have cancer, open surgery is preferred. Although simple cholecystectomy may be sufficient for T1 lesions, radical resection is needed for advanced cases. Postoperative external beam radiation therapy may reduce the rates of local recurrence. Concomitant 5-fluorouracil (5-FU), with or without mitomycin C, is part of adjuvant chemoradiotherapy. Improved survival attributed to adjuvant chemoradiotherapy has not been established. A biliary stent has been proposed as a palliative measure to relieve obstructive jaundice.

ADDITIONAL RESOURCES

Berberat PO, Kunzli BM, Gulbinas A, et al: An audit of outcomes of a series of periampullary carcinomas, *Eur J Surg Oncol* 35:187-191, 2009.

Gourgiotis S, Kocher HM, Solaini L, et al: Gallbladder cancer, *Am J Surg* 196:252-264, 2008.

Sheth S, Bedford A, Chopra S: Primary gallbladder cancer: recognition of risk factors and the role of prophylactic cholecystectomy, *Am J Gastroenterol* 95:1402-1409, 2000.

Wistuba LL, Gazdar AF: Gallbladder cancer: lessons from a rare tumor, *Nat Rev Cancer* 4:695-706, 2004.

Cholangiocarcinoma

208

C. S. Pitchumoni

Cholangiocarcinoma (CCA) is a malignancy originating from the epithelium of the intrahepatic or extrahepatic biliary duct system. Although rare in the United States, its incidence is increasing. The clinical presentation of CCA varies, depending on the location of the tumor and the level of obstruction in the biliary system (**Fig. 208-1**). Intrahepatic CCA arising from the small ducts or ductules resembles hepatoma. Extrahepatic tumor may be from the upper portion (perihilar, hilar, or Klatskin type) or lower portion of the common bile duct (CBD).

The incidence of CCA is not well defined; some CCAs are even grouped with liver or gallbladder cancer. CCA develops in almost 10% of patients with *primary sclerosing cholangitis* (PSC), a complication of ulcerative colitis. Cirrhosis of any etiology is associated with a higher-than-expected incidence of CCA. Also, CCA is a complication of choledochal cyst disease and Caroli disease (congenital biliary duct ectasia). Anomalous high insertion of the pancreatic duct with the CBD is another predisposing factor for CCA; **Box 208-1** lists other factors. Rare conditions associated with CCA include multiple biliary papillomatosis, bile duct adenoma, and exposure to thorium dioxide, a contrast agent no longer used in radiologic studies.

Box 208-1 Predisposing Factors for Cholangiocarcinoma

- Primary sclerosing cholangitis
- Ulcerative colitis
- Choledochal cysts
- Caroli disease (biliary duct ectasia)
- Intrahepatic stones
- Chronic viral hepatitis (HCV, HBV?)
- Long common channel of pancreatic and biliary duct
- Infections
  - *Opisthorchis viverrini* (Thailand, Laos, Malaysia)
  - *Clonorchis sinensis* (Japan, Korea, Vietnam)
- Miscellaneous
  - Human immunodeficiency virus (HIV)
  - Cirrhosis of any etiology
  - Alcoholism

CLINICAL PICTURE AND DIAGNOSIS

CCA presents classically as painless jaundice, clay-colored stools, and cola-colored urine. Pain, fatigue, malaise, and weight loss accompany advanced disease. Abnormal laboratory values indicate obstructive jaundice and demonstrate elevated alkaline phosphatase, bilirubin, and γ-glutamyltransferase levels. Serum CA 19-9 levels are often elevated above 100 U/mL; this test is particularly useful in diagnosing CCA in PSC. Useful radiologic studies are abdominal ultrasound, computed tomography of the abdomen, magnetic resonance cholangiopancreatography, endoscopic ultrasound, and endoscopic retrograde cholangiopancreatography (ERCP). ERCP and percutaneous cholangiography are effective at assessing the location of the tumor. A hilar location of the tumor (Klatskin tumor) has a classic cholangiographic appearance. Positron emission tomography (PET) is useful in diagnosis and staging. Brush cytology is possible during ERCP in 40% to 70% of patients. Angiography accurately shows vascular encasement and thrombosis of portal vein and hepatic artery. The differential diagnosis should include carcinoma of the head of the pancreas, gallbladder cancer, Mirizzi syndrome, and PSC.

TREATMENT AND MANAGEMENT

Treatment of cholangiocarcinoma is unsatisfactory. Surgical excision is the mainstay. Radiation therapy and chemotherapy provide symptomatic relief by removing the obstruction. Endoscopic placement of plastic or metal stents is another palliative procedure. All these measures have been essentially ineffective, and the prognosis is poor.

ADDITIONAL RESOURCES

De Groen PC: Cholangiocarcinoma: making the diagnosis, *Clin Perspect Gastroenterol* 4:77-89, 2001.

Yachimski P, Pratt DS: Cholangiocarcinoma: natural history, treatment, and strategies for surveillance in high risk patients, *J Clin Gastroenterol* 42:178-190, 2008.

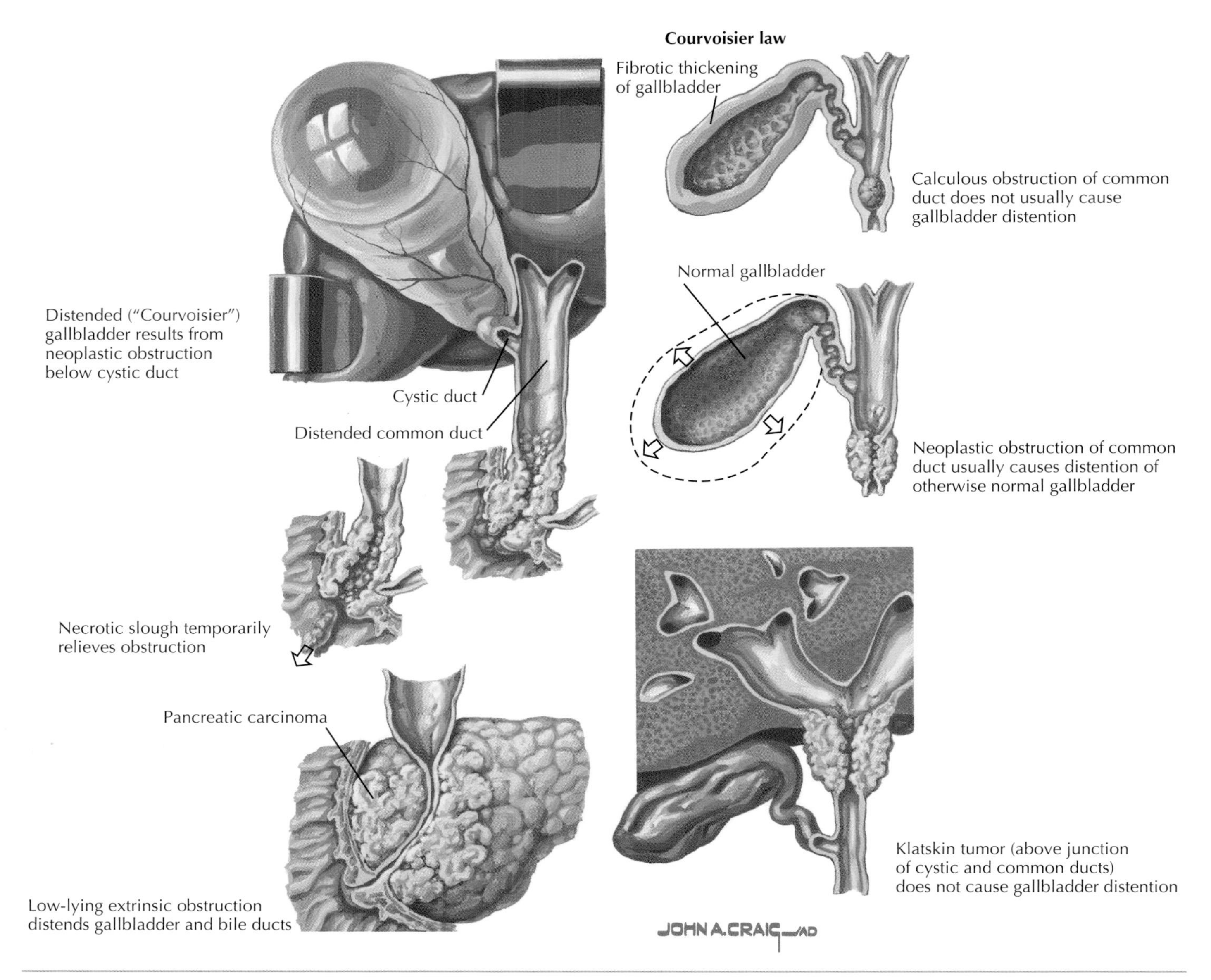

Figure 208-1 *Neoplastic Obstruction of the Bile Ducts.*

SECTION IX

Liver

Topography of the Liver

209

Kris V. Kowdley

The liver (Greek *hepar*) is located in the upper part of the abdomen, where it occupies the right hypochondriac and the greater part of the epigastric regions (**Fig. 209-1**). The left lobe of the liver extends into the left hypochondrium. The liver is the largest organ of the body and weighs 1400 to 1600 g in men and 1200 to 1400 g in women. In healthy persons, the *liver margin* extending below the thoracic cage is smooth and offers little resistance to the palpating finger. Downward displacement, enlargement, hardening, and formation of nodes or cysts produce impressive palpatory findings. Using percussion, the examiner must consider that the lungs overlay the upper portion of the liver and that the liver, in turn, overlays the intestines and the stomach.

The projections of the liver on the body surface have added significance with liver biopsy. The projections vary, depending on patient position and body build, especially the configuration of the thorax. The liver lies close to the diaphragm, and the upper pole of the right lobe projects as far as the level of the fourth intercostal space or the fifth rib; the highest point is 1 cm below the nipple, near the lateral body line. The upper limit of the left lobe projects to the upper border of the sixth rib. Here, the left tip of the liver is close to the diaphragm.

The ribs cover the greater part of the liver's right lobe, whereas a small part of its anterior surface is in contact with the anterior abdominal wall. In the *erect position*, the liver extends downward to the tenth or eleventh rib in the right midaxillary line. Here, the pleura projects down to the tenth rib, and the lung projects down to the eighth rib. The anterior margin of the liver crosses the costal arch in the right lateral body line, approximately on the level of the pylorus (*transpyloric line*). In the epigastrium, the thoracic cage does not cover the liver. It extends approximately three fingers below the base of the xiphoid process in the midline. Part of the left lobe is covered again by the rib cage.

Over the upper third of the right half of the liver, percussion indicates a *dull zone*, because here the diaphragm, pleura, and lung overlay the liver. Over the middle portion, percussion results in a *flat tone*. Over the lowest third of the liver, usually a flat tone is heard as well, except that sometimes *intestinal resonance* is produced by gas-filled intestinal loops. The border between dullness and flatness moves on respiration and is altered by enlargement or displacement of the liver and by conditions within the thoracic cage, which change the percussion qualities of the thoracic organs. In the *horizontal position*, the projection of the liver moves a little upward, and the area of flatness appears slightly enlarged. Measurement of the span of the flat sound, best percussed in the horizontal position, allows assessment of the size of the liver.

Projections of the liver are altered in some liver diseases (e.g., tumor infiltration, cirrhosis, syphilitic hepar lobatum) by displacement of the liver or, more often, by thoracic conditions pushing the liver downward. Subphrenic abscesses, depending on location and size, also displace the liver downward. Ascites, excessive dilatation of the colon, and abdominal tumors may push the liver upward, and retroperitoneal tumors may move it forward. Kyphoscoliosis or "barrel chest" alters the position of the liver. In some patients, the liver is abnormally movable *(hepatoptosis)*, causing peculiar palpatory findings.

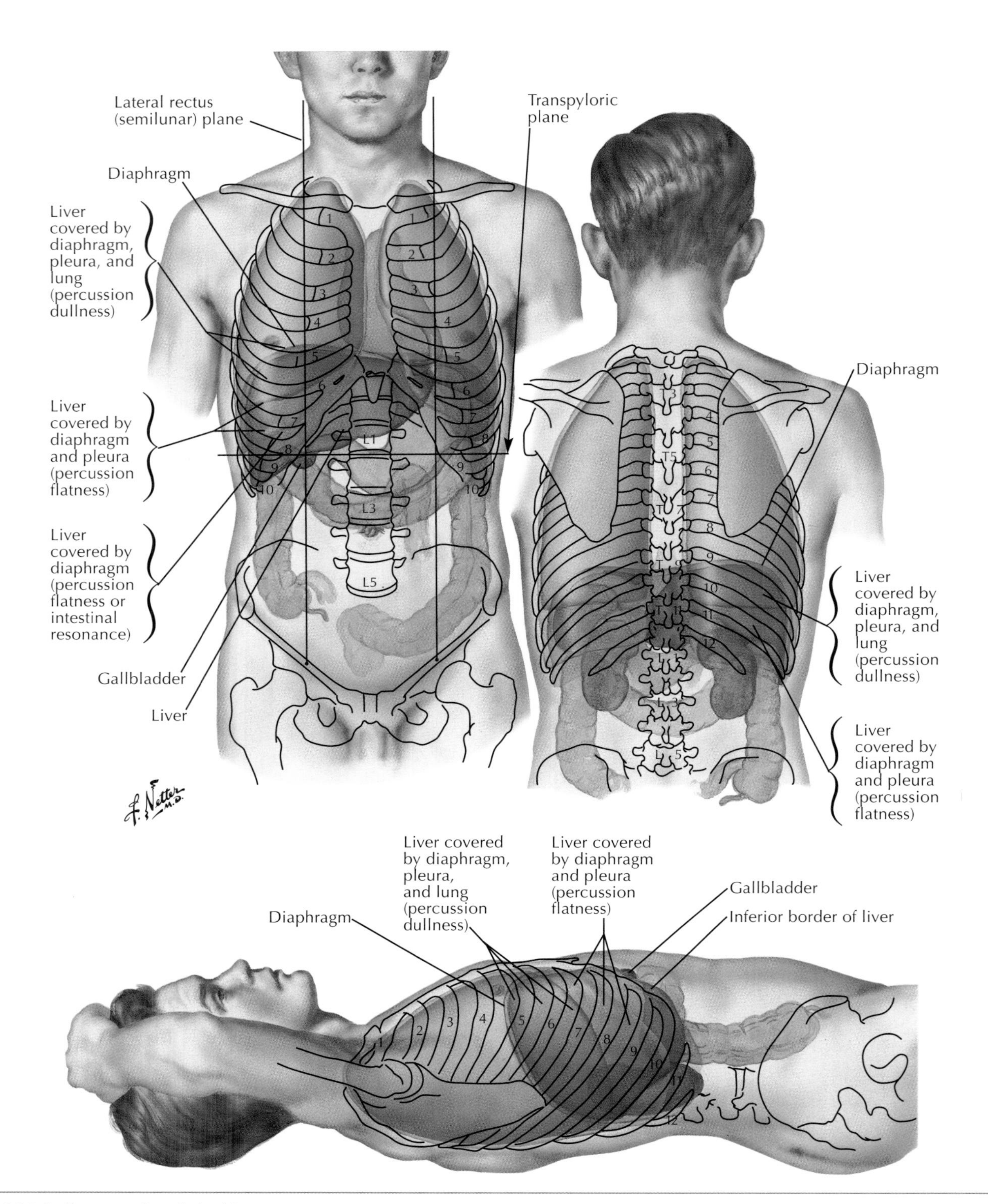

Figure 209-1 *Topography of the Liver.*

Surfaces and Bed of the Liver

210

Kris V. Kowdley

The liver is pyramid shaped, and its apex is formed by the thin, flattened left extremity of the left lobe (**Fig. 210-1**). Its base is seated on the right lateral surface, which rests on the diaphragm, and on the right thoracic cage, which produces the *costal impressions* on this surface. Its sides are formed by the anterior, posterior, and inferior surfaces. The border between the anterior and inferior surfaces is the *anterior margin*. The liver's consistency, sharpness of edge, smoothness of surface, and movement on respiration provide clinical information. On laparotomy, the anterior margin and the anterior surface are first exposed. Otherwise, the hepatic surfaces are not separated by distinct margins.

The liver is covered by peritoneum, except for the gallbladder bed, the hilus, adjacent parts surrounding the inferior vena cava (IVC), and a space to the right of the IVC called the *bare area*, which is in contact with the right adrenal gland (*adrenal impression*) and the right kidney (*renal impression*). *Peritoneal duplications*, which extend from the anterior abdominal wall and the diaphragm to the organ, form the ligaments of the liver. Formerly, it was thought that the ligaments maintain the liver in its position but probably add little to its fixation. It is now believed that the liver is kept in place by intraabdominal pressure.

The horizontal peritoneal duplication is the *coronary ligament*, the upper layer of which is exposed if the liver is pulled away from the diaphragm. The free right lateral margin of the coronary ligaments forms the *right triangular ligament*, whereas the *left triangular ligament* surrounds and merges with the left tip of the liver, the *appendix fibrosa hepatis* (fibrous appendix of liver). Over the right lobe, the space between the upper and lower layers of the coronary ligament is filled with areolar connective tissue. Below the insertion of the lower layer of the right coronary ligament, the hepatorenal space extends behind the liver.

From the middle portion of the coronary ligament originates another peritoneal duplication, the *falciform ligament*, which extends from the liver to the anterior abdominal wall between the diaphragm and the umbilicus. Its insertion on the liver divides the organ into a *right lobe* and *left lobe*. The inferior edge of the falciform ligament is enforced to form the round ligament (*ligamentum teres*), which extends to a point at which the longitudinal fissure of the liver crosses the inferior surface. With its anterior part, this fissure separates the quadrate lobe and the left lobe (tuber omentale) and forms a fossa for the umbilical vein or its remnant. The fissure proceeds toward the posterior surface, creating the fossa for the ductus venosus (*ligamentum venosum* in adult life). The two fossae may be regarded as the right limb of an H-shaped pattern, characteristic of the inferior surface of the liver. The left limb is formed by the gallbladder bed and the fossa for the IVC. The horizontal limb is marked by the *porta hepatis*, which contains the common hepatic duct, hepatic artery, portal vein, lymphatics, and nerves.

The *quadrate lobe*, between the gallbladder and the fossa for the umbilical vein, is in contact with the pylorus and the first portion of the duodenum *(duodenal impression)*. On the inferior and posterior surfaces lies the *caudate lobe*, between the fossa for the ligamentum venosum and the IVC; its anterior projection is the *papillary process*.

The inferior surface of the liver reveals further impressions of the organs it contacts: on the right lobe, impressions for the colon and right kidney, and on the left lobe, impressions for the esophagus and stomach. The superior surface is related to the diaphragm and forms the domes of the liver.

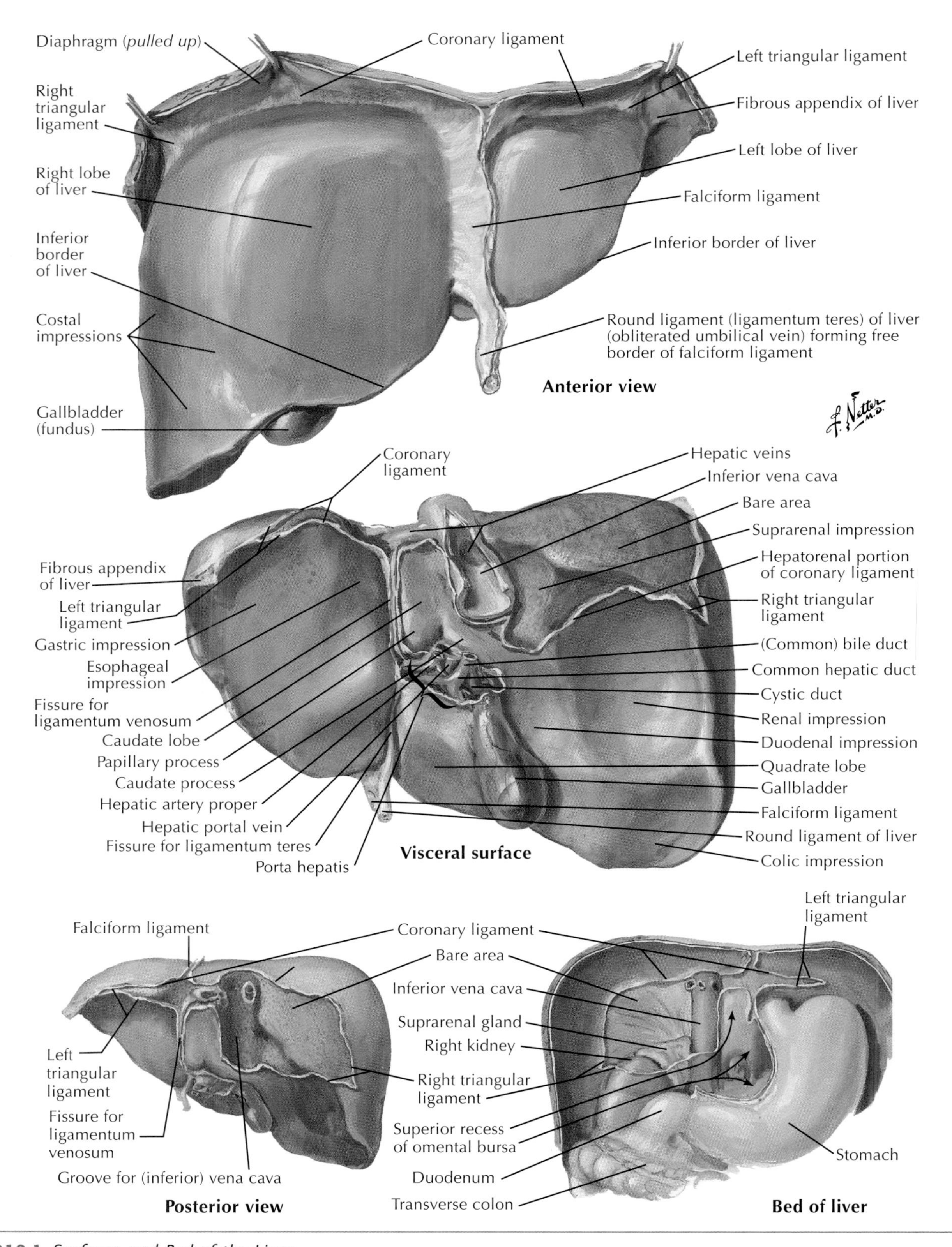

Figure 210-1 *Surfaces and Bed of the Liver.*

Lesser Omentum and Variations in Form of the Liver

211

Kris V. Kowdley

If the anterior margin of the liver is lifted, the *lesser omentum* is exposed. It represents a peritoneal fold that extends from the first portion of the duodenum and lesser curvature of the stomach and diaphragm to the liver, where the fold inserts at the fossa of the ligamentum venosum and continues to the porta hepatis (**Fig. 211-1**). There, the layers are separated to accommodate the structures running to and from the hilus of the liver.

On the free right edge of the lesser omentum, the reunited peritoneal layers are enforced to form the *hepatoduodenal ligament*. It is the anterior boundary of the *omental foramen* (foramen of Winslow, epiploic foramen), which is the entrance to the lesser abdominal cavity. The posterior wall of this cavity is formed by the inferior vena cava and the caudate lobe of the liver (see Chapter 210).

Near the right margin of the lesser omentum is the *common bile duct* (CBD), which divides into the cystic and common hepatic ducts. To the left of the CBD lies the *hepatic artery*, and behind both is the *portal vein*. The nerves and the lymph vessels of the liver accompany these structures. The hilus of the liver is anteriorly limited by the quadrate lobe and posteriorly by the caudate lobe. On the right side of the hilus, the right and left hepatic ducts branch from the main hepatic duct and enter the liver. To the left of the ducts, the hepatic artery enters the liver behind the ductal branches. The forking portal vein enters posteriorly to the ductal and arterial ramifications.

The shape of the liver varies. Its great regenerative ability and the plasticity of its tissue permit a wide variety of forms, depending in part on pressure exerted by neighboring organs and on disease processes or vascular alterations. A greatly reduced left lobe is offset by an enlarged right lobe, which reveals conspicuous and deep costal impressions. Occasionally, the left lobe is completely atrophic (see Fig. 211-1), with a wrinkled and thickened capsule and, microscopically, an impressive approximation of the portal triads, with almost no lobular parenchyma between them (see Chapter 212).

Vascular aberrations include partial obstruction of the lumen of the left branch of the portal vein by a dilated left hepatic duct, or obstruction of the bile ducts, considered the result of a local nutritional deficiency, especially because the nutritional condition of the left lobe initially is poor (see Chapter 216). In other situations, associated with a transverse position of the liver, the left lobe is unduly large.

In previous centuries, the liver frequently was disfigured by laced corsets or by tight belts or straps. Such physical forces may flatten and elongate the liver from above downward, with reduction of the superior diaphragmatic surface and sometimes with tonguelike extension of the right lobe (see Fig. 211-1). In other cases, the *corset liver* is displaced, and the renal impression is exaggerated. Clinical symptoms such as dyspepsia, cholelithiasis, and chlorosis have been ascribed to the corset liver, but it is questionable whether the corset liver actually caused the clinical manifestations, other than peculiar findings on palpation.

Indentations on the liver produced by the ribs, diaphragmatic insertions, and costal arch are normal. In kyphoscoliosis, the rib insertions may be prominent. Parallel sagittal furrows on the hepatic convexity have been designated *diaphragmatic grooves*.

Functionally, none of these variations currently is considered significant.

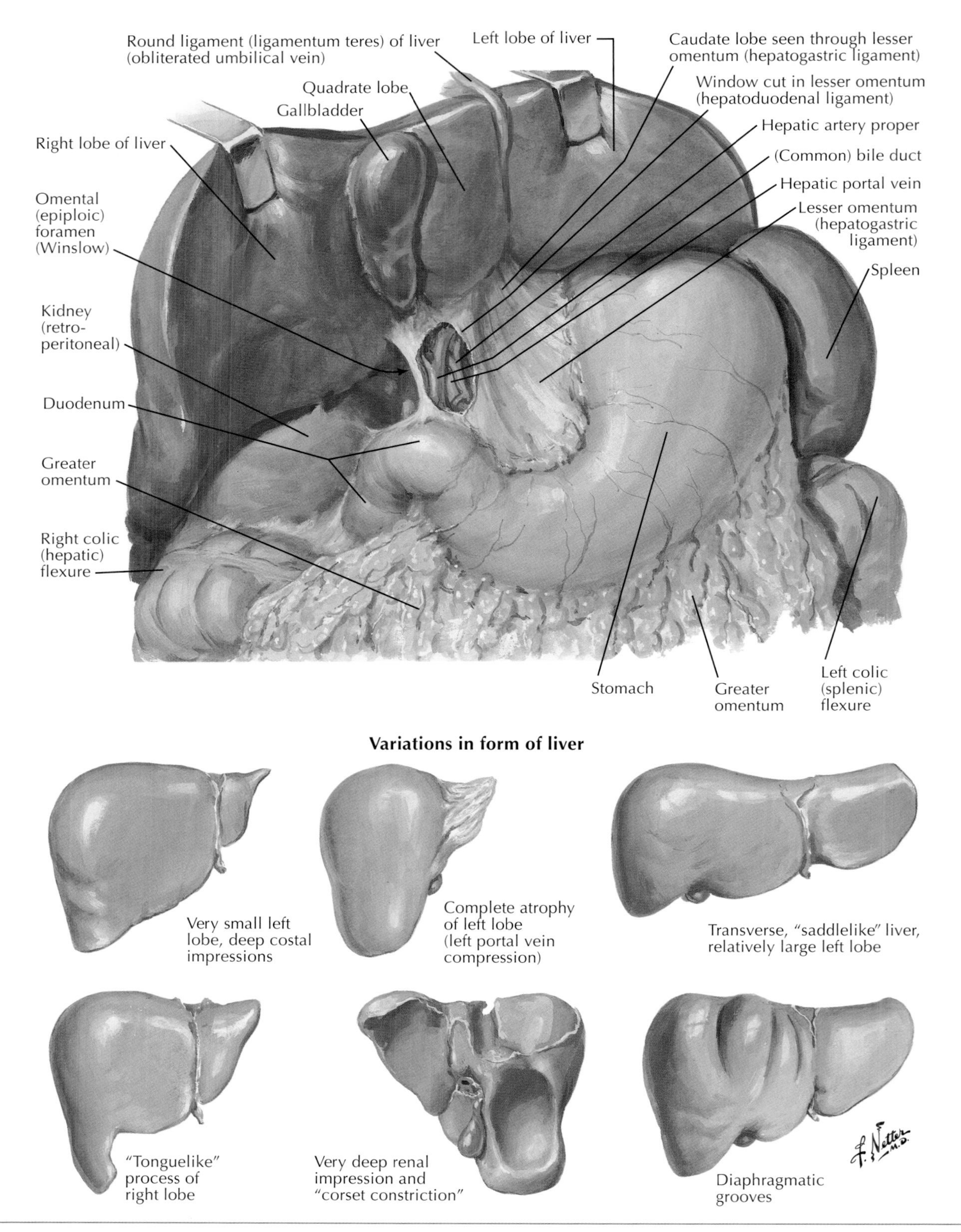

Figure 211-1 *Lesser Omentum and Variations in Form of the Liver.*

Cell Types within the Liver

212

Kris V. Kowdley

Major cell types unique to the liver include hepatocytes, bile duct cells, sinusoidal-lining cells (including Kupffer cells and endothelial cells), stellate cells, and immune cells (**Fig. 212-1**). *Hepatocytes* are the primary cell type within the liver and are responsible for a vast array of metabolic functions, including gluconeogenesis, fatty acid oxidation, synthesis of albumin and other plasma proteins, metabolism of drugs and toxins, synthesis of cholesterol, and bile acids.

Sinusoids, which are structurally different from capillaries, are of variable sizes and contain fenestrations. Sizes increase from zone 1 to zone 3. Sinusoids are lined by Kupffer cells and endothelial cells. *Kupffer cells* are large macrophages unique to the liver, and they have phagocytic activity. These cells clear endotoxin, bacteria, and senescent red blood cells (RBCs) from the circulation. Kupffer cells also store iron from erythrocytes. Therefore, Kupffer cell iron content is increased in conditions of increased RBC turnover, such as hemolysis or transfusional iron overload. By contrast, iron deposition in the liver in hereditary hemochromatosis is found primarily in periportal hepatocytes. *Endothelial cells* in the liver express receptors for several proteins and may be crucial for the maintenance of sinusoidal blood flow by producing vasoactive substances such as endothelin-1 and nitric oxide.

Stellate cells are unique to the liver and the subject of much study. Stellate cells (also called Ito cells, vitamin A–storing cells, and lipocytes) appear to play a central role in hepatic fibrogenesis and fibrinolysis. Stellate cells are transformed into fibroblasts after stimulation by cytokines, which may be released locally in response to injury. Activated stellate cells produce various types of collagen and may be responsible for the deposition of extracellular matrix. In addition, the contractility of stellate cells may be an important step in the development of portal hypertension. Stellate cells also appear to be involved in the degradation of extracellular matrix. Thus, stellate cells may play a critical role in two very important hepatic pathologic processes, the development of cirrhosis and portal hypertension. Manipulation of stellate cell activation may offer future therapeutic options for the treatment of cirrhosis and portal hypertension.

The liver also contains lymphatics. The normal production of *lymph* in the liver is approximately 2 L/day, but lymph production can increase greatly in patients with conditions leading to venous obstruction or with cirrhosis. Hepatic lymph has a high protein content and is collected from subendothelial *Disse* (perisinusoidal) *spaces*, adjacent to hepatic sinusoids. Hepatic lymph drains from the liver through the cisterna chyli and ultimately, it is thought, into the thoracic duct.

The liver also contains other immune cells, including T cells and natural killer (NK) cells, which are especially interesting because of *innate liver immunity* and its role in a variety of drug/toxin-induced liver injury.

The organization of cellular elements is controversial. Two models describe current concepts of microscopic hepatic organization. The *lobule model* focuses on the hepatic vein, with the portal areas organized around the points of a pentagon (**Fig. 212-2**). The more complex *acinus model* is based on the functional unit of the liver being an acinus organized around terminal portal venules, bile ductules, and lymph ducts. There is axial distribution around the portal venule with the acini clustered around the portal vein. The sinusoids, where blood leaves discrete vascular structures and freely contacts hepatocytes, radiate around the central veins (**Figs. 212-3** and **212-4**).

Blood enters the sinusoids from the terminal portal venules. End arterioles of the hepatic arteries also drain into the terminal portal venules. Therefore, the oxygen tension (Po_2) of the blood entering the sinusoid is richest around the portal area (zone 1 of Rappaport) and lowest in the region surrounding the central vein (zone 3 of Rappaport). Po_2 is intermediate between zone 1 and zone 3 (zone 2 of Rappaport). The centrilobular region of the lobule is most susceptible to toxic, hypoxic, and ischemic injury. Classic examples include acetaminophen toxicity or hepatic artery thrombosis after orthotopic liver transplantation. Each of these zones appears to have differentiated functions. For example, zone 1 hepatocytes play an important role in gluconeogenesis, whereas zone 3 hepatocytes are critical to lipid synthesis and glycolysis.

Bile ducts branch into smaller bile ductules that terminate in the bile canaliculus. Canaliculi are located between hepatocytes; bile drains cross the bile canalicular membrane in hepatocytes into bile ductules and subsequently into bile ducts.

ADDITIONAL RESOURCES

Rappaport AM, Wanless IR: Physioanatomic considerations. In Schiff L, Schiff ER, editors: *Diseases of the liver*, ed 7, Philadelphia, 1993, Lippincott, pp 1-41.

Saxena R, Theise ND, Crawford JM: Microanatomy of the human liver-exploring the hidden interfaces, *Hepatology* 30:1339-1346, 1999.

Yamamoto K, Sherman I, Phillips MJ, Fisher MM: Three-dimensional observations of the hepatic arterial terminations in rat, hamster and human liver by scanning electron microscopy of microvascular casts, *Hepatology* 5:452-456, 1985.

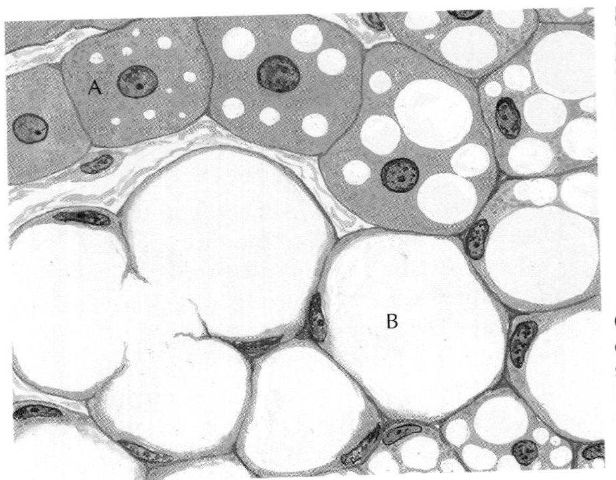

Liver cells with various degrees of fat accumulation, ranging from fine droplets *(A)* to large fatty cysts *(B)*

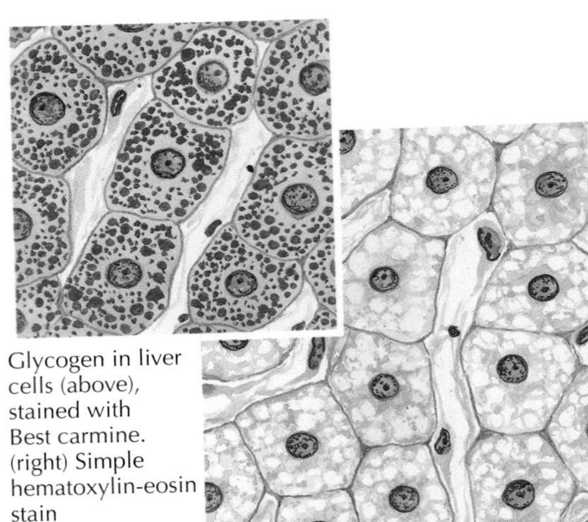
Glycogen in liver cells (above), stained with Best carmine. (right) Simple hematoxylin-eosin stain

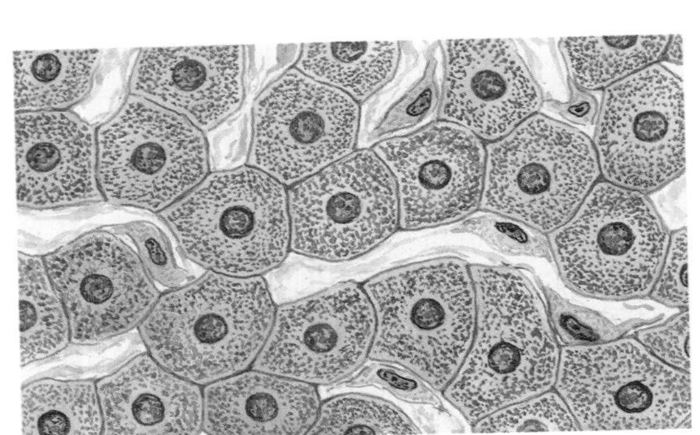
Liver cells with methyl green-pyronine stain (methyl green stains chromatin; pyronine stain cytoplasmic inclusions and nucleolus)

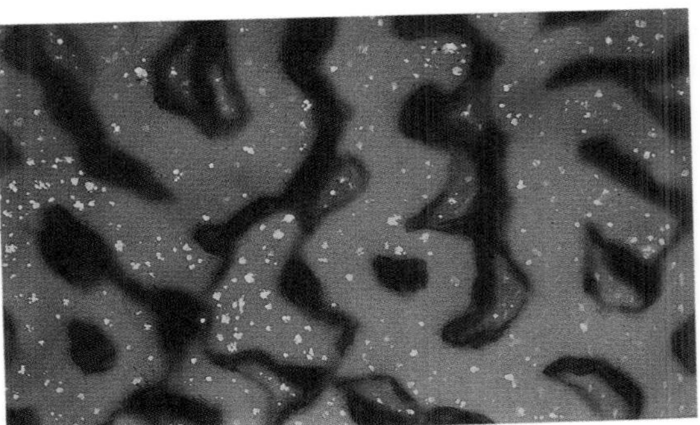
Vitamin A in liver cells and Kupffer cells made visible by fluorescence

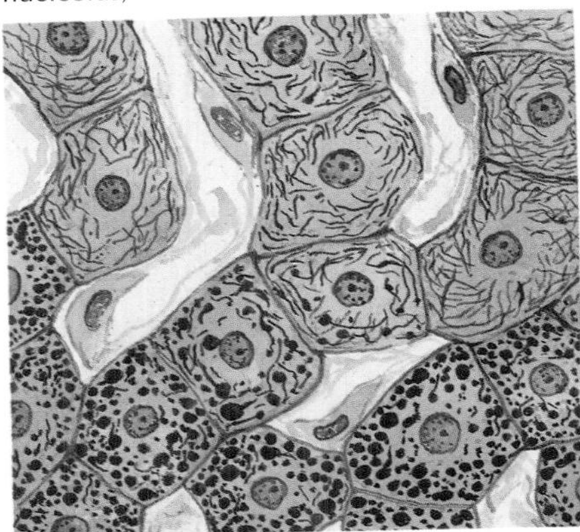
Variform mitochondria in liver cells reflecting differences in functional activity (Janus green stain)

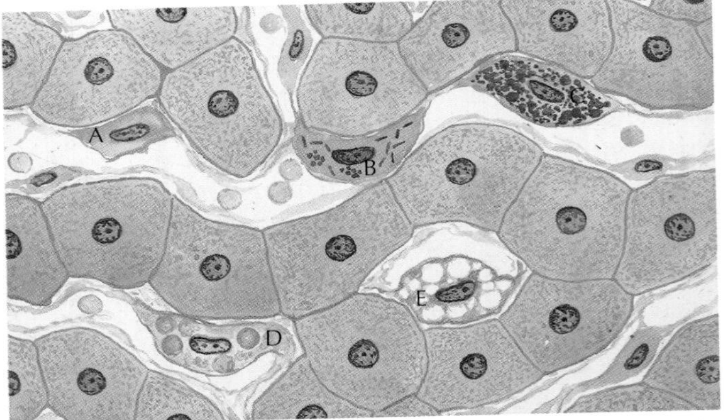

Kupffer cells in various stages. *(A)* Resting stage; *(B)* containing bacteria; *(C)* containing pigment; *(D)* containing red blood cells; *(E)* containing fat droplets

Figure 212-1 *Cell Types within the Liver.*

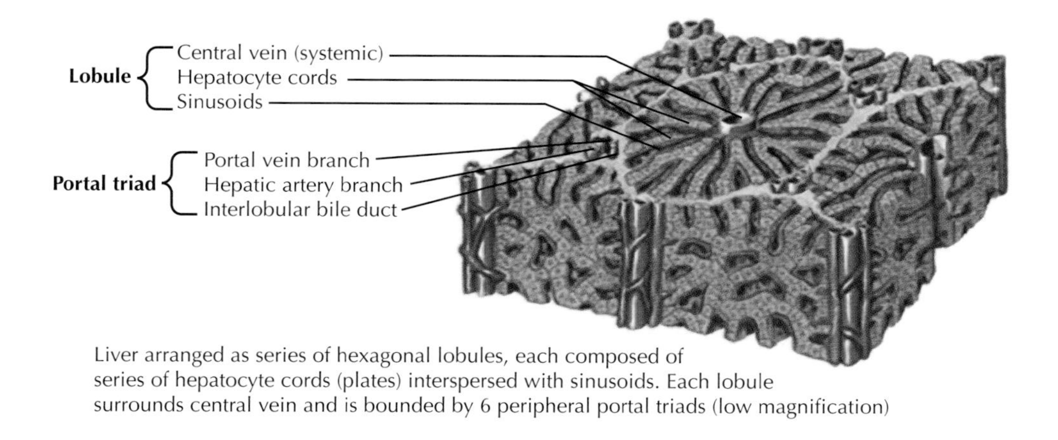

Liver arranged as series of hexagonal lobules, each composed of series of hepatocyte cords (plates) interspersed with sinusoids. Each lobule surrounds central vein and is bounded by 6 peripheral portal triads (low magnification)

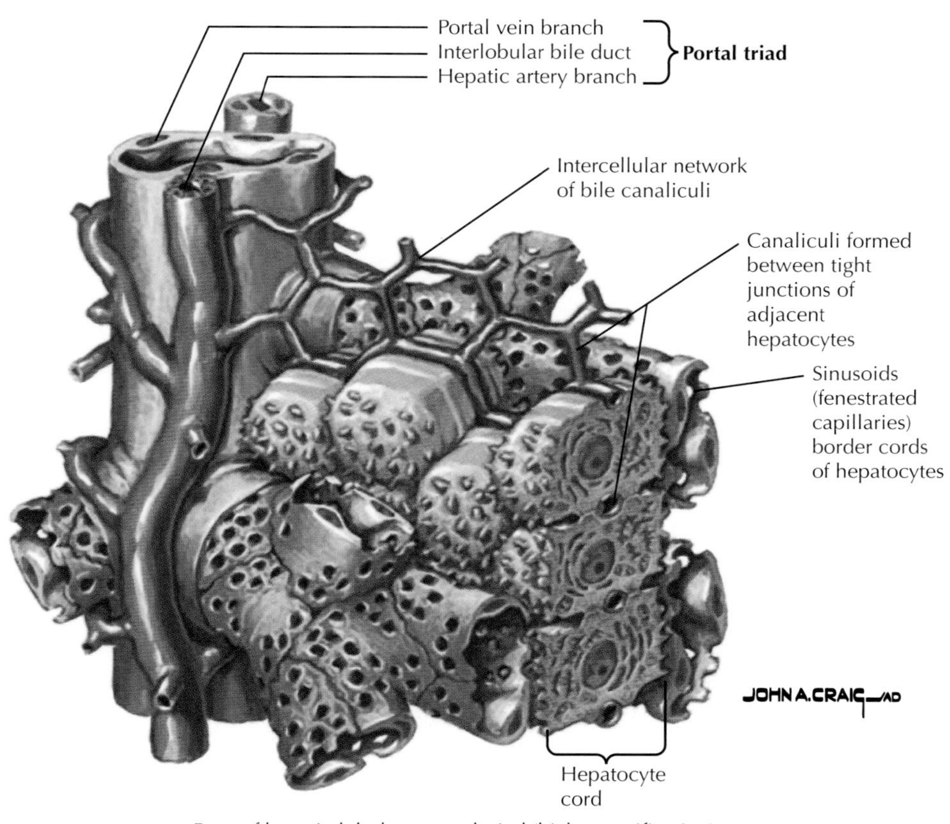

Parts of hepatic lobule at portal triad (high magnification)

Figure 212-2 *Hepatic Architecture.*

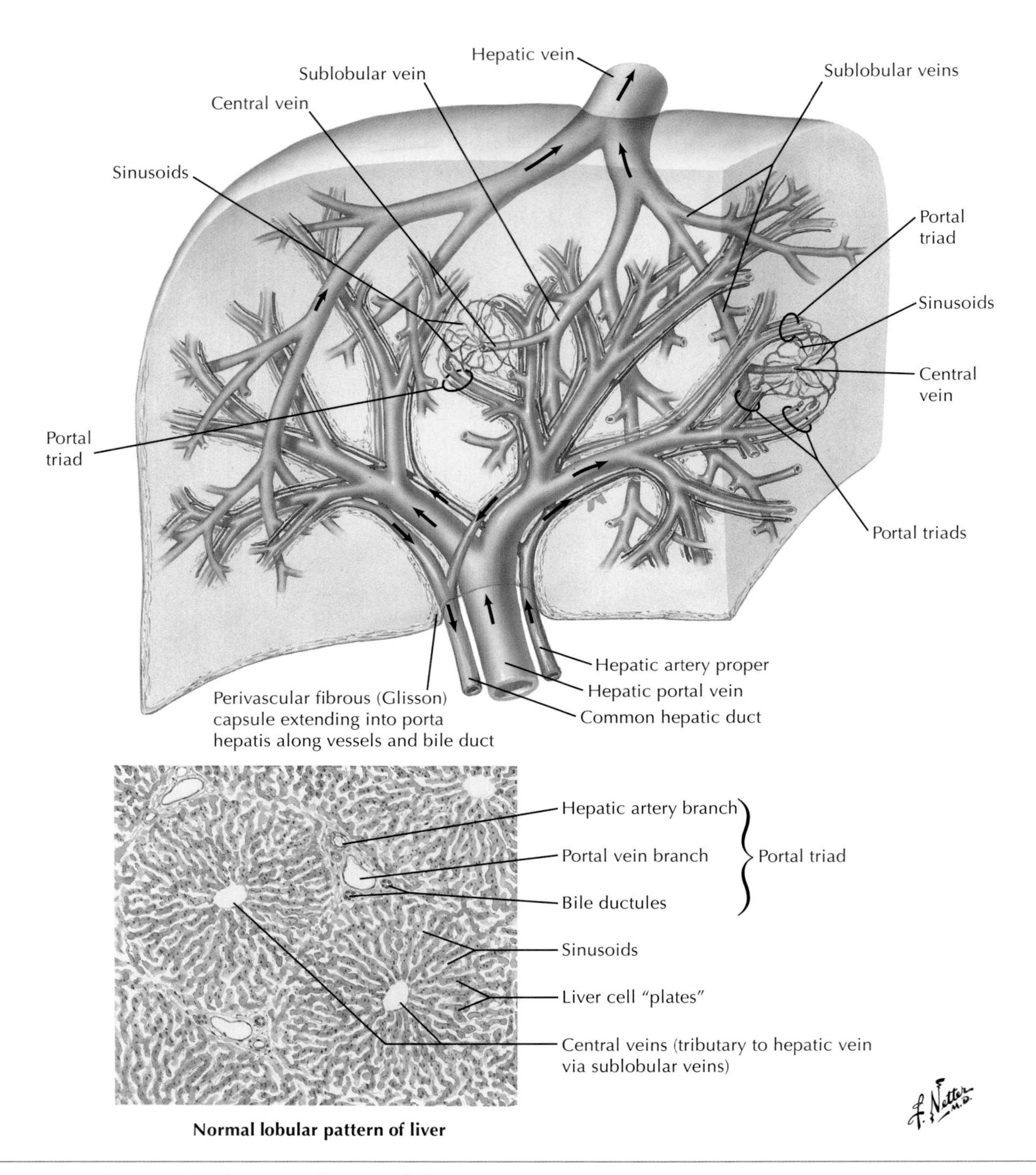

Figure 212-3 *Vascular Ductal Relations and Liver Lobules.*

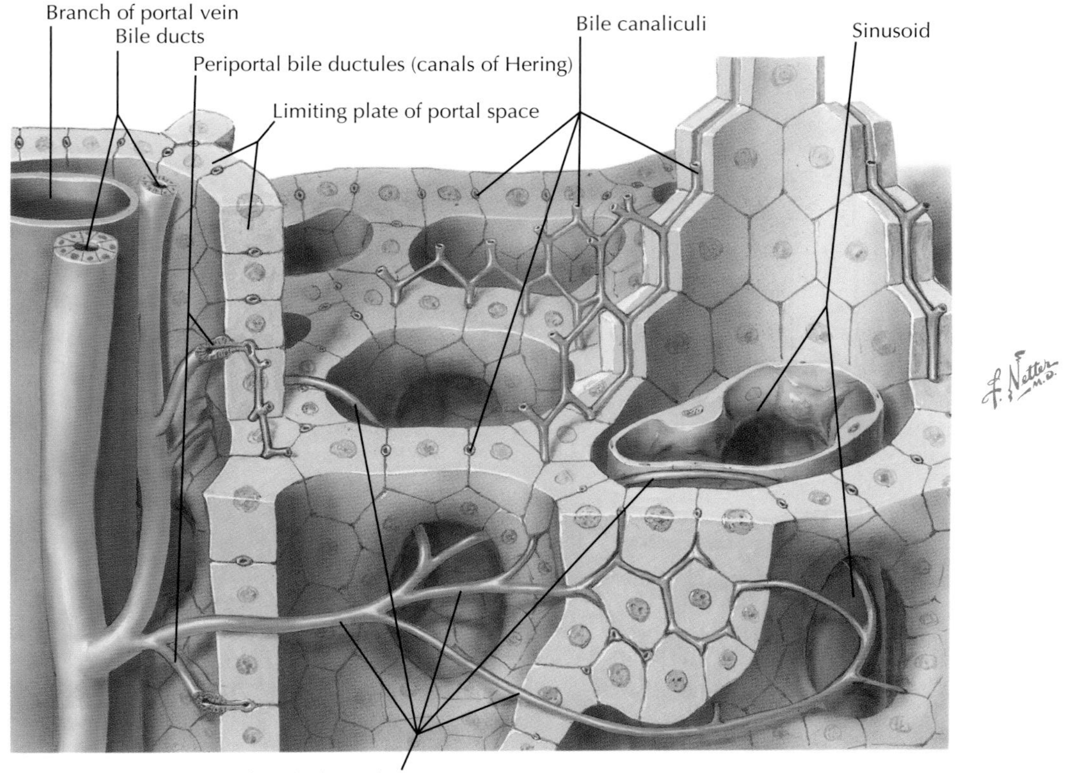

Low-power section of liver

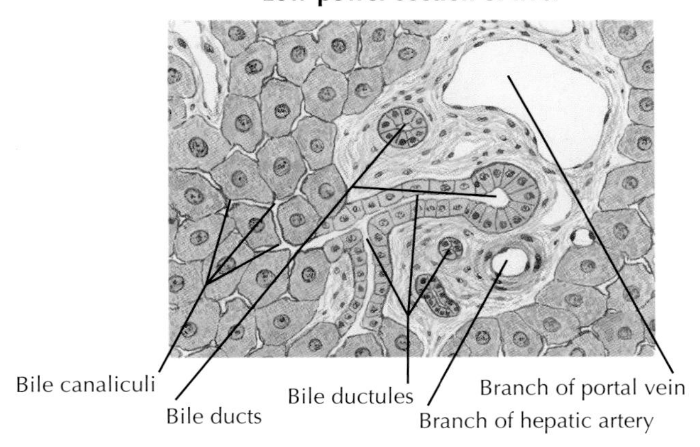

Note: In the top illustration, bile canaliculi appear as structures with walls of their own. However, as shown in histologic section, boundaries of canaliculi are actually a specialization of surface membranes of adjoining liver parenchymal cells

Figure 212-4 *Intrahepatic Biliary System.*

Vessel and Duct Distributions and Liver Segments

213

Kris V. Kowdley

Intraphepatic distribution of vessels and bile ducts was studied on casts prepared by injecting plastic into the vascular and biliary conduits before removing tissue. Besides being valuable for the cholangiographic demonstration of the vascular apparatus in vivo, this new recognition of the liver's segmental divisions, similar to those in the lungs, allowed partial hepatectomy or excision of single metastatic nodules.

Although the human liver, in contrast to the livers of some animals, fails to displace surface lobulation, the parallel course of the branches of the hepatic artery, portal vein, and bile ducts and the appearance of clefts in these preparations indicated a distinct lobular composition. A major lobar fissure extends obliquely downward from the fossa for the inferior vena cava to the gallbladder fossa (see Chapter 210). This does not coincide with the surface separation between the right and left lobes running along the insertion of the falciform ligament and the fossa for the ductus venosus. Through this fissure extends one of the main trunks of the hepatic vein, the tributaries of which never follow the distribution of the other vessels, but instead cross the portal vein branches in an interdigitated manner.

Each lobe of the liver is partitioned by a segmental division and is drained by a lobar bile duct of the first order (**Fig. 213-1**). The right division extends obliquely from the junction of the anterior and posterior surfaces downward toward the lower border of the liver and continues on the inferior surface toward the hilus, dividing the *right lobe* into anterior and posterior segments, each of which is drained by a bile duct of the second order. The left segmental cleft runs on the anterior surface along the attachments of the falciform ligament and on the visceral surface through the umbilical fossa and fossa for the ductus venosus, extending toward the hilus. This fissure divides the *left lobe* into medial and lateral segments, but in a significant number of cases, it is crossed by bile ducts and vessels. The *lateral segment* corresponds to the classic descriptions of the left lobe, whereas the aspect of the *medial segment* on the visceral live surface corresponds to the quadrate lobe. The four bile ducts of the second order form those of the third order, which drain either the superior or the inferior corresponding segment. Thus, the bile ducts and the accompanying vessels can be designated according to the lobes, segments, and areas to which they belong.

The anatomically distinct *caudate lobe* has a vascular arrangement that divides it into a left portion, drained by the left lobar duct, and a right portion, drained by the right lobar duct. The *caudate process*, connecting the caudate lobe with the right lobe of the liver, has a separate net of vessels, which usually communicates with branches of the right lobar duct.

Neither the caudate lobe nor other parts of the liver provide effective communication between the right and left lobar duct systems. Intrahepatic anastomoses between intraparenchymal branches of the arteries also have not been found, but in 25% of cases, interconnections between the right and left systems exist through small extrahepatic or subcapsular anastomosing vessels.

The distribution of draining bile ducts and afferent blood vessels, as schematically depicted here, is valid in most cases, but individual variations are abundant, especially in the lateral superior vessel and ducts for the appendix fibrosa. Rudimentary bile ducts are common in this region. Segmental bile duct variation occurs more often on the right side, whereas variation in segmental arteries is more common on the left side.

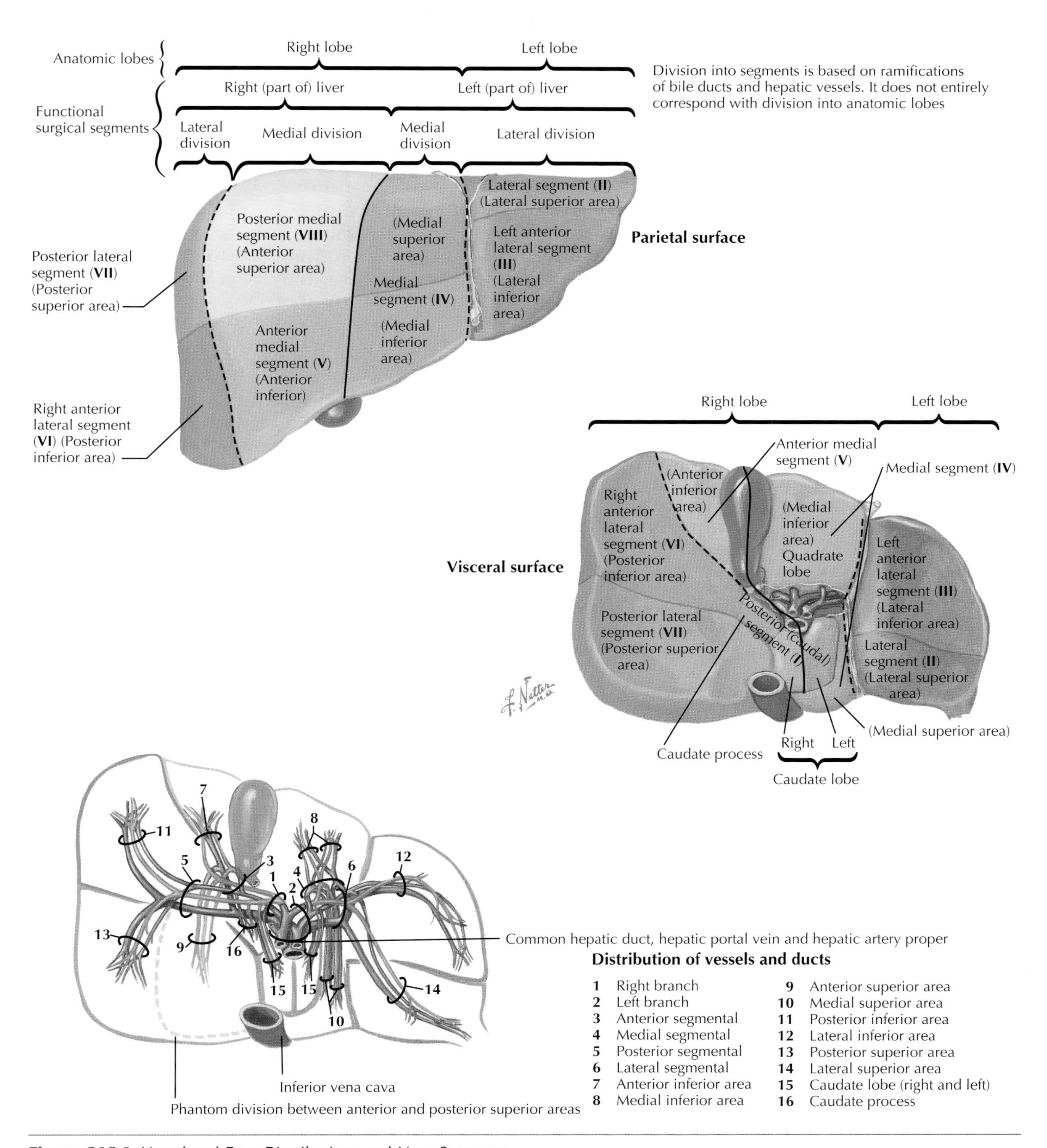

Figure 213-1 *Vessel and Duct Distributions and Liver Segments.*

Arterial Blood Supply of the Liver, Biliary System, and Pancreas

214

Kris V. Kowdley

Certain studies, especially the painstaking dissections of Michels, have disclosed considerable variations (see Chapter 215) in the arterial supply of the liver, biliary system, and pancreas (**Fig. 214-1**). According to the conventional description, found in only 55% of examined specimens, the *celiac artery* or *celiac axis* is a short, thick trunk that originates in the aorta just below the aortic hiatus in the diaphragm. It extends horizontally and forward above the pancreas and splits into the left gastric, hepatic, and splenic arteries. An inferior phrenic artery, usually starting from the aorta, or a dorsal pancreatic artery (see later), otherwise departing from the splenic artery, the hepatic artery, or the aorta, may exceptionally derive from the celiac axis. The *left gastric* (or coronary) artery, the smallest of the three celiac branches, starting at the cardia, extends along the lesser curvature of the stomach to anastomose with the *right gastric* artery.

The *splenic artery*, the largest of the three celiac branches (in the adult), takes a somewhat tortuous course to the left, along and behind the upper border of the pancreas. At a variable distance from the spleen, it breaks up into a number of terminal branches that enter the hilus of the spleen. The *left gastroepiploic* artery and the *short gastric* arteries usually have their origins in one of these terminal branches.

The *hepatic artery*, intermediate in size, passes forward and to the right to enter the right margin of the lesser omentum (see Chapter 211), in which it ascends, lying to the left of the common bile duct (CBD) and anterior to the portal vein. As the hepatic artery turns upward, it gives origin first to the gastroduodenal artery (see later), then usually to the supraduodenal artery, and finally to the right gastric artery. The *supraduodenal artery*, which may also originate from the right hepatic or the retroduodenal artery, descends to supply the anterior, superior, and posterior surfaces of the first inch (2.5 cm) of the duodenum. The right gastric artery passes to the left along the lesser curvature of the stomach to anastomose with the left gastric artery. The continuation of the hepatic artery beyond the origins of these vessels is known as the *common hepatic artery* (arteria hepatica propria). It ascends and divides into several branches, most often the *right hepatic* and *left hepatic* arteries; the *middle hepatic* artery usually arises from the left hepatic artery. The right hepatic artery generally passes behind the common hepatic duct to enter the cystic *Calot triangle*, formed by the cystic duct, the hepatic duct, and, cephalad, the liver. Occasionally, however, the right hepatic artery crosses in front of the bile duct (see Chapter 215). All terminal branches of the hepatic artery enter the liver at the porta hepatis. The *cystic artery* also has many variations.

In general, the arterial supply to the pancreas, common bile duct, and adjacent portions of the duodenum comes from branches of the gastroduodenal, superior mesenteric, and splenic arteries. The *gastroduodenal artery*, after its origin from the common hepatic artery, passes downward to course behind the first portion of the duodenum and in front of the head of the pancreas. Before or immediately after passing behind the duodenum, it gives origin to the *posterior superior pancreaticoduodenal* artery, which has been renamed (Michels) the *retroduodenal* artery. Its origin is often hidden by dense fibrous tissue, and passing to the right and downward over the CBD, it gives off a branch comprising the principal blood supply of that duct. The retroduodenal artery continues downward behind the head of the pancreas and between the duodenum and CBD, finally turning to the left to unite with the posterior branch of the *inferior pancreaticoduodenal* artery, also known as the *posterior inferior pancreaticoduodenal* artery.

At the lower border of the pylorus, the gastroduodenal artery divides into a larger right gastroepiploic artery and a smaller anterior superior pancreaticoduodenal artery. The *right gastroepiploic* enters the greater omentum to follow the greater curvature of the stomach. The *anterior superior pancreaticoduodenal* artery continues downward on the anterior surface of the head of the pancreas as far as its lower border, where it turns upward to unite with the *anterior branch* of the inferior pancreaticoduodenal artery, also known as the *anterior inferior pancreaticoduodenal* artery. In approximately 40% of the cases, no common inferior pancreaticoduodenal artery exists, and the anterior and posterior vessels originate separately from the superior mesenteric artery.

The head of the pancreas and the second and third portions of the duodenum are thus supplied by two arcades, an anterior and a posterior arch. The *posterior arch* is formed by the posterior superior pancreaticoduodenal (retroduodenal) artery uniting with the posterior inferior pancreaticoduodenal artery. The *anterior arch* is formed by the gastroduodenal and anterior superior pancreaticoduodenal arteries uniting with the anterior inferior pancreaticoduodenal artery. The posterior span is situated at a somewhat higher level than the anterior arch. Both give off branches that anastomose with each other through and around the pancreas, supplying that organ and the duodenum.

Branches of the splenic artery are the chief suppliers to the neck, body, and tail of the pancreas. Some of these are small twigs given off by the splenic artery as it courses along the upper border of the pancreas. Three branches, however, are usually larger than the others and have achieved the distinction of individual names. The *dorsal pancreatic* artery, also known as the *superior pancreatic* artery, although usually originating from the beginning of the splenic artery, may also arise from the hepatic or celiac artery or the aorta. It runs downward behind and in the substance of the pancreas, dividing into left and right branches. The left branch generally comprises the transverse

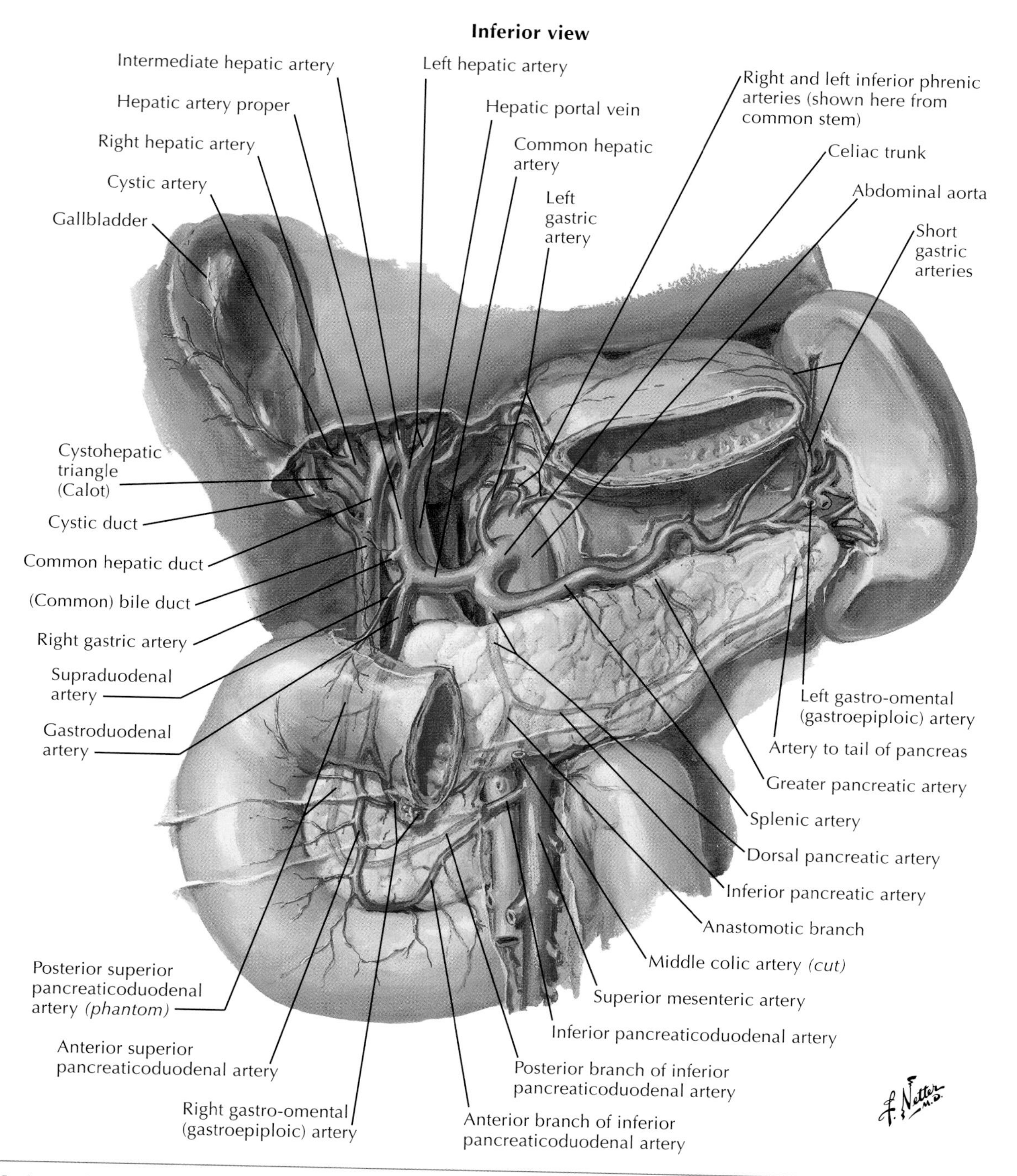

Figure 214-1 *Arterial Blood Supply of the Liver, Biliary System, and Pancreas.*

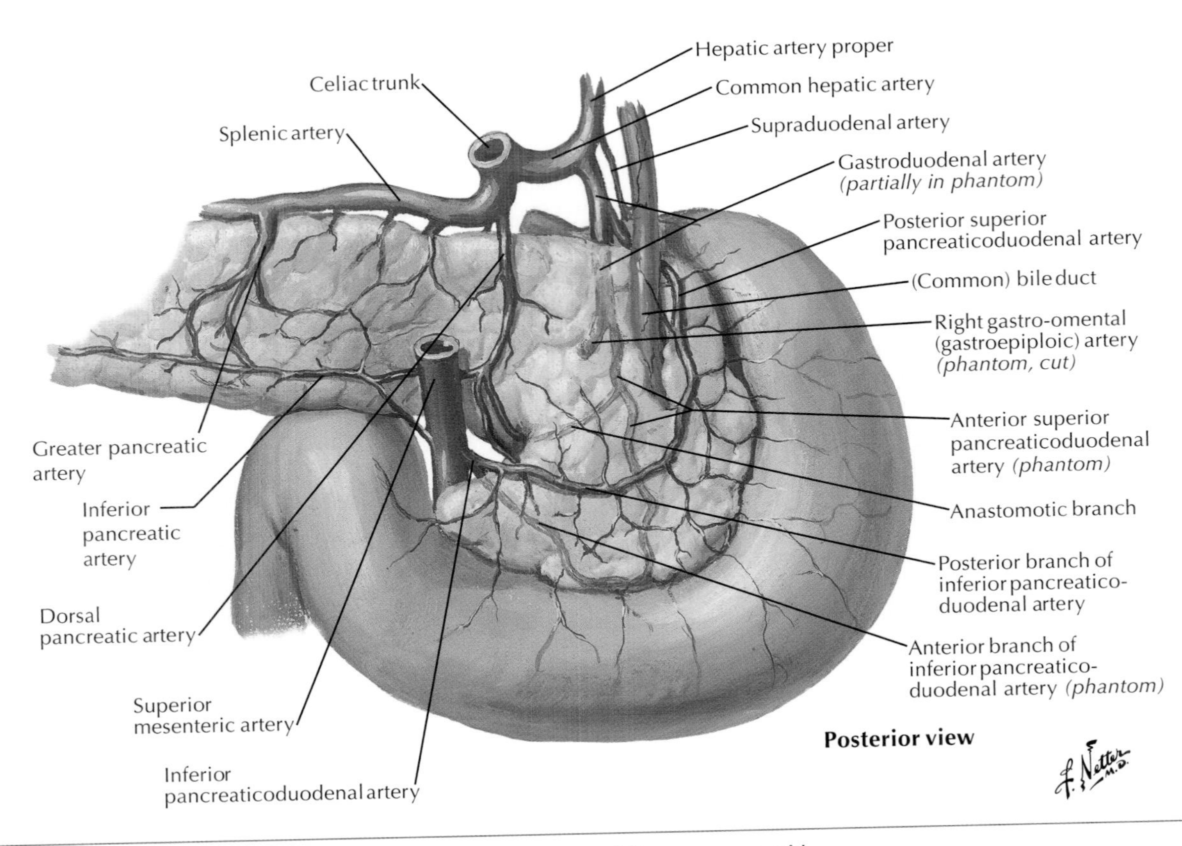

Figure 214-1 *Arterial Blood Supply of the Liver, Biliary System, and Pancreas—cont'd*

pancreatic artery. The right branches constitute an anastomotic vessel to the anterior pancreatic arch and also a branch to the pancreatic lingual. The *great pancreatic* artery originates from the splenic further to the left and passes downward, dividing into branches that anastomose with the transverse or inferior pancreatic artery. The *artery for tail of pancreas* (arteria caudae pancreatis) originates from the splenic artery, or from its terminal branches at the tail of the pancreas, and divides into branches that anastomose with the terminal twigs of the transverse pancreatic artery. The *transverse pancreatic* artery, usually the left branch of the dorsal pancreatic, courses behind the body and tail of the pancreas close to its lower border. It may originate from or communicate with the superior mesenteric artery.

The other branches of the splenic artery are variable terminal branches to the spleen, the left gastroepiploic artery, the short gastric arteries to the fundus of the stomach, and usually branches that anastomose with the left inferior phrenic artery.

Hepatic Artery Variations

215

Kris V. Kowdley

In more than 40% of dissections, variations were found in the origin and course of the hepatic artery or its branches (Michels) (**Fig. 215-1**). These involve, with equal incidence, the right and left hepatic arteries and are of more than passing surgical significance, primarily because their unintended ligation results in liver necrosis.

A *replaced* hepatic artery originates from a different source than in the standard description and substitutes for the typical vessel (see Chapter 214). An *accessory* artery is a vessel additional to those originating according to standard descriptions.

An example of replacement is the origin of the common hepatic artery from the superior mesenteric artery. The common hepatic artery passes through or behind the head of the pancreas, and its ligation during a pancreaticoduodenal resection deprives the liver of its arterial blood supply. Under these circumstances, only the left gastric and splenic arteries arise from the celiac axis.

Sometimes, right or left hepatic arteries originate independently from the celiac axis or fork from a short common hepatic artery. Under these conditions, the *gastroduodenal* artery originates from the right hepatic artery. Frequently, the right hepatic artery, giving off the gastroduodenal artery, originates from the *superior mesenteric* artery, whereas the left hepatic artery, in turn giving off the middle hepatic artery, derives from the *celiac axis*. Ligation of the replaced right hepatic artery, especially where it crosses the junction of the cystic and the common ducts, as during cholecystectomy, deprives the right lobe of the liver of its blood supply. In contrast, ligation of an accessory right hepatic artery that derives from the superior mesenteric is less significant because another right hepatic artery runs its typical course. Under these circumstances, two right hepatic arteries may be found in the Calot triangle.

A replaced right hepatic artery is more common than an accessory right hepatic artery.

An aberrant left hepatic artery, originating from the *left gastric* artery, is a replaced artery in 50% of patients and is an accessory in the other 50%. If replaced, only the right hepatic artery comes from the celiac axis, whereas in the presence of an accessory vessel, the common hepatic artery takes its usual course. Ligation of a replaced left hepatic artery, such as during gastrectomy, endangers the blood supply to the left lobe of the liver.

An accessory left hepatic artery may also come from the right hepatic artery. In approximately 12% of patients, the right hepatic artery, originating at its typical site of departure, crosses in front of the common hepatic duct instead of behind it, a variation worth remembering during exploration of the duct. The described variations are also significant in the formation of collaterals after obstruction or ligation of an artery.

Other variations shown in Figure 215-1, but not described, are less common, but their potential existence should not be ignored when operating in this field.

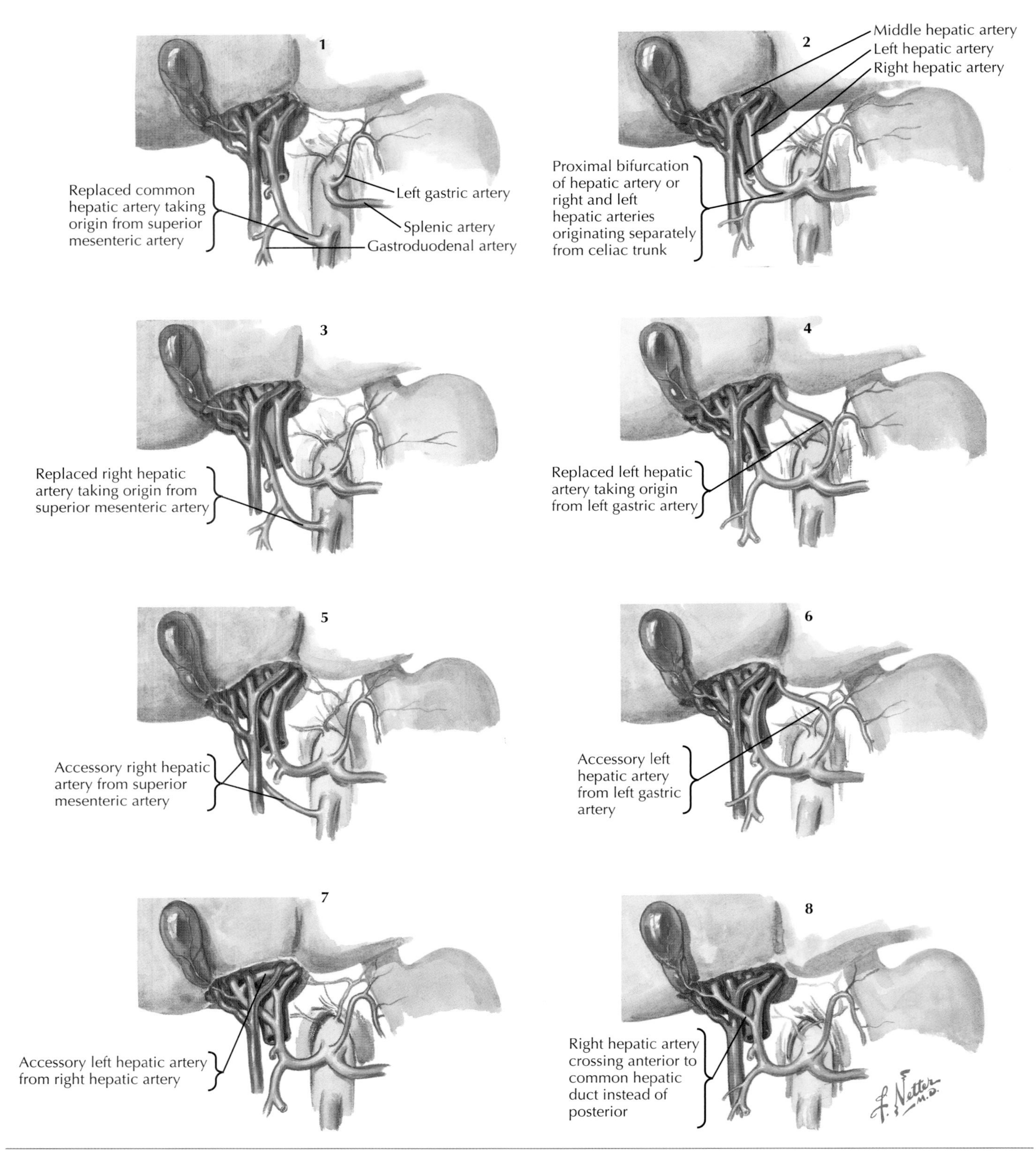

Figure 215-1 *Variations in the Origin and Course of the Hepatic Artery and Its Branches.*

Portal Vein Tributaries and Portacaval Anastomoses

Kris V. Kowdley

The portal vein forms behind the head of the pancreas at the height of the second lumbar vertebra (L2) through a confluence of the superior mesenteric and splenic veins (**Fig. 216-1**). It runs behind the first portion of the duodenum and then along the right border of the lesser omentum to the hilus of the liver, where it splints into its hepatic branches. The portal vein receives the *coronary vein*, which is the continuation of the left gastric vein and the esophageal venous plexus. The coronary vein, in turn, connects with the *short gastric* veins, the *azygos* and *hemiazygos* veins, in the lower and middle parts and with various branches of the superior vena cava, such as the innominate and inferior thyroid veins in the upper part of the esophageal region. The portal vein further accepts the *pyloric* vein, which, together with the coronary and gastric veins, forms a loop. The left main branch of the portal vein admits the *paraumbilical* veins and occasionally a persisting umbilical vein.

The *superior mesenteric* vein, one of the constituents of the vena portae, originates at the root of the mesentery, mainly from the *middle colic*, *right colic*, and *ileocolic* veins, receiving in addition many small veins. It further accepts the *inferior pancreaticoduodenal* vein, which runs in front of the third portion of the duodenum and the uncinate process of the pancreas. The *right gastroepiploic* vein, coming from the right aspects of the greater curvature of the stomach, enters the superior mesenteric vein before the latter unites with the splenic vein.

Splenic and inferior mesenteric veins usually have a common terminal end portion behind the body of the pancreas. The *inferior mesenteric* vein starts with the superior hemorrhoidal veins and continues in the posterior abdominal wall, receiving many tributaries, especially the *left colic* vein. The *splenic* vein begins at the hilus of the spleen and admits the *left gastroepiploic* vein, short gastric veins (both communicating with esophageal veins), and pancreatic veins, which anastomose with retroperitoneal veins and thus with the caval system.

The shortness of the main stem of the portal vein prevents complete mixing of the blood coming from its constituents, so the right extremity of the liver receives chiefly blood coming from the superior mesenteric vein. The left lobe receives blood from the coronary, inferior mesenteric, and splenic veins, whereas the left part of the right lobe, including the caudate and quadrate lobes, receives mixed blood. These "streamlines," demonstrated in animals, are not seen during portal venography, and it is uncertain whether they occur in humans. Their existence has been assumed, however, to explain the localization of tumor metastases and abscesses and also the predominance of massive necrosis in acute fatal viral hepatitis in the left lobe, which supposedly does not receive nutrient-rich protective blood from the small intestine.

Portacaval anastomoses have great clinical significance. They dilate when blood flow in the portal vein and through the liver is restrained. They relieve portal hypertension (see Chapters 224 and 227) and may be lifesaving in acute portal hypertension. As in chronic obstruction, however, portacaval anastomoses may shunt blood from the liver, depriving the organism of the liver's vital functions. Consequently, they contribute to hepatic insufficiency.

Dilatation of the hemorrhoidal veins results in hemorrhoidal piles, with the danger of hemorrhage, thrombosis, and inflammation. Varicosities of the esophageal veins (less with cardiac veins of stomach) may lead to *esophageal hemorrhage*, the most dangerous complication of portal hypertension (see Chapter 223). *Retroperitoneal varicose* portacaval anastomoses have less clinical significance. *Paraumbilical anastomoses* lead to marked dilatation of the veins in the anterior abdominal wall. If these veins converge toward the umbilicus, they form *caput medusae*.

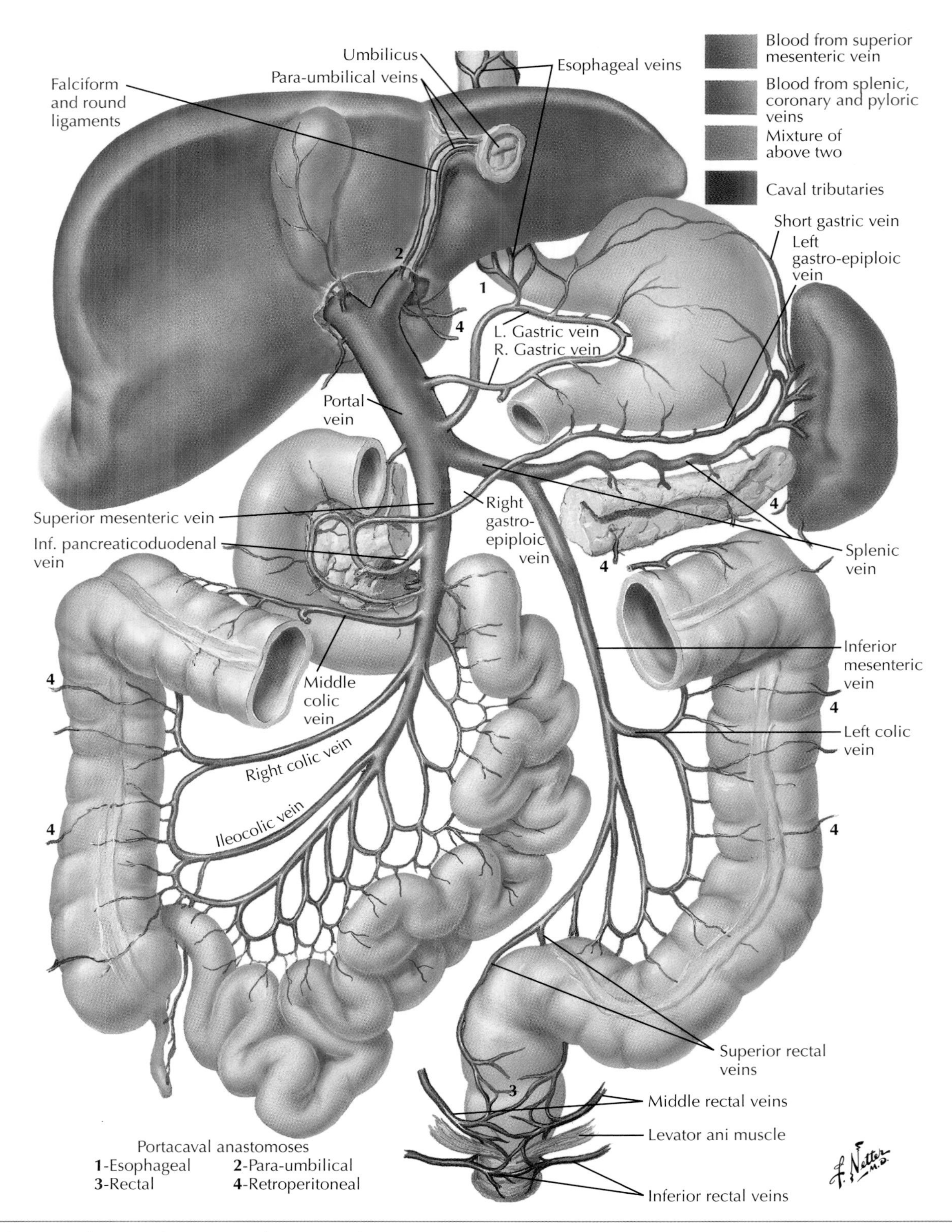

Figure 216-1 *Portal Vein Tributaries and Portacaval Anastomoses.*

Portal Vein Variations and Anomalies

217

Kris V. Kowdley

The anatomy of the portal vein system is said to reveal fewer major anatomic variations than the hepatic arterial system. Nevertheless, shunt procedures for portal hypertension have created considerable interest in the anatomy of the portal vein (see Chapter 227), and dissection studies indicate frequent minor variations of surgical importance (**Fig. 217-1**).

The length of the portal vein varies between 5.5 and 8 cm, with an average of approximately 6.5 cm (about 2½ inches); mean diameter is usually 1.09 cm. In cirrhosis, however, the diameter is considerably wider. It is of practical importance that in slightly more than 10% of the studied patients, no vessel enters the main stem of the portal vein, but in most patients, several veins are admitted that may be torn during the dissection for portacaval anastomoses. Dangerous hemorrhages may result, and their ligation may interfere with the size of the portal vein and the performance of the anastomosis.

In more than two thirds of patients, the *gastric coronary* vein is of major significance because portal drainage from esophageal varices enters the left aspect of the portal vein. Otherwise, portal drainage enters at the junction of the splenic and superior mesenteric veins, whereas in almost 25% of patients it joins the splenic vein. Under all these circumstances, the *pyloric* vein may enter the portal vein stem. On its right aspect, the portal vein may admit the superior pancreaticoduodenal vein, and close to the liver the cystic vein frequently joins the right branch of the portal vein.

The usual anatomic description of the formation of the portal vein is found in only about 50% of patients. In the other half, the *inferior mesenteric* vein enters the junction of the splenic and superior mesenteric veins, or it joins the superior mesenteric vein.

The size of the *splenic vein*, of major importance in splenorenal shunt, averages less than 0.5 cm between the splenic hilus and the junction with the inferior mesenteric vein. As a rule, the splenic vein is widened to a lesser degree in portal hypertension than the portal vein. Because the splenic vein is more or less embedded in the cephalad portion of the pancreas; the many pancreatic venous tributaries are so short that they may be easily torn during shunt procedures, and their ligation again creates technical problems.

Of rare *congenital anomalies* of the portal vein, the one of surgical significance involves an abnormal position anterior to the head of the pancreas and the duodenum. Another rare but physiologically interesting anomaly is the entrance of the portal vein into the inferior vena cava; this indicates that the morphologically normal-appearing liver can function without portal vein blood, and the hepatic artery is considerably enlarged. Extremely rare is an entrance of the pulmonary vein into the portal vein, probably caused by a disturbance in early fetal development of the venous systems. Again, congenital *strictures* of the portal vein at the hilus of the liver are rare but produce severe portal hypertension that may not be relieved by surgical anastomoses.

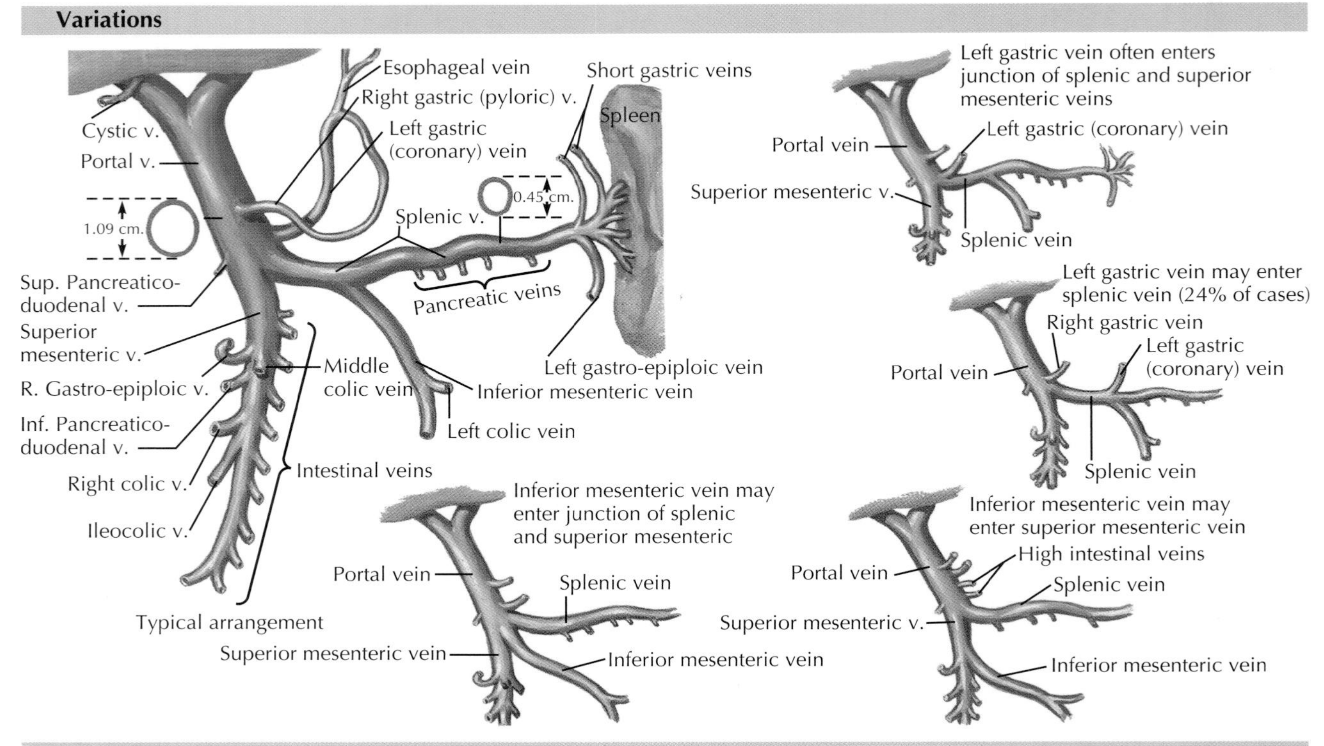

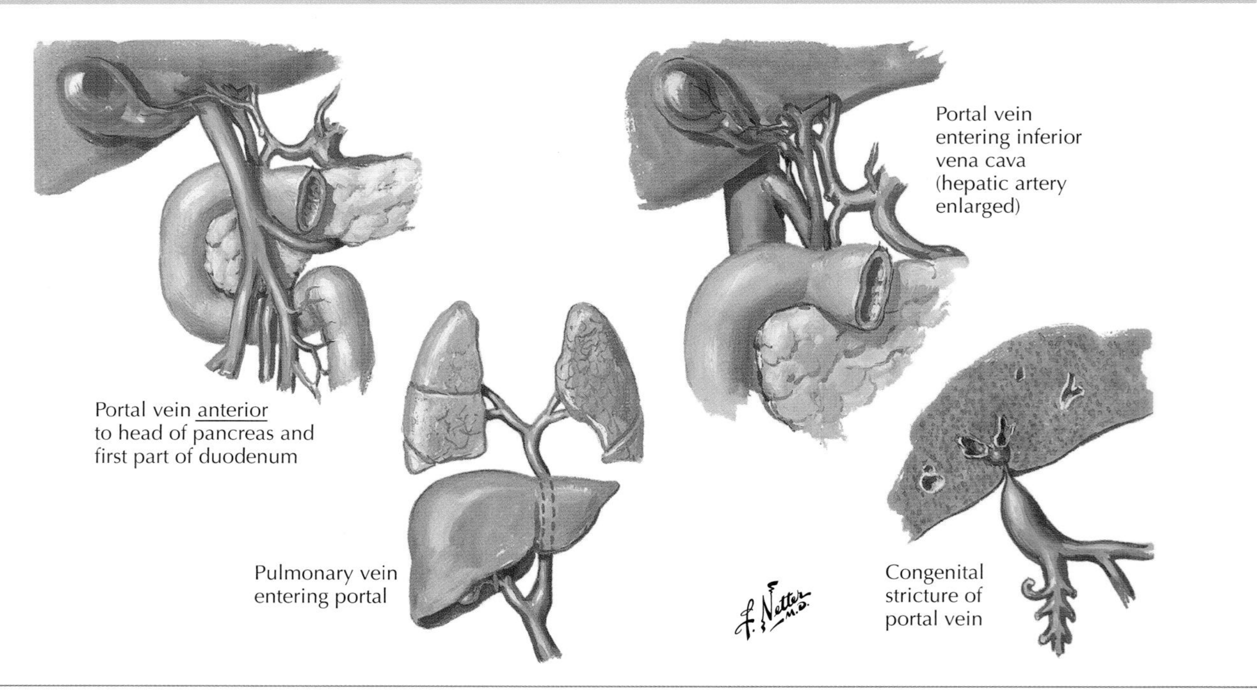

Figure 217-1 *Portal Vein Variations and Anomalies.*

Liver Function Tests

Kris V. Kowdley

Liver function tests (LFTs) refer to a panel of serum biochemical studies to screen for and monitor liver disease. This panel evaluates hepatic components ranging from hepatocellular necrosis and damaged hepatic synthetic capacity to the liver's ability to excrete breakdown products. LFTs also describe a battery of dynamic tests that provide real-time evaluation of liver function, generally by measuring hepatic blood flow, metabolic capacity, or excretory function.

Serum liver biochemical tests can be divided into three groups: (1) markers of hepatocellular injury, including alanine transaminase (ALT) and aspartate transaminase (AST); (2) markers of cholestasis, including alkaline phosphatase (ALP), 5′-nucleotidase, γ-glutamyltransferase (GGT), and bilirubin; and (3) markers of liver synthetic function, including serum albumin and prothrombin time.

In general, the hepatocellular liver enzymes reflect the degree or severity of hepatic necroinflammation. Serum AST and ALT are found in much higher concentrations within hepatocytes than in the circulation. Therefore, processes leading to necrosis or swelling of hepatocytes are associated with leakage of these enzymes into the plasma, resulting in increased concentrations. By contrast, elevated serum concentrations of the so-called cholestatic liver enzymes (e.g., ALP) may be attributed to release from damaged hepatocytes or to induction of these enzymes by processes that damage the biliary epithelia, such as obstruction by stones or cholestatic liver disease (e.g., primary biliary cirrhosis, primary sclerosing cholangitis).

Serum concentrations of albumin and bilirubin are more appropriately described as true LFTs; serum *albumin* is a measure of hepatic synthetic capacity, whereas *bilirubin* level reflects the uptake, conjugation, and biliary excretion of bilirubin, a breakdown product from senescent red blood cells. *Prothrombin time* (PT) is a measure of prothrombin synthesis, a vitamin K–dependent process.

The *serum bilirubin test* is the only blood test that measures the liver's excretory function, that is, the uptake, conjugation, and biliary excretion of this pigment. Other tests examine the liver's ability to perform specific metabolic functions and often involve administration of a compound, after which its metabolite can be measured in the serum, breath, or urine. These metabolic LFTs are more sensitive for hepatic dysfunction than the serum bilirubin test, but they may lack specificity. Nevertheless, these tests have proved useful in certain patients, such as those with compensated cirrhosis being considered for liver resection or surgery to decompress portal hypertension and whose hepatic functional reserve must be determined.

BREATH TESTS

Breath tests of liver function have generally been based on the principle that the rate of conversion of orally or intravenously radiolabeled carbon 14 (^{14}C), which is converted to $^{14}CO_2$ and is subsequently exhaled, can be collected and quantified as a measure of liver function. Breath tests are infrequently used in the United States, primarily because of the inconvenience of using radiolabeled ^{14}C. Current LFTs now measure galactose or caffeine clearance.

Human and animal studies have shown that the *aminopyrine breath test* is a quantitative measure of the mixed-function system. Patients with cirrhosis have a decreased clearance of aminopyrine, which correlates with a decreased rate of $^{14}CO_2$ appearance in breath, presumably because of the decreased mass of hepatic microsomal mass containing mixed-function oxidases. Aminopyrine breath test results also correlate with other markers of hepatic function, such as serum albumin and PT. The aminopyrine breath test is also helpful to determine the extent of severe hepatocellular disease, as in acute hepatitis, although it is rarely used clinically in this context.

Recently, a *methacetin breath test* was introduced to measure flow-dependent hepatic microsomal function. The substrate, ^{13}C methacetin, is metabolized in the liver by *O*-demethylation to $^{13}CO_2$ and acetaminophen.

CLEARANCE TESTS

Indocyanine green (ICG) is a dye with a high hepatic extraction ratio. Almost all the dye is excreted unchanged in bile; there is no significant enterohepatic circulation, and levels in the serum can be measured by atomic absorption spectrophotometry. Because ICG is excreted with little metabolic activity in the liver, the ICG clearance test has been used primarily to measure hepatic blood flow using the Fick equation. The proportion of ICG retained is inversely related to liver function; that is, higher retention correlates with decreased liver function.

Galactose clearance is reduced in patients with cirrhosis and chronic hepatitis but not in patients with biliary obstruction. However, it is unclear whether this test contributes increased sensitivity to standard LFTs, such as albumin and bilirubin.

Caffeine clearance is another test of hepatic metabolic activity and is measured by caffeine concentration in saliva. The caffeine clearance test has the advantages of noninvasive collection and avoidance of radioactivity. The main limitation is the confusion introduced by smoking, which increases caffeine clearance.

MEGX FORMATION

Monoethylglycinexylidide (MEGX) is a first-pass metabolite of lidocaine. *Lidocaine*, when administered intravenously at a subtherapeutic dose (1 mg/kg), is rapidly cleared and metabolized by the liver to MEGX. Thus, MEGX formation reflects hepatic blood flow and metabolic activity. In the early era of liver transplantation, the MEGX test engendered much enthusiasm as a useful test to determine hepatic reserve and optimal timing for liver transplantation. However, the rapid increase in demand for donor organs and the subsequent implementation of the Model for End-stage Liver Disease (MELD) scoring systems have rendered the MEGX test less clinically relevant, except when liver resection or portosystemic shunt is considered.

ADDITIONAL RESOURCES

Braden B, Lembcke B, Kuker W, Caspary WF: ^{13}C-breath tests: current state of the art and future directions, *Dig Liver Dis* 39(9):795-805, 2007.

Burke MD: Liver function: test selection and interpretation of results, *Clin Lab Med* 22:377-390, 2002.

Collier J, Bassendine M: How to respond to abnormal liver function tests, *Clin Med* 2:406-409, 2002.

Green RM, Flamm S: AGA technical review on the evaluation of liver chemistry tests, *Gastroenterology* 123:1367-1384, 2002.

Ilan Y: The assessment of liver function using breath tests (review), *Aliment Pharmacol Ther* 26(10):1293-1302, 2007.

Limdi JK, Hyde GM: Evaluation of abnormal liver function tests, *Postgrad Med J* 79:307-312, 2003.

Mallory MA, Lee SW, Kowdley KV: Abnormal liver test results on routine screening: how to evaluate, when to refer for a biopsy, *Postgrad Med* 115(3):53-56, 59-62, 66, 2004.

Morrison ED, Kowdley KV: Genetic liver disease in adults, *Postgrad Med* 107:147-152, 155, 158-159, 2000.

Nista EC, Fini L, Armuzzi A, et al: ^{13}C-breath tests in the study of microsomal liver function, *Eur Rev Med Pharmacol Sci* 8(1):33-46, 2004.

Pratt DS, Kaplan MM: Evaluation of abnormal liver-enzyme results in asymptomatic patients, *N Engl J Med* 342:1266-1271, 2000.

Prothrombin Formation

Kris V. Kowdley

The liver is the site of synthesis of several major proteins involved in coagulation, including factors I, II, V, VII, IX, X, XII, and XIII. The formation of factors II (prothrombin), VII, IX, and X is dependent on γ-carboxylation, which is a vitamin K–dependent step.

Prothrombin time (PT) is a measure of the time taken to convert prothrombin (factor II) to thrombin (activated factor II). This test is a simple yet useful measure of the body's coagulation. Causes of PT prolongation include the following (**Fig. 219-1**):

- Inadequate dietary intake of vitamin K, leading to decreased γ-carboxylation of proteins involved in coagulation.
- Inability to absorb vitamin K because of insufficient bile acid concentrations in the lumen of the intestine, such as from cholestatic liver disease or bacterial overgrowth.
- Intrinsic hepatocellular dysfunction, resulting in failure to synthesize prothrombin despite adequate vitamin K stores.
- Ingestion of drugs or toxins (e.g., warfarin) that interfere with prothrombin production.

Parenteral administration can help differentiate between inadequate vitamin K stores and liver dysfunction as the cause of a prolonged PT. Usually, three parenteral doses of vitamin K is sufficient to differentiate intrinsic liver disease from vitamin K deficiency. Patients with intrinsic liver disease fail to correct PT after vitamin K supplementation. This test can be particularly helpful in patients with cholestatic liver disease, in whom PT prolongation results from vitamin K deficiency and liver dysfunction.

ADDITIONAL RESOURCES

Bajaj SP, Joist JH: New insights into blood clots: implications for the use of APTT and PT as coagulation screening tests and in monitoring of anticoagulant therapy, *Semin Thromb Hemost* 25:407-418, 1999.

Nilsson IM: Coagulation and fibrinolysis, *Scand J Gastroenterol Suppl* 137: 11-18, 1987.

Uotila L: The metabolic functions and mechanism of action of vitamin K, *Scand J Clin Lab Invest Suppl* 201:109-117, 1990.

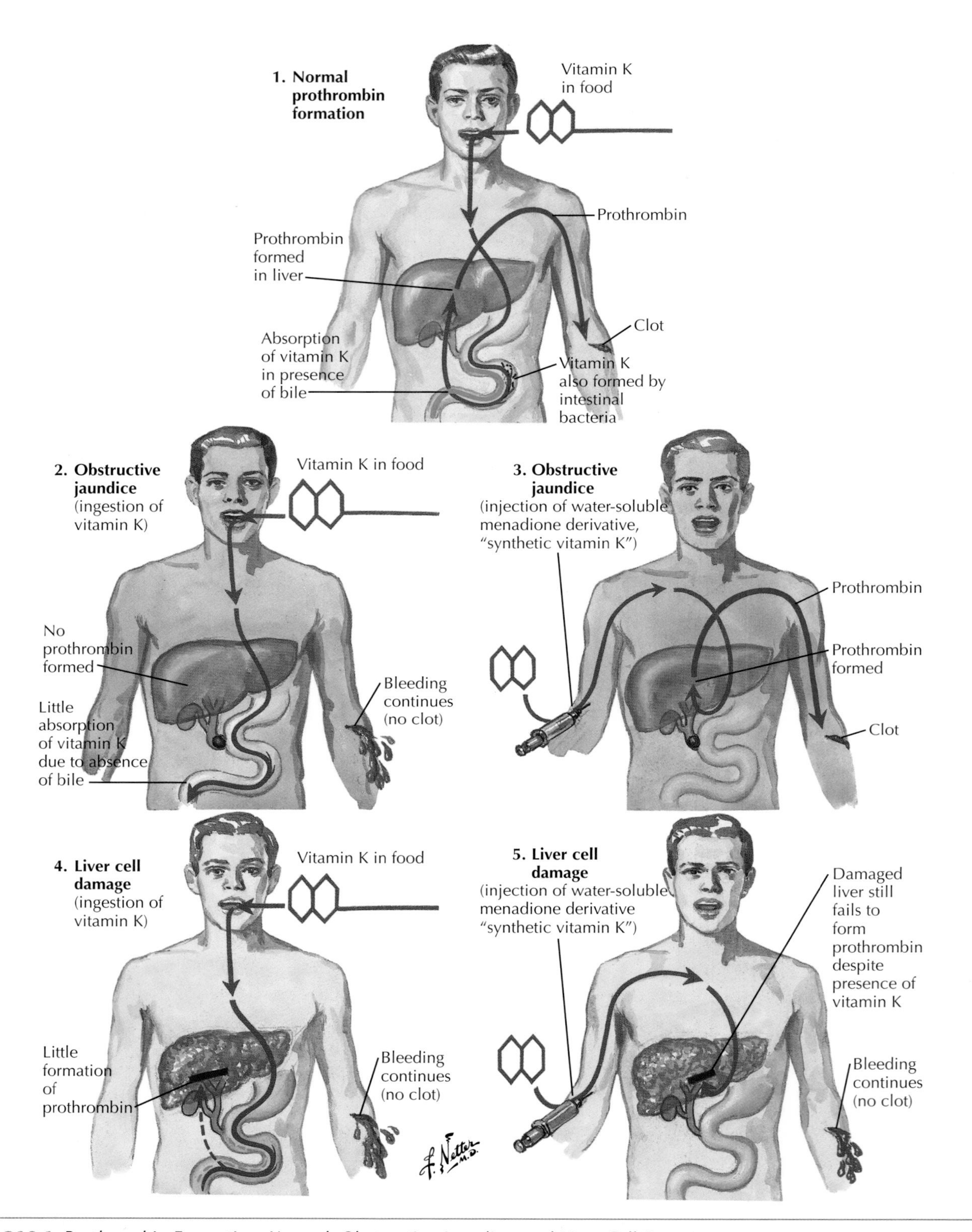

Figure 219-1 *Prothrombin Formation: Normal, Obstructive Jaundice, and Liver Cell Damage.*

Bilirubin and Bile Acid Metabolism

220

Kris V. Kowdley

Bile is secreted by the liver and serves many functions. *Hepatocytes* are the source of bile and secrete bile through specialized receptors. The biliary route is responsible for the elimination of lipid-soluble toxins, secretion of bile acids, transport of cholesterol into the gastrointestinal (GI) tract, and absorption of fats and fat-soluble vitamins, in addition to drugs, toxins, and heavy metals. Biliary epithelial cells further modify the bile secreted by hepatocytes through the addition of water, bicarbonate, and other compounds. Bile is stored in the gallbladder, where it is concentrated and then secreted into the lumen in response to hormonal and dietary signals.

The composition of bile has been studied in detail. Its predominant components are bile acids (**Figs. 220-1** and **220-2**). These compounds serve several functions, most importantly the absorption of fats and fat-soluble vitamins through the formation of *micelles*, which act as detergents. Micelles are formed by bile acids and by cholesterol, phosphatidyl choline, and lecithin.

The other major function of bile is the excretion of bilirubin, a breakdown product of red blood cells (RBCs). Bilirubin from senescent RBCs is conjugated by the liver through glucuronidation into bilirubin diglucuronide, thus rendering it water soluble and capable of transport in bile. Failure of the liver to conjugate or excrete bilirubin can result in jaundice and scleral icterus, caused by the retention of unconjugated bilirubin in the plasma. Other clinical signs of failure to transport bilirubin in the bile include dark urine and acholic stools.

Many other organic anions and cations are excreted in bile, including drugs and toxins. Other components of bile include steroid hormones, certain vitamins, cytokines such as tumor necrosis factor-α, and leukotrienes and divalent cations, most importantly *copper*. In fact, regulating body copper stores in the human occurs predominantly through the excretion of biliary copper. Chronic cholestatic disorders, such as primary biliary cirrhosis and primary sclerosing cholangitis, are often associated with excess copper accumulation in the liver. Hepatic copper content in these disorders can approach levels observed in patients with *Wilson disease*, a genetic disorder caused by a loss-of-function mutation in the *ATPB7* gene, which regulates copper excretion through the biliary tract. Copper accumulates in the liver over time, with subsequent copper overload in other organs, and is associated with disease in multiple organs through copper toxicity.

Other proteins found in bile include albumin, lysosomal enzymes, and haptoglobin. Bile also likely serves an important role in immune surveillance in the GI tract because secretory immunoglobulin A (sIgA) is an important component of bile.

Bile flow has bile acid–dependent and bile acid–independent components. Most bile flow depends on bile acid; furthermore, bile acids appear to have varying effects on bile flow, based on their physicochemical properties and other factors. By contrast, bile acid–independent bile flow generally results from an osmotic effect of anions and represents a smaller portion of the bile. Research has advanced knowledge of bilirubin and hepatic organic anion/cation transport by identifying and cloning specific transporters (e.g., mdr2/MDR3) in mice and humans responsible for biliary secretion and transport of these compounds (**Fig. 220-3**). For example, MDR3 is a member of a class of proteins that confer resistance to certain chemotherapeutic agents, and its normal function is to maintain biliary phospholipid transport. Biliary phospholipids may have a cytoprotective effect in biliary epithelia against the toxic effect of bile acids on cell membranes. Creation of a knockout mouse lacking the *mdr2* gene results in chronic cholestatic liver disease, analogous to chronic cholestatic disorders such as primary biliary cirrhosis in humans.

Several nuclear receptors are activated by ligands that regulate expression of hepatobiliary transporters at a transcriptional and posttranscriptional level. These are regulated, in turn, by bile acids, drugs, hormones, and cytokines. The clinical expression of many cholestatic liver disorders may be influenced by activation of these nuclear receptors, which include FXR, PXR, CAR, VDR, RAR, LRH, PPARα and GR.

ADDITIONAL RESOURCES

Elferink RP: Understanding and controlling hepatobiliary function, *Best Pract Res Clin Gastroenterol* 16:1025-1034, 2002.

Kullak-Ublick GA, Stieger B, Hagenbuch B, Meier PJ: Hepatic transport of bile salts, *Semin Liver Dis* 20:273-292, 2000.

Oude Elferink RP, Groen AK: Mechanisms of biliary lipid secretion and their role in lipid homeostasis, *Semin Liver Dis* 20:293-305, 2000.

Tomer G, Shneider BL: Disorders of bile formation and biliary transport, *Gastroenterol Clin North Am* 32:839-855, 2003.

Wagner M, Trauner M: Transcriptional regulation of hepatobiliary transport systems in health and disease: implications for a rationale approach to the treatment of intrahepatic cholestasis, *Ann Hepatol* 4(2):77-99, 2005.

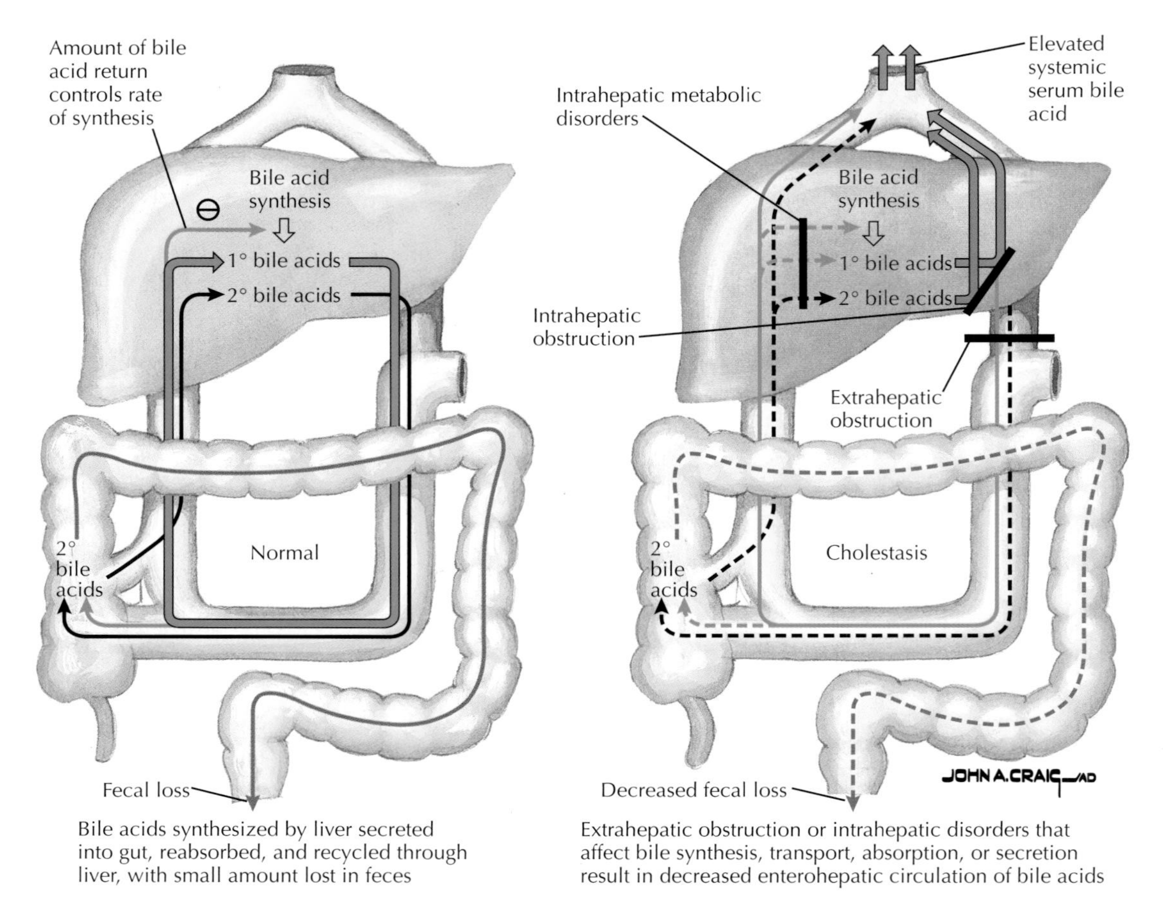

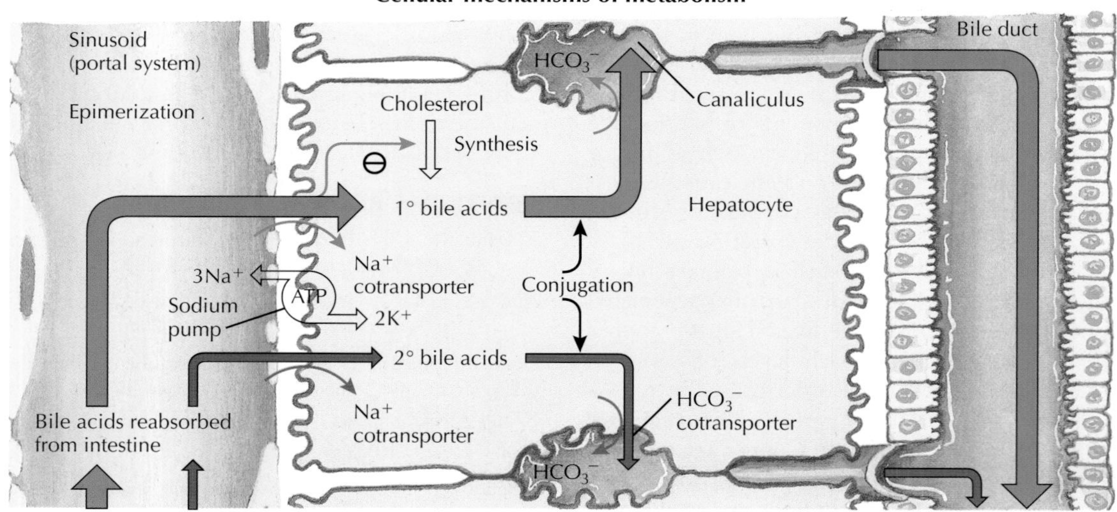

Figure 220-1 *Bile Acid Circulation and Metabolism: Enterohepatic Circulation and Cellular Mechanisms of Metabolism.*

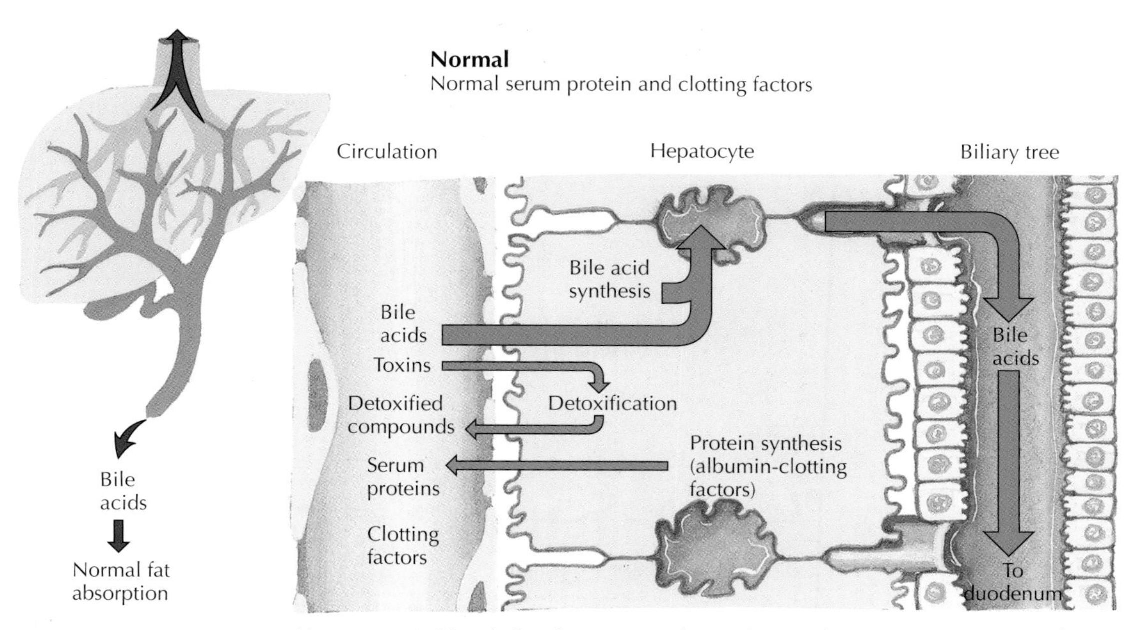

Hepatocytes (with polarity of transport and secretion) synthesize serum proteins and clotting factors and secrete them into bloodstream. Bile acids absorbed from circulation and secreted along with newly synthesized bile acids into biliary tree. Toxins absorbed from circulation, detoxified, and returned to circulation

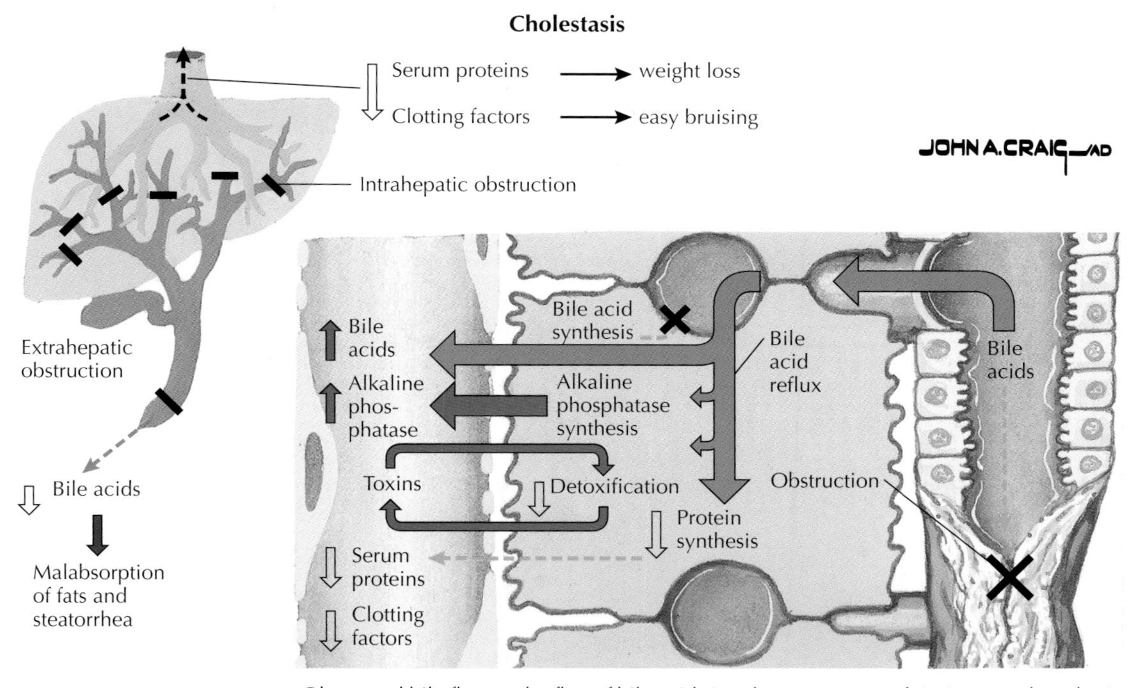

Obstructed bile flow and reflux of bile acids into hepatocytes result in increased synthesis and secretion of alkaline phosphatase. Resultant hepatocellular damage inhibits synthesis of proteins and clotting factors and limits detoxification

Figure 220-2 *Hepatic Protein and Bile Acid Metabolism: Normal Serum Protein and Clotting Factors and Cholestasis.*

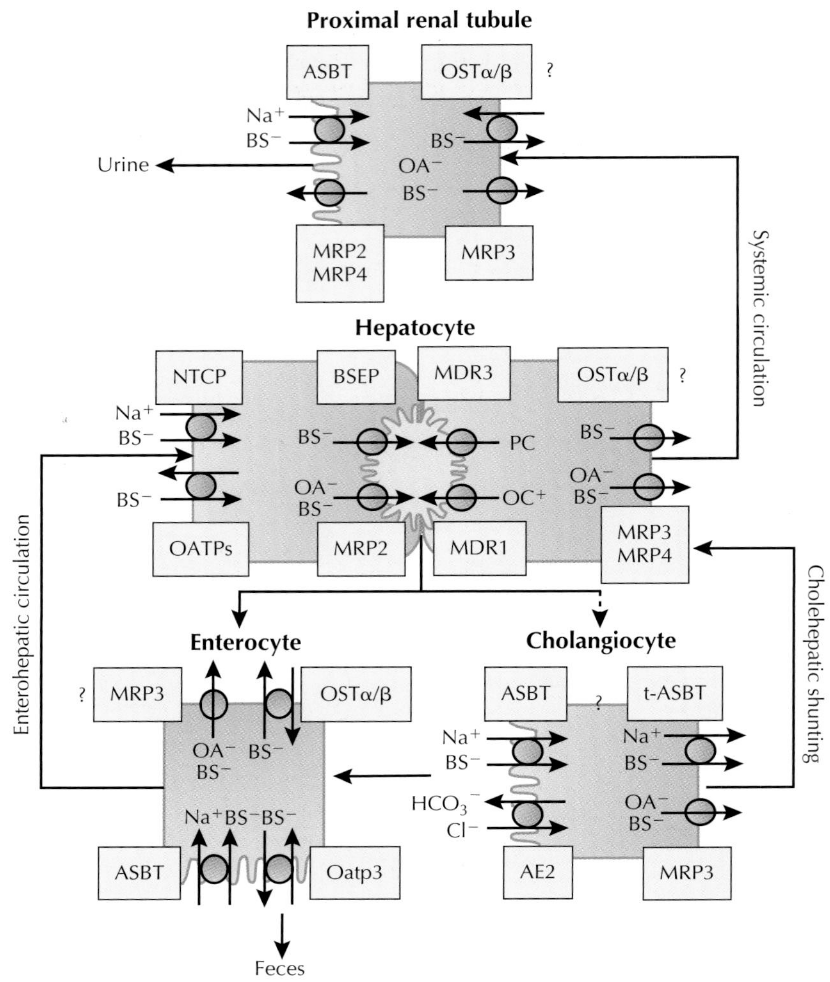

Bile acids are taken up by the hepatocytes via Na^+, the bile canaliculus via BSEP–determining bile salt–dependent bile flow, while divalent bile acids and anionic conjugates (e.g., bilirubin diglucuronide, glutathione) are excreted via MRP2–determining bile salt–independent bile flow. MDR3 mediates the canalicular secretion of phospholipids, which form mixed micelles together with bile acids and cholesterol. MDR1 excretes the bulky of organic cations. Basolateral bile acid export pumps (e.g., MRP3, MRP4) provide alternative excretory pathway for otherwise accumulated biliary constituents. Bile composition is further modified along the biliary passage by secretion of bicarbonate via AE2 and reabsorption of bile acids via luminal ASBT. A shortcut (called "cholehepatic shunting") between cholangiocytes and hepatocytes is proposed by basolateral export of bile acids from the cholangiocytes via MRP3 and possibly t-ASBT and hepatocellular reuptake. In the terminal ilieum bile acids are reabsorbed by ASBT and, to certain degrees, in rodents by Oatp3 and effluxed on the basolateral pole of enterocytes via OSTα/β and possibly to a lesser extent via MRP3 into portal circulation. Similar to the cholangiocyte and enterocyte, in the proximal renal tubules, bile acids are reabsorbed from the glomerular filtrate via ASBT to minimize bile salt loss. Possibly, renal MRP2 and MRP4 may be involved in the secretion of bile acids into the urine under cholestatic conditions.

Figure 220-3 *Hepatobiliary Transporters in Liver and Extrahepatic Tissues. From Wagner M, Trauner M:* Ann Hepatol *4(2):77-99, 2005.*

Clinical Manifestations of Cirrhosis

221

Kris V. Kowdley

Clinical manifestations of cirrhosis vary with its different forms (**Fig. 221-1**). Chills, high fever, and leukocytosis are induced by an infected extrahepatic biliary obstruction and are observed in secondary biliary cirrhosis. Peripheral neuropathy, as a symptom of malnutrition, is encountered in cirrhosis associated with malnutrition. Hepatic insufficiency, the result of liver cell damage, and loss of hepatic function because of the diversion of blood by extrahepatic portacaval collaterals and intrahepatic portohepatic venous anastomoses, explain many clinical symptoms. Palpation and percussion uncover a sometimes enlarged, sometimes shrunken, liver depending on the stage of disease. In most patients with cirrhosis, the organ is firm, and the granular character of the surface is rarely palpable.

Severe jaundice is seen in only a minority of patients; a subicteric hue is common. Central nervous system (CNS) manifestations vary from somnolence to precomatose manifestations—reflected in flapping tremor, mental confusion, and electroencephalographic changes—to frank coma. *Bleeding diathesis* is caused by a defect in the hepatic formation of serum proteins active in the processes of blood coagulation, especially factors II (prothrombin) and VII. Capillary damage also may be important.

Fibrinogen formation is usually not decreased in cirrhosis, but it may be increased. Hypoproteinemia and compression of the inferior vena cava by ascites may explain the appearance of ankle edema. Liver disease can result in anemia of the macrocytic or normochromic type, caused by hypersplenism and a nonspecific toxic effect on the bone marrow. Testicular atrophy, gynecomastia, female escutcheon, pectoral and axillary alopecia, and marked reddening of the thenar and hypothenar *(palmar erythema)* are all thought to be caused by excess circulating estrogen resulting from decreased clearance by the liver. Cutaneous *spider nevi* are found mainly on the upper half of the body, usually the neck, forearm, and dorsum but sometimes the mucous membranes, and consist of a central arteriole from which many small vessels radiate. Most manifestations of increased estrogenic activity appear more in men. In women, hair loss and spider nevi are seen most often, but some masculinizing effects, such as hirsutism, may be noted.

Liver failure combined with the effects of portal hypertension leads to ascites, esophageal varices, and the dilatation of abdominal veins *(caput medusae)*. Leukopenia, thrombocytopenia, and anemia result from enlargement of the spleen and signs of hypersplenism.

The pathogenesis and diagnosis of cirrhosis are discussed in Chapter 231.

ADDITIONAL RESOURCES

Rockey DC: The cell and molecular biology of hepatic fibrogenesis: clinical and therapeutic implications, *Clin Liver Dis* 4:319-355, 2000.

Schuppan D, Afdhal NH: Liver cirrhosis, *Lancet* 371(9615):838-851, 2008.

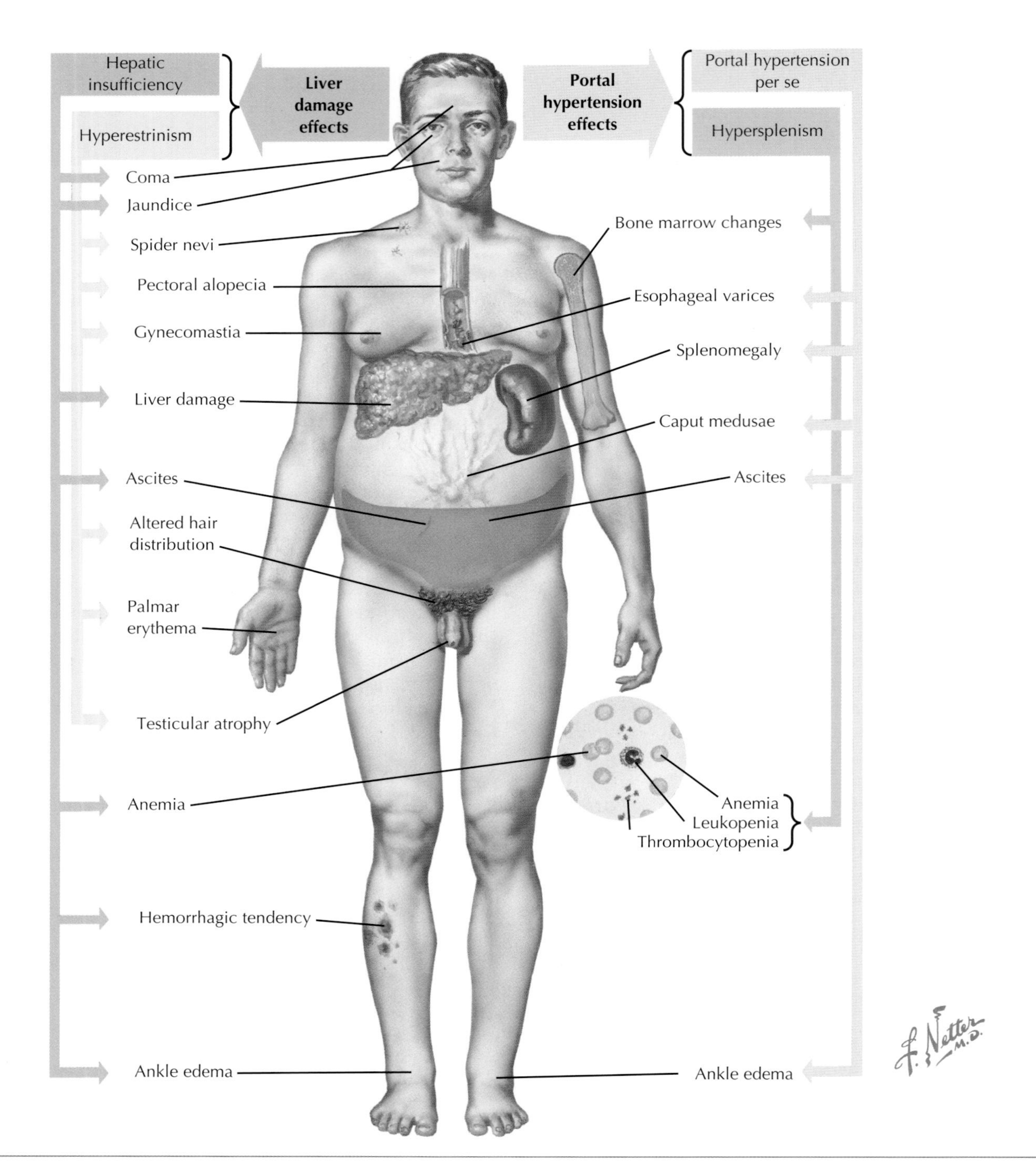

Figure 221-1 *Clinical Manifestations of Cirrhosis.*

Physical Diagnosis of Liver Disease

222

Kris V. Kowdley

The clinical diagnosis of liver disease is not difficult in patients with advanced hepatic decompensation. A history of deepening jaundice, dark urine, light stools, progressive increase in abdominal girth, and subjective symptoms of weakness, anorexia, and other digestive difficulties focuses attention on the liver.

Icterus, basically a deep staining of the skin, sclerae, and mucous membranes, may be present in extrahepatic obstructive jaundice and in hepatocellular injury (**Fig. 222-1**). The icterus present in *prehepatic* (hemolytic) jaundice, however, usually does not stain the tissues as deeply as in the other forms. In *hepatic* and *posthepatic* jaundice, the urine is dark and the feces are light, particularly if the jaundice is deep. In prehepatic jaundice, on the other hand, bilirubin does not appear in the urine, but the urine may be dark because of increased amounts of urobilinogen. For the same reason, the feces in prehepatic jaundice are also dark. It is important to remember that in certain advanced cases of liver disease, little or no jaundice may be apparent.

Detecting an enlarged or tender liver is the most striking indication of some hepatic disease, either primary or secondary. In patients with relaxed abdominal walls or in thin persons with low diaphragms, the liver may be palpable even in the absence of hepatic disease. In patients with biliary cirrhosis, fatty liver, or primary or secondary hepatic neoplasm, the liver may be massively enlarged and nodular. Patients with congestive heart failure (CHF) or constrictive pericarditis may also have an enlarged, tender liver. In some patients with far-advanced or rapidly progressing liver disease, the organ may be small and nonpalpable. An atrophic liver incapable of regeneration is usually an ominous finding in patients with known hepatic disease.

The presence of splenomegaly, ascites, and caput medusae raises the suspicion of portal hypertension, although the spleen may be enlarged in patients with parenchymal liver disease without portal hypertension (e.g., CHF patients).

Clubbing of the fingers and whitening of the nail beds are seen in some patients with cirrhosis. These signs are not specific for hepatic disease, but they serve to confirm the diagnosis. Severe *pruritus*, with or without jaundice, may be the outstanding symptom in patients with cholestatic liver disease and is frequently present in patients with posthepatic jaundice. Increased concentration of bile salts in the bloodstream is thought to cause pruritus. Elevated alkaline phosphatase and serum cholesterol concentrations are frequently seen in association with pruritus, comprising the outstanding features of so-called primary biliary cirrhosis.

Presacral and ankle edema, often notable in patients with advanced liver disease, results from decreased serum albumin and sodium retention. Intermittent fever may be observed in patients with cirrhosis. This febrile reaction is noted in approximately 25% of patients with chronic liver disease and may be caused by hepatic necroinflammation.

ADDITIONAL RESOURCES

Heidelbaugh JJ, Bruderly M: Cirrhosis and chronic liver failure: Part I—Diagnosis and evaluation, *Am Fam Physician* 74(5):756-762, 2006.

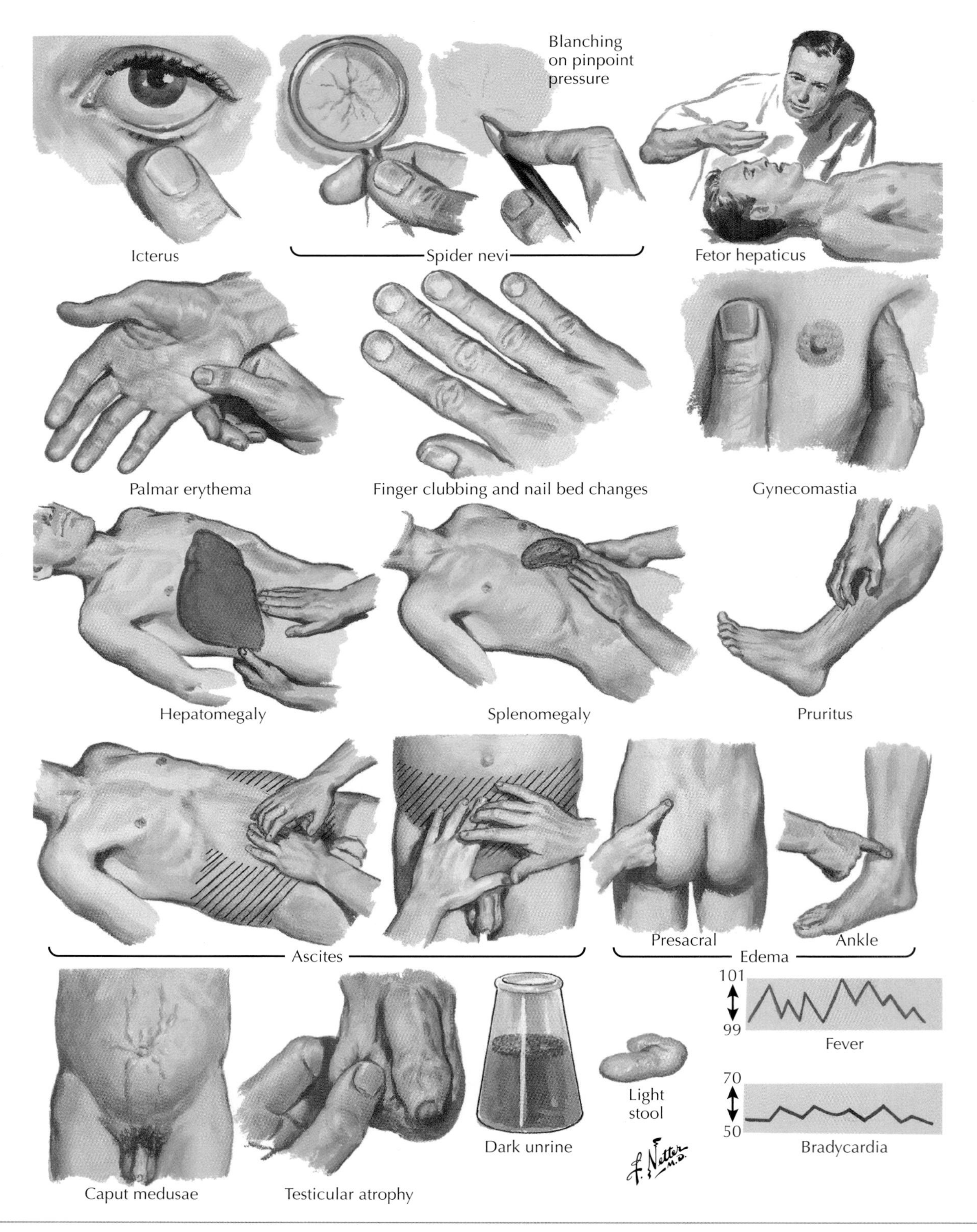

Figure 222-1 *Physical Diagnosis of Liver Disease.*

Causes of Portal Hypertension

Kris V. Kowdley

223

Portal venous pressure rises above the normal value of approximately 5 to 6 mm Hg because of (1) blockage in the intrahepatic portal venous system (sinusoidal portal hypertension), (2) impaired outflow of blood from the liver (postsinusoidal portal hypertension), or (3) increased pressure in the extrahepatic portal venous system, as from postal venous thrombosis (presinusoidal portal hypertension).

The *suprahepatic* (postsinusoidal) form of portal hypertension is most often caused by heart failure with passive congestion, particularly in patients with tricuspid regurgitation or constrictive pericarditis, or by obstruction of the main hepatic veins (**Fig. 223-1**). Occlusion of the hepatic veins, known as *Budd-Chiari syndrome*, is a rare condition resulting from spontaneous thrombosis associated with hypercoagulable states, congenital abnormalities, or mechanical obstruction. In suprahepatic portal hypertension, the liver is large and tender; ascites develops, and the spleen is slightly to moderately enlarged.

The most frequent type, *intrahepatic* (sinusoidal) portal hypertension, is caused by cirrhosis, although primary hepatic carcinoma or schistosomiasis may also lead to similar pathophysiologic changes. The spleen is greatly enlarged, and esophageal varices are often present.

In *infrahepatic* (presinusoidal) portal hypertension, the liver is of normal size, but the spleen is greatly enlarged, as are the esophageal veins. This form occurs more frequently in younger patients. The most important cause is portal vein thrombosis. Portal vein compression, by tumors or inflammatory masses, and congenital anomalies may also represent causative factors. On rare occasions, severe portal hypertension has been observed in children without detectable anatomic alterations.

ADDITIONAL RESOURCES

Baron F, Deprez M, Beguin Y: The veno-occlusive disease of the liver, *Haematologica* 82:718-725, 1997.

De Bruyn G, Graviss EA: A systematic review of the diagnostic accuracy of physical examination for the detection of cirrhosis, *BMC Med Inform Decis Mak* 1:6, 2001.

Hennenberg M, Trebicka J, Sauerbruch T, Heller J: Mechanisms of extrahepatic vasodilation in portal hypertension, *Gut* 57(9):1300-1314, 2008.

McGuire BM, Bloomer JR: Complications of cirrhosis: why they occur and what to do about them, *Postgrad Med* 103:209-212, 217-218, 223-224, 1998.

McHutchison JG: Differential diagnosis of ascites, *Semin Liver Dis* 17:191-202, 1997.

Menon KV, Gores GJ, Shah VH: Pathogenesis, diagnosis, and treatment of alcoholic liver disease, *Mayo Clin Proc* 76:1021-1029, 2001.

Sanyal AJ, Bosch J, Blei A, Arroyo V: Portal hypertension and its complications, *Gastroenterology* 134(6):1715-1728, 2008.

Schuppan D, Afdhal NH: Liver cirrhosis, *Lancet* 371(9615):838-851, 2008.

Simpson KJ, Finlayson ND: Clinical evaluation of liver disease, *Baillieres Clin Gastroenterol* 9:639-659, 1995.

Yu AS, Hu KQ: Management of ascites, *Clin Liver Dis* 5:541-568, viii, 2001.

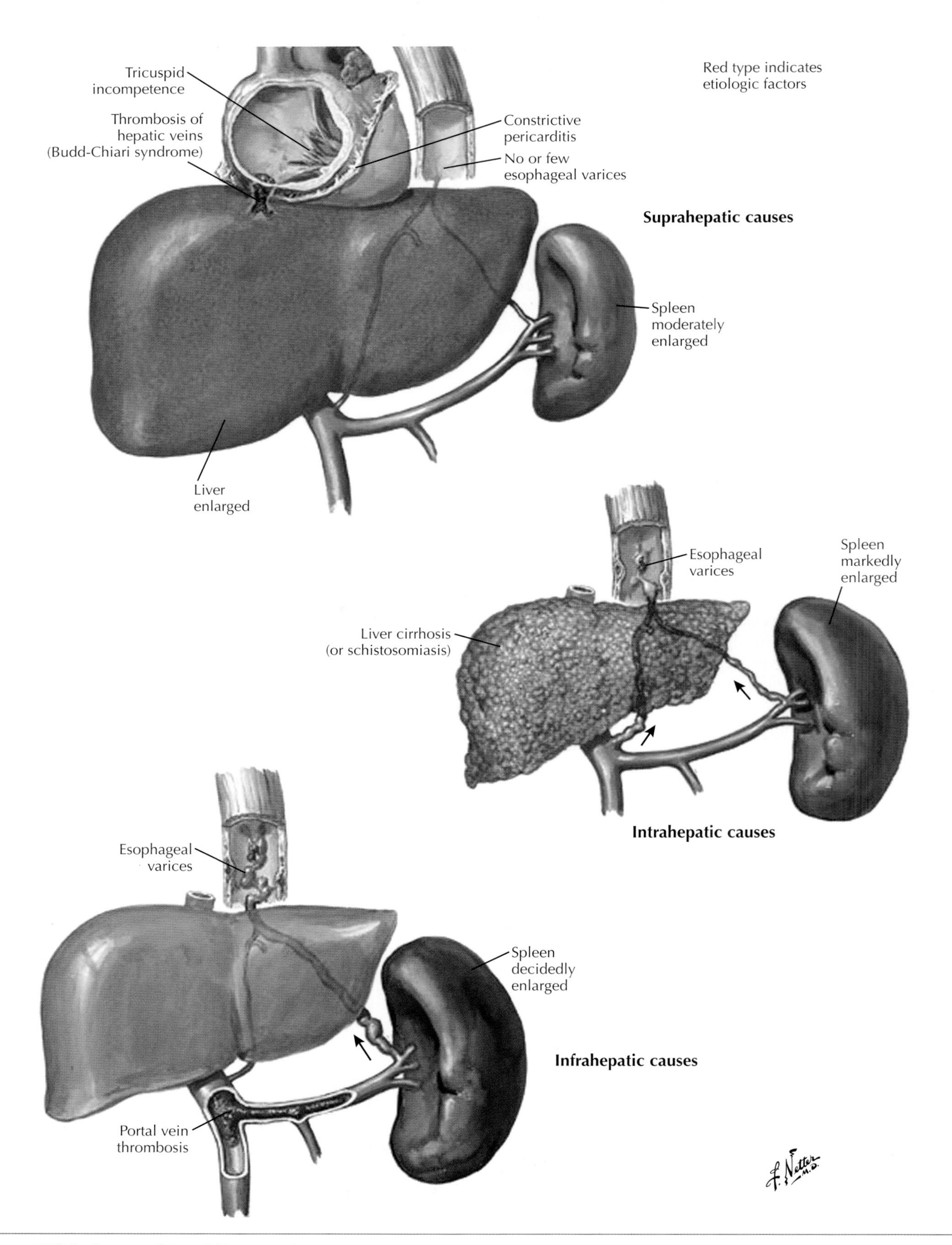

Figure 223-1 *Causes of Portal Hypertension.*

Ascites

224

Kris V. Kowdley

Ascites is a common and serious complication of cirrhosis. Long-term survival for patients with cirrhosis decreases after the development of ascites. Furthermore, the development of spontaneous bacterial peritonitis is associated with a 2-year survival rate of 50% to 60%, particularly in patients with advanced liver disease.

Ascites is an accumulation in the abdominal cavity of a transudative fluid that presumably emanates from the liver and possibly the peritoneum. The pathogenesis of ascites is complex and likely multifactorial (**Fig. 224-1**).

To explain ascites formation, the *underfill theory* states that extravasation of fluid into the abdominal cavity results in intravascular volume depletion, leading to renal salt and water retention and increased total body volume. This process may be compounded by peripheral vasodilatation, a hallmark of cirrhosis. The *overflow theory* states that ascites in cirrhosis results from primary renal salt and water retention, with leakage of fluid into the extravascular space because of increased plasma volume. The *hepatorenal hypothesis* states that ascites results from a hepatorenal reflex signaling the kidney to increase salt and water absorption in response to changes in sinusoidal blood flow or pressure, through neuronally and hormonally mediated mechanisms.

CLINICAL PICTURE AND DIAGNOSIS

Prostaglandins, the renin-angiotensin system, atrial natriuretic peptide, alcohol dehydrogenase, and many other hormones likely play a role in the altered hemodynamic state in patients with cirrhosis, which resembles sepsis with its increased cardiac output and decreased systemic vascular resistance. Portal hypertension and portosystemic shunts may exacerbate vasodilatation, which is likely mediated by nitric oxide.

Ascites may also develop from other causes of portal hypertension and from non–portal hypertension causes, such as peritoneal carcinomatosis, or inflammatory conditions involving the peritoneum.

The serum/ascites albumin gradient, calculated by subtracting the ascites albumin level from the serum albumin, is useful for differentiating ascites caused by portal hypertension from non–portal hypertension causes. The serum/ascites albumin gradient is usually greater than 1.0 in patients with portal hypertension but is less than 1.0 in patients with ascites from non–portal hypertension causes.

Determining the protein content of ascitic fluid is also helpful in the diagnosis of some patients. Ascites from cardiac cirrhosis and Budd-Chiari syndrome is often associated with a higher protein content, especially compared with patients with alcoholic cirrhosis, in whom ascitic fluid protein concentration is very low.

Spontaneous Bacterial Peritonitis

Measuring ascitic fluid white blood cell counts can also be clinically useful to identify the presence of spontaneous bacterial peritonitis (SBP). An ascitic fluid polymorphonuclear leukocyte (PMN) count greater than 250 is diagnostic of SBP. Occasionally, SBP may be diagnosed by the presence of elevated PMNs, but no bacteria can be cultured; such cases are described as *neutrocytic ascites. Bacterascites* describes infected peritoneal fluid not associated with an elevated PMN count, as seen in patients with neutropenia.

A serious complication of ascites, SBP may be associated with significant short-term morbidity. The pathogenesis is presumably the translocation of bacteria into otherwise sterile ascitic fluid. Usually, only one strain of bacteria is identified when ascitic fluid is cultured. The identification of polymicrobial infection should raise the suspicion of secondary peritonitis, such as diverticulitis or appendicitis.

TREATMENT AND MANAGEMENT

Prophylactic antibiotic therapy is indicated to prevent a recurrence of SBP. Primary prophylaxis has also been advocated in patients with ascites, especially in those with advanced cirrhosis. Ascites caused by cirrhosis is treated with diuretics. Most experts suggest initiating therapy with furosemide and spironolactone.

Peritoneovenous shunt procedures have been performed for patients with refractory ascites. However, these are seldom used now because of the high rates of shunt failure and occlusion and the infrequent but serious complication of infection or a disseminated intravascular coagulation–like syndrome. Rather, liver transplantation is the preferred therapy for patients with refractory ascites.

Transjugular intrahepatic portosystemic shunt (TIPS) has also been used to treat refractory ascites. The procedure is effective in relieving ascites compared with repeated large-volume paracentesis, but it improves survival only in patients with alcoholic cirrhosis. The high risk for hepatic encephalopathy and complications (e.g., renal insufficiency) should be considered before using TIPS in patients with advanced liver disease.

It is hoped that new agents that target elevated plasma antidiuretic hormone levels and that promote aquaeresis, such as the V_2-receptor antagonists, will be used in the future management of ascites in patients with cirrhosis.

ADDITIONAL RESOURCES

Angeli P, Merkel C: Pathogenesis and management of hepatorenal syndrome in patients with cirrhosis, *J Hepatol* 48(suppl 1):93-103, 2008.

Arroyo V, Colmenero J: Ascites and hepatorenal syndrome in cirrhosis: pathophysiological basis of therapy and current management, *J Hepatol* 38(suppl 1):69-89, 2003.

Dong MH, Saab S: Complications of cirrhosis, *Dis Mon* 54(7):445-456, 2008.

Ginès P, Cárdenas A, Arroyo V, Rodés J: Management of cirrhosis and ascites, *N Engl J Med* 350(16):1646-1654, 2004.

Gentilini P, Vizzutti F, Gentilini A, et al: Update on ascites and hepatorenal syndrome, *Dig Liver Dis* 34:592-605, 2002.

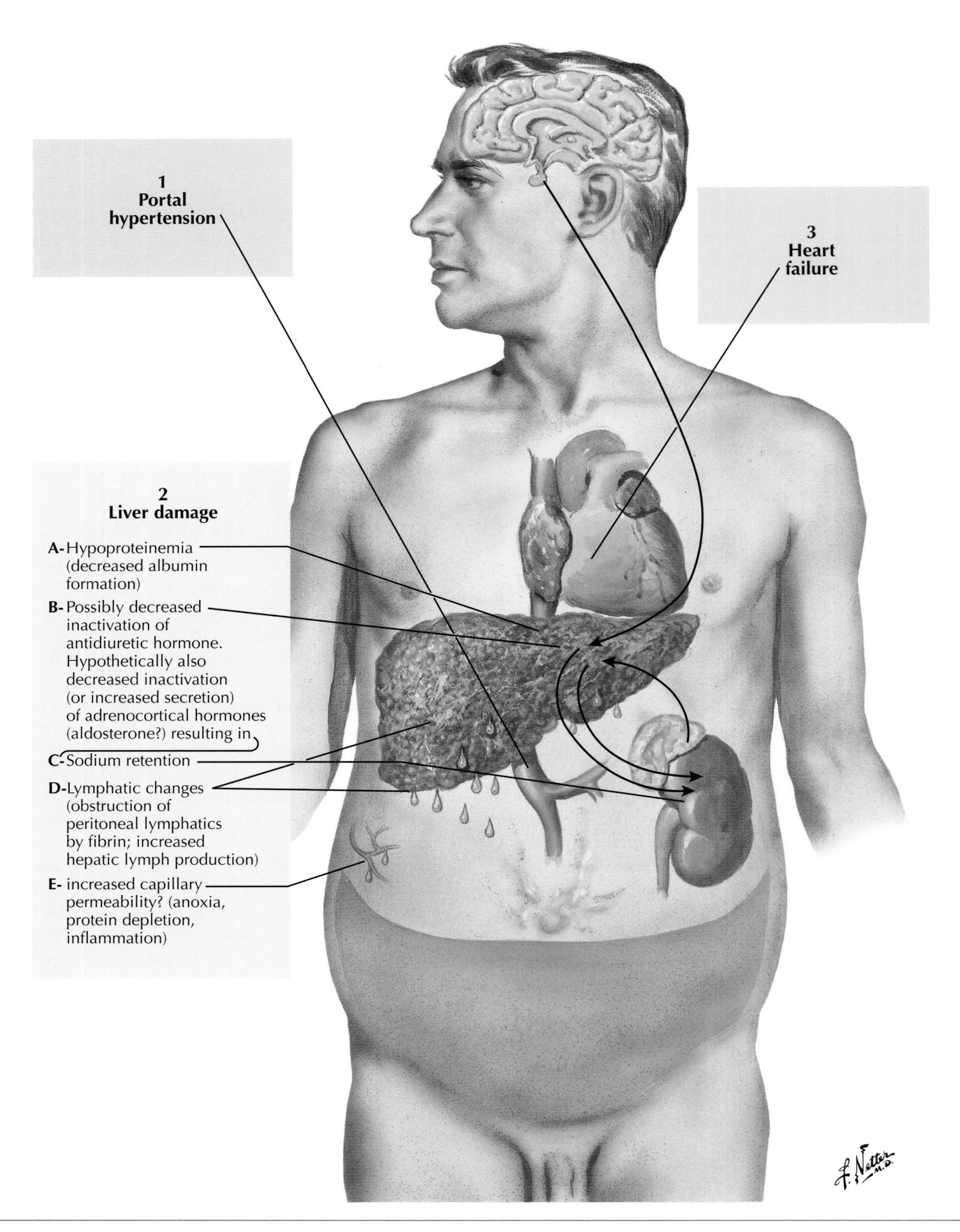

Figure 224-1 *Causes of Ascites: Heart Failure, Portal Hypertension, and Hepatic Insufficiency.*

Ginès P, Cárdenas A: The management of ascites and hyponatremia in cirrhosis, *Semin Liver Dis* 28(1):43-58, 2008.

Moore KP, Wong F, Ginès P, et al: The management of ascites in cirrhosis: report on the consensus conference of the International Ascites Club, *Hepatology* 38:258-266, 2003.

Runyon BA, Practice Guidelines Committee, American Association for the Study of Liver Diseases (AASLD): Management of adult patients with ascites due to cirrhosis, *Hepatology* 39:841, 2004.

Hepatic Encephalopathy

225

Kris V. Kowdley

Hepatic encephalopathy is characterized by abnormal mental status among patients with chronic liver disease. The pathophysiology of hepatic encephalopathy remains incompletely understood, but ongoing research for more than 50 years has resulted in several hypotheses to explain the mechanism for abnormal cognitive function in patients with chronic liver disease.

In one longstanding hypothesis, increased concentrations of ammonia in the central nervous system (CNS) are caused by failure of the liver to metabolize nitrogen-containing compounds and result in toxicity to neurons and consequent cognitive deficits. Altered ammonia metabolism clearly plays a role in hepatic encephalopathy, also known as *portosystemic encephalopathy* (PSE); patients with PSE often have significantly elevated ammonia concentrations, and treatment or spontaneous improvement is often associated with decreased blood ammonia concentrations. However, elevated serum ammonia concentration is not necessary for PSE to occur. Furthermore, patients with chronic liver disease often have elevated blood ammonia concentrations in the absence of demonstrable PSE. Other proposed causes for PSE are altered CNS levels of endogenous false neurotransmitters, such as excessive γ-aminobutyric acid (GABA), opiates, or GABA-induced benzodiazepine-like compounds. PSE may occur in patients with acute or chronic decompensation of chronic liver disease or may be caused by portosystemic shunting, which is usually (but not always) seen in patients with cirrhosis.

Recent studies suggest that hepatic encephalopathy is a consequence of early cerebral edema, astrocyte swelling, and resultant oxidative stress in the brain associated with production of reactive oxygen and nitrogen species. In addition, disruption of oscillatory networks in the brain has been demonstrated in PSE patients on magnetoencephalography and may underlie their cognitive and motor abnormalities.

Factors that can precipitate hepatic encephalopathy include dehydration, infection, gastrointestinal (GI) bleeding, constipation, electrolyte imbalance, or use of CNS depressants, particularly narcotics and benzodiazepines. GI bleeding is a particularly common cause of PSE because the combination of nitrogen-rich blood to the gut and volume depletion may exacerbate the condition.

CLINICAL PICTURE

Clinical features of hepatic encephalopathy include various states of altered mental status, ranging from mild alterations such as sleep/wake reversal, change in mood, and forgetfulness to progressive degrees of lethargy and obtundation. In the most serious stage of PSE, patients enter a deep coma, develop cerebral edema, and die of brainstem herniation (**Fig. 225-1**). Brainstem herniation is primarily observed in patients with acute or fulminant liver failure, although it has been described in patients with chronic liver disease who have experienced severe acute decompensation.

Therefore, PSE has been classified into four grades (**Table 225-1**). Grade 4 encephalopathy is associated with an increased risk for severe cerebral edema, and close clinical monitoring is required in such patients.

Clinical signs of PSE may be obvious or subtle. The most common clinical signs include *asterixis*, defined as the inability to maintain a particular position. This is typically examined by asking the patient to hold the arms out straight, with elbows unflexed and with the palms facing (extended) the examiner. A characteristic "flap" may be observed (see Fig. 225-1), sometimes called the *liver flap*. In patients unable to comply with this test, demonstrating diffuse clonus may substitute for asterixis. Additionally, the patient's breath may have a characteristic, musty, fruity odor, called *fetor hepaticus*.

The syndrome of *minimal hepatic encephalopathy*, previously described as "subclinical hepatic encephalopathy," has been the focus of several recent studies. Although changes in level of alertness or cognitive function may not be apparent, this syndrome may be associated with decreased attention, as in driving skill and reduced quality of life, and may be identified using psychometric testing. Lactulose therapy was shown to improve quality of life and cognitive function in one study. However, widely available standardized testing is needed, as well as additional data on efficacy of therapy.

DIAGNOSIS

The diagnosis of hepatic encephalopathy depends on a high index of clinical suspicion because of the absence of a pathognomonic sign or diagnostic test and the varied clinical presentations. In early stages, patients may be unaware of altered mental status, although family members may report alterations in the patient's personality, irritability level, or reversal of day-night cycles. Patients with more advanced disease stages show more obvious depression in the level of consciousness, with increased lethargy and confusion that may progress to stupor and coma.

Electroencephalographic abnormalities reported in PSE patients include diffuse, bilateral, high-voltage slow waves. Magnetic resonance imaging (MRI) with T1-weighted images may demonstrate increased signals in certain portions of the brain (globus pallidus). Abnormal evoked responses and abnormalities on positron emission tomography also have been noted. However, none of these findings is specific for hepatic encephalopathy.

In patients with acute liver failure, the speed of PSE progression is of prognostic importance. It is therefore critically important to complete the pretransplantation workup early after hospital admission of these patients, particularly the evaluation of psychosocial status, psychiatric conditions, and potential problems with drug or alcohol dependence.

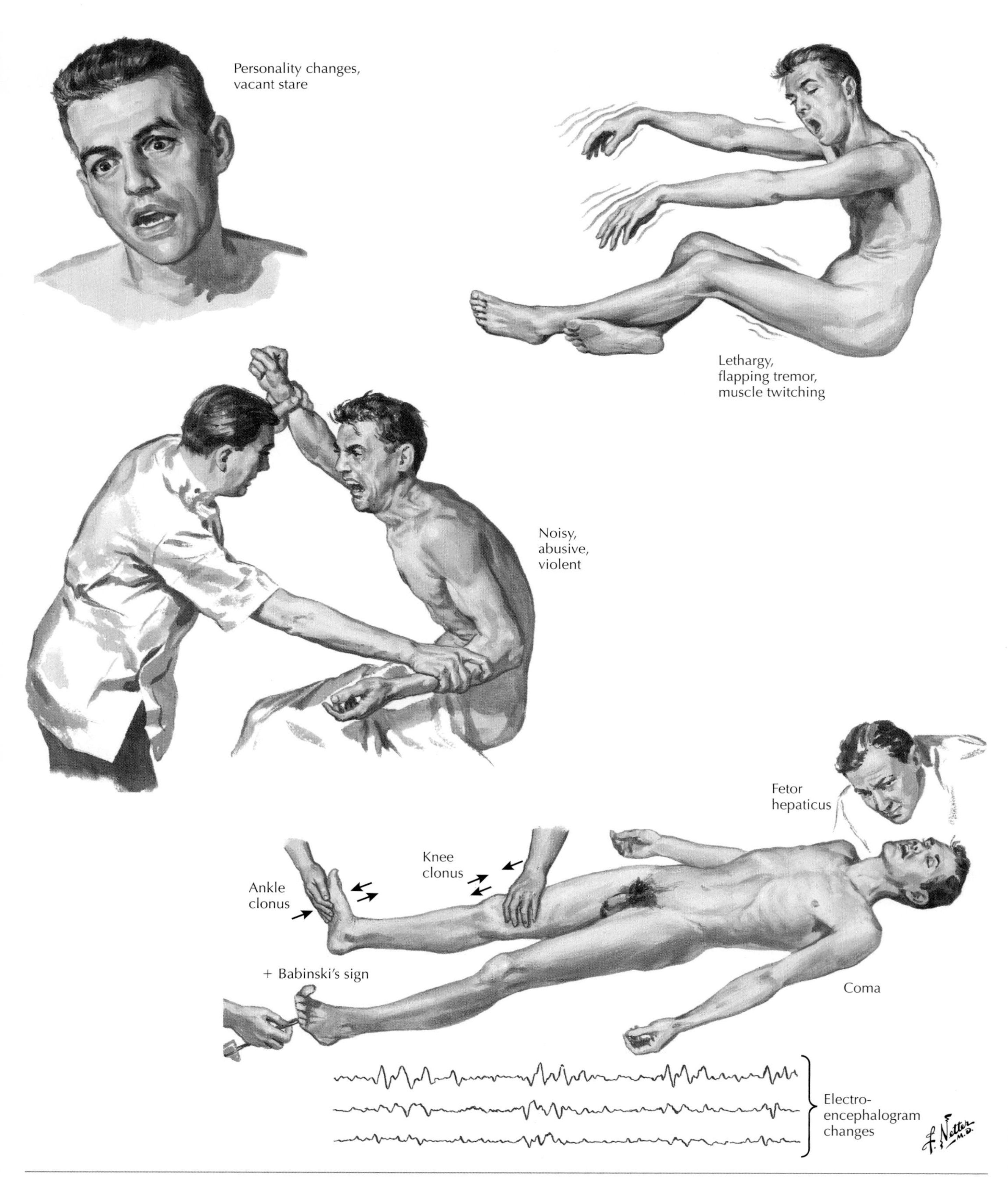

Figure 225-1 *Hepatic Encephalopathy: Clinical Manifestations.*

Table 225-1 Grading/Staging System for Hepatic Encephalopathy

Grade	Characteristics
1	Altered sleep patterns, altered mood, irritability, inability to maintain attention
2	Lethargy, altered speech, increased memory loss, dysarthria
3	Progressive stupor, decreased level of consciousness but responsive to stimuli
4	Coma, unresponsive to painful stimuli

TREATMENT AND MANAGEMENT

Treatment of hepatic encephalopathy consists primarily of removing the offending stimulus and administering nonabsorbable disaccharides and antibiotics. Although the mechanism of action of lactulose remains controversial, many investigators have proposed that the rationale for the use of nonabsorbable antibiotics (primarily lactulose) is based on the cathartic effect that may reduce the concentration of substrates for nitrogen formation and on the possible increased excretion by colonic bacteria of nitrogen or its metabolites in patients receiving lactulose. Nonabsorbable antibiotics, in particular neomycin, have also been used to treat PSE.

Rifaximin, a nonabsorbed derivative of rifamycin with a broad spectrum of activity, has been used for patients who cannot tolerate, or remain refractory to, lactulose or neomycin. Clinical trials suggest that rifaximin may be more effective and better tolerated than lactulose or neomycin. However, rifaximin is substantially more expensive and thus better reserved as a second-line agent.

In patients with acute liver failure, intracranial pressure monitoring is of benefit to monitor and treat intracranial hypertension. Protein restriction has also been advocated as a treatment for hepatic encephalopathy, though this is at best a short-term therapy for inpatients with moderate to severe PSE. Prolonged protein restriction is deleterious to patients with advanced liver disease because it may contribute to malnutrition and muscle wasting.

ADDITIONAL RESOURCES

Dhiman RK, Chawla YK: Minimal hepatic encephalopathy: time to recognise and treat, *Trop Gastroenterol* 29(1):6-12, 2008.

Ferenci P, Herneth A, Steindl P: Newer approaches to therapy of hepatic encephalopathy, *Semin Liver Dis* 16:329-338, 1996.

Häussinger D, Schliess F: Pathogenetic mechanisms of hepatic encephalopathy, *Gut* 57(8):1156-1165, 2008.

Lawrence KR, Klee JA: Rifaximin for the treatment of hepatic encephalopathy, *Pharmacotherapy* 28(8):1019-1032, 2008.

Prasad S, Dhiman RK, Duseja A, et al: Lactulose improves cognitive functions and health-related quality of life in patients with cirrhosis who have minimal hepatic encephalopathy, *Hepatology* 45(3):549-559, 2007.

Riordan SM, Williams R: Treatment of hepatic encephalopathy, *N Engl J Med* 337:473-479, 1997.

Hepatorenal Syndrome

Kris V. Kowdley

226

Hepatorenal syndrome (HRS) is defined as renal failure in patients with end-stage liver disease that is not associated with *intrinsic* renal disease. Recent series suggest that the incidence of HRS in patients with cirrhosis and ascites is approximately 18% at 1 year and 39% at 5 years.

The pathophysiology of HRS appears to be related to the underlying hemodynamic alterations in cirrhosis, characterized by splanchnic and systemic vasodilatation, decreased peripheral vascular resistance, and increased cardiac output in a pattern similar to that in patients with sepsis. The systemic vasodilatation is associated with renal vasoconstriction, which may be a physiologic response to decreased systemic blood pressure. Thus, the kidneys in these patients are exquisitely sensitive to changes in renal perfusion resulting from hypovolemia caused by gastrointestinal (GI) bleeding, overuse of diuretics, dehydration, or sepsis. It has been proposed that HRS is a sequela of unopposed renal vasoconstriction despite decreased renal perfusion. Once oliguria and azotemia develop, renal failure can progress rapidly and may be irreversible unless liver transplantation is performed promptly.

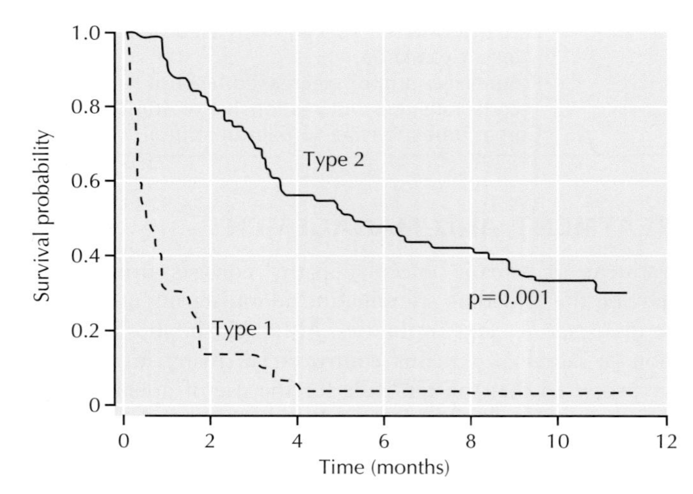

Reprinted with permission from Ginès P, Guevara M, Arroyo V, Rodés J. Hepatorenal syndrome. *Lancet* 2003;362:1819–1827.

Figure 226-1 *Hepatorenal Syndrome: Survival Comparison for Types 1 and 2. From Ginès P, Guevara M, Arroyo V, Rodés J:* Lancet *362:1819-1827, 2003.*

DIAGNOSIS AND TYPES

Specific diagnostic criteria have been established for HRS. The International Ascites Club defines HRS as low glomerular filtration rate (GFR) in the setting of acute or chronic liver disease with the absence of other causes of acute renal insufficiency (e.g., shock, nephrotoxic drugs, infection). Another criterion is low urine volume. However, the kidneys of HRS patients are morphologically and histologically normal.

Two types of HRS have been described; type 1 is much more life threatening than type 2 (**Fig. 226-1**). *Type 1* HRS is associated with rapidly decreasing renal function and rising serum creatinine over 2 weeks. This complication is often observed among patients with acute decompensation of liver function and severe acute alcoholic hepatitis. Clinical features include oliguria and hyponatremia. Type 1 HRS may develop in patients with type 2 HRS if there is an acute insult, such as a bacterial infection or arteriolar vasoconstriction within the kidneys, which could lead to ischemia, followed by progressive azotemia and oliguria.

Type 2 HRS is a more insidious and gradual type of renal insufficiency that develops in patients with advanced cirrhosis. It is frequently observed in patients with severe ascites that is often refractory to diuretic therapy; in fact, patients with type 2 HRS often do not tolerate diuretics because of renal insufficiency. Serum creatinine in these patients is usually between 1.5 and 2.5 mg/dL, although GFR may be greatly reduced because of a falsely low serum creatinine resulting from muscle wasting.

TREATMENT AND MANAGEMENT

Treatment of type 1 HRS is largely supportive. It is critically important to avoid nephrotoxic drugs, intravenous (IV) contrast, aminoglycosides, and hypovolemia in patients with type 2 because these may lead to type 1. Many advocate placing a central venous line to ensure that adequate filling pressures are maintained. Large-volume paracentesis with albumin has been advocated to decrease intraabdominal pressure, which may theoretically have a deleterious effect on renal blood flow. Diuretics should be used with great caution in these patients. Acute hemodialysis is appropriate if the patient is a candidate for liver transplantation because renal function frequently returns after successful liver transplantation, although hemodialysis is sometimes needed for several weeks after liver transplantation.

Specific therapies for HRS have included pharmacologic agents to increase splanchnic vasoconstriction and thus benefit the renin-angiotensin system, reduce renal sodium retention, and increase renal blood flow. The combination of *midodrine*, a selective α_1-adrenergic agonist associated with systemic vasoconstriction, and *octreotide*, a somatostatin analog that prevents splanchnic vasodilatation, is being increasingly used for the treatment of HRS. *Ornipressin*, a short-acting agent that stimulates splanchnic vasoconstriction, has been studied in Europe with some success. *Terlipressin*, a longer-acting vasopressin analog, has also been shown to benefit HRS patients and has few adverse effects, mainly associated with ischemia caused by arterial vasoconstriction. Both ornipressin and terlipressin appear to be most effective when administered with IV albumin, presumably because of the effects of plasma volume expansion.

Transjugular intrahepatic portosystemic shunt (TIPS) has been used in HRS patients to reduce the severity of portal hypertension and thus reduce hormonal and neuronal alterations that may lead to the circulatory and hemodynamic changes that initiate HRS. TIPS is effective in some patients with HRS, but careful patient selection is essential. Patients with advanced HRS may not tolerate this procedure well, and IV contrast must be used with great caution during the procedure. Furthermore, the high risk for hepatic encephalopathy and the small but serious risk for precipitating an acute decline in liver function after the procedure must be borne in mind before proceeding with TIPS. However, TIPS may be helpful in preserving renal function, reducing ascites, and maintaining the patient in a more stable state while awaiting liver transplantation.

ADDITIONAL RESOURCES

Arroyo V, Colmenero J: Ascites and hepatorenal syndrome in cirrhosis: pathophysiological basis of therapy and current management, *J Hepatol* 38(suppl 1):69-89, 2003.

Ginès P, Guevara M, Arroyo V, Rodés J: Hepatorenal syndrome, *Lancet* 362:1819-1827, 2003.

Kalambokis G, Economou M, Fotopoulos A, et al: The effects of chronic treatment with octreotide versus octreotide plus midodrine on systemic hemodynamics and renal hemodynamics and function in nonazotemic cirrhotic patients with ascites, *Am J Gastroenterol* 100(4):879-885, 2005.

Martín-Llahí M, Pépin MN, Guevara M, et al: Terlipressin and albumin vs albumin in patients with cirrhosis and hepatorenal syndrome: a randomized study, *Gastroenterology* 134(5):1352-1359, 2008.

Moreau R, Lebrec D: Acute renal failure in patients with cirrhosis: perspectives in the age of MELD, *Hepatology* 37:233-243, 2003.

Sanyal AJ, Boyer T, Garcia-Tsao G, et al: A randomized, prospective, double-blind, placebo-controlled trial of terlipressin for type 1 hepatorenal syndrome, *Gastroenterology* 134(5):1360-1368, 2008.

Variceal Bleeding

227

Kris V. Kowdley

The portal venous system is a high-flow, low-pressure system; normal portal pressure is less than 5 mm Hg. The *hepatic venous pressure gradient* (HVPG) is the difference in pressure between the portal vein and the inferior vena cava and is generally less than 6 mm Hg. Portal hypertension is defined as an elevation in the HVPG (**Fig. 227-1**).

Cirrhosis results in architectural and functional changes that lead to increased hepatic vascular resistance and increased portal blood inflow, consequently leading to increased portal venous pressure and elevated HVPG. Increased portal pressure results in the formation of gastroesophageal varices and return of blood into the superior vena cava through the azygous vein. Once the HVPG is greater than 12 mm Hg, risk for bleeding from esophageal varices increases. The incidence of bleeding from esophageal varices is 20% to 40% at 2 years of follow-up, and the mortality rate associated with the first bleeding episode remains at approximately 50%. The risk for recurrent bleeding approaches 80% at 2 years.

DIAGNOSIS: ROLE OF ENDOSCOPY

Endoscopy is useful to identify the presence of varices in patients with cirrhosis and to distinguish high-risk features associated with an increased bleeding risk. Endoscopy is also indicated for patients with acute bleeding to identify the cause of bleeding and for therapeutic intervention. Reports indicate that 10% to 47% of liver disease and upper gastrointestinal bleeding has a nonvariceal source.

The risk for bleeding is related to the severity of liver disease using Child criteria, the presence of red wales, and the size of the varices. The simplest of several grading systems classifies varices from grades 1 to 3. *Grade 3* varices are those that occlude the lumen; *grade 1* varices disappear completely with insufflation; and *grade 2* varices are between grades 1 and 3 in size. Using these scoring systems, patients can be stratified into categories ranging from 6% to 76% risk for bleeding during 1 year of follow-up. One study concluded that varices were most likely to be found during endoscopy in patients with compensated cirrhosis when prothrombin activity was less than 70%, platelet count less than 100×10^9/L, and ultrasonographic portal vein diameter greater than 13 mm.

PRIMARY PROPHYLAXIS

Nonselective *beta-adrenergic blocking agents* (β-blockers; e.g., propranolol, nadolol) are effective in reducing the risk for a first bleed by 40% to 50% in patients with esophageal varices. Treatment with propranolol begins at 40 mg daily, and doses up to 160 mg daily may be used. Treatment is associated with lower bleed-related mortality and may improve survival.

Endoscopic sclerotherapy has been extensively studied as primary prophylaxis for variceal bleeding, with more than 1500 patients enrolled in trials. With some exceptions, however, sclerotherapy has not been effective, largely because of the high rate of adverse effects, such as pulmonary complications, fever, chest pain, and esophageal ulceration.

Portosystemic shunts are effective in primary prevention but are associated with an unacceptably high incidence of hepatic encephalopathy. A significant proportion of patients fail to tolerate β-blockers or fail to experience a reduction in portal pressure with shunt therapy. Therefore, alternative therapies are needed.

Variceal band ligation has been studied as primary prevention (**Fig. 227-2**) and compared with β-blockers. The largest study showed reduced bleeding, from 43% in the propranolol group to 15% in the ligation group. Risk for bleeding in the propranolol group appeared unusually high, possibly because of underdosage. Band ligation does not appear to reduce overall or bleed-related mortality. Band ligation as primary prophylaxis should be reserved for patients with compensated cirrhosis with large varices or for patients with advanced liver disease with small or large varices who cannot tolerate β-blockers.

Isosorbide mononitrate at 40 mg twice daily can be used if neither ligation nor β-blocker therapy is an option.

ACUTE VARICEAL BLEEDING

Acute variceal bleeding is generally defined as visible bleeding from an esophageal varix or a gastric varix, or the presence of blood in the stomach of a patient with varices, with no other cause for bleeding. Clinically significant bleeding is defined as more than 2 U blood transfused over 24 hours, with blood pressure decreased to less than 100 mm Hg or pulse rate increased to greater than 100 beats/min. Mortality associated with acute variceal bleeding is greater than 30%, and the risk is increased in patients with decompensated liver disease.

Current management of acute variceal bleeding includes pharmacologic and endoscopic therapy. Key elements of management include airway protection, early use of endoscopy, and administration of antibiotics (e.g., norfloxacin, clavulanic acid–amoxicillin) for infection prophylaxis.

Historically, vasopressin and nitroglycerin have been used as pharmacologic therapy, but this regimen is infrequently used in the United States at present. Somatostatin and its analogs (octreotide, lanreotide, and vapreotide) are used widely. In the United States, *octreotide* is the only available agent in this category. Because of octreotide's long half-life, no loading dose is needed; a continuous infusion of 25 to 50 µg per hour is the usual dose. *Terlipressin*, an analog of vasopressin used in Europe, can be given as intermittent injection and has a favorable safety profile compared with vasopressin. A Cochrane review of terlipressin for variceal hemorrhage found that terlipressin was superior to placebo and comparable to somatostatin or sclerotherapy in mortality risk. A randomized trial of terlipressin versus sclerotherapy showed the two therapies equally effective compared with placebo in controlling acute bleeding and preventing early rebleeding.

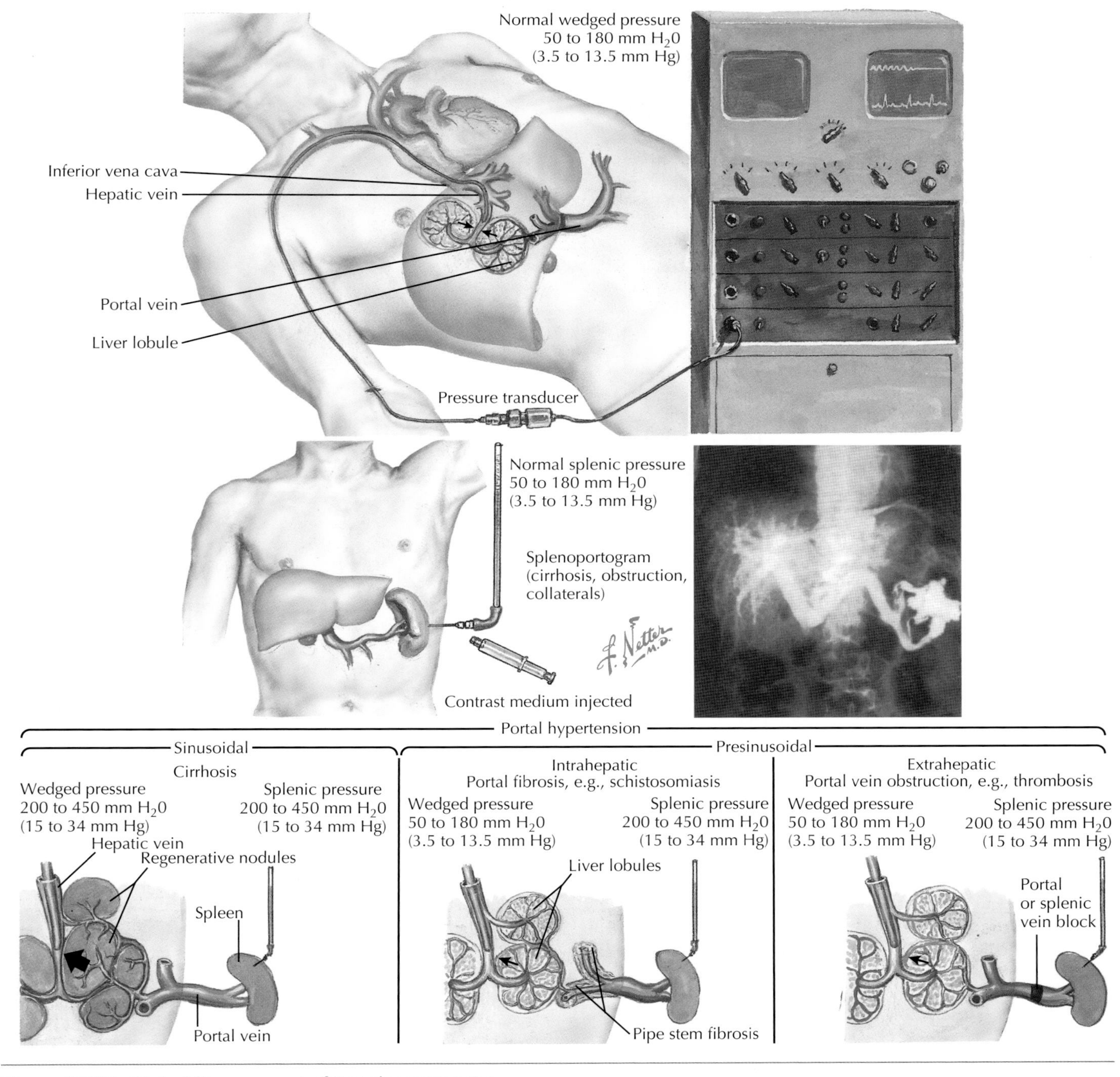

Figure 227-1 *Clinical Measurement of Portal Hypertension.*

Administering somatostatin analogs along with endoscopic therapy reduces transfusion requirements and is a reasonable adjunctive therapy given the low toxicity associated with these agents.

Endoscopic sclerotherapy has been used in acute variceal bleeding for two decades. A meta-analysis showed that sclerotherapy was more effective than balloon tamponade, no therapy, or vasopressin, with a 90% rate of bleeding control. However, the trials varied in type of sclerosant, technique, and follow-up. Sodium tetradecyl sulfate (15%), morrhuate sodium (5%), and ethanolamine (5%) are the most widely used sclerosants. No consensus exists as to the preferred sclerosant; variable rates of ulceration are reported with these agents.

Variceal band ligation is now the endoscopic treatment of choice because of the lower complication rate associated with this procedure (see **Fig. 227-2**). In one study, band ligation was shown to be equally or more effective in controlling acute bleeding than sclerotherapy and was associated with better survival.

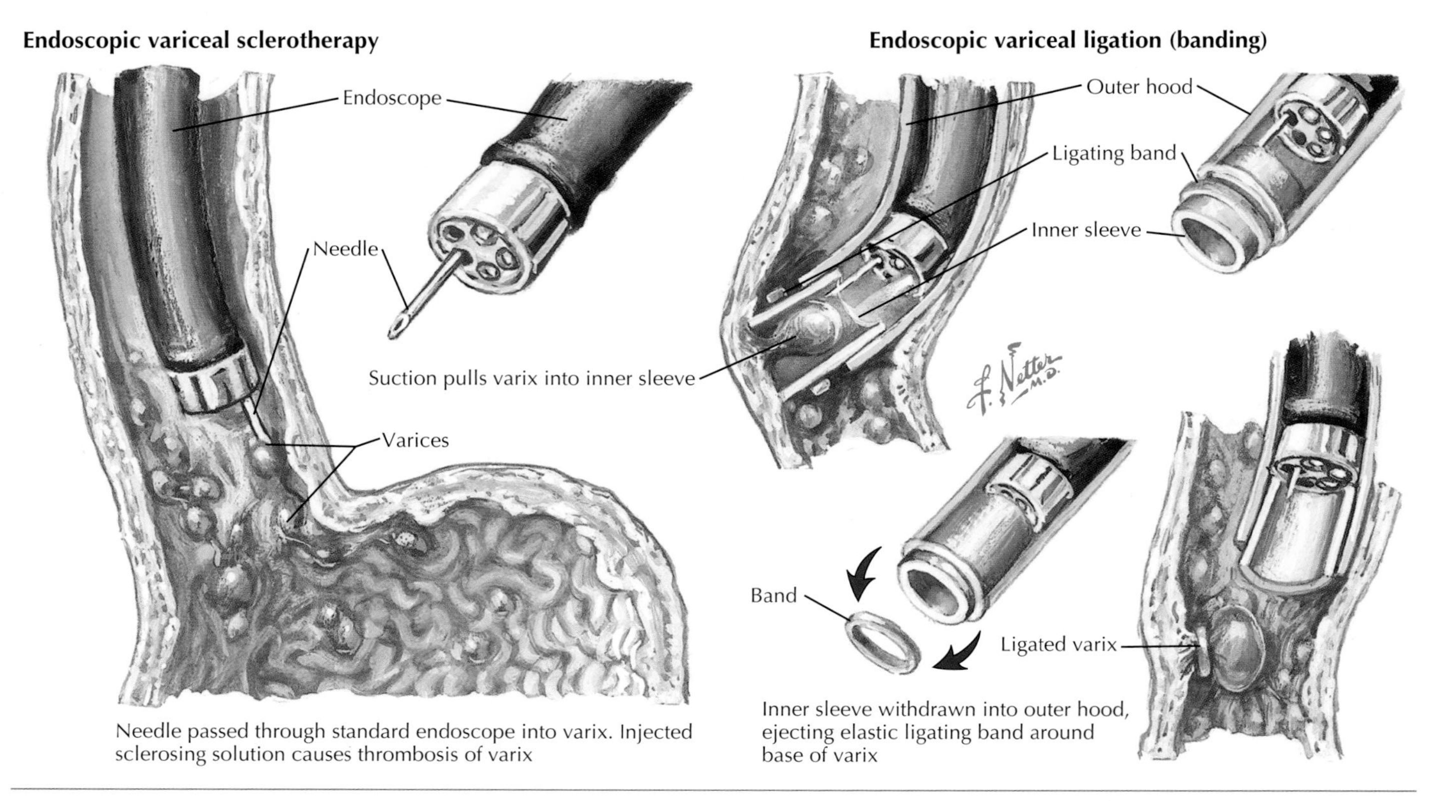

Figure 227-2 *Ligation Techniques.*

Acute rebleeding was reduced when band ligation was combined with octreotide. Although more expensive, multiband ligators should be used because of the risk for esophageal perforation associated with the overtubing necessary with the single-band ligator. The average number of bands placed ranges from 5 to 10. Patients should be treated with β-blockers or continued variceal band ligation, or a combination, for secondary prophylaxis against rebleeding.

Measuring HVPG is also predictive of early rebleeding and mortality. Patients with HVPG exceeding 20 mm Hg during the acute bleeding episode are at higher risk for early rebleeding and death at 1 year.

Therapy for Uncontrolled Bleeding

Patients whose bleeding is uncontrolled despite pharmacologic and endoscopic therapy should have a Sengstaken-Blakemore tube inserted and gastric balloon inflated. Endotracheal intubation should be strongly considered for these patients, although complications are common (10%-30%) and severe, ranging from esophageal perforation to aspiration pneumonia. Sengstaken-Blakemore tube placement is a temporizing measure and should be instituted while evaluating the patient for transjugular intrahepatic portosystemic shunt (TIPS), referral to a liver transplantation center, or consideration for a surgical portosystemic shunt or devascularization procedure.

Shunt Procedures

Shunt surgery is best reserved for patients with controlled bleeding who have Child class A cirrhosis, patients who may fail to comply with follow-up surveillance for TIPS stenosis, and patients with portal vein thrombosis. The preferred surgical shunt is a *splenorenal* shunt because of the lower risk for encephalopathy and because this procedure does not complicate future liver transplantation (**Fig. 227-3**).

GASTRIC VARICES

Gastric varices are classified as *gastroesophageal varices* (GEVs) and *isolated gastric varices* (IGVs). Type 1 GEVs are present in the cardia contiguous with esophageal varices, whereas type 2 GEVs extend beyond the cardia into the fundus. Type 1 IGVs are located in the fundus, whereas type 2 IGVs are located anywhere else in the stomach or duodenum. Mortality from IGV-related bleeding, especially in the fundus, is significantly higher than that from GEV-related bleeding.

Endoscopic therapy has been used to treat bleeding gastric varices. Endoscopic therapy appears to be equally effective for GEVs as for esophageal varices. However, the success rates are much lower for IGVs, and rebleed rates are high, approaching 90% in some series. Absolute ethanol may be more effective than other sclerosing agents. Injections of cyanoacrylate and thrombin have also been used. Cyanoacrylate appears to be the more promising agent and controls acute hemorrhage in most patients, but it is associated with a high rate of rebleeding.

Splenic vein thrombosis should always be suspected in patients with IGV in the fundus. Angiographic or surgical splenectomy should be considered as treatment. Given the lack of effective and widely available endoscopic therapy for IGVs, especially those in the gastric fundus, TIPS is emerging as an

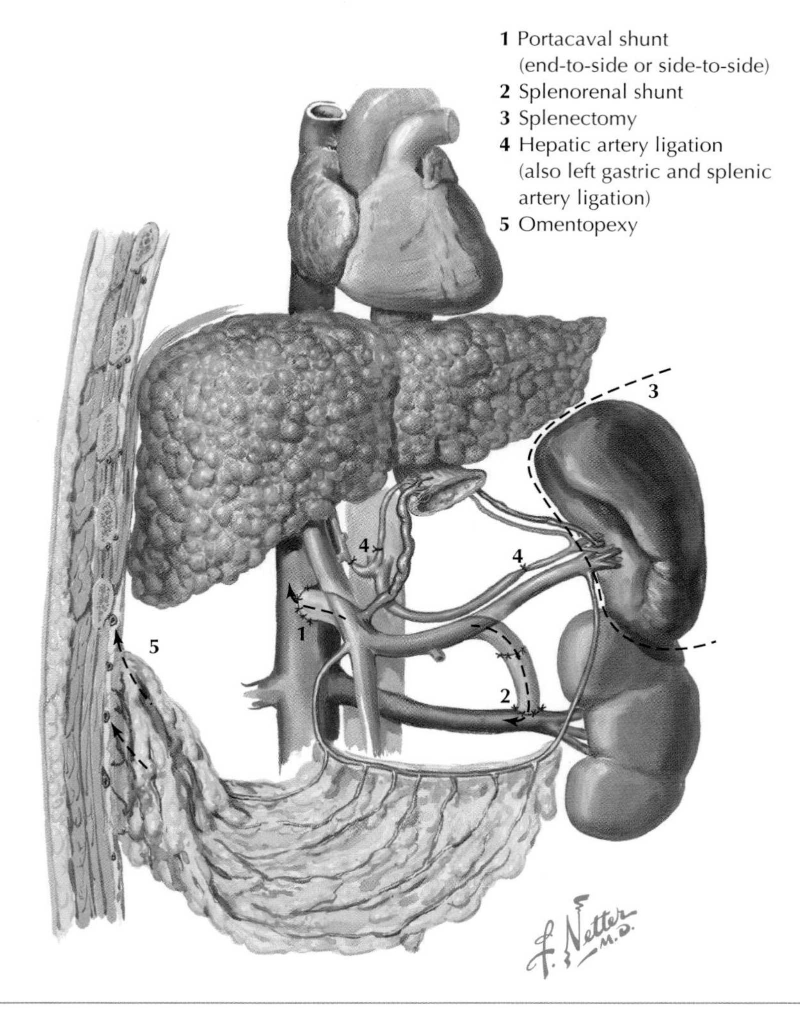

Figure 227-3 *Surgical Procedures to Relieve Portal Hypertension.*

important modality for such patients because mortality risk is low and rebleeding risk is acceptable.

ECTOPIC VARICES

Ectopic varices may form at many sites in the gastrointestinal tract, including the peritoneum, biliary tract, and genitourinary system. Ectopic varices are more common in patients with extrahepatic portal hypertension than in those with cirrhosis. Varices may also develop at enterostomy sites after abdominal surgery.

In a suggested management algorithm for patients with ectopic varices, angiographic embolization has been used with good initial control of bleeding. However, portal decompression is often necessary. Because bleeding from stomal varices can often be controlled by direct pressure, mortality has been estimated at less than 5%, and conservative measures have been advocated rather than surgery. Injection sclerotherapy of stomal varices should be undertaken with caution because of the high risk for ulceration and tissue injury.

ADDITIONAL RESOURCES

Bosch J, Garcia-Pagan JC: Complications of cirrhosis: I—Portal hypertension, *Hepatology* 32(suppl):141-156, 2000.

D'Amico G, Pagliaro L, Bosch J: The treatment of portal hypertension: a meta-analytic review, *Hepatology* 22: 332-354, 1995.

Garcia-Pagan JC, De Gottardi A, Bosch J: The modern management of portal hypertension: primary and secondary prophylaxis of variceal bleeding in cirrhotic patients (review), *Aliment Pharmacol Ther* 28(2):178-186, 2008.

Jalan R, Hayes PC: UK guidelines on the management of variceal haemorrhage in cirrhotic patients, *Gut* 6(suppl 3):iii1-iii15, 2000.

Vlachogiannakos J, Goulis J, Patch D, Burroughs AK: Primary prophylaxis for portal hypertensive bleeding in cirrhosis (review), *Aliment Pharmacol Ther* 14:851-860, 2000.

Transjugular Intrahepatic Portosystemic Shunt

228

Kris V. Kowdley

Transjugular intrahepatic portosystemic shunt (TIPS) represents another option for decompression of portal hypertension. This procedure was introduced in 1982, but long-term success was demonstrated only after development of expandable metal stents.

Interventional radiologists usually perform TIPS, cannulating the jugular vein and inserting a catheter in the right hepatic vein. A needle is then used to create a tract to the portal vein, usually the right portal vein, under fluoroscopic control. The tract is dilated and an expandable metal stent placed, bridging the hepatic vein to the portal vein.

Although essentially functioning as a side-to-side portacaval shunt, TIPS has several advantages, including lower mortality risk than a surgical shunt, especially in patients with advanced liver disease. It can be completely removed at liver transplantation, and unlike a surgically created portacaval shunt, TIPS does not increase the risks involved in liver transplantation.

INDICATIONS AND APPROACH

Current indications for TIPS include (1) esophageal variceal bleeding that is refractory to endoscopic band ligation or sclerotherapy, (2) refractory ascites or hepatic hydrothorax in patients awaiting liver transplantation, and (3) bleeding from gastric varices or other ectopic sites. TIPS is also effective for treating portal hypertension caused by Budd-Chiari syndrome and hepatic venoocclusive disease, which is most frequently observed after conditioning regimens for hematopoietic stem cell transplantation. TIPS has been performed to treat hepatorenal syndrome (HRS), but this indication should be considered experimental.

Figure 228-1 demonstrates patency of TIPS deployed with contrast. Hepatic venous pressure measurement before and after TIPS placement can be used to examine the efficacy of the procedure. Previous studies have shown that a portosystemic gradient (difference between portal and hepatic venous pressures) greater than 12 mm Hg is associated with an increased risk for bleeding varices. Variable-sized stents of 10 and 12 mm are available to increase the size of the shunt. Occasionally, side-by-side stents are placed to lower portal pressure further.

COMPLICATIONS

Complications of TIPS can be classified as immediate or late. Immediate complications include bleeding (hemoperitoneum, capsular hematoma, and hemobilia) and, in rare cases, cardiorespiratory failure, which may be related to hemodynamic alterations resulting from the procedure. Other immediate complications include fever (possibly associated with bacteremia), renal insufficiency, and shunt thrombosis. Acute shunt thrombosis is relatively uncommon but may be difficult to manage. Renal insufficiency may be multifactorial and primarily caused by radiographic contrast administration. Careful assessment of the intravascular fluid status and administration of mannitol may be appropriate for patients with mild renal insufficiency. There are reports of thrombotic and paradoxic emboli after TIPS therapy for HRS.

One of the most common clinical complications of TIPS is hepatic encephalopathy, reported in up to 20% to 30% of patients and occasionally resulting in a comatose state. Risk factors for hepatic encephalopathy include older age, larger TIPS diameter, and advanced liver disease. Therefore, TIPS should probably be avoided in patents with a previous history of recurrent severe hepatic encephalopathy.

ADDITIONAL RESOURCES

Jenkins RL: Defining the role of transjugular intrahepatic portosystemic shunts in the management of portal hypertension, *Liver Transplant Surg* 1:225-228, 1995.

Reichelderfer M: Bleeding stomal varices: case series and systematic review of the literature, *Clin Gastroenterol Hepatol* 6(3):346-352, 2008.

Shiffman ML, Jeffers L, Hoofnagle JH, Tralka TS: The role of transjugular intrahepatic portosystemic shunt for treatment of portal hypertension and its complications: a conference sponsored by the National Digestive Diseases Advisory Board, *Hepatology* 22:1591-1597, 1995.

Spier BJ, Fayyad AA, Lucey MR, et al: Portal hypertension and variceal hemorrhage, *Med Clin North Am* 92(3):551-574, viii, 2008.

Zheng M, Chen Y, Bai J, et al: Transjugular intrahepatic portosystemic shunt versus endoscopic therapy in the secondary prophylaxis of variceal rebleeding in cirrhotic patients: meta-analysis update, *J Clin Gastroenterol* 42(5):507-516, 2008.

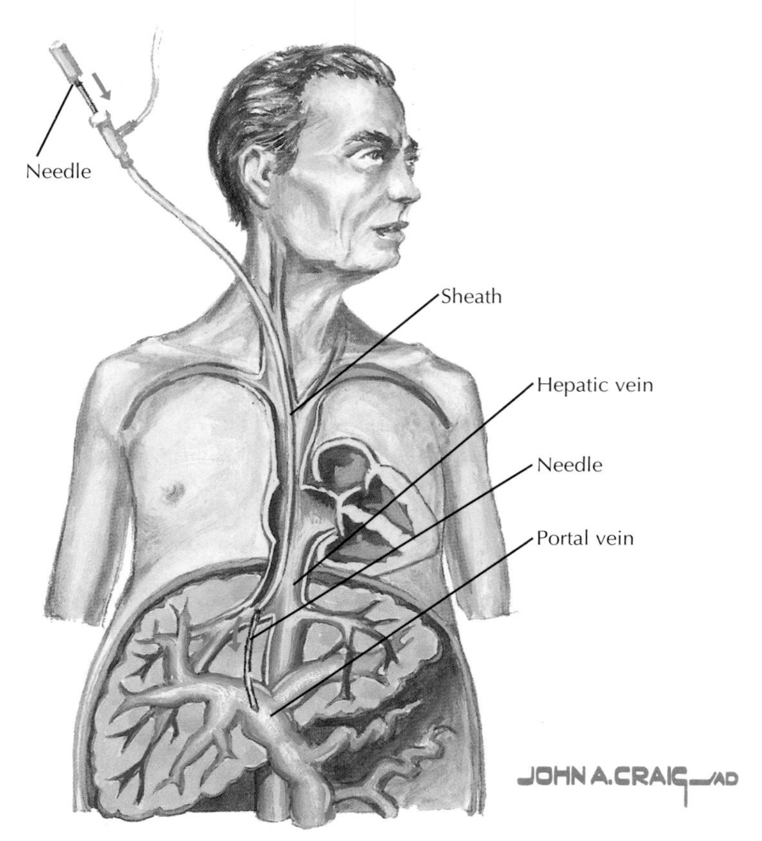

Jugular sheath passed into hepatic vein. Flexible needle inserted into sheath and advanced under fluoroscopic control across hepatic parenchyma into portal vein

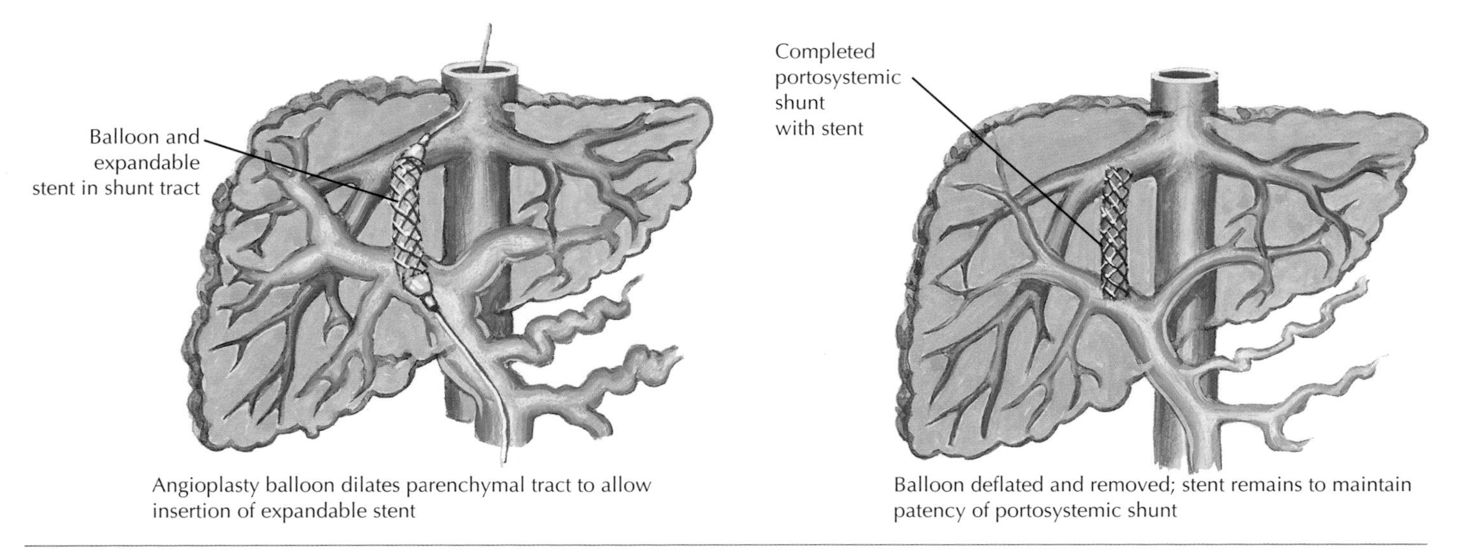

Angioplasty balloon dilates parenchymal tract to allow insertion of expandable stent

Balloon deflated and removed; stent remains to maintain patency of portosystemic shunt

Figure 228-1 *Transjugular Intrahepatic Portosystemic Shunt.*

Liver Biopsy

Kris V. Kowdley

Significant advances have increased the clinician's ability to image the liver using ultrasonography (US), computed tomography (CT), and magnetic resonance imaging (MRI) (see Chapter 232). However, liver biopsy remains an essential procedure for determining the diagnosis and prognosis of liver disease. In many patients, liver biopsy confirms the cause of liver disease.

More importantly, liver biopsy remains the best method for establishing the presence or absence of cirrhosis. In many liver diseases, including alcoholic liver disease, hemochromatosis, and hepatitis C, the risk for hepatocellular carcinoma is associated primarily with the presence of cirrhosis. Therefore, identification of cirrhosis can facilitate decisions about screening for liver cancer.

Furthermore, findings on liver biopsy are essential for making treatment decisions in many chronic liver diseases, including hepatitis B and hepatitis C. Liver biopsy remains the criterion for the diagnosis of chronic hepatic diseases such as Wilson disease, nonalcoholic steatohepatitis, and autoimmune hepatitis.

Liver biopsy occasionally identifies unexpected findings that may change management. Evaluating the minute structure of the liver is critical for assessing patients with drug-induced hepatitis or fulminant liver failure. In patients with liver failure, biopsy can yield important information about the liver's ability to regenerate from acute injury, informing decisions about liver transplantation. Managing patients after liver transplantation also largely depends on liver biopsy, particularly in conditions associated with a high rate of recurrence, such as hepatitis C and hepatitis B.

TECHNIQUES

Liver biopsy can be performed percutaneously, either blindly using percussion techniques or with US guidance (**Fig. 229-1**). US localizes an intercostal space with clear access to the liver, averting major vessels, gallbladder, and lungs. This location is marked, and the biopsy is subsequently performed at the bedside, usually in a day-surgery or ambulatory setting.

Liver biopsy can also be performed in the radiology suite under direct US or CT guidance. This technique is usually selected when biopsy is performed to study focal lesions, such as suspected hepatocellular carcinoma or adenoma. Liver biopsy can also be done using laparoscopy, although usually only when the patient is already undergoing laparoscopic surgery.

In addition to the percutaneous approach, liver biopsy can be performed using transjugular techniques, generally by interventional radiologists and for patients with a higher risk for bleeding, such as those with thrombocytopenia or coagulation abnormalities. A catheter is introduced into the right internal jugular vein and then passed into the liver through the hepatic vein. Biopsy forceps can then be used to obtain a tissue sample. The theoretic advantage of this approach is that any bleeding would occur within the vascular compartment and would be associated with a lower risk for intraabdominal bleeding.

Percutaneous liver biopsy is generally performed using local anesthesia alone, although many hepatologists have begun using a mild sedative to reduce anxiety. However, it is important that the patient be awake and alert during the performance of percutaneous liver biopsy because it is essential for patients to hold their breath, usually in the end-expiratory state. After local anesthesia with lidocaine, a trocar is used to create a tract to facilitate passage of a 16-gauge needle. A suction needle is then used to aspirate a core of liver tissue. Different types of needles are available for liver biopsy; the Klatskin needle is often used (see Fig. 229-1).

The biopsy specimen is usually placed in formalin or another fixative. A number of stains are used to evaluate the liver parenchyma. Hematoxylin and eosin stain is used to evaluate for inflammation and necrosis; trichrome stain is used to assess the presence and degree of fibrosis; and stains such as reticulin can be used to evaluate the architecture. In addition, special stains are useful to screen for specific liver diseases, such as periodic acid–Schiff with diastase for α_1-antitrypsin deficiency, Perls Prussian blue stain for iron, and a special stain for hepatitis B core antigen. Biochemical measurement of iron or copper can be performed from fresh or paraffin-embedded tissue to establish a specific diagnosis of hemochromatosis or Wilson disease, respectively.

COMPLICATIONS AND CONTRAINDICATIONS

The main risks involved in liver biopsy are pain and bleeding. Pain may be localized over the biopsy site, diffusely spread over the abdomen, or more often, referred to the right shoulder from irritation of the diaphragm. Bleeding is more serious and may result in hepatic capsular hematoma or even intraabdominal bleeding, although this is rare.

Other complications include infection, pneumothorax or hemothorax, and perforation of the gallbladder or the bile ducts. The risk for fatal complication of liver biopsy is approximately 1 in 10,000. Risk factors for complications of liver biopsy include coagulopathy and increased number of passes.

Most centers require that the patient lie on the right side for several hours after the biopsy to allow internal compression of the liver against the rib cage.

Contraindications to liver biopsy include lack of patient cooperation, bacterial cholangitis, extrahepatic bile duct obstruction, and significant coagulopathy or thrombocytopenia. Some advocate avoiding liver biopsy in patients with cystic liver lesions because of infection risk and in those with amyloidosis because of hemorrhage risk. Many experts believe that patients with large amounts of ascites should not undergo percutaneous biopsy because of the inability to compress the liver internally against the rib cage and the resulting increased risk for bleeding.

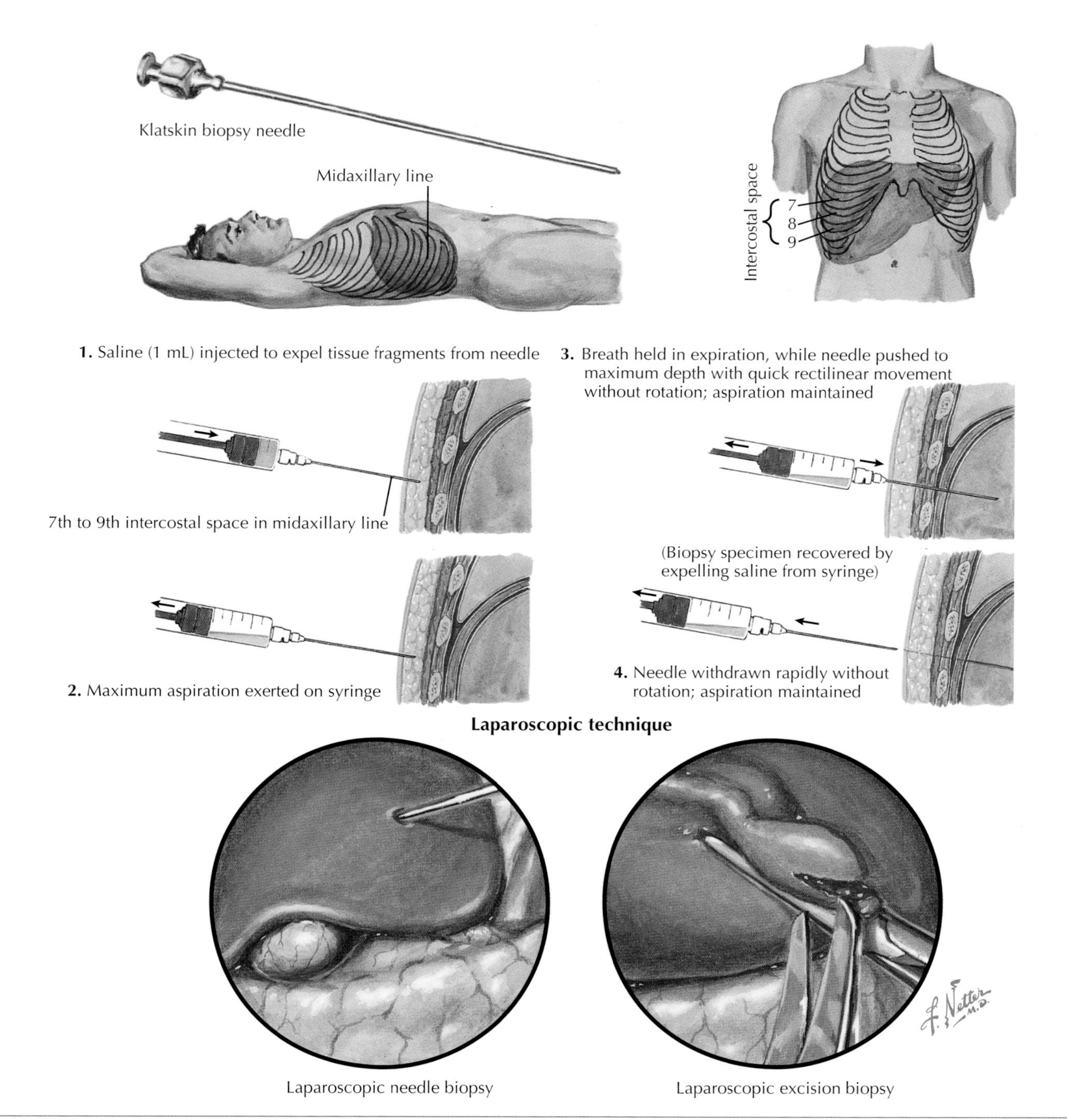

Figure 229-1 *Liver Biopsy: Percutaneous Suction and Laparoscopic Techniques.*

However, data from controlled trials are limited to support these recommendations.

Alternate Techniques

The most important limitation of liver biopsy is *sampling variability*, especially in liver diseases with focal involvement, such as cystic fibrosis and primary sclerosing cholangitis. This has led to active investigation into other methods to assess the presence or absence of cirrhosis. Imaging techniques used include transient elastography, MRI, and noninvasive serum markers of fibrogenesis (e.g., hyaluronic acid, fragments of collagen) that may form in the liver in response to injury. Although promising, these techniques have not yet achieved sufficient positive and negative predictive value to replace liver biopsy in the clinical setting.

In the future, the combination of structural imaging studies and dynamic serum or plasma markers of fibrosis may reach sufficient reliability to replace liver biopsy in the evaluation of cirrhosis. However, liver biopsy will likely remain essential for diagnostic purposes in some patients for whom serologic testing has not established the cause of liver disease.

ADDITIONAL RESOURCES

Buckley A, Petrunia D: Practice guidelines for liver biopsy: Canadian Association of Gastroenterology, *Can J Gastroenterol* 14:481-482, 2000.

Desmet VJ: Liver tissue examination, *J Hepatol* 39(suppl 1):43-49, 2003.

Grant A, Neuberger J: Guidelines on the use of liver biopsy in clinical practice: British Society of Gastroenterology, *Gut* 45(suppl 4):IV11, 1999.

Larson AM, Chan GC, Wartelle CF, et al: Infection complicating percutaneous liver biopsy in liver transplant recipients, *Hepatology* 26:1406-1409, 1997.

Rockey DC: Noninvasive assessment of liver fibrosis and portal hypertension with transient elastography, *Gastroenterology* 134(1):8-14, 2008.

Hepatic Necrosis

230

Kris V. Kowdley

Any acute or chronic process resulting in liver injury may be characterized by hepatocellular necrosis. Cell death may occur through apoptosis or necrosis. Apoptotic bodies are sometimes described as *acidophilic bodies*. The term *cell dropout* has also been used to describe some cases of hepatocellular necrosis. *Necrosis* implies not only the death of cells, but also the phenomena following cell death, namely, the disappearance of cells and frequently the accompanying inflammatory response. The final and irreversible stage of degeneration, hepatic necrosis, involves only the liver cells in most forms, whereas Kupffer cells and stroma remain intact. Kupffer cells respond to most types of hepatocellular degeneration and necrosis with reactive proliferation.

Acute viral, toxic, or drug-induced hepatitis is usually the cause of hepatocellular necrosis, but any process resulting in a systemic inflammation or a liver-specific injury can lead to hepatic necrosis, as can a variety of insults that cause hepatic ischemia or hypoxemia. Necrosis may be *focal;* that is, single cells or a small group of cells have been injured or have disappeared and are replaced by scavenger cells, usually neutrophilic segmented leukocytes but occasionally, especially in viral infections, histiocytes and lymphocytes (**Fig. 230-1**). Focal necrosis may also be the result of focal obstruction of the sinusoidal blood flow, such as by cellular debris or fibrin thrombi.

Zonal necrosis, in contrast, is characterized by its lobular distribution. In *central* necrosis, the destructive process takes place around the central vein and may extend toward the periphery of the lobule. Depending on the intensity of the damage and the age of the lesions, liver cell fragments may be recognizable, or the liver cells may entirely disappear, and red blood cells (RBCs) may engorge sinusoids and tissue spaces. In more advanced stages, the framework is collapsed, and only a few scavenger cells are found mixed with Kupffer cells and RBCs. Necrosis of the liver cells in the center of the lobule is often the result of ischemia (as seen in passive congestion or shock), hypoxemia (low atmospheric pressure), or both, because this part of the hepatic lobule is most sensitive to conditions of hypoxia.

Periportal or *peripheral* necrosis indicates damage to periportal hepatocytes and in the adjoining peripheral zone of the parenchymal lobule. Inflammatory cells accumulate, often with inflammation in the portal triads. Proliferation of bile ducts and cholangioles is also common. Typically, periportal necrosis results from inflammation in the portal triads that extend to the peripheral zone; thus it is seen in infections involving the portal triads, in chronic biliary obstruction, and in chronic viral hepatitis. Isolated *midzonal* necrosis is rare in humans.

Extensive zonal, mainly central, necrosis often results from exposure to various poisons, toxins, or drugs but is also observed after infection or shock. Because necrosis is also produced or aggravated by cardiac failure, it is sometimes difficult to determine how much primary liver cell damage and vascular factors are each responsible for the hepatic necrosis. If central necrosis becomes more extensive, bridges develop that connect the central zones or the portal and central zones (*bridging* necrosis). This may proceed further to almost complete loss of liver cells in a lobule (*massive* necrosis). Massive necrosis in a considerable part of the liver produces hepatic insufficiency, sometimes fatal, that historically has been termed *acute yellow atrophy* or *acute red atrophy* of the liver. The normal architecture of the liver may be difficult to recognize in massive necrosis.

ADDITIONAL RESOURCES

Akazawa Y, Gores GJ: Death receptor–mediated liver injury, *Semin Liver Dis* 27(4):327-338, 2007.

Canbay A, Friedman S, Gores GJ: Apoptosis: the nexus of liver injury and fibrosis, *Hepatology* 39:273-278, 2004.

Desmet VJ: Liver tissue examination, *J Hepatol* 39(suppl 1):43-49, 2003.

Rutherford A, Chung RT: Acute liver failure: mechanisms of hepatocyte injury and regeneration, *Semin Liver Dis* 28(2):167-174, 2008.

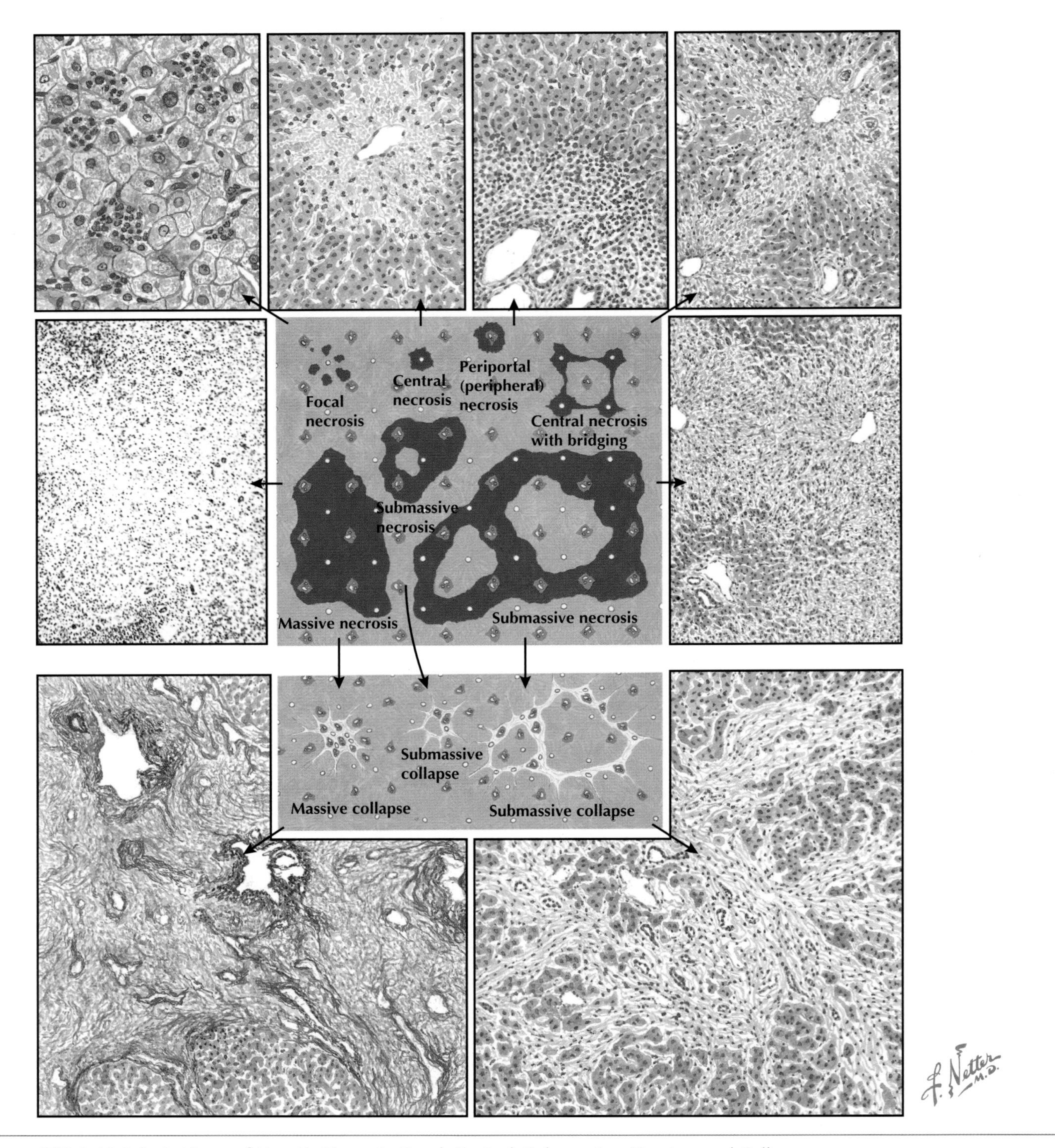

Figure 230-1 *Histologic Views of Hepatic Necrosis: Focal, Central, Submassive, Massive, and Collapse.*

Cirrhosis

Kris V. Kowdley

Cirrhosis represents the end stage of any process resulting in chronic injury to the liver parenchyma. Progressive fibrosis results in the alteration of hepatic architecture and the formation of small or large nodules. The associated circulatory and hemodynamic changes may result in portal hypertension with ascites, esophageal varices, and necrosis (**Figs. 231-1** and **231-2**).

PATHOGENESIS

The hepatic *stellate cell*, also referred to as the *Ito cell* or *lipocyte*, is the primary storage site for retinoids and is critical to the process of hepatic fibrogenesis. On stimulation by inflammation or other noxious factors, stellate cells become contractile and are activated, demonstrating features of myofibroblasts and releasing cytokines and other inflammatory mediators. Subsequently, a deposition of extracellular matrix results in fibrosis, development of scar tissue, vascular obstruction, and sinusoidal hypertension, followed by portal hypertension and eventually hepatic failure. This dynamic process may be accelerated by the addition of hepatotoxins (e.g., alcohol, drugs) or may be decelerated by the reduction of hepatic necroinflammation through immunomodulator therapy (e.g., corticosteroids) or elimination of the inciting agent (e.g., interferon-α for hepatitis C).

Other cells, such as fibroblasts, are also involved in the development of hepatic fibrosis. Over time, this process can result in the development of widespread scar tissue throughout the liver and ultimately leads to established cirrhosis.

Cytokines that play a key role in stellate cell proliferation and activation include transforming growth factor-β, interleukins, hepatocyte growth factor, and platelet-derived growth factor. These cytokines have multiple effects; some are fibrogenetic, whereas others may promote fibrinolysis. Some are primarily proinflammatory, and others are antiinflammatory.

The progression to cirrhosis in a patient with chronic liver injury is characterized by increased deposition of extracellular matrix, ongoing inflammation, and an imbalance in favor of fibrogenetic rather than fibrinolytic pathways. The current research to identify the key mediators responsible for hepatic fibrogenesis is the first step in the development of novel antifibrotic therapies potentially capable of halting or even reversing the process leading to cirrhosis.

CLINICAL PICTURE

Clinical manifestations associated with cirrhosis are described in Chapter 221. The most common signs of cirrhosis are thrombocytopenia caused by hypersplenism resulting from portal hypertension. In patients with chronic liver disease, the finding of a mildly to moderately decreased platelet count and associated splenomegaly can be considered diagnostic of cirrhosis.

DIAGNOSIS

The diagnosis of cirrhosis generally is made by liver biopsy or laparoscopy. Gross visualization of the liver by laparoscopy or surgery is probably the standard for the diagnosis of cirrhosis. Liver biopsy is useful to establish the presence of cirrhosis, but often the biopsy specimen is fragmented, and the diagnosis of cirrhosis cannot be made unequivocally based on the biopsy specimen.

Trichrome stain is the classic stain used to evaluate the degree or stage of fibrosis in the specimen. Most scoring systems classify the biopsy specimens into four histologic stages. Generally, stages 1 and 2 represent *periportal* or *septal fibrosis*, whereas stages 3 and 4 are used to describe *bridging fibrosis* and *cirrhosis*, respectively. Some have advocated the use of additional morphometric techniques that stain collagen to better classify patients into various stages. Recent studies have examined transient elastography and magnetic resonance elastography to identify the presence of cirrhosis (see Chapter 232).

The differential diagnosis of cirrhosis includes several conditions that may have a similar histologic appearance but that are not associated with hepatic synthetic dysfunction. Some examples include focal nodular hyperplasia, nodular regenerative hyperplasia, and congenital hepatic fibrosis.

Recent research has focused on the use of noninvasive serum markers of fibrogenesis to classify patients into those with a low or high probability of cirrhosis. It is hoped that in the next decade, clinicians will have the ability to stratify patients at low, medium, or high risk for cirrhosis using combinations of serum markers. However, at present, the so-called noninvasive markers of fibrosis should be considered investigative and are not yet ready for widespread clinical application. These noninvasive markers are generally enzymes involved in the production of extracellular matrix. Among the most promising of these markers are hyaluronic acid, collagen types IV and VI, propeptides of various collagens (e.g., PIIINP), propeptides of collagen types I and IV, and enzymes involved in the breakdown of matrix (e.g., matrix metalloproteinase-2).

ADDITIONAL RESOURCES

Friedman SL: Liver fibrosis: from bench to bedside, *J Hepatol* 38(suppl 1):38-53, 2003.

Friedman SL: Mechanisms of hepatic fibrogenesis, *Gastroenterology* 134(6):1655-1669, 2008.

Manning DS, Afdhal NH: Diagnosis and quantitation of fibrosis, *Gastroenterology* 134(6):1670-1681, 2008.

Oh S, Afdhal NH: Hepatic fibrosis: are any of the serum markers useful? *Curr Gastroenterol Rep* 3:12-18, 2001.

Rockey DC: The cell and molecular biology of hepatic fibrogenesis: clinical and therapeutic implications, *Clin Liver Dis* 4:319-355, 2000.

Rockey DC: Noninvasive assessment of liver fibrosis and portal hypertension with transient elastography, *Gastroenterology* 134(1):8-14, 2008.

Schuppan D, Afdhal NH: Liver cirrhosis, *Lancet* 371(9615):838-851, 2008.

Talwalkar JA, Yin M, Fidler JL, et al: Magnetic resonance imaging of hepatic fibrosis: emerging clinical applications, *Hepatology* 47(1):332-3342, 2008.

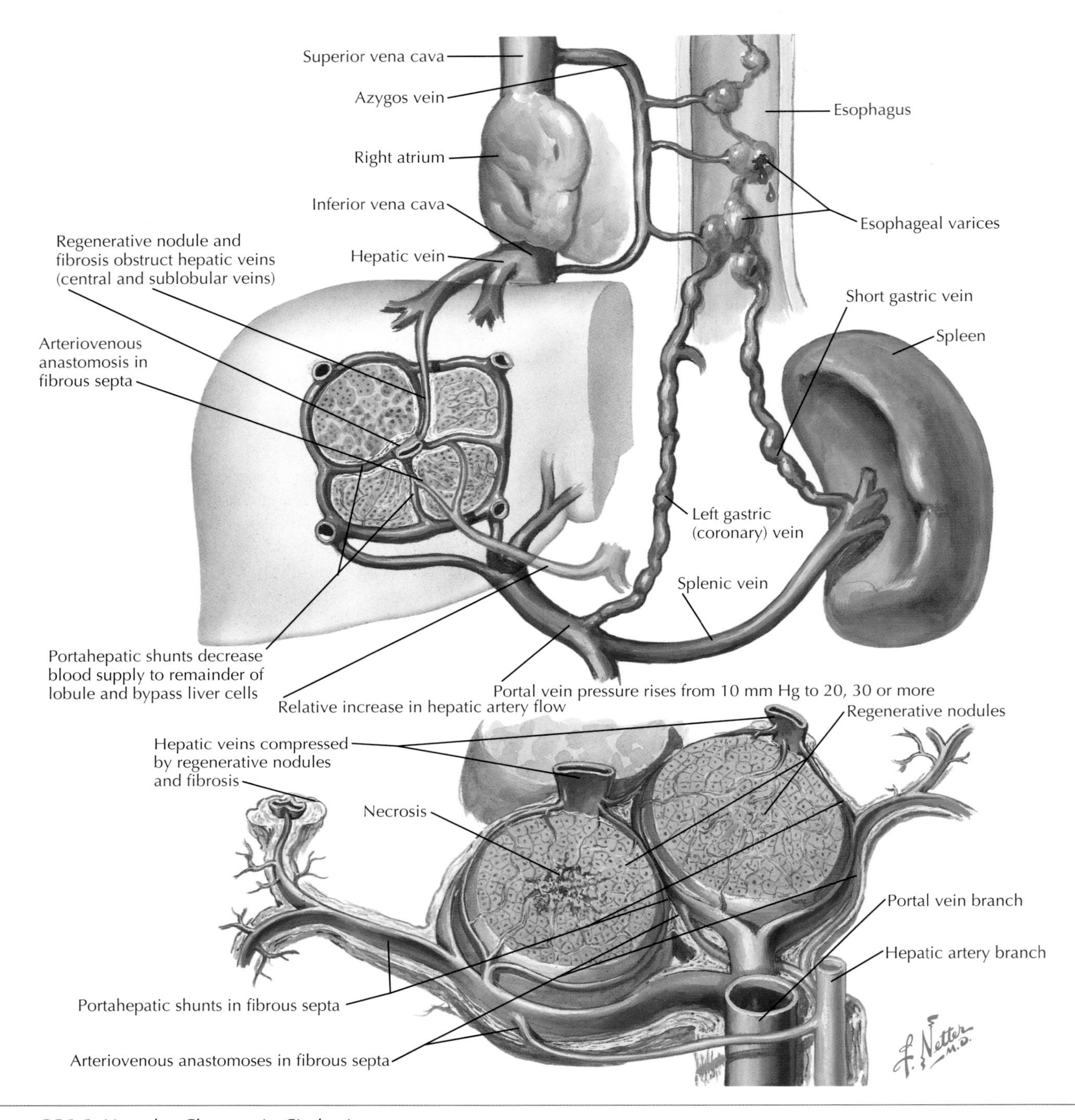

Figure 231-1 *Vascular Changes in Cirrhosis.*

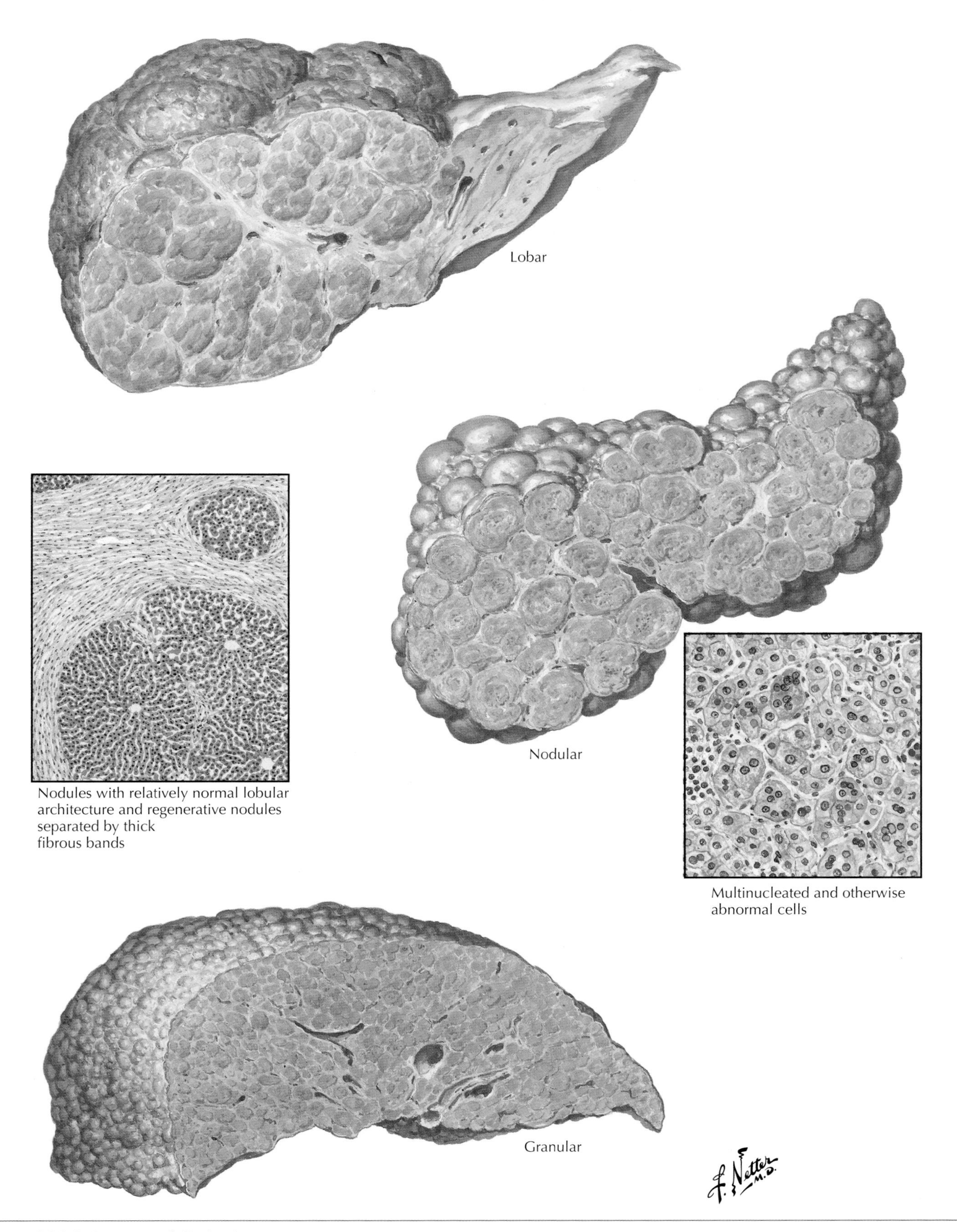

Figure 231-2 *Patterns of Cirrhosis.*

Imaging Studies of the Liver

232

Kris V. Kowdley

Great strides have been made in the past several decades in developing imaging modalities used to examine the liver. These include abdominal ultrasonography (US), computed tomography (CT), and magnetic resonance imaging (MRI). Recent advances in imaging have included positron emission tomography (PET) and magnetic resonance spectroscopy (MRS), but these remain primarily research tools and have had limited clinical application.

ABDOMINAL ULTRASONOGRAPHY

Abdominal US is often the first imaging modality used to evaluate the liver. US is easily performed, does not require intravenous (IV) access, and can provide a large amount of clinically relevant information in evaluating the liver and biliary tract. The use of Doppler technology also enables examination of the hepatic and portal venous system and the hepatic arterial flow. These latter developments have been particularly useful in the evaluation of patients after orthotopic liver transplantation and in those who have undergone placement of a transjugular intrahepatic portosystemic shunt (TIPS).

US is particularly helpful in evaluating suspected cystic lesions and in excluding the presence of dilated bile ducts. Liver cysts are often found during US and are usually asymptomatic. Based on US features, these can usually be classified as *simple cysts*. Simple cysts are anechoic and show posterior enhancement, which means there is an echogenic region behind the cystic lesion. In questionable cases, cyst aspiration can be performed to obtain cells for cytologic examination or to institute drainage if infection is a concern.

US is the most sensitive method for determining the presence of dilated bile ducts. Both intrahepatic and extrahepatic ductal dilatation can be identified, particularly in slender patients without fatty liver; thus, both the presence and the level of bile duct obstruction can be identified. In patients with acute biliary obstruction (e.g., acute choledocholithiasis), however, US may not reveal dilated ducts. A number of focal liver lesions may be identified, including bacterial, fungal, and parasitic abscesses and benign and malignant lesions (e.g., hepatic adenomas, hemangiomas, hepatocellular carcinoma).

However, CT and MRI are superior to US because of the ability to administer contrast and to obtain images in various phases—arterial, portal, and hepatic venous—which can further increase the specificity of the diagnosis.

Ultrasonography is also helpful to evaluate the hepatic parenchyma. In patients with fatty liver infiltration associated with obesity, hyperlipidemia, or type 2 diabetes, diffusely increased echogenicity may be observed in the liver. Increased echogenicity of the liver is also observed in patients with cirrhosis. In addition, US may reveal features of portal hypertension, such as splenomegaly, perisplenic varices, portosystemic collaterals, and reversal of flow (hepatofugal) in the portal vein. In some patients with cirrhosis, particularly resulting from alcohol, the left lobe may be enlarged.

COMPUTED TOMOGRAPHY

CT allows visualization of the liver with the addition of contrast, which has greatly improved the ability to differentiate hypervascular from hypovascular lesions. In addition, recent advances in CT have significantly enhanced radiologic diagnosis of liver lesions. Improvements in image acquisition techniques, such as helical and spiral CT, now allow rapid imaging of the liver and the opportunity to obtain arterial and portal venous phase images with the administration of IV contrast. These techniques have made so-called four-phase CT possible, permitting imaging of the liver during noncontrast, arterial, portal venous, and hepatic venous phases. Such improvements in technique have been helpful in differentiating vascular lesions, such as cavernous hemangiomas and hepatocellular carcinoma, which usually enhances in the arterial phases and may wash out during the portal venous phase.

CT has been invaluable in the evaluation of patients before liver transplantation. In addition to increased sensitivity for the screening of hepatocellular carcinoma in patients awaiting liver transplantation, three-dimensional (3D) reconstruction of the hepatic arterial system provides the surgeon with a useful road map, which is necessary given the great variability in normal hepatic arterial anatomy (see Chapter 215).

MAGNETIC RESONANCE IMAGING

MRI has become an increasingly common approach to liver imaging. MRI has been particularly useful in patients with renal insufficiency who may be unable to receive IV contrast. In such patients, MRI with magnetic resonance (MR) angiography can be used to evaluate liver texture, assess for liver masses, and delineate vascular anatomy. In addition, T1- and T2-weighted images can improve diagnostic sensitivity. Fat emits a bright signal on T1-weighted images, which may be helpful in identifying fat or blood. Fluid and pathologic lesions may be more visible on T2-weighted images.

Ongoing research is defining the role of MR contrast agents (e.g., gadolinium), and improvements in scanning time and 3D imaging techniques have allowed better visualization of the vessels and biliary tree. The paramagnetic properties of iron can be exploited using MRI because relaxation times are inversely related to hepatic iron content and can be quantified. Measuring hepatic iron content noninvasively will likely be possible in the near future using MRI.

CHOLANGIOPANCREATOGRAPHY

Endoscopic retrograde cholangiopancreatography (ERCP) was first introduced into clinical practice in 1968, although widespread use of this technique did not become established until the 1980s. ERCP has gradually evolved from a primarily diagnostic to a therapeutic procedure. Many therapeutic interventions are possible during ERCP, including placement of transpapillary stents to treat obstructive jaundice caused by benign or malignant strictures, bile fistulae, endoscopic sphincterotomy to remove common bile duct stones, and endoscopic balloon dilatation for benign strictures.

The ERCP technique involves the use of a specialized side-viewing endoscope with an elevator that allows placement of a 5-French plastic catheter into the biliary tree. Once deep cannulation into the bile duct has been achieved, radiographic contrast is injected, and fluoroscopic images are obtained. Excellent visualization of all major branches of the biliary tree and the intrahepatic ducts is possible with the injection of contrast under sufficient pressure.

Additionally, ERCP is invaluable for evaluating cholestatic liver disease. It is particularly useful in assessing patients with chronic cholestatic liver disease in the absence of an antimitochondrial antibody. ERCP has been considered the criterion for the ruling in (or ruling out) a diagnosis of primary sclerosing cholangitis (PSC). However, the newer technique of magnetic resonance cholangiopancreatography (MRCP) is increasingly used in the diagnosis of PSC.

Representative cross-sectional imaging techniques used in evaluation of liver disease are shown in **Figure 232-1**. Representative ERCP and MRCP images are shown in **Figure 232-2**. ERCP is also helpful for evaluating patients with cirrhosis who have abdominal pain of suspected biliary origin, because the serum liver enzymes may not be helpful in these patients, and US may not reveal dramatic biliary dilatation. Finally, ERCP has become an indispensable modality in the management of the patient after liver transplantation, especially now that T tubes are no longer frequently used in the context of biliary duct-to-duct anastomosis.

ELASTOGRAPHY

Magnetic resonance elastography is a newer technique that estimates liver stiffness using MRI technology (**Fig. 232-3**). *Transient elastography* is another technique that measures liver stiffness. Both techniques may be widely used in the future as an alternative to liver biopsy in determining the presence or absence of advanced hepatic fibrosis, which is associated with increased liver stiffness.

ADDITIONAL RESOURCES

Beckingham IJ, Ryder SD: ABC of diseases of liver, pancreas, and biliary system: investigation of liver and biliary disease, *BMJ* 322:33-36, 2001.

El Sherif A, McPherson SJ, Dixon AK: Spiral CT of the abdomen: increased diagnostic potential, *Eur J Radiol* 31:43-52, 1999.

Friedrich-Rust M, Ong MF, Martens S, et al: Performance of transient elastography for the staging of liver fibrosis: a meta-analysis, *Gastroenterology* 134(4):960-974, 2008.

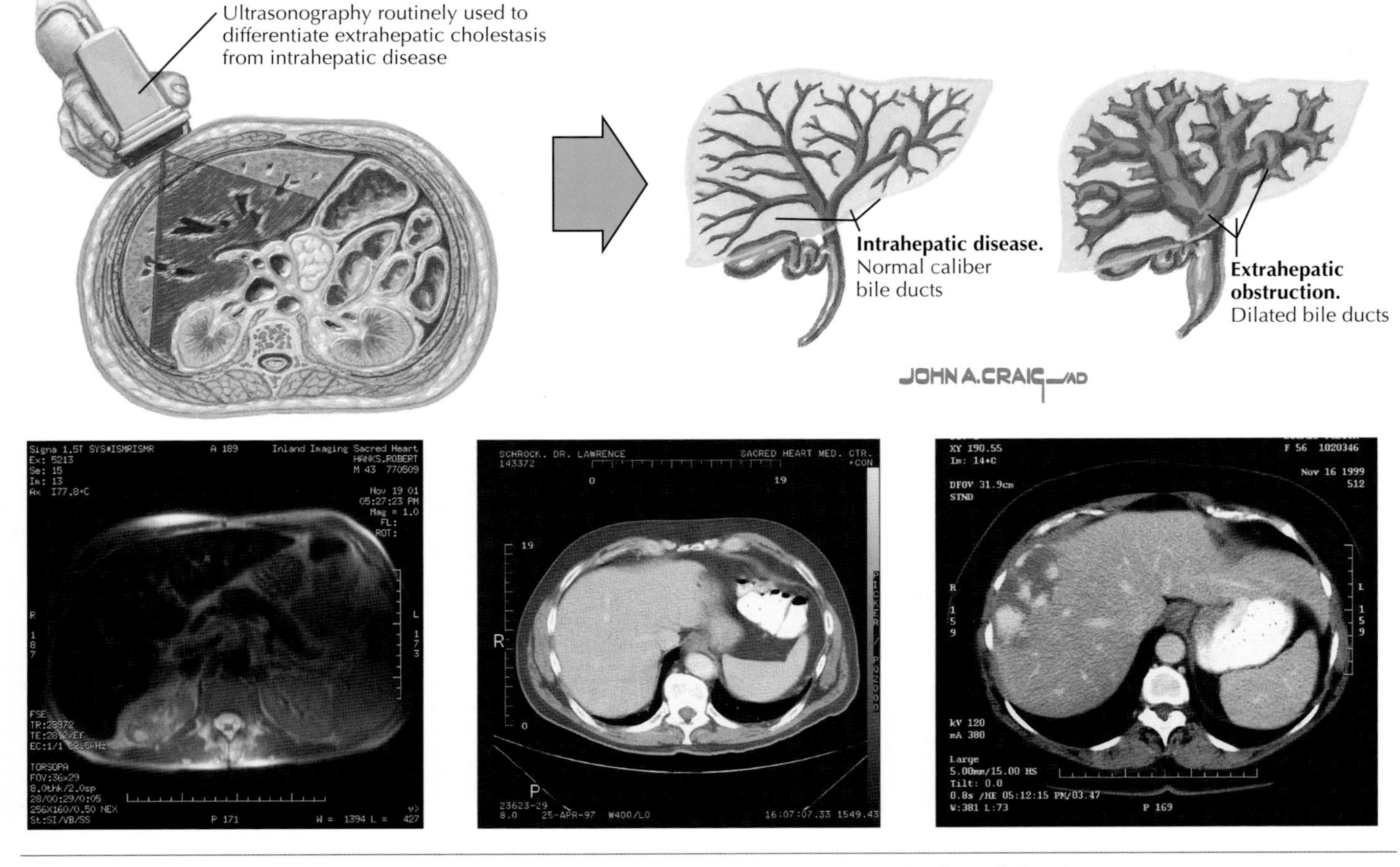

Figure 232-1 *Ultrasound, Computed Tomography, and Magnetic Resonance Imaging Studies of the Liver.*

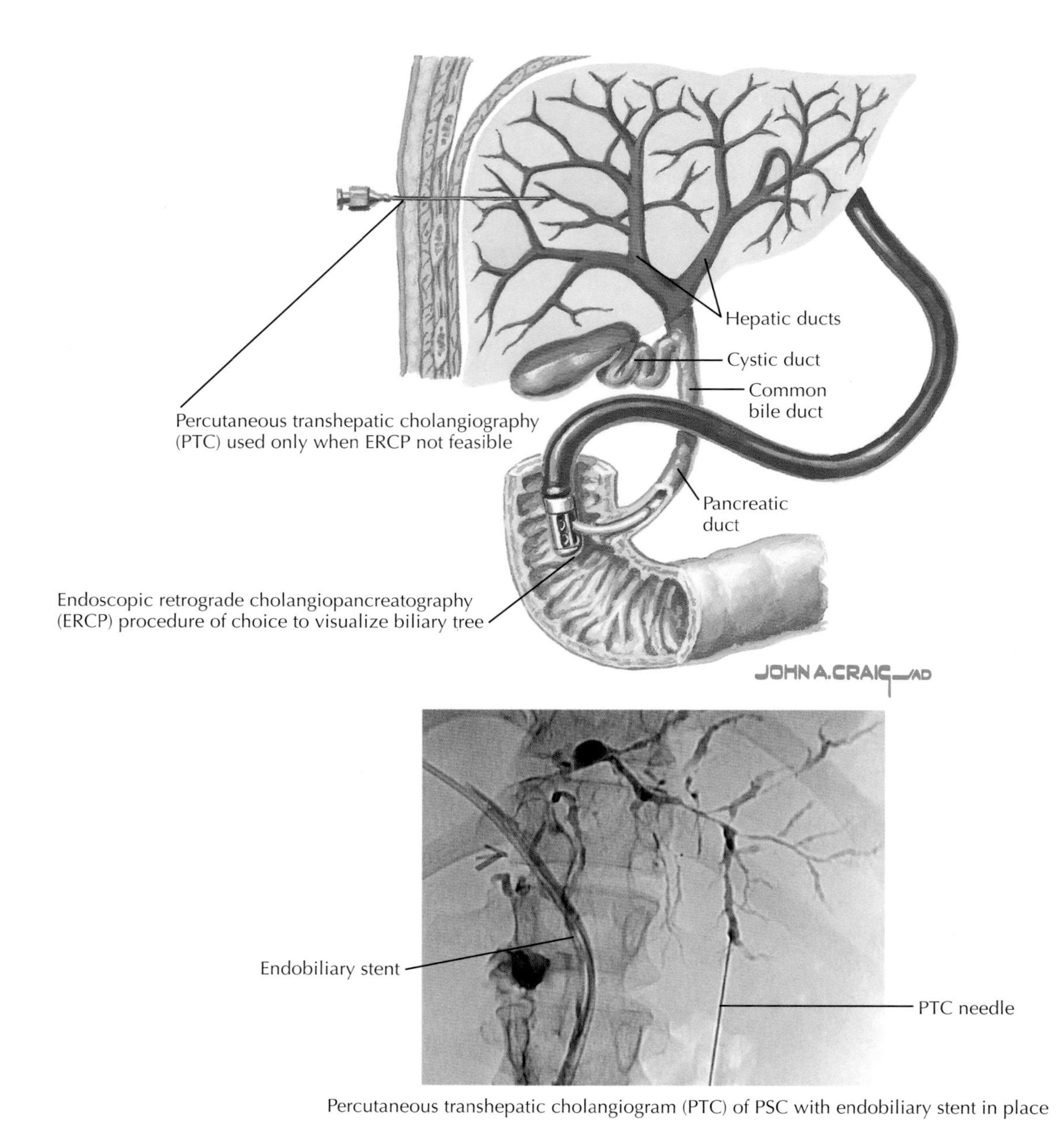

Percutaneous transhepatic cholangiogram (PTC) of PSC with endobiliary stent in place

Figure 232-2 *Percutaneous Cholangiography and Endoscopic and Magnetic Resonance Cholangiopancreatography Imaging Studies of the Liver.*

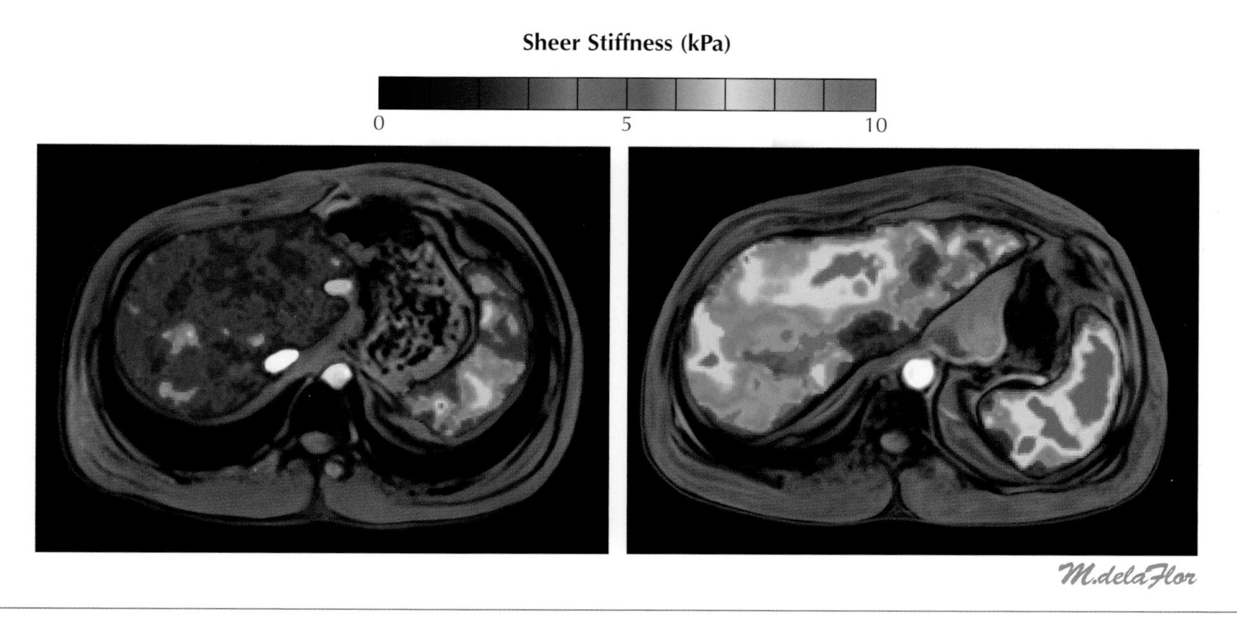

Figure 232-3 *Magnetic Resonance Elastography to Stage Hepatic Fibrosis.*

Levy AD: Noninvasive imaging approach to patients with suspected hepatobiliary disease, *Tech Vasc Interv Radiol* 4:132-140, 2001.

Noone TC, Semelka RC, Chaney DM, Reinhold C: Abdominal imaging studies: comparison of diagnostic accuracies resulting from ultrasound, computed tomography, and magnetic resonance imaging in the same individual, *Magn Reson Imaging* 22:19-24, 2004.

Ros PR, Mortele KJ: Hepatic imaging: an overview, *Clin Liver Dis* 6:1-16, 2002.

Talwalkar JA, Yin M, Fidler JL, et al: Magnetic resonance imaging of hepatic fibrosis: emerging clinical applications, *Hepatology* 47(1):332-342, 2008.

Alcoholic Liver Disease

233

Kris V. Kowdley

Alcoholic liver disease is the major cause of chronic liver disease in the United States and other Western countries. Furthermore, alcohol has been implicated as a major exacerbating factor in many other liver diseases, such as hepatitis C and hemochromatosis. Alcoholic liver disease may therefore play a role in most cases of liver disease.

Despite enormous progress in the past several decades, several key aspects of alcoholic liver disease remain unexplained, perhaps most importantly, the great variability in the relationship between quantity of alcohol consumed and risk for liver damage. However, it is generally recognized that the typical threshold level of alcohol intake associated with liver disease is 60 g daily over a 10-year period. The threshold for alcoholic liver disease is much lower among women than among men. Factors associated with this phenomenon may include the lighter body weight of women and decreased gastric alcohol dehydrogenase activity.

The pathophysiology of alcoholic liver disease is multifactorial. Numerous factors have been proposed, including genetic factors, toxic effects of alcohol, effect of prooxidant cytochromes (e.g., CYP 2E1), hypoxia, immune activation, and concomitant conditions (e.g., obesity). A key step in the metabolism of alcohol is the production of *acetaldehyde*, a hepatotoxin that mediates many steps in the evolution of hepatic necroinflammation in alcoholic liver disease. Stellate cell activation is central to the process of fibrogenesis. Over time, it can lead to the development of cirrhosis.

CLINICAL PICTURE

Clinical features of alcoholic liver disease are similar to other causes of liver disease, with some exceptions. Alcoholic liver disease is often associated with more prominent ascites in patients with otherwise-compensated liver disease. Serum liver biochemical tests may also point to a diagnosis of alcoholic liver disease because the AST/ALT ratio is often greater than 2:1 and is frequently greater than 3:1.

The role of liver biopsy remains controversial in alcoholic liver disease. Some argue that biopsy should be routinely performed because occasionally, other unexpected causes of liver disease can be found in patients given a presumptive diagnosis of alcoholic liver disease (**Fig. 233-1**). Histologic features of alcoholic liver disease may follow one of three patterns: fatty liver (**Fig. 233-2**), alcoholic hepatitis, and alcoholic cirrhosis (**Fig. 233-3**). Frequently, all three of these findings can be found in the same patient. In addition, there may be features of ballooning degeneration of hepatocytes, Mallory (hyaline) bodies, and variable degrees of fibrosis. In contrast to patients with nonalcoholic fatty liver disease, lobular damage, Mallory bodies, periportal or bridging fibrosis, and cirrhosis are observed in a much higher proportion of patients.

DIAGNOSIS

The diagnosis of alcoholic liver disease is based on the exclusion of other causes of liver disease and the appropriate history of alcohol use. It is useful to quantify the amount of alcohol consumed by the patient on a chronic basis. The quantity of alcohol consumed in grams per day can be estimated as follows: one 12-oz can of beer, one 4-oz glass of wine, and one 1-oz shot of spirits each contain approximately 11 g of alcohol. In the assessment of alcoholic liver disease, it is also useful to identify whether the patient is tolerant of or dependent on alcohol. Several standardized questionnaires, such as CAGE (cut down, annoyance, guilt, eye opener), are widely available for this purpose.

TREATMENT AND MANAGEMENT

Abstinence is the mainstay of therapy for alcoholic liver disease. Liver function can recover remarkably well after alcohol intake ceases. Some patients with advanced liver disease who are candidates for liver transplant recover to such a degree that liver transplantation is no longer needed.

Patients with cirrhosis are at long-term risk for hepatocellular carcinoma. Therefore, screening using ultrasonography or computed tomography is appropriate among patients with established cirrhosis. Chronic alcohol consumption can also lead to increased serum transferrin–iron saturation and ferritin levels and may lead to increased hepatic iron stores, sometimes mimicking hemochromatosis. However, some patients with a history of heavy alcohol consumption may indeed have hereditary hemochromatosis (see Chapter 246). Serum iron studies should be obtained in all patients with a history of alcoholic liver disease, and *HFE* gene testing should be performed in patients with elevated serum transferrin–iron saturation and ferritin levels.

COURSE AND PROGNOSIS

Long-term survival of patients with alcoholic liver disease is significantly lower than in patients with other causes of chronic liver disease, possibly as low as 7% at 10 years. Several variables may be associated with outcome in patients with alcoholic liver disease, including nutritional status, obesity, genetic factors, and concomitant use of hepatotoxic medications (e.g., acetaminophen) or infection with hepatitis C.

ALCOHOLIC HEPATITIS

Alcoholic hepatitis is a potentially life-threatening complication of alcoholic liver disease. It is characterized by jaundice and moderately to markedly elevated levels of serum transaminase. Patients often have fever, right upper quadrant pain, and tender hepatomegaly. Acute cholecystitis or choledocholithiasis is

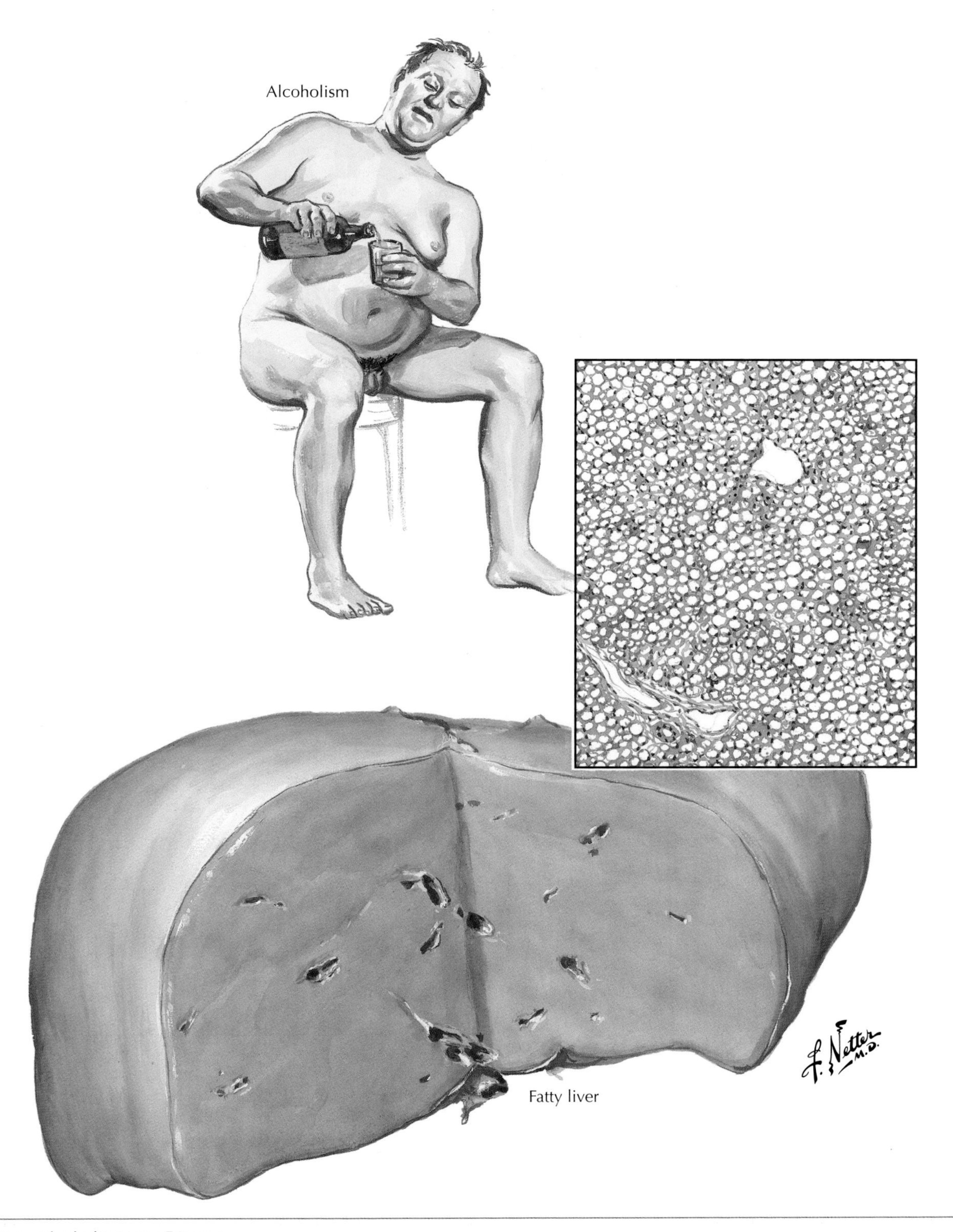

Figure 233-1 *Alcoholic Liver Disease.*

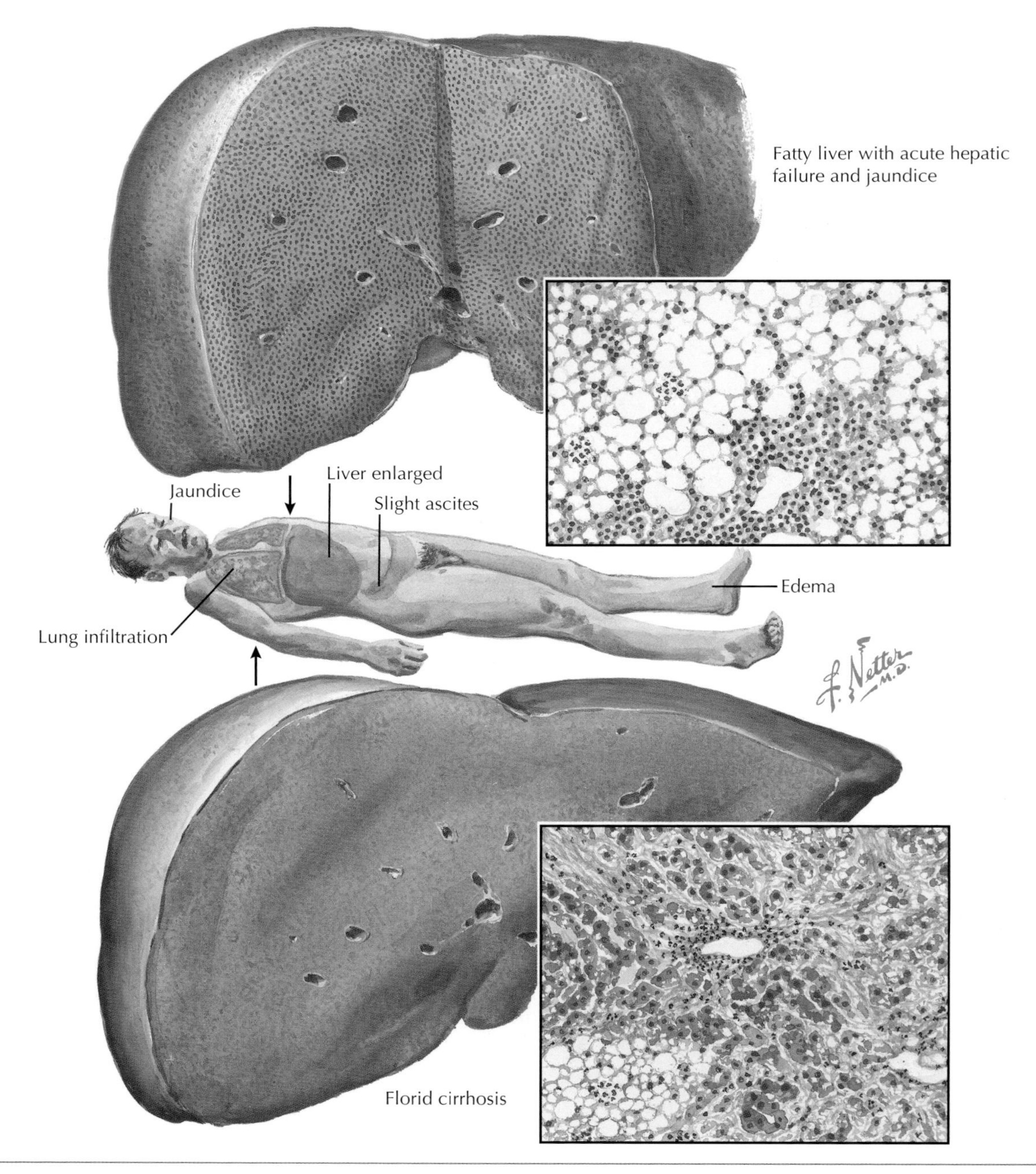

Figure 233-2 *Fatty Liver with Acute Hepatic Failure and Florid Cirrhosis.*

often suspected because of the presence of right upper quadrant tenderness and leukocytosis. Liver biopsy reveals a neutrophilic infiltrate accompanying steatohepatitis, with florid ballooning, degeneration of hepatocytes, and Mallory bodies.

A *discriminant function* has been developed to predict survival in patients with acute alcoholic hepatitis. It incorporates serum bilirubin, prothrombin time, and encephalopathy; a discriminant function ([4.6 × (prothrombin time/control [sec]) + bilirubin] > 52) or the presence of hepatic encephalopathy portend a poor prognosis in patients with acute alcoholic hepatitis.

Treatment with corticosteroids or pentoxifylline may improve survival of patients with alcoholic hepatitis.

ADDITIONAL RESOURCES

Diehl AM: Liver disease in alcohol abusers: clinical perspective, *Alcohol* 27:7-11, 2002.

Maher JJ: Alcoholic steatosis and steatohepatitis, *Semin Gastrointest Dis* 13:31-39. 2002.

Maher JJ: Treatment of alcoholic hepatitis, *J Gastroenterol Hepatol* 17:448-455, 2002.

Mathurin P, Louvet A, Dharancy S: Treatment of severe forms of alcoholic hepatitis: where are we going? *J Gastroenterol Hepatol* 23(suppl 1):60-62, 2008.

Menon KV, Cores CJ, Shah VH: Pathogenesis, diagnosis, and treatment of alcoholic liver disease, *Mayo Clin Proc* 76:1021-1029, 2001.

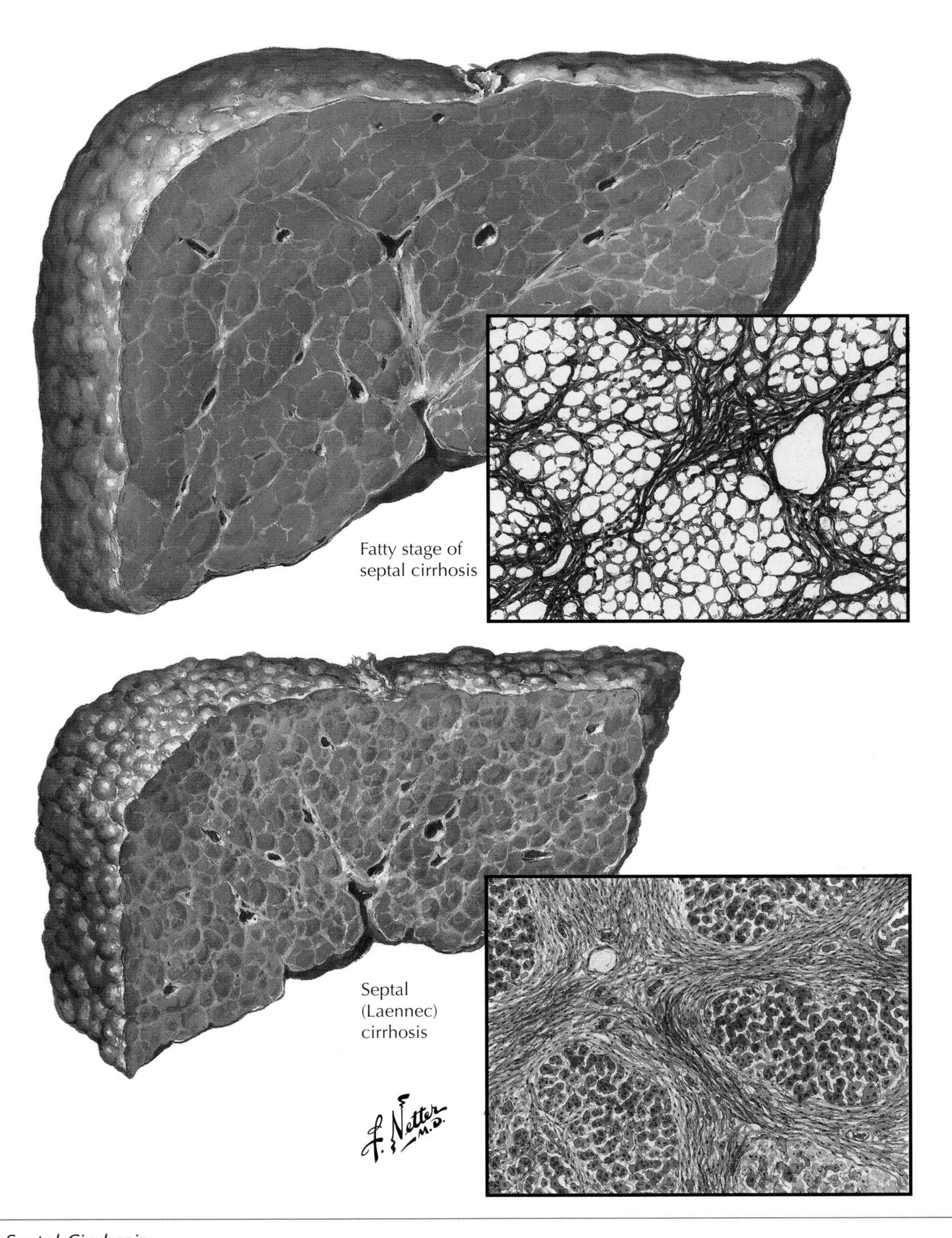

Figure 233-3 *Septal Cirrhosis.*

Nonalcoholic Fatty Liver Disease and Steatohepatitis

234

Kris V. Kowdley

The term "nonalcoholic steatohepatitis" was first used by Ludwig and colleagues to describe the liver histopathology in middle-aged women with diabetes who appeared to have alcoholic liver disease but did not drink alcohol. Great progress has been made in the past decade in understanding the clinical features of this liver disease and its epidemiologic features, pathophysiology, and natural history.

The spectrum of nonalcoholic liver disease ranges from simple fatty liver, or *nonalcoholic fatty liver disease* (NAFLD), to *nonalcoholic steatohepatitis* (NASH), characterized by cytolytic changes in hepatocytes, such as ballooning degeneration, Mallory (hyaline) bodies, and lobular inflammation (**Fig. 234-1**). Variable degrees of fibrosis may be present, ranging from pericellular and perivenular or perisinusoidal fibrosis in zone 3 of the liver lobule to bridging fibrosis and cirrhosis. In the late stages, the fatty infiltration may disappear, leaving a picture of cirrhosis of unclear etiology. Such patients are often classified as having *cryptogenic cirrhosis*.

Most patients with NAFLD or NASH have features of insulin resistance and may also have *syndrome X*, characterized by central or visceral obesity, type 2 diabetes, and dyslipidemia. The prevalence of NAFLD appears to be increasing rapidly, in parallel with the epidemic of obesity in Western countries. Multiple environmental factors likely play roles, including high-calorie and high-carbohydrate intake, sedentary lifestyle, and greater consumption of highly refined or processed sugars. In fact, NASH is increasingly observed in children, the population with the fastest-growing incidence of type 2 diabetes. Population-based studies suggest that the prevalence of NAFLD and NASH may be as high as 20% and 2% to 3%, respectively, in the general population. Prospective liver biopsy data from living donors in living-related liver transplantation have shown NASH in up to 25%.

The primary insult in NASH is thought to be *insulin resistance*, which may lead to increased circulating concentrations of free fatty acids. These may accumulate in the liver, leading to steatosis. A second insult, or second "hit," may lead to oxidative stress in the liver, resulting in progression to NASH. Many such possible second hits have been proposed, such as altered mitochondrial uncoupling proteins (UCP-2), cytochrome P450 2E1, excess iron, and activation of proinflammatory cytokines, such as tumor necrosis factor-α, nuclear factor kappa B, and interleukin-1, as well as other cascade pathways.

CLINICAL PICTURE

A detailed weight history should be obtained for all patients suspected to have fatty liver disease, including a history of obesity during childhood, glucose intolerance, gestational or type 2 diabetes, hypertriglyceridemia, low high-density lipoprotein level, and other disorders of lipid metabolism.

DIAGNOSIS

A diagnosis of NAFLD requires a careful and detailed history of alcohol use, potentially hepatotoxic medications, herbal supplements, and over-the-counter medications. Other factors known to cause steatohepatitis, such as jejunoileal bypass, total parenteral nutrition, and genetic or metabolic derangements, should be excluded. Severe malnutrition may also lead to NASH, manifested by *brown atrophy* of the liver (**Fig. 234-2**). Kwashiorkor may be associated with fatty liver disease among infants.

Steatosis and steatohepatitis are often discovered during routine medical screening. Most patients do not have symptoms. Some patients report fatigue, malaise, right upper quadrant fullness, or tenderness. Physical examination may reveal an android habitus, including central or truncal obesity and increased waist-to-hip ratio. Children with the disease are frequently obese and have *acanthosis nigricans*, a marker of insulin resistance that may be present in one third of children and adolescents. Occasionally, patients have symptoms of advanced liver disease, including ascites, variceal bleeding, and hepatic encephalopathy.

Blood tests frequently show mild to moderate elevations of serum liver enzymes, with twofold to fivefold increases in serum aspartate transaminase (AST) and alanine transaminase (ALT). The AST/ALT ratio may be a sign of cirrhosis, but it is usually lower in patients with cirrhosis than in those with alcoholic liver disease. Fasting serum glucose and triglyceride levels may be elevated; in addition, increased levels of serum ferritin, an acute-phase protein, may be present in up to 50% of patients, and increased levels of serum transferrin–iron saturation may be present in approximately 10%, although hepatic iron is not usually increased in the absence of *HFE* mutations.

Ultrasonography may reveal a so-called bright liver. Computed tomography often reveals a liver that is darker than the spleen. Magnetic resonance imaging may demonstrate fat as increased attenuation on T1-weighted images. However, noninvasive imaging modalities cannot distinguish simple steatosis from NASH.

TREATMENT AND MANAGEMENT

Weight loss through diet and exercise should be encouraged. Even a 10% decrease in body weight may greatly improve insulin resistance. Bariatric surgery with controlled and relatively gradual weight loss has recently been shown to reverse many of the histologic features of NASH. Additionally, some have advocated reduction in the carbohydrate content of the diet.

Numerous promising agents have been studied, although none is currently approved by the U.S. Food and Drug Administration for treatment of NASH. Although ursodeoxycholic acid was not found to be effective, metformin, vitamin E, and

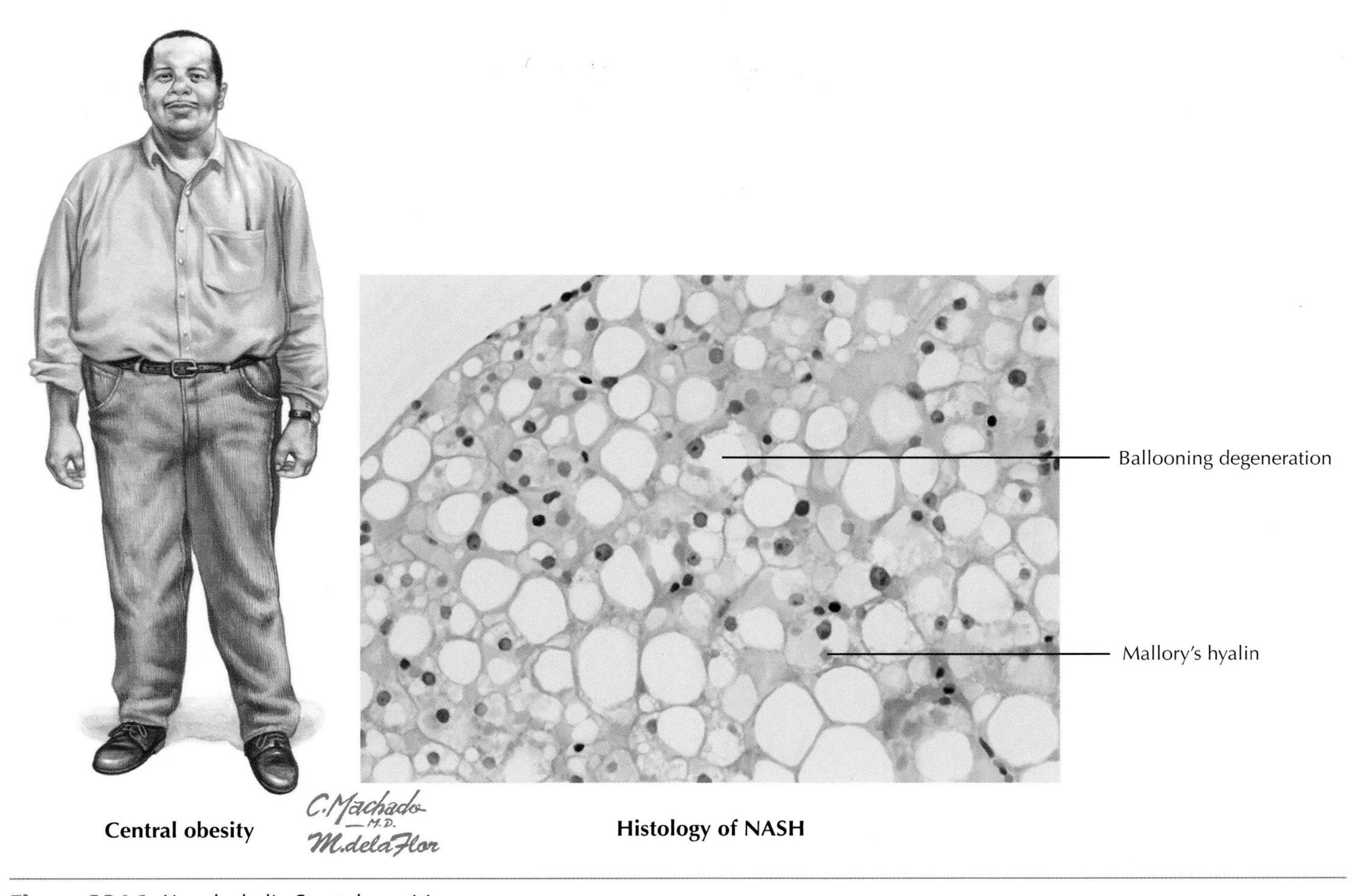

Figure 234-1 *Nonalcoholic Steatohepatitis.*

the newer thiazolidinediones (e.g., pioglitazone, rosiglitazone) are all being studied for the treatment of patients with NASH.

ADDITIONAL RESOURCES

Alba LM, Lindor K: Non-alcoholic fatty liver disease (review), *Aliment Pharmacol Ther* 17:977-986, 2003.

Li Z, Clark J, Diehl AM: The liver in obesity and type 2 diabetes mellitus, *Clin Liver Dis* 6:867-877, 2002.

Mofrad PS, Sanyal AJ: Nonalcoholic fatty liver disease, *MedGenMed* 5:14, 2003.

Mulhall BP, Ong JP, Younossi ZM: Non-alcoholic fatty liver disease: an overview, *J Gastroenterol Hepatol* 17:1136-1143, 2002.

Neuschwander-Tetri BA, Caldwell SH: Nonalcoholic steatohepatitis: summary of an AASLD Single Topic Conference, *Hepatology* 37:1202-1219, 2003.

Oh MK, Winn J, Poordad F: Diagnosis and treatment of non-alcoholic fatty liver disease (review), *Aliment Pharmacol Ther* 28(5):503-522, 2008.

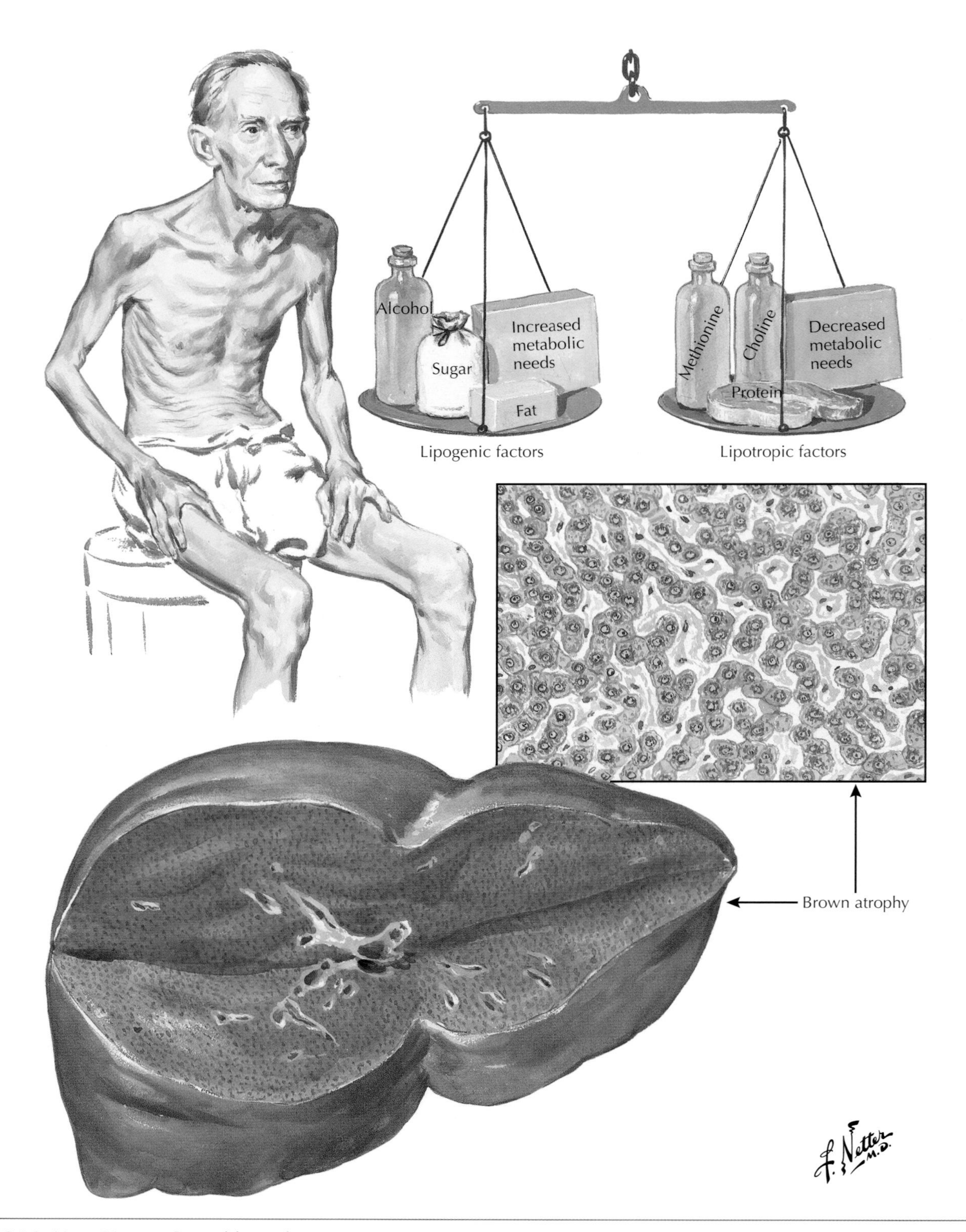

Figure 234-2 *Liver Disease Caused by Malnutrition.*

Extrahepatic Biliary Obstruction

Kris V. Kowdley

Extrahepatic biliary obstruction causes jaundice only if it is located between the confluence of the right and left hepatic ducts and the tip of the papilla of Vater. Obstruction of a single branch of the main hepatic duct does not produce jaundice because the unobstructed part of the liver compensates. The excretion of bile components other than bile pigments may not be as readily compensated, and serum alkaline phosphatase (ALP) or total serum cholesterol levels may be elevated in the absence of jaundice. Obstruction may be complete or incomplete; if incomplete, it is often intermittent.

Complete extrahepatic biliary obstruction is usually caused by tumors, which initially produce an incomplete obstruction but subsequently a permanent, complete occlusion (**Figs. 235-1** and **235-2**). Occasionally, regressive changes or hemorrhage into a tumor may result in the sloughing off of obstructive tissue, with temporary relief of a complete obstruction. This may be associated with at least chemical evidence of melena.

Intrinsic obstructive tumors are usually malignant and are represented by cancer of the biliary ducts or cancer of the papilla of Vater. Carcinoma of the pancreas (which may compress or kink the ducts), extension of carcinoma of the gallbladder, carcinomatous metastases to the hepatic lymph nodes, and other types of *extrinsic* carcinoma must invade the wall of the bile duct to produce obstruction. A tumor not fixed to the duct remains movable and cannot cause complete obstruction. For this reason, even large metastases to the hepatic lymph nodes, as seen in Hodgkin disease, leukemia, and reticulum cell sarcoma, seldom produce obstructive jaundice. Jaundice, if it appears in such conditions, with inflammatory swelling of the hepatic lymph nodes, is almost always attributed to intrahepatic causes.

Gallstones enter the biliary ducts and become impacted there, cause initial spasm and edema, and may be associated with complete obstruction that is usually transient, because spasm and edema subside quickly. If the stone is not expelled from the duct, incomplete obstruction persists. If a stone moves or acts like a ball valve, the obstruction may become intermittent. In *calculous* obstruction, a variable, usually short period of complete obstruction is followed by intermittent obstruction reflected in intermittent hyperbilirubinemia.

Strictures, whether produced by surgical injury to the biliary ducts or resulting from inflammatory lesions, may also cause biliary obstruction. Rarer causes of biliary obstruction are congenital atresia, inflammatory processes in neighboring organs (peptic ulcers, pancreatitis), duodenal diverticula, foreign bodies, and parasites.

Mechanical obstruction leads rapidly to dilatation of the biliary system above the sites of obstruction (**Fig. 235-3**). Obstruction of the *cystic duct* leads to dilatation of the gallbladder. If the obstruction involves the terminal portion of the cystic duct, the stone may bulge into the lumen of the common duct and produce jaundice and enlargement of the gallbladder (Mirizzi syndrome). Obstruction of the *common duct* by stones is usually associated with inflammation of the gallbladder, which may be fibrotic and does not dilate significantly. Obstruction of the duct by a tumor located near the papilla of Vater is usually associated with a normally expanding gallbladder that readily dilates and that may be palpated as a large, thin-walled cyst (Courvoisier gallbladder).

Hepatic effects of biliary obstruction, seen in biopsy specimens, develop more rapidly in complete than in incomplete obstruction. The first change is accumulation of bile pigment in the liver cells and Kupffer cells in the central zone of the lobule. Simultaneously, bile may amass in the form of ramified bile plugs in the dilated bile capillaries. The cytoplasm of some liver cells adjacent to the bile ductules is degenerated, and *pyknosis* (feathery degeneration) may be seen. At this stage, hyperbilirubinemia and bilirubinuria appear, and serum ALP activity is elevated. Subsequently, inflammatory infiltration of the portal triads develops, with proliferation of perilobular cholangioles and periportal ducts. At this stage, ALP elevation is more marked, and the total serum cholesterol may be elevated.

If the obstruction is prolonged, proliferation of the cholangioles increases, and bile casts may form even in peripheral ductules. Dilated cholangioles contain thick bile plugs, *microcalculi*, especially on the border between the lobular parenchyma and the portal triads, around which fibrosis often develops.

Although the features just described may be seen in intrahepatic and extrahepatic cholestasis, two features, the *extravasation of bile* and *bile infarcts*, are characteristically seen only in extrahepatic obstruction. Both appear after prolonged obstruction, when the obstruction is complete. Necroses of the epithelial lining of the interlobular bile ducts permit bile to escape into their walls, and granulation tissue appears around the golden-yellow bile in the portal triad. In circumscribed foci, the cytoplasm of liver cells is abnormal, and the bile is pigmented.

In the late stages of obstruction, secondary hepatocellular damage may be severe and may be reflected in marked abnormalities on liver function tests. The liver in the late stages of biliary obstruction is enlarged and dark green. On the cut surface, the bile ducts appear greatly dilated. Eventually, bands of fibrosis form and nodules regenerate, marking the beginning of cirrhosis.

ADDITIONAL RESOURCES

Abou-Saif A, Al-Kawas FH: Complications of gallstone disease: Mirizzi syndrome, cholecystocholedochal fistula, and gallstone ileus, *Am J Gastroenterol* 97:249-254, 2002.

Bezerra JA, Balistreri WF: Cholestatic syndromes of infancy and childhood, *Semin Gastrointest Dis* 12:54-65, 2001.

Brunt EM: Liver biopsy interpretation for the gastroenterologist, *Curr Gastroenterol Rep* 2:27-32, 2000.

Farah M, McLoughlin M, Byrne MF: Endoscopic retrograde cholangiopancreatography in the management of benign biliary strictures, *Curr Gastroenterol Rep* 10(2):150-156, 2008.

Heathcote EJ: Diagnosis and management of cholestatic liver disease, *Clin Gastroenterol Hepatol* 5(7):776-782, 2007.

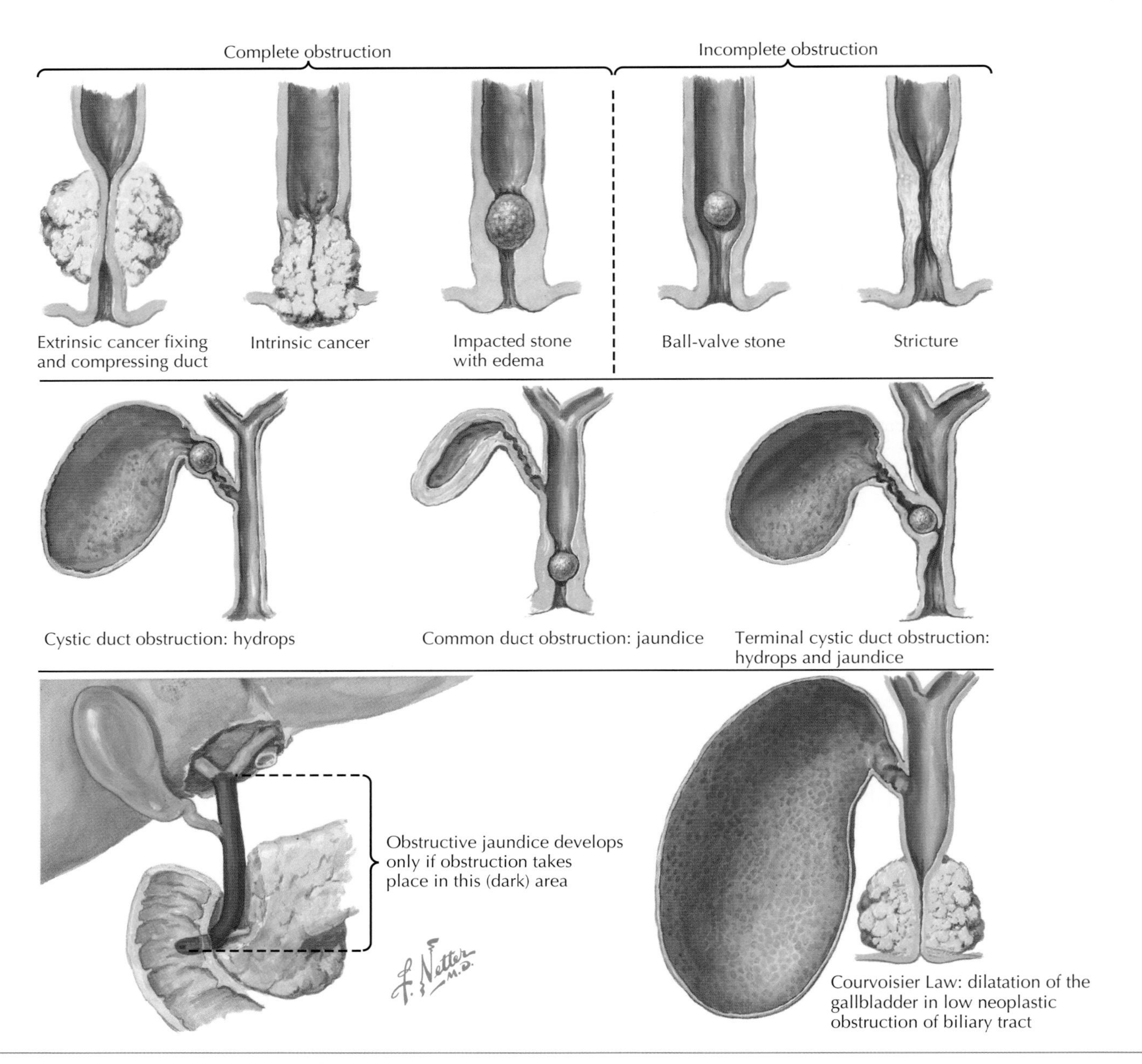

Figure 235-1 *Mechanism and Types of Extrahepatic Biliary Obstruction.*

Hofmann AF: Cholestatic liver disease: pathophysiology and therapeutic options, *Liver* 22(suppl 2):14-19, 2002.

Kim VVR, Ludwig J, Lindor KD: Variant forms of cholestatic diseases involving small bile ducts in adults, *Am J Gastroenterol* 95:1130-1138, 2000.

Pasha TM, Lindor KD: Diagnosis and therapy of cholestatic liver disease, *Med Clin North Am* 80:995-1019, 1996.

Rumalla A, Petersen BT: Diagnosis and therapy of biliary tract malignancy, *Semin Gastrointest Dis* 11:168-173, 2000.

Velayudham LS, Farrell GC: Drug-induced cholestasis, *Expert Opin Drug Safety* 2:287-304, 2003.

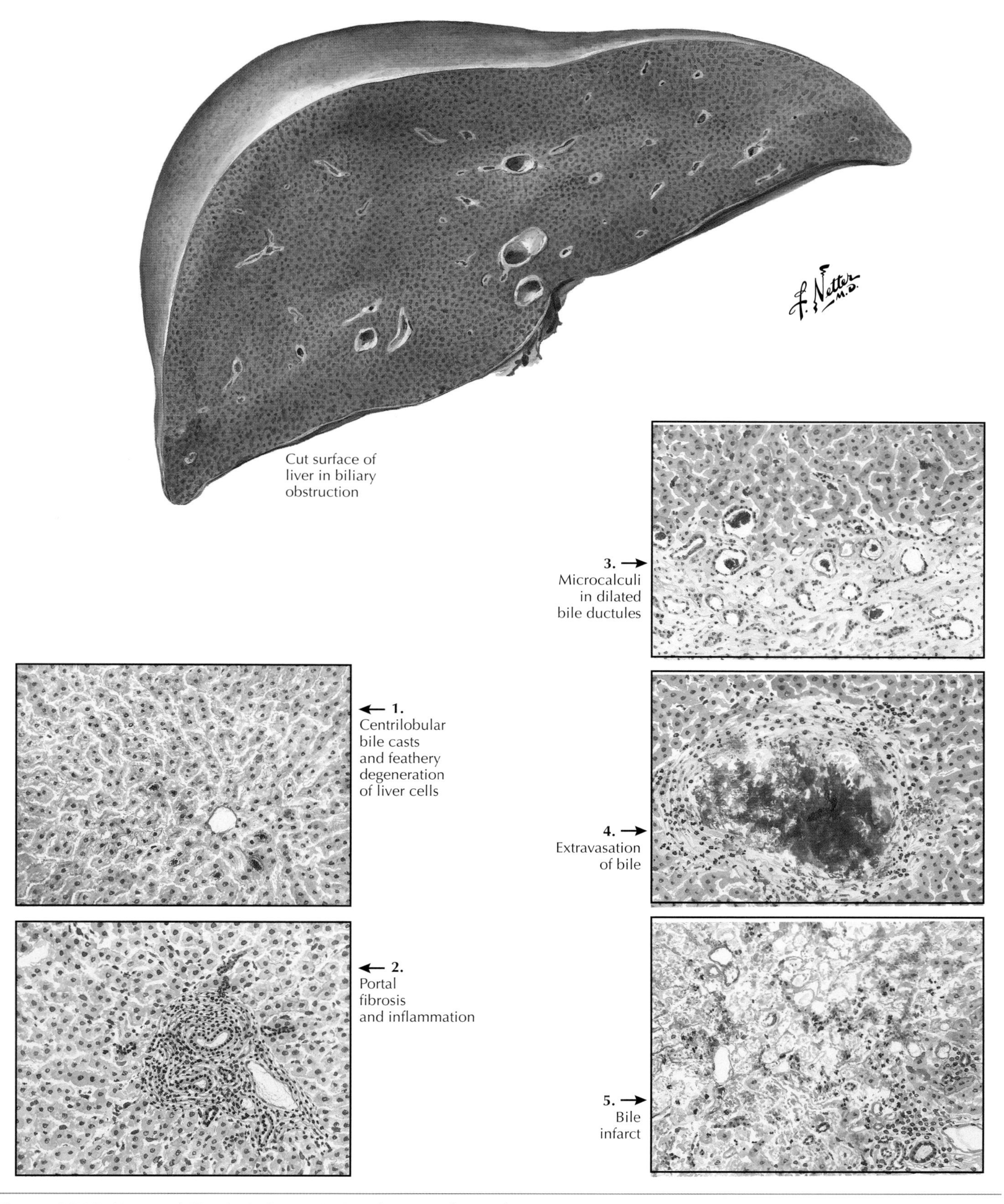

Figure 235-2 *Liver and Histologic Stages in Extrahepatic Biliary Obstruction.*

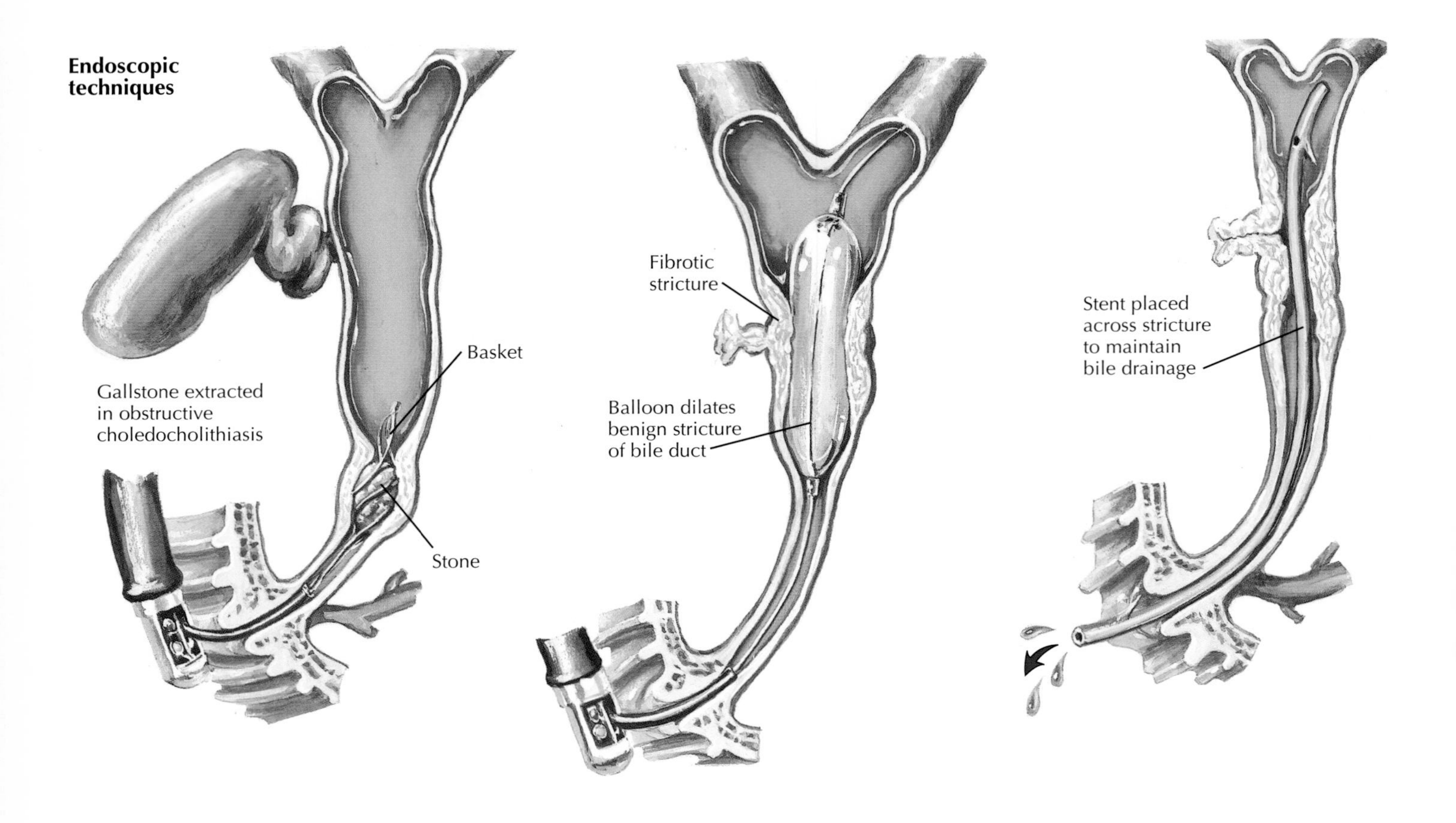

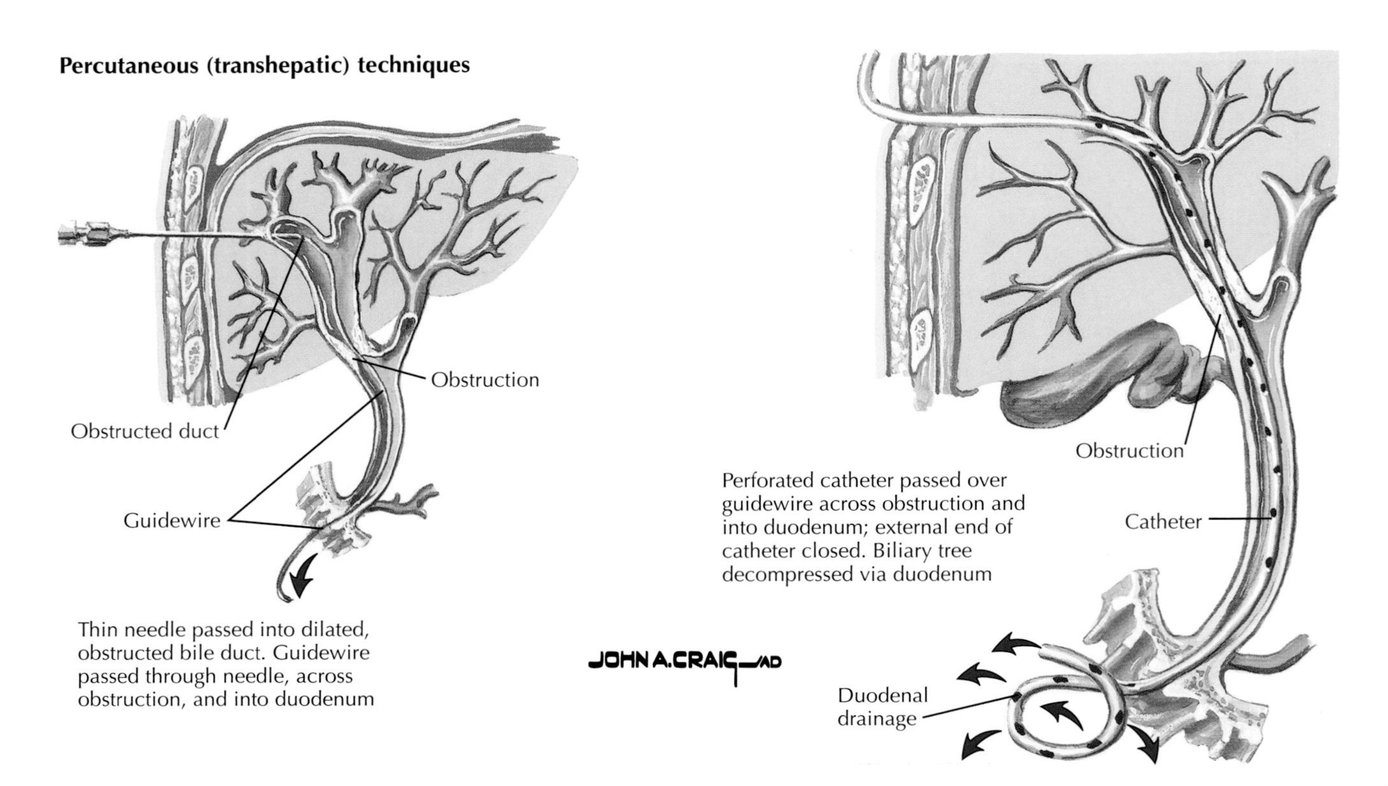

Figure 235-3 *Mechanical Decompression Techniques for Biliary Obstruction.*

Primary Biliary Cirrhosis

236

Kris V. Kowdley

Primary biliary cirrhosis (PBC) is a disorder of unknown cause characterized by lymphocytic cholangitis involving the small intralobular bile ducts within the liver (**Figs. 236-1** and **236-2**). This first step in PBC pathophysiology is followed by hepatotoxicity caused by the retention of toxic bile acids secondary to cholestasis and bile duct loss. The putative first "hit" is immunologic attack on intralobular bile ducts by activated T lymphocytes; this appears to be an autoimmune response in genetically predisposed persons. Information is limited, however, on the actual triggers initiating the T cell–mediated attack on bile duct cells. Although most patients with PBC have antibodies directed against the pyruvate dehydrogenase complex of mitochondria, it is unclear how this autoantibody leads to the attack of bile duct cells.

Recent studies suggest novel hypotheses for the pathogenesis of PBC. One study found high titers of an antibody against a retrovirus in patients with PBC. The presence of fetal cells in the maternal circulation (fetal microchimerism) has also been implicated. Others suggest that xenobiotic viruses or bacteria may induce the autoimmune reaction directed against bile ducts.

Approximately 95% of patients with PBC are women, and most are 25 to 85 years old. PBC is therefore rare in adolescents. A study from Rochester, Minnesota, found a prevalence of approximately 65 per 100,000 women and 12 per 100,000 men.

DIAGNOSIS

The diagnosis of PBC is relatively straightforward. PBC should be suspected in any patient with chronically elevated liver test findings in a cholestatic pattern. Almost all patients are positive for the antimitochondrial antibody. Liver biopsy characteristically reveals a biliary type of chronic injury with lymphocytic cholangitis, bile ductular proliferation, and variable degrees of fibrosis. Noncaseating granulomas are frequently present. Histologic stages include stage I (florid bile duct lesion), stage II (bile ductular proliferation), stage III (bridging fibrosis), and stage IV (cirrhosis) (**Figs. 236-3** and **236-4**). Interface hepatitis is usually absent.

Variants of PBC include the so-called autoimmune cholangiopathies. Such patients may have histologic features of PBC, but serologic findings suggest autoimmune hepatitis (antinuclear or anti–smooth muscle antibody positive).

The differential diagnosis of PBC includes any other cause of intrahepatic or extrahepatic cholestasis (**Fig. 236-5**). Drug-induced cholestasis is one of the most common liver diseases that may manifest similar to PBC. Increased estrogen levels, as might be observed in pregnancy and in women using oral contraceptives, may also lead to intrahepatic cholestasis. Extrahepatic biliary tract obstruction is generally recognized by the identification of dilated intrahepatic or extrahepatic bile ducts on imaging studies. Common causes include postoperative biliary strictures, bile duct cancer, and choledocholithiasis. Therefore, it is essential to rule out biliary obstruction in any patient with acute or chronic cholestasis, particularly because cholestasis resulting from biliary obstruction is often easily treated by cholangiographic or surgical intervention. Histologic features of intrahepatic cholestasis caused by drugs or obstruction may be different than those observed in PBC.

TREATMENT AND MANAGEMENT

Because PBC is presumed to be secondary to an autoimmune process and accumulation of toxic hydrophobic bile acids, treatment has focused on immunosuppressive agents or drugs that may reduce the toxicity associated with the retention of hydrophobic bile acids. Corticosteroids, azathioprine, cyclosporine, and other immunosuppressive therapies have been used without clear evidence of benefit. D-Penicillamine was used because patients with PBC had elevated hepatic copper levels that might have contributed to hepatotoxicity. Findings were negative. Subsequent studies showed that the elevated copper levels are secondary to cholestasis rather than a primary defect in copper metabolism. Corticosteroids are associated with improved serum biochemical tests and histology but also with an unacceptably high rate of bone loss, leading to accelerated osteoporosis; therefore they should not be used. One study showed that colchicine variably improves survival.

The mainstay of therapy for PBC is *ursodeoxycholic acid* (UDCA). Multicenter randomized trials have clearly shown that UDCA slows the progression of PBC and decreases liver transplantation and mortality. Patients with moderate or severe disease appeared to derive the greatest benefit from UDCA, whereas those with mild histologic disease and well-compensated liver function (serum bilirubin <1.4 mg/dL) did not benefit. By contrast, a U.S. multicenter study found that patients with mild disease were most likely to benefit from UDCA; however, this study was only 2 years in duration. UDCA has been shown to improve histologic findings and possibly to reduce the severity of portal hypertension. Based on these data, UDCA was approved for the treatment of PBC at a dose of 10 to 15 mg/kg body weight daily.

Nevertheless, it is controversial whether UDCA improves long-term outcomes in PBC. Differences in outcome with UDCA are likely related to patient selection and study design. Because PBC has a long natural history, studies of short duration or those that include patients with mild disease may show no benefit. A recent Cochrane review found that UDCA improved laboratory tests and some complications of PBC but did not reduce morbidity and mortality and liver transplantation.

In summary, UDCA is probably of greatest benefit for those with mild or moderate disease. Those with advanced disease should probably not be treated because of the possibility of exacerbating pruritus and liver disease. Furthermore, UDCA may lower serum bilirubin levels in these patients without improving survival and thus may lead to a delay in liver transplantation. Although UDCA therapy may normalize liver enzymes in many patients, some patients appear to have

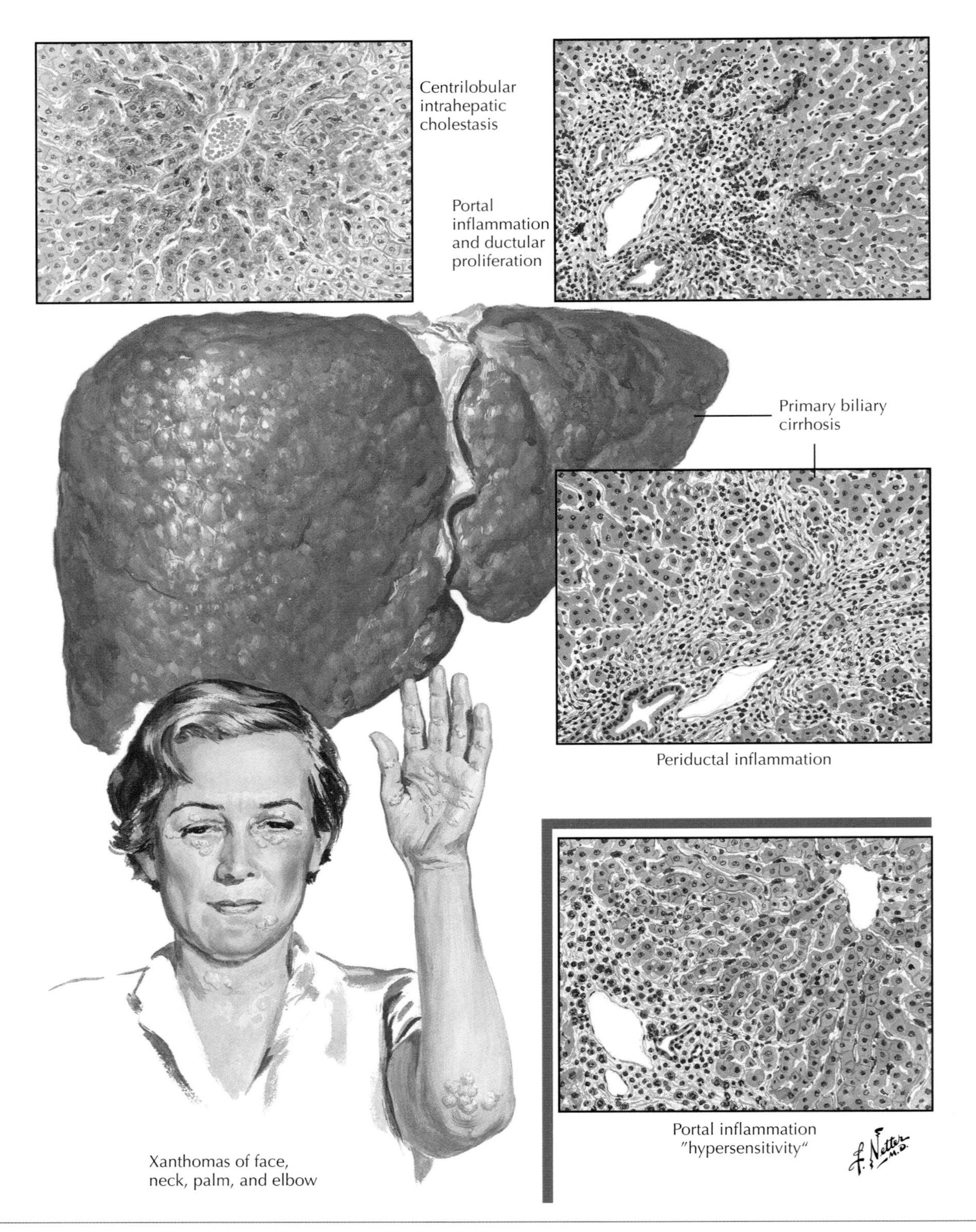

Figure 236-1 *Pathologic Features of Primary Biliary Cirrhosis.*

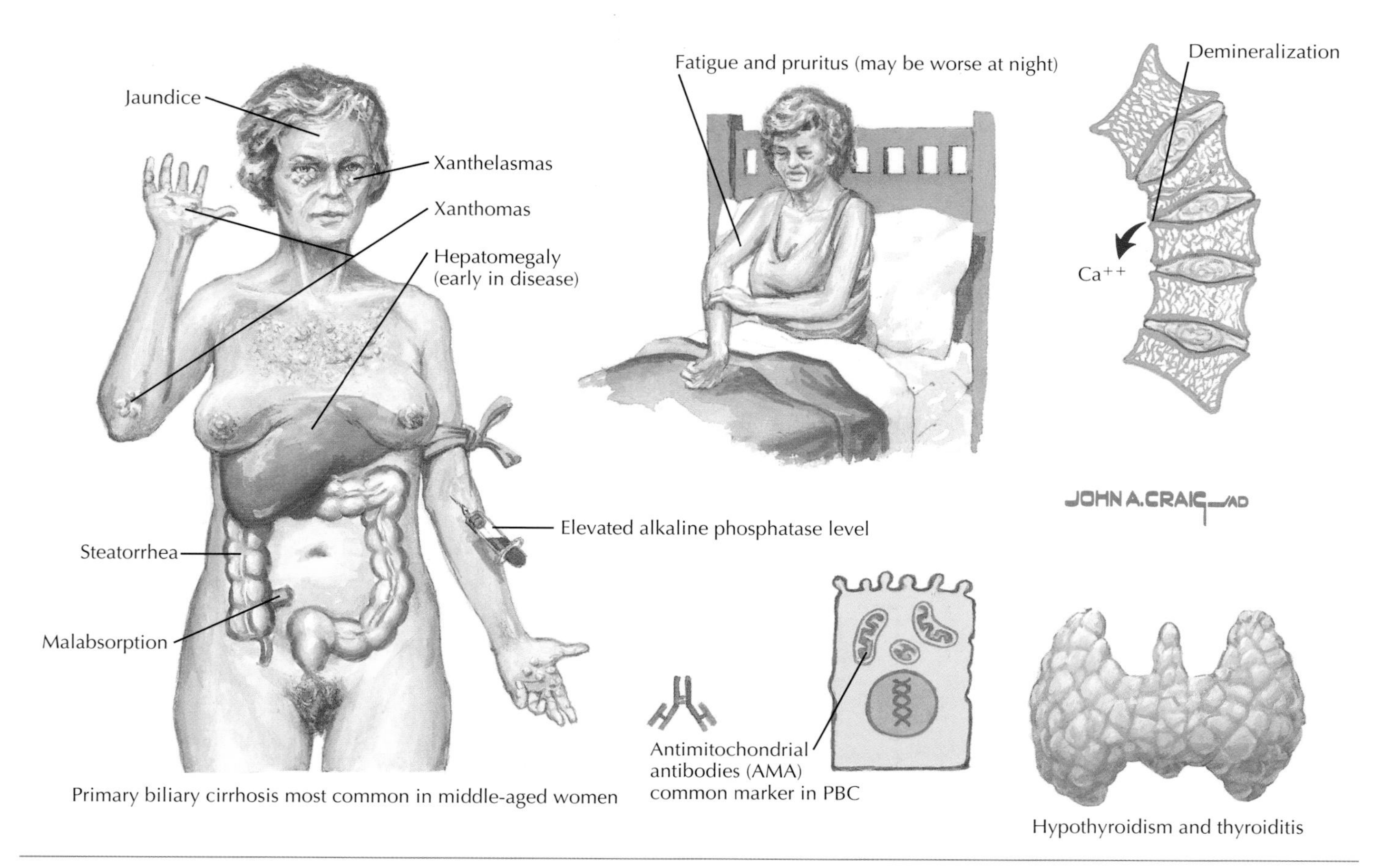

Figure 236-2 *Clinical Features of Primary Biliary Cirrhosis.*

marginal responses. Therefore, additional therapies for PBC are needed.

COURSE AND PROGNOSIS

Survival for patients with PBC is related to the stage of disease. Asymptomatic patients are not at increased risk for death from liver disease. Symptomatic patients are more likely to die of liver disease than are control subjects. One study, however, found that 90% of asymptomatic patients became symptomatic after a median follow-up of 7 years.

A prognostic model called the *Mayo model* has been developed for PBC. For variables, it uses serum albumin, bilirubin, prothrombin time, age, and degree of edema. The Mayo model can then predict survival for a patient with and without liver transplantation. Thus, the Mayo model has been useful for determining the optimal time to place a patient on the liver transplantation list.

ADDITIONAL RESOURCES

Burt AD: Primary biliary cirrhosis and other ductopenic diseases, *Clin Liver Dis* 6:363-380, 2002.

Gong Y, Huang ZB, Christensen E, Gluud C: Ursodeoxycholic acid for primary biliary cirrhosis, *Cochrane Database Syst Rev* (3):CD000551, 2008.

Kowdley KV: Ursodeoxycholic acid therapy in hepatobiliary disease, *Am J Med* 108:481-486, 2000.

Levy C, Lindor KD: Current management of primary biliary cirrhosis and primary sclerosing cholangitis, *J Hepatol* 38(suppl 1):24-37, 2003.

Lleo A, Invernizzi P, Mackay IR, et al: Etiopathogenesis of primary biliary cirrhosis, *World J Gastroenterol* 14(21):3328-3337, 2008.

Talwalkar JA, Lindor KD: Primary biliary cirrhosis, *Lancet* 362:53-61, 2003.

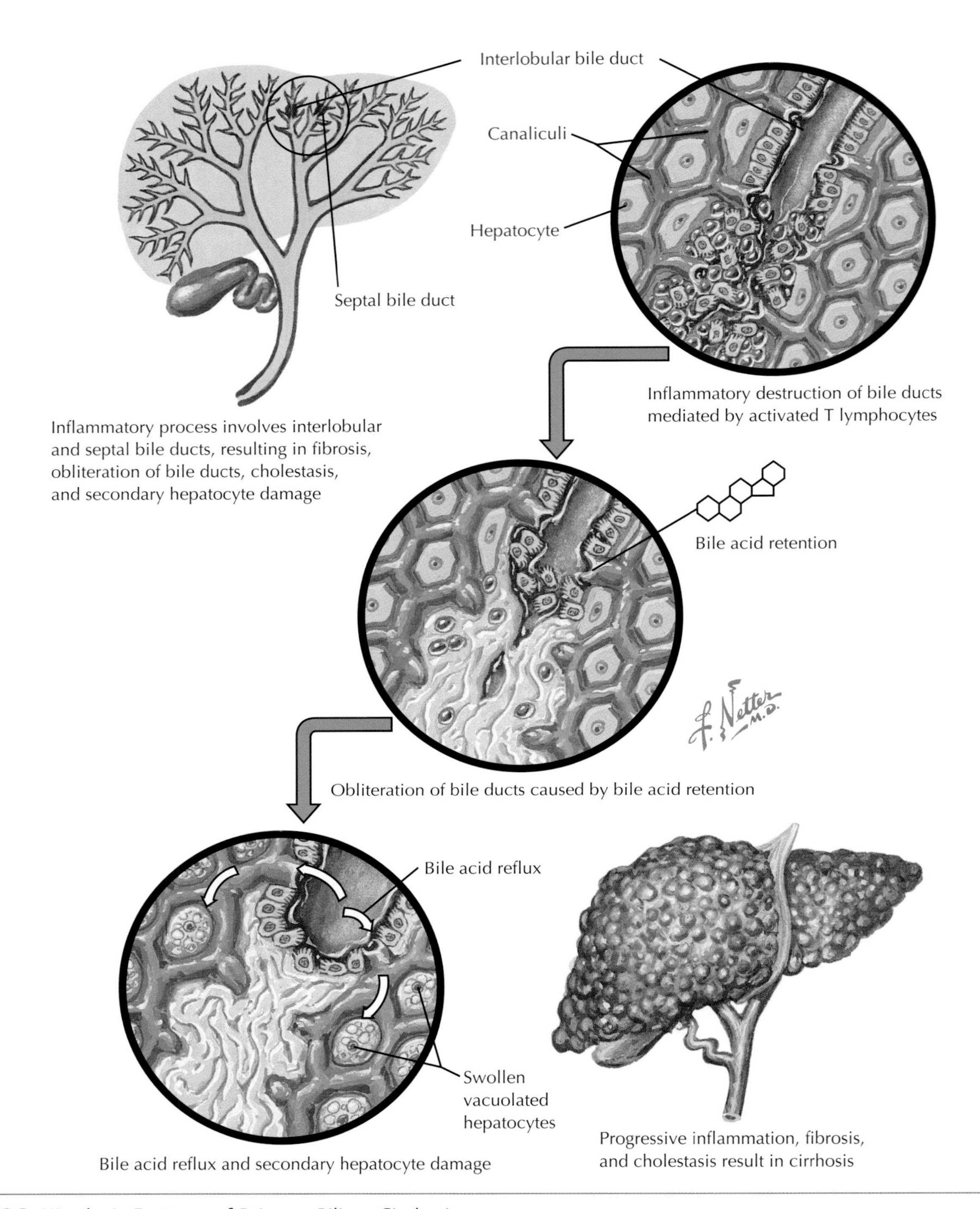

Figure 236-3 *Histologic Features of Primary Biliary Cirrhosis.*

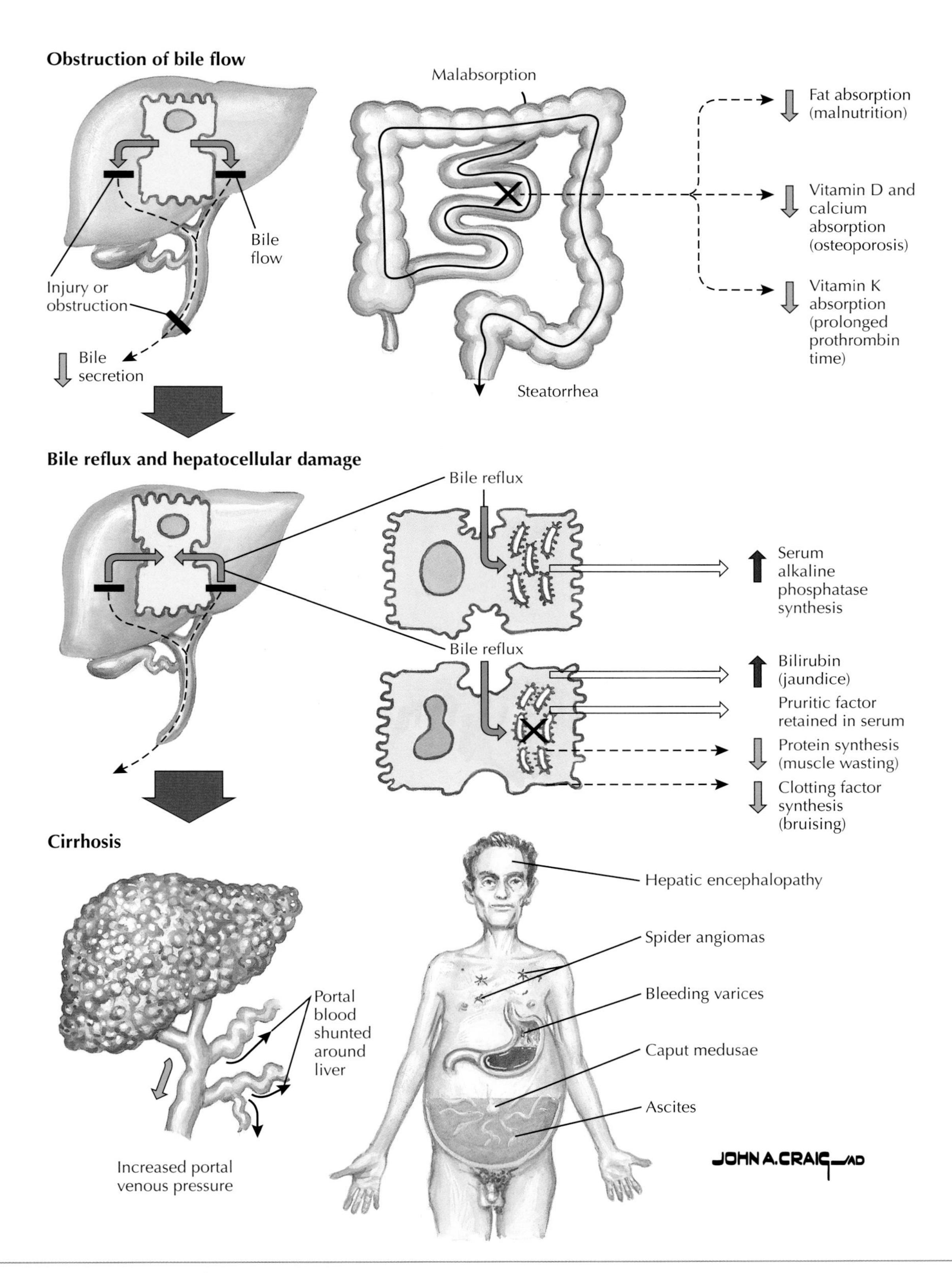

Figure 236-4 *Natural History of Cholestasis.*

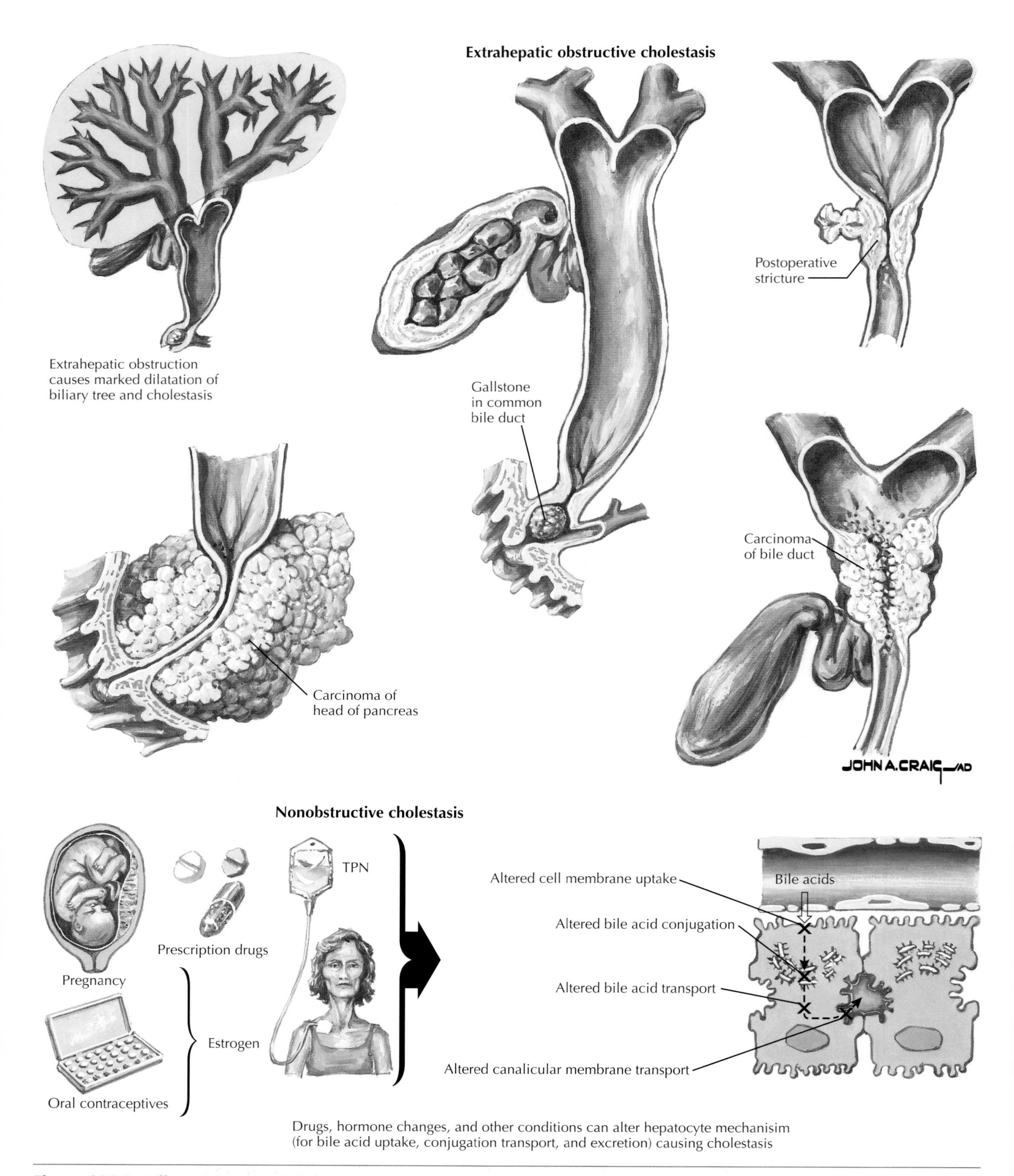

Figure 236-5 *Differential Diagnosis for Primary Biliary Cirrhosis.*

Primary Sclerosing Cholangitis

237

Kris V. Kowdley

Most patients with primary sclerosing cholangitis (PSC) have inflammatory bowel disease (IBD). This has led to speculation that PSC may have an autoimmune basis. Several abnormalities of humoral immunity have been described in PSC, including elevated levels of serum immunoglobulin M (IgM) and IgG, as well as a high prevalence of autoantibodies, such as anti–smooth muscle antibody and antineutrophil cytoplasmic antibody. Cellular immunity is also altered in PSC. Expression of CD4 T cells in the liver is increased in PSC. One study found an antibody (CEP) in patients with PSC that had cross-reactivity with colonic epithelial cells (**Fig. 237-1**).

CLINICAL PICTURE

Evidence shows that genetic predisposition to PSC may influence expression of the disease. HLA DRw52a is more prevalent among patients with PSC. HLA B8 also appears more common among PSC patients than among control subjects. Another possible explanation for the relationship between PSC and *ulcerative colitis* (UC) proposes that increased permeability of the colonic mucosa in PSC leads to bacterial infection of the biliary tree, resulting in chronic cholangitis.

Mutations in *CFTR*, the cystic fibrosis gene, may have a pathogenetic role in PSC. Cholangiographic and histologic features of PSC can resemble those of cystic fibrosis. Recent studies also suggest a relationship between mutations in *CFTR* and sclerosing pancreatitis. Therefore, it is possible that mutations in *CFTR* may cause biliary strictures and lead to the syndrome of PSC.

The exact of prevalence of PSC is unknown. Some studies report that approximately 5% of UC patients have PSC. The estimated prevalence of UC ranges from 40 to 225 cases per 100,000; therefore, 1 to 6 per 100,000 persons in the United States may have PSC. The challenge in deriving population-based estimates is the likelihood that many patients with PSC are asymptomatic. One study from Spain reported that the prevalence of PSC increased from 0.78 to 2.2 cases per 1 million over a 4-year period, 1984 to 1988. Approximately 1 in 6 patients was asymptomatic.

As stated, most patients with PSC have IBD. In one study from a tertiary care referral center, almost 90% of PSC patients were found to have UC on histologic evaluation of rectal biopsy specimens. A significant proportion of the patients did not have symptoms of colitis. Approximately two thirds of PSC patients are men, but among patients without IBD, the male/female ratio is closer to 1:1 (**Fig. 237-2**).

DIAGNOSIS

The standard tool for diagnosing PSC is cholangiography. Usually, endoscopic retrograde cholangiopancreatography (ERCP) is used. Percutaneous cholangiography can be considered if ERCP is unsuccessful, but it should be avoided, if possible, because of the greater technical difficulty and the increased risk for complications in the absence of intrahepatic dilatation. Several recent studies suggest that magnetic resonance cholangiopancreatography (MRCP) may be sensitive for detecting PSC. MRCP should be considered as an alternative to ERCP if this technology is available.

Adequate filling of the intrahepatic ducts at ERCP is necessary before PSC can be excluded based on cholangiography. Cholangiography with an occlusion balloon may improve the quality of the study, but it may be associated with a higher risk for complications. Routine use of parenteral antibiotics is associated with a low rate of infectious complications in patients with PSC undergoing ERCP. A small proportion of patients with PSC may have normal findings on cholangiography. PSC in these patients may be limited to smaller segmental bile ducts that may not be visible on standard fluoroscopic images obtained at cholangiography.

Liver biopsy has an important role for patients with PSC but is not necessary for diagnosis. Characteristic findings of *onion skinning*, or concentric fibrosis around medium-sized bile ducts, may be seen in only 20% of histologic specimens. However, biopsy is useful for determining the presence or absence of cirrhosis and thus may have important prognostic value.

Cholangiocarcinoma and Other Complications

The risk for cholangiocarcinoma is increased in patients with PSC (see Fig. 237-2). The true lifetime incidence of cholangiocarcinoma in PSC patients is unknown but is estimated at 10% to 15%. Unfortunately, no effective screening tests are available for detecting cholangiocarcinoma early, while it is treatable. Serum markers such as carcinoembryonic antigen (CEA) and cancer antigen 19-9 (CA 19-9) have been studied in PSC patients. One study suggested that a CA 19-9 level greater than 100 was highly sensitive and specific for the diagnosis of cholangiocarcinoma in patients with PSC. It is important to remember that cholangitis or cholestasis may lead to elevations in serum CA 19-9 in the absence of cholangiocarcinoma. Patients with PSC are also at higher risk for progression to end-stage liver disease and may need liver transplantation.

In addition, patients with PSC are at increased risk for cholangitis, both to bile stasis associated with stricture and from obstruction caused by sludge or calculi.

TREATMENT AND MANAGEMENT

A number of agents have been used to treat PSC, including corticosteroids and other immunosuppressive agents, methotrexate, and ursodeoxycholic acid (UDCA). In the largest randomized controlled trial, UDCA at 13 to 15 mg/kg/day was associated with a reduction in serum liver biochemistry values, but it did not improve clinical outcomes after up to 6 years of therapy. None of these agents was definitively shown to be effective in PSC. However, recent studies suggest that high-dose UDCA may be effective in PSC; 20 to 25 mg/kg/day was well tolerated and reduced serum liver enzyme levels much

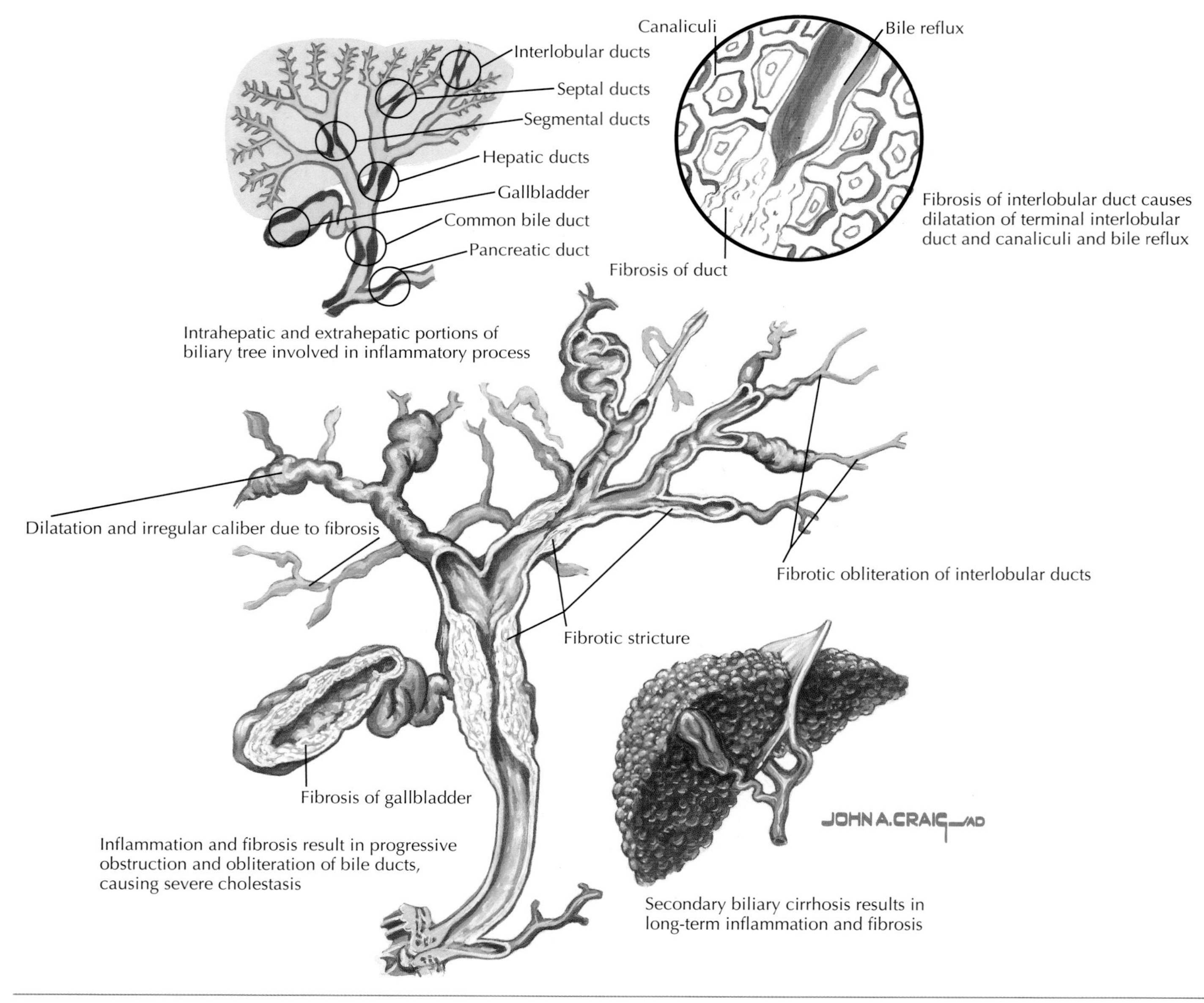

Figure 237-1 *Pathologic Features of Primary Sclerosing Cholangitis.*

more than lower doses. However, high-dose UDCA was not shown to improve survival or prevent cholangiocarcinoma in two large multicenter trials.

COURSE AND PROGNOSIS

Several studies have clearly shown that the risk for colon cancer is increased in patients with PSC and UC compared with patients with UC alone. Therefore, it is appropriate for patients with UC associated with PSC to undergo annual surveillance colonoscopy with multiple biopsies every 10 cm from cecum to rectum. Patients found to have high-grade dysplasia or dysplasia associated with a mass lesion should be offered colectomy. Recent data suggest that UDCA may have a chemopreventive role in reducing the risk for dysplasia in patients with ulcerative colitis and PSC.

Patients with PSC are at increased risk for fat-soluble vitamin deficiency. In one study, up to 25% of PSC patients had low plasma levels of vitamin K_1 (phylloquinone, phytonadione); serum levels of the other fat-soluble vitamins (vitamin A, ^{25}[OH]-vitamin D, vitamin E) were also frequently low. Therefore, it is recommended that serum levels of these vitamins be measured in patients with advanced cholestatic liver disease and that replacements be given if low levels are found. Vitamin A should be replaced with caution because of potential hepatotoxicity. Hyperlipidemia may overestimate vitamin E stores. Adjusting for plasma total lipids has been recommended when assessing vitamin E levels.

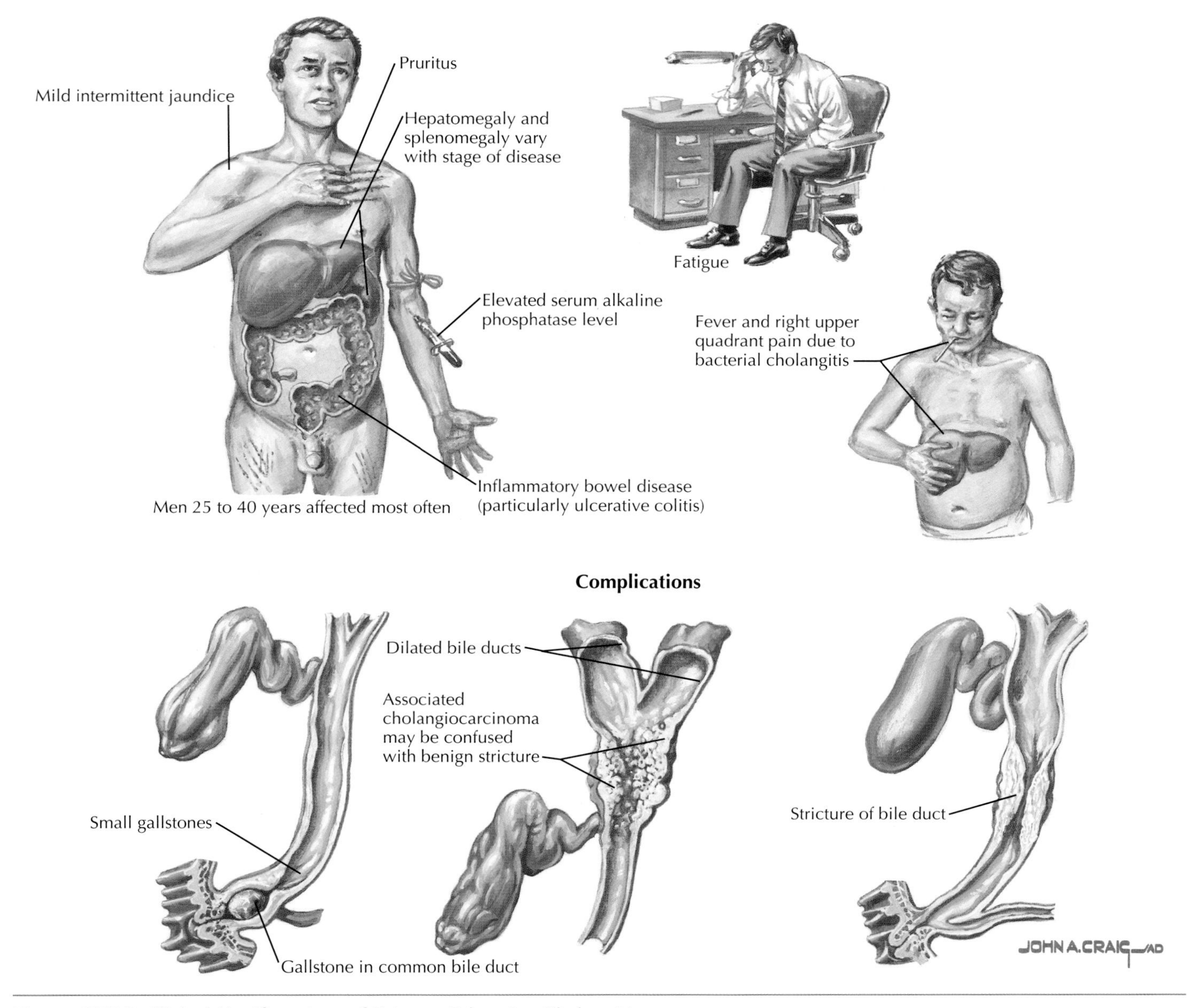

Figure 237-2 *Clinical Manifestations of Primary Sclerosing Cholangitis.*

Metabolic bone disease is also more common among patients with PSC. Although osteomalacia caused by vitamin D deficiency may occasionally be found, osteoporosis is much more common.

ADDITIONAL RESOURCES

Ahmad J, Slivka A: Hepatobiliary disease in inflammatory bowel disease, *Gastroenterol Clin North Am* 31:329-345, 2002.

Chapman RW: The management of primary sclerosing cholangitis, *Curr Gastroenterol Rep* 5:9-17, 2003.

Cullen S, Chapman R: Primary sclerosing cholangitis, *Autoimmun Rev* 2:305-312, 2003.

Cullen SN, Rust C, Fleming K, et al: High dose ursodeoxycholic acid for the treatment of primary sclerosing cholangitis is safe and effective, *J Hepatol* 48(5):792-800, 2008.

Lindor KD, Enders FB, Schmoll JA, et al: Randomized, double-blind controlled trial of high-dose ursodeoxycholic acid (UDCA) for primary sclerosing cholangitis, *Hepatology* 48(4; abstract LB2), 2008.

Olsson R, Boberg KM, de Muckadell OS, et al: High-dose ursodeoxycholic acid in primary sclerosing cholangitis: a 5-year multicenter, randomized, controlled study, *Gastroenterology* 129(5):1464-1472, 2005.

Schrumpf E, Boberg KM: Epidemiology of primary sclerosing cholangitis, *Best Pract Res Clin Gastroenterol* 15:553-562, 2001.

Silveira MG, Lindor KD: Primary sclerosing cholangitis, *Can J Gastroenterol* 22(8):689-698, 2008.

Talwalkar JA, Lindor KD: Natural history and prognostic models in primary sclerosing cholangitis, *Best Pract Res Clin Gastroenterol* 15:563-575, 2001.

Autoimmune Hepatitis

238

Kris V. Kowdley

Autoimmune hepatitis may represent a spectrum of liver disease starting with idiopathic liver disease, or it may be triggered by a number of antigens, including drugs, toxins, and viruses, in patients with an underlying genetic predisposition (**Fig. 238-1**). Autoimmune hepatitis is frequently a disease of young and middle-aged women. Patients often have a history of other autoimmune disease, such as autoimmune thyroid disease. Autoimmune hepatitis may manifest in acute form, with jaundice and marked elevation of serum aminotransferases (transaminases). Other modes of presentation include subfulminant or fulminant hepatitis and chronic hepatitis with or without cirrhosis.

Autoimmune hepatitis has been classified as types 1, 2, and 3. Laboratory features of *type 1* include elevated liver enzymes, associated with hypergammaglobulinemia and autoantibodies at high titer (>1:120). The most common autoantibodies are antinuclear antibodies and anti–smooth muscle antibodies. Other circulating autoantibodies (e.g., those directed against SLA/LP and atypical pANCA) are frequently found in type 1 autoimmune hepatitis. *Type 2* autoimmune hepatitis is associated with antibodies directed against LKM1 and LC-1, whereas liver biopsy for *type 3* reveals features of chronic hepatitis with an active chronic inflammatory infiltrate.

The hallmark of autoimmune hepatitis is *interface hepatitis* and may reveal *bridging necrosis.* The International Autoimmune Hepatitis Study Group has developed a classification system to facilitate the diagnosis of autoimmune hepatitis based on clinical, laboratory, and histologic features.

Seminal studies from the 1970s demonstrated that corticosteroid therapy improves survival among patients with severe autoimmune hepatitis. *Azathioprine* can be used as a steroid-sparing agent and may allow a reduction in steroid dosage. Initial treatment is associated with a high rate of biochemical response, but relapse is common with discontinuation and requires lifelong maintenance treatment. Liver transplantation can be performed in patients with liver failure and is associated with excellent long-term outcomes.

ADDITIONAL RESOURCES

Johnson PJ, McFarlane IG: Meeting report of the International Autoimmune Hepatitis Group, *Hepatology* 18:998-1005, 1993.

Krawitt EL: Autoimmune hepatitis, *N Engl J Med* 354(1):54-66, 2006.

Luxon BA: Autoimmune hepatitis: making sense of all those antibodies, *Postgrad Med* 114:79-82, 85-88, 2003.

McFarlane IG: Definition and classification of autoimmune hepatitis, *Semin Liver Dis* 22:317-324, 2002.

Medina J, Garcia-Buey L, Moreno-Otero R: Immunopathogenetic and therapeutic aspects of autoimmune hepatitis (review), *Aliment Pharmacol Ther* 17:1-16, 2003.

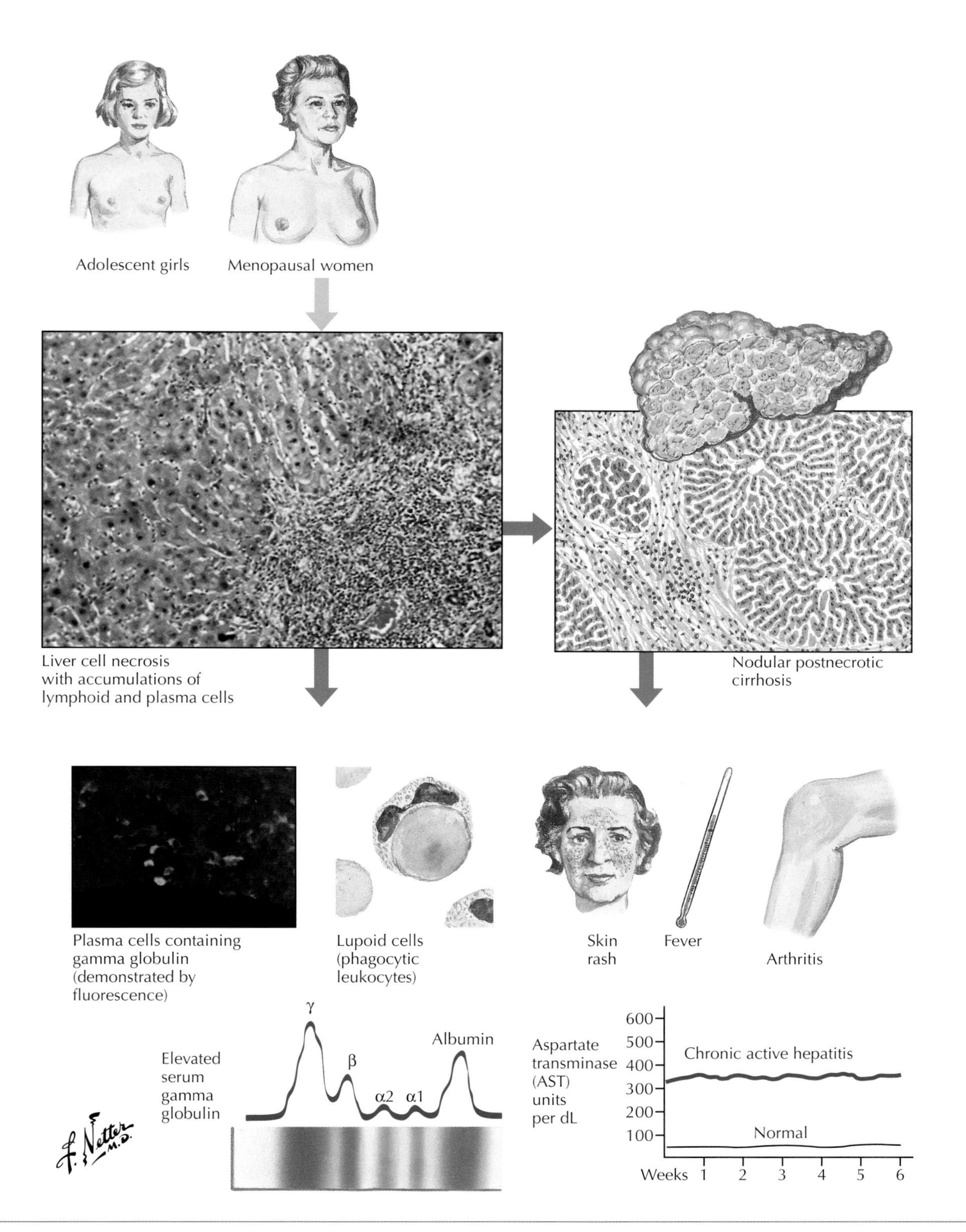

Figure 238-1 *Autoimmune Hepatitis.*

Acute Viral Hepatitis (Hepatitis A, D, E)

239

Kris V. Kowdley

Hepatitis A, hepatitis B (see Chapter 240), and hepatitis E viruses are the most common hepatotropic viruses associated with acute viral hepatitis. Hepatitis D requires hepatitis B to replicate. Other viruses associated with acute viral hepatitis include cytomegalovirus (CMV) and Epstein-Barr virus (EBV), which is less often associated with icteric hepatitis.

HEPATITIS A

Hepatitis A virus (HAV) is phylogenetically distinct from the other hepatitis viruses and belongs to a new genus described as *Hepatovirus.* HAV is a nonenveloped virus capable of surviving in a variety of external environments, such as dried feces and live oysters, for a prolonged time, and it can withstand relatively warm temperatures. Major modes of transmission are fecal-oral and person to person and through contaminated food and water. The incubation period is generally 1 to 2 weeks. Viral shedding in the stool often occurs before symptoms develop.

The typical course of acute HAV and other viral hepatitis infections is shown in **Figures 239-1** (acute form), **239-2** (acute massive necrosis), and **239-3** (subacute fatal form).

Clinical Picture

Acute HAV infection may present clinically in a manner similar to any other form of acute hepatitis. Moderate to marked elevation of serum transaminases (aminotransferases) may be present. The likelihood of jaundice increases with increasing length of exposure. Neonates and children are often asymptomatic. In fact, in many areas of the developing world, a large proportion of the population appears to have been previously exposed to HAV, based on the presence of positive anti-HAV antibody.

Several clinical patterns of HAV infection may be recognized. Acute infection may be completely asymptomatic, especially in very young patients. Many patients have symptomatic acute *icteric* hepatitis and may have all the characteristic symptoms, such as fatigue, lethargy, nausea, abdominal pain, and anorexia. Occasionally, a *cholestatic* variant may be observed, with prolonged jaundice and a highly cholestatic pattern of liver test abnormalities. *Relapsing* hepatitis has been reported in some patients, with apparent remissions and relapses that may last several months. Cholestasis and relapsing hepatitis do not increase mortality risk. In a small subset of patients, *fulminant* hepatitis may occur, necessitating urgent liver transplantation. Aplastic anemia is another rare but serious complication of acute HAV.

The incidence of these serious complications is less than 5%. Mortality risk in patients older than 50 is significantly greater than among younger patients. Chronic hepatitis is not observed among patients with acute HAV infection.

Diagnosis and Management

The diagnosis of acute HAV is established with the detection of immunoglobulin M (IgM) anti-HAV antibody in the serum. IgM anti-HAV antibody appears early in the course of infection and disappears after approximately 4 to 5 months, after which immunoglobulin G (IgG) antibody appears in the serum. IgG anti-HAV antibody is usually detectable years after infection and presumably confers lifelong immunity.

Vaccination against HAV is now widely available. In addition to its necessity for travel to areas where HAV is endemic, vaccination is recommended for all patients with chronic liver disease or immunosuppression and for homosexual men. Serum immunoglobulin injections are used for postexposure prophylaxis and should be given within 2 weeks of exposure.

HEPATITIS D

Hepatitis D virus (HDV) is a defective ribonucleic acid (RNA) virus that requires the presence of hepatitis B virus (HBV) for infectivity. The most common modes of transmission are *coinfection* with HBV in acute infection or *superinfection* in a chronic HBV patient. In Western countries, HDV is observed predominantly in patients who are intravenous (IV) drug users and those who have had multiple blood transfusions. Acute coinfection of HDV is associated with an increased risk of fulminant hepatitis, whereas HDV superinfection in the patient with chronic HBV infection may accelerate HBV progression.

Three common HDV genotypes (1-3) have been described. Interferon therapy has been used with success in patients with chronic HDV infection.

HEPATITIS E

Hepatitis E virus (HEV) is a nonenveloped RNA virus transmitted through the fecal-oral route. Large epidemics have been attributed to HEV in southeast and central Asia, the Middle East, Africa, and Mexico. Outbreaks in Mexico make this form of acute viral hepatitis clinically relevant in the United States.

Hepatitis E may be more common in developed countries than previously estimated, and HEV may be transmitted from other animals, particularly pigs. The common modes of transmission in endemic countries is usually contaminated water or water supply, although vertical transmission has been associated with acute, severe hepatitis in neonates in the the presence of HEV RNA. The incubation period may be as short as 2 weeks,

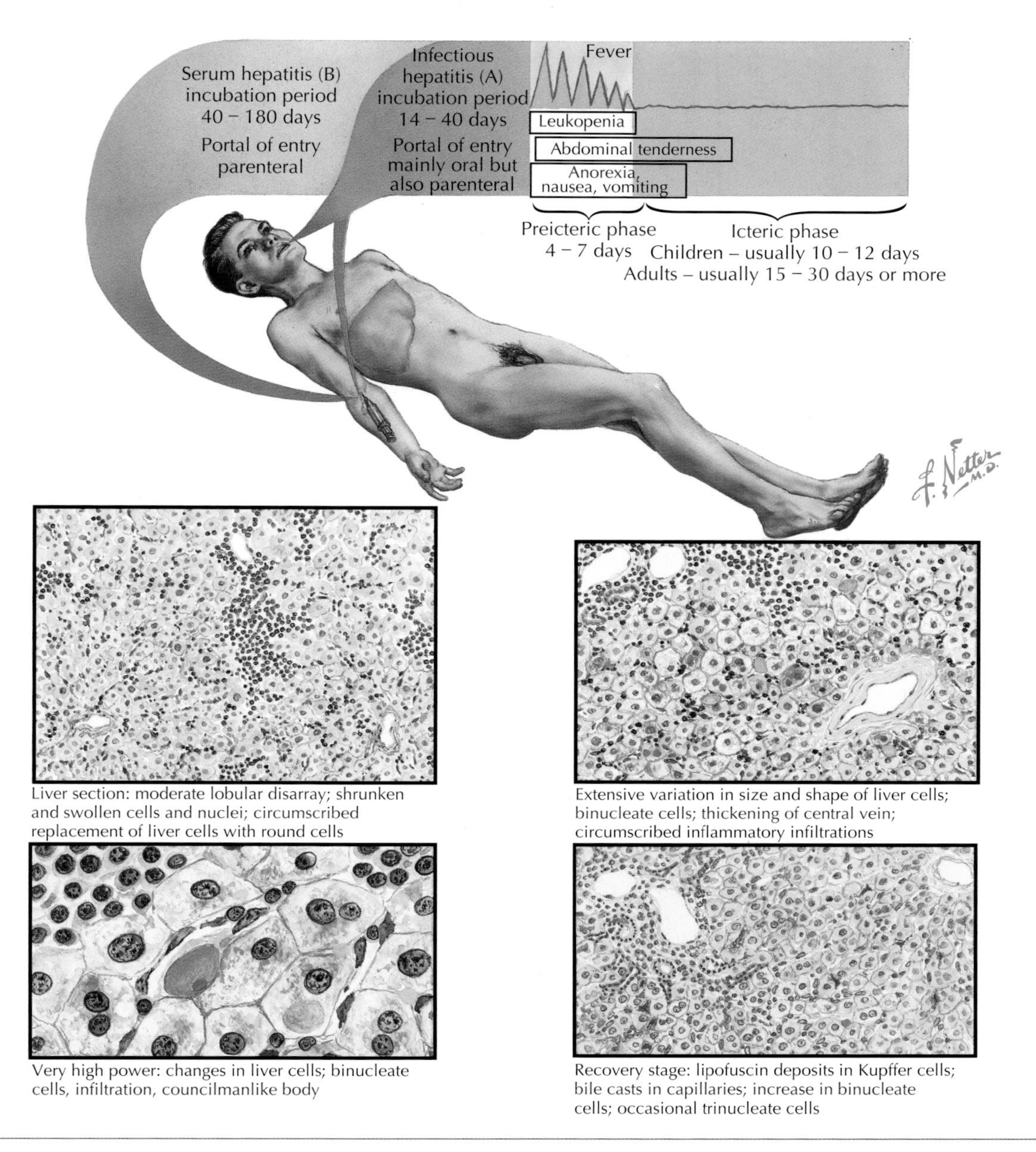

Figure 239-1 *Viral Hepatitis: Acute Form.*

but the average is 6 weeks. Viral shedding begins 1 week before the onset of symptoms and last 2 to 3 weeks.

Clinical Picture

The clinical features of HEV infection are distinct. Some have reported that gastrointestinal symptoms such as diarrhea are more common than with other causes of acute viral hepatitis. Young adults appear to be at greatest risk for clinical disease. The case-fatality ratio is particularly high among pregnant women, especially in the third trimester, with mortality rates as high as 25%. As with HAV infection, chronic disease does not occur, and HEV patients who recover do not develop chronic liver disease.

Diagnosis and Management

The diagnosis of HEV is made by detecting anti-HEV antibody in the serum. As with HAV, there is initially an IgM antibody to HEV that disappears over 4 to 5 months and is replaced by IgG antibody. Measuring HEV RNA in serum is possible and may be helpful early in the course of infection or in immunosuppressed persons.

No vaccines against HEV are available; therefore, the best form of prevention is to avoid possibly contaminated water.

ADDITIONAL RESOURCES

Dalton HR, Bendall R, Ijaz S, Banks M: Hepatitis E: an emerging infection in developed countries, *Lancet Infect Dis* 8(11):698-709, 2008.

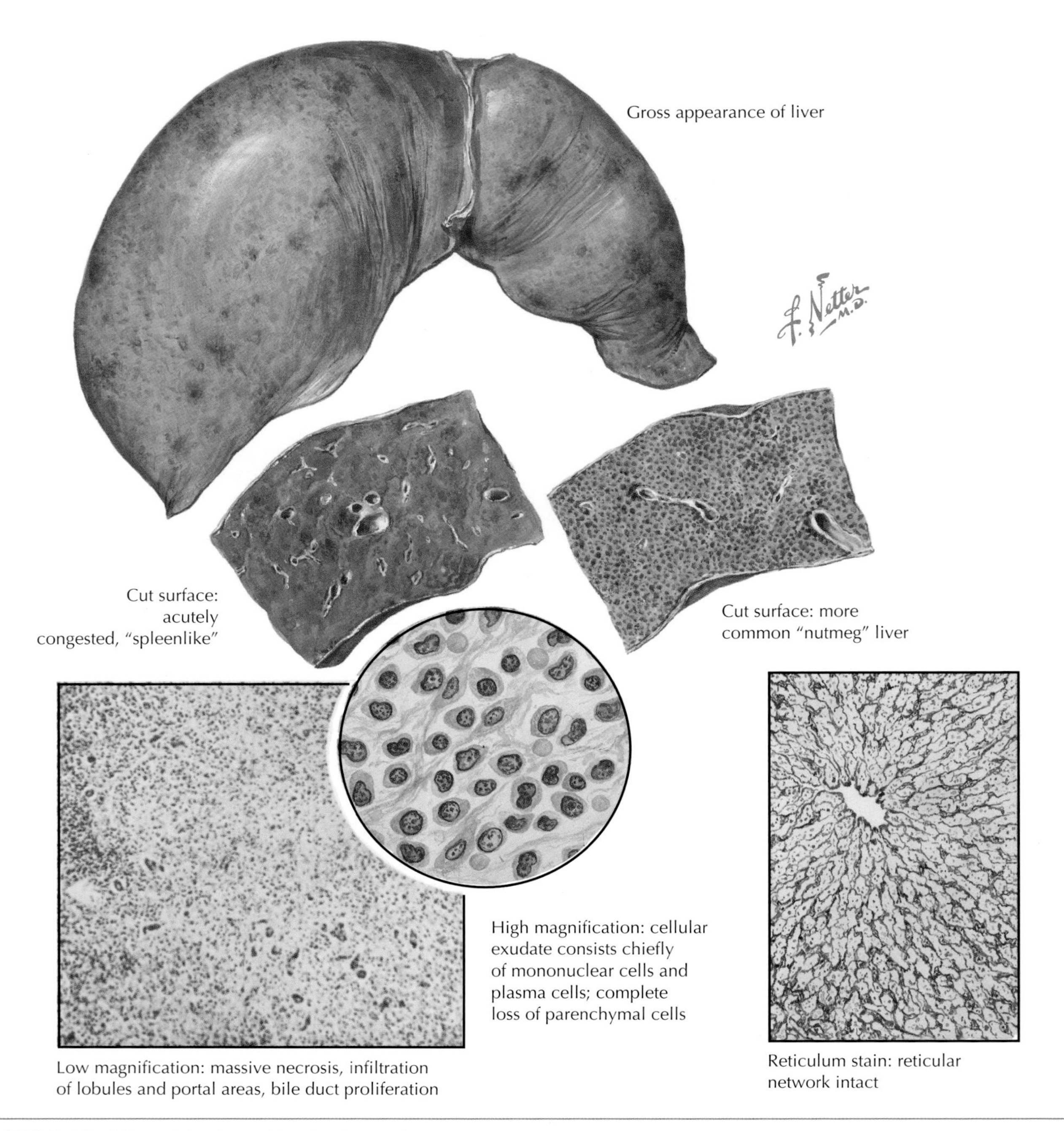

Figure 239-2 *Viral Hepatitis: Acute Massive Necrosis.*

Dény P: Hepatitis delta virus genetic variability: from genotypes I, II, III to eight major clades? *Curr Top Microbiol Immunol* 307:151-171, 2006.

Farci P, Chessa L, Balestrieri C, et al: Treatment of chronic hepatitis D, *J Viral Hepat* 14(suppl 1):58-63, 2007.

Hsieh TH, Liu CJ, Chen DS, Chen PJ: Natural course and treatment of hepatitis D virus infection, *J Formos Med Assoc* 105(11):869-881, 2006.

Purcell RH, Emerson SU: Hepatitis E: an emerging awareness of an old disease, *J Hepatol* 48(3):494-503, 2008.

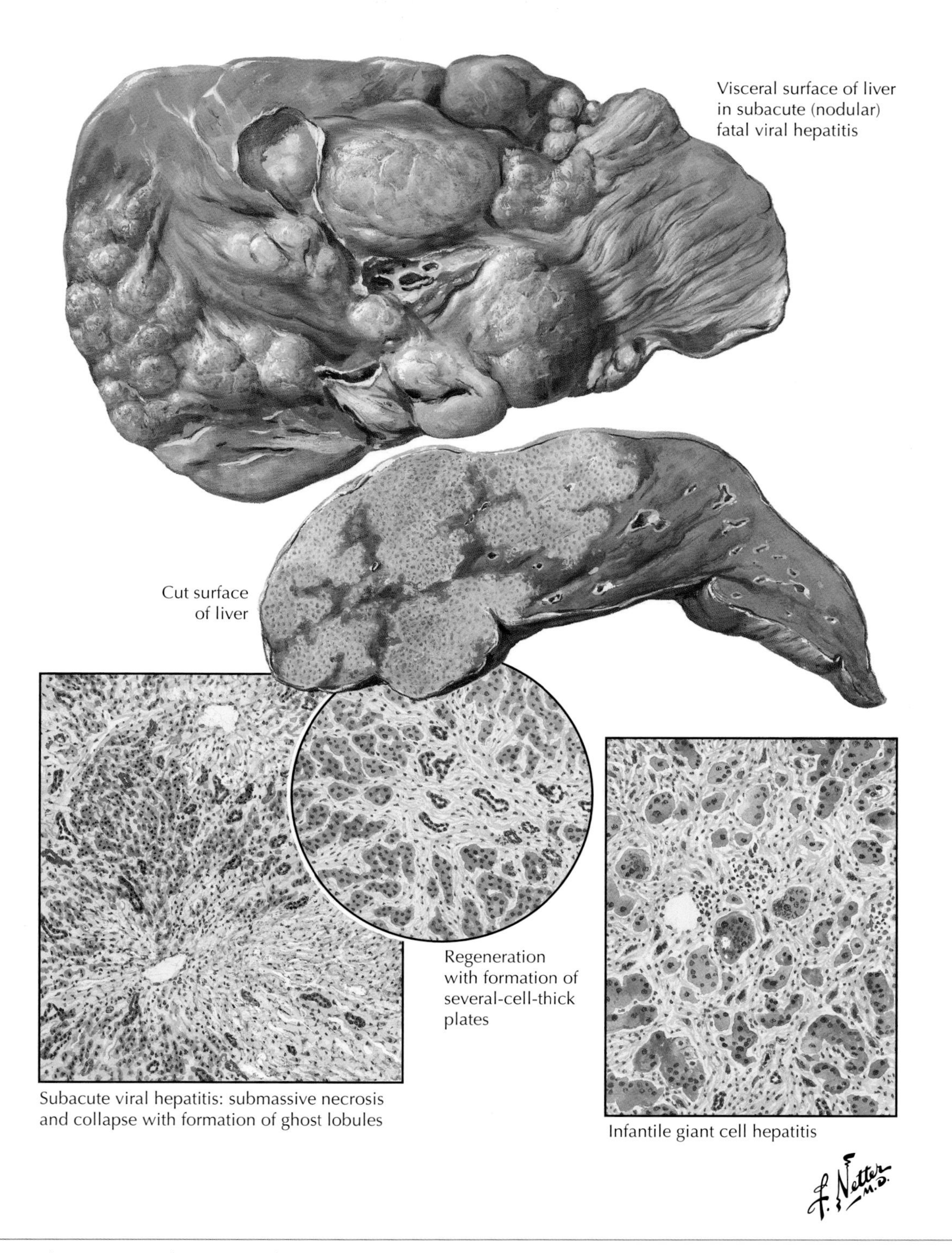

Figure 239-3 *Viral Hepatitis: Subacute Fatal Form.*

Hepatitis B

240

Kris V. Kowdley

Hepatitis B virus (HBV) is an enveloped, double-strand DNA virus that can cause acute and chronic hepatitis. In certain populations, including Southeast Asians, Alaskan natives, and sub-Saharan Africans, in areas where HBV is highly endemic, the prevalence of chronic hepatitis B is as high as 20%. In these populations, the virus is transmitted primarily through the maternal-neonatal route, and infection usually develops during infancy or early childhood. Most infected persons acquire chronic infection.

In parts of the world with low endemicity, including the United States, Canada, and Western Europe, HBV transmission occurs primarily through sexual contact in early adulthood. In this population, the clearance of hepatitis B surface antigen (HBsAg) and the development of immunity to HBV follow episodes of acute HBV infection. Fewer than 5% of these patients acquire chronic hepatitis B. However, a small proportion of patients with acute hepatitis B develops fulminant hepatitis and must undergo emergency liver transplantation.

CLINICAL PICTURE

The clinical features of acute hepatitis B are variable. Most patients do not have jaundice, and infection may be subclinical. Patients with neonatally acquired infection may have a long immunotolerant phase in the first several decades of life associated with high levels of HBV viremia but normal liver enzyme levels and minimal changes on biopsy. Over time, exacerbations and remissions of viral replication may be associated with elevated liver enzymes and symptoms, although symptoms occasionally may be absent or minimal.

Major long-term sequelae include cirrhosis and hepatocellular carcinoma (HCC). The risk for HCC is greatest in patients infected early in life with ongoing viral replication. In contrast to other types of liver disease, cirrhosis is not a prerequisite for development of HCC in chronic hepatitis B, although the risk is higher in patients with cirrhosis. Therefore, some experts suggest that HCC screening is appropriate for patients with or without cirrhosis and should be conducted after two or more decades of infection.

Extrahepatic Manifestations

Several extrahepatic manifestations of chronic hepatitis B have been reported, including vasculitis (particularly polyarteritis nodosa), glomerulonephritis, and essential mixed cryoglobulinemia.

DIAGNOSIS

The discovery of the so-called Australia antigen by Blumberg and colleagues in 1965 was a major advance in establishing a serologic diagnosis of acute and chronic hepatitis B. This test detects the HBsAg. Subsequently, the antibody to HBsAg and the antibody to the hepatitis B core antigen (HBcAb) and hepatitis B e antigen (HBeAg, HBeAb) were discovered. A panel of these tests now allows the clinician to identify acute versus chronic HBV infection, to assess the degree of viral replication in patients with chronic infection, and to determine prior HBV exposure and subsequent immunity (**Fig. 240-1**).

In addition, the widespread availability of tests for quantitative measurement of hepatitis B DNA using polymerase chain reaction (PCR)–based techniques also allows for the detection of active viremia. These can identify patients with active viral replication (serum HBV DNA–positive titer, usually >100,000 copies/mL, or >20,000 IU/mL), patients with HBeAg-

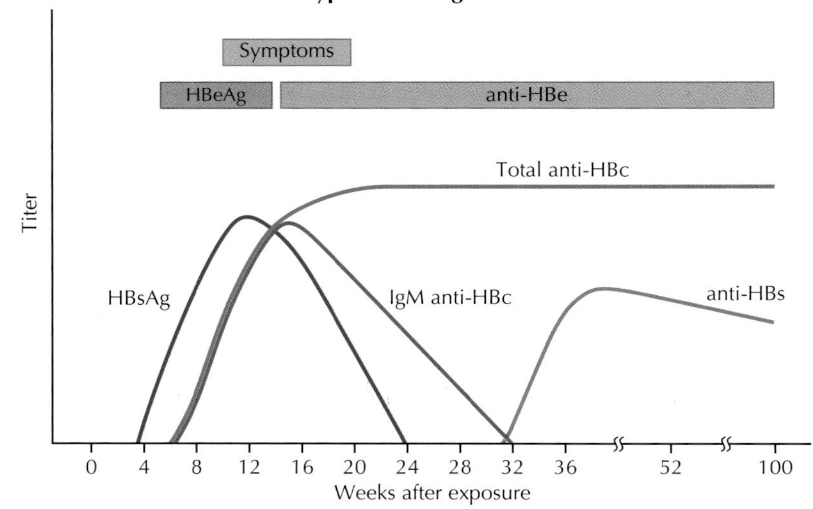

Note: Serologic markers of HBV infection vary depending on whether the infection is acute or chronic.

The first serologic marker to appear following acute infection is HBsAg, which can be detected as early as 1 or 2 weeks and as late as 11 or 12 weeks (mode, 30-60 days) after exposure to HBV. In persons who recover, HBsAg is no longer detectable in serum after an average period of about 3 months. HBeAg is generally detectable in patients with acute infection; the presence of HBeAg in serum correlates with higer titers of HBV and greater infectivity. A diagnosis of acute HBV infection can be made based on the detection of IgM class antibody to hepatitis B core antigen (IgM anti-HBc) in serum; IgM anti-HBc is generally detectable at the time of clinical onset and declines to sub-detectable levels within 6 months. IgG anti-HBc persists indefinitely as a marker of past infection. Anti-HBs becomes detectable during convalescence after the disappearance of HBsAg in patients who do not progress to chronic infection. The presence of anti-HBs following acute infection generally indicates recovery and immunity from reinfection.

Source: Centers for Disease Control.

Figure 240-1 *Typical Serologic Course of Acute Hepatitis B with Recovery.*

positive chronic hepatitis B (>10,000 copies/mL, or >2000 IU/mL), and patients with HBeAg-negative chronic hepatitis B, as well as patients with nonreplicative status.

The terms "chronic active hepatitis" and "chronic carrier" or "asymptomatic carrier" are no longer used. Similarly, "chronic persistent" and "chronic active" hepatitis are not currently used to classify patients. Rather, it is more appropriate to use *chronic hepatitis B* to describe patients who are chronically infected (i.e., HBsAg positive >6 months). Chronic hepatitis B with evidence of persistent HBV DNA in the serum is described as *replicative* ("chronic active hepatitis"), whereas chronic hepatitis B that is persistently HBV DNA negative (very low levels in serum using PCR-based assay) is classified as *nonreplicative* ("carrier hepatitis").

Thus, the most useful initial test to screen for chronic hepatitis B is the HBsAg. Patients who are HBsAg positive should have the titer of HBV DNA in the serum measured, preferably using a PCR-based assay. Patients with HBV DNA levels greater than 2000 IU/mL have replicative disease, although this cutoff level is arbitrary. The role of the HBeAg is to classify patients who are in the replicative phase as patients with a *wild-type* (HBeAg positive) or *precore mutant* profile. Most patients with high levels of HBV DNA in the blood but without measurable levels of HBeAg are thought to have a precore mutant variant of HBV (associated with stop codon in precore region of genome) and thus an inability to transcribe and translate the e protein. Alternatively, patients with HBeAg-negative chronic hepatitis B may have mutations in the basal core promoter, which downregulates eAg production.

Histologic findings in chronic hepatitis B may resemble any other type of chronic hepatitis, with features of hepatic necroinflammation, interface hepatitis, and variable amounts of fibrosis. Unique to HBV may be the finding of ground-glass hepatocytes, which represent HBsAg. Histologic scoring systems used to grade and stage chronic hepatitis B include the Knodell Histologic Activity Index and the Batts-Ludwig, Ishak, and METAVIR systems.

TREATMENT AND MANAGEMENT

The goal of treatment for patients with chronic hepatitis B is to suppress viral replication, enable seroconversion from the HBeAg-positive to the HBeAb-positive state, to normalize serum liver enzymes, and to improve or stabilize histology (**Fig. 240-2**). Seven treatments are currently approved for chronic hepatitis B: interferon-α (IFN-α), pegylated IFN alpha-2a (PEG-IFN), lamivudine, adefovir, entecavir, telbivudine, and tenofovir.

Interferon-α is given subcutaneously at 5 million units daily or 10 million units three times a week. PEG-IFN is given at 180 μg subcutaneously once a week for 48 weeks. The advantages of IFN therapy are the finite duration of therapy and the lack of resistance. However, IFN is associated with many adverse effects, such as flulike symptoms and bone marrow suppression. Patients frequently experience a "flare" of hepatitis during IFN therapy, which may be poorly tolerated by those with cirrhosis. Therefore, IFN should be avoided in patients with cirrhosis.

Lamivudine was the first oral nucleoside analog to be approved for treatment of chronic hepatitis B. Lamivudine is safe and effective in lowering serum HBV DNA levels and improving histologic abnormalities. However, resistance to lamivudine is increasingly observed with longer duration of therapy (30%-40% after 2 years; up to 70% after 5 years) and is associated with loss of therapeutic benefit over time. *Telbivudine* is another nucleoside analog that was shown to be superior to lamivudine. The limitation of telbivudine is the risk of resistance in patients who do not achieve complete viral suppression in the first several months of therapy. *Adefovir*, the second oral therapy approved for hepatitis B, is effective in patients with lamivudine resistance but is also associated with increased resistance after several years of therapy. *Tenofovir* and *entecavir* are potent oral agents with much lower resistance profiles and are safe and effective for long-term therapy.

Liver transplantation has been performed for end-stage liver disease caused by hepatitis B. Early reports suggested a high rate

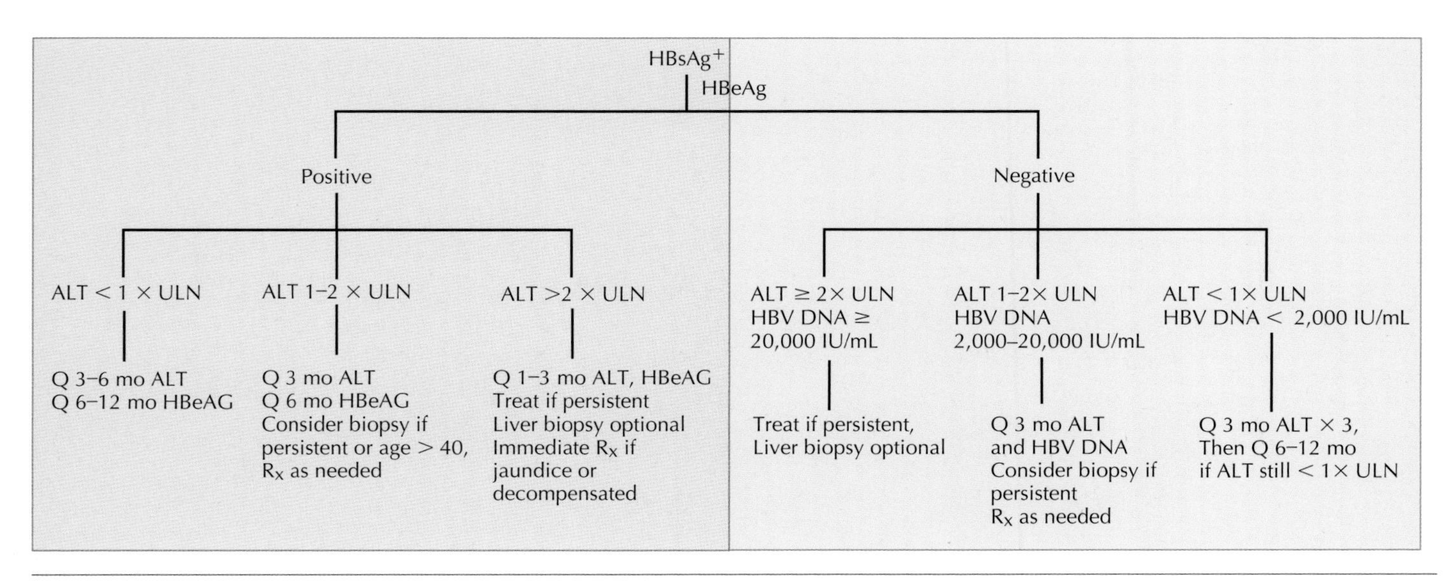

Figure 240-2 *Management of Chronic Hepatitis B Virus.*

of recurrence of hepatitis and associated graft failure and decreased survival. The use of high-dose parenteral hepatitis B immune globulin and, more recently, the addition of antiviral therapy have greatly reduced the risk for recurrent hepatitis and improved graft and patient survival.

ADDITIONAL RESOURCES

Alter MJ: Epidemiology of hepatitis B in Europe and worldwide, *J Hepatol* 39(suppl 1):64-69, 2003.

Ayoub WS, Keeffe EB: Current antiviral therapy of chronic hepatitis B (review), *Aliment Pharmacol Ther* 28(2):167-177, 2008.

De Franchis R, Hadengue A, Lau G, et al: EASE International Consensus Conference on Hepatitis B (2002, Geneva): consensus statement (long version), *J Hepatol* 39(suppl 1):3-25, 2003.

Fattovich G: Natural history of hepatitis B, *J Hepatol* 39(suppl 1):50-58, 2003.

Lai CL, Ratziu V, Yuen MF, Poynard T: Viral hepatitis B, *Lancet* 362: 2089-2094, 2003.

Singh NA, Reau N: Management of hepatitis B virus, *J Antimicrob Chemother* 62(2):224-228, 2008.

Hepatitis C

241

Kris V. Kowdley

Chronic hepatitis C was previously described as "non-A, non-B" hepatitis and was found to be transmitted through contaminated blood or blood products. This agent was first recognized in the mid-1970s, but an antibody to the virus was not identified until 1989. Since that time, it has become clear that hepatitis C is the major cause of non-A, non-B posttransfusion hepatitis.

Most patients infected with the hepatitis C virus (HCV) develop chronic hepatitis; 65% to 90% have been reported to have chronic hepatitis after acute exposure. Controversy surrounds the natural history of hepatitis C. Some studies indicate that a high percentage of patients may develop the major complication of cirrhosis, end-stage liver disease, or hepatocellular carcinoma (HCC). On the other hand, studies of patients who acquired hepatitis C after blood transfusion failed to show a significant increase in mortality in HCV-infected patients compared with control subjects.

CLINICAL PICTURE

Most patients with HCV have abnormal liver test findings, including alanine transaminase (ALT) and aspartate transaminase (AST). One third of patients are estimated to have normal liver test findings, and only about 20% of patients develop symptoms. Liver biopsy plays an important role in diagnosis and prognosis because histologic examination at diagnosis is the best predictor of progression. Overall, 20% of patients develop cirrhosis by 20 years. The risk for HCC is estimated at 1% to 5% per year after 20 years of disease, or 1% to 4% per year in patients with cirrhosis (**Fig 241-1**).

DIAGNOSIS

Chronic hepatitis C should be considered in any patient with elevated liver enzymes (ALT, AST). Hepatitis C should also be considered in patients at significant risk for HCV infection, such as blood transfusion before 1990, intravenous (IV) drug use, multiple sex partners, or multiple tattoos or skin piercings. Chronic hemodialysis is also a risk factor for hepatitis C, and 20% to 40% of patients receiving chronic hemodialysis may be infected with HCV.

The most widely used diagnostic test for hepatitis C is a second- or third-generation *enzyme-linked immunosorbent assay* (ELISA). *Recombinant immunoblot assay* is used to confirm ELISA results. The sensitivity of second-generation ELISA is estimated at 92% to 95%; the positive predictive value is 25% to 60%. If the patient has positive findings by immunoblot assay, the positive predictive value is 70% to 75%. Based on these data, the National Institutes of Health (NIH) Consensus Conference recommends that for patients at low risk for hepatitis C, recombinant immunoblot assay should be performed to confirm positive findings on ELISA. Conversely, immunoblot should not be performed in a patient with a well-established risk factor for hepatitis C, because such patients typically are positive for HCV ribonucleic acid (RNA) in the serum. For patients at low risk by history and with elevated ALT levels, either immunoblot or polymerase chain reaction (PCR)–based assay for direct measurement of HCV RNA in the serum should be used.

After exposure to hepatitis C, HCV RNA is present in the serum 1 to 3 weeks later in patients who experience seroconversion; the serum ALT is elevated by 50 days. However, only 25% to 35% of patients are estimated to have symptoms after acute hepatitis C infection. About 90% of patients have antibodies to hepatitis C by 3 months. As noted, only about 15% of patients have self-limited acute hepatitis C, and most develop chronic hepatitis C.

TREATMENT AND MANAGEMENT

Candidates for therapy of hepatitis C include patients with elevated serum ALT levels, HCV RNA in the serum, and liver biopsy showing fibrosis or significant necroinflammatory changes. Currently, synthetic interferon (IFN alpha-2a, IFN alpha-2b, IFN con-1) and combination therapy (IFN alpha-2b plus ribavirin) are approved for the treatment of chronic hepatitis C. A major advance has been the development of pegylated interferons such as IFN alpha-2a (Pegasys; Roche) and IFN alpha-2b (Peg-Intron; Schering-Plough). *Pegylation* involves the addition of a large polyethylene glycol residue to the interferon moiety. This results in significantly delayed clearance of the drug, leading to maintenance of high plasma concentrations for up to 10 days. Pegylated interferon is administered subcutaneously once a week.

Sustained virologic response rates (likelihood of remaining virus free 6 months after therapy) have improved dramatically with pegylated interferon and ribavirin. The best predictors of response with interferon and ribavirin include non–genotype 1 infection (specifically, genotype 2 or 3), absence of cirrhosis, low viral titer (<200,000 viral particles/mL), and low body weight. Several new oral agents, including the protease inhibitors *telaprevir* and *boceprevir*, are currently in late-stage clinical trials as part of combination therapy regimens with pegylated interferon and ribavirin.

Extrahepatic manifestations of hepatitis C are multiple and include porphyria cutanea tarda, focal lymphocytic sialadenitis, Mooren ulcers, type 2 cryoglobulinemia, membranoproliferative glomerulonephritis, and a variety of other autoantibodies. Treatment may occasionally be considered in these patients even if their liver disease is relatively mild, or if they have significant symptoms or morbidity from these extrahepatic manifestations.

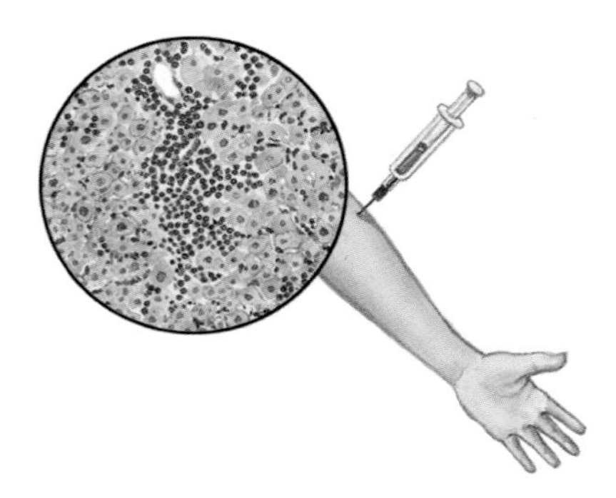

Most patients with HCV have abnormal liver (ALT, AST) findings.

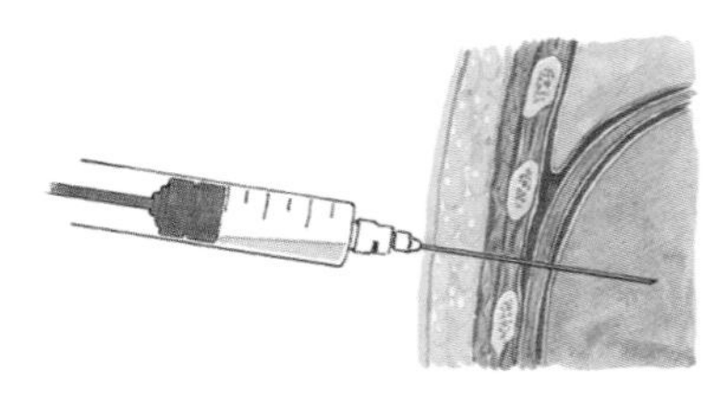

Liver biopsy plays an important role in diagnosis and prognosis.

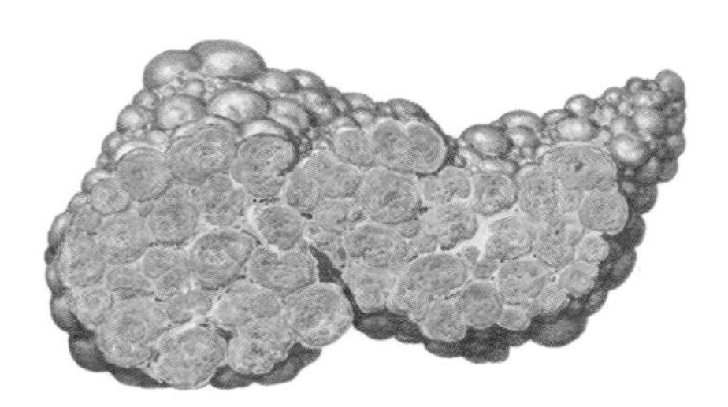

Twenty percent of patients develop cirrhosis by 20 years of age.

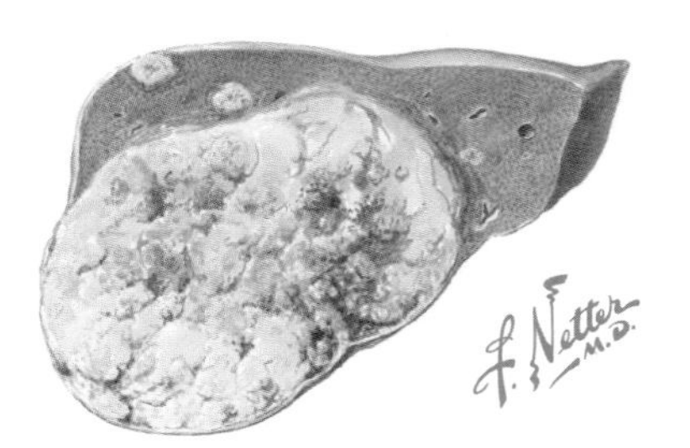

The risk for hepatocellular carcinoma is estimated to be 1% to 5% per year after 20 years of disease or 1% to 4% per year in patients with cirrhosis.

Figure 241-1 *Clinical Picture of Hepatitis C Infection.*

ADDITIONAL RESOURCES

Carey W: Tests and screening strategies for the diagnosis of hepatitis C, *Cleve Clin J Med* 70(suppl 4):7-13, 2003.

Jou JH, Muir AJ: In the clinic: hepatitis C, *Ann Intern Med* 148(11; ITC6): 1-16, 2008.

Kronenberger B, Welsch C, Forestier N, Zeuzem S: Novel hepatitis C drugs in current trials, *Clin Liver Dis* 12(3):529-555, viii, 2008.

National Institutes of Health: Consensus statement on management of hepatitis C, 19:1-46, 2002.

Poynard T, Yuen MF, Ratziu V, Lai CL: Viral hepatitis C, *Lancet* 362: 2095-2100, 2003.

Hepatitis Caused by Other Viruses

242

Kris V. Kowdley

INFECTIOUS MONONUCLEOSIS

Infectious mononucleosis is caused by the *Epstein-Barr virus* (EBV). This virus has been associated with a number of diseases involving B cells, including Burkitt lymphoma, nasopharyngeal carcinoma, and posttransplantation lymphoproliferative disorder.

Acute clinical features of infectious mononucleosis include fever, lymphadenopathy, severe pharyngitis, and atypical lymphocytosis (**Fig. 242-1**). Results of the heterophil antibody test are usually positive. Hepatic involvement resembles other causes of nonfatal acute viral hepatitis; jaundice is rare. Hepatosplenomegaly may be present in approximately 20% of patients. Serious complications include splenic rupture, meningitis, and pericarditis.

Liver histology findings may be similar to those in other forms of acute hepatitis. Portal and lobular inflammatory infiltrates, hepatocellular necrosis and acidophilic bodies, sinusoidal infiltration by monocytes, and atypical lymphocytes have been described.

The diagnosis of EBV infection is made by serologic testing for the heterophil antibody, agglutinating sheep cells, and immunoglobulin M (IgM) and IgG antibody to EBV. However, serologic diagnosis may be difficult because the presence of antibody may indicate prior exposure rather than acute infection, given the ubiquitous nature of EBV, and heterophil antibody may be negative early in infection. Direct measurement of EBV viral DNA is used in immunosuppressed patients, and antiviral prophylaxis is being considered.

Posttransplantation lymphoproliferative disorder is a serious complication that must be treated with a reduction in immunosuppression and, occasionally, with chemotherapy. EBV infection has also been associated with malignant transformation of lymphocytes, resulting in lymphomas.

YELLOW FEVER

Yellow fever is produced by a *Flavivirus* present in the blood in the first 3 days of the disease. The *urban* form is transmitted from person to person by the female mosquito, *Aedes aegypti* (in the jungles, white monkeys may serve as intermediate hosts). The mosquito becomes infective approximately 12 days after it has fed on infected blood. In humans, after an incubation period of 3 to 6 days, a self-limited disease develops, with headache, backache, photophobia, and some gastrointestinal symptoms, or a severe toxic condition evolves, usually with jaundice, that may be characterized by high fever, tachycardia, jaundice, altered mental status, oliguria, and possible multiorgan failure. This *toxic* form has a high mortality rate. The liver is large and is often yellow. Histologic changes are characteristic; hepatocellular degeneration develops and is associated with fatty metamorphosis and midzonal necrosis. Councilman bodies are often present. The incidence of yellow fever has been significantly reduced by eradication of the *Aedes* mosquito and by use of a live attenuated vaccine that confers long-lasting immunity.

CYTOMEGALOVIRUS INFECTION

Cytomegalovirus (CMV) is a member of a group of herpesviruses that produce a range of diseases, depending on a person's age and level of immune competence and the mode of exposure. The virus is present everywhere, and the risk for seropositivity increases with increasing age.

Although most infections are asymptomatic, CMV *hepatitis* can be associated with significant morbidity in the neonatal period. It is accompanied by jaundice, hepatosplenomegaly, central nervous system (CNS) involvement, and hemolytic anemia. In adults, CMV may cause acute hepatitis, which is indistinguishable from other causes of hepatitis. CMV viremia and CMV disease are important clinical issues in transplantation. The virus can be transmitted with the graft; a CMV-negative patient who receives a CMV-positive organ is at very high risk for CMV disease. Routine surveillance for CMV is performed in this population, and prophylactic therapy is given to patients at risk for CMV disease.

Histologic features are characteristic and consist of cytomegalic cells with inclusion bodies. The diagnosis of CMV is made by a combination of serologic tests and direct measurement of virus in the serum using polymerase chain reaction (PCR)–based techniques. CMV can be treated or prevented with several antiviral drugs, the most common of which is ganciclovir.

HERPESVIRUSES AND ADENOVIRUS

Other causes of acute viral hepatitis include herpesviruses and adenovirus. These viruses may cause severe hepatitis, particularly in the immunosuppressed population, and they may frequently be fatal. *Herpes simplex hepatitis* has been reported in pregnant women.

ADDITIONAL RESOURCES

Cainelli F, Vento S: Infections and solid organ transplant rejection: a cause-and-effect relationship? *Lancet Infect Dis* 2:539-549, 2002.

Carrigan DR: Adenovirus infections in immunocompromised patients, *Am J Med* 102:71-74, 1997.

Elgui de Oliveira D: DNA viruses in human cancer: an integrated overview on fundamental mechanisms of viral carcinogenesis, *Cancer Lett* 247(2):182-196, 2007.

Feranchak AP, Tyson RW, Narkewicz MR, et al: Fulminant Epstein-Barr viral hepatitis: orthotopic liver transplantation and review of the literature, *Liver Transpl Surg* 4:469-476, 1998.

Hjalgrim H, Engels EA: Infectious aetiology of Hodgkin and non-Hodgkin lymphomas: a review of the epidemiological evidence, *J Intern Med* 264(6):537-48, 2008.

Howard CR, Ellis DS, Simpson DI: Exotic viruses and the liver, *Semin Liver Dis* 4:361-374, 1984.

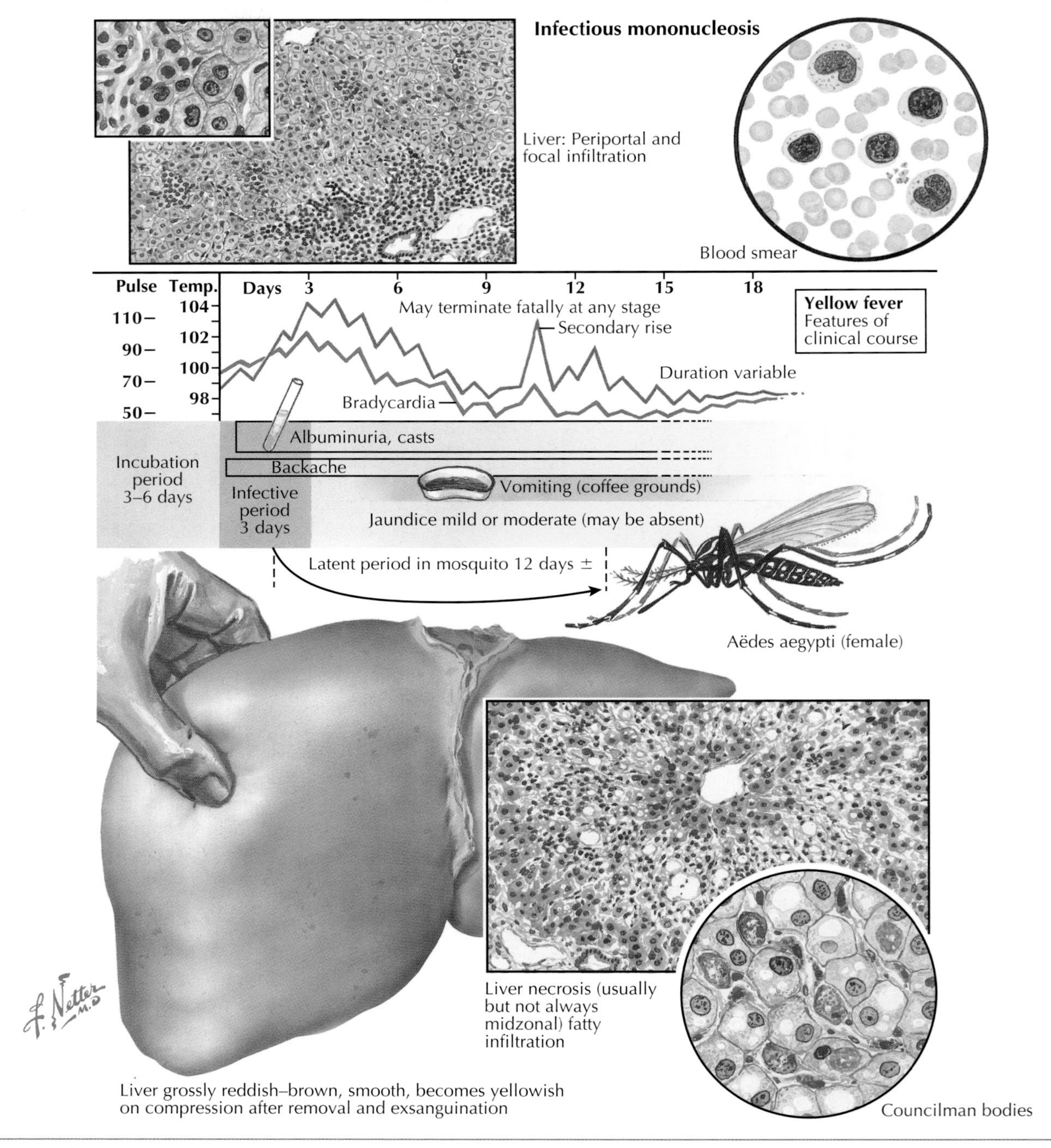

Figure 242-1 *Hepatitis Caused by Other Viruses: Infectious Mononucleosis and Yellow Fever.*

Markin RS: Manifestations of Epstein-Barr virus–associated disorders in liver, *Liver* 14:1-13, 1994.

Monath TP: Yellow fever: an update, *Lancet Infect Dis* 1:11-20, 2001.

Monath TP: Treatment of yellow fever, *Antiviral Res* 78(1):116-124, 2008.

Razonable RR: Cytomegalovirus infection after liver transplantation: current concepts and challenges, *World J Gastroenterol* 14(31):4849-4860, 2008.

Rubin RH: Cytomegalovirus in solid organ transplantation, *Transpl Infect Dis* 3(suppl 2):1-5, 2001.

Sayers MH: Transfusion-transmitted viral infections other than hepatitis and human immunodeficiency virus infection: cytomegalovirus, Epstein-Barr virus, human herpesvirus 6, and human parvovirus B19, *Arch Pathol Lab Med* 118:346-349, 1994.

Hepatotoxicity

243

Kris V. Kowdley

Numerous drugs and chemical agents can cause hepatotoxicity (**Fig. 243-1**). In addition, a growing list of herbal remedies and over-the-counter (OTC) supplements can lead to liver injury. Hepatotoxic reactions may be acute or chronic; occasionally, they can lead to severe or even fulminant liver injury and may require emergency liver transplantation.

CLASSIFICATION

Drug-induced liver injury has been classified as idiosyncratic or intrinsic. *Idiosyncratic* drug reactions are not predictable, can occur with different amounts of exposure, and are highly variable in clinical presentation. Age, gender, and genetic factors may influence drug metabolism and immune response, as may dietary factors such as amount of protein and alcohol intake. *Intrinsic* hepatotoxicity is generally a feature of agents that are inherently hepatotoxic or that produce toxic metabolites, which can then lead to liver injury.

Intrinsic Hepatotoxicity

Intrinsic hepatotoxins may cause liver injury by the production of free radicals or toxic metabolites. This type of hepatotoxicity is dose related and is common among exposed persons. However, many variables may influence the development of intrinsic hepatotoxicity. The agent may be directly toxic to cell membranes and may lead to hepatocellular necrosis; occasionally, there is accompanying steatosis. Cholestatic reactions have also been described. In some cases, the toxic agent is produced by metabolism in the liver. The classic example of an intrinsic hepatotoxin is *carbon tetrachloride* (CCl_4). It is believed that CCl_4 is converted to a toxic metabolite that leads to lipid peroxidation, thus causing damage to cell membranes (**Fig. 243-2**).

Idiosyncratic Hepatotoxicity

Idiosyncratic hepatotoxins may cause hepatotoxicity either through the unpredictable production of toxic metabolites or by induction of an autoimmune process that subsequently causes liver injury. In contrast to intrinsic hepatotoxicity, only a small number of idiosyncratically exposed persons develop liver injury.

CLINICAL PICTURE AND DIAGNOSIS

The clinical features of hepatotoxic reactions can be highly variable, ranging from asymptomatic elevation of serum liver enzyme levels to fulminant liver failure. Some drugs may induce a pattern of chronic injury. Serum liver biochemical tests often reveal the type of liver injury. Some drugs produce a predominantly hepatocellular pattern, with greatly increased alanine transaminase (ALT) and aspartate transaminase (AST) levels, whereas other agents reveal a cholestatic pattern of liver injury, with elevations primarily in serum alkaline phosphatase (ALP) and γ-glutamyltransferase (GGT) and less impressive increases in serum ALT and AST.

Some drug reactions also produce unique patterns of liver enzyme elevations. For example, *acetaminophen* hepatotoxicity is typically observed in chronic alcoholism in those who ingest as little as 6 g. This hepatotoxicity likely is related to the increased production of a toxic intermediate of acetaminophen caused by the upregulation of cytochrome P450 enzymes through chronic consumption of alcohol and through its reduced clearance because of glutathione depletion. Serum AST is often strikingly elevated (5000-10,000 IU/L) and may be increased several times compared with serum ALT.

Other drugs can produce clinical, biochemical, and serologic patterns identical to those of idiopathic chronic autoimmune hepatitis, with positive antinuclear antibody and elevated plasma globulins; the classic example is *methyldopa*. Pruritus is common with drugs that lead to cholestatic injury pattern; this symptom may be prolonged (weeks to months) and may be difficult to manage. Some drugs or toxins may be associated with the development of portal hypertension. This may be intrasinusoidal (alcohol), presinusoidal (nodular regenerative hyperplasia), or postsinusoidal (hepatic venoocclusive disease caused by alkylating agents as conditioning regimens for hematopoietic cell transplantation or pyrrolidine alkaloids).

Histologic Patterns

Drug-induced and toxin-induced liver injury can take any histologic form. Acute, severe liver injury may lead to massive hepatocellular necrosis in a diffuse zone 1 (periportal) or zone 3 (centrilobular) pattern. There may be features of an allergic response, such as eosinophilia. Variable amounts of steatosis may be present. Some drug reactions may cause diffuse steatosis, which may be macrovesicular or microvesicular. Classic examples of microvesicular steatosis are caused by *tetracycline* and *valproic acid*. Steatohepatitis, caused by *amiodarone*, may be indistinguishable from alcoholic liver disease.

Patients with chronic drug-induced liver disease may have features of *chronic hepatitis*, with portal-based inflammatory infiltrates, or *interface hepatitis* (piecemeal necrosis). Vascular diseases may manifest with sinusoidal dilatation, venous thrombi, or peliosis of the liver. Cholestatic drug reactions may be acute or chronic and may have features of hepatocellular damage as well (sulfa-containing compounds) or may demonstrate only cholestasis without necroinflammatory changes (anabolic steroids).

Industrial and Household Hepatotoxins

Many chemical agents are solvents capable of causing liver injury and have been used in industry and as household products. Occasional outbreaks of hepatotoxicity have been reported with these agents, including hydrocarbons and dimethylformamide (DMF), although such outbreaks are increasingly rare.

Benign and Malignant Neoplasms

Oral contraceptive use has been implicated in focal nodular hyperplasia and nodular regenerative hyperplasia, presumably because of the stimulation of estrogen receptors in the liver. Of

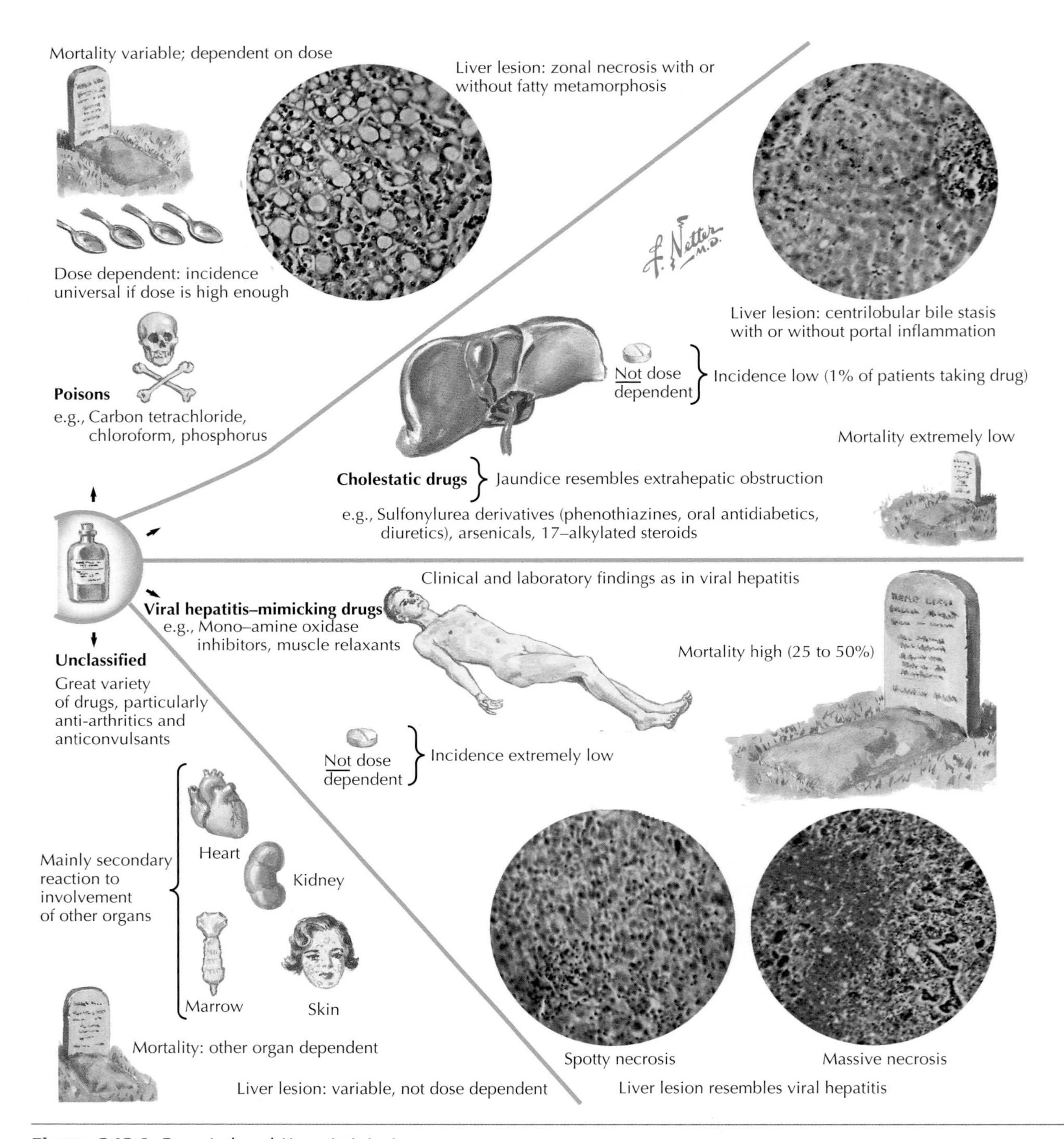

Figure 243-1 *Drug-Induced Hepatic Injuries.*

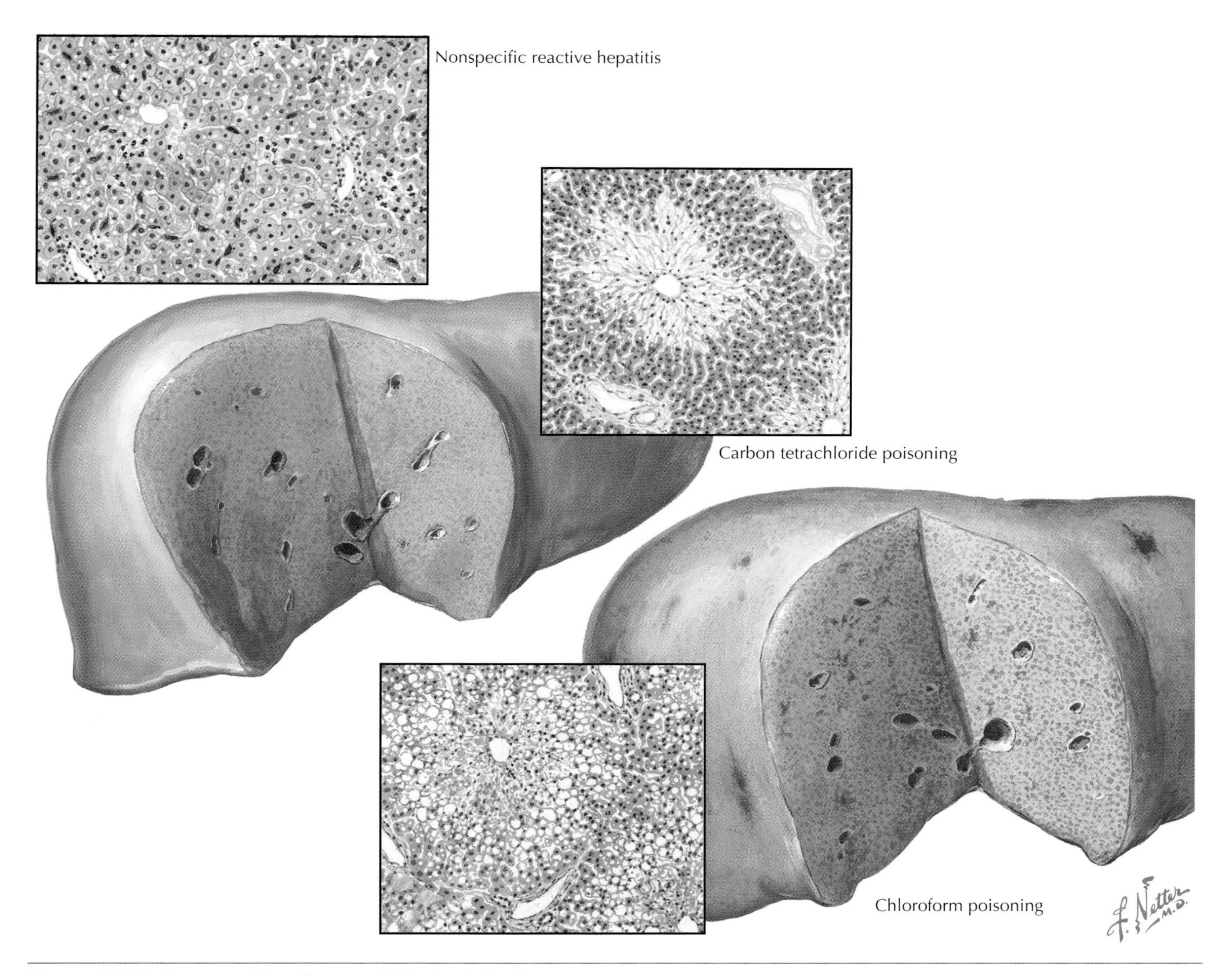

Figure 243-2 *Hepatotoxicity: Gross and Histologic Findings.*

greater concern is the possible development of malignant neoplasms, as noted with thorium dioxide (Thorotrast), vinyl chloride, and aflatoxin exposure.

ADDITIONAL RESOURCES

Fontana RJ: Acute liver failure due to drugs, *Semin Liver Dis* 28(2):175-187, 2008.

Lahoti S, Lee WM: Hepatotoxicity of anticholesterol, cardiovascular, and endocrine drugs and hormonal agents, *Gastroenterol Clin North Am* 24:907-922, 1995.

Lee WM: Drug-induced hepatotoxicity, *N Engl J Med* 349:474-485, 2003.

Liu ZX, Kaplowitz N: Immune-mediated drug-induced liver disease, *Clin Liver Dis* 6:467-486, 2002.

O'Connor N, Dargan PI, Jones AL: Hepatocellular damage from non-steroidal anti-inflammatory drugs, *Q J Med* 96:787-791, 2003.

Schiano TD: Hepatotoxicity and complementary and alternative medicines, *Clin Liver Dis* 7:453-473, 2003.

Velayudham LS, Farrell GC: Drug-induced cholestasis, *Expert Opin Drug Safety* 2:287-304, 2003.

Zapater P, Moreu R, Horga JF: The diagnosis of drug-induced liver disease, *Curr Clin Pharmacol* 1(2):207-217, 2006.

Disorders of Bilirubin Transport

244

Kris V. Kowdley

Bilirubin is a breakdown product of *heme*, which is derived from red blood cells (RBCs). Heme is initially cleaved by heme oxygenase into *biliverdin*, which is subsequently converted to bilirubin. In the circulation, bilirubin is bound to *albumin*. Bilirubin is then taken by the liver and conjugated through glucuronidation into bilirubin monoglucuronide or diglucuronide, which is mediated by a group of enzymes called uridine diphosphate glucuronyltransferase (UGT), of which UGT1A1 is the key isoform responsible for bilirubin glucuronidation. This step enables the nonpolar bilirubin to be converted into a water-soluble form that can be excreted in bile (**Fig. 244-1**).

Unconjugated hyperbilirubinemia is common in the early neonatal period and is generally benign. Serum bilirubin levels that are primarily unconjugated may rise to 6 mg/dL and, in a minority of patients, may rise threefold or fourfold. Such high levels may be toxic in the neonatal period. *Kernicterus* is the term used to describe encephalopathy associated with high levels of unconjugated hyperbilirubinemia in the neonatal period. Factors associated with increased risk for neurotoxicity from bilirubin include the use of drugs that displace albumin from bilirubin, thus increasing the exposure of free bilirubin to brain tissue. Clinical features of kernicterus include lack of muscle tone, abnormal reflexes, and possible progression to atony and death. Patients who recover from acute bilirubin toxicity may have long-term sequelae, such as hearing impairment resulting from cochlear damage, cerebellar abnormalities, and varying degrees of mental impairment. The mechanism behind bilirubin neurotoxicity may include impairment of DNA and RNA synthesis and protein and carbohydrate metabolism.

Disorders of bilirubin can be classified as those resulting from increased production, decreased hepatic uptake, decreased hepatic conjugation, and decreased biliary excretion. These are discussed briefly in the following text.

Often, increased bilirubin production results from increased RBC turnover, as may be observed in hemolytic conditions. ABO incompatibility was a common cause of neonatal hemolysis before anti-Rh immunoglobulin was used to treat mothers. Hereditary spherocytosis and sickle cell disease are examples of common conditions that may lead to increased serum bilirubin levels, particularly in the neonatal period. Serum bilirubin levels rarely rise above 3 to 5 mg/dL when unconjugated hyperbilirubinemia occurs in persons with normal liver function. However, if the liver's capacity to transport conjugated bilirubin is overwhelmed, conjugated and unconjugated hyperbilirubinemia may be present.

Inherited disorders of bilirubin conjugation are caused by mutations in the bilirubin-UGT1A1 enzyme, resulting in varying levels of deficiencies in its activity. *Crigler-Najjar syndrome* type 1 is defined by minimal to no activity of this enzyme, and it includes severe, indirect hyperbilirubinemia with a high incidence of kernicterus. The syndrome is a rare autosomal recessive disorder often found in consanguineous families. Phototherapy, plasmapheresis, and orthotopic liver transplantation have all been used to treat this life-threatening disorder. Crigler-Najjar syndrome *type 2* is a milder form of hepatic bilirubin-UGT deficiency characterized by elevated serum bilirubin levels but minimal clinically significant sequelae. A third form is the common *Gilbert syndrome*, which results in a mild decrease in enzyme activity and is associated with a mild elevation in serum bilirubin level (<3 mg/dL), although it may increase more in patients who are fasting or under stress. The mutation responsible is in the promoter region of the gene encoding UGT1A1, which results in reduced production of bilirubin-UGT. The prevalence of the homozygous form of the variant promoter mutation is 9% in Western populations.

In contrast to these disorders, *Rotor syndrome* and *Dubin-Johnson syndrome* are inherited disorders that lead to predominantly *conjugated* hyperbilirubinemia. Despite the presence of conjugated hyperbilirubinemia and jaundice, pruritus is notably absent in these patients because bile acid transport is not impaired. A dark, heavily pigmented liver characterizes Dubin-Johnson syndrome, whereas in Rotor syndrome, the liver is normal. The two conditions can be differentiated using oral cholecystography; the gallbladder is visualized in Rotor syndrome but not in Dubin-Johnson syndrome. In addition, the pattern of urinary coproporphyrin excretion can be useful to distinguish the two disorders; excretion is increased in Rotor syndrome (2.5-5 times normal) and the predominant form is coproporphyrin 1; by contrast, in Dubin-Johnson syndrome, excretion of urinary coproporphyrin is normal, although coproporphyrin 1 is also predominant. Each condition has a relatively benign clinical course and is not associated with life-threatening complications.

The molecular basis of these disorders has recently been characterized. Conjugated bilirubin is secreted into the biliary canaliculus from the hepatocyte by the multidrug-resistant–related protein (MRP-2), also known as canalicular multispecific organic anion transporter (cMOAT). Mutations in the gene encoding this transporter appear to be responsible for Dubin-Johnson syndrome. The molecular defect in Rotor syndrome is unknown, although an autosomal recessive form of inheritance is suspected.

ADDITIONAL RESOURCES

Blackburn S: Hyperbilirubinemia and neonatal jaundice, *Neonatal Netw* 14:15-25, 1995.

Bosma PJ: Inherited disorders of bilirubin metabolism, *J Hepatol* 38:107-117, 2003.

Bosma PJ, Seppen J, Goldhoorn B, et al: Bilirubin UDP-glucuronosyltransferase 1 is the only relevant bilirubin glucuronidating isoform in man, *J Biol Chem* 269(27):17960-17964, 1994.

Nowicki MJ, Poley JR: The hereditary hyperbilirubinaemias, *Baillieres Clin Gastroenterol* 12:355-367, 1998.

Shaffer EA: Cholestasis: the ABCs of cellular mechanisms for impaired bile secretion—transporters and genes, *Can J Gastroenterol* 16(6):380-389, 2002.

Trauner M, Fickert P, Zollner G: Genetic disorders and molecular mechanisms in cholestatic liver disease: a clinical approach, *Semin Gastrointest Dis* 12(2):66-88, 2001.

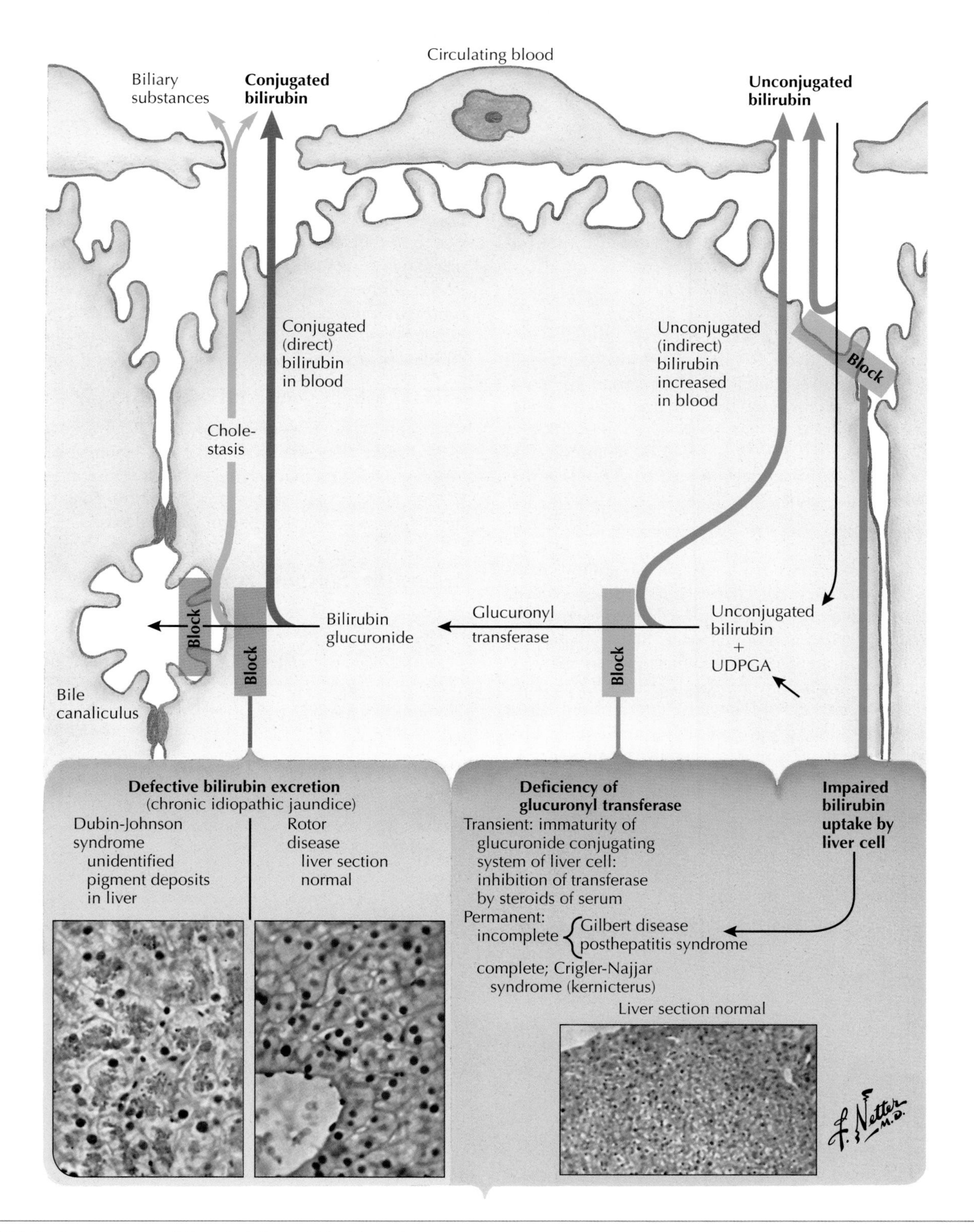

Figure 244-1 *Congenital and Familial Hyperbilirubinemia.*

α_1-Antitrypsin Deficiency

Kris V. Kowdley

Since its discovery in 1963, alpha$_1$-antitrypsin (α_1-AT) deficiency has been recognized as the most common genetic liver disease in infants and children, and it may cause cirrhosis and end-stage liver disease in adults. This deficiency affects approximately 1 in 1600 to 2800 live births in the United States and northern Europe.

Alpha$_1$-antitrypsin α_1-AT is a serine protease inhibitor that is made primarily in the liver. It inhibits trypsin, neutrophil elastase, collagenase, and chymotrypsin. More than 75 structural variants of this protein have been characterized. Alleles that lead to a reduction in the protein are referred to as *deficiency alleles;* Z variant (designated PiZZ) is the most common, and persons homozygous for this mutation have α_1-AT levels that are only 15% of normal value.

CLINICAL PICTURE

Persistent jaundice in the newborn is the most common manifestation of α_1-AT deficiency. Alternatively, liver disease caused by α_1-AT may not be diagnosed until adolescence or early adulthood, when it manifests as abdominal pain, hepatomegaly, or variceal hemorrhage. In adults, α_1-AT deficiency may manifest as emphysema, chronic hepatitis, cryptogenic cirrhosis, hepatocellular cancer, or symptoms of portal hypertension. Liver disease from α_1-AT in adults progresses slowly and is associated with a high risk for hepatocellular cancer in the presence of cirrhosis.

DIAGNOSIS

Determining the α_1-AT phenotype is the best means of diagnosing α_1-AT deficiency. Most persons with chronic liver disease are homozygous for PiZZ or are compound heterozygous for SZ (PiSZ). Serum α_1-AT levels are typically 10% to 15% of normal in PiZZ patients but may be normal in those with infection or other conditions that can lead to elevations in acute-phase reactants such as α_1-AT. Liver biopsy reveals characteristic PAS+, diastase-resistant globules within the hepatocytes that represent the retained protein. Some data suggest that PiMZ heterozygotes may develop liver disease in the presence of another liver disease, such as viral or autoimmune hepatitis.

TREATMENT AND PROGNOSIS

Orthotopic liver transplantation is the only proven therapy for liver disease associated with α_1-AT deficiency. Liver transplantation corrects the underlying defect because most of the α_1-AT is expressed in the hepatocytes. Liver transplantation in adults has a reported 65% long-term survival rate.

ADDITIONAL RESOURCES

Fairbanks KD, Tavill AS: Liver disease in alpha$_1$-antitrypsin deficiency: a review, *Am J Gastroenterol* 103(8):2136-2141, 2008.

Ishak KG: Inherited metabolic diseases of the liver, *Clin Liver Dis* 6:455-479, 2002.

Morrison ED, Kowdley KV: Genetic liver disease in adults: early recognition of the three most common causes, *Postgrad Med* 107:147-152, 155, 158-159, 2000.

Perlmutter DH: Liver injury in alpha$_1$-antitrypsin deficiency, *Clin Liver Dis* 4:387-408, 2000.

Stoller JK, Aboussouan LS: Alpha$_1$-antitrypsin deficiency, *Lancet* 365(9478):2225-2236, 2005.

Zhang KY, Tung BY, Kowdley KV: Liver transplantation for metabolic liver diseases, *Clin Liver Dis* 11(2):265-281, 2007.

Hereditary Hemochromatosis

Kris V. Kowdley

Hereditary hemochromatosis (HH) is an inherited disorder associated with parenchymal iron deposition in multiple organs caused by the excessive absorption of iron from a normal diet (**Fig. 246-1**). Iron deposition occurs in multiple organs, including the liver, heart, pancreas, skin, joints, and anterior pituitary. Consequences include cirrhosis and hepatocellular carcinoma, diabetes mellitus, cardiomyopathy, "bronzing" or hyperpigmentation of the skin, arthropathy involving the metacarpophalangeal (MCP) joints, and hypogonadotropic hypogonadism.

The term "hereditary hemochromatosis" is usually used to indicate *HLA-linked hemochromatosis*, a disorder found only in white populations. This disorder has been recognized since 1865, when Trousseau described the first case. Marcel Simon and colleagues discovered the association with the HLA-A haplotype and the autosomal recessive pattern of inheritance in the mid-1970s. The putative hemochromatosis gene was discovered in 1996. This gene, *HFE*, is mutated in most patients with phenotypic HH. Two common mutations were initially described, the *C282Y mutation*, indicating a cysteine-to-tyrosine substitution at amino acid 282, and the *H63D mutation*, a histidine-to-aspartate substitution at amino acid 63. Most patients with the phenotype of HH are homozygous for the C282Y mutation or are compound heterozygous for the C282Y and H63D mutations.

Hereditary hemochromatosis is the most common genetic disease among persons of northern European descent. Most studies have estimated a prevalence of 1:200 to 1:500. The highest allelic frequency for the C282Y mutation is in northern Europe (6.4%-9.5%); it has not been found in indigenous populations in the Americas, the Indian subcontinent, Africa, or the Middle East. The clinical significance of the H63D mutation is unclear.

The regulation of body iron stores in the human is through gastrointestinal (GI) absorption. This is accomplished by variable expression of different iron transporters in response to physiologic signals in the crypt cells of the small intestine. In iron deficiency, iron absorption in the proximal small intestine is upregulated. This is characterized by increased expression or activity of DMT1 (mucosal iron transporter), ferroportin (FPN1), basolateral iron transporter, and transferrin receptor, and by decreased mucosal ferritin content. A similar pattern was found in some patients with HH, leading to speculation that the crypt cell senses a state of iron deficiency. Recent studies have identified that *hepcidin*, a novel antimicrobial peptide produced by the liver, appears to have an inhibitory effect on iron absorption. Patients with *HFE*-associated HH appear to have inappropriately low hepcidin levels compared with healthy subjects. These data suggest that the *HFE* mutation may lead to decreased efficiency in signaling hepatic iron content to the crypt or intestinal absorptive cells, leading to greater iron absorption.

CLINICAL PICTURE

Early in the disease, iron accumulation begins in various tissues and may be associated with no symptoms or with nonspecific symptoms, such as fatigue, abdominal pain, arthralgias, and impotence. With progressive iron loading, diabetes, cirrhosis, cardiomyopathy, and arthritis may develop. Later signs typically do not develop until a significant amount of iron has accumulated, usually after the patient is 40 to 60 years of age. Phenotypic expression, however, is variable and depends on other factors, such as alcohol intake, diet, and pathologic or physiologic blood loss (e.g., through menstruation).

Since the advent of molecular diagnosis and increased awareness of the prevalence of HH, patients are being identified at an earlier, asymptomatic stage through screening serum iron studies or genetic testing. It has been clearly established that diagnosis of HH before the development of cirrhosis is associated with a normal life expectancy. However, patients who have cirrhosis at diagnosis are at increased risk for premature death from cirrhosis and hepatocellular carcinoma compared with an age/gender-matched cohort, even with iron depletion therapy. It was previously estimated that the risk for hepatocellular carcinoma is increased up to 200-fold in HH patients, but recent studies indicate the incidence may be lower.

DIAGNOSIS

Iron overload can be classified as primary or secondary. The term *primary iron overload* is used to indicate otherwise unexplained iron overload, usually caused by excessive GI absorption. The most common cause is *HFE*-associated HH, although other forms of primary iron overload have been described, including non–*HFE*-associated HH, juvenile HH, and African iron overload. These rare forms of hereditary iron overload are now recognized as being caused by mutations in several genes regulating iron metabolism (hepcidin, hemojuvelin, transferrin receptor 2, ferroportin). *Secondary iron overload* is used to describe overload secondary to iatrogenic iron administration, transfusion, or a compensatory increase in iron absorption caused by increased red blood cell turnover, alcoholism, dysmetabolic syndrome, or cirrhosis.

The diagnosis of HH has historically been made using phenotypic criteria because no genetic marker was available. Given the identification of the homozygous C282Y and C282Y/H63D compound heterozygous mutations in the *HFE* gene, the presence of these mutations has also been used for the diagnosis of HH. Patients with suspected HH based on elevated serum transferrin–iron saturation (>45%) should have confirmatory testing with liver biopsy or *HFE* gene testing. Magnetic resonance imaging (MRI) has been used to assess for hepatic iron overload, but it is insensitive unless specialized software is used.

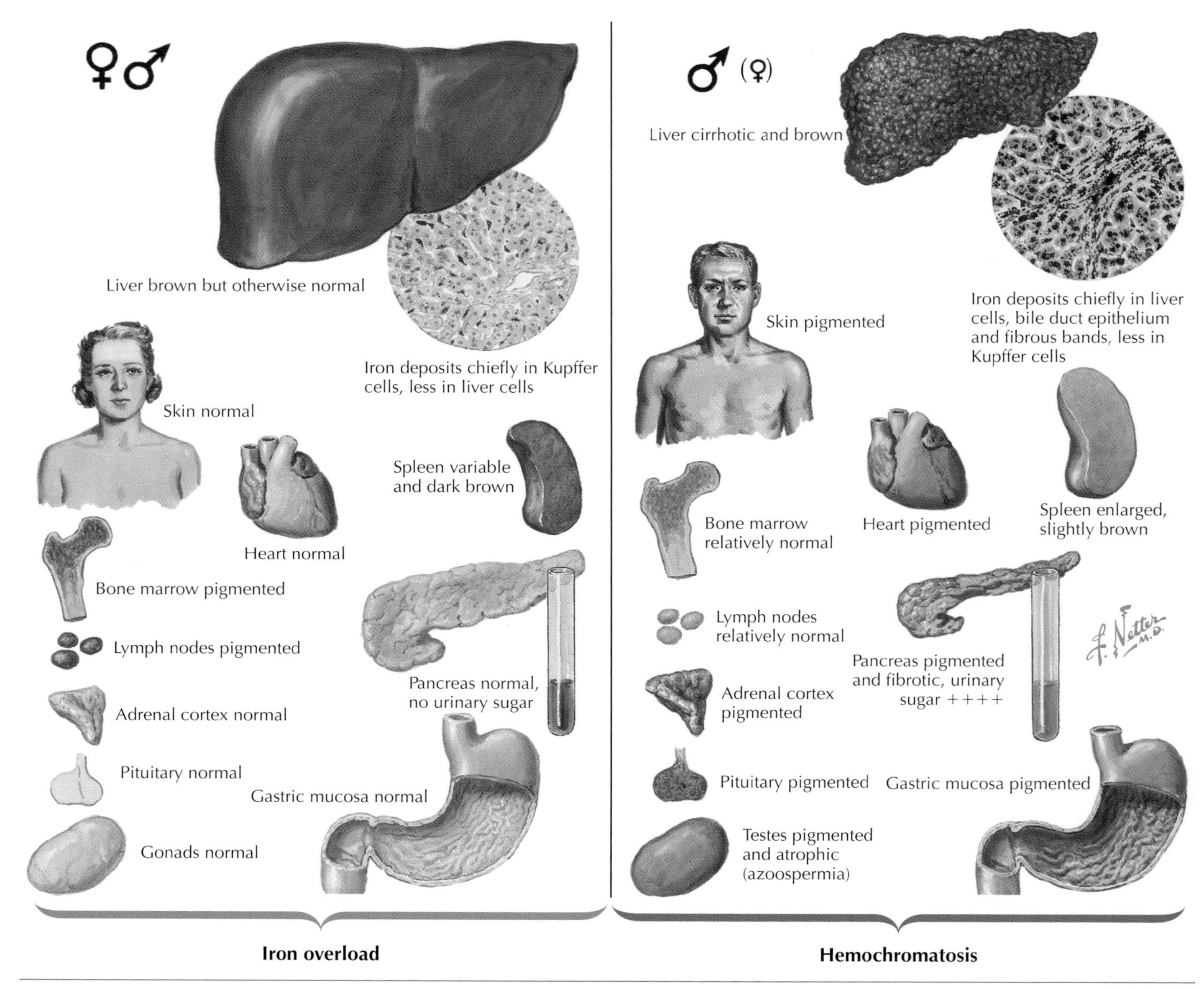

Figure 246-1 *Secondary Iron Overload and Hereditary Hemochromatosis.*

HFE *Genetic Testing*

From 65% to 100% of patients with typical phenotypic HH are homozygous for C282Y. In addition, compound heterozygosity (C282Y/H63D) occurs in approximately 5% of patients who have clinical evidence of HH, although the degree of iron overloading seems to be lower. Persons who are heterozygous for the C282Y mutation typically do not develop iron overload unless another disorder coexists, such as chronic hepatitis C infection or alcoholic liver disease. Therefore, only the C282Y homozygous and C282Y/H63D heterozygous genotypes are considered diagnostic for classic HH. Patients with suspected HH based on elevated serum transferrin–iron saturation who have either of these two mutation patterns can be confirmed to have HH. Liver biopsy is generally not required in these patients for diagnosis, but it may be needed to exclude the presence of cirrhosis. **Table 246-1** lists common mutations in the *HFE* gene with their clinical interpretations.

Liver Biopsy

Liver biopsy is of crucial importance to establish the presence or absence of cirrhosis. Patients with a diagnosis of HH confirmed by *HFE* gene testing (C282Y homozygous or compound heterozygous) who have normal liver enzyme levels and serum ferritin concentrations less than 1000 ng/mL and no other complication of liver disease (alcohol abuse, chronic hepatitis, steatohepatitis) are at low risk for cirrhosis and can avoid liver biopsy. Histologic features characteristic of HH include increased stainable iron in hepatocytes and bile duct cells, with a paucity of iron in Kupffer cells. There is a greater density of iron staining in periportal hepatocytes than in those around the central veins. Measuring the hepatic iron concentration may also be helpful.

Table 246-1 *HFE* Gene Testing in Hereditary Hemochromatosis

Genetic Test Result	Clinical Significance
C282Y homozygous mutation	Greatly increased risk for iron overload.
C282Y heterozygous mutation	Usually not associated with iron overload; serum transferrin—iron saturation may be elevated.
C282Y/H63D compound heterozygous mutation	Moderately increased risk for iron overload.
H63D homozygous mutation	Increased serum iron levels, but no increased risk for iron overload.
H63D heterozygous mutation	Not associated with iron overload.

TREATMENT AND MANAGEMENT

Iron reduction therapy by phlebotomy is the mainstay of treatment for HH. Phlebotomy therapy is safe, easy, and inexpensive and should be performed before symptoms develop. Weekly phlebotomy of 500 mL of whole blood is generally well tolerated. Phlebotomy is continued until iron depletion is confirmed by a mild anemia and ferritin concentrations less than 50 ng/mL. Cardiac function may improve if treatment is undertaken before dilated cardiomyopathy develops. However, joint symptoms may not respond to therapy.

Orthotopic liver transplantation remains an option for patients with advanced liver disease or hepatocellular carcinoma. However, outcomes after orthotopic liver transplantation are disappointing, with 1- and 5-year survival rates of 58% and 42%, respectively, although recent studies suggest that outcomes may be improving. The increased mortality after transplantation appears to be the result of infectious and cardiac complications. Iron reduction therapy before and after transplantation may reduce the mortality rate.

ADDITIONAL RESOURCES

Beutler E, Felitti VJ, Koziol JA, et al: Penetrance of 845G > A (C282Y) HFE hereditary haemochromatosis mutation in the USA, *Lancet* 359:211-218, 2002.

Fix OK, Kowdley KV: Hereditary hemochromatosis, *Minerva Med* 99(6):605-617, 2008.

Gordon SC, Galan MV, Tung BY, et al: Serum ferritin level predicts advanced hepatic fibrosis among U.S. patients with phenotypic hemochromatosis, *Ann Intern Med* 138:627-633, 2003.

Morrison ED, Brandhagen DJ, Phatak PD, et al: Genetic liver disease in adults, *Postgrad Med* 107:147-159, 2000.

Niederau C, Fischer R, Sonnenberg A, et al: Survival and causes of death in cirrhotic and noncirrhotic patients with primary hemochromatosis, *N Engl J Med* 313:1256-1262, 1985.

Whittington CA, Kowdley KV: Haemochromatosis (review), *Aliment Pharmacol Ther* 16:1963-1975, 2002.

Zhang KY, Tung BY, Kowdley KV: Liver transplantation for metabolic liver diseases, *Clin Liver Dis* 11(2):265-281, 2007.

Liver Disease in Pregnancy

247

Kris V. Kowdley

Hyperemesis gravidarum is characterized by severe nausea and vomiting in the first trimester of pregnancy. Its incidence is 0.35% to 0.8% of pregnancies. It is rare to observe this disorder after 20 weeks of gestation. Hyperthyroidism may occasionally accompany it, but its pathogenesis is unknown. Hospitalization for intravenous hydration and antiemetic therapy may be required. As many as 50% of patients with hyperemesis gravidarum have abnormal results on liver function tests. However, transaminase values rarely exceed 1000 IU/L. The differential diagnosis includes viral hepatitis, gastric outlet obstruction, and gastroenteritis. Treatment is supportive with hydration and antiemetics. Promethazine, ondansetron, and droperidol have all been used with success.

INTRAHEPATIC CHOLESTASIS OF PREGNANCY

Intrahepatic cholestasis of pregnancy (ICP) is a disorder of unknown etiology that typically develops in the second trimester. The prevalence is reported to be 0.7% (United States) to 6.5% (Chile). The pathogenesis is unknown but is thought to be related to genetic and hormonal factors. Patients with ICP appear to be more sensitive to the cholestatic effects of estrogen.

Characterized by severe pruritus and jaundice in the mother, ICP is associated with premature delivery and fetal death. The main symptom is pruritus involving the palms and soles and also perhaps the trunk and extremities. Serum levels of bile acids, transaminases, and bilirubin may be elevated. Jaundice may develop 1 to 4 weeks after the onset of itching in 20% to 60% of patients.

Laboratory tests may suggest a cholestatic or a hepatocellular process. The optimal test for ICP is the serum bile acid level, which is measured as cholylglycine. Liver biopsy reveals bland cholestasis with bile plugs predominantly in zone 3 (central vein region) and intact portal tracts. However, liver biopsy is generally not needed for diagnosis.

Management of ICP depends on symptoms. Early delivery has been advocated, either after 36 weeks for severe cases (if fetal lungs are mature) or at 38 weeks for less severe cases. Cholestyramine has been used but may exacerbate fat malabsorption. Vitamin K supplementation should be considered if cholestyramine is used. UDCA therapy has been studied for this disorder and appears promising. ICP carries an increased risk of fetal complications. Presumably, elevated bile acid concentrations in amniotic fluid and umbilical cord bile account for the observed fetal complications. Clinical and biochemical abnormalities usually resolve a few weeks after delivery. Some evidence indicates that ICP may be associated with progesterone therapy. Patients are at risk for recurrence with subsequent pregnancies or with oral contraceptive use.

HELLP SYNDROME

The acronym HELLP stands for *h*emolysis, *e*levated *l*iver tests, and *l*ow *p*latelet levels. This disorder usually develops in the third trimester of pregnancy. The incidence is 0.17% to 0.85% of all live births. Mean age at diagnosis is 25 years (range, 14-40 years); HEELP is usually diagnosed at 32 to 34 weeks of gestation (range, 22-40 weeks). HELLP syndrome is also associated with preeclampsia, which is characterized by hypertension, proteinuria, and edema. Preeclampsia is a common syndrome (5%-10% of all pregnancies). Ethnic factors affect the prevalence of preeclampsia; white, Chinese, and black women are at higher risk.

Patients may have no symptoms or may report abdominal pain, which is the most common symptom. Many diagnoses are made after delivery. Severe cases may be associated with renal failure or seizures (eclampsia) (**Fig. 247-1**). A universal feature of HELLP syndrome is the presence of abnormal liver enzymes; serum aspartate transaminase (AST) level may range from 70 to 6000 U/L (mean, 250 U/L). Prothrombin time (PT) is generally normal, except in patients with severe hemolysis and disseminated intravascular coagulation (DIC). Peripheral blood smear reveals features of hemolysis with schistocytes and burr cells. However, hemolytic anemia may be evanescent. Serum haptoglobin should be measured if the peripheral smear does not reveal obvious hemolysis. Platelet count is often less than 100,000. It should be emphasized that gestational thrombocytopenia occurs in up to 8% of uncomplicated pregnancies. Therefore, not all pregnant women with low platelet counts have HELLP syndrome. However, women with gestational thrombocytopenia are at seven times greater risk for HELLP syndrome. Liver enzyme levels are elevated before complications develop. Liver biopsy reveals fibrin deposition and hemorrhage localized to the periportal areas. Liver biopsy is usually not needed for diagnosis because other features of the disease can be ascertained from clinical and laboratory testing.

The differential diagnosis includes viral hepatitis, hemolytic uremic syndrome, thrombocytopenic purpura, and acute fatty liver of pregnancy, discussed next.

There is high rate of complications in HELLP syndrome, including DIC, placental abruption, renal failure, ascites, pulmonary or cerebral edema, adult respiratory distress syndrome, and hepatic rupture. The most serious complication of HELLP syndrome may be the development of hepatic infarction, characterized by abdominal pain, fever, and marked elevation of transaminases (>5000 U/L) and accompanied by subcapsular hematoma or intraperitoneal hemorrhage. Immediate surgical intervention may be needed, and there is a risk for maternal and fetal death. In addition, patients are at increased risk for recurrent HELLP syndrome with subsequent pregnancies. Overall maternal mortality in HELLP syndrome is as high as 8%. Fetal

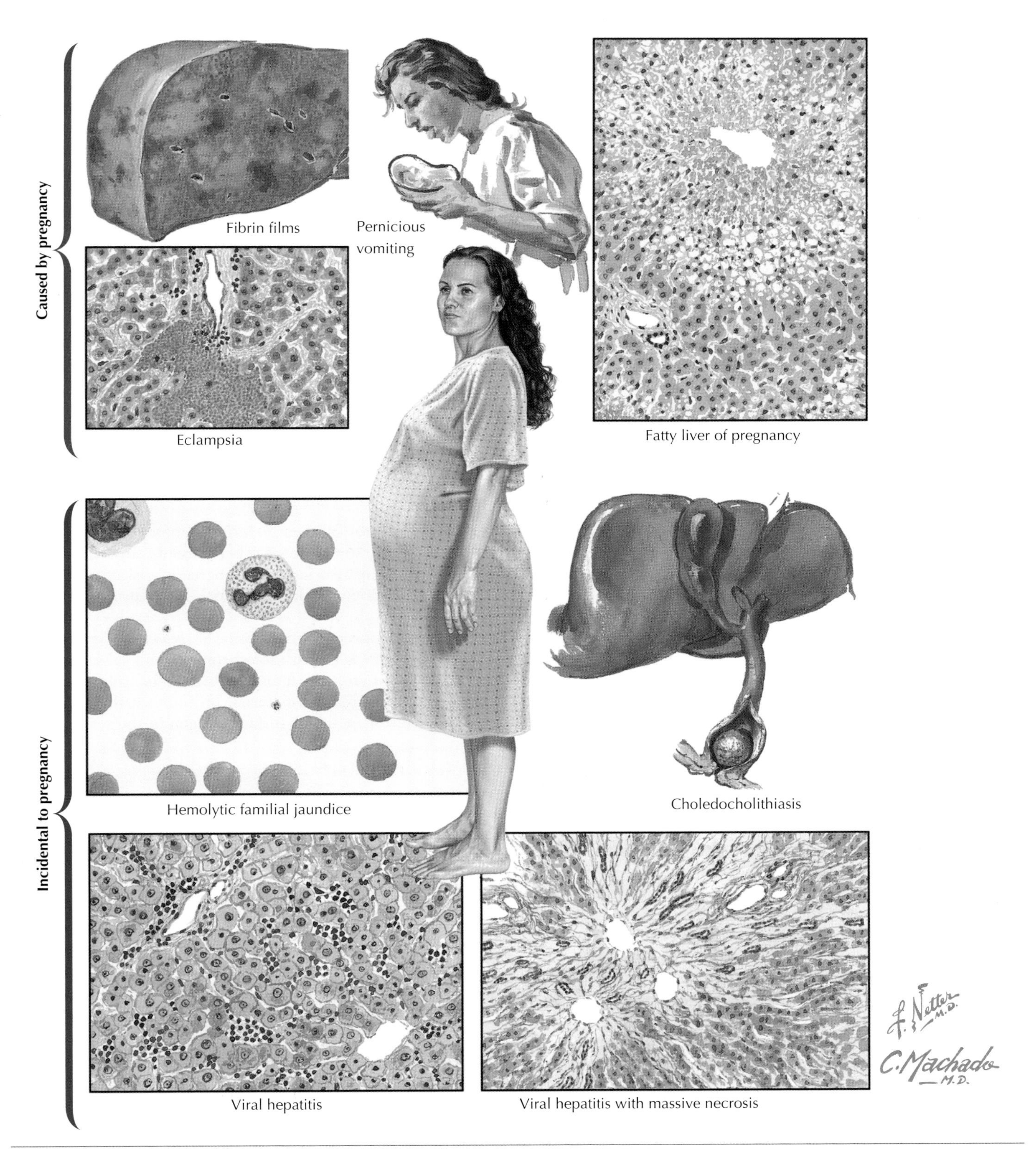

Figure 247-1 *Liver Disease in Pregnancy.*

mortality is as high as 35% to 37%. Rupture of hepatic hematoma may be associated with fetal and maternal mortality rates in excess of 50%.

Treatment is primarily obstetric. Careful fetal monitoring by an expert in high-risk obstetrics is appropriate, and prompt delivery is important. Clinical features of HELLP syndrome resolve within days of delivery. Plasmapheresis has been used in patients whose platelet counts continue to decrease after delivery. Liver enzymes generally return to normal after 3 to 5 days. Infants who survive have outcomes comparable to other infants of similar gestational age.

ACUTE FATTY LIVER OF PREGNANCY

Acute fatty liver of pregnancy is a rare condition, with an incidence of 1:13,000 to 1:16,000 deliveries. It may be associated with significant hepatic dysfunction, including PT prolongation, hyperammonemia, and hypoglycemia. Patients usually seek treatment during the third trimester of pregnancy. In its most severe form, fatty liver of pregnancy manifests as fulminant hepatic failure.

Patients with severe acute fatty liver of pregnancy have malaise, fatigue, anorexia, headache, nausea, and vomiting. Patients may be asymptomatic, and the diagnosis may be suspected based on incidentally discovered abnormalities on liver tests. Early in the disease course, right upper quadrant pain or epigastric pain may mimic acute cholecystitis or reflux esophagitis. Polydipsia is often an early symptom. Pruritus may develop in some patients, and jaundice eventually develops in most patients. The disease may rapidly worsen within days of initial presentation. Signs of acute liver failure, including hepatic encephalopathy, ascites, edema, and renal insufficiency, may develop. Signs of preeclampsia (hypertension, proteinuria) are present in more than 50% of patients.

Serum aminotransferase (transaminase) levels are elevated but are usually lower than 1000 IU/L. The level of transaminase elevation may underrepresent the severity of liver dysfunction. Marked jaundice and hyperbilirubinemia are common, and serum bilirubin levels are as high as 40 mg/dL. Extrahepatic complications have been reported, including upper gastrointestinal (GI) bleeding and renal dysfunction, which may require dialysis. DIC is common (up to 50% of patients). Pancreatitis may develop in up to 30% of patients. Severe hypoglycemia may be seen in 25% to 50% of patients and can occur at any stage of disease.

Liver biopsy is the criterion standard for diagnosis of acute fatty liver of pregnancy. Biopsy reveals vacuolization of hepatocytes and pallor in the central zone regions (see Fig. 247-1). Microvesicular steatosis is characteristic. The differential diagnosis includes acute viral hepatitis, acute toxic or drug-induced hepatitis, preeclampsia-related liver disease (including HELLP), drug-induced fatty liver, and biliary tract disorders.

Because patients may progress to fulminant liver failure and death or may require liver transplantation, acute fatty liver of pregnancy should be considered a medical and obstetric emergency. Patients should be promptly admitted to a liver failure unit for better monitoring, and some may need urgent liver transplantation. Prompt delivery is essential. Aggressive supportive care is required for the mother. Early diagnosis and management can minimize the severity of disease and the need for liver transplantation.

ACUTE LIVER DISEASE NOT SPECIFICALLY CAUSED BY PREGNANCY: VIRAL HEPATITIS

Viral hepatitis may be the most common liver disease during pregnancy, when hepatitis A, B, and E may all develop. *Hepatitis E* is associated with a case-fatality rate as high as 20%, especially if infection occurs in the third trimester. Most cases of acute hepatitis E are described among patients in the Indian subcontinent, northern Africa, and Mexico. Hepatitis E is often characterized by the presence of predominant GI symptoms such as diarrhea. Therefore, it is reasonable to caution pregnant women against travel to endemic areas, especially late in pregnancy.

Clinical features of acute *hepatitis A* infection during pregnancy are similar to those of other patients. Infection late in pregnancy may be associated with increased risk for premature delivery, but there is no increased risk for transmission to the fetus.

The major concern with exposure to *hepatitis B* virus (HBV) during pregnancy is the risk for chronicity and subsequent risk for transmission to the fetus. The risk for transmission is increased with increasing maternal HBV DNA titers and with the presence of hepatitis B surface antigen in the serum.

ADDITIONAL RESOURCES

Guntupalli SR, Steingrub J: Hepatic disease and pregnancy: an overview of diagnosis and management, *Crit Care Med* 33(10 suppl):332-339, 2005.

Hepburn IS, Schade RR: Pregnancy-associated liver disorders, *Dig Dis Sci* 53(9):2334-2258, 2008.

McDonald JA: Cholestasis of pregnancy, *J Gastroenterol Hepatol* 14:515-518, 1999.

Mishra L, Seeff LB: Viral hepatitis, A though E, complicating pregnancy, *Gastroenterol Clin North Am* 21:873-887, 1992.

Riely CA: Acute fatty liver of pregnancy, *Semin Liver Dis* 7:47-54, 1987.

Sandhu BS, Sanyal AJ: Pregnancy and liver disease, *Gastroenterol Clin North Am* 32:407-436, ix, 2003.

Simms J, Duff P: Viral hepatitis in pregnancy, *Semin Perinatol* 17:384-393, 1993.

Benign Liver Tumors

Kris V. Kowdley

248

Numerous nonmalignant conditions can result in the formation of nodular liver lesions (**Fig. 248-1**). These include regenerative nodules that develop as part of the liver's response to injury in patients with cirrhosis. Other common types of benign nodular lesions include focal nodular hyperplasia and nodular regenerative hyperplasia.

FOCAL NODULAR HYPERPLASIA

Focal nodular hyperplasia (FNH) is a benign liver condition described more than 100 years ago. It has been known by other terms, including hepatic hamartoma, focal cirrhosis, and hepatic pseudotumor.

The condition usually develops in women between 30 and 50 years of age. Many patients with FNH report a history of oral contraceptive (OC) use. Most patients are asymptomatic, and FNH is often an incidental finding or is discovered during evaluation after abnormal liver test findings. FNH has been reported to cause elevation of serum alkaline phosphatase (ALP) and γ-glutamyltransferase (GGT) levels in a minority of patients. In symptomatic patients, the diagnosis of FNH is usually made by imaging studies performed to evaluate abdominal pain. Some patients have hepatomegaly or a palpable mass over the right upper quadrant, although most patients have no specific abdominal finding on physical examination. Most patients have solitary lesions. Lesions usually measure 3 to 5 cm, often without a capsule. Although FNH can be present in any location within the liver, most lesions are subcapsular.

The histopathology of the liver in FNH is characteristic. Hepatocytes are normal and surround a central area of fibrosis (the so-called central scar). Nodules of liver parenchyma surround this central scar; in addition, there may be aberrant blood vessels. FNH may not be clearly seen on abdominal ultrasonography. Computed tomography (CT) may reveal a hypervascular lesion with enhancement during the arterial phase; the characteristic central scar may be evident during the portal venous phase. Dynamic magnetic resonance imaging (MRI) with contrast reveals hypervascularity and the central scar and has high specificity for the diagnosis of FNH. Sulfur colloid scintigraphy has been used to differentiate FNH from adenomas and other hypervascular lesions because the presence of Kupffer cells in FNH results in uptake by the lesion, whereas adenomas appear "cold" on scintigraphy. However, this test is not a criterion standard for the diagnosis of FNH (compared with adenoma) because some adenomas may contain Kupffer cells and therefore may appear "warm."

Overall, MRI may be the most useful study in FNH. Because the radiologic features are diagnostic and malignant transformation is not thought to be a concern in patients with typical presentation, histologic confirmation is unnecessary in patients with characteristic CT or MRI findings.

NODULAR REGENERATIVE HYPERPLASIA

Nodular regenerative hyperplasia (NRH) is used to describe regenerative nodules observed in patients without cirrhosis. The pattern of hepatic fibrosis surrounding these regenerative nodules is different from that observed in patients with cirrhosis. In contrast to FNH, NRH occurs in older persons and is often associated with various rheumatologic conditions. NRH is usually asymptomatic, but in some patients the large regenerative nodules lead to compression of the portal vein, resulting in portal hypertension and its complications, primarily variceal bleeding. NRH should be considered in patients with rheumatologic or autoimmune conditions who have portal hypertension, particularly those with preserved hepatic synthetic function. *Felty syndrome*, a condition associated with rheumatoid arthritis, is characterized by NRH and leukopenia because of the hypersplenism.

The diagnosis of NRH is best made by histologic examination. Ideally, wedge biopsy specimens removed through surgery are required to confirm the diagnosis. Radiologic features of NRH are nonspecific. Patients with bleeding from portal hypertension may benefit from endoscopic therapy. In patients with normal liver synthetic function and recurrent variceal bleeding, splenorenal shunt is an option.

HEPATIC ADENOMA

Hepatic adenomas are typically observed in young women, frequently those with a history of OC use. Approximately half the patients have symptoms such as abdominal mass or abdominal pain. Classic histologic features include normal-appearing hepatocytes with a paucity of bile ductular cells or normal liver lobules. Adenomas are often associated with the presence of a capsule.

Clinical complications include hemorrhage and malignant transformation into hepatocellular carcinoma. The risk for hemorrhage is related to the size of the lesion. Therefore, surgical resection is recommended for biopsy-proven adenomas, especially if they are large or symptomatic. Liver scintigraphy has been used to differentiate adenomas from FNH because adenomas generally do not contain Kupffer cells and thus may be "cold."

CAVERNOUS HEMANGIOMA

Cavernous hemangiomas are the most frequently identified nonmalignant liver tumors. Their prevalence has been estimated at 5% to 20%. These lesions are more common in women and may be related to OC use, because some cavernous hemangiomas are sensitive to estrogen. The diagnosis of cavernous hemangioma is often made incidentally; however, large lesions may cause abdominal pain and, rarely, hemorrhage from rupture

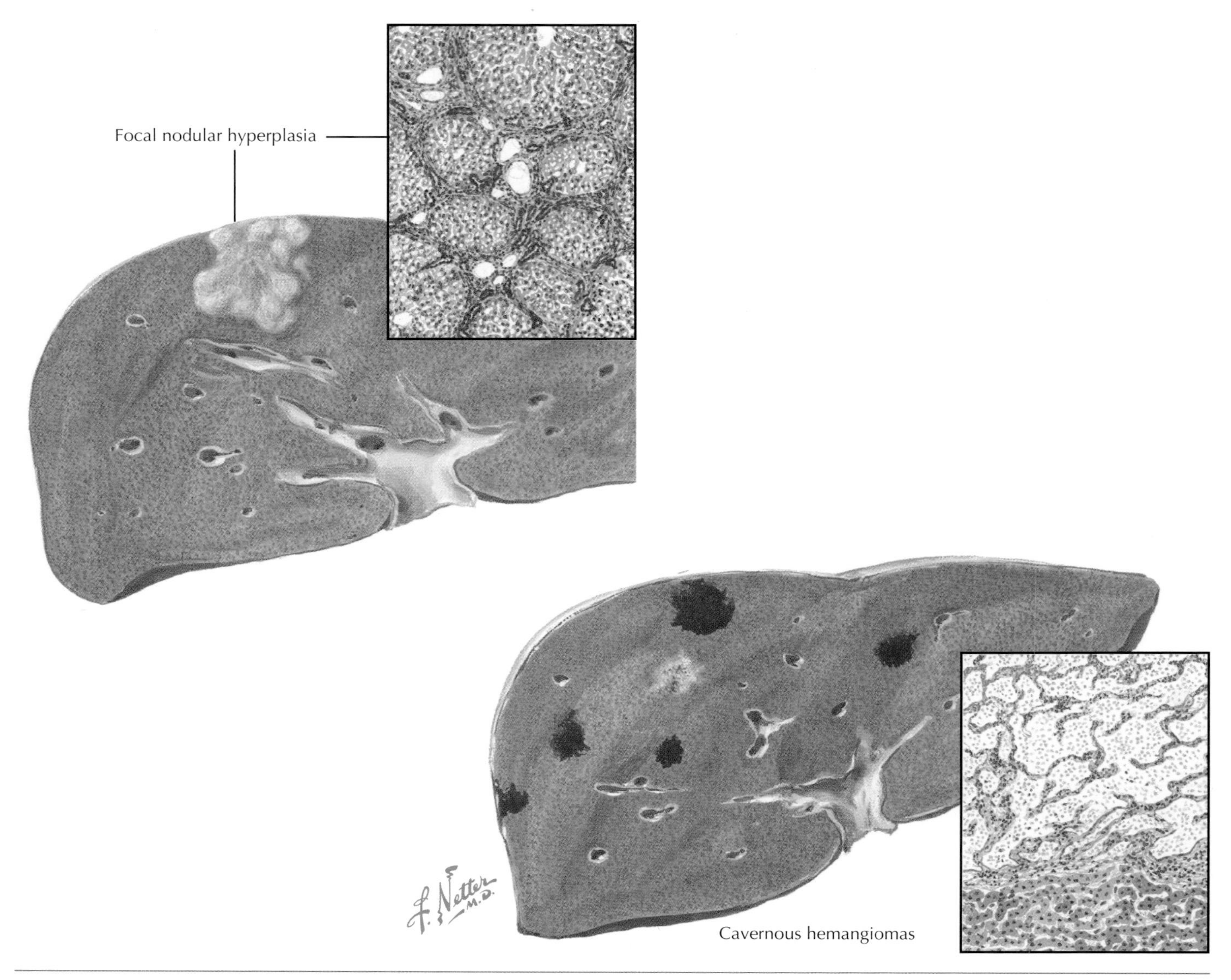

Figure 248-1 *Nodular Liver Lesions: Focal Nodular Hyperplasia and Cavernous Hemangioma.*

of the hemangioma. Occasionally, large cavernous hemangiomas are associated with disseminated intravascular coagulation, which is termed *Kasabach-Merritt syndrome.*

The diagnosis of cavernous hemangioma can be readily made by radiologic imaging. CT and MRI may reveal characteristic findings, including "puddling" of contrast within the tumor and increased intensity of signal during T2-weighted imaging.

Management of cavernous hemangiomas is largely expectant because there is minimal risk for malignant transformation. Surgery or angiographic embolization is contemplated primarily in symptomatic patients and those with evidence of bleeding. OC use should be discontinued. Liver transplantation has been performed in patients with very large lesions, especially in those with consumptive coagulopathy.

ADDITIONAL RESOURCES

Bartolozzi C, Cioni D, Donati F, Lencioni R: Focal liver lesions: MR imaging–pathologic correlation, *Eur Radiol* 11:1374-1388, 2001.

Gibbs JF, Litwin AM, Kahlenberg MS: Contemporary management of benign liver tumors, *Surg Clin North Am* 84(2):463-480, 2004.

Lin EC, Kuni CC: Radionuclide imaging of hepatic and biliary disease, *Semin Liver Dis* 21:179-194, 2001.

Trotter JF, Everson GT: Benign focal lesions of the liver, *Clin Liver Dis* 5:17-42, 2001.

Wanless IR: Benign liver tumors, *Clin Liver Dis* 6:513-526, 2002.

Granulomatous Liver Diseases

249

Kris V. Kowdley

Granulomatous liver diseases may be caused by infectious microbes or by systemic inflammatory or autoimmune processes. If granulomatous hepatitis is idiopathic, as it is in rare cases, the patient frequently has a fever of unknown origin. Granulomas are thought to originate from macrophages that are transformed into epithelioid cells after stimulation by antigens. Occasionally, granulomata are composed of multinucleated giant cells. Granulomas have been further classified into caseating and noncaseating granulomas. This pathologic distinction is clinically helpful because infectious processes such as tuberculosis often are associated with *caseating* granulomas, whereas autoimmune or inflammatory processes frequently cause *noncaseating* granulomas.

INFECTIOUS PROCESSES

The most common infectious diseases associated with granulomas are bacterial infections such as *Mycobacterium tuberculosis*, *Mycobacterium avium-intracellulare*, brucellosis, listeriosis, and tularemia. Other infections associated with granulomas include schistosomiasis, leishmaniasis, and visceral larva migrans. Rickettsial diseases and protozoal infections also may be associated with granulomas, as well as viruses (e.g., cytomegalovirus).

Hepatic Tuberculosis

Primary hepatic tuberculosis may occur in the extremely rare congenital form, but it is usually secondary to *miliary tuberculosis* (**Fig. 249-1**).

The most frequent lesion is the small miliary granuloma (tubercle), which may be scattered over the liver in all forms of active organ tuberculosis. Granuloma formation begins with a focal proliferation of Kupffer cells, which form small histiocytic nodules located throughout the parenchyma. Subsequently, liver cells surrounded by the histiocytes become necrotic, and in some cases, smaller or larger foci of hepatocellular necrosis with minimal mesenchymal reaction develop. In the nodules, some cells become larger and develop into epithelioid cells, the nuclei of which can divide without division of the cytoplasm, resulting in large giant cells (Langerhans). A lymphocytic infiltrate can be seen on the periphery of the granuloma. As the tubercle enlarges, central caseation necrosis may develop.

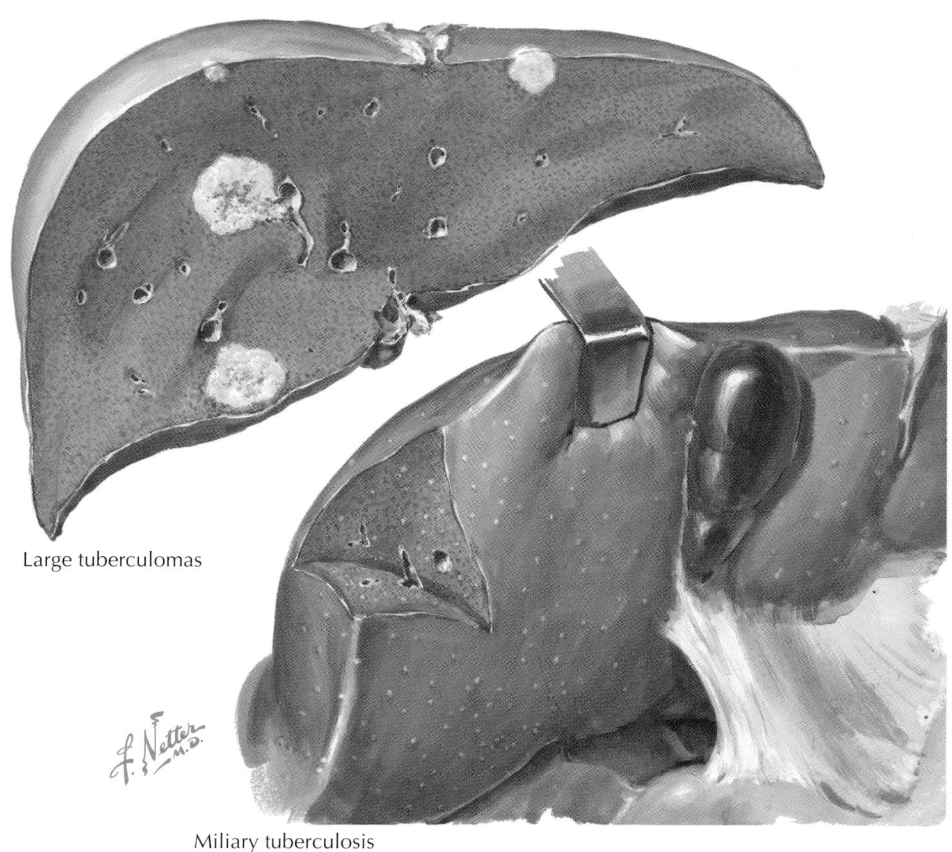

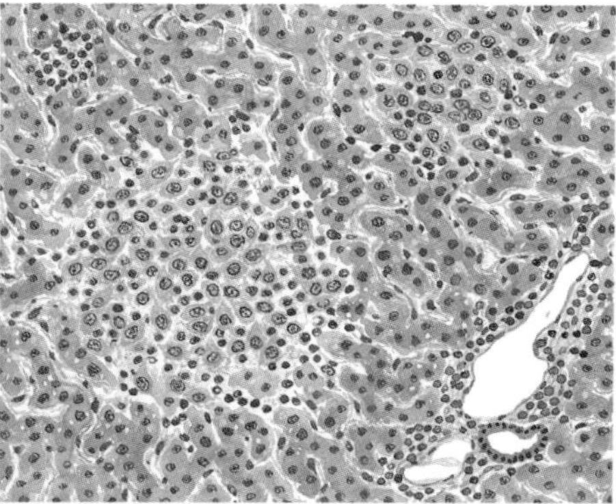

Acute miliary granulomas (soft tubercle – chiefly histiocytes)

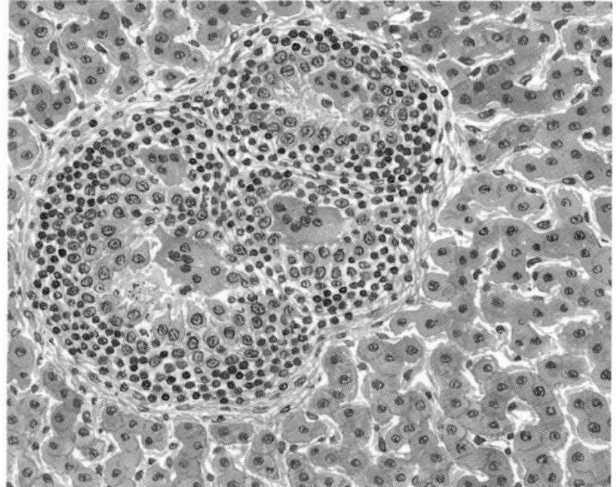

Subacute conglomerate tubercle (giant cells, caseation, histiocytes, surrounded by lymphocytes and fibrosis)

Figure 249-1 *Granulomatous Liver Diseases: Gross and Histologic Appearance.*

Eventually, the histiocytes may transform into fibroblasts and form a capsule around the tubercle. Finally, the entire lesion becomes transformed into a nodule of collagenous connective tissue.

Acid-fast bacilli are usually not seen in the scar, and tubercle bacilli are also difficult to culture from liver biopsy specimens. Additionally, the morphologic picture of the tubercles is not specific because other granulomatous diseases may produce similar lesions in the wall of the central vein. Tuberculous granulomas are spread all over the lobule, frequently close to the portal triads, where they tend to coalescence.

In miliary tuberculosis, the tubercles are densely spread, as readily seen on gross inspection, and they appear as white, pinhead-sized nodules best recognized through the capsule by inspection of the inferior surface of the left lobe. Hepatic miliary tuberculosis may precede pulmonary involvement, or the lungs may remain unaffected. Fever may be prolonged, and chest radiography findings may be negative; the diagnosis is made on liver biopsy.

Q Fever

Q fever is caused by *Coxiella burnetii*, a rickettsial agent that frequently infects cattle and other farm animals through oral or parenteral routes. The clinical features include fever, pulmonary infiltrates, and flulike symptoms.

SYSTEMIC AUTOIMMUNE OR IDIOPATHIC DISEASES

Granulomas can also be found in a variety of liver diseases localized to the liver and with liver involvement from systemic autoimmune or idiopathic diseases. Possibly the most common systemic disease associated with granulomas in the liver is *sarcoidosis* (**Fig. 249-2**). This is a relatively common disease of unknown cause, although dysregulation of the immune system is thought to play a role. Liver involvement in sarcoidosis is often manifested by abnormal liver test results in a cholestatic pattern, with a predominant elevation of serum alkaline phosphatase (ALP). Pruritus and fatigue may be common symptoms, but many patients are asymptomatic. The diagnosis is confirmed by an elevated serum angiotensin-converting enzyme (ACE) level, especially in the presence of other manifestations, such as mediastinal lymphadenopathy.

Idiopathic granulomatous hepatitis is an unusual disorder characterized by high fever and florid granulomatous infiltration in the liver. Liver biopsy is frequently performed in these patients because of an abnormal ALP level and a fever of unknown origin.

OTHER CONDITIONS CHARACTERIZED BY GRANULOMAS

Hepatic granulomas similar to those in tuberculosis and sarcoidosis are seen in many other conditions. In *brucellosis*, granulomas are irregularly spaced throughout the liver; they vary in size and degree of development and are accompanied by focal necrosis and portal inflammation (see Fig 249-2). In *histoplasmosis*, granulomas, resembling tubercles, occur together with a diffuse proliferation of the Kupffer cells, the cytoplasm of which is sometimes loaded with the fungus *Histoplasma capsulatum*. In other fungal diseases, such as *blastomycosis* or *coccidioidomycosis*, nonspecific reactive hepatitis is more frequent than granulomas in the liver. Tularemia, leprosy, and beryllium poisoning may also be associated with hepatic granulomas.

Primary biliary cirrhosis (PBC) is often characterized by multiple, small, noncaseating granulomas, and classic descriptions of PBC often include granulomas as a diagnostic criterion (see Chapter 236). Granulomas in PBC are primarily located in the periportal regions, whereas they may be found throughout the hepatic lobule with drug-induced or infectious causes.

DIAGNOSIS

Careful assessment of the patient with hepatic granulomas is essential because many infectious causes can be identified based on thorough history taking, with particular attention to travel, medication history, occupational history, social history (e.g., drug abuse) sexual history, and presence of systemic disorders, immunosuppression, or autoimmunity.

TREATMENT AND MANAGEMENT

Treatment of granulomatous liver diseases is based on the underlying cause. Discontinuation of any offending drug, such as allopurinol or phenylbutazone, is indicated. Idiopathic granulomatous hepatitis may be responsive to corticosteroids.

ADDITIONAL RESOURCES

Akritidis N, Tzivras M, Delladetsima I, et al: The liver in brucellosis, *Clin Gastroenterol Hepatol* 5:1109, 2007.

Drebber U, Kasper HU, Ratering J, et al: Hepatic granulomas: histological and molecular pathological approach to differential diagnosis—a study of 442 cases, *Liver Int* 28(6):828-834, 2008.

Guckian JC, Perry JE: Granulomatous hepatitis of unknown etiology: an etiologic and functional evaluation, *Am J Med* 44(2):207-215, 1968.

Ishak KG: Sarcoidosis of the liver and bile ducts, *Mayo Clin Proc* 73(5): 467-472, 1998.

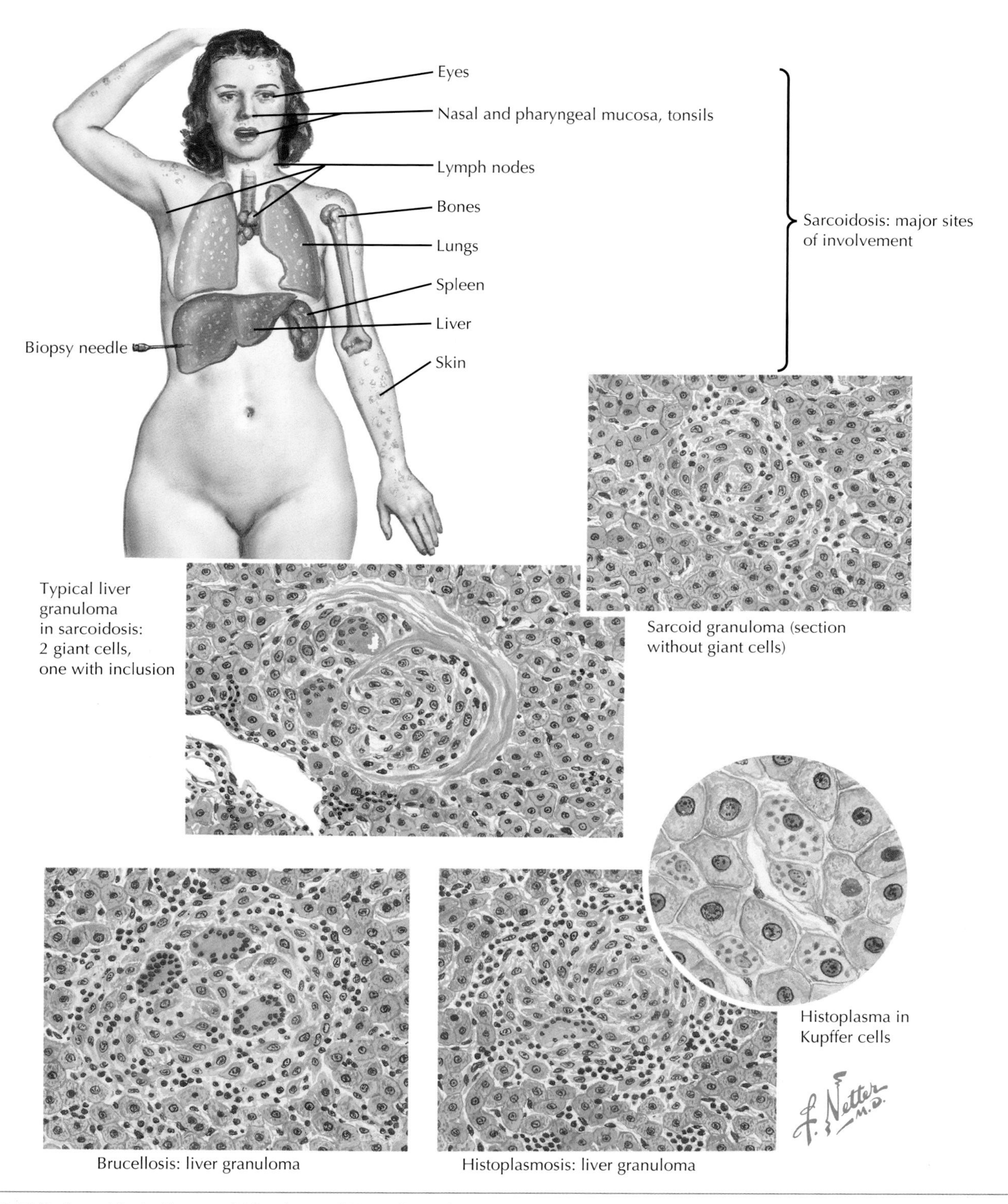

Figure 249-2 *Sarcoidosis Sites and Histology of Liver Granulomas.*

Hepatic Amebiasis

250

Kris V. Kowdley

Invasive amebiasis may be complicated by the development of an *amebic liver abscess* (**Fig. 250-1**). *Entamoeba histolytica* is associated with amebic liver abscess. This parasite is a commensal organism, and most exposed individuals pass cysts in the stool. However, invasive amebiasis can result in systemic illness, including fever, nausea, and diarrhea. Symptoms of amebic liver abscess include pain (often in right upper quadrant), fever, nonproductive cough, anorexia, nausea, vomiting, and diarrhea. *E. histolytica* is transmitted primarily through feces in contaminated water. The source of entry into the liver is presumed to be the portal vein. Young men 20 to 40 years of age are most often affected.

The typical amebic liver abscess is found in the right lobe of the liver and is a solitary lesion, although multiple lesions may be present. The most frequent method of initial diagnosis is ultrasonography or computed tomography. The diagnosis is best made by finding antiamebic antibodies using indirect hemagglutination and enzyme-linked immunosorbent assay. Antibody test findings may be negative very early after infection but will remain positive for several months after infection.

The most serious complications of amebic abscess include spread by rupture or fistulization into the chest, resulting in pulmonary or hepatopulmonary abscess, hepatobronchial fistula, or hematogenous dissemination into the brain or other organs. Abdominal ultrasonography is generally adequate for identifying amebic abscess and reveals a hypoechoic lesion. Percutaneous aspiration should be considered if there is concern about rupture or spread to adjacent organs. Stool tests to detect *E. histolytica* may not be helpful because patients generally do not have intestinal amebiasis at presentation with an amebic liver abscess. Treatment with metronidazole or dehydroemetine, as well as tinidazole, ornidazole, and nitazoxanide, are effective for amebic liver abscess. After therapy for the liver abscess, treatment for luminal infection is also recommended, even if the stool is not negative for *E. histolytica* (see Chapter 174).

ADDITIONAL RESOURCES

Fung HB, Doan TL: Tinidazole: a nitroimidazole antiprotozoal agent, *Clin Ther* 27(12):1859-1884, 2005.

Haque R, Huston CD, Hughes M, et al: Amebiasis, *N Engl J Med* 348(16):1565-1573, 2003.

Hughes MA, Petri WA Jr: Amebic liver abscess, *Infect Dis Clin North Am* 14:565-582, 2000.

Salles JM, Salles MJ, Moraes LA, Silva MC: Invasive amebiasis: an update on diagnosis and management, *Expert Rev Anti Infect Ther* 5(5):893-901, 2007.

Sharma MP, Ahuja V: Management of amebic and pyogenic liver abscess, *Indian J Gastroenterol* 20(suppl 1):C33-C36, 2001.

Stanley SL Jr: Amoebiasis, *Lancet* 361:1025-1034, 2003.

Wells CD, Arguedas M: Amebic liver abscess, *South Med J* 97(7):673-682, 2004.

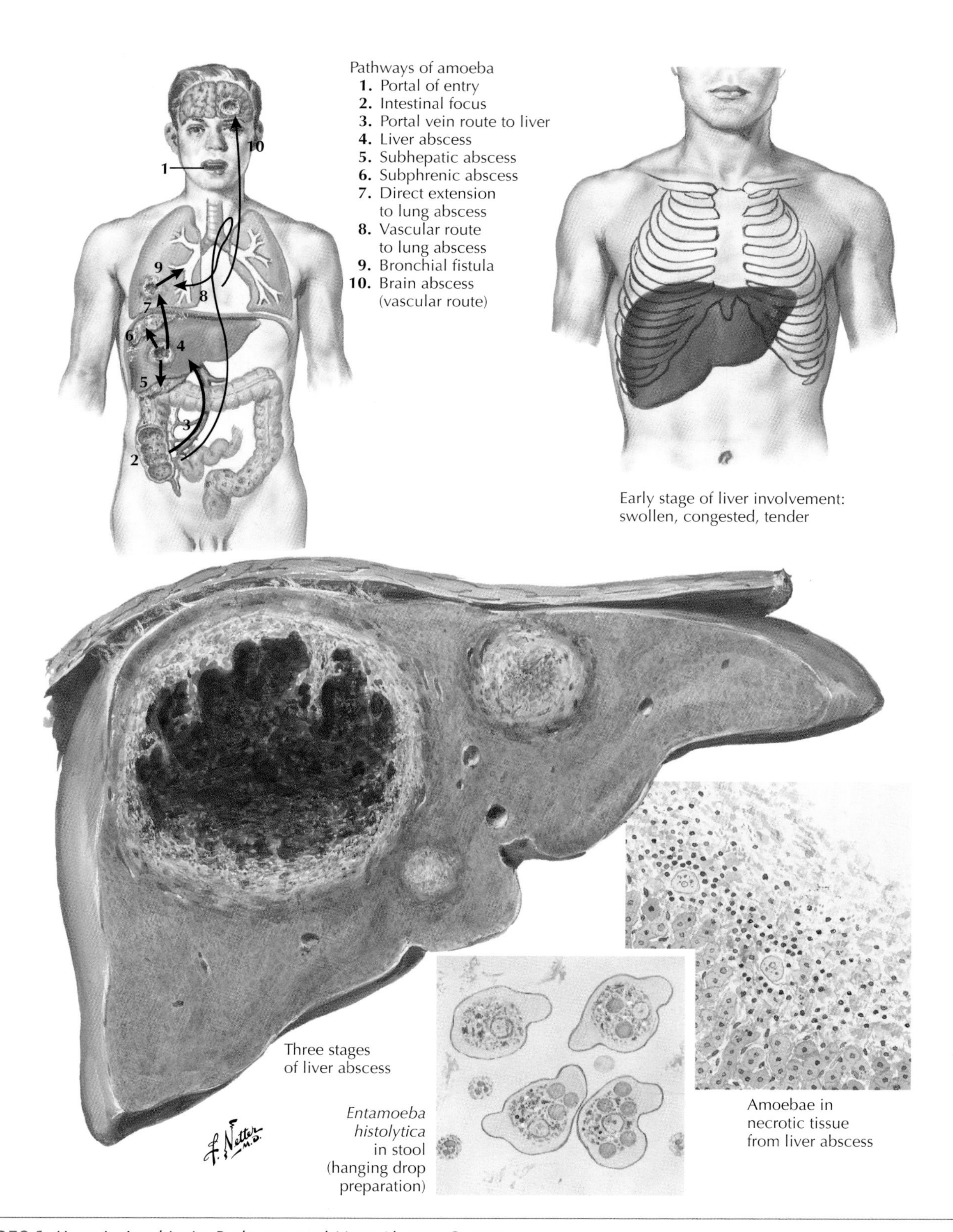

Figure 250-1 *Hepatic Amebiasis: Pathways and Liver Abscess Stages.*

Wilson Disease

251

Kris V. Kowdley

Wilson disease (WD) is named for the neurologist Samuel Kinnier Wilson, who first characterized the familial syndrome of lenticular degeneration associated with cirrhosis in 1912. It was not until 1948, however, that WD was recognized as a disorder of *copper* metabolism in which excess copper accumulates in the liver, central nervous system, and other tissues.

Wilson disease is an autosomal recessive disorder with an incidence in most populations of approximately 1 in 30,000. Gene frequency varies between 0.3% and 0.7%, leading to an estimated heterozygous carrier rate of 1 in 90. WD affects men and women equally and occurs in all races.

A gene for Wilson disease *(WND)* has been mapped to chromosome 13 (13q 14.3). The gene is expressed primarily in liver, kidney, and placenta but has also been found in other organs. The gene product is a P-type adenosine triphosphatase (ATPase), which transports copper into bile and incorporates it into ceruloplasmin.

Up to 60 mutations within the *WND* locus have been identified, although less than half are thought to have clinical significance. Point mutations are most common, but deletions, insertions, and splice-site mutations have all been reported. A specific mutation, His 1069Gln, appears more common in WD patients of European descent, although this mutation is present in only 15% to 25% of all patients with WD. Because of the genetic heterogeneity, genetic testing is not useful for a diagnosis of WD in probands. Genetic testing can be used, however, to identify affected family members once a proband has received a diagnosis.

Impaired biliary excretion of copper appears to be fundamental to the pathogenesis of WD. Body copper stores are normally regulated through biliary excretion. The *WND* gene product may be responsible for transporting copper from hepatocytes into the biliary system, and this function may be impaired in WD patients. However, it is unclear how the defect in ATP7B, the *WND* protein, alters this pathway.

The clinical sequelae of WD result from the accumulation of excess tissue copper. It is believed that tissue damage is caused by free radicals, along with depletion of glutathione and oxidative stress. Accumulation of copper in the liver results in chronic hepatitis, fibrosis, and cirrhosis. Fulminant hepatic failure may be a rare complication. Extrahepatic accumulation of copper in brain, joints, kidney, cornea, heart, and pancreas may occur. Neuropsychiatric manifestations are the result of copper overload in the brain. Kayser-Fleischer rings result from copper deposition in Descemet membrane of the cornea.

Abnormal liver histology may be seen in biopsy specimens of patients with asymptomatic WD, even in childhood. The earliest changes seen are fatty deposition, glycogenation of nuclei, and mitochondrial abnormalities. If the patient receives no treatment, WD progresses to cirrhosis.

CLINICAL PICTURE

The clinical presentation of WD is variable, but most patients seek treatment for hepatic or neurologic symptoms (**Fig. 251-1**). The biochemical defect is present at birth, but clinical symptoms rarely develop before 5 years. Age at presentation seems to correlate with the organ system involved. The average age for hepatic symptoms is 10 to 14 years, but for neurologic symptoms, 19 to 22 years; patients rarely present after age 40.

Neuropsychiatric manifestations are typically seen in adolescents or young adults and are often the presenting symptoms in symptomatic patients. Psychiatric symptoms include depression, mood disorders, and personality changes. The most common neurologic manifestations are tremor, drooling, hypertonicity, choreoathetosis, and parkinsonian-like findings. Other manifestations include osteopenia, distal renal tubular acidosis, hypercalciuria, arrhythmias, congestive heart failure, glucose intolerance, and amenorrhea caused by copper deposition in extrahepatic organs.

DIAGNOSIS

The diagnosis of WD is made using a combination of clinical, histologic, and biochemical data. The simplest screening tests are slit-lamp examination for Kayser-Fleischer rings and serum ceruloplasmin level. Although a low ceruloplasmin level (<20 mg/dL) suggests WD, it is not specific. Urinary copper excretion is usually greater than 100 μg daily, but it may not be elevated to this degree in some asymptomatic patients.

Liver biopsy with hepatic copper concentration is helpful for making the diagnosis of WD. Hepatic copper concentrations greater than 250 μg/g dry weight (normal, <35 μg/g dry weight) are often found in patients with untreated WD.

Genetic diagnosis of WD is not universally available at present and may not be definitive, although it is expected that genotyping for the common WD-associated mutations may be feasible in the near future.

TREATMENT AND MANAGEMENT

The goals of WD therapy include removing excess body copper and preventing reaccumulation. Copper-chelating agents such as *D-penicillamine* and *trientine* represent the mainstay of initial therapy. Trientine is increasingly used as first-line therapy because of adverse effects with D-penicillamine. *Zinc acetate* (Galzin) therapy has recently become popular as maintenance; the mechanism of action may be induction of metallothionein, which blocks copper absorption from the intestinal tract. *Ammonium tetrathiomolybdate* is an oral agent that binds to plasma copper and inhibits intestinal absorption; it has been advocated as better therapy for neurologic WD.

Patients with WD are advised to avoid copper-containing foods such as shellfish, chocolate, nuts, and liver.

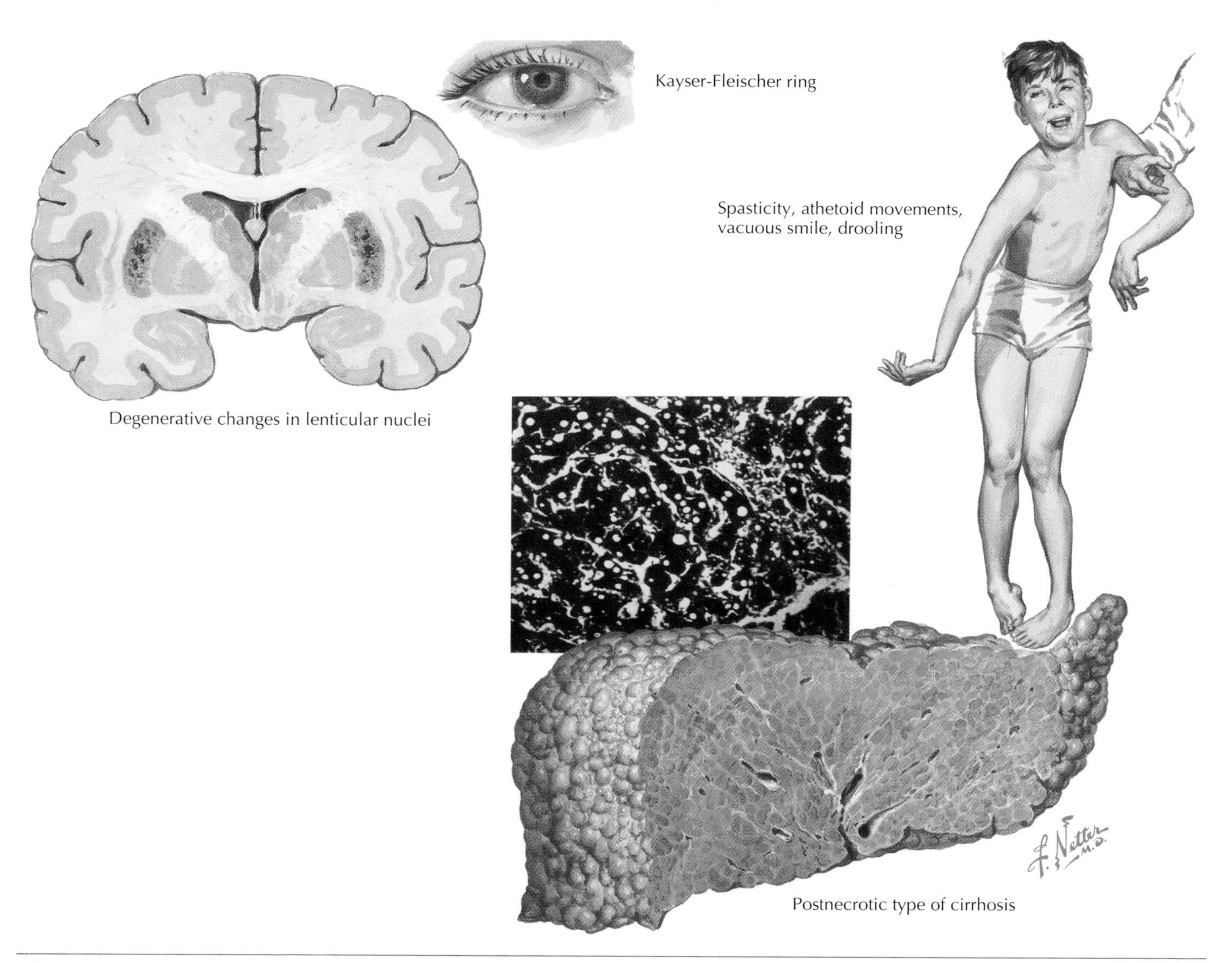

Figure 251-1 *Wilson Disease.*

Liver transplantation is effective therapy for patients who have progressive disease despite therapy or who have fulminant WD. It results in excellent posttransplantation outcomes. Because the primary defect resides within the liver, transplantation is curative.

ADDITIONAL RESOURCES

El-Youssef M: Wilson disease, *Mayo Clin Proc* 78:1126-1136, 2003.

Ferenci P, Caca K, Loudianos G, et al: Diagnosis and phenotypic classification of Wilson disease, *Liver Int* 23:139-142, 2003.

Gitlin JD: Wilson disease, *Gastroenterology* 125:1868-1877, 2003.

Roberts EA, Schilsky ML, American Association for Study of Liver Diseases (AASLD): Diagnosis and treatment of Wilson disease: an update, *Hepatology* 47(6):2089-2111, 2008.

Schisky ML: Diagnosis and treatment of Wilson's disease, *Pediatr Transpl* 6:15-19, 2002.

Hepatocellular Carcinoma

Kris V. Kowdley

Hepatocellular carcinoma (HCC) is potentially the most serious complication of chronic liver disease. Five-year survival of patients with HCC remains disappointing, despite dramatic advances in the epidemiology, diagnosis, and management of this disease. The epidemiology of HCC has been studied in detail recently, and its incidence appears to be increasing. Major risk factors associated with HCC worldwide include chronic hepatitis B and cirrhosis of any cause, but particular risk factors are chronic hepatitis C and alcoholic liver disease.

Histologic features of HCC are similar to those of other malignancies, with increased nuclear/cytoplasmic ratios and multiple mitotic figures (**Fig. 252-1**). Distinguishing features of hepatocytes may be retained, although normal sinusoidal arrangement is not seen. In addition, features of bile duct cells may be present. HCC has also been classified from "very well differentiated" to "poorly differentiated" cancer. Some types may be associated with extensive scar formation and are called *scirrhous* form. Other types may form structures similar to glands (*pseudoglandular* form).

Stage of HCC is important for predicting prognosis and directing therapy. Several classification schemes have been proposed.

CLINICAL PICTURE

Hepatocellular carcinoma should be suspected in any patient with evidence of acute or subacute decompensation of chronic liver disease. HCC should also be suspected in patients with increased abdominal pain, particularly over the right upper quadrant. A firm or "rock-hard" mass may be palpable; a vascular bruit may be present given the highly vascular nature of this tumor. Associated paraneoplastic phenomena may be present. Acute rupture of the tumor may lead to hemoperitoneum, shock, or even death.

DIAGNOSIS

The diagnosis of HCC may be suspected in a patient with high serum *α-fetoprotein* (AFP) level and/or a mass lesion seen on ultrasonography (US) or computed tomography (CT). Many patients with cirrhosis are now being diagnosed during biannual US or CT surveillance. Screening for HCC is recommended because the patients identified may be eligible for specific therapies, such as surgical resection, ablative therapy, or liver transplantation.

Serum AFP has been widely used to screen for HCC in populations at risk. Immature liver cells and HCC cells produce AFP. Patients with cirrhosis who have AFP levels above 300 ng/mL likely have HCC. However, it is being increasingly recognized that AFP is not sufficiently sensitive or specific for a diagnosis of HCC, and the most recent American Association for the Study of Liver Diseases (AASLD) guidelines do not include AFP as a preferred test for HCC surveillance and favor imaging techniques. Furthermore, approximately 30% of HCCs do not produce AFP. Serum AFP can be used if imaging studies are not available.

Ultrasound is a useful screening test for patients with cirrhosis. However, a significant proportion of nodular lesions found on US are regenerative nodules rather than HCCs. Helical CT permits rapid acquisition of liver images at various stages of contrast injection (noncontrast, hepatic arterial phase, portal venous phase) and has dramatically improved the ability to diagnose HCC. More importantly, helical CT can differentiate HCC nodules from regenerative nodules. The main disadvantage of CT screening is the cumulative risk of radiation over time and the possible nephrotoxicity of contrast agents used in CT and magnetic resonance imaging. Other means to enhance the visualization of HCC include administering oil-based contrast agents such as lipiodol. Delayed imaging can be performed several days after administration of the agent. Although widely used in Japan, this technique is seldom used in the United States.

TREATMENT AND MANAGEMENT

Treatment of HCC has evolved rapidly over the past several years. It was long believed that resection for cure was the only possible therapy offering long-term survival. However, several new modalities have been increasingly used in addition to resection, including radiofrequency ablation and chemoembolization using lipiodol and chemotherapeutic agents. Liver transplantation in appropriately selected candidates can be highly effective, with excellent long-term survival without recurrence. *Sorafenib* was recently approved as an oral chemotherapeutic agent for patients with advanced HCC and is the first such agent shown to be associated with improved survival in HCC patients.

ADDITIONAL RESOURCES

Bruix J, Sherman M: Practice Guidelines Committee, American Association for the Study of Liver Diseases: Management of hepatocellular carcinoma, *Hepatology* 42(5):1208-1236, 2005.

Di Bisceglie AM: Epidemiology and clinical presentation of hepatocellular carcinoma, *J Vasc Interv Radiol* 13(pt 2):S169-S171, 2002.

El-Serag HB: Hepatocellular carcinoma and hepatitis C in the United States, *Hepatology* 36(suppl 1):74-83, 2002.

Ijzermans JN, Bac DJ: Recent developments in screening, diagnosis and surgical treatment of hepatocellular carcinoma, *Scand J Gastroenterol Suppl* 223:50-54, 1997.

Llovet JM, Burroughs A, Bruix J: Hepatocellular carcinoma, *Lancet* 362:1907-1917, 2003.

Llovet JM, Ricci S, Mazzaferro V, et al: SHARP Investigators Study Group: Sorafenib in advanced hepatocellular carcinoma, *N Engl J Med* 359(4):378-390, 2008.

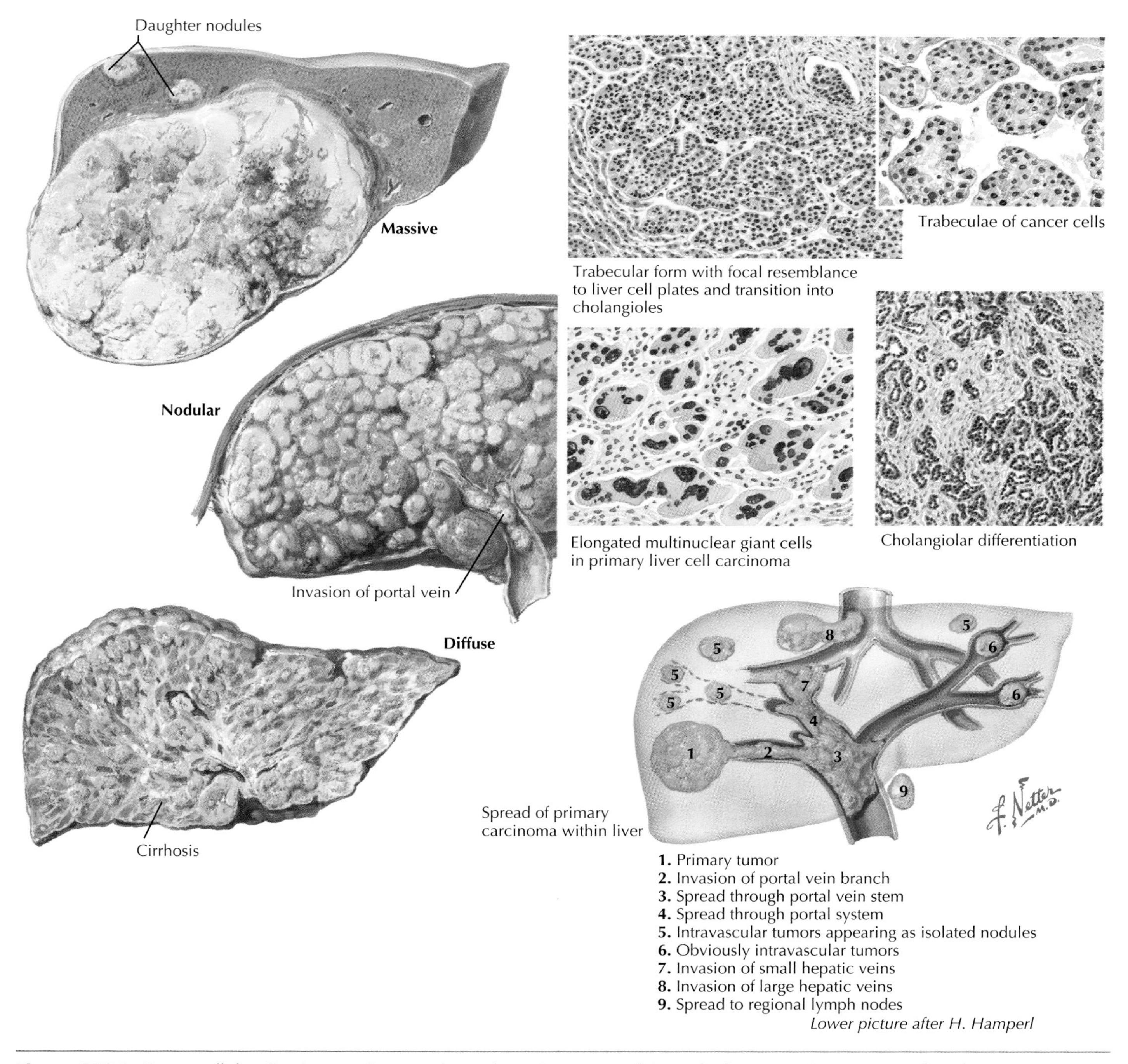

Figure 252-1 *Hepatocellular Carcinoma: Gross and Histologic Features and Spread of Primary Carcinoma Within the Liver.*

Ryder SD: British Society of Gastroenterology: guidelines for the diagnosis and treatment of hepatocellular carcinoma (HCC) in adults, *Gut* 52(suppl 3):1-8, 2003.

Sherman M: Screening for hepatocellular carcinoma, *Best Pract Res Clin Gastroenterol* 19(1):101-18, 2005.

Sherman M: Surveillance for hepatocellular carcinoma and early diagnosis, *Clin Liver Dis* 11(4):817-837, viii, 2007.

Trinchet JC, Ganne-Carrie N, Beaugrand M: Intra-arterial treatments in patients with hepatocellular carcinoma (review), *Aliment Pharmacol Ther* 17(suppl 2):111-118, 2003.

Liver Transplantation

Kris V. Kowdley

Liver transplantation has developed from an experimental procedure in the 1970s to a highly successful therapy at present. The success of liver transplantation and great demand for donor organs have resulted in priority allocation systems and increased live-donor transplantation, in which a portion of the liver from a donor is transplanted into a recipient.

Major current indications for liver transplantation are end-stage liver disease with significantly decreased short-term survival, acute fulminant liver failure, and unresectable hepatocellular carcinoma (HCC) (**Fig. 253-1**). Rare indications for liver transplantation include metabolic disease, large benign tumors, and severely impaired quality of life, as in patients with refractory pruritus from cholestatic liver disease.

The process of selection and assignment of priority for liver transplantation is managed by the United Network for Organ Sharing (UNOS), a national nonprofit organization that administers organ procurement and allocation with funding from the U.S. government through the Organ Procurement and Transplantation Network.

Major complications of chronic end-stage liver disease that should prompt referral to a liver transplantation center include evidence of decompensated liver disease, such as ascites, variceal bleeding, spontaneous bacterial peritonitis, or suspected hepatorenal syndrome. Known or suspected HCC is another indication for liver transplantation, if the lesion is within accepted UNOS criteria.

Because cadaveric donor livers are a scarce resource, with many more potential recipients than available donors, several allocation schemes have been used for determining priority for liver transplantation. The original system prioritized patients according to the Child-Pugh-Turcotte scoring system. Previous studies had demonstrated that survival was significantly worse in patients with Child C cirrhosis, followed by Child B cirrhosis. By contrast, patients with Child A cirrhosis had superior short and intermediate survival. Child-Pugh class is determined by assigning the patients 1, 2, or 3 points based on serum albumin, bilirubin, prothrombin time (PT), and presence or absence of encephalopathy and ascites. Patients were classified as having status 2b or 3 based on Child-Pugh-Turcotte score. Within each status, waiting time, blood type, and body weight determined priority. However, this system placed at a disadvantage those patients who were very ill but were referred late for transplantation evaluation, resulting in shorter times and lower priority status. Furthermore, the system did not have a mechanism for providing early transplantation for patients with HCC, for whom it is critical that transplantation be performed early (see Chapter 252).

These challenges led to the development of the Model for End-Stage Liver Disease (MELD) scoring system, which is currently used for determining priority for liver transplantation. The MELD system was originally designed to predict prognosis among patients who underwent transjugular intrahepatic portosystemic shunt (TIPS) placement. Subsequently, it was discovered that this model also was useful for predicting survival among patients with advanced liver disease. The MELD scoring system uses serum bilirubin, PT/international normalized ratio (INR), and creatinine to calculate a MELD score. Priority for liver transplantation in patients with chronic liver disease is based strictly on MELD score. In addition, patients with HCC are given additional MELD "points" to facilitate early transplantation.

These systems do not apply to patients with acute fulminant liver failure, as occurs in acute severe viral or drug-induced hepatitis or in acute severe Wilson disease. Patients in this category are assigned the highest priority for transplantation and are eligible to receive a donor organ not only from the local and UNOS regions, but also through a national network based on priority.

Detailed and thorough pretransplantation evaluation is conducted to identify any concurrent medical conditions that might preclude liver transplantation, including a detailed psychosocial assessment, with particular emphasis on adequate social support, need for counseling or recovery, support programs to maintain alcohol abstinence (in patients with history of chronic alcohol use), vaccinations, and detailed serologic and radiologic evaluation. HLA haplotyping and detailed immunologic assessment are not required.

Posttransplantation immunosuppression regimens include tacrolimus or cyclosporine, azathioprine or mycophenolate mofetil, with or without corticosteroids. In addition, many transplantation programs use antagonists to interleukin-2 as induction therapy to prevent early rejection.

Outcomes after liver transplantation are excellent among patients who undergo transplantation for chronic liver disease, and more than 95% of patients survive surgery. One-year survival rates exceed 80% in most established programs. Factors associated with lower survival after transplantation include fulminant liver failure, age older than 60, and renal failure requiring concomitant renal transplantation.

Long-term outcomes are strongly influenced by the underlying disease. Thus, hemochromatosis, hepatitis B, and HCC are associated with lower long-term survival; hepatitis C and alcoholic liver disease with intermediate survival; and autoimmune liver disease (autoimmune hepatitis, primary biliary cirrhosis, primary sclerosing cholangitis) with excellent long-term survival (85%). Improved selection of HCC patients and new antiviral therapies for hepatitis C have been associated with improved outcomes in these patients.

Developments to maximize use include split-liver transplantation, in which portions of one liver can be transplanted into more than one recipient, and live-donor liver transplantation, in which a portion of the liver (usually the right lobe in adults) is removed from a living donor (usually a family member) and is transplanted into a recipient. These procedures have rapidly become alternative means of liver replacement therapy and have been lifesaving treatments for many patients, especially those

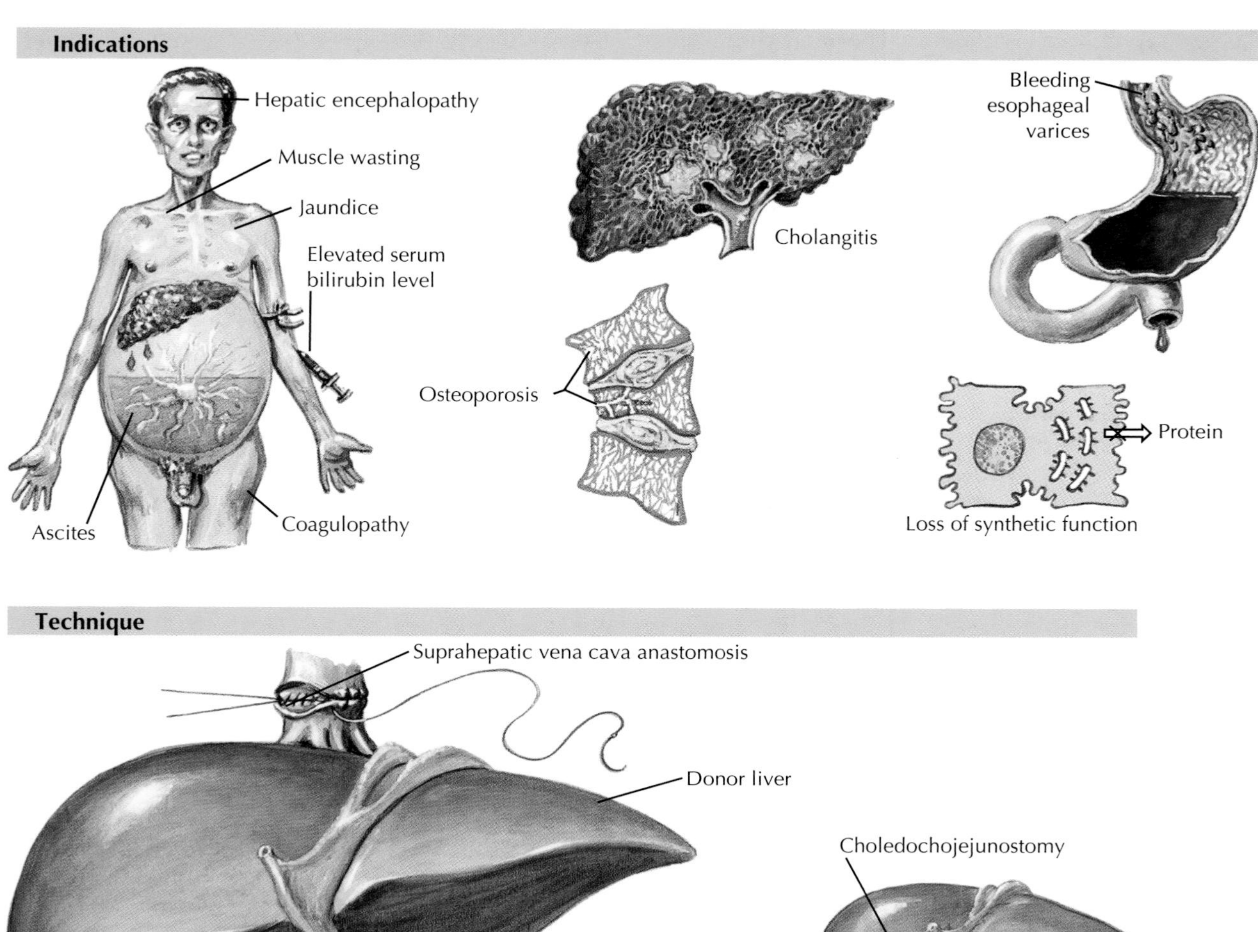

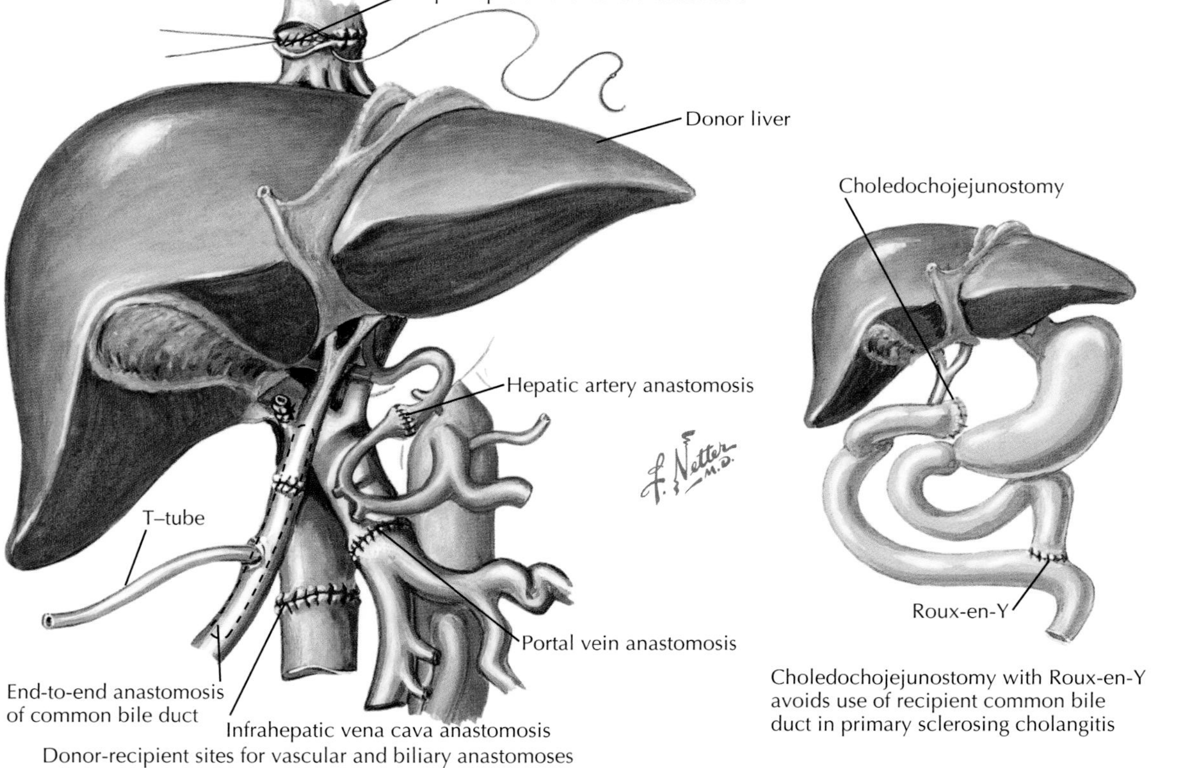

Figure 253-1 *Liver Transplantation: Indications and Technique.*

with HCC and severe liver insufficiency. However, ethical and psychosocial issues surround live-donor liver transplantation and are debated among ethicists and social scientists.

The long-term care of the liver transplant recipient requires a multidisciplinary approach. In addition to monitoring for rejection of the allograft, patients require surveillance for malignancy, opportunistic infection, osteoporosis, diabetes, hypertension, atherosclerosis, and renal insufficiency.

ADDITIONAL RESOURCES

Botero RC, Lucey MR: Organ allocation: model for end-stage liver disease, Child-Turcotte-Pugh, Mayo Risk Score, or something else, *Clin Liver Dis* 7:715-727, 2003.

Carthers RL Jr: Liver transplantation: American Association for the Study of Liver Diseases, *Liver Transpl* 6:122-135, 2000.

Devlin J, O'Grady J: Indications for referral and assessment in adult liver transplantation: a clinical guideline, British Society of Gastroenterology, *Gut* 45(suppl 6, VI):1-22, 1999.

Freeman RB: The impact of the model for end-stage liver disease on recipient selection for adult living liver donation, *Liver Transpl* 9(suppl 2):54-59, 2003.

Koffron A, Stein JA: Liver transplantation: indications, pretransplant evaluation, surgery, and posttransplant complications, *Med Clin North Am* 92(4):861-888, ix, 2008.

Lucey MR, Brown KA, Everson GT, et al: Minimal criteria for placement of adults on the liver transplant waiting list: a report of a national conference organized by the American Society of Transplant Physicians and the American Association for the Study of Liver Diseases, *Transplantation* 66:956-962, 1998.

O'Leary JG, Lepe R, Davis GL: Indications for liver transplantation, *Gastroenterology* 134(6):1764-1776, 2008.

Rosen HR: Transplantation immunology: what the clinician needs to know for immunotherapy, *Gastroenterology* 134(6):1789-801, 2008.

Spirochetal Infections

254

Kris V. Kowdley

Weil syndrome, also called infectious jaundice or spirochetal jaundice, is a severe form of leptospirosis caused by *Leptospira icterohaemorrhagiae* (**Fig. 254-1**). The disease has spread worldwide. Carriers are wild rats, dogs, and to a lesser degree, mice. These animals excrete leptospiras with the urine into stagnant water, where the organisms may survive for months. Human infection takes place either through skin abrasions or through the mouth.

Weil syndrome varies in severity. After an incubation period (6-12 days), high fever, headaches, abdominal pain, prostration, muscle pain, and conjunctivitis appear. At this stage, leptospiral organisms can be demonstrated in the blood or cerebrospinal fluid. About 10 days later, the fever subsides and a toxic stage develops in which renal manifestations (sometimes progressing to renal failure), meningitis, myocardial damage, dermal and conjunctival petechiae, epistaxis, and skin rashes are conspicuous. Liver involvement occurs in approximately 50% of patients. In this period, leptospiras are more readily found in urine than in blood. The fever may recur. After the third week, a slow convalescence begins, and serum antibody findings become positive.

Despite the frequency of hepatic involvement in the patient with spirochetal jaundice, the liver shows nonspecific changes such as centrilobular necrosis, a portal inflammatory infiltrate, and swollen Kupffer cells. The degree of jaundice is out of proportion to the liver dysfunction, partly because of hemolysis. Care is primarily supportive, including treatment of bleeding and renal failure. Death resulting from liver failure is rare. Antibiotics are effective only if given early in the course of Weil syndrome.

Liver disease caused by *syphilis* is infrequently observed now because of improvements in early diagnosis and the availability of effective therapy. Moreover, many cases of hepatic disease in persons with syphilis may be caused by hepatitis B or C. The liver lesion now recognized to be specifically caused by secondary syphilis is the scar formed after extensive specific coagulation necrosis (gumma), which leads to focal loss of hepatic tissue. The resultant irregular deformation of the liver frequently causes bizarre shapes and is designated *hepar lobatum*. Occasionally, however, deformation with enlargement of the left lobe and shrinkage of the right lobe causes unusual findings on liver palpation. Rarely, fresh yellow gummatous areas are found in the depths of the scars. Formerly, the now almost-extinct "brimstone" liver was frequently found in deeply jaundiced newborns as a characteristic of congenital syphilis, together with other syphilitic manifestations. Microscopic features are small miliary necroses (gummata), diffuse interstitial hepatitis, separated and distorted liver cell plates, increased interlobular connective tissue with intense inflammation, and demonstration of numerous spirochetes using a silver stain.

Lyme disease, a tick-borne disease caused by the spirochete *Borrelia burgdorferi*, can also be associated with liver test abnormalities in addition to myositis, fever, and splenic involvement.

ADDITIONAL RESOURCES

Aguero-Rosenfeld ME: Laboratory aspects of tick-borne diseases: Lyme, human granulocytic ehrlichiosis and babesiosis, *Mt Sinai J Med* 70(3):197-206, 2003.

Campisi D, Whitcomb C: Liver disease in early syphilis, *Arch Intern Med* 139:365-366, 1979.

Edwards GE, Domm BM: Human leptospirosis, *Medicine* 39:117, 1960.

Heath CW, Alexander AD, Gallon MM: Leptospirosis in the United States: analysis of 483 cases in man, 1949-1961, *N Engl J Med* 273:915-922, 1975.

Keisler DS, Starke W, Looney DJ, Mark WW: Early syphilis with liver involvement, *JAMA* 247:1999-2000, 1982.

Lee RV, Thornton GF, Conn HO: Liver disease associated with secondary syphilis, *N Engl J Med* 284:1423-1425, 1971.

Schlossberg D: Syphilitic hepatitis: a case report and review of the literature, *Am J Gastroenterol* 82:552-553, 1987.

Veeravahu M: Diagnosis of liver involvement in early syphilis, *Arch Intern Med* 145:132-134, 1985.

Zaidi SA, Singer C: Gastrointestinal and hepatic manifestations of tickborne diseases in the United States, *Clin Infect Dis* 34(9):1206-1212, 2002.

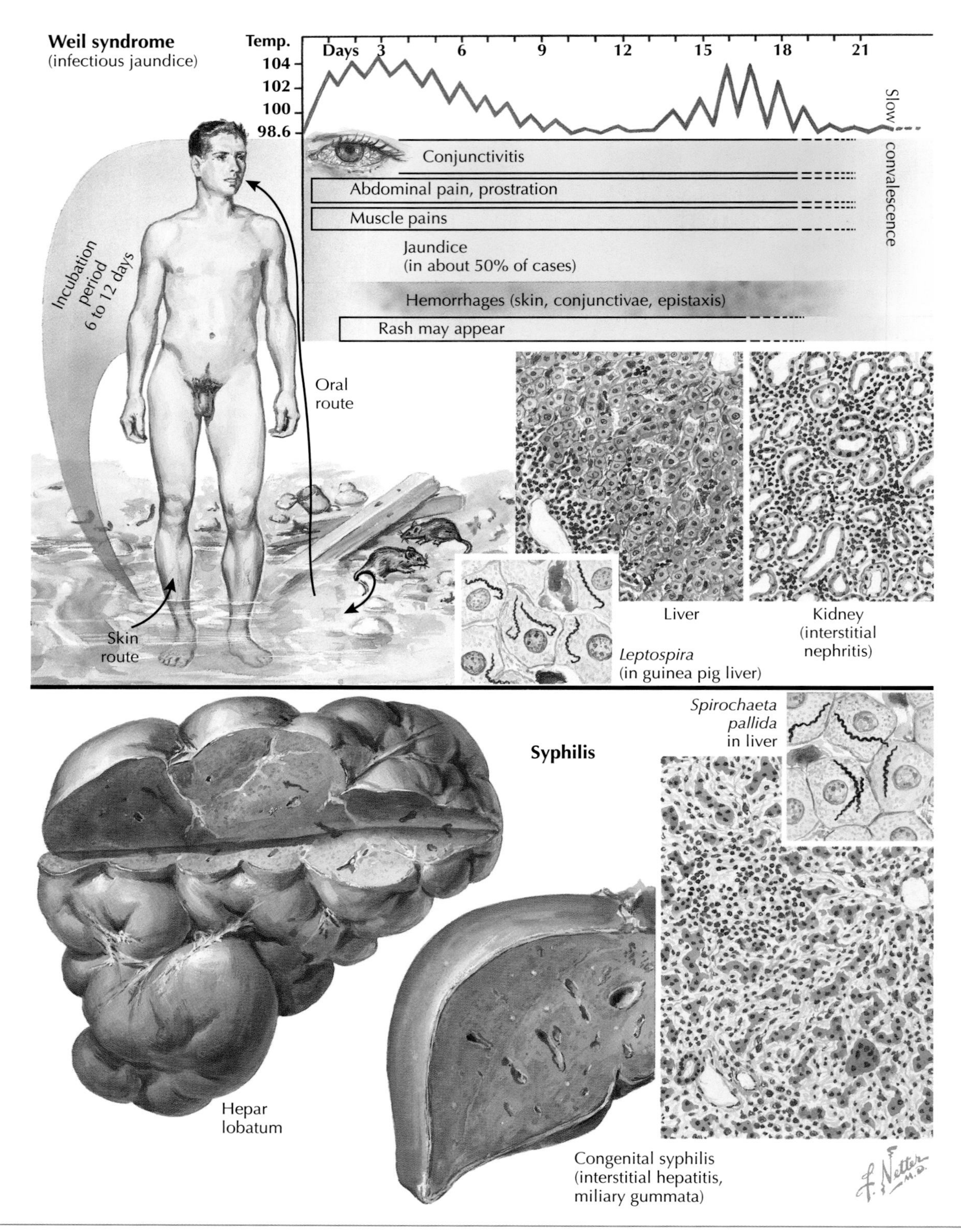

Figure 254-1 *Spirochetal Infections: Weil Syndrome and Syphilis.*

Actinomycosis

Kris V. Kowdley

Actinomycosis is an infection by an anaerobic fungus, *Actinomyces bovis*, which is found on many plants and also frequently as a harmful saprophyte in the oral cavity, especially in peridental structures and on the tonsils. The fungus, on rare occasions, enters the deeper tissues through a break in the mucosa or skin and produces suppuration. Typical initial localizations of the abscesses are the jaws, the lung, and the intestine, especially cecum and appendix. From the primary localization, the suppuration spreads through the vicinity.

Characteristically, actinomycotic abscesses do not respect the natural borders of the organs; they extend in all directions in the form of fistulae, which frequently extend from any original site to the body surface. Fistulous tracts are multiple, and the skin surface and the surfaces of involved organs assume a characteristic honeycomb appearance. Only rarely does the actinomycotic infection spread by the hematogenous route, and then metastatic abscesses develop; even endocarditis has been reported.

The liver is rarely the site of actinomycotic abscesses, but the primary focus is usually in the proximal colon, especially in the appendix. The liver is reached either by direct spread or through the portal vein. The *liver abscess* may also be a complication of a pulmonary actinomycosis, in which case a combined *pneumopleurohepatic abscess* frequently results (**Fig. 255-1**). Actinomycosis and amebiasis are the main causes for hepatobronchial fistulae. Again, the liver usually does not become involved by the hematogenous route. In cases of isolated hepatic actinomycosis, the original site of the infection may not be identified.

The smaller liver abscess represents a yellow focus, not sharply limited, that clearly reveals its development from the coalescence of even smaller abscesses. Central portions exhibit multiple, partially communicating cavities of different size. In the pus are small, yellow granules (sulfur granules) that consist of concentric, moderately basophilic branching filaments with eosinophilic-clubbed endings; the arrangement of these filaments, best seen in tissue sections, accounts for the name *ray fungus*. In cultures, the fungus grows in short, single-branched forms, simulating diphtheria bacilli, and in branching filaments. The ray fungus is surrounded by leukocytes that, in turn, are engulfed by granulation tissue earmarked by many fat-containing foam cells. This fat accumulation accounts for the bright yellow of the lesion.

The abscess grows by direct distention until it involves the hepatic capsule, with resultant subdiaphragmatic, subhepatic, or perinephritic perihepatitis. Eventually, perforation into the surrounding viscus or skin takes place. Diffuse peritonitis is rare. An extremely shaggy wall characterizes the large abscess cavity resulting from the expansion of the smaller lesion. Secondary infection by pyogenic bacteria is a dangerous complication.

Clinically, *hepatic actinomycosis* is a toxic wasting condition associated with fever, anemia, and leukocytosis. The liver is enlarged and tender, and abdominal pain develops. Ascites and jaundice are rare; involvement of the surrounding organs and multiple cutaneous fistulae contribute to the clinical picture. Hepatic function tests reveal no characteristic alterations except for manifestations of a space-occupying lesion. The prognosis is serious because of the high risk for suppuration and spread to other organs. Combined therapy with surgical intervention and antibiotics is often required.

Chapter 173 discusses abdominal actinomycosis.

ADDITIONAL RESOURCES

Christodoulou N, Papadakis I, Velegrakis M: Actinomycotic liver abscess: case report and review of the literature, *Chir Ital* 56(1):141-146, 2004.

Lai AT, Lam CM, Ng KK, et al: Hepatic actinomycosis presenting as a liver tumour: case report and literature review, *Asian J Surg* 27(4):345-347, 2004.

Miyamoto MI, Fang FC: Pyogenic liver abscess involving *Actinomyces:* case report and review, *Clin Infect Dis* 16:303-309, 1993.

Sharma M, Briski LE, Khatib R: Hepatic actinomycosis: an overview of salient features and outcome of therapy, *Scand J Infect Dis* 34:386-391, 2002.

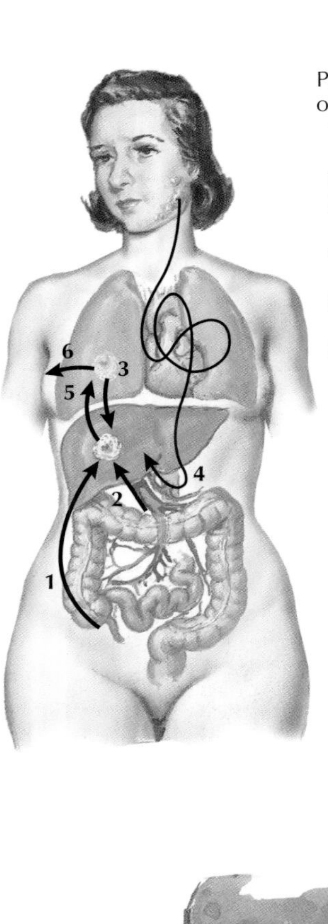

Possible routes of dissemination
1. Directly from gut (appendix) to liver
2. Via portal vein
3. Extension from lung to liver
4. Hematogenous route to liver
5. Extension from liver to lung
6. Cutaneous fistula

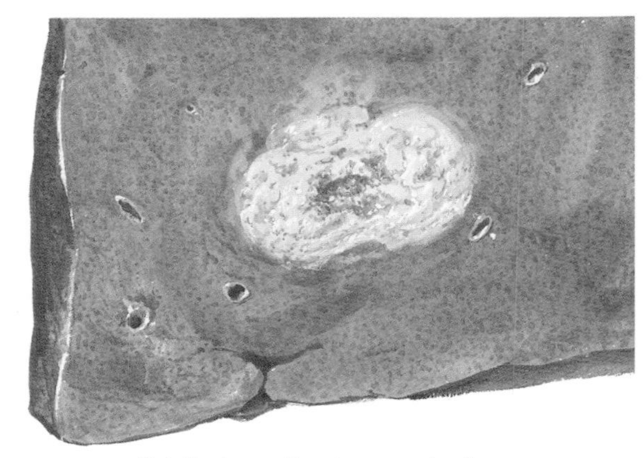

Relatively small actinomycotic abscess

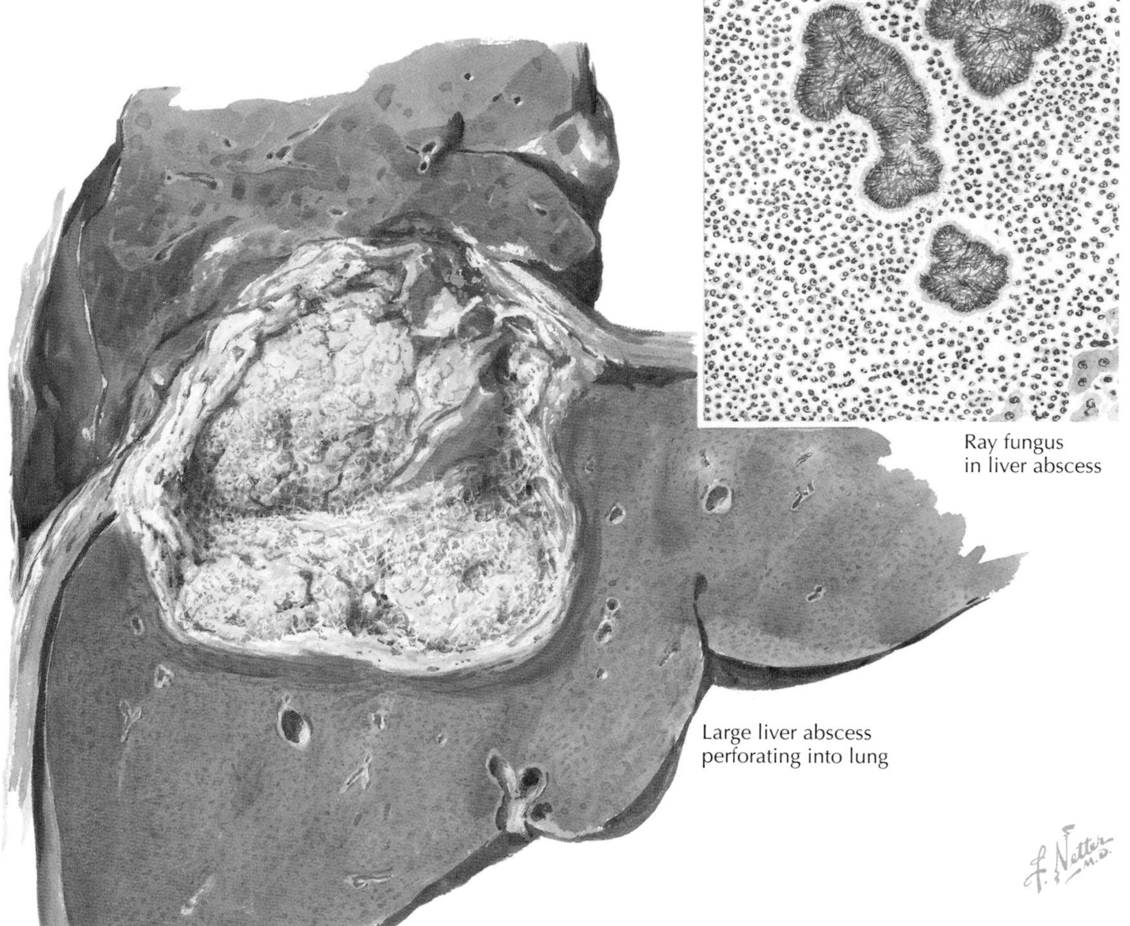

Ray fungus in liver abscess

Large liver abscess perforating into lung

Figure 255-1 *Actinomycosis.*

Echinococcosis (Hydatid Disease)

256

Kris V. Kowdley

Taenia echinococcus, or *Echinococcus granulosus*, is a tapeworm only 5 mm long in the adult stage. It lives in the small intestine of dogs and other canines that have been infected by ingestion of scolices containing viscera of other animals, mainly sheep (**Fig. 256-1**). In the canine intestine, the scolices develop into the adult *Taenia*, a piriform head with four suckers and numerous hooklets, a short neck, and only a few segments, of which the terminal (proglottis) releases the ova.

The ova are ingested by the larval or intermediate host (sheep, cattle, and hogs), but also by humans, mostly children. In the intestinal tract of the host, the larvae hatch from the egg and migrate into the liver and, much less often, into lungs, brain, and other organs, where the larvae develop into a cyst with an outer laminated and inner germinal layer, around which forms a capsule of collagenous tissue. From the cells of the germinal layer evolve embryonal scolices, either directly or after invaginations (brood capsules) form, and eventually become endogenous *daughter cysts.*

With successive invaginations and generations of cysts, the original unilocular main cyst is eventually filled by hundreds of daughter cysts of varying size. The main cyst grows through the years, initially symptomless, until it becomes 20 cm (8 inches) or larger in diameter. Daughter cysts are often discharged from the wall and float in the lumen containing the hydatid fluid. The fluid also contains the *hydatid sand* in which the scolices may be microscopically recognized. Daughter cysts may be seen as outpouchings on the wall of the main cyst or in the surrounding hepatic tissue and, occasionally, implanted in the peritoneal lining of the mesentery. When this asexual production of scolices in the cysts eventually stops, the capsule invades the cyst. The inner surface, formerly granular, becomes smooth; the wall becomes fibrotic and sometimes calcified, and thus radiographically visible. Inflammatory reactions in the vicinity of the cyst are rare.

Echinococcosis has its highest incidence in sheep-raising countries. Three types have been identified, caused by *E. granulosus*, *E. multilocularis*, and *E. vogeli.* Hydatid disease is most often associated with *E. granulosus*, which may be seen worldwide. *E. multilocularis* is most frequently observed in the Northern Hemisphere and is associated with alveolar hydatid disease.

Many hydatid infections are asymptomatic, and the cysts are mostly incidental findings during liver imaging studies. Clinical symptoms are caused by complications, most often, rupture of the cysts. Hydatid fluid entering the circulation can produce allergic manifestations and, rarely, anaphylactic shock. Rupture of daughter cysts into bile ducts or compression of the bile duct can lead to jaundice. Secondary bacterial infection of cysts causes fever and chills. Serum indirect hemagglutination test findings are usually positive.

Treatment with *albendazole*, a benzimidazole drug, is effective. Resection using open or laparoscopic approaches has been described, and ultrasonographic percutaneous drainage is being increasingly used. Care is taken to avoid spillage of the cyst contents because this can lead to systemic allergic reactions and possibly anaphylaxis.

ADDITIONAL RESOURCES

Ammann RW, Eckert J: Cestodes: *Echinococcus*, *Gastroenterol Clin North Am* 25:655-689, 1996.

Bastani B, Dehdashli F: Hepatic hydatid disease in Iran, with review of the literature, *Mt Sinai J Med* 62:62-69, 1995.

Czermak BV, Akhan O, Hiemetzberger R, et al: Echinococcosis of the liver, *Abdom Imaging* 33(2):133-143, 2008.

Kumar A, Chattopadhyay TK: Management of hydatid disease of the liver, *Postgrad Med J* 68:853-856, 1992.

Voros D, Katsarelias D, Polymeneas G, et al: Treatment of hydatid liver disease, *Surg Infect (Larchmt)* 8(6):621-627, 2007.

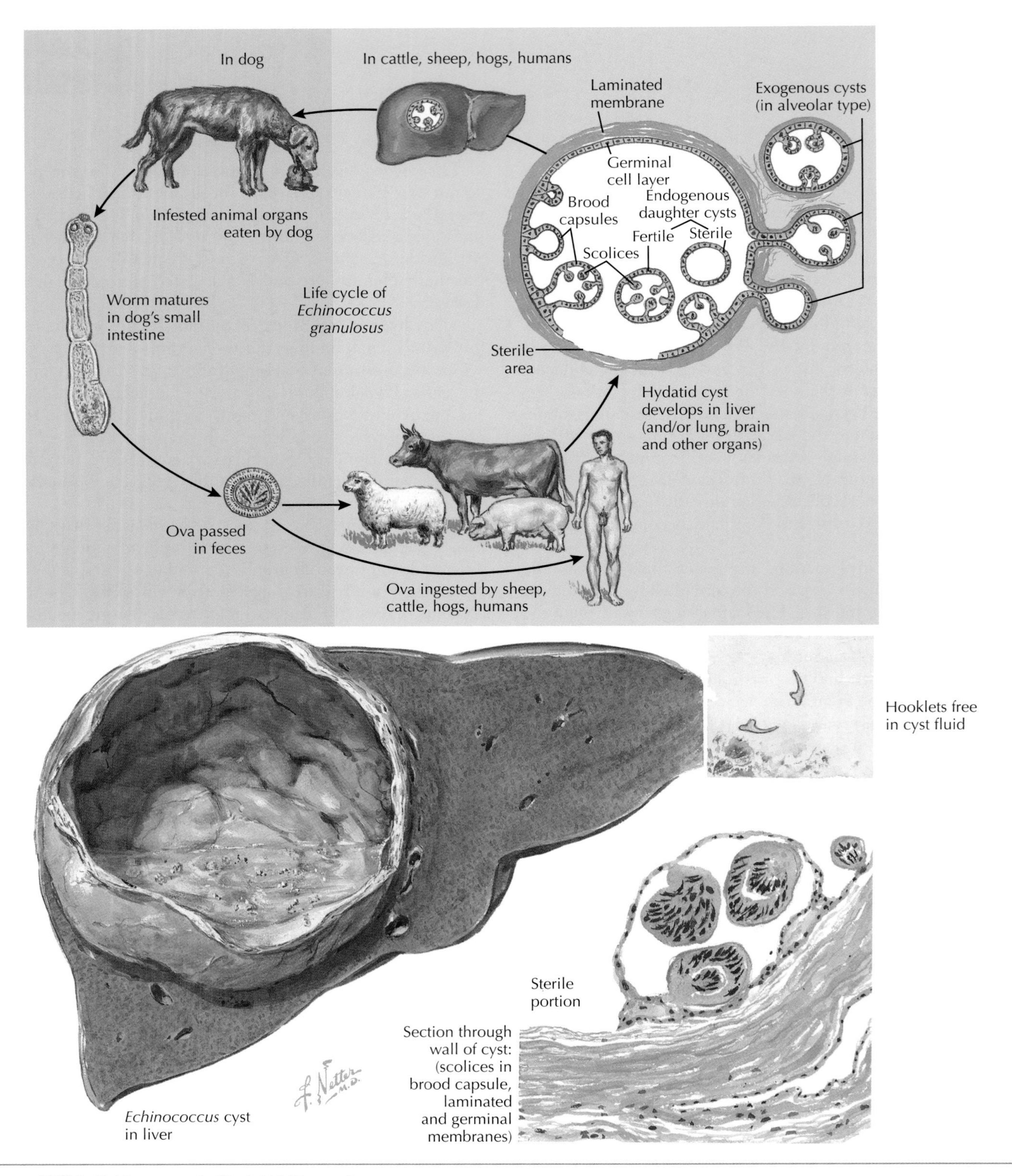

Figure 256-1 *Echinococcosis (Hydatid Disease).*

Schistosomiasis

257

Kris V. Kowdley

Schistosoma is a genus of trematode parasites or blood flukes, of which *Schistosoma mansoni*, *Schistosoma japonicum*, and *Schistosoma haematobium* are of importance in human pathology. *S. mansoni* is found mainly in Africa, parts of South America, and Puerto Rico, from where, with increasing emigration, it is imported to the United States. *S. japonicum* is common in the Far East. *S. haematobium* is found in Africa, especially Egypt, and in endemic foci in southern Europe and Asia. The life cycles of the three species are similar, but *S. haematobium* predominantly involves the vessels of the urinary bladder. Eggs of *S. mansoni* and *S. japonicum* are excreted with the feces of human carriers; those of *S. haematobium* are excreted with the urine.

Eggs of *S. mansoni*, approximately 140 μm long, exhibit a characteristic lateral spine (**Fig. 257-1**). They hatch when they fall into fresh water. The larvae, or *miracidia*, survive only a few hours, unless they can attach themselves to snails, which they penetrate. In the snail's digestive gland, the larvae pass through several stages (sporocysts) and develop into *cercariae*, which, having left the snail, propel themselves with a forked tail. They are most active in shallow water exposed to sunlight, where they may attach themselves to wading or swimming humans, whose unbroken skin or mucous membranes they enter. They eventually reach the extrahepatic tributaries and the intrahepatic branches of the portal vein, where they grow to full sexual maturity, depositing the fertilized eggs. Some eggs are extruded through the vascular wall into the intestinal lumen, where they pass with the feces, maintaining the life cycle. Other eggs are carried into the smallest portal radicles in the liver, where they are responsible for the clinical manifestations of *hepatic schistosomiasis.*

Immediately after infestation and during migration, localized or generalized skin reactions occur, accompanied by pruritus (swimmer's itch) and fever. The liver becomes enlarged. In the peripheral blood, granulocytosis and eosinophilia (14,000-20,000 cells) are found. Within about 6 weeks, symptoms may entirely subside, whereas in other patients, an acute toxic stage may develop. This stage is characterized by constant or intermittent fever, mild gastrointestinal discomfort to severe abdominal pain, nausea, vomiting, and occasionally persistent cough. The liver becomes large and tender, and splenomegaly develops.

After a variable interval, chronic colitis, mesenteric lymphadenitis, and pulmonary fibrosis may develop. However, the most dangerous manifestations occur in the portal system, where worms and ova obstruct portal venous blood flow, resulting in portal hypertension. The ova adhere to the lining endothelium, which grows over them. First, an inflammatory reaction develops, followed eventually by granuloma formation, with fibroblasts, epithelioid cells, and even giant cells. The ovum becomes necrotic, frequently calcified, and may entirely disappear, whereas the fibrosing *pseudotubercle* persists.

Granulomas are readily demonstrated by liver biopsy, and their etiology can frequently be established by demonstrating the ova or their remnants. If the changes to the hepatic parenchyma become severe, *pipestem cirrhosis* develops. More frequently, however, the cirrhosis is similar to other forms of cirrhosis, probably because of concomitant hepatitis C infection or alcoholic liver disease.

ADDITIONAL RESOURCES

Andrade ZA: Hepatic schistosomiasis: morphological aspects, *Prog Liver Dis* 2:228-242, 1965.

Bica I, Hamer DH, Stadecker MJ: Hepatic schistosomiasis, *Infect Dis Clin North Am* 14:583-604, viii, 2000.

Dunn MA, Kamel R: Hepatic schistosomiasis, *Hepatology* 1:653-661, 1981.

Elliott DE: Schistosomiasis: pathophysiology, diagnosis, and treatment, *Gastroenterol Clin North Am* 25:599-625, 1996.

Gryseels B, Polman K, Clerinx J, Kestens L: Human schistosomiasis, *Lancet* 368(9541):1106-1118, 2006.

Manzella A, Ohtomo K, Monzawa S, Lim JH: Schistosomiasis of the liver, *Abdom Imaging* 33(2):144-150, 2008.

McKerrow JH, Sun E: Hepatic schistosomiasis, *Prog Liver Dis* 12:121-135, 1994.

Nash TE, Cheever AW, Ottesen EA, Cook JA: Schistosome infections in humans: perspectives and recent findings—NIH conference, *Ann Intern Med* 97:740-754, 1982.

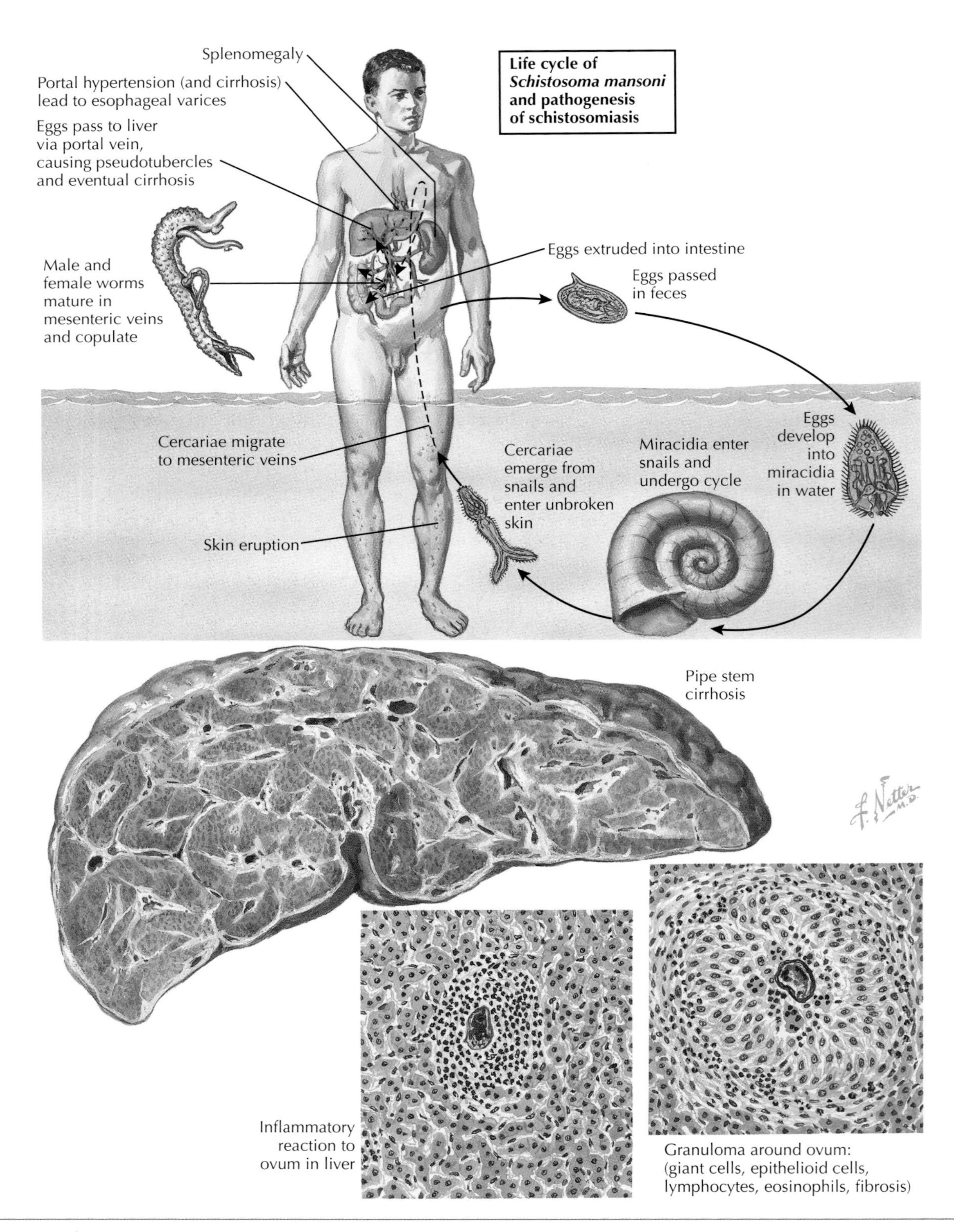

Figure 257-1 *Schistosomiasis.*

The Liver in Heart Failure

258

Kris V. Kowdley

Passive congestion of the liver may occur in a variety of clinical circumstances, including heart failure. However, the severity of passive liver congestion does not correlate well with the degree of hepatic involvement.

In *acute* passive congestion, the liver is greatly enlarged, the capsule is tense, the anterior edge is blunt, and on the cut surface, the lobular markings are much more distinct than usual (**Fig. 258-1**). On closer inspection, the zones around the central veins appear dark red and depressed, distinct from the intermediate and peripheral zones, which sometimes exhibit a yellow hue caused by fatty metamorphosis. The hepatic veins are extremely dilated. Histologically, the liver cells in the central zone have disappeared, and the sinusoids, as well as the tissue spaces, are crowded with red blood cells, as are the dilated branches of the hepatic veins. Central necrosis is more marked on autopsy than on biopsy specimens.

Therefore, *central necrosis* in a patient with heart failure may be a terminal or a preterminal event in end-stage or severe heart failure. Occasionally, only a small rim of parenchyma is preserved on the periphery of hepatic lobules in severe acute cardiac failure, such as after rupture of a chorda tendinea. Clinically, the liver is very large and exquisitely tender, especially in the gallbladder region.

In *chronic* passive congestion, the liver is smaller than in the acute stages and sometimes even smaller than normal. The surface is irregular and may be finely granular; the capsule often is thickened and covered by organized fibrin. The liver may be diffusely fibrotic, and regenerating nodules may be present. The hepatic veins appear wider than in acute stages. Fibrosis may initially surround the central veins, leading to a "reverse lobular" pattern.

With progressive disease, bridging fibrosis may develop between central veins and portal areas, finally leading to established cirrhosis. True *cardiac cirrhosis* is the result of severe and usually longstanding passive congestion, as occurs in severe tricuspid insufficiency or constrictive pericarditis. In *cardiac hepatic fibrosis*, the tender liver appears relatively small. Ascites may be present, and patients may be moderately jaundiced.

Patients with acute severe left ventricular heart failure may also develop acute ischemic hepatitis, characterized by high serum transaminase levels and possibly similar to *shock liver* associated with hypotension or hypoxemia. Jaundice and tender hepatomegaly may develop. Some patients may have evidence of synthetic dysfunction with hepatic encephalopathy. Liver biopsy may show centrilobular necrosis, which may be severe. With progressive damage, necrosis of hepatocytes in central areas may lead to sinusoidal congestion and a neutrophilic inflammatory infiltrate.

ADDITIONAL RESOURCES

Giallourakis CC, Rosenberg PM, Friedman LS: The liver in heart failure, *Clin Liver Dis* 6(4):947-967, viii-ix, 2002.

Wanless IR, Liu JJ, Butany J: Role of thrombosis in the pathogenesis of congestive hepatic fibrosis (cardiac cirrhosis), *Hepatology* 21(5):1232-1237, 1995.

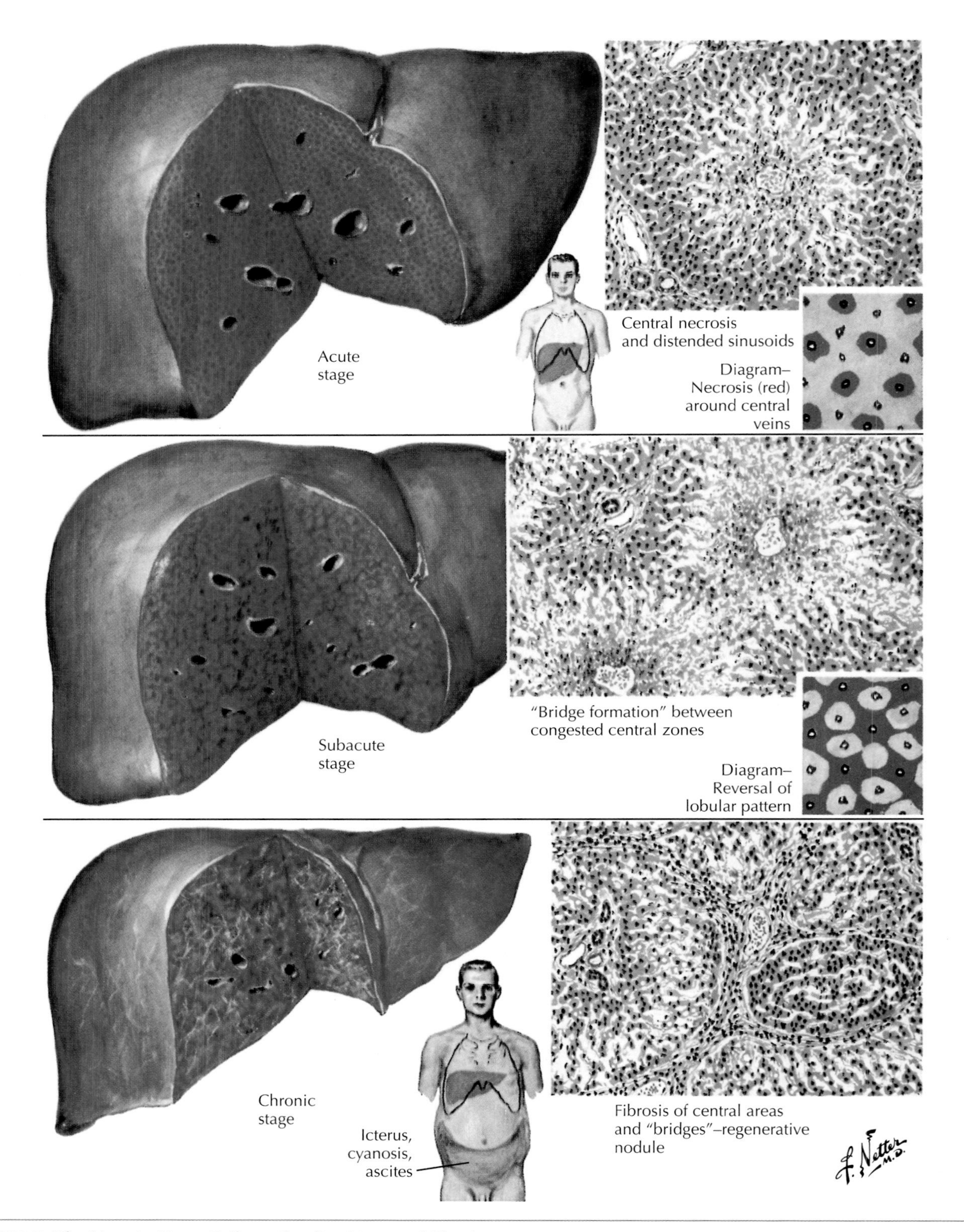

Figure 258-1 *The Liver in Heart Failure: Cardiac Liver and Shock.*

Portal Vein Obstruction

259

Kris V. Kowdley

Most cases of portal vein thrombosis are associated with underlying cirrhosis, malignancy, or pancreatitis. An acute decrease in systemic blood pressure may also lead to portal venous obstruction (**Fig. 259-1**).

In the absence of a known cause, many cases of portal vein thrombosis appear to be associated with hypercoagulable states such as polycythemia vera. Sudden, complete obstruction of the portal vein and its branches by a thrombus may lead, in rare cases, to a clinically dramatic picture dominated by hematemesis, melena with diarrhea, rapidly developing ascites, abdominal pain, peritonitis, ileus, and within a few days, coma and death. However, many of these patients also have associated thrombosis of the superior mesenteric vein. Jaundice is uncommon.

Precipitating factors include splenectomy or other procedures involving the portal system. Portal vein thrombosis may also develop in patients during the course of cirrhosis and may be a feature of hepatocellular carcinoma.

In *acute* portal vein thrombosis, the wall of the small intestine may also show changes, including edema and hemorrhage. The spleen is generally enlarged, but in patients without history of cirrhosis, the liver may be unremarkable. The thrombosis may originate from the portal vein itself or may extend into it from a splenic or a mesenteric vein thrombus or distally from thrombi in the intrahepatic branches of the portal vein.

A more gradual decrease in the portal circulation is well tolerated, possibly because of the development of collaterals. The thrombosis may result in cordlike shrinkage of the portal vein or a spongy cavernous transformation caused by recanalization of the thrombus itself. The liver shows minimal changes in isolated portal vein thrombosis. Again, jaundice is usually absent. The main complication, variceal bleeding, is generally better tolerated than in patients with cirrhosis. Ascites may be present.

Extrahepatic portal vein obstruction is common in developing countries, particularly India, and is a common cause of noncirrhotic portal hypertension. It is associated with extensive collateral circulation around the portal vein, leading to formation of ectopic varices in the peribiliary and peripancreatic regions. Variceal hemorrhage is the most common complication.

Treatment of portal vein obstruction focuses on managing the complications of portal hypertension. In patients without an underlying history of cirrhosis, a workup for a hypercoagulable state is indicated. Endoscopic therapy is effective for patients with variceal bleeding, and portal decompression is appropriate for patients with preserved hepatic synthetic function. Transjugular intrahepatic portosystemic shunt (TIPS) has been used in some patients with portal venous obstruction, especially when the acute portal vein thrombosis is associated with Budd-Chiari syndrome.

ADDITIONAL RESOURCES

Garcia-Pagán JC, Hernández-Guerra M, Bosch J: Extrahepatic portal vein thrombosis, *Semin Liver Dis* 28(3):282-292, 2008.

Janssen HL: Changing perspectives in portal vein thrombosis, *Scand J Gastroenterol Suppl* 232:69-73, 2000.

Moreau R, Lebrec D: Molecular and structural basis of portal hypertension, *Clin Liver Dis* 10(3):445-457, vii, 2006.

Sarin SK, Agarwal SR: Extrahepatic portal vein obstruction, *Semin Liver Dis* 22:43-58, 2002.

Sarin SK, Kumar A: Noncirrhotic portal hypertension, *Clin Liver Dis* 10(3):627-651, x, 2006.

Sobhonslidsuk A, Reddy KR: Portal vein thrombosis: a concise review, *Am J Gastroenterol* 97:535-541, 2002.

Valla DC, Condat B: Portal vein thrombosis in adults: pathophysiology, pathogenesis and management, *J Hepatol* 32:865-871, 2000.

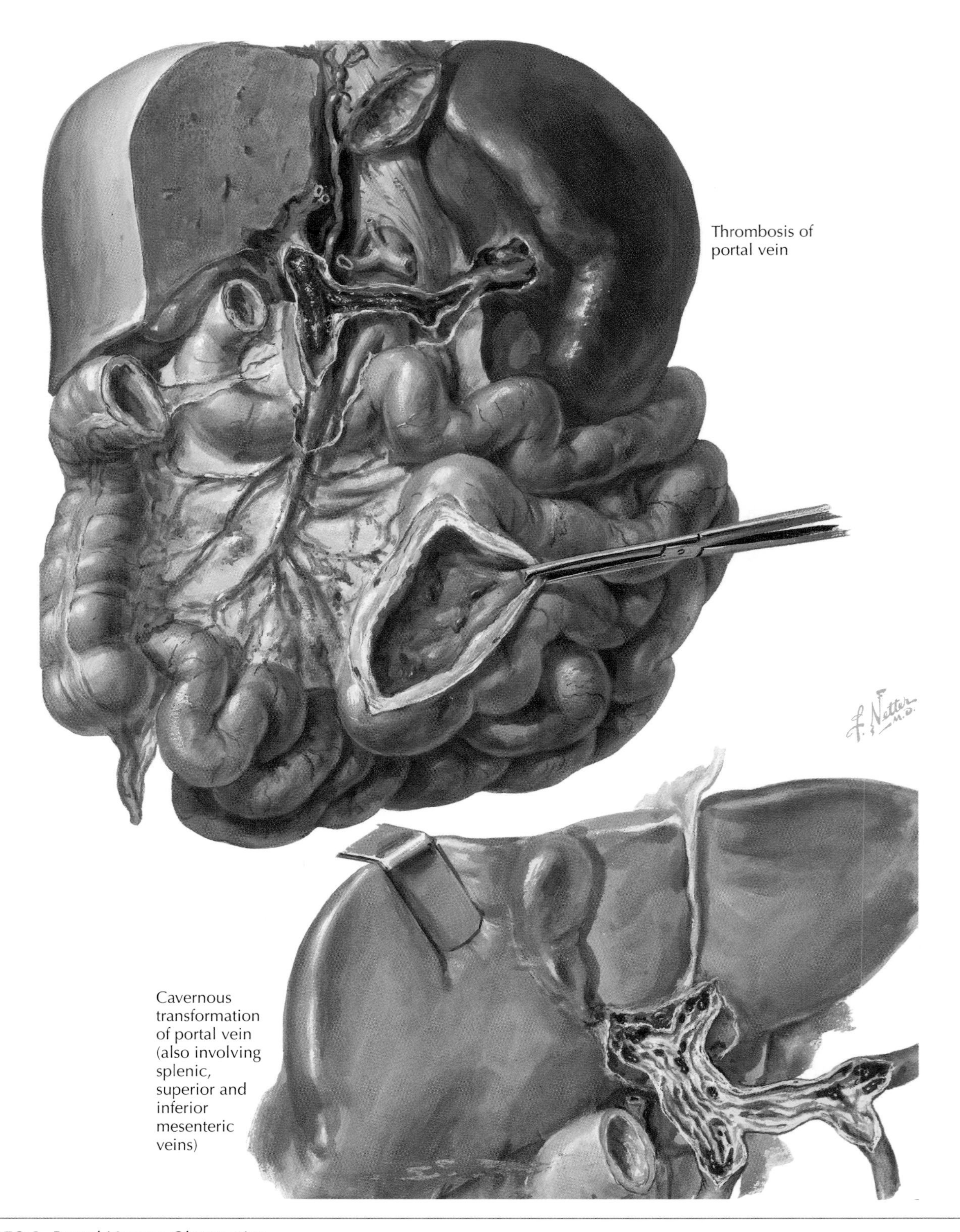

Figure 259-1 *Portal Venous Obstruction.*

Budd-Chiari Syndrome

Kris V. Kowdley

Budd-Chiari syndrome describes clinical features associated with hepatic venous outflow obstruction. Although this term is generally used to describe a specific entity, namely, hepatic venous obstruction associated with acute or chronic thrombosis, the level of outflow obstruction in the liver may be in any location from the suprahepatic cava to the right atrium.

CLINICAL PICTURE

Most cases of Budd-Chiari syndrome are observed in patients with hypercoagulable states. Acute or chronic outflow obstruction at the level of the hepatic vein results in portal hypertension and may lead to necrosis caused by decreased perfusion from impaired inflow into the sinusoids through the portal vein. Portal vein thrombosis may develop and may exacerbate this ischemic process.

Patients with Budd-Chiari syndrome may also have acute hepatic venous obstruction from thrombosis. This presentation may be associated with the development of rapidly worsening liver failure and may be considered a medical emergency. Acute thrombosis develops suddenly in the hepatic veins, with inadequate time for the development of collaterals to decompress the portal hypertension. The clinical presentation may vary depending on the chronicity of the thrombotic process and the formation of portosystemic collaterals.

Clinical manifestations of Budd-Chiari syndrome may include tender hepatomegaly, ascites, jaundice, and coagulopathy. Hepatic encephalopathy and liver failure are ominous signs.

DIAGNOSIS

Radiologic diagnosis using abdominal sonography and duplex examination of the hepatic veins is extremely sensitive and can be diagnostic in many patients. Direct venography using the transjugular technique can be used to confirm the diagnosis. In addition, transjugular liver biopsy can be performed to determine the presence or absence of cirrhosis. Often, computed tomography shows a prominent caudate lobe, which drains uniquely and directly into the vena cava, thus perhaps avoiding atrophy in classic cases of Budd-Chiari syndrome. In some patients, the prominent caudate lobe is mistaken for a mass on imaging studies.

The differential diagnosis of Budd-Chiari syndrome includes any cause of hepatic outflow obstruction, such as heart failure, severe tricuspid insufficiency, or constrictive pericarditis. Physical examination and cardiac echocardiography can usually differentiate these causes. Hepatic venoocclusive disease may seem similar on liver biopsy, but in Budd-Chiari syndrome, the hepatic veins are usually visualized and patent. Liver biopsy in Budd-Chiari syndrome may reveal centrilobular hemorrhage, congestion, and even organized thrombi in advanced cases.

The most common cause of a hypercoagulable state leading to Budd-Chiari syndrome is *polycythemia vera.* Other causes include protein C, protein S, and antithrombin III deficiency. Paroxysmal nocturnal hemoglobinuria is another cause of Budd-Chiari syndrome. Usually, an underlying cause can be identified with a complete workup.

TREATMENT AND MANAGEMENT

Management of Budd-Chiari syndrome is generally aimed at relieving portal hypertension in acute cases and in patients without cirrhosis. Surgical decompression and transjugular intrahepatic portosystemic shunt (TIPS) have been used with success. Medical therapy with thrombolytic agents is helpful for patients with acute thrombosis and possibly for other patients.

Liver transplantation is an option for patients with cirrhosis or with acute liver failure and if previous therapies have failed. Although there is concern about recurrent thrombosis after liver transplantation, long-term outcomes are good. However, long-term anticoagulation, which is generally used after liver transplantation, may be associated with an increased risk for bleeding complications.

ADDITIONAL RESOURCES

Hoekstra J, Janssen HL: Vascular liver disorders. I. Diagnosis, treatment and prognosis of Budd-Chiari syndrome, *Neth J Med* 66(8):334-339, 2008.

Janssen HL, Garcia-Pagan JC, Elias E, et al: Budd-Chiari syndrome: a review by an expert panel, *J Hepatol* 38:364-371, 2003.

Klein AS, Molmenti EP: Surgical treatment of Budd-Chiari syndrome, *Liver Transpl* 9:891-896, 2003.

Menon KV, Shah V, Kamath PS: The Budd-Chiari syndrome, *N Engl J Med* 350:578-585, 2004.

Plessier A, Valla DC: Budd-Chiari syndrome, *Semin Liver Dis* 28(3):259-269, 2008.

Valla DC: The diagnosis and management of the Budd-Chiari syndrome: consensus and controversies, *Hepatology* 38:793-803, 2003.

Bile Duct Cancer

261

Kris V. Kowdley

Bile duct cancer, also called *cholangiocarcinoma*, is the term used to describe malignancies arising from the epithelia of biliary ductal cells. The incidence of bile duct cancer is low, estimated at 8 per 1 million population. Major risk factors for bile duct cancer include primary sclerosing cholangitis (PSC), *Clonorchis sinensis* or *Opisthorchis viverrini* infection, older age, exposure to thorium dioxide (Thorotrast), congenital abnormalities of the bile ducts (e.g., Caroli disease), and possibly, prior biliary-enteric drainage procedures.

Cholangiocarcinomas are generally classified as hilar, intrahepatic, and distal. *Hilar* cholangiocarcinoma represents the most common variety, accounting for 50% to 60% of cases (**Figs. 261-1** and **261-2**). Most cholangiocarcinomas are adenocarcinomas, although different histologic patterns may be seen. Often, an intense desmoplastic reaction occurs around the tumor, facilitating tissue acquisition for diagnosis, particularly with endoscopic retrograde cholangiopancreatography (ERCP) or percutaneous techniques.

The diagnosis of cholangiocarcinoma can be difficult to establish early in the disease course, because local spread often has already occurred at clinical presentation, with jaundice or dilated ducts. Clinical manifestations may also include fever from associated cholangitis, weight loss, and abdominal pain. Jaundice is more common in patients with distal cholangiocarcinoma, whereas abdominal pain is more common in those with the peripheral or intrahepatic form of bile duct cancer.

Diagnosis is best made using ERCP, with confirmation by biopsy of the involved bile duct. Serum markers have been studied, including CA 19-9, carcinoembryonic antigen, and CA 125. However, their clinical usefulness remains unproven except in patients with PSC, in whom a CA 19-9 value greater than 100 is may indicate cholangiocarcinoma, especially in the absence of cholangitis.

The prognosis for patients with cholangiocarcinoma is poor, and 5-year survival is about 10%. Surgery for cure remains the best hope for long-term survival. For patients with unresectable lesions, palliative stenting using plastic or metal stents can provide relief of pruritus or can prevent cholangitis.

Liver transplantation has not been performed for patients with hilar cholangiocarcinoma because of the high risk for recurrent disease and the disappointing long-term survival. However, in carefully selected patients with PSC, long-term survival is possible in conjunction with adjuvant or neoadjuvant therapy.

Gemcitabine has been approved for treatment of cholangiocarcinoma and is frequently used in combination with other agents (e.g., 5-FU). Recent studies suggest photodynamic therapy may have role in treating hilar cholangiocarcinoma.

ADDITIONAL RESOURCES

Blechacz B, Gores GJ: Cholangiocarcinoma: advances in pathogenesis, diagnosis, and treatment, *Hepatology* 48(1):308-321, 2008.

Gores GJ: A spotlight on cholangiocarcinoma, *Gastroenterology* 125:1536-1538, 2003.

Gores GJ: Cholangiocarcinoma: current concepts and insights, *Hepatology* 37:961-969, 2003.

Khan SA, Davidson BR, Goldin R, et al: Guidelines for the diagnosis and treatment of cholangiocarcinoma: consensus document, *Gut* 51(suppl 6; VI):1-9, 2002.

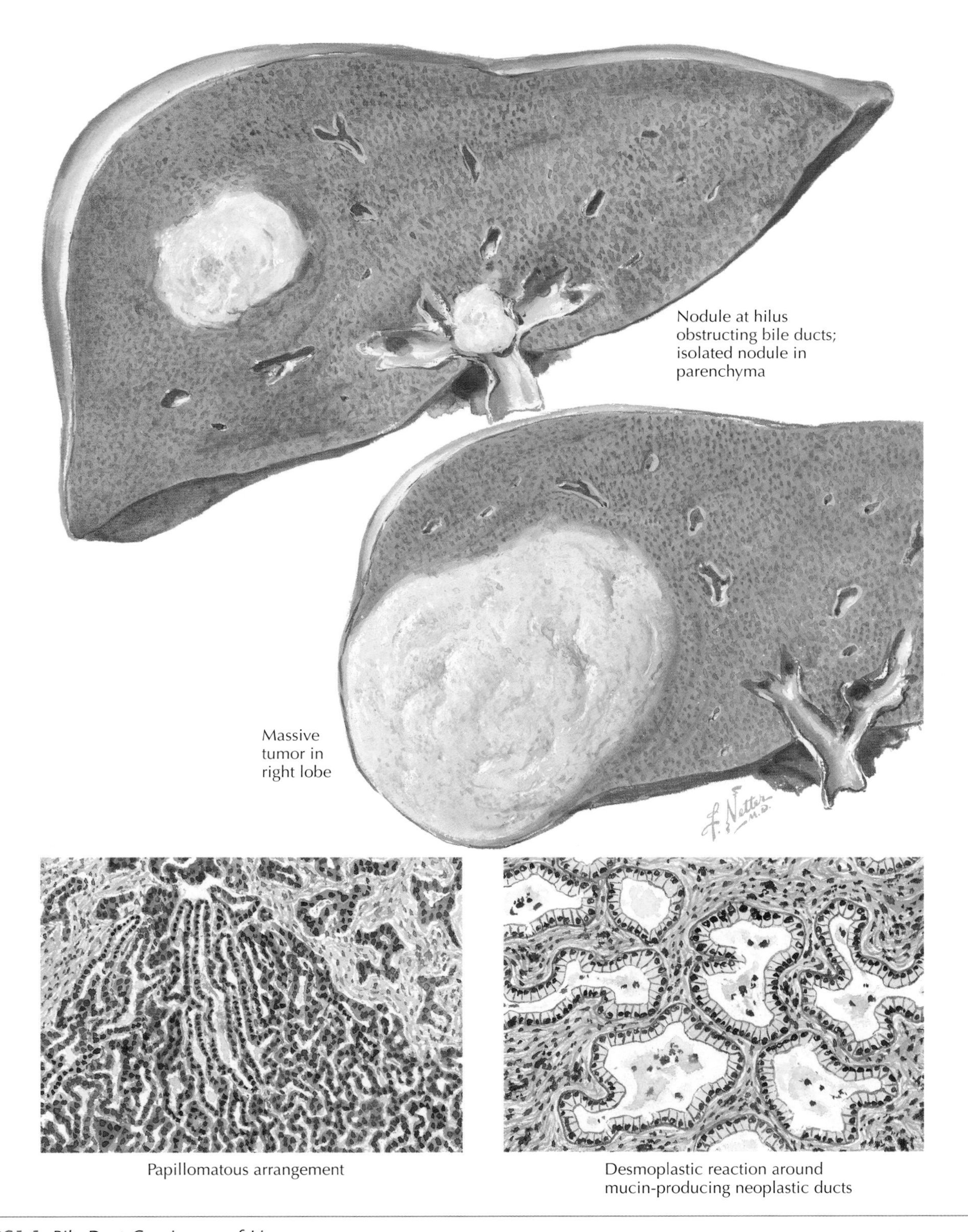

Figure 261-1 *Bile Duct Carcinoma of Liver.*

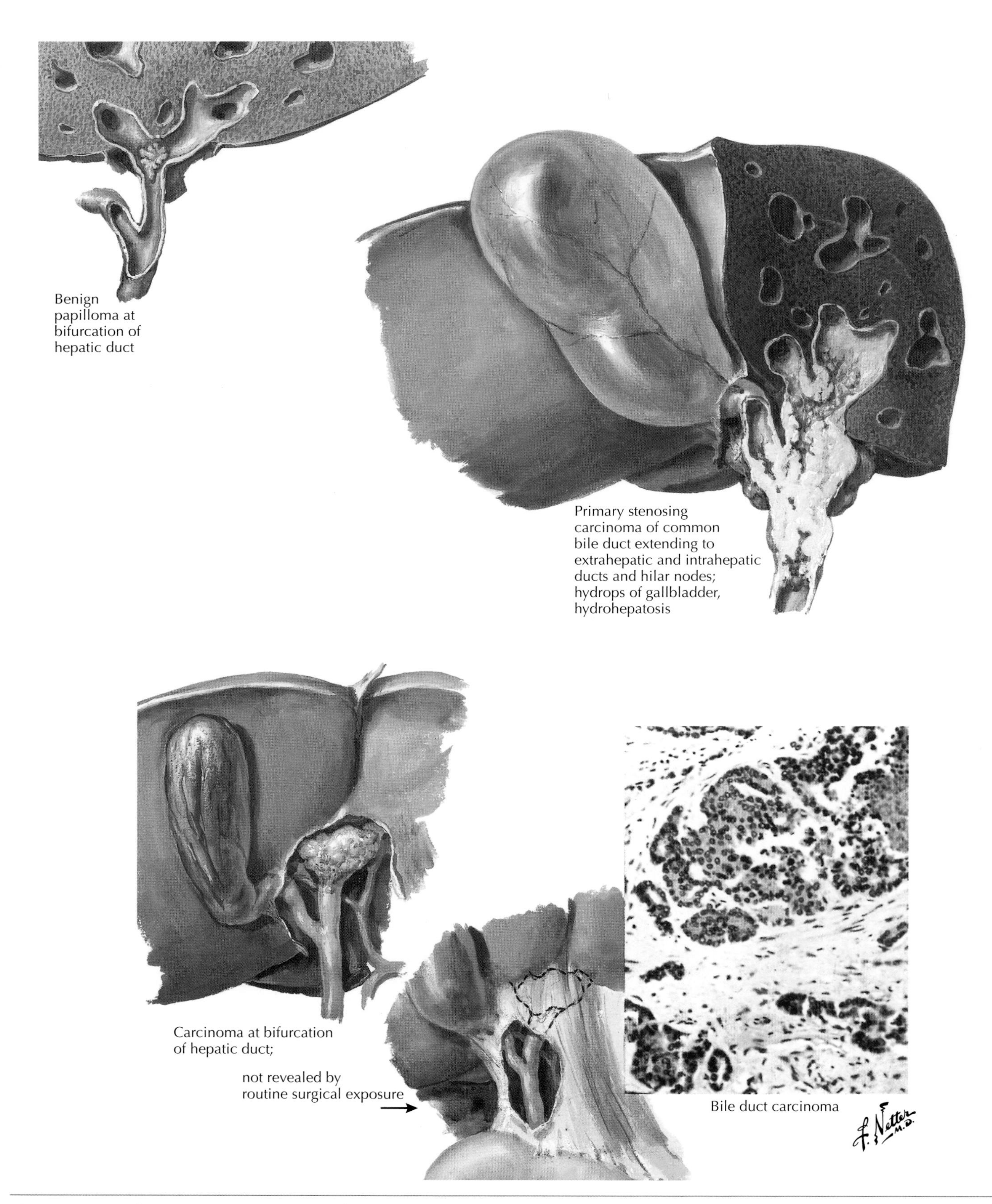

Figure 261-2 *Bile Duct Cancer: Benign Papilloma and Carcinomas.*

Metastatic Cancer

262

Kris V. Kowdley

Secondary liver cancer is more common than primary hepatocellular carcinoma (see Chapter 252). One of the primary sites from which carcinoma metastasizes to the liver is the gastrointestinal (GI) tract, particularly in colon cancer. Other common primary cancers with metastasis to the liver include stomach, esophagus, pancreas, and gallbladder. Cancers such as malignant melanoma (particularly ocular melanoma) and primary lung cancer can also spread to the liver. In most cases, liver metastases are detected through computed tomography (CT) or other cross-sectional imaging during staging or follow-up surveillance (**Fig. 262-1**).

Colon cancer may be the most common primary GI tract tumor with spread to the liver. Isolated liver metastases may occur years after the primary tumor has been treated. Therefore, long-term CT surveillance is indicated to monitor for evidence of hepatic involvement. Therapies for hepatic metastases from primary colon cancer include chemotherapy and surgery. Long-term remissions and even cures have been described after resection of isolated liver metastases from colorectal cancer.

Neuroendocrine tumors also frequently spread to the liver. Many of these tumors are functionally active and produce a variety of hormones, which can have clinical and biochemical effects. Islet cell tumors, glucagonomas, insulinomas, and carcinoid tumors that produce vasoactive intestinal peptide (VIP) can lead to clinical symptoms associated with the hormones they produce. Antihormonal therapies (e.g., octreotide) and surgical management have been advocated for such neuroendocrine tumors.

Other, rare tumors involving the liver include mesenchymal cell tumors (e.g., GI stromal tumors) and adenocarcinomas from unknown primary sites.

ADDITIONAL RESOURCES

Bhattacharya R, Rao S, Kowdley KV: Liver involvement in patients with solid tumors of nonhepatic origin, *Clin Liver Dis* 6:1033-1043, 2002.

Modlin IM, Kidd M, Drozdov I, et al: Pharmacotherapy of neuroendocrine cancers, *Expert Opin Pharmacother* 9(15):2617-2626, 2008.

Scholefield JH, Steele RJ: British Society for Gastroenterology; Association of Coloproctology for Great Britain and Ireland: Guidelines for follow-up after resection of colorectal cancer, *Gut* 51(suppl 5):V3-V5, 2002.

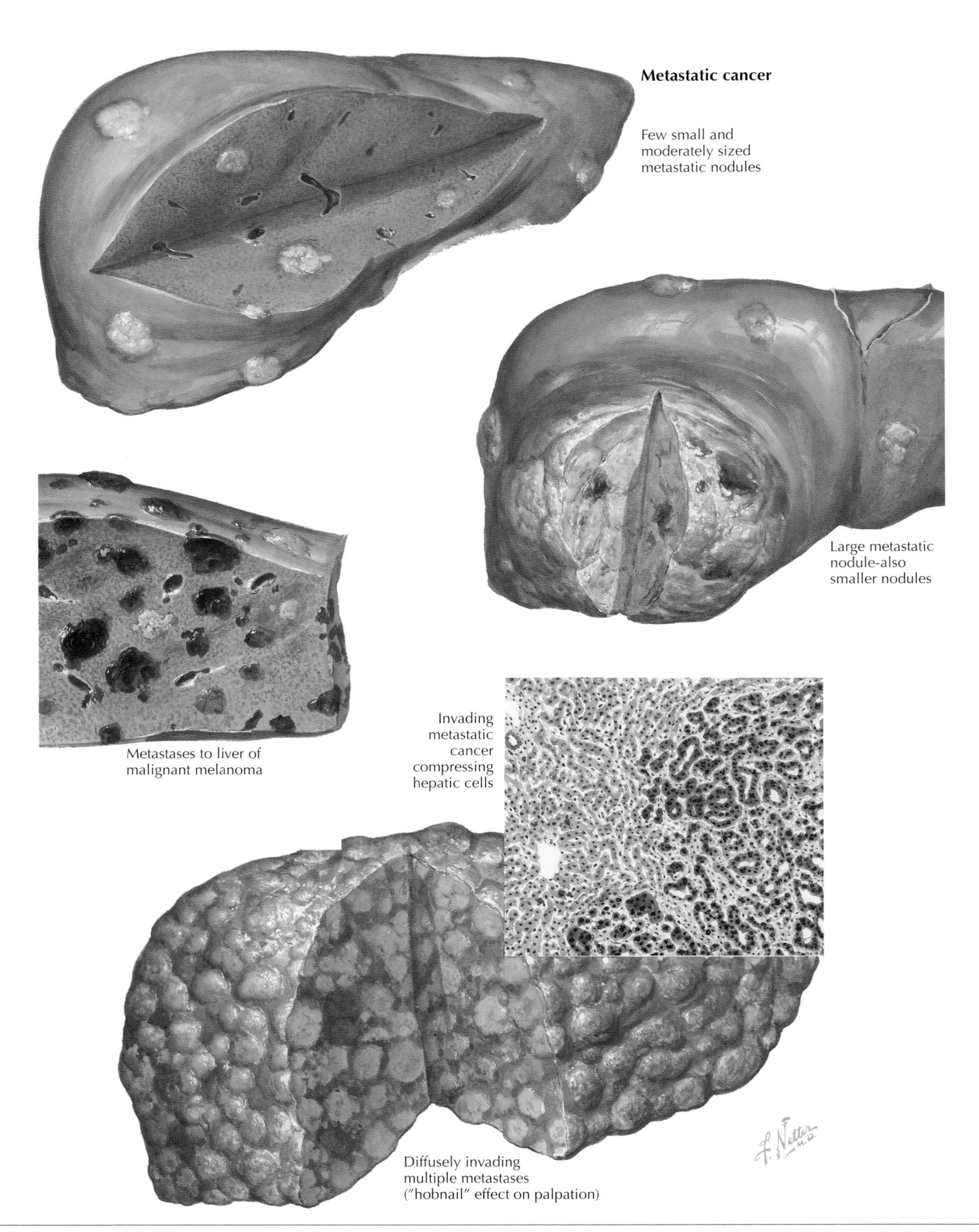

Figure 262-1 *Metastatic Liver Cancer.*

Hepatic Trauma

Kris V. Kowdley

Because of its size, location, and fixation, the liver is frequently subjected to trauma, which may be penetrating or blunt. After the brain, the liver is the organ that most often sustains blunt trauma.

Bullet or stab wounds penetrate to various depths and produce an intrahepatic canal with a ragged wall and a lumen filled with blood. In more than 25% of penetrating thoracoabdominal wounds, the liver is injured. Blunt injuries lead to *ruptures* or *lacerations*, which vary in size and number (**Fig. 263-1**). Blunt trauma usually results from automobile crashes or falls. The liver may be lacerated by broken ribs or may be crushed by the impact of the thoracic cage and the resisting spine. Internal stress of countercoup effects during a blunt injury may cause subcapsular or central lacerations, or only a subcapsular hematoma if the impact is mild.

Rupture from blunt injury is more likely to occur if the liver has become more friable, or if capsular tension has increased because of abscess, cyst, infectious disease (e.g., malaria), or hepatitis and fatty infiltration. In contrast to the spleen, so-called spontaneous ruptures of a minimally damaged liver are rare. Some claim that postprandial hyperemia may predispose to rupture of the liver; rupture during pregnancy has also been reported. There are rare reports of rupture of the liver in amyloidosis.

Except for temporary peritoneal irritation from blood oozing into the peritoneal cavity, subcapsular hematomas and small lacerations or ruptures usually heal with few clinical manifestations, leaving a pigmented or white subcapsular scar. If the hematoma becomes infected, intrahepatic, subphrenic, or subhepatic abscess may complicate the clinical course. Hepatic cysts and biliary fistulae also may develop, depending on the site of rupture. Infrequent complications include portal vein thrombosis (see Chapter 259) and arterial aneurysms.

Severe lacerations or rupture of the liver result in high mortality rates, particularly as a consequence of military injury. Early death is caused by hemorrhage that may be severe and that cannot be readily stopped, because the walls of the hepatic veins are thin, the liver is highly vascular, and the bile has anticoagulant effects. Late death is often caused by bile peritonitis and hepatorenal syndrome. However, completely detached liver tissue pieces are well tolerated within the peritoneal cavity and may be even organically attached in the lateral gutter (see Fig. 263-1).

Laboratory manifestations of hepatic trauma are surprisingly minimal. Jaundice is rare and occurs primarily if the gallbladder and the bile ducts are ruptured. In later stages, jaundice may be the result of liver abscesses or traumatic cholangitis. Foreign bodies (e.g., bullets) in the liver may eventually migrate into the biliary ducts and produce obstructive jaundice.

The liver is relatively insensitive to external ionizing radiation; even the effects of internal radiation by radioactive substances accumulating in the liver (e.g., phosphorus 32) are not severe.

ADDITIONAL RESOURCES

Carrillo EH, Richardson JD: The current management of hepatic trauma, *Adv Surg* 35:39-59, 2001.

Feliciano DV, Rozycki GS: Hepatic trauma, *Scand J Surg* 91:72-79, 2002.

Letoublon C, Arvieux C: Nonoperative management of blunt hepatic trauma, *Minerva Anestesiol* 68:132-137, 2002.

Maull KI: Current status of nonoperative management of liver injuries, *World J Surg* 25:1403-1404, 2001.

Richardson JD: Changes in the management of injuries to the liver and spleen, *J Am Coll Surg* 200(5):648-669, 2005.

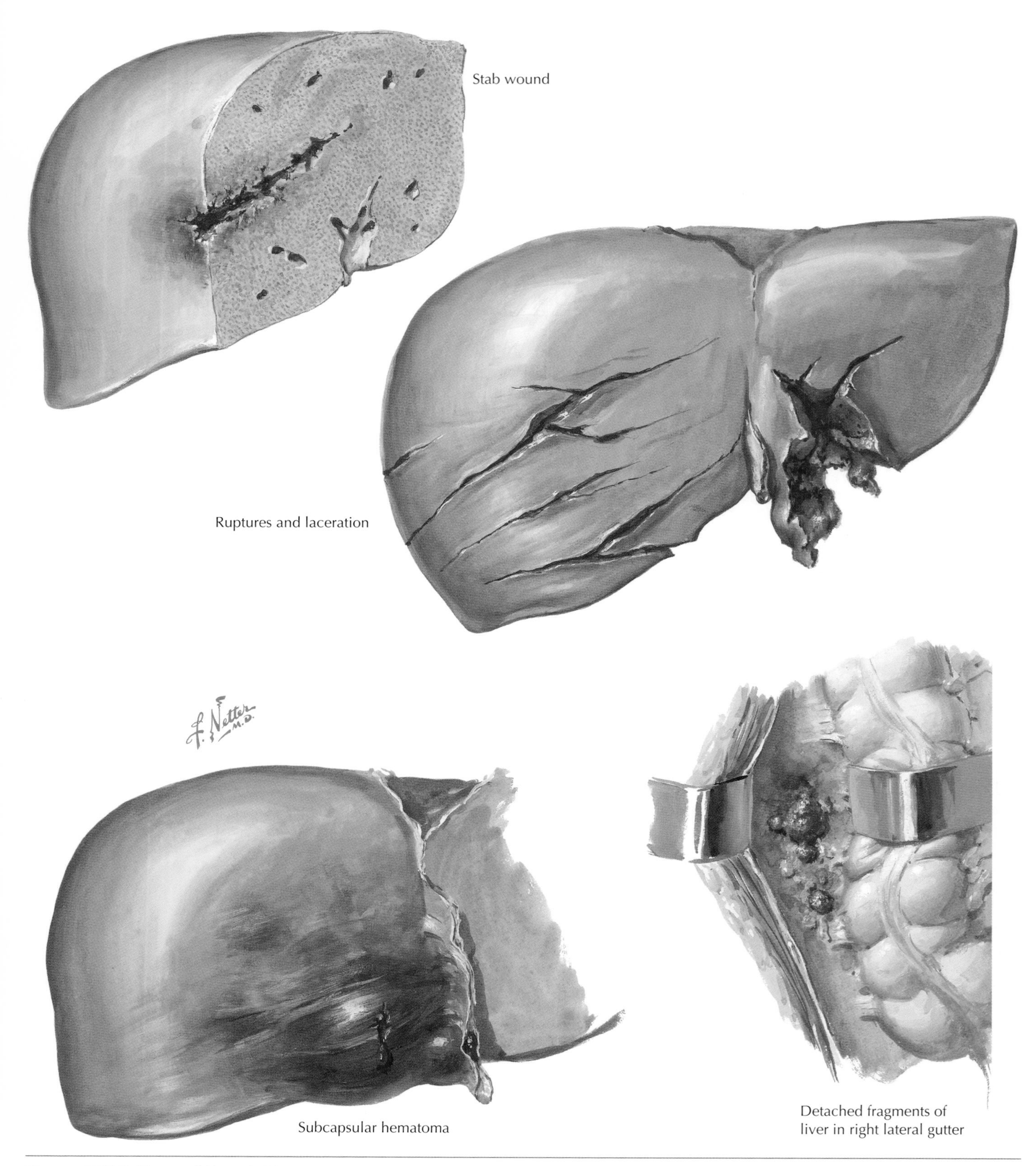

Figure 263-1 *Types of Hepatic Trauma.*

SECTION
X

Nutrition and Gastrointestinal Disease

Dietary and Nutritional Assessment

264

James S. Scolapio

Clinical history, including dietary history, laboratory values, and anthropometric measurements, remain the standards for assessing patients' nutritional status. No single laboratory test or piece of clinical information provides a definitive diagnosis of "malnutrition." For example, an isolated finding of low serum albumin does not necessarily imply the patient is malnourished. Knowledge of the patient's underlying disease process is fundamental to understanding and diagnosing malnutrition.

Assessment begins with screening and identifying persons at risk for malnutrition. For hospital patients, the Joint Commission on Accreditation of Healthcare Organizations requires nutritional screening within 24 hours of the hospitalization. A registered dietitian usually performs this initial screening. Dietary risk factors are easy to identify and provide clues that a patient may be malnourished. These risk factors include inadequate quantity and quality of food intake, presence of chronic disease, history of alcohol abuse, increased nutrient losses from diarrhea, increased nutrient requirements from infection and burns, and various psychosocial factors. A *24-hour diet recall* is one method to determine the dietary intake of a patient. Accuracy of the recall depends greatly on the patient's memory, and family members often need to be interviewed. Written *food diaries* and *calorie counting* are other methods used to collect information regarding food intake.

Nutritional assessment should also include anthropometric measurements, typically body weight, height, skinfold thickness, and midarm muscle circumference. Body weight is one of the most useful nutritional parameters to monitor in patients who are ill. Unintentional weight loss greater than 20% over less than 6 months is associated with *protein-energy malnutrition* (PEM) and functional abnormalities in most patients. *Body mass index* (BMI), calculated as weight (kg) divided by height (m^2), is used to characterize patients who have PEM or who are obese. Although all experts may not accept consistent BMI ranges, the normal BMI range is considered to be 18 to 25 (**Table 264-1**).

Table 264-1 Body Mass Index (BMI)

Grade	BMI
Obesity	
III	>40
II	30 to 40
I	25 to 29.9
Normal	>18.5 to <25
Protein-Energy Malnutrition	
I	17.0 to 18.4
II	16.0 to 16.9
III	<16

Measuring *skinfold thickness* is one of the easiest methods to estimate body fat stores. Measuring body fat requires a skinfold *caliper*. The *triceps skinfold* (TSF) thickness is representative of total body fat stores. Skinfold represents a double layer of subcutaneous tissue, including a small and relatively constant amount of skin and variable amounts of adipose tissue. For arm measurements, the most important factor is to use the midpoint of the upper arm and the same arm for repeat measurements. A thickness of less than 3 mm suggests severe depletion of fat. A TSF thickness greater than 8 mm is usually considered to represent adequate stores.

The *midarm muscle circumference* (MAMC) is a method to estimate skeletal muscle mass. A tape measure is used to determine the upper arm circumference at the same arm location used in the TSF thickness. The MAMC is calculated using the following equation:

$$\text{MAMC (cm)} = \text{Upper arm circumference (cm)} - [0.314 \times \text{TSF thickness (mm)}]$$

An MAMC less than 15 cm (6 inches) is considered severely depleted muscle reserve, and an MAMC greater than 21 cm is considered to represent adequate reserve.

A number of laboratory tests can be used to obtain information about a patient's nutritional status; however, none is specific for malnutrition. Assessing *nitrogen balance* is perhaps the best test for balance of food intake and total losses. Unfortunately, accurate nitrogen collection is cumbersome and requires an experienced laboratory. Difficulty in obtaining a complete 24-hour collection of urine and feces poses another potential clinical limitation of this test. Although low *serum albumin* level may be a marker of increased morbidity and mortality, serum levels can also be altered by the retention of extracellular fluid and the acute stress response. Therefore, serum albumin is not usually a true marker of PEM. Other laboratory tests that may be used to evaluate malnutrition and potential etiologies include a 72-hour fecal fat collection, creatinine-height index, total lymphocyte count, total iron-binding capacity, and isolated vitamin, mineral, and trace element deficiencies.

Subjective global assessment (SGA) has become an accepted method for evaluating the degree of malnutrition (**Box 264-1**). SGA provides a combination of the patient's medical history and physical findings. No laboratory testing is used. Studies have shown a good correlation between SGA score and more sophisticated laboratory tests. Specific questions from the medical history include amount of weight loss over the preceding 6 months, changes in dietary intake, gastrointestinal symptoms that may account for reduced food intake or malabsorption, the patient's functional status (bedridden or full capacity), and the stress response of the patient's underlying illness. Physical

Box 264-1 Subjective Global Assessment (SGA)

A. Medical history

1. Weight change

Overall loss in past 6 months: _______ kg

Change in past 2 weeks: _______ Increase

_______ No change

_______ Decrease

2. Dietary intake change (relative to normal)

_______ No change

_______ Change: duration = _______ weeks

Type: _______ Suboptimal solid diet _______ Full liquid diet

_______ Hypocaloric liquids _______ Starvation

3. Gastrointestinal symptoms that persisted longer than 2 weeks

_______ None _______ Anorexia _______ Nausea _______ Vomiting _______ Diarrhea

4. Functional capacity

_______ No dysfunction (full capacity)

_______ Dysfunction: duration = #_______ weeks

_______ Working suboptimally

_______ Ambulatory

_______ Bedridden

5. Disease and its relation to nutritional requirements

Primary diagnosis (specify): ___________________________________

Metabolic demand (stress): _______ None _______ Low _______ Moderate _______ High

B. Physical findings (for each trait, specify 0 = normal, 1+ = moderate, 3+ = severe)

_________ Loss of subcutaneous fat (triceps, chest)

_________ Muscle wasting (quadriceps, deltoids, temporals)

_________ Ankle edema, sacral edema

_________ Ascites

_________ Tongue or skin lesions suggesting nutrient deficiency

C. SGA rating (select one)

_______ A = Well nourished (minimal or no restriction of food intake or absorption, minimal change in function, weight stable or increasing)

_______ B = Moderately malnourished (food restriction, some function changes, little or no change in body mass)

_______ C = Severely malnourished (definitely decreased intake, function, and body mass)

SGA of nutritional status: _______

Reprinted with permission from Detsky AS, McLaughlin JR, Baker JP, et al: JPEN J Parenter Enteral Nutr *11:8-13, 1987; with permission from the American Society for Parenteral and Enteral Nutrition (A.S.P.E.N). A.S.P.E.N does not endorse the use of this material in any form other than its entirety.*

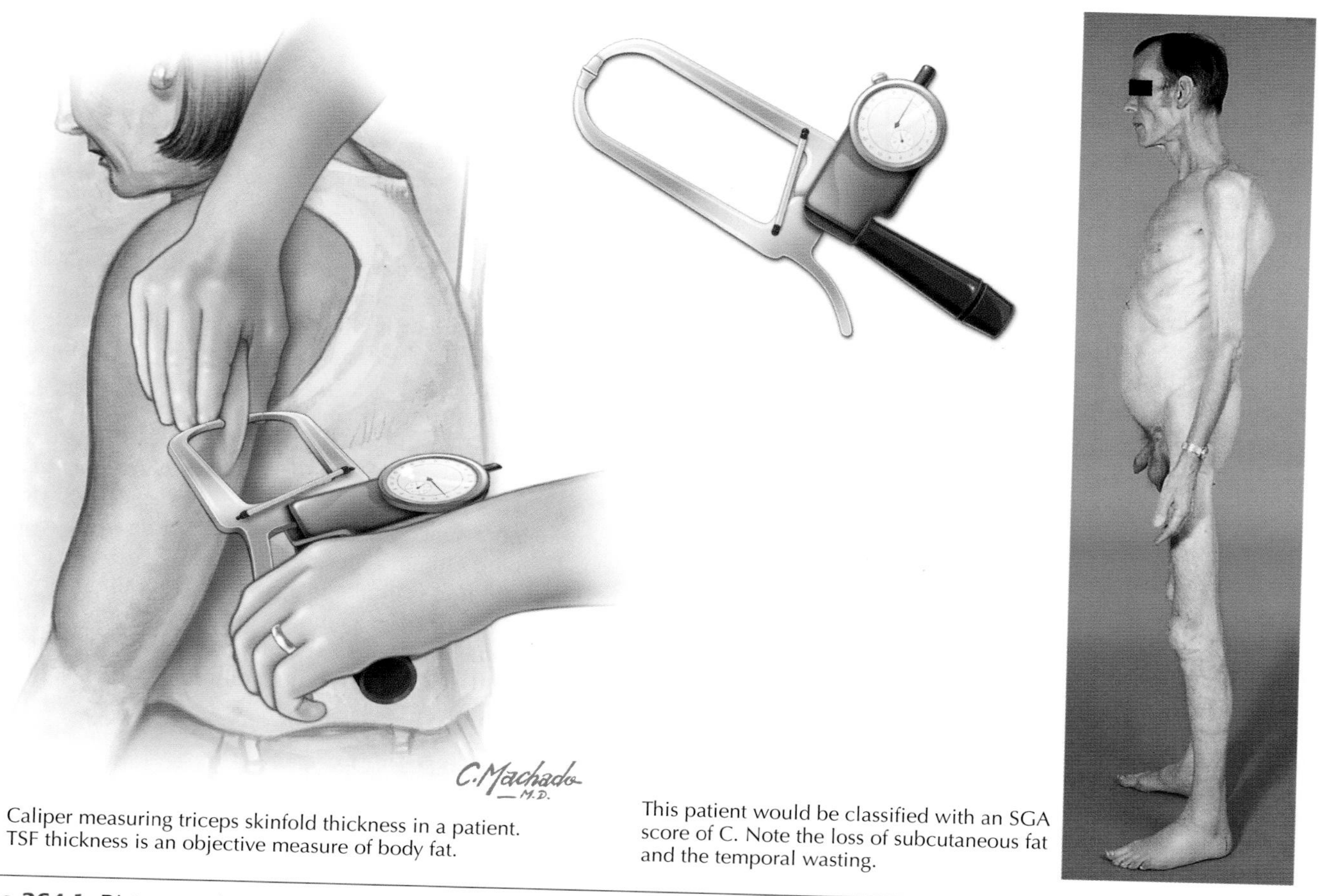

Caliper measuring triceps skinfold thickness in a patient. TSF thickness is an objective measure of body fat.

This patient would be classified with an SGA score of C. Note the loss of subcutaneous fat and the temporal wasting.

Figure 264-1 *Dietary and Nutritional Assessment. From Forbes CD, Jackson WF:* A color atlas and text of clinical medicine, *St Louis, 1993, Mosby.*

findings are scored as normal (0), mild (1+), or severe (3+) and include the degree of subcutaneous fat and muscle loss. Based on medical history and physical findings, patients are ranked in one of three categories: *A*, good nutrition; *B*, moderate malnutrition; or *C*, severe malnutrition. Patients with C scores demonstrate obvious physical signs of malnutrition and an overall loss of at least 10% of their usual weight (**Fig. 264-1**).

ADDITIONAL RESOURCES

Detsky AS, McLaughlin JR, Baker JP, et al: What is subjective global assessment of nutritional status? *JPEN J Parenter Enteral Nutr* 11:8-13, 1987.

Morgan SL, Weinsier RL: *Fundamentals of clinical nutrition*, ed 2, St Louis, 1998, Mosby–Year Book.

Macronutrients and Energy

James S. Scolapio

Energy requirements can be measured at bedside by indirect calorimetry or can be estimated using the Harris-Benedict equation. An estimate of *resting energy expenditure* (REE) can be derived from standard regression formulas based on various population studies. The most common formula used is the *Harris-Benedict equation*, as follows:

Women (kcal/day) = 655.10 + (9.46 × Weight, kg) + (1.86 × Height, cm) – (4.68 × Age, yr)

Men (kcal/day) = 66.47 + (13.75 × Weight, kg) + (5 × Height, cm) – (6.76 × Age, yr)

This regression equation was derived from studies on healthy subjects at rest and was not designed to address the stress and hypercatabolism seen in many disease states. Stress factors have been developed for certain clinical situations. The *stress factor* for patients after elective surgery is 1.2 times the resting REE and 1.5 times the REE for burn patients.

A reasonable correlation seems to exist between the measured REE by indirect calorimetry and that predicted from the Harris-Benedict equation. For most hospital patients, caloric needs are approximately REE × 1.2 to 1.5. *Indirect calorimetry* is based on the principle that energy expenditure is proportional to oxygen (O_2) consumption and carbon dioxide (CO_2) production. The *respiratory quotient* (RQ) is the ratio of CO_2 produced and O_2 consumed and provides information on substrate use. Each of the three major substrates has *n* RQ: 1.0 for carbohydrate, 0.8 for protein, and 0.7 for fat. An RQ less than 0.7 suggests that the patient is using fat as the primary fuel, whereas an RQ greater than 1.0 suggests the patient is being overfed with carbohydrate.

Nutrition substrates or macronutrients include protein, carbohydrate, and fat. The appropriate mix of substrate depends on the clinical state and the desired goals. In general, 1.0 to 1.5 g/kg/day of *protein*, 30% of total calories as *lipid* (fat), and the remaining substrate as *carbohydrate* is a desirable combination. In some circumstances, the substrate mixture may need modification. For example, in a patient with pulmonary disease and CO_2 retention for whom carbohydrates impose a greater demand on the respiratory system, carbohydrate intake should be minimized. Higher protein requirements are common in trauma and burn patients.

ADDITIONAL RESOURCES

Morgan SL, Weinsier RL: *Fundamentals of clinical nutrition*, ed 2, St Louis, 1998, Mosby–Year Book.

Micronutrient and Vitamin Deficiency

James S. Scolapio

266

Various vitamins and trace elements may be deficient in patients with gastrointestinal disease, who may present clinically with abnormal physical and laboratory findings (**Table 266-1**). The best guide to determine the dietary adequacy of vitamins and minerals is the U.S. *recommended daily allowance* (RDA).

Fat-soluble vitamin (e.g., A, D, E, K) deficiency can result in significant fat malabsorption, as occurs in short bowel syndrome, celiac disease, and chronic pancreatitis (**Fig. 266-1**). *Water-soluble vitamin* (e.g., niacin, thiamine) deficiency is less common in malabsorptive states and tends to occur more often in patients with poor dietary habits, as in chronic alcohol abuse. Vitamin B_{12} deficiency can occur after total gastrectomy, with terminal ileal disease, and with greater than 100-cm resection of the terminal ileum (**Fig. 266-2**).

Inorganic trace elements are essential for health and include iron, chromium, manganese, copper, zinc, and selenium. A deficiency of any of these trace elements may also result in abnormal clinical and laboratory findings (see Fig. 266-2).

Table 266-1 Vitamins and Trace Elements: Dietary Sources and Deficiencies

Vitamin	Dietary Source	Deficiency
Thiamine (B_1)	Cereals, grains, pork, legumes, wheat grain seeds, nuts	Wernicke-Korsakoff encephalopathy, high-output congestive heart failure, lactic acidosis, peripheral neuropathy, nystagmus
Niacin	Red meat, liver, milk, eggs, corn	Pellagra, dermatitis, diarrhea, dementia, stomatitis, Hartnup disease
Cobalamin (B_{12})	Meat, eggs, dairy	Megaloblastic anemia, subacute combined degeneration of spinal cord, pernicious anemia, progressive neuropathy
Folic acid	Yeast, liver, vegetables, fruits, nuts	Pancytopenia, megaloblastic anemia, glossitis, stomatitis
Ascorbic acid	Citrus fruit, tomatoes, green vegetables, peppers	Perifollicular hyperkeratosis (scurvy), hemorrhage
Biotin	Milk products, eggs, liver	Scaly dermatitis, alopecia, lethargy, hypotonia, lactic acidosis
Vitamin A	Fish oils, liver, egg yolk, fortified dairy products, carotenoids, green leafy vegetables	Night blindness, xerosis, Bitot spots, hyperkeratosis of skin
Vitamin D	Fortified milk, breads, fatty fish	Osteomalacia, rickets, reduced serum calcium
Vitamin E	Whole wheat, vegetable oils	Hemolytic anemia, spinocerebellar degeneration, neuropathy, ophthalmoplegia
Vitamin K	Green leafy vegetables, dairy products, cereals	Bleeding, increased prothrombin time
Iron	Red meat, fish, oysters, dried beans, fortified breads and cereals	Hypochromic microcytic anemia, cheilosis, weakness
Zinc	Shellfish, meat, eggs	Acrodermatitis enteropathica, diarrhea, apathy, impaired growth, hair loss, skin rash, dysgeusia, reduced wound healing
Copper	Liver, legumes, shellfish, nuts, seeds, whole grains	Microcytic hypochromic anemia, leukopenia, neutropenia, osteoporosis, Menkes syndrome
Chromium	Brewer's yeast, vegetable oils, liver, cereals	Glucose intolerance, peripheral neuropathy, metabolic encephalopathy
Selenium	Meat, poultry, fish, cereal, grains, seafood	Dilated cardiomyopathy, myositis, weakness, white nails, Keshan disease

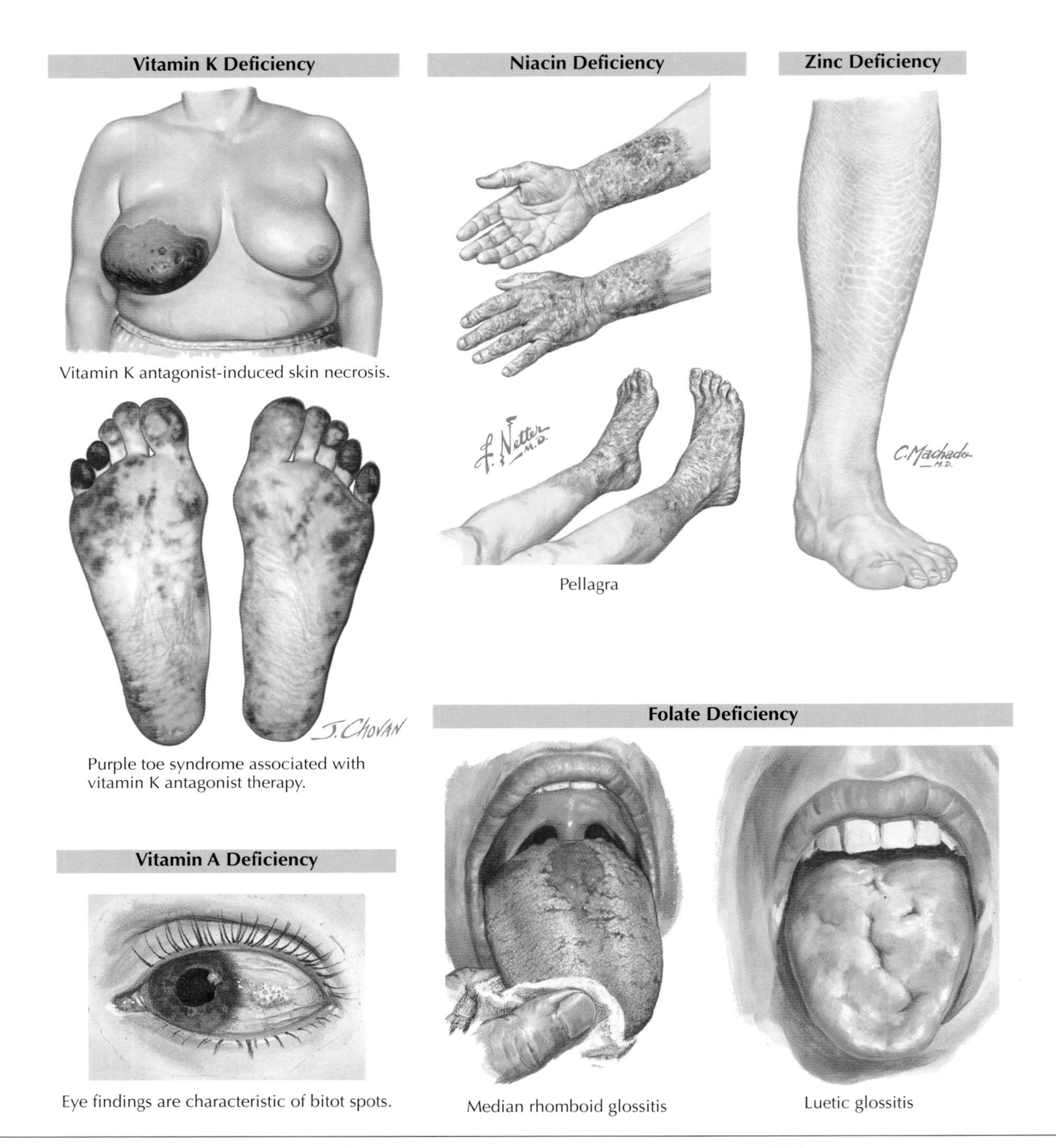

Figure 266-1 *Micronutrient and Vitamin Deficiencies: Vitamin K, Niacin, Zinc, Vitamin A, and Folate.*

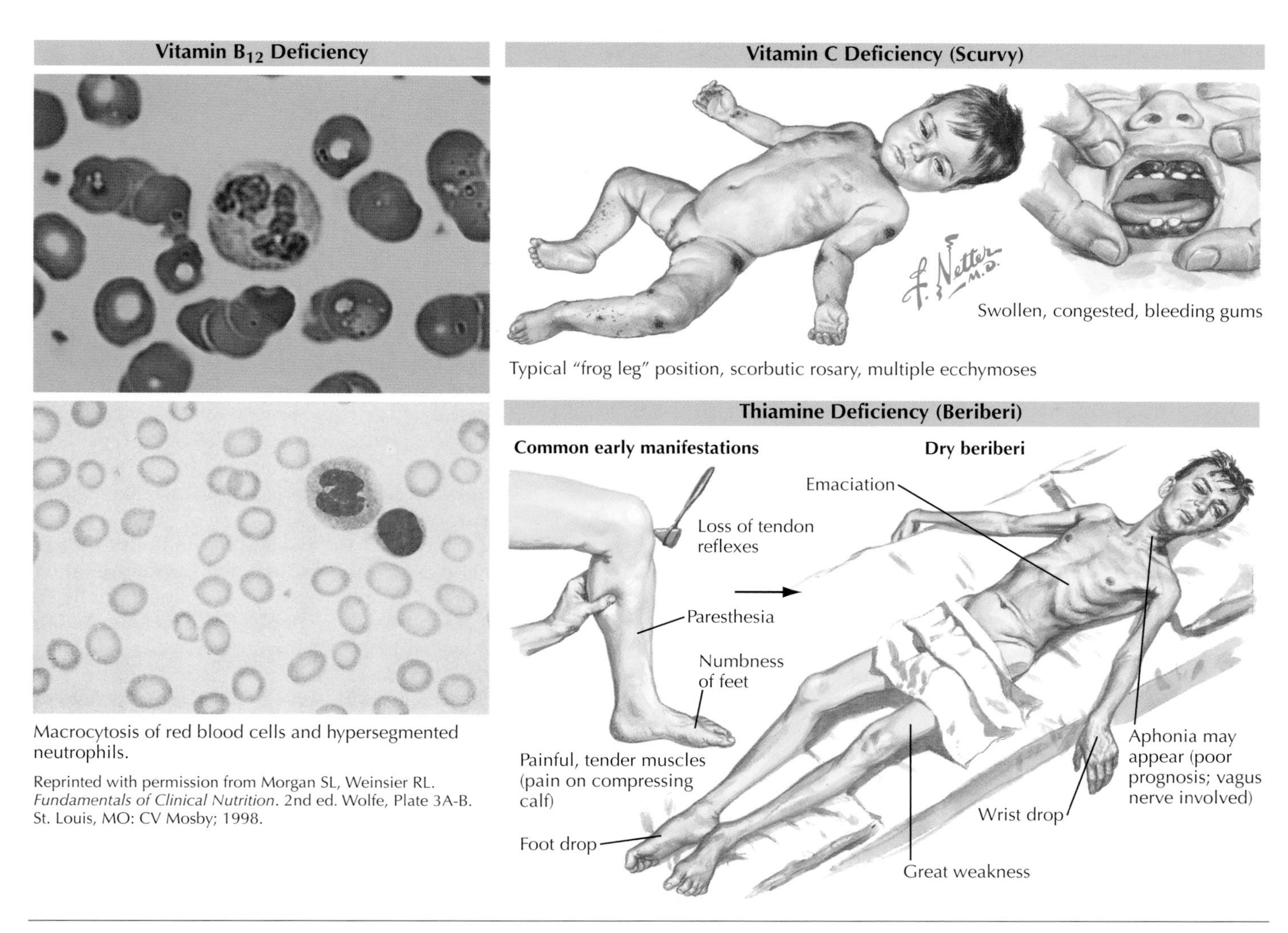

Macrocytosis of red blood cells and hypersegmented neutrophils.

Reprinted with permission from Morgan SL, Weinsier RL. *Fundamentals of Clinical Nutrition*. 2nd ed. Wolfe, Plate 3A-B. St. Louis, MO: CV Mosby; 1998.

Figure 266-2 *Micronutrient and Vitamin Deficiencies: Vitamin B_{12}, Vitamin C, and Thiamine.*

Enteral Nutrition

James S. Scolapio

267

Clinical guidelines for nutrition support, including enteral nutrition, have been published by the American Society of Parenteral and Enteral Nutrition.

For patients unable to take in sufficient calories by mouth for more than 10 to 14 days, *total enteral nutrition* (TEN), or tube feeding, is the preferred route to provide nutritional support. Compared with total parenteral nutrition (TPN), enteral feeding is associated with fewer infectious complications and is less expensive. Access for TEN can be through the nasal route or by a percutaneously placed gastric or small-bowel tube. In patients who require TEN for less than 4 weeks, the nasal route is preferred. Using a nasal tube beyond 4 weeks may cause significant nasal irritation and is uncomfortable for many patients.

A nasogastric or nasojejunal tube can usually be placed at beside. Nasojejunal tube placement may require the assistance of endoscopy or fluoroscopy by interventional radiology. Usually, a soft, 10-French to 12-French-diameter tube should be used. Tubes smaller than 10-French tend to become occluded, and tubes larger than 12-French are uncomfortable for most patients. Although limited studies are available, the current literature suggests that placing a nasal or a percutaneous jejunal tube may reduce the risk for aspiration compared with gastric tubes. Therefore, any patient at risk for gastric aspiration should have a feeding tube placed postpylorically, preferably beyond the ligament of Treitz. Similarly, any patient who has gastric dysmotility or is at increased risk for gastric reflux should have a percutaneous jejunal tube rather than a percutaneous gastric tube.

If tubes are placed at bedside without endoscopy or fluoroscopy, tube position should be confirmed with abdominal radiography. Data regarding the use of promotility agents to enhance the placement of feeding tubes are inconclusive.

With gastric feeding, the enteral formula can be given by gravity or by continuous pump feeding. Data comparing the two methods regarding gastric aspiration risk are inconclusive but suggest that risk may be reduced with pump feeding. When feeding into the small intestine, pump feeding is encouraged. Gravity feeding into the small intestine has been associated with bloating and diarrhea.

Feeding should be initiated at a rate of 20 to 25 mL per hour and should be advanced every 12 hours in 20-mL increments until the goal rate is reached. Gastric residual volumes should be checked in gastric-fed patients. A *residual volume* (RV) greater than 200 mL 2 hours after the last feeding indicates delayed gastric emptying and increases the risk for gastric aspiration. The patient must be carefully assessed and examined to determine whether it is appropriate to continue with feeding. Adding blue dye or methylene blue to the enteral formula should no longer be used as a marker of aspiration because deaths have been reported in patients with sepsis. Other complications of tube feeding besides aspiration include diarrhea, nasal ulceration, and hyperglycemia. Metabolic complications with TEN are less common than in patients receiving TPN.

Various enteral formulas are available for clinical use (**Table 267-1**). Hospitals usually have a set formulary. Most enteral formulas contain 1.0 kcal/mL. Defined formulas, including elemental, semi-elemental, disease-modifying, and immune-enhancing formulas, are specific for disease states. In general, data are limited regarding their clinical benefit compared with the standard polymeric formula. For example, immune-

Table 267-1 Enteral Formulas

Product	kcal/mL	mL/can	Protein/L (g)	Osmolality (mOsm/kg water)	Fiber	Fat, % (MCT + LCT of total calories)	MCT, % (from total fat)	Lactose
Osmolite*	1.06	237	37.1	300	No	29	20	No
Osmolite HN*	1.06	237	44.3	300	No	29	19	No
Osmolite HN Plus*	1.2	237	55.5	360	No	33	25	No
Nutren 1.0†	1.4	240	40.0	300	No	33	25	No
Isosource HN‡	1.2	250	53.0	330	No	30	50	No
Isocal§	1.06	240	34.0	270	No	37	20	No
Promote*	1.0	237	62.5	340	No	23	19	No
Jevity*	1.06	237	44.3	300	Yes	29	20	No
Peptamen†‖	1.0	240	40.0	270	No	33	70	No
Vivonex TEN‡¶	1.0	Powder 300 mL mix/H_2O	38.0	630	No	3	0	No

Product manufacturers: *Ross Laboratories Division/Abbott Laboratories; †Nestle Clinical Nutrition; ‡Novartis Nutrition; §Mead Johnson.
‖Semielemental.
¶Elemental.
MCT, Medium-chain triglycerides; *LCT,* long-chain triglycerides.

enhancing formulas have not been shown to improve mortality rates compared with traditional formulas.

Long-term enteral feeding is usually given in the home through a percutaneous endoscopic gastrostomy tube. The most common indications are neurologic disease, such as cerebrovascular accident (stroke), and after radiation therapy for head and neck cancer. Medicare and most insurance companies require that (1) enteral feeding will be needed for at least 3 months and (2) the patient will be unable to take sufficient calories by the oral route.

ADDITIONAL RESOURCES

ASPEN Board of Directors, Clinical Guideline Task Force: Guidelines for the use of parenteral and enteral nutrition in adult and pediatric patients, *JPEN J Parenter Enteral Nutr* 26(1 Suppl):1SA-138SA, 2002.

Dietary Fiber

James S. Scolapio

Dietary fiber is the *nonstarch polysaccharide* part of plant foods that is poorly digested by human enzymes. In the latter half of the twentieth century, physicians and scientists began to understand the importance of dietary fiber in maintaining health and dietary fiber deficiencies in causing disease. Cleave first noted the increased sugar intake that resulted from decreased dietary fiber intake. Epidemiologists and gastroenterologists such as Burkett, Trowell, Painter, Walker, Heaten, and Eastwood then began to stress the importance of dietary fiber in maintaining normal gastrointestinal (GI) function and in preventing disease.

CHEMISTRY AND PROPERTIES

Food chemists analyzed plant foods first for crude fiber content, then began to correlate fiber content with various parts of plants. Because dietary fibers are extremely complex chemical compounds, it has been difficult for food scientists to decide on simple methods to identify their components in all foods. However, understanding the properties of fiber in the GI tract has facilitated making the distinction between soluble and insoluble fibers. In Englyst's classic method, after extraction, substances can be broken down as *soluble*, *insoluble*, and *cellulose*, which constitute the nonstarch polysaccharide component. Some resistant starch is left over from the chemical process.

Components in the plant cell walls are cellulose, noncellulose polysaccharides (soluble and insoluble), lignin, waxes, protein, and ash. Major classes of *noncellulose polysaccharide* are rhamnogalacturonans, arabinogalactans, β-glucans, xylans, mannans, and xyloglucans. Gums and mucilages develop but are not strictly part of the plant cell walls; they are complex heteroglycans with branch structures. Bacterial fermentation clearly shows that bacterial enzymes almost completely ferment soluble fibers but poorly ferment insoluble fibers or cellulose.

PHYSICAL PROPERTIES

Important physical properties of dietary fiber and its components are particle size and polysaccharides. Depending on how food is cooked or processed, *particle size* will be large or greatly reduced. In some cases, reduction in size can completely disrupt the plant cell wall. Therefore, particle size is important in determining some of the properties discussed here.

Polysaccharides may be hydrophilic and may have a definite water-holding capacity. This varies with the food, and certainly, cellulose is limited to swelling property, which depends on polysaccharide type. Some polysaccharides are able to form gels, and some can become extremely viscous. These properties affect nutrient ion absorption, although the gels are completely fermented by bacterial action, freeing any substances that are trapped.

The water-holding property of insoluble fibers is particularly important for maintaining larger, softer stool. Insoluble fibers are poorly fermented by bacteria. Hence, this water-holding property is helpful in maintaining a larger stool bulk throughout the colon.

Ion Binding

Uronic acid–containing polysaccharides and the lignin components of dietary fiber have acidic functional groups that react with ions. Calcium, iron, and zinc can be bound, but they are readily freed. Similarly, bile salts can be bound. This appears to be an extremely dynamic process that may not necessarily interfere with absorption but may actually enhance it by bringing a particular substance to the site in the intestine where it is best absorbed.

FIBER INTAKE

The intake of dietary fiber varies greatly from society to society and within societies. Studies indicate that people from underdeveloped countries, such as Asia and Africa, eat as much as 60 to 80 g of dietary fiber daily, given that the main component of their diet is cereal fiber. In Western societies, the consumption is generally 5 to 10 g of dietary fiber daily. This great variation is attributed to selectivity and lifestyle. Diets are becoming more balanced worldwide as knowledge of the benefits of dietary fiber increases.

EFFECT ON THE GASTROINTESTINAL TRACT

Dietary fiber has specific effects on the GI tract. It should be remembered that *soluble* fiber may have different effects than *insoluble* fiber, but that most foods are mixed. Because fiber enters into the dynamics of intraluminal microecology, its effects vary with the fluid, pH, and lumen contents of other substances. Nevertheless, overall dietary fiber may slow gastric emptying, may have an inhibitory effect on pancreatic enzyme activity, and depending on the nutrients and particular foods involved, may slow absorption from the small intestine.

Fibers vary in their ability to slow oral-to-fecal transit time. It is a complicated interface, but guar gum can delay hydrogen-to-breath time, whereas bran and gum tragacanth have lesser effects. Pectin and cellulose seem to have no effect. When treating constipation, however, it is important knowing that transit time and bulk are increased. Pectin has been clearly shown to decrease the absorption of cholesterol.

COLON AS A FERMENTER

Bacterial flora of the distal ileum and colon ferment various fibers at various rates. The matrix that exists in the colon, with its vast bacterial population of aerobes and anaerobes, ferments soluble fibers at a 10-fold greater rate than it ferments insoluble fibers. It also produces short-chain fatty acids that are pivotal for the health of the colon and for the control of cholesterol metabolism through the enterohepatic circulation. *Butyric acid* is the main fuel of colonocytes; *acetic acid* is the building block for cholesterol; and *propionic acid* appears to have some controlling mechanism on cholesterol development. These short-chain fatty acids are produced in the colon and are absorbed through the enterohepatic circulation, if not directly used for fuel by colonocytes.

Depending on the amount of dietary fiber eaten, stool size varies greatly. In general, an 8- to 10-g diet produces approximately 100 mL of stool, whereas a 25- to 30-g diet produces as much as 300 mL of stool. Eventually, the amount of short-chain fatty acid produced will vary. Furthermore, if the dietary fiber intake is heavy in soluble fiber, such as from fruits and vegetables or psyllium seed, the bacterial flora will increase and flourish and will be larger than if the dietary fiber consists of cellulose and insoluble fibers from bran. In the latter case, the stool volume might still be large, but this results from the water-holding property of insoluble fiber rather than increased bacterial flora.

EFFECT ON DISEASE

It is now well accepted that dietary fiber ameliorates constipation and diarrhea (see Chapters 111 and 136). Also, colon diverticula are less common in those who eat high-fiber diets, and diverticular formation may be prevented in susceptible subjects. Diverticular disease may develop in persons who eat low-fiber diets for decades. It is recommended that the patient with diverticular disease maintain a high-fiber diet.

When acute diverticulitis develops, physicians usually restrict the amount of dietary fiber the patient can consume until the episodes have been resolved.

USE AND TREATMENT

Treatment and prevention of colonic polyps, colonic cancer, coronary artery disease, and stroke have remained extremely controversial. Meta-analysis has revealed conflicting results. Some large national studies reveal, however, that subjects on high-fiber diets experience less polyp formation and fewer coronary artery diseases. Consequently, most clinicians recommend a high-fiber diet to prevent diverticulosis of the colon, polyp formation, and atherosclerotic disease. This is a controversial issue, and the literature substantiates both positions.

Recent studies also indicate that subjects who eat high-fiber diets tend to experience less morbid obesity. Naturally, these subjects eat less fat and sugar; thus, their high-fiber diet decreases their nutrient energy intake.

The therapeutic recommendation for the intake of dietary fiber is 20 to 35 g daily, depending on the meal size and caloric intake. Some recommend approximately 10 to 12 g per 1000 calories. The recommendation also states that broad types of dietary fiber should be consumed, and that the intake should include a mixture of soluble and insoluble fiber.

Table 265-1 lists the dietary fiber content of the most common foods. A daily high-fiber cereal plus three to five portions (depending on size) of fruits or vegetables will usually

Table 268-1 Dietary Fiber Food Sources

	Serving Size	Soluble Fiber Content per Serving (g)	Insoluble Fiber Content per Serving (g)	Total Fiber Content per Serving (g)
Vegetables (cooked, unless otherwise noted)				
Asparagus	¾ cup	0.8	2.3	3.1
Bean sprouts, raw	½ cup	0.3	1.3	1.6
Beans				
Green	½ cup	0.5	1.6	2.1
Kidney	½ cup	2.5	3.3	5.8
Lima	½ cup	1.1	3.2	4.4
Pinto	½ cup	2.3	3.3	5.3
White	½ cup	1.4	3.6	5.0
Broccoli	½ cup	0.9	1.1	2.0
Brussels sprouts	½ cup	1.6	2.3	3.9
Cabbage	½ cup	0.9	1.1	2.0
Carrots	7 inch	1.1	1.2	2.3
Cauliflower	½ cup	0.4	0.6	1.0
Celery, raw	½ cup	0.4	0.9	1.3
Corn, kernels	½ cup	1.7	2.2	3.9
Eggplant	½ cup	0.8	1.2	2.0
Kale	½ cup	1.4	1.4	2.8
Lettuce, raw	½ cup	0.1	0.2	0.3
Okra	½ cup	1.0	3.1	4.1
Onions, raw	½ cup	0.8	1.8	2.6
Peas	½ cup	0.4	2.8	3.2
Potatoes				
Sweet, baked	½ large	0.7	1.0	1.7
White, baked	½ medium	1.0	1.0	1.9
Radishes, raw	5 medium	0.1	0.5	0.6

Continued

Table 268-1 Dietary Fiber Food Sources—cont'd

	Serving Size	Soluble Fiber Content per Serving (g)	Insoluble Fiber Content per Serving (g)	Total Fiber Content per Serving (g)
Squash				
Acorn	½ cup	0.5	3.8	4.3
Zucchini	½ cup	1.3	1.4	2.7
Tomato, raw	1 medium	0.2	0.6	0.8
Turnip	½ cup	0.8	0.9	1.7
Zucchini	½ cup	0.5	0.7	1.2
Fruits (raw)				
Apple, with skin	1	0.8	2.0	2.8
Apricots	2	0.7	0.8	1.5
Avocado	⅛ fresh	0.5	0.7	1.2
Banana	½ medium	0.3	0.7	1.0
Blackberries	½ cup	0.7	3.9	4.5
Cherries	10	0.3	0.9	1.2
Figs	1½	1.1	1.2	2.3
Grapefruit	½ medium	0.6	1.1	1.7
Grapes	12	0.1	0.4	0.5
Melon, cantaloupe	1 cup	0.3	0.8	1.1
Orange	1 small	0.3	0.9	1.2
Peach	1 medium	0.6	1.0	1.6
Pear	½ medium	0.5	2.0	2.5
Pineapple	½ cup	0.3	0.9	1.2
Plums	3 small	0.7	1.1	1.8
Raspberries	¾ cup	0.4	6.4	6.8
Strawberries	¾ cup	0.7	1.3	2.0
Grain Products				
Bread				
Bagel, plain	½	0.3	0.4	0.7
French	1 slice	0.3	0.7	1.0
Rye	1 slice	0.3	0.6	0.9
White enriched	1 slice	0.3	0.3	0.5
Whole wheat	1 slice	0.3	1.2	1.4
Cereal				
All-Bran (100%)	⅓ cup	1.7	7.0	8.6
Corn flakes	1 cup	0.2	0.3	0.4
Shredded wheat	1 biscuit	0.4	2.4	2.8
Fiber One	½ cup	0.8	11.1	11.9
Raisin bran	¾ cup	0.9	4.4	5.3
Oatmeal (oats)	⅓ cup	1.4	1.3	2.7
Crackers				
Graham	2 squares	0.5	2.3	2.8
Saltine	6 crackers	0.3	0.4	0.7
Rice				
Brown	½ cup	0.2	2.2	2.4
White	½ cup	0.01	0.09	0.1
Spaghetti	½ cup	0.3	0.5	0.8
Nuts				
Almonds	1 tbsp	0.1	1.0	1.1
Peanuts, roasted	10	0.2	0.4	0.6

satisfy the body's need for soluble and insoluble fiber. Dietitians and clinicians usually recommend five portions of fruits, vegetables, or grains, averaging 4 to 5 g each, and thus meeting the necessary requirement of 20 to 35 g daily.

ADDITIONAL RESOURCES

Bazzano LA, He J, Ogden LG, et al: Fruit and vegetable intake and risk of cardiovascular disease in U.S. adults: the first National Health and Nutrition Examination Survey epidemiologic follow-up study, *Am J Clin Nutr* 76:13-19, 2002.

Cleave TL: *The saccharine disease*, New Canaan, Conn, 1975, Keats Publishing.

Slattery ML, Curtin KP, Edwards SL, Schaffer DM: Plant foods, fiber, and rectal cancer, *Am J Clin Nutr* 79:274-281, 2004.

Trowell H, Burkett D, Heaton K: *Dietary fibre, fibre-depleted foods and disease*, London, 1985, Academic Press.

Parenteral Nutrition

James S. Scolapio

Parenteral nutrition, or *total parenteral nutrition* (TPN), is indicated for patients who are unable to take in sufficient calories by the oral route for more than 10 to 14 days, and for whom total enteral nutrition (TEN) is not possible (e.g., intestinal obstruction, severe malabsorption). Contraindications for TPN include a functional gastrointestinal (GI) tract, intended use less than 3 days, and imminent death from the underlying disease. Although patients usually prefer TPN to a nasogastric feeding tube, TPN is associated with a higher risk of infection from catheter sepsis and other metabolic complications.

Access for parenteral nutrition can be through a peripheral vein, if the dextrose concentration will be less than 5%, or through a central vein if the dextrose concentration will be greater than 5%. Administering a dextrose concentration greater than 5% through a peripheral vein can result in thrombophlebitis. When administering *central* parenteral nutrition, direct subclavian vein placement has traditionally been the method of placement. However, a peripherally inserted central catheter inserted through the brachial vein and advanced to the superior vena cava is the preferred method because it averts the risk for pneumothorax associated with the placement of subclavian lines. Before infusion is initiated, proper insertion of the catheter tip in the superior vena cava should be confirmed by chest radiograph.

Total parenteral nutrition is usually given as a 2-L, 3-in-1 (3:1) solution containing carbohydrate, protein, and lipid mixed together. Intravenous lipid is typically used to supply 20% to 40% of daily calories. Energy and protein requirements should be calculated as described in Chapter 265. TPN volume should be reduced in patients with congestive heart failure, significant renal disease, and fluid overload. The TPN formula also contains electrolytes, multivitamins, and trace elements.

Infusion rate should be based on a 24-hour period. The patient should be started on half the rate (40 mL/hr) for the first 24 hours; this can be increased to full rate (80 mL/hr) if tolerated by the patient. Serum electrolyte levels should be checked at least twice a week while the patient receives TPN in the hospital. Blood glucose levels should also be monitored closely while the patient receives TPN. If blood glucose levels are higher than 200 mg/dL, regular insulin may need to be added to the TPN solution to maintain glucose levels lower than 200 mg/dL, because hyperglycemia is a risk factor for infection. A sliding scale of regular insulin will also suffice without having to add insulin to the TPN solution. Serum triglyceride levels should be checked at least weekly to prevent serum levels from exceeding 500 mg/dL.

The most common complication of TPN is *catheter sepsis*. Nursing staff must follow strict sterile catheter techniques. To prevent bacterial contamination of the solution, the TPN solution should not be kept room temperature longer than 24 hours, and an in-line filter should always be used. Tapering the TPN infusion by 50% for 30 to 60 minutes before discontinuing it helps prevent symptomatic hypoglycemia.

When TPN is used in the home, it is referred to as *home parenteral nutrition*. Indications for home parenteral nutrition include short bowel syndrome, severe radiation enteritis, distal intestinal fistula, and mechanical intestinal obstruction when surgery is not immediately possible. Medicare will not reimburse patients with a functioning GI tract or in whom TPN will be required for less than 3 months. A certificate of medical necessity must be completed before hospital discharge, documenting that the TPN will be required for at least 3 months. Hospital case managers should be directly involved in arranging hospital discharge, ensuring all criteria are met for insurance reimbursement.

Home parenteral nutrition is associated with complications that must be recognized and addressed appropriately. These include catheter infection, liver disease, and metabolic bone disease. Multiple trace elements, which include manganese, copper, chromium, selenium, and zinc, should be checked at least every 6 months. Cholestatic liver disease increases the risk for manganese and copper toxicity because these two trace elements are excreted primarily through the hepatobiliary system.

ADDITIONAL RESOURCES

ASPEN Board of Directors, Clinical Guideline Task Force: Guidelines for the use of parenteral and enteral nutrition in adult and pediatric patients, *JPEN J Parenter Enteral Nutr* 26(1 Suppl):1SA-138SA, 2002.

Scolapio JS, Fleming CR, Kelly DG, et al: Survival of home parenteral nutrition–treated patients: 20 years of experience at the Mayo Clinic, *Mayo Clin Proc* 74:217-222, 1999.

Malnutrition

James S. Scolapio

Malnutrition can be classified as primary or secondary. *Primary malnutrition* is caused by inadequate food supply, which is common in many countries in the developing world, but less common in the United States. *Secondary malnutrition* results from an underlying chronic disease process. Anorexia nervosa is an example of a condition causing secondary malnutrition (**Fig. 270-1**).

Patients with gastrointestinal disease are most predisposed to secondary malnutrition because of reduced oral intake of food, intestinal obstruction, altered absorption, and digestion of nutrients. *Malnutrition*, which can be defined as unintentional weight loss of more than 10% of usual body weight over 3 months, may occur in hospital patients. Malnutrition has been associated with increased infections, impaired wound healing, increased postoperative complications, longer hospital stay, and higher mortality. Dietary energy and protein deficiencies usually occur together, although one form may predominate.

Kwashiorkor is the term used when severe protein deficiency is the primary cause of malnutrition, and *marasmus* is the term used when severe energy or calorie deficiency is the primary cause of the malnutrition. *Marasmic kwashiorkor* is the term used to describe the combination of chronic energy and protein deficiency. *Protein-energy malnutrition* (PEM) and *protein-calorie malnutrition* are more common terms for malnutrition. A body mass index (BMI) less than 16 represents severe PEM. Worldwide, socioeconomic and environmental factors are leading causes of PEM.

The diagnosis of marasmus includes a clinical history of inadequate calorie intake, usually from a chronic illness, and physical findings of severe wasting of muscle and subcutaneous fat. Patients with marasmus frequently are at 60% of their expected weight for height, and children often have greatly impaired longitudinal growth. Hair thinning and hair loss with dry, flaky skin are also common. Diminished skinfold thickness and reduced midarm muscle circumference and temporal wasting illustrate the loss of fat and skeletal muscle, respectively, in these patients. Serum albumin levels are usually normal in patients with marasmus.

Diagnostic features of kwashiorkor include pitting edema of the feet and legs, skin ulceration, and epidermal sloughing. In contrast to marasmus, kwashiorkor occurs in the United States predominantly in patients with acute, highly metabolic illness, such as trauma and burns. Subcutaneous fat and muscle mass are often preserved, although careful inspection usually reveals more muscle wasting. Delayed wound healing, skin breakdown, and infection are also common. The abdomen may protrude because of an edematous stomach and intestinal loops. The most common biochemical findings in kwashiorkor are reduced serum albumin level, lymphopenia, and anemia.

Treating PEM requires replacing fluids, macronutrients, and micronutrients. Caution must be used to avoid rapid repletion and when reintroducing feeding in patients with severe PEM. Refeeding should be done slowly to prevent *refeeding syndrome.* Rapid refeeding with oral, enteral, or parenteral nutrition can result in acute decreases in serum phosphorus, potassium, and magnesium levels, resulting in cardiac arrhythmia and death. Acute thiamine deficiency is also a potential concern during refeeding in patients with severe PEM.

ADDITIONAL RESOURCES

Shils ME: *Modern nutrition in health and disease*, ed 9, Baltimore, 1999, Williams & Wilkins.

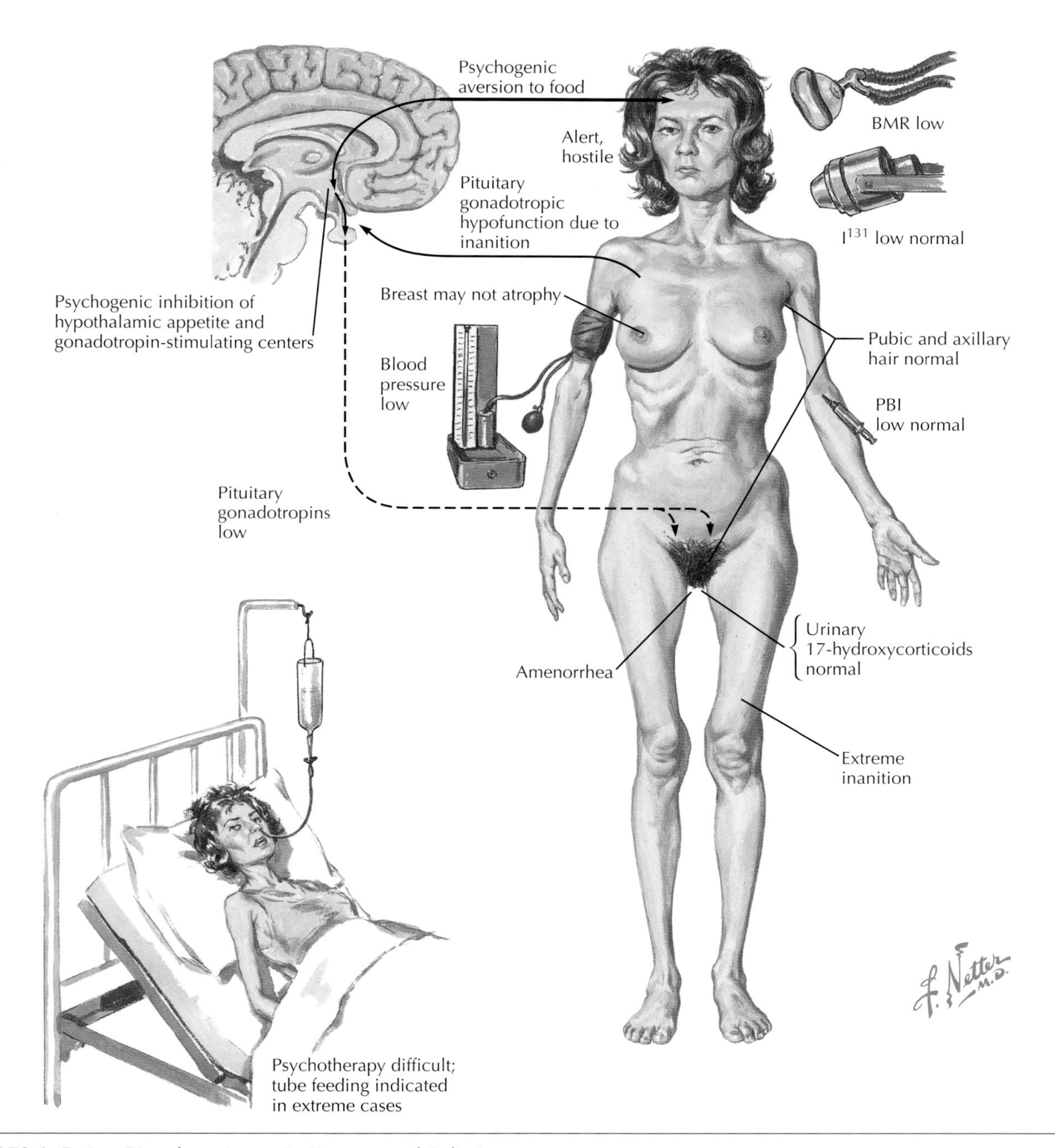

Figure 270-1 *Eating Disorders: Anorexia Nervosa and Bulimia.*

Obesity

271

James S. Scolapio

Obesity is reaching epidemic proportions in the United States. In the last 30 years, the prevalence of obesity has more than doubled, from 13% to 27%. It is estimated that 300,000 deaths per year are caused by obesity. Patients with a body mass index (BMI) between 25 and 29.9 are considered overweight, and a BMI greater than 30 is considered obese. Other factors, such a fat distribution and weight gain, modify the risk within each BMI category.

Persons with increased abdominal fat are at increased risk for diabetes, hypertension, hyperlipidemia, and ischemic heart disease (metabolic syndrome) compared with those with increased gluteal and femoral fat (**Fig. 271-1**). In all persons, obesity is caused by the ingestion of more calories than are expended. Excess calories are stored as fat. Genetic and environmental factors contribute to obesity. The marked increase in obesity in the last 20 years cannot be attributed to genetic factors alone and are most likely caused by changes in the environment.

Treatment of obesity centers on modifying behavior, including diet and physical activity. *Diet modification* should encourage patients to eat three meals daily, to avoid snacking between meals, to avoid energy-dense and high-fat foods, and to increase the intake of fruits and vegetables. *Physical activity* is also important for weight reduction and overall health. Aerobic exercise has additional health benefits independent of weight loss itself.

Pharmacotherapy can help selected patients maintain long-term weight loss. Patients receiving pharmacotherapy for obesity should also be involved in efforts to change their lifestyles, including developing the habits of healthy eating and adequate exercise. Only sibutramine and orlistat are approved for long-term use. *Sibutramine* is a monoamine reuptake inhibitor that inhibits the reuptake of norepinephrine, serotonin, and dopamine. It can cause increases in blood pressure and therefore should not be given to patients with uncontrollable hypertension. *Orlistat* inhibits intestinal lipases and results in intestinal malabsorption. The most common adverse effect with orlistat is oily stool.

Surgery is the most effective approach for achieving weight loss in the extremely obese patient (BMI >40). Indications for surgery include BMI greater than 40 or BMI between 35 and 40 and severe obesity-related diseases, such as diabetes and obstructive sleep apnea. Serious psychiatric disorders are absolute contraindications for surgery. Of the various types of obesity surgery, the *gastric bypass* procedure, also known as the Roux-en-Y, is the most accepted. A small pouch is created in the cardia of the stomach, which drains into a segment of bypassed jejunum. Bypassing the duodenum can cause malabsorption of iron, calcium, and vitamin B_{12}. Serum levels of these micronutrients should be monitored at least twice a year. In addition, gastric bypass surgery can cause dumping syndrome, and postoperative dietary changes may be necessary.

ADDITIONAL RESOURCES

DeMaria EJ: Bariatric surgery for morbid obesity, *N Engl J Med* 356:2176-2183, 2007.

Glenny AM, O'Meara S, Melville A, Wilson C: The treatment and prevention of obesity: a systemic review of the literature, *Int J Obes Relat Metab Disord* 21:715-737, 1997.

Ogden CL, Yanovski SZ, Carroll MD, Flegal KM: The epidemiology of obesity, *Gastroenterology* 132:2087-2102, 2007.

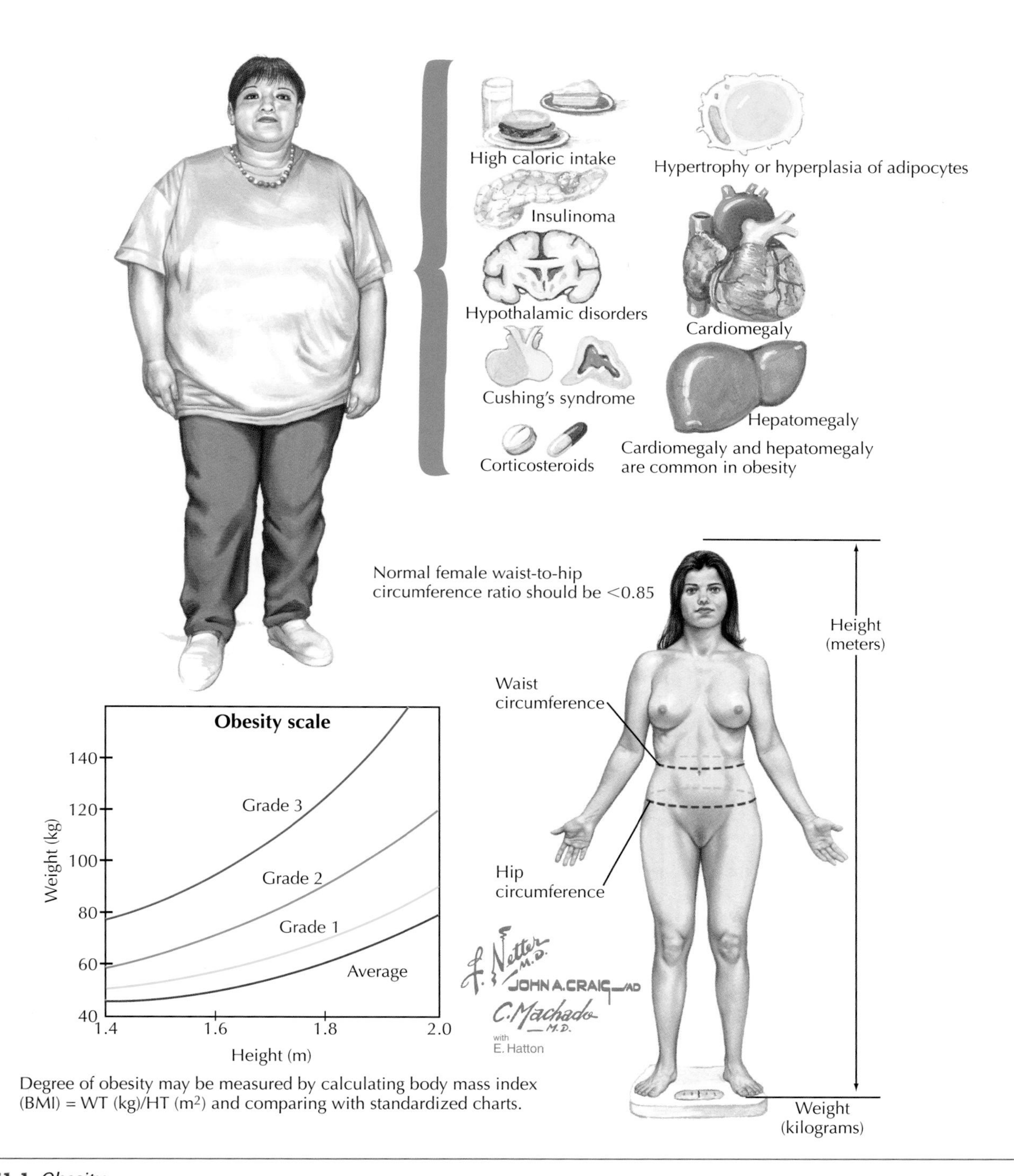

Degree of obesity may be measured by calculating body mass index (BMI) = WT (kg)/HT (m^2) and comparing with standardized charts.

Figure 271-1 *Obesity.*

Gastrointestinal Diseases Related to Nutrition

272

James S. Scolapio

PANCREATITIS

Acute pancreatitis is an example of a disease process possibly affected by method of feeding (total parenteral nutrition [TPN] vs. total enteral nutrition [TEN]). In a prospective study of 54 patients with acute mild pancreatitis, catheter sepsis was 10 times higher in the TPN group than in control subjects who received intravenous fluids only. In the same study, pneumothorax occurred only in the TPN group, and length of hospital stay was longer in the TPN group. This study suggests a significant risk associated with giving TPN to patients with mild pancreatitis.

Based on recent recommendations from the American Society of Parenteral and Enteral Nutrition, patients with mild pancreatitis, who typically resume eating within 7 to 10 days of leaving the hospital, should not receive TEN or TPN unless they have lost significant weight, defined as greater than 10% of usual weight, in the previous 6 months. A more recent study compared complications in 38 patients with severe acute pancreatitis treated with either TEN or TPN. Septic complications, hyperglycemia, peripancreatic necrosis, and cost were all greater in those patients who had received TPN. Mortality rate was similar in the two groups.

Another prospective study of 34 patients with severe pancreatitis compared outcomes with TEN and TPN and found that multiorgan failure, abdominal sepsis, and mortality were greater in the TPN group. In addition, markers of disease severity, including C-reactive protein, Acute Physiology and Chronic Health Evaluation II score, and immunoglobulin M endotoxin, were all higher in patients receiving TPN. Although bacterial translocation was not formally evaluated, TEN might have reduced bacterial translocation compared with TPN, explaining why outcomes were better with TEN. Another study of 156 patients with pancreatitis reported that hypocaloric feeding with TEN resulted in fewer septic and metabolic complications than TPN feeding. More than 50% of the TPN patients were hyperglycemic, versus only 15% of TEN patients. Despite fewer complications in the TEN group, mortality rates were the same in both groups. Current data support the superiority of TEN over TPN in managing patients with acute severe pancreatitis.

In patients with chronic pancreatitis, the cause of malnutrition is multifactorial, including fear of postprandial abdominal pain (sitophobia), steatorrhea, anorexia, and often coexistent alcoholism. Steatorrhea and azotorrhea (fecal protein loss) occur when lipase and trypsin secretion are reduced by 90%. Nutritional management first begins with appropriate management of a patient's abdominal pain. Analgesics should be given at least 30 minutes before meals to prevent postprandial exacerbation of pain. Meta-analysis failed to show a beneficial effect of exogenous pancreatic enzyme replacement in relieving abdominal pain. Treatment of exocrine insufficiency focuses on giving adequate pancreatic enzymes. The minimal dose of lipase required is 28,000 IU per meal. Enzymes should be given with meals to ensure adequate mixing with food. Weight maintenance, symptomatic improvement of diarrhea, and decreased 72-hour fecal fat excretion are the goals of therapy. Dietary fat intake should not be restricted. Fat-soluble vitamins and vitamin B_{12} should be replaced if necessary.

INFLAMMATORY BOWEL DISEASE

The outcomes of TEN and TPN have also been evaluated in patients with inflammatory bowel disease (IBD). Whether the combination of complete bowel rest and TPN can be used successfully as primary therapy in patients with acute IBD, with or without other medical therapy, is controversial.

The literature suggests that patients with Crohn's *enteritis* might achieve clinical remission with the combination of bowel rest and TPN. Indirect evidence, however, suggests that TPN is less effective than steroid therapy in active Crohn's disease. Results suggest that nothing by mouth and TPN for 3 to 6 weeks results in a clinical response rate of 64% in patients with acute Crohn's disease. In most of these studies, however, prednisone was given simultaneously with TPN, making it difficult to discern whether the positive effects observed resulted completely from bowel rest and TPN or from the combined effects of prednisone and TPN. Data also suggest that administering TEN with an elemental, peptide-based, or polymeric formula for 3 to 6 weeks results in a remission rate of approximately 68%, which is similar to the remission rate reported with TPN and bowel rest. The positive benefit is most likely related to the lipid composition of the enteral formulas. Monounsaturated fatty acid (oleic acid), which is present in most enteral formulas, is not a precursor to inflammatory mediators (arachidonic acid and eicosanoic synthesis), which may explain the beneficial effect observed.

On the other hand, current literature suggests that patients with Crohn's colitis and idiopathic ulcerative colitis do not respond any better to TPN and bowel rest (with or without prednisone) than patients treated with prednisone and oral diet. In these studies, TPN resulted in 10% more complications than enteral feeding. Complications included pneumothorax from central catheter placement, catheter sepsis, and various metabolic complications. Enteral feedings using elemental formulas appear to have limited benefit in Crohn's colitis and idiopathic ulcerative colitis. Therefore, surgery should not be delayed for TPN or TEN in patients with refractory colitis. Indications for bowel rest and TPN include patients with mechanical bowel obstruction, distal small bowel fistula, and toxic megacolon who are not judged to be surgical candidates and who have been without food for 10 to 14 days.

LIVER DISEASE

Cirrhosis is perhaps one of the best examples of protein-calorie malnutrition. Poor dietary intake and altered substrate metabolism are the primary reasons for malnutrition in this group of patients. Malnutrition has a negative impact on survival.

Oral dietary therapy is the mainstay of nutritional treatment. Although nasogastric feeding tubes can be safely placed in most patients, use beyond 4 weeks is difficult because of nasal discomfort and risk for bleeding from thrombocytopenia and impaired coagulation. Gastrostomy feeding tubes cannot be safely placed in patients with ascites given the risk for peritonitis. In most patients, except those with refractory hepatic encephalopathy, protein intake should not be restricted and should provide 1.5 g/kg/day of total calories. Vegetable protein is better tolerated than animal protein because it produces fewer aromatic amino acids, thought to be the responsible mediators of hepatic encephalopathy.

Branched-chain amino acids may be helpful in select patients with refractory disease. Fat-soluble vitamin levels should be checked and augmented if they are low. Serum vitamin A concentration levels may not reflect hepatic vitamin A concentrations; therefore, replacement should be administered with caution given the risk for vitamin A toxicity to the liver.

Orthotopic liver transplantation is the principal treatment for malnutrition in patients with cirrhosis. Within 4 months of transplantation, increased muscle mass and improved functional status are clinically evident.

SHORT BOWEL SYNDROME

Short bowel syndrome is a collection of signs and symptoms used to describe the nutritional and metabolic consequences after major resection of the small intestine. In adults, resection for Crohn's disease and intestinal infarction are primary reasons for short bowel syndrome.

Diarrhea, fluid and electrolyte losses, and weight loss characterize short bowel syndrome. Patients with less than 100 to 150 cm of remaining small intestine often require TPN for survival. Patients with enough remaining colon and 50 cm or more of residual small bowel can usually survive on an oral diet without TPN. After surgical resection, the remaining intestine undergoes a process of adaptation, including structural and functional changes that tend to maximize nutrient and fluid absorption. Changes include villus cell hyperplasia and increased brush border enzyme activity. These changes can occur up to 1 year after intestinal resection.

Dietary management of short bowel syndrome depends on whether a patient has retained part of the colon. The patient who has retained colon benefits from a diet high in complex carbohydrates and low in fat. The colon can convert complex carbohydrates to short-chain fatty acids (acetate, propionate, butyrate). The production of short-chain fatty acids stimulates sodium and water absorption and provides additional calories for intestinal absorption. Oxalate should be restricted in the diets of patients with short bowel syndrome and partial colon because oxalate is absorbed principally from the colon, resulting in oxalate nephropathy. Vitamin B_{12} should be replaced if more than 100 cm of a patient's terminal ileum has been resected or if it is severely diseased.

Patients with large volumes of diarrhea (2 L/day) should be administered antimotility agents to reduce the amount of volume loss. Oral rehydration solutions are also beneficial for some patients with short bowel syndrome. Octreotide should be reserved for patients with high stool output despite the use of antimotility agents and oral rehydration therapy. Exogenous trophic factors, including growth hormone, glutamine, and glucagon-like peptide-2, have been experimentally used to increase nutrient absorption in patients with short bowel syndrome. However, results have not been clinically significant.

Parenteral nutrition remains the treatment of choice for the patient with refractory disease who cannot survive on an oral diet alone. Small-bowel transplantation is indicated for patients with recurrent line sepsis, lack of venous access, and progressive TPN-induced liver disease.

ADDITIONAL RESOURCES

ASPEN Board of Directors, Clinical Guideline Task Force: Guidelines for the use of parenteral and enteral nutrition in adult and pediatric patients, *JPEN J Parenter Enteral Nutr* 26(1 Suppl):1SA-138SA, 2002.

Florez DA, Aranda-Michel J: Nutritional management of acute and chronic liver disease, *Semin Gastrointest Dis* 13:169-178, 2002.

Greenberg GR, Fleming CR, Jeejeebhoy KN, et al: Controlled trial of bowel rest and nutritional support in the management of Crohn's disease, *Gut* 29:1309-1315, 1988.

McClave SA, Chang WK, Dhaliwal R, Heyland DK: Nutrition support in acute pancreatitis: a systemic review of the literature, *JPEN J Parenter Enteral Nutr* 30: 143-156, 2006.

Scolapio JS, Raimondo M, Lankisch M: Nutritional support in pancreatitis, *Scand Gastroenterol* 35:1010-1015, 2000.

Scolapio JS, Ukleja A: Short-bowel syndrome, *Curr Opin Clin Nutr Metab Care* 1:391-394, 1998.

Index

F

G

J

O

P

S

U